DIMENSION	METRIC	METRIC/ENGLISH
Power, heat transfer rate	$1\ W = 1\ J/s$ $1\ kW = 1000\ W = 1.341\ hp$ $1\ hp^{\ddagger} = 745.7\ W$	$1\ kW = 3412.14\ Btu/h$ $\quad = 737.56\ lbf \cdot ft/s$ $1\ hp = 550\ lbf \cdot ft/s = 0.7068\ Btu/s$ $\quad = 42.41\ Btu/min = 2544.5\ Btu/h$ $\quad = 0.74570\ kW$ $1\ Btu/h = 1.055056\ kJ/h$ $1\ ton\ of\ refrigeration = 200\ Btu/min$
Pressure	$1\ Pa = 1\ N/m^2$ $1\ kPa = 10^3\ Pa = 10^{-3}\ MPa$ $1\ atm = 101.325\ kPa = 1.01325\ bars$ $\quad = 760\ mmHg\ at\ 0°C$ $\quad = 1.03323\ kgf/cm^2$ $1\ mmHg = 0.1333\ kPa$	$1\ Pa = 1.4504 \times 10^{-4}\ psia$ $\quad = 0.020886\ lbf/ft^2$ $1\ psia = 144\ lbf/ft^2 = 6.894757\ kPa$ $1\ atm = 14.696\ psia = 29.92\ inHg\ at\ 30°F$ $1\ inHg = 3.387\ kPa$
Specific heat	$1\ kJ/kg \cdot °C = 1\ kJ/kg \cdot K$ $\quad = 1\ J/g \cdot °C$	$1\ Btu/lbm \cdot °F = 4.1868\ kJ/kg \cdot °C$ $1\ Btu/lbmol \cdot R = 4.1868\ kJ/kmol \cdot K$ $1\ kJ/kg \cdot °C = 0.23885\ Btu/lbm \cdot °F$ $\quad = 0.23885\ Btu/lbm \cdot R$
Specific volume	$1\ m^3/kg = 1000\ L/kg$ $\quad = 1000\ cm^3/g$	$1\ m^3/kg = 16.02\ ft^3/lbm$ $1\ ft^3/lbm = 0.062428\ m^3/kg$
Temperature	$T(K) = T(°C) + 273.15$ $\Delta T(K) = \Delta T(°C)$	$T(R) = T(°F) + 459.67 = 1.8T(K)$ $T(°F) = 1.8\ T(°C) + 32$ $\Delta T(°F) = \Delta T(R) = 1.8^* \Delta T(K)$
Thermal conductivity	$1\ W/m \cdot °C = 1\ W/m \cdot K$	$1\ W/m \cdot °C = 0.57782\ Btu/h \cdot ft \cdot °F$
Thermal resistance	$1°C/W = 1\ K/W$	$1\ K/W = 0.52750°F/h \cdot Btu$
Velocity	$1\ m/s = 3.60\ km/h$	$1\ m/s = 3.2808\ ft/s = 2.237\ mi/h$ $1\ mi/h = 1.609\ km/h$
Viscosity, dynamic	$1\ N \cdot s/m^2 = 1\ kg/m \cdot s$	$1\ kg/m \cdot s = 2419.1\ lbm/ft \cdot h$ $\quad = 5.8016 \times 10^{-6}\ lbf \cdot h/ft^2$
Viscosity, kinematic	$1\ m^2/s = 10^4\ cm^2/s$	$1\ m^2/s = 10.764\ ft^2/s = 3.875 \times 10^4\ ft^2/h$
Volume	$1\ m^3 = 1000\ L = 10^6\ cm^3\ (cc)$	$1\ m^3 = 6.1024 \times 10^4\ in^3 = 35.315\ ft^3$ $\quad = 264.17\ gal\ (U.S.)$ $1\ U.S.\ gallon = 231\ in^3 = 3.7854\ L$

*Exact conversion factor between metric and English units.

‡Mechanical horsepower. The electrical horsepower is taken to be exactly 746 W.

HEAT TRANSFER
A PRACTICAL APPROACH

WCB/McGRAW-HILL SERIES IN MECHANICAL ENGINEERING

CONSULTING EDITORS

Jack P. Holman, *Southern Methodist University*

John R. Lloyd *Michigan State University*

Also available from McGraw-Hill

SCHAUM'S OUTLINE SERIES
IN MECHANICAL ENGINEERING

Most outlines include basic theory, definitions, hundreds of example problems solved in step-by-step detail, and supplementary problems with answers.

Related titles on the Current List Include:

Acoustics
Continuum Mechanics
Elementary Statics & Strength of Materials
Engineering Economics
Engineering Mechanics
Engineering Thermodynamics
Fluid Dynamics
Fluid Mechanics & Hydraulics
Heat Transfer
Lagrangian Dynamics
Machine Design
Mathematical Handbook of Formulas & Tables
Mechanical Vibrations
Operations Research
Statics & Mechanics of Materials
Strength of Materials
Theoretical Mechanics
Thermodynamics with Chemical Applications

SCHAUM'S SOLVED PROBLEMS BOOKS

Each title in this series is a complete and expert source of solved problems with solutions worked out in step-by-step detail.

Related Titles on the Current List Include:

3000 Solved Problems in Calculus
2500 Solved Problems in Differential Equations
2500 Solved Problems in Fluid Mechanics & Hydraulics
1000 Solved Problems in Heat Transfer
3000 Solved Problems in Linear Algebra
2000 Solved Problems in Mechanical Engineering Thermodynamics
2000 Solved Problems in Numerical Analysis
700   Solved Problems in Vector Mechanics for Engineers: Dynamics
800   Solved Problems in Vector Mechanics for Engineers: Statics

Available at most college bookstores, or for a complete list of titles and prices, write to: Schaum Division
      McGraw-Hill, Inc.
      Princeton Road, S-1
      Hightstown, NJ 08520

HEAT TRANSFER

A PRACTICAL APPROACH

Yunus A. Çengel
University of Nevada, Reno

Boston, Massachusetts   Burr Ridge, Illinois   Dubuque, Iowa
Madison, Wisconsin   New York, New York   San Francisco, California   St. Louis, Missouri

McGraw-Hill Higher Education

*A Division of The **McGraw-Hill** Companies*

HEAT TRANSFER: A PRACTICAL APPROACH

This book is printed on acid-free paper.

domestic   3 4 5 6 7 8 9 0 DOC/DOC 0 9 8 7 6 5 4 3 2
international   1 2 3 4 5 6 7 8 9 0 DOC/DOC 0 9 8 7 6 5 4 3 2 1

ISBN 0-07-011505-2

Vice president and editorial director: *Kevin T. Kane*
Publisher: *Tom Casson*
Senior sponsoring editor: *Debra Riegert*
Marketing manager: *John Wannemacher*
Project manager: *Gladys True*
Production supervisor: *Heather D. Burbridge*
Designer: *Matthew Baldwin*
Cover designer: *Ellen Pettengell, Ellen Pettengell Design*
Compositor: *GAC/Indianapolis*
Typeface: *10.5/12 Times Roman*
Printer: *R. R. Donnelley & Sons Company*

Library of Congress Cataloging-in-Publication Data
Çengel, Yunus A.
    Heat transfer: a practical approach / Yunus A. Çengel
       p.   cm.
    Includes index.
    ISBN 0-07-01105-2
    1. Heat—Transmission.   2. Heat—Transmission—Industrial
applications.   I. Title.
QC320.C46 1997
621.402′2—dc21   97-6640

When ordering the title, use ISBN -0-7-115223-7

If interested in ordering the book with the EES problems disk use ISBN 0-07-561176-7 since the book will be available in two different versions.

http://www.mhhe.com

About the Author

Yunus A. Çengel received his Ph.D. in mechanical engineering from North Carolina State University and joined the faculty of mechanical engineering at the University of Nevada, Reno, where he has been teaching undergraduate and graduate courses in thermodynamics and heat transfer while conducting research. He has published primarily in the areas of thermodynamics, radiation heat transfer, natural convection, solar energy, geothermal energy, energy conservation, and engineering education. Dr. Çengel is the author of the textbook, *Introduction to Thermodynamics and Heat Transfer,* and the coauthor of *Thermodynamics: An Engineering Approach,* both published by McGraw-Hill. He has led teams of engineering students to numerous manufacturing facilities in northern Nevada and California to conduct energy audits and has prepared energy conservation reports for them. Dr. Çengel has been voted the outstanding teacher by the ASME student sections in both North Carolina State University and the University of Nevada, Reno. He is a member of the American Society of Mechanical Engineers (ASME) and the American Society for Engineering Education (ASEE). Dr. Çengel is also the recipient of the ASEE Meriam/Wiley Distinguished Author Award.

Contents

Table
of Examples

Chapter 3 ■ STEADY HEAT CONDUCTION   **129**

Chapter 14 ■ **COOLING OF ELECTRONIC EQUIPMENT**   863

Preface

GENERAL APPROACH

This introductory text is intended for use in a first course in heat transfer for undergraduate engineering students in their junior or senior year, and as a reference book for practicing engineers. The text covers the *basic principles* of heat transfer with a broad range of *engineering applications*. It contains sufficient material to give instructors flexibility and to accommodate their preferences on the right blend of fundamentals and applications.

The students are assumed to have completed their basic physics and calculus sequence. The completion of first courses in thermodynamics, fluid mechanics, and differential equations prior to taking heat transfer is desirable. The relevant concepts from these topics are introduced and reviewed as needed. The emphasis throughout the text is kept on the *physics* and the *physical arguments* in order to develop an *intuitive understanding* of the subject matter.

There are several textbooks in heat transfer currently available, and one cannot help wondering if there is a need for another one. After all, heat transfer is a mature science, and the topics of heat transfer are well established. However, it is often stated that the current education system needs to be modernized, and major research programs have been undertaken in recent years to come up with a better match between engineering education and engineering practice. The author has long felt that current engineering education is academically oriented and geared toward bringing up academicians rather than practicing engineers. This text is the outcome of an attempt to have a suitable textbook for a practically oriented heat transfer course for engineering students. The text covers all the standard topics in heat transfer with an emphasis on physical mechanisms and practical applications, while de-emphasizing heavy mathematical aspects, which are being left to computers.

In engineering practice, an understanding of the mechanisms of heat transfer is becoming increasingly important since heat transfer plays a crucial role in the design of vehicles, power plants, refrigerators, electronic devices, buildings, and bridges, among other things. Even a chef needs to have an intuitive understanding of the heat transfer mechanism in order to cook the food "right" every time. We may not be aware of it, but we always use the principles of heat transfer when seeking thermal comfort. We insulate our bodies by putting on heavy coats in winter, and we minimize heat gain by radiation by staying in the shadow in summer. We speed up the cooling of hot food by blowing on it and keep warm in cold weather by cuddling up and thus minimizing the exposed surface area. That is, we already use heat transfer whether we realize it or not. So we might as well go ahead and learn what we already practice.

The philosophy that contributed to the popularity of the thermodynamics book I coauthored with Dr. Boles has remained unchanged in this text: talk directly to the minds of tomorrow's engineers in a simple yet precise manner, and encourage *creative thinking* and development of a *deeper understanding* of the subject matter. The goal throughout this project has been to offer an engineering textbook that is *read* by the students with *interest* and *enthusiasm* instead of one that is used as a reference book to solve problems. Special effort is made to touch the curious minds and take them on a pleasant journey in the wonderful world of thermodynamics and heat transfer and explore the wonders of these exciting subjects.

Yesterday's engineer spent a major portion of his or her time substituting values into the formulas and obtaining numerical results. But all the formula manipulations and number crunching are being left to the computers. Tomorrow's engineer will have to have a clear understanding and a firm grasp of the *basic principles* so that he or she can understand even the most complex problems, formulate them, and interpret the results. A conscious effort is made to lead students in this direction.

CONTENTS

In Chapter 1 we introduce the *basic concepts of thermodynamics and heat transfer,* and the first law of thermodynamics. We also review the ideal gas equation and the specific heat relations since they are commonly used in the heating and cooling of gases. In this chapter we also introduce the basic mechanisms of heat transfer. In Chapter 2 we derive the *heat conduction equation,* discuss the boundary conditions, and solve some steady one-dimensional heat conduction problems. We also discuss heat generation and variable conductivity.

Chapters 3 and 4 deal with *steady* and *transient* heat conduction, respectively, and Chapter 5 with *numerical methods.* A practical approach is used in these chapters and the thermal resistance concept is emphasized. Extensive discussions are given on *thermal insulations* and the *optimum thickness of insulation* because of the widespread use of insulations in industry and the key role they play in any energy conservation project. Chapters 6 and 7 deal with *forced and natural convection,* respectively. Again, a practical engineering approach is used with a wealth of physical explanations and empirical correlations. The velocity and thermal boundary layers are discussed without resorting to the boundary layer equations. *Boiling and condensation heat transfer* is presented in Chapter 8, followed by *radiation heat transfer* in Chapter 9, *heat exchangers* in Chapter 10, and *mass transfer* in Chapter 11.

Chapters 12, 13, and 14 deal with common application areas of heat transfer in engineering practice. They are intended to show students how heat transfer is used in *real-world* situations and to give them a chance *to practice* what they have been mastering during much of the course. In Chapter 12 we discuss *heat transfer in residential and commercial buildings,* thermal comfort, heating and cooling loads, infiltration losses, and thermal design using the local weather data. In Chapter 13 we discuss heat transfer associated with the *cooling and freezing of foods* such as vegetables, fruits, fish, and meat products. Finally, in Chapter 14, we discuss the cooling techniques currently used in the *thermal control of electronic equipment.* These chapters are independent of each other and can be covered in any order. Or they can be skipped all together to allow more time on fundamentals in earlier chapters. Most students will find the material in these chapters very interesting and highly informative, and they will probably read them anyway even if they are not covered. The material in these chapters is based primarily on *handbooks* such as the authoritative series of handbooks by the American Society for Heating, Refrigerating, and Air-Conditioning Engineers (ASHRAE), Inc.

LEARNING TOOLS

A distinctive feature of this book is the absence of heavy mathematical and theoretical aspects of subject matter such as the separation of variables and the Navier-Stokes equations. The author believes that such material has little practical value, and it is better suited for graduate-level courses. The emphasis in undergraduate education should remain on *developing a sense of underlying physical mechanism* and a *mastery of solving practical problems* an engineer is likely to face in the real world. The absence of such time-consuming and intimidating material should free the instructor to cover more material on the fundamentals and applications of heat transfer. This should also make heat transfer a more pleasant and worthwhile experience for the students.

An observant mind should have no difficulty understanding heat transfer. After all, the principles of heat transfer are based on our *everyday experiences* and *experimental observations.* A more physical, intuitive approach is used throughout this text. Frequently, *parallels are drawn* between the subject matter and students' everyday experiences so that they can relate the subject matter to what they already know. The process of cooking, for example, serves as an excellent vehicle to demonstrate the basic principles of heat transfer.

The material in the text is introduced at a level that an average student can follow comfortably. It speaks to the students, not over the students. In fact, it is *self-instructive*. Noting that the principles of heat transfer are based on experimental observations, all the derivations in this text are based on physical arguments, and thus they are easy to follow and understand.

Figures are important learning tools that help the students "get the picture." The text makes effective use of graphics. It probably contains more figures and illustrations than any other heat transfer book. Figures attract attention and stimulate curiosity and interest. Some of the figures in this text are intended to serve as a means of emphasizing some key concepts that would otherwise go unnoticed, or as paragraph summaries.

Each chapter contains numerous worked-out *examples* that clarify the material and illustrate the use of the basic principles. An *intuitive* and *systematic* approach is used in the solution of the example problems, with particular attention to the proper use of *units*. A *summary* is included at the end of each chapter for a quick overview of basic concepts and important relations.

The *end-of-chapter problems* are grouped under specific topics in the order they are covered to make problem selection easier for both instructors and students. The problems within each group start with concept questions, indicated by "C," to check the students' level of understanding of basic concepts. The problems under *Review Problems* are more comprehensive in nature and are not directly tied to any specific section of a chapter. The problems under the *Computer, Design, and Essay Problems* title are intended to encourage students to make engineering judgments, to conduct independent searches on topics of interest, and to communicate their findings in a professional manner. Several economics- and safety-related problems are incorporated throughout to enhance cost and safety awareness among engineering students. Answers to selected problems are listed immediately following the problem for convenience to the students.

In recognition of the fact that English units are still widely used in some industries, both SI and English units are used in this text, with an emphasis on SI. The material in this text can be covered using combined SI/English units or SI units alone, depending on the preference of the instructor. The property tables and charts in the appendix are presented in both units, except the ones that involve dimensionless quantities. Problems, tables, and charts in English units are designated by "E" after the number for easy recognition. Frequently used conversion factors and the physical constants are listed on the inner cover pages of the text for easy reference.

SUPPLEMENTS

Solutions Manual

This manual features detailed typed solutions with illustrations suitable for posting.

EES Software

Developed by Sandy Klein and Bill Beckman from the University of Wisconsin—Madison, this software program allows students to solve problems, especially design problems, and to ask "what if" questions. EES (pronounced "ease") is an acronym for Engineering Equation Solver. EES is very easy to master, and equations can be entered in any form and in any order. The combination of equation-solving capability and engineering property data makes EES an extremely powerful tool for students.

EES can do optimization, parametric analysis, linear and nonlinear regression, and provide publication-quality plotting capability. Equations can be entered in any form and in any order. EES automatically rearranges the equations to solve them in the most efficient manner.

EES is particularly useful for heat transfer problems since most of the property data needed for solving heat transfer problems are provided in the program. For example, the steam tables are implemented such that any thermodynamic property can be obtained from a built-in function call in terms of any two properties. Similar capability is provided for many organic refrigerants, ammonia, methane, carbon dioxide, and many other fliuds. Air tables are built-in, as are psychometric functions and JANAF table data for many common gases. Transport properties are also provided for all substances. EES also allows the user to enter property data or functional relationships with lookup tables, with internal functions written with EES, or with externally compiled functions written in Pascal, C, C++, or Fortran.

The EES engine is available to those who adopt the text with the problems disk, or it will be available via a password protected system on the Web to departments who wish to use EES without a McGraw-Hill text. A license for EES is provided to departments of educational institutions that adopt *Heat Transfer: A Practical Approach* from WCB/McGraw-Hill. If you would like this option, be sure to order your book using ISBN 0-07-561176-7. If you need more information, contact your local WCB/McGraw-Hill representative, call 1-800-338-3987, or visit our Web site at www.mhhe.com.

EES Software Problems Disk contains the EES programs that have been developed to solve some of the problems in this text. Problems solved on the EES problems disk are denoted in the text with a disk symbol. Each program provides detailed comments and on-line help. These programs should help the student master the important concepts without the calculational burden that has been previously required. A Windows-based Software Demo is available for your review.

ACKNOWLEDGMENTS

I would like to acknowledge with appreciation the numerous and valuable comments, suggestions, criticisms, and praise of the following academic reviewers:

Edward Anderson, *Texas Tech University*

Bahman Litkouhi, *Manhattan College*

Hameed Metghalchi, *Northeastern University*

William Moses, *Mercer University*

Sidney Roberts, *Old Dominion University*

Ramendra P. Roy, *Arizona State University*

Brian Vick, *Virginia Polytechnic Institute & State University*

Their suggestions have greatly helped to improve the quality of this text. I also would like to thank my students at the University of Nevada, Reno, who provided plenty of feedback from students' perspectives while class-testing the material. Finally, I would like to express my appreciation to my wife Zehra and my children for their continued patience, understanding, and support throughout the preparation of this text.

Yunus A. Çengel

Nomenclature

A	Area, m^2
A_c	Cross-sectional area
ACH	Air changes per hour
Bi	Biot number
C	Specific heat, $kJ/kg \cdot K$; capacity ratio
C	Molar concentration, $kmol/m^3$
C_h, C_c	Heat capacity rate, $W/°C$
C_D	Drag coefficient
C_f	Friction coefficient
C_p	Constant pressure specific heat, $kJ/kg \cdot K$
C_v	Constant volume specific heat, $kJ/kg \cdot K$
COP	Coefficient of performance
d, D	Diameter, m
D_{AB}	Diffusion coefficient
D_h	Hydraulic diameter, m
e	Specific total energy, kJ/kg
erfc	Complimentary error function
E	Total energy, kJ
E_b	Blackbody emissive flux

f	Friction factor
f_λ	Blackbody radiation function
F	Force, N
F_D	Drag coefficient
F_{ij}	View factor
g	Gravitational acceleration, m/s^2
$\dot{g}$	Heat generation rate, W/m^3
G	Incident radiation, W/m^2
$\overline{Gr}$	Grashof number
h	Enthalpy, $u + Pv$, kJ/kg
h	Convection heat transfer coefficient, W/m$^2 \cdot$ °C
h_c	Thermal contact conductance, W/m$^2 \cdot$ °C
h_{fg}	Latent heat of vaporization, kJ/kg
h_{if}	Latent heat of fusion, kJ/kg
j	Diffusive mass flux, kg/s $\cdot$ m^2
J	Radiosity, W/m^2; Bessel function
k	Thermal conductivity
k_{eff}	Effective thermal conductivity, W/m $\cdot$ °C
L	Length; half thickness of a plane wall
L_c	Characteristic or corrected length
L_h	Hydrodynamic entry length
L_t	Thermal entry length
Le	Lewis number
m	Mass, kg
$\dot{m}$	Mass flow rate, kg/s
M	Molar mass, kg/kmol
N	Number of moles, kmol
NTU	Number of transfer units
Nu	Nusselt number
p	Perimeter, m
P	Pressure, kPa
P_v	Vapor pressure
Pr	Prandtl number
Q	Total amount heat transfer, kJ
$\dot{q}$	Heat flux, W/m^2
$\dot{Q}$	Heat transfer rate, kW
r	Radius, m
r_{cr}	Critical radius of insulation

R	Gas constant, kJ/kg · K
R	Thermal resistance, °C/W
Ra	Rayleigh number
R_c	Thermal contact resistance, m^2 · °C/W
R_f	Fouling factor
R_u	Universal gas constant, kJ/kmol · K
R-value	R-value of insulation
Re	Reynolds number
S	Conduction shape factor
Sc	Schmidt number
Sh	Sherwood number
St	Stanton number
SC	Shading coefficient
SHGC	Solar heat gain coefficient
t	Thickness, m
t	Time, s
T	Temperature, °C or K
T_i	Initial temperature, °C or K
T_f	Film temperature, °C
T_{sat}	Saturation temperature, °C
T_{sky}	Sky temperature, K
u	Internal energy, kJ/kg
U	Overall heat transfer coefficient, W/m^2 · °C
v	Specific volume, m^3/kg
V	Total volume, m^3
$\dot{V}$	Volume flow rate, m^3/s
$\mathcal{V}$	Velocity, m/s
$\mathcal{V}_\infty$	Free stream velocity
w	Mass fraction
$\dot{W}$	Power, kW
y	Mole fraction

Greek Letters

α	Absorptivity; thermal diffusivity, m^2/s
α_s	Solar absorptivity
β	Volume expansion coefficient, 1/K
δ	Characteristic length

ΔT_{lm}	Log mean temperature difference
ε	Emissivity; heat exchanger or fin effectiveness
η_{fin}	Fin efficiency
μ	Dynamic viscosity, kg/m · s
ν	Kinematic viscosity, m^2/s; frequency, 1/s
ρ	Density, kg/m^3
σ	Stefan–Boltzmann constant
τ	Transmissivity; Fourier number
ϕ	Relative humidity
θ	Dimensionless temperature

Subscripts

atm	Atmospheric
av	Average
b	Bulk fluid
cyl	Cylinder
e	Electrical
e	Exit conditions
f	Saturated liquid; fluid
gen	Generation
i	Inlet, initial, or indoor conditions
i	ith component
l	Liquid
m	Mixture
o	Outdoor conditions
rad	Radiation
s	Surface
sat	Saturated
semi-inf	Semi-infinite medium
sph	Sphere
surr	Surrounding surfaces
sys	System
v	Vapor
∞	Ambient conditions

Superscripts

· (over dot)	Quantity per unit time
⁻ (over bar)	Quantity per unit mole

Basic Concepts of Thermodynamics and Heat Transfer

The science of *thermodynamics* deals with the *amount* of heat transfer as a system undergoes a process from one equilibrium state to another, and makes no reference to *how long* the process will take. But in engineering, we are often interested in the *rate* of heat transfer, which is the topic of the science of *heat transfer*.

We start this chapter with a review of the fundamental concepts of thermodynamics that form the framework for heat transfer. We first present the relation of heat to other forms of energy and review the first law of thermodynamics. We then present the three basic mechanisms of heat transfer, which are conduction, convection, and radiation, and discuss thermal conductivity. *Conduction* is the transfer of energy from the more energetic particles of a substance to the adjacent, less energetic ones as a result of interactions between the particles. *Convection* is the mode of heat transfer between a solid surface and the adjacent liquid or gas that is in motion, and it involves the combined effects of conduction and fluid motion. *Radiation* is the energy emitted by matter in the form of electromagnetic waves (or photons) as a result of the changes in the electronic configurations of the atoms or molecules. We close this chapter with a discussion of simultaneous heat transfer.

1-1 ■ THERMODYNAMICS AND HEAT TRANSFER

We all know from experience that a cold canned drink left in a room warms up and a warm canned drink left in a refrigerator cools down. This is accomplished by the transfer of *energy* from the warm medium to the cold one. The energy transfer is always from the higher temperature medium to the lower temperature one, and the energy transfer stops when the two mediums reach the same temperature.

You will recall from thermodynamics that energy exists in various forms. In this text we are primarily interested in **heat,** which is *the form of energy that can be transferred from one system to another as a result of temperature difference*. The science that deals with the determination of the *rates* of such energy transfers is **heat transfer.**

You may be wondering why we need to undertake a detailed study on heat transfer. After all, we can determine the amount of heat transfer for any system undergoing any process using a thermodynamic analysis alone. The reason is that thermodynamics is concerned with the *amount* of heat transfer as a system undergoes a process from one equilibrium state to another, and it gives no indication about *how long* the process will take. A thermodynamic analysis simply tells us how much heat must be transferred to realize a specified change of state to satisfy the conservation of energy principle.

In practice we are more concerned about the rate of heat transfer (heat transfer per unit time) than we are with the amount of it. For example, we can determine the amount of heat transferred from a thermos bottle as the hot coffee inside cools from 90°C to 80°C by a thermodynamic analysis alone. But a typical user or designer of a thermos is primarily interested in *how long* it will be before the hot coffee inside cools to 80°C, and a thermodynamic analysis cannot answer this question. Determining the rates of heat transfer to or from a system and thus the times of cooling or heating, as well as the variation of the temperature, is the subject of *heat transfer* (Fig. 1-1).

Thermodynamics deals with equilibrium states and changes from one equilibrium state to another. Heat transfer, on the other hand, deals with systems that lack thermal equilibrium, and thus it is a *nonequilibrium* phenomenon. Therefore, the study of heat transfer cannot be based on the principles of thermodynamics alone. However, the laws of thermodynamics lay the framework for the science of heat transfer. The *first law* requires that the rate of energy transfer into a system be equal to the rate of increase of the energy of that system. The *second law* requires that heat be transferred in the direction of decreasing temperature (Fig. 1-2). This is like a car parked on an inclined road that must go downhill in the direction of decreasing elevation when its brakes are released. It is also analogous to the electric current flowing in the direction of decreasing voltage or the fluid flowing in the direction of decreasing pressure.

The basic requirement for heat transfer is the presence of a *temperature difference*. There can be no net heat transfer between two mediums that are at the same temperature. The temperature difference is the *driving force* for heat transfer, just as the *voltage difference* is the driving force for electric

Thermos
bottle

Hot
coffee

Insulation

FIGURE 1-1

We are normally interested in how long it takes for the hot coffee in a thermos to cool to a certain temperature, which cannot be determined from a thermodynamic analysis alone.

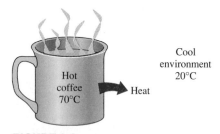

Cool
environment
20°C

Hot
coffee
70°C

Heat

FIGURE 1-2

Heat flows in the direction of decreasing temperature.

current flow and *pressure difference* is the driving force for fluid flow. The rate of heat transfer in a certain direction depends on the magnitude of the *temperature gradient* (the temperature difference per unit length or the rate of change of temperature) in that direction. The larger the temperature gradient, the higher the rate of heat transfer.

Application Areas of Heat Transfer

Heat transfer is commonly encountered in engineering systems and other aspects of life, and one does not need to go very far to see some application areas of heat transfer. In fact, one does not need to go anywhere. The human body is constantly rejecting heat to its surroundings, and human comfort is closely tied to the rate of this heat rejection. We try to control this heat transfer rate by adjusting our clothing to the environmental conditions.

Many ordinary household appliances are designed, in whole or in part, by using the principles of heat transfer. Some examples include the electric or gas range, the heating and air-conditioning system, the refrigerator and freezer, the water heater, the iron, and even the computer, the TV, and the VCR. Of course, energy-efficient homes are designed on the basis of minimizing heat loss in winter and heat gain in summer. Heat transfer plays a major role in the design of many other devices, such as car radiators, solar collectors, various components of power plants, and even spacecraft. The optimal insulation thickness in the walls and roofs of the houses, on hot water or steam pipes, or on water heaters is again determined on the basis of a heat transfer analysis with economic consideration (Fig. 1-3).

FIGURE 1-3

Some application areas of heat transfer.

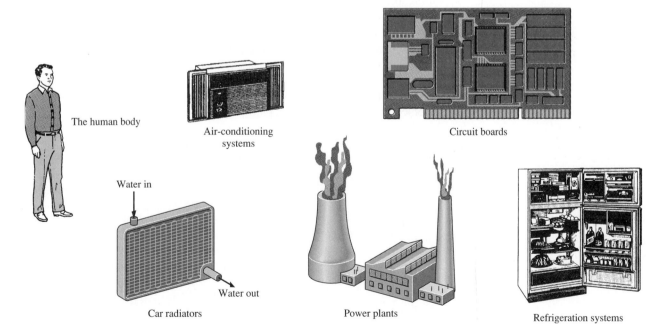

The human body

Air-conditioning
systems

Circuit boards

Water in

Water out

Car radiators

Power plants

Refrigeration systems

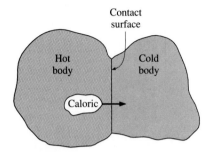

FIGURE 1-4

In the early 19th century, heat was thought to be an invisible fluid called the *caloric* that flowed from warmer bodies to the cooler ones.

Historical Background

Heat has always been perceived to be something that produces in us a sensation of warmth, and one would think that the nature of heat is one of the first things understood by mankind. But it was only in the middle of the nineteenth century that we had a true physical understanding of the nature of heat, thanks to the development at that time of the **kinetic theory,** which treats molecules as tiny balls that are in motion and thus possess kinetic energy. Heat is then defined as the energy associated with the random motion of atoms and molecules. Although it was suggested in the eighteenth and early nineteenth centuries that heat is the manifestation of motion at the molecular level (called the *live force*), the prevailing view of heat until the middle of the nineteenth century was based on the **caloric theory** proposed by the French chemist Antoine Lavoisier (1743–1794) in 1789. The caloric theory asserts that heat is a fluid-like substance called the **caloric** that is a massless, colorless, odorless, and tasteless substance that can be poured from one body into another (Fig. 1-4). When caloric was added to a body, its temperature increased; and when caloric was removed from a body, its temperature decreased. When a body could not contain any more caloric, much the same way as when a glass of water could not dissolve any more salt or sugar, the body was said to be saturated with caloric. This interpretation gave rise to the terms *saturated liquid* and *saturated vapor* that are still in use today.

The caloric theory came under attack soon after its introduction. It maintained that heat is a substance that could not be created or destroyed. Yet it was known that heat can be generated indefinitely by rubbing one's hands together or rubbing two pieces of wood together. In 1798, the American Benjamin Thompson (Count Rumford) (1753–1814) showed in his papers that heat can be generated continuously through friction. The validity of the caloric theory was also challenged by several others. But it was the careful experiments of the Englishman James P. Joule (1818–1889) published in 1843 that finally convinced the skeptics that heat was not a substance after all, and thus put the caloric theory to rest. Although the caloric theory was totally abandoned in the middle of the nineteenth century, it contributed greatly to the development of thermodynamics and heat transfer.

1-2 ■ ENGINEERING HEAT TRANSFER

Heat transfer equipment such as heat exchangers, boilers, condensers, radiators, heaters, furnaces, refrigerators, and solar collectors are designed primarily on the basis of heat transfer analysis. The heat transfer problems encountered in practice can be considered in two groups: (1) *rating* and (2) *sizing* problems. The rating problems deal with the determination of the heat transfer rate for an existing system at a specified temperature difference. The sizing problems deal with the determination of the size of a system in order to transfer heat at a specified rate for a specified temperature difference.

A heat transfer process or equipment can be studied either *experimentally* (testing and taking measurements) or *analytically* (by analysis or cal-

Fundamentals

culations). The experimental approach has the advantage that we deal with the actual physical system, and what we get is what it is, within the limits of experimental error. However, this approach is expensive, time-consuming, and often impractical. Besides, the system we are analyzing may not even exist. For example, the size of a heating system of a building must usually be determined *before* the building is actually built on the basis of the dimensions and specifications given. The analytical approach (including numerical approach) has the advantage that it is fast and inexpensive, but the results obtained are subject to the accuracy of the assumptions and idealizations made in the analysis. In heat transfer studies, often a good compromise is reached by reducing the choices to just a few by analysis, and then verifying the findings experimentally and doing some fine-tuning.

Modeling in Heat Transfer

The descriptions of most scientific problems involve relations that relate the changes in some key variables to each other. Usually the smaller the increment chosen in the changing variables, the more general and accurate the description. In the limiting case of infinitesimal or differential changes in variables, we obtain *differential equations* that provide precise mathematical formulations for the physical principles and laws by representing the rates of changes as *derivatives*. Therefore, differential equations are used to investigate a wide variety of problems in sciences and engineering, including heat transfer. However, most heat transfer problems encountered in practice can be solved without resorting to differential equations and the complications associated with them.

The study of physical phenomena involves two important steps. In the first step, all the variables that affect the phenomena are identified, reasonable assumptions and approximations are made, and the interdependence of these variables is studied. The relevant physical laws and principles are invoked, and the problem is formulated mathematically. The equation itself is very instructive as it shows the degree of dependence of some variables on others, and the relative importance of various terms. In the second step, the problem is solved using an appropriate approach, and the results are interpreted.

Many processes that seem to occur in nature randomly and without any order are, in fact, being governed by some visible or not-so-visible physical laws. Whether we notice them or not, these laws are there, governing consistently and predictably what seem to be ordinary events. Most of these laws are well defined and well understood by scientists. This makes it possible to predict the course of an event before it actually occurs, or to study various aspects of an event mathematically without actually running expensive and time-consuming experiments. This is where the power of analysis lies. Very accurate results to meaningful practical problems can be obtained with relatively little effort by using a suitable and realistic mathematical model. The preparation of such models requires an adequate knowledge of the natural phenomena involved and the relevant laws, as well as a sound judgment. An unrealistic model will obviously give inaccurate and thus unacceptable results.

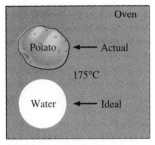

FIGURE 1-5

Modeling is a powerful engineering tool
that provides great insight and
simplicity at the expense of
some accuracy.

An analyst working on an engineering problem often finds himself or herself in a position to make a choice between a very accurate but complex model, and a simple but not-so-accurate model. The right choice depends on the situation at hand. The right choice is usually the simplest model that yields adequate results. For example, the process of baking potatoes or roasting a round chunk of beef in an oven can be studied analytically in a simple way by modeling the potato or the roast as a spherical solid ball that has the properties of water (Fig. 1-5). The model is quite simple, but the results obtained are sufficiently accurate for most practical purposes. As another example, when we analyze the heat losses from a building in order to select the right size for a heater, we determine the heat losses under anticipated worst conditions and select a furnace that will provide sufficient heat to make up for those losses. Often we tend to choose a larger furnace in anticipation of some future expansion, or just to provide a factor of safety. A very simple analysis will be adequate in this case.

When selecting heat transfer equipment, it is important to consider the actual operating conditions. For example, when purchasing a heat exchanger that will handle hard water, we must consider that some calcium deposits will form on the heat transfer surfaces over time, causing fouling and thus a gradual decline in performance. The heat exchanger must be selected on the basis of operation under these adverse conditions instead of under new conditions.

Preparing very accurate but complex models is usually not so difficult. But such models are not much use to an analyst if they are very difficult and time-consuming to solve. At the minimum, the model should reflect the essential features of the physical problem it represents. There are many significant real-world problems that can be analyzed with a simple model. But it should always be kept in mind that the results obtained from an analysis are as accurate as the assumptions made in simplifying the problem. Therefore, the solution obtained should not be applied to situations for which the original assumptions do not hold.

A solution that is not quite consistent with the observed nature of the problem indicates that the mathematical model used is too crude. In that case, a more realistic model should be prepared by eliminating one or more of the questionable assumptions. This will result in a more complex problem that, of course, is more difficult to solve. Thus any solution to a problem should be interpreted within the context of how that problem is formulated.

Engineering Software Packages

Perhaps you are wondering why we are about to undertake a painstaking study of the fundamentals of heat transfer. After all, almost all the heat transfer problems we are likely to encounter in practice can be solved using one of several sophisticated software packages readily available in the market today. These software packages not only give the desired numerical results, but also supply the outputs in colorful graphical form for impressive presentations. It is unthinkable to practice engineering today without using some of these packages. This tremendous computing power available to us at the touch of a button is both a blessing and a curse. It certainly enables engineers to solve problems easily and quickly, but it also opens the door for abuses and misinformation. In the hands of poorly educated people, these software packages are as dangerous as sophisticated powerful weapons in the hands of poorly trained soldiers.

Thinking that a person who can use the engineering software packages without proper training on fundamentals can practice engineering is like thinking that a person who can use a wrench can work as a car mechanic. If it were true that the engineering students do not need all these fundamental courses they are taking because practically everything can be done by computers quickly and easily, then it would also be true that the employers would no longer need high-salaried engineers since any person who knows how to use a word-processing program can also learn how to use those software packages. But the statistics show that the need for engineers is on the rise, not on the decline, despite the availability of these powerful packages.

We should always remember that all the computing power and the engineering software packages available today are just *tools,* and tools have meaning only in the hands of masters. Having the best word-processing program does not make a person a good writer, but it certainly makes the job of a good writer much easier and makes the writer more productive (Fig. 1-6). Hand calculators did not eliminate the need to teach our children how to add or subtract, and the sophisticated medical software packages did not take the place of medical school training. Neither will engineering software packages replace the traditional engineering courses. They will simply cause a shift in emphasis in the courses from mathematics to physics. That is, more time will be spent in the classrooms discussing the physical aspects of the problems in greater detail, and less time on the mechanics of solution procedures.

All these marvelous and powerful tools available today put an extra burden on today's engineers. They must still have a thorough understanding of the fundamentals, develop a "feel" of the physical phenomena, be able to put the data into proper perspective, and make sound engineering judgments, just like their predecessors. But they must do it much better and much faster because of the powerful tools available today. The engineers in the past had to rely on hand calculations, slide rules, and later hand calculators and computers. Today they rely on software packages. The easy access to such power and the possibility of a simple misunderstanding or misinterpretation causing great damage make it more important today than ever to have a solid training in the fundamentals of engineering. In this text we make an extra effort to put the emphasis on developing an intuitive and physical understanding of heat transfer phenomena instead of on the mathematical details of solution procedures.

1-3 ■ HEAT AND OTHER FORMS OF ENERGY

Energy can exist in numerous forms such as thermal, mechanical, kinetic, potential, electrical, magnetic, chemical, and nuclear, and their sum constitutes the **total energy** E (or e on a unit mass basis) of a system. The forms of energy related to the molecular structure of a system and the degree of the

FIGURE 1-6

An excellent word-processing program does not make a person a good writer; it simply makes a good writer a better and more efficient writer.

molecular activity are referred to as the *microscopic energy*. The sum of all microscopic forms of energy is called the **internal energy** of a system, and is denoted by U (or u on a unit mass basis).

The international unit of energy is *joule* (J) or *kilojoule* (1 kJ = 1000 J). In the English system, the unit of energy is the *British thermal unit* (Btu), which is defined as the energy needed to raise the temperature of 1 lbm of water at 68°F by 1°F. The magnitudes of kJ and Btu are almost identical (1 Btu = 1.055056 kJ). Another well-known unit of energy is the *calorie* (1 cal = 4.1868 J), which is defined as the energy needed to raise the temperature of 1 gram of water at 15°C by 1°C.

Internal energy may be viewed as the sum of the kinetic and potential energies of the molecules. The portion of the internal energy of a system associated with the kinetic energy of the molecules is called **sensible energy** or **sensible heat.** The average velocity and the degree of activity of the molecules are proportional to the temperature. Thus, at higher temperatures the molecules will possess higher kinetic energy, and as a result, the system will have a higher internal energy.

The internal energy is also associated with the intermolecular forces between the molecules of a system. These are the forces that bind the molecules to each other, and, as one would expect, they are strongest in solids and weakest in gases. If sufficient energy is added to the molecules of a solid or liquid, they will overcome these molecular forces and simply break away, turning the system to a gas. This is a *phase change* process and because of this added energy, a system in the gas phase is at a higher internal energy level than it is in the solid or the liquid phase. The internal energy associated with the phase of a system is called **latent energy** or **latent heat.**

The changes mentioned above can occur without a change in the chemical composition of a system. Most heat transfer problems fall into this category, and one does not need to pay any attention to the forces binding the atoms in a molecule together. The internal energy associated with the atomic bonds in a molecule is called **chemical** (or **bond**) **energy,** whereas the internal energy associated with the bonds within the nucleus of the atom itself is called **nuclear energy.** The chemical and nuclear energies are absorbed or released during chemical or nuclear reactions, respectively.

In the analysis of systems that involve fluid flow, we frequently encounter the combination of properties u and Pv. For the sake of simplicity and convenience, this combination is defined as **enthalpy** h. That is, $h = u + Pv$ where the term Pv represents the *flow energy* of the fluid (also called the *flow work*), which is the energy needed to push a fluid and to maintain flow. In the energy analysis of flowing fluids, it is convenient to treat the flow energy as part of the energy of the fluid and to represent the microscopic energy of a fluid stream by enthalpy h (Fig. 1-7).

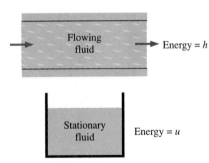

FIGURE 1-7

The *internal energy u* represents the microscopic energy of a nonflowing fluid, whereas *enthalpy h* represents the microscopic energy of a flowing fluid.

Specific Heats of Gases, Liquids, and Solids

You may recall that an **ideal gas** is defined as a gas that obeys the relation

$$Pv = RT \qquad \text{or} \qquad P = \rho RT \qquad (1\text{-}1)$$

where P is the absolute pressure, v is the specific volume, T is the absolute temperature, ρ is the density, and R is the gas constant. It has been experimentally observed that the ideal gas relation given above closely approximates the P-v-T behavior of real gases at low densities. At low pressures and high temperatures, the density of a gas decreases and the gas behaves like an ideal gas. In the range of practical interest, many familiar gases such as air, nitrogen, oxygen, hydrogen, helium, argon, neon, and krypton and even heavier gases such as carbon dioxide can be treated as ideal gases with negligible error (often less than one percent). Dense gases such as water vapor in steam power plants and refrigerant vapor in refrigerators, however, should not always be treated as ideal gases since they usually exist at a state near saturation.

You may also recall that **specific heat** is defined as *the energy required to raise the temperature of a unit mass of a substance by one degree* (Fig. 1-8). In general, this energy depends on how the process is executed. In thermodynamics, we are interested in two kinds of specific heats: specific heat at constant volume C_v and specific heat at constant pressure C_p. The **specific heat at constant volume** C_v can be viewed as the energy required to raise the temperature of a unit mass of a substance by one degree as the volume is held constant. The energy required to do the same as the pressure is held constant is the **specific heat at constant pressure** C_p. The specific heat at constant pressure C_p is always greater than C_v because at constant pressure the system is allowed to expand and the energy for this expansion work must also be supplied to the system. For ideal gases, these two specific heats are related to each other by $C_p = C_v + R$.

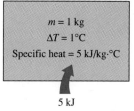

FIGURE 1-8

Specific heat is the energy required to raise the temperature of a unit mass of a substance by one degree in a specified way.

A common unit for specific heats is kJ/kg · °C or kJ/kg · K. Notice that these two units are *identical* since $\Delta T(°C) = \Delta T(K)$, and 1°C change in temperature is equivalent to a change of 1 K. Also,

$$1 \text{ kJ/kg} \cdot °C \equiv 1 \text{ J/g} \cdot °C \equiv 1 \text{ kJ/kg} \cdot K \equiv 1 \text{ J/g} \cdot K$$

The specific heats of a substance, in general, depend on two independent properties such as temperature and pressure. For an *ideal gas,* however, they depend on *temperature* only (Fig. 1-9). At low pressures all real gases approach ideal gas behavior, and therefore their specific heats depend on temperature only.

The differential changes in the internal energy u and enthalpy h of an ideal gas can be expressed in terms of the specific heats as

$$du = C_v dT \quad \text{and} \quad dh = C_p dT \quad (1\text{-}2)$$

The finite changes in the internal energy and enthalpy of an ideal gas during a process can be expressed approximately by using specific heat values at the average temperature as

$$\Delta u = C_{v,\text{ ave}}\Delta T \quad \text{and} \quad \Delta h = C_{p,\text{ ave}}\Delta T \quad \text{(J/kg)} \quad (1\text{-}3)$$

or

$$\Delta U = mC_{v,\text{ ave}}\Delta T \quad \text{and} \quad \Delta H = mC_{p,\text{ ave}}\Delta T \quad \text{(J)} \quad (1\text{-}4)$$

where m is the mass of the system.

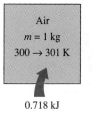 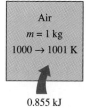

FIGURE 1-9

The specific heat of a substance changes with temperature.

FIGURE 1-10
The C_v and C_p values of incompressible substances are identical and are denoted by C.

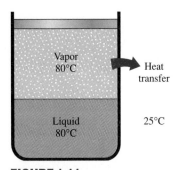

FIGURE 1-11
The sensible and latent forms of internal energy can be transferred as a result of a temperature difference, and they are referred to as *heat* or *thermal energy.*

A substance whose specific volume (or density) does not change with temperature or pressure is called an **incompressible substance.** The specific volumes of solids and liquids essentially remain constant during a process, and thus they can be approximated as incompressible substances without sacrificing much in accuracy.

The constant-volume and constant-pressure specific heats are identical for incompressible substances (Fig. 1-10). Therefore, for solids and liquids the subscripts on C_v and C_p can be dropped and both specific heats can be represented by a single symbol, C. That is, $C_p \cong C_v \cong C$. This result could also be deduced from the physical definitions of constant-volume and constant-pressure specific heats. Specific heats of several common gases, liquids, and solids are given in the Appendix.

The specific heats of incompressible substances depend on temperature only. Therefore, the change in the internal energy of solids and liquids can be expressed as

$$\Delta U = mC_{ave}\Delta T \qquad \text{(J)} \qquad (1\text{-}5)$$

where C_{ave} is the average specific heat evaluated at the average temperature. Note that the internal energy change of the systems that remain in a single phase (liquid, solid, or gas) during the process can be determined very easily using average specific heats.

Energy Transfer

Energy can be transferred to or from a given mass by two mechanisms: *heat Q* and *work W.* An energy interaction is heat transfer if its driving force is a temperature difference. Otherwise, it is work. A rising piston, a rotating shaft, and an electrical wire crossing the system boundaries are all associated with work interactions. Work done *per unit time* is called **power,** and is denoted by $\dot{W}$. The unit of power is W or hp (1 hp = 746 W). Car engines and hydraulic, steam, and gas turbines produce work; compressors, pumps, and mixers consume work. Notice that the energy of a system decreases as it does work, and increases as work is done on the system.

In daily life, we frequently refer to the sensible and latent forms of internal energy as **heat,** and we talk about the heat content of bodies (Fig. 1-11). In thermodynamics, however, those forms of energy are usually referred to as **thermal energy** to prevent any confusion with *heat transfer.*

The term *heat* and the associated phrases such as *heat flow, heat addition, heat rejection, heat absorption, heat gain, heat loss, heat storage, heat generation, electrical heating, latent heat, body heat,* and *heat source* are in common use today, and the attempt to replace *heat* in these phrases by *thermal energy* had only limited success. These phrases are deeply rooted in our vocabulary and they are used by both the ordinary people and scientists without causing any misunderstanding. For example, the phrase *body heat* is understood to mean the *thermal energy content* of a body. Likewise, *heat flow* is understood to mean the *transfer of thermal energy,* not the flow of a fluid-like substance called *heat,* although the latter incorrect interpretation, based on the caloric theory, is the origin of this phrase. Also, the transfer of heat into a system is frequently referred to as *heat addition* and the transfer of heat out of a system as *heat rejection*.

Perhaps there are thermodynamic reasons for being so reluctant to replace *heat* by *thermal energy*: it takes less time and energy to say, write, and comprehend *heat* than it does *thermal energy*. Keeping in line with current practice, we will refer to the thermal energy as *heat* and the transfer of thermal energy as *heat flow* or *heat transfer*. The amount of heat transferred during the process is denoted by Q. The amount of heat transferred per unit time is called **heat transfer rate,** and is denoted by $\dot{Q}$. The overdot stands for the time derivative, or "per unit time." The heat transfer rate $\dot{Q}$ has the unit J/s, which is equivalent to W.

When the *rate* of heat transfer $\dot{Q}$ is available, then the total amount of heat transfer Q during a time interval Δt can be determined from

$$Q = \int_0^{\Delta t} \dot{Q} dt \qquad \text{(J)} \qquad (1\text{-}6)$$

provided that the variation of $\dot{Q}$ with time is known. For the special case of $\dot{Q} = $ constant, the equation above reduces to

$$Q = \dot{Q}\Delta t \qquad \text{(J)} \qquad (1\text{-}7)$$

The rate of heat transfer per unit area normal to the direction of heat transfer is called **heat flux,** and the average heat flux on a surface is expressed as (Fig. 1-12)

$$\dot{q} = \frac{\dot{Q}}{A} \qquad \text{(W/m}^2\text{)} \qquad (1\text{-}8)$$

where A is the heat transfer area normal to the direction of heat transfer. The unit of heat flux in English units is $\text{Btu/h} \cdot \text{ft}^2$. Note that heat flux may vary with time as well as position on a surface.

EXAMPLE 1-1   Heating of a Copper Ball

A 10-cm diameter copper ball is to be heated from 100°C to an average temperature of 150°C in 30 minutes (Fig. 1-13). Taking the average density and specific heat of copper in this temperature range to be $\rho = 8950 \text{ kg/m}^3$ and $C_p = 0.395$ kJ/kg $\cdot$ °C, respectively, determine (*a*) the total amount of heat transfer to the copper ball, (*b*) the average rate of heat transfer to the ball, and (*c*) the average heat flux.

Solution   The copper ball is to be heated from 100°C to 150°C. The total heat transfer, the average rate of heat transfer, and the average heat flux are to be determined.

Assumptions   Constant properties can be used for copper at the average temperature.

Properties   The average density and specific heat of copper are given to be $\rho = 8950 \text{ kg/m}^3$ and $C_p = 0.395$ kJ/kg $\cdot$ °C.

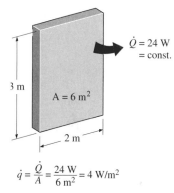

$$\dot{q} = \frac{\dot{Q}}{A} = \frac{24 \text{ W}}{6 \text{ m}^2} = 4 \text{ W/m}^2$$

FIGURE 1-12

Heat flux is heat transfer *per unit time* and *per unit area*, and is equal to $\dot{q} = \dot{Q}/A$ when $\dot{Q}$ is uniform over the area A.

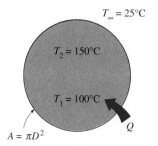

FIGURE 1-13

Schematic for Example 1-1.

Analysis (*a*) The amount of heat transferred to the copper ball is simply the change in its internal energy, and is determined from

Energy transfer to the system = Energy increase of the system

$$Q = \Delta U = mC_{ave} (T_2 - T_1)$$

where

$$m = \rho V = \frac{\pi}{6}\rho D^3 = \frac{\pi}{6}(8950 \text{ kg/m}^3)(0.1 \text{ m})^3 = 4.69 \text{ kg}$$

Substituting,

$$Q = (4.69 \text{ kg})(0.395 \text{ kJ/kg} \cdot {}°\text{C})(150 - 100){}°\text{C} = \textbf{92.6 kJ}$$

Therefore, 92.6 kJ of heat needs to be transferred to the copper ball to heat it from 100°C to 150°C.

(*b*) The rate of heat transfer normally changes during a process with time the same way the velocity of a car changes with time in normal driving. However, we can determine the *average* rate of heat transfer by dividing the total amount of heat transfer by the time interval, the same way we can determine the average velocity of a car by dividing the distance traveled by the driving time. Therefore,

$$\dot{Q}_{ave} = \frac{Q}{\Delta t} = \frac{92.6 \text{ kJ}}{1800 \text{ s}} = 0.0514 \text{ kJ/s} = \textbf{51.4 W}$$

(*c*) Heat flux is defined as the heat transfer per unit time per unit area, or the rate of heat transfer per unit area. Therefore, the average heat flux in this case is

$$\dot{q}_{ave} = \frac{\dot{Q}_{ave}}{A} = \frac{\dot{Q}_{ave}}{\pi D^2} = \frac{51.4 \text{ W}}{\pi(0.1 \text{ m})^2} = \textbf{1636 W/m}^2$$

Discussion Note that heat flux may vary with location on a surface. The value calculated above is the *average* heat flux over the entire surface of the ball.

1-4 ◾ THE FIRST LAW OF THERMODYNAMICS

The **first law of thermodynamics,** also known as the **conservation of energy principle,** states that *energy can neither be created nor destroyed; it can only change forms.* Therefore, every bit of energy must be accounted for during a process. The conservation of energy principle (or the energy balance) for *any system* undergoing *any process* may be expressed as follows: *The net change (increase or decrease) in the total energy of the system during a process is equal to the difference between the total energy entering and the total energy leaving the system during that process.* That is,

$$\begin{pmatrix} \text{Total energy} \\ \text{entering the} \\ \text{system} \end{pmatrix} - \begin{pmatrix} \text{Total energy} \\ \text{leaving the} \\ \text{system} \end{pmatrix} = \begin{pmatrix} \text{Change in the} \\ \text{total energy of} \\ \text{the system} \end{pmatrix} \quad (1\text{-}9)$$

Noting that energy can be transferred to or from a system by *heat, work,* and *mass flow,* and that the total energy of a simple compressible system consists of internal, kinetic, and potential energies, the **energy balance** for any system undergoing any process can be expressed more explicitly as

General: $\underbrace{E_{in} - E_{out}}_{\substack{\text{Net energy transfer} \\ \text{by heat, work, and mass}}}$ $=$ $\underbrace{\Delta E_{system}}_{\substack{\text{Change in internal, kinetic,} \\ \text{potential, etc., energies}}}$ (J)   (1-10)

or, in the **rate form,** as

$\underbrace{\dot{E}_{in} - \dot{E}_{out}}_{\substack{\text{Rate of net energy transfer} \\ \text{by heat, work, and mass}}}$ $=$ $\underbrace{dE_{system}/dt}_{\substack{\text{Rate of change in internal} \\ \text{kinetic, potential, etc., energies}}}$ (W)   (1-11)

Energy is a property, and the value of a property does not change unless the state of the system changes. Therefore, the energy change of a system is zero ($\Delta E_{system} = 0$) if the state of the system does not change during the process, that is, the process is steady. The energy balance in this case reduces to (Fig. 1-14)

Steady, rate form: $\underbrace{\dot{E}_{in}}_{\substack{\text{Rate of net energy transfer in} \\ \text{by heat, work, and mass}}}$ $=$ $\underbrace{\dot{E}_{out}}_{\substack{\text{Rate of net energy transfer out} \\ \text{by heat, work, and mass}}}$ (1-12)

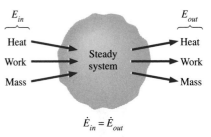

FIGURE 1-14

In steady operation, the rate of energy transfer to a system is equal to the rate of energy transfer from the system.

In the absence of significant electric, magnetic, motion, gravity, and surface tension effects (i.e., for stationary simple compressible systems), the change in the *total energy* of a system during a process is simply the change in its *internal energy.* That is, $\Delta E_{system} = \Delta U_{system}$.

In heat transfer analysis, we are usually interested only in the forms of energy that can be transferred as a result of a temperature difference, that is, heat or thermal energy. In such cases it is convenient to write a **heat balance** and to treat the conversion of nuclear, chemical, electrical, and so on, energies into thermal energy as *heat generation.* The *energy balance* in that case can be expressed as

$\underbrace{Q_{in} - Q_{out}}_{\substack{\text{Net heat} \\ \text{transfer}}}$ $+$ $\underbrace{E_{gen}}_{\substack{\text{Heat} \\ \text{generation}}}$ $=$ $\underbrace{\Delta E_{thermal, system}}_{\substack{\text{Change in thermal} \\ \text{energy of the system}}}$ (J)   (1-13)

Energy Balance for Closed Systems (*Fixed Mass*)

A closed system consists of a *fixed mass.* The total energy E for most systems encountered in practice consists of the internal energy U. This is especially the case for stationary systems since they don't involve any changes in their velocity or elevation during a process. The energy balance relation in that case reduces to

Stationary closed system: $E_{in} - E_{out} = \Delta U = mC_v\Delta T$ (J)   (1-14)

where we expressed the internal energy change in terms of mass m, the specific heat at constant volume C_v, and the temperature change ΔT of the system. When the system involves heat transfer only and no work interactions across its boundary, the energy balance relation further reduces to (Fig. 1-15)

$$\text{Stationary closed system, no work:} \qquad Q = mC_v\Delta T \qquad \text{(J)} \qquad (1\text{-}15)$$

where Q is the net amount of heat transfer to or from the system. This is the form of the energy balance relation we will use most often when dealing with a fixed mass.

Specific heat = C_v
Mass = m
Initial temp = T_1
Final temp = T_2

$$Q = mC_v(T_1 - T_2)$$

FIGURE 1-15

In the absence of any work interactions, the change in the energy content of a closed system is equal to the net heat transfer.

Energy Balance for Steady-Flow Systems

A large number of engineering devices such as water heaters and car radiators involve mass flow in and out of a system, and are modeled as *control volumes*. Most control volumes are analyzed under steady operating conditions. The term *steady* means *no change with time* at a specified location. The opposite of steady is *unsteady* or *transient*. Also, the term *uniform* implies *no change with position* throughout a surface or region at a specified time. These meanings are consistent with their everyday usage (steady girl-friend, uniform distribution, etc.). The total energy content of a control volume during a *steady-flow process* remains constant (E_{CV} = constant). That is, the change in the total energy of the control volume during such a process is zero ($\Delta E_{CV} = 0$). Thus the amount of energy entering a control volume in all forms (heat, work, mass transfer) for a steady-flow process must be equal to the amount of energy leaving it.

The amount of mass flowing through a cross-section of a flow device per unit time is called the **mass flow rate,** and is denoted by $\dot{m}$. A fluid flows in and out of a control volume through pipes or ducts. The mass flow rate of a fluid flowing in a pipe or duct is proportional to the cross-sectional area A_c of the pipe or duct, the density ρ, and the velocity $\mathcal{V}$ of the fluid. The mass flow rate through a differential area dA_c can be expressed as $\delta\dot{m} = \rho\mathcal{V}_n dA_c$ where $\mathcal{V}_n$ is the velocity component normal to dA_c. The mass flow rate through the entire cross-sectional area is obtained by integration over A_c.

The flow of a fluid through a pipe or duct can often be approximated to be *one-dimensional*. That is, the properties can be assumed to vary in one direction only (the direction of flow). As a result, all properties are assumed to be uniform at any cross-section normal to the flow direction, and the properties are assumed to have *bulk average values* over the entire cross-section. Under the one-dimensional flow approximation, the mass flow rate of a fluid flowing in a pipe or duct can be expressed as (Fig. 1-16)

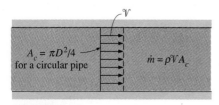

$A_c = \pi D^2/4$
for a circular pipe

$\dot{m} = \rho\mathcal{V}A_c$

FIGURE 1-16

The mass flow rate of a fluid at a cross-section is equal to the product of the fluid density, average fluid velocity, and the cross-sectional area.

$$\dot{m} = \rho\mathcal{V}A_c \qquad \text{(kg/s)} \qquad (1\text{-}16)$$

where ρ is the fluid density, $\mathcal{V}$ is the average fluid velocity in the flow direction, and A_c is the cross-sectional area of the pipe or duct.

The volume of a fluid flowing through a pipe or duct per unit time is called the **volume flow rate** $\dot{V}$, and is expressed as

$$\dot{V} = \mathcal{V}A_c = \frac{\dot{m}}{\rho} \qquad \text{(m}^3\text{/s)} \qquad (1\text{-}17)$$

Note that the mass flow rate of a fluid through a pipe or duct remains constant during steady flow. This is not the case for the volume flow rate, however, unless the density of the fluid remains constant.

For a steady-flow system with one inlet and one exit, the rate of mass flow into the control volume must be equal to the rate of mass flow out of it. That is, $\dot{m}_{in} = \dot{m}_{out} = \dot{m}$. When the changes in kinetic and potential energies are negligible, which is usually the case, and there is no work interaction, the energy balance for such a steady-flow system reduces to (Fig. 1-17)

$$\dot{Q} = \dot{m}\Delta h = \dot{m}C_p\Delta T \qquad \text{(kJ/s)} \qquad (1\text{-}18)$$

where $\dot{Q}$ is the rate of net heat transfer into or out of the control volume. This is the form of the energy balance relation we will use most often for steady-flow systems.

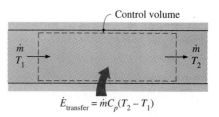

$$\dot{E}_{transfer} = \dot{m}C_p(T_2 - T_1)$$

FIGURE 1-17

Under steady conditions, the net rate of energy transfer to a fluid in a control volume is equal to the rate of increase in the energy of the fluid stream flowing through the control volume.

EXAMPLE 1-2   Heating of Water in an Electric Teapot

1.2 kg of liquid water initially at 15°C is to be heated to 95°C in a teapot equipped with a 1200 W electric heating element inside (Fig. 1-18). The teapot is 0.5 kg and has an average specific heat of 0.7 kJ/kg · °C. Taking the specific heat of water to be 4.18 kJ/kg · °C and disregarding any heat loss from the teapot, determine how long it will take for the water to be heated.

Solution   Liquid water is to be heated from 15°C to 95°C in an electric teapot. The heating time is to be determined.

Assumptions   **1** Heat loss from the teapot is negligible. **2** Constant properties can be used for both the teapot and the water.

Properties   The average specific heats are given to be 0.7 kJ/kg · °C for the teapot and 4.18 kJ/kg · °C for water.

Analysis   We take the teapot and the water in it as our system, which is a closed system (fixed mass). The energy balance in this case can be expressed as

$$E_{in} - E_{out} = \Delta E_{system}$$
$$E_{in} = \Delta U_{system} = \Delta U_{water} + \Delta U_{teapot}$$

Then the amount of energy needed to raise the temperature of water and the teapot from 15°C to 95°C is

$$E_{in} = (mC\Delta T)_{water} + (mC\Delta T)_{teapot}$$
$$= (1.2 \text{ kg})(4.18 \text{ kJ/kg} \cdot °\text{C})(95 - 15)°\text{C} + (0.5 \text{ kg})(0.7 \text{ kJ/kg} \cdot °\text{C})(95 - 15)°\text{C}$$
$$= 429.3 \text{ kJ}$$

The 1200 W electric heating unit will supply energy at a rate of 1.2 kW or 1.2 kJ per second. Therefore, the time needed for this heater to supply 429.3 kJ of heat is determined from

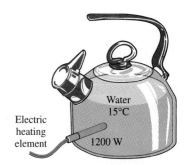

Water
15°C

Electric
heating
element

1200 W

FIGURE 1-18

Schematic for Example 1-2.

$$\Delta t = \frac{\text{Total energy transferred}}{\text{Rate of energy transfer}} = \frac{E_{in}}{\dot{E}_{transfer}} = \frac{429.3 \text{ kJ}}{1.2 \text{ kJ/s}} = 358 \text{ s} = \textbf{6.0 min}$$

Discussion   In reality, it will take more than 6 minutes to accomplish this heating process since some heat loss is inevitable during heating.

EXAMPLE 1-3   Heat Loss from Heating Ducts in a Basement

A 5-m-long section of an air heating system of a house passes through an un-heated space in the basement (Fig. 1-19). The cross-section of the rectangular duct of the heating system is 20 cm × 25 cm. Hot air enters the duct at 100 kPa and 60°C at an average velocity of 5 m/s. The temperature of the air in the duct drops to 54°C as a result of heat loss to the cool space in the basement. Deter-mine the rate of heat loss from the air in the duct to the basement under steady conditions. Also, determine the cost of this heat loss per hour if the house is heated by a natural gas furnace that has an efficiency of 80 percent, and the cost of the natural gas in that area is $0.60/therm (1 therm = 100,000 Btu = 105,500 kJ).

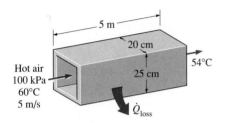

FIGURE 1-19
Schematic for Example 1-3.

Solution   The temperature of the air in the heating duct of a house drops from 60°C to 54°C as a result of heat losses to the cool space in the basement. The rate of heat loss from the hot air and its cost are to be determined.

Assumptions   **1** Steady operating conditions exist. **2** Air can be treated as an ideal gas with constant properties at room temperature.

Properties   The constant pressure specific heat of air at room temperature is 1.005 kJ/kg · °C (Table A-11).

Analysis   We take the basement section of the heating system as our system, which is a steady-flow system. The rate of heat loss from the air in the duct can be determined from

$$\dot{Q} = \dot{m}C_p\Delta T$$

where $\dot{m}$ is the mass flow rate and ΔT is the temperature drop. The density of air at the inlet conditions is determined to be

$$\rho = \frac{P}{RT} = \frac{100 \text{ kPa}}{(0.287 \text{ kPa} \cdot \text{m}^3/\text{kg} \cdot \text{K})(60 + 273)\text{K}} = 1.046 \text{ kg/m}^3$$

The cross-sectional area of the duct is

$$A_c = (0.20 \text{ m})(0.25 \text{ m}) = 0.05 \text{ m}^2$$

Then the mass flow rate of air through the duct and the rate of heat loss become

$$\dot{m} = \rho \mathcal{V} A_c = (1.046 \text{ kg/m}^3)(5 \text{ m/s})(0.05 \text{ m}^2) = 0.2615 \text{ kg/s}$$

and

$$\dot{Q}_{loss} = \dot{m}C_p(T_{in} - T_{out})$$
$$= (0.2615 \text{ kg/s})(1.005 \text{ kJ/kg} \cdot °\text{C})(60 - 54)°\text{C}$$
$$= \textbf{1.577 kJ/s}$$

or 5677 kJ/h. The cost of this heat loss to the home owner is

$$\text{Cost of heat loss} = \frac{(\text{Rate of heat loss})(\text{Unit cost of energy input})}{\text{Furnace efficiency}}$$

$$= \frac{(5677 \text{ kJ/h})(\$0.60/\text{therm})}{0.80}\left(\frac{1 \text{ therm}}{105,500 \text{ kJ}}\right)$$

$$= \$0.040/\text{h}$$

Discussion The heat loss from the heating ducts in the basement is costing the home owner 4 cents per hour. Assuming the heater operates 2000 hours during a heating season, the annual cost of this heat loss adds up to $80. Most of this money can be saved by insulating the heating ducts in the unheated areas.

EXAMPLE 1-4   Electric Heating of a House at High Elevation

Consider a house that has a floor space of 2000 ft² and an average height of 9 ft at 5000 ft elevation where the standard atmospheric pressure is 12.2 psia (Fig. 1-20). Initially the house is at a uniform temperature of 50°F. Now the electric heater is turned on, and the heater runs until the air temperature in the house rises to an average value of 70°F. Determine the amount of energy transferred to the air assuming (a) the house is air-tight and thus no air escapes during the heating process and (b) some air escapes through the cracks as the heated air in the house expands at constant pressure. Also determine the cost of this heat for each case if the cost of electricity in that area is $0.075/kWh.

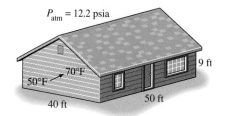

FIGURE 1-20

Schematic for Example 1-4.

Solution   The air in the house is heated from 50°F to 70°F by an electric heater. The amount and cost of the energy transferred to the air are to be determined for constant-volume and constant-pressure cases.

Assumptions   1 Air can be treated as an ideal gas with constant properties at room temperature. 2 Heat loss from the house during heating is negligible. 3 The volume occupied by the furniture and other belongings is negligible.

Properties   The specific heats of air at room temperature are $C_p = 0.240$ Btu/lbm · °F and $C_v = C_p - R = 0.171$ Btu/lbm · °F (Tables A-1E and A-11E).

Analysis   The volume and the mass of the air in the house are

$$V = (\text{Floor area})(\text{Height}) = (2000 \text{ ft}^2)(9 \text{ ft}) = 18,000 \text{ ft}^3$$

$$m = \frac{PV}{RT} = \frac{(12.2 \text{ psia})(18,000 \text{ ft}^3)}{(0.3704 \text{ psia} \cdot \text{ft}^3/\text{lbm} \cdot \text{R})(50 + 460)\text{R}} = 1162 \text{ lbm}$$

(a) The amount of energy transferred to air at constant volume is simply the change in its internal energy, and is determined from

$$E_{in} - E_{out} = \Delta E_{system}$$

$$E_{in, \text{ constant volume}} = \Delta U_{air} = mC_v\Delta T$$

$$= (1162 \text{ lbm})(0.171 \text{ Btu/lbm} \cdot \text{°F})(70 - 50)\text{°F}$$

$$= 3974 \text{ Btu}$$

At a unit cost of $0.075/kWh, the total cost of this energy is

Cost of energy = (Amount of energy)(Unit cost of energy)

$$= (3974 \text{ Btu})(\$0.075/\text{kWh})\left(\frac{1 \text{ kWh}}{3412 \text{ Btu}}\right)$$

$$= \mathbf{\$0.087}$$

(*b*) The amount of energy transferred to air at constant pressure is the change in its enthalpy, and is determined from

$$E_{in, \text{ constant pressure}} = \Delta H_{air} = mC_p \Delta T$$

$$= (1162 \text{ lbm})(0.240 \text{ Btu/lbm} \cdot °F)(70 - 50)°F$$

$$= \mathbf{5578 \text{ Btu}}$$

At a unit cost of $0.075/kWh, the total cost of this energy is

Cost of energy = (Amount of energy)(Unit cost of energy)

$$= (5578 \text{ Btu})(\$0.075/\text{kWh})\left(\frac{1 \text{ kWh}}{3412 \text{ Btu}}\right)$$

$$= \mathbf{\$0.123}$$

Discussion   It will cost about 12 cents to raise the temperature of the air in this house from 50°F to 70°F. The second answer is more realistic since every house has cracks, especially around the doors and windows, and the pressure in the house remains essentially constant during a heating process. Therefore, the second approach is used in practice. This conservative approach somewhat overpredicts the amount of energy used, however, since some of the air will escape through the cracks before it is heated to 70°F.

1-5 ■ HEAT TRANSFER MECHANISMS

Heat can be transferred in three different ways: *conduction, convection,* and *radiation.* All modes of heat transfer require the existence of a temperature difference, and all modes of heat transfer are from the high temperature medium to a lower temperature one. Below we give a brief description of each mode.

Conduction

Conduction is the transfer of energy from the more energetic particles of a substance to the adjacent less energetic ones as a result of interactions between the particles. Conduction can take place in solids, liquids, or gases. In *gases* and *liquids,* conduction is due to the *collisions* and *diffusion* of the molecules during their random motion. In *solids,* it is due to the combination of *vibrations* of the molecules in a lattice and the energy transport by *free electrons.* A cold canned drink in a warm room, for example, eventually warms up to the room temperature as a result of heat transfer from the room to the drink through the aluminum can by conduction.

The *rate* of heat conduction through a medium depends on the *geometry* of the medium, its *thickness,* and the *material* of the medium, as well as the temperature *difference* across the medium. We know that wrapping a hot water tank with glass wool (an insulating material) reduces the rate of heat loss from the tank. The thicker the insulation, the smaller the heat loss. We also know that a hot water tank will lose heat at a higher rate when the temperature of the room housing the tank is lowered. Further, the larger the tank, the larger the surface area and thus the rate of heat loss.

Consider steady heat conduction through a large plane wall of thickness $\Delta x = L$ and surface area A, as shown in Figure 1-21. The temperature difference across the wall is $\Delta T = T_2 - T_1$. Experiments have shown that the rate of heat transfer $\dot{Q}$ through the wall is *doubled* when the temperature difference ΔT across the wall or the area A normal to the direction of heat transfer is doubled, but is *halved* when the wall thickness L is doubled. Thus we conclude that *the rate of heat conduction through a plane layer is proportional to the temperature difference across the layer and the heat transfer area, but is inversely proportional to the thickness of the layer.* That is,

$$\text{Rate of heat conduction} \propto \frac{(\text{Area})(\text{Temperature difference})}{\text{Thickness}} \qquad (1\text{-}19)$$

or,

$$\dot{Q}_{\text{cond}} = kA\frac{\Delta T}{\Delta x} \qquad (\text{W}) \qquad (1\text{-}20)$$

where the constant of proportionality k is the **thermal conductivity** of the material, which is a *measure of the ability of a material to conduct heat* (Fig. 1-22). In the limiting case of $\Delta x \rightarrow 0$, the equation above reduces to the differential form

$$\dot{Q}_{\text{cond}} = -kA\frac{dT}{dx} \qquad (\text{W}) \qquad (1\text{-}21)$$

which is called **Fourier's law of heat conduction** after J. Fourier, who expressed it first in his heat transfer text in 1822. Here dT/dx is the **temperature gradient,** which is the slope of the temperature curve on a *T-x* diagram (the rate of change of T with x), at location x. The relation above indicates that the rate of heat conduction in a direction is proportional to the temperature gradient in that direction. Heat is conducted in the direction of decreasing temperature, and the temperature gradient becomes negative when temperature decreases with increasing x. Therefore, a *negative sign* is added to Equation 1-21 to make heat transfer in the positive x direction a positive quantity.

The heat transfer area A is always *normal* to the direction of heat transfer. For heat loss through a 5-m-long, 3-m-high, and 25-cm-thick wall, for example, the heat transfer area is $A = 15$ m². Note that the thickness of the wall has no effect on A (Fig. 1-23).

EXAMPLE 1-5   The Cost of Heat Loss through the Roof

The roof of an electrically heated home is 6 m long, 8 m wide, and 0.25 m thick, and is made of a flat layer of concrete whose thermal conductivity is

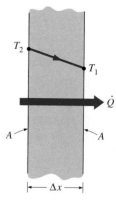

FIGURE 1-21

Heat conduction through a large plane wall of thickness Δx and area A.

FIGURE 1-22

The rate of heat conduction through a solid is directly proportional to its thermal conductivity.

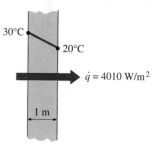

(a) Copper ($k = 401$ W/m·°C)

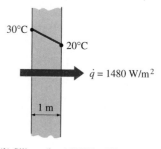

(b) Silicon ($k = 148$ W/m·°C)

$k = 0.8$ W/m · °C (Fig. 1-24). On a certain winter night, the temperatures of the inner and the outer surfaces of the roof are measured to be about 15°C and 4°C, respectively, for a period of 10 hours. Determine (a) the rate of heat loss through the roof that night and (b) the cost of that heat loss to the home owner if the cost of electricity is $0.08/kWh.

Solution The inner and outer surfaces of the flat concrete roof of an electrically heated home are maintained at 15°C and 4°C, respectively, during a night. The heat loss through the roof and its cost that night are to be determined.

Assumptions **1** Steady operating conditions exist during the entire night since the surface temperatures of the roof remain constant at the specified values. **2** Constant properties can be used for the roof.

Properties The thermal conductivity of the roof is given to be $k = 0.8$ W/m · °C.

Analysis (a) Noting that heat transfer through the roof is by conduction and the area of the roof is $A = 6$ m $\times$ 8 m $= 48$ m², the steady rate of heat transfer through the roof can be determined from Equation 1-20 to be

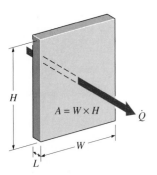

FIGURE 1-23

In heat conduction analysis, A represents the area *normal* to the direction of heat transfer.

$$\dot{Q} = kA\frac{T_1 - T_2}{L} = (0.8 \text{ W/m} \cdot \text{°C})(48 \text{ m}^2)\frac{(15 - 4)\text{°C}}{0.25 \text{ m}} = \textbf{1690 W} = \textbf{1.690 kW}$$

(b) The amount of heat lost through the roof during a 10-hour period and its cost are determined from

$$Q = \dot{Q}\Delta t = (1.690 \text{ kW})(10 \text{ h}) = 16.90 \text{ kWh}$$
$$\text{Cost} = (\text{Amount of energy})(\text{Unit cost of energy})$$
$$= (16.90 \text{ kWh})(\$0.08/\text{kWh}) = \textbf{\$1.35}$$

Discussion The cost to the home owner of the heat loss through the roof that night was $1.35. The total heating bill of the house will be much larger since the heat losses through the walls are not considered in the above calculations.

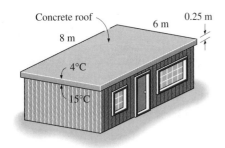

FIGURE 1-24

Schematic for Example 1-5.

Thermal Conductivity

We have seen in Section 1-3 that different materials store heat differently, and we have defined the property specific heat C_p as a measure of a material's ability to store heat. For example, $C_p = 4.18$ kJ/kg · °C for water and $C_p = 0.45$ kJ/kg · °C for iron at room temperature, which indicates that water can store almost 10 times the energy that iron can per unit mass. Likewise, the thermal conductivity k is a measure of a material's ability to conduct heat. For example, $k = 0.608$ W/m · °C for water and $k = 80.2$ W/m · °C for iron at room temperature, which indicates that iron conducts heat more than 100 times faster than water can. Thus we say that water is a poor heat conductor relative to iron, although water is an excellent medium to store heat.

Equation 1-20 for the rate of conduction heat transfer under steady conditions can also be viewed as the defining equation for thermal conductivity. Thus the **thermal conductivity** of a material can be defined as *the rate of* *heat transfer through a unit thickness of the material per unit area per unit temperature difference.* The thermal conductivity of a material is a measure of how fast heat will flow in that material. A large value for thermal conductivity indicates that the material is a good heat conductor, and a low value indicates that the material is a poor heat conductor or *insulator.* The thermal conductivities of some common materials at room temperature are given in Table 1-1. The thermal conductivity of pure copper at room temperature is $k = 401$ W/m · °C, which indicates that a 1-m-thick copper wall will conduct heat at a rate of 401 W per m^2 area per °C temperature difference across the wall. Note that materials such as copper and silver that are good electric conductors are also good heat conductors, and have high values of thermal conductivity. Materials such as rubber, wood, and Styrofoam are poor conductors of heat and have low conductivity values.

A layer of material of known thickness and area can be heated from one side by an electric resistance heater of known output. If the outer surfaces of the heater are well insulated, all the heat generated by the resistance heater will be transferred through the material whose conductivity is to be determined. Then measuring the two surface temperatures of the material when steady heat transfer is reached and substituting them into Equation 1-20 together with other known quantities give the thermal conductivity (Fig. 1-25).

The thermal conductivities of materials vary over a wide range, as shown in Figure 1-26. The thermal conductivities of gases such as air vary by a factor of 10^4 from those of pure metals such as copper. Note that pure crystals and metals have the highest thermal conductivities, and gases and insulating materials the lowest.

Temperature is a measure of the kinetic energies of the particles such as the molecules or atoms of a substance. In a liquid or gas, the kinetic energy of the molecules is due to their random translational motion as well as their vibrational and rotational motions. When two molecules possessing different kinetic energies collide, part of the kinetic energy of the more energetic (higher-temperature) molecule is transferred to the less energetic (lower-temperature) molecule, much the same as when two elastic balls of the same mass at different velocities collide, part of the kinetic energy of the faster ball is transferred to the slower one. The higher the temperature, the faster the molecules move and the higher the number of such collisions, and the better the heat transfer.

The *kinetic theory* of gases predicts and the experiments confirm that the thermal conductivity of gases is proportional to the *square root of the absolute temperature T,* and inversely proportional to the *square root of the molar mass M.* Therefore, the thermal conductivity of a gas increases with increasing temperature and decreasing molar mass. So it is not surprising that the thermal conductivity of helium ($M = 4$) is much higher than those of air ($M = 29$) and argon ($M = 40$).

The thermal conductivities of *gases* at 1 atm pressure are listed in Table A-11. However, they can also be used at pressures other than 1 atm, since the thermal conductivity of gases is *independent of pressure* in a wide range of pressures encountered in practice.

TABLE 1-1

The thermal conductivities of some materials at room temperature

Material	k, W/m · °C*
Diamond	2300
Silver	429
Copper	401
Gold	317
Aluminum	237
Iron	80.2
Mercury (l)	8.54
Glass	0.78
Brick	0.72
Water (l)	0.613
Human skin	0.37
Wood (oak)	0.17
Helium (g)	0.152
Soft rubber	0.13
Refrigerant-12	0.072
Glass fiber	0.043
Air (g)	0.026
Urethane, rigid foam	0.026

*Multiply by 0.5778 to convert to Btu/h · ft · °F.

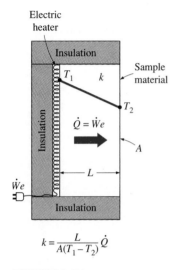

Electric
heater

$$k = \frac{L}{A(T_1 - T_2)} \dot{Q}$$

FIGURE 1-25

A simple experimental setup to
determine the thermal conductivity
of a material.

The mechanism of heat conduction in a *liquid* is complicated by the fact that the molecules are more closely spaced, and they exert a stronger intermolecular force field. The thermal conductivities of liquids usually lie between those of solids and gases. The thermal conductivity of a substance is normally highest in the solid phase and lowest in the gas phase. Unlike gases, the thermal conductivities of most liquids decrease with increasing temperature, with water being a notable exception. Like gases, the conductivity of liquids decreases with increasing molar mass. Liquid metals such as mercury and sodium have high thermal conductivities and are very suitable for use in applications where a high heat transfer rate to a liquid is desired, as in nuclear power plants.

In *solids,* heat conduction is due to two effects: the *lattice vibrational waves* induced by the vibrational motions of the molecules positioned at relatively fixed positions in a periodic manner called a lattice, and the energy transported via the *free flow of electrons* in the solid (Fig. 1-27). The thermal conductivity of a solid is obtained by adding the lattice and electronic components. The relatively high thermal conductivities of pure metals are primarily due to the electronic component. The lattice component of thermal conductivity strongly depends on the way the molecules are arranged. For example, diamond, which is a highly ordered crystalline solid, has the highest known thermal conductivity at room temperature.

Unlike metals, which are good electrical and heat conductors, *crystalline solids* such as diamond and semiconductors such as silicon are good heat conductors but poor electrical conductors. As a result, such materials find widespread use in the electronics industry. Despite their higher price, diamond heat sinks are used in the cooling of sensitive electronic components because of the excellent thermal conductivity of diamond. Silicon oils and gaskets are commonly used in the packaging of electronic components because they provide both good thermal contact and good electrical insulation.

Pure metals have high thermal conductivities, and one would think that *metal alloys* should also have high conductivities. One would expect an alloy made of two metals of thermal conductivities k_1 and k_2 to have a conductivity k between k_1 and k_2. But this turns out not to be the case. The thermal conductivity of an alloy of two metals is usually much lower than that of either metal, as shown in Table 1-2. Even small amounts in a pure metal of "foreign" molecules that are good conductors themselves seriously disrupt the flow of heat in that metal. For example, the thermal conductivity of steel containing just 1 percent of chrome is 62 W/m · °C, while the thermal conductivities of iron and chromium are 83 and 95 W/m · °C, respectively.

The thermal conductivities of materials vary with temperature (Table 1-3). The variation of thermal conductivity over certain temperature ranges is negligible for some materials, but significant for others, as shown in Figure 1-28. The thermal conductivities of certain solids exhibit dramatic increases at temperatures near absolute zero, when these solids become *superconductors.* For example, the conductivity of copper reaches a maximum value of about 20,000 W/m · °C at 20 K, which is about 50 times the conductivity at room temperature. The thermal conductivities and other thermal properties of various materials are given in Tables A-3 to A-12.

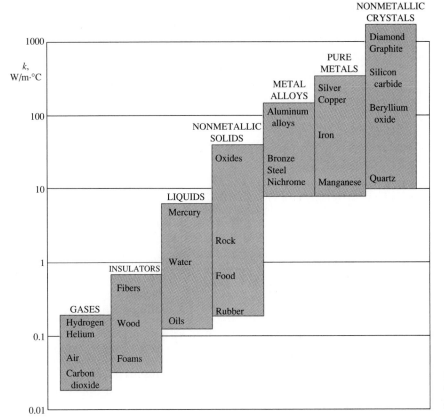

FIGURE 1-26
The range of thermal conductivity of
various materials at room temperature.

The temperature dependence of thermal conductivity causes considerable complexity in conduction analysis. Therefore, it is common practice to evaluate the thermal conductivity k at the *average temperature* and treat it as a *constant* in calculations.

In heat transfer analysis, a material is normally assumed to be *isotropic;* that is, to have uniform properties in all directions. This assumption is realistic for most materials, except those that exhibit different structural characteristics in different directions, such as laminated composite materials and wood. The thermal conductivity of wood across the grain, for example, is different than that parallel to the grain.

Thermal Diffusivity

The product ρC_p, which is frequently encountered in heat transfer analysis, is called the **heat capacity** of a material. Both the specific heat C_p and the heat capacity ρC_p represent the heat storage capability of a material. But C_p expresses it *per unit mass* whereas ρC_p expresses it *per unit volume,* as can be noticed from their units J/kg · °C and J/m^3 · °C, respectively.

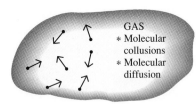

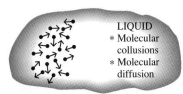

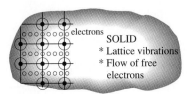

FIGURE 1-27

The mechanisms of heat conduction in different phases of a substance.

TABLE 1-2

The thermal conductivity of an alloy is usually much lower than the thermal conductivity of either metal of which it is composed

Pure metal or alloy	k, W/m · °C, at 300 K
Copper	401
Nickel	91
Constantan (55% Cu, 45% Ni)	23
Copper	401
Aluminum	237
Commercial bronze (90% Cu, 10% Al)	52

Another material property that appears in the transient heat conduction analysis is the **thermal diffusivity,** which represents how fast heat diffuses through a material and is defined as

$$\alpha = \frac{\text{Heat conducted}}{\text{Heat stored}} = \frac{k}{\rho C_p} \quad (\text{m}^2/\text{s}) \qquad (1\text{-}22)$$

Note that the thermal conductivity k represents how well a material conducts heat, and the heat capacity ρC_p represents how much energy a material stores per unit volume. Therefore, the thermal diffusivity of a material can be viewed as the ratio of the *heat conducted* through the material to the *heat stored* per unit volume. A material that has a high thermal conductivity or a low heat capacity will obviously have a large thermal diffusivity. The larger the thermal diffusivity, the faster the propagation of heat into the medium. A small value of thermal diffusivity means that heat is mostly absorbed by the material and a small amount of heat will be conducted further.

The thermal diffusivities of some common materials at 20°C are given in Table 1-4. Note that the thermal diffusivity ranges from $\alpha = 0.14 \times 10^{-6}$ m²/s for water to 174×10^{-6} m²/s for silver, which is a difference of more than a thousand times. Also note that the thermal diffusivities of beef and water are the same. This is not surprising, since meat as well as fresh vegetables and fruits are mostly water, and thus they possess the thermal properties of water.

EXAMPLE 1-6   Measuring the Thermal Conductivity of a Material

A common way of measuring the thermal conductivity of a material is to sandwich an electric thermofoil heater between two identical samples of the material, as shown in Figure 1-29. The thickness of the resistance heater, including its cover, which is made of thin silicon rubber, is usually less than 0.5 mm. A circulating fluid such as tap water keeps the exposed ends of the samples at constant temperature. The lateral surfaces of the samples are well insulated to ensure that heat transfer through the samples is one-dimensional. Two thermocouples are embedded into each sample some distance L apart, and a differential thermometer reads the temperature drop ΔT across this distance along each sample. When steady operating conditions are reached, the total rate of heat transfer through both samples becomes equal to the electric power drawn by the heater, which is determined by multiplying the electric current by the voltage.

In a certain experiment, cylindrical samples of diameter 5 cm and length 10 cm are used. The two thermocouples in each sample are placed 3 cm apart. After initial transients, the electric heater is observed to draw 0.4 A at 110 V, and both differential thermometers read a temperature difference of 15°C. Determine the thermal conductivity of the sample.

Solution   The thermal conductivity of a material is to be determined by ensuring one-dimensional heat conduction, and by measuring temperatures when steady operating conditions are reached.

Assumptions   **1** Steady operating conditions exist since the temperature readings do not change with time. **2** Heat losses through the lateral surfaces of the apparatus are negligible since those surfaces are well insulated, and thus the entire heat generated by the heater is conducted through the samples. **3** The apparatus possesses thermal symmetry.

Analysis The electrical power consumed by the resistance heater and converted to heat is

$$\dot{W}_e = VI = (110 \text{ V})(0.4 \text{ A}) = 44 \text{ W}$$

The rate of heat flow through each sample is

$$\dot{Q} = \tfrac{1}{2}\dot{W}_e = \tfrac{1}{2} \times 44 \text{ W} = 22 \text{ W}$$

since only half of the heat generated will flow through each sample because of symmetry. Reading the same temperature difference across the same distance in each sample also confirms that the apparatus possesses thermal symmetry. The heat transfer area is the area normal to the direction of heat flow, which is the cross-sectional area of the cylinder in this case:

$$A = \tfrac{1}{4}\pi D^2 = \tfrac{1}{4}\pi(0.05 \text{ m})^2 = 0.00196 \text{ m}^2$$

Noting that the temperature drops by 15°C within 3 cm in the direction of heat flow, the thermal conductivity of the sample is determined to be

$$\dot{Q} = kA\frac{\Delta T}{L} \rightarrow k = \frac{\dot{Q}L}{A\Delta T} = \frac{(22 \text{ W})(0.03 \text{ m})}{(0.00196 \text{ m}^2)(15°C)} = \textbf{22.4 W/m} \cdot \textbf{°C}$$

Discussion Perhaps you are wondering if we really need to use two samples in the apparatus, since the measurements on the second sample do not give any

TABLE 1-3

**Thermal conductivities of materials
vary with temperature**

T, K	Copper	Aluminum
100	482	302
200	413	237
300	401	237
400	393	240
600	379	231
800	366	218

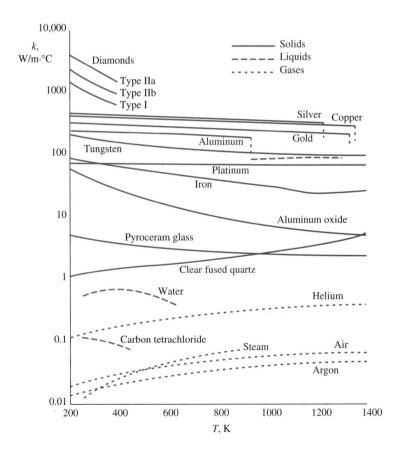

FIGURE 1-28

The variation of the thermal conductivity of various solids, liquids, and gases with temperature.

TABLE 1-4

The thermal diffusivities of some materials at room temperature

Material	α, m²/s*
Silver	149×10^{-6}
Gold	127×10^{-6}
Copper	113×10^{-6}
Aluminum	97.5×10^{-6}
Iron	22.8×10^{-6}
Mercury (l)	4.7×10^{-6}
Marble	1.2×10^{-6}
Ice	1.2×10^{-6}
Concrete	0.75×10^{-6}
Brick	0.52×10^{-6}
Heavy soil (dry)	0.52×10^{-6}
Glass	0.34×10^{-6}
Glass wool	0.23×10^{-6}
Water (l)	0.14×10^{-6}
Beef	0.14×10^{-6}
Wood (oak)	0.13×10^{-6}

*Multiply by 10.76 to convert to ft²/s.

FIGURE 1-29

Apparatus to measure the thermal conductivity of a material using two identical samples and a thin resistance heater (Example 1-6).

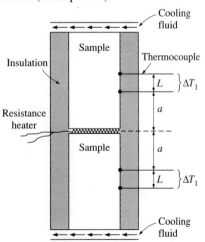

additional information. It seems like we can replace the second sample by insulation. Indeed, we do not need the second sample; however, it enables us to verify the temperature measurements on the first sample and provides thermal symmetry, which reduces experimental error.

EXAMPLE 1-7   Conversion between SI and English Units

An engineer who is working on the heat transfer analysis of a brick building in English units needs the thermal conductivity of brick. But the only value he can find from his handbooks is 0.72 W/m · °C, which is in SI units. To make matters worse, the engineer does not have a direct conversion factor between the two unit systems for thermal conductivity (he should have kept his heat transfer textbook instead of selling it back to the bookstore). Can you help him out?

Solution   The situation this engineer is facing is not unique, and most engineers often find themselves in a similar position. A person must be very careful during unit conversion in order not to fall into some common pitfalls and to avoid some costly mistakes. Although unit conversion is a simple process, it requires utmost care and careful reasoning.

The conversion factors for W and m are straightforward and are given in conversion tables to be

$$1 \text{ W} = 3.41214 \text{ Btu/h}$$
$$1 \text{ m} = 3.2808 \text{ ft}$$

But the conversion of °C into °F is not so simple, and it can be a source of error if one is not careful. Perhaps the first thought that comes to mind is to replace °C by (°F − 32)/1.8 since $T(°C) = [T(°F) − 32]/1.8$. But this will be wrong since the °C in the unit W/m · °C represents *per °C change in temperature*. Noting that 1°C change in temperature corresponds to 1.8°F, the proper conversion factor to be used is

$$1°C = 1.8°F$$

Substituting, we get

$$1 \text{ W/m} \cdot °C = \frac{3.41214 \text{ Btu/h}}{(3.2808 \text{ ft})(1.8°F)} = 0.5778 \text{ Btu/h} \cdot \text{ft} \cdot °F$$

which is the desired conversion factor. Therefore, the thermal conductivity of the brick in English units is

$$k_{\text{brick}} = 0.72 \text{ W/m} \cdot °C$$
$$= 0.72 \times 0.5778 \text{ Btu/h} \cdot \text{ft} \cdot °F$$
$$= \textbf{0.42 Btu/h} \cdot \textbf{ft} \cdot \textbf{°F}$$

Discussion   Note that the thermal conductivity value of a material in English units is about half that in SI units (Fig. 1-30). Also note that we rounded the result to two significant digits (the same number in the original value) since expressing the result in more significant digits (such as 0.4160 instead of 0.42) would falsely imply a more accurate value than the original one.

Convection

Convection is the mode of energy transfer between a solid surface and the adjacent liquid or gas that is in motion, and it involves the combined effects of *conduction* and *fluid motion*. The faster the fluid motion, the greater the convection heat transfer. In the absence of any bulk fluid motion, heat transfer between a solid surface and the adjacent fluid is by pure conduction. The presence of bulk motion of the fluid enhances the heat transfer between the solid surface and the fluid, but it also complicates the determination of heat transfer rates.

Consider the cooling of a hot block by blowing cool air over its top surface (Fig. 1-31). Energy is first transferred to the air layer adjacent to the block by conduction. This energy is then carried away from the surface by convection; that is, by the combined effects of conduction within the air that is due to random motion of air molecules and the bulk or macroscopic motion of the air that removes the heated air near the surface and replaces it by the cooler air.

Convection is called **forced convection** if the fluid is forced to flow over the surface by external means such as a fan, pump, or the wind. In contrast, convection is called **natural** (or **free**) **convection** if the fluid motion is caused by buoyancy forces that are induced by density differences due to the variation of temperature in the fluid (Fig. 1-32). For example, in the absence of a fan, heat transfer from the surface of the hot block in Figure 1-31 will be by natural convection since any motion in the air in this case will be due to the rise of the warmer (and thus lighter) air near the surface and the fall of the cooler (and thus heavier) air to fill its place. Heat transfer between the block and the surrounding air will be by conduction if the temperature difference between the air and the block is not large enough to overcome the resistance of air to movement and thus to initiate natural convection currents.

Heat transfer processes that involve *change of phase* of a fluid are also considered to be convection because of the fluid motion induced during the process, such as the rise of the vapor bubbles during boiling or the fall of the liquid droplets during condensation.

Despite the complexity of convection, the rate of *convection heat transfer* is observed to be proportional to the temperature difference, and is conveniently expressed by **Newton's law of cooling** as

$$\dot{Q}_{convection} = hA\,(T_s - T_\infty) \qquad (\text{W}) \qquad (1\text{-}23)$$

where h is the *convection heat transfer coefficient* in $\text{W/m}^2 \cdot \text{°C}$ or $\text{Btu/h} \cdot \text{ft}^2 \cdot \text{°F}$, A is the surface area through which convection heat transfer takes place, T_s is the surface temperature, and T_∞ is the temperature of the fluid sufficiently far from the surface. Note that at the surface, the fluid temperature equals the surface temperature of the solid.

The convection heat transfer coefficient h is not a property of the fluid. It is an experimentally determined parameter whose value depends on all the variables influencing convection such as the surface geometry, the nature of fluid motion, the properties of the fluid, and the bulk fluid velocity. Typical values of h are given in Table 1-5.

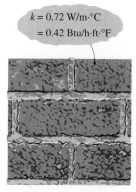

$k = 0.72 \text{ W/m·°C}$
$= 0.42 \text{ Btu/h·ft·°F}$

FIGURE 1-30
The thermal conductivity value in English units is obtained by multiplying the value in SI units by 0.5778.

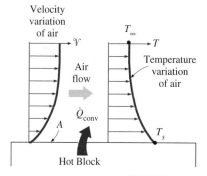

FIGURE 1-31
Heat transfer from a hot surface to air by convection.

Some people do not consider convection to be a fundamental mechanism of heat transfer since it is essentially heat conduction in the presence of fluid motion. But we still need to give this combined phenomenon a name, unless we are willing to keep referring to it as "conduction with fluid motion." Thus, it is practical to recognize convection as a separate heat transfer mechanism despite the valid arguments to the contrary.

EXAMPLE 1-8 Measuring Convection Heat Transfer Coefficient

A 2-m-long, 0.3-cm-diameter electrical wire extends across a room at 15°C, as shown in Figure 1-33. Heat is generated in the wire as a result of resistance heating, and the surface temperature of the wire is measured to be 152°C in steady operation. Also, the voltage drop and electric current through the wire are measured to be 60 V and 1.5 A, respectively. Disregarding any heat transfer by radiation, determine the convection heat transfer coefficient for heat transfer between the outer surface of the wire and the air in the room.

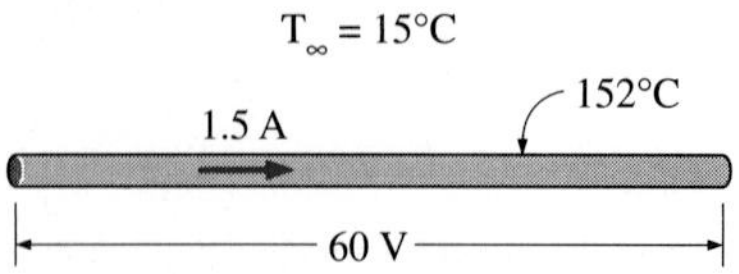

Solution The convection heat transfer coefficient for heat transfer from an electrically heated wire to air is to be determined by measuring temperatures when steady operating conditions are reached and the electric power consumed.

Assumptions 1 Steady operating conditions exist since the temperature readings do not change with time. 2 Radiation heat transfer is negligible.

Analysis When steady operating conditions are reached, the rate of heat loss from the wire will equal the rate of heat generation in the wire as a result of resistance heating. That is,

$$\dot{Q} = \dot{E}_{generated} = VI = (60 \text{ V})(1.5 \text{ A}) = 90 \text{ W}$$

The surface area of the wire is

$$A = (\pi D)L = \pi(0.003 \text{ m})(2 \text{ m}) = 0.01885 \text{ m}^2$$

Newton's law of cooling for convection heat transfer is expressed as

$$\dot{Q} = hA(T_s - T_\infty)$$

Disregarding any heat transfer by radiation and thus assuming all the heat loss from the wire to occur by convection, the convection heat transfer coefficient is determined to be

$$h = \frac{\dot{Q}}{A(T_s - T_\infty)} = \frac{90 \text{ W}}{(0.01885 \text{ m}^2)(152 - 15)°C} = \textbf{34.9 W/m}^2 \cdot \textbf{°C}$$

Discussion Note that the simple setup described above can be used to determine the average heat transfer coefficients from a variety of surfaces in air. Also, heat transfer by radiation can be eliminated by keeping the surrounding surfaces at the temperature of the wire.

Radiation

Radiation is the energy emitted by matter in the form of *electromagnetic waves* (or *photons*) as a result of the changes in the electronic configurations

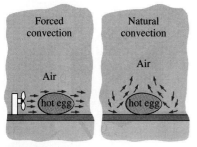

FIGURE 1-32

The cooling of a boiled egg by forced and natural convection.

TABLE 1-5

Typical values of convection heat transfer coefficient

Type of convection	h, W/m$^2 \cdot$ °C*
Free convection of gases	2–25
Free convection of liquids	10–1000
Forced convection of gases	25–250
Forced convection of liquids	50–20,000
Boiling and condensation	2500–100,000

*Multiply by 0.176 to convert to Btu/h · ft^2 · °F.

FIGURE 1-33

Schematic for Example 1-8.

$T_\infty = 15°C$

1.5 A   152°C

60 V

of the atoms or molecules. Unlike conduction and convection, the transfer of energy by radiation does not require the presence of an *intervening medium*. In fact, energy transfer by radiation is fastest (at the speed of light) and it suffers no attenuation in a vacuum. This is exactly how the energy of the sun reaches the earth.

In heat transfer studies we are interested in *thermal radiation*, which is the form of radiation emitted by bodies because of their temperature. It differs from other forms of electromagnetic radiation such as x-rays, gamma rays, microwaves, radio waves, and television waves that are not related to temperature. All bodies at a temperature above absolute zero emit thermal radiation.

Radiation is a *volumetric phenomenon*, and all solids, liquids, and gases emit, absorb, or transmit radiation to varying degrees. However, radiation is usually considered to be a *surface phenomenon* for solids that are opaque to thermal radiation such as metals, wood, and rocks since the radiation emitted by the interior regions of such material can never reach the surface, and the radiation incident on such bodies is usually absorbed within a few microns from the surface.

The maximum rate of radiation that can be emitted from a surface at an absolute temperature T_s (in K or R) is given by the **Stefan-Boltzmann law** as

$$\dot{Q}_{emit, max} = \sigma A T_s^4 \qquad \text{(W)} \qquad (1\text{-}24)$$

where $\sigma = 5.67 \times 10^{-8} \text{ W/m}^2 \cdot \text{K}^4$ or $0.1714 \times 10^{-8} \text{ Btu/h} \cdot \text{ft}^2 \cdot \text{R}^4$ is the *Stefan-Boltzmann constant*. The idealized surface that emits radiation at this maximum rate is called a **blackbody,** and the radiation emitted by a blackbody is called **blackbody radiation** (Fig. 1-34). The radiation emitted by all real surfaces is less than the radiation emitted by a blackbody at the same temperature, and is expressed as

$$\dot{Q}_{emit} = \varepsilon \sigma A T_s^4 \qquad \text{(W)} \qquad (1\text{-}25)$$

where ε is the *emissivity* of the surface. The property emissivity, whose value is in the range $0 \le \varepsilon \le 1$, is a measure of how closely a surface approximates a blackbody for which $\varepsilon = 1$. The emissivities of some surfaces are given in Table 1-6.

Another important radiation property of a surface is its *absorptivity* α, which is the fraction of the radiation energy incident on a surface that is absorbed by the surface. Like emissivity, its value is in the range $0 \le \alpha \le 1$. A blackbody absorbs the entire radiation incident on it. That is, a blackbody is a perfect absorber ($\alpha = 1$) as it is a perfect emitter.

In general, both ε and α of a surface depend on the temperature and the wavelength of the radiation. **Kirchhoff's law** of radiation states that the emissivity and the absorptivity of a surface are equal at the same temperature and wavelength. In most practical applications, the dependence of ε and α on the temperature and wavelength is ignored, and the average absorptivity of a surface is taken to be equal to its average emissivity. The rate at which a surface absorbs radiation is determined from (Fig. 1-35)

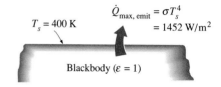

FIGURE 1-34

Blackbody radiation represents the *maximum amount of radiation that can be emitted from a surface at a specified temperature.*

TABLE 1-6

Emissivities of some materials at 300 K

Material	Emissivity
Aluminum foil	0.07
Anodized aluminum	0.82
Polished copper	0.03
Polished gold	0.03
Polished silver	0.02
Polished stainless steel	0.17
Black paint	0.98
White paint	0.90
White paper	0.92–0.97
Asphalt pavement	0.85–0.93
Red brick	0.93–0.96
Human skin	0.95
Wood	0.82–0.92
Soil	0.93–0.96
Water	0.96
Vegetation	0.92–0.96

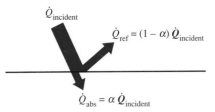

FIGURE 1-35

The absorption of radiation incident on an opaque surface of absorptivity α.

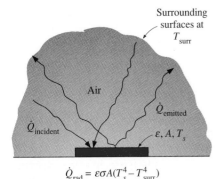

$$\dot{Q}_{rad} = \varepsilon\sigma A(T_s^4 - T_{surr}^4)$$

FIGURE 1-36

Radiation heat transfer between a surface and the surfaces surrounding it.

$$\dot{Q}_{absorbed} = \alpha\dot{Q}_{incident} \qquad (W) \qquad (1\text{-}26)$$

where $\dot{Q}_{incident}$ is the rate at which radiation is incident on the surface and α is the absorptivity of the surface. For opaque (nontransparent) surfaces, the portion of incident radiation not absorbed by the surface is reflected back.

The difference between the rates of radiation emitted by the surface and the radiation absorbed is the *net* radiation heat transfer. If the rate of radiation absorption is greater than the rate of radiation emission, the surface is said to be *gaining* energy by radiation. Otherwise, the surface is said to be *losing* energy by radiation. In general, the determination of the net rate of heat transfer by radiation between two surfaces is a complicated matter since it depends on the properties of the surfaces, their orientation relative to each other, and the interaction of the medium between the surfaces with radiation.

When a surface of emissivity ε and surface area A at an *absolute temperature* T_s is *completely enclosed* by a much larger (or black) surface at absolute temperature T_{surr} separated by a gas (such as air) that does not intervene with radiation, the net rate of radiation heat transfer between these two surfaces is given by (Fig. 1-36)

$$\dot{Q}_{rad} = \varepsilon\sigma A\ (T_s^4 - T_{surr}^4) \qquad (W) \qquad (1\text{-}27)$$

In this special case, the emissivity and the surface area of the surrounding surface do not have any effect on the net radiation heat transfer.

Radiation heat transfer to or from a surface surrounded by a gas such as air occurs *parallel* to conduction (or convection, if there is bulk gas motion) between the surface and the gas. Thus the total heat transfer is determined by *adding* the contributions of both heat transfer mechanisms. For simplicity and convenience, this is often done by defining a **combined heat transfer coefficient** $h_{combined}$ that includes the effects of both convection and radiation. Then the *total* heat transfer to or from a surface by convection and radiation is expressed as

$$\dot{Q}_{total} = h_{combined}A(T_s - T_\infty) \qquad (W) \qquad (1\text{-}28)$$

Note that the combined heat transfer coefficient is essentially a convection heat transfer coefficient modified to include the effects of radiation.

Radiation is usually significant relative to conduction or natural convection, but negligible relative to forced convection. Thus radiation in forced convection applications is normally disregarded, especially when the surfaces involved have low emissivities and low to moderate temperatures.

EXAMPLE 1-9 Radiation Effect on Thermal Comfort

It is a common experience to feel "chilly" in winter and "warm" in summer in our homes even when the thermostat setting is kept the same. This is due to the so called "radiation effect" resulting from radiation heat exchange between our bodies and the surrounding surfaces of the walls and the ceiling.

Consider a person standing in a room maintained at 22°C at all times. The inner surfaces of the walls, floors, and the ceiling of the house are observed to be at an average temperature of 10°C in winter and 25°C in summer. Determine the rate of radiation heat transfer between this person and the surrounding surfaces if the exposed surface area and the average outer surface temperature of the person are 1.4 m² and 30°C, respectively (Fig. 1-37).

Solution The rate of radiation heat transfer between a person and the surrounding surfaces at specified temperatures are to be determined in summer and winter.

Assumptions **1** Steady operating conditions exist. **2** Heat transfer by convection is not considered. **3** The person is completely surrounded by the interior surfaces of the room. **4** The surrounding surfaces are at a uniform temperature.

Properties The emissivity of a person is $\varepsilon = 0.95$ (Table A-14).

Analysis The net rates of radiation heat transfer from the body to the surrounding walls, ceiling, and floor in winter and summer are

$$\dot{Q}_{\text{rad, winter}} = \varepsilon\sigma A(T_s^4 - T_{\text{surr, winter}}^4)$$
$$= (0.95)(5.67 \times 10^{-8} \text{ W/m}^2 \cdot \text{K}^4)(1.4 \text{ m}^2)$$
$$\times [(30 + 273)^4 - (10 + 273)^4] \text{ K}^4$$
$$= \mathbf{152 \text{ W}}$$

and

$$\dot{Q}_{\text{rad, summer}} = \varepsilon\sigma A (T_s^4 - T_{\text{surr, summer}}^4)$$
$$= (0.95)(5.67 \times 10^{-8} \text{ W/m}^2 \cdot \text{K}^4)(1.4 \text{ m}^2)$$
$$\times [(30 + 273)^4 - (25 + 273)^4] \text{ K}^4$$
$$= \mathbf{40.9 \text{ W}}$$

Discussion Note that we must use *absolute temperatures* in radiation calculations. Also note that the rate of heat loss from the person by radiation is almost four times as large in winter than it is in summer, which explains the "chill" we feel in winter even if the thermostat setting is kept the same.

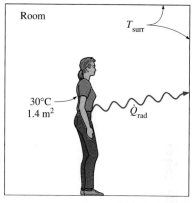

FIGURE 1-37
Schematic for Example 1-9.

1-6 ■ SIMULTANEOUS HEAT TRANSFER MECHANISMS

We mentioned that there are three mechanisms of heat transfer, but not all three can exist simultaneously in a medium. For example, heat transfer is only by conduction in *opaque solids,* but by conduction and radiation in *semitransparent solids.* Thus, a solid may involve conduction and radiation but not convection. However, a solid may involve heat transfer by convection and/or radiation on its surfaces exposed to a fluid or other surfaces. For example, the outer surfaces of a cold piece of rock will warm up in a warmer environment as a result of heat gain by convection (from the air) and radiation (from the sun or the warmer surrounding surfaces). But the inner parts of the rock will warm up as this heat is transferred to the inner region of the rock by conduction.

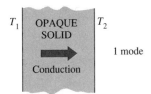

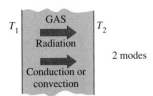

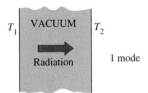

FIGURE 1-38
Although there are three mechanisms of heat transfer, a medium may involve only two of them simultaneously.

FIGURE 1-39
Heat transfer from the person described in Example 1-10.

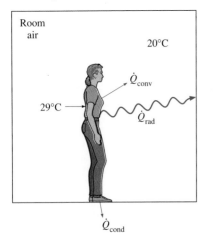

Heat transfer is by conduction and possibly by radiation in a *still fluid* (no bulk fluid motion) and by convection and radiation in a *flowing fluid*. In the absence of radiation, heat transfer through a fluid is either by conduction or convection, depending on the presence of any bulk fluid motion. Convection can be viewed as combined conduction and fluid motion, and conduction in a fluid can be viewed as a special case of convection in the absence of any fluid motion (Fig. 1-38).

Thus, when we deal with heat transfer through a *fluid,* we have either *conduction* or *convection*, but not both. Also, gases are practically transparent to radiation, except that some gases are known to absorb radiation strongly at certain wavelengths. Ozone, for example, strongly absorbs ultraviolet radiation. But in most cases, a gas between two solid surfaces does not interfere with radiation and acts effectively as a vacuum. Liquids, on the other hand, are usually strong absorbers of radiation.

Finally, heat transfer through a *vacuum* is by radiation only since conduction or convection requires the presence of a material medium.

EXAMPLE 1-10   Heat Loss from a Person

Consider a person standing in a breezy room at 20°C. Determine the total rate of heat transfer from this person if the exposed surface area and the average outer surface temperature of the person are 1.6 m² and 29°C, respectively, and the convection heat transfer coefficient is 6 W/m² · °C (Fig. 1-39).

Solution   The total rate of heat transfer from a person by both convection and radiation to the surrounding air and surfaces at specified temperatures is to be determined.

Assumptions   **1** Steady operating conditions exist. **2** The person is completely surrounded by the interior surfaces of the room. **3** The surrounding surfaces are at the same temperature as the air in the room. **4** Heat conduction to the floor through the feet is negligible.

Properties   The emissivity of a person is $\varepsilon = 0.95$ (Table A-14).

Anslysis   The heat transfer between the person and the air in the room will be by convection (instead of conduction) since it is conceivable that the air in the vicinity of the skin or clothing will warm up and rise as a result of heat transfer from the body, initiating natural convection currents. It appears that the experimentally determined value for the rate of convection heat transfer in this case is 6 W per unit surface area (m²) per unit temperature difference (in K or °C) between the person and the air away from the person. Thus, the rate of convection heat transfer from the person to the air in the room is

$$\dot{Q}_{conv} = hA(T_s - T_\infty)$$
$$= (6 \text{ W/m}^2 \cdot °\text{C})(1.6 \text{ m}^2)(29 - 20)°\text{C}$$
$$= 86.4 \text{ W}$$

The person will also lose heat by radiation to the surrounding wall surfaces. We take the temperature of the surfaces of the walls, ceiling, and floor to be equal to the air temperature in this case for simplicity, but we recognize that this does not need to be the case. These surfaces may be at a higher or lower temperature than the average temperature of the room air, depending on the outdoor conditions and the structure of the walls. Considering that air does not intervene with radiation and the person is completely enclosed by the surrounding surfaces, the net rate of radiation heat transfer from the person to the surrounding walls, ceiling, and floor is

$$\dot{Q}_{rad} = \varepsilon \sigma A (T_s^4 - T_{surr}^4)$$
$$= (0.95)(5.67 \times 10^{-8} \, \text{W/m}^2 \cdot \text{K}^4)(1.6 \, \text{m}^2)$$
$$\times [(29 + 273)^4 - (20 + 273)^4] \, \text{K}^4$$
$$= 81.7 \, \text{W}$$

Note that we must use *absolute* temperatures in radiation calculations. Also note that we used the emissivity value for the skin and clothing at room temperature since the emissivity is not expected to change significantly at a slightly higher temperature.

Then the rate of total heat transfer from the body is determined by adding these two quantities:

$$\dot{Q}_{total} = \dot{Q}_{conv} + \dot{Q}_{rad} = (86.4 + 81.7) \, \text{W} = \textbf{168.1 W}$$

Discussion The heat transfer would be much higher if the person were not dressed since the exposed surface temperature would be higher. Thus, an important function of the clothes is to serve as a barrier against heat transfer.

In the above calculations, heat transfer through the feet to the floor by conduction, which is usually very small, is neglected. Heat transfer from the skin by perspiration, which is the dominant mode of heat transfer in hot environments, is not considered here.

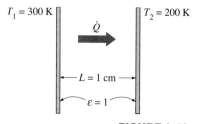

FIGURE 1-40
Schematic for Example 1-11.

EXAMPLE 1-11   Heat Transfer between Two Isothermal Plates

Consider steady heat transfer between two large parallel plates at constant temperatures of $T_1 = 300$ K and $T_2 = 200$ K that are $L = 1$ cm apart, as shown in Figure 1-40. Assuming the surfaces to be black (emissivity $\varepsilon = 1$), determine the rate of heat transfer between the plates per unit surface area assuming the gap between the plates is (a) filled with atmospheric air, (b) evacuated, (c) filled with urethane insulation, and (d) filled with superinsulation that has an apparent thermal conductivity of 0.00002 W/m · °C.

Solution   The total rate of heat transfer between two large parallel plates at specified temperatures is to be determined for four different cases.

Assumptions   **1** Steady operating conditions exist. **2** There are no natural convection currents in the air between the plates. **3** The surfaces are black and thus $\varepsilon = 1$.

Properties   The thermal conductivity at the average temperature of 250 K is $k = 0.0223$ W/m · °C for air (Table A-11), 0.026 W/m · °C for urethane insulation (Table A-6), and 0.00002 W/m · °C for the superinsulation.

Analysis   (a) The rates of conduction and radiation heat transfer between the plates through the air layer are

$$\dot{Q}_{cond} = kA\frac{T_1 - T_2}{L} = (0.0223 \text{ W/m} \cdot {}^\circ\text{C})(1 \text{ m}^2)\frac{(300 - 200){}^\circ\text{C}}{0.01 \text{ m}} = 223 \text{ W}$$

and

$$\dot{Q}_{rad} = \varepsilon\sigma A(T_1^4 - T_2^4)$$
$$= (1)(5.67 \times 10^{-8} \text{ W/m}^2 \cdot \text{K}^4)(1 \text{ m}^2)[(300 \text{ K})^4 - (200 \text{ K})^4] = 368 \text{ W}$$

Therefore,

$$\dot{Q}_{total} = \dot{Q}_{cond} + \dot{Q}_{rad} = 223 + 368 = \mathbf{591 \text{ W}}$$

The heat transfer rate in reality will be higher because of the natural convection currents that are likely to occur in the air space between the plates.

(*b*) When the air space between the plates is evacuated, there will be no conduction or convection, and the only heat transfer between the plates will be by radiation. Therefore,

$$\dot{Q}_{total} = \dot{Q}_{rad} = \mathbf{368 \text{ W}}$$

(*c*) An opaque solid material placed between two plates blocks direct radiation heat transfer between the plates. Also, the thermal conductivity of an insulating material accounts for the radiation heat transfer that may be occurring through the voids in the insulating material. The rate of heat transfer through the urethane insulation is

$$\dot{Q}_{total} = \dot{Q}_{cond} = kA\frac{T_1 - T_2}{L} = (0.026 \text{ W/m} \cdot {}^\circ\text{C})(1 \text{ m}^2)\frac{(300 - 200){}^\circ\text{C}}{0.01 \text{ m}} = \mathbf{260 \text{ W}}$$

Note that heat transfer through the urethane material is less than the heat transfer through the air determined in (*a*), although the thermal conductivity of the insulation is higher than that of air. This is because the insulation blocks the radiation whereas air transmits it.

(*d*) The layers of the superinsulation prevents any direct radiation heat transfer between the plates. However, radiation heat transfer between the sheets of superinsulation does occur, and the apparent thermal conductivity of the superinsulation accounts for this effect. Therefore,

$$\dot{Q}_{total} = kA\frac{T_1 - T_2}{L} = (0.00002 \text{ W/m} \cdot {}^\circ\text{C})(1 \text{ m}^2)\frac{(300 - 200){}^\circ\text{C}}{0.01 \text{ m}} = \mathbf{0.2 \text{ W}}$$

which is $\frac{1}{1840}$ of the heat transfer through the vacuum. The results of this example are summarized in Figure 1-41 to put them into perspective.

Discussion   This example demonstrates the effectiveness of superinsulations, which are discussed in the next chapter, and explains why they are the insulation of choice in critical applications despite their high cost.

EXAMPLE 1-12   Heat Transfer in Conventional and Microwave Ovens
The fast and efficient cooking of microwave ovens made them one of the essential appliances in modern kitchens (Fig. 1-42). Discuss the heat transfer mechanisms associated with the cooking of a chicken in microwave and conventional ovens, and explain why cooking in a microwave oven is more efficient.

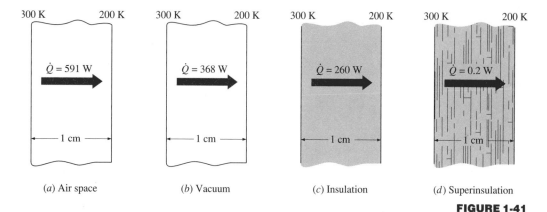

(a) Air space   (b) Vacuum   (c) Insulation   (d) Superinsulation

FIGURE 1-41

Different ways of reducing heat transfer between two isothermal plates, and their effectiveness.

Solution Food is cooked in a microwave oven by absorbing the electromagnetic radiation energy generated by the microwave tube, called the magnetron. The radiation emitted by the magnetron is not thermal radiation, since its emission is not due to the temperature of the magnetron; rather, it is due to the conversion of electrical energy into electromagnetic radiation at a specified wavelength. The wavelength of the microwave radiation is such that it is *reflected* by metal surfaces; *transmitted* by the cookware made of glass, ceramic, or plastic; and *absorbed* and converted to internal energy by food (especially the water, sugar, and fat) molecules.

In a microwave oven, the *radiation* that strikes the chicken is absorbed by the skin of the chicken and the outer parts. As a result, the temperature of the chicken at and near the skin rises. Heat is then *conducted* toward the inner parts of the chicken from its outer parts. Of course, some of the heat absorbed by the outer surface of the chicken is lost to the air in the oven by *convection*.

In a conventional oven, the air in the oven is first heated to the desired temperature by the electric or gas heating element. This preheating may take several minutes. The heat is then transferred from the air to the skin of the chicken by *natural convection* in most ovens or by *forced convection* in the newer convection ovens that utilize an air mover. The air motion in convection ovens increases the convection heat transfer coefficient and thus decreases the cooking time. Heat is then *conducted* toward the inner parts of the chicken from its outer parts as in microwave ovens.

Microwave ovens replace the slow convection heat transfer process in conventional ovens by the instantaneous radiation heat transfer. As a result, microwave ovens transfer energy to the food at full capacity the moment they are turned on, and thus they cook faster while consuming less energy.

FIGURE 1-42

A chicken being cooked in a microwave oven (Example 1-12).

FIGURE 1-43

Schematic for Example 1-13.

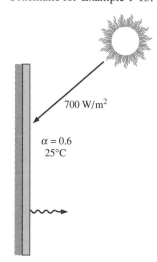

EXAMPLE 1-13   Heating of a Plate by Solar Energy

A thin metal plate is insulated on the back and exposed to solar radiation at the front surface (Fig. 1-43). The exposed surface of the plate has an absorptivity of 0.6 for solar radiation. If solar radiation is incident on the plate at a rate of 700 W/m² and the surrounding air temperature is 25°C, determine the surface temperature of the plate when the heat loss by convection and radiation equals the solar energy absorbed by the plate. Assume the combined convection and radiation heat transfer coefficient to be 50 W/m² · °C.

Solution The back side of the thin metal plate is insulated and the front side is exposed to solar radiation. The surface temperature of the plate is to be determined when it stabilizes.

Assumptions **1** Steady operating conditions exist. **2** Heat transfer through the insulated side of the plate is negligible. **3** The heat transfer coefficient remains constant.

Properties The solar absorptivity of the plate is given to be $\alpha = 0.6$.

Analysis The absorptivity of the plate is 0.6, and thus 60 percent of the solar radiation incident on the plate will be absorbed continuously. As a result, the temperature of the plate will rise, and the temperature difference between the plate and the surroundings will increase. This increasing temperature difference will cause the rate of heat loss from the plate to the surroundings to increase. At some point, the rate of heat loss from the plate will equal the rate of solar energy absorbed, and the temperature of the plate will no longer change. The temperature of the plate when steady operation is established is determined from

$$\dot{E}_{\text{gained}} = \dot{E}_{\text{lost}} \quad \text{or} \quad \alpha A \, \dot{q}_{\text{incident, solar}} = h_{\text{combined}} A \, (T_s - T_\infty)$$

Solving for T_s and substituting, the plate surface temperature is determined to be

$$T_s = T_\infty + \frac{\alpha \dot{q}_{\text{incident, solar}}}{h_{\text{combined}}} = 25°\text{C} + \frac{0.6 \times (700 \text{ W/m}^2)}{(50 \text{ W/m}^2 \cdot °\text{C})} = \mathbf{33.4°C}$$

Discussion Note that the heat losses will prevent the plate temperature from rising above 33.4°C. Also, the combined heat transfer coefficient accounts for the effects of both convection and radiation, and thus it is very convenient to use in heat transfer calculations when its value is known with reasonable accuracy.

1-7 ■ SUMMARY

In this chapter, the basic concepts of thermodynamics and heat transfer are introduced and discussed. The science of *thermodynamics* deals with the amount of heat transfer as a system undergoes a process from one equilibrium state to another, whereas the science of *heat transfer* deals with the rate of heat transfer, which is the main quantity of interest in the design and evaluation of heat transfer equipment. The sum of all forms of energy of a system is called *total energy,* and it includes the internal, kinetic, and potential energies. The *internal energy* represents the molecular energy of a system, and it consists of sensible, latent, chemical, and nuclear forms. The sensible and latent forms of internal energy can be transferred from one medium to another as a result of temperature difference, and are referred to as *heat* or *thermal energy.* Thus, *heat transfer* is the exchange of the sensible and latent forms of internal energy between two mediums as a result of temperature difference. The amount of heat transferred per unit time is called *heat transfer rate* and is denoted by $\dot{Q}$. The rate of heat transfer per unit surface area is called *heat flux, $\dot{q}$*.

The *specific heat* is defined as the energy required to raise the temperature of a unit mass of a substance by one degree. The *specific heat at constant volume C_v* can be viewed as the energy required to raise the temperature of a unit mass of a substance by one degree as the volume is held constant. The energy required to do the same as the pressure is held constant is the *specific heat at constant pressure C_p*. The changes in the internal energy and enthalpy of an ideal gas can be expressed in terms of the specific heats as

$$\Delta u = C_{v,\text{ave}}\Delta T \qquad \text{and} \qquad \Delta h = C_{p,\text{ave}}\Delta T \qquad \text{(J/kg)}$$

The specific heats of an ideal gas are related to each other by $C_p = C_v + R$. For an incompressible substance (solid or liquid), these two specific heats are identical and are denoted by C. That is, $C_p \cong C_v \cong C$.

A system of fixed mass is called a *closed system* and a system that involves mass transfer across its boundaries is called an *open system* or *control volume*. The *first law of thermodynamics* or the *energy balance* for any system undergoing any process can be expressed as

$$E_{in} - E_{out} = \Delta E_{\text{system}} \qquad \text{(J)}$$

When a stationary closed system involves heat transfer only and no work interactions across its boundary, the energy balance relation reduces to

$$Q = mC_v\Delta T \qquad \text{(J)}$$

where Q is the amount of net heat transfer to or from the system. When heat is transferred at a constant rate of $\dot{Q}$, the amount of heat transfer during a time interval Δt can be determined from $Q = \dot{Q}\Delta t$.

Under steady conditions and in the absence of any work interactions, the conservation of energy relation for a control volume with one inlet and one exit with negligible changes in kinetic and potential energies can be expressed as

$$\dot{Q} = \dot{m}C_p\Delta T \qquad \text{(W)}$$

where $\dot{m} = \rho \mathcal{V} A_c$ is the mass flow rate and $\dot{Q}$ is the rate of net heat transfer into or out of the control volume.

Heat can be transferred in three different ways: Conduction, convection, and radiation. *Conduction* is the transfer of energy from the more energetic particles of a substance to the adjacent less energetic ones as a result of interactions between the particles, and is expressed by *Fourier's law of heat conduction* as

$$\dot{Q}_{\text{cond}} = -kA\frac{dT}{dx} \qquad \text{(W)}$$

where k is the *thermal conductivity* of the material, A is the *area* normal to the direction of heat transfer, and dT/dx is the *temperature gradient*. The rate of heat conduction across a plane layer of thickness L is given by

$$\dot{Q}_{cond} = kA\frac{\Delta T}{L} \qquad \text{(W)}$$

where $\Delta T = T_2 - T_1$ is the temperature difference across the layer.

Convection is the mode of heat transfer between a solid surface and the adjacent liquid or gas that is in motion, and involves the combined effects of conduction and fluid motion. The rate of convection heat transfer is expressed by *Newton's law of cooling* as

$$\dot{Q}_{convection} = hA(T_s - T_\infty) \qquad \text{(W)}$$

where h is the *convection heat transfer coefficient* in $W/m^2 \cdot °C$ or $Btu/h \cdot ft^2 \cdot °F$, A is the *surface area* through which convection heat transfer takes place, T_s is the *surface temperature,* and T_∞ is the *temperature of the fluid* sufficiently far from the surface.

Radiation is the energy emitted by matter in the form of electromagnetic waves (or photons) as a result of the changes in the electronic configurations of the atoms or molecules. The maximum rate of radiation that can be emitted from a surface at an absolute temperature T_s is given by the *Stefan-Boltzmann law* as $\dot{Q}_{emit, max} = \sigma AT_s^4$, where $\sigma = 5.67 \times 10^{-8} \ W/m^2 \cdot K^4$ or 0.1714×10^{-8} $Btu/h \cdot ft^2 \cdot R^4$ is the *Stefan-Boltzmann constant.*

When a surface of emissivity ε and surface area A at an absolute temperature T_s is completely enclosed by a much larger (or black) surface at absolute temperature T_{surr} separated by a gas (such as air) that does not intervene with radiation, the net rate of radiation heat transfer between these two surfaces is given by

$$\dot{Q}_{rad} = \varepsilon \sigma A(T_s^4 - T_{surr}^4) \qquad \text{(W)}$$

In this special case, the emissivity and the surface area of the surrounding surface do not have any effect on the net radiation heat transfer.

The rate at which a surface absorbs radiation is determined from $\dot{Q}_{absorbed} = \alpha \dot{Q}_{incident}$ where $\dot{Q}_{incident}$ is the rate at which radiation is incident on the surface and α is the absorptivity of the surface.

REFERENCES AND SUGGESTED READING

1. Y. A. Çengel. *Introduction to Thermodynamics and Heat Transfer.* New York: McGraw-Hill, 1997.

2. Y. A. Çengel and M. A. Boles. *Thermodynamics—An Engineering Approach.* 2nd ed. New York: McGraw-Hill, 1994.

3. J. P. Holman. *Heat Transfer.* 8th ed. New York: McGraw-Hill, 1997.

4. F. P. Incropera and D. P. DeWitt. *Introduction to Heat Transfer.* 2nd ed. New York: John Wiley & Sons, 1990.

5. F. Kreith and M. S. Bohn. *Principles of Heat Transfer.* 5th ed. St. Paul, MN: West Publishing, 1993.

6. A. F. Mills. *Basic Heat and Mass Transfer.* Burr Ridge, IL: Richard D. Irwin, 1995.

7. M. N. Ozisik. *Heat Transfer—A Basic Approach.* New York: McGraw-Hill, 1985.

PROBLEMS*

Thermodynamics and Heat Transfer

1-1C How does the science of heat transfer differ from the science of thermodynamics?

1-2C What is the driving force for (*a*) heat transfer, (*b*) electric current flow, and (*c*) fluid flow?

1-3C Why is heat transfer a nonequilibrium phenomenon?

1-4C What is the caloric theory? When and why was it abandoned?

1-5C How do rating problems in heat transfer differ from the sizing problems?

1-6C What is the difference between the analytical and experimental approach to heat transfer? Discuss the advantages and disadvantages of each approach.

1-7C What is the importance of modeling in engineering? How are the mathematical models for engineering processes prepared?

1-8C When modeling an engineering process, how is the right choice made between a simple but crude and a complex but accurate model? Is the complex model necessarily a better choice since it is more accurate?

1-9C How do the differential equations in the study of physical problems arise?

1-10C What is the value of the engineering software packages in (*a*) engineering education and (*b*) engineering practice?

Heat and Other Forms of Energy

1-11C What is heat flux? How is it related to the heat transfer rate?

1-12C What are the mechanisms of energy transfer to a closed system? How is heat transfer distinguished from the other forms of energy transfer?

1-13C How are heat, internal energy, and thermal energy related to each other?

---

*Students are encouraged to answer all the concept "C" questions.

1-14C When is the energy crossing the boundaries of a system heat, and when is it work?

1-15C Is the energy required to heat air from 295 to 305 K the same as the energy required to heat it from 345 to 355 K? Assume the pressure remains constant in both cases.

1-16C An ideal gas is heated from 50°C to 80°C at a constant pressure of (*a*) 1 atm and (*b*) 3 atm. For which case do you think the energy required will be greater? Why?

1-17C An ideal gas is heated from 50°C to 80°C (*a*) at constant volume and (*b*) at constant pressure. For which case do you think the energy required will be greater? Why?

1-18 A cylindrical resistor element on a circuit board dissipates 0.2 W of power. The resistor is 1.5 cm long, and has a diameter of 0.4 cm. Assuming heat to be transferred uniformly from all surfaces, determine (*a*) the amount of heat this resistor dissipates during a 24-hour period, (*b*) the heat flux, and (*c*) the fraction of heat dissipated from the top and bottom surfaces.

1-19E A logic chip used in a computer dissipates 3 W of power in an environment at 120°F, and has a heat transfer surface area of 0.08 in². Assuming the heat transfer from the surface to be uniform, determine (*a*) the amount of heat this chip dissipates during an eight-hour work day, in kWh, and (*b*) the heat flux on the surface of the chip, in W/in².

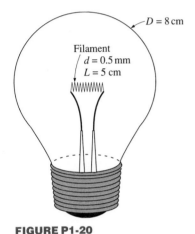

Filament
$d = 0.5$ mm
$L = 5$ cm

$D = 8$ cm

FIGURE P1-20

1-20 Consider a 150-W incandescent lamp. The filament of the lamp is 5 cm long and has a diameter of 0.5 mm. The diameter of the glass bulb of the lamp is 8 cm. Determine the heat flux, in W/m², (*a*) on the surface of the filament and (*b*) on the surface of the glass bulb, and (*c*) calculate how much it will cost per year to keep that lamp on for eight hours a day every day if the unit cost of electricity is $0.08/kWh.
Answers: (*a*) 1.91×10^6 W/m², (*b*) 7500 W/m², (*c*) $35.04/yr

1-21 A 1200-W iron is left on the ironing board with its base exposed to the air. About 90 percent of the heat generated in the iron is dissipated through its base whose surface area is 150 cm², and the remaining 10 percent through other surfaces. Assuming the heat transfer from the surface to be uniform, determine (*a*) the amount of heat the iron dissipates during a two-hour period, in kWh, (*b*) the heat flux on the surface of the iron base, in W/m², and (*c*) the total cost of the electrical energy consumed during this two-hour period. Take the unit cost of electricity to be $0.07/kWh.

FIGURE P1-22

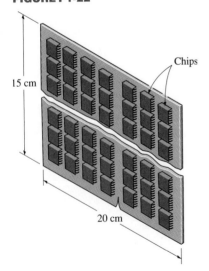

15 cm

Chips

20 cm

1-22 A 15-cm × 20-cm circuit board houses on its surface 120 closely spaced logic chips, each dissipating 0.1 W. If the heat transfer from the back surface of the board is negligible, determine (*a*) the amount of heat this circuit board dissipates during a 10-hour period, in kWh, and (*b*) the heat flux on the surface of the circuit board, in W/m².

1-23 A 15-cm-diameter aluminum ball is to be heated from 80°C to an average temperature of 200°C. Taking the average density and specific heat of aluminum in this temperature range to be $\rho = 2700 \text{ kg/m}^3$ and $C_p = 0.90 \text{ kJ/kg} \cdot \text{°C}$, respectively, determine the amount of energy that needs to be transferred to the aluminum ball. *Answer:* 515 kJ

1-24 The average specific heat of the human body is 3.6 kJ/kg · °C. If the body temperature of an 80-kg man rises from 37°C to 39°C during strenuous exercise, determine the increase in the thermal energy content of the body as a result of this rise in body temperature.

1-25 Infiltration of cold air into a warm house during winter through the cracks around doors, windows, and other openings is a major source of energy loss since the cold air that enters needs to be heated to the room temperature. The infiltration is often expressed in terms of ACH (air changes per hour). An ACH of 2 indicates that the entire air in the house is replaced twice every hour by the cold air outside.

Consider an electrically heated house that has a floor space of 200 m² and an average height of 3 m at 1000 m elevation, where the standard atmospheric pressure is 89.6 kPa. The house is maintained at a temperature of 22°C, and the infiltration losses are estimated to amount to 0.7 ACH. Assuming the pressure and the temperature in the house remain constant, determine the amount of energy loss from the house due to infiltration for a day during which the average outdoor temperature is 5°C. Also, determine the cost of this energy loss for that day if the unit cost of electricity in that area is $0.082/kWh. *Answers:* 53.7 kWh/day, $4.40/day

1-26 Consider a house with a floor space of 200 m² and an average height of 3 m at sea level, where the standard atmospheric pressure is 101.3 kPa. Initially the house is at a uniform temperature of 10°C. Now the electric heater is turned on, and the heater runs until the air temperature in the house rises to an average value of 22°C. Determine how much heat is absorbed by the air assuming some air escapes through the cracks as the heated air in the house expands at constant pressure. Also, determine the cost of this heat if the unit cost of electricity in that area is $0.075/kWh.

1-27E Consider a 40-gallon water heater that is initially filled with water at 45°F. Determine how much energy needs to be transferred to the water to raise its temperature to 140°F. Take the density and specific heat of water to be 62 lbm/ft³ and 1.0 Btu/lbm · °F, respectively.

The First Law of Thermodynamics

1-28C On a hot summer day, a student turns his fan on when he leaves his room in the morning. When he returns in the evening, will his room be warmer or cooler than the neighboring rooms? Why? Assume all the doors and windows are kept closed.

1-29C Consider two identical rooms, one with a refrigerator in it and the other without one. If all the doors and windows are closed, will the room that contains the refrigerator be cooler or warmer than the other room? Why?

1-30C Define mass and volume flow rates. How are they related to each other?

1-31 Two 800-kg cars moving at a velocity of 90 km/h have a head-on collision on a road. Both cars come to a complete rest after the crash. Assuming all the kinetic energy of cars is converted to thermal energy, determine the average temperature rise of the remains of the cars immediately after the crash. Take the average specific heat of the cars to be 0.45 kJ/kg · °C.

1-32 A classroom that normally contains 40 people is to be air-conditioned using window air-conditioning units of 5-kW cooling capacity. A person at rest may be assumed to dissipate heat at a rate of about 360 kJ/h. There are 10 light bulbs in the room, each with a rating of 100 W. The rate of heat transfer to the classroom through the walls and the windows is estimated to be 15,000 kJ/h. If the room air is to be maintained at a constant temperature of 21°C, determine the number of window air-conditioning units required.
Answer: 2 units

1-33E A rigid tank contains 20 lbm of air at 50 psia and 80°F. The air is now heated until its pressure doubles. Determine (*a*) the volume of the tank and (*b*) the amount of heat transfer.   *Answers:* (*a*) 80 ft³, (*b*) 2035 Btu

1-34 A 1-m³ rigid tank contains hydrogen at 250 kPa and 500 K. The gas is now cooled until its temperature drops to 300 K. Determine (*a*) the final pressure in the tank and (*b*) the amount of heat transfer from the tank.

1-35 A 4-m × 5-m × 6-m room is to be heated by a baseboard resistance heater. It is desired that the resistance heater be able to raise the air temperature in the room from 7 to 25°C within 15 minutes. Assuming no heat losses from the room and an atmospheric pressure of 100 kPa, determine the required power rating of the resistance heater. Assume constant specific heats at room temperature.   *Answer:* 2.70 kW

1-36 A 4-m × 5-m × 7-m room is heated by the radiator of a steam heating system. The steam radiator transfers heat at a rate of 10,000 kJ/h and a 100-W fan is used to distribute the warm air in the room. The heat losses from the room are estimated to be at a rate of about 5000 kJ/h. If the initial temperature of the room air is 10°C, determine how long it will take for the air temperature to rise to 20°C. Assume constant specific heats at room temperature.

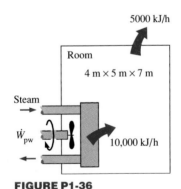

FIGURE P1-36

1-37 A student living in a 4-m × 6-m × 6-m dormitory room turns his 150-W fan on before she leaves her room on a summer day hoping that the room will be cooler when she comes back in the evening. Assuming all the doors and windows are tightly closed and disregarding any heat transfer through the walls and the windows, determine the temperature in the room when she comes back 10 hours later. Use specific heat values at room temperature and assume the room to be at 100 kPa and 15°C in the morning when she leaves.   *Answer:* 58.2°C

FIGURE P1-37

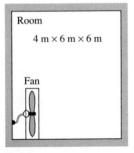

1-38E A 10-ft³ tank contains oxygen initially at 14.7 psia and 27°C. A paddle wheel within the tank is rotated until the pressure inside rises to 20 psia. During the process 20 Btu of heat is lost to the surroundings.

Neglecting the energy stored in the paddle wheel, determine the work done by the paddle wheel.

1-39 A room is heated by a baseboard resistance heater. When the heat losses from the room on a winter day amount to 8000 kJ/h, it is observed that the air temperature in the room remains constant even though the heater operates continuously. Determine the power rating of the heater, in kW.

1-40 A 50-kg mass of copper at 70°C is dropped into an insulated tank containing 80 kg of water at 25°C. Determine the final equilibrium temperature in the tank.

1-41 A 20-kg mass of iron at 100°C is brought into contact with 20 kg of aluminum at 200°C in an insulated enclosure. Determine the final equilibrium temperature of the combined system.   *Answer:* 168°C

1-42 An unknown mass of iron at 90°C is dropped into an insulated tank that contains 80 L of water at 20°C. At the same time, a paddle wheel driven by a 200-W motor is activated to stir the water. Thermal equilibrium is established after 25 minutes with a final temperature of 27°C. Determine the mass of the iron. Neglect the energy stored in the paddle wheel, and take the density of water to be 1000 kg/m^3.   *Answer:* 72.1 kg

1-43E A 90-lbm mass of copper at 160°F and a 50-lbm mass of iron at 200°F are dropped into a tank containing 180 lbm of water at 70°F. If 600 Btu of heat is lost to the surroundings during the process, determine the final equilibrium temperature.

1-44 A 5-m × 6-m × 8-m room is to be heated by an electrical resistance heater placed in a short duct in the room. Initially, the room is at 15°C, and the local atmospheric pressure is 98 kPa. The room is losing heat steadily to the outside at a rate of 200 kJ/min. A 200-W fan circulates the air steadily through the duct and the electric heater at an average mass flow rate of 50 kg/min. The duct can be assumed to be adiabatic, and there is no air leaking in or out of the room. If it takes 15 minutes for the room air to reach an average temperature of 25°C, find (*a*) the power rating of the electric heater and (*b*) the temperature rise that the air experiences each time it passes through the heater.

1-45 A house has an electric heating system that consists of a 300-W fan and an electric resistance heating element placed in a duct. Air flows steadily through the duct at a rate of 0.6 kg/s and experiences a temperature rise of 5°C. The rate of heat loss from the air in the duct is estimated to be 400 W. Determine the power rating of the electric resistance heating element.

1-46 A hair dryer is basically a duct in which a few layers of electric resistors are placed. A small fan pulls the air in and forces it to flow over the resistors where it is heated. Air enters a 1200-W hair dryer at 100 kPa and 22°C, and leaves at 47°C. The cross-sectional area of the hair dryer at the exit is 60 cm^2. Neglecting the power consumed by the fan and the heat losses through the walls of the hair dryer, determine (*a*) the volume flow rate of air at the inlet and (*b*) the velocity of the air at the exit.

Answers: (*a*) 0.0404 m^3/s, (*b*) 7.31 m/s

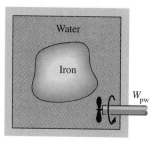

FIGURE P1-42

FIGURE P1-46

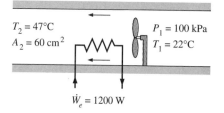

FIGURE P1-49

1-47 The ducts of an air heating system pass through an unheated area. As a result of heat losses, the temperature of the air in the duct drops by 4°C. If the mass flow rate of air is 120 kg/min, determine the rate of heat loss from the air to the cold environment.

1-48E Air enters the duct of an air-conditioning system at 15 psia and 50°F at a volume flow rate of 450 ft³/min. The diameter of the duct is 10 inches and heat is transferred to the air in the duct from the surroundings at a rate of 2 Btu/s. Determine (*a*) the velocity of the air at the duct inlet and (*b*) the temperature of the air at the exit. *Answers:* (*a*) 825 ft/min, (*b*) 64°F

1-49 Water is heated in an insulated, constant diameter tube by a 7-kW electric resistance heater. If the water enters the heater steadily at 15°C and leaves at 70°C, determine the mass flow rate of water.

Heat Transfer Mechanisms

1-50C Define thermal conductivity and explain its significance in heat transfer.

1-51C What are the mechanisms of heat transfer? How are they distinguished from each other?

1-52C What is the physical mechanism of heat conduction in a solid, a liquid, and a gas?

1-53C Consider heat transfer through a windowless wall of a house in a winter day. Discuss the parameters that affect the rate of heat conduction through the wall.

1-54C Write down the expressions for the physical laws that govern each mode of heat transfer, and identify the variables involved in each relation.

1-55C How does heat conduction differ from convection?

1-56C Does any of the energy of the sun reach the earth by conduction or convection?

1-57C How does forced convection differ from natural convection?

1-58C Define emissivity and absorptivity. What is Kirchhoff's law of radiation?

1-59C What is a blackbody? How do real bodies differ from blackbody?

1-60C Judging from its unit W/m · °C, can we define thermal conductivity of a material as the rate of heat transfer through the material per unit thickness per unit temperature difference? Explain.

1-61C Consider heat loss through the two walls of a house on a winter night. The walls are identical, except that one of them has a tightly fit glass window. Through which wall will the house lose more heat? Explain.

1-62C Consider two walls of a house that are identical except that one is made of 10-cm-thick wood, while the other is made of 25-cm-thick brick. Through which wall will the house lose more heat in winter?

1-63C How do the thermal conductivity of gases and liquids vary with temperature?

1-64C Why is the thermal conductivity of superinsulation orders of magnitude lower than the thermal conductivity of ordinary insulation?

1-65C Why do we characterize the heat conduction ability of insulators in terms of their apparent thermal conductivity instead of the ordinary thermal conductivity?

1-66C Consider an alloy of two metals whose thermal conductivities are k_1 and k_2. Will the thermal conductivity of the alloy be less than k_1, greater than k_2, or between k_1 and k_2?

1-67 The inner and outer surfaces of a 5-m × 6-m brick wall of thickness 30 cm and thermal conductivity 0.69 W/m · °C are maintained at temperatures of 20°C and 5°C, respectively. Determine the rate of heat transfer through the wall, in W.   *Answer:* 1035 W

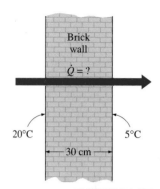

FIGURE P1-67

1-68 The inner and outer surfaces of a 0.5-cm-thick 2-m × 2-m window glass in winter are 10°C and 3°C, respectively. If the thermal conductivity of the glass is 0.78 W/m · °C, determine the amount of heat loss, in kJ, through the glass over a period of 5 hours. What would your answer be if the glass were 1 cm thick?   *Answers:* 78,624 kJ,   39,312 kJ

1-69 An aluminum pan whose thermal conductivity is 237 W/m · °C has a flat bottom with diameter 20 cm and thickness 0.4 cm. Heat is transferred steadily to boiling water in the pan through its bottom at a rate of 500 W. If the inner surface of the bottom of the pan is at 105°C, determine the temperature of the outer surface of the bottom of the pan.

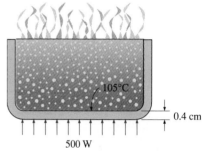

FIGURE P1-69

1-70E The north wall of an electrically heated home is 20 ft long, 10 ft high, and 1 ft thick, and is made of brick whose thermal conductivity is $k = 0.42$ Btu/h · ft · °F. On a certain winter night, the temperatures of the inner and the outer surfaces of the wall are measured to be at about 62°F and 25°F, respectively, for a period of 8 hours. Determine (*a*) the rate of heat loss through the wall that night and (*b*) the cost of that heat loss to the home owner if the cost of electricity is $0.07/kWh.

1-71 In a certain experiment, cylindrical samples of diameter 4 cm and length 7 cm are used. The two thermocouples in each sample are placed 3 cm apart. After initial transients, the electric heater is observed to draw 0.6 A at 110 V, and both differential thermometers read a temperature difference of 10°C. Determine the thermal conductivity of the sample.
    Answer: 78.8 W/m · °C

1-72 One way of measuring the thermal conductivity of a material is to sandwich an electric thermofoil heater between two identical rectangular samples of the material and to heavily insulate the four outer edges, as shown

FIGURE P1-72

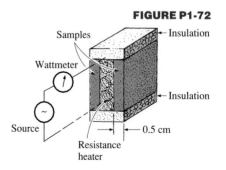

in the figure. Thermocouples attached to the inner and outer surfaces of the samples record the temperatures.

During an experiment, two 0.5-cm-thick samples 10 cm × 10 cm in size are used. When steady operation is reached, the heater is observed to draw 35 W of electric power, and the temperature of each sample is observed to drop from 82°C at the inner surface to 74°C at the outer surface. Determine the thermal conductivity of the material at the average temperature.

1-73   Repeat Problem 1-72 for an electric power consumption of 20 W.

1-74   A heat flux meter attached to the inner surface of a 3-cm-thick refrigerator door indicates a heat flux of 25 W/m² through the door. Also, the temperatures of the inner and the outer surfaces of the door are measured to be 7°C and 15°C, respectively. Determine the average thermal conductivity of the refrigerator door.   *Answer:* 0.0938 W/m · °C

1-75   Consider a person standing in a room maintained at 20°C at all times. The inner surfaces of the walls, floors, and ceiling of the house are observed to be at an average temperature of 12°C in winter and 23°C in summer. Determine the rates of radiation heat transfer between this person and the surrounding surfaces in both summer and winter if the exposed surface area, emissivity, and the average outer surface temperature of the person are 1.6 m², 0.95, and 32°C, respectively.

1-76   For heat transfer purposes, a standing man can be modeled as a 30-cm-diameter, 170-cm-long vertical cylinder with both the top and bottom surfaces insulated and with the side surface at an average temperature of 34°C. For a convection heat transfer coefficient of 15 W/m² · °C, determine the rate of heat loss from this man by convection in an environment at 20°C.
      Answer: 336 W

1-77   Hot air at 80°C is blown over a 2-m × 4-m flat surface at 30°C. If the average convection heat transfer coefficient is 55 W/m² · °C, determine the rate of heat transfer from the air to the plate, in kW.   *Answer:* 22,000 W

1-78   The heat generated in the circuitry on the surface of a silicon chip (k = 130 W/m · °C) is conducted to the ceramic substrate to which it is attached. The chip is 6 mm × 6 mm in size and 0.5 mm thick and dissipates 3 W of power. Disregarding any heat transfer through the 0.5-mm-high side surfaces, determine the temperature difference between the front and back surfaces of the chip in steady operation.

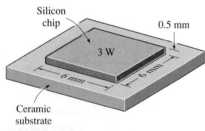

Silicon chip

0.5 mm

3 W

6 mm

6 mm

Ceramic substrate

FIGURE P1-78

1-79   A 50-cm-long, 800-W electric resistance heating element with diameter 0.5 cm and surface temperature 120°C is immersed in 40 kg of water initially at 20°C. Determine how long it will take for this heater to raise the water temperature to 80°C. Also, determine the convection heat transfer coefficients at the beginning and at the end of the heating process.

1-80   A 5-cm-external-diameter, 10-m-long hot water pipe at 80°C is losing heat to the surrounding air at 5°C by natural convection with a heat transfer coefficient of 25 W/m² · °C. Determine the rate of heat loss from the pipe by natural convection, in W.   *Answer:* 2945 W

1-81 A hollow spherical iron container with outer diameter 20 cm and thickness 0.4 cm is filled with iced water at 0°C. If the outer surface temperature is 5°C, determine the approximate rate of heat loss from the sphere, in kW, and the rate at which ice melts in the container. The heat from fusion of water is 333.7 kJ/kg.

1-82E The inner and outer glasses of a 6-ft × 6-ft double-pane window are at 60°F and 42°F, respectively. If the 0.5-in. space between the two glasses is filled with still air, determine the rate of heat transfer through the window, in Btu/h. *Answer:* 224 Btu/h

1-83 Two surfaces of a 2-cm-thick plate are maintained at 0°C and 100°C, respectively. If it is determined that heat is transferred through the plate at a rate of 500 W/m², determine its thermal conductivity.

1-84 Four power transistors, each dissipating 15 W, are mounted on a thin vertical aluminum plate 22 cm × 22 cm in size. The heat generated by the transistors is to be dissipated by both surfaces of the plate to the surrounding air at 25°C, which is blown over the plate by a fan. The entire plate can be assumed to be nearly isothermal, and the exposed surface area of the transistor can be taken to be equal to its base area. If the average convection heat transfer coefficient is 25 W/m² · °C, determine the temperature of the aluminum plate. Disregard any radiation effects.

1-85 An ice chest whose outer dimensions are 30 cm × 40 cm × 40 cm is made of 3-cm-thick Styrofoam ($k = 0.033$ W/m · °C). Initially, the chest is filled with 40 kg of ice at 0°C, and the inner surface temperature of the ice chest can be taken to be 0°C at all times. The heat of fusion of ice at 0°C is 333.7 kJ/kg, and the surrounding ambient air is at 30°C. Disregarding any heat transfer from the 40-cm × 40-cm base of the ice chest, determine how long it will take for the ice in the chest to melt completely if the outer surfaces of the ice chest are at 8°C. *Answer:* 32.7 days

1-86 A transistor with a height of 0.4 cm and a diameter of 0.6 cm is mounted on a circuit board. The transistor is cooled by air flowing over it with an average heat transfer coefficient of 30 W/m² · °C. If the air temperature is 55°C and the transistor case temperature is not to exceed 70°C, determine the amount of power this transistor can dissipate safely. Disregard any heat transfer from the transistor base.

1-87E A 200-ft-long section of a steam pipe whose outer diameter is 4 inches passes through an open space at 50°F. The average temperature of the outer surface of the pipe is measured to be 280°F, and the average heat transfer coefficient on that surface is determined to be 6 Btu/h · ft² · °F. Determine (*a*) the rate of heat loss from the steam pipe and (*b*) the annual cost of this energy loss if steam is generated in a natural gas furnace having an efficiency of 86 percent, and the price of natural gas is $0.58/therm (1 therm = 100,000 Btu). *Answers:* (*a*) 289,000 Btu/h, (*b*) $17,074/yr

1-88 The boiling temperature of nitrogen at atmospheric pressure at sea level (1 atm) is −196°C. Therefore, nitrogen is commonly used in low

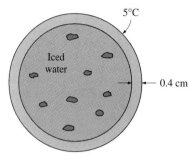

FIGURE P1-81

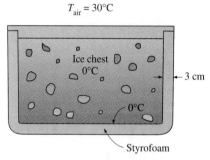

FIGURE P1-85

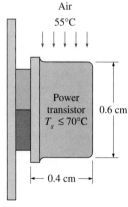

FIGURE P1-86

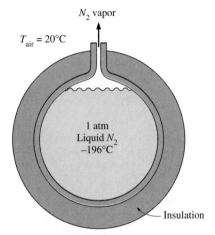

FIGURE P1-88

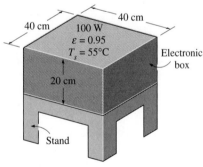

FIGURE P1-92

temperature scientific studies since the temperature of liquid nitrogen in a tank open to the atmosphere will remain constant at $-196°C$ until the liquid nitrogen in the tank is depleted. Any heat transfer to the tank will result in the evaporation of some liquid nitrogen, which has a heat of vaporization of 198 kJ/kg and a density of 810 kg/m^3 at 1 atm.

Consider a 4-m-diameter spherical tank initially filled with liquid nitrogen at 1 atm and $-196°C$. The tank is exposed to $20°C$ ambient air with a heat transfer coefficient of 25 W/m$^2 \cdot °C$. The temperature of the thin-shelled spherical tank is observed to be almost the same as the temperature of the nitrogen inside. Disregarding any radiation heat exchange, determine the rate of evaporation of the liquid nitrogen in the tank as a result of the heat transfer from the ambient air.

1-89 Repeat Problem 1-88 for liquid oxygen, which has a boiling temperature of $-183°C$, a heat of vaporization of 213 kJ/kg, and a density of 1140 kg/m^3 at 1 atm pressure.

1-90 Consider a person whose exposed surface area is 1.7 m^2, emissivity is 0.7, and surface temperature is $32°C$. Determine the rate of heat loss from that person by radiation in a large room having walls at a temperature of (*a*) 300 K and (*b*) 280 K.   *Answers:* (*a*) 37.4 W, (*b*) 169.2 W

1-91 A 0.3-cm-thick, 12-cm-high, and 18-cm-long circuit board houses 80 closely spaced logic chips on one side, each dissipating 0.04 W. The board is impregnated with copper fillings and has an effective thermal conductivity of 16 W/m $\cdot °C$. All the heat generated in the chips is conducted across the circuit board and is dissipated from the back side of the board to the ambient air. Determine the temperature difference between the two sides of the circuit board.   *Answer:* $0.028°C$

1-92 Consider a sealed 20-cm-high electronic box whose base dimensions are 40 cm $\times$ 40 cm placed in a vacuum chamber. The emissivity of the outer surface of the box is 0.95. If the electronic components in the box dissipate a total of 100 W of power and the outer surface temperature of the box is not to exceed $55°C$, determine the temperature at which the surrounding surfaces must be kept if this box is to be cooled by radiation alone. Assume the heat transfer from the bottom surface of the box to the stand to be negligible.

1-93 Using the conversion factors between W and Btu/h, m and ft, and K and R, express the Stefan-Boltzmann constant $\sigma = 5.67 \times 10^{-8}$ W/m$^2 \cdot$ K^4 in the English unit Btu/h $\cdot$ ft$^2 \cdot$ R^4.

1-94 An engineer who is working on the heat transfer analysis of a house in English units needs the convection heat transfer coefficient on the outer surface of the house. But the only value he can find from his handbooks is 20 W/m$^2 \cdot °C$, which is in SI units. The engineer does not have a direct conversion factor between the two unit systems for the convection heat transfer coefficient. Using the conversion factors between W and Btu/h, m and ft, and $°C$ and $°F$, express the given convection heat transfer coefficient in Btu/h $\cdot$ ft$^2 \cdot °F$.   *Answer:* 3.52 Btu/h $\cdot$ ft$^2 \cdot °C$

1-95C Can all three modes of heat transfer occur simultaneously (in parallel) in a medium?

1-96C Can a medium involve (*a*) conduction and convection, (*b*) conduction and radiation, or (*c*) convection and radiation simultaneously? Give examples for the "yes" answers.

1-97C The deep human body temperature of a healthy person remains constant at 37°C while the temperature and the humidity of the environment change with time. Discuss the heat transfer mechanisms between the human body and the environment both in summer and winter, and explain how a person can keep cooler in summer and warmer in winter.

1-98C We often turn the fan on in summer to help us cool. Explain how a fan makes us feel cooler in the summer. Also explain why some people use ceiling fans also in winter.

1-99 Consider a person standing in a room at 23°C. Determine the total rate of heat transfer from this person if the exposed surface area and the skin temperature of the person are 1.7 m^2 and 32°C, respectively, and the convection heat transfer coefficient is 5 W/m$^2 \cdot$ °C. Take the emissivity of the skin and the clothes to be 0.9, and assume the temperature of the inner surfaces of the room to be the same as the air temperature.   *Answer:* 161.3 W

1-100 Consider steady heat transfer between two large parallel plates at constant temperatures of $T_1 = 290$ K and $T_2 = 150$ K that are $L = 2$ cm apart. Assuming the surfaces to be black (emissivity $\epsilon = 1$), determine the rate of heat transfer between the plates per unit surface area assuming the gap between the plates is (*a*) filled with atmospheric air, (*b*) evacuated, (*c*) filled with fiberglass insulation, and (*d*) filled with superinsulation having an apparent thermal conductivity of 0.00015 W/m $\cdot$ °C.

1-101 A 1.4-m-long, 0.2-cm-diameter electrical wire extends across a room that is maintained at 20°C. Heat is generated in the wire as a result of resistance heating, and the surface temperature of the wire is measured to be 240°C in steady operation. Also, the voltage drop and electric current through the wire are measured to be 110 V and 3 A, respectively. Disregarding any heat transfer by radiation, determine the convection heat transfer coefficient for heat transfer between the outer surface of the wire and the air in the room.   *Answer:* 170.5 W/m$^2 \cdot$ °C

1-102E A 2-in-diameter spherical ball whose surface is maintained at a temperature of 170°F is suspended in the middle of a room at 70°F. If the convection heat transfer coefficient is 3 Btu/h $\cdot$ ft$^2 \cdot$ °F and the emissivity of the surface is 0.8, determine the total rate of heat transfer from the ball.

1-103 A 1000-W iron is left on the iron board with its base exposed to the air at 20°C. The convection heat transfer coefficient between the base surface and the surrounding air is 35 W/m$^2 \cdot$ °C. If the base has an emissivity of 0.6

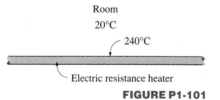

Room
20°C

240°C

Electric resistance heater

FIGURE P1-101

FIGURE P1-103

Iron
1000 W

20°C

and a surface area of 0.02 m², determine the temperature of the base of the iron.   *Answer:* 674°C

1-104 The outer surface of a spacecraft in space has an emissivity of 0.8 and an absorptivity of 0.3 for solar radiation. If solar radiation is incident on the spacecraft at a rate of 1000 W/m², determine the surface temperature of the spacecraft when the radiation emitted equals the solar energy absorbed.

1-105 A 3-m-internal-diameter spherical tank made of 1-cm-thick stainless steel is used to store iced water at 0°C. The tank is located outdoors at 25°C. Assuming the entire steel tank to be at 0°C and thus the thermal resistance of the tank to be negligible, determine (*a*) the rate of heat transfer to the iced water in the tank and (*b*) the amount of ice at 0°C that melts during a 24-hour period. The heat of fusion of water at atmospheric pressure is h_{if} = 333.7 kJ/kg. The emissivity of the outer surface of the tank is 0.6, and the convection heat transfer coefficient on the outer surface can be taken to be 30 W/m² · °C. Assume the average surrounding surface temperature for radiation exchange to be 15°C.   *Answer:* 5898 kg

1-106 The roof of a house consists of a 15-cm-thick concrete slab (k = 2 W/m · °C) that is 15 m wide and 20 m long. The emissivity of the outer surface of the roof is 0.9, and the convection heat transfer coefficient on that surface is estimated to be 15 W/m² · °C. The inner surface of the roof is maintained at 15°C. On a clear winter night, the ambient air is reported to be at 10°C while the night sky temperature for radiation heat transfer is 255 K. Considering both radiation and convection heat transfer, determine the outer surface temperature and the rate of heat transfer through the roof.
   If the house is heated by a furnace burning natural gas with an efficiency of 85 percent, and the unit cost of natural gas is $0.60/therm (1 therm = 105,500 kJ of energy content), determine the money lost through the roof that night during a 14-hour period.

1-107E Consider a flat plate solar collector placed horizontally on the flat roof of a house. The collector is 5 ft wide and 15 ft long, and the average temperature of the exposed surface of the collector is 100°F. The emissivity of the exposed surface of the collector is 0.9. Determine the rate of heat loss from the collector by convection and radiation during a calm day when the ambient air temperature is 70°F and the effective sky temperature for radiation exchange is 50°F. Take the convection heat transfer coefficient on the exposed surface to be 2.5 Btu/h · ft² · °F.

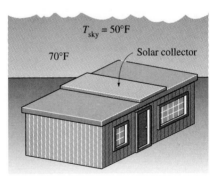

T_{sky} = 50°F

70°F   Solar collector

FIGURE P1-107E

Review Problems

1-108 The average atmospheric temperature and pressure in the world is approximated as a function of altitude by the relations

$$T_{atm} = 288.15 - 65z \quad \text{and} \quad P_{atm} = 101.325(1 - 0.02256z)^{5.256}$$

where T_{atm} is the temperature of the atmosphere in K, P_{atm} is the atmospheric pressure in kPa, and z is the altitude in km, with z = 0 at sea level. Determine the approximate atmospheric temperatures and pressures at Atlanta ($z = 306$ m), Denver ($z = 1610$ m), Mexico City ($z = 2309$ m), and the top of Mount Everest ($z = 8848$ m).

1-109 2.5 kg of liquid water initially at 18°C is to be heated to 96°C in a teapot equipped with a 1500-W electric heating element inside. The teapot is 0.8 kg and has an average specific heat of 0.6 kJ/kg · °C. Taking the specific heat of water to be 4.18 kJ/kg · °C and disregarding any heat loss from the teapot, determine how long it will take for the water to be heated.

1-110 A 4-m-long section of an air heating system of a house passes through an unheated space in the attic. The inner diameter of the circular duct of the heating system is 20 cm. Hot air enters the duct at 100 kPa and 65°C at an average velocity of 3 m/s. The temperature of the air in the duct drops to 60°C as a result of heat loss to the cool space in the attic. Determine the rate of heat loss from the air in the duct to the attic under steady conditions. Also, determine the cost of this heat loss per hour if the house is heated by a natural gas furnace having an efficiency of 82 percent, and the cost of the natural gas in that area is $0.58/therm (1 therm = 105,500 kJ).

 Answers: 0.488 kJ/s, $0.012/h

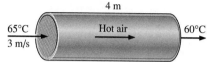

FIGURE P1-110

1-111E A 15-in-diameter watermelon is to be cooled from 75°F to 50°F in a refrigerator. Previous observations indicate that heat is removed from the watermelon at an average rate of 200 Btu/h. Using the properties of water for the watermelon, determine (*a*) the average heat flux on the surface of the watermelon, in Btu/h · ft², and (*b*) how long it will take to cool the watermelon.   *Answers:* (*a*) 40.8 Btu/h · ft², (*b*) 7.9 h

1-112 Water flows through a shower head steadily at a rate of 10 L/min. An electric resistance heater placed in the water pipe heats the water from 16°C to 43°C. Taking the density of water to be 1 kg/L, determine the electric power input to the heater, in kW.

 In an effort to conserve energy, it is proposed to pass the drained warm water at a temperature of 39°C through a heat exchanger to preheat the incoming cold water. If the heat exchanger has an effectiveness of 0.50 (that is, it recovers only half of the energy that can possibly be transferred from the drained water to incoming cold water), determine the electric power input required in this case. If the price of the electric energy is 8.5 ¢/kWh, determine how much money is saved during a 10-minute shower as a result of installing this heat exchanger.   *Answers:* 18.8 kW, 10.8 kW, $0.0113

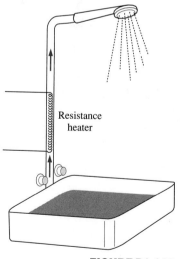

FIGURE P1-112

1-113 It is proposed to have a water heater that consists of an insulated pipe of 5 cm diameter and an electrical resistor inside. Cold water at 15°C enters the heating section steadily at a rate of 30 L/min. If water is to be heated to 50°C, determine (*a*) the power rating of the resistance heater and (*b*) the average velocity of the water in the pipe.

1-114 A passive solar house that is losing heat to the outdoors at an average rate of 50,000 kJ/h is maintained at 22°C at all times during a winter night for 10 hours. The house is to be heated by 50 glass containers each containing 20 L of water heated to 80°C during the day by absorbing solar energy. A

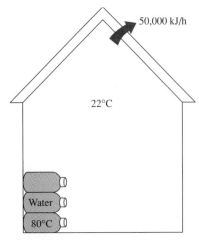

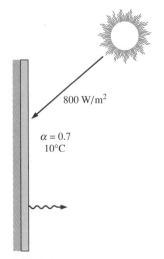

FIGURE P1-114

FIGURE P1-116

thermostat-controlled 15-kW back-up electric resistance heater turns on whenever necessary to keep the house at 22°C. (*a*) How long did the electric heating system run that night? (*b*) How long would the electric heater have run that night if the house incorporated no solar heating?

Answers: (*a*) 4.77 h, (*b*) 9.26 h

1-115 It is well known that wind makes the cold air feel much colder as a result of the *windchill* effect that is due to the increase in the convection heat transfer coefficient with increasing air velocity. The windchill effect is usually expressed in terms of the *windchill factor*, which is the difference between the actual air temperature and the equivalent calm-air temperature. For example, a windchill factor of 20°C for an actual air temperature of 5°C means that the windy air at 5°C feels as cold as the still air at −15°C. In other words, a person will lose as much heat to air at 5°C with a windchill factor of 20°C as he or she would in calm air at −15°C.

For heat transfer purposes, a standing man can be modeled as a 30-cm-diameter, 170-cm-long vertical cylinder with both the top and bottom surfaces insulated and with the side surface at an average temperature of 34°C. For a convection heat transfer coefficient of 15 W/m² · °C, determine the rate of heat loss from this man by convection in still air at 20°C. What would your answer be if the convection heat transfer coefficient is increased to 50 W/m² · °C as a result of winds? What is the windchill factor in this case?

Answers: 336 W, 1120 W, 32.7°C

1-116 A thin metal plate is insulated on the back and exposed to solar radiation on the front surface. The exposed surface of the plate has an absorptivity of 0.7 for solar radiation. If solar radiation is incident on the plate at a rate of 800 W/m² and the surrounding air temperature is 10°C, determine the surface temperature of the plate when the heat loss by convection equals the solar energy absorbed by the plate. Take the convection heat transfer coefficient to be 30 W/m² · °C, and disregard any heat loss by radiation.

1-117E One short ton (2000 lbm) of liquid water at 180°F is brought into a well-insulated and well-sealed 10-ft × 10-ft × 15-ft room initially at 72°F and 14.6 psia. Assuming constant specific heats for both air and water at room temperature, determine the final equilibrium temperature in the room.

1-118 A 4-m × 5-m × 6-m room is to be heated by one ton (1000 kg) of liquid water contained in a tank placed in the room. The room is losing heat to the outside at an average rate of 10,000 kJ/h. The room is initially at 20°C and 100 kPa, and is maintained at an average temperature of 20°C at all times. If the hot water is to meet the heating requirements of this room for a 24-hour period, determine the minimum temperature of the water when it is first brought into the room. Assume constant specific heats for both air and water at room temperature. *Answer:* 77.4°C

1-119 Consider a 3-m × 3-m × 3-m cubical furnace whose top and side surfaces closely approximate black surfaces at a temperature of 1200 K. The base surface has an emissivity of $\epsilon = 0.7$, and is maintained at 800 K. Determine the net rate of radiation heat transfer to the base surface from the top and side surfaces. *Answer:* 594,400 W

1-120   Consider a refrigerator whose dimensions are 1.8 m × 1.2 m × 0.8 m and whose walls are 3 cm thick. The refrigerator consumes 600 W of power when operating and has a COP of 2.5. It is observed that the motor of the refrigerator remains on for 5 minutes and then is off for 15 minutes periodically. If the average temperatures at the inner and outer surfaces of the refrigerator are 6°C and 17°C, respectively, determine the average thermal conductivity of the refrigerator walls. Also, determine the annual cost of operating this refrigerator if the unit cost of electricity is $0.06/kWh.

1-121   The energy content of a certain food is to be determined in a bomb calorimeter containing 3 kg of water by burning a 2-g sample of it in the presence of 100 g of air in the reaction chamber. If the water temperature rises by 3.2°C when equilibrium is established, determine the energy content of the food, in kJ/kg, by neglecting the thermal energy stored in the reaction chamber and the energy supplied by the mixer. What is a rough estimate of the error involved in neglecting the thermal energy stored in the reaction chamber?   *Answer:* 20,083 kJ/kg

1-122   A 68-kg man whose average body temperature is 38°C drinks one liter of cold water at 3°C in an effort to cool down. Taking the average specific heat of the human body to be 3.6 kJ/kg · °C, determine the drop in the average body temperature of this person under the influence of this cold water.

1-123   A 0.2-L glass of water at 20°C is to be cooled with ice to 5°C. Determine how much ice needs to be added to the water, in grams, if the ice is at 0°C. Also, determine how much water would be needed if the cooling is to be done with cold water at 0°C. The melting temperature and the heat of fusion of ice at atmospheric pressure are 0°C and 333.7 kJ/kg, respectively, and the density of water is 1 kg/L.

1-124E   In order to cool 1 short ton (2000 lbm) of water at 70°F in a tank, a person pours 160 lbm of ice at 25°F into the water. Determine the final equilibrium temperature in the tank. The melting temperature and the heat of fusion of ice at atmospheric pressure are 32°F and 143.5 Btu/lbm, respectively.   *Answer:* 56.3°F

Computer, Design, and Essay Problems

1-125   Find out how the specific heats of gases, liquids, and solids are determined in national laboratories. Describe the experimental apparatus and the procedures used.

1-126   Using information from the utility bills for the coldest month last year, estimate the average rate of heat loss from your house for that month. In your analysis, consider the contribution of the internal heat sources such as people, lights, and appliances. Identify the primary sources of heat loss from your house and propose ways of improving the energy efficiency of your house.

1-127   Design an experiment complete with instrumentation to determine the specific heats of a gas using a resistance heater. Discuss how the

17°C →
6°C →

FIGURE P1-120

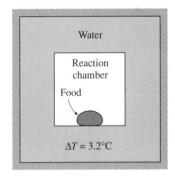

Water

Reaction
chamber

Food

ΔT = 3.2°C

FIGURE P1-121

FIGURE P1-123

Ice, 0°C

Water
0.2 L
20°C

experiment will be conducted, what measurements need to taken, and how the specific heats will be determined. What are the sources of error in your system? How can you minimize the experimental error?

1-128　Design an experiment complete with instrumentation to determine the specific heat of a liquid using a resistance heater. Discuss how the experiment will be conducted, what measurements need to taken, and how the specific heats will be determined. What are the sources of error in your system? How can you minimize the experimental error? How would you modify this system to determine the specific heat of a solid?

1-129　Write an essay on how microwave ovens work, and explain how they cook much faster than conventional ovens. Discuss whether conventional electric or microwave ovens consume more electricity for the same task.

1-130　Design a 1200-W electric hair dryer such that the air temperature and velocity in the dryer will not exceed 50°C and 3 m/s, respectively.

1-131　Design an electric hot water heater for a family of four in your area. The maximum water temperature in the tank and the power consumption are not to exceed 60°C and 4 kW, respectively. There are two showers in the house, and the flow rate of water through each of the shower heads is about 10 L/min. Each family member takes a 5-minute shower every morning. Explain why a hot water tank is necessary, and determine the proper size of the tank for this family.

Heat Conduction Equation

Heat transfer has *direction* as well as *magnitude*. The rate of heat conduction in a specified direction is proportional to the *temperature gradient,* which is the change in temperature per unit length in that direction. Heat conduction in a medium, in general, is three-dimensional and time dependent. That is, $T = T(x, y, z, t)$ and the temperature in a medium varies with position as well as time. Heat conduction in a medium is said to be *steady* when the temperature does not vary with time, and *unsteady* or *transient* when it does. Heat conduction in a medium is said to be *one-dimensional* when conduction is significant in one dimension only and negligible in the other two dimensions, *two-dimensional* when conduction in the third dimension is negligible, and *three-dimensional* when conduction in all dimensions is significant.

We start this chapter with a description of steady, unsteady, and multi-dimensional heat conduction. After a brief review of differential equations, we derive the differential equation that governs heat conduction in a large plane wall, a long cylinder, and a sphere, and generalize the results to three-dimensional cases in rectangular, cylindrical, and spherical coordinates. Following a discussion of the boundary conditions, we present the formulation of heat conduction problems and their solutions. Finally, we consider heat conduction problems with variable thermal conductivity.

This chapter deals with the theoretical and mathematical aspects of heat conduction, and it can be covered selectively, if desired, without causing a significant loss in continuity. The more practical aspects of heat conduction are covered in the following two chapters.

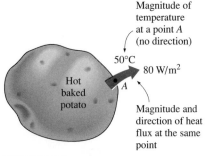

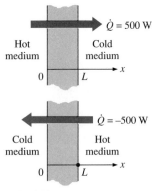

FIGURE 2-1

Heat transfer has direction as well as magnitude, and thus it is a *vector* quantity.

FIGURE 2-2

Sign convention for heat transfer (positive in the positive direction; negative in the negative direction).

2-1 ■ INTRODUCTION

In Chapter 1 heat conduction was defined as the transfer of thermal energy from the more energetic particles of a medium to the adjacent less energetic ones as a result of interactions between the particles. It was stated that conduction can take place in liquids and gases as well as solids provided that there is no bulk motion involved in the liquid or the gas.

Although heat transfer and temperature are closely related, they are of a different nature. Unlike temperature, heat transfer has direction as well as magnitude, and thus it is a *vector* quantity (Fig. 2-1). Therefore, we must specify both direction and magnitude in order to describe heat transfer completely at a point. For example, saying that the temperature on the inner surface of a wall is 18°C describes the temperature at that location fully. But saying that the heat flux on that surface is 50 W/m² immediately prompts the question "in what direction?" We can answer this question by saying that heat conduction is toward the inside (indicating heat gain for the house) or toward the outside (indicating heat loss for the house).

To avoid such questions, we can work with a coordinate system and adopt a **sign convention** to indicate direction with plus or minus signs. The universally accepted sign convention for heat transfer is that heat transfer in the positive direction of a coordinate axis is positive and in the opposite direction it is negative. Therefore, a positive quantity indicates heat transfer in the positive direction and a negative quantity indicates heat transfer in the negative direction (Fig. 2-2).

The driving force for any form of heat transfer is the *temperature difference,* and the larger the temperature difference, the larger the rate of heat transfer. Some heat transfer problems in engineering require the determination of the *temperature distribution* (the variation of temperature) throughout the medium in order to calculate some quantities of interest such as the local heat transfer rate, thermal expansion, and thermal stress at some critical locations at specified times. The specification of the *temperature* at a point in a medium first requires the specification of the *location* of that point. This can be done by choosing a suitable coordinate system such as the *rectangular, cylindrical,* or *spherical* coordinates, depending on the geometry involved, and a convenient reference point (the origin).

The *location* of a point is specified as (x, y, z) in rectangular coordinates, as (r, ϕ, z) in cylindrical coordinates, and as (r, ϕ, θ) in spherical coordinates, where the distances x, y, z, and r and the angles ϕ and θ are as shown in Figure 2-3. Then the temperature at a point (x, y, z) at time t in rectangular coordinates is expressed as $T(x, y, z, t)$. The best coordinate system for a given geometry is the one that describes the surfaces of the geometry best. For example, a parallelepiped is best described in rectangular coordinates since each surface can be described by a constant value of the x, y, or z coordinates. A cylinder is best suited for cylindrical coordinates since its lateral surface can be described by a constant value of the radius. Similarly, the entire outer surface of a spherical body can best be described by a constant value of the radius in spherical coordinates. For an arbitrarily shaped

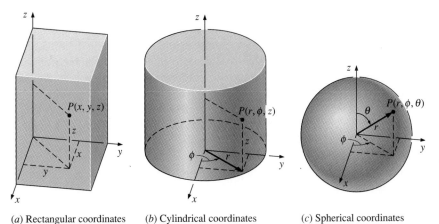

(a) Rectangular coordinates (b) Cylindrical coordinates (c) Spherical coordinates

FIGURE 2-3

The various distances and angles involved when describing the location of a point in different coordinate systems.

body, we normally use rectangular coordinates since it is easier to deal with distances than with angles.

The notation described above is also used to identify the variables involved in a heat transfer problem. For example, the notation $T(x, y, z, t)$ implies that the temperature varies with the space variables x, y, and z as well as time. The notation $T(x)$, on the other hand, indicates that the temperature varies in the x direction only and there is no variation with the other two space coordinates or time.

Steady versus Transient Heat Transfer

Heat transfer problems are often classified as being **steady** (also called *steady-state*) or **transient** (also called *unsteady*). The term *steady* implies *no change* with time at any point within the medium, while *transient* implies *variation with time* or *time dependence*. Therefore, the temperature or heat flux remains unchanged with time during steady heat transfer through a medium at any location, although both quantities may vary from one location to another (Fig. 2-4). For example, heat transfer through the walls of a house will be steady when the conditions inside the house and the outdoors remain constant for several hours. But even in this case, the temperatures on the inner and outer surfaces of the wall will be different unless the temperatures inside and outside the house are the same. The cooling of an apple in a refrigerator, on the other hand, is a transient heat transfer process since the temperature at any fixed point within the apple will change with time during cooling. During transient heat transfer, the temperature normally varies with time as well as position. In the special case of variation with time but not with position, the temperature of the medium changes *uniformly* with time. Such heat transfer systems are called **lumped systems.** A small metal object such as a thermocouple junction or a thin copper wire, for example, can be analyzed as a lumped system during a heating or cooling process.

FIGURE 2-4

Steady and transient heat conduction in a plane wall.

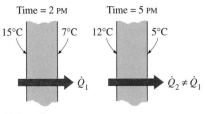

(a) Transient

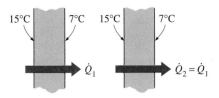

(b) Steady-state

Most heat transfer problems encountered in practice are *transient* in nature, but they are usually analyzed under some presumed *steady* conditions since steady processes are easier to analyze, and they provide the answers to our questions. For example, heat transfer through the walls and ceiling of a typical house is never steady since the outdoor conditions such as the temperature, the speed and direction of the wind, the location of the sun, and so on, change constantly. The conditions in a typical house are not so steady either. Therefore, it is almost impossible to perform a heat transfer analysis of a house accurately. But then, do we really need an in-depth heat transfer analysis? If the purpose of a heat transfer analysis of a house is to determine the proper size of a heater, which is usually the case, we need to know the *maximum* rate of heat loss from the house, which is determined by considering the heat loss from the house under *worst* conditions for an extended period of time, that is, during *steady* operation under worst conditions. Therefore, we can get the answer to our question by doing a heat transfer analysis under steady conditions. If the heater is large enough to keep the house warm under the presumed worst conditions, it is large enough for all conditions. The approach described above is a common practice in engineering.

Multidimensional Heat Transfer

Heat transfer problems are also classified as being *one-dimensional, two-dimensional,* or *three-dimensional,* depending on the relative magnitudes of heat transfer rates in different directions and the level of accuracy desired. In the most general case, heat transfer through a medium is **three-dimensional.** That is, the temperature varies along all three primary directions within the medium during the heat transfer process. The temperature distribution throughout the medium at a specified time as well as the heat transfer rate at any location in this general case can be described by a set of three coordinates such as the x, y, and z in the rectangular (or Cartesian) coordinate system; the r, ϕ, and z in the cylindrical coordinate system; and the r, ϕ, and θ in the spherical (or polar) coordinate system. The temperature distribution in this general case is expressed as $T(x, y, z, t)$, $T(r, \phi, z, t)$, and $T(r, \phi, \theta, t)$ in the respective coordinate systems.

The temperature in a medium, in some cases, varies mainly in two primary directions, and the variation of temperature in the third direction (and thus heat transfer in that direction) is negligible. A heat transfer problem in that case is said to be **two-dimensional.** For example, the steady temperature distribution in a long bar of rectangular cross-section can be expressed as $T(x,y)$ if the temperature variation in the z direction (along the bar) is negligible and there is no change with time (Fig. 2-5).

A heat transfer problem is said to be **one-dimensional** if the temperature in the medium varies in one direction only and thus heat is transferred in one direction, and the variation of temperature and thus heat transfer in other directions are negligible or zero. For example, heat transfer through the glass of a window can be considered to be one-dimensional since heat transfer through the glass will occur predominantly in one direction (the direction

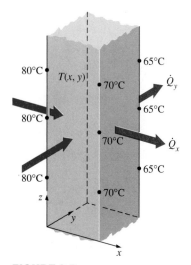

FIGURE 2-5

Two-dimensional heat transfer in a long rectangular bar.

normal to the surface of the glass) and heat transfer in other directions (from one side edge to the other and from the top edge to the bottom) is negligible (Fig. 2-6). Likewise, heat transfer through a hot water pipe can be considered to be one-dimensional since heat transfer through the pipe occurs predominantly in the radial direction from the hot water to the ambient, and heat transfer along the pipe and along the circumference of a cross-section (z and ϕ directions) is typically negligible. Heat transfer to an egg dropped into boiling water is also one-dimensional because of symmetry. Heat will be transferred to the egg in this case in the radial direction, that is, along straight lines passing through the midpoint of the egg.

We also mentioned in Chapter 1 that the rate of heat conduction through a medium in a specified direction (say, in the x direction) is proportional to the temperature difference across the medium and the area normal to the direction of heat transfer, but is inversely proportional to the distance in that direction. This was expressed in the differential form by **Fourier's law of heat conduction** for one-dimensional heat conduction as

$$\dot{Q}_{cond} = -kA\frac{dT}{dx} \qquad \text{(W)} \qquad (2\text{-}1)$$

where k is the *thermal conductivity* of the material, which is a measure of the ability of a material to conduct heat, and dT/dx is the *temperature gradient,* which is the slope of the temperature curve on a T-x diagram (Fig. 2-7). The thermal conductivity of a material, in general, varies with temperature. But sufficiently accurate results can be obtained by using a constant value for thermal conductivity at the *average* temperature.

Heat is conducted in the direction of decreasing temperature, and thus the temperature gradient is negative when heat is conducted in the positive x direction. Therefore, a *negative sign* is added to Equation 2-1 to make heat transfer in the positive x direction a positive quantity.

To obtain a general relation for Fourier's law of heat conduction, consider a medium in which the temperature distribution is three-dimensional. Figure 2-8 shows an isothermal surface in that medium. The heat flux vector at a point P on this surface must be perpendicular to the surface, and it must point in the direction of decreasing temperature. If n is the normal of the isothermal surface at point P, the rate of heat conduction at that point can be expressed from Fourier's law as

$$\dot{Q}_n = -kA\frac{dT}{dn} \qquad \text{(W)} \qquad (2\text{-}2)$$

In rectangular coordinates, the heat conduction vector can be expressed in terms of its components as

$$\vec{\dot{Q}}_n = \dot{Q}_x\,\vec{i} + \dot{Q}_y\,\vec{j} + \dot{Q}_z\,\vec{k} \qquad (2\text{-}3)$$

where $\vec{i}$, $\vec{j}$, and $\vec{k}$ are the unit vectors, and $\dot{Q}_x$, $\dot{Q}_y$, and $\dot{Q}_z$ are the magnitudes of the heat transfer rates in the x, y, and z directions, which again can be determined from Fourier's law as

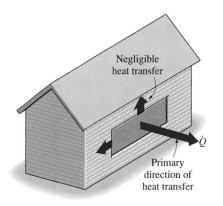

FIGURE 2-6
Heat transfer through the window of a house can be taken to be one-dimensional.

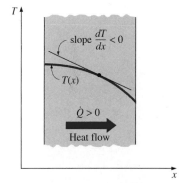

FIGURE 2-7
The temperature gradient dT/dx is simply the slope of the temperature curve on a T-x diagram.

$$\dot{Q}_x = -kA_x\frac{\partial T}{\partial x}, \qquad \dot{Q}_y = -kA_y\frac{\partial T}{\partial y}, \qquad \text{and} \qquad \dot{Q}_z = kA_z\frac{\partial T}{\partial z} \qquad (2\text{-}4)$$

Here A_x, A_y and A_z are heat conduction areas normal to the x, y, and z directions, respectively (Fig. 2-8).

Most engineering materials are *isotropic* in nature, and thus they have the same properties in all directions. For such materials we do not need to be concerned about the variation of properties with direction. But in *anisotropic* materials such as the fibrous or composite materials, the properties may change with direction. For example, some of the properties of wood along the grain are different than those in the direction normal to the grain. In such cases the thermal conductivity may need to be expressed as a tensor quantity to account for the variation with direction. The treatment of such advance topics is beyond the scope of this text, and we will assume the thermal conductivity of a material to be independent of direction.

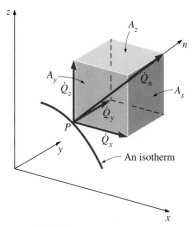

FIGURE 2-8

The heat transfer vector is always normal to an isothermal surface and can be resolved into its components like any other vector.

Heat Generation

A medium through which heat is conducted may involve the conversion of electrical, nuclear, or chemical energy into heat (or thermal) energy. In heat conduction analysis, such conversion processes are characterized as **heat generation.**

For example, the temperature of a resistance wire rises rapidly when electric current passes through it as a result of the electrical energy being converted to heat at a rate of I^2R, where I is the current and R is the electrical resistance of the wire (Fig. 2-9). The safe and effective removal of this heat away from the sites of heat generation (the electronic circuits) is the subject of *electronics cooling,* which is one of the major application areas of heat transfer.

Likewise, a large amount of heat is generated in the fuel elements of nuclear reactors as a result of nuclear fission that serves as the *heat source* for the nuclear power plants. The natural disintegration of radioactive elements in nuclear waste or other radioactive material also results in the generation of heat throughout the body. The heat generated in the sun as a result of the fusion of hydrogen into helium makes the sun a large nuclear reactor that supplies heat to the heavens and the earth.

Another source of heat generation in a medium is exothermic chemical reactions that may occur throughout the medium. The chemical reaction in this case serves as a *heat source* for the medium. In the case of endothermic reactions, however, heat is absorbed instead of being released during reaction, and thus the chemical reaction serves as a *heat sink*. The heat generation term becomes a negative quantity in this case.

Often it is also convenient to model the absorption of radiation such as solar energy or gamma rays as heat generation when these rays penetrate deep into the body while being absorbed gradually. For example, the absorption of solar energy in large bodies of water can be treated as heat generation throughout the water at a rate equal to the rate of absorption, which varies

FIGURE 2-9

Heat is generated in the heating coils of an electric range as a result of the conversion of electrical energy to heat.

with depth (Fig. 2-10). But the absorption of solar energy by an opaque body occurs within a few microns of the surface, and the solar energy that penetrates into the medium in this case can be treated as specified heat flux on the surface.

Note that heat generation is a *volumetric phenomenon*. That is, it occurs throughout the body of a medium. Therefore, the rate of heat generation in a medium is usually specified *per unit volume* and is denoted by $\dot{g}$, whose unit is W/m^3 or Btu/h $\cdot$ ft^3.

The rate of heat generation in a medium may vary with time as well as position within the medium. When the variation of heat generation with position is known, the *total* rate of heat generation in a medium of volume V can be determined from

$$\dot{G} = \int_V \dot{g}dV \qquad \text{(W)} \qquad (2\text{-}5)$$

In the special case of *uniform* heat generation, as in the case of electric resistance heating throughout a homogeneous material, the relation above reduces to $\dot{G} = \dot{g}V$ where $\dot{g}$ is the constant rate of heat generation per unit volume.

EXAMPLE 2-1   Heat Gain by a Refrigerator

In order to size the compressor of a new refrigerator, it is desired to determine the rate of heat transfer from the kitchen air into the refrigerated space through the walls, door, and the top and bottom section of the refrigerator (Fig. 2-11). In your analysis, would you treat this as a transient or steady-state heat transfer problem? Also, would you consider the heat transfer to be one-dimensional or multi-dimensional? Explain.

Solution   The heat transfer process from the kitchen air to the refrigerated space is transient in nature since the thermal conditions in the kitchen and the refrigerator, in general, change with time. However, we would analyze this problem as a steady heat transfer problem under the worst anticipated conditions such as the lowest thermostat setting for the refrigerated space, and the anticipated highest temperature in the kitchen (the so-called design conditions). If the compressor is large enough to keep the refrigerated space at the desired temperature setting under the presumed worst conditions, then it is large enough to do so under all conditions by cycling on and off.

Heat transfer into the refrigerated space is three-dimensional in nature since heat will be entering through all six sides of the refrigerator. However, heat transfer through any wall or floor takes place in the direction normal to the surface, and thus it can be analyzed as being one-dimensional. Therefore, this problem can be simplified greatly by considering the heat transfer to be one-dimensional at each of the four sides as well as the top and bottom sections, and then by adding the calculated values of heat transfer at each surface.

EXAMPLE 2-2   Heat Generation in a Hair Dryer

The resistance wire of a 1200-W hair dryer is 80 cm long and has a diameter of $D = 0.3$ cm (Fig. 2-12). Determine the rate of heat generation in the wire per unit

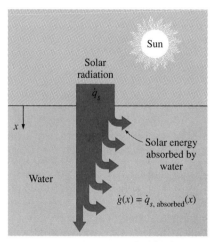

FIGURE 2-10
The absorption of solar radiation by water can be treated as heat generation.

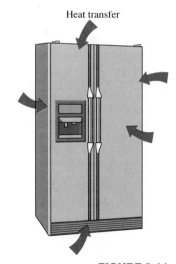

FIGURE 2-11
Schematic for Example 2-1.

Hair dryer
1200 W

FIGURE 2-12
Schematic for Example 2-2.

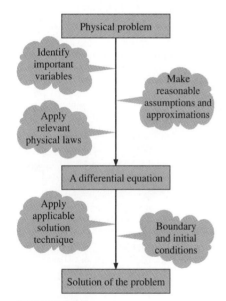

FIGURE 2-13
Mathematical modeling of
physical problems.

volume, in W/cm³, and the heat flux on the outer surface of the wire as a result of this heat generation.

Solution The power consumed by the resistance wire of a hair dryer is given. The heat generation and the heat flux are to be determined.

Assumptions Heat is generated uniformly in the resistance wire.

Analysis A 1200-W hair dryer will convert electrical energy into heat in the wire at a rate of 1200 W. Therefore, the rate of heat generation in a resistance wire is simply equal to the power rating of a resistance heater. Then the rate of heat generation in the wire per unit volume is determined by dividing the total rate of heat generation by the volume of the wire,

$$\dot{g} = \frac{\dot{G}}{V_{wire}} = \frac{\dot{G}}{(\pi D^2/4)L} = \frac{1200 \text{ W}}{[\pi(0.3 \text{ cm})^2/4](80 \text{ cm})} = \textbf{212.2 W/cm}^3$$

Similarly, heat flux on the outer surface of the wire as a result of this heat generation is determined by dividing the total rate of heat generation by the surface area of the wire,

$$\dot{q} = \frac{\dot{G}}{A_{wire}} = \frac{\dot{G}}{\pi DL} = \frac{1200 \text{ W}}{\pi(0.3 \text{ cm})(80 \text{ cm})} = \textbf{15.9 W/cm}^2$$

Discussion Note that heat generation is expressed per unit volume in W/cm³ or Btu/h · ft³, whereas heat flux is expressed per unit surface area in W/cm² or Btu/h · ft².

2-2 ■ A BRIEF REVIEW OF DIFFERENTIAL EQUATIONS*

As we mentioned in Chapter 1, the description of most scientific problems involves relations that involve the changes in some key variables with respect to each other. Usually the smaller the increment chosen in the changing variables, the more general and accurate the description. In the limiting case of infinitesimal or differential changes in variables, we obtain *differential equations,* which provide precise mathematical formulations for the physical principles and laws by representing the rates of change as *derivatives.* Therefore, differential equations are used to investigate a wide variety of problems in science and engineering, including heat transfer.

Differential equations arise when relevant *physical laws* and *principles* are applied to a problem by considering infinitesimal changes in the variables of interest. Therefore, obtaining the governing differential equation for a specific problem requires an adequate knowledge of the nature of the problem, the variables involved, appropriate simplifying assumptions, and the applicable physical laws and principles involved, as well as a careful analysis (Fig. 2-13).

*This section is given as a review for the students, and can be skipped if desired without a loss in continuity.

An equation, in general, may involve one or more variables. As the name implies, a **variable** is a quantity that may assume various values during a study. A quantity whose value is fixed during a study is called a **constant.** Constants are usually denoted by the earlier letters of the alphabet such as a, b, c, and d, whereas variables are usually denoted by the later ones such as t, x, y, and z. A variable whose value can be changed arbitrarily is called an **independent variable** (or argument). A variable whose value depends on the value of other variables and thus cannot be varied independently is called a **dependent variable** (or a function).

A dependent variable y that depends on a variable x is usually denoted as $y(x)$ for clarity. However, this notation becomes very inconvenient and cumbersome when y is repeated several times in an expression. In such cases it is desirable to denote $y(x)$ simply as y when it is clear that y is a function of x. This shortcut in notation improves the appearance and the readability of the equations. The value of y at a fixed number a is denoted by $y(a)$.

The **derivative** of a function $y(x)$ at a point is equivalent to the *slope* of the tangent line to the graph of the function at that point and is defined as (Fig. 2-14)

$$y'(x) = \frac{dy(x)}{dx} = \lim_{\Delta x \to 0} \frac{\Delta y}{\Delta x} = \lim_{\Delta x \to 0} \frac{y(x + \Delta x) - y(x)}{\Delta x} \qquad (2\text{-}6)$$

Here Δx represents a (small) change in the independent variable x and is called an *increment* of x. The corresponding change in the function y is called an increment of y and is denoted by Δy. Therefore, the derivative of a function can be viewed as the ratio of the increment Δy of the function to the increment Δx of the independent variable for very small Δx. Note that Δy and thus $y'(x)$ will be zero if the function y does not change with x.

Most problems encountered in practice involve quantities that change with time t, and their first derivatives with respect to time represent the rate of change of those quantities with time. For example, if $N(t)$ denotes the population of a bacteria colony at time t, then the first derivative $N' = dN/dt$ represents the rate of change of the population, which is the amount the population increases or decreases per unit time.

The derivative of the first derivative of a function y is called the *second derivative* of y, and is denoted by y'' or d^2y/dy^2. In general, the derivative of the $(n - 1)$st derivative of y is called the nth derivative of y and is denoted by $y^{(n)}$ or d^ny/dx^n. Here, n is a positive integer and is called the **order** of the derivative. The order n should not be confused with the *degree* of a derivative. For example, y''' is the third order derivative of y, but $(y')^3$ is the third degree of the first derivative of y. Note that the first derivative of a function represents the *slope* or the *rate of change* of the function with the independent variable, and the second derivative represents the *rate of change of the slope* of the function with the independent variable.

When a function y depends on two or more independent variables such as x and t, it is sometimes of interest to examine the dependence of the function on one of the variables only. This is done by taking the derivative of

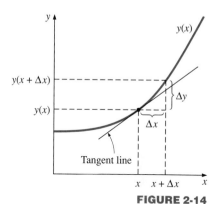

FIGURE 2-14

The derivative of a function at a point represents the slope of the tangent line of the function at that point.

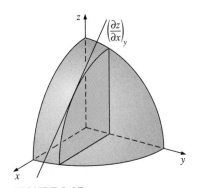

FIGURE 2-15

Graphical representation of partial derivative $\partial z/\partial x$.

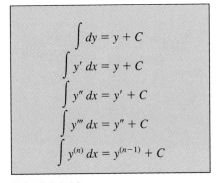

FIGURE 2-16

Some indefinite integrals that involve derivatives.

the function with respect to that variable while holding the other variables constant. Such derivatives are called **partial derivatives.** The first partial derivatives of the function $y(x, t)$ with respect to x and t are defined as (Fig. 2-15)

$$\frac{\partial y}{\partial x} = \lim_{\Delta x \to 0} \frac{y(x + \Delta x, t) - y(x, t)}{\Delta x} \qquad (2\text{-}7)$$

$$\frac{\partial y}{\partial t} = \lim_{\Delta t \to 0} \frac{y(x, t + \Delta t) - y(x, t)}{\Delta t} \qquad (2\text{-}8)$$

Note that when finding $\partial y/\partial x$ we treat t as a constant and differentiate y with respect to x. Likewise, when finding $\partial y/\partial t$ we treat x as a constant and differentiate y with respect to t.

Integration can be viewed as the inverse process of differentiation. Integration is commonly used in solving differential equations since solving a differential equation is essentially a process of removing the derivatives from the equation. Differentiation is the process of finding $y'(x)$ when a function $y(x)$ is given, whereas integration is the process of finding the function $y(x)$ when its derivative $y'(x)$ is given. The integral of this derivative is expressed as

$$\int y'(x)dx = \int dy = y(x) + C \qquad (2\text{-}9)$$

since $y'(x)dx = dy$ and the integral of the differential of a function is the function itself (plus a constant, of course). In the equation above, x is the integration variable and C is an arbitrary constant called the **integration constant.**

The derivative of $y(x) + C$ is $y'(x)$ no matter what the value of the constant C is. Therefore, two functions that differ by a constant have the same derivative, and we always add a constant C during integration to recover this constant that is lost during differentiation. The integral in Equation 2-9 is called an **indefinite integral** since the value of the arbitrary constant C is indefinite. The procedure described above can be extended to higher order derivatives (Fig. 2-16). For example,

$$\int y''(x)dx = y'(x) + C \qquad (2\text{-}10)$$

This can be proved by defining a new variable $u(x) = y'(x)$, differentiating it to obtain $u'(x) = y''(x)$, and then applying Equation 2-9. Therefore, the order of a derivative decreases by one each time it is integrated.

Classification of Differential Equations

A differential equation that involves only ordinary derivatives is called an **ordinary differential equation,** and a differential equation that involves partial derivatives is called a **partial differential equation.** Then it follows that problems that involve a single independent variable result in ordinary differential equations, and problems that involve two or more independent variables result in partial differential equations. A differential equation may involve several derivatives of various orders of an unknown function. The order of the highest derivative in a differential equation is the order of the equation. For example, the order of $y''' + (y'')^4 = 7x^5$ is 3 since it contains no fourth or higher order derivatives.

You will remember from algebra that the equation $3x - 5 = 0$ is much easier to solve than the equation $x^4 + 3x - 5 = 0$ because the first equation is linear whereas the second one is nonlinear. This is also true for differential equations. Therefore, before we start solving a differential equation, we usually check for linearity. A differential equation is said to be **linear** if the dependent variable and all of its derivatives are of the first degree and their coefficients depend on the independent variable only. In other words, a differential equation is linear if it can be written in a form that does not involve (1) any powers of the dependent variable or its derivatives such as y^3 or $(y')^2$, (2) any products of the dependent variable or its derivatives such as yy' or $y'y'''$, and (3) any other nonlinear functions of the dependent variable such as $\sin y$ or e^y. If any of these conditions apply, it is **nonlinear** (Fig. 2-17).

A linear differential equation, however, may contain (1) powers or nonlinear functions of the independent variable, such as and x^2 and $\cos x$ and (2) products of the dependent variable (or its derivatives) and functions of the independent variable, such as x^3y', x^2y, and $e^{-2x}y''$. A linear differential equation of order n can be expressed in the most general form as

$$y^{(n)} + f_1(x)y^{(n-1)} + \cdots + f_{n-1}(x)y' + f_n(x)y = R(x) \qquad (2\text{-}11)$$

A differential equation that cannot be put into this form is nonlinear. A linear differential equation in y is said to be **homogeneous** as well if $R(x) = 0$. Otherwise, it is nonhomogeneous. That is, each term in a linear homogeneous equation contains the dependent variable or one of its derivatives after the equation is cleared of any common factors. The term $R(x)$ is called the *nonhomogeneous term*.

Differential equations are also classified by the nature of the coefficients of the dependent variable and its derivatives. A differential equation is said to have **constant coefficients** if the coefficients of all the terms that involve the dependent variable or its derivatives are constants. If, after clearing any common factors, any of the terms with the dependent variable or its derivatives involve the independent variable as a coefficient, that equation is said to have **variable coefficients** (Fig. 2-18). Differential equations with constant coefficients are usually much easier to solve than those with variable coefficients.

Solutions of Differential Equations

Solving a differential equation can be as easy as taking one or more integrations; but such simple differential equations are usually the exception rather than the rule. There is no single general solution method applicable to all differential equations. There are different solution techniques, each being applicable to different classes of differential equations. Sometimes solving a differential equation requires the use of two or more techniques as well as

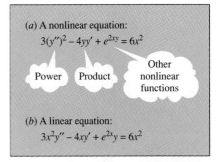

(a) A nonlinear equation:
$$3(y'')^2 - 4yy' + e^{2xy} = 6x^2$$

Power   Product   Other nonlinear functions

(b) A linear equation:
$$3x^2y'' - 4xy' + e^{2x}y = 6x^2$$

FIGURE 2-17

A differential equation that is (a) nonlinear and (b) linear. When checking for linearity, we examine the dependent variable only.

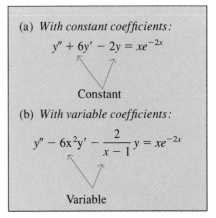

(a) *With constant coefficients:*
$$y'' + 6y' - 2y = xe^{-2x}$$

Constant

(b) *With variable coefficients:*
$$y'' - 6x^2y' - \frac{2}{x-1}y = xe^{-2x}$$

Variable

FIGURE 2-18

A differential equation with (a) constant coefficients and (b) variable coefficients.

ingenuity and mastery of solution methods. Some differential equations can be solved only by using some very clever tricks. Some cannot be solved analytically at all.

In algebra, we usually seek discrete values that satisfy an algebraic equation such as $x^2 - 7x - 10 = 0$. When dealing with differential equations, however, we seek functions that satisfy the equation in a specified interval. For example, the algebraic equation $x^2 - 7x - 10 = 0$ is satisfied by two numbers only: 2 and 5. But the differential equation $y' - 7y = 0$ is satisfied by the function e^{7x} for any value of x (Fig. 2-19).

Consider the algebraic equation $x^3 - 6x^2 + 11x - 6 = 0$. Obviously, $x = 1$ satisfies this equation, and thus it is a solution. However, it is not the only solution of this equation. We can easily show by direct substitution that $x = 2$ and $x = 3$ also satisfy this equation, and thus they are solutions as well. But there are no other solutions to this equation. Therefore, we say that the set 1, 2, and 3 forms the complete solution to this algebraic equation.

The same line of reasoning also applies to differential equations. Typically, differential equations have multiple solutions that contain at least one arbitrary constant. Any function that satisfies the differential equation on an interval is called a *solution* of that differential equation in that interval. A solution that involves one or more arbitrary constants represents a family of functions that satisfy the differential equation and is called **a general solution** of that equation. Not surprisingly, a differential equation may have more than one general solution. A general solution is usually referred to as **the general solution** or the **complete solution** if every solution of the equation can be obtained from it as a special case. A solution that can be obtained from a general solution by assigning particular values to the arbitrary constants is called a **specific solution.**

You will recall from algebra that a number is a solution of an algebraic equation if it satisfies the equation. For example, 2 is a solution of the equation $x^3 - 8 = 0$ because the substitution of 2 for x yields identically zero. Likewise, a function is a solution of a differential equation if that function satisfies the differential equation. In other words, a solution function yields identity when substituted into the differential equation. For example, it can be shown by direct substitution that the function $3e^{-2x}$ is a solution of $y'' - 4y = 0$ (Fig. 2-20).

(a) *An algebraic equation:*

$$y^2 - 7y - 10 = 0$$

Solution: $y = 2$ and $y = 5$

(b) *A differential equation:*

$$y' - 7y = 0$$

Solution: $y = e^{7x}$

FIGURE 2-19

Unlike those of algebraic equations, the solutions of differential equations are typically functions instead of discrete values.

Function: $f = 3e^{-2x}$

Differential equation: $y'' - 4y = 0$

Derivatives of f:

$$f' = -6e^{-2x}$$
$$f'' = 12e^{-2x}$$

Substituting into $y'' - 4y = 0$:

$$f'' - 4f \overset{?}{=} 0$$
$$12e^{-2x} - 4 \times 3e^{-2x} \overset{?}{=} 0$$
$$0 = 0$$

Therefore, the function $3e^{-2x}$ is a solution of the differential equation $y'' - 4y = 0$.

FIGURE 2-20

Verifying that a given function is a solution of a differential equation.

2-3 ■ ONE-DIMENSIONAL HEAT CONDUCTION EQUATION

Consider heat conduction through a large plane wall such as the wall of a house, the glass of a single pane window, the metal plate at the bottom of a pressing iron, a cast iron steam pipe, a cylindrical nuclear fuel element, an electrical resistance wire, the wall of a spherical container, or a spherical metal ball that is being quenched or tempered. Heat conduction in these and many other geometries can be approximated as being *one-dimensional* since heat conduction through these geometries will be dominant in one direction and negligible in other directions. Below we will develop the one-dimensional heat conduction equation in rectangular, cylindrical, and spherical coordinates.

Heat Conduction Equation in a Large Plane Wall

Consider a thin element of thickness Δx in a large plane wall, as shown in Figure 2-21. Assume the density of the wall is ρ, the specific heat is C, and the area of the wall normal to the direction of heat transfer is A. An *energy balance* on this thin element during a small time interval Δt can be expressed as

$$\begin{pmatrix} \text{Rate of heat} \\ \text{conduction} \\ \text{at } x \end{pmatrix} - \begin{pmatrix} \text{Rate of heat} \\ \text{conduction} \\ \text{at } x + \Delta x \end{pmatrix} + \begin{pmatrix} \text{Rate of heat} \\ \text{generation} \\ \text{inside the} \\ \text{element} \end{pmatrix} = \begin{pmatrix} \text{Rate of change} \\ \text{of the energy} \\ \text{content of the} \\ \text{element} \end{pmatrix}$$

or

$$\dot{Q}_x - \dot{Q}_{x+\Delta x} + \dot{G}_{\text{element}} = \frac{\Delta E_{\text{element}}}{\Delta t} \tag{2-12}$$

But the change in the energy content of the element and the rate of heat generation within the element can be expressed as

$$\Delta E_{\text{element}} = E_{t+\Delta t} - E_t = mC(T_{t+\Delta t} - T_t) = \rho CA\Delta x(T_{t+\Delta t} - T_t)$$

$$\dot{G}_{\text{element}} = \dot{g}V_{\text{element}} = \dot{g}A\Delta x$$

Substituting into Equation 2-12, we get

$$\dot{Q}_x - \dot{Q}_{x+\Delta x} + \dot{g}A\Delta x = \rho CA\Delta x \frac{T_{t+\Delta t} - T_t}{\Delta t}$$

Dividing by $A\Delta x$ gives

$$-\frac{1}{A}\frac{\dot{Q}_{x+\Delta x} - \dot{Q}_x}{\Delta x} + \dot{g} = \rho C \frac{T_{t+\Delta t} - T_t}{\Delta t} \tag{2-13}$$

Taking the limit as $\Delta x \to 0$ and $\Delta t \to 0$ yields

$$\frac{1}{A}\frac{\partial}{\partial x}\left(kA\frac{\partial T}{\partial x}\right) + \dot{g} = \rho C \frac{\partial T}{\partial t} \tag{2-14}$$

since, from the definition of the derivative and Fourier's law of heat conduction,

$$\lim_{\Delta x \to 0} \frac{\dot{Q}_{x+\Delta x} - \dot{Q}_x}{\Delta x} = \frac{\partial \dot{Q}}{\partial x} = \frac{\partial}{\partial x}\left(-kA\frac{\partial T}{\partial x}\right) \tag{2-15}$$

Noting that the area A is constant for a plane wall, the one-dimensional transient heat conduction equation in a plane wall becomes

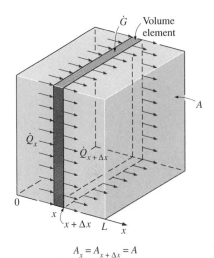

$$A_x = A_{x+\Delta x} = A$$

FIGURE 2-21

One-dimensional heat conduction through a volume element in a large plane wall.

$$\text{Variable conductivity:} \qquad \frac{\partial}{\partial x}\left(k\frac{\partial T}{\partial x}\right) + \dot{g} = \rho C \frac{\partial T}{\partial t} \qquad (2\text{-}16)$$

The thermal conductivity k of a material, in general, depends on the temperature T (and therefore x), and thus it cannot be taken out of the derivative. However, the *thermal conductivity* in most practical applications can be assumed to remain *constant* at some average value. The equation above in that case reduces to

$$\text{Constant conductivity:} \qquad \frac{\partial^2 T}{\partial x^2} + \frac{\dot{g}}{k} = \frac{1}{\alpha}\frac{\partial T}{\partial t} \qquad (2\text{-}17)$$

where the property $\alpha = k/\rho C$ is the **thermal diffusivity** of the material and represents how fast heat propagates through a material. It reduces to the following forms under specified conditions (Fig. 2-22):

(1) *Steady-state:*
($\partial/\partial t = 0$)
$$\frac{d^2 T}{dx^2} + \frac{\dot{g}}{k} = 0 \qquad (2\text{-}18)$$

(2) *Transient, no heat generation:*
($\dot{g} = 0$)
$$\frac{\partial^2 T}{\partial x^2} = \frac{1}{\alpha}\frac{\partial T}{\partial t} \qquad (2\text{-}19)$$

(3) *Steady-state, no heat generation:*
($\partial/\partial t = 0$ and $\dot{g} = 0$)
$$\frac{d^2 T}{dx^2} = 0 \qquad (2\text{-}20)$$

Note that we replaced the partial derivatives by the ordinary derivatives in the one-dimensional steady heat conduction case since the partial and ordinary derivatives of a function are identical when the function depends on a single variable only ($T = T(x)$ in this case).

General, one dimensional:

No generation	Steady-state

$$\frac{\partial^2 T}{\partial x^2} + \overset{0}{\cancel{\frac{\dot{g}}{k}}} = \overset{0}{\cancel{\frac{1}{\alpha}\frac{\partial T}{\partial t}}}$$

Steady, one-dimensional:

$$\frac{d^2 T}{dx^2} = 0$$

FIGURE 2-22

The simplification of the one-dimensional heat conduction equation in a plane wall for the case of constant conductivity for steady conduction with no heat generation.

FIGURE 2-23

One-dimensional heat conduction through a volume element in a long cylinder.

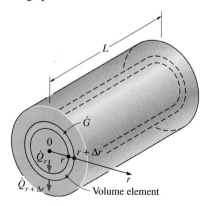
Volume element

Heat Conduction Equation in a Long Cylinder

Now consider a thin cylindrical shell element of thickness Δr in a long cylinder, as shown in Figure 2-23. Assume the density of the cylinder is ρ, the specific heat is C, and the length is L. The area of the cylinder normal to the direction of heat transfer at any location is $A = 2\pi r L$ where r is the value of the radius at that location. Note that the heat transfer area A depends on r in this case, and thus it varies with location. An *energy balance* on this thin cylindrical shell element during a small time interval Δt can be expressed as

$$\begin{pmatrix}\text{Rate of heat}\\\text{conduction}\\\text{at } r\end{pmatrix} - \begin{pmatrix}\text{Rate of heat}\\\text{conduction}\\\text{at } r + \Delta r\end{pmatrix} + \begin{pmatrix}\text{Rate of heat}\\\text{generation}\\\text{inside the}\\\text{element}\end{pmatrix} = \begin{pmatrix}\text{Rate of change}\\\text{of the energy}\\\text{content of the}\\\text{element}\end{pmatrix}$$

or

$$\dot{Q}_r - \dot{Q}_{r+\Delta r} + \dot{G}_{\text{element}} = \frac{\Delta E_{\text{element}}}{\Delta t} \qquad (2\text{-}21)$$

The change in the energy content of the element and the rate of heat generation within the element can be expressed as

$$\Delta E_{\text{element}} = E_{t+\Delta t} - E_t = mC(T_{t+\Delta t} - T_t) = \rho CA\Delta r(T_{t+\Delta t} - T_t)$$

$$\dot{G}_{\text{element}} = \dot{g}V_{\text{element}} = \dot{g}A\Delta r$$

Substituting into Equation 2-21, we get

$$\dot{Q}_r - \dot{Q}_{r+\Delta r} + \dot{g}A\Delta r = \rho CA\Delta r \frac{T_{t+\Delta t} - T_t}{\Delta t} \qquad (2\text{-}22)$$

where $A = 2\pi r L$. You may be tempted to express the area at the *middle* of the element using the *average* radius as $A = 2\pi(r + \Delta r/2)L$. But there is nothing we can gain from this complication since later in the analysis we will take the limit as $\Delta r \to 0$ and thus the term $\Delta r/2$ will drop out. Now dividing the equation above by $A\Delta r$ gives

$$-\frac{1}{A}\frac{\dot{Q}_{r+\Delta r} - \dot{Q}_r}{\Delta r} + \dot{g} = \rho C \frac{T_{t+\Delta t} - T_t}{\Delta t} \qquad (2\text{-}23)$$

Taking the limit as $\Delta r \to 0$ and $\Delta t \to 0$ yields

$$\frac{1}{A}\frac{\partial}{\partial r}\left(kA\frac{\partial T}{\partial r}\right) + \dot{g} = \rho C \frac{\partial T}{\partial t} \qquad (2\text{-}24)$$

since, from the definition of the derivative and Fourier's law of heat conduction,

$$\lim_{\Delta r \to 0} \frac{\dot{Q}_{r+\Delta r} - \dot{Q}_r}{\Delta r} = \frac{\partial \dot{Q}}{\partial r} = \frac{\partial}{\partial r}\left(-kA\frac{\partial T}{\partial r}\right) \qquad (2\text{-}25)$$

Noting that the heat transfer area in this case is $A = 2\pi r L$, the one-dimensional transient heat conduction equation in a cylinder becomes

Variable conductivity:	$\dfrac{1}{r}\dfrac{\partial}{\partial r}\left(rk\dfrac{\partial T}{\partial r}\right) + \dot{g} = \rho C \dfrac{\partial T}{\partial t}$	(2-26)

For the case of constant thermal conductivity, the equation above reduces to

Constant conductivity:	$\dfrac{1}{r}\dfrac{\partial}{\partial r}\left(r\dfrac{\partial T}{\partial r}\right) + \dfrac{\dot{g}}{k} = \dfrac{1}{\alpha}\dfrac{\partial T}{\partial t}$	(2-27)

where again the property $\alpha = k/\rho C$ is the thermal diffusivity of the material, which represents how fast heat propagates through the material. It reduces to the following forms under specified conditions (Fig. 2-24):

(a) The form that is ready to integrate

$$\frac{d}{dr}\left(r\frac{dT}{dr}\right) = 0$$

(b) The equivalent alternative form

$$r\frac{d^2T}{dr^2} + \frac{dT}{dr} = 0$$

FIGURE 2-24

Two equivalent forms of the differential equation for the one-dimensional steady heat conduction in a cylinder with no heat generation.

(1) *Steady-state:*
   $(\partial/\partial t = 0)$
$$\frac{1}{r}\frac{d}{dr}\left(r\frac{dT}{dr}\right) + \frac{\dot{g}}{k} = 0 \qquad (2\text{-}28)$$

(2) *Transient, no heat generation:*
   $(\dot{g} = 0)$
$$\frac{1}{r}\frac{\partial}{\partial r}\left(r\frac{\partial T}{\partial r}\right) = \frac{1}{\alpha}\frac{\partial T}{\partial t} \qquad (2\text{-}29)$$

(3) *Steady-state, no heat generation:*
   $(\partial/\partial t = 0 \text{ and } \dot{g} = 0)$
$$\frac{d}{dr}\left(r\frac{dT}{dr}\right) = 0 \qquad (2\text{-}30)$$

Note that we again replaced the partial derivatives by the ordinary derivatives in the one-dimensional steady heat conduction case since the partial and ordinary derivatives of a function are identical when the function depends on a single variable only ($T = T(r)$ in this case).

Heat Conduction Equation in a Sphere

Now consider a sphere with density ρ, specific heat C, and outer radius R. The area of the sphere normal to the direction of heat transfer at any location is $A = 4\pi r^2$ where r is the value of the radius at that location. Note that the heat transfer area A depends on r in this case also, and thus it varies with location. By considering a thin spherical shell element of thickness Δr and repeating the approach described above for the cylinder by using $A = 4\pi r^2$ instead of $A = 2\pi rL$, the one-dimensional transient heat conduction equation for a sphere is determined to be (Fig. 2-25)

Variable conductivity:
$$\frac{1}{r^2}\frac{\partial}{\partial r}\left(r^2 k\frac{\partial T}{\partial r}\right) + \dot{g} = \rho C\frac{\partial T}{\partial t} \qquad (2\text{-}31)$$

which, in the case of constant thermal conductivity, reduces to

Constant conductivity:
$$\frac{1}{r^2}\frac{\partial}{\partial r}\left(r^2\frac{\partial T}{\partial r}\right) + \frac{\dot{g}}{k} = \frac{1}{\alpha}\frac{\partial T}{\partial t} \qquad (2\text{-}32)$$

where again the property $\alpha = k/\rho C$ is the thermal diffusivity of the material. It reduces to the following forms under specified conditions:

(1) *Steady-state:*
   $(\partial/\partial t = 0)$
$$\frac{1}{r^2}\frac{d}{dr}\left(r^2\frac{dT}{dr}\right) + \frac{\dot{g}}{k} = 0 \qquad (2\text{-}33)$$

(2) *Transient,*
   no heat generation:
   $(\dot{g} = 0)$
$$\frac{1}{r^2}\frac{\partial}{\partial r}\left(r^2\frac{\partial T}{\partial r}\right) = \frac{1}{\alpha}\frac{\partial T}{\partial t} \qquad (2\text{-}34)$$

(3) *Steady-state,*
   no heat generation:
   $(\partial/\partial t = 0 \text{ and } \dot{g} = 0)$
$$\frac{d}{dr}\left(r^2\frac{dT}{dr}\right) = 0 \quad \text{or} \quad r\frac{d^2T}{dr^2} + 2\frac{dT}{dr} \qquad (2\text{-}35)$$

where again we replaced the partial derivatives by the ordinary derivatives in the one-dimensional steady heat conduction case.

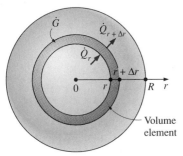

FIGURE 2-25
One-dimensional heat conduction through a volume element in a sphere.

Combined One-Dimensional Heat Conduction Equation

An examination of the one-dimensional transient heat conduction equations for the plane wall, cylinder, and sphere reveals that all three equations can be expressed in a compact form as

$$\frac{1}{r^n}\frac{\partial}{\partial r}\left(r^n k \frac{\partial T}{\partial r}\right) + \dot{g} = \rho C \frac{\partial T}{\partial t} \qquad (2\text{-}36)$$

where $n = 0$ for a plane wall, $n = 1$ for a cylinder, and $n = 2$ for a sphere. In the case of a plane wall, it is customary to replace the variable r by x. This equation can be simplified for steady-state or no heat generation cases as described before.

EXAMPLE 2-3 Heat Conduction through the Bottom of a Pan

Consider a steel pan placed on top of an electric range to cook spaghetti (Fig. 2-26). The bottom section of the pan is $L = 0.4$ cm thick and has a diameter of $D = 18$ cm. The electric heating unit on the range top consumes 800 W of power during cooking, and 80 percent of the heat generated in the heating element is transferred uniformly to the pan. Assuming constant thermal conductivity, obtain the differential equation that describes the variation of the temperature in the bottom section of the pan during steady operation.

Solution The bottom section of the pan has a large surface area relative to its thickness and can be approximated as a large plane wall. Heat flux is applied to the bottom surface of the pan uniformly, and the conditions on the inner surface are also uniform. Therefore, we expect the heat transfer through the bottom section of the pan to be from the bottom surface toward the top, and heat transfer in this case can reasonably be approximated as being one-dimensional. Taking the direction normal to the bottom surface of the pan to be the x axis, we will have $T = T(x)$ during steady operation since the temperature in this case will depend on x only.

The thermal conductivity is given to be constant, and there is no heat generation in the medium (within the bottom section of the pan). Therefore, the differential equation governing the variation of temperature in the bottom section of the pan in this case is simply Equation 2-20,

$$\frac{d^2 T}{dx^2} = 0$$

which is the steady one-dimensional heat conduction equation in rectangular coordinates under the conditions of constant thermal conductivity and no heat generation. Note that the conditions at the surface of the medium have no effect on the differential equation.

800 W

FIGURE 2-26

Schematic for Example 2-3.

FIGURE 2-27

Schematic for Example 2-4.

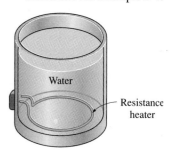

Water

Resistance
heater

EXAMPLE 2-4 Heat Conduction in a Resistance Heater

A 2-kW resistance heater wire with thermal conductivity $k = 15$ W/m·°C, diameter $D = 0.4$ cm, and length $L = 50$ cm is used to boil water by immersing it in water (Fig. 2-27). Assuming the variation of the thermal conductivity of the wire with temperature to be negligible, obtain the differential equation that describes the variation of the temperature in the wire during steady operation.

Solution The resistance wire can be considered to be a very long cylinder since its length is more than 100 times its diameter. Also, heat is generated uniformly in the wire and the conditions on the outer surface of the wire are uniform. Therefore, it is reasonable to expect the temperature in the wire to vary in the radial r direction only and thus the heat transfer to be one-dimensional. Then we will have $T = T(r)$ during steady operation since the temperature in this case will depend on r only.

The rate of heat generation in the wire per unit volume can be determined from

$$\dot{g} = \frac{\dot{G}}{V_{wire}} = \frac{\dot{G}}{(\pi D^2/4)L} = \frac{2000 \text{ W}}{[\pi(0.004 \text{ m})^2/4](0.5 \text{ m})} = 0.318 \times 10^9 \text{ W/m}^3$$

Noting that the thermal conductivity is given to be constant, the differential equation that governs the variation of temperature in the wire is simply Equation 2-28,

$$\frac{1}{r}\frac{d}{dr}\left(r\frac{dT}{dr}\right) + \frac{\dot{g}}{k} = 0$$

which is the steady one-dimensional heat conduction equation in cylindrical coordinates for the case of constant thermal conductivity. Note again that the conditions at the surface of the wire have no effect on the differential equation.

EXAMPLE 2-5 Cooling of a Hot Metal Ball in Air

A spherical metal ball of radius R is heated in an oven to a temperature of 600°F throughout and is then taken out of the oven and allowed to cool in ambient air at $T_\infty = 75°F$ by convection and radiation (Fig. 2-28). The thermal conductivity of the ball material is known to vary linearly with temperature. Assuming the ball is cooled uniformly from the entire outer surface, obtain the differential equation that describes the variation of the temperature in the ball during cooling.

Solution The ball is initially at a uniform temperature and is cooled uniformly from the entire outer surface. Also, the temperature at any point in the ball will change with time during cooling. Therefore, this is a one-dimensional transient heat conduction problem since the temperature within the ball will change with the radial distance r and the time t. That is, $T = T(r, t)$.

The thermal conductivity is given to be variable, and there is no heat generation in the ball. Therefore, the differential equation that governs the variation of temperature in the ball in this case is obtained from Equation 2-31 by setting the heat generation term equal to zero. We obtain

$$\frac{1}{r^2}\frac{\partial}{\partial r}\left(r^2 k \frac{\partial T}{\partial r}\right) = \rho C \frac{\partial T}{\partial t}$$

which is the one-dimensional transient heat conduction equation in spherical coordinates under the conditions of variable thermal conductivity and no heat generation. Note again that the conditions at the outer surface of the ball have no effect on the differential equation.

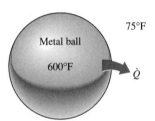

75°F

Metal ball

600°F

$\dot{Q}$

FIGURE 2-28
Schematic for Example 2-5.

In the last section we considered one-dimensional heat conduction and assumed heat conduction in other directions to be negligible. Most heat transfer problems encountered in practice can be approximated as being one-dimensional, and we will mostly deal with such problems in this text. However, this is not always the case, and sometimes we need to consider heat transfer in other directions as well. In such cases heat conduction is said to be *multidimensional,* and in this section we will develop the governing differential equation in such systems in rectangular, cylindrical, and spherical coordinate systems.

Rectangular Coordinates

Consider a small rectangular element of length Δx, width Δy, and height Δz, as shown in Figure 2-29. Assume the density of the body is ρ and the specific heat is C. An *energy balance* on this element during a small time interval Δt can be expressed as

$$\begin{pmatrix} \text{Rate of heat} \\ \text{conduction at} \\ x, y, \text{and } z \end{pmatrix} - \begin{pmatrix} \text{Rate of heat} \\ \text{conduction} \\ \text{at } x + \Delta x, \\ y + \Delta y, \text{and } z + \Delta z \end{pmatrix} + \begin{pmatrix} \text{Rate of heat} \\ \text{generation} \\ \text{inside the} \\ \text{element} \end{pmatrix} = \begin{pmatrix} \text{Rate of change} \\ \text{of the energy} \\ \text{content of} \\ \text{the element} \end{pmatrix}$$

or

$$\dot{Q}_x + \dot{Q}_y + \dot{Q}_z - \dot{Q}_{x + \Delta x} - \dot{Q}_{y + \Delta y} - \dot{Q}_{z + \Delta z} + \dot{G}_{\text{element}} = \frac{\Delta E_{\text{element}}}{\Delta t} \quad (2\text{-}37)$$

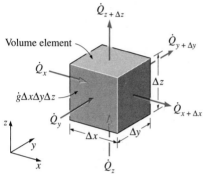

FIGURE 2-29

Three-dimensional heat conduction through a rectangular volume element.

Noting that the volume of the element is $V_{\text{element}} = \Delta x \Delta y \Delta z$, the change in the energy content of the element and the rate of heat generation within the element can be expressed as

$$\Delta E_{\text{element}} = E_{t + \Delta t} - E_t = mC(T_{t + \Delta t} - T_t) = \rho C \Delta x \Delta y \Delta z(T_{t + \Delta t} - T_t)$$

$$\dot{G}_{\text{element}} = \dot{g} V_{\text{element}} = \dot{g} \Delta x \Delta y \Delta z$$

Substituting into Equation 2-37, we get

$$\dot{Q}_x + \dot{Q}_y + \dot{Q}_z - \dot{Q}_{x + \Delta x} - \dot{Q}_{y + \Delta y}$$

$$- \dot{Q}_{z + \Delta z} + \dot{g} \Delta x \Delta y \Delta z = \rho C \Delta x \Delta y \Delta z \frac{T_{t + \Delta t} - T_t}{\Delta t}$$

Dividing by $\Delta x \Delta y \Delta z$ gives

$$-\frac{1}{\Delta y \Delta z} \frac{\dot{Q}_{x + \Delta x} - \dot{Q}_x}{\Delta x} - \frac{1}{\Delta x \Delta z} \frac{\dot{Q}_{y + \Delta y} - \dot{Q}_y}{\Delta y}$$

$$-\frac{1}{\Delta x \Delta y} \frac{\dot{Q}_{z + \Delta z} - \dot{Q}_z}{\Delta z} + \dot{g} = \rho C \frac{T_{t + \Delta t} - T_t}{\Delta_t}$$

Noting that the heat transfer areas of the element for heat conduction in the x, y, and z directions are $A_x = \Delta y \Delta z$, $A_y = \Delta x \Delta z$, and $A_z = \Delta x \Delta y$, respectively, and taking the limit as Δx, Δy, Δz and $\Delta t \to 0$ yields

$$\frac{\partial}{\partial x}\left(k\frac{\partial T}{\partial x}\right) + \frac{\partial}{\partial y}\left(k\frac{\partial T}{\partial y}\right) + \frac{\partial}{\partial z}\left(k\frac{\partial T}{\partial z}\right) + \dot{g} = \rho C \frac{\partial T}{\partial t} \qquad (2\text{-}38)$$

since, from the definition of the derivative and Fourier's law of heat conduction,

$$\lim_{\Delta x \to 0} \frac{1}{\Delta y \Delta z} \frac{\dot{Q}_{x+\Delta x} - \dot{Q}_x}{\Delta x} = \frac{1}{\Delta y \Delta z} \frac{\partial \dot{Q}_x}{\partial x} = \frac{1}{\Delta y \Delta z} \frac{\partial}{\partial x}\left(-k\Delta y \Delta z \frac{\partial T}{\partial x}\right) = -\frac{\partial}{\partial x}\left(k\frac{\partial T}{\partial x}\right)$$

$$\lim_{\Delta y \to 0} \frac{1}{\Delta x \Delta z} \frac{\dot{Q}_{y+\Delta y} - \dot{Q}_y}{\Delta y} = \frac{1}{\Delta x \Delta z} \frac{\partial \dot{Q}_y}{\partial y} = \frac{1}{\Delta x \Delta z} \frac{\partial}{\partial y}\left(-k\Delta x \Delta z \frac{\partial T}{\partial y}\right) = -\frac{\partial}{\partial y}\left(k\frac{\partial T}{\partial y}\right)$$

$$\lim_{\Delta z \to 0} \frac{1}{\Delta x \Delta y} \frac{\dot{Q}_{z+\Delta z} - \dot{Q}_z}{\Delta z} = \frac{1}{\Delta x \Delta y} \frac{\partial \dot{Q}_z}{\partial z} = \frac{1}{\Delta x \Delta y} \frac{\partial}{\partial z}\left(-k\Delta x \Delta y \frac{\partial T}{\partial z}\right) = -\frac{\partial}{\partial z}\left(k\frac{\partial T}{\partial z}\right)$$

Equation 2-38 is the general heat conduction equation in rectangular coordinates. In the case of constant thermal conductivity, it reduces to

$$\frac{\partial^2 T}{\partial x^2} + \frac{\partial^2 T}{\partial y^2} + \frac{\partial^2 T}{\partial z^2} + \frac{\dot{g}}{k} = \frac{1}{\alpha}\frac{\partial T}{\partial t} \qquad (2\text{-}39)$$

where the property $\alpha = k/\rho C$ is again the *thermal diffusivity* of the material. The equation above is known as the **Fourier-Biot equation,** and it reduces to the following forms under specified conditions:

(1) *Steady-state:*
(called the **Poisson equation**)
$$\frac{\partial^2 T}{\partial x^2} + \frac{\partial^2 T}{\partial y^2} + \frac{\partial^2 T}{\partial z^2} + \frac{\dot{g}}{k} = 0 \quad (2\text{-}40)$$

(2) *Transient, no heat generation*
(called the **diffusion equation**)
$$\frac{\partial^2 T}{\partial x^2} + \frac{\partial^2 T}{\partial y^2} + \frac{\partial^2 T}{\partial z^2} = \frac{1}{\alpha}\frac{\partial T}{\partial t} \quad (2\text{-}41)$$

(3) *Steady-state, no heat generation:*
(called the **Laplace equation**)
$$\frac{\partial^2 T}{\partial x^2} + \frac{\partial^2 T}{\partial y^2} + \frac{\partial^2 T}{\partial z^2} = 0 \quad (2\text{-}42)$$

Note that in the special case of one-dimensional heat transfer in the x direction, the derivatives with respect to y and z drop out and the equations above reduce to the ones developed in the previous section for a plane wall (Fig. 2-30).

Cylindrical Coordinates

The general heat conduction equation in cylindrical coordinates can be obtained from an energy balance on a volume element in cylindrical coordinates, shown in Figure 2-31, by following the steps outlined above. It can also be obtained directly from Equation 2-38 by coordinate transformation using the following relations between the coordinates of a point in rectangular and cylindrical coordinate systems:

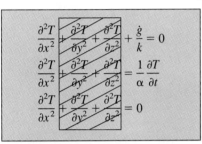

$$\frac{\partial^2 T}{\partial x^2} + \frac{\partial^2 T}{\partial y^2} + \frac{\partial^2 T}{\partial z^2} + \frac{\dot{g}}{k} = 0$$

$$\frac{\partial^2 T}{\partial x^2} + \frac{\partial^2 T}{\partial y^2} + \frac{\partial^2 T}{\partial z^2} = \frac{1}{\alpha}\frac{\partial T}{\partial t}$$

$$\frac{\partial^2 T}{\partial x^2} + \frac{\partial^2 T}{\partial y^2} + \frac{\partial^2 T}{\partial z^2} = 0$$

FIGURE 2-30

The three-dimensional heat conduction equations reduce to the one-dimensional ones when the temperature varies in one dimension only.

FIGURE 2-31

A differential volume element in cylindrical coordinates.

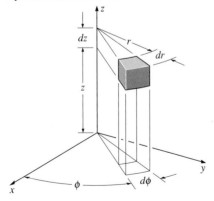

$$x = r \cos \phi, \qquad y = r \sin \phi, \qquad \text{and} \qquad z = z$$

After lengthy manipulations, we obtain

$$\frac{1}{r} \frac{\partial}{\partial r} \left(kr \frac{\partial T}{\partial r} \right) + \frac{1}{r^2} \frac{\partial}{\partial \phi} \left(k \frac{\partial T}{\partial \phi} \right) + \frac{\partial}{\partial z} \left(k \frac{\partial T}{\partial z} \right) + \dot{g} = \rho C \frac{\partial T}{\partial t} \qquad (2\text{-}43)$$

Spherical Coordinates

The general heat conduction equations in spherical coordinates can be obtained from an energy balance on a volume element in spherical coordinates, shown in Figure 2-32, by following the steps outlined above. It can also be obtained directly from Equation 2-38 by coordinate transformation using the following relations between the coordinates of a point in rectangular and spherical coordinate systems:

$$x = r \cos \phi \sin \theta, \qquad y = r \sin \phi \sin \theta, \qquad \text{and} \qquad z = \cos \theta$$

Again after lengthy manipulations, we obtain

$$\frac{1}{r^2} \frac{\partial}{\partial r} \left(kr^2 \frac{\partial T}{\partial r} \right) + \frac{1}{r^2 \sin^2 \theta} \frac{\partial}{\partial \phi} \left(k \frac{\partial T}{\partial \phi} \right)$$
$$+ \frac{1}{r^2 \sin \theta} \frac{\partial}{\partial \theta} \left(k \sin \theta \frac{\partial T}{\partial \theta} \right) + \dot{g} = \rho C \frac{\partial T}{\partial t} \qquad (2\text{-}44)$$

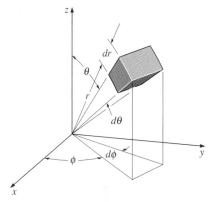

FIGURE 2-32

A differential volume element
in spherical coordinates.

Obtaining analytical solutions to the differential equations above requires a knowledge of the solution techniques of partial differential equations, which is beyond the scope of this introductory text. Here we limit our consideration to one-dimensional steady-state cases or lumped systems since they result in ordinary differential equations.

EXAMPLE 2-6 Heat Conduction in a Short Cylinder

A short cylindrical metal billet of radius R and height h is heated in an oven to a temperature of 600°F throughout and is then taken out of the oven and allowed to cool in ambient air at $T_\infty = 65$°F by convection and radiation. Assuming the billet is cooled uniformly from all outer surfaces and the variation of the thermal conductivity of the material with temperature is negligible, obtain the differential equation that describes the variation of the temperature in the billet during this cooling process.

Solution   The billet shown in Figure 2-33 is initially at a uniform temperature and is cooled uniformly from the top and bottom surfaces in the z direction as well as the lateral surface in the radial r direction. Also, the temperature at any point in the ball will change with time during cooling. Therefore, this is a two-dimensional transient heat conduction problem since the temperature within the billet will change with the radial and axial distances r and z and with time t. That is, $T = T(r,z,t)$.

The thermal conductivity is given to be constant, and there is no heat generation in the billet. Therefore, the differential equation that governs the variation of temperature in the billet in this case is obtained from Equation 2-43 by setting

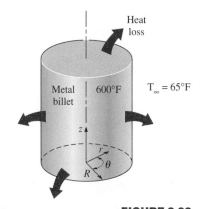

FIGURE 2-33

Schematic for Example 2-6.

the heat generation term and the derivatives with respect to ϕ equal to zero. We obtain

$$\frac{1}{r}\frac{\partial}{\partial r}\left(kr\frac{\partial T}{\partial r}\right) + \frac{\partial}{\partial z}\left(k\frac{\partial T}{\partial z}\right) = \rho C \frac{\partial T}{\partial t}$$

In the case of constant thermal conductivity, it reduces to

$$\frac{1}{r}\frac{\partial}{\partial r}\left(r\frac{\partial T}{\partial r}\right) + \frac{\partial^2 T}{\partial z^2} = \frac{1}{\alpha}\frac{\partial T}{\partial t}$$

which is the desired equation.

2-5 ■ BOUNDARY AND INITIAL CONDITIONS

The heat conduction equations above were developed using an energy balance on a differential element inside the medium, and they remain the same regardless of the *thermal conditions* on the *surfaces* of the medium. That is, the differential equations do not incorporate any information related to the conditions on the surfaces such as the surface temperature or a specified heat flux. Yet we know that the heat flux and the temperature distribution in a medium depend on the conditions at the surfaces, and the description of a heat transfer problem in a medium is not complete without a full description of the thermal conditions at the bounding surfaces of the medium. The *mathematical expressions* of the thermal conditions at the boundaries are called the **boundary conditions.**

From a mathematical point of view, solving a differential equation is essentially a process of *removing derivatives,* or an *integration* process, and thus the solution of a differential equation typically involves arbitrary constants (Fig. 2-34). It follows that to obtain a unique solution to a problem, we need to specify more than just the governing differential equation. We need to specify some conditions (such as the value of the function or its derivatives at some value of the independent variable) so that forcing the solution to satisfy these conditions at specified points will result in unique values for the arbitrary constants and thus a *unique solution*. But since the differential equation has no place for the additional information or conditions, we need to supply them separately in the form of boundary or initial conditions.

Consider the variation of temperature along the wall of a brick house in winter. The temperature at any point in the wall depends on, among other things, the conditions at the two surfaces of the wall such as the air temperature of the house, the velocity and direction of the winds, and the solar energy incident on the outer surface. That is, the temperature distribution in a medium depends on the conditions at the boundaries of the medium as well as the heat transfer mechanism inside the medium. To describe a heat transfer problem completely, *two boundary conditions* must be given for *each direction* of the coordinate system along which heat transfer is significant (Fig. 2-35). Therefore, we need to specify *two boundary conditions* for one-dimensional problems, *four boundary conditions* for two-dimensional

FIGURE 2-34

The general solution of a typical differential equation involves arbitrary constants, and thus an infinite number of solutions.

The differential equation:

$$\frac{d^2 T}{dx^2} = 0$$

General solution:

$$T(x) = C_1 x + C_2$$

Arbitrary constants

Some specific solutions:

$$T(x) = 2x + 5$$
$$T(x) = -x + 12$$
$$T(x) = -3$$
$$T(x) = 6.2x$$
$$\vdots$$

FIGURE 2-35

To describe a heat transfer problem completely, two boundary conditions must be given for each direction along which heat transfer is significant.

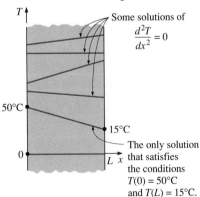

Some solutions of

$$\frac{d^2 T}{dx^2} = 0$$

50°C

15°C

0   L   x

The only solution that satisfies the conditions $T(0) = 50°C$ and $T(L) = 15°C$.

problems, and *six boundary conditions* for three-dimensional problems. In the case of the wall of a house, for example, we need to specify the conditions at two locations (the inner and the outer surfaces) of the wall since heat transfer in this case is one-dimensional. But in the case of a parallelepiped, we need to specify six boundary conditions (one at each face) when heat transfer in all three dimensions is significant.

The physical argument presented above is consistent with the mathematical nature of the problem since the heat conduction equation is second order (i.e., involves second derivatives with respect to the space variables) in all directions along which heat conduction is significant, and the general solution of a second order linear differential equation involves two arbitrary constants for each direction. That is, the number of boundary conditions that needs to be specified in a direction is equal to the order of the differential equation in that direction.

Reconsider the brick wall discussed above. The temperature at any point on the wall at a specified time also depends on the condition of the wall at the beginning of the heat conduction process. Such a condition, which is usually specified at time $t = 0$, is called the **initial condition,** which is a mathematical expression for the temperature distribution of the medium initially. Note that we need only one initial condition for a heat conduction problem regardless of the dimension since the conduction equation is first order in time (it involves the first derivative of temperature with respect to time).

In rectangular coordinates, the initial condition can be specified in the general form as

$$T(x, y, z, 0) = f(x, y, z) \qquad (2\text{-}45)$$

where the function $f(x, y, z)$ represents the temperature distribution throughout the medium at time $t = 0$. When the medium is initially at a uniform temperature of T_i, the initial condition above can be expressed as $T(x, y, z, 0) = T_i$. Note that under *steady* conditions, the heat conduction equation does not involve any time derivatives, and thus we do not need to specify an initial condition.

The heat conduction equation is first order in time, and thus the initial condition cannot involve any derivatives (it is limited to a specified temperature). However, the heat conduction equation is second order in space coordinates, and thus a boundary condition may involve first derivatives at the boundaries as well as specified values of temperature. Boundary conditions most commonly encountered in practice are the *specified temperature, specified heat flux, convection,* and *radiation* boundary conditions.

1 Specified Temperature Boundary Condition

The *temperature* of an exposed surface can usually be measured directly and easily. Therefore, one of the easiest ways to specify the thermal conditions on a surface is to specify the temperature. For one-dimensional heat transfer through a plane wall of thickness L, for example, the specified temperature boundary conditions can be expressed as (Fig. 2-36)

FIGURE 2-36

Specified temperature boundary conditions on both surfaces of a plane wall.

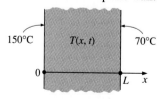

$T(0, t) = 150°C$

$T(L, t) = 70°C$

$$T(0, t) = T_1$$
$$T(L, t) = T_2$$
(2-46)

where T_1 and T_2 are the specified temperatures at surfaces at $x = 0$ and $x = L$, respectively. The specified temperatures can be constant, which is the case during steady heat conduction, or may vary with time.

2 Specified Heat Flux Boundary Condition

When there is sufficient information about energy interactions at a surface, it may be possible to determine the rate of heat transfer and thus the *heat flux* $\dot{q}$ (heat transfer rate per unit surface area, W/m^2) on that surface, and this information can be used as one of the boundary conditions. The heat flux in the positive x direction anywhere in the medium, including the boundaries, can be expressed by *Fourier's law* of heat conduction as

$$\dot{q} = -k\frac{\partial T}{\partial x} = \left(\begin{array}{c}\text{Heat flux in the} \\ \text{positive } x \text{ direction}\end{array}\right) \qquad \text{(W/m}^2\text{)} \qquad (2\text{-}47)$$

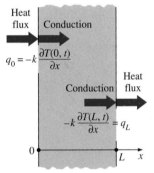

$$q_0 = -k\frac{\partial T(0, t)}{\partial x}$$

$$-k\frac{\partial T(L, t)}{\partial x} = q_L$$

FIGURE 2-37

Specified heat flux boundary conditions on both surfaces of a plane wall.

Then the boundary condition at a boundary is obtained by setting the specified heat flux equal to $-k(\partial T/\partial x)$ at that boundary. The sign of the specified heat flux is determined by inspection: *positive* if the heat flux is in the positive direction of the coordinate axis, and *negative* if it is in the opposite direction. Note that it is extremely important to have the *correct sign* for the specified heat flux since the wrong sign will invert the direction of heat transfer and cause the heat gain to be interpreted as heat loss.

For a plate of thickness L subjected to heat flux of 50 W/m^2 into the medium from both sides, for example, the specified heat flux boundary conditions can be expressed as (Fig. 2-37)

$$-k\frac{\partial T(0, t)}{\partial x} = 50 \qquad \text{and} \qquad -k\frac{\partial T(L, t)}{\partial x} = -50 \qquad (2\text{-}48)$$

Note that the heat flux at the surface at $x = L$ is in the *negative x* direction, and thus it is -50 W/m^2.

Special Case: Insulated Boundary

Some surfaces are commonly insulated in practice in order to minimize heat loss (or heat gain) through them. Insulation reduces heat transfer but does not totally eliminate it unless its thickness is infinity. However, heat transfer through a properly insulated surface can be taken to be zero since adequate insulation reduces the heat transfer through a surface to negligible levels. Therefore, a well-insulated surface can be modeled as a surface with a specified heat flux of zero. Then the boundary condition on a perfectly insulated surface (at $x = 0$, for example) can be expressed as (Fig. 2-38)

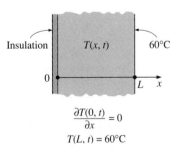

$$\frac{\partial T(0, t)}{\partial x} = 0$$

$$T(L, t) = 60°C$$

FIGURE 2-38

A plane wall with insulation and specified temperature boundary conditions.

$$k\frac{\partial T(0, t)}{\partial x} = 0 \qquad \text{or} \qquad \frac{\partial T(0, t)}{\partial t} = 0 \qquad (2\text{-}49)$$

That is, *on an insulated surface, the first derivative of temperature with respect to the space variable (the temperature gradient) in the direction normal to the insulated surface is zero*. This also means that the temperature function must be perpendicular to an insulated surface since the slope of temperature at the surface must be zero.

Another Special Case: Thermal Symmetry

Some heat transfer problems possess *thermal symmetry* as a result of the symmetry in imposed thermal conditions. For example, the two surfaces of a large hot plate of thickness L suspended vertically in air will be subjected to the same thermal conditions, and thus the temperature distribution in one half of the plate will be the same as that in the other half. That is, the heat transfer problem in this plate will possess thermal symmetry about the center plane at $x = L/2$. Also, the direction of heat flow at any point in the plate will be toward the surface closer to the point, and there will be no heat flow across the center plane. Therefore, the center plane can be viewed as an insulated surface, and the thermal condition at this plane of symmetry can be expressed as (Fig. 2-39)

$$\frac{\partial T(L/2, t)}{\partial x} = 0 \qquad (2\text{-}50)$$

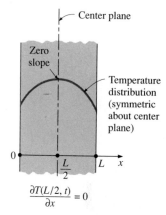

$$\frac{\partial T(L/2, t)}{\partial x} = 0$$

FIGURE 2-39

Thermal symmetry boundary condition at the center plane of a plane wall.

which resembles the *insulation* or *zero heat flux* boundary condition. This result can also be deduced from a plot of temperature distribution with a maximum, and thus zero slope, at the center plane.

In the case of cylindrical (or spherical) bodies having thermal symmetry about the center line (or midpoint), the thermal symmetry boundary condition requires that the first derivative of temperature with respect to r (the radial variable) be zero at the centerline (or the midpoint).

EXAMPLE 2-7   Heat Flux Boundary Condition

Consider an aluminum pan used to cook beef stew on top of an electric range. The bottom section of the pan is $L = 0.3$ cm thick and has a diameter of $D = 20$ cm. The electric heating unit on the range top consumes 800 W of power during cooking, and 90 percent of the heat generated in the heating element is transferred to the pan. During steady operation, the temperature of the inner surface of the pan is measured to be 110°C. Express the boundary conditions for the bottom section of the pan during this cooking process.

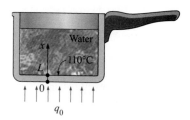

FIGURE 2-40

Schematic for Example 2-7.

Solution The heat transfer through the bottom section of the pan is from the bottom surface toward the top and can reasonably be approximated as being one-dimensional. We take the direction normal to the bottom surfaces of the pan as the x axis with the origin at the outer surface, as shown in Figure 2-40. Then the inner and outer surfaces of the bottom section of the pan can be represented by $x = 0$ and $x = L$, respectively. During steady operation, the temperature will depend on x only and thus $T = T(x)$.

The boundary condition on the outer surface of the bottom of the pan at $x = 0$ can be approximated as being specified heat flux since it is stated that 90 percent of the 800 W (i.e., 720 W) is transferred to the pan at that surface.

Therefore,

$$-k\frac{dT(0)}{dx} = q_0$$

where

$$q_0 = \frac{\text{Heat transfer rate}}{\text{Bottom surface area}} = \frac{0.720 \text{ kW}}{\pi(0.1 \text{ m})^2} = 22.9 \text{ kW/m}^2$$

The temperature at the inner surface of the bottom of the pan is specified to be 110°C. Then the boundary condition on this surface can be expressed as

$$T(L) = 110°C$$

where $L = 0.003$ m. Note that the determination of the boundary conditions may require some reasoning and approximations.

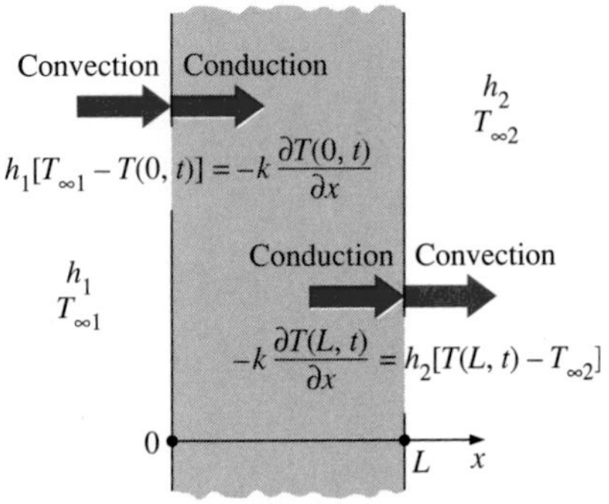

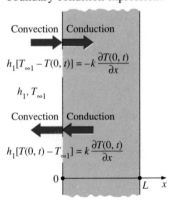

FIGURE 2-41

Convection boundary conditions on both surfaces of a plane wall.

3 Convection Boundary Condition

Convection is probably the most common boundary condition encountered in practice since most heat transfer surfaces are exposed to an environment at a specified temperature. The convection boundary condition is based on an *energy balance on the surface* and can be expressed as

$$\begin{pmatrix} \text{Heat conduction} \\ \text{at the surface in a} \\ \text{selected direction} \end{pmatrix} = \begin{pmatrix} \text{Heat convection} \\ \text{at the surface in} \\ \text{the same direction} \end{pmatrix}$$

For one-dimensional heat transfer in the x direction in a plate of thickness L, the convection boundary conditions on both surfaces can be expressed as

$$-k\frac{\partial T(0, t)}{\partial x} = h_1[T_{\infty 1} - T(0, t)] \tag{2-51a}$$

FIGURE 2-42

The assumed direction of heat transfer at a boundary has no effect on the boundary condition expression.

and

$$-k\frac{\partial T(L, t)}{\partial x} = h_2[T(L, t) - T_{\infty 2}] \tag{2-51b}$$

where h_1 and h_2 are the convection heat transfer coefficients and $T_{\infty 1}$ and $T_{\infty 2}$ are the temperatures of the surrounding mediums on the two sides of the plate, as shown in Figure 2-41.

In writing Equations 2-51 for convection boundary conditions, we have selected the direction of heat transfer to be the positive x direction at both surfaces. But those expressions are equally applicable when heat transfer is in the opposite direction at one or both surfaces since reversing the direction of heat transfer at a surface simply reverses the signs of *both* conduction and convection terms at that surface. This is equivalent to multiplying an equation by -1, which has no effect on the equality (Fig. 2-42). Being able to select either direction as the direction of heat transfer is certainly a relief since often we do not know the surface temperature and thus the direction of heat transfer at a surface in advance. This argument is also valid for other boundary conditions such as the radiation and combined boundary conditions discussed below.

Note that a surface has zero thickness and thus no mass, and it cannot store any energy. Therefore, the entire net heat entering the surface from one side must leave the surface from the other side. The convection boundary condition simply states that heat continues to flow from a body to the surrounding medium at the same rate, and it just changes vehicles at the surface from conduction to convection (or vice versa in the other direction). This is analogous to people traveling on buses on land and transferring to the ships at the shore. If the passengers are not allowed to wander around at the shore, then the rate at which the people are unloaded at the shore from the buses must equal the rate at which they board the ships. We may call this the conservation of "people" principle.

Also note that the surface temperatures $T(0, t)$ and $T(L, t)$ are not known (if they were known, we would simply use them as the specified temperature boundary condition and not bother with convection). But a surface temperature can be determined once the solution $T(x, t)$ is obtained by substituting the value of x at that surface into the solution.

EXAMPLE 2-8 Convection and Insulation Boundary Conditions

Steam flows through a pipe shown in Figure 2-43 at an average temperature of $T_\infty = 200°C$. The inner and outer radii of the pipe are $r_1 = 8$ cm and $r_2 = 8.5$ cm, respectively, and the outer surface of the pipe is heavily insulated. If the convection heat transfer coefficient on the inner surface of the pipe is $h = 65$ W/m² · °C, express the boundary conditions on the inner and outer surfaces of the pipe during transient periods.

Solution During initial transient periods, heat transfer through the pipe material will predominantly be in the radial direction, and thus can be approximated as being one-dimensional. Then the temperature within the pipe material will change with the radial distance r and the time t. That is, $T = T(r, t)$.

It is stated that heat transfer between the steam and the pipe at the inner surface is by convection. Then taking the direction of heat transfer to be the positive r direction, the boundary condition on that surface can be expressed as

$$-k\frac{\partial T(r_1, t)}{\partial r} = h[T_\infty - T(r_1)]$$

FIGURE 2-43
Schematic for Example 2-8.

The pipe is said to be well insulated on the outside, and thus heat loss through the outer surface of the pipe can be assumed to be negligible. Then the boundary condition at the outer surface can be expressed as

$$\frac{\partial T(r_2, t)}{\partial r} = 0$$

That is, the temperature gradient must be zero on the outer surface of the pipe at all times.

4 Radiation Boundary Condition

In some cases, such as those encountered in space and cryogenic applications, a heat transfer surface is surrounded by an evacuated space and thus there is no convection heat transfer between a surface and the surrounding medium. In such cases, *radiation* becomes the only mechanism of heat transfer between the surface under consideration and the surroundings. Using an energy balance, the radiation boundary condition on a surface can be expressed as

$$\begin{pmatrix} \text{Heat conduction} \\ \text{at the surface in a} \\ \text{selected direction} \end{pmatrix} = \begin{pmatrix} \text{Radiation exchange} \\ \text{at the surface in} \\ \text{the same direction} \end{pmatrix}$$

For one-dimensional heat transfer in the x direction in a plate of thickness L, the radiation boundary conditions on both surfaces can be expressed as (Fig. 2-44)

$$-k\frac{\partial T(0, t)}{\partial x} = \varepsilon_1 \sigma [T_{surr, 1}^4 - T(0, t)^4] \qquad (2\text{-}52a)$$

and

$$-k\frac{\partial T(L, t)}{\partial x} = \varepsilon_2 \sigma [T(L, t)^4 - T_{surr, 2}^4] \qquad (2\text{-}52b)$$

where ε_1 and ε_2 are the emissivities of the boundary surfaces, $\sigma = 5.67 \times 10^{-8}$ W/m$^2 \cdot$ K^4 is the Stefan-Boltzmann constant, and $T_{surr, 1}$ and $T_{surr, 2}$ are the average temperatures of the surfaces surrounding the two sides of the plate, respectively. Note that the temperatures in radiation calculations must be expressed in K or R (not in °C or °F).

The radiation boundary condition involves the fourth power of temperature, and thus it is a *nonlinear* condition. As a result, the application of this boundary condition results in powers of the unknown coefficients, which makes it difficult to determine them. Therefore, it is tempting to ignore radiation exchange at a surface during a heat transfer analysis in order to avoid the complications associated with nonlinearity. This is especially the case when heat transfer at the surface is dominated by convection, and the role of radiation is minor.

5 Interface Boundary Conditions

Some bodies are made up of layers of different materials, and the solution of a heat transfer problem in such a medium requires the solution of the heat transfer problem in each layer. This, in turn, requires the specification of the boundary conditions at each *interface*.

The boundary conditions at an interface are based on the requirements that (1) two bodies in contact must have the *same temperature* at the area of contact and (2) an interface (which is a surface) cannot store any energy, and thus the *heat flux* on the two sides of an interface *must be the same*. The

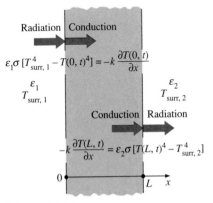

FIGURE 2-44

Radiation boundary conditions on both surfaces of a plane wall.

boundary conditions at the interface of two bodies A and B in perfect contact at $x = x_0$ can be expressed as (Fig. 2-45)

$$T_A(x_0, t) = T_B(x_0, t) \tag{2-53}$$

and

$$-k_A \frac{\partial T_A(x_0, t)}{\partial x} = -k_B \frac{\partial T_B(x_0, t)}{\partial x} \tag{2-54}$$

where k_A and k_B are the thermal conductivities of the layers A and B, respectively. The case of imperfect contact results in thermal contact resistance, which is considered in the next chapter.

6 Generalized Boundary Conditions

So far we have considered surfaces subjected to *single mode* heat transfer, such as the specified heat flux, convection, or radiation for simplicity. In general, however, a surface may involve convection, radiation, *and* specified heat flux simultaneously. The boundary condition in such cases is again obtained from an energy balance on the surface, which can be expressed as

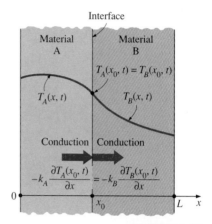

FIGURE 2-45

Boundary conditions at the interface of two bodies in perfect contact.

$$\begin{pmatrix} \text{Heat transfer} \\ \text{to the surface} \\ \text{in all modes} \end{pmatrix} = \begin{pmatrix} \text{Heat transfer} \\ \text{from the surface} \\ \text{in all modes} \end{pmatrix} \tag{2-55}$$

This is illustrated below with examples.

EXAMPLE 2-9   Combined Convection and Radiation Condition

A spherical metal ball of radius r_0 is heated in an oven to a temperature of 600°F throughout and is then taken out of the oven and allowed to cool in ambient air at $T_\infty = 78°F$, as shown in Figure 2-46. The thermal conductivity of the ball material is $k = 8.3$ Btu/h · ft · °F, and the average convection heat transfer coefficient on the outer surface of the ball is evaluated to be $h = 4.5$ Btu/h · ft² · °F. The emissivity of the outer surface of the ball is $\varepsilon = 0.6$, and the average temperature of the surrounding surfaces is $T_{surr} = 525$ R. Assuming the ball is cooled uniformly from the entire outer surface, express the initial and boundary conditions for the cooling process of the ball.

Solution   The ball is initially at a uniform temperature and is cooled uniformly from the entire outer surface. Therefore, this is a one-dimensional transient heat transfer problem since the temperature within the ball will change with the radial distance r and the time t. That is, $T = T(r, t)$. Taking the moment the ball is removed from the oven to be $t = 0$, the initial condition can be expressed as

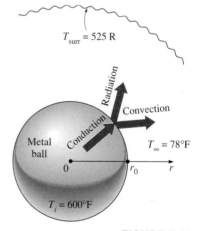

FIGURE 2-46

Schematic for Example 2-9.

$$T(r, 0) = T_i = 600°F$$

The problem possesses symmetry about the midpoint ($r = 0$) since the isotherms in this case will be concentric spheres, and thus no heat will be crossing the midpoint of the ball. Then the boundary condition at the midpoint can be expressed as

$$\frac{\partial T(0, t)}{\partial r} = 0$$

The heat conducted to the outer surface of the ball is lost to the environment by convection and radiation. Then taking the direction of heat transfer to be the positive r direction, the boundary condition on the outer surface can be expressed as

$$-k\frac{\partial T(r_0, t)}{\partial r} = h[T(r_0) - T_\infty] + \varepsilon\sigma[T(r_0)^4 - T_{surr}^4]$$

All the quantities in the above relations are known except the temperatures and their derivatives at $r = 0$ and r_0. Also, the radiation part of the boundary condition is often ignored for simplicity by modifying the convection heat transfer coefficient to account for the contribution of radiation. The convection coefficient h in that case becomes the combined heat transfer coefficient.

EXAMPLE 2-10   Combined Convection, Radiation, and Heat Flux

Consider the south wall of a house that is $L = 0.2$ m thick. The outer surface of the wall is exposed to solar radiation and has an absorptivity of $\alpha = 0.5$ for solar energy. The interior of the house is maintained at $T_{\infty 1} = 20°C$, while the ambient air temperature outside remains at $T_{\infty 2} = 5°C$. The sky, the ground, and the surfaces of the surrounding structures at this location can be modeled as a surface at an effective temperature of $T_{sky} = 255$ K for radiation exchange on the outer surface. The radiation exchange between the inner surface of the wall and the surfaces of the walls, floor, and ceiling it faces is negligible. The convection heat transfer coefficients on the inner and the outer surfaces of the wall are $h_1 = 6$ W/m^2 · °C and $h_2 = 25$ W/m^2 · °C, respectively. The thermal conductivity of the wall material is $k = 0.7$ W/m · °C, and the emissivity of the outer surface is $\varepsilon_1 = 0.9$. Assuming the heat transfer through the wall to be steady and one-dimensional, express the boundary conditions on the inner and the outer surfaces of the wall.

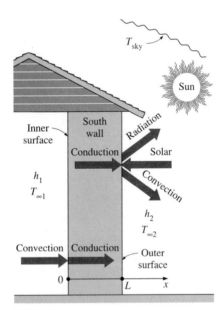

FIGURE 2-47
Schematic for Example 2-10.

Solution   We take the direction normal to the wall surfaces as the x axis with the origin at the inner surface of the wall, as shown in Figure 2-47. Then the inner and outer surfaces of the wall can be represented by $x = 0$ and $x = L$, respectively. The heat transfer through the wall is given to be steady and one-dimensional, and thus the temperature depends on x only and not on time. That is, $T = T(x)$.

The boundary condition on the inner surface of the wall at $x = 0$ is a typical convection condition since it does not involve any radiation or specified heat flux. Taking the direction of heat transfer to be the positive x direction, the boundary condition on the inner surface can be expressed as (see Eq. 2-51)

$$-k\frac{dT(0)}{dx} = h_1[T_{\infty 1} - T(0)]$$

Note that we have replaced the partial derivative by the ordinary derivative since the temperature depends on a single variable only.

The boundary condition on the outer surface at $x = L$ is as general as it gets: it involves conduction, convection, radiation, and specified heat flux. Again taking the direction of heat transfer to be the positive x direction, the boundary condition on the outer surface can be expressed as

$$-k\frac{dT(L)}{dx} = h_1[T(L) - T_{\infty 2}] + \varepsilon_2\sigma[T(L)^4 - T_{sky}^4] - \alpha\dot{q}_{solar}$$

where $\dot{q}_{solar}$ is the incident solar heat flux. Assuming the opposite direction for heat transfer would give the same result multiplied by -1, which is equivalent to the relation above. All the quantities in the above relations are known except the temperatures and their derivatives at the two boundaries.

Note that a heat transfer problem may involve different kinds of boundary conditions on different surfaces. For example, a plate may be subject to *heat flux* on one surface while losing or gaining heat by *convection* from the other surface. Also, the two boundary conditions in a direction may be specified *at the same boundary* while no condition is imposed on the other boundary. For example, specifying the temperature and heat flux at $x = 0$ of a plate of thickness L will result in a unique solution for the one-dimensional steady temperature distribution in the plate, including the value of temperature at the surface $x = L$. Although not necessary, there is nothing wrong with specifying more than two boundary conditions in a specified direction, provided that there is no contradiction. The extra conditions in this case can be used to verify the results.

2-6 ■ SOLUTION OF STEADY ONE-DIMENSIONAL HEAT CONDUCTION PROBLEMS

So far we have derived the differential equations for heat conduction in various coordinate systems and discussed the possible boundary conditions. A heat conduction problem can be formulated by specifying the applicable differential equation and a set of proper boundary conditions.

In this section we will solve a wide range of heat conduction problems in rectangular, cylindrical, and spherical geometries. We will limit our attention to problems that result in *ordinary differential equations* such as the *steady one-dimensional* heat conduction problems. We will also assume *constant thermal conductivity,* but will consider variable conductivity later in this chapter. If you feel rusty on differential equations or haven't taken differential equations yet, no need to panic. *Simple integration* is all you need to solve the steady one-dimensional heat conduction problems.

The solution procedure for solving heat conduction problems can be summarized as (1) *formulate* the problem by obtaining the applicable differential equation in its simplest form and specifying the boundary conditions, (2) obtain the *general solution* of the differential equation, and (3) apply the *boundary conditions* and determine the arbitrary constants in the general solution (Fig. 2-48). This is demonstrated below with several representative examples.

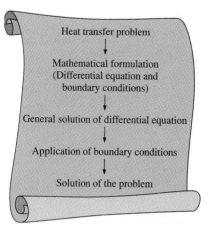

FIGURE 2-48

Basic steps involved in the solution of heat transfer problems.

EXAMPLE 2-11 Heat Conduction in a Plane Wall

Consider a large plane wall of thickness $L = 0.2$ m, thermal conductivity $k = 1.2$ W/m · °C, and surface area $A = 15$ m². The two sides of the wall are

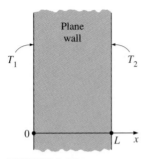

FIGURE 2-49

Schematic for Example 2-11.

maintained at constant temperatures of $T_1 = 120°C$ and $T_2 = 50°C$, respectively, as shown in Figure 2-49. Determine (a) the variation of temperature within the wall and the value of temperature at $x = 0.1$ m and (b) the rate of heat conduction through the wall under steady conditions.

Solution A plane wall with specified surface temperatures is given. The variation of temperature and the rate of heat transfer are to be determined.

Assumptions **1** Heat conduction is steady. **2** Heat conduction is one-dimensional since the wall is large relative to its thickness and the thermal conditions on both sides are uniform. **3** Thermal conductivity is constant. **4** There is no heat generation.

Properties The thermal conductivity is given to be $k = 1.2$ W/m · °C.

Analysis (a) Taking the direction normal to the surface of the wall to be the x direction, the differential equation for this problem can be expressed as

$$\frac{d^2T}{dx^2} = 0$$

with boundary conditions

$$T(0) = T_1 = 120°C$$
$$T(L) = T_2 = 50°C$$

The differential equation is linear and second order, and a quick inspection of this differential equation reveals that it has a single term involving derivatives and no terms involving the unknown function T as a factor. Thus, it can be solved by direct integration. Noting that an integration reduces the order of a derivative by one, the general solution of the differential equation above can be obtained by two simple successive integrations, each of which introduces an integration constant.

Integrating the differential equation once with respect to x yields

$$\frac{dT}{dx} = C_1$$

where C_1 is an arbitrary constant. Notice that the order of the derivative went down by one as a result of integration. As a check, if we take the derivative of this equation, we will obtain the original differential equation. This equation is not the solution yet since it involves a derivative.

Integrating one more time, we obtain

$$T(x) = C_1x + C_2$$

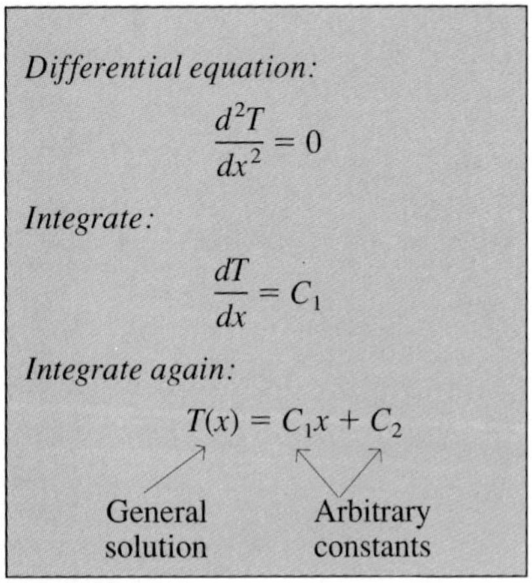

which is the general solution of the differential equation (Fig. 2-50). The general solution in this case resembles the general formula of a straight line whose slope is C_1 and whose value at $x = 0$ is C_2. This is not surprising since the second derivative represents the change in the slope of a function, and a zero second derivative indicates that the slope of the function remains constant. Therefore, *any straight line* is a solution of the differential equation above.

The general solution contains two unknown constants C_1 and C_2, and thus we need two equations to determine them uniquely and obtain the specific solution. These equations are obtained by forcing the general solution to satisfy the

---

Differential equation:

$$\frac{d^2T}{dx^2} = 0$$

Integrate:

$$\frac{dT}{dx} = C_1$$

Integrate again:

$$T(x) = C_1x + C_2$$

General solution   Arbitrary constants

FIGURE 2-50

Obtaining the general solution of a simple second order differential equation by integration.

specified boundary conditions. The application of each condition yields one equation, and thus we need to specify two conditions to determine the constants C_1 and C_2.

When applying a boundary condition to an equation, *all occurrences of the dependent and independent variables and any derivatives are replaced by the specified values.* Thus the only unknowns in the resulting equations are the arbitrary constants.

The first boundary condition can be interpreted as *in the general solution, replace all the x's by zero and T(x) by T_1.* That is (Fig. 2-51),

$$T(0) = C_1 \times 0 + C_2 \quad \rightarrow \quad C_2 = T_1$$

The second boundary condition can be interpreted as *in the general solution, replace all the x's by L and T(x) by T_2.* That is,

$$T(L) = C_1 L + C_2 \quad \rightarrow \quad T_2 = C_1 L + T_1 \quad \rightarrow \quad C_1 = \frac{T_2 - T_1}{L}$$

Substituting the C_1 and C_2 expressions into the general solution, we obtain

$$T(x) = \frac{T_2 - T_1}{L} x + T_1 \tag{2-56}$$

which is the desired solution since it satisfies not only the differential equation but also the two specified boundary conditions. That is, differentiating Equation 2-56 with respect to x twice will give d^2T/dx^2, which is the given differential equation, and substituting $x = 0$ and $x = L$ into Equation 2-56 gives $T(0) = T_1$ and $T(L) = T_2$, respectively, which are the specified conditions at the boundaries.

Substituting the given information, the value of the temperature at $x = 0.1$ m is determined to be

$$T(0.1 \text{ m}) = \frac{(50 - 120)°C}{0.2 \text{ m}} (0.1 \text{ m}) + 120°C = \textbf{85°C}$$

(b) The rate of heat conduction anywhere in the wall is determined from Fourier's law to be

$$\dot{Q}_{\text{wall}} = -kA \frac{dT}{dx} = -kAC_1 = -kA \frac{T_2 - T_1}{L} = kA \frac{T_1 - T_2}{L} \tag{2-57}$$

The numerical value of the rate of heat conduction through the wall is determined by substituting the given values to be

$$\dot{Q} = kA \frac{T_1 - T_2}{L} = (1.2 \text{ W/m} \cdot °C)(15 \text{ m}^2) \frac{(120 - 50)°C}{0.2 \text{ m}} = \textbf{6300 W}$$

Discussion Note that under steady conditions, the rate of heat conduction through a plain wall is constant.

EXAMPLE 2-12   A Wall with Various Sets of Boundary Conditions

Consider steady one-dimensional heat conduction in a large plane wall of thickness L and constant thermal conductivity k with no heat generation. Obtain expressions for the variation of temperature within the wall for the following pairs of boundary conditions (Fig. 2-52):

Boundary condition:
$$T(0) = T_1$$
General solution:
$$T(x) = C_1 x + C_2$$
Applying the boundary condition:
$$T(x) = C_1 x + C_2$$
$$\uparrow \qquad \uparrow$$
$$0 \qquad 0$$
$$T_1$$
Substituting:
$$T_1 = C_1 \times 0 + C_2 \rightarrow C_2 = T_1$$
It cannot involve x or T(x) after the boundary condition is applied.

FIGURE 2-51
When applying a boundary condition to the general solution at a specified point, all occurrences of the dependent and independent variables should be replaced by their specified values at that point.

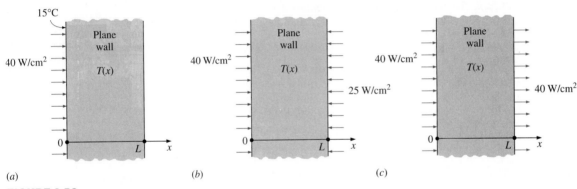

(a)

(b)

(c)

FIGURE 2-52
Schematic for Example 2-12.

$$(a) \quad -k\frac{dT(0)}{dx} = \dot{q}_0 = 40 \text{ W/cm}^2 \quad \text{and} \quad T(0) = T_0 = 15°C$$

$$(b) \quad -k\frac{dT(0)}{dx} = \dot{q}_0 = 40 \text{ W/cm}^2 \quad \text{and} \quad -k\frac{dT(L)}{dx} = \dot{q}_L = -25 \text{ W/cm}^2$$

$$(c) \quad -k\frac{dT(0)}{dx} = \dot{q}_0 = 40 \text{ W/cm}^2 \quad \text{and} \quad -k\frac{dT(L)}{dx} = \dot{q}_0 = 40 \text{ W/cm}^2$$

Solution This is a steady one-dimensional heat conduction problem with constant thermal conductivity and no heat generation in the medium, and the heat conduction equation in this case can be expressed as (Eq. 2-20)

$$\frac{d^2T}{dx^2} = 0$$

whose general solution was determined in the previous example by direct integration to be

$$T(x) = C_1 x + C_2$$

where C_1 and C_2 are two arbitrary integration constants. The specific solutions corresponding to each specified pair of boundary conditions are determined as follows.

(a) In this case, both boundary conditions are specified at the same boundary at $x = 0$, and no boundary condition is specified at the other boundary at $x = L$. Noting that

$$\frac{dT}{dx} = C_1$$

the application of the boundary conditions gives

$$-k\frac{dT(0)}{dx} = \dot{q}_0 \quad \rightarrow \quad -kC_1 = \dot{q}_0 \quad \rightarrow \quad C_1 = -\frac{\dot{q}_0}{k}$$

and

$$T(0) = T_0 \quad \rightarrow \quad T_0 = C_1 \times 0 + C_2 \quad \rightarrow \quad C_2 = T_0$$

Substituting, the specific solution in this case is determined to be

$$T(x) = -\frac{\dot{q}_0}{k} x + T_0$$

Therefore, the two boundary conditions can be specified at the same boundary, and it is not necessary to specify them at different locations. In fact, the fundamental theorem of linear ordinary differential equations guarantees that a unique solution exists when both conditions are specified at the same location. But no such guarantee exists when the two conditions are specified at different boundaries, as you will see below.

(b) In this case different heat fluxes are specified at the two boundaries. The application of the boundary conditions gives

$$-k\frac{dT(0)}{dx} = \dot{q}_0 \quad \rightarrow \quad -kC_1 = \dot{q}_0 \quad \rightarrow \quad C_1 = -\frac{\dot{q}_0}{k}$$

and

$$-k\frac{dT(L)}{dx} = \dot{q}_L \quad \rightarrow \quad -kC_1 = \dot{q}_L \quad \rightarrow \quad C_1 = -\frac{\dot{q}_L}{k}$$

Since $\dot{q}_0 \neq \dot{q}_L$ and the constant C_1 cannot be equal to two different things at the same time, there is no solution in this case. This is not surprising since this case corresponds to supplying heat to the plane wall from both sides and expecting the temperature of the wall to remain steady (not to change with time). This is impossible.

(c) In this case, the same values for heat flux are specified at the two boundaries. The application of the boundary conditions gives

$$-k\frac{dT(0)}{dx} = \dot{q}_0 \quad \rightarrow \quad -kC_1 = \dot{q}_0 \quad \rightarrow \quad C_1 = -\frac{\dot{q}_0}{k}$$

and

$$-k\frac{dT(L)}{dx} = \dot{q}_0 \quad \rightarrow \quad -kC_1 = \dot{q}_0 \quad \rightarrow \quad C_1 = -\frac{\dot{q}_0}{k}$$

Thus, both conditions result in the same value for the constant C_1, but no value for C_2. Substituting, the specific solution in this case is determined to be

$$T(x) = -\frac{\dot{q}_0}{k} x + C_2$$

which is not a unique solution since C_2 is arbitrary. This solution represents a family of straight lines whose slope is $-\dot{q}_0/k$. Physically, this problem corresponds to requiring the rate of heat supplied to the wall at $x = 0$ be equal to the rate of heat removal from the other side of the wall at $x = L$. But this is a consequence of the heat conduction through the wall being steady, and thus the second boundary condition does not provide any new information. So it is not surprising that the solution of this problem is not unique. The three cases discussed above are summarized in Figure 2-53.

Differential equation:
$$T''(x) = 0$$

General solution:
$$T(x) = C_1 x + C_2$$

(a) *Unique solution:*
$$\left. \begin{array}{r} -kT'(0) = \dot{q}_0 \\ T(0) = T_0 \end{array} \right\} T(x) = -\frac{\dot{q}_0}{k} x + T_0$$

(b) *No solution:*
$$\left. \begin{array}{r} -kT'(0) = \dot{q}_0 \\ -kT'(L) = \dot{q}_L \end{array} \right\} T(x) = \text{None}$$

(c) *Multiple solutions:*
$$\left. \begin{array}{r} -kT'(0) = \dot{q}_0 \\ -kT'(L) = \dot{q}_0 \end{array} \right\} T(x) = -\frac{\dot{q}_0}{k} x + C_2$$
$$\uparrow$$
Arbitrary

FIGURE 2-53

A boundary-value problem may have a unique solution, infinitely many solutions, or no solutions at all.

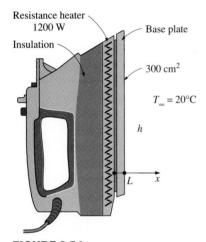

FIGURE 2-54

Schematic for Example 2-13.

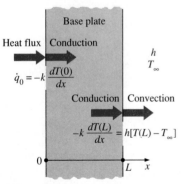

FIGURE 2-55

The boundary conditions on the base plate of the iron discussed in Example 2-13.

EXAMPLE 2-13   Heat Conduction in the Base Plate of an Iron

Consider the base plate of a 1200-W household iron that has a thickness of $L = 0.5$ cm, base area of $A = 300$ cm^2, and thermal conductivity of $k = 15$ W/m $\cdot$ °C. The inner surface of the base plate is subjected to uniform heat flux generated by the resistance heaters inside, and the outer surface loses heat to the surroundings at $T_\infty = 20$°C by convection, as shown in Figure 2-54. Taking the convection heat transfer coefficient to be $h = 80$ W/m$^2 \cdot$ °C and disregarding heat loss by radiation, obtain an expression for the variation of temperature in the base plate, and evaluate the temperatures at the inner and the outer surfaces.

Solution   The base plate of an iron is considered. The variation of temperature in the plate and the surface temperatures are to be determined.

Assumptions   **1** Heat transfer is steady since there is no change with time. **2** Heat transfer is one-dimensional since the surface area of the base plate is large relative to its thickness, and the thermal conditions on both sides are uniform. **3** Thermal conductivity is constant. **4** There is no heat generation in the medium. **5** Heat transfer by radiation is negligible. **6** The upper part of the iron is well insulated so that the entire heat generated in the resistance wires is transferred to the base plate through its inner surface.

Properties   The thermal conductivity is given to be $k = 15$ W/m $\cdot$ °C.

Analysis   The inner surface of the base plate is subjected to uniform heat flux at a rate of

$$\dot{q}_0 = \frac{\dot{Q}_0}{A_{\text{base}}} = \frac{1200 \text{ W}}{0.03 \text{ m}^2} = 40{,}000 \text{ W/m}^2$$

The outer side of the plate is subjected to the convection condition. Taking the direction normal to the surface of the wall as the x direction with its origin on the inner surface, the differential equation for this problem can be expressed as (Fig. 2-55)

$$\frac{d^2T}{dx^2} = 0$$

with the boundary conditions

$$-k\frac{dT(0)}{dx} = \dot{q}_0 = 40{,}000 \text{ W/m}^2$$

$$-k\frac{dT(L)}{dx} = h[T(L) - T_\infty]$$

The general solution of the differential equation is again obtained by two successive integrations to be

$$\frac{dT}{dx} = C_1$$

and

$$T(x) = C_1x + C_2 \tag{a}$$

where C_1 and C_2 are arbitrary constants. Applying the first boundary condition,

$$-k\frac{dT(0)}{dx} = \dot{q}_0 \quad \rightarrow \quad -kC_1 = \dot{q}_0 \quad \rightarrow \quad C_1 = -\frac{\dot{q}_0}{k}$$

Noting that $dT/dx = C_1$ and $T(L) = C_1L + C_2$, applying the second boundary condition gives

$$-k\frac{dT(L)}{dx} = h[T(L) - T_\infty] \quad \rightarrow \quad -kC_1 = h[(C_1L + C_2) - T_\infty]$$

Substituting $C_1 = -\dot{q}_0/k$ and solving for C_2, we obtain

$$C_2 = T_\infty - \frac{C_1k}{h} - C_1L = T_\infty + \frac{\dot{q}_0}{h} + \frac{\dot{q}_0}{k}L$$

Now substituting C_1 and C_2 into the general solution (a), we obtain

$$T(x) = T_\infty + \dot{q}_0\left(\frac{L-x}{k} + \frac{1}{h}\right) \qquad (b)$$

which is the solution for the variation of the temperature in the plate. The temperatures at the inner and outer surfaces of the plate are determined by substituting $x = 0$ and $x = L$, respectively, into the relation above:

$$T(0) = T_\infty + \dot{q}_0\left(\frac{L}{k} + \frac{1}{h}\right)$$

$$= 20°C + (40{,}000 \text{ W/m}^2)\left(\frac{0.005 \text{ m}}{15 \text{ W/m}\cdot°C} + \frac{1}{80 \text{ W/m}^2\cdot°C}\right) = \mathbf{533°C}$$

and

$$T(L) = T_\infty + \dot{q}_0\left(0 + \frac{1}{h}\right) = 20°C + \frac{40{,}000 \text{ W/m}^2}{80 \text{ W/m}^2\cdot°C} = \mathbf{520°C}$$

Discussion Note that the temperature of the inner surface of the base plate will be 13°C higher than the temperature of the outer surface when steady operating conditions are reached. Also note that the heat transfer analysis above enables us to calculate the temperatures of surfaces that we cannot even reach. This example demonstrates how the heat flux and convection boundary conditions are applied to heat transfer problems.

EXAMPLE 2-14   Heat Conduction in a Solar Heated Wall

Consider a large plane wall of thickness $L = 0.06$ m and thermal conductivity $k = 1.2$ W/m·°C in space. The wall is covered with white porcelain tiles that have an emissivity of $\varepsilon = 0.85$ and a solar absorptivity of $\alpha = 0.26$, as shown in Figure 2-56. The inner surface of the wall is maintained at $T_1 = 300$ K at all times, while the outer surface is exposed to solar radiation that is incident at a rate of $\dot{q}_{solar} = 800$ W/m². The outer surface is also losing heat by radiation to deep space at 0 K. Determine the temperature of the outer surface of the wall and the rate of heat transfer through the wall when steady operating conditions are reached. What would your response be if no solar radiation was incident on the surface?

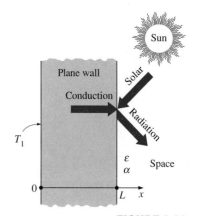

FIGURE 2-56

Schematic for Example 2-14.

Solution A plane wall in space is subjected to specified temperature on one side and solar radiation on the other side. The outer surface temperature and the rate of heat transfer are to be determined.

Assumptions **1** Heat transfer is steady since there is no change with time. **2** Heat transfer is one-dimensional since the wall is large relative to its thickness, and the thermal conditions on both sides are uniform. **3** Thermal conductivity is constant. **4** There is no heat generation.

Properties The thermal conductivity is given to be $k = 1.2$ W/m $\cdot$ °C.

Analysis Taking the direction normal to the surface of the wall as the x direction with its origin on the inner surface, the differential equation for this problem can be expressed as

$$\frac{d^2T}{dx^2} = 0$$

with boundary conditions

$$T(0) = T_1 = 300 \text{ K}$$

$$-k\frac{dT(L)}{dx} = \varepsilon\sigma[T(L)^4 - T_{\text{space}}^4] - \alpha\dot{q}_{\text{solar}}$$

where $T_{\text{space}} = 0$. The general solution of the differential equation is again obtained by two successive integrations to be

$$T(x) = C_1 x + C_2 \qquad (a)$$

where C_1 and C_2 are arbitrary constants. Applying the first boundary condition yields

$$T(0) = C_1 \times 0 + C_2 \quad \rightarrow \quad C_2 = T_1$$

Noting that $dT/dx = C_1$ and $T(L) = C_1 L + C_2 = C_1 L + T_1$, the application of the second boundary conditions gives

$$-k\frac{dT(L)}{dx} = \varepsilon\sigma T(L)^4 + \alpha\dot{q}_{\text{solar}} \quad \rightarrow \quad -kC_1 = \varepsilon\sigma(C_1 L + T_1)^4 - \alpha\dot{q}_{\text{solar}}$$

Well, it looks like we have hit a wall. Although C_1 is the only unknown in the equation above, we cannot get an explicit expression for it because the equation is nonlinear, and thus we cannot get a closed form expression for the temperature distribution. This should explain why we do our best to avoid nonlinearities in the analysis, such as those associated with radiation.

But there is no reason to give up. Let us back up a little bit and denote the outer surface temperature by $T(L) = T_L$ instead of $T(L) = C_1 L + T_1$. The application of the second boundary condition in this case gives

$$-k\frac{dT(L)}{dx} = \varepsilon\sigma T(L)^4 + \alpha\dot{q}_{\text{solar}} \quad \rightarrow \quad kC_1 = \varepsilon\sigma T_L^4 - \alpha\dot{q}_{\text{solar}}$$

Solving for C_1 gives

$$C_1 = \frac{\alpha\dot{q}_{\text{solar}} - \varepsilon\sigma T_L^4}{k} \qquad (b)$$

Now substituting C_1 and C_2 into the general solution (a), we obtain

$$T(x) = \frac{\alpha \dot{q}_{solar} - \varepsilon \sigma T_L^4}{k} x + T_1 \qquad (c)$$

which is the solution for the variation of the temperature in the wall in terms of the unknown outer surface temperature T_L. At $x = L$ it becomes

$$T_L = \frac{\alpha \dot{q}_{solar} - \varepsilon \sigma T_L^4}{k} L + T_1 \qquad (d)$$

which is an implicit relation for the outer surface temperature T_L. Substituting the given values, we get

$$T_L = \frac{0.26 \times (800 \text{ W/m}^2) - 0.85 \times (5.67 \times 10^{-8} \text{ W/m}^2 \cdot \text{K}^4)\, T_L^4}{1.2 \text{ W/m} \cdot \text{K}} (0.06 \text{ m}) + 300 \text{ K}$$

which simplifies to

$$T_L = 310.4 - 0.240975 \left(\frac{T_L}{100}\right)^4$$

This equation can be solved by one of the several nonlinear equation solvers available (or by the old fashioned trial and error method) to give (Fig. 2-57)

$$T_L = 292.7 \text{ K}$$

Knowing the outer surface temperature and knowing that it must remain constant under steady conditions, the temperature distribution in the wall can be determined by substituting the T_L value above into Equation (c):

$$T(x) = \frac{0.26 \times (800 \text{ W/m}^2) - 0.85 \times (5.67 \times 10^{-8} \text{ W/m}^2 \cdot \text{K}^4)(292.7 \text{ K})^4}{1.2 \text{ W/m} \cdot \text{K}} x + 300 \text{K}$$

which simplifies to

$$T(x) = (-121.5 \text{ m}^{-1})x + 300 \text{ K}$$

Note that the outer surface temperature turned out to be lower than the inner surface temperature. Therefore, the heat transfer through the wall will be toward the outside despite the absorption of solar radiation by the outer surface. Knowing both the inner and outer surface temperatures of the wall, the steady rate of heat conduction through the wall can be determined from

$$\dot{q} = k \frac{T_0 - T_L}{L} = (1.2 \text{ W/m} \cdot \text{K}) \frac{(300 - 292.7) \text{ K}}{0.06 \text{ m}} = 146 \text{ W/m}^2$$

Discussion   In the case of no incident solar radiation, the outer surface temperature, determined from Equation (d) by setting $\dot{q}_{solar} = 0$, will be $TL = 284.3$ K. It is interesting to note that the solar energy incident on the surface causes the surface temperature to increase by about 8 K only when the inner surface temperature of the wall is maintained at 300 K.

(1) *Rearrange the equation to be solved:*

$$T_L = 310.4 - 0.240975 \left(\frac{T_L}{100}\right)^4$$

The equation is in the proper form since the left side consists of T_L only.

(2) *Guess the value of T_L, say 300 K, and substitute into the right side of the equation. It gives*

$$T_L = 290.2 \text{ K}$$

(3) *Now substitute this value of T_L into the right side of the equation and get*

$$T_L = 293.1 \text{ K}$$

(4) *Repeat step (3) until convergence to desired accuracy is achieved. The subsequent iterations give*

$$T_L = 292.6 \text{ K}$$
$$T_L = 292.7 \text{ K}$$
$$T_L = 292.7 \text{ K}$$

Therefore, the solution is $T_L = 292.7$ K. The result is independent of the initial guess.

FIGURE 2-57

A simple method of solving a nonlinear equation is to arrange the equation such that the unknown is alone on the left side while everything else is on the right side, and to iterate after an initial guess until convergence.

EXAMPLE 2-15   Heat Loss through a Steam Pipe

Consider a steam pipe of length $L = 20$ m, inner radius $r_1 = 6$ cm, outer radius $r_2 = 8$ cm, and thermal conductivity $k = 20$ W/m · °C, as shown in Figure 2-58. The inner and outer surfaces of the pipe are maintained at average temperatures of $T_1 = 150$°C and $T_2 = 60$°C, respectively. Obtain a general relation for the

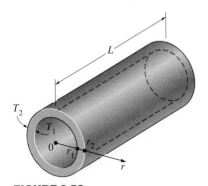

FIGURE 2-58
Schematic for Example 2-15.

Differential equation:

$$\frac{d}{dr}\left(r\frac{dT}{dr}\right) = 0$$

Integrate:

$$r\frac{dT}{dr} = C_1$$

Divide by r (r ≠ 0):

$$\frac{dT}{dr} = \frac{C_1}{r}$$

Integrate again:

$$T(r) = C_1 \ln r + C_2$$

which is the general solution.

FIGURE 2-59
Basic steps involved in the solution of
the steady one-dimensional
heat conduction equation in
cylindrical coordinates.

temperature distribution inside the pipe under steady conditions, and determine the rate of heat loss from the steam through the pipe.

Solution A steam pipe is subjected to specified temperatures on its surfaces. The variation of temperature and the rate of heat transfer are to be determined.

Assumptions **1** Heat transfer is steady since there is no change with time. **2** Heat transfer is one-dimensional since there is thermal symmetry about the centerline and no variation in the axial direction, and thus $T = T(r)$. **3** Thermal conductivity is constant. **4** There is no heat generation.

Properties The thermal conductivity is given to be $k = 20$ W/m · °C.

Analysis The mathematical formulation of this problem can be expressed as

$$\frac{d}{dr}\left(r\frac{dT}{dr}\right) = 0$$

with boundary conditions

$$T(r_1) = T_1 = 150°C$$

$$T(r_2) = T_2 = 60°C$$

Integrating the differential equation once with respect to r gives

$$r\frac{dT}{dr} = C_1$$

where C_1 is an arbitrary constant. We now divide both sides of the equation above by r to bring it to a readily integrable form,

$$\frac{dT}{dr} = \frac{C_1}{r}$$

Again integrating with respect to r gives (Fig. 2-59)

$$T(r) = C_1 \ln r + C_2 \tag{a}$$

We now apply both boundary conditions by replacing all occurrences of r and $T(r)$ in the relation above with the specified values at the boundaries. We get

$$T(r_1) = T_1 \quad \rightarrow \quad C_1 \ln r_1 + C_2 = T_1$$

$$T(r_2) = T_2 \quad \rightarrow \quad C_1 \ln r_2 + C_2 = T_2$$

which are two equations in two unknowns, C_1 and C_2. Solving them simultaneously gives

$$C_1 = \frac{T_2 - T_1}{\ln(r_2/r_1)} \quad \text{and} \quad C_2 = T_1 - \frac{T_2 - T_1}{\ln(r_2/r_1)}\ln r_1$$

Substituting them into Eq. (*a*) and rearranging, the variation of temperature within the pipe is determined to be

$$T(r) = \frac{\ln(r/r_1)}{\ln(r_2/r_1)}(T_2 - T_1) + T_1 \tag{2-58}$$

The rate of heat loss from the steam is simply the total rate of heat conduction through the pipe, and is determined from Fourier's law to be

$$\dot{Q}_{cylinder} = -kA\frac{dT}{dr} = -k(2\pi rL)\frac{C_1}{r} = -2\pi kHC_1 = 2\pi kH\frac{T_1 - T_2}{\ln(r_2/r_1)} \quad (2\text{-}59)$$

The numerical value of the rate of heat conduction through the pipe is determined by substituting the given values

$$\dot{Q} = 2\pi(20 \text{ W/m} \cdot °C)(20 \text{ m})\frac{(150 - 60)°C}{\ln(0.08/0.06)} = \textbf{786.3 kW}$$

Discussion Note that the total rate of heat transfer through a pipe is constant, but the heat flux is not since it decreases in the direction of heat transfer with increasing radius since $\dot{q} = \dot{Q}/(2\pi rL)$.

EXAMPLE 2-16 Heat Conduction through a Spherical Shell

Consider a spherical container of inner radius $r_1 = 8$ cm, outer radius $r_2 = 10$ cm, and thermal conductivity $k = 45$ W/m · °C, as shown in Figure 2-60. The inner and outer surfaces of the container are maintained at constant temperatures of $T_1 = 200°C$ and $T_2 = 80°C$, respectively, as a result of some chemical reactions occurring inside. Obtain a general relation for the temperature distribution inside the shell under steady conditions, and determine the rate of heat loss from the container.

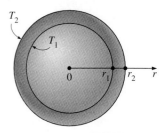

FIGURE 2-60

Schematic for Example 2-16.

Solution A spherical container is subjected to specified temperatures on its surfaces. The variation of temperature and the rate of heat transfer are to be determined.

Assumptions **1** Heat transfer is steady since there is no change with time. **2** Heat transfer is one-dimensional since there is thermal symmetry about the midpoint, and thus $T = T(r)$. **3** Thermal conductivity is constant. **4** There is no heat generation.

Properties The thermal conductivity is given to be $k = 45$ W/m · °C.

Analysis The mathematical formulation of this problem can be expressed as

$$\frac{d}{dr}\left(r^2\frac{dT}{dr}\right) = 0$$

with boundary conditions

$$T(r_1) = T_1 = 200°C$$
$$T(r_2) = T_2 = 80°C$$

Integrating the differential equation once with respect to r yields

$$r^2\frac{dT}{dr} = C_1$$

where C_1 is an arbitrary constant. We now divide both sides of the equation above by r^2 to bring it to a readily integrable form,

$$\frac{dT}{dr} = \frac{C_1}{r^2}$$

Again integrating with respect to r gives

$$T(r) = -\frac{C_1}{r} + C_2 \qquad (a)$$

We now apply both boundary conditions by replacing all occurrences of r and $T(r)$ in the relation above by the specified values at the boundaries. We get

$$T(r_1) = T_1 \quad \rightarrow \quad -\frac{C_1}{r_1} + C_2 = T_1$$

$$T(r_2) = T_2 \quad \rightarrow \quad -\frac{C_1}{r_2} + C_2 = T_2$$

which are two equations in two unknowns, C_1 and C_2. Solving them simultaneously gives

$$C_1 = -\frac{r_1 r_2}{r_2 - r_1}(T_1 - T_2) \quad \text{and} \quad C_2 = \frac{r_2 T_2 - r_1 T_1}{r_2 - r_1}$$

Substituting into Equation (a), the variation of temperature within the spherical shell is determined to be

$$T(r) = \frac{r_1 r_2}{r(r_2 - r_1)}(T_1 - T_2) + \frac{r_2 T_2 - r_1 T_1}{r_2 - r_1} \qquad (2\text{-}60)$$

The rate of heat loss from the container is simply the total rate of heat conduction through the container wall and is determined from Fourier's law

$$\dot{Q}_{sphere} = -kA\frac{dT}{dr} = -k(4\pi r^2)\frac{C_1}{r^2} = -4\pi kC_1 = 4\pi kr_1 r_2 \frac{T_1 - T_2}{r_2 - r_1} \qquad (2\text{-}61)$$

The numerical value of the rate of heat conduction through the wall is determined by substituting the given values to be

$$\dot{Q} = 4\pi(45 \text{ W/m} \cdot °\text{C})(0.08 \text{ m})(0.10 \text{ m})\frac{(200 - 80)°\text{C}}{(0.10 - 0.08) \text{ m}} = \textbf{27,140 W}$$

Discussion Note that the total rate of heat transfer through a spherical shell is constant, but the heat flux, $\dot{q} = \dot{Q}/4\pi r^2$, is not since it decreases in the direction of heat transfer with increasing radius as shown in Figure 2-61.

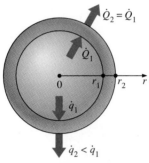

$$\dot{q}_1 = \frac{\dot{Q}_1}{A_1} = \frac{27.14 \text{ kW}}{4\pi(0.08 \text{ m})^2} = 337.5 \text{ kW/m}^2$$

$$\dot{q}_2 = \frac{\dot{Q}_2}{A_2} = \frac{27.14 \text{ kW}}{4\pi(0.10 \text{ m})^2} = 216.0 \text{ kW/m}^2$$

FIGURE 2-61

During steady one-dimensional heat conduction in a spherical (or cylindrical) container, the total rate of heat transfer remains constant, but the heat flux decreases with increasing radius.

FIGURE 2-62

Heat generation in solids is commonly encountered in practice.

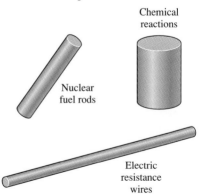

2-7 ■ HEAT GENERATION IN A SOLID

Many practical heat transfer applications involve the conversion of some form of energy into *heat* energy in the medium. Such mediums are said to involve internal *heat generation,* which manifests itself as a rise in temperature throughout the medium. Some examples of heat generation are *resistance heating* in wires, exothermic *chemical reactions* in a solid, and *nuclear reactions* in nuclear fuel rods where electrical, chemical, and nuclear energies are converted to heat, respectively (Fig. 2-62). The absorption of radiation throughout the volume of a semitransparent medium such as water can also be considered as heat generation within the medium, as explained earlier.

Heat generation is usually expressed *per unit volume* of the medium, and is denoted by $\dot{g}$, whose unit is W/m^3. For example, heat generation in an electrical wire of outer radius r_0 and length L can be expressed as

$$\dot{g} = \frac{\dot{E}_{g.\text{electric}}}{V_{\text{wire}}} = \frac{I^2 R_e}{\pi r_o^2 L} \qquad (\text{W/m}^3) \qquad (2\text{-}62)$$

where I is the electric current and R_e is the electrical resistance of the wire.

The temperature of a medium *rises* during heat generation as a result of the absorption of the generated heat by the medium during transient start-up period. As the temperature of the medium increases, so does the heat transfer from the medium to its surroundings. This continues until steady operating conditions are reached and the rate of heat generation equals the rate of heat transfer to the surroundings. Once steady operation has been established, the temperature of the medium at any point no longer changes.

The *maximum temperature* T_{max} in a solid that involves uniform heat generation will occur at a location *furthest away* from the outer surface when the outer surface of the solid is maintained at a constant temperature T_s. For example, the maximum temperature occurs at the *midplane* in a plane wall, at the *centerline* in a long cylinder, and at the *midpoint* in a sphere. The temperature distribution within the solid in these cases will be *symmetrical* about the center of symmetry.

The quantities of major interest in a medium with heat generation are the surface temperature T_s and the maximum temperature T_{max} that occurs in the medium in *steady* operation. Below we develop expressions for these two quantities for common geometries for the case of *uniform* heat generation ($\dot{g}$ = constant) within the medium.

Consider a solid medium of surface area A, volume V, and constant thermal conductivity k, where heat is generated at a constant rate of $\dot{g}$ per unit volume. Heat is transferred from the solid to the surrounding medium at T_∞, with a constant heat transfer coefficient of h. All the surfaces of the solid are maintained at a common temperature T_s. Under *steady* conditions, the energy balance for this solid can be expressed as (Fig. 2-63)

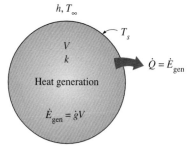

$$\begin{pmatrix} \text{Rate of} \\ \textit{heat transfer} \\ \text{from the solid} \end{pmatrix} = \begin{pmatrix} \text{Rate of} \\ \textit{energy generation} \\ \text{within the solid} \end{pmatrix} \qquad (2\text{-}63)$$

FIGURE 2-63
At steady conditions, the entire heat generated in a solid must leave the solid through its outer surface.

or

$$\dot{Q} = \dot{g}V \qquad (\text{W}) \qquad (2\text{-}64)$$

Disregarding radiation (or incorporating it in the heat transfer coefficient h), the heat transfer rate can also be expressed from Newton's law of cooling as

$$\dot{Q} = hA(T_s - T_\infty) \qquad (\text{W}) \qquad (2\text{-}65)$$

Combining Equations 2-64 and 2-65 and solving for the surface temperature T_s gives

$$T_s = T_\infty + \frac{\dot{g}V}{hA} \qquad (2\text{-}66)$$

For a large *plane wall* of thickness $2L$ ($A = 2A_{wall}$ and $V = 2LA_{wall}$), a long solid *cylinder* of radius r_o ($A = 2\pi r_o L$ and $V = \pi r_o^2 L$), and a solid *sphere* of radius r_0 ($A = 4\pi r_o^2$ and $V = \frac{4}{3}\pi r_o^3$), Equation 2-66 reduces to

$$T_{s,\text{ plane wall}} = T_\infty + \frac{\dot{g}L}{h} \tag{2-67}$$

$$T_{s,\text{ cylinder}} = T_\infty + \frac{\dot{g}r_o}{2h} \tag{2-68}$$

$$T_{s,\text{ sphere}} = T_\infty + \frac{\dot{g}r_o}{3h} \tag{2-69}$$

Note that the rise in surface temperature T_s is due to heat generation in the solid.

Reconsider heat transfer from a long solid cylinder with heat generation. We mentioned above that, under *steady* conditions, the entire heat generated within the medium is conducted through the outer surface of the cylinder. Now consider an imaginary inner cylinder of radius r within the cylinder (Fig. 2-64). Again the *heat generated* within this inner cylinder must be equal to the *heat conducted* through the outer surface of this inner cylinder. That is, from Fourier's law of heat conduction,

$$-kA_r \frac{dT}{dr} = \dot{g}V_r \tag{2-70}$$

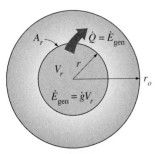

FIGURE 2-64

Heat conducted through a cylindrical shell of radius r is equal to the heat generated within a shell.

where $A_r = 2\pi r L$ and $V_r = \pi r^2 L$ at any location r. Substituting these expressions into the above relation and separating the variables, we get

$$-k(2\pi r L)\frac{dT}{dr} = \dot{g}(\pi r^2 L) \quad \rightarrow \quad dT = -\frac{\dot{g}}{2k}r\,dr$$

Integrating from $r = 0$ where $T(0) = T_0$ to $r = r_o$ where $T(r_o) = T_s$ yields

$$\Delta T_{\text{max, cylinder}} = T_o - T_s = \frac{\dot{g}r_o^2}{4k} \tag{2-71}$$

where T_o is the centerline temperature of the cylinder, which is the *maximum temperature*, and ΔT_{max} is the difference between the centerline and the surface temperatures of the cylinder, which is the *maximum temperature rise* in the cylinder above the surface temperature. Once ΔT_{max} is available, the centerline temperature can easily be determined from (Fig. 2-65)

$$T_{\text{center}} = T_o = T_s + \Delta T_{max} \tag{2-72}$$

The approach outlined above can also be used to determine the *maximum temperature rise* in a plane wall of thickness $2L$ and a solid sphere of radius r_0, with the following results:

$$\Delta T_{\text{max, plane wall}} = \frac{\dot{g}L^2}{2k} \tag{2-73}$$

$$\Delta T_{\text{max, sphere}} = \frac{\dot{g}r_o^2}{6k} \tag{2-74}$$

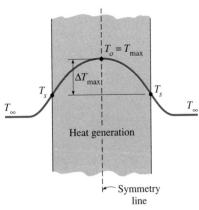

FIGURE 2-65

The maximum temperature in a symmetrical solid with uniform heat generation occurs at its center.

Again the maximum temperature at the center can be determined from Equation 2-72 by adding the maximum temperature rise to the surface temperature of the solid.

Note that the thermal resistance concept discussed earlier cannot be used when there is heat generation in the medium since the heat transfer rate through the medium in this case is no longer constant.

EXAMPLE 2-17   Centerline Temperature of a Resistance Heater

A 2-kW resistance heater wire whose thermal conductivity is $k = 15$ W/m·°C has a diameter of $D = 4$ mm and a length of $L = 0.5$ m, and is used to boil water (Fig. 2-66). If the outer surface temperature of the resistance wire is $T_s = 105$°C, determine the temperature at the center of the wire.

Solution   The surface temperature of a resistance heater submerged in water is to be determined.

Assumptions   **1** Heat transfer is steady since there is no change with time. **2** Heat transfer is one-dimensional since there is thermal symmetry about the centerline and no change in the axial direction. **3** Thermal conductivity is constant. **4** Heat generation in the heater is uniform.

Properties   The thermal conductivity is given to be $k = 15$ W/m·°C.

Analysis   The 2-kW resistance heater converts electric energy into heat at a rate of 2 kW. The heat generation per unit volume of the wire is

$$\dot{g} = \frac{\dot{Q}_{gen}}{V_{wire}} = \frac{\dot{Q}_{gen}}{\pi r_o^2 L} = \frac{2000 \text{ W}}{\pi (0.002 \text{ m})^2 (0.5 \text{ m})} = 0.318 \times 10^9 \text{ W/m}^3$$

Then the center temperature of the wire is determined from Equation 2-71 to be

$$T_o = T_s + \frac{\dot{g} r_o^2}{4k} = 105°C + \frac{(0.318 \times 10^9 \text{ W/m}^3)(0.002 \text{ m})^2}{4 \times (15 \text{ W/m} \cdot °C)} = \mathbf{126.2°C}$$

Discussion   Note that the temperature difference between the center and the surface of the wire is 21.2°C.

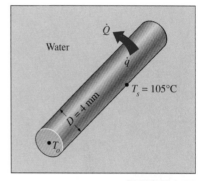

FIGURE 2-66
Schematic for Example 2-17.

We have developed the relations above using the intuitive *energy balance* approach. However, we could have obtained the same relations by setting up the appropriate *differential equations* and solving them, as illustrated in the examples below.

EXAMPLE 2-18   Variation of Temperature in a Resistance Heater

A long homogeneous resistance wire of radius $r_0 = 0.2$ in. and thermal conductivity $k = 7.8$ Btu/h·ft.°F is being used to boil water at atmospheric pressure by the passage of electric current, as shown in Figure 2-67. Heat is generated in the wire uniformly as a result of resistance heating at a rate of $\dot{g} = 2400$ Btu/h·in³. If the outer surface temperature of the wire is measured to be $T_s = 226$°F, obtain a relation for the temperature distribution, and determine the temperature at the centerline of the wire when steady operating conditions are reached.

Solution   This heat transfer problem is similar to the problem above, except that we need to obtain a relation for the variation of temperature within the wire with r. Differential equations are well suited for this purpose.

FIGURE 2-67
Schematic for Example 2-18.

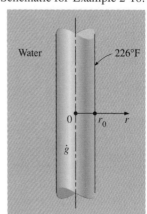

Assumptions   **1** Heat transfer is steady since there is no change with time. **2** Heat transfer is one-dimensional since there is no thermal symmetry about the centerline and no change in the axial direction. **3** Thermal conductivity is constant. **4** Heat generation in the wire is uniform.

Properties   The thermal conductivity is given to be $k = 7.8$ Btu/h · ft · °F.

Analysis   The differential equation which governs the variation of temperature in the wire is simply Equation 2-28,

$$\frac{1}{r}\frac{d}{dr}\left(r\frac{dT}{dr}\right) + \frac{\dot{g}}{k} = 0$$

This is a second order linear ordinary differential equation, and thus its general solution will contain two arbitrary constants. The determination of these constants requires the specification of two boundary conditions, which can be taken to be

$$T(r_0) = T_s = 226°F$$

and

$$\frac{dT(0)}{dr} = 0$$

The first boundary condition simply states that the temperature of the outer surface of the wire is 226°F. The second boundary condition is the symmetry condition at the centerline, and states that the maximum temperature in the wire will occur at the centerline, and thus the slope of the temperature at $r = 0$ must be zero (Fig. 2-68). This completes the mathematical formulation of the problem. Now we will try to solve it.

Although not immediately obvious, the differential equation is in a form that can be solved by direct integration. Multiplying both sides of the equation by r and rearranging, we obtain

$$\frac{d}{dr}\left(r\frac{dT}{dr}\right) = -\frac{\dot{g}}{k}r$$

Integrating with respect to r gives

$$r\frac{dT}{dr} = -\frac{\dot{g}}{k}\frac{r^2}{2} + C_1 \qquad (a)$$

since the heat generation is constant, and the integral of a derivative of a function is the function itself. That is, integration removes a derivative. It is convenient at this point to apply the second boundary condition, since it is related to the first derivative of the temperature, by replacing all occurrences of r and dT/dr in the equation above by zero. It yields

$$0 \times \frac{dT(0)}{dr} = -\frac{\dot{g}}{2k} \times 0 + C_1 \quad \rightarrow \quad C_1 = 0$$

Thus C_1 cancels from the solution. We now divide Equation (a) by r to bring it to a readily integrable form,

$$\frac{dT}{dr} = -\frac{\dot{g}}{2k}r$$

Again integrating with respect to r gives

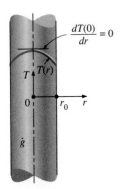

FIGURE 2-68

The thermal symmetry condition at the centerline of a wire in which heat is generated uniformly.

$$T(r) = -\frac{\dot{g}}{4k} r^2 + C_2 \qquad (b)$$

We now apply the first boundary condition by replacing all occurrences of r by r_0 and all occurrences of T by T_s. We get

$$T_s = -\frac{\dot{g}}{4k} r_0^2 + C_2 \quad \rightarrow \quad C_2 = T_s + \frac{\dot{g}}{4k} r_0^2$$

Substituting this C_2 relation into Eq. (b) and rearranging give

$$T(r) = T_s + \frac{\dot{g}}{4k} (r_0^2 - r^2) \qquad (c)$$

which is the desired solution for the temperature distribution in the wire as a function of r. The temperature at the centerline $(r = 0)$ is obtained by replacing r in Equation (c) by zero and substituting the known quantities,

$$T(0) = T_s + \frac{\dot{g}}{4k} r_0^2 = 226°F + \frac{2400 \text{ Btu/h} \cdot \text{in}^3}{4 \times (7.8 \text{ Btu/h} \cdot \text{ft} \cdot °F)} \left(\frac{12 \text{ in.}}{1 \text{ ft}} \right)(0.2 \text{ in.})^2 = \mathbf{263°F}$$

Discussion The temperature of the centerline will be 37°F above the temperature of the outer surface of the wire. Note that the expression above for the centerline temperature is identical to Equation 2-71, which was obtained using an energy balance on a control volume.

EXAMPLE 2-19 Heat Conduction in a Two-Layer Medium

Consider a long resistance wire of radius $r_1 = 0.2$ cm and thermal conductivity $k_{\text{wire}} = 15$ W/m · °C in which heat is generated uniformly as a result of resistance heating at a constant rate of $\dot{g} = 50$ W/cm³ (Fig. 2-69). The wire is embedded in a 0.5-cm-thick layer of ceramic whose thermal conductivity is $k_{\text{ceramic}} = 1.2$ W/m · °C. If the outer surface temperature of the ceramic layer is measured to be $T_s = 45°C$, determine the temperatures at the center of the resistance wire and the interface of the wire and the ceramic layer under steady conditions.

Solution The surface and interface temperatures of a resistance wire covered with a ceramic layer are to be determined.

Assumptions **1** Heat transfer is steady since there is no change with time. **2** Heat transfer is one-dimensional since this two-layer heat transfer problem possesses symmetry about the centerline and involves no change in the axial direction, and thus $T = T(r)$. **3** Thermal conductivities are constant. **4** Heat generation in the wire is uniform.

Properties It is given that $k_{\text{wire}} = 15$ W/m · °C and $k_{\text{ceramic}} = 1.2$ W/m · °C.

Analysis Letting T_I denote the unknown interface temperature, the heat transfer problem in the wire can be formulated as

$$\frac{1}{r} \frac{d}{dr} \left(r \frac{dT_{\text{wire}}}{dr} \right) + \frac{\dot{g}}{k} = 0$$

with

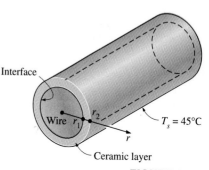

FIGURE 2-69

Schematic for Example 2-19.

$$T_{wire}(r_1) = T_I$$

$$\frac{dT_{wire}(0)}{dr} = 0$$

This problem was solved in Example 2-18, and its solution was determined to be

$$T_{wire}(r) = T_I + \frac{\dot{g}}{4k_{wire}}(r_1^2 - r^2) \tag{a}$$

Noting that the ceramic layer does not involve any heat generation and its outer surface temperature is specified, the heat conduction problem in that layer can be expressed as

$$\frac{d}{dr}\left(r\frac{dT_{ceramic}}{dr}\right) = 0$$

with

$$T_{ceramic}(r_1) = T_I$$

$$T_{ceramic}(r_2) = T_s = 45°C$$

This problem was solved in Example 2-15, and its solution was determined to be

$$T_{ceramic}(r) = \frac{\ln(r/r_1)}{\ln(r_2/r_1)}(T_s - T_I) + T_I \tag{b}$$

We have already utilized the first interface condition by setting the wire and ceramic layer temperatures equal to T_I at the interface $r = r_1$. The interface temperature T_1 is determined from the second interface condition that the heat flux in the wire and the ceramic layer at $r = r_1$ must be the same:

$$-k_{wire}\frac{dT_{wire}(r_1)}{dr} = -k_{ceramic}\frac{dT_{ceramic}(r_1)}{dr} \rightarrow \frac{\dot{g}r_1}{2} = -k_{ceramic}\frac{T_s - T_I}{\ln(r_2/r_1)}\frac{1}{r_1}$$

Solving for T_I and substituting the given values, the interface temperature is determined to be

$$T_I = \frac{\dot{g}r_1^2}{2k_{ceramic}}\ln\frac{r_2}{r_1} + T_s$$

$$= \frac{(50 \times 10^6 \text{ W/m}^3)(0.002 \text{ m})^2}{2(1.2 \text{ W/m} \cdot °C)}\ln\frac{0.007 \text{ m}}{0.002 \text{ m}} + 45° C = \textbf{149.4°C}$$

Knowing the interface temperature, the temperature at the centerline ($r = 0$) is obtained by substituting the known quantities into Equation (a),

$$T_{wire}(0) = T_I + \frac{\dot{g}r_1^2}{4k_{wire}} = 149.4°C + \frac{(50 \times 10^6 \text{ W/m}^3)(0.002 \text{ m})^2}{4 \times (15 \text{ W/m} \cdot °C)} = \textbf{152.7°C}$$

Thus the temperature of the centerline will be slightly above the interface temperature.

Discussion This example demonstrates how steady one-dimensional heat conduction problems in composite mediums can be solved. We could also solve this problem by determining the heat flux at the interface by dividing the total heat generated in the wire by the surface area of the wire, and then using this value as the specifed heat flux boundary condition for both the wire and the ceramic layer. This way the two problems are decoupled and can be solved separately.

You will recall from Chapter 1 that the thermal conductivity of a material, in general, varies with temperature (Fig. 2-70). However, this variation is mild for many materials in the range of practical interest and can be disregarded. In such cases we can use an average value for the thermal conductivity and treat it as a constant, as we have been doing so far. This is also common practice for other temperature-dependent properties such as the density and specific heat.

When the variation of thermal conductivity with temperature in a specified temperature interval is large, however, it may be necessary to account for this variation to minimize the error. Accounting for the variation of the thermal conductivity with temperature, in general, complicates the analysis. But in the case of simple one-dimensional cases, we can obtain heat transfer relations in a straightforward manner.

When the variation of thermal conductivity with temperature $k(T)$ is known, the average value of the thermal conductivity in the temperature range between T_1 and T_2 can be determined from

$$k_{ave} = \frac{\displaystyle\int_{T_1}^{T_2} k(T)\,dT}{T_2 - T_1} \tag{2-75}$$

This relation is based on the requirement that the rate of heat transfer through a medium with constant average thermal conductivity k_{ave} equals the rate of heat transfer through the same medium with variable conductivity $k(T)$. Note that in the case of constant thermal conductivity $k(T) = k$, the relation above reduces to $k_{ave} = k$, as expected.

Then the rate of steady heat transfer through a plane wall, cylindrical layer, or spherical layer for the case of variable thermal conductivity can be determined by replacing the constant thermal conductivity k in Equations 2-57, 2-59, and 2-61 by the k_{ave} expression (or value) from the above relation:

$$\dot{Q}_{plane\ wall} = k_{ave} A \frac{T_1 - T_2}{L} = \frac{A}{L}\int_{T_2}^{T_1} k(T)\,dT \tag{2-76}$$

$$\dot{Q}_{cylinder} = 2\pi k_{ave} L \frac{T_1 - T_2}{\ln(r_2/r_1)} = \frac{2\pi L}{\ln(r_2/r_1)}\int_{T_2}^{T_1} k(T)\,dT \tag{2-77}$$

$$\dot{Q}_{sphere} = 4\pi k_{ave} r_1 r_2 \frac{T_1 - T_2}{r_2 - r_1} = \frac{4\pi r_1 r_2}{r_2 - r_1}\int_{T_2}^{T_1} k(T)\,dT \tag{2-78}$$

The variation in thermal conductivity of a material with temperature in the temperature range of interest can often be approximated as a linear function and expressed as

$$k(T) = k_0(1 + \beta T) \tag{2-79}$$

where β is called the **temperature coefficient of thermal conductivity.** The *average* value of thermal conductivity in the temperature range T_1 to T_2 in this case can be determined from

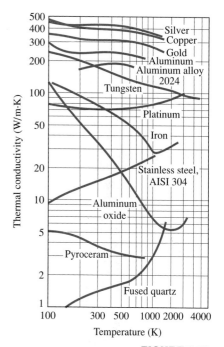

Variable Thermal
Conductivity, $k(T)$

FIGURE 2-70

Variation of the thermal conductivity of some solids with temperature.

$$k_{ave} = \frac{\int_{T_1}^{T_2} k_0(1 + \beta T)dT}{T_2 - T_1} = k_0\left(1 + \beta \frac{T_2 + T_1}{2}\right) = k(T_{ave}) \quad (2\text{-}80)$$

Note that the *average thermal conductivity* in this case is equal to the thermal conductivity value at the *average temperature*.

We have mentioned earlier that in a plane wall the temperature varies linearly during steady one-dimensional heat conduction when the thermal conductivity is constant. But this is no longer the case when the thermal conductivity changes with temperature, even linearly, as shown in Figure 2-71.

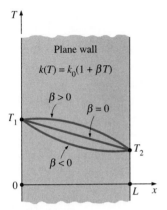

FIGURE 2-71

The variation of temperature in a plane wall during steady one-dimensional heat conduction for the cases of constant and variable thermal conductivity.

EXAMPLE 2-20   Variation of Temperature in a Wall with $k(T)$

Consider a plane wall of thickness L whose thermal conductivity varies linearly in a specified temperature range as $k(T) = k_0(1 + \beta T)$ where k_0 and β are constants. The wall surface at $x = 0$ is maintained at a constant temperature of T_1 while the surface at $x = L$ is maintained at T_2, as shown in Figure 2-72. Assuming steady one-dimensional heat transfer, obtain a relation for (a) the heat transfer rate through the wall and (b) the temperature distribution $T(x)$ in the wall.

Solution   A plate with variable conductivity is subjected to specified temperatures on both sides. The variation of temperature and the rate of heat transfer are to be determined.

Assumptions   **1** Heat transfer is given to be steady and one-dimensional. **2** Thermal conductivity varies linearly. **3** There is no heat generation.

Properties   The thermal conductivity is given to be $k(T) = k_0(1 + \beta T)$.

Analysis   (a) The rate of heat transfer through the wall can be determined from

$$\dot{Q} = k_{ave} A \frac{T_1 - T_2}{L}$$

where A is the heat conduction area of the wall and

$$k_{ave} = k(T_{ave}) = k_0\left(1 + \beta \frac{T_2 + T_1}{2}\right)$$

is the average thermal conductivity (Eq. 2-80).

(b) To determine the temperature distribution in the wall, we begin with Fourier's law of heat conduction, expressed as

$$\dot{Q} = -k(T) A \frac{dT}{dx}$$

where the rate of conduction heat transfer $\dot{Q}$ and the conduction area A are constant. Separating the variables in the above equation and integrating from $x = 0$ where $T(0) = T_1$ to any x where $T(x) = T$, we get

$$\int_0^x \dot{Q}dx = -A\int_{T_1}^T k(T)dT$$

Substituting $k(T) = k_0(1 + \beta T)$ and performing the integrations we obtain

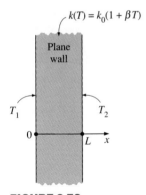

FIGURE 2-72

Schematic for Example 2-20.

$$\dot{Q}x = -Ak_0[(T - T_1) + \beta(T^2 - T_1^2)/2]$$

Substituting the $\dot{Q}$ expression from part (a) and rearranging give

$$T^2 + \frac{2}{\beta}T + \frac{2k_{ave}}{\beta k_0}\frac{x}{L}(T_1 - T_2) - T_1^2 - \frac{2}{\beta}T_1 = 0$$

which is a *quadratic* equation in the unknown temperature T. Using the quadratic formula, the temperature distribution $T(x)$ in the wall is determined to be

$$T(x) = -\frac{1}{\beta} \pm \sqrt{\frac{1}{\beta^2} - \frac{2k_{ave}}{\beta k_0}\frac{x}{L}(T_1 - T_2) + T_1^2 + \frac{2}{\beta}T_1}$$

The proper sign of the square root term (+ or −) is determined from the requirement that the temperature at any point within the medium must remain between T_1 and T_2. The result above explains why the temperature distribution in a plane wall is no longer a straight line when the thermal conductivity varies with temperature.

EXAMPLE 2-21   Heat Conduction through a Wall with $k(T)$

Consider a 2-m-high and 0.7-m-wide bronze plate whose thickness is 0.1 m. One side of the plate is maintained at a constant temperature of 600 K while the other side is maintained at 400 K, as shown in Figure 2-73. The thermal conductivity of the bronze plate can be assumed to vary linearly in that temperature range as $k(T) = k_0(1 + \beta T)$ where $k_0 = 38$ W/m · K and $\beta = 9.21 \times 10^{-4}$ K^{-1}. Disregarding the edge effects and assuming steady one-dimensional heat transfer, determine the rate of heat conduction through the plate.

Solution   A plate with variable conductivity is subjected to specified temperatures on both sides. The rate of heat transfer is to be determined.

Assumptions   **1** Heat transfer is given to be steady and one-dimensional. **2** Thermal conductivity varies linearly. **3** There is no heat generation.

Properties   The thermal conductivity is given to be $k(T) = k_0(1 + \beta T)$.

Analysis   The average thermal conductivity of the medium in this case is simply the value at the average temperature and is determined from

$$k_{ave} = k(T_{ave}) = k_0\left(1 + \beta\frac{T_2 + T_1}{2}\right)$$

$$= (38 \text{ W/m} \cdot \text{K})\left[1 + (9.21 \times 10^{-4} \text{ K}^{-1})\frac{(600 + 400)\text{ K}}{2}\right]$$

$$= 55.5 \text{ W/m} \cdot \text{K}$$

Then the rate of heat conduction through the plate can be determined from Equation 2-76 to be

$$\dot{Q} = k_{ave} A\frac{T_1 - T_2}{L}$$

$$= (55.5 \text{ W/m} \cdot \text{K})(2 \text{ m} \times 0.7 \text{ m})\frac{(600 - 400)\text{K}}{0.1 \text{ m}} = \mathbf{155{,}400 \text{ W}}$$

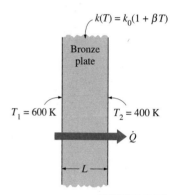

$k(T) = k_0(1 + \beta T)$

Bronze
plate

$T_1 = 600$ K $T_2 = 400$ K

$\dot{Q}$

L

FIGURE 2-73
Schematic for Example 2-21.

Discussion   We would have obtained the same result by substituting the given $k(T)$ relation into the second part of Equation 2-76 and performing the indicated integration.

2-9 ■ SUMMARY

In this chapter we have studied the heat conduction equation and its solutions. Heat conduction in a medium is said to be *steady* when the temperature does not vary with time and *unsteady* or *transient* when it does. Heat conduction in a medium is said to be *one-dimensional* when conduction is significant in one dimension only and negligible in the other two dimensions. It is said to be *two-dimensional* when conduction in the third dimension is negligible and *three-dimensional* when conduction in all dimensions is significant. In heat transfer analysis, the conversion of electrical, chemical, or nuclear energy into heat (or thermal) energy is characterized as *heat generation*.

The heat conduction equation can be derived by performing an energy balance on a differential volume element. The one-dimensional heat conduction equation in rectangular, cylindrical, and spherical coordinate systems for the case of constant thermal conductivities are expressed as

$$\frac{\partial^2 T}{\partial x^2} + \frac{\dot{g}}{k} = \frac{1}{\alpha}\frac{\partial T}{\partial t}$$

$$\frac{1}{r}\frac{\partial}{\partial r}\left(r\frac{\partial T}{\partial r}\right) + \frac{\dot{g}}{k} = \frac{1}{\alpha}\frac{\partial T}{\partial t}$$

$$\frac{1}{r^2}\frac{\partial}{\partial r}\left(r^2\frac{\partial T}{\partial r}\right) + \frac{\dot{g}}{k} = \frac{1}{\alpha}\frac{\partial T}{\partial t}$$

where the property $\alpha = k/\rho C$ is the *thermal diffusivity* of the material.

The solution of a heat conduction problem depends on the conditions at the surfaces, and the mathematical expressions for the thermal conditions at the boundaries are called the *boundary conditions*. The solution of transient heat conduction problems also depends on the condition of the medium at the beginning of the heat conduction process. Such a condition, which is usually specified at time $t = 0$, is called the *initial condition,* which is a mathematical expression for the temperature distribution of the medium initially. Complete mathematical description of a heat conduction problem requires the specification of two boundary conditions for each dimension along which heat conduction is significant, and an initial condition when the problem is transient. The most common boundary conditions are the *specified temperature, specified heat flux, convection,* and *radiation* boundary conditions. A boundary surface, in general, may involve specified heat flux, convection, and radiation at the same time.

For steady one-dimensional heat transfer through a plate of thickness L, the various types of boundary conditions at the surfaces at $x = 0$ and $x = L$ can be expressed as follows:

Specified temperature:

$$T(0) = T_1 \quad \text{and} \quad T(L) = T_2$$

where T_1 and T_2 are the specified temperatures at surfaces at $x = 0$ and $x = L$.

Specified heat flux:

$$-k\frac{dT(0)}{dx} = \dot{q}_0 \quad \text{and} \quad -k\frac{dT(L)}{dx} = \dot{q}_L$$

where $\dot{q}_0$ and $\dot{q}_L$ are the specified heat fluxes at surfaces at $x = 0$ and $x = L$.

Insulation or thermal symmetry:

$$\frac{dT(0)}{dx} = 0 \quad \text{and} \quad \frac{dT(L)}{dx} = 0$$

Convection:

$$-k\frac{dT(0)}{dx} = h_1[T_{\infty 1} - T(0)] \quad \text{and} \quad -k\frac{dT(L)}{dx} = h_2[T(L) - T_{\infty 2}]$$

where h_1 and h_2 are the convection heat transfer coefficients and $T_{\infty 1}$ and $T_{\infty 2}$ are the temperatures of the surrounding mediums on the two sides of the plate.

Radiation:

$$-k\frac{dT(0)}{dx} = \varepsilon_1\sigma[T_{\text{surr, 1}}^4 - T(0)^4] \quad \text{and} \quad -k\frac{dT(L)}{dx} = \varepsilon_2\sigma[T(L)^4 - T_{\text{surr, 2}}^4]$$

where ε_1 and ε_2 are the emissivities of the boundary surfaces, $\sigma = 5.67 \times 10^{-8}$ W/m$^2 \cdot$ K^4 is the Stefan-Boltzmann constant, and $T_{\text{surr, 1}}$ and $T_{\text{surr, 2}}$ are the average temperatures of the surfaces surrounding the two sides of the plate. In radiation calculations, the temperatures must be in K or R.

Interface of two bodies A *and* B *in perfect contact at* $x = x_0$:

$$T_A(x_0) = T_B(x_0) \quad \text{and} \quad -k_A\frac{dT_A(x_0)}{dx} = -k_B\frac{dT_B(x_0)}{dx}$$

where k_A and k_B are the thermal conductivities of the layers A and B.

Heat generation is usually expressed *per unit volume* of the medium and is denoted by $\dot{g}$, whose unit is W/m^3. Under steady conditions, the surface temperature T_s of a plane wall, a cylinder, and a sphere in which heat is generated at a constant rate of $\dot{g}$ per unit volume in a surrounding medium at T_∞ can be expressed as

$$T_{s,\text{ plane wall}} = T_\infty + \frac{\dot{g}L}{h}$$

$$T_{s,\text{ cylinder}} = T_\infty + \frac{\dot{g}r_o}{2h}$$

$$T_{s,\text{ sphere}} = T_\infty + \frac{\dot{g}r_o}{3h}$$

where h is the convection heat transfer coefficient. The maximum temperature rise between the surface and the midsection of a medium is given by

$$\Delta T_{\text{max, plane wall}} = \frac{\dot{g}L^2}{2k}$$

$$\Delta T_{\text{max, cylinder}} = \frac{\dot{g}r_o^2}{4k}$$

$$\Delta T_{\text{max, sphere}} = \frac{\dot{g}r_o^2}{6k}$$

where T_0 is the midsection temperature of the medium, which is the highest temperature.

When the variation of thermal conductivity with temperature $k(T)$ is known, the average value of the thermal conductivity in the temperature range between T_1 and T_2 can be determined from

$$k_{\text{ave}} = \frac{\displaystyle\int_{T_1}^{T_2} k(T)dT}{T_2 - T_1}$$

Then the rate of steady heat transfer through a plane wall, cylindrical layer, or spherical layer can be expressed as

$$\dot{Q}_{\text{plane wall}} = k_{\text{ave}}A\frac{T_1 - T_2}{L} = \frac{A}{L}\int_{T_2}^{T_1} k(T)dT$$

$$\dot{Q}_{\text{cylinder}} = 2\pi k_{\text{ave}}L\frac{T_1 - T_2}{\ln(r_2/r_1)} = \frac{2\pi L}{\ln(r_2/r_1)}\int_{T_2}^{T_1} k(T)dT$$

$$\dot{Q}_{\text{sphere}} = 4\pi k_{\text{ave}}r_1r_2\frac{T_1 - T_2}{r_2 - r_1} = \frac{4\pi r_1r_2}{r_2 - r_1}\int_{T_2}^{T_1} k(T)dT$$

The variation of thermal conductivity of a material with temperature can often be approximated as a linear function and expressed as

$$k(T) = k_0(1 + \beta T)$$

where β is called the *temperature coefficient of thermal conductivity*.

REFERENCES AND SUGGESTED READING

1. W. E. Boyce and R. C. Diprima. *Elementary Differential Equations and Boundary Value Problems.* 4th ed. New York: John Wiley & Sons, 1986.

2. J. P. Holman. *Heat Transfer.* 8th ed. New York: McGraw-Hill, 1997.

3. F. P. Incropera and D. P. DeWitt. *Introduction to Heat Transfer.* 2nd ed. New York: John Wiley & Sons, 1990.

4. F. Kreith and M. S. Bohn. *Principles of Heat Transfer.* 5th ed. St. Paul, MN: West Publishing, 1993.

5. S. S. Kutateladze. *Fundamentals of Heat Transfer.* New York: Academic Press, 1963.

6. A. F. Mills. *Basic Heat and Mass Transfer.* Burr Ridge, IL: Richard D. Irwin, 1995.

7. M. N. Ozisik. *Heat Transfer—A Basic Approach.* New York: McGraw-Hill, 1985.

8. L. C. Thomas. *Heat Transfer.* Englewood Cliffs, NJ: Prentice Hall, 1992.

9. F. M. White. *Heat and Mass Transfer.* Reading, MA: Addison-Wesley, 1988.

PROBLEMS

Introduction

2-1C   Is heat transfer a scalar or vector quantity? Explain. Answer the same question for temperature.

2-2C   What is the sign convention for heat transfer? Why do we use a coordinate system when solving a heat transfer problem to determine the temperature distribution in a medium?

2-3C   How does transient heat transfer differ from steady heat transfer? How does one-dimensional heat transfer differ from two-dimensional heat transfer?

2-4C   Consider a cold canned drink left on a dinner table. Would you model the heat transfer to the drink as one-, two-, or three-dimensional? Would the heat transfer be steady or transient? Also, which coordinate system would you use to analyze this heat transfer problem, and where would you place the origin? Explain.

2-5C   Consider a round potato being baked in an oven. Would you model the heat transfer to the potato as one-, two-, or three-dimensional? Would the heat transfer be steady or transient? Also, which coordinate system would you use to solve this problem, and where would you place the origin? Explain.

FIGURE P2-5

2-6C   Consider an egg being cooked in boiling water in a pan. Would you model the heat transfer to the egg as one-, two-, or three-dimensional? Would the heat transfer be steady or transient? Also, which coordinate system would you use to solve this problem, and where would you place the origin? Explain.

2-7C   Consider a hot dog being cooked in boiling water in a pan. Would you model the heat transfer to the hot dog as one-, two-, or three-dimensional? Would the heat transfer be steady or transient? Also, which coordinate system would you use to solve this problem, and where would you place the origin? Explain.

FIGURE P2-7

2-8C   What is a lumped system? How does heat transfer in a lumped system differ from steady heat transfer?

2-9C Consider the cooking process of a roast beef in an oven. Would you consider this to be a steady or transient heat transfer problem? Also, would you consider this to be one-, two-, or three-dimensional? Explain.

2-10C Consider heat loss from a 200-L cylindrical hot water tank in a house to the surrounding medium. Would you consider this to be a steady or transient heat transfer problem? Also, would you consider this heat transfer problem to be one-, two-, or three-dimensional? Explain.

2-11C Does a heat flux vector at a point P on an isothermal surface of a medium have to be perpendicular to the surface at that point? Explain.

2-12C From a heat transfer point of view, what is the difference between isotropic and unisotropic materials?

2-13C What is heat generation in a solid? Give examples.

2-14C Heat generation is also referred to as energy generation or thermal energy generation. What do you think of these phrases?

2-15 In order to determine the size of the heating element of a new oven, it is desired to determine the rate of heat transfer through the walls, door, and the top and bottom section of the oven. In your analysis, would you consider this to be a steady or transient heat transfer problem? Also, would you consider the heat transfer to be one-dimensional or multidimensional? Explain.

2-16E The resistance wire of a 1000-W iron is 20 in. long and has a diameter of $D = 0.08$ in. Determine the rate of heat generation in the wire per unit volume, in $Btu/h \cdot ft^3$, and the heat flux on the outer surface of the wire, in $Btu/h \cdot ft^2$, as a result of this heat generation.

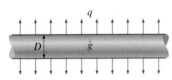

FIGURE P2-16E

2-17 In a nuclear reactor, heat is generated uniformly in the 5-cm- diameter cylindrical uranium rods at a rate of 7×10^7 W/m^3. If the length of the rods is 1 m, determine the rate of heat generation in each rod.
Answer: 137.4 kW

2-18 In a solar pond, the absorption of solar energy can be modeled as heat generation and can be approximated by $\dot{g} = \dot{g}_0 e^{-bx}$ where $\dot{g}_0$ is the rate of heat absorption at the top surface per unit volume and b is a constant. Obtain a relation for the total rate of heat generation in a water layer of surface area A and thickness L at the top of the pond.

2-19 Consider a large 3-cm-thick stainless steel plate in which heat is generated uniformly at a rate of 5×10^6 W/m^3. Assuming the plate is losing heat from both sides, determine the heat flux on the surface of the plate during steady operation. *Answer:* 75,000 W/m^2

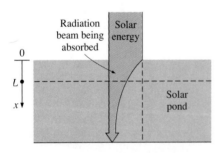

FIGURE P2-18

Review of Differential Equations

2-20C Why do we often utilize simplifying assumptions when we derive differential equations?

(a) Is heat transfer steady or transient?
(b) Is heat transfer one-, two-, or three-dimensional?
(c) Is there heat generation in the medium?
(d) Is the thermal conductivity of the medium constant or variable?

2-45 Starting with an energy balance on a volume element, derive the two-dimensional transient heat conduction equation in rectangular coordinates for $T(x, y, t)$ for the case of constant thermal conductivity and no heat generation.

2-46 Starting with an energy balance on a ring-shaped volume element, derive the two-dimensional steady heat conduction equation in cylindrical coordinates for $T(r, z)$ for the case of constant thermal conductivity and no heat generation.

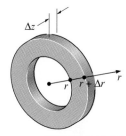

FIGURE P2-46

2-47 Starting with an energy balance on a disk volume element, derive the one-dimensional transient heat conduction equation for $T(z, t)$ in a cylinder of diameter D with an insulated side surface for the case of constant thermal conductivity with heat generation.

2-48 Consider a medium in which the heat conduction equation is given in its simplest form as

$$\frac{\partial^2 T}{\partial x^2} + \frac{\partial^2 T}{\partial y^2} = \frac{1}{\alpha} \frac{\partial T}{\partial t}$$

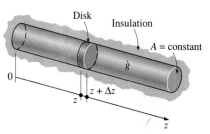

FIGURE P2-47

(a) Is heat transfer steady or transient?
(b) Is heat transfer one-, two-, or three-dimensional?
(c) Is there heat generation in the medium?
(d) Is the thermal conductivity of the medium constant or variable?

2-49 Consider a medium in which the heat conduction equation is given in its simplest form as

$$\frac{1}{r} \frac{\partial}{\partial r}\left(kr \frac{\partial T}{\partial r}\right) + \frac{\partial}{\partial z}\left(k \frac{\partial T}{\partial z}\right) + \dot{g} = 0$$

(a) Is heat transfer steady or transient?
(b) Is heat transfer one-, two-, or three-dimensional?
(c) Is there heat generation in the medium?
(d) Is the thermal conductivity of the medium constant or variable?

2-50 Consider a medium in which the heat conduction equation is given in its simplest form as

$$\frac{1}{r^2} \frac{\partial}{\partial r}\left(r^2 \frac{\partial T}{\partial t}\right) + \frac{1}{r^2 \sin^2 \theta} \frac{\partial^2 T}{\partial \phi^2} = \frac{1}{\alpha} \frac{\partial T}{\partial t}$$

(a) Is heat transfer steady or transient?
(b) Is heat transfer one-, two-, or three-dimensional?
(c) Is there heat generation in the medium?
(d) Is the thermal conductivity of the medium constant or variable?

**Boundary and Initial Conditions; Formulation of
Heat Conduction Problems**

2-51C What is a boundary condition? How many boundary conditions do
we need to specify for a two-dimensional heat transfer problem?

2-52C What is an initial condition? How many initial conditions do we need
to specify for a two-dimensional heat transfer problem?

2-53C What is a thermal symmetry boundary condition? How is it expressed mathematically?

2-54C How is the boundary condition on an insulated surface expressed
mathematically?

2-55C It is claimed that the temperature profile in a medium must be perpendicular to an insulated surface. Is this a valid claim? Explain.

2-56C Why do we try to avoid the radiation boundary conditions in heat
transfer analysis?

2-57 Consider a spherical container of inner radius r_1, outer radius r_2, and
thermal conductivity k. Express the boundary condition on the inner surface
of the container for steady one-dimensional conduction for the following
cases: (a) specified temperature of 50°C, (b) specified heat flux of 30 W/m^2
toward the center, (c) convection to a medium at T_∞ with a heat transfer
coefficient of h.

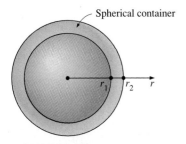

Spherical container

FIGURE P2-57

2-58 Heat is generated in a long wire of radius r_0 at a constant rate of $\dot{g}_0$ per
unit volume. The wire is covered with a plastic insulation layer. Express the
heat flux boundary condition at the interface in terms of the heat generated.

2-59 Consider a long pipe of inner radius r_1, outer radius r_2, and thermal
conductivity k. The outer surface of the pipe is subjected to convection to a
medium at T_∞ with a heat transfer coefficient of h, but the direction of heat
transfer is not known. Express the convection boundary condition on the
outer surface of the pipe.

2-60 Consider a spherical shell of inner radius r_1, outer radius r_2, thermal
conductivity k, and emissivity ε. The outer surface of the shell is subjected
to radiation to surrounding surfaces at T_{surr}, but the direction of heat transfer
is not known. Express the radiation boundary condition on the outer surface
of the shell.

FIGURE P2-62

Steel pan

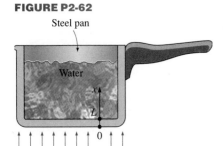

Water

2-61 A container consists of two spherical layers, A and B, that are in
perfect contact. If the radius of the interface is r_0, express the boundary
conditions at the interface.

2-62 Consider a steel pan used to boil water on top of an electric range.
The bottom section of the pan is $L = 0.3$ cm thick and has a diameter of
$D = 20$ cm. The electric heating unit on the range top consumes 1000 W of
power during cooking, and 85 percent of the heat generated in the heating

element is transferred uniformly to the pan. Heat transfer from the top surface of the bottom section to the water is by convection with a heat transfer coefficient of h. Assuming constant thermal conductivity and one-dimensional heat transfer, express the mathematical formulation (the differential equation and the boundary conditions) of this heat conduction problem during steady operation. Do not solve.

2-63E A 1.5-kW resistance heater wire whose thermal conductivity is $k = 10.4$ Btu/h · ft · °F has a radius of $r_0 = 0.06$ in. and a length of $L = 15$ in., and is used for space heating. Assuming constant thermal conductivity and one-dimensional heat transfer, express the mathematical formulation (the differential equation and the boundary conditions) of this heat conduction problem during steady operation. Do not solve.

2-64 Consider an aluminum pan used to cook stew on top of an electric range. The bottom section of the pan is $L = 0.25$ cm thick and has a diameter of $D = 18$ cm. The electric heating unit on the range top consumes 900 W of power during cooking, and 90 percent of the heat generated in the heating element is transferred to the pan. During steady operation, the temperature of the inner surface of the pan is measured to be 108°C. Assuming temperature-dependent thermal conductivity and one-dimensional heat transfer, express the mathematical formulation (the differential equation and the boundary conditions) of this heat conduction problem during steady operation. Do not solve.

2-65 Water flows through a pipe at an average temperature of $T_\infty = 50$°C. The inner and outer radii of the pipe are $r_1 = 6$ cm and $r_2 = 6.5$ cm, respectively. The outer surface of the pipe is wrapped with a thin electric heater that consumes 300 W per m length of the pipe. The exposed surface of the heater is heavily insulated so that the entire heat generated in the heater is transferred to the pipe. Heat is transferred from the inner surface of the pipe to the water by convection with a heat transfer coefficient of $h = 40$ W/m² · °C. Assuming constant thermal conductivity and one-dimensional heat transfer, express the mathematical formulation (the differential equation and the boundary conditions) of the heat conduction in the pipe during steady operation. Do not solve.

2-66 A spherical metal ball of radius $r_0 = 7$ cm is heated in an oven to a temperature of T_i throughout and is then taken out of the oven and dropped into a large body of water at T_∞ where it is cooled by convection with an average convection heat transfer coefficient of h. Assuming constant thermal conductivity and transient one-dimensional heat transfer, express the mathematical formulation (the differential equation and the boundary and initial conditions) of this heat conduction problem. Do not solve.

2-67 A spherical metal ball of radius r_0 is heated in an oven to a temperature of T_i throughout and is then taken out of the oven and allowed to cool in ambient air at T_∞ by convection and radiation. The emissivity of the outer surface of the cylinder is ε, and the temperature of the surrounding surfaces

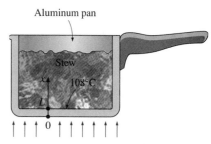

Aluminum pan

Stew

108°C

FIGURE P2-64

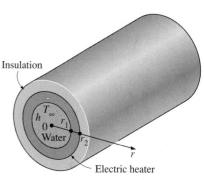

Insulation

Water

Electric heater

FIGURE P2-65

FIGURE P2-67

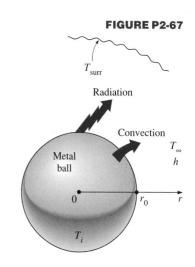

T_{surr}

Radiation

Convection

T_∞

h

Metal ball

r_0

T_i

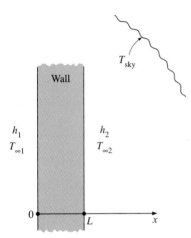

FIGURE P2-68

is T_{surr}. The average convection heat transfer coefficient is estimated to be h. Assuming variable thermal conductivity and transient one-dimensional heat transfer, express the mathematical formulation (the differential equation and the boundary and initial conditions) of this heat conduction problem. Do not solve.

2-68 Consider the north wall of a house that is $L = 0.25$ m thick. The outer surface of the wall exchanges heat by both convection and radiation. The interior of the house is maintained at $T_{\infty 1}$, while the ambient air temperature outside remains at $T_{\infty 2}$. The sky, the ground, and the surfaces of the surrounding structures at this location can be modeled as a surface at an effective temperature of T_{sky} for radiation exchange on the outer surface. The radiation exchange between the inner surface of the wall and the surfaces of the walls, floor, and ceiling it faces is negligible. The convection heat transfer coefficients on the inner and outer surfaces of the wall are h_1 and h_2, respectively. The thermal conductivity of the wall material is $k = 0.7$ W/m $\cdot$ °C and the emissivity of the outer surface is ε_2. Assuming the heat transfer through the wall to be steady and one-dimensional, express the mathematical formulation (the differential equation and the boundary and initial conditions) of this heat conduction problem. Do not solve.

Solution of Steady One-Dimensional Heat Conduction Problems

2-69C Consider one-dimensional heat conduction through a large plane wall with no heat generation that is perfectly insulated on one side and is subjected to convection and radiation on the other side. It is claimed that under steady conditions, the temperature in a plane wall must be uniform (the same everywhere). Do you agree with this claim? Why?

2-70C It is stated that the temperature in a plane wall with constant thermal conductivity and no heat generation varies linearly during steady one-dimensional heat conduction. Will this still be the case when the wall loses heat by radiation from its surfaces?

2-71C Consider a solid cylindrical rod whose ends are maintained at constant but different temperatures while the side surface is perfectly insulated. There is no heat generation. It is claimed that the temperature along the axis of the rod varies linearly during steady heat conduction. Do you agree with this claim? Why?

2-72C Consider a solid cylindrical rod whose side surface is maintained at a constant temperature while the end surfaces are perfectly insulated. The thermal conductivity of the rod material is constant and there is no heat generation. It is claimed that the temperature in the radial direction within the rod will not vary during steady heat conduction. Do you agree with this claim? Why?

2-73 Consider a large plane wall of thickness $L = 0.4$ m, thermal conductivity $k = 2.3$ W/m $\cdot$ °C, and surface area $A = 20$ m^2. The left side of the wall is maintained at a constant temperature of $T_1 = 80$°C while the right side loses heat by convection to the surrounding air at $T_\infty = 15°C$ with a heat transfer coefficient of $h = 24$ W/m$^2 \cdot$ °C. Assuming constant thermal conductivity and no heat generation in the wall, (*a*) express the differential equation and the boundary conditions for steady one-dimensional heat conduction through the wall, (*b*) obtain a relation for the variation of temperature in the wall by solving the differential equation, and (*c*) evaluate the rate of heat transfer through the wall. *Answer:* (*c*) 6031 W

2-74 Consider a solid cylindrical rod of length 0.15 m and diameter 0.05 m. The top and bottom surfaces of the rod are maintained at constant temperatures of 20°C and 95°C, respectively, while the side surface is perfectly insulated. Determine the rate of heat transfer through the rod if it is made of (*a*) copper, $k = 380$ W/m $\cdot$ °C, (*b*) steel, $k = 18$ W/m $\cdot$ °C, and (*c*) granite, $k = 1.2$ W/m $\cdot$ °C.

2-75 Consider the base plate of a 800-W household iron with a thickness of $L = 0.6$ cm, base area of $A = 160$ cm^2, and thermal conductivity of $k = 20$ W/m $\cdot$ °C. The inner surface of the base plate is subjected to uniform heat flux generated by the resistance heaters inside. When steady operating conditions are reached, the outer surface temperature of the plate is measured to be 85°C. Disregarding any heat loss through the upper part of the iron, (*a*) express the differential equation and the boundary conditions for steady one-dimensional heat conduction through the plate, (*b*) obtain a relation for the variation of temperature in the base plate by solving the differential equation, and (*c*) evaluate the inner surface temperature. *Answer:* (*c*) 100°C

2-76 Repeat Problem 2-75 for a 1000-W iron.

2-77E Consider a steam pipe of length $L = 15$ ft, inner radius $r_1 = 2$ in., outer radius $r_2 = 2.4$ in., and thermal conductivity $k = 7.2$ Btu/h $\cdot$ ft $\cdot$ °F. Steam is flowing through the pipe at an average temperature of 250°F, and the average convection heat transfer coefficient on the inner surface is given to be $h = 1.25$ Btu/h $\cdot$ ft $\cdot$ °F . If the average temperature on the outer surfaces of the pipe is $T_2 = 160$°F, (*a*) express the differential equation and the boundary conditions for steady one-dimensional heat conduction through the pipe, (*b*) obtain a relation for the variation of temperature in the pipe by solving the differential equation, and (*c*) evaluate the rate of heat loss from the steam through the pipe. *Answer:* (*c*) 16,800 Btu/h

2-78 A spherical container of inner radius $r_1 = 2$ m, outer radius $r_2 = 2.1$ m, and thermal conductivity $k = 30$ W/m $\cdot$ °C is filled with iced water at 0°C. The container is gaining heat by convection from the surrounding air at $T_\infty = 25$°C with a heat transfer coefficient of $h = 18$ W/m$^2 \cdot$ °C. Assuming the inner surface temperature of the container to be 0°C, (*a*) express the differential equation and the boundary conditions for steady one-dimensional heat conduction through the container, (*b*) obtain a relation for the variation of temperature in the container by solving the differential equation, and (*c*) evaluate the rate of heat gain to the iced water.

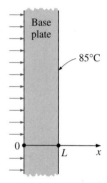

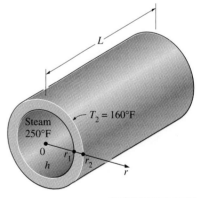

FIGURE P2-75

FIGURE P2-77E

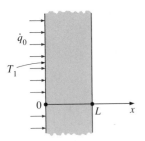

$\dot{q}_0$

T_1

0   L   x

FIGURE P2-79

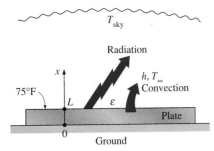

T_{sky}

Radiation

h, T_∞
Convection

75°F

L ε

Plate

0

Ground

FIGURE P2-81E

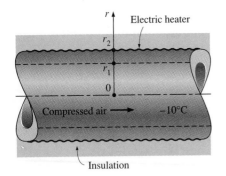

r   Electric heater

r_2

r_1

0

Compressed air ⟶ −10°C

Insulation

FIGURE P2-83

2-79 Consider a large plane wall of thickness $L = 0.3$ m, thermal conductivity $k = 2.5$ W/m · °C, and surface area $A = 12$ m². The left side of the wall at $x = 0$ is subjected to a net heat flux of $\dot{q}_0 = 700$ W/m² while the temperature at that surface is measured to be $T_1 = 80°C$. Assuming constant thermal conductivity and no heat generation in the wall, (a) express the differential equation and the boundary conditions for steady one-dimensional heat conduction through the wall, (b) obtain a relation for the variation of temperature in the wall by solving the differential equation, and (c) evaluate the temperature of the right surface of the wall at $x = L$.   Answer: (c) −4°C

2-80 Repeat Prob. 2-79 for a heat flux of 800 W/m² and a surface temperature of 85°C at the left surface at $x = 0$.

2-81E A large steel plate having a thickness of $L = 4$ in., thermal conductivity of $k = 7.2$ Btu/h · ft · °F, and an emissivity of $\varepsilon = 0.6$ is lying on the ground. The exposed surface of the plate at $x = L$ is known to exchange heat by convection with the ambient air at $T_\infty = 90°F$ with an average heat transfer coefficient of $h = 12$ Btu/h · ft² · °F as well as by radiation with the open sky with an equivalent sky temperature of $T_{sky} = 510$ R. Also, the temperature of the upper surface of the plate is measured to be 75°F. Assuming steady one-dimensional heat transfer, (a) express the differential equation and the boundary conditions for heat conduction through the plate, (b) obtain a relation for the variation of temperature in the plate by solving the differential equation, and (c) determine the value of the lower surface temperature of the plate at $x = 0$.

2-82E Repeat Problem 2-81E by disregarding radiation heat transfer.

2-83 When a long section of a compressed air line passes through the outdoors, it is observed that the moisture in the compressed air freezes in cold weather, disrupting and even completely blocking the air flow in the pipe. To avoid this problem, the outer surface of the pipe is wrapped with electric strip heaters and then insulated.

Consider a compressed air pipe of length $L = 6$ m, inner radius $r_1 = 3.7$ cm, outer radius $r_2 = 4.0$ cm, and thermal conductivity $k = 14$ W/m · °C equipped with a 300-W strip heater. Air is flowing through the pipe at an average temperature of −10°C, and the average convection heat transfer coefficient on the inner surface is $h = 30$ W/m² · °C. Assuming 15 percent of the heat generated in the strip heater is lost through the insulation, (a) express the differential equation and the boundary conditions for steady one-dimensional heat conduction through the pipe, (b) obtain a relation for the variation of temperature in the pipe material by solving the differential equation, and (c) evaluate the inner and outer surface temperatures of the pipe.   Answers: (c) −3.91°C, −3.87°C

2-84 In a food processing facility, a spherical container of inner radius $r_1 = 40$ cm, outer radius $r_2 = 41$ cm, and thermal conductivity $k = 1.5$ W/m · °C is used to store hot water and to keep it at 100°C at all times. To accomplish this, the outer surface of the container is wrapped with a 500-W electric strip heater and then insulated. The temperature of the inner surface of the container is observed to be nearly 100°C at all times. Assuming 10 percent of the heat generated in the heater is lost through the insulation, (a) express the differential equation and the boundary conditions for steady one-dimensional heat conduction through the container, (b) obtain a relation for the variation of temperature in the container material by solving the differential equation, and (c) evaluate the outer surface temperature of the container. Also determine how much water at 100°C this tank can supply steadily if the cold water enters at 20°C.

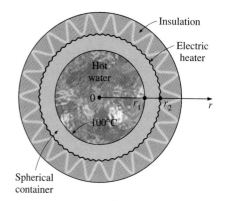

FIGURE P2-84

Heat Generation in a Solid

2-85C Does heat generation in a solid violate the first law of thermodynamics, which states that energy cannot be created or destroyed? Explain.

2-86C What is heat generation? Give some examples.

2-87C An iron is left unattended and its base temperature rises as a result of resistance heating inside. When will the rate of heat generation inside the iron be equal to the rate of heat loss from the iron?

2-88C Consider the uniform heating of a plate in an environment at a constant temperature. Is it possible for part of the heat generated in the left half of the plate to leave the plate through the right surface? Explain.

2-89C Consider uniform heat generation in a cylinder and a sphere of equal radius made of the same material in the same environment. Which geometry will have a higher temperature at its center? Why?

2-90 A 1.5-kW resistance heater wire with thermal conductivity of $k = 20$ W/m · °C, a diameter of $D = 5$ mm, and a length of $L = 0.7$ m is used to boil water. If the outer surface temperature of the resistance wire is $T_s = 110$°C, determine the temperature at the center of the wire.

2-91 Consider a long solid cylinder of radius $r_0 = 4$ cm and thermal conductivity $k = 25$ W/m · °C. Heat is generated in the cylinder uniformly at a rate of $\dot{g}_0 = 35$ W/cm³. The side surface of the cylinder is maintained at a constant temperature of $T_s = 80$°C. The variation of temperature in the cylinder is given by

$$T(r) = \frac{\dot{g} r_0^2}{k}\left[1 - \left(\frac{r}{r_0}\right)^2\right] + T_s$$

Based on this relation, determine (a) if the heat conduction is steady or transient, (b) if it is one-, two-, or three-dimensional, and (c) the value of heat flux on the side surface of the cylinder at $r = r_0$.

2-92E A long homogeneous resistance wire of radius $r_0 = 0.25$ in. and thermal conductivity $k = 8.6$ Btu/h · ft · °F is being used to boil water at atmospheric pressure by the passage of electric current. Heat is generated in the wire uniformly as a result of resistance heating at a rate of $\dot{g} = 1800$ Btu/h · in³. The heat generated is transferred to water at 212°F by convection with an average heat transfer coefficient of $h = 820$ Btu/h · ft² · °F.

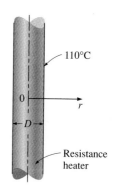

FIGURE P-90

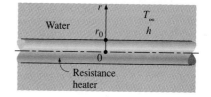

FIGURE P2-92E

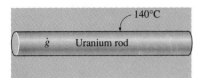

FIGURE P2-93

FIGURE P2-95

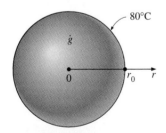

FIGURE P2-99

FIGURE P2-100

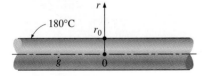

Assuming steady one-dimensional heat transfer, (a) express the differential equation and the boundary conditions for heat conduction through the wire, (b) obtain a relation for the variation of temperature in the wire by solving the differential equation, and (c) determine the temperature at the centerline of the wire.   Answer: (c) 290.8°F

2-93   In a nuclear reactor, 5-cm-diameter cylindrical uranium rods cooled by water from outside serve as the fuel. Heat is generated uniformly in the rods (k = 29.5 W/m · °C) at a rate of 7×10^7 W/m^3. If the outer surface temperature of rods is 140°C, determine the temperature at their center.

2-94   Consider a large 3-cm-thick stainless steel plate (k = 15.1 W/m · °C) in which heat is generated uniformly at a rate of 5×10^6 W/m^3. Both sides of the plate are exposed to an environment at 30°C with a heat transfer coefficient of 60 W/m^2 · °C. Explain where in the plate the highest and the lowest temperatures will occur, and determine their values.

2-95   Consider a large 5-cm-thick brass plate (k = 111 W/m · °C) in which heat is generated uniformly at a rate of 2×10^5 W/m^3. One side of the plate is insulated while the other side is exposed to an environment at 25°C with a heat transfer coefficient of 44 W/m^2 · °C. Explain where in the plate the highest and the lowest temperatures will occur, and determine their values.

2-96   A 6-m-long 2-kW electrical resistance wire is made of 0.2-mm-diameter stainless steel (k = 15.1 W/m · °C). The resistance wire operates in an environment at 30°C with a heat transfer coefficient of 140 W/m^2 · °C at the outer surface. Determine the surface temperature of the wire (a) by using the applicable relation and (b) by setting up the proper differential equation and solving it.   Answers: (a) 409°C, (b) 409°C

2-97E   Heat is generated uniformly at a rate of 5 kW per m length in a 0.08-in-diameter electric resistance wire made of nickel steel (k = 5.8 Btu/h · ft · °F). Determine the temperature difference between the centerline and the surface of the wire.

2-98E   Repeat Problem 2-97E for a manganese wire (k = 4.5 Btu/h · ft · °F).

2-99   Consider a homogeneous spherical radioactive material of radius r_0 = 0.04 m that is generating heat at a constant rate of $\dot{g}$ = 4×10^7 W/m^3. The heat generated is dissipated to the environment steadily. The outer surface of the sphere is maintained at a uniform temperature of 80°C and the thermal conductivity of the sphere is k = 15 W/m · °C. Assuming steady one-dimensional heat transfer, (a) express the differential equation and the boundary conditions for heat conduction through the sphere, (b) obtain a relation for the variation of temperature in the sphere by solving the differential equation, and (c) determine the temperature at the center of the sphere.

2-100   A long homogeneous resistance wire of radius r_0 = 5 mm is being used to heat the air in a room by the passage of electric current. Heat is generated in the wire uniformly at a rate of $\dot{g}$ = 5×10^7 W/m^3 as a result of resistance heating. If the temperature of the outer surface of the wire remains at 180°C, determine the temperature at r = 2 mm after steady operation conditions are reached. Take the thermal conductivity of the wire to be $k = 8$ W/m · °C.   *Answer:* 212.8°C

2-101   Consider a large plane wall of thickness $L = 0.05$ m. The wall surface at $x = 0$ is insulated while the surface at $x = L$ is maintained at a temperature of 30°C. The thermal conductivity of the wall is $k = 30$ W/m · °C, and heat is generated in the wall at a rate of $\dot{g} = \dot{g}_0 e^{-0.5x/L}$ W/m³ where $\dot{g}_0 = 8 \times 10^6$ W/m³. Assuming steady one-dimensional heat transfer, (*a*) express the differential equation and the boundary conditions for heat conduction through the wall, (*b*) obtain a relation for the variation of temperature in the wall by solving the differential equation, and (*c*) determine the temperature of the insulated surface of the wall.   *Answer:* (*c*) 314.1°C

Variable Thermal Conductivity, $k(T)$

2-102C   Consider steady one-dimensional heat conduction in a plane wall, long cylinder, and sphere with constant thermal conductivity and no heat generation. Will the temperature in any of these mediums vary linearly? Explain.

2-103C   Is the thermal conductivity of a medium, in general, constant or does it vary with temperature?

2-104C   Consider steady one-dimensional heat conduction in a plane wall in which the thermal conductivity varies linearly. The error involved in heat transfer calculations by assuming constant thermal conductivity at the average temperature is (*a*) none, (*b*) small, or (*c*) significant.

2-105C   The temperature of a plane wall during steady one-dimensional heat conduction varies linearly when the thermal conductivity is constant. Is this still the case when the thermal conductivity varies linearly with temperature?

2-106C   When the thermal conductivity of a medium varies linearly with temperature, is the average thermal conductivity always equivalent to the conductivity value at the average temperature?

2-107   Consider a plane wall of thickness L whose thermal conductivity varies in a specified temperature range as $k(T) = k_0(1 + \beta T^2)$ where k_0 and β are two specified constants. The wall surface at $x = 0$ is maintained at a constant temperature of T_1, while the surface at $x = L$ is maintained at T_2. Assuming steady one-dimensional heat transfer, obtain a relation for the heat transfer rate through the wall.

2-108   Consider a cylindrical shell of length L, inner radius r_1, and outer radius r_2 whose thermal conductivity varies linearly in a specified temperature range as $k(T) = k_0(1 + \beta T)$ where k_0 and β are two specified constants. The inner surface of the shell is maintained at a constant temperature of T_1, while the outer surface is maintained at T_2. Assuming steady one-dimensional heat transfer, obtain a relation for (*a*) the heat transfer rate through the wall and (*b*) the temperature distribution $T(r)$ in the shell.

FIGURE P2-108

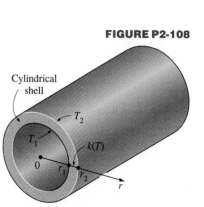

2-109 Consider a spherical shell of inner radius r_1 and outer radius r_2 whose thermal conductivity varies linearly in a specified temperature range as $k(T) = k_0(1 + \beta T)$ where k_0 and β are two specified constants. The inner surface of the shell is maintained at a constant temperature of T_1 while the outer surface is maintained at T_2. Assuming steady one-dimensional heat transfer, obtain a relation for (a) the heat transfer rate through the shell and (b) the temperature distribution $T(r)$ in the shell.

2-110 Consider a 1.5-m-high and 0.6-m-wide plate whose thickness is 0.15 m. One side of the plate is maintained at a constant temperature of 500 K while the other side is maintained at 350 K. The thermal conductivity of the plate can be assumed to vary linearly in that temperature range as $k(T) = k_0(1 + \beta T)$ where $k_0 = 25$ W/m $\cdot$ K and $\beta = 8.7 \times 10^{-4}$ K^{-1}. Disregarding the edge effects and assuming steady one-dimensional heat transfer, determine the rate of heat conduction through the plate.
Answer: 30,780 W

Review Problems

2-111 Consider a small hot metal object of mass m and specific heat C that is initially at a temperature of T_i. Now the object is allowed to cool in an environment at T_∞ by convection with a heat transfer coefficient of h. The temperature of the metal object is observed to vary uniformly with time during cooling. Writing an energy balance on the entire metal object, derive the differential equation that describes the variation of temperature of the ball with time, $T(t)$. Assume constant thermal conductivity and no heat generation in the object. Do not solve.

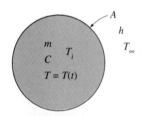

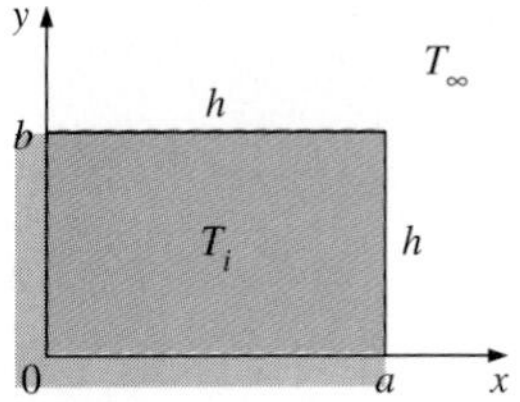

FIGURE P2-111

2-112 Consider a long rectangular bar of length a in the x direction and width b in the y direction that is initially at a uniform temperature of T_i. The surfaces of the bar at $x = 0$ and $y = 0$ are insulated, while heat is lost from the other two surfaces by convection to the surrounding medium at temperature T_∞ with a heat transfer coefficient of h. Assuming constant thermal conductivity and transient two-dimensional heat transfer with no heat generation, express the mathematical formulation (the differential equation and the boundary and initial conditions) of this heat conduction problem. Do not solve.

FIGURE P2-112

2-113 Consider a short cylinder of radius r_0 and height h in which heat is generated at a constant rate of $\dot{g}_0$. Heat is lost from the cylindrical surface at $r = r_0$ by convection to the surrounding medium at temperature T_∞ with a heat transfer coefficient of h. The bottom surface of the cylinder at $z = 0$ is insulated, while the top surface at $z = H$ is subjected to uniform heat flux $\dot{q}_h$. Assuming constant thermal conductivity and steady two-dimensional heat transfer, express the mathematical formulation (the differential equation and the boundary conditions) of this heat conduction problem. Do not solve.

2-114E Consider a large plane wall of thickness $L = 0.5$ ft and thermal conductivity $k = 1.2$ Btu/h $\cdot$ ft $\cdot$ °F in space. The wall is covered with a material which has an emissivity of $\varepsilon = 0.80$ and a solar absorptivity of $\alpha = 0.45$. The inner surface of the wall is maintained at $T_1 = 520$ R at all times, while the outer surface is exposed to solar radiation that is incident at a rate of $\dot{q}_{solar} = 300$ Btu/h $\cdot$ ft^2. The outer surface is also losing heat by radiation to deep space at 0 K. Determine the temperature of the outer surface of the wall and the rate of heat transfer through the wall when steady operating conditions are reached.   *Answers:* 530.9 R, 26.2 Btu/h $\cdot$ ft^2

2-115E Repeat Problem 2-114E for the case of no solar radiation incident on the surface.

2-116 Consider a steam pipe of length L, inner radius r_1, outer radius r_2, and constant thermal conductivity k. Steam flows inside the pipe at an average temperature of T_i with a convection heat transfer coefficient of h_i. The outer surface of the pipe is exposed to convection to the surrounding air at a temperature of T_0 with a heat transfer coefficient of h_o. Assuming steady one-dimensional heat conduction through the pipe, (*a*) express the differential equation and the boundary conditions for heat conduction through the pipe material, (*b*) obtain a relation for the variation of temperature in the pipe material by solving the differential equation, and (*c*) obtain a relation for the temperature of the outer surface of the pipe.

2-117 The boiling temperature of nitrogen at atmospheric pressure at sea level (1 atm pressure) is $-196°C$. Therefore, nitrogen is commonly used in low temperature scientific studies since the temperature of liquid nitrogen in a tank open to the atmosphere will remain constant at $-196°C$ until the liquid nitrogen in the tank is depleted. Any heat transfer to the tank will result in the evaporation of some liquid nitrogen, which has a heat of vaporization of 198 kJ/kg and a density of 810 kg/m^3 at 1 atm.

Consider a thick-walled spherical tank of inner radius $r_1 = 2$ m, outer radius $r_2 = 2.1$ m , and constant thermal conductivity $k = 18$ W/m $\cdot$ °C. The tank is initially filled with liquid nitrogen at 1 atm and $-196°C$, and is exposed to ambient air at $T_\infty = 20°C$ with a heat transfer coefficient of $h = 25$ W/m^2 $\cdot$ °C. The inner surface temperature of the spherical tank is observed to be almost the same as the temperature of the nitrogen inside. Assuming steady one-dimensional heat transfer, (*a*) express the differential equation and the boundary conditions for heat conduction through the tank, (*b*) obtain a relation for the variation of temperature in the tank material by solving the differential equation, and (*c*) determine the rate of evaporation of the liquid nitrogen in the tank as a result of the heat transfer from the ambient air.   *Answer:* (*c*) 1.32 kg/s

2-118 Repeat Problem 2-117 for liquid oxygen, which has a boiling temperature of $-183°C$, a heat of vaporization of 213 kJ/kg, and a density of 1140 kg/m^3 at 1 atm.

2-119 Consider a large plane wall of thickness $L = 0.4$ m and thermal conductivity $k = 8.4$ W/m $\cdot$ °C. There is no access to the inner side of the wall at $x = 0$ and thus the thermal conditions on that surface are not known. However, the outer surface of the wall at $x = L$, whose emissivity is $\varepsilon = 0.7$,

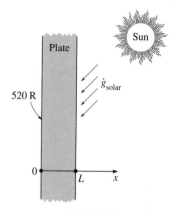

FIGURE P2-114E

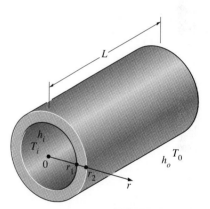

FIGURE P2-116

FIGURE P2-119

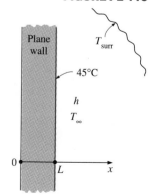

is known to exchange heat by convection with ambient air at $T_\infty = 25°C$ with an average heat transfer coefficient of $h = 14$ W/m² · °C as well as by radiation with the surrounding surfaces at an average temperature of $T_{surr} = 290$ K. Further, the temperature of the outer surface is measured to be $T_2 = 45°C$. Assuming steady one-dimensional heat transfer, (a) express the differential equation and the boundary conditions for heat conduction through the plate, (b) obtain a relation for the temperature of the outer surface of the plate by solving the differential equation, and (c) evaluate the inner surface temperature of the wall at $x = 0$.   Answer: (c) 64.3°C

2-120   A 1000-W iron is left on the iron board with its base exposed to ambient air. The base plate of the iron has a thickness of $L = 0.5$ cm, base area of $A = 150$ cm², and thermal conductivity of $k = 18$ W/m · °C. The inner surface of the base plate is subjected to uniform heat flux generated by the resistance heaters inside. The outer surface of the base plate whose emissivity is $\varepsilon = 0.7$, loses heat by convection to ambient air at $T_\infty = 22°$ C with an average heat transfer coefficient of $h = 30$ W/m² · °C as well as by radiation to the surrounding surfaces at an average temperature of $T_{surr} = 290$ K. Disregarding any heat loss through the upper part of the iron, (a) express the differential equation and the boundary conditions for steady one-dimensional heat conduction through the plate, (b) obtain a relation for the temperature of the outer surface of the plate by solving the differential equation, and (c) evaluate the outer surface temperature.

2-121   Repeat Problem 2-120 for a 1200-W iron.

2-122E   The roof of a house consists of a 0.8-ft-thick concrete slab ($k = 1.1$ Btu/h · ft · °F) that is 25 ft wide and 35 ft long. The emissivity of the outer surface of the roof is 0.9, and the convection heat transfer coefficient on that surface is estimated to be 3.2 Btu/h · ft² · °F. On a clear winter night, the ambient air is reported to be at 50°F, while the night sky temperature for radiation heat transfer is 310 R. If the inner surface temperature of the roof is $T_1 = 62°F$, determine the outer surface temperature of the roof and the rate of heat loss through the roof when steady operating conditions are reached.

2-123   Consider a long resistance wire of radius $r_1 = 0.3$ cm and thermal conductivity $k_{wire} = 18$ W/m · °C in which heat is generated uniformly at a constant rate of $\dot{g} = 1.5$ W/cm³ as a result of resistance heating. The wire is embedded in a 0.4-cm-thick layer of plastic whose thermal conductivity is $k_{plastic} = 1.8$ W/m · °C. The outer surface of the plastic cover loses heat by convection to the ambient air at $T_\infty = 25°C$ with an average combined heat transfer coefficient of $h = 14$ W/m² · °C. Assuming one-dimensional heat transfer, determine the temperatures at the center of the resistance wire and the wire-plastic layer interface under steady conditions.
     Answers: 97.1°C, 97.3°C

2-124   Consider a cylindrical shell of length L, inner radius r_1, and outer radius r_2 whose thermal conductivity varies in a specified temperature range as $k(T) = k_0(1 + \beta T^2)$ where k_0 and β are two specified constants. The inner surface of the shell is maintained at a constant temperature of T_1 while the

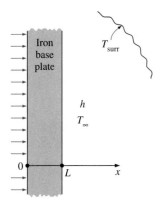

FIGURE P2-120

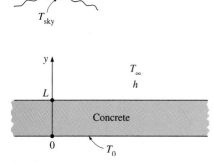

FIGURE P2-122E

FIGURE P2-123

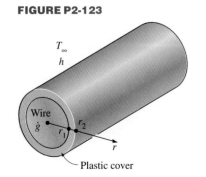

outer surface is maintained at T_2. Assuming steady one-dimensional heat transfer, obtain a relation for the heat transfer rate through the shell.

2-125 In a nuclear reactor, heat is generated in 2.5-cm-diameter cylindrical uranium fuel rods at a rate of 4×10^7 W/m^3. Determine the temperature difference between the center and the surface of the fuel rod.

  Answer: 56.6°C

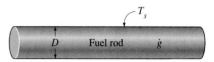

FIGURE P2-125

Computer, Design, and Essay Problems

2-126 Write an essay on heat generation in nuclear fuel rods. Obtain information on the ranges of heat generation, the variation of heat generation with position in the rods, and the absorption of emitted radiation by the cooling medium.

2-127 Write an interactive computer program to calculate the heat transfer rate and the value of temperature anywhere in the medium for steady one-dimensional heat conduction in a plane wall for any combination of specified temperature, specified heat flux, and convection boundary condition. Run the program for five different sets of specified boundary conditions.

2-128 Write an interactive computer program to calculate the heat transfer rate and the value of temperature anywhere in the medium for steady one-dimensional heat conduction in a long cylindrical shell for any combination of specified temperature, specified heat flux, and convection boundary conditions. Run the program for five different sets of specified boundary conditions.

2-129 Write an interactive computer program to calculate the heat transfer rate and the value of temperature anywhere in the medium for steady one-dimensional heat conduction in a spherical shell for any combination of specified temperature, specified heat flux, and convection boundary conditions. Run the program for five different sets of specified boundary conditions.

2-130 Write an interactive computer program to calculate the heat transfer rate and the value of temperature anywhere in the medium for steady one-dimensional heat conduction in a plane wall whose thermal conductivity varies linearly as $k(T) = k_0(1 + \beta T)$ where the constants k_0 and β are specified by the user for specified temperature boundary conditions.

Steady Heat Conduction

3

In heat transfer analysis, we are often interested in the rate of heat transfer through a medium under steady conditions, and perhaps surface temperatures. Such problems can be solved easily without involving any differential equations by the introduction of *thermal resistance concepts* in an analogous manner to electrical circuit problems. In this case, the thermal resistance corresponds to electrical resistance, temperature difference corresponds to voltage, and the heat transfer rate corresponds to electric current.

We start this chapter with *one-dimensional steady heat conduction* in a plane wall, a cylinder, and a sphere, and develop relations for *thermal resistances* in these geometries. We also develop thermal resistance relations for convection and radiation conditions at the boundaries. We apply this concept to heat conduction problems in *multilayer* plane walls, cylinders, and spheres and generalize it to systems that involve heat transfer in two or three dimensions. We also discuss the *thermal contact resistance* and the *overall heat transfer coefficient* and develop relations for the critical radius of insulation for a cylinder and a sphere. We then present extensive discussions on *thermal insulations* because of their importance and widespread use, including the optimum thickness of insulation. Finally, we discuss steady heat transfer from *finned surfaces* and some complex geometries commonly encountered in practice through the use of *conduction shape factors*.

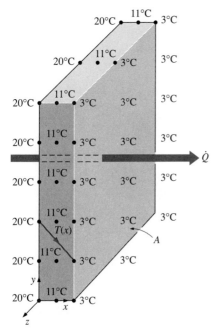

FIGURE 3-1

Heat flow through a wall is one-dimensional when the temperature of the wall varies in one direction only.

3-1 ■ STEADY HEAT CONDUCTION IN PLANE WALLS

Consider steady heat conduction through the walls of a house during a winter day. We know that heat is continuously lost to the outdoors through the wall. We intuitively feel that heat transfer through the wall is in the *normal direction* to the wall surface, and no significant heat transfer takes place in the wall in other directions (Fig. 3-1).

Recall that heat transfer in a certain direction is driven by the *temperature gradient* in that direction. There will be no heat transfer in a direction in which there is no change in temperature. Temperature measurements at several locations on the inner or outer wall surface will confirm that a wall surface is nearly *isothermal*. That is, the temperatures at the top and bottom of a wall surface as well as at the right or left ends are almost the same. Therefore, there will be no heat transfer through the wall from the top to the bottom, or from left to right, but there will be considerable temperature difference between the inner and the outer surfaces of the wall, and thus significant heat transfer in the direction from the inner surface to the outer one.

The small thickness of the wall causes the temperature gradient in that direction to be large. Further, if the air temperatures in and outside the house remain constant, then heat transfer through the wall of a house can be modeled as *steady* and *one-dimensional*. The temperature of the wall in this case will depend on one direction only (say the x-direction) and can be expressed as $T(x)$.

Noting that heat transfer is the only energy interaction involved in this case and there is no heat generation, the *energy balance* for the wall can be expressed as

$$
\left(\begin{array}{c} \text{Rate of} \\ \text{heat transfer} \\ \text{into the wall} \end{array} \right) - \left(\begin{array}{c} \text{Rate of} \\ \text{heat transfer} \\ \text{out of the wall} \end{array} \right) = \left(\begin{array}{c} \text{Rate of change} \\ \text{of the energy} \\ \text{of the wall} \end{array} \right)
$$

or

$$
\dot{Q}_{\text{in}} - \dot{Q}_{\text{out}} = \frac{dE_{\text{wall}}}{dt} \tag{3-1}
$$

But $dE_{\text{wall}}/dt = 0$ for *steady* operation, since there is no change in the temperature of the wall with time at any point. Therefore, the rate of heat transfer into the wall must be equal to the rate of heat transfer out of it. In other words, *the rate of heat transfer through the wall must be constant,* $\dot{Q}_{\text{cond, wall}} = \text{constant}$.

Consider a plane wall of thickness L and average thermal conductivity k. The two surfaces of the wall are maintained at constant temperatures of T_1 and T_2. For one-dimensional steady heat conduction through the wall, we have $T(x)$. Then Fourier's law of heat conduction for the wall can be expressed as

$$
\dot{Q}_{\text{cond, wall}} = -kA\frac{dT}{dx} \qquad \text{(W)} \tag{3-2}
$$

where the rate of conduction heat transfer $\dot{Q}_{\text{cond, wall}}$ and the surface area A are constant. Thus we have dT/dx = constant, which means that *the temperature through the wall varies linearly with x*. That is, the temperature distribution in the wall under steady conditions is a *straight line* (Fig. 3-2).

Separating the variables in the above equation and integrating from $x = 0$, where $T(0) = T_1$, to $x = L$, where $T(L) = T_2$, we get

$$\int_{x=0}^{L} \dot{Q}_{\text{cond, wall}} dx = -\int_{T=T_1}^{T_2} kA \, dT$$

Performing the integration and rearranging gives

$$\dot{Q}_{\text{cond, wall}} = kA\frac{T_1 - T_2}{L} \qquad \text{(W)} \qquad (3\text{-}3)$$

which is identical to Equation 1-20. Again, *the rate of heat conduction through a plane wall is proportional to the average thermal conductivity, the wall area, and the temperature difference, but is inversely proportional to the wall thickness*. Also, once the rate of heat conduction is available, the temperature $T(x)$ at any location x can be determined by replacing T_2 in Equation 3-3 by T, and L by x.

The Thermal Resistance Concept

Equation 3-3 for heat conduction through a plane wall can be rearranged as

$$\dot{Q}_{\text{cond, wall}} = \frac{T_1 - T_2}{R_{\text{wall}}} \qquad \text{(W)} \qquad (3\text{-}4)$$

where

$$R_{\text{wall}} = \frac{L}{kA} \qquad \text{(°C/W)} \qquad (3\text{-}5)$$

is the *thermal resistance* of the wall against heat conduction or simply the **conduction resistance** of the wall. Note that the thermal resistance of a medium depends on the *geometry* and the *thermal properties* of the medium.

The equation above for heat flow is analogous to the relation for *electric current flow I*, expressed as

$$I = \frac{\mathcal{V}_1 - \mathcal{V}_2}{R_e} \qquad (3\text{-}6)$$

where $R_e = L/\sigma_e A$ is the *electric resistance* and $\mathcal{V}_1 - \mathcal{V}_2$ is the *voltage difference* across the resistance (σ_e is the electrical conductivity). Thus, the *rate of heat transfer* through a layer corresponds to the *electric current*, the *thermal resistance* corresponds to *electrical resistance*, and the *temperature difference* corresponds to *voltage difference* across the layer (Fig. 3-3).

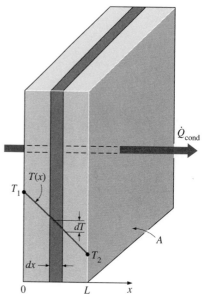

FIGURE 3-2

Under steady conditions, the temperature distribution in a plane wall is a straight line.

FIGURE 3-3

Analogy between thermal and electrical resistance concepts.

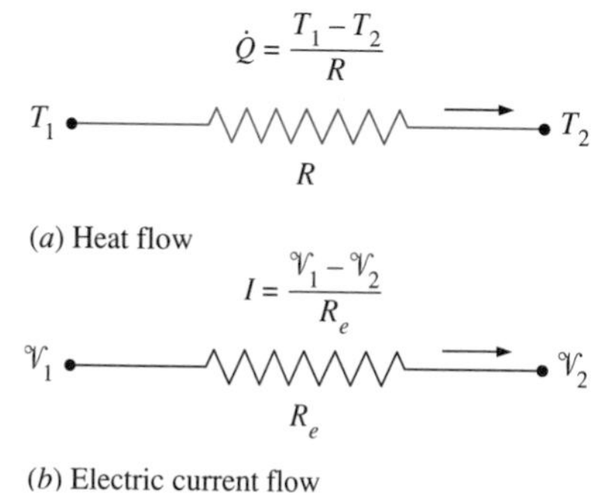

$$\dot{Q} = \frac{T_1 - T_2}{R}$$

$T_1 \bullet\!\!\!-\!\!\!\wedge\!\!\wedge\!\!\wedge\!\!\wedge\!\!\!-\!\!\!\rightarrow\!\bullet T_2$

R

(a) Heat flow

$$I = \frac{\mathcal{V}_1 - \mathcal{V}_2}{R_e}$$

$\mathcal{V}_1 \bullet\!\!\!-\!\!\!\wedge\!\!\wedge\!\!\wedge\!\!\wedge\!\!\!-\!\!\!\rightarrow\!\bullet \mathcal{V}_2$

R_e

(b) Electric current flow

Consider convection heat transfer from a solid surface of area A and temperature T_s to a fluid whose temperature sufficiently far from the surface is T_∞, with a convection heat transfer coefficient h. Newton's law of cooling for convection heat transfer rate $\dot{Q}_{conv} = hA(T_s - T_\infty)$ can be rearranged as

$$\dot{Q}_{conv} = \frac{T_s - T_\infty}{R_{conv}} \qquad \text{(W)} \qquad (3\text{-}7)$$

where

$$R_{conv} = \frac{1}{hA} \qquad \text{(°C/W)} \qquad (3\text{-}8)$$

is the *thermal resistance* of the surface against heat convection, or simply the **convection resistance** of the surface (Fig. 3-4). Note that when the convection heat transfer coefficient is very large ($h \rightarrow \infty$), the convection resistance becomes *zero* and $T_s \approx T_\infty$. That is, the surface offers *no resistance to convection,* and thus it does not slow down the heat transfer process. This situation is approached in practice at surfaces where boiling and condensation occur. Also note that the surface does not have to be a plane surface. Equation 3-8 for convection resistance is valid for surfaces of any shape, provided that the assumption of $h = $ constant and uniform is reasonable.

When the wall is surrounded by a gas, the *radiation effects,* which we have ignored so far, can be significant and may need to be considered. The rate of radiation heat transfer between a surface of emissivity ε and area A at temperature T_s and the surrounding surfaces at some average temperature T_{surr} can be expressed as

$$\dot{Q}_{rad} = \varepsilon \sigma A(T_s^4 - T_{surr}^4) = h_{rad}A(T_s - T_{surr}) = \frac{T_s - T_{surr}}{R_{rad}} \qquad \text{(W)} \quad (3\text{-}9)$$

where

$$R_{rad} = \frac{1}{h_{rad}A} \qquad \text{(K/W)} \qquad (3\text{-}10)$$

is the *thermal resistance* of a surface against radiation, or the *radiation resistance,* and

$$h_{rad} = \frac{\dot{Q}_{rad}}{A(T_s - T_{surr})} = \varepsilon \sigma (T_s^2 + T_{surr}^2)(T_s + T_{surr}) \qquad \text{(W/m}^2 \cdot \text{K)} \quad (3\text{-}11)$$

is the **radiation heat transfer coefficient.** Note that both T_s and T_{surr} *must* be in K in the evaluation of h_{rad}. The definition of the radiation heat transfer coefficient enables us to express radiation conveniently in an analogous manner to convection in terms of a temperature difference. But h_{rad} depends strongly on temperature while h_{conv} usually does not.

A surface exposed to the surrounding air involves convection and radiation simultaneously, and the total heat transfer at the surface is determined

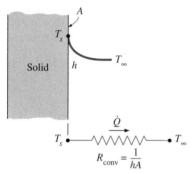

FIGURE 3-4

Schematic for convection resistance at a surface.

by adding (or subtracting, if in the opposite direction) the radiation and convection components. The convection and radiation resistances are parallel to each other, as shown in Figure 3-5, and may cause some complication in the thermal resistance network. When $T_{surr} \approx T_\infty$, the radiation effect can properly be accounted for by replacing h in the convection resistance relation by

$$h_{combined} = h_{conv} + h_{rad} \qquad (W/m^2 \cdot K) \qquad (3\text{-}12)$$

where $h_{combined}$ is the **combined heat transfer coefficient.** This way all the complications associated with radiation are avoided.

Thermal Resistance Network

Now consider steady one-dimensional heat flow through a plane wall of thickness L and thermal conductivity k that is exposed to convection on both sides to fluids at temperatures $T_{\infty 1}$ and $T_{\infty 2}$ with heat transfer coefficients h_1 and h_2, respectively, as shown in Figure 3-6. Assuming $T_{\infty 2} < T_{\infty 1}$, the variation of temperature will be as shown in the figure. Note that the temperature varies linearly in the wall, and asymptotically approaches $T_{\infty 1}$ and $T_{\infty 2}$ in the fluids as we move away from the wall.

Under steady conditions we have

$$\begin{pmatrix} Rate\ of \\ heat\ convection \\ into\ the\ wall \end{pmatrix} = \begin{pmatrix} Rate\ of \\ heat\ conduction \\ through\ the\ wall \end{pmatrix} = \begin{pmatrix} Rate\ of \\ heat\ convection \\ from\ the\ wall \end{pmatrix}$$

or

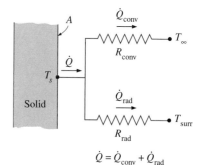

$$\dot{Q} = \dot{Q}_{conv} + \dot{Q}_{rad}$$

FIGURE 3-5

Schematic for convection and radiation resistances at a surface.

FIGURE 3-6

The thermal resistance network for heat transfer through a plane wall subjected to convection on both sides, and the electrical analogy.

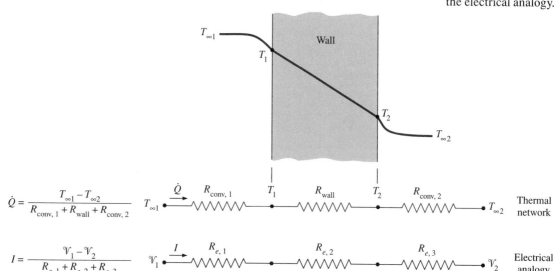

$$\dot{Q} = \frac{T_{\infty 1} - T_{\infty 2}}{R_{conv,1} + R_{wall} + R_{conv,2}}$$

$$I = \frac{\mathcal{V}_1 - \mathcal{V}_2}{R_{e,1} + R_{e,2} + R_{e,3}}$$

$$\dot{Q} = h_1 A(T_{\infty 1} - T_1) = kA\frac{T_1 - T_2}{L} = h_2 A(T_2 - T_{\infty 2}) \qquad (3\text{-}13)$$

which can be rearranged as

$$\dot{Q} = \frac{T_{\infty 1} - T_1}{1/h_1 A} = \frac{T_1 - T_2}{L/kA} = \frac{T_2 - T_{\infty 2}}{1/h_2 A}$$

$$= \frac{T_{\infty 1} - T_1}{R_{conv,1}} = \frac{T_1 - T_2}{R_{wall}} = \frac{T_2 - T_{\infty 2}}{R_{conv,2}} \qquad (3\text{-}14)$$

Adding the numerators and denominators yields (Fig. 3-7)

$$\dot{Q} = \frac{T_{\infty 1} - T_{\infty 2}}{R_{total}} \qquad (W) \qquad (3\text{-}15)$$

where

$$R_{total} = R_{conv,1} + R_{wall} + R_{conv,2} = \frac{1}{h_1 A} + \frac{L}{kA} + \frac{1}{h_2 A} \qquad (°C/W) \quad (3\text{-}16)$$

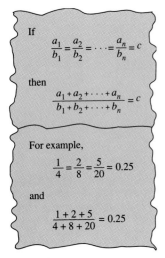

FIGURE 3-7

A useful mathematical identity.

Note that the heat transfer area A is constant for a plane wall, and the rate of heat transfer through a wall separating two mediums is equal to the temperature difference divided by the total thermal resistance between the mediums. Also note that the thermal resistances are in *series,* and the equivalent thermal resistance is determined by simply *adding* the individual resistances, just like the electrical resistances connected in series. Thus, the electrical analogy still applies. We summarize this as *the rate of steady heat transfer between two surfaces is equal to the temperature difference divided by the total thermal resistance between those two surfaces.*

Another observation that can be made from Equation 3-15 is that the ratio of the temperature drop to the thermal resistance across any layer is constant, and thus the temperature drop across any layer is proportional to the thermal resistance of the layer. The larger the resistance, the larger the temperature drop. In fact, the equation $\dot{Q} = \Delta T/R$ can be rearranged as

$$\Delta T = \dot{Q}R \qquad (°C) \qquad (3\text{-}17)$$

which indicates that the *temperature drop* across any layer is equal to the *rate of heat transfer* times the *thermal resistance* across that layer (Fig. 3-8). You may recall that this is also true for voltage drop across an electrical resistance when the electric current is constant.

It is sometimes convenient to express heat transfer through a medium in an analogous manner to Newton's law of cooling as

$$\dot{Q} = UA\Delta T \qquad (W) \qquad (3\text{-}18)$$

where U is the **overall heat transfer coefficient.** A comparison of Equations 3-15 and 3-18 reveals that

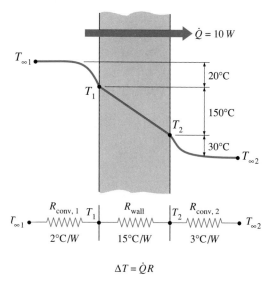

$$\dot{Q} = 10 \text{ W}$$

$$\Delta T = \dot{Q} R$$

FIGURE 3-8

The temperature drop across a layer is proportional to its thermal resistance.

$$UA = \frac{1}{R_{\text{total}}} \qquad (3\text{-}19)$$

Therefore, for a unit area, the overall heat transfer coefficient is equal to the inverse of the total thermal resistance.

Note that we do not need to know the surface temperatures of the wall in order to evaluate the rate of steady heat transfer through it. All we need to know is the convection heat transfer coefficients and the fluid temperatures on both sides of the wall. The *surface temperature* of the wall can be determined as described above using the thermal resistance concept, but by taking the surface at which the temperature is to be determined as one of the terminal surfaces. For example, once $\dot{Q}$ is evaluated, the surface temperature T_1 can be determined from

$$\dot{Q} = \frac{T_{\infty 1} - T_1}{R_{\text{conv}, 1}} = \frac{T_{\infty 1} - T_1}{1/h_1 A} \qquad (3\text{-}20)$$

Multilayer Plane Walls

In practice we often encounter plane walls that consist of several layers of different materials. The thermal resistance concept can still be used to determine the rate of steady heat transfer through such *composite* walls. As you may have already guessed, this is done by simply noting that the conduction resistance of each wall is L/kA connected in series, and using the electrical analogy; that is, by dividing the *temperature difference* between two surfaces at known temperatures by the *total thermal resistance* between them.

Consider a plane wall that consists of two layers (such as a brick wall with a layer of insulation). The rate of steady heat transfer through this two-layer composite wall can be expressed as (Fig. 3-9)

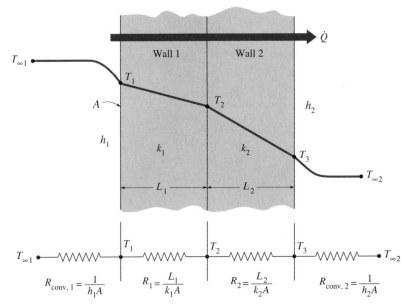

FIGURE 3-9

The thermal resistance network for heat transfer through a two-layer plane wall subjected to convection on both sides.

$$\dot{Q} = \frac{T_{\infty 1} - T_{\infty 2}}{R_{\text{total}}} \tag{3-21}$$

where R_{total} is the *total thermal resistance*, expressed as

$$R_{\text{total}} = R_{\text{conv}, 1} + R_{\text{wall}, 1} + R_{\text{wall}, 2} + R_{\text{conv}, 2} \tag{3-22}$$

$$= \frac{1}{h_1 A} + \frac{L_1}{k_1 A} + \frac{L_2}{k_2 A} + \frac{1}{h_2 A}$$

The subscripts 1 and 2 in the R_{wall} relations above indicate the first and the second layers, respectively. We could also obtain this result by following the approach used above for the single-layer case by noting that the rate of steady heat transfer $\dot{Q}$ through a multilayer medium is constant, and thus it must be the same through each layer. Note from the thermal resistance network that the resistances are *in series,* and thus the *total thermal resistance* is simply the *arithmetic sum* of the individual thermal resistances in the path of heat flow.

The result above for the *two-layer* case is analogous to the *single-layer* case, except that an *additional resistance* is added for the *additional layer.* This result can be extended to plane walls that consist of *three* or *more layers* by adding an *additional resistance* for each *additional layer.*

Once $\dot{Q}$ is *known,* an unknown surface temperature T_j at any surface or interface j can be determined from

$$\dot{Q} = \frac{T_i - T_j}{R_{\text{total}, i-j}} \tag{3-23}$$

where T_i is a *known* temperature at location i and $R_{\text{total}, i-j}$ is the total thermal resistance between locations i and j. For example, when the fluid temperatures $T_{\infty 1}$ and $T_{\infty 2}$ for the two-layer case shown in Figure 3-9 are available and $\dot{Q}$ is calculated from Equation 3-23, the interface temperature T_2 between the two walls can be determined from (Fig. 3-10)

$$\dot{Q} = \frac{T_{\infty 1} - T_2}{R_{\text{conv}, 1} + R_{\text{wall}, 1}} = \frac{T_{\infty 1} - T_2}{\dfrac{1}{h_1 A} + \dfrac{L_1}{k_1 A}} \tag{3-24}$$

The temperature drop across a layer is easily determined from Equation 3-17 by multiplying $\dot{Q}$ by the thermal resistance of that layer.

The thermal resistance concept is widely used in practice because it is intuitively easy to understand and it has proven to be a powerful tool in the solution of a wide range of heat transfer problems. But its use is limited to systems through which the rate of heat transfer $\dot{Q}$ remains *constant*; that is, to systems involving *steady* heat transfer with *no heat generation* (such as resistance heating or chemical reactions) within the medium.

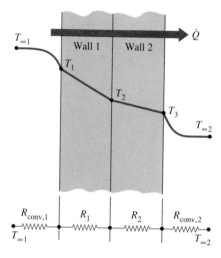

EXAMPLE 3-1   Heat Loss through a Wall

Consider a 3-m-high, 5-m-wide, and 0.3-m-thick wall whose thermal conductivity is $k = 0.9 \text{ W/m} \cdot \text{°C}$ (Fig. 3-11). On a certain day, the temperatures of the inner and the outer surfaces of the wall are measured to be 16°C and 2°C, respectively. Determine the rate of heat loss through the wall on that day.

To find T_1: $\quad \dot{Q} = \dfrac{T_{\infty 1} - T_1}{R_{\text{conv}, 1}}$

To find T_2: $\quad \dot{Q} = \dfrac{T_{\infty 1} - T_2}{R_{\text{conv}, 1} + R_1}$

To find T_3: $\quad \dot{Q} = \dfrac{T_3 - T_{\infty 2}}{R_{\text{conv}, 2}}$

FIGURE 3-10

The evaluation of the surface and interface temperatures when $T_{\infty 1}$ and $T_{\infty 2}$ are given and $\dot{Q}$ is calculated.

Solution   The two surfaces of a wall are maintained at specified temperatures. The rate of heat loss through the wall is to be determined.

Assumptions   **1** Heat transfer through the wall is steady since the surface temperatures remain constant at the specified values. **2** Heat transfer is one-dimensional since any significant temperature gradients will exist in the direction from the indoors to the outdoors. **3** Thermal conductivity is constant.

Properties   The thermal conductivity is given to be $k = 0.9 \text{ W/m} \cdot \text{°C}$.

FIGURE 3-11

Schematic for Example 3-1.

Analysis   Noting that the heat transfer through the wall is by conduction and the surface area of the wall is $A = 3 \text{ m} \times 5 \text{ m} = 15 \text{ m}^2$, the steady rate of heat transfer through the wall can be determined from Equation 3-3 to be

$$\dot{Q} = kA \frac{T_1 - T_2}{L} = (0.9 \text{ W/m} \cdot \text{°C})(15 \text{ m}^2) \frac{(16 - 2)\text{° C}}{0.3 \text{ m}} = \textbf{630 W}$$

We could also determine the steady rate of heat transfer through the wall by making use of the thermal resistance concept from

$$\dot{Q} = \frac{\Delta T_{\text{wall}}}{R_{\text{wall}}}$$

where

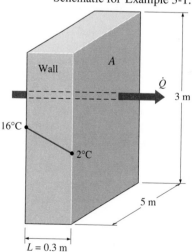

$$R_{\text{wall}} = \frac{L}{kA} = \frac{0.3 \text{ m}}{(0.9 \text{ W/m} \cdot \text{°C})(15 \text{ m}^2)} = 0.02222\text{°C/W}$$

Substituting, we get

$$\dot{Q} = \frac{(16 - 2)°C}{0.02222°C/W} = 630 \text{ W}$$

Discussion This is the same result obtained earlier. Note that heat conduction through a plane wall with specified surface temperatures can be determined directly and easily without utilizing the thermal resistance concept. However, the thermal resistance concept serves as a valuable tool in more complex heat transfer problems, as you will see in the following examples.

EXAMPLE 3-2   Heat Loss through a Single-Pane Window

Consider a 0.8-m-high and 1.5-m-wide glass window with a thickness of 8 mm and a thermal conductivity of $k = 0.78$ W/m $\cdot$ °C. Determine the steady rate of heat transfer through this glass window and the temperature of its inner surface for a day during which the room is maintained at 20°C while the temperature of the outdoors is -10°C. Take the heat transfer coefficients on the inner and outer surfaces of the window to be $h_1 = 10$ W/m² $\cdot$ °C and $h_2 = 40$ W/m² $\cdot$ °C, which includes the effects of radiation.

Solution The two surfaces of a window are maintained at specified temperatures. The rate of heat loss through the window and the inner surface temperature are to be determined.

Assumptions **1** Heat transfer through the window is steady since the surface temperatures remain constant at the specified values. **2** Heat transfer through the wall is one-dimensional since any significant temperature gradients will exist in the direction from the indoors to the outdoors. **3** Thermal conductivity is constant. **4** Heat transfer by radiation is negligible.

Properties The thermal conductivity is given to be $k = 0.78$ W/m $\cdot$ °C.

Analysis This problem involves conduction through the glass window and convection at its surfaces, and can best be handled by making use of the thermal resistance concept and drawing the thermal resistance network, as shown in Figure 3-12. Noting that the surface area of the window is $A = 0.8$ m $\times$ 1.5 m = 1.2 m², the individual resistances are evaluated from their definitions to be

$$R_i = R_{\text{conv},1} = \frac{1}{h_1 A} = \frac{1}{(10 \text{ W/m}^2 \cdot °C)(1.2 \text{ m}^2)} = 0.08333°C/W$$

$$R_{\text{glass}} = \frac{L}{kA} = \frac{0.008 \text{ m}}{(0.78 \text{ W/m} \cdot °C)(1.2 \text{ m}^2)} = 0.00855°C/W$$

$$R_o = R_{\text{conv},2} = \frac{1}{h_2 A} = \frac{1}{(40 \text{ W/m}^2 \cdot °C)(1.2 \text{ m}^2)} = 0.02083°C/W$$

Noting that all three resistances are in series, the total resistance is determined to be

$$R_{\text{total}} = R_{\text{conv},1} + R_{\text{glass}} + R_{\text{conv},2} = 0.08333 + 0.00855 + 0.02083$$
$$= 0.1127°C/W$$

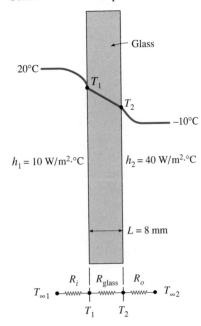

FIGURE 3-12

Schematic for Example 3-2.

Then the steady rate of heat transfer through the window becomes

$$\dot{Q} = \frac{T_{\infty 1} - T_{\infty 2}}{R_{total}} = \frac{[20 - (-10)]°C}{0.1127°C/W} = \textbf{266 W}$$

Knowing the rate of heat transfer, the inner surface temperature of the window glass can be determined from

$$\dot{Q} = \frac{T_{\infty 1} - T_1}{R_{conv,1}} \longrightarrow T_1 = T_{\infty 1} - \dot{Q}R_{conv,1}$$

$$= 20°C - (266\ W)(0.08333°C/W)$$

$$= \textbf{-2.2°C}$$

Discussion Note that the inner surface temperature of the window glass will be −2.2°C even though the temperature of the air in the room is maintained at 20°C. Such low surface temperatures are highly undesirable since they cause the formation of fog or even frost on the inner surfaces of the glass when the humidity in the room is high.

EXAMPLE 3-3   Heat Loss through Double-Pane Windows

Consider a 0.8-m-high and 1.5-m-wide double-pane window consisting of two 4-mm-thick layers of glass ($k = 0.78$ W/m · °C) separated by a 10-mm-wide stagnant air space ($k = 0.026$ W/m · °C). Determine the steady rate of heat transfer through this double-pane window and the temperature of its inner surface for a day during which the room is maintained at 20°C while the temperature of the outdoors is −10°C. Take the convection heat transfer coefficients on the inner and outer surfaces of the window to be $h_1 = 10$ W/m² · °C and $h_2 = 40$ W/m² · °C, which includes the effects of radiation.

Solution This example problem is identical to the previous one except that the single 8-mm-thick window glass is replaced by two 4-mm-thick glasses that enclose a 10-mm-wide stagnant air space. Therefore, the thermal resistance network of this problem will involve two additional conduction resistances corresponding to the two additional layers, as shown in Figure 3-13. Noting that the surface area of the window is again $A = 0.8\ m \times 1.5\ m = 1.2\ m^2$, the individual resistances are evaluated from their definitions to be

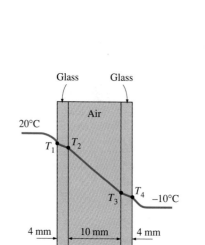

$$R_i = R_{conv,1} = \frac{1}{h_1 A} = \frac{1}{(10\ W/m^2 \cdot °C)(1.2\ m^2)} = 0.08333°C/W$$

$$R_1 = R_3 = R_{glass} = \frac{L_1}{k_1 A} = \frac{0.004\ m}{(0.78\ W/m \cdot °C)(1.2\ m^2)} = 0.00427°C/W$$

$$R_2 = R_{air} = \frac{L_2}{k_2 A} = \frac{0.01\ m}{(0.026\ W/m \cdot °C)(1.2\ m^2)} = 0.3205°C/W$$

$$R_o = R_{conv,2} = \frac{1}{h_2 A} = \frac{1}{(40\ W/m^2 \cdot °C)(1.2\ m^2)} = 0.02083°C/W$$

FIGURE 3-13
Schematic for Example 3-3.

Noting that all three resistances are in series, the total resistance is determined to be

$$R_{total} = R_{conv,1} + R_{glass,1} + R_{air} + R_{glass,2} + R_{conv,2}$$

$$= 0.08333 + 0.00427 + 0.3205 + 0.00427 + 0.02083$$

$$= 0.4332°C/W$$

Then the steady rate of heat transfer through the window becomes

$$\dot{Q} = \frac{T_{\infty 1} - T_{\infty 2}}{R_{\text{total}}} = \frac{[20 - (-10)]°C}{0.4332°C/W} = \textbf{69.2 W}$$

which is about one-fourth of the result obtained in the previous example. This explains the popularity of the double- and even triple-pane windows in cold climates. The drastic reduction in the heat transfer rate in this case is due to the large thermal resistance of the air layer between the glasses. In reality, the thermal resistance of the air layer will be somewhat lower because of the natural convection currents that are likely to occur in the air space.

The inner surface temperature of the window in this case will be

$$T_1 = T_{\infty 1} - \dot{Q}R_{\text{conv, 1}} = 20°C - (69.2 \text{ W})(0.08333°C/W) = \textbf{14.2°C}$$

which is considerably higher than the −2.2°C obtained in the previous example. Therefore, a double-pane window will rarely get fogged. A double-pane window will also reduce the heat gain in summer, and thus reduce the air-conditioning costs.

3-2 ■ THERMAL CONTACT RESISTANCE

In the analysis of heat conduction through multilayer solids, we assumed "perfect contact" at the interface of two layers, and thus no temperature drop at the interface. This would be the case when the surfaces are perfectly smooth and they produce a perfect contact at each point. In reality, however, even flat surfaces that appear smooth to the eye turn out to be rather rough when examined under a microscope, as shown in Figure 3-14, with numerous peaks and valleys. That is, a surface is *microscopically rough* no matter how smooth it appears to be.

FIGURE 3-14

Temperature distribution and heat flow lines along two solid plates pressed against each other for the case of perfect and imperfect contact.

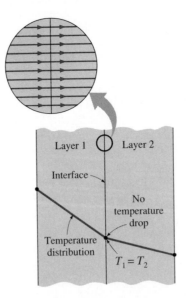

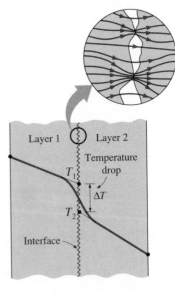

(a) Ideal (perfect) thermal contact

(b) Actual (imperfect) thermal contact

When two such surfaces are pressed against each other, the peaks will form good material contact but the valleys will form voids filled with air. As a result, an interface will contain numerous *air gaps* of varying sizes that act as *insulation* because of the low thermal conductivity of air. Thus, an interface offers some resistance to heat transfer, and this resistance per unit interface area is called the **thermal contact resistance, R_c.** The value of R_c is determined experimentally using a setup like the one shown in Figure 3–15, and as expected, there is considerable scatter of data because of the difficulty in characterizing the surfaces.

Consider heat transfer through two metal rods of cross-sectional area A that are pressed against each other. Heat transfer through the interface of these two rods is the sum of the heat transfers through the *solid contact spots* and the *gaps* in the noncontact areas and can be expressed as

$$\dot{Q} = \dot{Q}_{\text{contact}} + \dot{Q}_{\text{gap}} \tag{3-25}$$

It can also be expressed in an analogous manner to Newton's law of cooling as

$$\dot{Q} = h_c A \Delta T_{\text{interface}} \tag{3-26}$$

where A is the apparent interface area (which is the same as the cross-sectional area of the rods) and $\Delta T_{\text{interface}}$ is the effective temperature difference at the interface. The quantity h_c, which corresponds to the convection heat transfer coefficient, is called the **thermal contact conductance** and is expressed as

$$h_c = \frac{\dot{Q}/A}{\Delta T_{\text{interface}}} \quad (\text{W/m}^2 \cdot {}^\circ\text{C}) \tag{3-27}$$

It is related to thermal contact resistance by

$$R_c = \frac{1}{h_c} = \frac{\Delta T_{\text{interface}}}{\dot{Q}/A} \quad (\text{m}^2 \cdot {}^\circ\text{C/W}) \tag{3-28}$$

That is, thermal contact resistance is the inverse of thermal contact conductance. Usually, thermal contact conductance is reported in the literature, but the concept of thermal contact resistance serves as a better vehicle for explaining the effect of interface on heat transfer. Note that R_c represents thermal contact resistance *per unit area*. The thermal resistance for the entire interface is obtained by dividing R_c by the apparent interface area A.

The thermal contact resistance can be determined from Equation 3–28 by measuring the temperature drop at the interface and dividing it by the heat flux under steady conditions. The value of thermal contact resistance depends on the *surface roughness* and the *material properties* as well as the *temperature* and *pressure* at the interface and the *type of fluid* trapped at the interface. The situation becomes more complex when plates are fastened by bolts, screws, or rivets since the interface pressure in this case is nonuniform. The thermal contact resistance in that case also depends on the plate thickness, the bolt radius, and the size of the contact zone. Thermal contact

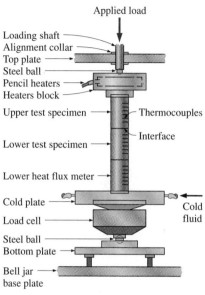

Applied load

Loading shaft
Alignment collar
Top plate
Steel ball
Pencil heaters
Heaters block

Upper test specimen — Thermocouples

— Interface
Lower test specimen

Lower heat flux meter

Cold plate
Load cell
Cold fluid

Steel ball
Bottom plate

Bell jar
base plate

FIGURE 3–15

A typical experimental setup for the determination of thermal contact resistance (from Song et al., Ref. 23).

TABLE 3-1

Thermal contact conductance for aluminum plates with different fluids at the interface for a surface roughness of 10 μm and interface pressure of 1 atm (from Fried, Ref. 9)

Fluid at the interface	Contact conductance, h_c, W/m² · °C
Air	3640
Helium	9520
Hydrogen	13,900
Silicone oil	19,000
Glycerin	37,700

FIGURE 3-16

Effect of metallic coatings on thermal contact conductance (from Peterson, Ref. 21).

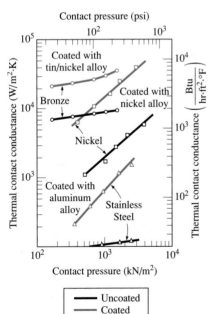

resistance is observed to *decrease* with *decreasing surface roughness* and *increasing interface pressure,* as expected. Most experimentally determined values of the thermal contact resistance fall between 0.000005 and 0.0005 m² · °C/W (the corresponding range of thermal contact conductance is 2000 to 200,000 W/m² · °C).

When we analyze heat transfer in a medium consisting of two or more layers, the first thing we need to know is whether the thermal contact resistance is *significant* or not. We can answer this question by comparing the magnitudes of the thermal resistances of the layers with typical values of thermal contact resistance. For example, the thermal resistance of a 1-cm-thick layer of an insulating material per unit surface area is

$$R_{c,\,\text{insulation}} = \frac{L}{k} = \frac{0.01 \text{ m}}{0.04 \text{ W/m} \cdot {}^\circ\text{C}} = 0.25 \text{ m}^2 \cdot {}^\circ\text{C/W}$$

whereas for a 1-cm-thick layer of copper, it is

$$R_{c,\,\text{copper}} = \frac{L}{k} = \frac{0.01 \text{ m}}{386 \text{ W/m} \cdot {}^\circ\text{C}} = 0.000026 \text{ m}^2 \cdot {}^\circ\text{C/W}$$

Comparing the values above with typical values of thermal contact resistance, we conclude that thermal contact resistance is significant and can even dominate the heat transfer for good heat conductors such as metals, but can be disregarded for poor heat conductors such as insulations. This is not surprising since insulating materials consist mostly of air space just like the interface itself.

The thermal contact resistance can be minimized by applying a thermally conducting liquid called a *thermal grease* such as silicon oil on the surfaces before they are pressed against each other. This is commonly done when attaching electronic components such as power transistors to heat sinks. The thermal contact resistance can also be reduced by replacing the air at the interface by a *better conducting gas* such as helium or hydrogen, as shown in Table 3-1.

Another way to minimize the contact resistance is to insert a *soft metallic foil* such as tin, silver, copper, nickel, or aluminum between the two surfaces. Experimental studies show that the thermal contact resistance can be reduced by a factor of up to 7 by a metallic foil at the interface. For maximum effectiveness, the foils must be very thin. The effect of metallic coatings on thermal contact conductance is shown in Figure 3-16 for various metal surfaces.

There is considerable uncertainty in the contact conductance data reported in the literature, and care should be exercised when using them. In Table 3-2 some experimental results are given for the contact conductance between similar and dissimilar metal surfaces for use in preliminary design calculations. Note that the *thermal contact conductance* is *highest* (and thus the contact resistance is lowest) for *soft metals* with *smooth surfaces* at *high pressure.*

TABLE 3-2   143

Thermal contact conductance of some metal surfaces in air (from Holman, Ref. 12, and Kreith and Bohn, Ref. 16)

Material	Surface condition	Rough-ness, μm	Tempera-ture, °C	Pressure, MPa	h_c,* W/m² · °C
Identical Metal Pairs					
416 Stainless steel	Ground	2.54	90–200	0.3–2.5	3800
304 Stainless steel	Ground	1.14	20	4–7	1900
Aluminum	Ground	2.54	150	1.2–2.5	11,400
Copper	Ground	1.27	20	1.2–20	143,000
Copper	Milled	3.81	20	1–5	55,500
Copper (vacuum)	Milled	0.25	30	0.7–7	11,400
Dissimilar Metal Pairs					
Stainless steel–		20–30	20	10	2900
Aluminum				20	3600
Stainless steel–		1.0–2.0	20	10	16,400
Aluminum				20	20,800
Steel Ct-30–	Ground	1.4–2.0	20	10	50,000
Aluminum				15–35	59,000
Steel Ct-30–	Milled	4.5–7.2	20	10	4800
Aluminum				30	8300
Aluminum-Copper	Ground	1.3–1.4	20	5	42,000
				15	56,000
Aluminum-Copper	Milled	4.4–4.5	20	10	12,000
				20–35	22,000

*Divide the given values by 5.678 to convert to Btu/h · ft² · °F.

EXAMPLE 3-4   Equivalent Thickness for Contact Resistance

The thermal contact conductance at the interface of two 1-cm-thick aluminum plates is measured to be 11,000 W/m² · °C. Determine the thickness of the aluminum plate whose thermal resistance is equal to the thermal resistance of the interface between the plates (Fig. 3-17).

Solution   The thickness of aluminum plate whose thermal resistance is equal to the thermal contact resistance is to be determined.

Analysis   Noting that thermal contact resistance is the inverse of thermal contact conductance, the thermal contact resistance is determined to be

$$R_c = \frac{1}{h_c} = \frac{1}{11{,}000 \text{ W/m}^2 \cdot °\text{C}} = 0.909 \times 10^{-4} \text{ m}^2 \cdot °\text{C/W}$$

For a unit surface area, the thermal resistance of a flat plate is defined as

$$R = \frac{L}{k}$$

where L is the thickness of the plate and k is the thermal conductivity. Setting $R = R_c$, the equivalent thickness is determined from the relation above to be

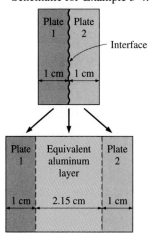

FIGURE 3-17
Schematic for Example 3-4.

$$L = kR_c = (237 \text{ W/m} \cdot °C)(0.909 \times 10^{-4} \text{ m}^2 \cdot °C/W) = 0.0215 \text{ m} = \textbf{2.15 cm}$$

Discussion Note that the interface between the two plates offers as much resistance to heat transfer as a 2.15-cm-thick aluminum plate. Note that the thermal contact resistance in this case is greater than the sum of the thermal resistances of both plates.

EXAMPLE 3-5 Contact Resistance of Transistors

Four identical power transistors with aluminum casing are attached on one side of a 1-cm-thick 20-cm × 20-cm square copper plate ($k = 386$ W/m · °C) by screws that exert an average pressure of 6 MPa (Fig. 3-18). The base area of each transistor is 8 cm², and each transistor is placed at the center of a 10-cm × 10-cm quarter section of the plate. The interface roughness is estimated to be about 1.5 μm. All transistors are covered by a thick Plexiglas layer, which is a poor conductor of heat, and thus all the heat generated at the junction of the transistor must be dissipated to the ambient at 20°C through the back surface of the copper plate. The combined convection/radiation heat transfer coefficient at the back surface can be taken to be 25 W/m² · °C. If the case temperature of the transistor is not to exceed 70°C, determine the maximum power each transistor can dissipate safely, and the temperature jump at the case-plate interface.

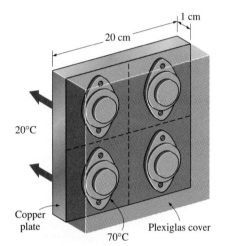

FIGURE 3-18
Schematic for Example 3-5.

Solution Four identical power transistors are attached on a copper plate. For a maximum case temperature of 70°C, the maximum power dissipation and the temperature jump at the interface are to be determined.

Assumptions **1** Steady operating conditions exist. **2** Heat transfer can be approximated as being one-dimensional, although it is recognized that heat conduction in some parts of the plate will be two-dimensional since the plate area is much larger than the base area of the transistor. But the large thermal conductivity of copper will minimize this effect. **3** All the heat generated at the junction is dissipated through the back surface of the plate since the transistors are covered by a thick Plexiglas layer. **4** Thermal conductivities are constant.

Properties The thermal conductivity of copper is given to be $k = 386$ W/m · °C. The contact conductance is obtained from Table 3-2 to be $h_c = 42{,}000$ W/m² · °C, which corresponds to copper-aluminum interface for the case of 1.3–1.4 μm roughness and 5 MPa pressure, which is sufficiently close to what we have.

Analysis The contact area between the case and the plate is given to be 8 cm², and the plate area for each transistor is 100 cm². The thermal resistance network of this problem consists of three resistances in series (interface, plate, and convection), which are determined to be

$$R_{\text{interface}} = \frac{1}{h_c A_c} = \frac{1}{(42{,}000 \text{ W/m}^2 \cdot °C)(8 \times 10^{-4} \text{ m}^2)} = 0.030°C/W$$

$$R_{\text{plate}} = \frac{L}{kA} = \frac{0.01 \text{ m}}{(386 \text{ W/m} \cdot °C)(0.01 \text{ m}^2)} = 0.0026°C/W$$

$$R_{\text{conv}} = \frac{1}{h_o A} = \frac{1}{(25 \text{ W/m}^2 \cdot °C)(0.01 \text{ m}^2)} = 4.0°C/W$$

The total thermal resistance is then

$$R_{total} = R_{interface} + R_{plate} + R_{ambient} = 0.030 + 0.0026 + 4.0 = 4.0326°C/W$$

Note that the thermal resistance of a copper plate is very small and can be ignored altogether. Then the rate of heat transfer is determined to be

$$\dot{Q} = \frac{\Delta T}{R_{total}} = \frac{(70 - 20)°C}{4.0326°C/W} = \textbf{12.4 W}$$

Therefore, the power transistor should not be operated at power levels greater than 12.4 W if the case temperature is not to exceed 70°C.

The temperature jump at the interface is determined from

$$\Delta T_{interface} = \dot{Q}R_{interface} = (12.4 \text{ W})(0.030°C/W) = \textbf{0.37°C}$$

which is not very large. Therefore, even if we eliminate the thermal contact resistance at the interface completely, we will lower the operating temperature of the transistor in this case by less than 0.4°C.

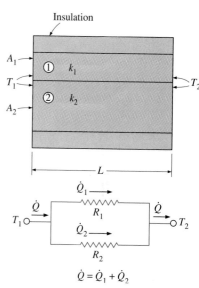

3-3 ■ GENERALIZED THERMAL RESISTANCE NETWORKS

The *thermal resistance* concept or the *electrical analogy* can also be used to solve steady heat transfer problems that involve parallel layers or combined series-parallel arrangements. Although such problems are often two- or even three-dimensional, approximate solutions can be obtained by assuming one-dimensional heat transfer and using the thermal resistance network.

Consider the composite wall shown in Figure 3-19, which consists of two parallel layers. The thermal resistance network, which consists of two parallel resistances, can be represented as shown in the figure. Noting that the total heat transfer is the sum of the heat transfers through each layer, we have

$$\dot{Q} = \dot{Q}_1 + \dot{Q}_2 = \frac{T_1 - T_2}{R_1} + \frac{T_1 - T_2}{R_2} = (T_1 - T_2)\left(\frac{1}{R_1} + \frac{1}{R_2}\right) \quad (3\text{-}29)$$

Utilizing electrical analogy, we get

$$\dot{Q} = \frac{T_1 - T_2}{R_{total}} \quad (3\text{-}30)$$

where

$$\frac{1}{R_{total}} = \frac{1}{R_1} + \frac{1}{R_2} \longrightarrow R_{total} = \frac{R_1 R_2}{R_1 + R_2} \quad (3\text{-}31)$$

since the resistances are in parallel.

Now consider the combined series-parallel arrangement shown in Figure 3-20. The total rate of heat transfer through this composite system can again be expressed as

$$\dot{Q} = \frac{T_1 - T_\infty}{R_{total}} \quad (3\text{-}32)$$

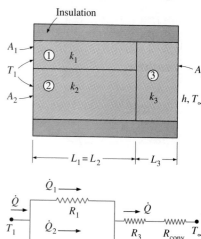

FIGURE 3-19

Thermal resistance network for two parallel layers.

FIGURE 3-20

Thermal resistance network for combined series-parallel arrangement.

where

$$R_{\text{total}} = R_{12} + R_3 + R_{\text{conv}} = \frac{R_1 R_2}{R_1 + R_2} + R_3 + R_{\text{conv}} \qquad (3\text{-}33)$$

and

$$R_1 = \frac{L_1}{k_1 A_1}, \qquad R_2 = \frac{L_2}{k_2 A_2}, \qquad R_3 = \frac{L_3}{k_3 A_3}, \qquad R_{\text{conv}} = \frac{1}{h A_3} \qquad (3\text{-}34)$$

Once the individual thermal resistances are evaluated, the total resistance and the total rate of heat transfer can easily be determined from the relations above.

The result obtained will be somewhat approximate, since the surfaces of the third layer will probably not be isothermal, and heat transfer between the first two layers is likely to occur.

Two assumptions commonly used in solving complex multidimensional heat transfer problems by treating them as one-dimensional (say, in the *x*-direction) using the thermal resistance network are (1) any plane wall normal to the *x*-axis is *isothermal* (i.e., to assume the temperature to vary in the *x*-direction only) and (2) any plane parallel to the *x*-axis is *adiabatic* (i.e., to assume heat transfer to occur in the *x*-direction only). These two assumptions result in different resistance networks, and thus different (but usually close) values for the total thermal resistance and thus heat transfer. The actual result lies between these two values. In geometries in which heat transfer occurs predominantly in one direction, either approach gives satisfactory results.

FIGURE 3-21

Schematic for Example 3-6.

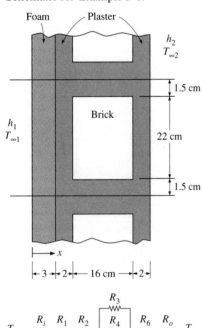

EXAMPLE 3-6   Heat Loss through a Composite Wall

A 3-m-high and 5-m-wide wall consists of long 16-cm × 22-cm cross-section horizontal bricks ($k = 0.72$ W/m · °C) separated by 3-cm-thick plaster layers ($k = 0.22$ W/m · °C). There are also 2-cm-thick plaster layers on each side of the brick and a 3-cm-thick rigid foam ($k = 0.026$ W/m · °C) on the inner side of the wall, as shown in Figure 3-21. The indoor and the outdoor temperatures are 20°C and −10°C, and the convection heat transfer coefficients on the inner and the outer sides are $h_1 = 10$ W/m² · °C and $h_2 = 25$ W/m² · °C, respectively. Assuming one-dimensional heat transfer and disregarding radiation, determine the rate of heat transfer through the wall.

Solution   The composition of a composite wall is given. The rate of heat transfer through the wall is to be determined.

Assumptions   **1** Heat transfer is steady since there is no indication of change with time. **2** Heat transfer can be approximated as being one-dimensional since it is predominantly in the *x*-direction. **3** Thermal conductivities are constant. **4** Heat transfer by radiation is negligible.

Properties   The thermal conductivities are given to be $k = 0.72$ W/m · °C for bricks, $k = 0.22$ W/m · °C for plaster layers, and $k = 0.026$ W/m · °C for the rigid foam.

Analysis There is a pattern in the construction of this wall that repeats itself every 25-cm distance in the vertical direction. There is no variation in the horizontal direction. Therefore, we consider a 1-m-deep and 0.25-m-high portion of the wall, since it is representative of the entire wall.

Assuming any cross-section of the wall normal to the *x*-direction to be *isothermal*, the thermal resistance network for the representative section of the wall becomes as shown in Figure 3-21. The individual resistances are evaluated as follows:

$$R_i = R_{conv,\,1} = \frac{1}{h_1 A} = \frac{1}{(10\ W/m^2 \cdot {}^\circ C)(0.25 \times 1\ m^2)} = 0.4{}^\circ C/W$$

$$R_1 = R_{foam} = \frac{L}{kA} = \frac{0.03\ m}{(0.026\ W/m \cdot {}^\circ C)(0.25 \times 1\ m^2)} = 4.6{}^\circ C/W$$

$$R_2 = R_6 = R_{plaster,\,side} = \frac{L}{kA} = \frac{0.02\ m}{(0.22\ W/m \cdot {}^\circ C)(0.25 \times 1\ m^2)}$$

$$= 0.36{}^\circ C/W$$

$$R_3 = R_5 = R_{plaster,\,center} = \frac{L}{kA} = \frac{0.16\ m}{(0.22\ W/m \cdot {}^\circ C)(0.015 \times 1\ m^2)}$$

$$= 48.48{}^\circ C/W$$

$$R_4 = R_{brick} = \frac{L}{kA} = \frac{0.16\ m}{(0.72\ W/m \cdot {}^\circ C)(0.22 \times 1\ m^2)} = 1.01{}^\circ C/W$$

$$R_o = R_{conv,\,2} = \frac{1}{h_2 A} = \frac{1}{(25\ W/m^2 \cdot {}^\circ C)(0.25 \times 1\ m^2)} = 0.16{}^\circ C/W$$

The three resistances R_3, R_4, and R_5 in the middle are parallel, and their equivalent resistance is determined from

$$\frac{1}{R_{mid}} = \frac{1}{R_3} + \frac{1}{R_4} + \frac{1}{R_5} = \frac{1}{48.48} + \frac{1}{1.01} + \frac{1}{48.48} = 1.03\ W/{}^\circ C$$

which gives

$$R_{mid} = 0.97{}^\circ C/W$$

Now all the resistances are in series, and the total resistance is determined to be

$$R_{total} = R_i + R_1 + R_2 + R_{mid} + R_6 + R_o$$

$$= 0.4 + 4.6 + 0.36 + 0.97 + 0.36 + 0.16$$

$$= 6.85{}^\circ C/W$$

Then the steady rate of heat transfer through the wall becomes

$$\dot{Q} = \frac{T_{\infty 1} - T_{\infty 2}}{R_{total}} = \frac{[20 - (-10)]{}^\circ C}{6.85{}^\circ C/W} = 4.38W \qquad (per\ 0.25\ m^2\ surface\ area)$$

or 4.38/0.25 = 17.5 W per m² surface area. The total surface area of the wall is $A = 3\ m \times 5\ m = 15\ m^2$. Then the rate of heat transfer through the entire wall becomes

$$\dot{Q}_{total} = (17.5\ W/m^2)(15\ m^2) = \mathbf{262.5\ W}$$

Of course, this result is approximate, since we assumed the temperature within the wall to vary in one direction only and ignored any temperature change (and thus heat transfer) in the other two directions.

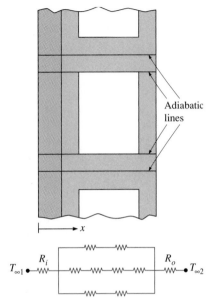

FIGURE 3-22

Alternative thermal resistance network for Example 3-6 for the case of surfaces parallel to the primary direction of heat transfer being adiabatic.

FIGURE 3-23

Heat is lost from a hot water pipe to the air outside in the radial direction, and thus heat transfer from a long pipe is very nearly one-dimensional.

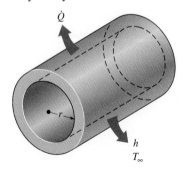

Discussion In the above solution, we assumed the temperature at any cross-section of the wall normal to the *x*-direction to be *isothermal.* We could also solve this problem by going to the other extreme and assuming the surfaces parallel to the *x*-direction to be *adiabatic.* The thermal resistance network in this case will be as shown in Figure 3-22. By following the approach outlined above, the total thermal resistance in this case is determined to be R_{total} = 6.97°C/W, which is almost identical to the value 6.85°C/W obtained before. Thus either approach would give roughly the same result in this case. This example demonstrates that either approach can be used in practice to obtain satisfactory results.

3-4 ■ HEAT CONDUCTION IN CYLINDERS AND SPHERES

Consider steady heat conduction through a hot water pipe. Heat is continuously lost to the outdoors through the wall of the pipe, and we intuitively feel that heat transfer through the pipe is in the normal direction to the wall surface and no significant heat transfer takes place in the pipe in other directions (Fig. 3-23). The wall of the pipe, whose thickness is rather small, separates two fluids at different temperatures, and thus the temperature gradient in the radial direction will be relatively large. Further, if the fluid temperatures inside and outside the pipe remain constant, then heat transfer through the pipe can be modeled as *steady.* Thus heat transfer through the pipe can be modeled as *steady* and *one-dimensional.* The temperature of the pipe in this case will depend on one-direction only (the radial *r*-direction) and can be expressed as $T = T(r)$. The temperature is independent of the azimuthal angle or the axial distance. This situation is approximated in practice in long cylindrical pipes and spherical containers.

In *steady* operation, there is no change in the temperature of the pipe with time at any point. Therefore, the rate of heat transfer into the pipe must be equal to the rate of heat transfer out of it. In other words, heat transfer through the pipe must be constant, $\dot{Q}_{cond,cyl}$ = constant.

Consider a long cylindrical layer (such as a circular pipe) of inner radius r_1, outer radius r_2, length L, and average thermal conductivity k (Fig. 3-24). The two surfaces of the cylindrical layer are maintained at constant temperatures T_1 and T_2. There is no heat generation in the layer and the thermal conductivity is constant. For one-dimensional heat conduction through the cylindrical layer, we have $T(r)$. Then Fourier's law of heat conduction for heat transfer through the cylindrical layer can be expressed as

$$\dot{Q}_{cond,cyl} = -kA \frac{dT}{dr} \quad \text{(W)} \qquad (3\text{-}35)$$

where $A = 2\pi rL$ is the heat transfer surface area at location *r*. Note that A depends on *r*, and thus it *varies* in the direction of heat transfer. Separating the variables in the above equation and integrating from $r = r_1$, where $T(r_1) = T_1$, to $r = r_2$, where $T(r_2) = T_2$, gives

$$\int_{r=r_1}^{r_2} \frac{\dot{Q}_{cond,cyl}}{A} \, dr = -\int_{T=T_1}^{T_2} k \, dT \qquad (3\text{-}36)$$

Substituting $A = 2\pi r L$ and performing the integrations give

$$\dot{Q}_{\text{cond,cyl}} = 2\pi L k \frac{T_1 - T_2}{\ln(r_2/r_1)} \qquad (\text{W}) \qquad (3\text{-}37)$$

since $\dot{Q}_{\text{cond,cyl}} = $ constant. This equation can be rearranged as

$$\dot{Q}_{\text{cond,cyl}} = \frac{T_1 - T_2}{R_{\text{cyl}}} \qquad (\text{W}) \qquad (3\text{-}38)$$

where

$$R_{\text{cyl}} = \frac{\ln(r_2/r_1)}{2\pi L k} = \frac{\ln(\text{Outer radius/Inner radius})}{2\pi \times (\text{Length}) \times (\text{Thermal conductivity})} \qquad (3\text{-}39)$$

is the *thermal resistance* of the cylindrical layer against heat conduction, or simply the **conduction resistance** of the cylinder layer.

We can repeat the analysis above for a *spherical layer* by taking $A = 4\pi r^2$ and performing the integrations in Equation 3-36. The result can be expressed as

$$\dot{Q}_{\text{cond,sph}} = \frac{T_1 - T_2}{R_{\text{sph}}} \qquad (3\text{-}40)$$

where

$$R_{\text{sph}} = \frac{r_2 - r_1}{4\pi r_1 r_2 k}$$
$$= \frac{\text{Outer radius} - \text{Inner radius}}{4\pi(\text{Outer radius})(\text{Inner radius})(\text{Thermal conductivity})} \qquad (3\text{-}41)$$

is the *thermal resistance* of the spherical layer against heat conduction, or simply the **conduction resistance** of the spherical layer.

Now consider steady one-dimensional heat flow through a cylindrical or spherical layer that is exposed to convection on both sides to fluids at temperatures $T_{\infty 1}$ and $T_{\infty 2}$ with heat transfer coefficients h_1 and h_2, respectively, as shown in Figure 3-25. The thermal resistance network in this case consists of one conduction and two convection resistances in series, just like the one for the plane wall, and the rate of heat transfer under steady conditions can be expressed as

$$\dot{Q} = \frac{T_{\infty 1} - T_{\infty 2}}{R_{\text{total}}} \qquad (3\text{-}42)$$

where

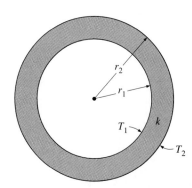

FIGURE 3-24

A long cylindrical pipe (or spherical shell) with specified inner and outer surface temperatures T_1 and T_2.

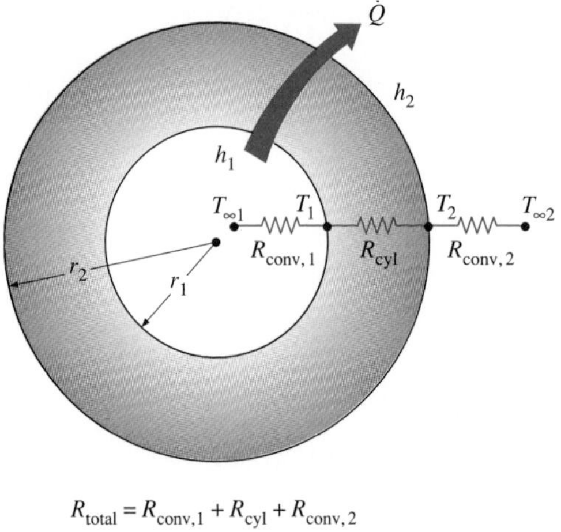

FIGURE 3-25

The thermal resistance network for a cylindrical (or spherical) shell subjected to convection from both the inner and the outer sides.

$$R_{total} = R_{conv,1} + R_{cyl} + R_{conv,2}$$

$$R_{total} = R_{conv,1} + R_{cyl} + R_{conv,2} \tag{3-43}$$
$$= \frac{1}{(2\pi r_1 L)h_1} + \frac{\ln(r_2/r_1)}{2\pi L k} + \frac{1}{(2\pi r_2 L)h_2}$$

for a *cylindrical* layer, and

$$R_{total} = R_{conv,1} + R_{sph} + R_{conv,2} \tag{3-44}$$
$$= \frac{1}{(4\pi r_1^2)h_1} + \frac{r_2 - r_1}{4\pi r_1 r_2 k} + \frac{1}{(4\pi r_2^2)h_2}$$

for a *spherical* layer. Note that A in the convection resistance relation $R_{conv} = 1/hA$ is the *surface area at which convection occurs*. It is equal to $A = 2\pi r L$ for a cylindrical surface and $A = 4\pi r^2$ for a spherical surface of radius r. Also note that the thermal resistances are in series, and thus the total thermal resistance is determined by simply adding the individual resistances, just like the electrical resistances connected in series.

Multilayered Cylinders and Spheres

Steady heat transfer through multilayered cylindrical or spherical shells can be handled just like multilayered plane walls discussed earlier by simply adding an *additional resistance* in series for each *additional layer*. For example, the steady heat transfer rate through the three-layered composite cylinder of length L shown in Figure 3-26 with convection on both sides can be expressed as

$$\dot{Q} = \frac{T_{\infty 1} - T_{\infty 2}}{R_{total}} \tag{3-45}$$

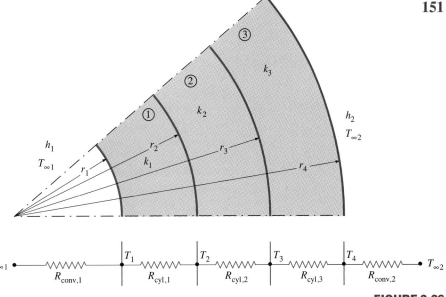

FIGURE 3-26

The thermal resistance network for heat transfer through a three-layered composite cylinder subjected to convection on both sides.

where R_{total} is the *total thermal resistance*, expressed as

$$R_{\text{total}} = R_{\text{conv},1} + R_{\text{cyl},1} + R_{\text{cyl},2} + R_{\text{cyl},3} + R_{\text{conv},2} \qquad (3\text{-}46)$$

$$= \frac{1}{h_1 A_1} + \frac{\ln(r_2/r_1)}{2\pi L k_1} + \frac{\ln(r_3/r_2)}{2\pi L k_2} + \frac{\ln(r_4/r_3)}{2\pi L k_3} + \frac{1}{h_2 A_4}$$

where $A_1 = 2\pi r_1 L$ and $A_4 = 2\pi r_4 L$. Equation 3-46 can also be used for a three-layered spherical shell by replacing the thermal resistances of cylindrical layers by the corresponding spherical ones. Again, note from the thermal resistance network that the resistances are in series, and thus the total thermal resistance is simply the *arithmetic sum* of the individual thermal resistances in the path of heat flow.

Once $\dot{Q}$ is known, we can determine any intermediate temperature T_j by applying the relation $\dot{Q} = (T_i - T_j)/R_{\text{total},i-j}$ across any layer or layers such that T_i is a *known* temperature at location i and $R_{\text{total},i-j}$ is the total thermal resistance between locations i and j (Fig. 3-27). For example, once $\dot{Q}$ has been calculated, the interface temperature T_2 between the first and second cylindrical layers can be determined from

$$\dot{Q} = \frac{T_{\infty 1} - T_2}{R_{\text{conv},1} + R_{\text{cyl},1}} = \frac{T_{\infty 1} - T_2}{\dfrac{1}{h_1 (2\pi r_1 L)} + \dfrac{\ln(r_2/r_1)}{2\pi L k_1}} \qquad (3\text{-}47)$$

We could also calculate T_2 from

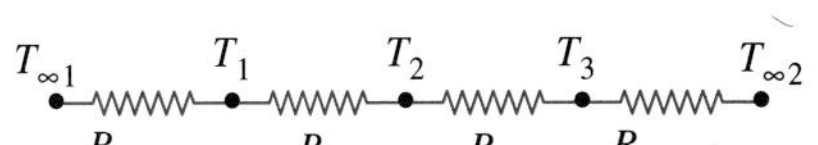

$$\dot{Q} = \frac{T_{\infty 1} - T_1}{R_{\text{conv},1}}$$

$$= \frac{T_{\infty 1} - T_2}{R_{\text{conv},1} + R_1}$$

$$= \frac{T_1 - T_3}{R_1 + R_2}$$

$$= \frac{T_2 - T_3}{R_2}$$

$$= \frac{T_2 - T_{\infty 2}}{R_2 + R_{\text{conv},2}}$$

$$= \cdots$$

FIGURE 3-27

The ratio $\Delta T/R$ across any layer is equal to $\dot{Q}$, which remains constant in one-dimensional steady conduction.

$$\dot{Q} = \frac{T_2 - T_{\infty 2}}{R_2 + R_3 + R_{conv,2}} = \frac{T_2 - T_{\infty 2}}{\frac{\ln(r_3/r_2)}{2\pi L k_2} + \frac{\ln(r_4/r_3)}{2\pi L k_3} + \frac{1}{h_o(2\pi r_4 L)}} \quad (3\text{-}48)$$

Although both relations will give the same result, we prefer the first one since it involves fewer terms and thus less work.

The thermal resistance concept can also be used for *other geometries*, provided that the proper conduction resistances and the proper surface areas in convection resistances are used.

EXAMPLE 3-7 Heat Transfer to a Spherical Container

A 3-m internal diameter spherical tank made of 2-cm-thick stainless steel ($k = 15$ W/m · °C) is used to store iced water at $T_{\infty 1} = 0$°C. The tank is located in a room whose temperature is $T_{\infty 2} = 22$°C. The walls of the room are also at 22°C. The outer surface of the tank is black and heat transfer between the outer surface of the tank and the surroundings is by natural convection and radiation. The convection heat transfer coefficients at the inner and the outer surfaces of the tank are $h_1 = 80$ W/m² · °C and $h_2 = 10$ W/m² · °C, respectively. Determine (a) the rate of heat transfer to the iced water in the tank and (b) the amount of ice at 0°C that melts during a 24-h period.

Solution A spherical container filled with iced water is subjected to convection and radiation heat transfer at its outer surface. The rate of heat transfer and the amount of ice that melts per day are to be determined.

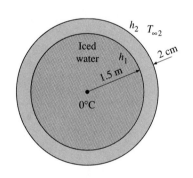

Assumptions 1 Heat transfer is steady since the specified thermal conditions at the boundaries do not change with time. 2 Heat transfer is one-dimensional since there is thermal symmetry about the midpoint. 3 Thermal conductivity is constant.

Properties The thermal conductivity of steel is given to be $k = 15$ W/m · °C. The heat of fusion of water at atmospheric pressure is $h_{if} = 333.7$ kJ/kg. The outer surface of the tank is black and thus its emissivity is $\varepsilon = 1$.

Analysis (a) The thermal resistance network for this problem is given in Figure 3-28. Noting that the inner diameter of the tank is $D_1 = 3$ m and the outer diameter is $D_2 = 3.04$ m, the inner and the outer surface areas of the tank are

$$A_1 = \pi D_1^2 = \pi(3 \text{ m})^2 = 28.3 \text{ m}^2$$
$$A_2 = \pi D_2^2 = \pi(3.04 \text{ m})^2 = 29.0 \text{ m}^2$$

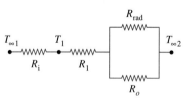

FIGURE 3-28

Schematic for Example 3-7.

Also, the radiation heat transfer coefficient is given by

$$h_{rad} = \varepsilon \sigma (T_2^2 + T_{\infty 2}^2)(T_2 + T_{\infty 2})$$

But we do not know the outer surface temperature T_2 of the tank, and thus we cannot calculate h_{rad}. Therefore, we need to assume a T_2 value now and check the accuracy of this assumption later. We will repeat the calculations if necessary using a revised value for T_2.

We note that T_2 must be between 0°C and 22°C, but it must be closer to 0°C, since the heat transfer coefficient inside the tank is much larger. Taking

$T_2 = 5°C = 278$ K, the radiation heat transfer coefficient is determined to be

$$h_{rad} = (1)(5.67 \times 10^{-8} \text{ W/m}^2 \cdot \text{K}^4)[(295 \text{ K})^2 + (278 \text{ K})^2][(295 + 278) \text{ K}]$$
$$= 5.34 \text{ W/m}^2 \cdot \text{K} = 5.34 \text{ W/m}^2 \cdot °C$$

Then the individual thermal resistances become

$$R_i = R_{conv, 1} = \frac{1}{h_1 A_1} = \frac{1}{(80 \text{ W/m}^2 \cdot °C)(28 \cdot 3 \text{ m}^2)} = 0.000442°C/W$$

$$R_1 = R_{sphere} = \frac{r_2 - r_1}{4\pi k r_1 r_2} = \frac{(1.52 - 1.50) \text{ m}}{4\pi (15 \text{ W/m} \cdot °C)(1.52 \text{ m})(1.50 \text{m})}$$
$$= 0.000047°C/W$$

$$R_o = R_{conv, 2} = \frac{1}{h_2 A_2} = \frac{1}{(10 \text{ W/m}^2 \cdot °C)(29.0 \text{ m}^2)} = 0.00345°C/W$$

$$R_{rad} = \frac{1}{h_{rad} A_2} = \frac{1}{(5.34 \text{ W/m}^2 \cdot °C)(29.0 \text{ m}^2)} = 0.00646°C/W$$

The two parallel resistances R_o and R_{rad} can be replaced by an equivalent resistance R_{equiv} determined from

$$\frac{1}{R_{equiv}} = \frac{1}{R_o} + \frac{1}{R_{rad}} = \frac{1}{0.00345} + \frac{1}{0.00646} = 444.7 \text{ W/°C}$$

which gives

$$R_{equiv} = 0.00225°C/W$$

Now all the resistances are in series, and the total resistance is determined to be

$$R_{total} = R_i + R_1 + R_{equiv} = 0.000442 + 0.000047 + 0.00225 = 0.00274°C/W$$

Then the steady rate of heat transfer to the iced water becomes

$$\dot{Q} = \frac{T_{\infty 2} - T_{\infty 1}}{R_{total}} = \frac{(22 - 0)°C}{0.00274°C/W} = \textbf{8029 W} \qquad (\text{or } \dot{Q} = 8.027 \text{ kJ/s}).$$

To check the validity of our original assumption, we now determine the outer surface temperature from

$$\dot{Q} = \frac{T_\infty - T_2}{R_{equiv}} \longrightarrow T_2 = T_\infty - \dot{Q} R_{equiv}$$
$$= 22°C - (8029 \text{ W})(0.00225°C/W) = 4°C$$

which is sufficiently close to the 5°C assumed in the determination of the radiation heat transfer coefficient. Therefore, there is no need to repeat the calculations using 4°C for T_2.

(b) The total amount of heat transfer during a 24-h period is

$$Q = \dot{Q} \Delta t = (8.029 \text{ kJ/s})(24 \times 3600 \text{ s}) = 673,700 \text{ kJ}$$

Noting that it takes 333.7 kJ of energy to melt 1 kg of ice at 0°C, the amount of ice which will melt during a 24-h period is

$$m_{ice} = \frac{Q}{h_{if}} = \frac{673,700 \text{ kJ}}{333.7 \text{ kJ/kg}} = \textbf{2079 kg}$$

Therefore, about 2 metric tons of ice will melt in the tank every day.

Discussion An easier way to deal with combined convection and radiation at a surface when the surrounding medium and surfaces are at the same temperature is to add the radiation and convection heat transfer coefficients and to treat the result as the convection heat transfer coefficient. That is, to take $h = 10 + 5.34 = 15.34$ W/m² · °C in this case. This way, we can ignore radiation since its contribution is accounted for in the convection heat transfer coefficient. The convection resistance of the outer surface in this case would be

$$R_{combined} = \frac{1}{h_{combined} A_2} = \frac{1}{(15.34 \text{ W/m}^2 \cdot °C)(29.0 \text{ m}^2)} = 0.00225°C/W$$

which is identical to the value obtained for the equivalent resistance for the parallel convection and the radiation resistances.

EXAMPLE 3-8   Heat Loss through an Insulated Steam Pipe

Steam at $T_{\infty 1} = 320°C$ flows in a cast iron pipe ($k = 80$ W/m · °C) whose inner and outer diameters are $D_1 = 5$ cm and $D_2 = 5.5$ cm, respectively. The pipe is covered with 3-cm-thick glass wool insulation with $k = 0.05$ W/m · °C. Heat is lost to the surroundings at $T_\infty = 5°C$ by natural convection and radiation, with a combined heat transfer coefficient of $h_2 = 18$ W/m² · °C. Taking the heat transfer coefficient inside the pipe to be $h_1 = 60$ W/m² · °C, determine the rate of heat loss from the steam per unit length of the pipe. Also determine the temperature drops across the pipe shell and the insulation.

Solution A steam pipe covered with glass wool insulation is subjected to convection on its surfaces. The rate of heat transfer per unit length and the temperature drops across the pipe and the insulation are to be determined.

FIGURE 3-29

Schematic for Example 3-8.

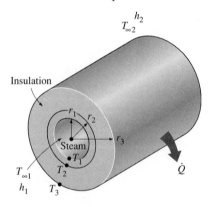

Assumptions **1** Heat transfer is steady since there is no indication of any change with time. **2** Heat transfer is one-dimensional since there is thermal symmetry about the centerline and no variation in the axial direction. **3** Thermal conductivities are constant. **4** The thermal contact resistance at the interface is negligible.

Properties The thermal conductivities are given to be $k = 80$ W/m · °C for cast iron and $k = 0.05$ W/m · °C for glass wool insulation.

Analysis The thermal resistance network for this problem involves four resistances in series and is given in Figure 3-29. Taking $L = 1$ m, the areas of the surfaces exposed to convection are determined to be

$$A_1 = 2\pi r_1 L = 2\pi(0.025 \text{ m})(1 \text{ m}) = 0.157 \text{ m}^2$$
$$A_3 = 2\pi r_3 L = 2\pi(0.0575 \text{ m})(1\text{m}) = 0.361 \text{ m}^2$$

Then the individual thermal resistances become

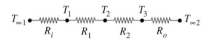

$$R_i = R_{conv,1} = \frac{1}{h_1 A} = \frac{1}{(60 \text{ W/m}^2 \cdot °C)(0.157 \text{ m}^2)} = 0.106°C/W$$

$$R_1 = R_{pipe} = \frac{\ln(r_2/r_1)}{2\pi k_1 L} = \frac{\ln(2.75/2.5)}{2\pi(80 \text{ W/m} \cdot °C)(1 \text{ m})} = 0.0002°C/W$$

$$R_2 = R_{insulation} = \frac{\ln(r_3/r_2)}{2\pi k_2 L} = \frac{\ln(5.75/2.75)}{2\pi(0.05 \text{ W/m} \cdot °C)(1 \text{ m})} = 2.35°C/W$$

$$R_o = R_{conv,2} = \frac{1}{h_2 A_3} = \frac{1}{(18 \text{ W/m}^2 \cdot °C)(0.361 \text{ m}^2)} = 0.154°C/W$$

Noting that all resistances are in series, the total resistance is determined to be

$$R_{total} = R_i + R_1 + R_2 + R_o = 0.106 + 0.0002 + 2.35 + 0.154 = 2.61°C/W$$

Then the steady rate of heat loss from the steam becomes

$$\dot{Q} = \frac{T_{\infty 1} - T_{\infty 2}}{R_{total}} = \frac{(320 - 5)°C}{2.61°C/W} = \textbf{120.7 W} \qquad \text{(per m pipe length)}$$

The heat loss for a given pipe length can be determined by multiplying the above quantity by the pipe length L.

The temperature drops across the pipe and the insulation are determined from Equation 3-17 to be

$$\Delta T_{pipe} = \dot{Q} R_{pipe} = (120.7 \text{ W})(0.0002°C/W) = \textbf{0.02°C}$$

$$\Delta T_{insulation} = \dot{Q} R_{insulation} = (120.7 \text{ W})(2.35°C/W) = \textbf{284°C}$$

That is, the temperatures between the inner and the outer surfaces of the pipe differ by 0.02°C, whereas the temperatures between the inner and the outer surfaces of the insulation differ by 284°C.

Discussion Note that the thermal resistance of the pipe is too small relative to the other resistances and can be neglected without causing any significant error. Also note that the temperature drop across the pipe is practically zero, and thus the pipe can be assumed to be isothermal. The resistance to heat flow in insulated pipes is primarily due to the insulation.

3-5 ■ CRITICAL RADIUS OF INSULATION

We know that adding more insulation to a wall or to the attic always decreases heat transfer. The thicker the insulation, the lower the heat transfer rate. This is expected, since the heat transfer area A is constant, and adding insulation always increases the thermal resistance of the wall without affecting the convection resistance.

Adding insulation to a cylindrical pipe or a spherical shell, however, is a different matter. The additional insulation increases the conduction resistance of the insulation layer but decreases the convection resistance of the surface because of the increase in the outer surface area for convection. The heat transfer from the pipe may increase or decrease, depending on which effect dominates.

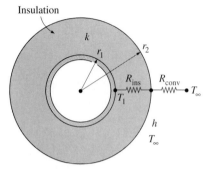

Insulation

FIGURE 3-30

An insulated cylindrical pipe exposed to convection from the outer surface and the thermal resistance network associated with it.

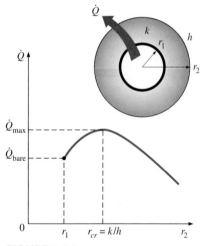

FIGURE 3-31

Consider a cylindrical pipe of outer radius r_1 whose outer surface temperature T_1 is maintained constant (Fig. 3-30). The pipe is now insulated with a material whose thermal conductivity is k and outer radius is r_2. Heat is lost from the pipe to the surrounding medium at temperature T_∞, with a convection heat transfer coefficient h. The rate of heat transfer from the insulated pipe to the surrounding air can be expressed as (Fig. 3-31)

$$\dot{Q} = \frac{T_1 - T_\infty}{R_{ins} + R_{conv}} = \frac{T_1 - T_\infty}{\dfrac{\ln(r_2/r_1)}{2\pi Lk} + \dfrac{1}{h(2\pi r_2 L)}} \qquad (3\text{-}49)$$

The variation of $\dot{Q}$ with the outer radius of the insulation r_2 is plotted in Figure 3-31. The value of r_2 at which $\dot{Q}$ reaches a maximum is determined from the requirement that $d\dot{Q}/dr_2 = 0$ (zero slope). Performing the differentiation and solving for r_2 yields the **critical radius of insulation** for a cylindrical body to be

$$r_{cr,\,cylinder} = \frac{k}{h} \qquad \text{(m)} \qquad (3\text{-}50)$$

Note that the critical radius of insulation depends on the thermal conductivity of the insulation k and the external convection heat transfer coefficient h. The rate of heat transfer from the cylinder increases with the addition of insulation for $r_2 < r_{cr}$, reaches a maximum when $r_2 = r_{cr}$, and starts to decrease for $r_2 > r_{cr}$. Thus, insulating the pipe may actually increase the rate of heat transfer from the pipe instead of decreasing it when $r_2 < r_{cr}$.

The important question to answer at this point is whether we need to be concerned about the critical radius of insulation when insulating hot water pipes or even hot water tanks. Should we always check and make sure that the outer radius of insulation exceeds the critical radius before we install any insulation? Probably not, as explained below.

The value of the critical radius r_{cr} will be the largest when k is large and h is small. Noting that the lowest value of h encountered in practice is about 5 W/m² · °C for the case of natural convection of gases, and that the thermal conductivity of common insulating materials is about 0.05 W/m² · °C, the largest value of the critical radius we are likely to encounter is

$$r_{cr,\,max} = \frac{k_{max,\,insulation}}{h_{min}} \approx \frac{0.05 \text{ W/m} \cdot °\text{C}}{5 \text{ W/m}^2 \cdot °\text{C}} = 0.01 \text{ m} = 1 \text{ cm}$$

This value would be even smaller when the radiation effects are considered. The critical radius would be much less in forced convection, often less than 1 mm, because of much larger h values associated with forced convection. Therefore, we can insulate hot water or steam pipes freely without worrying about the possibility of increasing the heat transfer by insulating the pipes.

The radius of electric wires may be smaller than the critical radius. Therefore, the plastic electrical insulation may actually *enhance* the heat transfer from electric wires and thus keep their steady operating temperatures at lower and thus safer levels.

The discussions above can be repeated for a sphere, and it can be shown in a similar manner that the critical radius of insulation for a spherical shell is

$$r_{cr, sphere} = \frac{2k}{h} \qquad (3\text{-}51)$$

where k is the thermal conductivity of the insulation and h is the convection heat transfer coefficient on the outer surface.

EXAMPLE 3-9 Heat Loss from an Insulated Electric Wire

A 3-mm-diameter and 5-m-long electric wire is tightly wrapped with a 2-mm-thick plastic cover whose thermal conductivity is $k = 0.15$ W/m·°C. Electrical measurements indicate that a current of 10 A passes through the wire and there is a voltage drop of 8 V along the wire. If the insulated wire is exposed to a medium at $T_\infty = 30$°C with a heat transfer coefficient of $h = 12$ W/m^2·°C, determine the temperature at the interface of the wire and the plastic cover in steady operation. Also determine whether doubling the thickness of the plastic cover will increase or decrease this interface temperature.

Solution An electric wire is tightly wrapped with a plastic cover. The interface temperature and the effect of doubling the thickness of the plastic cover on the interface temperature are to be determined.

Assumptions **1** Heat transfer is steady since there is no indication of any change with time. **2** Heat transfer is one-dimensional since there is thermal symmetry about the centerline and no variation in the axial direction. **3** Thermal conductivities are constant. **4** The thermal contact resistance at the interface is negligible. **5** Heat transfer coefficient incorporates the radiation effects, if any.

Properties The thermal conductivity is given to be $k = 0.15$ W/m·°C.

Analysis Heat is generated in the wire and its temperature rises as a result of resistance heating. We assume heat is generated uniformly throughout the wire and is transferred to the surrounding medium in the radial direction. In steady operation, the rate of heat transfer becomes equal to the heat generated within the wire, which is determined from

$$\dot{Q} = \dot{W}_e = VI = (8 \text{ V})(10 \text{ A}) = 80 \text{ W}$$

The thermal resistance network for this problem involves a conduction resistance for the plastic cover and a convection resistance for the outer surface in series, as shown in Figure 3-32. The values of these two resistances are determined to be

$$A_2 = (2\pi r_2)L = 2\pi(0.0035 \text{ m})(5 \text{ m}) = 0.110 \text{ m}^2$$

$$R_{conv} = \frac{1}{hA_2} = \frac{1}{(12 \text{ W/m}^2 \cdot °\text{C})(0.110 \text{ m}^2)} = 0.76°\text{C/W}$$

$$R_{plastic} = \frac{\ln(r_2/r_1)}{2\pi kL} = \frac{\ln(3.5/1.5)}{2\pi(0.15 \text{ W/m} \cdot °\text{C})(5 \text{ m})} = 0.18°\text{C/W}$$

FIGURE 3-32
Schematic for Example 3-9.

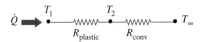

and therefore

$$R_{\text{total}} = R_{\text{plastic}} + R_{\text{conv}} = 0.76 + 0.18 = 0.94°C/W$$

Then the interface temperature can be determined from

$$\dot{Q} = \frac{T_1 - T_\infty}{R_{\text{total}}} \quad \longrightarrow \quad T_1 = T_\infty + \dot{Q}R_{\text{total}}$$

$$= 30°C + (80 \text{ W})(0.94°C/W) = \textbf{105°C}$$

Note that we did not involve the electrical wire directly in the thermal resistance network, since the wire involves heat generation.

To answer the second part of the question, we need to know the critical radius of insulation of the plastic cover. It is determined from Equation 3-50 to be

$$r_{\text{cr}} = \frac{k}{h} = \frac{0.15 \text{ W/m} \cdot °C}{12 \text{ W/m}^2 \cdot °C} = 0.0125 \text{ m} = 12.5 \text{ mm}$$

which is larger than the radius of the plastic cover. Therefore, increasing the thickness of the plastic cover will *enhance* heat transfer until the outer radius of the cover reaches 12.5 mm. As a result, the rate of heat transfer $\dot{Q}$ will *increase* when the interface temperature T_1 is held constant, or T_1 will *decrease* when $\dot{Q}$ is held constant, which is the case here.

Discussion It can be shown by repeating the calculations above for a 4-mm-thick plastic cover that the interface temperature drops to 90.6°C when the thickness of the plastic cover is doubled. It can also be shown in a similar manner that the interface reaches a minimum temperature of 83°C when the outer radius of the plastic cover equals the critical radius.

3-6 ■ THERMAL INSULATION

Thermal insulations are *materials or combinations of materials that are used primarily to provide resistance to heat flow* (Fig. 3-33). You are probably familiar with several kinds of insulation available in the market. Most insulations are heterogeneous materials made of low thermal conductivity materials, and they involve air pockets. This is not surprising since air has one of the lowest thermal conductivities and is readily available. The *Styrofoam* commonly used as a packaging material for TVs, VCRs, computers, and just about anything because of its light weight is also an excellent insulator.

Temperature difference is the driving force for heat flow, and the greater the temperature difference, the larger the rate of heat transfer. We can slow down the heat flow between two mediums at different temperatures by putting "barriers" on the path of heat flow. Thermal insulations serve as such barriers, and they play a major role in the design and manufacture of all energy-efficient devices or systems, and they are usually the cornerstone of all energy conservation projects. A 1991 Drexel University study of the energy-intensive U.S. industries revealed that insulation saves the U.S. industry nearly 2 billion barrels of oil per year, valued at $60 billion a year in energy costs, and more can be saved by practicing better insulation techniques and retrofitting the older industrial facilities.

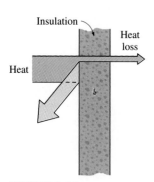

FIGURE 3-33

Thermal insulation retards heat transfer by acting as a barrier in the path of heat flow.

Heat is generated in *furnaces* or *heaters* by burning a fuel such as coal, oil, or natural gas or by passing electric current through a *resistance heater.* Electricity is rarely used for heating purposes since its unit cost is much higher. The heat generated is absorbed by the medium in the furnace and its surfaces, causing a temperature rise above the ambient temperature. This temperature difference drives heat transfer from the hot medium to the ambient, and insulation reduces the amount of heat loss and thus saves fuel and money. Therefore, insulation *pays for itself* from the energy it saves. Insulating properly requires a one-time capital investment, but its effects are dramatic and long term. The payback period of insulation is usually under two years. That is, the money insulation saves during the first two years is usually greater than its initial material and installation costs. On a broader perspective, insulation also helps the environment and fights air pollution and the greenhouse effect by reducing the amount of fuel burned and thus the amount of CO_2 and other gases released into the atmosphere (Fig. 3-34).

Saving energy with insulation is not limited to hot surfaces. We can also save energy and money by insulating *cold surfaces* (surfaces whose temperature is below the ambient temperature) such as chilled water lines, cryogenic storage tanks, refrigerated trucks, and air-conditioning ducts. The source of "coldness" is *refrigeration,* which requires energy input, usually electricity. In this case, heat is transferred from the surroundings to the cold surfaces, and the refrigeration unit must now work harder and longer to make up for this heat gain and thus it must consume more electrical energy. A cold canned drink can be kept cold much longer by wrapping it in a blanket. A refrigerator with well-insulated walls will consume much less electricity than a similar refrigerator with little or no insulation. Insulating a house well will result in reduced cooling load, and thus reduced electricity consumption for air-conditioning.

Whether we realize it or not, we have an *intuitive* understanding and appreciation of thermal insulation. As babies we feel much better in our blankies, and as children we know we should wear a sweater or coat when going outside in cold weather (Fig. 3-35). When getting out of a pool after swimming on a windy day, we quickly wrap in a towel to stop shivering. Similarly, early man used animal furs to keep warm and built shelters using mud bricks and wood. Cork was used as a roof covering for centuries. The need for effective thermal insulation became evident with the development of mechanical refrigeration later in the 19th century, and a great deal of work was done at universities and government and private laboratories in the 1910s and 1920s to identify and characterize thermal insulation (Powell, Ref. 22).

Thermal insulation in the form *mud, clay, straw, rags,* and *wood strips* was first used in the 18th century on steam engines to keep workmen from being burned by hot surfaces. As a result, boiler room temperatures dropped and it was noticed that fuel consumption was also reduced. The realization of improved engine efficiency and energy savings prompted the search for materials with improved thermal efficiency. One of the first such material was *mineral wool* insulation, which, like many materials, was discovered by accident. About 1840, an iron producer in Wales aimed a stream of high pressure steam at the slag flowing from a blast furnace, and manufactured mineral

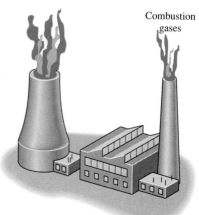

Combustion gases

FIGURE 3-34

Insulation also helps the environment by reducing the amount of fuel burned and the air pollutants released.

FIGURE 3-35

In cold weather, we minimize heat loss from our bodies by putting on thick layers of insulation (coats or furs).

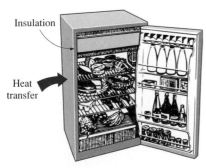

Insulation

Heat
transfer

FIGURE 3-36
The insulation layers in the walls of a refrigerator reduce the amount of heat flow into the refrigerator and thus the running time of the refrigerator, saving electricity.

wool was born. In the early 1860s, this slag wool was a by-product of manufacturing cannons for the Civil War and quickly found its way into many industrial uses. By 1880, builders began installing mineral wool in houses, with one of the most notable applications being General Grant's house. The insulation of this house was described in an article: "it keeps the house cool in summer and warm in winter; it prevents the spread of fire; and it deadens the sound between floors" (Edmunds, Ref. 7). An article published in 1887 in *Scientific American* detailing the benefits of insulating the entire house gave a major boost to the use of insulation in residential buildings.

The energy crisis of the 1970s had a tremendous impact on the public awareness of energy and limited energy reserves and brought an emphasis on *energy conservation*. We have also seen the development of new and more effective insulation materials since then, and a considerable increase in the use of insulation. Thermal insulation is used in more places than you may be aware of. The walls of your house are probably filled with some kind of insulation, and the roof is likely to have a thick layer of insulation. The "thickness" of the walls of your refrigerator is due to the insulation layer sandwiched between two layers of sheet metal (Fig. 3-36). The walls of your range are also insulated to conserve energy, and your hot water tank contains less water than you think because of the 2- to 4-cm-thick insulation in the walls of the tank. Also, your hot water pipe may look much thicker than the cold water pipe because of insulation.

1 Reasons for Insulating

If you examine the engine compartment of your car, you will notice that the firewall between the engine and the passenger compartment as well as the inner surface of the hood are insulated. The reason for insulating the hood is not to conserve the waste heat from the engine but to protect people from burning themselves by touching the hood surface, which will be too hot if not insulated. As this example shows, the use of insulation is not limited to energy conservation. Various reasons for using insulation can be summarized as follows:

- **Energy Conservation** Conserving energy by reducing the rate of heat flow is the primary reason for insulating surfaces. Insulation materials that will perform satisfactorily in the temperature range of $-268°C$ to $1000°C$ ($-450°F$ to $1800°F$) are widely available.
- **Personnel Protection and Comfort** A surface that is too hot poses a danger to people who are working in that area of accidentally touching the hot surface and burning themselves (Fig. 3-37). To prevent this danger and to comply with the OSHA (Occupational Safety and Health Administration) standards, the temperatures of hot surfaces should be reduced to below $60°C$ ($140°F$) by insulating them. Also, the excessive heat coming off the hot surfaces creates an unpleasant environment in which to work, which adversely affects the performance or productivity of the workers, especially in summer months.
- **Maintaining Process Temperature** Some processes in chemical industry are temperature-sensitive, and it may become necessary to

insulate the process tanks and flow sections heavily to maintain the same temperature throughout.

- **Reducing Temperature Variation and Fluctuations** The temperature in an enclosure may vary greatly between the midsection and the edges if the enclosure is not insulated. For example, the temperature near the walls of a poorly insulated house is much lower than the temperature at the midsections. Also, the temperature in an uninsulated enclosure will follow the temperature changes in the environment closely and fluctuate. Insulation minimizes temperature nonuniformity in an enclosure and slows down fluctuations.

- **Condensation and Corrosion Prevention** Water vapor in the air condenses on surfaces whose temperature is below the dew point, and the outer surfaces of the tanks or pipes that contain a cold fluid frequently fall below the dew-point temperature unless they have adequate insulation. The liquid water on exposed surfaces of the metal tanks or pipes may promote corrosion as well as algae growth.

- **Fire Protection** Damage during a fire may be minimized by keeping valuable combustibles in a safety box that is well insulated. Insulation may lower the rate of heat flow to such levels that the temperature in the box never rises to unsafe levels during fire.

- **Freezing Protection** Prolonged exposure to subfreezing temperatures may cause water in pipes or storage vessels to freeze and burst as a result of heat transfer from the water to the cold ambient. The bursting of pipes as a result of freezing can cause considerable damage. Adequate insulation will slow down the heat loss from the water and prevent freezing during limited exposure to subfreezing temperatures. For example, covering vegetables during a cold night will protect them from freezing, and burying water pipes in the ground at a sufficient depth will keep them from freezing during the entire winter. Wearing thick gloves will protect the fingers from possible frostbite. Also, a molten metal or plastic in a container will solidify on the inner surface if the container is not properly insulated.

- **Reducing Noise and Vibration** An added benefit of thermal insulation is its ability to dampen noise and vibrations (Fig. 3-38). The insulation materials differ in their ability to reduce noise and vibration, and the proper kind can be selected if noise reduction is an important consideration.

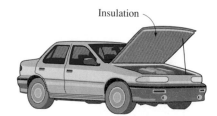

FIGURE 3-37

The hood of the engine compartment of a car is insulated to reduce its temperature and to protect people from burning themselves.

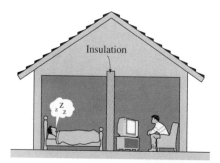

FIGURE 3-38

Insulation materials absorb vibration and sound waves, and are used to minimize sound transmission.

2 Classification of Thermal Insulation

There are numerous types of insulation available in the market, and sometimes selecting the right kind of insulation can become a confusing job. Therefore, it is helpful to classify the insulations in some ways to have a better perspective of them.

Insulation materials can be classified broadly as **capacitive, reflective,** and **resistive** materials. The *thick masonry walls* used in the past relied on heat storage capacity of the walls and can be viewed as capacitive insulation. They slowed down the flow of heat from one medium to another by simply

FIGURE 3-39

A thick layer of ground serves as capacitive insulation because of its large thermal mass.

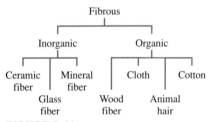

FIGURE 3-40

Classification of fibrous insulations.

FIGURE 3-41

A magnified view of the structures of the air cells in cork insulation.

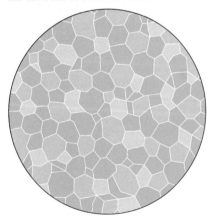

absorbing the heat and then releasing it slowly. Today it is no longer practical to build such massive structures, and thus capacitive insulations are not used in buildings. However, thick layers of earth can be viewed as capacitive insulation and can be used to advantage by burying water mains in the ground to protect them from freezing. Also, the lower part of the north wall of a house can be insulated effectively by covering it with a thick layer of dirt (Fig. 3-39).

Reflective insulation such as a piece of *aluminum foil* placed on a window glass is based on reflecting the incident radiation back by using highly reflective surfaces. Reflective insulations are effective against radiation but not conduction or convection. Therefore, they are mostly used in evacuated spaces. Superinsulations are constructed of several layers of reflective insulations placed in an evacuated enclosure.

When we say insulation, we normally mean *resistive insulation* that is made of a material of low thermal conductivity and offers effective resistance to heat flow despite its small thickness. Insulations exhibit considerable variations in their structure, but they can be classified into four main groups: *fibrous insulation, cellular insulation, granular insulation,* and *reflective insulation.* Each kind is explained below.

• **Fibrous Insulation**   As the name implies, *fibrous insulation* is composed of small-diameter fibers that fill an air space. The fibers can be *organic,* such as wool or other animal hair, cotton, wood, cloth, cane, and dry vegetable fibers, or *inorganic,* such as mineral wool, glass fiber, and ceramic fibers (Fig. 3-40).

    Mineral wool was first obtained in 1836 from volcanic craters in Hawaii, but now it is made from molten slag, rock, or other minerals. Mineral wool insulations are manufactured by bonding mineral wool fibers with a medium or high temperature binder or by enclosing the mineral wool between various facings. Mineral wool insulation is well suited for high temperature applications and can be used at temperatures up to about 1100°C (2000°F). *Glass fiber,* which is also a mineral fiber, was discovered in the late 1930s and found a widespread use as an insulating material during World War II when the first national program to conserve energy was established. Glass fiber insulation is relatively inexpensive and is suitable for use from −30 to 450°C (−20 to 850°F), depending on the internal structure and the binder used. *Ceramic fiber,* which is an alumina-silica compound, can be used at temperatures as high as 1750°C (3200°F).

• **Cellular Insulation**   Cellular insulation is characterized by cellular-like structures with closed cells and is made of cellular materials such as cork, foamed plastics, glass, polystyrene, polyurethane, and other polymers (Fig. 3-41). *Cellular glass* is impermeable and noncombustible and can be used indoors, outdoors, and even underground in the temperature range of −180 to 650°C (−290 to 1200°F). Common *foamed plastics* include polyurethane, polystyrene, and polyisocyanurate, which have upper service temperatures of 105°C (220°F), 135°C

(275°F), and 150°C (300°F), respectively. The use of cork has declined considerably over the years.

- **Granular Insulation** Granular insulations are characterized by small nodules with voids. Calcium silicate, vermiculate, and perlite are the best known granular insulations, and they are noncombustible. *Perlite* is a volcanic rock glass that occurs as small pearl-like masses, and it can be used in the temperature range of 15 to 815°C (60 to 1500°F). *Calcium silicate* reinforced with organic and inorganic fibers is molded into rigid forms and can be used in the temperature range of 15 to 815°C (60 to 1500°F). Calcium silicate may absorb water when it gets wet, but it dries easily without any deterioration in its performance.

- **Reflective Insulation** Reflective insulations are based on reflecting the thermal radiation incident on the surface back by using highly reflective surfaces. Reflective insulation can be used by itself to minimize heat flow by radiation, or it can be used as a covering on the exposed surfaces of resistive insulations to combat both radiation and conduction.

3 Structure and Form of Insulations

Fibrous, cellular, granular, and reflective insulations can be manufactured in a variety of physical forms. Below we describe some common forms encountered in practice.

- **Loose-Fill Insulations** These consist of fibers, powders, or granules that are poured or blown into walls or other cavities (Fig. 3-42).
- **Blankets and Batts** These flexible or semirigid insulations consist of organic and inorganic materials with or without binders. One or both sides of this insulation may have covering and facings such as wiremesh, metal lath, cloth, plastic, paper, or laminated foil (Fig. 3-43).
- **Rigid Insulations** These are preformed during manufacture in the form of rectangular blocks, boards, or sheets at standard dimensions. Insulation for pipes and curved surfaces is supplied in sections for a large range of nominal dimensions.
- **Insulating Cements** These are obtained by mixing a loose insulating material with water or a suitable binder. The prepared insulating cement is then directly applied or blown on a surface while still wet.
- **Formed-in-Place Insulations** A small amount of liquid mixture of insulation under pressure such as polyisocyanurate is poured into the cavity to be insulated or sprayed on the surface to form a rigid foam insulation as it expands and solidifies.
- **Reflective Insulations** These consist of single- or multilayer metallic sheets or rolls with shiny surfaces characterized by their high reflectivity. Some contain air bubbles to retard conduction as well as radiation.

There are a wide variety of insulation materials available in the market, but most are primarily made of fiberglass, mineral wool, polyethylene, foam, or calcium silicate. They come in various trade names such as

FIGURE 3-42

Loose-fill insulation can be poured or blown into place.

FIGURE 3-43

Blanket insulation comes in rolls and can be used to completely wrap surfaces.

Ethafoam Polyethylene Foam Sheeting, Solimide Polimide Foam Sheets, FPC Fiberglass Reinforced Silicone Foam Sheeting, Silicone Sponge Rubber Sheets, fiberglass/mineral wool insulation blankets, wire-reinforced mineral wool insulation, Reflect-All Insulation, granulated bulk mineral wool insulation, cork insulation sheets, foil-faced fiberglass insulation, blended sponge rubber sheeting, and numerous others. The unit costs of various insulations in 1995 dollars for 2.54-cm (1-in)-thick blankets are about $9/m^2$ for foil-faced fiberglass, $12/m^2$ for wire-reinforced mineral wool, and $13/m^2$ for cork.

Today various forms of **fiberglass insulation** are widely used in process industries and heating and air-conditioning applications because of their low cost, light weight, resiliency, and versatility. But they are not suitable for some applications because of their low resistance to moisture and fire and their limited maximum service temperature. Fiberglass insulations come in various forms such as unfaced fiberglass insulation, vinyl-faced fiberglass insulation, foil-faced fiberglass insulation, and fiberglass insulation sheets. The reflective foil-faced fiberglass insulation resists vapor penetration and retards radiation because of the aluminum foil on it and is suitable for use on pipes, ducts, and other surfaces.

Mineral wool is resilient, lightweight, fibrous, wool-like, thermally efficient, fire resistant up to 1100°C (2000°F), and forms a sound barrier. Mineral wool insulation comes in the form of blankets, rolls, or blocks. It is also available in composite forms such as fiberglass/mineral wool insulation blankets, wire-reinforced mineral wool insulation blankets, high temperature mineral wool insulation, and granulated bulk mineral wool insulation. **Calcium silicate** is a solid material that is suitable for use at high temperatures, but it is more expensive. Also, it needs to be cut with a saw during installation, and thus it takes longer to install and there is more waste. A comparison of fiberglass, mineral wool, and calcium silicate insulations is given in Table 3-3.

Cork is obtained from the outer bark of an oak tree, and cork insulation sheets consist of firmly bonded cork particles (Fig. 3-44). Cork has many uses because of its unique cellular structure. A small piece of natural cork contains more than 12 million air cells in a volume of 1 cm^3, which explains its light weight and low thermal conductivity. The important properties of cork are its low thermal conductivity, light weight, buoyancy, high coefficient of friction, resistance to moisture and liquid penetration, resilience, compressibility, and noise and vibration absorption. These desirable properties make it very suitable for use in a wide range of products such as shoe soles, floats and life preservers, gasketing, sealing, padding, sound absorption, and thermal insulation. Cork can be sawed, nailed, and trimmed like wood and can be used in the temperature range from −180°C (−292°F) to 110°C (230°F). Cork was a popular insulation material used on roofs, floors, walls, and many other places. Because of its higher cost, however, it has largely been replaced by fiberglass insulation.

An insulation job may be as easy as installing the insulation in a single step. But it may also involve vapor and air retarder coatings, sealants, jackets

Removed bark

FIGURE 3-44
Cork oak tree and a piece of removed bark 2 to 7 cm thick.

TABLE 3-3   165

Comparison of various aspects of insulations (from Danish, Ref. 6)

Quantity	Calcium silicate (at a density of 11 to 14 lbm/ft³)	Mineral wool (at a density of 11 to 14 lbm/ft³)	Fiberglass (at a density of 3 to 6 lbm/ft³)
Thermal conductivity in Btu/h · ft · °F:			
100°F	0.029	0.028	0.023
200°F	0.036	0.032	0.027
300°F	0.042	0.038	0.033
400°F	0.046	0.044	0.040
500°F	0.050	—	0.048
Service temperature limit*	1200°F	1200°F	850°F
Resistance to moisture damage	Absorbs moisture with structural damage	Absorbs moisture with structural damage	Negligible absorption; dries without damage
Shrinkage	1.5 to 2 percent	4 percent	Negligible
Abuse resistance	High compression resistance; breaks	Low compression resistance; does not recover	Low compression resistance; recovers
Resistance to hostile environment	Good	Good	Good
Fire resistance	Noncombustible	Noncombustible	Noncombustible
Corrosion of metals resistance	Contains inhibitor	Does not cause	Does not cause
Health aspects	Dusty; noncarcinogenic	Dusty; skin irritating; noncarcinogenic	Skin irritating; noncarcinogenic
Breakage resistance	10 to 20 percent loss	Breaks	No breakage loss
Ease of cutting and fitting	Difficult; needs saw	Knife cutting	Knife cutting
Total installed cost	High	High	Low

*May vary greatly depending on the temperature range of the binder used.

and weather coatings, and surface finishing. Attention to detail and specifications during installation will ensure high performance and long life.

4 Superinsulators

You may be tempted to think that the most effective way to reduce heat transfer is to use insulating materials that are known to have very low thermal conductivities such as urethane or rigid foam ($k = 0.026$ W/m · °C) or fiberglass ($k = 0.035$ W/m · °C). After all, they are widely available, inexpensive, and easy to install. Looking at the thermal conductivities of materials, you may also notice that the thermal conductivity of air at room temperature is 0.026 W/m · °C, which is lower than the conductivities of practically all of the ordinary insulating materials. Thus you may think that a layer of enclosed air space is as effective as any of the common insulating materials of the same thickness. Of course, heat transfer through the air will probably be higher than what a pure conduction analysis alone would indicate because of

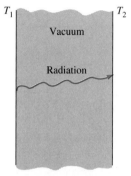

FIGURE 3-45

Evacuating the space between two surfaces completely eliminates heat transfer by conduction or convection but leaves the door wide open for radiation.

FIGURE 3-46

Superinsulators are built by closely packing layers of highly reflective thin metal sheets and evacuating the space between them.

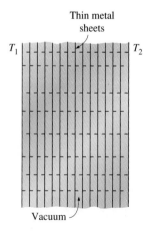

the natural convection currents that are likely to occur in the air layer. Besides, air is transparent to radiation, and thus heat will also be lost from the surface by radiation. The thermal conductivity of air is practically independent of pressure unless the pressure is extremely high or extremely low. Therefore, we can reduce the thermal conductivity of air and thus the conduction heat transfer through the air by evacuating the air space. In the limiting case of absolute vacuum, the thermal conductivity will be zero since there will be no particles in this case to "conduct" heat from one surface to the other, and thus the conduction heat transfer will be zero. Noting that the thermal conductivity cannot be negative, an absolute vacuum must be the ultimate insulator, right? Well, not quite.

The purpose of insulation is to reduce "total" heat transfer from a surface, not just conduction. A vacuum totally eliminates conduction but offers zero resistance to radiation, whose magnitude can be comparable to conduction or natural convection in gases (Fig. 3-45). Thus, a vacuum is no more effective in reducing heat transfer than sealing off one of the lanes of a two-lane road is in reducing the flow of traffic in a one-way road.

Insulation against radiation heat transfer between two surfaces is achieved by placing "barriers" between the two surfaces, which are highly reflective thin metal sheets. Radiation heat transfer between two surfaces is inversely proportional to the number of such sheets placed between the surfaces. Very effective insulations are obtained by using closely packed layers of highly reflective thin metal sheets such as aluminum foil (usually 25 sheets per cm) separated by fibers made of insulating material such as glass fiber (Fig. 3-46). Further, the space between the layers is evacuated to form a vacuum under 0.000001 atm pressure to minimize conduction or convection heat transfer through the air space between the layers. The result is an insulating material whose apparent thermal conductivity is below 2×10^{-5} W/m · °C, which is one thousand times less than the conductivity of air or any common insulating material. These specially built insulators are called **superinsulators,** and they are commonly used in space applications and cryogenics, which is the branch of heat transfer dealing with temperatures below 100 K (-173°C) such as those encountered in the liquefaction, storage, and transportation of gases, with helium, hydrogen, nitrogen, and oxygen being the most common ones.

5 Properties of Insulation

By definition, insulations are materials that offer resistance to heat flow, and thus you probably think that the best insulations are the ones with the lowest thermal conductivities. Well, having a low thermal conductivity is certainly desirable, but an application may require the insulation to possess certain properties before it can even be considered for the job. Therefore, we may need to examine a whole range of *characteristics* when selecting a suitable insulation. Below we give an overview of the relevant properties of insulation.

• **Thermal Conductivity**   This is a measure of the insulating ability of a material. The lower the thermal conductivity, the more effective the insulation for a specified thickness. Other things being equal, the best insulation is the one with the lowest conductivity. Some handbooks list the *R*-value of insulation or the thermal resistance, which is the inverse of thermal conductivity for a flat layer of unit thickness. The higher the *R*-value, the more effective the insulation. The thermal conductivity of insulation, in general, increases with temperature (Fig. 3-47). Therefore, results obtained using thermal conductivity values at room temperature for insulations utilized at high temperatures may be grossly misleading.

Most insulation materials have a porous structure, and they consist of a combination of solid matter and cavities filled with air or another gas. The properties of such loose insulations depend on the volumetric fraction of air space. Heat transfer through such insulations is by *conduction* through the solid material and *conduction* or *convection* through the air space as well as *radiation*. Insulations are *inhomogeneous* materials, and it is more proper to speak of **"effective"** or **"apparent" thermal conductivity** of insulation to account for all mechanisms of heat transfer (Fig. 3-48). The thermal conductivity of an insulation material depends on *the form; the density, diameter, and arrangement of the fibers or cells; the thickness of the cell walls; the radiation properties of the cell surfaces; the type of bonding material; the facing materials;* and *the type of gas contained in the cells.* For a specified insulation, it also varies with temperature. Radiation heat transfer plays a larger role in low-density fibrous and cellular insulations.

At temperatures below 200 to 300°C, the primary mode of heat transfer through the air (or other gas) cells is by conduction. Therefore, the conductivity of the insulation can be reduced by replacing the air in the cells by a gas with smaller conductivity. For example, the apparent thermal conductivity of an insulation can be reduced by 50 percent by replacing the air in the cells by a fluorinated hydrocarbon gas. It is difficult to retain this improved performance, however, since the air eventually diffuses into the cells and drives the gas out. The diffusion of air can be minimized by covering the insulation with continuous sheets that are impermeable to gases.

The *unit* of thermal conductivity in SI is W/m · °C, but it is more instructive to view it as W · m/m² · °C, which has the interpretation that thermal conductivity is the rate of heat transfer through the unit thickness of a material per unit surface area per unit temperature difference. Alternately, we can express it as W/m² · °C/m, which is the rate of heat transfer per unit surface area per unit temperature gradient. In the English system, the thermal conductivity is normally given in Btu/h · ft · °F, but it is also expressed in Btu · in/h · ft² · °F with the conversion relation 1 Btu/h · ft · °F = 12 Btu · in/h · ft² · °F since 1 ft = 12 in. Thermal conductivities in SI and the English system are related to each other by

$$1 \text{ W/m} \cdot °\text{C} = 0.5778 \text{ Btu/h} \cdot \text{ft} \cdot °\text{F}$$

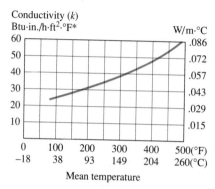

* Divide by 12 to convert to Btu/h·ft·°F

FIGURE 3-47

Variation of the thermal conductivity of rigid fiberglass pipe insulation (courtesy of Manville Mechanical Insulations, Denver, CO).

FIGURE 3-48

The *apparent thermal conductivity* of an insulating material accounts for the conduction through the solid material and conduction or convection through the air space as well as radiation.

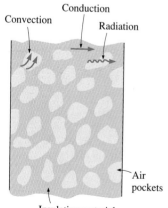

Apparent thermal
conductivity
W/m·°C

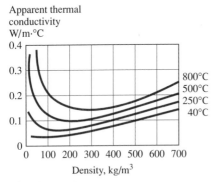

FIGURE 3-49

The variation of the apparent thermal
conductivity of some insulations with
density (from ASHRAE *Handbook of
Fundamentals,* p. 20.4, Fig. 1).

FIGURE 3-50

Typical variation of the apparent
thermal conductivity of fibrous
insulations with fiber diameter and
density (from Lotz, Ref. 19).

Note: Heat flow is perpendicular
to the fiber orientation at 24°C
mean temperature.

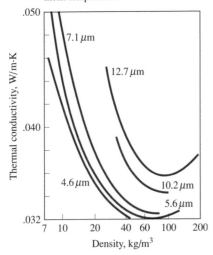

Therefore, the thermal conductivity values in English units are roughly half the values in SI units.

- **Density** The density of insulation is commonly listed in property tables since it affects weight as well as thermal properties. The variation of the apparent thermal conductivity of some insulation materials is shown in Figure 3-49. Note that the thermal conductivity, in general, decreases with density, reaches a minimum, and then increases. The value of density that corresponds to minimum thermal conductivity depends on the temperature, the type of material, the size of the fibers or cores, and their orientation. The variation of apparent thermal conductivity with fiber diameter and density is shown in Figure 3-50.

- **Service Temperature** Insulations are designed to perform in a specified range of temperature, and a surface temperature that is too high or too low will limit the choices considerably. For example, a surface temperature above 450°C (or 850°F) will eliminate all fiberglass insulations from consideration and will concentrate the search on mineral wool and other refractory insulations. Polyethylene pipe insulation that has a maximum operating temperature of 80°C, for example, cannot be used to insulate a steam pipe at 120°C.

- **Form** Insulations come in various forms such as rigid boards, preformed rigid structures, flexible blankets, loose-fill fibers or powders that can be poured in, cement mixtures, sprays, and formed-in-place expandable pellets or liquids. An irregular surface will rule out the use of rigid board insulation, and formed-in-place insulations may be the best choice for narrow cavities in hard-to-reach places.

- **Structural Strength** Insulations that are used in self-supporting partitions or other load-bearing positions such as floors and stepped-on pipes must have sufficient compressive strength or puncture resistance to withstand applied forces without suffering permanent deformation or even destruction. Some insulation materials are sandwiched between metal, plastic, or plywood layers to maximize their structural strength at the expense of added cost and bulk (Fig. 3-51). In applications with large temperature swings, it may also be desirable to have a low coefficient of thermal expansion.

- **Surface Reflectivity and Emissivity** When there is a large temperature difference between the exposed surface of the insulation and the surrounding surfaces, heat transfer by radiation becomes important. In such cases, radiation can be minimized by using an insulation covered with a surface that has a high reflectivity and low emissivity. An ordinary surface has a reflectivity of 0.1 and emissivity of 0.9, but a surface covered with a shiny metal foil may have an emissivity of 0.1 and reflectivity of 0.9. To impede radiation further, multiple layers of reflective materials separated by air space or vacuum can be used.

- **Health and Safety** Insulations should not pose a serious health or safety hazard to people exposed to them during transportation, storage, installation, or service. Insulations containing asbestos must be avoided because they have been shown to be carcinogenic. Insulations that cause skin irritations on some people should be handled with care.

Insulations are rated for combustibility, flame spread, and the amount of harmful gases released during fires, and they must meet the federal, state, or local codes for fire safety.

- **Acoustics** Insulation materials also have varying degrees of sound absorption capabilities, depending on their densities, physical structures, and surface porosities. Sound-absorbing insulations are best suited for interior walls and floors. Sound transmission in a stud wall can be reduced considerably by placing a sound absorption thermal insulation in the wall cavity. Flexible, semirigid, and formed-in-place fibrous insulations of various sizes are widely available for sound insulation.

- **Corrosiveness** Some insulations are corrosive, and they may corrode the metal surfaces on which they are installed. Therefore, corrosiveness of insulation materials is commonly listed in property tables. Mineral wool and fiberglass insulations do not cause corrosion, but some others may.

- **Water Vapor Transmission** The porous structure of insulation materials makes them prone to diffusion of water vapor and other gases, and insulations differ greatly with respect to moisture migration, moisture damage, and the amount of moisture or water retention. The thermal conductivity of water is about 25 times that of air, and the moisture that condenses in the insulation and replaces the air spaces in the cells may drastically reduce its insulating ability. Therefore, in applications involving temperatures below the dew point, such as chilled water pipelines and air-conditioning ducts, moisture condensation is a serious concern. Some insulations may be soaked in water without suffering any permanent damage, while others may be destroyed by it. In the first case, the insulation can simply be dried by removing the jacketing, whereas in the second case it must be replaced. To prevent moisture migration, insulations are often sealed with vapor retarders, which are sufficiently thick plastic sheets impermeable to water vapor (Fig. 3-52). Also, a weather protective jacket is commonly used in outdoor applications.

Other properties of insulation that can be important are resistance to settling, permanence, ease of handling, finishing, dimensional uniformity, resistance to chemical reactions and changes, reusability, the sizes and thicknesses available, and cost.

6   The *R*-value of Insulation

The effectiveness of insulation materials is given by some manufacturers in terms of their **R-value,** which is the *thermal resistance* of the material *per unit surface area*. For *flat insulation* the R-value is obtained by simply dividing the thickness of the insulation by its thermal conductivity. That is,

$$R\text{-value} = \frac{L}{k} \quad \text{(flat insulation)} \qquad (3\text{-}52)$$

FIGURE 3-51

Pipe-in-pipe insulation is obtained by dispensing into the annulus between two pipes a liquid that expands and solidifies, filling the entire space.

FIGURE 3-52

The migration and condensation of moisture in insulation should be prevented by covering the insulation with vapor retarders.

where L is the thickness and k is the thermal conductivity of the material. Note that doubling the thickness L doubles the R-value of flat insulation. For *pipe insulation*, the R-value is determined using the thermal resistance relation from

$$R\text{-value} = \frac{r_2}{k} \ln \frac{r_2}{r_1} \qquad \text{(pipe insulation)} \qquad (3\text{-}53)$$

where r_1 is the inside radius of insulation and r_2 is the outside radius of insulation. Once the R-value is available, the rate of heat transfer through the insulation can be determined from

$$\dot{Q} = \frac{\Delta T}{R\text{-value}} \times \text{Area} \qquad (3\text{-}54)$$

where ΔT is the temperature difference across the insulation and Area is the outer surface area for a cylinder.

In the United States, the R-values of insulation are expressed without any units, such as R-19 and R-30. These R-values are obtained by dividing the thickness of the material in *feet* by its thermal conductivity in the unit Btu/h · ft · °F so that the R-values actually have the unit h · ft^2 · °F/Btu. For example, the R-value of 6-in.-thick glass fiber insulation whose thermal conductivity is 0.025 Btu/h · ft · °F is (Fig. 3-53)

$$R\text{-value} = \frac{L}{k} = \frac{0.5 \text{ ft}}{0.025 \text{ Btu/h} \cdot \text{ft} \cdot \text{°F}} = 20 \text{ h} \cdot \text{ft}^2 \cdot \text{°F/Btu}$$

Thus, this 6-in.-thick glass fiber insulation would be referred to as R-20 insulation by the builders. The unit of R-value is m^2 · °C/W in SI units, with the conversion relation 1 m^2 · °C/W = 5.678 h · ft^2 · °F/Btu. Therefore, a small R-value in SI corresponds to a large R-value in English units.

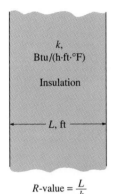

k,
Btu/(h·ft·°F)

Insulation

L, ft

$R\text{-value} = \frac{L}{k}$

FIGURE 3-53
The R-value of an insulating material is simply the ratio of the thickness of the material to its thermal conductivity in proper units.

7 Selecting the Proper Insulation

There are several considerations in the selection of insulation for an application, and the first step in selecting the right insulation is *understanding* the nature of the application. Once the requirements of the application are determined, the task becomes selecting the most economical insulation among those that meet these requirements. Here are the main considerations in the selection of insulation:

- **Purpose** First we need to know the primary reason for insulating a surface. Most likely it is to conserve energy, but the purpose may also be to reduce the surface temperature for safety reasons, to prevent freezing or solidification on surfaces, to protect from fire, to prevent condensation on surfaces, to prevent sound transmission, or to control process temperature by minimizing temperature variations or fluctuations, among other things. If the purpose is to reduce the temperature of hot surfaces below 60°C (140°F) to meet a safety requirement, for example, we will use as much insulation as it takes to satisfy the requirement whether it is economical to do so or not. If the reason for insulating is energy conservation, we should use the kind and thickness of insulation that results in the best return on investment. If there are two or more reasons for insulating, the selection must be made on the basis of *most restrictive* reason. For example, if insulation is to provide personnel protection, requiring an insulation thickness of 2 cm, and to conserve energy, requiring a thickness of 3 cm, installing 3-cm-thick insulation will satisfy both criteria since it is greater than what is needed for personnel protection.

- **Special Requirements**   Each insulation job has its own requirements, and they must be identified before a selection can be made. For example, rigid boards cannot be used for pipe insulation. Here the choice is between flexible insulation and preformed rigid pipe insulation at a fixed diameter. Formed-in-place or poured-in types come to mind first for hard-to-reach places, especially when it is desired that the insulation form a bond with the surface and carry some load.

- **Environment**   The environment to which the insulation will be exposed may severely limit the choices. The insulation for the underground steam pipes will be quite different than the insulation for the steam pipes hanging inside the production facilities (Fig. 3-54). An insulation that absorbs vapors should not be used in a facility where flammable vapors are released. An insulation that settles when agitated should not be used in places that may be subjected to vibration. An insulation that is likely to be stepped on should have a high compressive strength. For moist environments, the choice is limited to insulations with a high resistance to moisture penetration such as cork, foam, and polyethylene insulation.

- **Ease of Handling and Installation**   Some insulations have special requirements during storage prior to installation; others do not. Some can be installed by the plant personnel; others may require specialists. Some permit easy one-step installation; others require cutting, wrapping, painting, and so forth, and are more involved.

- **Cost**   As you might expect, the final selection among the insulations that meet the requirements is made on the basis of lowest cost. Once the choices are narrowed to a few, an economic analysis is performed to identify the one with the minimum total cost. The thickness of insulation is also determined on the basis of minimum cost, as discussed below.

Other considerations for the selection of insulation include freedom from harmful chemicals (such as asbestos, which is no longer used because it is carcinogenic) and local availability.

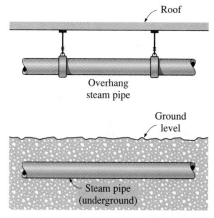

FIGURE 3-54

The environment in which the insulation will be used is an important consideration in the selection of insulation.

8   Optimum Thickness of Insulation

It should be realized that insulation does not eliminate heat transfer; it merely reduces it. The thicker the insulation, the lower the rate of heat transfer but also the higher the cost of insulation. Therefore, there should be an *optimum*

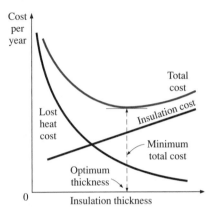

FIGURE 3-55

Determination of the optimum
thickness of insulation on the basis of
minimum total cost.

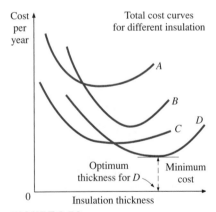

FIGURE 3-56

Determination of the most economical
type of insulation and its
optimum thickness.

thickness of insulation that corresponds to a minimum combined cost of insulation and heat lost. The determination of the optimum thickness of insulation is illustrated in Figure 3-55. Notice that the cost of insulation increases roughly linearly with thickness while the cost of heat loss decreases exponentially. The total cost, which is the sum of the insulation cost and the lost heat cost, decreases first, reaches a minimum, and then increases. The thickness corresponding to the minimum total cost is the optimum thickness of insulation, and this is the recommended thickness of insulation to be installed.

If you are mathematically inclined, you can determine the *optimum thickness* by obtaining an expression for the *total cost,* which is the sum of the expressions for the lost heat cost and insulation cost as a function of thickness; *differentiating* the total cost expression with respect to the thickness; and *setting* it equal to zero. The thickness value satisfying the resulting equation is the optimum thickness. The cost values can be determined from an annualized lifetime analysis or simply from the requirement that the insulation pay for itself within two or three years. Note that the optimum thickness of insulation depends on the fuel cost, and the higher the fuel cost, the larger the optimum thickness of insulation. Considering that insulation will be in service for many years and the fuel prices are likely to escalate, a reasonable increase in fuel prices must be assumed in calculations. Otherwise, what is optimum insulation today will be inadequate insulation in the years to come, and we may have to face the possibility of costly retrofitting projects. This is what happened in the 1970s and 1980s to insulations installed in the 1960s.

The discussion above on optimum thickness is valid when the type and manufacturer of insulation are already selected, and the only thing to be determined is the most economical thickness. But often there are several suitable insulations for a job, and the selection process can be rather confusing since each insulation can have a different thermal conductivity, different installation cost, and different service life. In such cases, a selection can be made by preparing an annualized cost versus thickness chart like Figure 3-56 for each insulation, and determining the one having the *lowest* minimum cost. The insulation with the lowest annual cost is obviously the most economical insulation, and the insulation thickness corresponding to the *minimum total cost* is the *optimum thickness.* When the optimum thickness falls between two commercially available thicknesses, it is a good practice to be conservative and choose the thicker insulation. The extra thickness will provide a little safety cushion for any possible decline in performance over time and will help the environment by reducing the production of greenhouse gases such as CO_2.

The determination of the optimum thickness of insulation requires a heat transfer and economic analysis, which can be tedious and time-consuming. But a selection can be made in a few minutes using the tables and charts prepared by TIMA (Thermal Insulation Manufacturers Association) and member companies. The primary inputs required for using these tables or charts are the operating and ambient temperatures, pipe diameter (in the case of pipe insulation), and the unit fuel cost. Recommended insulation thicknesses for hot surfaces at specified temperatures are given in Table 3-4. Recommended thicknesses of *pipe insulations* as a function of service temperatures are 0.5 to 1 in. for 150°F, 1 to 2 in. for 250°F, 1.5 to 3 in. for 350°F, 2 to 4.5 in. for 450°F, 2.5 to 5.5 in. for 550°F, and 3 to 6 in. for 650°F for nominal pipe diameters of 0.5 to 36 in. The lower recommended insulation thicknesses are for pipes with small diameters, and the larger ones are for pipes with large diameters.

9 Retrofitting Existing Insulation

When plants are first built or buildings are first constructed, they are usually adequately insulated using the standards prevailing at the time of construction. However, what was once adequate insulation may no longer be adequate, and the increases in the cost of energy and the development of cheaper and more effective easy-to-install insulation materials may justify retrofitting those plants or buildings with new insulation.

There are two reasons for considering a retrofitting project: the existing insulation is *no longer adequate* and needs to be upgraded or the insulation is *damaged* and needs to be replaced. Indications of damaged insulation can be *visible,* such as tears, cracks, sags, bare spots, discolorization, and rotting, or *invisible,* such as being saturated with water. An increase in heat rate of a boiler for the same operating conditions or declines in the operating temperatures of processing equipment are also indications of possible damage or deterioration of insulation. A temperature rise on the outer surface of insulation should also alert us that the effectiveness of insulation is declining. As a rule of thumb, temperatures 15°C (or 25°F) above ambient air temperature indicate damaged or inadequate insulation (Danish, Ref. 6). Also, a surface we cannot touch for more than five seconds with a bare hand calls for insulation.

Water droplets forming on air-conditioning ducts or chilled water pipes are indications that the vapor barrier on the existing insulation is damaged and no longer protecting the insulation. Moisture that condenses in insulation diminishes the effectiveness of insulation, and wet insulation must be replaced immediately if it cannot be dried out by removing its protective jacket.

The critical question in retrofitting projects is whether to *add* new insulation on top of the existing insulation or to *replace* the existing insulation entirely with the new one. The answer depends on the situation at hand. When space permits, it may be advisable to install a second layer of insulation over the existing one if it is in good shape. But inadequate space may require the original insulation to be stripped off all surfaces and replaced by the new one, especially when the existing insulation is damaged. New insulation should never be added on existing insulation that is badly deteriorated (Fig. 3-57). The labor cost of removing existing insulation versus fitting the new insulation on top of the existing one should also be considered.

Finally, care should be taken when installing a second layer of insulation over existing insulation that is covered with a facing material. Added insulation may cause the interface temperature of the two insulations to rise above the safe temperature of the cover for hot surfaces, and to drop below the

TABLE 3-4

Recommended insulation thicknesses for flat hot surfaces as a function of surface temperature (from TIMA *Energy Savings Guide*)

Surface temperature	Insulation thickness
150°F (66°C)	2″ (5.1 cm)
250°F (121°C)	3″ (7.6 cm)
350°F (177°C)	4″ (10.2 cm)
550°F (288°C)	6″ (15.2 cm)
750°F (400°C)	9″ (22.9 cm)
950°F (510°C)	10″ (25.44 cm)

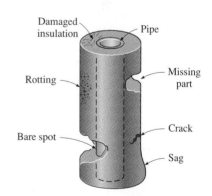

FIGURE 3-57

New insulation should never be added on damaged insulation.

dew-point temperature for cold surfaces. Therefore, it is a good practice to remove the jacketing from existing insulation before adding more insulation.

EXAMPLE 3-10 Effect of Insulation on Surface Temperature

Hot water at $T_i = 120°C$ flows in a stainless steel pipe ($k = 15$ W/m · °C) whose inner diameter is 1.6 cm and thickness is 0.2 cm. The pipe is to be covered with adequate insulation so that the temperature of the outer surface of the insulation does not exceed 40°C when the ambient temperature is $T_o = 25°C$. Taking the heat transfer coefficients inside and outside the pipe to be $h_i = 70$ W/m² · °C and $h_o = 20$ W/m² · °C, respectively, determine the thickness of fiberglass insulation ($k = 0.038$ W/m · °C) that needs to be installed on the pipe.

Solution A steam pipe is to be covered with enough insulation to reduce the exposed surface temperature to 40°C. The thickness of insulation that needs to be installed is to be determined.

Assumptions **1** Heat transfer is steady since there is no indication of any change with time. **2** Heat transfer is one-dimensional since there is thermal symmetry about the centerline and no variation in the axial direction. **3** Thermal conductivities are constant. **4** The thermal contact resistance at the interface is negligible.

Properties The thermal conductivities are given to be $k = 15$ W/m · °C for the steel pipe and $k = 0.038$ W/m · °C for fiberglass insulation.

Analysis The thermal resistance network for this problem involves four resistances in series and is given in Figure 3-58. The inner radius of the pipe is $r_1 = 0.8$ cm and the outer radius of the pipe and thus the inner radius of the insulation is $r_2 = 1.0$ cm. Letting r_3 represent the outer radius of the insulation, the areas of the surfaces exposed to convection for an $L = 1$-m-long section of the pipe become

$$A_1 = 2\pi r_1 L = 2\pi(0.008 \text{ m})(1 \text{ m}) = 0.0503 \text{ m}^2$$

$$A_3 = 2\pi r_3 L = 2\pi r_3 (1 \text{ m}) = 6.28 r_3 \text{ m}^2$$

Then the individual thermal resistances are determined to be

$$R_i = R_{conv,1} = \frac{1}{h_i A_1} = \frac{1}{(70 \text{ W/m}^2 \cdot °C)(0.0503 \text{ m}^2)} = 0.284°C/W$$

$$R_1 = R_{pipe} = \frac{\ln(r_2/r_1)}{2\pi k_1 L} = \frac{\ln(0.01/0.008)}{2\pi(15 \text{ W/m} \cdot °C)(1 \text{ m})} = 0.0024°C/W$$

$$R_2 = R_{insulation} = \frac{\ln(r_3/r_2)}{2\pi k_2 L} = \frac{\ln(r_3/0.01)}{2\pi(0.038 \text{ W/m} \cdot °C)(1 \text{ m})}$$

$$= 4.188 \ln(r_3/0.01)°C/W$$

$$R_o = R_{conv,2} = \frac{1}{h_o A_3} = \frac{1}{(20 \text{ W/m}^2 \cdot °C)(6.28 r_3 \text{ m}^2)} = \frac{1}{125.6 r_3}°C/W$$

Noting that all resistances are in series, the total resistance is determined to be

$$R_{total} = R_i + R_1 + R_2 + R_o$$

$$= [0.284 + 0.0024 + 4.188 \ln(r_3/0.01) + 1/125.6 r_3]°C/W$$

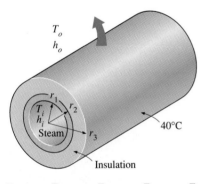

FIGURE 3-58

Schematic for Example 3-10.

Then the steady rate of heat loss from the steam becomes

$$\dot{Q} = \frac{T_i - T_o}{R_{total}} = \frac{(120 - 125)°C}{[0.284 + 0.0024 + 4.188 \ln(r_3/0.01) + 1/125.6r_3]°C/W}$$

Noting that the outer surface temperature of insulation is specified to be 40°C, the rate of heat loss can also be expressed as

$$\dot{Q} = \frac{T_3 - T_o}{R_o} = \frac{(40 - 25)°C}{(1/125.6r_3)°C/W} = 1884r_3$$

Setting the two relations above equal to each other and solving for r_3 gives $r_3 = 0.0170$ m. Then the minimum thickness of fiberglass insulation required is

$$t = r_3 - r_2 = 0.0170 - 0.0100 = 0.0070 \text{ m} = \textbf{0.70 cm}$$

Discussion   Insulating the pipe with at least 0.70-cm-thick fiberglass insulation will ensure that the outer surface temperature of the pipe will be at 40°C or below.

EXAMPLE 3-11   Optimum Thickness of Insulation

During a plant visit, you notice that the outer surface of a cylindrical curing oven is very hot, and your measurements indicate that the average temperature of the exposed surface of the oven is 180°F when the surrounding air temperature is 75°F. You suggest to the plant manager that the oven should be insulated, but the manager does not think it is worth the expense. Then you propose to the manager to pay for the insulation yourself if he lets you keep the savings from the fuel bill for one year. That is, if the fuel bill is $5000/yr before insulation and drops to $2000/yr after insulation, you will get paid $3000. The manager agrees since he has nothing to lose, and a lot to gain. Is this a smart bet on your part?

The oven is 12 ft long and 8 ft in diameter, as shown in Figure 3-59. The plant operates 16 h a day 365 days a year, and thus 5840 h/yr. The insulation to be used is fiberglass ($k_{ins} = 0.024$ Btu/h · ft · °F), whose cost is $0.70/ft^2 per inch of thickness for materials, plus $2.00/ft^2 for labor regardless of thickness. The combined heat transfer coefficient on the outer surface is estimated to be $h_o = 3.5$ Btu/h · ft^2 · °F. The oven uses natural gas, whose unit cost is $0.75/therm input (1 therm = 100,000 Btu), and the efficiency of the oven is 80 percent. Disregarding any inflation or interest, determine how much money you will make out of this venture, if any, and the thickness of insulation (in whole inches) that will maximize your earnings.

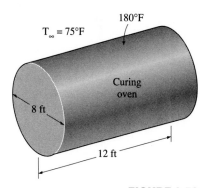

$T_\infty = 75°F$

180°F

Curing oven

8 ft

12 ft

FIGURE 3-59
Schematic for Example 3-11.

Solution   A cylindrical oven is to be insulated to reduce heat losses. The optimum thickness of insulation and the amount of money made by the vendor are to be determined.

Assumptions   **1** Steady operating conditions exist. **2** Heat transfer through the insulation is one-dimensional. **3** Thermal conductivities are constant. **4** The thermal contact resistance at the interface is negligible. **5** The surfaces of the cylindrical oven can be treated as plain surfaces since its diameter is greater than 3 ft.

Properties   The thermal conductivity is given to be $k = 0.024$ Btu/h · ft · °F.

Analysis The exposed surface area of the oven is

$$A = 2A_{base} + A_{side} = 2\pi r^2 + 2\pi rL = 2\pi(4\text{ ft})^2 + 2\pi(4\text{ ft})(12\text{ ft}) = 402\text{ ft}^2$$

The rate of heat loss from the oven before the insulation is installed is determined from

$$\dot{Q} = h_o A(T_s - T_\infty) = (3.5\text{ Btu/h}\cdot\text{ft}^2\cdot\text{°F})(402\text{ ft}^2)(180 - 75)\text{°F} = 147{,}700\text{ Btu/h}$$

Noting that the plant operates 5840 h/yr, the total amount of heat loss from the oven per year is

$$Q = \dot{Q}\Delta t = (147{,}700\text{ Btu/h})(5840\text{ h/yr}) = 0.863 \times 10^9\text{ Btu/yr}$$

The efficiency of the oven is given to be 80 percent. Therefore, to generate this much heat, the oven must consume energy (in the form of natural gas) at a rate of

$$Q_{in} = Q/\eta_{oven} = (0.863 \times 10^9\text{ Btu/yr})/0.80 = 1.079 \times 10^9\text{ Btu/yr}$$
$$= 10{,}790\text{ therms}$$

since 1 therm = 100,000 Btu. Then the annual fuel cost of this oven before insulation becomes

$$\text{Annual cost} = Q_{in} \times \text{Unit cost}$$
$$= (10{,}790\text{ therm/yr})(\$0.75/\text{therm}) = \$8093/\text{yr}$$

That is, the heat losses from the exposed surfaces of the oven are currently costing the plant over \$8,000.

When insulation is installed, the rate of heat transfer from the oven can be determined from

$$\dot{Q}_{ins} = \frac{T_s - T_\infty}{R_{total}} = \frac{T_s - T_\infty}{R_{ins} + R_{conv}} = A\frac{(T_s - T_\infty)}{\dfrac{t_{ins}}{k_{ins}} + \dfrac{1}{h_o}}$$

We expect the surface temperature of the oven to increase and the heat transfer coefficient to decrease somewhat when insulation is installed. We assume these two effects to counteract each other. Then the relation above for 1-in.-thick insulation gives the rate of heat loss to be

$$\dot{Q}_{ins} = \frac{A(T_s - T_\infty)}{\dfrac{t_{ins}}{k_{ins}} + \dfrac{1}{h_o}} = \frac{(402\text{ ft}^2)(180 - 75)\text{°F}}{\dfrac{1/12\text{ ft}}{0.024\text{ Btu/h}\cdot\text{ft}\cdot\text{°F}} + \dfrac{1}{3.5\text{ Btu/h}\cdot\text{ft}^2\cdot\text{°F}}}$$
$$= 11{,}230\text{ Btu/h}$$

Also, the total amount of heat loss from the oven per year and the amount and cost of energy consumption of the oven become

$$Q_{ins} = \dot{Q}_{ins}\Delta t = (11{,}230\text{ Btu/h})(5840\text{ h/yr}) = 0.6558 \times 10^8\text{ Btu/yr}$$

$$Q_{in,\,ins} = Q_{ins}/\eta_{oven} = (0.6558 \times 10^8\text{ Btu/yr})/0.80 = 0.820 \times 10^8\text{ Btu/yr}$$
$$= 820\text{ therms}$$

$$\text{Annual cost} = Q_{in,\,ins} \times \text{Unit cost}$$
$$= (820\text{ therm/yr})(\$0.75/\text{therm}) = \$615/\text{yr}$$

Therefore, insulating the oven by 1-in.-thick fiberglass insulation will reduce the fuel bill by $8093 − $615 = $7362 per year. The unit cost of insulation is given to be $2.70/ft². Then the installation cost of insulation becomes

Insulation cost = (Unit cost)(Surface area) = ($2.70/ft²)(402 ft²) = $1085

The sum of the insulation and heat loss costs is

Total cost = Insulation cost + Heat loss cost = $1085 + $615 = $1700

Then the net earnings will be

Earnings = Income − Expenses = $8093 − $1700 = $6393

To determine the thickness of insulation that maximizes your earnings, we repeat the calculations above for 2-, 3-, 4-, and 5-in.-thick insulations, and list the results in Table 3-5. Note that the total cost of insulation decreases first with increasing insulation thickness, reaches a minimum, and then starts to increase.

TABLE 3-5

The variation of total insulation cost with insulation thickness

Insulation thickness	Heat loss, Btu/h	Lost fuel, therms/yr	Lost fuel cost, $/yr	Insulation cost, $	Total cost, $
1 in.	11,230	820	615	1085	1700
2 in.	5838	426	320	1367	1687
3 in.	3944	288	216	1648	1864
4 in.	2978	217	163	1930	2093
5 in.	2392	175	131	2211	2342

We observe that the total insulation cost is a minimum at $1687 for the case of **2-in-thick** insulation. The earnings in this case are

Maximum earnings = Income − Minimum expenses

= $8093 − $1687 = **$6406**

which is not bad for a day's worth of work. The plant manager is also a big winner in this venture since the heat losses will cost him only $320/yr during the second and consequent years instead of $8093/yr. A thicker insulation could probably be justified in this case if the cost of insulation is annualized over the lifetime of insulation, say 20 years. Several energy conservation measures are being marketed as explained above by several power companies and private firms.

3-7 ■ HEAT TRANSFER FROM FINNED SURFACES

The rate of heat transfer from a surface at a temperature T_s to the surrounding medium at T_∞ is given by Newton's law of cooling as

$$\dot{Q}_{conv} = hA(T_s - T_\infty)$$

where A is the heat transfer surface area and h is the convection heat transfer coefficient. When the temperatures T_s and T_∞ are fixed by design considerations, as is often the case, there are *two ways* to increase the rate of heat

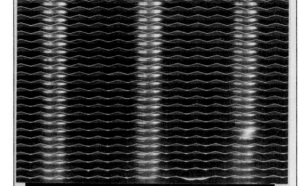

FIGURE 3-60
The thin plate fins of a car radiator
greatly increase the rate of heat transfer
to the air (courtesy of James Kleiser).

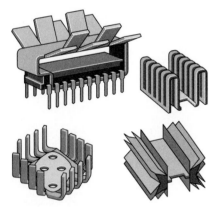

FIGURE 3-61
Some innovative fin designs.

transfer: to increase the *convection heat transfer coefficient h* or to increase the *surface area A*. Increasing *h* may require the installation of a pump or fan, or replacing the existing one with a larger one, but this approach may or may not be practical. Besides, it may not be adequate. The alternative is to increase the surface area by attaching to the surface *extended surfaces* called *fins* made of highly conductive materials such as aluminum. Finned surfaces are manufactured by extruding, welding, or wrapping a thin metal sheet on a surface. Fins enhance heat transfer from a surface by exposing a larger surface area to convection and radiation.

Finned surfaces are commonly used in practice to enhance heat transfer, and they often increase the rate of heat transfer from a surface severalfold. The *car radiator* shown in Figure 3-60 is an example of a finned surface. The closely packed thin metal sheets attached to the hot water tubes increase the surface area for convection and thus the rate of convection heat transfer from the tubes to the air many times. There are a variety of innovative fin designs available in the market, and they seem to be limited only by imagination (Fig. 3-61).

In the analysis of the fins, we consider *steady* operation with *no heat generation* in the fin, and we assume the thermal conductivity *k* of the material to remain constant. We also assume the convection heat transfer coefficient *h* to be *constant* and *uniform* over the entire surface of the fin for convenience in the analysis. We recognize that the convection heat transfer coefficient *h*, in general, varies along the fin as well as its circumference, and its value at a point is a strong function of the *fluid motion* at that point. The value of *h* is usually much lower at the *fin base* than it is at the *fin tip* because the fluid is surrounded by solid surfaces near the base, which seriously disrupt its motion to the point of "suffocating" it, while the fluid near the fin tip has little contact with a solid surface and thus encounters little resistance to flow. Therefore, adding too many fins on a surface may actually decrease the overall heat transfer when the decrease in *h* offsets any gain resulting from the increase in the surface area.

Fin Equation

Consider a volume element of a fin at location x having a length of Δx, cross-sectional area of A_c, and a perimeter of p, as shown in Figure 3-62. Under steady conditions, the energy balance on this volume element can be expressed as

$$\begin{pmatrix} \text{Rate of } heat \\ conduction \text{ into} \\ \text{the element at } x \end{pmatrix} = \begin{pmatrix} \text{Rate of } heat \\ conduction \text{ from the} \\ \text{element at } x + \Delta x \end{pmatrix} + \begin{pmatrix} \text{Rate of } heat \\ convection \text{ from} \\ \text{the element} \end{pmatrix}$$

or

$$\dot{Q}_{\text{cond}, x} = \dot{Q}_{\text{cond}, x + \Delta x} + \dot{Q}_{\text{conv}}$$

where

$$\dot{Q}_{\text{conv}} = h(p\Delta x)(T - T_\infty)$$

Substituting and dividing by Δx, we obtain

$$\frac{\dot{Q}_{\text{cond}, x + \Delta x} - \dot{Q}_{\text{cond}, x}}{\Delta x} + hp(T - T_\infty) = 0$$

Taking the limit as $\Delta x \to 0$ and using the definition of the derivative gives

$$\frac{d\dot{Q}_{\text{cond}}}{dx} + hp(T - T_\infty) = 0 \qquad (3\text{-}55)$$

From Fourier's law of heat conduction we have

$$\dot{Q}_{\text{cond}} = -kA_c \frac{dT}{dx}$$

where A_c is the cross-sectional area of the fin at location x. Substitution of this relation into Equation 3-55 gives the differential equation governing heat transfer in fins,

$$\frac{d}{dx}\left(kA_c \frac{dT}{dx}\right) - hp(T - T_\infty) = 0 \qquad (3\text{-}56)$$

In general, the cross-sectional area A_c and the perimeter p of a fin vary with x, which makes this differential equation difficult to solve. In the special case of *constant cross-section* and *constant thermal conductivity*, the differential equation above reduces to

$$\frac{d^2\theta}{dx^2} - a^2\theta = 0 \qquad (3\text{-}57)$$

where

$$a^2 = \frac{hp}{kA_c}$$

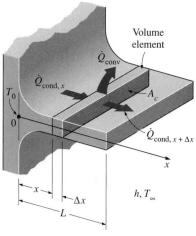

FIGURE 3-62

Volume element of a fin at location x having a length of Δx, cross-sectional area of A_c, and perimeter of p.

and $\theta = T - T_\infty$ is the *temperature excess*. At the fin base we have $\theta_b = T_b - T_\infty$.

Equation 3-57 is a linear, homogeneous, second-order differential equation with constant coefficients. A fundamental theory of differential equations states that such an equation has two linearly independent solution functions, and its general solution is the linear combination of those two solution functions. A careful examination of the differential equation reveals that subtracting a constant multiple of the solution function θ from its second derivative yields zero. Thus we conclude that the function θ and its second derivative must be *constant multiples* of each other. The only functions whose derivatives are constant multiples of the functions themselves are the *exponential functions* (or a linear combination of exponential functions such as sine and cosine hyperbolic functions). Therefore, the solution functions of the differential equation above are the exponential functions e^{-ax} or e^{ax} or constant multiples of them. This can be verified by direct substitution. For example, the second derivative of e^{-ax} is $a^2 e^{-ax}$, and its substitution into Equation 3-57 yields zero. Therefore, the general solution of the differential equation Eq. 3-57 is

$$\theta(x) = C_1 e^{ax} + C_2 e^{-ax} \tag{3-58}$$

where C_1 and C_2 are arbitrary constants whose values are to be determined from the boundary conditions at the base and at the tip of the fin. Note that we need only two conditions to determine C_1 and C_2 uniquely.

The temperature of the plate to which the fins are attached is normally known in advance. Therefore, at the fin base we have a *specified temperature* boundary condition, expressed as

Boundary condition at fin base: $\qquad \theta(0) = \theta_b = T_b - T_\infty \tag{3-59}$

At the fin tip we have several possibilities, including specified temperature, negligible heat loss (idealized as an insulated tip), convection, and combined convection and radiation (Fig. 3-63). Below we consider each case separately.

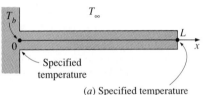

(a) Specified temperature
(b) Negligible heat loss
(c) Convection
(d) Convection and radiation

FIGURE 3-63

Boundary conditions at the fin base and the fin tip.

1 Infinitely Long Fin ($T_{\text{fin tip}} = T_\infty$)

For a sufficiently long fin of *uniform* cross-section (A_c = constant), the temperature of the fin at the fin tip will approach the environment temperature T_∞ and thus θ will approach zero. That is,

Boundary condition at fin tip: $\quad \theta(L) = T(L) - T_\infty = 0 \qquad$ as $\qquad L \to \infty$

This condition will be satisfied by the function e^{-ax}, but not by the other prospective solution function e^{ax} since it tends to infinity as x gets larger. Therefore, the general solution in this case will consist of a constant multiple of e^{-ax}. The value of the constant multiple is determined from the requirement that at the fin base where $x = 0$ the value of θ will be θ_b. Noting that $e^{-ax} = e^0 = 1$, the proper value of the constant is θ_b, and the solution function we are looking for is $\theta(x) = \theta_b e^{-ax}$. This function satisfies the differential equation as well as the requirements that the solution reduce to θ_b at the fin base and approach zero at the fin tip for large x. Noting that $\theta = T - T_\infty$ and

$\boxed{a = \sqrt{hp/kA_c}}$, the variation of temperature along the fin in this case can be expressed as

$$\text{Infinitely long fin:} \qquad \frac{T(x) - T_\infty}{T_b - T_\infty} = e^{-ax} = e^{-x\sqrt{hp/kA_c}} \qquad (3\text{-}60)$$

Note that the temperature along the fin in this case decreases *exponentially* from T_b to T_∞, as shown in Figure 3-64. The steady rate of *heat transfer* from the entire fin can be determined from Fourier's law of heat conduction

$$\text{Infinitely long fin:} \qquad \dot{Q}_{\text{long fin}} = -kA_c \frac{dT}{dx}\bigg|_{x=0} = \sqrt{hpkA_c}\,(T_b - T_\infty) \qquad (3\text{-}61)$$

where p is the perimeter, A_c is the cross-sectional area of the fin, and x is the distance from the fin base. Alternately, the rate of heat transfer from the fin could also be determined by considering heat transfer from a differential volume element of the fin and integrating it over the entire surface of the fin. That is,

$$\dot{Q}_{\text{fin}} = \int_{A_{\text{fin}}} h[T(x) - T_\infty]dA_{\text{fin}} = \int_{A_{\text{fin}}} h\theta(x)dA_{\text{fin}} \qquad (3\text{-}62)$$

The two approaches described above are equivalent and give the same result since, under steady conditions, the heat transfer from the exposed surfaces of the fin is equal to the heat transfer to the fin at the base (Fig. 3-65).

2 Negligible Heat Loss from the Fin Tip (Insulated fin tip, $\dot{Q}_{\text{fin tip}} = 0$)

Fins are not likely to be so long that their temperature approaches the surrounding temperature at the tip. A more realistic situation is for heat transfer from the fin tip to be negligible since the heat transfer from the fin is proportional to its surface area, and the surface area of the fin tip is usually a negligible fraction of the total fin area. Then the fin tip can be assumed to be insulated, and the condition at the fin tip can be expressed as

$$\text{Boundary condition at fin tip:} \qquad \frac{d\theta}{dx}\bigg|_{x=L} = 0 \qquad (3\text{-}63)$$

The condition at the fin base remains the same as expressed in Equation 3-59. The application of these two conditions on the general solution (Eq. 3-58) yields, after some manipulations, the following relation for the temperature distribution:

$$\text{Insulated fin tip:} \qquad \frac{T(x) - T_\infty}{T_b - T_\infty} = \frac{\cosh a(L - x)}{\cosh aL} \qquad (3\text{-}64)$$

The rate of *heat transfer* from the fin can be determined again from Fourier's law of heat conduction:

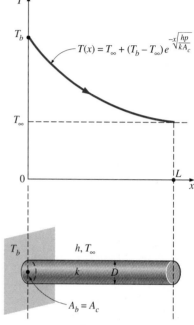

$(p = \pi D, A_c = \pi D^2/4$ for a cylindrical fin$)$

FIGURE 3-64

A long circular fin of uniform cross-section and the variation of temperature along it.

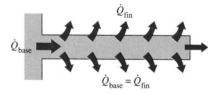

FIGURE 3-65

Under steady conditions, heat transfer from the exposed surfaces of the fin is equal to heat conduction to the fin at the base.

$$\text{Insulated fin tip:} \qquad \dot{Q}_{\text{insulated tip}} = -kA_c \frac{dT}{dx}\bigg|_{x=0} \tag{3-65}$$
$$= \sqrt{hpkA_c}\,(T_b - T_\infty) \tanh aL$$

Note that the heat transfer relations for the very long fin and the fin with negligible heat loss at the tip differ by the factor tanh aL, which approaches 1 as L becomes very large.

3 Convection (or Combined Convection and Radiation) from Fin Tip

The fin tips, in practice, are exposed to the surroundings, and thus the proper boundary condition for the fin tip is convection that also includes the effects of radiation. The fin equation can still be solved in this case using the convection at the fin tip as the second boundary condition, but the analysis becomes more involved, and it results in rather lengthy expressions for the temperature distribution and the heat transfer. Yet, in general, the fin tip area is a small fraction of the total fin surface area, and thus the complexities involved can hardly justify the improvement in accuracy.

A practical way of accounting for the heat loss from the fin tip is to replace the *fin length L* in the relation for the *insulated tip* case by a **corrected length** defined as (Fig. 3-66)

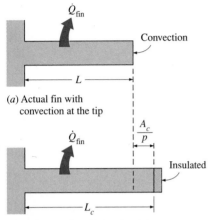

(a) Actual fin with convection at the tip

(b) Equivalent fin with insulated tip

FIGURE 3-66

Corrected fin length L_c is defined such that heat transfer from a fin of length L_c with insulated tip is equal to heat transfer from the actual fin of length L with convection at the fin tip.

$$\text{Corrected fin length:} \qquad L_c = L + \frac{A_c}{p} \tag{3-66}$$

where A_c is the cross-sectional area and p is the perimeter of the fin at the tip. Multiplying the relation above by the perimeter gives $A_{\text{corrected}} = A_{\text{fin (lateral)}} + A_{\text{tip}}$, which indicates that the fin area determined using the corrected length is equivalent to the sum of the lateral fin area plus the fin tip area.

The corrected length approximation gives very good results when the variation of temperature near the fin tip is small (which is the case when $aL \geq 1$) and the heat transfer coefficient at the fin tip is about the same as that at the lateral surface of the fin. Therefore, *fins subjected to convection at their tips can be treated as fins with insulated tips by replacing the actual fin length by the corrected length in Equations 3-64 and 3-65.*

Using the proper relations for A_c and p, the corrected lengths for rectangular and cylindrical fins are easily determined to be

$$L_{c,\,\text{rectangular fin}} = L + \frac{t}{2} \qquad \text{and} \qquad L_{c,\,\text{cylindrical fin}} = L + \frac{D}{4}$$

where t is the thickness of the rectangular fins and D is the diameter of the cylindrical fins.

Fin Efficiency

Consider the surface of a *plane wall* at temperature T_b exposed to a medium at temperature T_∞. Heat is lost from the surface to the surrounding medium by convection with a heat transfer coefficient of h. Disregarding radiation or accounting for its contribution in the convection coefficient h, heat transfer from a surface area A is expressed as $\dot{Q} = hA(T_s - T_\infty)$.

Now let us consider a fin of constant cross-sectional area $A_c = A_b$ and length L that is attached to the surface with a perfect contact (Fig. 3-67). This time heat will flow from the surface to the fin *by conduction* and from the fin to the surrounding medium *by convection* with the same heat transfer coefficient h. The temperature of the fin will be T_b at the fin base and gradually decrease toward the fin tip. Convection from the fin surface causes the temperature at any cross-section to drop somewhat from the midsection toward the outer surfaces. However, the cross-sectional area of the fins is usually very small, and thus the temperature at any cross-section can be considered to be uniform. Also, the fin tip can be assumed for convenience and simplicity to be insulated by using the corrected length for the fin instead of the actual length.

In the limiting case of *zero thermal resistance* or *infinite thermal conductivity* ($k \rightarrow \infty$), the temperature of the fin will be uniform at the base value of T_b. The heat transfer from the fin will be *maximum* in this case and can be expressed as

$$\dot{Q}_{\text{fin, max}} = hA_{\text{fin}} (T_b - T_\infty) \qquad (3\text{-}67)$$

In reality, however, the temperature of the fin will drop along the fin, and thus the heat transfer from the fin will be less because of the decreasing temperature difference $T(x) - T_\infty$ toward the fin tip, as shown in Figure 3-68. To account for the effect of this decrease in temperature on heat transfer, we define a **fin efficiency** as

$$\eta_{\text{fin}} = \frac{\dot{Q}_{\text{fin}}}{\dot{Q}_{\text{fin, max}}} = \frac{\text{Actual heat transfer rate from the fin}}{\substack{\text{Ideal heat transfer rate from the fin} \\ \text{if the entire fin were at base temperature}}} \qquad (3\text{-}68)$$

or

$$\dot{Q}_{\text{fin}} = \eta_{\text{fin}}\dot{Q}_{\text{fin, max}} = \eta_{\text{fin}}hA_{\text{fin}} (T_b - T_\infty) \qquad (3\text{-}69)$$

where A_{fin} is the total surface area of the fin. This relation enables us to determine the heat transfer from a fin when its efficiency is known. For the cases of constant cross-section of *very long fins* and *fins with insulated tips*, the fin efficiency can be expressed as

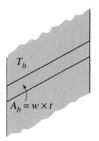

(*a*) Surface without fins

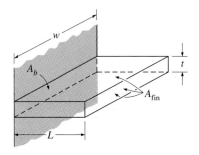

(*b*) Surface with a fin

$$A_{\text{fin}} = 2 \times w \times L + w \times t$$
$$\cong 2 \times w \times L$$

FIGURE 3-67

Fins enhance heat transfer from a surface by enhancing surface area.

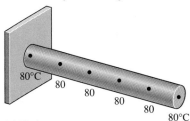

(*a*) Ideal

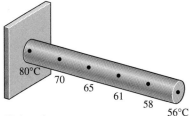

(*b*) Actual

FIGURE 3-68

Ideal and actual temperature distribution in a fin.

$$\eta_{\text{long fin}} = \frac{\dot{Q}_{\text{fin}}}{\dot{Q}_{\text{fin, max}}} = \frac{\sqrt{hpkA_c}\,(T_b - T_\infty)}{hA_{\text{fin}}\,(T_b - T_\infty)} = \frac{1}{L}\sqrt{\frac{kA_c}{hp}} = \frac{1}{aL} \quad (3\text{-}70)$$

and

$$\eta_{\text{insulated tip}} = \frac{\dot{Q}_{\text{fin}}}{\dot{Q}_{\text{fin, max}}} = \frac{\sqrt{hpkA_c}\,(T_b - T_\infty)\tanh aL}{hA_{\text{fin}}\,(T_b - T_\infty)}$$

$$= \frac{\tanh aL}{L}\sqrt{\frac{kp}{hA_c}} = \frac{\tanh aL}{aL} \quad (3\text{-}71)$$

since $A_{\text{fin}} = pL$ for fins with constant cross-section. Equation 3-71 can also be used for fins subjected to convection provided that the fin length L is replaced by the corrected length L_c.

Fin efficiency relations are developed for fins of various profiles and are plotted in Figure 3-69 for fins on a *plain surface* and in Figure 3-70 for *circular fins* of constant thickness. The fin surface area associated with each profile is also given on each figure. For most fins of constant thickness encountered in practice, the fin thickness t is too small relative to the fin length L, and thus the fin tip area is negligible.

Note that fins with triangular and parabolic profiles contain less material and are more efficient than the ones with rectangular profiles, and thus are more suitable for applications requiring minimum weight such as space applications.

An important consideration in the design of finned surfaces is the selection of the proper *fin length L*. Normally the *longer* the fin, the *larger* the heat transfer area and thus the *higher* the rate of heat transfer from the fin. But also the larger the fin, the bigger the mass, the higher the price, and the larger the fluid friction. Therefore, increasing the length of the fin beyond a certain value cannot be justified unless the added benefits outweigh the added cost. Also, the fin efficiency decreases with increasing fin length because of the decrease in fin temperature with length. Fin lengths that cause the fin efficiency to drop below 60 percent usually cannot be justified economically and should be avoided. The efficiency of most fins used in practice is above 90 percent.

Fin Effectiveness

Fins are used to *enhance* heat transfer, and the use of fins on a surface cannot be recommended unless the enhancement in heat transfer justifies the added cost and complexity associated with the fins. In fact, there is no assurance that adding fins on a surface will *enhance* heat transfer. The performance of the fins is judged on the basis of the enhancement in heat transfer relative to

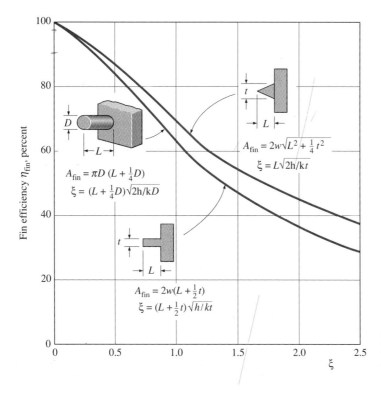

FIGURE 3-69

Efficiency of circular, rectangular, and triangular fins on a plain surface of width w (from Gardner, Ref. 10).

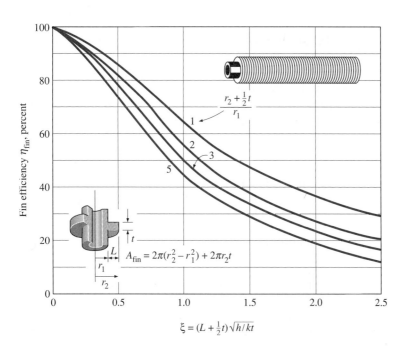

FIGURE 3-70

Efficiency of circular fins of length L and constant thickness t (from Gardner, Ref. 10).

the no-fin case. The performance of fins expressed in terms of the *fin effectiveness* ε_{fin} defined as (Fig. 3-71)

$$\varepsilon_{\text{fin}} = \frac{\dot{Q}_{\text{fin}}}{\dot{Q}_{\text{no fin}}} = \frac{\dot{Q}_{\text{fin}}}{hA_b\,(T_b - T_\infty)} = \frac{\text{Heat transfer rate from the fin of } \textit{base area } A_b}{\text{Heat transfer rate from the surface of } \textit{area } A_b} \quad (3\text{-}72)$$

Here, A_b is the cross-sectional area of the fin at the base and $\dot{Q}_{\text{no fin}}$ represents the rate of heat transfer from this area if no fins are attached to the surface. An effectiveness of $\varepsilon_{\text{fin}} = 1$ indicates that the addition of fins to the surface does not affect heat transfer at all. That is, heat conducted to the fin through the base area A_b is equal to the heat transferred from the same area A_b to the surrounding medium. An effectiveness of $\varepsilon_{\text{fin}} < 1$ indicates that the fin actually acts as *insulation,* slowing down the heat transfer from the surface. This situation can occur when fins made of low thermal conductivity materials are used. An effectiveness of $\varepsilon_{\text{fin}} > 1$ indicates that fins are *enhancing* heat transfer from the surface, as they should. However, the use of fins cannot be justified unless ε_{fin} is sufficiently larger than 1. Finned surfaces are designed on the basis of *maximizing* effectiveness for a specified cost or *minimizing* cost for a desired effectiveness.

Note that both the fin efficiency and fin effectiveness are related to the performance of the fin, but they are different quantities. However, they are related to each other by

$$\varepsilon_{\text{fin}} = \frac{\dot{Q}_{\text{fin}}}{\dot{Q}_{\text{no fin}}} = \frac{\dot{Q}_{\text{fin}}}{hA_b\,(T_b - T_\infty)} = \frac{\eta_{\text{fin}}hA_{\text{fin}}\,(T_b - T_\infty)}{hA_b\,(T_b - T_\infty)} = \frac{A_{\text{fin}}}{A_b}\,\eta_{\text{fin}} \quad (3\text{-}73)$$

Therefore, the fin effectiveness can be determined easily when the fin efficiency is known, or vice versa.

The rate of heat transfer from a sufficiently *long* fin of *uniform* cross-section under steady conditions is given by Equation 3-61. Substituting this relation into Equation 3-72, the effectiveness of such a long fin is determined to be

$$\varepsilon_{\text{long fin}} = \frac{\dot{Q}_{\text{fin}}}{\dot{Q}_{\text{no fin}}} = \frac{\sqrt{hpkA_c}\,(T_b - T_\infty)}{hA_b\,(T_b - T_\infty)} = \sqrt{\frac{kp}{hA_c}} \quad (3\text{-}74)$$

since $A_c = A_b$ in this case. We can draw several important conclusions from the fin effectiveness relation above for consideration in the design and selection of the fins:

- The *thermal conductivity k* of the fin material should be as high as possible. Thus it is no coincidence that fins are made from metals, with copper, aluminum, and iron being the most common ones. Perhaps the most widely used fins are made of aluminum because of its low cost and weight and its resistance to corrosion.

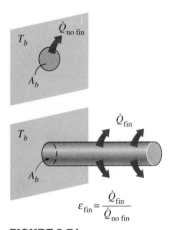

FIGURE 3-71

The effectiveness of a fin.

- The ratio of the *perimeter* to the *cross-sectional area* of the fin p/A_c should be as high as possible. This criterion is satisfied by *thin* plate fins or *slender* pin fins.
- The use of fins is *most effective* in applications involving a *low convection heat transfer coefficient*. Thus, the use of fins is more easily justified when the medium is a *gas* instead of a liquid and the heat transfer is by *natural convection* instead of by forced convection. Therefore, it is no coincidence that in liquid-to-gas heat exchangers such as the car radiator, fins are placed on the *gas* side.

When determining the rate of heat transfer from a finned surface, we must consider the *unfinned portion* of the surface as well as the *fins*. Therefore, the rate of heat transfer for a surface containing n fins can be expressed as

$$\dot{Q}_{\text{total, fin}} = \dot{Q}_{\text{unfin}} + \dot{Q}_{\text{fin}}$$
$$= hA_{\text{unfin}}(T_b - T_\infty) + \eta_{\text{fin}}hA_{\text{fin}}(T_b - T_\infty) \qquad (3\text{-}75)$$
$$= h(A_{\text{unfin}} + \eta_{\text{fin}}A_{\text{fin}})(T_b - T_\infty)$$

We can also define an **overall effectiveness** for a finned surface as the ratio of the total heat transfer from the finned surface to the heat transfer from the same surface if there were no fins,

$$\varepsilon_{\text{fin, overall}} = \frac{\dot{Q}_{\text{total, fin}}}{\dot{Q}_{\text{total, no fin}}} = \frac{h(A_{\text{unfin}} + \eta_{\text{fin}}A_{\text{fin}})(T_b - T_\infty)}{hA_{\text{no fin}}(T_b - T_\infty)} \qquad (3\text{-}76)$$

where $A_{\text{no fin}}$ is the area of the surface when there are no fins, A_{fin} is the total surface area of all the fins on the surface, and A_{unfin} is the area of the unfinned portion of the surface (Fig. 3-72). Note that the overall fin effectiveness depends on the fin density (number of fins per unit length) as well as the effectiveness of the individual fins. The overall effectiveness is a better measure of the performance of a finned surface than the effectiveness of the individual fins.

Proper Length of a Fin

An important step in the design of a fin is the determination of the appropriate length of the fin once the fin material and the fin cross-section are specified. You may be tempted to think that the longer the fin, the larger the surface area and thus the higher the rate of heat transfer. Therefore, for maximum heat transfer, the fin should be infinitely long. However, the temperature drops along the fin exponentially and reaches the environment temperature at some length. The part of the fin beyond this length does not contribute to heat transfer since it is at the temperature of the environment, as shown in Figure 3-73. Therefore, designing such an "extra long" fin is out of question since it results in material waste, excessive weight, and increased size and thus increased cost with no benefit in return (in fact, such a long fin will hurt performance since it will suppress fluid motion and thus reduce the

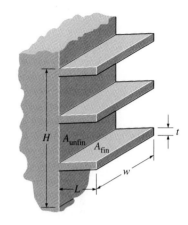

$A_{\text{no fin}} = w \times H$
$A_{\text{unfin}} = w \times H - 3 \times (t \times w)$
$A_{\text{fin}} = 2 \times L \times w + t \times w$ (one fin
$\qquad \approx 2 \times L \times w$

FIGURE 3-72

Various surface areas associated with a rectangular surface with three fins.

FIGURE 3-73

Because of the gradual temperature drop along the fin, the region near the fin tip makes little or no contribution to heat transfer.

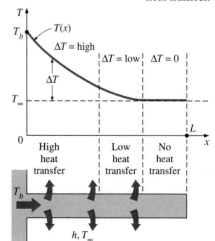

convection heat transfer coefficient). Fins that are so long that the temperature approaches the environment temperature cannot be recommended either since the little increase in heat transfer at the tip region cannot justify the large increase in the weight and cost.

To get a sense of the proper length of a fin, we compare heat transfer from a fin of finite length to heat transfer from an infinitely long fin under the same conditions. The ratio of these two heat transfers is

$$\text{Heat transfer ratio:} \quad \frac{\dot{Q}_{\text{fin}}}{\dot{Q}_{\text{long fin}}} = \frac{\sqrt{hpkA_c}\,(T_b - T_\infty)\,\tanh aL}{\sqrt{hpkA_c}\,(T_b - T_\infty)} = \tanh aL \quad (3\text{-}77)$$

TABLE 3-6

The variation of heat transfer from a fin relative to that from an infinitely long fin

aL	$\dfrac{\dot{Q}_{\text{fin}}}{\dot{Q}_{\text{long fin}}} = \tanh aL$
0.1	0.100
0.2	0.197
0.5	0.462
1.0	0.762
1.5	0.905
2.0	0.964
2.5	0.987
3.0	0.995
4.0	0.999
5.0	1.000

Using a hand calculator, the values of $\tanh aL$ are evaluated for some values of aL and the results are given in Table 3-6. We observe from the table that heat transfer from a fin increases with aL almost linearly at first, but the curve reaches a plateau later and reaches a value for the infinitely long fin at about $aL = 5$. Therefore, a fin whose length is $L = \frac{1}{5}a$ can be considered to be an infinitely long fin. We also observe that reducing the fin length by half in that case (from $aL = 5$ to $aL = 2.5$) causes a drop of just 1 percent in heat transfer. We certainly would not hesitate sacrificing 1 percent in heat transfer performance in return for 50 percent reduction in the size and possibly the cost of the fin. In practice, a fin length that corresponds to about $aL = 1$ will transfer 76.2 percent of the heat that can be transferred by an infinitely long fin, and thus it should offer a good compromise between heat transfer performance and the fin size.

A common approximation used in the analysis of fins is to assume the fin temperature varies in one direction only (along the fin length) and the temperature variation along other directions is negligible. Perhaps you are wondering if this one-dimensional approximation is a reasonable one. This is certainly the case for fins made of thin metal sheets such as the fins on a car radiator, but we wouldn't be so sure for fins made of thick materials. Studies have shown that the error involved in one-dimensional fin analysis is negligible (less than about 1 percent) when

$$\frac{h\delta}{k} < 0.2$$

where δ is the characteristic thickness of the fin, which is taken to be the plate thickness t for rectangular fins and the diameter D for cylindrical ones.

Specially designed finned surfaces called *heat sinks,* which are commonly used in the cooling of electronic equipment, involve one-of-a-kind complex geometries, as shown in Table 3-7. The heat transfer performance of heat sinks is usually expressed in terms of their *thermal resistances* R in °C/W, which is defined as

$$\dot{Q}_{\text{fin}} = \frac{T_b - T_\infty}{R} = hA_{\text{fin}}\,\eta_{\text{fin}}\,(T_b - T_\infty) \quad (3\text{-}78)$$

TABLE 3-7

Combined natural convection and radiation thermal resistance of various heat sinks used in the cooling of electronic devices between the heat sink and the surroundings. All fins are made of aluminum 6063T-5, are black anodized, and are 76 mm (3 in.) long (courtesy of Vemaline Products, Inc.).

HS 5030

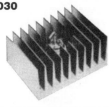

R = 0.9°C/W (vertical)
R = 1.2°C/W (horizontal)

Dimensions: 76 mm $\times$ 105 mm $\times$ 44 mm
Surface area: 677 cm^2

HS 6065

R = 5°C/W

Dimensions: 76 mm $\times$ 38 mm $\times$ 24 mm
Surface area: 387 cm^2

HS 6071

R = 1.4°C/W (vertical)
R = 1.8°C/W (horizontal)

Dimensions: 76 mm $\times$ 92 mm $\times$ 26 mm
Surface area: 968 cm^2

HS 6105

R = 1.8°C/W (vertical)
R = 2.1°C/W (horizontal)

Dimensions: 76 mm $\times$ 127 mm $\times$ 91 mm
Surface area: 677 cm^2

HS 6115

R = 1.1°C/W (vertical)
R = 1.3°C/W (horizontal)

Dimensions: 76 mm $\times$ 102 mm $\times$ 25 mm
Surface area: 929 cm^2

HS 7030

R = 2.9°C/W (vertical)
R = 3.1°C/W (horizontal)

Dimensions: 76 mm $\times$ 97 mm $\times$ 19 mm
Surface area: 290 cm^2

A small value of thermal resistance indicates a small temperature drop across the heat sink, and thus a high fin efficiency.

EXAMPLE 3-12 Maximum Power Dissipation of a Transistor

Power transistors that are commonly used in electronic devices consume large amounts of electric power. The failure rate of electronic components increases almost exponentially with operating temperature. As a rule of thumb, the failure rate of electronic components is halved for each 10°C reduction in the junction operating temperature. Therefore, the operating temperature of electronic components is kept below a safe level to minimize the risk of failure.

The sensitive electronic circuitry of a power transistor at the junction is protected by its case, which is a rigid metal enclosure. Heat transfer characteristics of a power transistor is usually specified by the manufacturer in terms of the case-to-ambient thermal resistance, which accounts for both the natural convection and radiation heat transfers.

The case-to-ambient thermal resistance of a power transistor that has a maximum power rating of 10 W is given to be 20°C/W. If the case temperature of the transistor is not to exceed 85°C, determine the power at which this transistor can be operated safely in an environment at 25°C.

Solution The maximum power rating of transistor whose case temperature is not to exceed 85°C is to be determined.

Assumptions **1** Steady operating conditions exist. **2** The transistor case is isothermal at 85°C.

Properties The case-to-ambient thermal resistance is given to be 20°C/W.

Analysis The power transistor and the thermal resistance network associated with it are shown in Figure 3-74. We notice from the thermal resistance network that there is a single resistance of 20°C/W between the case at T_c = 85°C and the ambient at T_∞ = 25°C, and thus the rate of heat transfer is

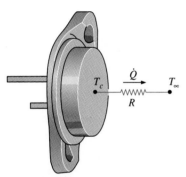

FIGURE 3-74
Schematic for Example 3-12.

$$\dot{Q} = \left(\frac{\Delta T}{R}\right)_{\text{case-ambient}} = \frac{T_c - T_\infty}{R_{\text{case-ambient}}} = \frac{(85 - 25)°C}{20°C/W} = 3 \text{ W}$$

Therefore, this power transistor should not be operated at power levels above 3 W if its case temperature is not to exceed 85°C.

Discussion This transistor can be used at higher power levels by attaching it to a heat sink (which lowers the thermal resistance by increasing the heat transfer surface area, as discussed in the next example) or by using a fan (which lowers the thermal resistance by increasing the convection heat transfer coefficient).

EXAMPLE 3-13 Selecting a Heat Sink for a Transistor

A 60-W power transistor is to be cooled by attaching it to one of the commercially available heat sinks shown in Table 3-7. Select a heat sink that will allow the case temperature of the transistor not to exceed 90°C in the ambient air at 30°C.

Solution A commercially available heat sink from Table 3-7 is to be selected to keep the case temperature of a transistor below 90°C.

Assumptions **1** Steady operating conditions exist. **2** The transistor case is isothermal at 90°C. **3** The contact resistance between the transistor and the heat sink is negligible.

Analysis The rate of heat transfer from a 60-W transistor at full power is $\dot{Q}$ = 60 W. The thermal resistance between the transistor attached to the heat sink and the ambient air for the specified temperature difference is determined to be

$$\dot{Q} = \frac{\Delta T}{R} \quad\longrightarrow\quad R = \frac{\Delta T}{\dot{Q}} = \frac{(90 - 30)°C}{60\ W} = 1.0°C/W$$

Therefore, the thermal resistance of the heat sink should be below 1.0°C/W. An examination of Table 3-7 reveals that the HS 5030, whose thermal resistance is 0.9°C/W in the vertical position, is the only heat sink that will meet this requirement.

EXAMPLE 3-14 Effect of Fins on Heat Transfer from Steam Pipes

Steam in a heating system flows through tubes whose outer diameter is D_1 = 3 cm and whose walls are maintained at a temperature of 120°C. Circular aluminum fins (k = 180 W/m · °C) of outer diameter D_2 = 6 cm and constant thickness t = 2 mm are attached to the tube, as shown in Figure 3-75. The space between the fins is 3 mm, and thus there are 200 fins per meter length of the tube. Heat is transferred to the surrounding air at T_∞ = 25°C, with a combined heat transfer coefficient of h = 60 W/m² · °C. Determine the increase in heat transfer from the tube per meter of its length as a result of adding fins.

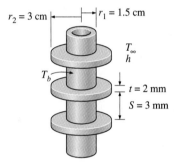

FIGURE 3-75
Schematic for Example 3-14.

Solution Circular aluminum fins are to be attached to the tubes of a heating system. The increase in heat transfer from the tubes per unit length as a result of adding fins is to be determined.

Assumptions **1** Steady operating conditions exist. **2** The heat transfer coefficient is uniform over the entire fin surfaces. **3** Thermal conductivity is constant. **4** Heat transfer by radiation is negligible.

Properties The thermal conductivity of the fins is given to be k = 180 W/m · °C.

Analysis In the case of no fins, heat transfer from the tube per meter of its length is determined from Newton's law of cooling to be

$$A_{no\ fin} = \pi D_1 L = \pi(0.03\ m)(1\ m) = 0.0942\ m^2$$
$$\dot{Q}_{no\ fin} = hA_{no\ fin}(T_b - T_\infty)$$
$$= (60\ W/m^2 \cdot °C)(0.0942\ m^2)(120 - 25)°C$$
$$= 537\ W$$

The efficiency of the circular fins attached to a circular tube is plotted in Figure 3-70. Noting that $L = \frac{1}{2}(D_2 - D_1) = \frac{1}{2}(0.06 - 0.03) = 0.015$ m in this case, we have

$$\frac{r_2 + \frac{1}{2}t}{r_1} = \frac{(0.03 + \frac{1}{2} \times 0.002)\ \text{m}}{0.015\ \text{m}} = 2.07$$

$$(L + \tfrac{1}{2}t) \sqrt{\frac{h}{kt}} = (0.015 + \tfrac{1}{2} \times 0.002)\ \text{m} \times \sqrt{\frac{60\ \text{W/m}^2 \cdot °\text{C}}{(180\ \text{W/m} \cdot °\text{C})(0.002\ \text{m})}} = 0.207$$

$\Bigg\}\ \eta_{\text{fin}} = 0.95$

$$A_{\text{fin}} = 2\pi(r_2^2 - r_1^2) + 2\pi r_2 t$$
$$= 2\pi[(0.03\ \text{m})^2 - (0.015\ \text{m})^2] + 2\pi(0.03\ \text{m})(0.002\ \text{m})$$
$$= 0.00462\ \text{m}^2$$

$$\dot{Q}_{\text{fin}} = \eta_{\text{fin}}\dot{Q}_{\text{fin, max}} = \eta_{\text{fin}}hA_{\text{fin}}(T_b - T_\infty)$$
$$= 0.95(60\ \text{W/m}^2 \cdot °\text{C})(0.00462\ \text{m}^2)(120 - 25)°\text{C}$$
$$= 25.0\ \text{W}$$

Noting that the space between the two fins is 3 mm, heat transfer from the un-finned portion of the tube is

$$A_{\text{unfin}} = \pi D_1 S = \pi(0.03\ \text{m})(0.003\ \text{m}) = 0.000283\ \text{m}^2$$
$$\dot{Q}_{\text{unfin}} = hA_{\text{unfin}}(T_b - T_\infty)$$
$$= (60\ \text{W/m}^2 \cdot °\text{C})(0.000283\ \text{m}^2)(120 - 25)°\text{C}$$
$$= 1.60\ \text{W}$$

Noting that there are 200 fins and thus 200 interfin spacings per meter length of the tube, the total heat transfer from the finned tube becomes

$$\dot{Q}_{\text{total, fin}} = n(\dot{Q}_{\text{fin}} + \dot{Q}_{\text{unfin}}) = 200(25.0 + 1.6)\ \text{W} = 5320\ \text{W}$$

Therefore, the increase in heat transfer from the tube per meter of its length as a result of the addition of fins is

$$\dot{Q}_{\text{increase}} = \dot{Q}_{\text{total, fin}} - \dot{Q}_{\text{no fin}} = 5320 - 537 = \textbf{4783 W} \qquad \text{(per m tube length)}$$

Discussion   The overall effectiveness of the finned tube is

$$\varepsilon_{\text{fin, overall}} = \frac{\dot{Q}_{\text{total, fin}}}{\dot{Q}_{\text{total, no fin}}} = \frac{5320\ \text{W}}{537\ \text{W}} = 9.9$$

That is, the rate of heat transfer from the steam tube increases by a factor of almost 10 as a result of adding fins. This explains the widespread use of the finned surfaces.

3-8 ■ HEAT TRANSFER IN COMMON CONFIGURATIONS

So far, we have considered heat transfer in *simple* geometries such as large plane walls, long cylinders, and spheres. This is because heat transfer in such geometries can be approximated as *one-dimensional,* and simple analytical solutions can be obtained easily. But most problems encountered in practice are two- or three-dimensional and involve rather complicated geometries for which no simple solutions are available.

An important class of heat transfer problems for which simple solutions are obtained encompasses those involving two surfaces maintained at *con-*

stant temperatures T_1 and T_2. The steady rate of heat transfer between these two surfaces is expressed as

$$\dot{Q} = Sk(T_1 - T_2) \qquad (3\text{-}79)$$

where S is the **conduction shape factor,** which has the dimension of *length*, and k is the thermal conductivity of the medium between the surfaces. The conduction shape factor depends on the *geometry* of the system only.

Conduction shape factors have been determined for a number of configurations encountered in practice and are given in Table 3-8 for some common cases. More comprehensive tables are available in the literature. Once the value of the shape factor is known for a specific geometry, the total steady heat transfer rate can be determined from the equation above using the specified two constant temperatures of the two surfaces and the thermal conductivity of the medium between them. Note that conduction shape factors are applicable only when heat transfer between the two surfaces is by *conduction*. Therefore, they cannot be used when the medium between the surfaces is a liquid or gas, which involves natural or forced convection currents.

A comparison of Equations 3-4 and 3-79 reveals that the conduction shape factor S is related to the thermal resistance R by $R = 1/kS$ or $S = 1/kR$. Thus, these two quantities are the inverse of each other when the thermal conductivity of the medium is unity. The use of the conduction shape factors is illustrated below with examples.

EXAMPLE 3-15   Heat Loss from Buried Steam Pipes

A 30-m-long, 10-cm-diameter hot water pipe of a district heating system is buried in the soil 50 cm below the ground surface, as shown in Figure 3-76. The outer surface temperature of the pipe is 80°C. Taking the surface temperature of the earth to be 10°C and the thermal conductivity of the soil at that location to be 0.9 W/m · °C, determine the rate of heat loss from the pipe.

Solution   The hot water pipe of a district heating system is buried in the soil. The rate of heat loss from the pipe is to be determined.

Assumptions   **1** Steady operating conditions exist. **2** Heat transfer is two-dimensional (no change in the axial direction). **3** Thermal conductivity of the soil is constant.

Properties   The thermal conductivity of the soil is given to be $k = 0.9$ W/m · °C.

Analysis   The shape factor for this configuration is given in Table 3-8 to be

$$S = \frac{2\pi L}{\ln(4z/D)}$$

since $z > 1.5D$, where z is the distance of the pipe from the ground surface, and D is the diameter of the pipe. Substituting,

$$S = \frac{2\pi \times (30 \text{ m})}{\ln(4 \times 0.5/0.1)} = 62.9 \text{ m}$$

Then the steady rate of heat transfer from the pipe becomes

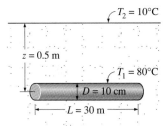

FIGURE 3-76
Schematic for Example 3-15.

TABLE 3-8

Conduction shape factors S for several configurations for use in $\dot{Q} = kS(T_1 - T_2)$ to determine the steady rate of heat transfer through a medium of thermal conductivity k between the surfaces at temperatures T_1 and T_2

(1) Isothermal cylinder of length L buried in a semi-infinite medium ($L \gg D$ and $z > 1.5D$) $$S = \frac{2\pi L}{\ln(4z/D)}$$ 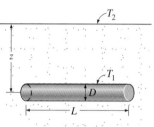	(2) Vertical isothermal cylinder of length L buried in a semi-infinite medium ($L \gg D$) $$S = \frac{2\pi L}{\ln(4L/D)}$$ 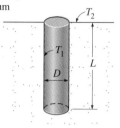
(3) Two parallel isothermal cylinders placed in an infinite medium ($L \gg D_1, D_2, z$) $$S = \frac{2\pi L}{\cosh^{-1}\left(\dfrac{4z^2 - D_1^2 - D_2^2}{2 D_1 D_2}\right)}$$ 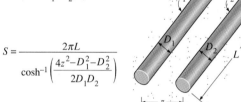	(4) A row of equally spaced parallel isothermal cylinders buried in a semi-infinite medium ($L \gg D$, z and $w > 1.5D$) $$S = \frac{2\pi L}{\ln\left(\dfrac{2w}{\pi D} \sinh \dfrac{2\pi z}{w}\right)}$$ (per cylinder) 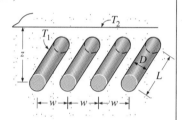
(5) Circular isothermal cylinder of length L in the midplane of an infinite wall ($z > 0.5D$) $$S = \frac{2\pi L}{\ln(8z/\pi D)}$$ 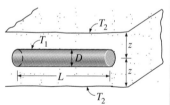	(6) Circular isothermal cylinder of length L at the center of a square solid bar of the same length $$S = \frac{2\pi L}{\ln(1.08 w/D)}$$ 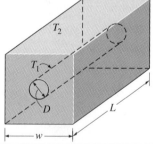
(7) Eccentric circular isothermal cylinder of length L in a cylinder of the same length ($L > D_2$) $$S = \frac{2\pi L}{\cosh^{-1}\left(\dfrac{D_1^2 + D_2^2 - 4z^2}{2 D_1 D_2}\right)}$$ 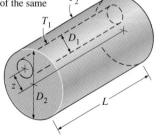	(8) Large plane wall $$S = \frac{A}{L}$$ 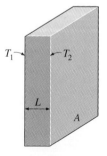

TABLE 3-8 (Concluded)

(9) A long cylindrical layer

$$S = \frac{2\pi L}{\ln(D_2/D_1)}$$

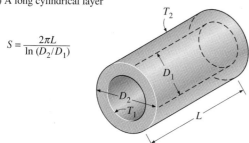

(10) A square flow passage

(a) For $a/b > 1.4$,

$$S = \frac{2\pi L}{0.93 \ln(0.948 a/b)}$$

(b) For $a/b < 1.41$,

$$S = \frac{2\pi L}{0.785 \ln(a/b)}$$

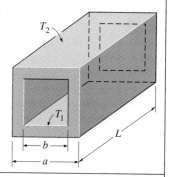

(11) A spherical layer

$$S = \frac{2\pi D_1 D_2}{D_2 - D_1}$$

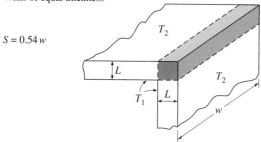

(12) Disk buried parallel to the surface in a semi-infinite medium $(z \gg D)$

$$S = 4D$$

$(S = 2D$ when $z = 0)$

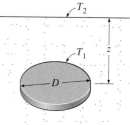

(13) The edge of two adjoining walls of equal thickness

$$S = 0.54 w$$

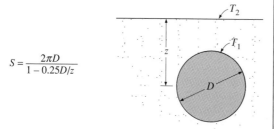

(14) Corner of three walls of equal thickness

$$S = 0.15 L$$

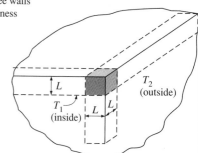

(15) Isothermal sphere buried in a semi-infinite medium

$$S = \frac{2\pi D}{1 - 0.25 D/z}$$

(16) Isothermal sphere buried in a semi-infinite medium at T_2 whose surface is insulated

$$S = \frac{2\pi D}{1 + 0.25 D/z}$$

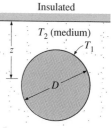

$$\dot{Q} = Sk(T_1 - T_2) = (62.9 \text{ m})(0.9 \text{ W/m} \cdot °C)(80 - 10)°C = \textbf{3963 W}$$

Discussion Note that this heat is conducted from the pipe surface to the surface of the earth through the soil and then transferred to the atmosphere by convection and radiation.

EXAMPLE 3-16 Heat Transfer between Hot and Cold Water Pipes

A 5-m-long section of hot and cold water pipes run parallel to each other in a thick concrete layer, as shown in Figure 3-77. The diameters of both pipes are 5 cm, and the distance between the centerline of the pipes is 30 cm. The surface temperatures of the hot and cold pipes are 70°C and 15°C, respectively. Taking the thermal conductivity of the concrete to be $k = 0.75$ W/m · °C, determine the rate of heat transfer between the pipes.

Solution Hot and cold water pipes run parallel to each other in a thick concrete layer. The rate of heat transfer between the pipes is to be determined.

Assumptions **1** Steady operating conditions exist. **2** Heat transfer is two-dimensional (no change in the axial direction). **3** Thermal conductivity of the concrete is constant.

Properties The thermal conductivity of concrete is given to be $k = 0.75$ W/m · °C.

Analysis The shape factor for this configuration is given in Table 3-8 to be

$$S = \frac{2\pi L}{\cosh^{-1}\left(\dfrac{4z^2 - D_1^2 - D_2^2}{2D_1 D_2}\right)}$$

where z is the distance between the centerlines of the pipes and L is their length. Substituting,

$$S = \frac{2\pi \times (5 \text{ m})}{\cosh^{-1}\left(\dfrac{4 \times 0.3^2 - 0.05^2 - 0.05^2}{2 \times 0.05 \times 0.05}\right)} = 6.34 \text{ m}$$

Then the steady rate of heat transfer between the pipes becomes

$$\dot{Q} = Sk(T_1 - T_2) = (6.34 \text{ m})(0.75 \text{ W/m} \cdot °C)(70 - 15°)C = \textbf{262 W}$$

Discussion We can reduce this heat loss by placing the hot and cold water pipes further away from each other.

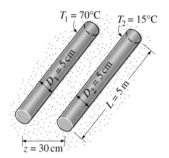

$T_1 = 70°C$ $T_2 = 15°C$

$D_1 = 5$ cm

$D_2 = 5$ cm

$L = 5$ m

$z = 30$ cm

FIGURE 3-77
Schematic for Example 3-16.

It is well known that insulation reduces heat transfer and saves energy and money. Decisions on the right amount of insulation are based on a heat transfer analysis, followed by an economic analysis to determine the "monetary value" of energy loss. This is illustrated below with an example.

EXAMPLE 3-17 Cost of Heat Loss through Walls in Winter

Consider an electrically heated house whose walls are 9 ft high and have an *R*-value of insulation of 13 (i.e., a thickness-to-thermal conductivity ratio of

$L/k = 13 \text{ h} \cdot \text{ft}^2 \cdot °\text{F/Btu}$). Two of the walls of the house are 40 ft long and the others are 30 ft long. The house is maintained at 75°F at all times, while the temperature of the outdoors varies. Determine the amount of heat lost through the walls of the house on a certain day during which the average temperature of the outdoors is 45°F. Also, determine the cost of this heat loss to the homeowner if the unit cost of electricity is \$0.075/kWh. For combined convection and radiation heat transfer coefficients, use the ASHRAE (American Society of Heating, Refrigeration, and Air Conditioning Engineers) recommended values of $h_i = 1.46 \text{ Btu/h} \cdot \text{ft}^2 \cdot °\text{F}$ for the inner surface of the walls and $h_o = 4.0 \text{ Btu/h} \cdot \text{ft}^2 \cdot °\text{F}$ for the outer surface of the walls under 15 mph wind conditions in winter.

Solution An electrically heated house with R-13 insulation is considered. The amount of heat lost through the walls and its cost are to be determined.

Assumptions **1** The indoor and outdoor air temperatures have remained at the given values for the entire day so that heat transfer through the walls is steady. **2** Heat transfer through the walls is one-dimensional since any significant temperature gradients in this case will exist in the direction from the indoors to the outdoors. **3** The radiation effects are accounted for in the heat transfer coefficients.

Analysis This problem involves conduction through the wall and convection at its surfaces and can best be handled by making use of the thermal resistance concept and drawing the thermal resistance network, as shown in Figure 3-78. The heat transfer surface area of the walls is

$$A = \text{Circumference} \times \text{Height} = (2 \times 30 \text{ ft} + 2 \times 40 \text{ ft})(9 \text{ ft}) = 1260 \text{ ft}^2$$

Then the individual resistances are evaluated from their definitions to be

$$R_i = R_{\text{conv}, i} = \frac{1}{h_i A} = \frac{1}{(1.46 \text{ Btu/h} \cdot \text{ft}^2 \cdot °\text{F})(1260 \text{ ft}^2)} = 0.00054 \text{ h} \cdot °\text{F/Btu}$$

$$R_{\text{wall}} = \frac{L}{kA} = \frac{13 \text{ h} \cdot \text{ft}^2 \cdot °\text{F/Btu}}{1260 \text{ ft}^2} = 0.01032 \text{ h} \cdot °\text{F/Btu}$$

$$R_o = R_{\text{conv}, o} = \frac{1}{h_c A} = \frac{1}{(4.0 \text{ Btu/h} \cdot \text{ft}^2 \cdot °\text{F})(1260 \text{ ft}^2)} = 0.00020 \text{ h} \cdot °\text{F/Btu}$$

Noting that all three resistances are in series, the total resistance is determined to be

$$R_{\text{total}} = R_i + R_{\text{wall}} + R_o = 0.00054 + 0.01032 + 0.00020 = 0.01106 \text{ h} \cdot °\text{F/Btu}$$

Then the steady rate of heat transfer through the walls of the house becomes

$$\dot{Q} = \frac{T_{\infty 1} - T_{\infty 2}}{R_{\text{total}}} = \frac{(75 - 45)°\text{F}}{0.01106 \text{ h} \cdot °\text{F/Btu}} = 2712 \text{ Btu/h}$$

Finally, the total amount of heat lost through the walls during a 24-h period and its cost to the home owner are

$$Q = \dot{Q} \, \Delta t = (2712 \text{ Btu/h})(24\text{-h/day}) = \mathbf{65,099 \ Btu/day = 19.1 \ kWh/day}$$

since 1 kWh = 3412 Btu, and

Heating cost = (Energy lost)(Cost of energy) = (19.1 kWh/day)(\$0.075/kWh)
$$= \mathbf{\$1.43/day}$$

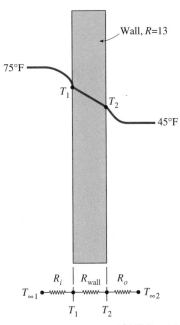

FIGURE 3-78
Schematic for Example 3-17.

Discussion   The heat losses through the walls of the house will cost the home owner that day $1.43 worth of electricity.

3-9 ■ SUMMARY

One-dimensional heat transfer through a simple or composite body exposed to convection from both sides to mediums at temperatures $T_{\infty 1}$ and $T_{\infty 2}$ can be expressed as

$$\dot{Q} = \frac{T_{\infty 1} - T_{\infty 2}}{R_{\text{total}}} \qquad \text{(W)}$$

where R_{total} is the total thermal resistance between the two mediums. For a plane wall exposed to convection on both sides, the total resistance is expressed as

$$R_{\text{total}} = R_{\text{conv, 1}} + R_{\text{wall}} + R_{\text{conv, 2}} = \frac{1}{h_1 A} + \frac{L}{kA} + \frac{1}{h_2 A} \qquad \text{(°C/W)}$$

This relation can be extended to plane walls that consist of two or more layers by adding an additional resistance for each additional layer. The elementary thermal resistance relations can be expressed as follows:

$$\text{Conduction resistance (plane wall):} \qquad R_{\text{wall}} = \frac{L}{kA}$$

$$\text{Conduction resistance (cylinder):} \qquad R_{\text{cyl}} = \frac{\ln(r_2/r_1)}{2\pi L k}$$

$$\text{Conduction resistance (sphere):} \qquad R_{\text{sph}} = \frac{r_2 - r_1}{4\pi r_1 r_2 k}$$

$$\text{Convection resistance:} \qquad R_{\text{conv}} = \frac{1}{hA}$$

$$\text{Interface resistance:} \qquad R_{\text{interface}} = \frac{1}{h_c A} = \frac{R_c}{A}$$

$$\text{Radiation resistance:} \qquad R_{\text{rad}} = \frac{1}{h_{\text{rad}} A}$$

where h_c is the thermal contact conductance, R_c is the thermal contact resistance, and the radiation heat transfer coefficient is defined as

$$h_{\text{rad}} = \varepsilon\sigma(T_s^2 + T_{\text{surr}}^2)(T_s + T_{\text{surr}}) \qquad \text{(W/m}^2 \cdot \text{K)}$$

Once the rate of heat transfer is available, the *temperature drop* across any layer can be determined from

$$\Delta T = \dot{Q} R \qquad \text{(°C)}$$

The thermal resistance concept can also be used to solve steady heat transfer problems involving parallel layers or combined series-parallel arrangements.

Adding insulation to a cylindrical pipe or a spherical shell will increase the rate of heat transfer if the outer radius of the insulation is less than the

$$r_{\text{cr, cylinder}} = \frac{k_{\text{ins}}}{h}$$

$$r_{\text{cr, sphere}} = \frac{2k_{\text{ins}}}{h}$$

Thermal insulations are materials or a combination of materials that are used primarily to provide resistance to heat flow. Thermal insulations are used for various reasons, such as energy conservation, personnel protection and comfort, maintenance of process temperature, reduction of temperature variation and fluctuations, condensation and corrosion prevention, fire protection, freezing protection, and reduction of noise and vibration. Insulation materials are classified as fibrous, cellular, granular, and reflective, and they are available in the form of loose-fill, blankets and batts, rigid insulations, insulating cements, formed-in-place insulations, and reflective insulations. Important considerations in the selection of insulations are the purpose, environment, ease of handling and installation, and cost. Optimum thickness of insulation is usually determined on the basis of minimum combined cost of insulation and heat lost.

The effectiveness of an insulation is often given in terms of its *R-value,* the thermal resistance of the material per unit surface area, expressed as

$$R\text{-value} = \frac{L}{k} \qquad \text{(flat insulation)}$$

where L is the thickness and k is the thermal conductivity of the material.

Finned surfaces are commonly used in practice to enhance heat transfer. Fins enhance heat transfer from a surface by exposing a larger surface area to convection. The temperature distribution along the fin for very long fins and for fins with negligible heat transfer at the fin are given by

Infinitely long fin: $\qquad \dfrac{T(x) - T_\infty}{T_b - T_\infty} = e^{-x\sqrt{hp/kA_c}}$

Insulated fin tip: $\qquad \dfrac{T(x) - T_\infty}{T_b - T_\infty} = \dfrac{\cosh a(L - x)}{\cosh aL}$

where $a = \sqrt{hp/kA_c}$, p is perimeter, and A_c is the cross-sectional area of the fin. The rates of heat transfer for both cases are given to be

Infinitely long fin: $\qquad \dot{Q}_{\text{long fin}} = -kA_c \left.\dfrac{dT}{dx}\right|_{x=0} = \sqrt{hpkA_c}\,(T_b - T_\infty)$

Insulated fin tip: $\qquad \dot{Q}_{\text{insulated tip}} = -kA_c \left.\dfrac{dT}{dx}\right|_{x=0} = \sqrt{hpkA_c}\,(T_b - T_\infty)\tanh aL$

Fins exposed to convection at their tips can be treated as fins with insulated tips by using the corrected length $L_c = L + A_c/p$ instead of the actual fin length.

The temperature of a fin drops along the fin, and thus the heat transfer from the fin will be less because of the decreasing temperature difference toward the fin tip. To account for the effect of this decrease in temperature on heat transfer, we define *fin efficiency* as

$$\eta_{fin} = \frac{\dot{Q}_{fin}}{\dot{Q}_{fin,\,max}} = \frac{\text{Actual heat transfer rate from the fin}}{\text{Ideal heat transfer rate from the fin if}}$$
$$\text{the entire fin were at base temperature}$$

When the fin efficiency is available, the rate of heat transfer from a fin can be determined from

$$\dot{Q}_{fin} = \eta_{fin}\dot{Q}_{fin,\,max} = \eta_{fin}hA_{fin}\,(T_b - T_\infty)$$

The performance of the fins is judged on the basis of the enhancement in heat transfer relative to the no-fin case and is expressed in terms of the *fin effectiveness* ε_{fin}, defined as

$$\varepsilon_{fin} = \frac{\dot{Q}_{fin}}{\dot{Q}_{no\;fin}} = \frac{\dot{Q}_{fin}}{hA_b\,(T_b - T_\infty)} = \frac{\begin{array}{c}\text{Heat transfer rate from}\\ \text{the fin of } base\ area\ A_b\end{array}}{\begin{array}{c}\text{Heat transfer rate from}\\ \text{the surface of } area\ A_b\end{array}}$$

Here, A_b is the cross-sectional area of the fin at the base and $\dot{Q}_{no\;fin}$ represents the rate of heat transfer from this area if no fins are attached to the surface. The *overall effectiveness* for a finned surface is defined as the ratio of the total heat transfer from the finned surface to the heat transfer from the same surface if there were no fins,

$$\varepsilon_{fin,\,overall} = \frac{\dot{Q}_{total,\,fin}}{\dot{Q}_{total,\,no\;fin}} = \frac{h(A_{unfin} + \eta_{fin}A_{fin})(T_b - T_\infty)}{hA_{no\;fin}\,(T_b - T_\infty)}$$

Fin efficiency and fin effectiveness are related to each other by

$$\varepsilon_{fin} = \frac{A_{fin}}{A_b}\,\eta_{fin}$$

Certain multidimensional heat transfer problems involve two surfaces maintained at *constant* temperatures T_1 and T_2. The steady rate of heat transfer between these two surfaces is expressed as

$$\dot{Q} = Sk(T_1 - T_2)$$

where S is the *conduction shape factor* that has the dimension of *length* and k is the thermal conductivity of the medium between the surfaces.

REFERENCES AND SUGGESTED READING

1. American Society of Heating, Refrigeration, and Air Conditioning Engineers. *Handbook of Fundamentals*. Atlanta: ASHRAE, 1993.

2. R. V. Andrews. "Solving Conductive Heat Transfer Problems with Electrical-Analogue Shape Factors." *Chemical Engineering Progress* 5 (1955), p. 67.

3. R. Barron. *Cryogenic Systems.* New York: McGraw-Hill, 1967.

4. Y. Bayazitoglu and M. N. Özişik. *Elements of Heat Transfer.* New York: McGraw-Hill, 1988.

5. H. S. Carslaw and J. C. Jaeger. *Conduction of Heat in Solids.* London: Oxford University Press, 1959.

6. C. Danish. "Factors to Consider in Planning Reinsulation Projects." *Plant Engineering,* November 1985, pp. 81–83.

7. W. M. Edmunds. "Residential Insulation." *ASTM Standardization News,* January 1989, pp. 36–39.

8. L. S. Fletcher. "Recent Developments in Contact Conductance Heat Transfer." *Journal of Heat Transfer* 110, no. 4B (1988), pp. 1059–79.

9. E. Fried. "Thermal Conduction Contribution to Heat Transfer at Contacts." *Thermal Conductivity,* vol. 2, ed. R. P. Tye. London: Academic Press, 1969.

10. K. A. Gardner. "Efficiency of Extended Surfaces." *Trans. ASME* 67 (1945), pp. 621–31.

11. E. Hahne and U. Grigull. "Formfactor und Formwiderstand der stationaren mehrdimensionalen Warmeleiteitung." *International Journal of Heat Mass Transfer* 18 (1975), p. 75.

12. J. P. Holman. *Heat Transfer.* 7th ed. New York: McGraw-Hill, 1990.

13. F. P. Incropera and D. P. DeWitt. *Introduction to Heat Transfer.* 2nd ed. New York: John Wiley & Sons, 1990.

14. W. I. Irwin. "Insulate Intelligently." *Chemical Engineering Progress,* May 1991, pp. 51–55.

15. D. Q. Kern and A. D. Kraus. *Extended Surface Heat Transfer.* New York: McGraw-Hill, 1972.

16. F. Kreith and M. S. Bohn. *Principles of Heat Transfer.* 5th ed. St. Paul, MN: West Publishing, 1993.

17. S. S. Kutateladze. *Fundamentals of Heat Transfer.* New York: Academic Press, 1963.

18. V. M. Liss. "Selecting Thermal Insulation." *Chemical Engineering,* May 26, 1986, pp. 103–5.

19. W. A. Lotz. "Facts About Thermal Insulation." *ASHRAE Journal,* June 1969, pp. 83–84.

20. M. N. Özişik. *Heat Transfer—A Basic Approach.* New York: McGraw-Hill, 1985.

21. G. P. Peterson. "Thermal Contact Resistance in Waste Heat Recovery Systems." *Proceedings of the 18th ASME/ETCE Hydrocarbon Processing Symposium.* Dallas, TX, 1987, pp. 45–51.

22. F. J. Powell. "Thermal Insulation—Still Number One." *ASTM Standardization News,* January 1989, pp. 32–35.

23. S. Song, M. M. Yovanovich, and F. O. Goodman. "Thermal Gap Conductance of Conforming Surfaces in Contact." *Journal of Heat Transfer* 115 (1993), p. 533.

24. J. E. Sunderland and K. R. Johnson. "Shape Factors for Heat Conduction through Bodies with Isothermal or Convective Boundary Conditions." *Trans. ASHRAE* 10 (1964), pp. 237–41.

25. N. V. Suryanarayana. *Engineering Heat Transfer.* St. Paul, MN: West Publishing, 1995.

26. L. C. Thomas. *Heat Transfer.* Englewood Cliffs, NJ: Prentice Hall, 1992.

27. F. M. White. *Heat and Mass Transfer.* Reading, MA: Addison-Wesley, 1988.

PROBLEMS

Steady Heat Conduction in Plane Walls

3-1C Consider one-dimensional heat conduction through a cylindrical rod of diameter D and length L. What is the heat transfer area of the rod if (a) the lateral surfaces of the rod are insulated and (b) the top and bottom surfaces of the rod are insulated?

3-2C Consider a 1.5-m $\times$ 2-m glass window whose thickness is 0.01 m. What is the heat transfer area of the window?

3-3C Consider heat conduction through a plane wall. Does the energy content of the wall change during steady heat conduction? How about during transient conduction? Explain.

3-4C Consider heat conduction through a wall of thickness L and area A. Under what conditions will the temperature distributions in the wall be a straight line?

3-5C What does the thermal resistance of a medium represent?

3-6C How is the combined heat transfer coefficient defined? What convenience does it offer in heat transfer calculations?

3-7C Can we define the convection resistance per unit surface area as the inverse of the convection heat transfer coefficient?

3-8C Why are the convection and the radiation resistances at a surface in parallel instead of being in series?

3-9C Consider a surface of area A at which the convection and radiation heat transfer coefficients are h_{conv} and h_{rad}, respectively. Explain how you would determine (a) the single equivalent heat transfer coefficient, and (b) the equivalent thermal resistance. Assume the medium and the surrounding surfaces are at the same temperature.

3-10C How does the thermal resistance network associated with a single-layer plane wall differ from the one associated with a five-layer composite wall?

3-11C Consider steady one-dimensional heat transfer through a multilayer medium. If the rate of heat transfer $\dot{Q}$ is known, explain how you would determine the temperature drop across each layer.

3-12C Consider steady one-dimensional heat transfer through a plane wall exposed to convection from both sides to environments at known temperatures $T_{\infty 1}$ and $T_{\infty 2}$ with known heat transfer coefficients h_1 and h_2. Once the rate of heat transfer $\dot{Q}$ has been evaluated, explain how you would determine the temperature of each surface.

3-13C Someone comments that a microwave oven can be viewed as a conventional oven with zero convection resistance at the surface of the food. Is this an accurate statement?

3-14C Consider a window glass consisting of two 4-mm-thick glass sheets pressed tightly against each other. Compare the heat transfer rate through this window with that of one consisting of a single 8-mm-thick glass sheet under identical conditions.

3-15C Consider steady heat transfer through the wall of a room in winter. The convection heat transfer coefficient at the outer surface of the wall is three times that of the inner surface as a result of the winds. On which surface of the wall do you think the temperature will be closer to the surrounding air temperature? Explain.

3-16C The bottom of a pan is made of a 4-mm-thick aluminum layer. In order to increase the rate of heat transfer through the bottom of the pan, someone proposes a design for the bottom that consists of a 3-mm-thick copper layer sandwiched between two 2-mm-thick aluminum layers. Will the new design conduct heat better? Explain. Assume perfect contact between the layers.

3-17C Consider two cold canned drinks, one wrapped in a blanket and the other placed on a table in the same room. Which drink will warm up faster?

3-18 Consider a 4-m-high, 6-m-wide, and 0.3-m-thick brick wall whose thermal conductivity is $k = 0.8 \text{W/m} \cdot {}^\circ\text{C}$. On a certain day, the temperatures of the inner and the outer surfaces of the wall are measured to be 14°C and 6°C, respectively. Determine the rate of heat loss through the wall on that day.

3-19 Consider a 1.2-m-high and 2-m-wide glass window whose thickness is 6 mm and thermal conductivity is $k = 0.78 \text{ W/m} \cdot {}^\circ\text{C}$. Determine the steady rate of heat transfer through this glass window and the temperature of its inner surface for a day during which the room is maintained at 24°C while the temperature of the outdoors is -5°C. Take the convection heat

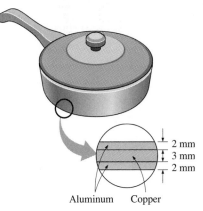

Aluminum   Copper

FIGURE P3-16C

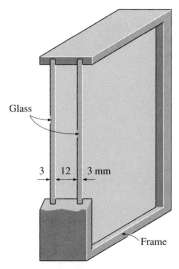

Glass

3 | 12 | 3 mm

Frame

FIGURE P3-20

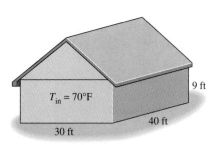

$T_{in} = 70°F$

30 ft

40 ft

9 ft

FIGURE P3-22E

FIGURE P3-24

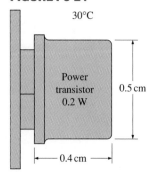

30°C

Power
transistor
0.2 W

0.5 cm

0.4 cm

transfer coefficients on the inner and outer surfaces of the window to be $h_1 = 10$ W/m$^2 \cdot$ °C and $h_2 = 25$ W/m$^2 \cdot$ °C, and disregard any heat transfer by radiation.

3-20 Consider a 1.2-m-high and 2-m-wide double-pane window consisting of two 3-mm-thick layers of glass ($k = 0.78$ W/m $\cdot$ °C) separated by a 12-mm-wide stagnant air space ($k = 0.026$ W/m $\cdot$ °C). Determine the steady rate of heat transfer through this double-pane window and the temperature of its inner surface for a day during which the room is maintained at 24°C while the temperature of the outdoors is −5°C. Take the convection heat transfer coefficients on the inner and outer surfaces of the window to be $h_1 = 10$ W/m$^2 \cdot$ °C and $h_2 = 25$ W/m$^2 \cdot$ °C, and disregard any heat transfer by radiation. *Answers:* 114 W, 19.2°C

3-21 Repeat Problem 3-20, assuming the space between the two glass layers is evacuated.

3-22E Consider an electrically heated brick house ($k = 0.40$ Btu/h $\cdot$ ft $\cdot$ °F) whose walls are 9 ft high and 1 ft thick. Two of the walls of the house are 40 ft long and the others are 30 ft long. The house is maintained at 70°F at all times while the temperature of the outdoors varies. On a certain day, the temperature of the inner surface of the walls is measured to be at 55°F while the average temperature of the outer surface is observed to remain at 45°F during the day for 10 h and at 35°F at night for 14 h. Determine the amount of heat lost from the house that day. Also determine the cost of that heat loss to the homeowner for an electricity price of $0.075/kWh.

3-23 A cylindrical resistor element on a circuit board dissipates 0.15 W of power in an environment at 40°C. The resistor is 1.2 cm long, and has a diameter of 0.3 cm. Assuming heat to be transferred uniformly from all surfaces, determine (*a*) the amount of heat this resistor dissipates during a 24-h period, (*b*) the heat flux on the surface of the resistor, in W/m^2, and (*c*) the surface temperature of the resistor for a combined convection and radiation heat transfer coefficient of 9 W/m$^2 \cdot$ °C.

3-24 Consider a power transistor that dissipates 0.2 W of power in an environment at 30°C. The transistor is 0.4 cm long and has a diameter of 0.5 cm. Assuming heat to be transferred uniformly from all surfaces, determine (*a*) the amount of heat this resistor dissipates during a 24-h period, in kWh; (*b*) the heat flux on the surface of the transistor, in W/m^2; and (*c*) the surface temperature of the resistor for a combined convection and radiation heat transfer coefficient of 12 W/m$^2 \cdot$ °C.

3-25 A 12-cm × 18-cm circuit board houses on its surface 100 closely spaced logic chips, each dissipating 0.07 W. The heat transfer from the back surface of the board is negligible. If the heat transfer coefficient on the surface of the board is 10 W/m$^2 \cdot$ °C, determine (*a*) the heat flux on the surface of the circuit board, in W/m^2; (*b*) the surface temperature of the chips; and (*c*) the thermal resistance between the surface of the circuit board and the cooling medium, in °C/W.

3-26 Consider a person standing in a room at 20°C with an exposed surface area of 1.7 m^2. The deep body temperature of the human body is 37°C, and the thermal conductivity of the human tissue near the skin is about 0.3 W/m · °C. The body is losing heat at a rate of 150 W by natural convection and radiation to the surroundings. Taking the body temperature 0.5 cm beneath the skin to be 37°C, determine the skin temperature of the person. *Answer:* 35.5° C

3-27 Water is boiling in a 25-cm-diameter aluminum pan ($k = 237$ W/m · °C) at 95°C. Heat is transferred steadily to the boiling water in the pan through its 0.5-cm-thick flat bottom at a rate of 600 W. If the inner surface temperature of the bottom of the pan is 108°C, determine (*a*) the boiling heat transfer coefficient on the inner surface of the pan, and (*b*) the outer surface temperature of the bottom of the pan.

3-28E A wall is constructed of two layers of 0.5-in-thick sheetrock ($k = 0.10$ Btu/h · ft · °F), which is a plasterboard made of two layers of heavy paper separated by a layer of gypsum, placed 5 in. apart. The space between the sheetrocks is filled with fiberglass insulation ($k = 0.020$ Btu/h · ft · °F). Determine (*a*) the thermal resistance of the wall, and (*b*) its *R*-value of insulation in English units.

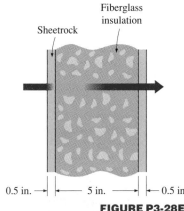

FIGURE P3-28E

3-29 The roof of a house consists of a 15-cm-thick concrete slab ($k = 2$ W/m · °C) that is 15 m wide and 20 m long. The convection heat transfer coefficients on the inner and outer surfaces of the roof are 5 and 12 W/m² · °C, respectively. In a clear winter night, the ambient air is reported to be at 10°C, while the night sky temperature is 100 K. The house and the interior surfaces of the wall are maintained at a constant temperature of 20°C. The emissivity of both surfaces of the concrete roof is 0.9. Considering both radiation and convection heat transfers, determine the rate of heat transfer through the roof, and the inner surface temperature of the roof.

If the house is heated by a furnace burning natural gas with an efficiency of 80 percent, and the price of natural gas is $0.60/therm (1 therm = 105,500 kJ of energy content), determine the money lost through the roof that night during a 14-h period.   *Answers:* 16.0°C, $4.37

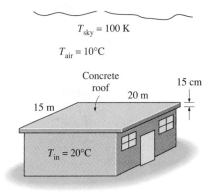

FIGURE P3-29

3-30 The heat generated in the circuitry on the surface of a silicon chip ($k = 130$ W/m · °C) is conducted to the ceramic substrate to which it is attached. The chip is 6 mm × 6 mm in size and 0.5 mm thick and dissipates 3 W of power. Determine the temperature difference between the front and back surfaces of the chip in steady operation.

3-31 A 2-m × 1.5-m section of wall of an industrial furnace burning natural gas is not insulated, and the temperature at the outer surface of this section is measured to be 80°C. The temperature of the furnace room is 30°C, and the combined convection and radiation heat transfer coefficient at the surface of the outer furnace is 10 W/m² · °C. It is proposed to insulate this section of the furnace wall with glass wool insulation ($k = 0.038$ W/m · °C) in order to reduce the heat loss by 90 percent. Assuming the outer surface temperature of the metal section still remains at about 80°C, determine the thickness of the insulation that needs to be used.

The furnace operates continuously and has an efficiency of 78 percent. The price of the natural gas is $0.55/therm (1 therm = 105,500 kJ of energy

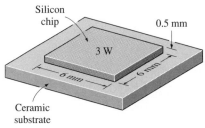

FIGURE P3-30

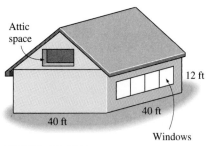

Attic
space

12 ft

40 ft

40 ft

Windows

FIGURE P3-32E

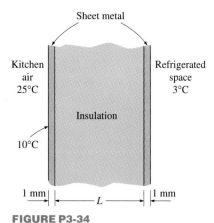

Sheet metal

Kitchen
air
25°C

Refrigerated
space
3°C

Insulation

10°C

1 mm

L

1 mm

FIGURE P3-34

FIGURE P3-36E

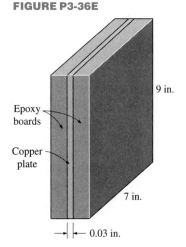

9 in.

Epoxy
boards

Copper
plate

7 in.

0.03 in.

content). If the installation of the insulation will cost $250 for materials and labor, determine how long it will take for the insulation to pay for itself from the energy it saves.

3-32E Consider a house whose walls are 12 ft high and 40 ft long. Two of the walls of the house have no windows, while each of the other two walls has four windows made of 0.25-in.-thick glass ($k = 0.45$ Btu/h · ft · °F), 3 ft × 5 ft in size. The walls are certified to have an R-value of 19 (i.e., an L/k value of 19 h · ft^2 · °F/Btu). Disregarding any direct radiation gain or loss through the windows and taking the heat transfer coefficients at the inner and outer surfaces of the house to be 2 and 4 Btu/h · ft^2 · °F, respectively, determine the ratio of the heat transfer through the walls with and without windows.

3-33 Consider a house that has a 10-m × 20-m base and a 4-m-high wall. All four walls of the house have an R-value of 2.31 m^2 · °C/W. The two 10-m × 4-m walls have no windows. The third wall has five windows made of 0.5-cm-thick glass ($k = 0.78$ W/m · °C), 1.2 m × 1.8 m in size. The fourth wall has the same size and number of windows, but they are double-paned with a 1.5-cm-thick stagnant air space ($k = 0.026$ W/m · °C) enclosed between two 0.5-cm-thick glass layers. The thermostat in the house is set at 22°C and the average temperature outside at that location is 5°C during the seven-month-long heating season. Disregarding any direct radiation gain or loss through the windows and taking the heat transfer coefficients at the inner and outer surfaces of the house to be 7 and 15 W/m^2 · °C, respectively, determine the average rate of heat transfer through each wall.

If the house is electrically heated and the price of electricity is $0.09/kWh, determine the amount of money this household will save per heating season by converting the single-pane windows to double-pane windows.

3-34 The wall of a refrigerator is constructed of fiberglass insulation ($k = 0.035$ W/m · °C) sandwiched between two layers of 1-mm-thick sheet metal ($k = 15.1$ W/m · °C). The refrigerated space is maintained at 3°C, and the average heat transfer coefficients at the inner and outer surfaces of the wall are 4 W/m^2 · °C and 9 W/m^2 · °C, respectively. The kitchen temperature averages 25°C. It is observed that condensation occurs on the outer surfaces of the refrigerator when the temperature of the outer surface drops to 20°C. Determine the minimum thickness of fiberglass insulation that needs to be used in the wall in order to avoid condensation on the outer surfaces.

3-35 Heat is to be conducted along a circuit board that has a copper layer on one side. The circuit board is 15 cm long and 15 cm wide, and the thicknesses of the copper and epoxy layers are 0.1 mm and 1.2 mm, respectively. Disregarding heat transfer from side surfaces, determine the percentages of heat conduction along the copper ($k = 386$ W/m · °C) and epoxy ($k = 0.26$ W/m · °C) layers. Also determine the effective thermal conductivity of the board. *Answers:* 0.8 percent, 99.2 percent, and 29.9 W/m · °C

3-36E A 0.03-in-thick copper plate ($k = 223$ Btu/h · ft · °F) is sandwiched between two 0.1-in.-thick epoxy boards ($k = 0.15$ Btu/h · ft · °F) that are

7 in. × 9 in. in size. Determine the effective thermal conductivity of the board along its 9-in.-long side. What fraction of the heat conducted along that side is conducted through copper?

Thermal Contact Resistance

3-37C What is thermal contact resistance? How is it related to thermal contact conductance?

3-38C Will the thermal contact resistance be greater for smooth or rough plain surfaces?

3-39C A wall consists of two layers of insulation pressed against each other. Do we need to be concerned about the thermal contact resistance at the interface in a heat transfer analysis or can we just ignore it?

3-40C A plate consists of two thin metal layers pressed against each other. Do we need to be concerned about the thermal contact resistance at the interface in a heat transfer analysis or can we just ignore it?

3-41C Consider two surfaces pressed against each other. Now the air at the interface is evacuated. Will the thermal contact resistance at the interface increase or decrease as a result?

3-42C Explain how the thermal contact resistance can be minimized.

3-43 The thermal contact conductance at the interface of two 1-cm-thick copper plates is measured to be 18,000 W/m² · °C. Determine the thickness of the copper plate whose thermal resistance is equal to the thermal resistance of the interface between the plates.

3-44 Six identical power transistors with aluminum casing are attached on one side of a 1.2-cm-thick 20-cm × 30-cm copper plate ($k = 386$ W/m · °C) by screws that exert an average pressure of 10 MPa. The base area of each transistor is 9 cm², and each transistor is placed at the center of 10-cm × 10-cm section of the plate. The interface roughness is estimated to be about 1.4 μm. All transistors are covered by a thick Plexiglas layer, which is a poor conductor of heat, and thus all the heat generated at the junction of the transistor must be dissipated to the ambient at 15°C through the back surface of the copper plate. The combined convection/radiation heat transfer coefficient at the back surface can be taken to be 30 W/m² · °C. If the case temperature of the transistor is not to exceed 85°C, determine the maximum power each transistor can dissipate safely, and the temperature jump at the case-plate interface.

3-45 Two 5-cm-diameter, 15-cm-long aluminum bars ($k = 176$ W/m · °C) with ground surfaces are pressed against each other with a pressure of 20 atm. The bars are enclosed in an insulation sleeve and, thus, heat transfer from the lateral surfaces is negligible. If the top and bottom surfaces of the two-bar system are maintained at temperatures of 150°C and 20°C, respectively,

FIGURE P3-44

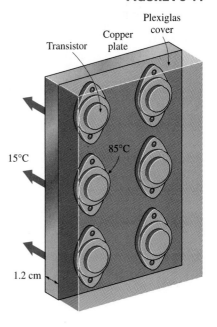

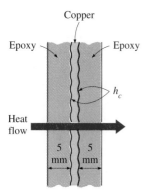

Copper

Epoxy

Epoxy

h_c

Heat
flow

5
mm

5
mm

FIGURE P3-46

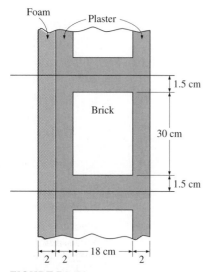

Foam

Plaster

1.5 cm

Brick

30 cm

1.5 cm

2   2

18 cm

2

FIGURE P3-50

determine (*a*) the rate of heat transfer along the cylinders under steady conditions and (*b*) the temperature drop at the interface.
   Answers: (*a*) 142.4 W, (*b*) 6.4°C

3-46   A 1-mm-thick copper plate ($k = 386$ W/m · °C) is sandwiched between two 5-mm-thick epoxy boards ($k = 0.26$ W/m · °C) that are 15 cm × 20 cm in size. If the thermal contact conductance on both sides of the copper plate is estimated to be 4000 W/m · °C, determine the error involved in the total thermal resistance of the plate if the thermal contact conductances are ignored.

Generalized Thermal Resistance Networks

3-47C   When plotting the thermal resistance network associated with a heat transfer problem, explain when two resistances are in series and when they are in parallel.

3-48C   The thermal resistance networks can also be used approximately for multidimensional problems. For what kind of multidimensional problems will the thermal resistance approach give adequate results?

3-49C   What are the two approaches used in the development of the thermal resistance network for two-dimensional problems?

3-50   A 4-m-high and 6-m-wide wall consists of a long 18-cm × 30-cm cross-section of horizontal bricks ($k = 0.72$ W/m · °C) separated by 3-cm-thick plaster layers ($k = 0.22$ W/m · °C). There are also 2-cm-thick plaster layers on each side of the wall, and a 2-cm-thick rigid foam ($k = 0.026$ W/m · °C) on the inner side of the wall. The indoor and the outdoor temperatures are 22°C and −4°C, and the convection heat transfer coefficients on the inner and the outer sides are $h_1 = 10$ W/m² · °C and $h_2 = 20$ W/m² · °C, respectively. Assuming one-dimensional heat transfer and disregarding radiation, determine the rate of heat transfer through the wall.

3-51   A 10-cm-thick wall is to be constructed with 2.5-m-long wood studs ($k = 0.11$ W/m · °C) that have a cross-section of 10 cm × 10 cm. At some point the builder ran out of those studs and started using pairs of 2.5-m-long wood studs that have a cross-section of 5 cm × 10 cm nailed to each other instead. The manganese steel nails ($k = 50$ W/m · °C) are 10 cm long and have a diameter of 0.4 cm. A total of 50 nails are used to connect the two studs, which are mounted to the wall such that the nails cross the wall. The temperature difference between the inner and outer surfaces of the wall is 15°C. Assuming the thermal contact resistance between the two layers to be negligible, determine the rate of heat transfer (*a*) through a solid stud and (*b*) through a stud pair of equal length and width nailed to each other. (c) Also determine the effective conductivity of the nailed stud pair.

3-52 A 12-m-long and 5-m-high wall is constructed of two layers of 1-cm-thick sheetrock ($k = 0.17$ W/m · °C) spaced 12 cm by wood studs ($k = 0.11$ W/m · °C) whose cross-section is 12 cm × 5 cm. The studs are placed vertically 60 cm apart, and the space between them is filled with fiberglass insulation ($k = 0.034$ W/m · °C). The house is maintained at 20°C and the ambient temperature outside is −5°C. Taking the heat transfer coefficients at the inner and outer surfaces of the house to be 8.3 and 34 W/m² · °C, respectively, determine (a) the thermal resistance of the wall considering a representative section of it and (b) the rate of heat transfer through the wall.

3-53E A 10-in.-thick, 30-ft-long, and 10-ft-high wall is to be constructed using 9-in.-long solid bricks ($k = 0.40$ Btu/h · ft · °F) of cross-section 7 in. × 7 in., or identical size bricks with nine square air holes ($k = 0.015$ Btu/h · ft · °F) that are 9 in. long and have a cross-section of 1.5 in. × 1.5 in. There is a 0.5-in.-thick plaster layer ($k = 0.10$ Btu/h · ft · °F) between two adjacent bricks on all four sides and on both sides of the wall. The house is maintained at 75°F and the ambient temperature outside is 35°F. Taking the heat transfer coefficients at the inner and outer surfaces of the wall to be 1.5 and 4 Btu/h · ft² · °F, respectively, determine the rate of heat transfer through the wall constructed of (a) solid bricks and (b) bricks with air holes.

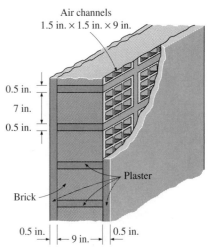

FIGURE P3-53E

3-54 Consider a 5-m-high, 8-m-long, and 0.22-m-thick wall whose representative section is as given in Figure P3-54. The thermal conductivities of various materials used, in W/m · °C, are $k_A = k_F = 2$, $k_B = 8$, $k_C = 20$, $k_D = 15$, and $k_E = 35$. The left and right surfaces of the wall are maintained at uniform temperatures of 300°C and 100°C, respectively. Assuming heat transfer through the wall to be one-dimensional, determine (a) the rate of heat transfer through the wall; (b) the temperature at the point where the sections B, D, and E meet; and (c) the temperature drop across the section F. Disregard any contact resistances at the interfaces.

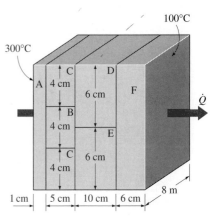

FIGURE P3-54

3-55 Repeat Problem 3-54 assuming that the thermal contact resistance at the interfaces D-F and E-F is 0.00012 m² · °C/W.

3-56 Clothing made of several thin layers of fabric with trapped air in between, often called ski clothing, is commonly used in cold climates because it is light, fashionable, and a very effective thermal insulator. So it is no surprise that such clothing has largely replaced thick and heavy old-fashioned coats.

Consider a jacket made of five layers of 0.1-mm-thick synthetic fabric ($k = 0.13$ W/m · °C) with 1.5-mm-thick air space ($k = 0.026$ W/m · °C) between the layers. Assuming the inner surface temperature of the jacket to be 28°C and the surface area to be 1.1 m², determine the rate of heat loss

FIGURE P3-56

Multilayered ski jacket

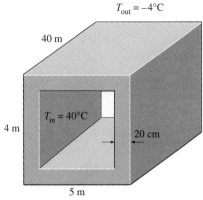

$T_{out} = -4°C$

40 m

4 m

$T_{in} = 40°C$

20 cm

5 m

FIGURE P3-58

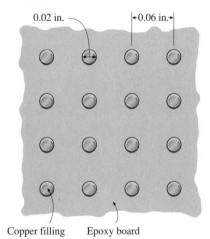

0.02 in.

0.06 in.

Copper filling   Epoxy board

FIGURE P3-59E

FIGURE P3-64

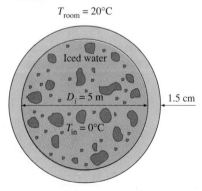

$T_{room} = 20°C$

Iced water

$D_i = 5$ m

1.5 cm

$T_{in} = 0°C$

through the jacket when the temperature of the outdoors is −5°C and the heat transfer coefficient at the outer surface is 25 W/m² · °C.

What would your response be if the jacket is made of a single layer of 0.5-mm-thick synthetic fabric? What should be the thickness of a wool fabric ($k = 0.035$ W/m · °C) if the person is to achieve the same level of thermal comfort wearing a thick wool coat instead of a ski jacket?

3-57 Repeat Problem 3-56 assuming the layers of the jacket are made of cotton fabric ($k = 0.06$ W/m · °C).

3-58 A 5-m-wide, 4-m-high, and 40-m-long kiln used to cure concrete pipes is made of 20-cm-thick concrete walls and ceiling ($k = 0.9$ W/m · °C). The kiln is maintained at 40°C by injecting hot steam into it. The two ends of the kiln, 4 m × 5 m in size, are made of a 3-mm-thick sheet metal covered with 2-cm-thick Styrofoam ($k = 0.033$ W/m · °C). The convection heat transfer coefficients on the inner and the outer surfaces of the kiln are 3000 W/m² · °C and 25 W/m² · °C, respectively. Disregarding any heat loss through the floor, determine the rate of heat loss from the kiln when the ambient air is at −4°C.

3-59E Consider a 6-in. × 8-in. epoxy glass laminate ($k = 0.10$ Btu/h · ft · °F) whose thickness is 0.05 in. In order to reduce the thermal resistance across its thickness, cylindrical copper fillings ($k = 223$ Btu/h · ft · °F) of 0.02 in. diameter are to be planted throughout the board, with a center-to-center distance of 0.06 in. Determine the new value of the thermal resistance of the epoxy board for heat conduction across its thickness as a result of this modification.   *Answer:* 0.0099 h · °F/Btu

Heat Conduction in Cylinders and Spheres

3-60C Consider one-dimensional heat conduction through a plane wall, a long cylinder, and a sphere. For which of these geometries is the heat transfer area constant, and for which of them is it variable? Explain.

3-61C What is an infinitely long cylinder? When is it proper to treat an actual cylinder as being infinitely long, and when is it not?

3-62C Consider a short cylinder whose top and bottom surfaces are insulated. The cylinder is initially at a uniform temperature T_i and is subjected to convection from its side surface to a medium at temperature T_∞, with a heat transfer coefficient of h. Is the heat transfer in this short cylinder one- or two-dimensional? Explain.

3-63C Can the thermal resistance concept be used for a solid cylinder or sphere in steady operation? Explain.

3-64 A 5-m-internal-diameter spherical tank made of 1.5-cm-thick stainless steel ($k = 15$ W/m · °C) is used to store iced water at 0°C. The tank is located in a room whose temperature is 20°C. The walls of the room are also at 20°C. The outer surface of the tank is black (emissivity $\varepsilon = 1$), and heat transfer between the outer surface of the tank and the surroundings is by natural convection and radiation. The convection heat transfer coefficients at the inner and the outer surfaces of the tank are 80 W/m² · °C and 10 W/m² · °C, respectively. Determine (a) the rate of heat transfer to the iced water in the tank and (b) the amount of ice at 0°C that melts during a 24-h period. The heat of fusion of water at atmospheric pressure is h_{if} = 333.7 kJ/kg.

3-65 Steam at 320°C flows in a stainless steel pipe (k = 15 W/m · °C) whose inner and outer diameters are 5 cm and 5.5 cm, respectively. The pipe is covered with 3-cm-thick glass wool insulation (k = 0.038 W/m · °C). Heat is lost to the surroundings at 5°C by natural convection and radiation, with a combined natural convection and radiation heat transfer coefficient of 15 W/m² · °C. Taking the heat transfer coefficient inside the pipe to be 80 W/m² · °C, determine the rate of heat loss from the steam per unit length of the pipe. Also determine the temperature drops across the pipe shell and the insulation.

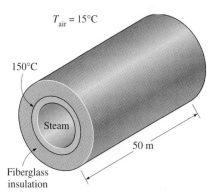

FIGURE P3-66

3-66 A 50-m-long section of a steam pipe whose outer diameter is 10 cm passes through an open space at 15°C. The average temperature of the outer surface of the pipe is measured to be 150°C. If the combined heat transfer coefficient on the outer surface of the pipe is 20 W/m² · °C, determine (a) the rate of heat loss from the steam pipe, (b) the annual cost of this energy lost if steam is generated in a natural gas furnace that has an efficiency of 75 percent and the price of natural gas is $0.52/therm (1 therm = 105,500 kJ), and (c) the thickness of fiberglass insulation (k = 0.035 W/m · °C) needed in order to save 90 percent of the heat lost. Assume the pipe temperature to remain constant at 150°C.

3-67 Consider a 2-m-high electric hot water heater that has a diameter of 40 cm and maintains the hot water at 55°C. The tank is located in a small room whose average temperature is 27°C, and the heat transfer coefficients on the inner and outer surfaces of the heater are 50 and 12 W/m² · °C, respectively. The tank is placed in another 46-cm-diameter sheet metal tank of negligible thickness, and the space between the two tanks is filled with foam insulation (k = 0.03 W/m · °C). The thermal resistances of the water tank and the outer thin sheet metal shell are very small and can be neglected. The price of electricity is $0.08/kWh, and the home owner pays $280 a year for water heating. Determine the fraction of the hot water energy cost of this household that is due to the heat loss from the tank.

Hot water tank insulation kits consisting of 3-cm-thick fiberglass insulation (k = 0.035 W/m · °C) large enough to wrap the entire tank are available in the market for about $30. If such an insulation is installed on this water tank by the home owner himself, how long will it take for this additional insulation to pay for itself?   *Answers:* 17.5 percent, 1.5 year

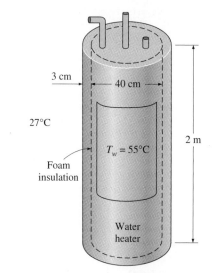

FIGURE P3-67

FIGURE P3-68

3-68 Consider a cold aluminum canned drink that is initially at a uniform temperature of 3°C. The can is 12.5 cm high and has a diameter of 6 cm. If the combined convection/radiation heat transfer coefficient between the can and the surrounding air at 25°C is 10 W/m² · °C, determine how long it will take for the average temperature of the drink to rise to 10°C.

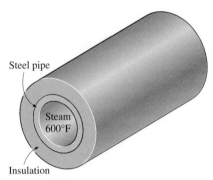

Steel pipe

Steam
600°F

Insulation

FIGURE P3-70E

In an effort to slow down the warming of the cold drink, a person puts the can in a perfectly fitting 1-cm-thick cylindrical rubber insulation ($k = 0.13$ W/m · °C). Now how long will it take for the average temperature of the drink to rise to 10°C? Assume the top of the can is not covered.

3-69 Repeat Problem 3-68, assuming a thermal contact resistance of 0.00008 m² · °C/W between the can and the insulation.

3-70E Steam at 600°F is flowing through a steel pipe ($k = 8.7$ Btu/h · ft · °F) whose inner and outer diameters are 3.5 in. and 4.0 in., respectively, in an environment at 60°F. The pipe is insulated with 2-in.-thick fiberglass insulation ($k = 0.020$ Btu/h · ft · °F). If the heat transfer coefficients on the inside and the outside of the pipe are 30 and 5 Btu/h · ft² · °F, respectively, determine the rate of heat loss from the steam per foot length of the pipe. What is the error involved in neglecting the thermal resistance of the steel pipe in calculations?

3-71 Hot water at an average temperature of 90°C is flowing through a 15-m section of a cast iron pipe ($k = 52$ W/m · °C) whose inner and outer diameters are 4 cm and 4.6 cm, respectively. The outer surface of the pipe, whose emissivity is 0.7, is exposed to the cold air at 10°C in the basement, with a heat transfer coefficient of 15 W/m² · °C. The heat transfer coefficient at the inner surface of the pipe is 120 W/m² · °C. Taking the walls of the basement to be at 10°C also, determine the rate of heat loss from the hot water. Also, determine the average velocity of the water in the pipe if the temperature of the water drops by 3°C as it passes through the basement.

3-72 Repeat Problem 3-71 for a pipe made of copper ($k = 386$ W/m · °C) instead of cast iron.

3-73E Steam exiting the turbine of a steam power plant at 100°F is to be condensed in a large condenser by cooling water flowing through copper pipes ($k = 223$ Btu/h · ft · °F) of inner diameter 0.4 in. and outer diameter 0.6 in. at an average temperature of 70°F. The heat of vaporization of water at 100°F is 1037 Btu/lbm. The heat transfer coefficients are 1500 Btu/h · ft² · °F on the steam side and 35 Btu/h · ft² · °F on the water side. Determine the length of the tube required to condense steam at a rate of 400 lbm/h.

Answer: 3830 ft

3-74E Repeat Problem 3-73E, assuming that a 0.01-in.-thick layer of mineral deposit ($k = 0.5$ Btu/h · ft · °F) has formed on the inner surface of the pipe.

3-75 The boiling temperature of nitrogen at atmospheric pressure at sea level (1 atm pressure) is −196°C. Therefore, nitrogen is commonly used in low-temperature scientific studies since the temperature of liquid nitrogen in a tank open to the atmosphere will remain constant at −196°C until it is depleted. Any heat transfer to the tank will result in the evaporation of some liquid nitrogen, which has a heat of vaporization of 198 kJ/kg and a density of 810 kg/m³ at 1 atm.

Consider a 3-m-diameter spherical tank that is initially filled with liquid nitrogen at 1 atm and −196°C. The tank is exposed to ambient air at

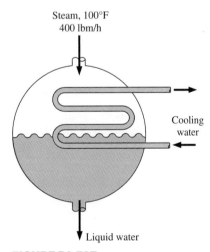

Steam, 100°F
400 lbm/h

Cooling
water

Liquid water

FIGURE P3-73E

15°C, with a combined convection and radiation heat transfer coefficient of 35 W/m² · °C. The temperature of the thin-shelled spherical tank is observed to be almost the same as the temperature of the nitrogen inside. Determine the rate of evaporation of the liquid nitrogen in the tank as a result of the heat transfer from the ambient air if the tank is (a) not insulated, (b) insulated with 5-cm-thick fiberglass insulation (k = 0.035 W/m · °C), and (c) insulated with 2-cm-thick superinsulation which has an effective thermal conductivity of 0.00005 W/m · °C.

3-76 Repeat Problem 3-75 for liquid oxygen, which has a boiling temperature of −183°C, a heat of vaporization of 213 kJ/kg, and a density of 1140 kg/m³ at 1 atm pressure.

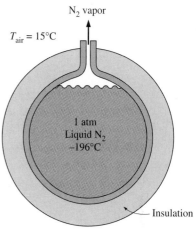

$T_{air} = 15°C$

N₂ vapor

1 atm
Liquid N₂
−196°C

Insulation

FIGURE P3-75

Critical Radius of Insulation

3-77C What is the critical radius of insulation? How is it defined for a cylindrical layer?

3-78C A pipe is insulated such that the outer radius of the insulation is less than the critical radius. Now the insulation is taken off. Will the rate of heat transfer from the pipe increase or decrease for the same pipe surface temperature?

3-79C A pipe is insulated to reduce the heat loss from it. However, measurements indicate that the rate of heat loss has increased instead of decreasing. Can the measurements be right?

3-80C Consider a pipe at a constant temperature whose radius is greater than the critical radius of insulation. Someone claims that the rate of heat loss from the pipe has increased when some insulation is added to the pipe. Is this claim valid?

3-81C Consider an insulated pipe exposed to the atmosphere. Will the critical radius of insulation be greater on calm days or on windy days? Why?

3-82 A 2-mm-diameter and 10-m-long electric wire is tightly wrapped with a 1-mm-thick plastic cover whose thermal conductivity is k = 0.15 W/m · °C. Electrical measurements indicate that a current of 10 A passes through the wire and there is a voltage drop of 8 V along the wire. If the insulated wire is exposed to a medium at $T_\infty = 30°C$ with a heat transfer coefficient of h = 18 W/m² · °C, determine the temperature at the interface of the wire and the plastic cover in steady operation. Also determine if doubling the thickness of the plastic cover will increase or decrease this interface temperature.

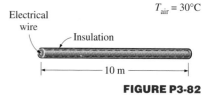

$T_{air} = 30°C$

Electrical
wire

Insulation

10 m

FIGURE P3-82

3-83E A 0.083-in.-diameter electrical wire at 115°F is covered by 0.02-in.-thick plastic insulation (k = 0.075 Btu/h · ft · °F). The wire is exposed to a medium at 50°F, with a combined convection and radiation heat transfer coefficient of 2.5 Btu/h · ft² · °F. Determine if the plastic insulation on the wire will increase or decrease heat transfer from the wire.

Answer: It helps

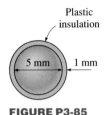

Plastic
insulation

5 mm   1 mm

FIGURE P3-85

3-84E Repeat Problem 3-83E, assuming a thermal contact resistance of 0.001 h · ft^2 · °F/Btu at the interface of the wire and the insulation.

3-85 A 5-mm-diameter spherical ball at 50°C is covered by a 1-mm-thick plastic insulation ($k = 0.13$ W/m · °C). The ball is exposed to a medium at 15°C, with a combined convection and radiation heat transfer coefficient of 20 W/m^2 · °C. Determine if the plastic insulation on the ball will help or hurt heat transfer from the ball.

Thermal Insulation

3-86C What is thermal insulation? How does a thermal insulator differ in purpose from an electrical insulator and from a sound insulator?

3-87C Does insulating cold surfaces save energy? Explain.

3-88C What is the R-value of insulation? How is it determined? Will doubling the thickness of flat insulation double its R-value?

3-89C How does the R-value of an insulation differ from its thermal resistance?

3-90C Why is the thermal conductivity of superinsulation orders of magnitude lower than the thermal conductivities of ordinary insulations?

3-91C Why do we characterize the heat conduction ability of insulators in terms of their apparent thermal conductivity instead of the ordinary thermal conductivity?

3-92C Someone suggests that one function of hair is to insulate the head. Do you agree with this suggestion?

3-93C Name five different reasons for using insulation in industrial facilities.

3-94C Define capacitive, resistive, and reflective insulations and give an example for each kind.

3-95C How do fibrous, cellular, and granular insulations differ from each other? Give an example for each kind.

3-96C How do loose fill, rigid, and formed-in-place insulations differ from each other? Give an example for each type.

3-97C Name five properties of insulations that are most important in the selection of insulation.

3-98C What is optimum thickness of insulation? How is it determined?

3-99C What are two common reasons for retrofitting of insulation? When should the existing insulation be replaced and when should new insulation be added on top of existing insulation?

3-100 What is the thickness of flat R-8 (in SI units) insulation whose thermal conductivity is 0.04 W/m · °C?

3-101E What is the thickness of flat R-20 (in English units) insulation whose thermal conductivity is 0.02 Btu/h · ft · °F?

3-102 Hot water at 110°C flows in a cast iron pipe ($k = 52$ W/m · °C) whose inner radius is 2.0 cm and thickness is 0.3 cm. The pipe is to be covered with adequate insulation so that the temperature of the outer surface of the insulation does not exceed 30°C when the ambient temperature is 22°C. Taking the heat transfer coefficients inside and outside the pipe to be $h_i = 80$ W/m² · °C and $h_o = 22$ W/m² · °C, respectively, determine the thickness of fiber glass insulation ($k = 0.038$ W/m · °C) that needs to be installed on the pipe.   *Answer:* 0.84 cm

3-103 Consider a furnace whose average outer surface temperature is measured to be 90°C when the average surrounding air temperature is 27°C. The furnace is 6 m long and 3 m in diameter. The plant operates 80 hours per week for 52 weeks per year. You are to insulate the furnace using fiberglass insulation ($k_{ins} = 0.038$ W/m · °C) whose cost is $10/m² per cm of thickness for materials, plus $30/m² for labor regardless of thickness. The combined heat transfer coefficient on the outer surface is estimated to be $h_o = 30$ W/m² · °C. The furnace uses natural gas whose unit cost is $0.50/therm input (1 therm = 105,500 kJ), and the efficiency of the furnace is 78 percent. The management is willing to authorize the installation of the thickest insulation (in whole cm) that will pay for itself (materials and labor) in one year. That is, the total cost of insulation should be roughly equal to the drop in the fuel cost of the furnace for one year. Determine the thickness of insulation to be used and the money saved per year. Assume the surface temperature of the furnace and the heat transfer coefficient are to remain constant.   *Answer:* 14 cm

3-104 Repeat Problem 3-103 for an outer surface temperature of 75°C for the furnace.

3-105E Steam at 400°F is flowing through a steel pipe ($k = 8.7$ Btu/h · ft · °F) whose inner and outer diameters are 3.5 in. and 4.0 in., respectively, in an environment at 60°F. The pipe is insulated with 1-in.-thick fiberglass insulation ($k = 0.020$ Btu/h · ft · °F), and the heat transfer coefficients on the inside and the outside of the pipe are 30 Btu/h · ft² · °F and 5 Btu/h · ft² · °F, respectively. It is proposed to add another 1-in.-thick layer of fiberglass insulation on top of the existing one to reduce the heat losses further and to save energy and money. The total cost of new insulation is $7 per ft length of the pipe, and the net fuel cost of energy in the steam is $0.01 per 1000 Btu (therefore, each 1000 Btu reduction in the heat loss will save the plant $0.01). The policy of the plant is to implement energy conservation measures that pay for themselves within two years. Assuming continuous operation (8760 h/year), determine if the proposed additional insulation is justified.

3-106 The plumbing system of a plant involves some section of a plastic pipe ($k = 0.16$ W/m · °C) of inner diameter 6 cm and outer diameter 6.6 cm exposed to the ambient air. You are to insulate the pipe with adequate weather-jacketed fiberglass insulation ($k = 0.035$ W/m · °C) to prevent freezing of water in the pipe. The plant is closed for the weekends for a period of 60 h, and the water in the pipe remains still during that period. The ambient temperature in the area gets as low as $-10°C$ in winter, and the high winds can cause heat transfer coefficients as high as 30 $W/m^2 \cdot °C$. Also, the water temperature in the pipe can be as cold as 15°C, and water starts freezing when its temperature drops to 0°C. Disregarding the convection resistance inside the pipe, determine the thickness of insulation that will protect the water from freezing under worst conditions.

3-107 Repeat Problem 3-106 assuming 20 percent of the water in the pipe is allowed to freeze without jeopardizing safety.   *Answer:* 27.9 cm

Heat Transfer from Finned Surfaces

3-108C What is the reason for the widespread use of fins on surfaces?

3-109C What is the difference between the fin effectiveness and the fin efficiency?

3-110C The fins attached to a surface are determined to have an effectiveness of 0.9. Do you think the rate of heat transfer from the surface has increased or decreased as a result of the addition of these fins?

3-111C Explain how the fins enhance heat transfer from a surface. Also, explain how the addition of fins may actually decrease heat transfer from a surface.

3-112C How does the overall effectiveness of a finned surface differ from the effectiveness of a single fin?

3-113C Hot water is to be cooled as it flows through the tubes exposed to atmospheric air. Fins are to be attached in order to enhance heat transfer. Would you recommend attaching the fins inside or outside the tubes? Why?

3-114C Hot air is to be cooled as it is forced to flow through the tubes exposed to atmospheric air. Fins are to be added in order to enhance heat transfer. Would you recommend attaching the fins inside or outside the tubes? Why? When would you recommend attaching fins both inside and outside the tubes?

3-115C Consider two finned surfaces that are identical except that the fins on the first surface are formed by casting or extrusion, whereas they are attached to the second surface afterwards by welding or tight fitting. For which case do you think the fins will provide greater enhancement in heat transfer? Explain.

3-116C The heat transfer surface area of a fin is equal to the sum of all surfaces of the fin exposed to the surrounding medium, including the surface area of the fin tip. Under what conditions can we neglect heat transfer from the fin tip?

3-117C Does the (*a*) efficiency and (*b*) effectiveness of a fin increase or decrease as the fin length is increased?

3-118C Two pin fins are identical, except that the diameter of one of them is twice the diameter of the other. For which fin will the (a) fin effectiveness and (b) fin efficiency be higher? Explain.

3-119C Two plate fins of constant rectangular cross-section are identical, except that the thickness of one of them is twice the thickness of the other. For which fin will the (a) fin effectiveness and (b) fin efficiency be higher? Explain.

3-120C Two finned surfaces are identical, except that the convection heat transfer coefficient of one of them is twice that of the other. For which finned surface will the (a) fin effectiveness and (b) fin efficiency be higher? Explain.

3-121 Obtain a relation for the fin efficiency for a fin of constant cross-sectional area A_c, perimeter p, length L, and thermal conductivity k exposed to convection to a medium at T_∞ with a heat transfer coefficient h. Assume the fins are sufficiently long so that the temperature of the fin at the tip is nearly T_∞. Take the temperature of the fin at the base to be T_b and neglect heat transfer from the fin tips. Simplify the relation for (a) a circular fin of diameter D and (b) rectangular fins of thickness t.

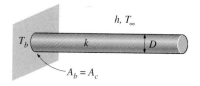

$$p = \pi D, \quad A_c = \pi D^2/4$$

FIGURE P3-121

3-122 The case-to-ambient thermal resistance of a power transistor that has a maximum power rating of 15 W is given to be 25°C/W. If the case temperature of the transistor is not to exceed 80°C, determine the power at which this transistor can be operated safely in an environment at 30°C.

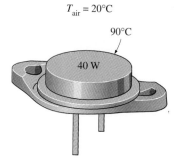

FIGURE P3-123

3-123 A 40-W power transistor is to be cooled by attaching it to one of the commercially available heat sinks shown in Table 3-7. Select a heat sink that will allow the case temperature of the transistor not to exceed 90° in the ambient air at 20°.

3-124 A 30-W power transistor is to be cooled by attaching it to one of the commercially available heat sinks shown in Table 3-7. Select a heat sink that will allow the case temperature of the transistor not to exceed 80°C in the ambient air at 35°C.

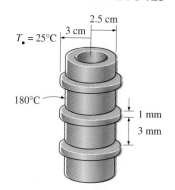

FIGURE P3-125

FIGURE P3-126E

3-125 Steam in a heating system flows through tubes whose outer diameter is 5 cm and whose walls are maintained at a temperature of 180°C. Circular aluminum alloy 2024-T6 fins ($k = 186$ W/m · °C) of outer diameter 6 cm and constant thickness 1 mm are attached to the tube. The space between the fins is 3 mm, and thus there are 250 fins per meter length of the tube. Heat is transferred to the surrounding air at $T_\infty = 25°C$, with a heat transfer coefficient of 40 W/m² · °C. Determine the increase in heat transfer from the tube per meter of its length as a result of adding fins. *Answer:* 2274 W

3-126E Consider a stainless steel spoon ($k = 8.7$ Btu/h · ft · °F) partially immersed in boiling water at 200°F in a kitchen at 75°F. The handle of the spoon has a cross-section of 0.08 in. × 0.5 in., and extends 7 in. in the air from the free surface of the water. If the heat transfer coefficient at the exposed surfaces of the spoon handle is 3 Btu/h · ft² · °F, determine the temperature difference across the exposed surface of the spoon handle. State your assumptions. *Answer:* 124.8°F

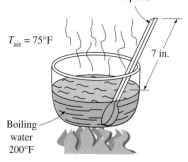

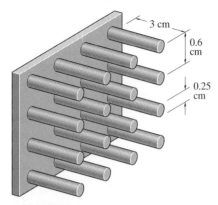

FIGURE P3-130

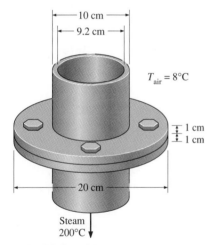

FIGURE P3-132

FIGURE P3-135

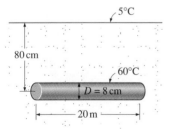

3-127E Repeat Problem 3-126 for a silver spoon ($k = 247$ Btu/h · ft · °F).

3-128 A 0.3-cm-thick, 12-cm-high, and 18-cm-long circuit board houses 80 closely spaced logic chips on one side, each dissipating 0.04 W. The board is impregnated with copper fillings and has an effective thermal conductivity of 20 W/m · °C. All the heat generated in the chips is conducted across the circuit board and is dissipated from the back side of the board to a medium at 40°C, with a heat transfer coefficient of 50 W/m² · °C. (a) Determine the temperatures on the two sides of the circuit board. (b) Now a 0.2-cm-thick, 12-cm-high, and 18-cm-long aluminum plate ($k = 237$ W/m · °C) with 864 2-cm-long aluminum pin fins of diameter 0.25 cm is attached to the back side of the circuit board with a 0.02-cm-thick epoxy adhesive ($k = 1.8$ W/m · °C). Determine the new temperatures on the two sides of the circuit board.

3-129 Repeat Problem 3-128 using a copper plate with copper fins ($k = 386$ W/m · °C) instead of aluminum ones.

3-130 A hot surface at 100°C is to be cooled by attaching 3-cm-long, 0.25-cm-diameter aluminum pin fins ($k = 237$ W/m · °C) to it, with a center-to-center distance of 0.6 cm. The temperature of the surrounding medium is 30°C, and the heat transfer coefficient on the surfaces is 35 W/m² · °C. Determine the rate of heat transfer from the surface for a 1-m × 1-m section of the plate. Also determine the overall effectiveness of the fins.

3-131 Repeat Problem 3-130 using copper fins ($k = 386$ W/m · °C) instead of aluminum ones.

3-132 Two 3-m-long and 0.4-cm-thick cast iron ($k = 52$ W/m · °C) steam pipes of outer diameter 10 cm are connected to each other through two 1-cm-thick flanges of outer diameter 20 cm. The steam flows inside the pipe at an average temperature of 200°C with a heat transfer coefficient of 180 W/m² · °C. The outer surface of the pipe is exposed to an ambient at 8°C, with a heat transfer coefficient of 25 W/m² · °C. (a) Disregarding the flanges, determine the average outer surface temperature of the pipe. (b) Using this temperature for the base of the flange and treating the flanges as the fins, determine the fin efficiency and the rate of heat transfer from the flanges. (c) What length of pipe is the flange section equivalent to for heat transfer purposes?

Heat Transfer in Common Configurations

3-133C What is a conduction shape factor? How is it related to the thermal resistance?

3-134C What is the value of conduction shape factors in engineering?

3-135 A 20-m-long and 8-cm-diameter hot water pipe of a district heating system is buried in the soil 80 cm below the ground surface. The outer surface temperature of the pipe is 60°C. Taking the surface temperature of the earth to be 5°C and the thermal conductivity of the soil at that location to be 0.9 W/m · °C, determine the rate of heat loss from the pipe.

3-136 Hot and cold water pipes 8 m long run parallel to each other in a thick concrete layer. The diameters of both pipes are 5 cm, and the distance between the centerlines of the pipes is 40 cm. The surface temperatures of the hot and cold pipes are 60°C and 15°C, respectively. Taking the thermal conductivity of the concrete to be $k = 0.75$ W/m · °C, determine the rate of heat transfer between the pipes.   *Answer:* ~~411 W~~ 306 w

3-137E A row of 3-ft-long and 1-in.-diameter used uranium fuel rods that are still radioactive are buried in the ground parallel to each other with a center-to-center distance of 8 in. at a depth 15 ft from the ground surface at a location where the thermal conductivity of the soil is 0.6 Btu/h · ft · °F. If the surface temperature of the rods and the ground are 400°F and 50°F, respectively, determine the rate of heat transfer from the fuel rods to the atmosphere through the soil.

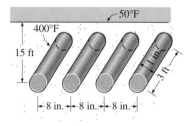

FIGURE P3-137E

3-138 Hot water at an average temperature of 60°C and an average velocity of 0.6 m/s is flowing through a 5-m section of a thin-walled hot water pipe that has an outer diameter of 2.5 cm. The pipe passes through the center of a 14-cm-thick wall filled with fiberglass insulation ($k = 0.035$ W/m · °C). If the surfaces of the wall are at 18°C, determine (*a*) the rate of heat transfer from the pipe to the air in the rooms and (*b*) the temperature drop of the hot water as it flows through this 5-m-long section of the wall.
    Answers: 23.5 W, 0.02°C

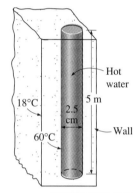

FIGURE P3-138

3-139 Hot water at an average temperature of 80°C and an average velocity of 1.5 m/s is flowing through a 25-m section of a pipe that has an outer diameter of 5 cm. The pipe extends 2 m in the ambient air above the ground, dips into the ground ($k = 1.5$ W/m · °C) vertically for 3 m, and continues horizontally at this depth for 20 m more before it enters the next building. The first section of the pipe is exposed to the ambient air at 8°C, with a heat transfer coefficient of 22 W/m² · °C. If the surface of the ground is covered with snow at 0°C, determine (*a*) the total rate of heat loss from the hot water and (*b*) the temperature drop of the hot water as it flows through this 25-m-long section of the pipe.

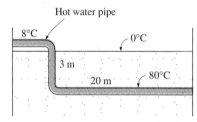

FIGURE P3-139

3-140 Consider a house with a flat roof whose outer dimensions are 12 m × 12 m. The outer walls of the house are 6 m high. The walls and the roof of the house are made of 20-cm-thick concrete ($k = 0.75$ W/m · °C). The temperatures of the inner and outer surfaces of the house are 15°C and 3°C, respectively. Accounting for the effects of the edges of adjoining surfaces, determine the rate of heat loss from the house through its walls and the roof. What is the error involved in ignoring the effects of the edges and corners and treating the roof as a 12 m × 12 m surface and the walls as 6 m × 12 m surfaces for simplicity?

FIGURE P3-141

3-141 Consider a 10-m-long thick-walled concrete duct ($k = 0.75$ W/m · °C) of square cross-section. The outer dimensions of the duct are 20 cm × 20 cm, and the thickness of the duct wall is 2 cm. If the inner and outer surfaces of the duct are at 100°C and 15°C, respectively, determine the rate of heat transfer through the walls of the duct.   *Answer:* 22.9 kW

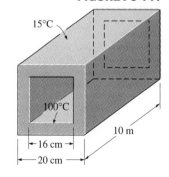

3-142 A 3-m-diameter spherical tank containing some radioactive material is buried in the ground ($k = 1.4$ W/m · °C). The distance between the top surface of the tank and the ground surface is 4 m. If the surface temperatures of the tank and the ground are 170°C and 15°C, respectively, determine the rate of heat transfer from the tank.

3-143 Hot water at an average temperature of 85°C passes through a row of eight parallel pipes that are 4 m long and have an outer diameter of 3 cm, located vertically in the middle of a concrete wall ($k = 0.75$ W/m · °C) that is 4 m high, 8 m long, and 15 cm thick. If the surfaces of the concrete walls are exposed to a medium at 20°C, with a heat transfer coefficient of 8 W/m² · °C, determine the rate of heat loss from the hot water and the surface temperature of the wall.

Review Problems

3-144E Steam is produced in the copper tubes ($k = 223$ Btu/h · ft · °F) of a heat exchanger at a temperature of 250°F by another fluid condensing on the outside surfaces of the tubes at 350°F. The inner and outer diameters of the tube are 1 in. and 1.3 in., respectively. When the heat exchanger was new, the rate of heat transfer per foot length of the tube was 2×10^4 Btu/h. Determine the rate of heat transfer per foot length of the tube when a 0.01-in.-thick layer of limestone ($k = 1.7$ Btu/h · ft · °F) has formed on the inner surface of the tube after extended use.

3-145E Repeat Problem 3-144E, assuming that a 0.01-in.-thick limestone layer has formed on both the inner and outer surfaces of the tube.

3-146 A 1.2 m-diameter and 6-m-long cylindrical propane tank is initially filled with liquid propane whose density is 581 kg/m³. The tank is exposed to the ambient air at 15°C, with a heat transfer coefficient of 20 W/m² · °C. Now a crack develops at the top of the tank and the pressure inside drops to 1 atm while the temperature drops to −42°C, which is the boiling temperature of propane at 1 atm. The heat of vaporization of propane at 1 atm is 425 kJ/kg. The propane is slowly vaporized as a result of the heat transfer from the ambient air into the tank, and the propane vapor escapes the tank at −42°C through the crack. Assuming the propane tank to be at about the same temperature as the propane inside at all times, determine how long it will take for the propane tank to empty if the tank is (*a*) not insulated and (*b*) insulated with 7.5-cm-thick glass wool insulation ($k = 0.038$ W/m · °C).

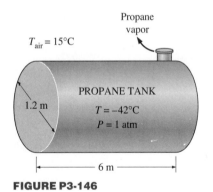

Propane
vapor

$T_{air} = 15°C$

PROPANE TANK

1.2 m

$T = -42°C$
$P = 1$ atm

6 m

FIGURE P3-146

3-147 Hot water is flowing at an average velocity of 1.5 m/s through a cast iron pipe ($k = 52$ W/m · °C) whose inner and outer diameters are 3 cm and 3.5 cm, respectively. The pipe passes through a 15-m-long section of a basement whose temperature is 15°C. If the temperature of the water drops from 70°C to 67°C as it passes through the basement and the heat transfer coefficient on the inner surface of the pipe is 400 W/m² · °C, determine the combined convection and radiation heat transfer coefficient at the outer surface of the pipe.   *Answer:* 272.5 W/m² · °C

3-148 Newly formed concrete pipes are usually cured first overnight by steam in a curing kiln maintained at a temperature of 45°C before the pipes are cured for several days outside. The heat and moisture to the kiln is provided by steam flowing in a pipe whose outer diameter is 12 cm. During a plant inspection, it was noticed that the pipe passes through a 10-m section that is completely exposed to the ambient air before it reaches the kiln. The temperature measurements indicate that the average temperature of the outer surface of the steam pipe is 82°C when the ambient temperature is 5°C. The combined convection and radiation heat transfer coefficient at the outer surface of the pipe is estimated to be 25 W/m$^2 \cdot$ °C. Determine the amount of heat lost from the steam during a 10-h curing process that night.

Steam is supplied by a gas-fired steam generator that has an efficiency of 80 percent, and the plant pays \$0.60/therm of natural gas (1 therm = 105,500 kJ). If the pipe is insulated and 90 percent of the heat loss is saved as a result, determine the amount of money this facility will save a year as a result of insulating the steam pipes. Assume that the concrete pipes are cured 110 nights a year. State your assumptions.

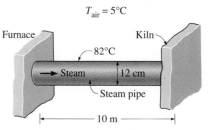

FIGURE P3-148

3-149 Consider an 18-cm × 18-cm multilayer circuit board dissipating 27 W of heat. The board consists of four layers of 0.2-mm-thick copper (k = 386 W/m $\cdot$ °C) and three layers of 1.5-mm-thick epoxy glass (k = 0.26 W/m $\cdot$ °C) sandwiched together, as shown in the figure. The circuit board is attached to a heat sink from both ends, and the temperature of the board at those ends is 35°C. Heat is considered to be uniformly generated in the epoxy layers of the board at a rate of 0.5 W per 1-cm × 18-cm epoxy laminate strip (or 1.5 W per 1-cm × 18-cm strip of the board). Considering only a portion of the board because of symmetry, determine the magnitude and location of the maximum temperature that occurs in the board. Assume heat transfer from the top and bottom faces of the board to be negligible.

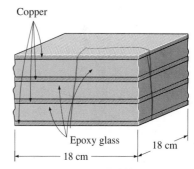

FIGURE P3-149

3-150 The plumbing system of a house involves 0.5-m section of a plastic pipe (k = 0.16 W/m $\cdot$ °C) of inner diameter 2 cm and outer diameter 2.4 cm exposed to the ambient air. During a cold and windy night, the ambient air temperature remains at about −5°C for a period of 14 h. The combined convection and radiation heat transfer coefficient on the outer surface of the pipe is estimated to be 40 W/m$^2 \cdot$ °C, and the heat of fusion of water is 333.7 kJ/kg. Assuming the pipe to contain stationary water initially at 0°C, determine if the water in that section of the pipe will completely freeze that night.

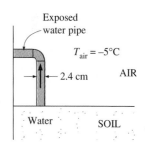

FIGURE P3-150

3-151 Repeat Problem 3-150 for the case of a heat transfer coefficient of 10 W/m$^2 \cdot$ °C on the outer surface as a result of putting a fence around the pipe that blocks the wind.

3-152E The surface temperature of an 3-in.-diameter baked potato is observed to drop from 300°F to 200°F in 5 minutes in an environment at 70°F. Determine the average heat transfer coefficient between the potato and its surroundings. Using this heat transfer coefficient and the same surface temperature, determine how long it will take for the potato to experience the same temperature drop if it is wrapped completely in a 0.12-in.-thick towel (k = 0.035 Btu/h $\cdot$ ft $\cdot$ °F). You may use the properties of water for potato.

3-153E Repeat Problem 3-152E assuming there is a 0.02-in.-thick air space (k = 0.015 Btu/h $\cdot$ ft $\cdot$ °F) between the potato and the towel.

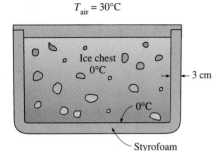

$T_{\text{air}} = 30°C$

Ice chest
0°C

3 cm

0°C

Styrofoam

FIGURE P3-154

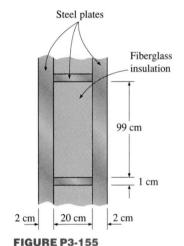

Steel plates

Fiberglass insulation

99 cm

1 cm

2 cm   20 cm   2 cm

FIGURE P3-155

3-154   An ice chest whose outer dimensions are 30 cm × 40 cm × 50 cm is made of 3-cm-thick Styrofoam ($k = 0.033$ W/m · °C). Initially, the chest is filled with 45 kg of ice at 0°C, and the inner surface temperature of the ice chest can be taken to be 0°C at all times. The heat of fusion of ice at 0°C is 333.7 kJ/kg, and the heat transfer coefficient between the outer surface of the ice chest and surrounding air at 30°C is 20 W/m² · °C. Disregarding any heat transfer from the 40-cm × 50-cm base of the ice chest, determine how long it will take for the ice in the chest to melt completely.

3-155   A 4-m-high and 6-m-long wall is constructed of two large 2-cm-thick steel plates ($k = 15$ W/m · °C) separated by 1-cm-thick and 20-cm-wide steel bars placed 99 cm apart. The remaining space between the steel plates is filled with fiberglass insulation ($k = 0.035$ W/m · °C). If the temperature difference between the inner and the outer surfaces of the walls is 15°C, determine the rate of heat transfer through the wall. Can we ignore the steel bars between the plates in heat transfer analysis since they occupy only 1 percent of the heat transfer surface area?

3-156   A 0.2-cm-thick, 10-cm-high, and 15-cm-long circuit board houses electronic components on one side that dissipate a total of 15 W of heat uniformly. The board is impregnated with conducting metal fillings and has an effective thermal conductivity of 12 W/m · °C. All the heat generated in the components is conducted across the circuit board and is dissipated from the back side of the board to a medium at 37°C, with a heat transfer coefficient of 45 W/m² · °C. (*a*) Determine the surface temperatures on the two sides of the circuit board. (*b*) Now a 0.1-cm-thick, 10-cm-high, and 15-cm-long aluminum plate ($k = 237$ W/m · °C) with 20 0.2-cm-thick, 2-cm-long, and 15-cm-wide aluminum fins of rectangular profile are attached to the back side of the circuit board with a 0.015-cm-thick epoxy adhesive ($k = 1.8$ W/m · °C). Determine the new temperatures on the two sides of the circuit board.

FIGURE P3-156

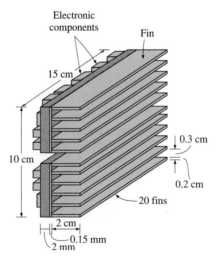

Electronic components

Fin

15 cm

10 cm

0.3 cm

0.2 cm

20 fins

2 cm

0.15 mm

2 mm

3-157   Repeat Problem 3-156 using a copper plate with copper fins ($k = 386 \ W/m \cdot °C$) instead of aluminum ones.

3-158   A row of 10 parallel pipes that are 5 m long and have an outer diameter of 6 cm are used to transport steam at 150°C through the concrete floor ($k = 0.75 \ W/m \cdot °C$) of a 10-m × 5-m room that is maintained at 25°C. The combined convection and radiation heat transfer coefficient at the floor is 12 $W/m^2 \cdot °C$. If the surface temperature of the concrete floor is not to exceed 40°C, determine how deep the steam pipes should be buried below the surface of the concrete floor.

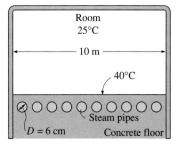

FIGURE P3-158

Computer, Design, and Essay Problems

3-159   The temperature in deep space is close to absolute zero, which presents thermal challenges for the astronauts who do space walks. Propose a design for the clothing of the astronauts that will be most suitable for the thermal environment in space. Defend the selections in your design.

3-160   In the design of electronic components, it is very desirable to attach the electronic circuitry to a substrate material that is a very good thermal conductor but also a very effective electrical insulator. If the high cost is not a major concern, what material would you propose for the substrate?

3-161   Using cylindrical samples of the same material, devise an experiment to determine the thermal contact resistance. Cylindrical samples are available at any length, and the thermal conductivity of the material is known.

3-162   What are the considerations in determining the proper length and right number of fins attached to a surface?

3-163   Find out about the wall construction of the cabins of large commercial airplanes, the range of ambient conditions under which they operate, typical heat transfer coefficients on the inner and outer surfaces of the wall, and the heat generation rates inside. Determine the size of the heating and air-conditioning system that will be able to maintain the cabin at 20°C at all times for an airplane capable of carrying 400 people.

3-164   Repeat Problem 3-163 for a submarine with a crew of 60 people.

3-165   A house with 200-m^2 floor space is to be heated with geothermal water flowing through pipes laid in the ground under the floor. The walls of the house are 4 m high, and there are 10 single-paned windows in the house that are 1.2 m wide and 1.8 m high. The house has R-19 (in h · ft^2 · °F/Btu) insulation in the walls and R-30 on the ceiling. The floor temperature is not to exceed 40°C. Hot geothermal water is available at 90°C, and the inner and outer diameter of the pipes to be used are 2.4 cm and 3.0 cm. Design such a heating system for this house in your area.

Transient Heat Conduction

The temperature of a body, in general, varies with time as well as position. In rectangular coordinates, this variation is expressed as $T(x, y, z, t)$, where (x, y, z) indicates variation in the x, y, and z directions, respectively, and t indicates variation with time. In the preceding chapter, we considered heat conduction under *steady* conditions, for which the temperature of a body at any point does not change with time. This certainly simplified the analysis, especially when the temperature varied in one direction only, and we were able to obtain analytical solutions. In this chapter, we consider the variation of temperature with *time* as well as *position* in one- and multidimensional systems.

We start this chapter with the analysis of *lumped systems* in which the temperature of a solid varies with time but remains uniform throughout the solid at any time. Then we consider the variation of temperature with time as well as position for one-dimensional heat conduction problems such as those associated with a large plane wall, a long cylinder, a sphere, and a semi-infinite medium using *transient temperature charts* and analytical solutions. Finally, we consider transient heat conduction in multidimensional systems by utilizing the *product solution*.

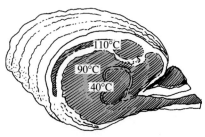

(a) Copper ball

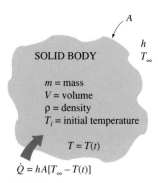

(b) Roast beef

FIGURE 4-1

A small copper ball can be modeled as a lumped system, but a roast beef cannot.

A

SOLID BODY

h
T_∞

m = mass
V = volume
ρ = density
T_i = initial temperature

$T = T(t)$

$\dot{Q} = hA[T_\infty - T(t)]$

FIGURE 4-2

The geometry and parameters involved in the lumped system analysis.

4-1 ■ LUMPED SYSTEM ANALYSIS

In heat transfer analysis, some bodies are observed to behave like a "lump" whose interior temperature remains essentially uniform at all times during a heat transfer process. The temperature of such bodies can be taken to be a function of time only, $T(t)$. Heat transfer analysis that utilizes this idealization is known as **lumped system analysis,** which provides great simplification in certain classes of heat transfer problems without much sacrifice from accuracy.

Consider a small hot copper ball coming out of an oven (Fig. 4-1). Measurements indicate that the temperature of the copper ball changes with time, but it does not change much with position at any given time. Thus the temperature of the ball remains uniform at all times, and we can talk about the temperature of the ball with no reference to a specific location.

Now let us go to the other extreme and consider a large roast in an oven. If you have done any roasting, you must have noticed that the temperature distribution within the roast is not even close to being uniform. You can easily verify this by taking the roast out before it is completely done and cutting it in half. You will see that the outer parts of the roast are well done while the center part is barely warm. Thus, lumped system analysis is not applicable in this case. Before presenting a criterion about applicability of lumped system analysis, we develop the formulation associated with it.

Consider a body of arbitrary shape of mass m, volume V, surface area A, density ρ, and specific heat C_p initially at a uniform temperature T_i (Fig. 4-2). At time $t = 0$, the body is placed into a medium at temperature T_∞, and heat transfer takes place between the body and its environment, with a heat transfer coefficient h. For the sake of discussion, we will assume that $T_\infty > T_i$, but the analysis is equally valid for the opposite case. We assume lumped system analysis to be applicable, so that the temperature remains uniform within the body at all times and changes with time only, $T = T(t)$.

During a differential time interval dt, the temperature of the body rises by a differential amount dT. An energy balance of the solid for the time interval dt can be expressed as

$$\begin{pmatrix} \text{Heat transfer into the body} \\ \text{during } dt \end{pmatrix} = \begin{pmatrix} \text{The increase in the} \\ \text{energy of the body} \\ \text{during } dt \end{pmatrix}$$

or

$$hA(T_\infty - T)\,dt = mC_p\,dT \tag{4-1}$$

Noting that $m = \rho V$ and $dT = d(T - T_\infty)$ since T_∞ = constant, Equation 4-1 can be rearranged as

$$\frac{d(T - T_\infty)}{T - T_\infty} = -\frac{hA}{\rho VC_p}\,dt \tag{4-2}$$

Integrating from $t = 0$, at which $T = T_i$, to any time t, at which $T = T(t)$, gives

$$\ln \frac{T(t) - T_\infty}{T_i - T_\infty} = -\frac{hA}{\rho V C_p} t \qquad (4\text{-}3)$$

Taking the exponential of both sides and rearranging, we obtain

$$\frac{T(t) - T_\infty}{T_i - T_\infty} = e^{-bt} \qquad (4\text{-}4)$$

where

$$b = \frac{hA}{\rho V C_p} \qquad (1/\text{s}) \qquad (4\text{-}5)$$

is a positive quantity whose dimension is $(\text{time})^{-1}$. Equation 4-4 is plotted in Figure 4-3 for different values of b. There are two observations that can be made from this figure and the relation above:

1. Equation 4-4 enables us to determine the temperature $T(t)$ of a body at time t, or alternatively, the time t required for the temperature to reach a specified value $T(t)$.

2. The temperature of a body approaches the ambient temperature T_∞ exponentially. The temperature of the body changes rapidly at the beginning, but rather slowly later on. A large value of b indicates that the body will approach the environment temperature in a short time. The larger the value of the exponent b, the higher the rate of decay in temperature. Note that b is proportional to the surface area, but inversely proportional to the mass and the specific heat of the body. This is not surprising since it takes longer to heat or cool a larger mass, especially when it has a large specific heat.

Once the temperature $T(t)$ at time t is available from Equation 4-4, the *rate* of convection heat transfer between the body and its environment at that time can be determined from Newton's law of cooling as

$$\dot{Q}(t) = hA[T(t) - T_\infty] \qquad (\text{W}) \qquad (4\text{-}6)$$

The *total amount* of heat transfer between the body and the surrounding medium over the time interval $t = 0$ to t is simply the change in the energy content of the body:

$$Q = mC_p[T(t) - T_i] \qquad (\text{kJ}) \qquad (4\text{-}7)$$

The amount of heat transfer reaches its *upper limit* when the body reaches the surrounding temperature T_∞. Therefore, the *maximum* heat transfer between the body and its surroundings is (Fig. 4-4)

$$Q_{\text{max}} = mC_p(T_\infty - T_i) \qquad (\text{kJ}) \qquad (4\text{-}8)$$

We could also obtain this equation by substituting the $T(t)$ relation from Equation 4-4 into the $\dot{Q}(t)$ relation in Equation 4-6 and integrating it from $t = 0$ to $t \rightarrow \infty$.

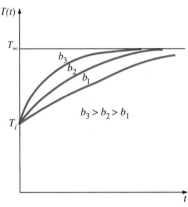

FIGURE 4-3

The temperature of a lumped system approaches the environment temperature as time gets larger.

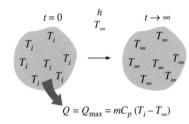

$$Q = Q_{\text{max}} = mC_p(T_i - T_\infty)$$

FIGURE 4-4

Heat transfer to or from a body reaches its maximum value when the body reaches the environment temperature.

Criteria for Lumped System Analysis

The lumped system analysis certainly provides great convenience in heat transfer analysis, and naturally we would like to know when it is appropriate to use it. The first step in establishing a criterion for the applicability of the lumped system analysis is to define a **characteristic length** as

$$L_c = \frac{V}{A}$$

and a **Biot number** Bi as

$$\text{Bi} = \frac{hL_c}{k} \tag{4-9}$$

It can also be expressed as (Fig. 4-5)

$$\text{Bi} = \frac{h}{k/L_c} \frac{\Delta T}{\Delta T} = \frac{\text{Convection at the surface of the body}}{\text{Conduction within the body}}$$

or

$$\text{Bi} = \frac{L_c/k}{1/h} = \frac{\text{Conduction resistance within the body}}{\text{Convection resistance at the surface of the body}}$$

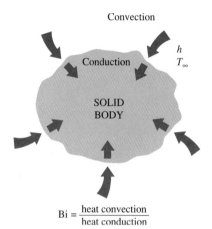

$$\text{Bi} = \frac{\text{heat convection}}{\text{heat conduction}}$$

FIGURE 4-5
The Biot number can be viewed as the ratio of the convection at the surface to conduction within the body.

When a solid body is being heated by the hotter fluid surrounding it (such as a potato being baked in an oven), heat is first *convected* to the body and subsequently *conducted* within the body. The Biot number is the *ratio* of the internal resistance of a body to *heat conduction* to its external resistance to *heat convection*. Therefore, a small Biot number represents small resistance to heat conduction, and thus small temperature gradients within the body.

Lumped system analysis assumes a *uniform* temperature distribution throughout the body, which will be the case only when the thermal resistance of the body to heat conduction (the *conduction resistance*) is zero. Thus, lumped system analysis is *exact* when Bi = 0 and *approximate* when Bi > 0. Of course, the smaller the Bi number, the more accurate the lumped system analysis. Then the question we must answer is, How much accuracy are we willing to sacrifice for the convenience of the lumped system analysis?

Before answering this question, we should mention that a 20 percent uncertainty in the convection heat transfer coefficient h in most cases is considered "normal" and "expected." Assuming h to be *constant* and *uniform* is also an approximation of questionable validity, especially for irregular geometries. Therefore, in the absence of sufficient experimental data for the specific geometry under consideration, we cannot claim our results to be better than ±20 percent, even when Bi = 0. This being the case, introducing another source of uncertainty in the problem will hardly have any effect on the overall uncertainty, provided that it is minor. It is generally accepted that lumped system analysis is *applicable* if

$$\text{Bi} \leq 0.1$$

When this criterion is satisfied, the temperatures within the body relative to the surroundings (i.e., $T - T_\infty$) remain within 5 percent of each other even for well-rounded geometries such as a spherical ball. Thus, when $\text{Bi} < 0.1$, the variation of temperature with location within the body will be slight and can reasonably be approximated as being uniform.

The first step in the application of lumped system analysis is the calculation of the *Biot number,* and the assessment of the applicability of this approach. One may still wish to use lumped system analysis even when the criterion $\text{Bi} < 0.1$ is not satisfied, if high accuracy is not a major concern.

Note that the Biot number is the ratio of the *convection* at the surface to *conduction* within the body, and this number should be as small as possible for lumped system analysis to be applicable. Therefore, *small bodies* with *high thermal conductivity* are good candidates for lumped system analysis, especially when they are in a medium that is a poor conductor of heat (such as air or another gas) and motionless. Thus, the hot small copper ball placed in quiescent air, discussed earlier, is most likely to satisfy the criterion for lumped system analysis (Fig. 4-6).

Some Remarks on Heat Transfer in Lumped Systems

To understand the heat transfer mechanism during the heating or cooling of a solid by the fluid surrounding it, and the criterion for lumped system analysis, consider the following analogy (Fig. 4-7). People from the mainland are to go *by boat* to an island whose entire shore is a harbor, and from the harbor to their destinations on the island *by bus.* The overcrowding of people at the harbor depends on the boat traffic to the island and the ground transportation system on the island. If there is an excellent ground transportation system with plenty of buses, there will be no overcrowding at the harbor, especially when the boat traffic is light. But when the opposite is true, there will be a huge overcrowding at the harbor, creating a large difference between the populations at the harbor and inland. The chance of overcrowding is much lower in a small island with plenty of fast buses.

In heat transfer, a poor ground transportation system corresponds to poor heat conduction in a body, and overcrowding at the harbor to the accumulation of heat and the subsequent rise in temperature near the surface of the body relative to its inner parts. Lumped system analysis is obviously not applicable when there is overcrowding at the surface. Of course, we have disregarded radiation in this analogy and thus the air traffic to the island. Like passengers at the harbor, heat changes *vehicles* at the surface from *convection* to *conduction.* Noting that a surface has zero thickness and thus cannot store any energy, heat reaching the surface of a body by convection must continue its journey within the body by conduction.

Consider heat transfer from a hot body to its cooler surroundings. Heat will be transferred from the body to the surrounding fluid as a result of a temperature difference. But this energy will come from the region near the surface, and thus the temperature of the body near the surface will drop. This creates a *temperature gradient* between the inner and outer regions of the

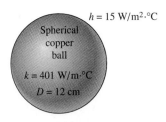

$$L_c = \frac{V}{A} = \frac{\frac{1}{6}\pi D^3}{\pi D^2} = \frac{1}{6}D = 0.02 \text{ m}$$

$$\text{Bi} = \frac{hL_c}{k} = \frac{15 \times 0.02}{401} = 0.00075 < 0.1$$

FIGURE 4-6

Small bodies with high thermal conductivities and low convection coefficients are most likely to satisfy the criterion for lumped system analysis.

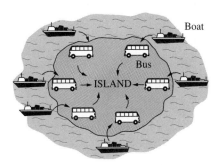

FIGURE 4-7

Analogy between heat transfer to a solid and passenger traffic to an island.

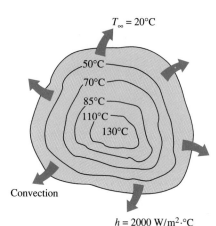

$T_\infty = 20°C$

50°C
70°C
85°C
110°C
130°C

Convection

$h = 2000 \ W/m^2 \cdot °C$

FIGURE 4-8
When the convection coefficient h is high and k is low, large temperature differences occur between the inner and outer regions of a large solid.

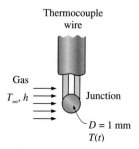

Thermocouple
wire

Gas
T_∞, h → Junction

$D = 1$ mm
$T(t)$

FIGURE 4-9
Schematic for Example 4-1.

body and initiates heat flow by conduction from the interior of the body toward the outer surface.

When the convection heat transfer coefficient h and thus convection heat transfer from the body are high, the temperature of the body near the surface will drop quickly (Fig. 4-8). This will create a larger temperature difference between the inner and outer regions unless the body is able to transfer heat from the inner to the outer regions just as fast. Thus, the magnitude of the maximum temperature difference within the body depends strongly on the ability of a body to conduct heat toward its surface relative to the ability of the surrounding medium to convect this heat away from the surface. The Biot number is a measure of the relative magnitudes of these two competing effects.

Recall that heat conduction in a specified direction n per unit surface area is expressed as $\dot{q} = -k\partial T/\partial n$, where $\partial T/\partial n$ is the temperature gradient and k is the thermal conductivity of the solid. Thus, the temperature distribution in the body will be *uniform* only when its thermal conductivity is *infinite*, and no such material is known to exist. Therefore, temperature gradients and thus temperature differences must exist within the body, no matter how small, in order for heat conduction to take place. Of course, the temperature gradient and the thermal conductivity are inversely proportional for a given heat flux. Therefore, the larger the thermal conductivity, the smaller the temperature gradient.

EXAMPLE 4-1 Temperature Measurement by Thermocouples

The temperature of a gas stream is to be measured by a thermocouple whose junction can be approximated as a 1-mm-diameter sphere, as shown in Figure 4-9. The properties of the junction are $k = 35$ W/m · °C, $\rho = 8500$ kg/m³, and $C_p = 320$ J/kg · °C, and the convection heat transfer coefficient between the junction and the gas is $h = 210$ W/m² · °C. Determine how long it will take for the thermocouple to read 99 percent of the initial temperature difference.

Solution The temperature of a gas stream is to be measured by a thermocouple. The time it takes to register 99 percent of the initial ΔT is to be determined.

Assumptions **1** The junction is spherical in shape with a diameter of $D = 0.001$ m. **2** The thermal properties of the junction and the heat transfer coefficient are constant. **3** Radiation effects are negligible.

Properties The properties of the junction are given in the problem statement.

Analysis The characteristic length of the junction is

$$L_c = \frac{V}{A} = \frac{\frac{1}{6}\pi D^3}{\pi D^2} = \frac{1}{6} D = \frac{1}{6}(0.001 \ m) = 1.67 \times 10^{-4} \ m$$

Then the Biot number becomes

$$Bi = \frac{hL_c}{k} = \frac{(210 \ W/m^2 \cdot °C)(1.67 \times 10^{-4} \ m)}{35 \ W/m \cdot °C} = 0.001 < 0.1$$

Therefore, lumped system analysis is applicable, and the error involved in this approximation is negligible.

In order to read 99 percent of the initial temperature difference $T_i - T_\infty$ between the junction and the gas, we must have

$$\frac{T(t) - T_\infty}{T_i - T_\infty} = 0.01$$

For example, when $T_i = 0°C$ and $T_\infty = 100°C$, a thermocouple is considered to have read 99 percent of this applied temperature difference when its reading indicates $T(t) = 99°C$.

The value of the exponent b is

$$b = \frac{hA_s}{\rho C_p V} = \frac{h}{\rho C_p L_c} = \frac{210 \text{ W/m}^2 \cdot °C}{(8500 \text{ kg/m}^3)(320 \text{ J/kg} \cdot °C)(1.67 \times 10^{-4} \text{ m})} = 0.462 \text{ s}^{-1}$$

We now substitute these values into Equation 4-4 and obtain

$$\frac{T(t) - T_\infty}{T_i - T_\infty} = e^{-bt} \longrightarrow 0.01 = e^{-(0.462 \text{ s}^{-1})t}$$

which yields

$$t = 10 \text{ s}$$

Therefore, we must wait at least 10 s for the temperature of the thermocouple junction to approach within 1 percent of the initial junction-gas temperature difference.

EXAMPLE 4-2   Predicting the Time of Death

A person is found dead at 5 PM in a room whose temperature is 20°C. The temperature of the body is measured to be 25°C when found, and the heat transfer coefficient is estimated to be $h = 8$ W/m$^2 \cdot$ °C. Modeling the body as a 30-cm-diameter, 1.70-m-long cylinder, estimate the time of death of that person (Fig. 4-10).

Solution   A body is found while still warm. The time of death is to be estimated.

Assumptions   **1** The body can be modeled as a 30-cm-diameter, 1.70-m-long cylinder. **2** The thermal properties of the body and the heat transfer coefficient are constant. **3** The radiation effects are negligible. **4** The person was healthy(!) when he or she died with a body temperature of 37°C.

Properties   The average human body is 72 percent water by mass, and thus we can assume the body to have the properties of water at the average temperature of $(37 + 25)/2 = 31°C$; $k = 0.617$ W/m $\cdot$ °C, $\rho = 996$ kg/m^3, and $C_p = 4178$ J/kg $\cdot$ °C (Table A-9).

Analysis   The characteristic length of the body is

$$L_c = \frac{V}{A} = \frac{\pi r_o^2 L}{2\pi r_o L + 2\pi r_o^2} = \frac{\pi(0.15 \text{ m})^2(1.7 \text{ m})}{2\pi(0.15 \text{ m})(1.7 \text{ m}) + 2\pi(0.15 \text{ m})^2} = 0.0689 \text{ m}$$

Then the Biot number becomes

FIGURE 4-10
Schematic for Example 4-2.

$$\text{Bi} = \frac{hL_c}{k} = \frac{(8 \text{ W/m}^2 \cdot {}^\circ\text{C})(0.0689 \text{ m})}{0.617 \text{ W/m} \cdot {}^\circ\text{C}} = 0.89 > 0.1$$

Therefore, lumped system analysis is *not* applicable. However, we can still use it to get a "rough" estimate of the time of death. The exponent b in this case is

$$b = \frac{hA}{\rho C_p V} = \frac{h}{\rho C_p L_c} = \frac{8 \text{ W/m}^2 \cdot {}^\circ\text{C}}{(996 \text{ kg/m}^3)(4178 \text{ J/kg} \cdot {}^\circ\text{C})(0.0689 \text{ m})}$$

$$= 2.79 \times 10^{-5} \text{ s}^{-1}$$

We now substitute these values into Equation 4-4,

$$\frac{T(t) - T_\infty}{T_i - T_\infty} = e^{-bt} \longrightarrow \frac{25 - 20}{37 - 20} = e^{-(2.79 \times 10^{-5} \text{ s}^{-1})t}$$

which yields

$$t = 43,860 \text{ s} = \textbf{12.2 h}$$

Therefore, as a rough estimate, the person died about 12 h before the body was found, and thus the time of death is 5 AM. This example demonstrates how to obtain "ball park" values using a simple analysis.

4-2 ■ TRANSIENT HEAT CONDUCTION IN LARGE PLANE WALLS, LONG CYLINDERS, AND SPHERES

In the preceding section, we considered bodies in which the variation of temperature within the body was negligible; that is, bodies that remain nearly *isothermal* during a process. Relatively *small* bodies of *highly conductive* materials approximate this behavior. In general, however, the temperature within a body will change from point to point as well as with time. In this section, we consider the variation of temperature with *time* and *position* in one-dimensional problems such as those associated with a large plane wall, a long cylinder, and a sphere.

Consider a plane wall of thickness $2L$, a long cylinder of radius r_o, and a sphere of radius r_o initially at a *uniform temperature T_i*, as shown in Figure 4-11. At time $t = 0$, each geometry is placed in a large medium that is at

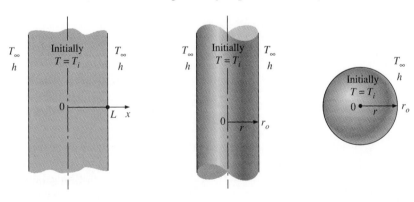

FIGURE 4-11

Schematic of the simple geometries in which heat transfer is one-dimensional.

(a) A large plane wall   (b) A long cylinder   (c) A sphere a constant temperature T_∞ and kept in that medium for $t > 0$. Heat transfer takes place between these bodies and their environments by convection with a *uniform* and *constant* heat transfer coefficient h. Note that all three cases possess geometric and thermal symmetry: the plane wall is symmetric about its *center plane* ($x = 0$), the cylinder is symmetric about its *centerline* ($r = 0$), and the sphere is symmetric about its *center point* ($r = 0$). We neglect *radiation* heat transfer between these bodies and their surrounding surfaces, or incorporate the radiation effect into the convection heat transfer coefficient h.

The variation of the temperature profile with *time* in the plane wall is illustrated in Figure 4-12. When the wall is first exposed to the surrounding medium at $T_\infty < T_i$ at $t = 0$, the entire wall is at its initial temperature T_i. But the wall temperature at and near the surfaces starts to drop as a result of heat transfer from the wall to the surrounding medium. This creates a *temperature gradient* in the wall and initiates heat conduction from the inner parts of the wall toward its outer surfaces. Note that the temperature at the center of the wall remains at T_i until $t = t_2$, and that the temperature profile within the wall remains symmetric at all times about the center plane. The temperature profile gets flatter and flatter as time passes as a result of heat transfer, and eventually becomes uniform at $T = T_\infty$. That is, the wall reaches *thermal equilibrium* with its surroundings. At that point, the heat transfer stops since there is no longer a temperature difference. Similar discussions can be given for the long cylinder or sphere.

The formulation of the problems for the determination of the one-dimensional transient temperature distribution $T(x, t)$ in a wall results in a partial differential equation, which can be solved using advanced mathematical techniques. The solution, however, normally involves infinite series, which are inconvenient and time-consuming to evaluate. Therefore, there is clear motivation to present the solution in *tabular* or *graphical* form. However, the solution involves the parameters x, L, t, k, α, h, T_i, and T_∞, which are too many to make any graphical presentation of the results practical. In order to reduce the number of parameters, we nondimensionalize the problem by defining the following dimensionless quantities:

Dimensionless temperature:
$$\theta(x, t) = \frac{T(x, t) - T_\infty}{T_i - T_\infty}$$

Dimensionless distance from the center: $\quad X = \frac{x}{L}$

Dimensionless heat transfer coefficient: $\quad \text{Bi} = \frac{hL}{k} \qquad \textbf{(Biot number)}$

Dimensionless time: $\quad \tau = \frac{\alpha t}{L^2} \qquad \textbf{(Fourier number)}$

The nondimensionalization enables us to present the temperature in terms of three parameters only: X, Bi, and τ. This makes it practical to present the solution in graphical form. The dimensionless quantities defined above for a plane wall can also be used for a *cylinder* or *sphere* by replacing the space variable x by r and the half-thickness L by the outer radius r_o. Note that the

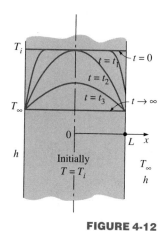

FIGURE 4-12

Transient temperature profiles in a plane wall exposed to convection from its surfaces for $T_i > T_\infty$.

characteristic length in the definition of the Biot number is taken to be the *half-thickness L* for the plane wall, and the *radius r_o* for the long cylinder and sphere instead of *V/A* used in lumped system analysis.

The one-dimensional transient heat conduction problem described above can be solved exactly for any of the three geometries, but the solution involves infinite series, which are difficult to deal with. However, the terms in the solutions converge rapidly with increasing time, and for $\tau > 0.2$, keeping the first term and neglecting all the remaining terms in the series results in an error under 2 percent. We are usually interested in the solution for times with $\tau > 0.2$, and thus it is very convenient to express the solution using this **one-term approximation,** given as

$$\text{Plane wall:} \quad \theta(x, t)_{\text{wall}} = \frac{T(x, t) - T_\infty}{T_i - T_\infty} = A_1 e^{-\lambda_1^2 \tau} \cos(\lambda_1 x/L), \quad \tau > 0.2 \quad (4\text{-}10)$$

$$\text{Cylinder:} \quad \theta(x, t)_{\text{cyl}} = \frac{T(r, t) - T_\infty}{T_i - T_\infty} = A_1 e^{-\lambda_1^2 \tau} J_0(\lambda_1 r/r_o), \quad \tau > 0.2 \quad (4\text{-}11)$$

$$\text{Sphere:} \quad \theta(x, t)_{\text{sph}} = \frac{T(r, t) - T_\infty}{T_i - T_\infty} = A_1 e^{-\lambda_1^2 \tau} \frac{\sin(\lambda_1 r/r_o)}{\lambda_1 r/r_o}, \quad \tau > 0.2 \quad (4\text{-}12)$$

where the constants A_1 and λ_1 are functions of the Bi number only, and their values are listed in Table 4-1 against the Bi number for all three geometries. The function J_0 is the zeroth-order Bessel function of the first kind, whose value can be determined from Table 4-2. Noting that $\cos(0) = J_0(0) = 1$ and the limit of $(\sin x)/x$ is also 1, the above relations simplify to the following at the center of a plane wall, cylinder, or sphere:

$$\text{Center of plane wall } (x = 0): \quad \theta_{0,\,\text{wall}} = \frac{T_o - T_\infty}{T_i - T_\infty} = A_1 e^{-\lambda_1^2 \tau} \quad (4\text{-}13)$$

$$\text{Center of cylinder } (r = 0): \quad \theta_{0,\,\text{cyl}} = \frac{T_o - T_\infty}{T_i - T_\infty} = A_1 e^{-\lambda_1^2 \tau} \quad (4\text{-}14)$$

$$\text{Center of sphere } (r = 0): \quad \theta_{0,\,\text{sph}} = \frac{T_o - T_\infty}{T_i - T_\infty} = A_1 e^{-\lambda_1^2 \tau} \quad (4\text{-}15)$$

Once the Bi number is known, the above relations can be used to determine the temperature anywhere in the medium. The determination of the constants A_1 and λ_1 usually requires interpolation. For those who prefer reading charts to interpolating, the relations above are plotted and the one-term approximation solutions are presented in graphical form, known as the *transient temperature charts*. Note that the charts are sometimes difficult to read, and they are subject to reading errors. Therefore, the relations above should be preferred to the charts.

The transient temperature charts in Figures 4-13, 4-14, and 4-15 for a large plane wall, long cylinder, and sphere were presented by M. P. Heisler in 1947 and are called **Heisler charts.** They were supplemented in 1961 with transient heat transfer charts by H. Gröber. There are *three* charts associated

TABLE 4-1

235

Coefficients used in the one-term approximate solution of transient
one-dimensional heat conduction in plane walls, cylinders, and spheres
($Bi = hL/k$ for a plane wall of thickness $2L$, and $Bi = hr_o/k$ for a cylinder or
sphere of radius r_o)

Bi	Plane wall λ_1	Plane wall A_1	Cylinder λ_1	Cylinder A_1	Sphere λ_1	Sphere A_1
0.01	0.0998	1.0017	0.1412	1.0025	0.1730	1.0030
0.02	0.1410	1.0033	0.1995	1.0050	0.2445	1.0060
0.04	0.1987	1.0066	0.2814	1.0099	0.3450	1.0120
0.06	0.2425	1.0098	0.3438	1.0148	0.4217	1.0179
0.08	0.2791	1.0130	0.3960	1.0197	0.4860	1.0239
0.1	0.3111	1.0161	0.4417	1.0246	0.5423	1.0298
0.2	0.4328	1.0311	0.6170	1.0483	0.7593	1.0592
0.3	0.5218	1.0450	0.7465	1.0712	0.9208	1.0880
0.4	0.5932	1.0580	0.8516	1.0931	1.0528	1.1164
0.5	0.6533	1.0701	0.9408	1.1143	1.1656	1.1441
0.6	0.7051	1.0814	1.0184	1.1345	1.2644	1.1713
0.7	0.7506	1.0918	1.0873	1.1539	1.3525	1.1978
0.8	0.7910	1.1016	1.1490	1.1724	1.4320	1.2236
0.9	0.8274	1.1107	1.2048	1.1902	1.5044	1.2488
1.0	0.8603	1.1191	1.2558	1.2071	1.5708	1.2732
2.0	1.0769	1.1785	1.5995	1.3384	2.0288	1.4793
3.0	1.1925	1.2102	1.7887	1.4191	2.2889	1.6227
4.0	1.2646	1.2287	1.9081	1.4698	2.4556	1.7202
5.0	1.3138	1.2403	1.9898	1.5029	2.5704	1.7870
6.0	1.3496	1.2479	2.0490	1.5253	2.6537	1.8338
7.0	1.3766	1.2532	2.0937	1.5411	2.7165	1.8673
8.0	1.3978	1.2570	2.1286	1.5526	2.7654	1.8920
9.0	1.4149	1.2598	2.1566	1.5611	2.8044	1.9106
10.0	1.4289	1.2620	2.1795	1.5677	2.8363	1.9249
20.0	1.4961	1.2699	2.2880	1.5919	2.9857	1.9781
30.0	1.5202	1.2717	2.3261	1.5973	3.0372	1.9898
40.0	1.5325	1.2723	2.3455	1.5993	3.0632	1.9942
50.0	1.5400	1.2727	2.3572	1.6002	3.0788	1.9962
100.0	1.5552	1.2731	2.3809	1.6015	3.1102	1.9990
∞	1.5708	1.2732	2.4048	1.6021	3.1416	2.0000

ξ	$J_o(\xi)$	$J_1(\xi)$
0.0	1.0000	0.0000
0.1	0.9975	0.0499
0.2	0.9900	0.0995
0.3	0.9776	0.1483
0.4	0.9604	0.1960
0.5	0.9385	0.2423
0.6	0.9120	0.2867
0.7	0.8812	0.3290
0.8	0.8463	0.3688
0.9	0.8075	0.4059
1.0	0.7652	0.4400
1.1	0.7196	0.4709
1.2	0.6711	0.4983
1.3	0.6201	0.5220
1.4	0.5669	0.5419
1.5	0.5118	0.5579
1.6	0.4554	0.5699
1.7	0.3980	0.5778
1.8	0.3400	0.5815
1.9	0.2818	0.5812
2.0	0.2239	0.5767
2.1	0.1666	0.5683
2.2	0.1104	0.5560
2.3	0.0555	0.5399
2.4	0.0025	0.5202
2.6	−0.0968	−0.4708
2.8	−0.1850	−0.4097
3.0	−0.2601	−0.3391
3.2	−0.3202	−0.2613

with each geometry: the first chart is to determine the temperature T_o at the
center of the geometry at a given time t. The second chart is to determine the
temperature at *other locations* at the same time in terms of T_o. The third
chart is to determine the total amount of *heat transfer* up to the time t. These
plots are valid for $\tau > 0.2$.

Note that the case $1/Bi = k/hL = 0$ corresponds to $h \rightarrow \infty$, which corresponds to the case of *specified surface temperature* T_∞. That is, the case in
which the surfaces of the body are suddenly brought to the temperature T_∞ at

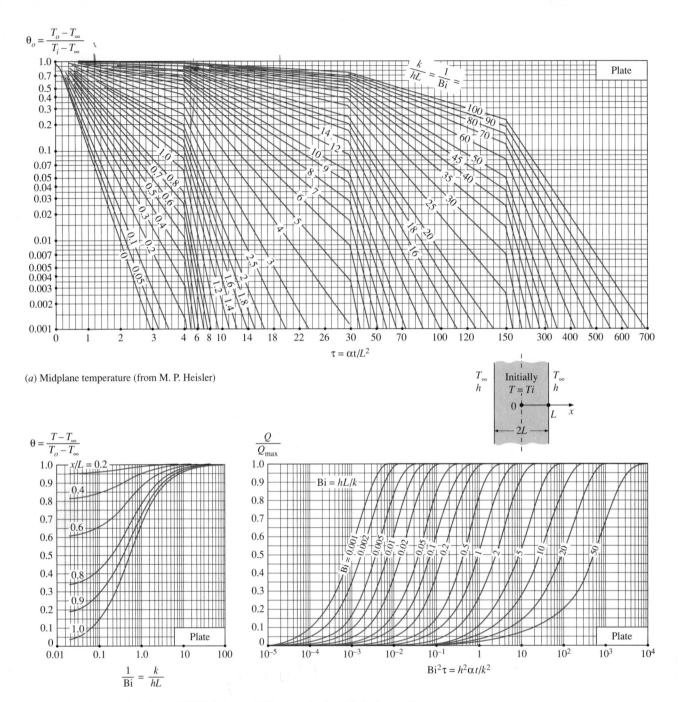

(a) Midplane temperature (from M. P. Heisler)

(b) Temperature distribution (from M. P. Heisler)   (c) Heat transfer (from H. Gröber *et al.*)

FIGURE 4-13

Transient temperature and heat transfer charts for a plane wall of thickness $2L$ initially at a uniform temperature T_i subjected to convection from both sides to an environment at temperature T_∞ with a convection coefficient of h.

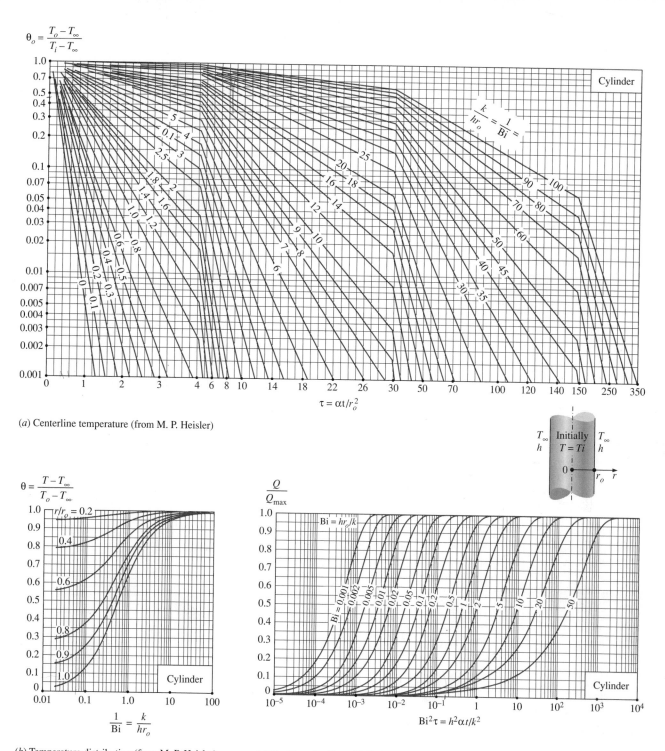

$\theta_o = \dfrac{T_o - T_\infty}{T_i - T_\infty}$

(a) Centerline temperature (from M. P. Heisler)

$\theta = \dfrac{T - T_\infty}{T_o - T_\infty}$

(b) Temperature distribution (from M. P. Heisler)

$\dfrac{Q}{Q_{max}}$

$Bi = hr_o/k$

(c) Heat transfer (from H. Gröber et al.)

FIGURE 4-14

Transient temperature and heat transfer charts for a long cylinder of radius r_o initially at a uniform temperature T_i subjected to convection from all sides to an environment at temperature T_∞ with a convection coefficient of h.

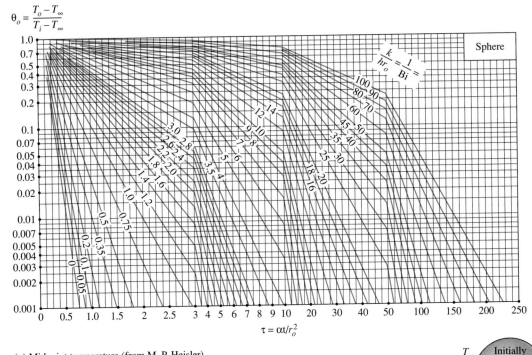

$$\theta_o = \frac{T_o - T_\infty}{T_i - T_\infty}$$

(a) Midpoint temperature (from M. P. Heisler)

$$\theta = \frac{T - T_\infty}{T_o - T_\infty}$$

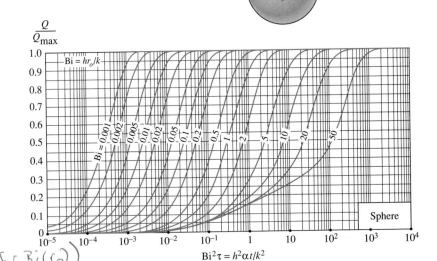

(b) Temperature distribution (from M. P. Heisler)

(c) Heat transfer (from H. Gröber et al.)

FIGURE 4-15

Transient temperature and heat transfer charts for a sphere of radius r_o initially at a uniform temperature T_i subjected to convection from all sides to an environment at temperature T_∞ with a convection coefficient of h.

$t = 0$ and kept at T_∞ at all times can be handled by setting h to infinity (Fig. 4-16).

The temperature of the body changes from the initial temperature T_i to the temperature of the surroundings T_∞ at the end of the transient heat conduction process. Thus, the *maximum* amount of heat that a body can gain (or lose if $T_i > T_\infty$) is simply the *change* in the *energy content* of the body. That is,

$$Q_{max} = mC_p(T_\infty - T_i) = \rho V C_p(T_\infty - T_i) \qquad \text{(kJ)} \qquad (4\text{-}16)$$

where m is the mass, V is the volume, ρ is the density, and C_p is the specific heat of the body. Thus, Q_{max} represents the amount of heat transfer for $t \to \infty$. The amount of heat transfer Q at a finite time t will obviously be less than this maximum. The ratio Q/Q_{max} is plotted in Figures 4-13c, 4-14c, and 4-15c against the variables Bi and $h^2\alpha t/k^2$ for the large plane wall, long cylinder, and sphere, respectively. Note that once the *fraction* of heat transfer Q/Q_{max} has been determined from these charts for the given t, the actual amount of heat transfer by that time can be evaluated by multiplying this fraction by Q_{max}. A *negative* sign for Q_{max} indicates that heat is *leaving* the body (Fig. 4-17).

The fraction of heat transfer can also be determined from the following relations, which are based on the one-term approximations discussed above:

Plane wall: $\qquad \left(\dfrac{Q}{Q_{max}}\right)_{wall} = 1 - \theta_{0,\,wall}\dfrac{\sin\lambda_1}{\lambda_1} \qquad (4\text{-}17)$

Cylinder: $\qquad \left(\dfrac{Q}{Q_{max}}\right)_{cyl} = 1 - 2\theta_{0,\,cyl}\dfrac{J_1(\lambda_1)}{\lambda_1} \qquad (4\text{-}18)$

Sphere: $\qquad \left(\dfrac{Q}{Q_{max}}\right)_{sph} = 1 - 3\theta_{0,\,sph}\dfrac{\sin\lambda_1 - \lambda_1\cos\lambda_1}{\lambda_1^3} \qquad (4\text{-}19)$

The use of the Heisler/Gröber charts and the one-term solutions discussed above is limited to the conditions specified at the beginning of this section: the body is initially at a *uniform* temperature, the temperature of the medium surrounding the body and the convection heat transfer coefficient are *constant* and *uniform*, and there is no *energy generation* in the body.

We discussed the physical significance of the *Biot number* earlier and indicated that it is a measure of the relative magnitudes of the two heat transfer mechanisms: *convection* at the surface and *conduction* through the solid. A *small* value of Bi indicates that the inner resistance of the body to heat conduction is *small* relative to the resistance to convection between the surface and the fluid. As a result, the temperature distribution within the solid becomes fairly uniform, and lumped system analysis becomes applicable. Recall that when Bi < 0.1, the error in assuming the temperature within the body to be *uniform* is negligible.

239

Transient Heat
Conduction in
Large Plane Walls,
Long Cylinders,
and Sphers

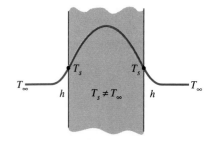

(a) Finite convection coefficient

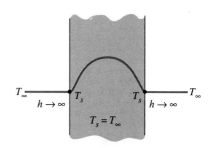

(b) Infinite convection coefficient

FIGURE 4-16

The specified surface temperature corresponds to the case of convection to an environment at T_∞ with a convection coefficient h that is *infinite*.

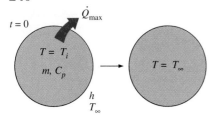

$t = 0$

$T = T_i$
m, C_p

h
T_∞

$\dot{Q}_{max}$

$T = T_\infty$

(a) Maximum heat transfer ($t \to \infty$)

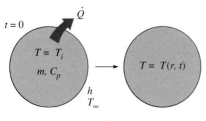

$t = 0$

$T = T_i$
m, C_p

h
T_∞

$\dot{Q}$

$T = T(r, t)$

$\text{Bi} = \dots$
$\left. \dfrac{h^2\alpha t}{k^2} = \text{Bi}^2\tau = \dots \right\}$ $\dfrac{Q}{Q_{max}} = \dots$
(Gröber chart)

(b) Actual heat transfer for time t

FIGURE 4-17

The fraction of total heat transfer Q/Q_{max} up to a specified time t is determined using the Gröber charts.

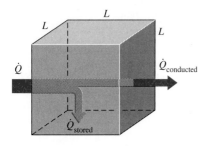

L
L
L
L

$\dot{Q}$

$\dot{Q}_{conducted}$

$\dot{Q}_{stored}$

Fourier number: $\tau = \dfrac{\alpha t}{L^2} = \dfrac{\dot{Q}_{conducted}}{\dot{Q}_{stored}}$

FIGURE 4-18

Fourier number at time t can be viewed as the ratio of the rate of heat conducted to the rate of heat stored at that time.

To understand the physical significance of the *Fourier number* τ, we express it as (Fig. 4-18)

$$\tau = \frac{\alpha t}{L^2} = \frac{kL^2\,(1/L)\,\Delta T}{\rho C_p L^3/t\,\Delta T} = \frac{\begin{array}{c}\text{The rate at which heat is } \textit{conducted}\\ \text{across } L \text{ of a body of volume } L^3\end{array}}{\begin{array}{c}\text{The rate at which heat is } \textit{stored}\\ \text{in a body of volume } L^3\end{array}} \quad (4\text{-}20)$$

Therefore, the Fourier number is a measure of *heat conducted* through a body relative to *heat stored*. Thus, a large value of the Fourier number indicates faster propagation of heat through a body.

Perhaps you are wondering about what constitutes an infinitely large plate or an infinitely long cylinder. After all, nothing in this world is infinite. A plate whose thickness is small relative to the other dimensions can be modeled as an infinitely large plate, except very near the outer edges. But the edge effects on large bodies are usually negligible, and thus a large plane wall such as the wall of a house can be modeled as an infinitely large wall for heat transfer purposes. Similarly, a long cylinder whose diameter is small relative to its length can be analyzed as an infinitely long cylinder. The use of the transient temperature charts and the one-term solutions is illustrated in the following examples.

EXAMPLE 4-3 Boiling Eggs

An ordinary egg can be approximated as a 5-cm-diameter sphere (Fig. 4-19). The egg is initially at a uniform temperature of 5°C and is dropped into boiling water at 95°C. Taking the convection heat transfer coefficient to be $h = 1200$ W/m$^2 \cdot$ °C, determine how long it will take for the center of the egg to reach 70°C.

Solution An egg is cooked in boiling water. The cooking time of the egg is to be determined.

Assumptions **1** The egg is spherical in shape with a radius of $r_0 = 2.5$ cm. **2** Heat conduction in the egg is one-dimensional because of thermal symmetry about the midpoint. **3** The thermal properties of the egg and the heat transfer coefficient are constant. **4** The Fourier number is $\tau > 0.2$ so that the one-term approximate solutions are applicable.

Properties The water content of eggs is about 74 percent, and thus the thermal conductivity and diffusivity of eggs can be approximated by those of water at the average temperature of $(5 + 70)/2 = 37.5$°C; $k = 0.627$ W/m $\cdot$ °C and $\alpha = k/\rho C_p = 0.151 \times 10^{-6}$ m^2/s (Table A-9).

Analysis The temperature within the egg varies with radial distance as well as time, and the temperature at a specified location at a given time can be determined from the Heisler charts or the one-term solutions. Here we will use the latter to demonstrate their use. The Biot number for this problem is

$$\text{Bi} = \frac{hr_0}{k} = \frac{(1200 \text{ W/m}^2 \cdot \text{°C})(0.025 \text{ m})}{0.627 \text{ W/m} \cdot \text{°C}} = 47.8$$

which is (Actually, the Biot number for use in lumped system analysis should be determined from $Bi = hLc/k = h(r/3)/k$.) much greater than 0.1, and thus the lumped system analysis is not applicable. The coefficients λ_1 and A_1 for a sphere corresponding to this Bi are, from Table 4-1,

$$\lambda_1 = 3.0753, \qquad A_1 = 1.9958$$

Substituting these and other values into Equation 4-15 and solving for τ gives

$$\frac{T_o - T_\infty}{T_i - T_\infty} = A_1 e^{-\lambda_1^2 \tau} \quad\longrightarrow\quad \frac{70 - 95}{5 - 95} = 1.9958 e^{-(3.0753)^2 \tau} \quad\longrightarrow\quad \tau = 0.209$$

which is greater than 0.2, and thus the one-term solution is applicable with an error of less than 2 percent. Then the cooking time is determined from the definition of the Fourier number to be

$$t = \frac{\tau r_o^2}{\alpha} = \frac{(0.209)(0.025 \text{ m})^2}{0.151 \times 10^{-6} \text{ m}^2/\text{s}} = 865 \text{ s} \approx \textbf{14.4 min}$$

Therefore, it will take about 15 min for the center of the egg to be heated from 5°C to 70°C.

241

Transient Heat
Conduction in
Large Plane Walls,
Long Cylinders,
and Spheres

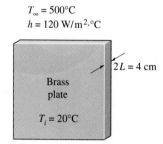

Egg

$T_i = 5°C$

$h = 1200 \text{ W/m}^2 \cdot °C$
$T_\infty = 95°C$

FIGURE 4-19

Schematic for Example 4-3.

EXAMPLE 4-4 Heating of Large Brass Plates in an Oven

In a production facility, large brass plates of 4 cm thickness that are initially at a uniform temperature of 20°C are heated by passing them through an oven that is maintained at 500°C (Fig. 4-20). The plates remain in the oven for a period of 7 min. Taking the combined convection and radiation heat transfer coefficient to be $h = 120 \text{ W/m}^2 \cdot °C$, determine the surface temperature of the plates when they come out of the oven.

Solution Large brass plates are heated in an oven. The surface temperature of the plates leaving the oven is to be determined.

Assumptions **1** Heat conduction in the plate is one-dimensional since the plate is large relative to its thickness and there is thermal symmetry about the center plane. **2** The thermal properties of the plate and the heat transfer coefficient are constant. **3** The Fourier number is $\tau > 0.2$ so that the one-term approximate solutions are applicable.

Properties The properties of brass at room temperature are $k = 110 \text{ W/m} \cdot °C$, $\rho = 8530 \text{ kg/m}^3$, $C_p = 380 \text{ J/kg} \cdot °C$, and $\alpha = 33.9 \times 10^{-6} \text{ m}^2/\text{s}$ (Table A-3). More accurate results are obtained by using properties at average temperature.

Analysis The temperature at a specified location at a given time can be determined from the Heisler charts or one-term solutions. Here we will use the charts to demonstrate their use. Noting that the half-thickness of the plate is $L = 0.02\text{m}$, from Figure 4-13 we have

$$\left.\begin{array}{l} \dfrac{1}{Bi} = \dfrac{k}{hL} = \dfrac{100 \text{ W/m} \cdot °C}{(120 \text{ W/m}^2 \cdot °C)(0.02 \text{ m})} = 45.8 \\[3mm] \tau = \dfrac{\alpha t}{L^2} = \dfrac{(33.9 \times 10^{-6} \text{ m}^2/\text{s})(7 \times 60 \text{ s})}{(0.02 \text{ m})^2} = 35.6 \end{array}\right\} \dfrac{T_o - T_\infty}{T_i - T_\infty} = 0.46$$

Also,

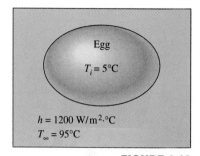

$T_\infty = 500°C$
$h = 120 \text{ W/m}^2 \cdot °C$

$2L = 4$ cm

Brass
plate

$T_i = 20°C$

FIGURE 4-20

Schematic for Example 4-4.

$$\left.\begin{array}{l} \dfrac{1}{\text{Bi}} = \dfrac{k}{hL} = 45.8 \\[2mm] \dfrac{x}{L} = \dfrac{L}{L} = 1 \end{array}\right\} \quad \dfrac{T - T_\infty}{T_o - T_\infty} = 0.99$$

Therefore,

$$\frac{T - T_\infty}{T_i - T_\infty} = \frac{T - T_\infty}{T_o - T_\infty}\frac{T_o - T_\infty}{T_i - T_\infty} = 0.46 \times 0.99 = 0.455$$

and

$$T = T_\infty + 0.455(T_i - T_\infty) = 500 + 0.455(20 - 500) = \textbf{282°C}$$

Therefore, the surface temperature of the plates will be 282°C when they leave the oven.

Discussion   We notice that the Biot number in this case is Bi = 1/45.8 = 0.022, which is much less than 0.1. Therefore, we expect the lumped system analysis to be applicable. This is also evident from $(T - T_\infty)/(T_o - T_\infty) = 0.99$, which indicates that the temperatures at the center and the surface of the plate relative to the surrounding temperature are within 1 percent of each other. Noting that the error involved in reading the Heisler charts is typically at least a few percent, the lumped system analysis in this case may yield just as accurate results with less effort.

   The heat transfer surface area of the plate is 2A, where A is the face area of the plate (the plate transfers heat through both of its surfaces), and the volume of the plate is V = (2L)A, where L is the half-thickness of the plate. The exponent b used in the lumped system analysis is determined to be

$$b = \frac{hA}{\rho C_p V} = \frac{h(2A)}{\rho C_p(2LA)} = \frac{h}{\rho C_p L}$$

$$= \frac{120 \text{ W/m}^2 \cdot \text{°C}}{(8530 \text{ kg/m}^3)(380 \text{ J/kg} \cdot \text{°C})(0.02 \text{ m})} = 0.00185 \text{ s}^{-1}$$

Then the temperature of the plate at t = 7 min = 420 s is determined from

$$\frac{T(t) - T_\infty}{T_i - T_\infty} = e^{-bt} \longrightarrow \frac{T(t) - 500}{20 - 500} = e^{-(0.00185 \text{ s}^{-1})(420 \text{ s})}$$

It yields

$$T(t) = 279°C$$

which is practically identical to the result obtained above using the Heisler charts. Therefore, we can use lumped system analysis with confidence when the Biot number is sufficiently small.

EXAMPLE 4-5   Cooling of a Long Stainless Steel Cylindrical Shaft

A long 20-cm-diameter cylindrical shaft made of stainless steel 304 comes out of an oven at a uniform temperature of 600°C (Fig. 4-21). The shaft is then allowed to cool slowly in an environment chamber at 200°C with an average heat transfer coefficient of h = 80 W/m² · °C. Determine the temperature at the center of the

FIGURE 4-21

Schematic for Example 4-5.

$T_\infty = 200°C$
$h = 80 \text{ W/m}^2 \cdot °C$

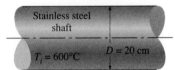

Stainless steel
shaft

$T_i = 600°C$ $D = 20$ cm

shaft 45 min after the start of the cooling process. Also, determine the heat transfer per unit length of the shaft during this time period.

Solution A long cylindrical shaft at 600°C is allowed to cool slowly. The center temperature and the heat transfer per unit length are to be determined.

Assumptions **1** Heat conduction in the shaft is one-dimensional since it is long and it has thermal symmetry about the centerline. **2** The thermal properties of the shaft and the heat transfer coefficient are constant. **3** The Fourier number is $\tau > 0.2$ so that the one-term approximate solutions are applicable.

Properties The properties of stainless steel 304 at room temperature are $k = 14.9$ W/m $\cdot$ °C, $\rho = 7900$ kg/m^3, $C_p = 477$ J/kg $\cdot$ °C, and $\alpha = 3.95 \times 10^{-6}$ m^2/s (Table A-3). More accurate results can be obtained by using properties at average temperature.

Analysis The temperature within the shaft may vary with the radial distance r as well as time, and the temperature at a specified location at a given time can be determined from the Heisler charts. Noting that the radius of the shaft is $r_o = 0.1$ m, from Figure 4-14 we have

$$\left. \begin{array}{l} \dfrac{1}{\text{Bi}} = \dfrac{k}{hr_o} = \dfrac{14.9 \text{ W/m} \cdot °\text{C}}{(80 \text{ W/m}^2 \cdot °\text{C})(0.1 \text{ m})} = 1.86 \\[4mm] \tau = \dfrac{\alpha t}{r_o^2} = \dfrac{(3.95 \times 10^{-6} \text{ m}^2/\text{s})(45 \times 60 \text{ s})}{(0.1 \text{ m})^2} = 1.07 \end{array} \right\} \dfrac{T_o - T_\infty}{T_i - T_\infty} = 0.40$$

and

$$T_o = T_\infty + 0.4(T_i - T_\infty) = 200 + 0.4(600 - 200) = \textbf{360°C}$$

Therefore, the center temperature of the shaft will drop from 600°C to 360°C in 45 min.

To determine the actual heat transfer, we first need to calculate the maximum heat that can be transferred from the cylinder, which is the sensible energy of the cylinder relative to its environment. Taking $L = 1$ m,

$$m = \rho V = \rho \pi r_o^2 L = (7900 \text{ kg/m}^3)\pi(0.1 \text{ m})^2(1 \text{ m}) = 248.2 \text{ kg}$$
$$Q_{max} = mC_p(T_\infty - T_i) = (248.2 \text{ kg})(0.477 \text{ kJ/kg} \cdot °\text{C})(600 - 200)°\text{C}$$
$$= 47,354 \text{ kJ}$$

The dimensionless heat transfer ratio is determined from Figure 4-14c for a long cylinder to be

$$\left. \begin{array}{l} \text{Bi} = \dfrac{1}{1/\text{Bi}} = \dfrac{1}{1.86} = 0.537 \\[4mm] \dfrac{h^2 \alpha t}{k^2} = \text{Bi}^2\tau = (0.537)^2(1.07) = 0.309 \end{array} \right\} \dfrac{Q}{Q_{max}} = 0.62$$

Therefore,

$$Q = 0.62 Q_{max} = 0.62 \times (47,354 \text{ kJ}) = \textbf{29,360 kJ}$$

which is the total heat transfer from the shaft during the first 45 min of the cooling.

Alternative solution   We could also solve this problem using the one-term solution relation instead of the transient charts. First we find the Biot number

$$\text{Bi} = \frac{hr_o}{k} = \frac{(80 \text{ W/m}^2 \cdot {}^\circ\text{C})(0.1 \text{ m})}{14.9 \text{ W/m} \cdot {}^\circ\text{C}} = 0.537$$

The coefficients λ_1 and A_1 for a cylinder corresponding to this Bi are determined from Table 4-1 to be

$$\lambda_1 = 0.970, \qquad A_1 = 1.122$$

Substituting these values into Equation 4-14 gives

$$\theta_0 = \frac{T_o - T_\infty}{T_i - T_\infty} = A_1 e^{-\lambda_1^2 \tau} = 1.122 e^{-(0.970)^2(1.07)} = 0.41$$

and thus

$$T_o = T_\infty + 0.41(T_i - T_\infty) = 200 + 041(600 - 200) = \textbf{364}^\circ\textbf{C}$$

The value of $J_1(\lambda_1)$ for $\lambda_1 = 0.970$ is determined from Table 4-2 to be 0.430. Then the fractional heat transfer is determined from Equation 4-18 to be

$$\frac{Q}{Q_{max}} = 1 - 2\theta_0 \frac{J_1(\lambda_1)}{\lambda_1} = 1 - 2 \times 0.41 \frac{0.430}{0.970} = 0.636$$

and thus

$$Q = 0.636 Q_{max} = 0.636 \times (47{,}354 \text{ kJ}) = \textbf{30{,}120 kJ}$$

Discussion   The slight difference between the two results is due to the reading error of the charts.

4-3 ■ TRANSIENT HEAT CONDUCTION IN SEMI-INFINITE SOLIDS

A semi-infinite solid is an idealized body that has a *single plane surface* and extends to infinity in all directions, as shown in Figure 4-22. This idealized body is used to indicate that the temperature change in the part of the body in which we are interested (the region close to the surface) is due to the thermal conditions on a single surface. The earth, for example, can be considered to be a semi-infinite medium in determining the variation of temperature near its surface. Also, a thick wall can be modeled as a semi-infinite medium if all we are interested in is the variation of temperature in the region near one of the surfaces, and the other surface is too far to have any impact on the region of interest during the time of observation.

Consider a semi-infinite solid that is at a uniform temperature T_i. At time $t = 0$, the surface of the solid at $x = 0$ is exposed to convection by a fluid at a constant temperature T_∞, with a heat transfer coefficient h. This problem can be formulated as a partial differential equation, which can be solved analytically for the transient temperature distribution $T(x, t)$. The solution obtained

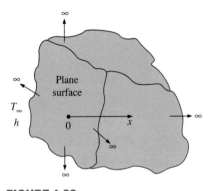

FIGURE 4-22
Schematic of a semi-infinite body.

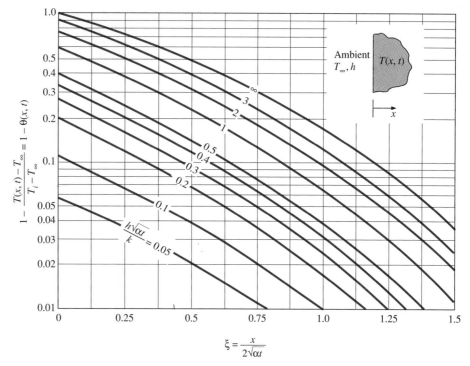

FIGURE 4-23

Variation of temperature with position and time in a semi-infinite solid initially at T_i subjected to convection to an environment at T_∞ with a convection heat transfer coefficient of h (from P. J. Schneider, Ref. 11).

is presented in Figure 4-23 graphically for the *nondimensionalized temperature* defined as

$$1 - \theta(x, t) = 1 - \frac{T(x, t) - T_\infty}{T_i - T_\infty} = \frac{T(x, t) - T_i}{T_\infty - T_i} \qquad (4\text{-}21)$$

against the dimensionless variable $x/(2\sqrt{\alpha t})$ for various values of the parameter $h\sqrt{\alpha t}/k$.

Note that the values on the vertical axis correspond to $x = 0$, and thus represent the surface temperature. The curve $h\sqrt{\alpha t}/k = \infty$ corresponds to $h \to \infty$, which corresponds to the case of *specified temperature* T_∞ at the surface at $x = 0$. That is, the case in which the surface of the semi-infinite body is suddenly brought to temperature T_∞ at $t = 0$ and kept at T_∞ at all times can be handled by setting h to infinity. The specified surface temperature case is closely approximated in practice when condensation or boiling takes place on the surface. For a *finite* heat transfer coefficient h, the surface temperature approaches the fluid temperature T_∞ as the time t approaches infinity.

The exact solution of the transient one-dimensional heat conduction problem in a semi-infinite medium that is initially at a uniform temperature of T_i and is suddenly subjected to convection at time $t = 0$ has been obtained, and is expressed as

$$\frac{T(x, t) - T_i}{T_\infty - T_i} = \text{erfc}\left(\frac{x}{2\sqrt{\alpha t}}\right) - \exp\left(\frac{hx}{k} + \frac{h^2\alpha t}{k^2}\right)\left[\text{erfc}\left(\frac{x}{2\sqrt{\alpha t}} + \frac{h\sqrt{\alpha t}}{k}\right)\right] \quad (4\text{-}22)$$

where the quantity erfc (ξ) is the **complementary error function,** defined as

$$\text{erfc}(\xi) = 1 - \frac{2}{\sqrt{\pi}}\int_0^\xi e^{-u^2}\, du \quad (4\text{-}23)$$

Despite its simple appearance, the integral that appears in the above relation cannot be performed analytically. Therefore, it is evaluated numerically for different values of ξ, and the results are listed in Table 4-3. For the special case of $h \to \infty$, the surface temperature T_s becomes equal to the fluid temperature T_∞, and Equation 4-22 reduces to

$$\frac{T(x, t) - T_i}{T_s - T_i} = \text{erfc}\left(\frac{x}{2\sqrt{\alpha t}}\right) \quad (4\text{-}24)$$

This solution corresponds to the case when the temperature of the exposed surface of the medium is suddenly raised (or lowered) to T_s at $t = 0$ and is maintained at that value at all times. Although the graphical solution given in Figure 4-23 is simply a plot of the exact analytical solution given by Equation 4-23, it is subject to reading errors, and thus is of limited accuracy.

TABLE 4-3

The complementary error function

ξ	erfc (ξ)	ξ	erfc (ξ)	ξ	erfc (ξ)	ξ	erfc (ξ)	ξ	erfc (ξ)	ξ	erfc (ξ)
0.00	1.00000	0.38	0.5910	0.76	0.2825	1.14	0.1069	1.52	0.03159	1.90	0.00721
0.02	0.9774	0.40	0.5716	0.78	0.2700	1.16	0.10090	1.54	0.02941	1.92	0.00662
0.04	0.9549	0.42	0.5525	0.80	0.2579	1.18	0.09516	1.56	0.02737	1.94	0.00608
0.06	0.9324	0.44	0.5338	0.82	0.2462	1.20	0.08969	1.58	0.02545	1.96	0.00557
0.08	0.9099	0.46	0.5153	0.84	0.2349	1.22	0.08447	1.60	0.02365	1.98	0.00511
0.10	0.8875	0.48	0.4973	0.86	0.2239	1.24	0.07950	1.62	0.02196	2.00	0.00468
0.12	0.8652	0.50	0.4795	0.88	0.2133	1.26	0.07476	1.64	0.02038	2.10	0.00298
0.14	0.8431	0.52	0.4621	0.90	0.2031	1.28	0.07027	1.66	0.01890	2.20	0.00186
0.16	0.8210	0.54	0.4451	0.92	0.1932	1.30	0.06599	1.68	0.01751	2.30	0.00114
0.18	0.7991	0.56	0.4284	0.94	0.1837	1.32	0.06194	1.70	0.01612	2.40	0.00069
0.20	0.7773	0.58	0.4121	0.96	0.1746	1.34	0.05809	1.72	0.01500	2.50	0.00041
0.22	0.7557	0.60	0.3961	0.98	0.1658	1.36	0.05444	1.74	0.01387	2.60	0.00024
0.24	0.7343	0.62	0.3806	1.00	0.1573	1.38	0.05098	1.76	0.01281	2.70	0.00013
0.26	0.7131	0.64	0.3654	1.02	0.1492	1.40	0.04772	1.78	0.01183	2.80	0.00008
0.28	0.6921	0.66	0.3506	1.04	0.1413	1.42	0.04462	1.80	0.01091	2.90	0.00004
0.30	0.6714	0.68	0.3362	1.06	0.1339	1.44	0.04170	1.82	0.01006	3.00	0.00002
0.32	0.6509	0.70	0.3222	1.08	0.1267	1.46	0.03895	1.84	0.00926	3.20	0.00001
0.34	0.6306	0.72	0.3086	1.10	0.1198	1.48	0.03635	1.86	0.00853	3.40	0.00000
0.36	0.6107	0.74	0.2953	1.12	0.1132	1.50	0.03390	1.88	0.00784	3.60	0.00000

EXAMPLE 4-6   Minimum Burial Depth of Water Pipes to Avoid Freezing

In areas where the air temperature remains below 0°C for prolonged periods of time, the freezing of water in underground pipes is a major concern. Fortunately, the soil remains relatively warm during those periods, and it takes weeks for the subfreezing temperatures to reach the water mains in the ground. Thus, the soil effectively serves as an insulation to protect the water from subfreezing temperatures in winter.

The ground at a particular location is covered with snow pack at −10°C for a continuous period of three months, and the average soil properties at that location are $k = 0.4$ W/m · °C and $\alpha = 0.15 \times 10^{-6}$ m^2/s (Fig. 4-24). Assuming an initial uniform temperature of 15°C for the ground, determine the minimum burial depth to prevent the water pipes from freezing.

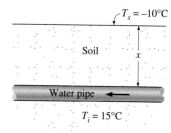

FIGURE 4-24
Schematic for Example 4-6.

Solution   The water pipes are buried in the ground to prevent freezing. The minimum burial depth at a particular location is to be determined.

Assumptions   **1** The temperature in the soil is affected by the thermal conditions at one surface only, and thus the soil can be considered to be a semi-infinite medium with a specified surface temperature of −10°C. **2** The thermal properties of the soil are constant.

Properties   The properties of the soil are as given in the problem statement.

Analysis   The temperature of the soil surrounding the pipes will be 0°C after three months in the case of minimum burial depth. Therefore, from Figure 4-23, we have

$$
\left.
\begin{aligned}
\frac{h\sqrt{\alpha t}}{k} &= \infty \quad \text{(since } h \to \infty\text{)} \\[2mm]
1 - \frac{T(x, t) - T_\infty}{T_i - T_\infty} &= 1 - \frac{0 - (-10)}{15 - (-10)} = 0.6
\end{aligned}
\right\}
\xi = \frac{x}{2\sqrt{\alpha t}} = 0.36
$$

We note that

$$t = (90 \text{ days})(24 \text{ h/day})(3600 \text{ s/h}) = 7.78 \times 10^6 \text{ s}$$

and thus

$$x = 2\xi\sqrt{\alpha t} = 2 \times 0.36\sqrt{(0.15 \times 10^{-6} \text{ m}^2/\text{s})(7.78 \times 10^6 \text{ s})} = \textbf{0.77 m}$$

Therefore, the water pipes must be buried to a depth of at least 77 cm to avoid freezing under the specified harsh winter conditions.

Alternative solution   The solution of this problem could also be determined from Equation 4-24:

$$\frac{T(x, t) - T_i}{T_s - T_i} = \text{erfc}\left(\frac{x}{2\sqrt{\alpha t}}\right) \longrightarrow \frac{0 - 15}{-10 - 15} = \text{erfc}\left(\frac{x}{2\sqrt{\alpha t}}\right) = 0.60$$

The argument that corresponds to this value of the complementary error function is determined from Table 4-3 to be $\xi = 0.37$. Therefore,

$$x = 2\xi\sqrt{\alpha t} = 2 \times 0.37\sqrt{(0.15 \times 10^{-6} \text{ m}^2/\text{s})(7.78 \times 10^{-6} \text{ s})} = \textbf{0.80 m}$$

Again, the slight difference is due to the reading error of the chart.

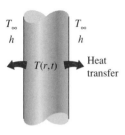

T_∞ T_∞
h h

$T(r,t)$ Heat transfer

(a) Long cylinder

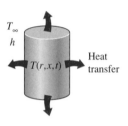

T_∞
h

$T(r,x,t)$ Heat transfer

(b) Short cylinder (two-dimensional)

FIGURE 4-25

The temperature in a short cylinder exposed to convection from all surfaces varies in both the radial and axial directions, and thus heat is transferred in both directions.

FIGURE 4-26

A short cylinder of radius r_o and height a is the *intersection* of a long cylinder of radius r_o and a plane wall of thickness a.

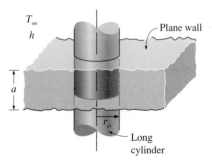

T_∞
h

Plane wall

a

r_o

Long cylinder

4-4 ■ TRANSIENT HEAT CONDUCTION IN MULTIDIMENSIONAL SYSTEMS

The transient temperature charts presented earlier can be used to determine the temperature distribution and heat transfer in *one-dimensional* heat conduction problems associated with a large plane wall, a long cylinder, a sphere, and a semi-infinite medium. Using a clever superposition principle called the **product solution,** these charts can also be used to construct solutions for the *two-dimensional* transient heat conduction problems encountered in geometries such as a short cylinder, a long rectangular bar, or a semi-infinite cylinder or plate, and even *three-dimensional* problems associated with geometries such as a rectangular prism or a semi-infinite rectangular bar, provided that *all* surfaces of the solid are subjected to convection to the *same* fluid at temperature T_∞, with the *same* heat transfer coefficient h, and the body involves no heat generation (Fig. 4-25). The solution in such multidimensional geometries can be expressed as the *product* of the solutions for the one-dimensional geometries whose intersection is the multidimensional geometry.

Consider a *short cylinder* of height a and radius r_o initially at a uniform temperature T_i. There is no heat generation in the cylinder. At time $t = 0$, the cylinder is subjected to convection from all surfaces to a medium at temperature T_∞ with a heat transfer coefficient h. The temperature within the cylinder will change with x as well as r and time t since heat transfer will occur from the top and bottom of the cylinder as well as its side surfaces. That is, $T = T(r, x, t)$ and thus this is a two-dimensional transient heat conduction problem. When the properties are assumed to be constant, it can be shown that the solution of this two-dimensional problem can be expressed as

$$\left(\frac{T(r, x, t) - T_\infty}{T_i - T_\infty}\right)_{\substack{\text{short} \\ \text{cylinder}}} = \left(\frac{T(x, t) - T_\infty}{T_i - T_\infty}\right)_{\substack{\text{plane} \\ \text{wall}}} \left(\frac{T(r, t) - T_\infty}{T_i - T_\infty}\right)_{\substack{\text{infinite} \\ \text{cylinder}}} \quad (4\text{-}25)$$

That is, the solution for the two-dimensional short cylinder of height a and radius r_o is equal to the *product* of the nondimensionalized solutions for the one-dimensional plane wall of thickness a and the long cylinder of radius r_o, which are the two geometries whose intersection is the short cylinder, as shown in Figure 4-26. We generalize this as follows: *the solution for a multidimensional geometry is the product of the solutions of the one-dimensional geometries whose intersection is the multidimensional body.*

For convenience, the one-dimensional solutions are denoted by

$$\theta_{\text{wall}}(x, t) = \left(\frac{T(x, t) - T_\infty}{T_i - T_\infty}\right)_{\substack{\text{plane} \\ \text{wall}}}$$

$$\theta_{\text{cyl}}(r, t) = \left(\frac{T(r, t) - T_\infty}{T_i - T_\infty}\right)_{\substack{\text{infinite} \\ \text{cylinder}}} \quad (4\text{-}26)$$

$$\theta_{\text{semi-inf}}(x, t) = \left(\frac{T(x, t) - T_\infty}{T_i - T_\infty}\right)_{\substack{\text{semi-infinite} \\ \text{solid}}}$$

For example, the solution for a long solid bar whose cross section is an $a \times b$ rectangle is the intersection of the two infinite plane walls of thicknesses a and b, as shown in Figure 4-27, and thus the transient temperature distribution for this rectangular bar can be expressed as

$$\left(\frac{T(x, y, t) - T_\infty}{T_i - T_\infty}\right)_{\substack{\text{rectangular} \\ \text{bar}}} = \theta_{\text{wall}}(x, t)\theta_{\text{wall}}(y, t) \quad (4\text{-}27)$$

The proper forms of the product solutions for some other geometries are given in Table 4-4. It is important to note that the x-coordinate is measured from the *surface* in a semi-infinite solid, and from the *midplane* in a plane wall. The radial distance r is always measured from the centerline.

Note that the solution of a *two-dimensional* problem involves the product of *two* one-dimensional solutions, whereas the solution of a *three-dimensional* problem involves the product of *three* one-dimensional solutions.

A modified form of the product solution can also be used to determine the total transient heat transfer to or from a multidimensional geometry by using the one-dimensional values, as shown by L. S. Langston in 1982. The transient heat transfer for a two-dimensional geometry formed by the intersection of two one-dimensional geometries 1 and 2 is

$$\left(\frac{Q}{Q_{\text{max}}}\right)_{\text{total, 2D}} = \left(\frac{Q}{Q_{\text{max}}}\right)_1 + \left(\frac{Q}{Q_{\text{max}}}\right)_2\left[1 - \left(\frac{Q}{Q_{\text{max}}}\right)_1\right] \quad (4\text{-}28)$$

Transient heat transfer for a three-dimensional body formed by the intersection of three one-dimensional bodies 1, 2, and 3 is given by

$$\left(\frac{Q}{Q_{\text{max}}}\right)_{\text{total, 3D}} = \left(\frac{Q}{Q_{\text{max}}}\right)_1 + \left(\frac{Q}{Q_{\text{max}}}\right)_2\left[1 - \left(\frac{Q}{Q_{\text{max}}}\right)_1\right]$$
$$+ \left(\frac{Q}{Q_{\text{max}}}\right)_3\left[1 - \left(\frac{Q}{Q_{\text{max}}}\right)_1\right]\left[1 - \left(\frac{Q}{Q_{\text{max}}}\right)_2\right] \quad (4\text{-}29)$$

The use of the product solution in transient two- and three-dimensional heat conduction problems is illustrated in the following examples.

EXAMPLE 4-7 Cooling of a Short Brass Cylinder

A short brass cylinder of diameter $D = 10$ cm and height $H = 12$ cm is initially at a uniform temperature $T_i = 120°C$. The cylinder is now placed in atmospheric air at 25°C, where heat transfer takes place by convection, with a heat transfer coefficient of $h = 60$ W/m² · °C. Calculate the temperature at (a) the center of the cylinder and (b) the center of the top surface of the cylinder 15 min after the start of the cooling.

Solution   A short cylinder is allowed to cool in atmospheric air. The temperatures at the centers of the cylinder and the top surface are to be determined.

Assumptions   **1** Heat conduction in the short cylinder is two-dimensional, and thus the temperature varies in both the axial x- and the radial r-directions. **2** The

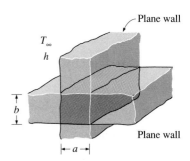

T_∞
h
b
a
Plane wall
Plane wall

FIGURE 4-27

A long solid bar of rectangular profile $a \times b$ is the *intersection* of two plane walls of thicknesses a and b.

TABLE 4-4

Multidimensional solutions expressed as products of one-dimensional solutions for bodies that are initially at a uniform temperature T_i and exposed to convection from all surfaces to a medium at T_∞

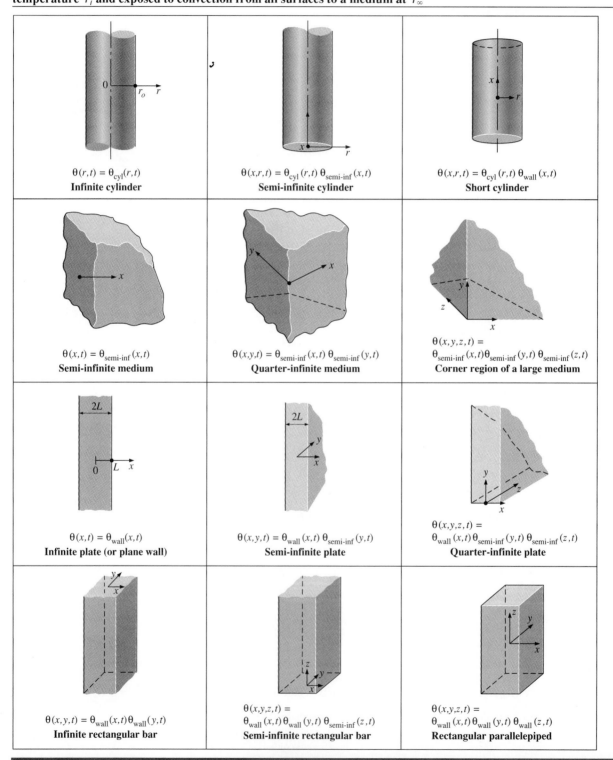

$\theta(r,t) = \theta_{cyl}(r,t)$
Infinite cylinder

$\theta(x,r,t) = \theta_{cyl}(r,t)\,\theta_{semi\text{-}inf}(x,t)$
Semi-infinite cylinder

$\theta(x,r,t) = \theta_{cyl}(r,t)\,\theta_{wall}(x,t)$
Short cylinder

$\theta(x,t) = \theta_{semi\text{-}inf}(x,t)$
Semi-infinite medium

$\theta(x,y,t) = \theta_{semi\text{-}inf}(x,t)\,\theta_{semi\text{-}inf}(y,t)$
Quarter-infinite medium

$\theta(x,y,z,t) =$
$\theta_{semi\text{-}inf}(x,t)\theta_{semi\text{-}inf}(y,t)\,\theta_{semi\text{-}inf}(z,t)$
Corner region of a large medium

$\theta(x,t) = \theta_{wall}(x,t)$
Infinite plate (or plane wall)

$\theta(x,y,t) = \theta_{wall}(x,t)\,\theta_{semi\text{-}inf}(y,t)$
Semi-infinite plate

$\theta(x,y,z,t) =$
$\theta_{wall}(x,t)\theta_{semi\text{-}inf}(y,t)\,\theta_{semi\text{-}inf}(z,t)$
Quarter-infinite plate

$\theta(x,y,t) = \theta_{wall}(x,t)\theta_{wall}(y,t)$
Infinite rectangular bar

$\theta(x,y,z,t) =$
$\theta_{wall}(x,t)\,\theta_{wall}(y,t)\,\theta_{semi\text{-}inf}(z,t)$
Semi-infinite rectangular bar

$\theta(x,y,z,t) =$
$\theta_{wall}(x,t)\,\theta_{wall}(y,t)\,\theta_{wall}(z,t)$
Rectangular parallelepiped

thermal properties of the cylinder and the heat transfer coefficient are constant.
3 The Fourier number is $\tau > 0.2$ so that the one-term approximate solutions are applicable.

Properties The properties of brass at room temperature are $k = 110$ W/m · °C and $\alpha = 33.9 \times 10^{-6}$ m²/s (Table A-3). More accurate results can be obtained by using properties at average temperature.

Analysis (a) This short cylinder can physically be formed by the intersection of a long cylinder of radius $r_o = 5$ cm and a plane wall of thickness $2L = 12$ cm, as shown in Figure 4-28. The dimensionless temperature at the center of the plane wall is determined from Figure 4-13a to be

$$\left.\begin{array}{l} \tau = \dfrac{\alpha t}{L^2} = \dfrac{(3.39 \times 10^{-5}\,\text{m}^2/\text{s})(900\,\text{s})}{(0.06\,\text{m})^2} = 8.48 \\[4mm] \dfrac{1}{Bi} = \dfrac{k}{hL} = \dfrac{110\,\text{W/m}\cdot°\text{C}}{(60\,\text{W/m}^2\cdot°\text{C})(0.06\,\text{m})} = 30.6 \end{array}\right\} \theta_{\text{wall}}(0,\,t) = \dfrac{T(0,\,t) - T_\infty}{T_i - T_\infty} = 0.8$$

Similarly, at the center of the cylinder, we have

$$\left.\begin{array}{l} \tau = \dfrac{\alpha t}{r_o^2} = \dfrac{(3.39 \times 10^{-5}\,\text{m}^2/\text{s})(900\,\text{s})}{(0.05\,\text{m})^2} = 12.2 \\[4mm] \dfrac{1}{Bi} = \dfrac{k}{hr_o} = \dfrac{110\,\text{W/m}\cdot°\text{C}}{(60\,\text{W/m}^2\cdot°\text{C})(0.05\,\text{m})} = 36.7 \end{array}\right\} \theta_{\text{cyl}}(0,\,t) = \dfrac{T(0,\,t) - T_\infty}{T_i - T_\infty} = 0.5$$

Therefore,

$$\left(\dfrac{T(0,\,0,\,t) - T_\infty}{T_i - T_\infty}\right)_{\substack{\text{short} \\ \text{cylinder}}} = \theta_{\text{wall}}(0,\,t) \times \theta_{\text{cyl}}(0,\,t) = 0.8 \times 0.5 = 0.4$$

and

$$T(0,\,0,\,t) = T_\infty + 0.4(T_i - T_\infty) = 25 + 0.4(120 - 25) = \textbf{63°C}$$

This is the temperature at the center of the short cylinder, which is also the center of both the long cylinder and the plate.

(b) The center of the top surface of the cylinder is still at the center of the long cylinder ($r = 0$), but at the outer surface of the plane wall ($x = L$). Therefore, we first need to find the surface temperature of the wall. Noting that $x = L = 0.06$ m,

$$\left.\begin{array}{l} \dfrac{x}{L} = \dfrac{0.06\,\text{m}}{0.06\,\text{m}} = 1 \\[4mm] \dfrac{1}{Bi} = \dfrac{k}{hL} = \dfrac{110\,\text{W/m}\cdot°\text{C}}{(60\,\text{W/m}^2\cdot°\text{C})(0.06\,\text{m})} = 30.6 \end{array}\right\} \dfrac{T(L,\,t) - T_\infty}{T_o - T_\infty} = 0.98$$

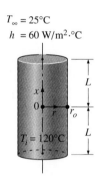

$T_\infty = 25°C$
$h = 60$ W/m²·°C

$T_i = 120°C$

FIGURE 4-28
Schematic for Example 4-7.

Then

$$\theta_{wall}(L, t) = \frac{T(L, t) - T_\infty}{T_i - T_\infty} = \left(\frac{T(L, t) - T_\infty}{T_o - T_\infty}\right)\left(\frac{T_o - T_\infty}{T_i - T_\infty}\right) = 0.98 \times 0.8 = 0.784$$

Therefore,

$$\left(\frac{T(L, 0, t) - T_\infty}{T_i - T_\infty}\right)_{\substack{short \\ cylinder}} = \theta_{wall}(L, t)\theta_{cyl}(0, t) = 0.784 \times 0.5 = 0.392$$

and

$$T(L, 0, t) = T_\infty + 0.392(T_i - T_\infty) = 25 + 0.392(120 - 25) = \mathbf{62.2°C}$$

which is the temperature at the center of the top surface of the cylinder.

EXAMPLE 4-8 Heat Transfer from a Short Cylinder

Determine the total heat transfer from the short brass cylinder ($\rho = 8530$ kg/m^3, $C_p = 0.380$ kJ/(kg $\cdot$ °C) discussed in Example 4-7.

Solution We first determine the maximum heat that can be transferred from the cylinder, which is the sensible energy content of the cylinder relative to its environment:

$$m = \rho V = \rho \pi r_o^2 L = (8530 \text{ kg/m}^3)\pi(0.05 \text{ m})^2(0.06 \text{ m}) = 4.02 \text{ kg}$$

$$Q_{max} = mC_p(T_i - T_\infty) = (4.02 \text{ kg})(0.380 \text{ kJ/kg} \cdot °C)(120 - 25)°C = 145.1 \text{ kJ}$$

Then we determine the dimensionless heat transfer ratios for both geometries. For the plane wall, it is determined from Figure 4-13c to be

$$\left.\begin{array}{l} Bi = \dfrac{1}{1/Bi} = \dfrac{1}{30.6} = 0.0327 \\[2mm] \dfrac{h^2\alpha t}{k^2} = Bi^2\tau = (0.0327)^2(8.48) = 0.0091 \end{array}\right\} \left(\frac{Q}{Q_{max}}\right)_{\substack{plane \\ wall}} = 0.23$$

Similarly, for the cylinder, we have

$$\left.\begin{array}{l} Bi = \dfrac{1}{1/Bi} = \dfrac{1}{36.7} = 0.0272 \\[2mm] \dfrac{h^2\alpha t}{k^2} = Bi^2\tau = (0.0272)^2(12.2) = 0.0090 \end{array}\right\} \left(\frac{Q}{Q_{max}}\right)_{\substack{infinite \\ cylinder}} = 0.47$$

Then the heat transfer ratio for the short cylinder is, from Equation 4-28,

$$\left(\frac{Q}{Q_{max}}\right)_{short\,cyl} = \left(\frac{Q}{Q_{max}}\right)_1 + \left(\frac{Q}{Q_{max}}\right)_2\left[1 - \left(\frac{Q}{Q_{max}}\right)_1\right]$$

$$= 0.23 + 0.47(1 - 0.23) = 0.592$$

Therefore, the total heat transfer from the cylinder during the first 15 min of cooling is

$$Q = 0.592 Q_{max} = 0.592 \times (145.1 \text{ kJ}) = \mathbf{85.9 \text{ kJ}}$$

EXAMPLE 4-9   Cooling of a Long Cylinder by Water

A semi-infinite aluminum cylinder of diameter $D = 20$ cm is initially at a uniform temperature $T_i = 200°C$. The cylinder is now placed in water at 15°C where heat transfer takes place by convection, with a heat transfer coefficient of $h = 120$ W/m² · °C. Determine the temperature at the center of the cylinder 15 cm from the end surface 5 min after the start of the cooling.

Solution   A semi-infinite aluminum cylinder is cooled by water. The temperature at the center of the cylinder 15 cm from the end surface is to be determined.

Assumptions   **1** Heat conduction in the semi-infinite cylinder is two-dimensional, and thus the temperature varies in both the axial x- and the radial r-directions. **2** The thermal properties of the cylinder and the heat transfer coefficient are constant. **3** The Fourier number is $\tau > 0.2$ so that the one-term approximate solutions are applicable.

Properties   The properties of aluminum at room temperature are $k = 237$ W/m · °C and $\alpha = 9.71 \times 10^{-6}$ m²/s (Table A-3). More accurate results can be obtained by using properties at average temperature.

Analysis   This semi-infinite cylinder can physically be formed by the intersection of an infinite cylinder of radius $r_o = 10$ cm and a semi-infinite medium, as shown in Figure 4-29.

We will solve this problem using the one-term solution relation for the cylinder and the analytic solution for the semi-infinite medium. First we consider the infinitely long cylinder and evaluate the Biot number:

$$Bi = \frac{hr_o}{k} = \frac{(120 \text{ W/m}^2 \cdot °C)(0.1 \text{ m})}{237 \text{ W/m} \cdot °C} = 0.05$$

The coefficients λ_1 and A_1 for a cylinder corresponding to this Bi are determined from Table 4-1 to be $\lambda_1 = 0.3126$ and $A_1 = 1.0124$. Fourier number in this case is

$$\tau = \frac{\alpha t}{r_o^2} = \frac{(9.71 \times 10^{-5} \text{ m}^2/\text{s})(5 \times 60 \text{ s})}{(0.1 \text{ m})^2} = 2.91 > 0.2$$

$T_\infty = 15°C$
$h = 120$ W/m²·°C
$T_i = 200°C$
$D = 20$ cm
$x = 15$ cm

FIGURE 4-29
Schematic for Example 4-9.

and thus the one-term approximation is applicable. Substituting these values into Equation 4-14 gives

$$\theta_0 = \theta_{cyl}(0, t) = A_1 e^{-\lambda_1^2 \tau} = 1.0124 e^{-(0.3126)^2(2.91)} = 0.762$$

The solution for the semi-infinite solid can be determined from

$$1 - \theta_{semi-inf}(x, t) = \text{erfc}\left(\frac{x}{2\sqrt{\alpha t}}\right) - \exp\left(\frac{hx}{k} + \frac{h^2 \alpha t}{k^2}\right)\left[\text{erfc}\left(\frac{x}{2\sqrt{\alpha t}} + \frac{h\sqrt{\alpha t}}{k}\right)\right]$$

First we determine the various quantities in parentheses:

$$\xi = \frac{x}{2\sqrt{\alpha t}} = \frac{0.15\,\text{m}}{2\sqrt{(9.71 \times 10^{-5}\,\text{m}^2/\text{s})(5 \times 60\,\text{s})}} = 0.44$$

$$\frac{h\sqrt{\alpha t}}{k} = \frac{(120\,\text{W/m}^2 \cdot {}^\circ\text{C})\sqrt{(9.71 \times 10^{-5}\,\text{m}^2/\text{s})(300\,\text{s})}}{237\,\text{W/m} \cdot {}^\circ\text{C}} = 0.086$$

$$\frac{hx}{k} = \frac{(120\,\text{W/m}^2 \cdot \text{C})(0.15\,\text{m})}{237\,\text{W/m} \cdot {}^\circ\text{C}} = 0.0759$$

$$\frac{h^2 \alpha t}{k^2} = \left(\frac{h\sqrt{\alpha t}}{k}\right)^2 = (0.086)^2 = 0.0074$$

Substituting and evaluating the complementary error functions from Table 4-3,

$$\theta_{\text{semi-inf}}(x, t) = 1 - \text{erfc}\,(0.44) + \exp\,(0.0759 + 0.0074)\,\text{erfc}\,(0.44 + 0.086)$$
$$= 1 - 0.5338 + \exp\,(0.0833) \times 0.457$$
$$= 0.963$$

Now we apply the product solution to get

$$\left(\frac{T(x, 0, t) - T_\infty}{T_i - T_\infty}\right)_{\substack{\text{semi-infinite} \\ \text{cylinder}}} = \theta_{\text{semi-inf}}(x, t)\theta_{\text{cyl}}(0, t) = 0.963 \times 0.762 = 0.734$$

and

$$T(x, 0, t) = T_\infty + 0.734(T_i - T_\infty) = 15 + 0.734(200 - 15) = \mathbf{151^\circ C}$$

which is the temperature at the center of the cylinder 15 cm from the exposed bottom surface.

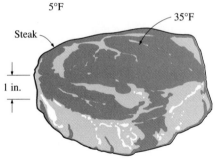

5°F

35°F

Steak

1 in.

FIGURE 4-30
Schematic for Example 4-10.

EXAMPLE 4-10 **Refrigerating Steaks while Avoiding Frostbite**

In a meat processing plant, 1-in.-thick steaks initially at 75°F are to be cooled in the racks of a large refrigerator that is maintained at 5°F (Fig. 4-30). The steaks are placed close to each other, so that heat transfer from the 1-in.-thick edges is negligible. The entire steak is to be cooled below 45°F, but its temperature is not to drop below 35°F at any point during refrigeration to avoid "frostbite." The convection heat transfer coefficient and thus the rate of heat transfer from the steak can be controlled by varying the speed of a circulating fan inside. Determine the heat transfer coefficient h that will enable us to meet both temperature constraints while keeping the refrigeration time to a minimum. The steak can be treated as a homogeneous layer having the properties $\rho = 74.9\,\text{lbm/ft}^3$, $C_p = 0.98\,\text{Btu/lbm} \cdot {}^\circ\text{F}$, $k = 0.26\,\text{Btu/h} \cdot \text{ft} \cdot {}^\circ\text{F}$, and $\alpha = 0.0035\,\text{ft}^2/\text{h}$.

Solution   Steaks are to be cooled in a refrigerator maintained at 5°F. The heat transfer coefficient that will allow cooling the steaks below 45°F while avoiding frostbite is to be determined.

Assumptions   **1** Heat conduction through the steaks is one-dimensional since the steaks form a large layer relative to their thickness and there is thermal symmetry about the center plane. **2** The thermal properties of the steaks and the heat transfer coefficient are constant. **3** The Fourier number is $\tau > 0.2$ so that the one-term approximate solutions are applicable.

Properties   The properties of the steaks are as given in the problem statement.

Analysis   The lowest temperature in the steak will occur at the surfaces and the highest temperature at the center at a given time, since the inner part will be the last place to be cooled. In the limiting case, the surface temperature at $x = L = 0.5$ in. from the center will be 35°F, while the midplane temperature is 45°F in an environment at 5°F. Then, from Figure 4-13b, we obtain

$$\left. \begin{array}{c} \dfrac{x}{L} = \dfrac{0.5 \text{ in.}}{0.5 \text{ in.}} = 1 \\[2mm] \dfrac{T(L, t) - T_\infty}{T_o - T_\infty} = \dfrac{35 - 5}{45 - 5} = 0.75 \end{array} \right\} \quad \dfrac{1}{\text{Bi}} = \dfrac{k}{hL} = 1.5$$

which gives

$$h = \frac{1}{1.5} \frac{k}{L} = \frac{0.26 \text{ Btu/h} \cdot \text{ft} \cdot °\text{F}}{1.5(0.5/12 \text{ ft})} = 4.16 \text{ Btu/h} \cdot \text{ft}^2 \cdot °\text{F}$$

Therefore, the convection heat transfer coefficient should be kept below this value to satisfy the constraints on the temperature of the steak during refrigeration. We can also meet the constraints by using a lower heat transfer coefficient, but doing so would extend the refrigeration time unnecessarily.

4-5 ■ SUMMARY

In this chapter we considered the variation of temperature with time as well as position in one- or multidimensional systems. We first considered the *lumped systems* in which the temperature varies with time but remains uniform throughout the system at any time. The temperature of a lumped body of arbitrary shape of mass m, volume V, surface area A_s, density ρ, and specific heat C_p initially at a uniform temperature T_i that is exposed to convection at time $t = 0$ in a medium at temperature T_∞ with a heat transfer coefficient h is expressed as

$$\frac{T(t) - T_\infty}{T_i - T_\infty} = e^{-bt}$$

where

$$b = \frac{hA}{\rho C_p V} = \frac{h}{\rho C_p L_c} \qquad (1/s)$$

is a positive quantity whose dimension is $(\text{time})^{-1}$. This relation can be used to determine the temperature $T(t)$ of a body at time t or, alternately, the time t required for the temperature to reach a specified value $T(t)$. Once the temperature $T(t)$ at time t is available, the *rate* of convection heat transfer between the body and its environment at that time can be determined from Newton's law of cooling as

$$\dot{Q}(t) = hA[T(t) - T_\infty] \qquad (\text{W})$$

The *total amount* of heat transfer between the body and the surrounding medium over the time interval $t = 0$ to t is simply the change in the energy content of the body,

$$Q = mC_p[T(t) - T_i] \qquad \text{(kJ)}$$

The amount of heat transfer reaches its upper limit when the body reaches the surrounding temperature T_∞. Therefore, the *maximum* heat transfer between the body and its surroundings is

$$Q_{max} = mC_p\,(T_\infty - T_i) \qquad \text{(kJ)}$$

The error involved in lumped system analysis is negligible when

$$\text{Bi} = \frac{hL_c}{k} < 0.1$$

where Bi is the *Biot number* and $L_c = V/A$ is the *characteristic length*.

When the lumped system analysis is not applicable, the variation of temperature with position as well as time can be determined using the *transient temperature charts* given in Figures 4-13, 4-14, 4-15, and 4-23 for a large plane wall, a long cylinder, a sphere, and a semi-infinite medium, respectively. These charts are applicable for one-dimensional heat transfer in those geometries. Therefore, their use is limited to situations in which the body is initially at a uniform temperature, all surfaces are subjected to the same thermal conditions, and the body does not involve any heat generation. These charts can also be used to determine the total heat transfer from the body up to a specified time t.

Using a *one-term approximation*, the solutions of one-dimensional transient heat conduction problems are expressed analytically as

Plane wall: $\quad \theta(x, t)_{\text{wall}} = \dfrac{T(x, t) - T_\infty}{T_i - T_\infty} = A_1 e^{-\lambda_1^2 \tau} \cos{(\lambda_1 x/L)}, \qquad \tau > 0.2$

Cylinder: $\quad \theta(x, t)_{\text{cyl}} = \dfrac{T(r, t) - T_\infty}{T_i - T_\infty} = A_1 e^{-\lambda_1^2 \tau} J_0(\lambda_1 r/r_o), \qquad \tau > 0.2$

Sphere: $\quad \theta(x, t)_{\text{sph}} = \dfrac{T(r, t) - T_\infty}{T_i - T_\infty} = A_1 e^{-\lambda_1^2 \tau} \dfrac{\sin{(\lambda_1 r/r_o)}}{\lambda_1 r/r_o}, \qquad \tau > 0.2$

where the constants A_1 and λ_1 are functions of the Bi number only, and their values are listed in Table 4-1 against the Bi number for all three geometries. The error involved in one-term solutions is less than 2 percent when $\tau > 0.2$.

Using the one-term solutions, the fractional heat transfers in different geometries are expressed as

Plane wall: $\quad \left(\dfrac{Q}{Q_{max}}\right)_{\text{wall}} = 1 - \theta_{0,\,\text{wall}} \dfrac{\sin \lambda_1}{\lambda_1}$

Cylinder: $\quad \left(\dfrac{Q}{Q_{max}}\right)_{\text{cyl}} = 1 - 2\theta_{0,\,\text{cyl}} \dfrac{J_1(\lambda_1)}{\lambda_1}$

Sphere: $\quad \left(\dfrac{Q}{Q_{max}}\right)_{\text{sph}} = 1 - 3\theta_{0,\,\text{sph}} \dfrac{\sin \lambda_1 - \lambda_1 \cos \lambda_1}{\lambda_1^3}$

The analytic solution for one-dimensional transient heat conduction in a semi-infinite solid subjected to convection is given by

$$\frac{T(x, t) - T_i}{T_\infty - T_i} = \text{erfc}\left(\frac{x}{2\sqrt{\alpha t}}\right) - \exp\left(\frac{hx}{k} + \frac{h^2 \alpha t}{k^2}\right)\left[\text{erfc}\left(\frac{x}{2\sqrt{\alpha t}} + \frac{h\sqrt{\alpha t}}{k}\right)\right]$$

where the quantity erfc (ξ) is the *complementary error function*. For the special case of $h \to \infty$, the surface temperature T_s becomes equal to the fluid temperature T_∞, and the above equation reduces to

$$\frac{T(x, t) - T_i}{T_s - T_i} = \text{erfc}\left(\frac{x}{2\sqrt{\alpha t}}\right) \qquad (T_s = \text{constant})$$

Using a clever superposition principle called the *product solution* these charts can also be used to construct solutions for the *two-dimensional* transient heat conduction problems encountered in geometries such as a short cylinder, a long rectangular bar, or a semi-infinite cylinder or plate, and even *three-dimensional* problems associated with geometries such as a rectangular prism or a semi-infinite rectangular bar, provided that all surfaces of the solid are subjected to convection to the same fluid at temperature T_∞, with the same convection heat transfer coefficient h, and the body involves no heat generation. The solution in such multidimensional geometries can be expressed as the product of the solutions for the one-dimensional geometries whose intersection is the multidimensional geometry.

The total heat transfer to or from a multidimensional geometry can also be determined by using the one-dimensional values. The transient heat transfer for a two-dimensional geometry formed by the intersection of two one-dimensional geometries 1 and 2 is

$$\left(\frac{Q}{Q_{max}}\right)_{total, 2D} = \left(\frac{Q}{Q_{max}}\right)_1 + \left(\frac{Q}{Q_{max}}\right)_2\left[1 - \left(\frac{Q}{Q_{max}}\right)_1\right]$$

Transient heat transfer for a three-dimensional body formed by the intersection of three one-dimensional bodies 1, 2, and 3 is given by

$$\left(\frac{Q}{Q_{max}}\right)_{total, 3D} = \left(\frac{Q}{Q_{max}}\right)_1 + \left(\frac{Q}{Q_{max}}\right)_2\left[1 - \left(\frac{Q}{Q_{max}}\right)_1\right]$$
$$+ \left(\frac{Q}{Q_{max}}\right)_3\left[1 - \left(\frac{Q}{Q_{max}}\right)_1\right]\left[1 - \left(\frac{Q}{Q_{max}}\right)_2\right]$$

REFERENCES AND SUGGESTED READING

1. Y. Bayazitoglu and M. N. Özişik. *Elements of Heat Transfer.* New York: McGraw-Hill, 1988.

2. H. S. Carslaw and J. C. Jaeger. *Conduction of Heat in Solids.* 2nd ed. London: Oxford University Press, 1959.

3. H. Gröber, S. Erk, and U. Grigull. *Fundamentals of Heat Transfer.* New York: McGraw-Hill, 1961.

4. M. P. Heisler. "Temperature Charts for Induction and Constant Temperature Heating." *ASME Transactions* 69 (1947), pp. 227–36.

5. J. P. Holman. *Heat Transfer.* 7th ed. New York: McGraw-Hill, 1990.

6. F. P. Incropera and D. P. DeWitt. *Introduction to Heat Transfer.* 2nd ed. New York: John Wiley & Sons, 1990.

7. M. Jakob. *Heat Transfer.* Vol. 1. New York: John Wiley & Sons, 1949.

8. F. Kreith and M. S. Bohn. *Principles of Heat Transfer.* 5th ed. St. Paul, MN: West Publishing, 1993.

9. L. S. Langston. "Heat Transfer from Multidimensional Objects Using One-Dimensional Solutions for Heat Loss." *International Journal of Heat and Mass Transfer* 25 (1982), pp. 149–50.

10. M. N. Özişik, *Heat Transfer—A Basic Approach.* New York: McGraw-Hill, 1985.

11. P. J. Schneider. *Conduction Heat Transfer.* Reading, MA: Addison-Wesley, 1955.

12. L. C. Thomas. *Heat Transfer.* Englewood Cliffs, NJ: Prentice Hall, 1992.

13. F. M. White. *Heat and Mass Transfer.* Reading, MA: Addison-Wesley, 1988.

PROBLEMS*

Lumped System Analysis

4-1C How does transient heat conduction differ from steady conduction? How do two-dimensional heat transfer problems differ from one-dimensional ones?

4-2C What is lumped system analysis? When is it applicable?

4-3C Consider heat transfer between two identical hot solid bodies and the air surrounding them. The first solid is being cooled by a fan while the second one is allowed to cool naturally. For which solid is the lumped system analysis more likely to be applicable? Why?

4-4C Consider heat transfer between two identical hot solid bodies and their environments. The first solid is dropped in a large container filled with water, while the second one is allowed to cool naturally in the air. For which solid is the lumped system analysis more likely to be applicable? Why?

FIGURE P4-5C

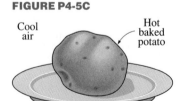

Cool air

Hot baked potato

4-5C Consider a hot baked potato on a plate. The temperature of the potato is observed to drop by 4°C during the first minute. Will the temperature drop during the second minute be less than, equal to, or more than 4°C? Why?

---

*Students are encouraged to answer *all* the concept "C" questions.

4-6C Consider a potato being baked in an oven that is maintained at a constant temperature. The temperature of the potato is observed to rise by 5°C during the first minute. Will the temperature rise during the second minute be less than, equal to, or more than 5°C? Why?

4-7C What is the physical significance of the Biot number? Is the Biot number more likely to be larger for highly conducting solids or poorly conducting ones?

4-8C Consider two identical 4-kg pieces of roast beef. The first piece is baked as a whole, while the second is baked after being cut into two equal pieces in the same oven. Will there be any difference between the cooking times of the whole and cut roasts? Why?

4-9C Consider a sphere and a cylinder of equal volume made of copper. Both the sphere and the cylinder are initially at the same temperature and are exposed to convection in the same environment. Which do you think will cool faster, the cylinder or the sphere? Why?

4-10C In what medium is the lumped system analysis more likely to be applicable: in water or in air? Why?

4-11C For which solid is the lumped system analysis more likely to be applicable: an actual apple or a golden apple of the same size? Why?

4-12C For which kind of bodies made of the same material is the lumped system analysis more likely to be applicable: slender ones or well-rounded ones of the same volume? Why?

4-13 Obtain relations for the characteristic lengths of a large plane wall of thickness $2L$, a very long cylinder of radius r_o, and a sphere of radius r_o.

4-14 Obtain a relation for the time required for a lumped system to reach the average temperature $\frac{1}{2}(T_i + T_\infty)$, where T_i is the initial temperature and T_∞ is the temperature of the environment.

4-15 The temperature of a gas stream is to be measured by a thermocouple whose junction can be approximated as a 1.2-mm-diameter sphere. The properties of the junction are $k = 35$ W/m · °C, $\rho = 8500$ kg/m³, and $C_p = 320$ J/kg · °C, and the heat transfer coefficient between the junction and the gas is $h = 65$ W/m² · °C. Determine how long it will take for the thermocouple to read 99 percent of the initial temperature difference.
Answer: 38.5 s

4-16E In a manufacturing facility, 2-in.-diameter brass balls ($k = 64.1$ Btu/h · ft · °F, $\rho = 532$ lbm/ft³, and $C_p = 0.092$ Btu/lbm · °F) initially at 250°F are quenched in a water bath at 120°F for a period of 2 min at a rate of 100 balls per minute. If the convection heat transfer coefficient is 42 Btu/h · ft² · °F, determine (*a*) the temperature of the balls after quenching and (*b*) the rate at which heat needs to be removed from the water in order to keep its temperature constant at 120°F.

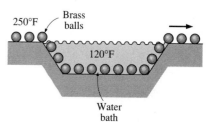

FIGURE P4-16E

4-17E Repeat Problem 4-16E for aluminum balls.

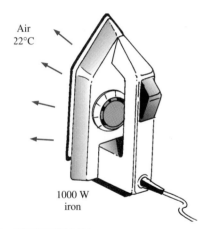

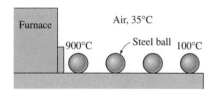

4-18 To warm up some milk for a baby, a mother pours milk into a thin-walled glass whose diameter is 6 cm. The height of the milk in the glass is 7 cm. She then places the glass into a large pan filled with hot water at 60°C. The milk is stirred constantly, so that its temperature is uniform at all times. If the heat transfer coefficient between the water and the glass is 120 W/m² · °C, determine how long it will take for the milk to warm up from 3°C to 38°C. Take the properties of the milk to be the same as those of water. Can the milk in this case be treated as a lumped system? Why?   *Answer:* 5.9 min

4-19 Repeat Problem 4-18 for the case of water also being stirred, so that the heat transfer coefficient is doubled to 240 W/m² · °C.

4-20E During a picnic on a hot summer day, all the cold drinks disappeared quickly, and the only available drinks were those at the ambient temperature of 75°F. In an effort to cool a 12-fluid-oz drink in a can, which is 5 in. high and has a diameter of 2.5 in., a person grabs the can and starts shaking it in the iced water of the chest at 32°F. The temperature of the drink can be assumed to be uniform at all times, and the heat transfer coefficient between the iced water and the aluminum can is 30 Btu/h · ft² · °F. Using the properties of water for the drink, estimate how long it will take for the canned drink to cool to 45°F.

4-21 Consider a 1000-W iron whose base plate is made of 0.5-cm-thick aluminum alloy 2024-T6 (ρ = 2770 kg/m³, C_p = 875 J/kg · °C, α = 7.3 × 10⁻⁵ m²/s). The base plate has a surface area of 0.03 m². Initially, the iron is in thermal equilibrium with the ambient air at 22°C. Taking the heat transfer coefficient at the surface of the base plate to be 12 W/m² · °C and assuming 85 percent of the heat generated in the resistance wires is transferred to the plate, determine how long it will take for the plate temperature to reach 140°C. Is it realistic to assume the plate temperature to be uniform at all times?

4-22 Stainless steel ball bearings (ρ = 8085 kg/m³, k = 15.1 W/m · °C, C_p = 0.480 kJ/kg · °C, and α = 3.91 × 10⁻⁶ m²/s) having a diameter of 1.2 cm are to be quenched in water. The balls leave the oven at a uniform temperature of 900°C and are exposed to air at 30°C for a while before they are dropped into the water. If the temperature of the balls is not to fall below 850°C prior to quenching and the heat transfer coefficient in the air is 125 W/m² · °C, determine how long they can stand in the air before being dropped into the water.   *Answer:* 3.7 s

4-23 Carbon steel balls (ρ = 7833 kg/m³, k = 54 W/m · °C, C_p = 0.465 kJ/kg · °C, and α = 1.474 × 10⁻⁶ m²/s) 8 mm in diameter are annealed by heating them first to 900°C in a furnace and then allowing them to cool slowly to 100°C in ambient air at 35°C. If the average heat transfer coefficient is 75 W/m² · °C, determine how long the annealing process will take. If 2500 balls are to be annealed per hour, determine the total rate of heat transfer from the balls to the ambient air.

4-24 An electronic device dissipating 30 W has a mass of 20 g, a specific heat of 850 J/kg · °C, and a surface area of 5 cm². The device is lightly used, and it is on for 5 min and then off for several hours, during which it cools to the ambient temperature of 25°C. Taking the heat transfer coefficient to be 12 W/m² · °C, determine the temperature of the device at the end of the 5-min operating period. What would your answer be if the device were attached to an aluminum heat sink having a mass of 200 g and a surface area of 50 cm²? Assume the device and the heat sink to be nearly isothermal.

Transient Heat Conduction in Large Plane Walls, Long Cylinders, and Spheres

4-25C What is an infinitely long cylinder? When is it proper to treat an actual cylinder as being infinitely long, and when is it not? For example, is it proper to use this model when finding the temperatures near the bottom or top surfaces of a cylinder? Explain.

4-26C Can the transient temperature charts in Figure 4-13 for a plane wall exposed to convection on both sides be used for a plane wall with one side exposed to convection while the other side is insulated? Explain.

4-27C Why are the transient temperature charts prepared using non-dimensionalized quantities such as the Biot and Fourier numbers instead of the actual variables such as thermal conductivity and time?

4-28C What is the physical significance of the Fourier number? Will the Fourier number for a specified heat transfer problem double when the time is doubled?

4-29C How can we use the transient temperature charts when the surface temperature of the geometry is specified instead of the temperature of the surrounding medium and the convection heat transfer coefficient?

4-30C A body at an initial temperature of T_i is brought into a medium at a constant temperature of T_∞. How can you determine the maximum possible amount of heat transfer between the body and the surrounding medium?

4-31C The Biot number during a heat transfer process between a sphere and its surroundings is determined to be 0.02. Would you use lumped system analysis or the transient temperature charts when determining the midpoint temperature of the sphere? Why?

4-32 A student calculates that the total heat transfer from a spherical copper ball of diameter 15 cm initially at 200°C and its environment at a constant temperature of 25°C during the first 20 min of cooling is 4200 kJ. Is this result reasonable? Why?

4-33 An ordinary egg can be approximated as a 5.5-cm-diameter sphere whose properties are roughly those of water at room temperature ($k = 0.6$ W/m · °C and $\alpha = 0.14 \times 10^{-6}$ m²/s). The egg is initially at a uniform temperature of 8°C and is dropped into boiling water at 97°C. Taking the convection heat transfer coefficient to be $h = 1400$ W/m² · °C, determine how long it will take for the center of the egg to reach 70°C.

FIGURE P4-33

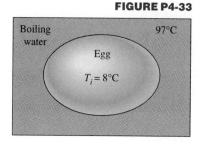

Boiling water   97°C

Egg

$T_i = 8°C$

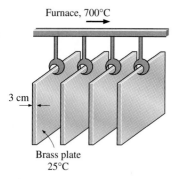

Furnace, 700°C

3 cm

Brass plate
25°C

FIGURE P4-34

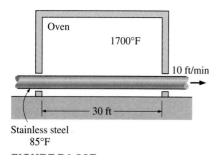

Oven

1700°F

10 ft/min

30 ft

Stainless steel
85°F

FIGURE P4-36E

FIGURE P4-39

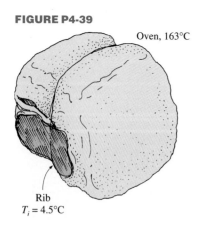

Oven, 163°C

Rib
$T_i = 4.5°C$

4-34 In a production facility, 3-cm-thick large brass plates ($k = 110$ W/m · °C, $\rho = 8530$ kg/m³, $C_p = 380$ J/kg · °C, and $\alpha = 33.9 \times 10^{-6}$ m²/s) that are initially at a uniform temperature of 25°C are heated by passing them through an oven maintained at 700°C. The plates remain in the oven for a period of 10 min. Taking the convection heat transfer coefficient to be $h = 80$ W/m² · °C, determine the surface temperature of the plates when they come out of the oven.

4-35 A long 35-cm-diameter cylindrical shaft made of stainless steel 304 ($k = 14.9$ W/m · °C, $\rho = 7900$ kg/m³, $C_p = 477$ J/kg · °C, and $\alpha = 3.95 \times 10^{-6}$ m²/s) comes out of an oven at a uniform temperature of 400°C. The shaft is then allowed to cool slowly in a chamber at 150°C with an average convection heat transfer coefficient of $h = 60$ W/m² · °C. Determine the temperature at the center of the shaft 20 min after the start of the cooling process. Also, determine the heat transfer per unit length of the shaft during this time period.   *Answers:* 390°C, 15,680 kJ

4-36E Long cylindrical AISI stainless steel rods ($k = 7.74$ Btu/h · ft · °F and $\alpha = 0.135$ ft²/h) of 4-in. diameter are heat-treated by drawing them at a velocity of 10 ft/min through a 30-ft-long oven maintained at 1700°F. The heat transfer coefficient in the oven is 20 Btu/h · ft² · °F. If the rods enter the oven at 85°F, determine their centerline temperature when they leave.

4-37 In a meat processing plant, 2-cm-thick steaks ($k = 0.45$ W/m · °C and $\alpha = 0.91 \times 10^{-7}$ m²/s) that are initially at 25°C are to be cooled by passing them through a refrigeration room at -10°C. The heat transfer coefficient on both sides of the steaks is 9 W/m² · °C. If both surfaces of the steaks are to be cooled to 3°C, determine how long the steaks should be kept in the refrigeration room.

4-38 A long cylindrical wood log ($k = 0.17$ W/m · °C and $\alpha = 1.28 \times 10^{-7}$ m²/s) is 10 cm in diameter and is initially at a uniform temperature of 10°C. It is exposed to hot gases at 500°C in a fireplace with a heat transfer coefficient of 13.6 W/m² · °C on the surface. If the ignition temperature of the wood is 420°C, determine how long it will be before the log ignites.

4-39 In *Betty Crocker's Cookbook,* it is stated that it takes 2 h 45 min to roast a 3.2-kg rib initially at 4.5°C "rare" in an oven maintained at 163°C. It is recommended that a meat thermometer be used to monitor the cooking, and the rib is considered rare done when the thermometer inserted into the center of the thickest part of the meat registers 60°C. The rib can be treated as homogeneous spherical object with the properties $\rho = 1200$ kg/m³, $C_p = 4.1$ kJ/kg · °C, $k = 0.45$ W/m · °C, and $\alpha = 0.91 \times 10^{-7}$ m²/s. Determine (*a*) the heat transfer coefficient at the surface of the rib, (*b*) the temperature of the outer surface of the rib when it is done, and (*c*) the amount of heat transferred to the rib. (*d*) Using the values obtained, predict how long it will take to roast this rib to "medium" level, which occurs when the innermost temperature of the rib reaches 71°C. Compare your result to the listed value of 3 h 20 min.

If the roast rib is to be set on the counter for about 15 min before it is sliced, it is recommended that the rib be taken out of the oven when the thermometer registers about 4°C below the indicated value because the rib will continue cooking even after it is taken out of the oven. Do you agree with this recommendation?

Answers: (*a*) 156.9 W/m² · °C, (*b*) 159.5°C, (*c*) 1829 kJ, (*d*) 3 h 1 min

4-40 Repeat Problem 4-39 for a roast rib that is to be "well-done" instead of "rare." A rib is considered to be well-done when its center temperature reaches 77°C, and the roasting in this case takes about 4 h 15 min.

4-41 For heat transfer purposes, an egg can be considered to be a 5.5-cm-diameter sphere having the properties of water. An egg that is initially at 8°C is dropped into the boiling water at 100°C. The heat transfer coefficient at the surface of the egg is estimated to be 500 W/m² · °C. If the egg is considered cooked when its center temperature reaches 60°C, determine how long the egg should be kept in the boiling water.

4-42 Repeat Problem 4-41 for a location at 1610-m elevation such as Denver, Colorado, where the boiling temperature of water is 94.4°C.

4-43 The author and his 6-year-old son have conducted the following experiment to determine the thermal conductivity of a hot dog. They first boiled water in a large pan and measured the temperature of the boiling water to be 94°C, which is not surprising, since they live at an elevation of about 1650 m in Reno, Nevada. They then took a hot dog that is 12.5 cm long and 2.2 cm in diameter and inserted a thermocouple into the midpoint of the hot dog and another thermocouple just under the skin. They waited until both thermocouples read 20°C, which is the ambient temperature. They then dropped the hot dog into boiling water and observed the changes in both temperatures. Exactly 2 min after the hot dog was dropped into the boiling water, they recorded the center and the surface temperatures to be 59°C and 88°C, respectively. The density of the hot dog can be taken to be 980 kg/m³, which is slightly less than the density of water, since the hot dog was observed to be floating in water while being almost completely immersed. The specific heat of a hot dog can be taken to be 3900 J/kg · °C, which is slightly less than that of water, since a hot dog is mostly water. Using transient temperature charts, determine (*a*) the thermal diffusivity of the hot dog (*b*) the thermal conductivity of the hot dog, and (*c*) the convection heat transfer coefficient.

Answers: (*a*) 2×10^{-7} m²/s, (*b*) 0.76 W/m · °C, (*c*) 658 W/m² · °C.

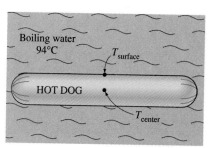

FIGURE P4-43

4-44 Using the data and the answers given in Problem 4-43, determine the center and the surface temperatures of the hot dog 4 min after the start of the cooking. Also determine the amount of heat transferred to the hot dog.

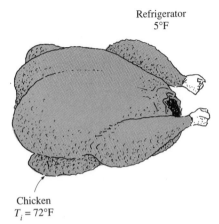

Refrigerator
5°F

Chicken
$T_i = 72°F$

FIGURE P4-45E

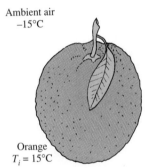

Ambient air
−15°C

Orange
$T_i = 15°C$

FIGURE P4-47

4-45E In a chicken processing plant, whole chickens averaging 5 lb each and initially at 72°F are to be cooled in the racks of a large refrigerator that is maintained at 5°F. The entire chicken is to be cooled below 45°F, but the temperature of the chicken is not to drop below 35°F at any point during refrigeration. The convection heat transfer coefficient and thus the rate of heat transfer from the chicken can be controlled by varying the speed of a circulating fan inside. Determine the heat transfer coefficient that will enable us to meet both temperature constraints while keeping the refrigeration time to a minimum. The chicken can be treated as a homogeneous spherical object having the properties $\rho = 74.9$ lbm/ft^3, $C_p = 0.98$ Btu/lbm $\cdot$ °F, $k = 0.26$ Btu/h $\cdot$ ft $\cdot$ °F, and $\alpha = 0.0035$ ft^2/h.

4-46 A person puts a few apples into the freezer at −15°C to cool them quickly for guests who are about to arrive. Initially, the apples are at a uniform temperature of 20°C, and the heat transfer coefficient on the surfaces is 8 W/m^2 $\cdot$ °C. Treating the apples as 9-cm-diameter spheres and taking their properties to be $\rho = 840$ kg/m^3, $C_p = 3.6$ kJ/kg $\cdot$ °C, $k = 0.513$ W/m $\cdot$ °C, and $\alpha = 1.3 \times 10^{-7}$ m^2/s, determine the center and surface temperatures of the apples in 1 h. Also, determine the amount of heat transfer from each apple.

4-47 Citrus fruits are very susceptible to cold weather, and extended exposure to subfreezing temperatures can destroy them. Consider an 8-cm-diameter orange that is initially at 15°C. A cold front moves in one night, and the ambient temperature suddenly drops to −6°C, with a heat transfer coefficient of 15 W/m^2 $\cdot$ °C. Using the properties of water for the orange and assuming the ambient conditions to remain constant for 4 h before the cold front moves out, determine if any part of the orange will freeze that night.

4-48 An 8-cm-diameter potato ($\rho = 1100$ kg/m^3, $C_p = 3900$ J/kg $\cdot$ °C, $k = 0.6$ W/m $\cdot$ °C, and $\alpha = 1.4 \times 10^{-7}$ m^2/s) that is initially at a uniform temperature of 25°C is baked in an oven at 170°C until a temperature sensor inserted to the center of the potato indicates a reading of 70°C. The potato is then taken out of the oven and wrapped in thick towels so that almost no heat is lost from the baked potato. Assuming the heat transfer coefficient in the oven to be 25 W/m^2 $\cdot$ °C, determine (*a*) how long the potato is baked in the oven and (*b*) the final equilibrium temperature of the potato after it is wrapped.

Transient Heat Conduction in Semi-Infinite Solids

4-49C What is a semi-infinite medium? Give examples of solid bodies that can be treated as semi-infinite mediums for heat transfer purposes.

4-50C Under what conditions can a plane wall be treated as a semi-infinite medium?

4-51C Consider a hot semi-infinite solid at an initial temperature of T_i that is exposed to convection to a cooler medium at a constant temperature of T_∞, with a heat transfer coefficient of h. Explain how you can determine the total amount of heat transfer from the solid up to a specified time t_o.

4-52   In areas where the air temperature remains below 0°C for prolonged periods of time, the freezing of water in underground pipes is a major concern. Fortunately, the soil remains relatively warm during those periods, and it takes weeks for the subfreezing temperatures to reach the water mains in the ground. Thus, the soil effectively serves as an insulation to protect the water from the freezing atmospheric temperatures in winter.

The ground at a particular location is covered with snow pack at $-8°C$ for a continuous period of 60 days, and the average soil properties at that location are $k = 0.4$ W/m · °C and $\alpha = 0.15 \times 10^{-6}$ m²/s. Assuming an initial uniform temperature of 10°C for the ground, determine the minimum burial depth to prevent the water pipes from freezing.

4-53   The soil temperature in the upper layers of the earth varies with the variations in the atmospheric conditions. Before a cold front moves in, the earth at a location is initially at a uniform temperature of 10°C. Then the area is subjected to a temperature of $-10°C$ and high winds that resulted in a convection heat transfer coefficient of 40 W/m² · °C on the earth's surface for a period of 10 h. Taking the properties of the soil at that location to be $k = 0.9$ W/m · °C and $\alpha = 1.6 \times 10^{-5}$ m²/s, determine the soil temperature at distances 0, 10, 20 and 50 cm from the earth's surface at the end of this 10-h period.

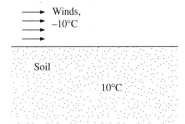

FIGURE P4-53

4-54E   The walls of a furnace are made of 1.5-ft-thick concrete ($k = 0.64$ Btu/h · ft · °F and $\alpha = 0.023$ ft²/h). Initially, the furnace and the surrounding air are in thermal equilibrium at 70°F. The furnace is then fired, and the inner surfaces of the furnace are subjected to hot gases at 1800°F with a very large heat transfer coefficient. Determine how long it will take for the temperature of the outer surface of the furnace walls to rise to 70.1°F.

Answer: 181 min

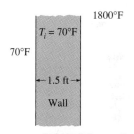

FIGURE P4-54E

4-55   A thick wood slab ($k = 0.17$ W/m · °C and $\alpha = 1.28 \times 10^{-7}$ m²/s) that is initially at a uniform temperature of 25°C is exposed to hot gases at 550°C for a period of 5 minutes. The heat transfer coefficient between the gases and the wood slab is 35 W/m² · °C. If the ignition temperature of the wood is 420°C, determine if the wood will ignite.

4-56   A large cast iron container ($k = 52$ W/m · °C and $\alpha = 1.70 \times 10^{-5}$ m²/s) with 5-cm-thick walls is initially at a uniform temperature of 0°C and is filled with ice at 0°C. Now the outer surfaces of the container are exposed to hot water at 60°C with a very large heat transfer coefficient. Determine how long it will be before the ice inside the container starts melting. Also, taking the heat transfer coefficient on the inner surface of the container to be 250 W/m² · °C, determine the rate of heat transfer to the ice through a 1.2-m-wide and 2-m-high section of the wall when steady operating conditions are reached. Assume the ice starts melting when its inner surface temperature rises to 0.1°C.

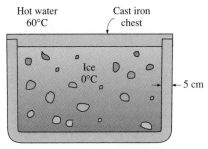

FIGURE P4-56

Transient Heat Conduction in Multidimensional Systems

4-57C   What is the product solution method? How is it used to determine the transient temperature distribution in a two-dimensional system?

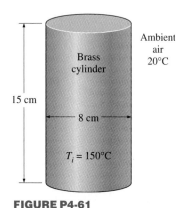

FIGURE P4-61

FIGURE P4-65

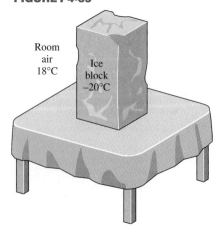

4-58C How is the product solution used to determine the variation of temperature with time and position in three-dimensional systems?

4-59C A short cylinder initially at a uniform temperature T_i is subjected to convection from all of its surfaces to a medium at temperature T_∞. Explain how you can determine the temperature of the midpoint of the cylinder at a specified time t.

4-60C Consider a short cylinder whose top and bottom surfaces are insulated. The cylinder is initially at a uniform temperature T_i and is subjected to convection from its side surface to a medium at temperature T_∞ with a heat transfer coefficient of h. Is the heat transfer in this short cylinder one- or two-dimensional? Explain.

4-61 A short brass cylinder ($\rho = 8530$ kg/m^3, $C_p = 0.389$ kJ/kg · °C, $k = 110$ W/m · °C, and $\alpha = 3.39 \times 10^{-5}$ m^2/s) of diameter $D = 8$ cm and height $H = 15$ cm is initially at a uniform temperature of $T_i = 150$°C. The cylinder is now placed in atmospheric air at 20°C, where heat transfer takes place by convection with a heat transfer coefficient of $h = 40$ W/m^2 · °C. Calculate (a) the center temperature of the cylinder, (b) the center temperature of the top surface of the cylinder, and (c) the total heat transfer from the cylinder 15 min after the start of the cooling.

4-62 A semi-infinite aluminum cylinder ($k = 237$ W/m · °C, $\alpha = 9.71 \times 10^{-5}$ m^2/s) of diameter $D = 15$ cm is initially at a uniform temperature of $T_i = 150$°C. The cylinder is now placed in water at 10°C, where heat transfer takes place by convection with a heat transfer coefficient of $h = 140$ W/m^2 · °C. Determine the temperature at the center of the cylinder 10 cm from the end surface 8 min after the start of the cooling.

4-63E A hot dog can be considered to be a cylinder 5 in. long and 0.8 in. in diameter whose properties are $\rho = 61.2$ lbm/ft^3, $C_p = 0.93$ Btu/lbm · °F, $k = 0.44$ Btu/h · ft · °F, and $\alpha = 0.0077$ ft^2/h. A hot dog initially at 40°F is dropped into boiling water at 212°F. If the heat transfer coefficient at the surface of the hot dog is estimated to be 120 Btu/h · ft^2 · °F, determine the center temperature of the hot dog after 5, 10, and 15 min by treating the hot dog as (a) a finite cylinder and (b) an infinitely long cylinder.

4-64E Repeat Problem 4-63E for a location at 5300 ft elevation such as Denver, Colorado, where the boiling temperature of water is 202°F.

4-65 A 5-cm-high rectangular ice block ($k = 2.22$ W/m · °C and $\alpha = 0.124 \times 10^{-7}$ m^2/s) initially at -20°C is placed on a table on its square base 4 cm $\times$ 4 cm in size in a room at 18°C. The heat transfer coefficient on the exposed surfaces of the ice block is 12 W/m^2 · °C. Disregarding any heat transfer from the base to the table, determine how long it will be before the ice block starts melting. Where on the ice block will the first liquid droplets appear?

4-66 A 2-cm-high cylindrical ice block ($k = 2.22$ W/m · °C and $\alpha = 0.124 \times 10^{-7}$ m^2/s) is placed on a table on its base of diameter 2 cm in a room at 20°C. The heat transfer coefficient on the exposed surfaces of the ice block is 13 W/m² · °C, and heat transfer from the base of the ice block to the table is negligible. If the ice block is not to start melting at any point for at least 2 h, determine what the initial temperature of the ice block should be.

4-67 Consider a cubic block whose sides are 5 cm long and a cylindrical block whose height and diameter are also 5 cm. Both blocks are initially at 20°C and are made of granite ($k = 2.5$ W/m · °C and $\alpha = 1.15 \times 10^{-6}$ m²/s). Now both blocks are exposed to hot gases at 500°C in a furnace on all of their surfaces with a heat transfer coefficient of 40 W/m² · °C. Determine the center temperature of each geometry after 10, 20, and 60 min.

4-68 Repeat Problem 4-67 with the heat transfer coefficient at the top and the bottom surfaces of each block being doubled to 80 W/m² · °C.

4-69 A 20-cm-long cylindrical aluminum block ($\rho = 2702$ kg/m³, $C_p = 0.896$ kJ/kg · °C, $k = 236$ W/m · °C, and $\alpha = 9.75 \times 10^{-5}$ m²/s), 15 cm in diameter, is initially at a uniform temperature of 20°C. The block is to be heated in a furnace at 1200°C until its center temperature rises to 300°C. If the heat transfer coefficient on all surfaces of the block is 50 W/m² · °C, determine how long the block should be kept in the furnace. Also, determine the amount of heat transfer from the aluminum block if it is allowed to cool in the room until its temperature drops to 20°C throughout.

4-70 Repeat Problem 4-69 for the case where the aluminum block is inserted into the furnace on a low-conductivity material so that the heat transfer to or from the bottom surface of the block is negligible.

Review Problems

4-71 Consider two 2-cm-thick large steel plates ($k = 43$ W/m · °C and $\alpha = 1.17 \times 10^{-5}$ m²/s) that were put on top of each other while wet and left outside during a cold winter night at -15°C. The next day, a worker needs one of the plates, but the plates are stuck together because the freezing of the water between the two plates has bonded them together. In an effort to melt the ice between the plates and separate them, the worker takes a large hair-dryer and blows hot air at 50°C all over the exposed surface of the plate on the top. The convection heat transfer coefficient at the top surface is estimated to be 40 W/m² · °C. Determine how long the worker must keep blowing hot air before the two plates separate. *Answer:* 507 s

4-72 Consider a curing kiln whose walls are made of 30-cm-thick concrete whose properties are $k = 0.9$ W/m · °C and $\alpha = 0.23 \times 10^{-5}$ m²/s. Initially, the kiln and its walls are in equilibrium with the surroundings at 5°C. Then all the doors are closed and the kiln is heated by steam so that the temperature of the inner surface of the walls is raised to 45°C and is maintained at that level for 3 h. The curing kiln is then opened and exposed to the atmospheric air after the stream flow is turned off. If the outer surfaces of the walls of the kiln were insulated, would it save any energy that day during the period the kiln was used for curing for 3 h only, or would it make no difference? Base your answer on calculations.

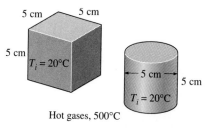

5 cm   5 cm

5 cm $T_i = 20°C$

5 cm   5 cm

$T_i = 20°C$

Hot gases, 500°C

FIGURE P4-67

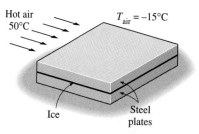

Hot air 50°C $T_{air} = -15°C$

Ice   Steel plates

FIGURE P4-71

FIGURE P4-72

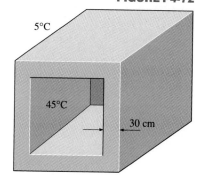

5°C

45°C

30 cm

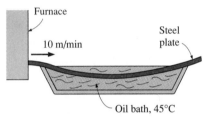

FIGURE P4-74

FIGURE P4-75

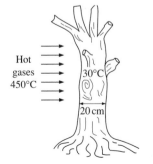

FIGURE P4-76E

FIGURE P4-77

4-73 The water main in the cities must be placed at sufficient depth below the earth's surface to avoid freezing during extended periods of subfreezing temperatures. Determine the minimum depth at which the water main must be placed at a location where the soil is initially at 15°C and the earth's surface temperature under the worst conditions is expected to remain at −10°C for a period of 75 days. Take the properties of soil at that location to be $k = 0.7$ W/m · °C and $\alpha = 1.4 \times 10^{-5}$ m²/s.   *Answer:* 7.05 m

4-74 A hot dog can be considered to be a 12-cm-long cylinder whose diameter is 2 cm and whose properties are $\rho = 980$ kg/m³, $C_p = 3.9$ kJ/kg · °C, $k = 0.76$ W/m · °C, and $\alpha = 2 \times 10^{-7}$ m²/s. A hot dog initially at 5°C is dropped into boiling water at 100°C. The heat transfer coefficient at the surface of the hot dog is estimated to be 600 W/m² · °C. If the hot dog is considered cooked when its center temperature reaches 80°C, determine how long it will take to cook it in the boiling water.

4-75 A long roll of 2-m-wide a 0.5-cm-thick 1-Mn manganese steel plate coming off a furnace at 820°C is to be quenched in an oil bath ($C_p = 2.0$ kJ/kg · °C) at 45°C. The metal sheet is moving at a steady velocity of 10 m/min, and the oil bath is 8 m long. Taking the convection heat transfer coefficient on both sides of the plate to be 860 W/m² · °C, determine the temperature of the sheet metal when it leaves the oil bath. Also, determine the required rate of heat removal from the oil to keep its temperature constant at 45°C.

4-76E In *Betty Crocker's Cookbook,* it is stated that it takes 5 h to roast a 14-lb stuffed turkey initially at 40°F in an oven maintained at 325°F. It is recommended that a meat thermometer be used to monitor the cooking, and the turkey is considered done when the thermometer inserted deep into the thickest part of the breast or thigh without touching the bone registers 185°F. The turkey can be treated as a homogeneous spherical object with the properties $\rho = 75$ lbm/ft³, $C_p = 0.98$ Btu/lbm · °F, $k = 0.26$ Btu/h · ft · °F, and $\alpha = 0.0035$ ft²/h. Assuming the tip of the thermometer is at one-third radial distance from the center of the turkey, determine (*a*) the average heat transfer coefficient at the surface of the turkey, (*b*) the temperature of the skin of the turkey when it is done, and (*c*) the total amount of heat transferred to the turkey in the oven. Will the reading of the thermometer be more or less than 185°F 5 min after the turkey is taken out of the oven?

4-77 During a fire, the trunks of some dry oak trees ($k = 0.17$ W/m × °C and $\alpha = 1.28 \times 10^{-7}$ m²/s) that are initially at a uniform temperature of 30°C are exposed to hot gases at 450°C for a period of 4 h, with a heat transfer coefficient of 65 W/m² · °C on the surface. The ignition temperature of the trees is 410°C. Treating the trunks of the trees as long cylindrical rods of diameter 20 cm, determine if these dry trees will ignite as the fire sweeps through them.

4-78 We often cut a watermelon in half and put it into the freezer to cool it quickly. But usually we forget to check on it and end up having a watermelon with a frozen layer on the top. To avoid this potential problem a person wants to set the timer such that it will go off when the temperature of the exposed surface of the watermelon drops to 3°C.

Consider a 30-cm-diameter spherical watermelon that is cut into two equal parts and put into a freezer at −12°C. Initially, the entire watermelon is at a uniform temperature of 25°C, and the heat transfer coefficient on the surfaces is 30 W/m² · °C. Assuming the watermelon to have the properties of water, determine how long it will take for the center of the exposed cut surfaces of the watermelon to drop to 3°C.

4-79 The thermal conductivity of a solid whose density and specific heat are known can be determined from the relation $k = \alpha/\rho C_p$ after evaluating the thermal diffusivity α.

Consider a 2-cm-diameter cylindrical rod made of a sample material whose density and specific heat are 3700 kg/m³ and 920 J/kg · °C, respectively. The sample is initially at a uniform temperature of 25°C. In order to measure the temperatures of the sample at its surface and its center, a thermocouple is inserted to the center of the sample along the centerline, and another thermocouple is welded into a small hole drilled on the surface. The sample is dropped into boiling water at 100°C. After 3 min, the surface and the center temperatures are recorded to be 93°C and 75°C, respectively. Determine the thermal diffusivity and the thermal conductivity of the material.

4-80 In desert climates, rainfall is not a common occurrence since the rain droplets formed in the upper layer of the atmosphere often evaporate before they reach the ground. Consider a raindrop that is initially at a temperature of 5°C and has a diameter of 5 mm. Determine how long it will take for the diameter of the raindrop to reduce to 3 mm as it falls through ambient air at 25°C with a heat transfer coefficient of 400 W/m² · °C. The water temperature can be assumed to remain constant and uniform at 5°C at all times, and the heat of vaporization of water at 5°C is 2490 kJ/kg.

4-81E Consider a plate of thickness 1 in., a long cylinder of diameter 1 in., and a sphere of diameter 1 in., all initially at 400°F and all made of bronze ($k = 15.0$ Btu/h · ft · °F and $\alpha = 0.333$ ft²/h). Now all three of these geometries are exposed to cool air at 75°F on all of their surfaces, with a heat transfer coefficient of 7 Btu/h · ft² · °F. Determine the center temperature of each geometry after 5, 10, and 30 min. Explain why the center temperature of the sphere is always the lowest.

4-82E Repeat Problem 4-81E for cast iron geometries ($k = 29$ Btu/h · ft · °F and $\alpha = 0.61$ ft²/h).

4-83 Long aluminum wires of diameter 3 mm ($\rho = 2702$ kg/m³, $C_p = 0.896$ kJ/kg · °C, $k = 236$ W/m · °C, and $\alpha = 9.75 \times 10^{-5}$ m²/s) are extruded at a temperature of 350°C and exposed to atmospheric air at 30°C with a heat transfer coefficient of 35 W/m² · °C. (a) Determine how long it will take for the wire temperature to drop to 50°C. (b) If the wire is extruded at a velocity of 10 m/min, determine how far the wire travels after extrusion by the time

Watermelon, 25°C

FIGURE P4-78

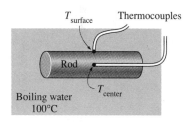

$T_{surface}$ · · · Thermocouples

Rod

T_{center}

Boiling water 100°C

FIGURE P4-79

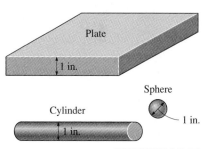

Plate

1 in.

Cylinder

1 in.

Sphere

1 in.

FIGURE P4-81E

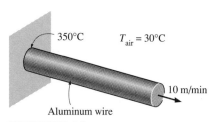

350°C · · · $T_{air} = 30$°C

10 m/min

Aluminum wire

FIGURE P4-83

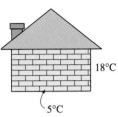

18°C

5°C

FIGURE P4-85

its temperature drops to 50°C. What change in the cooling process would you propose to shorten this distance? (c) Assuming the aluminum wire leaves the extrusion room at 50°C, determine the rate of heat transfer from the wire to the extrusion room.   *Answers:* (a) 144 s, (b) 24 m, (c) 855 W

4-84 Repeat Problem 4-83 for a copper wire (ρ = 8950 kg/m^3, C_p = 0.383 kJ/kg · °C, k = 386 W/m · °C, and α = 1.13 × 10^{-4} m^2/s).

4-85 Consider a brick house (k = 0.72 W/m · °C and α = 0.45 × 10^{-6} m^2/s) whose walls are 10 m long, 3 m high, and 0.3 m thick. The heater of the house broke down one night, and the entire house, including its walls, was observed to be 5°C throughout in the morning. The outdoors warmed up as the day progressed, but no change was felt in the house, which was tightly sealed. Assuming the outer surface temperature of the house to remain constant at 18°C, determine how long it would take for the temperature of the inner surfaces of the walls to rise to 5.1°C.

Computer, Design, and Essay Problems

4-86 Conduct the following experiment at home to determine the combined convection and radiation heat transfer coefficient at the surface of an apple exposed to the room air. You will need two thermometers and a clock.

First, weigh the apple and measure its diameter. You may measure its volume by placing it in a large measuring cup halfway filled with water, and measuring the change in volume when it is completely immersed in the water. Refrigerate the apple overnight so that it is at a uniform temperature in the morning and measure the air temperature in the kitchen. Then take the apple out and stick one of the thermometers to its middle and the other just under the skin. Record both temperatures every 5 min for an hour. Using these two temperatures, calculate the heat transfer coefficient for each interval and take their average. The result is the combined convection and radiation heat transfer coefficient for this heat transfer process. Using your experimental data, also calculate the thermal conductivity and thermal diffusivity of the apple and compare them to the values given above.

4-87 Repeat Problem 4-86 using a banana instead of an apple. The thermal properties of bananas are practically the same as those of apples.

4-88 Conduct the following experiment to determine the lumped exponent b in Equation 4-5 for a can of soda and then predict the temperature of the soda at different times. Leave the soda in the refrigerator overnight. Measure the air temperature in the kitchen and the temperature of the soda while it is still in the refrigerator by taping the sensor of the thermometer to the outer surface of the can. Then take the soda out and measure its temperature again in 5 min. Using these values, calculate the exponent b. Using this b-value, predict the temperatures of the soda in 10, 15, 20, 30, and 60 min and compare the results with the actual temperature measurements. Do you think the lumped system analysis is valid in this case?

4-89   Whole ready-to-cook turkeys range from about 2 to 11 kg. Roasted turkeys are considered done when a thermometer inserted deep into the turkey registers 85°C. From your favorite cookbook, obtain the instructions to bake a large stuffed turkey and evaluate the average heat transfer coefficient during baking. Using this heat transfer coefficient, estimate the baking time for a turkey that is only half as large.

4-90   Citrus trees are very susceptible to cold weather, and extended exposure to subfreezing temperatures can destroy the crop. In order to protect the trees from occasional cold fronts with subfreezing temperatures, tree growers in Florida usually install water sprinklers on the trees. When the temperature drops below a certain level, the sprinklers spray water on the trees and their fruits to protect them against the damage the subfreezing temperatures can cause. Explain the basic mechanism behind this protection measure and write an essay on how the system works in practice.

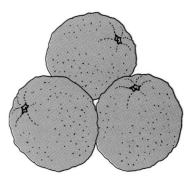

FIGURE P4-90

Numerical Methods in Heat Conduction

So far we have mostly considered relatively simple heat conduction problems involving *simple geometries* with simple boundary conditions because only such simple problems can be solved *analytically*. But many problems encountered in practice involve *complicated geometries* with complex boundary conditions or variable properties and cannot be solved analytically. In such cases, sufficiently accurate approximate solutions can be obtained by computers using a *numerical method*.

Analytical solution methods such as those presented in Chapter 2 are based on solving the governing differential equation together with the boundary conditions. They result in solution functions for the temperature at *every point* in the medium. Numerical methods, on the other hand, are based on replacing the differential equation by a set of *n* algebraic equations for the unknown temperatures at *n* selected points in the medium, and the simultaneous solution of these equations results in the temperature values at those *discrete points*.

There are several ways of obtaining the numerical formulation of a heat conduction problem, such as the *finite difference* method, the *finite element* method, the *boundary element* method, and the *energy balance* (or control volume) method. Each method has its own advantages and disadvantages, and each is used in practice. In this chapter we will use primarily the *energy balance* approach since it is based on the familiar energy balances on control volumes instead of heavy mathematical formulations, and thus it gives a better physical feel for the problem. Besides, it results in the same set of algebraic equations as the finite difference method. In this chapter, the numerical formulation and solution of heat conduction problems are demonstrated for both steady and transient cases in various geometries.

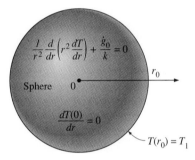

FIGURE 5-1

The analytical solution of a problem requires solving the governing differential equation and applying the boundary conditions.

Solution:

$$T(r) = T_1 + \frac{\dot{g}_0}{6k}(r_0{}^2 - r^2)$$

$$\dot{Q}(r) = -kA\frac{dT}{dr} = \frac{4\pi \dot{g}_0 r^3}{3}$$

5-1 ■ WHY NUMERICAL METHODS?

The ready availability of *high speed computers* and easy-to-use *powerful software packages* has had a major impact on engineering education and practice in recent years. Engineers in the past had to rely on *analytical skills* to solve significant engineering problems, and thus they had to undergo a rigorous training in mathematics. Today's engineers, on the other hand, have access to a tremendous amount of *computation power* under their fingertips, and they mostly need to understand the physical nature of the problem and interpret the results. But they also need to understand how calculations are performed by the computers to develop an awareness of the processes involved and the limitations, while avoiding any possible pitfalls.

In Chapter 2 we solved various heat conduction problems in various geometries in a systematic but highly mathematical manner by (1) deriving the governing differential equation by performing an energy balance on a differential volume element, (2) expressing the boundary conditions in the proper mathematical form, and (3) solving the differential equation and applying the boundary conditions to determine the integration constants. This resulted in a solution function for the temperature distribution in the medium, and the solution obtained in this manner is called the **analytical solution** of the problem. For example, the mathematical formulation of one-dimensional steady heat conduction in a sphere of radius r_0 whose outer surface is maintained at a uniform temperature of T_1 with uniform heat generation at a rate of $\dot{g}_0$ was expressed as (Fig. 5-1)

$$\frac{1}{r^2}\frac{d}{dr}\left(r^2\frac{dT}{dr}\right) + \frac{\dot{g}_0}{k} = 0 \tag{5-1}$$

$$\frac{dT(0)}{dr} = 0 \quad \text{and} \quad T(r_0) = T_1$$

whose (analytical) solution is

$$T(r) = T_1 + \frac{\dot{g}_0}{6k}(r_0^2 - r^2) \tag{5-2}$$

This is certainly a very desirable form of solution since the temperature at any point within the sphere can be determined simply by substituting the *r*-coordinate of the point into the analytical solution function above. The analytical solution of a problem is also referred to as the **exact solution** since it satisfies the differential equation and the boundary conditions. This can be verified by substituting the solution function into the differential equation and the boundary conditions. Further, the *rate of heat flow* at any location within the sphere or its surface can be determined by taking the derivative of the solution function $T(r)$ and substituting it into Fourier's law as

$$\dot{Q}(r) = -kA\frac{dT}{dr} = -k(4\pi r^2)\left(-\frac{\dot{g}_0 r}{3k}\right) = \frac{4\pi \dot{g}_0 r^3}{3} \tag{5-3}$$

The analysis above did not require any mathematical sophistication beyond the level of *simple integration,* and you are probably wondering why anyone would ask for something else. After all, the solutions obtained are exact and easy to use. Besides, they are instructive since they show clearly the functional dependence of temperature and heat transfer on the independent variable r. Well, there are several reasons for searching for alternative solution methods.

1 Limitations

Analytical solution methods are limited to *highly simplified problems* in *simple geometries* (Fig. 5-2). The geometry must be such that its entire surface can be described mathematically in a coordinate system by setting the variables equal to constants. That is, it must fit into a coordinate system *perfectly* with nothing sticking out or in. In the case of one-dimensional heat conduction in a solid sphere of radius r_0, for example, the entire outer surface can be described by $r = r_0$. Likewise, the surfaces of a finite solid cylinder of radius r_0 and height H can be described by $r = r_0$ for the side surface and $z = 0$ and $z = H$ for the bottom and top surfaces, respectively. Even minor complications in geometry can make an analytical solution impossible. For example, a spherical object with an extrusion like a *handle* at some location is impossible to handle analytically since the boundary conditions in this case cannot be expressed in any familiar coordinate system.

Even in simple geometries, heat transfer problems cannot be solved analytically if the *thermal conditions* are not sufficiently simple. For example, the consideration of the variation of thermal conductivity with temperature, the variation of the heat transfer coefficient over the surface, or the radiation heat transfer on the surfaces can make it impossible to obtain an analytical solution. Therefore, analytical solutions are limited to problems that are simple or can be simplified with reasonable approximations.

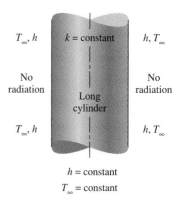

FIGURE 5-2

Analytical solution methods are limited to simplified problems in simple geometries.

2 Better Modeling

We mentioned earlier that analytical solutions are exact solutions since they do not involve any approximations. But this statement needs some clarification. Distinction should be made between an *actual real-world problem* and the *mathematical model* that is an idealized representation of it. The solutions we get are the solutions of mathematical models, and the degree of applicability of these solutions to the actual physical problems depends on the accuracy of the model. An "approximate" solution of a realistic model of a physical problem is usually more accurate than the "exact" solution of a crude mathematical model (Fig. 5-3).

When attempting to get an analytical solution to a physical problem, there is always the tendency to *oversimplify* the problem to make the mathematical model sufficiently simple to warrant an analytical solution. Therefore, it is common practice to ignore any effects that cause mathematical complications such as nonlinearities in the differential equation or the boundary conditions. So it comes as no surprise that *nonlinearities* such as temperature dependence of thermal conductivity and the radiation boundary conditions are seldom considered in analytical solutions. A mathematical model intended for a numerical solution is likely to represent the actual

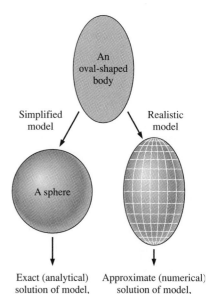

FIGURE 5-3

The approximate numerical solution of a real-world problem may be more accurate than the exact (analytical) solution of an oversimplified model of that problem.

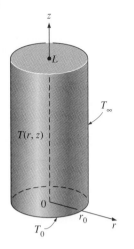

Analytical solution:

$$\frac{T(r,z) - T_\infty}{T_0 - T_\infty} = \sum_{n=1}^{\infty} \frac{J_0(\lambda_n r)}{\lambda_n J_1(\lambda_n r_0)} \frac{\sinh \lambda_n(L-z)}{\sinh (\lambda_n L)}$$

where λ_n's are roots of $J_0(\lambda_n r_0) = 0$

FIGURE 5-4

Some analytical solutions are so complex that one would hesitate calling them solutions.

FIGURE 5-5

The ready availability of high-powered computers with sophisticated software packages has made numerical solution the norm rather than the exception.

problem better. Therefore, the numerical solution of engineering problems has now become the norm rather than the exception even when analytical solutions are available.

3 Flexibility

Engineering problems often require extensive *parametric studies* to understand the influence of some variables on the solution in order to choose the right set of variables and to answer some "what-if" questions. This is an *iterative process* that is extremely tedious and time-consuming if done by hand. Computers and numerical methods are ideally suited for such calculations, and a wide range of related problems can be solved by minor modifications in the code or input variables. Today it is almost unthinkable to perform any significant optimization studies in engineering without the power and flexibility of computers and numerical methods.

4 Complications

Some problems can be solved analytically, but the solution procedure is so complex and the resulting solution expressions so complicated that it is not worth all that effort. With the exception of steady one-dimensional or transient lumped system problems, all heat conduction problems result in *partial* differential equations. Solving such equations usually requires mathematical sophistication beyond that acquired at the undergraduate level, such as orthogonality, eigenvalues, Fourier and Laplace transforms, Bessel and Legendre functions, and infinite series. In such cases, the evaluation of the solution, which often involves double or triple summations of infinite series at a specified point, is a challenge in itself (Fig. 5-4). Therefore, even when the solutions are available in some handbooks, they are intimidating enough to scare prospective users away.

5 Human Nature

As human beings, we like to sit back and make wishes, and we like our wishes to come true without much effort. The invention of TV remote controls made us feel like kings in our homes since the commands we give in our comfortable chairs by pressing buttons are immediately carried out by the obedient TV sets. After all, what good is cable TV without a remote control. We certainly would love to continue being the king in our little cubicle in the engineering office by solving problems at the press of a button on a computer (until they invent a remote control for the computers, of course). Well, this might have been a fantasy yesterday, but it is a reality today. Practically all engineering offices today are equipped with *high-powered computers* with *sophisticated software packages,* with impressive presentation-style colorful output in graphical and tabular form (Fig. 5-5). Besides, the results are as accurate as the analytical results for all practical purposes. The computers have certainly changed the way engineering is practiced.

The discussions above should not lead you to believe that analytical solutions are unnecessary and that they should be discarded from the engineering curriculum. On the contrary, insight to the *physical phenomena* and *engineering wisdom* is gained primarily through analysis. The "feel" that engineers develop during the analysis of simple but fundamental problems serves as an invaluable tool when interpreting a huge pile of results obtained from a computer when solving a complex problem. A simple analysis by hand for a limiting case can be used to check if the results are in the proper range. Also, nothing can take the place of getting "ball park" results on a piece of paper during preliminary discussions. The calculators made the basic arithmetic operations by hand a thing of the past, but they did not eliminate the need for instructing grade school children how to add or multiply.

In this chapter, you will learn how to *formulate* and *solve* heat transfer problems numerically using one or more approaches. In your professional life, you will probably solve the heat transfer problems you come across using a professional software package, and you are highly unlikely to write your own programs to solve such problems. (Besides, people will be highly skeptical of the results obtained using your own program instead of using a well-established commercial software package that has stood the test of time.) The insight you will gain in this chapter by formulating and solving some heat transfer problems will help you better understand the available software packages and be an informed and responsible user.

5-2 ■ FINITE DIFFERENCE FORMULATION OF DIFFERENTIAL EQUATIONS

The numerical methods for solving differential equations are based on replacing the *differential equations* by *algebraic equations*. In the case of the popular **finite difference** method, this is done by replacing the *derivatives* by *differences*. Below we will demonstrate this with both first- and second-order derivatives. But first we give a motivational example.

Consider a man who deposits his money in the amount of $A_0 = \$100$ in a savings account at an annual interest rate of 18 percent, and let us try to determine the amount of money he will have after one year if interest is compounded continuously (or instantaneously). In the case of simple interest, the money will earn $18 interest, and the man will have $100 + 100 \times 0.18 = \$118.00$ in his account after one year. But in the case of compounding, the interest earned during a compounding period will also earn interest for the remaining part of the year, and the year-end balance will be greater than $118. For example, if the money is compounded twice a year, the balance will be $100 + 100 \times (0.18/2) = \$109$ after six months, and $109 + 109 \times (0.18/2) = \$118.81$ at the end of the year. We could also determine the balance A directly from

$$A = A_0(1 + i)^n = (\$100)(1 + 0.09)^2 = \$118.81 \qquad (5\text{-}4)$$

where i is the interest rate for the compounding period and n is the number of periods. Using the same formula, the year-end balance is determined for monthly, daily, hourly, minutely, and even secondly compounding, and the results are given in Table 5-1.

TABLE 5-1

Year-end balance of a $100 account earning interest at an annual rate of 18 percent for various compounding periods

Compounding period	Number of periods, n	Year-end balance
1 year	1	$118.00
6 months	2	118.81
1 month	12	119.56
1 week	52	119.68
1 day	365	119.72
1 hour	8760	119.72
1 minute	525,600	119.72
1 second	31,536,000	119.72
Instantaneous	∞	119.72

Note that in the case of daily compounding, the year-end balance will be $119.72, which is $1.72 more than the simple interest case. (So it is no wonder that the credit card companies usually charge interest compounded daily when determining the balance.) Also note that compounding at smaller time intervals, even at the end of each second, does not change the result, and we suspect that instantaneous compounding using "differential" time intervals dt will give the same result. This suspicion is confirmed by obtaining the differential equation $dA/dt = iA$ for the balance A, whose solution is $A = A_0 \exp(it)$. Substitution yields

$$A = (\$100)\exp(0.18 \times 1) = \$119.72$$

which is identical to the result for daily compounding. Therefore, replacing a differential time interval dt by a finite time interval of $\Delta t = 1$ day gave the same result, which leads us into believing that *reasonably accurate results can be obtained by replacing differential quantities by sufficiently small differences*. Next, we develop the finite difference formulation of heat conduction problems by replacing the derivatives in the differential equations by differences. In the following section we will do it using the energy balance method, which does not require any knowledge of differential equations.

Derivatives are the building blocks of differential equations, and thus we first give a brief review of derivatives. Consider a function f that depends on x, as shown in Figure 5-6. The **first derivative** of $f(x)$ at a point is equivalent to the *slope* of a line tangent to the curve at that point and is defined as

$$\frac{df(x)}{dx} = \lim_{\Delta x \to 0} \frac{\Delta f}{\Delta x} = \lim_{\Delta x \to 0} \frac{f(x + \Delta x) - f(x)}{\Delta x} \tag{5-5}$$

which is the ratio of the increment Δf of the function to the increment Δx of the independent variable as $\Delta x \to 0$. If we don't take the indicated limit, we will have the following *approximate* relation for the derivative:

$$\frac{df(x)}{dx} \cong \frac{f(x + \Delta x) - f(x)}{\Delta x} \tag{5-6}$$

This approximate expression of the derivative in terms of differences is the **finite difference form** of the first derivative. The equation above can also be obtained by writing the *Taylor series expansion* of the function f about the point x,

$$f(x + \Delta x) = f(x) + \Delta x \frac{df(x)}{dx} + \frac{1}{2} \Delta x^2 \frac{d^2 f(x)}{dx^2} + \cdots \tag{5-7}$$

and neglecting all the terms in the expansion except the first two. The first term neglected is proportional to Δx^2, and thus the *error* involved in each step of this approximation is also proportional to Δx^2. However, the *commutative error* involved after M steps in the direction of length L is proportional to Δx since $M \Delta x^2 = (L/\Delta x)\Delta x^2 = L \Delta x$. Therefore, the smaller the Δx, the smaller the error, and thus the more accurate the approximation.

Now consider steady one-dimensional heat transfer in a plane wall of thickness L with heat generation. The wall is subdivided into M equal sec-

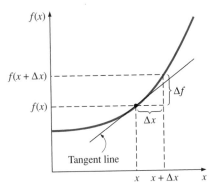

FIGURE 5-6

The derivative of a function at a point represents the slope of the function at that point.

tions of thickness $\Delta x = L/M$ in the x-direction, separated by $M + 1$ points $0, 1, 2, \ldots, m - 1, m, m + 1, \ldots, M$ called **nodes** or **nodal points,** as shown in Figure 5-7. The x-coordinate of any point m is simply $x_m = m\Delta x$, and the temperature at that point is simply $T(x_m) = T_m$.

The heat conduction equation involves the second derivatives of temperature with respect to the space variables, such as d^2T/dx^2, and the finite difference formulation is based on replacing the second derivatives by appropriate differences. But we need to start the process with first derivatives. Using Equation 5-6, the first derivative of temperature dT/dx at the midpoints $m - \frac{1}{2}$ and $m + \frac{1}{2}$ of the sections surrounding the node m can be expressed as

$$\frac{dT}{dx}\bigg|_{m-\frac{1}{2}} \cong \frac{T_m - T_{m-1}}{\Delta x} \quad \text{and} \quad \frac{dT}{dx}\bigg|_{m+\frac{1}{2}} \cong \frac{T_{m+1} - T_m}{\Delta x} \quad (5\text{-}8)$$

Noting that the second derivative is simply the derivative of the first derivative, the second derivative of temperature at node m can be expressed as

$$\frac{d^2T}{dx^2}\bigg|_m \cong \frac{\dfrac{dT}{dx}\bigg|_{m+\frac{1}{2}} - \dfrac{dT}{dx}\bigg|_{m-\frac{1}{2}}}{\Delta x} = \frac{\dfrac{T_{m+1} - T_m}{\Delta x} - \dfrac{T_m - T_{m-1}}{\Delta x}}{\Delta x}$$
$$= \frac{T_{m-1} - 2T_m + T_{m+1}}{\Delta x^2} \quad (5\text{-}9)$$

which is the *finite difference representation* of the *second derivative* at a general internal node m. Note that the second derivative of temperature at a node m is expressed in terms of the temperatures at node m and its two neighboring nodes. Then the differential equation

$$\frac{d^2T}{dx^2} + \frac{\dot{g}}{k} = 0 \quad (5\text{-}10)$$

which is the governing equation for *steady one-dimensional* heat transfer in a plane wall with heat generation and constant thermal conductivity, can be expressed in the *finite difference* form as (Fig. 5-8)

$$\frac{T_{m-1} - 2T_m + T_{m+1}}{\Delta x^2} + \frac{\dot{g}_m}{k} = 0, \quad m = 1, 2, 3, \ldots, M - 1 \quad (5\text{-}11)$$

where $\dot{g}_m$ is the rate of heat generation at node m. If the surface temperatures T_0 and T_M are specified, the application of this equation to each of the $M - 1$ interior nodes results in $M - 1$ equations for the determination of $M - 1$ unknown temperatures at the interior nodes. Solving these equations simultaneously gives the temperature values at the nodes. If the temperatures at the outer surfaces are not known, then we need to obtain two more equations in a similar manner using the specified boundary conditions. Then the

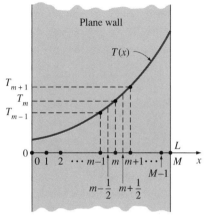

FIGURE 5-7

Schematic of the nodes and the nodal temperatures used in the development of the finite difference formulation of heat transfer in a plane wall.

FIGURE 5-8

The differential equation is valid at every point of a medium, whereas the finite difference equation is valid at discrete points (the nodes) only.

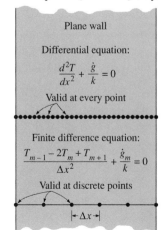

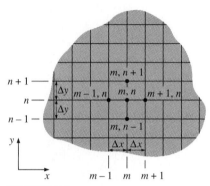

FIGURE 5-9

Finite difference mesh for
two-dimensional conduction in
rectangular coordinates.

unknown temperatures at $M + 1$ nodes are determined by solving the resulting system of $M + 1$ equations in $M + 1$ unknowns simultaneously.

Note that the *boundary conditions* have no effect on the finite difference formulation of interior nodes of the medium. This is not surprising since the control volume used in the development of the formulation does not involve any part of the boundary. You may recall that the boundary conditions had no effect on the differential equation of heat conduction in the medium either.

The finite difference formulation above can easily be extended to two- or three-dimensional heat transfer problems by replacing each second derivative by a difference equation in that direction. For example, the *finite difference formulation* for **steady two-dimensional heat conduction** in a region with heat generation and constant thermal conductivity can be expressed in rectangular coordinates as (Fig. 5-9)

$$\frac{T_{m+1,n} - 2T_{m,n} + T_{m-1,n}}{\Delta x^2} + \frac{T_{m,n+1} - 2T_{m,n} + T_{m,n-1}}{\Delta y^2} + \frac{\dot{g}_{m,n}}{k} = 0 \quad (5\text{-}12)$$

for $m = 1, 2, 3, \ldots, M-1$ and $n = 1, 2, 3, \ldots, N-1$ at any interior node (m, n). Note that a rectangular region that is divided into M equal subregions in the x-direction and N equal subregions in the y-direction has a total of $(M + 1)(N + 1)$ nodes, and Equation 5-12 can be used to obtain the finite difference equations at $(M - 1)(N - 1)$ of these nodes (i.e., all nodes except those at the boundaries).

The finite difference formulation is given above to demonstrate how difference equations are obtained from differential equations. However, we will use the *energy balance approach* in the following sections to obtain the numerical formulation because it is more *intuitive* and can handle *boundary conditions* more easily. Besides, the energy balance approach does not require having the differential equation before the analysis.

5-3 ■ ONE-DIMENSIONAL STEADY HEAT CONDUCTION

In this section we will develop the finite difference formulation of heat conduction in a plane wall using the energy balance approach and discuss how to solve the resulting equations. The **energy balance method** is based on *subdividing* the medium into a sufficient number of volume elements and then applying an *energy balance* on each element. This is done by first *selecting* the nodal points (or nodes) at which the temperatures are to be determined and then *forming elements* (or control volumes) over the nodes by drawing lines through the midpoints between the nodes. This way, the interior nodes remain at the middle of the elements, and the properties *at the node* such as the temperature and the rate of heat generation represent the *average* properties of the element. Sometimes it is convenient to think of temperature as varying *linearly* between the nodes, especially when expressing heat conduction between the elements using Fourier's law.

To demonstrate the approach, again consider steady one-dimensional heat transfer in a plane wall of thickness L with heat generation $\dot{g}(x)$ and constant conductivity k. The wall is now subdivided into M equal regions of thickness $\Delta x = L/M$ in the x-direction, and the divisions between the regions are selected as the nodes. Therefore, we have $M + 1$ nodes labeled $0, 1, 2, \ldots, m - 1, m, m + 1, \ldots, M$, as shown in Figure 5-10. The x-coordinate of any node m is simply $x_m = m\Delta x$, and the temperature at that point is $T(x_m) = T_m$. Elements are formed by drawing vertical lines through the midpoints between the nodes. Note that all interior elements represented by interior nodes are full-size elements (they have a thickness of Δx), whereas the two elements at the boundaries are half-sized.

To obtain a general difference equation for the interior nodes, consider the element represented by node m and the two neighboring nodes $m - 1$ and $m + 1$. Assuming the heat conduction to be *into* the element on all surfaces, an *energy balance* on the element can be expressed as

$$\begin{pmatrix} \text{Rate of heat} \\ \text{conduction} \\ \text{at the left} \\ \text{surface} \end{pmatrix} + \begin{pmatrix} \text{Rate of heat} \\ \text{conduction} \\ \text{at the right} \\ \text{surface} \end{pmatrix} + \begin{pmatrix} \text{Rate of heat} \\ \text{generation} \\ \text{inside the} \\ \text{element} \end{pmatrix} = \begin{pmatrix} \text{Rate of change} \\ \text{of the energy} \\ \text{content of} \\ \text{the element} \end{pmatrix}$$

or

$$\dot{Q}_{\text{cond, left}} + \dot{Q}_{\text{cond, right}} + \dot{G}_{\text{element}} = \frac{\Delta E_{\text{element}}}{\Delta t} = 0 \qquad (5\text{-}13)$$

since the energy content of a medium (or any part of it) does not change under *steady* conditions and thus $\Delta E_{\text{element}} = 0$. The rate of *heat generation* within the element can be expressed as

$$\dot{G}_{\text{element}} = \dot{g}_m V_{\text{element}} = \dot{g}_m A\Delta x \qquad (5\text{-}14)$$

where $\dot{g}_m$ is the rate of heat generation per unit volume in W/m^3 evaluated at node m and treated as a constant for the entire element, and A is heat transfer area, which is simply the inner (or outer) surface area of the wall.

Recall that when temperature varies *linearly,* the steady rate of heat conduction across a plane wall of thickness L can be expressed as

$$\dot{Q}_{\text{cond}} = kA\frac{\Delta T}{L} \qquad (5\text{-}15)$$

where ΔT is the temperature change across the wall and the direction of heat transfer is from the high temperature side to the low temperature. In the case of a plane wall with heat generation, the variation of temperature is not linear and thus the relation above is not applicable. However, the variation of temperature between the nodes can be *approximated* as being *linear* in the determination of heat conduction across a thin layer of thickness Δx between two nodes (Fig. 5-11). Obviously the smaller the distance Δx between two nodes, the more accurate is this approximation. (In fact, such approximations are the reason for classifying the numerical methods as approximate solution

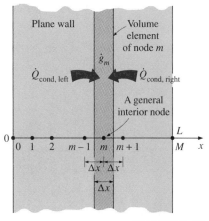

FIGURE 5-10

The nodal points and volume elements for the finite difference formulation of one-dimensional conduction in a plane wall.

FIGURE 5-11

In finite difference formulation, the temperature is assumed to vary linearly between the nodes.

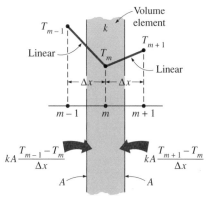

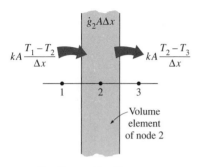

$$kA\frac{T_1 - T_2}{\Delta x} - kA\frac{T_2 - T_3}{\Delta x} + \dot{g}_2 A\Delta x = 0$$

or

$$T_1 - 2T_2 + T_3 + \dot{g}_2 A\Delta x^2 / k = 0$$

(a) Assuming heat transfer to be out of the
volume element at the right surface.

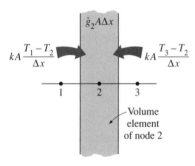

$$kA\frac{T_1 - T_2}{\Delta x} + kA\frac{T_3 - T_2}{\Delta x} + \dot{g}_2 A\Delta x = 0$$

or

$$T_1 - 2T_2 + T_3 + \dot{g}_2 A\Delta x^2 / k = 0$$

(b) Assuming heat transfer to be into the
volume element at all surfaces.

FIGURE 5-12

The assumed direction of heat transfer
at surfaces of a volume element has
no effect on the finite difference
formulation.

methods. In the limiting case of Δx approaching zero, the formulation becomes exact and we obtain a differential equation.) Noting that the direction of heat transfer on both surfaces of the element is assumed to be *toward* the node m, the rate of heat conduction at the left and right surfaces can be expressed as

$$\dot{Q}_{\text{cond, left}} = kA\frac{T_{m-1} - T_m}{\Delta x} \quad \text{and} \quad \dot{Q}_{\text{cond, right}} = kA\frac{T_{m+1} - T_m}{\Delta x} \quad (5\text{-}16)$$

Substituting Equations 5-14 and 5-16 into Equation 5-13 gives

$$kA\frac{T_{m-1} - T_m}{\Delta x} + kA\frac{T_{m+1} - T_m}{\Delta x} + \dot{g}_m A\Delta x = 0 \quad (5\text{-}17)$$

which simplifies to

$$\frac{T_{m-1} - 2T_m + T_{m+1}}{\Delta x^2} + \frac{\dot{g}_m}{k} = 0, \qquad m = 1, 2, 3, \ldots, M - 1 \quad (5\text{-}18)$$

which is *identical* to the difference equation (Eq. 5-11) obtained earlier. Again, this equation is applicable to each of the $M - 1$ interior nodes, and its application gives $M - 1$ equations for the determination of temperatures at $M + 1$ nodes. The two additional equations needed to solve for the $M + 1$ unknown nodal temperatures are obtained by applying the energy balance on the two elements at the boundaries (unless, of course, the boundary temperatures are specified).

You are probably thinking that if heat is conducted into the element from both sides, as assumed in the formulation, the temperature of the medium will have to rise and thus heat conduction cannot be steady. Perhaps a more realistic approach would be to assume the heat conduction to be *into* the element on the left side and *out of* the element on the right side. If you repeat the formulation using this assumption, you will again obtain the same result since the heat conduction term on the right side in this case will involve $T_m - T_{m+1}$ instead of $T_{m+1} - T_m$, which is subtracted instead of being added. Therefore, the assumed direction of heat conduction at the surfaces of the volume elements has no effect on the formulation, as shown in Figure 5-12. (Besides, the actual direction of heat transfer is usually not known.) However, it is convenient to assume heat conduction to be into the element at all surfaces and not worry about the sign of the conduction terms. Then all temperature differences in conduction relations are expressed as the temperature of the neighboring node minus the temperature of the node under consideration, and all conduction terms are added.

Boundary Conditions

Above we have developed a general relation for obtaining the finite difference equation for each interior node of a plane wall. This relation is not applicable to the nodes on the boundaries, however, since it requires the presence of nodes on both sides of the node under consideration, and a boundary node does not have a neighboring node on at least one side. Therefore, we need to obtain the finite difference equations of boundary nodes separately. This is best done by applying an *energy balance* on the volume elements of boundary nodes.

Boundary conditions most commonly encountered in practice are the *specified temperature, specified heat flux, convection,* and *radiation* boundary conditions, and below we develop the finite difference formulations for them for the case of steady one-dimensional heat conduction in a plane wall of thickness L as an example. The node number at the left surface at $x = 0$ is 0, and at the right surface at $x = L$ it is M. Note that the width of the volume element for either boundary node is $\Delta x/2$.

The **specified temperature** boundary condition is the simplest boundary condition to deal with. For one-dimensional heat transfer through a plane wall of thickness L, the *specified temperature boundary conditions* on both the left and right surfaces can be expressed as (Fig. 5-13)

$$
\begin{aligned}
T(0) &= T_0 = \text{Specified value} \\
T(L) &= T_M = \text{Specified value}
\end{aligned} \tag{5-19}
$$

where T_0 and T_m are the specified temperatures at surfaces at $x = 0$ and $x = L$, respectively. Therefore, the specified temperature boundary conditions are incorporated by simply assigning the given surface temperatures to the boundary nodes. We do not need to write an energy balance in this case unless we decide to determine the rate of heat transfer into or out of the medium after the temperatures at the interior nodes are determined.

When other boundary conditions such as the *specified heat flux, convection, radiation,* or *combined convection and radiation* conditions are specified at a boundary, the finite difference equation for the node at that boundary is obtained by writing an *energy balance* on the volume element at that boundary. The energy balance is again expressed as

$$
\sum_{\text{all sides}} \dot{Q} + \dot{G}_{\text{element}} = 0 \tag{5-20}
$$

for heat transfer under *steady* conditions. Again we assume all heat transfer to be *into* the volume element from all surfaces for convenience in formulation, except for specified heat flux since its direction is already specified. Specified heat flux is taken to be a *positive* quantity if into the medium and a *negative* quantity if out of the medium. Then the finite difference formulation at the node $m = 0$ (at the left boundary where $x = 0$) of a plane wall of thickness L during steady one-dimensional heat conduction can be expressed as (Fig. 5-14)

$$
\dot{Q}_{\text{left surface}} + kA \frac{T_1 - T_0}{\Delta x} + \dot{g}_0(A\Delta x/2) = 0 \tag{5-21}
$$

where $A\Delta x/2$ is the *volume* of the volume element (note that the boundary element has half thickness), $\dot{g}_0$ is the rate of heat generation per unit volume

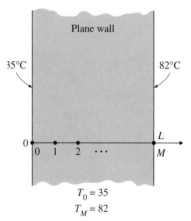

$T_0 = 35$
$T_M = 82$

FIGURE 5-13

Finite difference formulation of specified temperature boundary conditions on both surfaces of a plane wall.

FIGURE 5-14

Schematic for the finite difference formulation of the left boundary node of a plane wall.

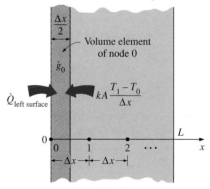

$$\dot{Q}_{\text{left surface}} + kA\frac{T_1 - T_0}{\Delta x} + \dot{g}_0 A\frac{\Delta x}{2} = 0$$

(in W/m^3) at $x = 0$, and A is the heat transfer area, which is constant for a plane wall. Note that we have Δx in the denominator of the second term instead of $\Delta x/2$. This is because the ratio in that term involves the temperature difference between nodes 0 and 1, and thus we must use the distance between those two nodes, which is Δx.

The finite difference form of various boundary conditions can be obtained from Equation 5-21 by replacing $\dot{Q}_{\text{left surface}}$ by a suitable expression. Below this is done for various boundary conditions at the left boundary.

1. Specified Heat Flux Boundary Condition

$$\dot{q}_0 A + kA \frac{T_1 - T_0}{\Delta x} + \dot{g}_0(A\Delta x/2) = 0 \qquad (5\text{-}22)$$

Special case: **Insulated Boundary** ($\dot{q}_0 = 0$)

$$kA \frac{T_1 - T_0}{\Delta x} + \dot{g}_0(A\Delta x/2) = 0 \qquad (5\text{-}23)$$

2. Convection Boundary Condition

$$hA(T_\infty - T_0) + kA \frac{T_1 - T_0}{\Delta x} + \dot{g}_0(A\Delta x/2) = 0 \qquad (5\text{-}24)$$

3. Radiation Boundary Condition

$$\varepsilon\sigma A(T_{\text{surr}}^4 - T_0^4) + kA \frac{T_1 - T_0}{\Delta x} + \dot{g}_0(A\Delta x/2) = 0 \qquad (5\text{-}25)$$

4. Combined Convection and Radiation Boundary Condition (Fig. 5-15)

$$hA(T_\infty - T_0) + \varepsilon\sigma A(T_{\text{surr}}^4 - T_0^4) + kA \frac{T_1 - T_0}{\Delta x} + \dot{g}_0(A\Delta x/2) = 0 \qquad (5\text{-}26)$$

or

$$h_{\text{combined}} A(T_\infty - T_0) + kA \frac{T_1 - T_0}{\Delta x} + \dot{g}_0(A\Delta x/2) = 0 \qquad (5\text{-}27)$$

5. Combined Convection, Radiation, and Heat Flux Boundary Condition

$$\dot{q}_0 A + hA(T_\infty - T_0) + \varepsilon\sigma A(T_{\text{surr}}^4 - T_0^4) + kA \frac{T_1 - T_0}{\Delta x} + \dot{g}_0(A\Delta x/2) = 0$$

$$(5\text{-}28)$$

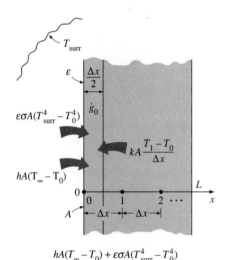

$$hA(T_\infty - T_0) + \varepsilon\sigma A(T_{\text{surr}}^4 - T_0^4)$$
$$+ kA\frac{T_1 - T_0}{\Delta x} + \dot{g}_0 A\frac{\Delta x}{2} = 0$$

FIGURE 5-15

Schematic for the finite difference formulation of combined convection and radiation on the left boundary of a plane wall.

6. Interface Boundary Condition Two different solid media A and B are assumed to be in perfect contact, and thus at the same temperature at the interface at node m (Fig. 5-16). Subscripts A and B indicate properties of media A *and* B, respectively.

$$k_A A \frac{T_{m-1} - T_m}{\Delta x} + k_B A \frac{T_{m+1} - T_m}{\Delta x} + \dot{g}_{A,m}(A\Delta x/2) + \dot{g}_{B,m}(A\Delta x/2) = 0$$

$$(5\text{-}29)$$

In the above relations, $\dot{q}_0$ is the specified heat flux in W/m^2, h is the convection coefficient, h_{combined} is the combined convection and radiation coefficient, T_∞ is the temperature of the surrounding medium, T_{surr} is the temperature of the surrounding surfaces, ε is the emissivity of the surface, and σ is the Stefan-Boltzman constant. The relations above can also be used for node M on the right boundary by replacing the subscript "0" by "M" and the subscript "1" by "$M-1$".

Note that *absolute temperatures* must be used in radiation heat transfer calculations, and all temperatures should be expressed in K or R when a boundary condition involves radiation to avoid mistakes. We usually try to avoid the *radiation boundary condition* even in numerical solutions since it causes the finite difference equations to be *nonlinear*, which are more difficult to solve.

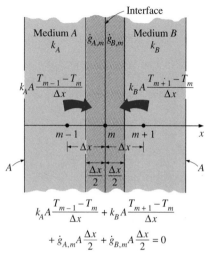

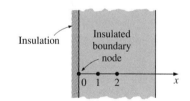

FIGURE 5-16

Schematic for the finite difference formulation of the interface boundary condition for two mediums A and B that are in perfect thermal contact.

Treating Insulated Boundary Nodes as Interior Nodes: The Mirror Image Concept

One way of obtaining the finite difference formulation of a node on an insulated boundary is to treat insulation as "zero" heat flux and to write an energy balance, as done in Equation 5-23. Another and more practical way is to treat the node on an insulated boundary as an interior node. Conceptually this is done by replacing the insulation on the boundary by a *mirror* and considering the reflection of the medium as its extension (Fig. 5-17). This way the node next to the boundary node appears on both sides of the boundary node because of symmetry, converting it into an interior node. Then using the general formula (Eq. 5-18) for an interior node, which involves the sum of the temperatures of the adjoining nodes minus twice the node temperature, the finite difference formulation of a node $m = 0$ on an insulated boundary of a plane wall can be expressed as

$$\frac{T_{m+1} - 2T_m + T_{m-1}}{\Delta x^2} + \frac{\dot{g}_m}{k} = 0 \quad \rightarrow \quad \frac{T_1 - 2T_0 + T_1}{\Delta x^2} + \frac{\dot{g}_0}{k} = 0$$

which is equivalent to Equation 5-23 obtained by the energy balance approach.

The mirror image approach can also be used for problems that possess thermal symmetry by replacing the plane of symmetry by a mirror. Alternately, we can replace the plane of symmetry by insulation and consider only half of the medium in the solution. The solution in the other half of the medium is simply the mirror image of the solution obtained.

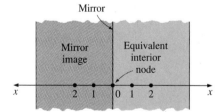

FIGURE 5-17

A node on an insulated boundary can be treated as an interior node by replacing the insulation by a mirror.

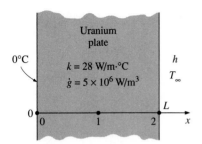

FIGURE 5-18

Schematic for Example 5-1.

EXAMPLE 5-1   Steady Heat Conduction in a Large Uranium Plate

Consider a large uranium plate of thickness $L = 4$ cm and thermal conductivity $k = 28$ W/m · °C in which heat is generated uniformly at a constant rate of $\dot{g} = 5 \times 10^6$ W/m³. One side of the plate is maintained at 0°C by iced water while the other side is subjected to convection to an environment at $T_\infty = 30$°C with a heat transfer coefficient of $h = 45$ W/m² · °C, as shown in Figure 5-18. Considering a total of three equally spaced nodes in the medium, two at the boundaries and one at the middle, estimate the exposed surface temperature of the plate under steady conditions using the finite difference approach.

Solution   A uranium plate is subjected to specified temperature on one side and convection on the other. The unknown surface temperature of the plate is to be determined numerically using three equally spaced nodes.

Assumptions   **1** Heat transfer through the wall is steady since there is no indication of any change with time. **2** Heat transfer is one-dimensional since the plate is large relative to its thickness. **3** Thermal conductivity is constant. **4** Radiation heat transfer is negligible.

Properties   The thermal conductivity is given to be $k = 28$ W/m · °C.

Analysis   The number of nodes is specified to be $M = 3$, and they are chosen to be at the two surfaces of the plate and the midpoint, as shown in the figure. Then the nodal spacing Δx becomes

$$\Delta x = \frac{L}{M-1} = \frac{0.04 \text{ m}}{3-1} = 0.02 \text{ m}$$

We number the nodes 0, 1, and 2. The temperature at node 0 is given to be $T_0 = 0$°C, and the temperatures at nodes 1 and 2 are to be determined. This problem involves only two unknown nodal temperatures, and thus we need to have only two equations to determine them uniquely. These equations are obtained by applying the finite difference method to nodes 1 and 2.

Node 1 is an interior node, and the finite difference formulation at that node is obtained directly from Equation 5-18 by setting $m = 1$:

$$\frac{T_0 - 2T_1 + T_2}{\Delta x^2} + \frac{\dot{g}_1}{k} = 0 \quad \rightarrow \quad \frac{0 - 2T_1 + T_2}{\Delta x^2} + \frac{\dot{g}_1}{k} = 0 \quad \rightarrow \quad 2T_1 - T_2 = \frac{\dot{g}_1 \Delta x^2}{k}$$

$$(1)$$

Node 2 is a boundary node subjected to convection, and the finite difference formulation at that node is obtained by writing an energy balance on the volume element of thickness $\Delta x/2$ at that boundary by assuming heat transfer to be into the medium at all sides:

$$hA(T_\infty - T_2) + kA\frac{T_1 - T_2}{\Delta x} + \dot{g}_2(A\Delta x/2) = 0$$

Canceling the heat transfer area A and rearranging give

$$T_1 - \left(1 + \frac{h\Delta x}{k}\right)T_2 = -\frac{h\Delta x}{k}T_\infty - \frac{\dot{g}_2\Delta x^2}{2k} \qquad (2)$$

Equations (1) and (2) form a system of two equations in two unknowns T_1 and T_2. Substituting the given quantities and simplifying give

$$2T_1 - T_2 = 71.43 \qquad \text{(in °C)}$$
$$T_1 - 1.032T_2 = -36.68 \qquad \text{(in °C)}$$

This is a system of two algebraic equations in two unknowns and can be solved easily by the elimination method. Solving the first equation for T_1 and substituting into the second equation result in an equation in T_2 whose solution is

$$T_2 = \textbf{136.1°C}$$

This is the temperature of the surface exposed to convection, which is the desired result. Substitution of this result into the first equation gives $T_1 = 103.8°C$, which is the temperature at the middle of the plate.

Discussion The purpose of this example is to demonstrate the use of the finite difference method with minimal calculations, and the accuracy of the result was not a major concern. But you might still be wondering how accurate the result obtained above is. After all, we used a mesh of only three nodes for the entire plate, which seems to be rather crude. This problem can be solved analytically as described in Chapter 2, and the analytical (exact) solution can be shown to be

$$T(x) = \frac{0.5\dot{g}hL^2/k + \dot{g}L + T_\infty h}{hL + k} x - \frac{\dot{g}x^2}{2k}$$

Substituting the given quantities, the temperature of the exposed surface of the plate at $x = L = 0.04$ m is determined to be 136.0°C, which is almost identical to the result obtained above with the approximate finite difference method (Fig. 5-19). Therefore, highly accurate results can be obtained with numerical methods by using a limited number of nodes.

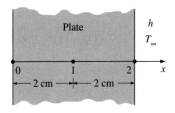

Finite difference solution:

$$T_2 = 136.1°C$$

Exact solution:

$$T_2 = 136.0°C$$

FIGURE 5-19

Despite being approximate in nature, highly accurate results can be obtained by numerical methods.

FIGURE 5-20

Schematic for Example 5-2 and the volume element of a general interior node of the fin.

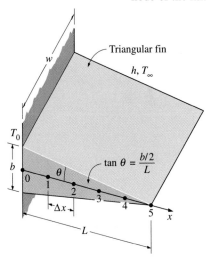

EXAMPLE 5-2   Heat Transfer from Triangular Fins

Consider an aluminum alloy fin ($k = 180$ W/m·°C) of triangular cross-section with length $L = 5$ cm, base thickness $b = 1$ cm, and very large width w in the direction normal to the plane of paper, as shown in Figure 5-20. The base of the fin is maintained at a temperature of $T_0 = 200°C$. The fin is losing heat to the surrounding medium at $T_\infty = 25°C$ with a heat transfer coefficient of $h = 15$ W/m²·°C. Using the finite difference method with six equally spaced nodes along the fin in the x-direction, determine (a) the temperatures at the nodes, (b) the rate of heat transfer from the fin for $w = 1$ m, and (c) the fin efficiency.

Solution A long triangular fin attached to a surface is considered. The nodal temperatures, the rate of heat transfer, and the fin efficiency are to be determined numerically using six equally spaced nodes.

Assumptions **1** Heat transfer is steady since there is no indication of any change with time. **2** The temperature along the fin varies in the x direction only. **3** Thermal conductivity is constant. **4** Radiation heat transfer is negligible.

Properties The thermal conductivity is given to be $k = 180$ W/m·°C.

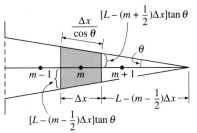

Analysis (a) The number of nodes in the fin is specified to be $M = 6$, and their location is as shown in the figure. Then the nodal spacing Δx becomes

$$\Delta x = \frac{L}{M-1} = \frac{0.05 \text{ m}}{6-1} = 0.01 \text{ m}$$

The temperature at node 0 is given to be $T_0 = 200°C$, and the temperatures at the remaining five nodes are to be determined. Therefore, we need to have five equations to determine them uniquely. Nodes 1, 2, 3, and 4 are interior nodes, and the finite difference formulation for a general interior node m is obtained by applying an energy balance on the volume element of this node. Noting that heat transfer is steady and there is no heat generation in the fin and assuming heat transfer to be into the medium at all sides, the energy balance can be expressed as

$$\sum_{\text{all sides}} \dot{Q} = 0 \quad \rightarrow \quad kA_{\text{left}} \frac{T_{m-1} - T_m}{\Delta x} + kA_{\text{right}} \frac{T_{m+1} - T_m}{\Delta x} + hA_{\text{conv}}(T_\infty - T_m) = 0$$

Note that heat transfer areas are different for each node in this case, and using geometrical relations, they can be expressed as

$$A_{\text{left}} = (\text{Height} \times \text{Width})_{@m - \frac{1}{2}} = 2w[L - (m - \tfrac{1}{2})\Delta x]\tan \theta$$

$$A_{\text{right}} = (\text{Height} \times \text{Width})_{@m + \frac{1}{2}} = 2w[L - (m + \tfrac{1}{2})\Delta x]\tan \theta$$

$$A_{\text{conv}} = 2 \times \text{Length} \times \text{Width} = 2w(\Delta x/\cos\theta)$$

Substituting,

$$2kw[L - (m - \tfrac{1}{2})\Delta x]\tan\theta \frac{T_{m-1} - T_m}{\Delta x}$$

$$+ 2kw[L - (m + \tfrac{1}{2})\Delta x]\tan\theta \frac{T_{m+1} - T_m}{\Delta x} + h\frac{2w\Delta x}{\cos\theta}(T_\infty - T_m) = 0$$

Dividing each term by $2kwL \tan\theta/\Delta x$ gives

$$\left[1 - (m - \tfrac{1}{2})\frac{\Delta x}{L}\right](T_{m-1} - T_m) + \left[1 - (m + \tfrac{1}{2})\frac{\Delta x}{L}\right](T_{m+1} - T_m)$$

$$+ \frac{h(\Delta x)^2}{kL \sin\theta}(T_\infty - T_m) = 0$$

Note that

$$\tan\theta = \frac{b/2}{L} = \frac{0.5 \text{ cm}}{5 \text{ cm}} = 0.1 \quad \rightarrow \quad \theta = \tan^{-1}0.1 = 5.71°$$

Also, $\sin 5.71° = 0.0995$. Then the substitution of known quantities gives

$$(5.5 - m)T_{m-1} - (10.00838 - 2m)T_m + (4.5 - m)T_{m+1} = -0.209$$

Now substituting 1, 2, 3, and 4 for m results in the following four finite difference equations for the interior nodes:

$m = 1$:	$-8.00838T_1 + 3.5T_2 = -900.209$	(1)
$m = 2$:	$3.5T_1 - 6.00838T_2 + 2.5T_3 = -0.209$	(2)
$m = 3$:	$2.5T_2 - 4.00838T_3 + 1.5T_4 = -0.209$	(3)
$m = 4$:	$1.5T_3 - 2.00838T_4 + 0.5T_5 = -0.209$	(4)

The finite difference equation for the boundary node 5 is obtained by writing an energy balance on the volume element of length $\Delta x/2$ at that boundary, again by assuming heat transfer to be into the medium at all sides (Fig. 5-21):

$$kA_{left}\frac{T_4 - T_5}{\Delta x} + hA_{conv}(T_\infty - T_5) = 0$$

where

$$A_{left} = 2w\frac{\Delta x}{2}\tan\theta \quad \text{and} \quad A_{conv} = 2w\frac{\Delta x/2}{\cos\theta}$$

Canceling w in all terms and substituting the known quantities give

$$T_4 - 1.00838T_5 = -0.209 \tag{5}$$

Equations (1) through (5) form a linear system of five algebraic equations in five unknowns. Solving them simultaneously using an equation solver or one of the solution methods discussed in the next section gives

$$T_1 = \textbf{198.6°C}, \qquad T_2 = \textbf{197.1°C}, \qquad T_3 = \textbf{195.7°C},$$
$$T_4 = \textbf{194.3°C}, \qquad T_5 = \textbf{192.9°C}$$

which is the desired solution for the nodal temperatures.

(b) The total rate of heat transfer from the fin is simply the sum of the heat transfer from each volume element to the ambient, and for $w = 1$ m it is determined from

$$\dot{Q}_{fin} = \sum_{m=0}^{5}\dot{Q}_{element, m} = \sum_{m=0}^{5}hA_{conv, m}(T_m - T_\infty)$$

Noting that the heat transfer surface area is $w\Delta x/\cos\theta$ for the boundary nodes 0 and 5, and twice as large for the interior nodes 1, 2, 3, and 4, we have

$$\dot{Q}_{fin} = h\frac{w\Delta x}{\cos\theta}[(T_0 - T_\infty) + 2(T_1 - T_\infty) + 2(T_2 - T_\infty) + 2(T_3 - T_\infty)$$
$$\quad + 2(T_4 - T_\infty) + (T_5 - T_\infty)]$$
$$= h\frac{w\Delta x}{\cos\theta}[T_0 + 2(T_1 + T_2 + T_3 + T_4) + T_5 - 10T_\infty]$$
$$= (15\ \text{W/m}^2 \cdot \text{°C})\frac{(1\ \text{m})(0.01\ \text{m})}{\cos 5.71°}[200 + 2 \times 785.7 + 192.9 - 10 \times 25]$$
$$= \textbf{258.4 W}$$

(c) If the entire fin were at the base temperature of $T_0 = 200°C$, the total rate of heat transfer from the fin for $w = 1$ m would be

$$\dot{Q}_{max} = hA_{fin, total}(T_0 - T_\infty) = h(2wL/\cos\theta)(T_0 - T_\infty)$$
$$= (15\ \text{W/m}^2 \cdot \text{°C})[2(1\ \text{m})(0.05\ \text{m})/\cos 5.71°](200 - 25)°C$$
$$= 263.8\ \text{W}$$

Then the fin efficiency is determined from

$$\eta_{fin} = \frac{\dot{Q}_{fin}}{\dot{Q}_{max}} = \frac{258.4\ \text{W}}{263.8\ \text{W}} = \textbf{0.98}$$

which is less than 1, as expected. We could also determine the fin efficiency in this case from the proper fin efficiency curve in Chapter 3, which is based on the

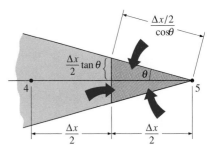

FIGURE 5-21
Schematic of the volume element of node 5 at the tip of a triangular fin.

analytical solution. We would read 0.98 for the fin efficiency, which is identical to the value determined above numerically.

5-4 ■ SOLUTION METHODS FOR SYSTEMS OF ALGEBRAIC EQUATIONS

Direct methods:
Solve in a systematic manner following a series of well-defined steps.

Iterative methods:
Start with an initial guess for the solution, and iterate until solution converges.

FIGURE 5-22

Two general categories of solution methods for solving systems of algebraic equations.

The finite difference formulation of steady heat conduction problems usually results in a system of N algebraic equations in N unknown nodal temperatures that need to be solved simultaneously. When N is small (such as 2 or 3), we can use the elementary *elimination method* to eliminate all unknowns except one and then solve for that unknown (see Example 5-1). The other unknowns are then determined by back substitution. When N is large, which is usually the case, the elimination method is not practical and we need to use a more systematic approach that can be adapted to computers. There are numerous systematic approaches available in the literature, and they are broadly classified as **direct** and **iterative** methods. The direct methods are based on a fixed number of well-defined steps that result in the solution in a systematic manner. The iterative methods, on the other hand, are based on an initial guess for the solution that is refined by iteration until a specified convergence criterion is satisfied (Fig. 5-22). The direct methods usually require a large amount of computer memory and computation time, and they are more suitable for systems with a relatively small number of equations. The computer memory requirements for iterative methods are minimal, and thus they are usually preferred for large systems. The convergence of iterative methods to the desired solution, however, may pose a problem.

Below we discuss two methods that are representatives of the two broad categories discussed above: the direct *Gaussian elimination* method and the iterative *Gauss-Seidel* method. Both approaches can be programmed easily to solve any number of algebraic equations simultaneously. Full details of these and other, less popular approaches such as the *matrix inversion* method can be found in numerical methods books.

You probably have access to an *equation solver* that solves a system of equations at the press of a button, but it is still worthwhile to go through the methods presented below to understand the procedures involved.

Gaussian Elimination Method

The Gaussian elimination method is a direct solution method based on a systematic elimination process during which one of the unknowns in a system of *linear* algebraic equations is eliminated during each step. The last equation of the system involves only one unknown at the end of the elimination process and is solved for that unknown, with the remaining unknowns determined by a series of back substitution processes. Below we demonstrate the method with an example.

Consider the following system of three linear algebraic equations with three unknowns:

$$x_1 - 3x_2 + x_3 = 10$$
$$-x_1 + x_2 - 2x_3 = -13 \qquad (5\text{-}30)$$
$$2x_1 + 5x_2 + x_3 = 4$$

It can be shown by direct substitution that the solution of this system is $x_1 = 2$, $x_2 = -1$, and $x_3 = 5$. Below we will obtain this solution in a systematic way using the Gaussian elimination method:

Step 1 Using the first equation, eliminate the first unknown x_1 from all other equations. This is done by multiplying the first equation by a suitable constant (by 1 in this case), and adding the result to the second equation. The process is repeated for the third equation. We get

$$x_1 - 3x_2 + x_3 = 10$$
$$-2x_2 - x_3 = -3$$
$$11x_2 - x_3 = -16$$

Step 2 Similarly, using the second equation, eliminate the second unknown x_2 from the third equation:

$$x_1 - 3x_2 + x_3 = 10$$
$$-2x_2 - x_3 = -3$$
$$-6.5x_3 = -32.5$$

Step 3 Solve the last equation for the last unknown:

$$-6.5x_3 = -32.5 \quad \rightarrow \quad x_3 = -32.5/(-6.5) = 5$$

Step 4 Using the value of x_3, solve the second equation for x_2:

$$-2x_2 - x_3 = -3 \quad \rightarrow \quad x_2 = (-3 + x_3)/(-2) = (-3 + 5)/(-2) = -1$$

Step 5 Using the values of x_2 and x_3, solve the first equation for x_1:

$$x_1 - 3x_2 + x_3 = 10 \quad \rightarrow \quad x_1 = 10 + 3x_2 - x_3$$
$$= 10 + [3 \times (-1)] - 5 = 2$$

Therefore, the solution of the given system of three equations with three unknowns is $x_1 = 2$, $x_2 = -1$, and $x_3 = 5$, as expected.

The application of the Gaussian elimination method to large systems of equations is best done using the matrix notation. In this case, the coefficients of the equations are stored in a square matrix, called the *coefficient matrix,* while the constant terms are stored in a vector called the *right-hand side vector.* Computationally it is convenient to combine the two into a single *augmented matrix* whose last column is the right-hand side matrix. Viewing each row of this matrix as an equation, the procedures discussed above are used to eliminate the terms below the main diagonal and to transform the augmented matrix into an *upper diagonal form.* The solution is then obtained by back substitution.

For example, the augmented matrix of the system given above is

$$\begin{bmatrix} 1 & -3 & 1 & 10 \\ -1 & 1 & -2 & -13 \\ 2 & 5 & 1 & 4 \end{bmatrix}$$

Once the elements under the main diagonal are eliminated by legitimate row operations (such as switching two rows in the matrix, multiplying all elements of a row by a constant, and multiplying one row by a constant and adding it to another row), the upper diagonal form of the augmented matrix is determined to be

$$\begin{bmatrix} 1 & -3 & 1 & 10 \\ 0 & -2 & -1 & -3 \\ 0 & 0 & -6.5 & -32.5 \end{bmatrix}$$

Now the last row can be interpreted as $-6.5x_3 = -32.5$, whose solution is $x_3 = 5$. The other unknowns are determined by back substitution as before. The matrix notation and the row operations make it very convenient to program the Gaussian elimination method for a general system of N linear algebraic equations with N unknowns, and to use it as a general purpose equation solver.

Solve for unknowns
in symbolic form

↓

Guess a solution
to start the process

↓

Iterate until the solution converges

FIGURE 5-23

The basic steps involved in the Gauss-Seidel iteration method.

Gauss-Seidel Iteration Method

This is one of the simplest iteration techniques available for solving a system of linear algebraic equations simultaneously. Despite its slow convergence, it remains a popular method because of its simplicity. The method is based on the following steps (Fig. 5-23):

1. Solve each equation in the system for one of the unknowns (preferably for the one with the largest coefficient) so that each unknown is expressed in terms of the other unknowns (in heat transfer, solve the equation for the mth node for T_m).

2. Make initial guesses for all unknowns (the same value can be chosen as the initial value for all unknowns for simplicity).

3. Using the equations in Step 1, calculate the values of the unknowns (always use the most recently computed values for unknowns when available).

4. Repeat Step 3 until a specified convergence criterion is satisfied.

We demonstrate the method with an example.

EXAMPLE 5-3 Gauss-Seidel Iteration Method

Solve the following system of three linear algebraic equations with three unknowns using the Gauss-Seidel iteration method.

$$x_1 - 3x_2 + x_3 = 10$$
$$2x_1 + 5x_2 + x_3 = 4$$
$$-x_1 + x_2 - 2x_3 = -13$$

Solution We start the solution process by solving the first equation for x_1, the second equation for x_2, and the third equation for x_3:

$$x_1 = 10 + 3x_2 - x_3$$
$$x_2 = (4 - 2x_1 - x_3)/5$$
$$x_3 = (-13 + x_1 - x_2)/(-2)$$

Next we make initial guesses for the unknowns. We can choose any arbitrary values we want for initial guesses, but the solution will converge faster (in fewer iterations) if we guess values that are close to the actual solution. In a fin problem, for example, we know that the nodal temperatures will be between the base temperature and the ambient temperature, and they will be decreasing toward the fin tip. Thus we can make the initial guesses accordingly.

In this case we have no idea about the solution range, and thus we take $x_1 = 0$, $x_2 = 0$, and $x_3 = 0$ for simplicity. We call these the "zeroth iteration" values. We use these guesses to start the iteration process, and replace them by the calculated values as they become available,

$$x_1 = 10 + 3x_2 - x_3 = 10 + 3 \times 0 - 0 = 10$$
$$x_2 = (4 - 2x_1 - x_3)/5 = [4 - 2 \times 10 - 0]/5 = -3.2$$
$$x_3 = (-13 + x_1 - x_2)/(-2) = [-13 + 10 - (-3.2)/(-2) = -0.1$$

Therefore, the values of the unknowns become $x_1 = 10$, $x_2 = -3.2$, and $x_3 = -0.1$ after the first iteration. Repeating the step above, the values of the unknowns after the second iteration become

$$x_1 = 10 + 3x_2 - x_3 = 10 + 3 \times (-3.2) - (-0.1) = 0.5$$
$$x_2 = (4 - 2x_1 - x_3)/5 = [4 - 2 \times 0.5 + 0.1]/5 = 0.62$$
$$x_3 = (-13 + x_1 - x_2)/(-2) = (-13 + 0.5 - 0.62)/(-2) = 6.56$$

The process is repeated for 100 iterations using the FORTRAN program given in Figure 5-24, and the results are listed in Table 5-2. Notice that the results do not change much after the 70th iteration, and thus the results after 70 iterations can be taken as the solution of the system. Note that we would have received the exact results of 2, -1, and 5 if we continued the process for 90 iterations. Needless to say, problems of iterative nature are best solved with a computer, and we can specify the iteration to terminate when the difference in the unknowns between two successive iterations is less than a specified value.

Despite the apparent simplicity of the iterative methods, stability and convergence remain as major concerns. It can be shown that for *steady* heat conduction problems, the Gauss-Seidel iteration method is *inherently stable* and *always* converges uniformly to the desired solution. The convergence can be rather slow for large systems of equations, however, but can be speeded up by using some modifications such as the successive overrelaxation (SOR) scheme.

```
PROGRAM GAUSS
DATA X1,X2,X3/.0,.0,.0/
DO 20 K = 1,100
X1 = 10 + 3*X2 - X3
X2 = (4 - 2*X1 - X3)/5.
X3 = (-13 + X1 - X2)/(-2.)
WRITE (*,10)K,X1,X2,X3
10 FORMAT (' ', I3, 3(F8.1))
20 CONTINUE
END
```

FIGURE 5-24

The complete FORTRAN program used to solve the three equations in Example 5-3 with the Gauss-Seidel iteration method.

TABLE 5-2

The results of the Gauss-Seidel iteration for the solution of the system of equations in Example 5-3 for various iterations

Iteration number	Unknowns		
	x_1	x_2	x_3
0	0.00	0.00	0.00
1	10.00	-3.20	-0.10
2	0.50	0.62	6.56
3	5.30	-2.63	2.53
4	-0.43	0.47	6.95
5	4.45	-2.37	3.09
10	0.38	-0.08	6.27
20	1.26	-0.58	5.58
30	1.66	-0.81	5.26
40	1.85	-0.91	5.12
50	1.93	-0.96	5.05
60	1.97	-0.98	5.03
70	1.99	-0.99	5.01
80	1.99	-1.00	5.01
90	2.00	-1.00	5.00
100	2.00	-1.00	5.00

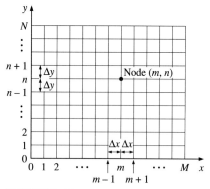

FIGURE 5-25

The nodal network for the finite
difference formulation of
two-dimensional conduction
in rectangular coordinates.

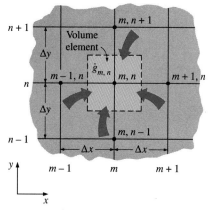

FIGURE 5-26

The volume element of a general
interior node (m, n) for two-dimensional
conduction in rectangular coordinates.

5-5 ■ TWO-DIMENSIONAL STEADY HEAT CONDUCTION

In Section 5-3 we considered one-dimensional heat conduction and assumed
heat conduction in other directions to be negligible. Many heat transfer problems encountered in practice can be approximated as being one-dimensional,
but this is not always the case. Sometimes we need to consider heat transfer
in other directions as well when the variation of temperature in other directions is significant. In this section we will consider the numerical formulation
and solution of *two-dimensional steady* heat conduction in rectangular coordinates using the finite difference method. The approach presented below can
be extended to three-dimensional cases.

Consider a *rectangular region* in which heat conduction is significant in
the x- and y-directions. Now divide the x-y plane of the region into a rectangular mesh of nodal points spaced Δx and Δy apart in the x- and y-directions,
respectively, as shown in Figure 5-25, and consider a unit depth of $\Delta z = 1$ in
the z-direction. Our goal is to determine the temperatures at the nodes, and
it is convenient to number the nodes and describe their position by the
numbers instead of actual coordinates. A logical numbering scheme for
two-dimensional problems is the *double subscript notation* (m, n) where
$m = 0, 1, 2, \ldots, M$ is the node count in the x-direction and $n = 0, 1, 2, \ldots,$
N is the node count in the y-direction. The coordinates of the node (m, n) are
simply $x = m\Delta x$ and $y = n\Delta y$, and the temperature at the node (m, n) is
denoted by $T_{m,n}$.

Now consider a *volume element* of size $\Delta x \times \Delta y \times 1$ centered about a
general interior node (m, n) in a region in which heat is generated at a rate of
$\dot{g}$ and the thermal conductivity k is constant, as shown in Figure 5-26. Again
assuming the direction of heat conduction to be *toward* the node under consideration at all surfaces, the energy balance on the volume element can be
expressed as

$$
\begin{pmatrix} \text{Rate of heat conduction} \\ \text{at the left, top, right,} \\ \text{and bottom surfaces} \end{pmatrix} + \begin{pmatrix} \text{Rate of heat} \\ \text{generation inside} \\ \text{the element} \end{pmatrix} = \begin{pmatrix} \text{Rate of change of} \\ \text{the energy content} \\ \text{of the element} \end{pmatrix}
$$

or

$$
\dot{Q}_{\text{cond, left}} + \dot{Q}_{\text{cond, top}} + \dot{Q}_{\text{cond, right}} + \dot{Q}_{\text{cond, bottom}} + \dot{G}_{\text{element}} = \frac{\Delta E_{\text{element}}}{\Delta t} = 0
$$

$$(5\text{-}31)$$

for the *steady* case. Again assuming the temperatures between the adjacent nodes to vary linearly and noting that the heat transfer area is
$A_x = \Delta y \times 1 = \Delta y$ in the x-direction and $A_y = \Delta x \times 1 = \Delta x$ in the
y-direction, the energy balance relation above becomes

$$kΔy \frac{T_{m-1,n} - T_{m,n}}{Δx} + kΔx \frac{T_{m,n+1} - T_{m,n}}{Δy} + kΔy \frac{T_{m+1,n} - T_{m,n}}{Δx}$$

$$+ kΔx \frac{T_{m,n-1} - T_{m,n}}{Δy} + \dot{g}_{m,n} Δx Δy = 0 \tag{5-32}$$

Dividing each term by $Δx × Δy$ and simplifying give

$$\frac{T_{m-1,n} - 2T_{m,n} + T_{m+1,n}}{Δx^2} + \frac{T_{m,n-1} - 2T_{m,n} + T_{m,n+1}}{Δy^2} + \frac{\dot{g}_{m,n}}{k} = 0$$

$$\tag{5-33}$$

for $m = 1, 2, 3, \ldots, M - 1$ and $n = 1, 2, 3, \ldots, N - 1$. This equation is identical to Equation 5-12 obtained earlier by replacing the derivatives in the differential equation by differences for an interior node (m, n). Again a rectangular region M equally spaced nodes in the x-direction and N equally spaced nodes in the y-direction has a total of $(M + 1)(N + 1)$ nodes, and Equation 5-33 can be used to obtain the finite difference equations at all interior nodes.

In finite difference analysis, usually a **square mesh** is used for simplicity (except when the magnitudes of temperature gradients in the x- and y-directions are very different), and thus $Δx$ and $Δy$ are taken to be the same. Then $Δx = Δy = l$, and the relation above simplifies to

$$T_{m-1,n} + T_{m+1,n} + T_{m,n+1} + T_{m,n-1} - 4T_{m,n} + \frac{\dot{g}_{m,n}l^2}{k} = 0 \tag{5-34}$$

That is, the finite difference formulation of an interior node is obtained by *adding the temperatures of the four nearest neighbors of the node, subtracting four times the temperature of the node itself, and adding the heat generation term*. It can also be expressed in the following form, which is easy to remember:

$$T_{\text{left}} + T_{\text{top}} + T_{\text{right}} + T_{\text{bottom}} - 4T_{\text{node}} + \frac{\dot{g}_{\text{node}}l^2}{k} = 0 \tag{5-35}$$

When there is no heat generation in the medium, the finite difference equation for an interior node further simplifies to $T_{\text{node}} = (T_{\text{left}} + T_{\text{top}} + T_{\text{right}} + T_{\text{bottom}})/4$, which has the interesting interpretation that *the temperature of each interior node is the arithmetic average of the temperatures of the four neighboring nodes*. This statement is also true for the three-dimensional problems except that the interior nodes in that case will have six neighboring nodes instead of four.

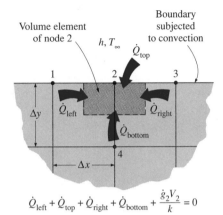

$$\dot{Q}_{\text{left}} + \dot{Q}_{\text{top}} + \dot{Q}_{\text{right}} + \dot{Q}_{\text{bottom}} + \frac{\dot{g}_2 V_2}{k} = 0$$

FIGURE 5-27

The finite difference formulation
of a boundary node is obtained
by writing an energy balance on
its volume element.

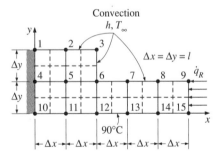

FIGURE 5-28

Schematic for Example 5-4 and the
nodal network (the boundaries of
valume elements of the nodes are
indicated by dashed lines).

Boundary Nodes

The development of finite difference formulation of *boundary nodes* in two- (or three-) dimensional problems is similar to the development in the one-dimensional case discussed earlier. Again, the region is partitioned between the nodes by forming *volume elements* around the nodes, and an *energy balance* is written for each boundary node. Various boundary conditions can be handled as discussed for a plane wall, except that the volume elements in the two-dimensional case involve heat transfer in the *y-direction* as well as the *x-direction*. Insulated surfaces can still be viewed as "mirrors," and the mirror image concept can be used to treat nodes on insulated boundaries as interior nodes.

For heat transfer under *steady* conditions, the basic equation to keep in mind when writing an *energy balance* on a volume element is (Fig. 5-27)

$$\sum_{\text{all sides}} \dot{Q} + \dot{g} V_{\text{element}} = 0 \qquad (5\text{-}36)$$

whether the problem is one-, two-, or three-dimensional. Again we assume, for convenience in formulation, all heat transfer to be *into* the volume element from all surfaces except for specified heat flux, whose direction is already specified. This is demonstrated in the example below for various boundary conditions.

EXAMPLE 5-4 Steady Two-Dimensional Heat Conduction in L-Bars

Consider steady heat transfer in an L-shaped solid body whose cross-section is given in Figure 5-28. Heat transfer in the direction normal to the plane of the paper is negligible, and thus heat transfer in the body is two-dimensional. The thermal conductivity of the body is $k = 15$ W/m · °C, and heat is generated in the body at a rate of $\dot{g} = 2 \times 10^6$ W/m³. The left surface of the body is insulated, and the bottom surface is maintained at a uniform temperature of 90°C. The entire top surface is subjected to convection to ambient air at $T_\infty = 25$°C with a convection coefficient of $h = 80$ W/m² · °C, and the right surface is subjected to heat flux at a uniform rate of $\dot{q}_R = 5000$ W/m². The nodal network of the problem consists of 15 equally spaced nodes with $\Delta x = \Delta y = 1.2$ cm, as shown in the figure. Five of the nodes are at the bottom surface, and thus their temperatures are known. Obtain the finite difference equations at the remaining nine nodes and determine the nodal temperatures by solving them.

Solution   Heat transfer in a long L-shaped solid bar with specified boundary conditions is considered. The nine unknown nodal temperatures are to be determined with the finite difference method.

Assumptions   **1** Heat transfer is steady and two-dimensional, as stated. **2** Thermal conductivity is constant. **3** Heat generation is uniform. **4** Radiation heat transfer is negligible.

Properties   The thermal conductivity is given to be $k = 15$ W/m · °C.

Analysis   We observe that all nodes are boundary nodes except node 5, which is an interior node. Therefore, we will have to rely on energy balances to obtain the finite difference equations. But first we form the volume elements by partitioning the region among the nodes equitably by drawing dashed lines between the nodes. If we consider the volume element represented by an interior node to be *full size* (i.e., $\Delta x \times \Delta y \times 1$), then the element represented by a regular boundary node such as node 2 becomes *half size* (i.e., $\Delta x \times \Delta y/2 \times 1$), and a corner node such as node 1 is *quarter size* (i.e., $\Delta x/2 \times \Delta y/2 \times 1$). Keeping Equation 5-36 in mind for the energy balance, the finite difference equations for each of the nine nodes are obtained as follows:

(a) *Node 1*. The volume element of this corner node is insulated on the left and subjected to convection at the top and to conduction at the right and bottom surfaces. An energy balance on this element gives [Fig. 5-29a]

$$0 + h\frac{\Delta x}{2}(T_\infty - T_1) + k\frac{\Delta y}{2}\frac{T_2 - T_1}{\Delta x} + k\frac{\Delta x}{2}\frac{T_4 - T_1}{\Delta y} + \dot{g}_1\frac{\Delta x}{2}\frac{\Delta y}{2} = 0$$

Taking $\Delta x = \Delta y = l$, it simplifies to

$$-\left(2 + \frac{hl}{k}\right)T_1 + T_2 + T_4 = -\frac{hl}{k}T_\infty - \frac{\dot{g}_1 l^2}{2k}$$

(b) *Node 2*. The volume element of this boundary node is subjected to convection at the top and to conduction at the right, bottom, and left surfaces. An energy balance on this element gives [Fig. 5-29b]

$$h\Delta x(T_\infty - T_2) + k\frac{\Delta y}{2}\frac{T_3 - T_2}{\Delta x} + k\Delta x\frac{T_5 - T_2}{\Delta y} + k\frac{\Delta y}{2}\frac{T_1 - T_2}{\Delta x} + \dot{g}_2\Delta x\frac{\Delta y}{2} = 0$$

Taking $\Delta x = \Delta y = l$, it simplifies to

$$T_1 - \left(4 + \frac{2hl}{k}\right)T_2 + T_3 + 2T_5 = -\frac{2hl}{k}T_\infty - \frac{\dot{g}_2 l^2}{k}$$

(c) *Node 3*. The volume element of this corner node is subjected to convection at the top and right surfaces and to conduction at the bottom and left surfaces. An energy balance on this element gives [Fig. 5-30a]

$$h\left(\frac{\Delta x}{2} + \frac{\Delta y}{2}\right)(T_\infty - T_3) + k\frac{\Delta x}{2}\frac{T_6 - T_3}{\Delta y} + k\frac{\Delta y}{2}\frac{T_2 - T_3}{\Delta x} + \dot{g}_3\frac{\Delta x}{2}\frac{\Delta y}{2} = 0$$

Taking $\Delta x = \Delta y = l$, it simplifies to

$$T_2 - \left(2 + \frac{2hl}{k}\right)T_3 + T_6 = -\frac{2hl}{k}T_\infty - \frac{\dot{g}_1 l^2}{2k}$$

(d) *Node 4*. This node is on the insulated boundary and can be treated as an interior node by replacing the insulation by a mirror. This puts a reflected image of node 5 to the left of node 4. Noting that $\Delta x = \Delta y = l$, the general interior node relation for the steady two-dimensional case (Eq. 5-35) gives [Fig. 5-30b]

$$T_5 + T_1 + T_5 + T_{10} - 4T_4 + \frac{\dot{g}_4 l^2}{k} = 0$$

or, noting that $T_{10} = 90°$ C,

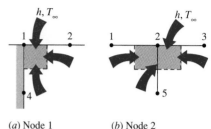

(a) Node 1   (b) Node 2

FIGURE 5-29

Schematics for energy balances on the volume elements of nodes 1 and 2.

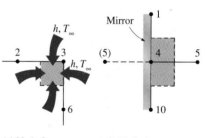

(a) Node 3   (b) Node 4

FIGURE 5-30

Schematics for energy balances on the volume elements of nodes 3 and 4.

$$T_1 - 4T_4 + 2T_5 = -90 - \frac{\dot{g}_4 l^2}{k}$$

(e) Node 5. This is an interior node, and noting that $\Delta x = \Delta y = l$, the finite difference formulation of this node is obtained directly from Equation 5-35 to be [Fig. 5-31a]

$$T_4 + T_2 + T_6 + T_{11} - 4T_5 + \frac{\dot{g}_5 l^2}{k} = 0$$

or, noting that $T_{11} = 90°C$,

$$T_2 + T_4 - 4T_5 + T_6 = -90 - \frac{\dot{g}_4 l^2}{k}$$

(f) Node 6. The volume element of this inner corner node is subjected to convection at the L-shaped exposed surface and to conduction at other surfaces. An energy balance on this element gives [Fig. 5-31b]

$$h\left(\frac{\Delta x}{2} + \frac{\Delta y}{2}\right)(T_\infty - T_6) + k\frac{\Delta y}{2}\frac{T_7 - T_6}{\Delta x} + k\Delta x\frac{T_{12} - T_6}{\Delta y}$$
$$+ k\Delta y\frac{T_5 - T_6}{\Delta x} + k\frac{\Delta x}{2}\frac{T_3 - T_6}{\Delta y} + \dot{g}_6\frac{3\Delta x \Delta y}{4} = 0$$

Taking $\Delta x = \Delta y = l$ and noting that $T_{12} = 90°C$, it simplifies to

$$T_3 + 2T_5 - \left(6 + \frac{2hl}{k}\right)T_6 + T_7 = -180 - \frac{2hl}{k}T_\infty - \frac{3\dot{g}_6 l^2}{2k}$$

(g) Node 7. The volume element of this boundary node is subjected to convection at the top and to conduction at the right, bottom, and left surfaces. An energy balance on this element gives [Fig. 5-32a]

$$h\Delta x(T_\infty - T_7) + k\frac{\Delta y}{2}\frac{T_8 - T_7}{\Delta x} + k\Delta x\frac{T_{13} - T_7}{\Delta y}$$
$$+ k\frac{\Delta y}{2}\frac{T_6 - T_7}{\Delta x} + \dot{g}_7\Delta x\frac{\Delta y}{2} = 0$$

Taking $\Delta x = \Delta y = l$ and noting that $T_{13} = 90°C$, it simplifies to

$$T_6 - \left(4 + \frac{2hl}{k}\right)T_7 + T_8 = -180 - \frac{2hl}{k}T_\infty - \frac{\dot{g}_7 l^2}{k}$$

(h) Node 8. This node is identical to Node 7, and the finite difference formulation of this node can be obtained from that of Node 7 by shifting the node numbers by 1 (i.e., replacing subscript m by $m + 1$). It gives

$$T_7 - \left(4 + \frac{2hl}{k}\right)T_8 + T_9 = -180 - \frac{2hl}{k}T_\infty - \frac{\dot{g}_8 l^2}{k}$$

(i) Node 9. The volume element of this corner node is subjected to convection at the top surface, to heat flux at the right surface, and to conduction at the bottom and left surfaces. An energy balance on this element gives [Fig. 5-32b]

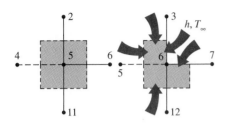

(a) Node 5   (b) Node 6

FIGURE 5-31

Schematics for energy balances on the volume elements of nodes 5 and 6.

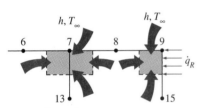

FIGURE 5-32

Schematics for energy balances on the volume elements of nodes 7 and 9.

$$h\frac{\Delta x}{2}(T_\infty - T_9) + \dot{q}_R\frac{\Delta y}{2} + k\frac{\Delta x}{2}\frac{T_{15}-T_9}{\Delta y} + k\frac{\Delta y}{2}\frac{T_8-T_9}{\Delta x} + \dot{g}_9\frac{\Delta x}{2}\frac{\Delta y}{2} = 0$$

Taking $\Delta x = \Delta y = l$ and noting that $T_{15} = 90°C$, it simplifies to

$$T_8 - \left(2 + \frac{hl}{k}\right)T_9 = -90 - \frac{\dot{q}_R l}{k} - \frac{hl}{k}T_\infty - \frac{\dot{g}_9 l^2}{2k}$$

This completes the development of finite difference formulation for this problem. Substituting the given quantities, the system of nine equations for the determination of nine unknown nodal temperatures becomes

$$-2.064T_1 + T_2 + T_4 = -11.2$$
$$T_1 - 4.128T_2 + T_3 + 2T_5 = -22.4$$
$$T_2 - 2.128T_3 + T_6 = -12.8$$
$$T_1 - 4T_4 + 2T_5 = -109.2$$
$$T_2 + T_4 - 4T_5 + T_6 = -109.2$$
$$T_3 + 2T_5 - 6.128T_6 + T_7 = -212.0$$
$$T_6 - 4.128T_7 + T_8 = -202.4$$
$$T_7 - 4.128T_8 + T_9 = -202.4$$
$$T_8 - 2.064T_9 = -105.2$$

which is a system of nine algebraic equations with nine unknowns. Starting with an initial guess of 90°C for all unknowns and using the computer program given in Figure 5-33, the solution accurate to the first decimal place is determined after 20 iterations with Gauss-Seidel iteration method to be

$$T_1 = 112.1°C \qquad T_2 = 110.8°C \qquad T_3 = 106.6°C$$
$$T_4 = 109.4°C \qquad T_5 = 108.1°C \qquad T_6 = 103.2°C$$
$$T_7 = \ \ 97.3°C \qquad T_8 = \ \ 96.3°C \qquad T_9 = \ \ 97.6°C$$

Note that the temperature is the highest at node 1 and the lowest at node 8. This is consistent with our expectations since node 1 is the farthest away from the bottom surface, which is maintained at 90°C and has one side insulated, and node 8 has the largest exposed area relative to its volume while being close to the surface at 90°C.

```
PROGRAM GAUSS
DATA T1,T2,T3,T4,T5/5*90./
DATA T6,T7,T8,T9/4*90./
DO 20 K=1,40
T1=(T2+T4+11.2)/2.064
T2=(T1+T3+2*T5+22.4)/4.128
T3=(T2+T6+12.8)/2.128
T4=(T1+2*T5+109.2)/4.
T5=(T2+T4+T6+109.2)/4.
T6=(T3+2*T5+T7+212)/6.128
T7=(T6+T8+202.4)/4.128
T8=(T7+T9+202.4)/4.128
T9=(T8+105.2)/2.064
WRITE (*,10)K,T1,T2,
1T3,T4,T5,T6,T7,T8,T9
10 FORMAT (' ', I3, 9(F8.1))
20 CONTINUE
END
```

FIGURE 5-33

The complete FORTRAN program used to solve the nine equations in Example 5-4 with the Gauss-Seidel iteration method.

Irregular Boundaries

In problems with simple geometries, we can fill the entire region using simple volume elements such as strips for a plane wall and rectangular elements for two-dimensional conduction in a rectangular region. We can also use cylindrical or spherical shell elements to cover the cylindrical and spherical bodies entirely. However, many geometries encountered in practice such as turbine blades or engine blocks do not have simple shapes, and it is difficult to fill such geometries having irregular boundaries with simple volume elements. A practical way of dealing with such geometries is to replace the irregular

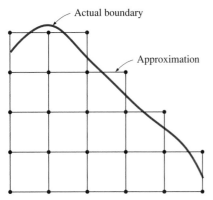

FIGURE 5-34

Approximating an irregular boundary
with a rectangular mesh.

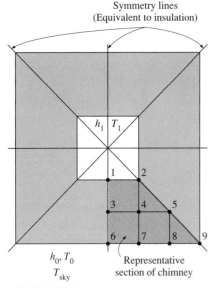

FIGURE 5-35

Schematic of the chimney discussed in
Example 5-5 and the nodal network for
a representative section.

geometry by a series of simple volume elements, as shown in Figure 5-34. This simple approach is often satisfactory for practical purposes, especially when the nodes are closely spaced near the boundary. More sophisticated approaches are available for handling irregular boundaries, and they are commonly incorporated into the commercial software packages.

EXAMPLE 5-5   Heat Loss through Chimneys

Hot combustion gases of a furnace are flowing through a square chimney made of concrete ($k = 1.4$ W/m · °C). The flow section of the chimney is 20 cm × 20 cm, and the thickness of the wall is 20 cm. The average temperature of the hot gases in the chimney is $T_i = 300$°C, and the average convection heat transfer coefficient inside the chimney is $h_i = 70$ W/m² · °C. The chimney is losing heat from its outer surface to the ambient air at $T_o = 20$°C by convection with a heat transfer coefficient of $h_o = 21$ W/m² · °C and to the sky by radiation. The emissivity of the outer surface of the wall is $\varepsilon = 0.9$, and the effective sky temperature is estimated to be 260 K. Using the finite difference method with $\Delta x = \Delta y = 10$ cm and taking full advantage of symmetry, determine the temperatures at the nodal points of a cross section and the rate of heat loss for a 1-m-long section of the chimney.

Solution   Heat transfer through a square chimney is considered. The nodal temperatures and the rate of heat loss per unit length are to be determined with the finite difference method.

Assumptions   **1** Heat transfer is steady since there is no indication of change with time. **2** Heat transfer through the chimney is two-dimensional since the height of the chimney is large relative to its cross-section, and thus heat conduction through the chimney in the axial direction is negligible. It is tempting to simplify the problem further by considering heat transfer in each wall to be one-dimensional, which would be the case if the walls were thin and thus the corner effects were negligible. This assumption cannot be justified in this case since the walls are very thick and the corner sections constitute a considerable portion of the chimney structure. **3** Thermal conductivity is constant.

Properties   The properties of chimney are given to be $k = 1.4$ W/m · °C and $\varepsilon = 0.9$.

Analysis   The cross-section of the chimney is given in Figure 5-35. The most striking aspect of this problem is the apparent symmetry about the horizontal and vertical lines passing through the midpoint of the chimney as well as the diagonal axes, as indicated on the figure. Therefore, we need to consider only one-eighth of the geometry in the solution whose nodal network consists of nine equally spaced nodes.
    No heat can cross a symmetry line, and thus symmetry lines can be treated as insulated surfaces and thus "mirrors" in the finite-difference formulation. Then the nodes in the middle of the symmetry lines can be treated as interior nodes by using mirror images. But still six of the nodes are boundary nodes, so we will have to write energy balances to obtain their finite difference formulations. But first we partition the region among the nodes equitably by drawing dashed lines between the nodes through the middle. Then the region around a node surrounded by the boundary or the dashed lines represents the volume element of the node. Considering a unit depth and using the energy balance approach for the boundary nodes (again assuming all heat transfer into the volume element

for convenience) and the formula for the interior nodes, the finite difference equations for the nine nodes are determined as follows:

(a) *Node 1.* On the inner boundary, subjected to convection, Figure 5-36a

$$0 + h_i \frac{\Delta x}{2}(T_i - T_1) + k \frac{\Delta y}{2} \frac{T_2 - T_1}{\Delta x} + k \frac{\Delta x}{2} \frac{T_3 - T_1}{\Delta y} + 0 = 0$$

Taking $\Delta x = \Delta y = l$, it simplifies to

$$-\left(2 + \frac{h_i l}{k}\right) T_1 + T_2 + T_3 = -\frac{h_i l}{k} T_i$$

(b) *Node 2.* On the inner boundary, subjected to convection, Figure 5-36b

$$k \frac{\Delta y}{2} \frac{T_1 - T_2}{\Delta x} + h_i \frac{\Delta x}{2}(T_i - T_2) + 0 + k\Delta x \frac{T_4 - T_2}{\Delta y} = 0$$

Taking $\Delta x = \Delta y = l$, it simplifies to

$$T_1 - \left(3 + \frac{h_i l}{k}\right) T_2 + 2T_4 = -\frac{h_i l}{k} T_i$$

(c) *Nodes 3, 4, and 5.* (Interior nodes, Fig. 5-37)

$$\text{Node 3:} \quad T_4 + T_1 + T_4 + T_6 - 4T_3 = 0$$
$$\text{Node 4:} \quad T_3 + T_2 + T_5 + T_7 - 4T_4 = 0$$
$$\text{Node 5:} \quad T_4 + T_4 + T_8 + T_8 - 4T_5 = 0$$

(d) *Node 6.* (On the outer boundary, subjected to convection and radiation)

$$0 + k \frac{\Delta x}{2} \frac{T_3 - T_6}{\Delta y} + k \frac{\Delta y}{2} \frac{T_7 - T_6}{\Delta x}$$
$$+ h_o \frac{\Delta x}{2}(T_o - T_6) + \varepsilon\sigma \frac{\Delta x}{2}(T_{sky}^4 - T_6^4) = 0$$

Taking $\Delta x = \Delta y = l$, it simplifies to

$$T_2 + T_3 - \left(2 + \frac{h_o l}{k}\right) T_6 = -\frac{h_o l}{k} T_o - \frac{\varepsilon\sigma l}{k}(T_{sky}^4 - T_6^4)$$

(e) *Node 7.* (On the outer boundary, subjected to convection and radiation, Fig. 5-38)

$$k \frac{\Delta y}{2} \frac{T_6 - T_7}{\Delta x} + k\Delta x \frac{T_4 - T_7}{\Delta y} + k \frac{\Delta y}{2} \frac{T_8 - T_7}{\Delta x}$$
$$+ h_o\Delta x(T_o - T_7) + \varepsilon\sigma\Delta x(T_{sky}^4 - T_7^4) = 0$$

Taking $\Delta x = \Delta y = l$, it simplifies to

$$2T_4 + T_6 - \left(4 + \frac{2h_o l}{k}\right) T_7 + T_8 = -\frac{2h_o l}{k} T_o - \frac{2\varepsilon\sigma l}{k}(T_{sky}^4 - T_7^4)$$

(f) *Node 8.* Same as Node 7, except shift the node numbers up by 1 (replace 4 by 5, 6 by 7, 7 by 8, and 8 by 9 in the above relation)

$$2T_5 + T_7 - \left(4 + \frac{2h_o l}{k}\right) T_8 + T_9 = -\frac{2h_o l}{k} T_o - \frac{2\varepsilon\sigma l}{k}(T_{sky}^4 - T_8^4)$$

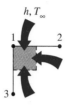

(a) Node 1   (b) Node 2

FIGURE 5-36

Schematics for energy balances on the volume elements of nodes 1 and 2.

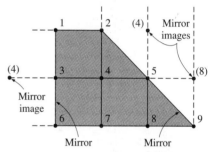

FIGURE 5-37

Converting the boundary nodes 3 and 5 on symmetry lines to interior nodes by using mirror images.

FIGURE 5-38

Schematics for energy balances on the volume elements of nodes 7 and 9.

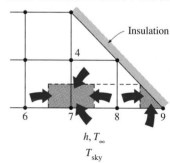

(g) Node 9. (On the outer boundary, subjected to convection and radiation, Fig. 5-38)

$$k\frac{\Delta y}{2}\frac{T_8 - T_9}{\Delta x} + 0 + h_o\frac{\Delta x}{2}(T_o - T_9) + \varepsilon\sigma\frac{\Delta x}{2}(T_{sky}^4 - T_9^4) = 0$$

Taking $\Delta x = \Delta y = l$, it simplifies to

$$T_8 - \left(1 + \frac{h_o l}{k}\right)T_9 = -\frac{h_o l}{k}T_o - \frac{\varepsilon\sigma l}{k}(T_{sky}^4 - T_9^4)$$

This problem involves radiation, which requires the use of absolute temperature, and thus all temperatures should be expressed in Kelvin. Alternately, we could use °C for all temperatures provided that the four temperatures in the radiation terms are expressed in the form $(T + 273)^4$. Substituting the given quantities, the system of nine equations for the determination of nine unknown nodal temperatures in a form suitable for use with the Gauss-Seidel iteration method becomes

$$T_1 = (T_2 + T_3 + 2865)/7$$
$$T_2 = (T_1 + 2T_4 + 2865)/8$$
$$T_3 = (T_1 + 2T_4 + T_6)/4$$
$$T_4 = (T_2 + T_3 + T_5 + T_7)/4$$
$$T_5 = (2T_4 + 2T_8)/4$$
$$T_6 = (T_2 + T_3 + 456.2 - 0.3645 \times 10^{-9}\ T_6^4)/3.5$$
$$T_7 = (2T_4 + T_6 + T_8 + 912.4 - 0.729 \times 10^{-9}\ T_7^4)/7$$
$$T_8 = (2T_5 + T_7 + T_9 + 912.4 - 0.729 \times 10^{-9}\ T_8^4)/7$$
$$T_9 = (T_8 + 456.2 - 0.3645 \times 10^{-9}\ T_9^4)/2.5$$

which is a system of *nonlinear* equations. Starting with initial guesses of 573 K at the inner boundary nodes, 293 K at the outer boundary nodes, and 400 K at the interior nodes for the Gauss-Seidel iteration, the solution, accurate to the first decimal place, is determined after seven iterations to be

$T_1 = 545.7$ K $= 272.6$°C	$T_2 = 529.2$ K $= 256.1$°C	$T_3 = 425.2$ K $= 152.1$°C
$T_4 = 411.2$ K $= 138.0$°C	$T_5 = 362.1$ K $= 89.0$°C	$T_6 = 332.9$ K $= 59.7$°C
$T_7 = 328.1$ K $= 54.9$°C	$T_8 = 313.1$ K $= 39.9$°C	$T_9 = 296.5$ K $= 23.4$°C

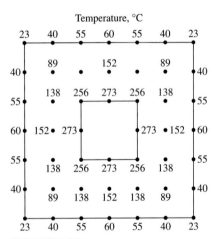

Temperature, °C

FIGURE 5-39

The variation of temperature in the chimney.

The variation of temperature in the chimney is shown in Figure 5-39.

Note that we can solve even nonlinear systems of equations easily with the Gauss-Seidel iteration method. The Gaussian elimination method cannot be used in this case since it is limited to linear equations. Also note that the temperatures are highest at the inner wall (but less than 300°C) and lowest at the outer wall (but more that 260 K), as expected.

The average temperature at the outer surface of the chimney weighed by the surface area is

$$T_{wall, out} = \frac{(0.5T_6 + T_7 + T_8 + 0.5T_9)}{(0.5 + 1 + 1 + 0.5)}$$

$$= \frac{0.5 \times 332.8 + 328.1 + 313.1 + 0.5 \times 296.5}{3} = 318.6\ K$$

Then the rate of heat loss through the 1-m-long section of the chimney can be determined approximately from

$$\dot{Q}_{\text{chimney}} = h_o A_o (T_{\text{wall, out}} - T_o) + \varepsilon \sigma A_o (T^4_{\text{wall, out}} - T^4_{\text{sky}})$$

$$= (21 \text{ W/m}^2 \cdot \text{K})[4 \times (0.6 \text{ m})(1 \text{ m})](318.6 - 293)\text{K}$$

$$+ 0.9(5.67 \times 10^{-8} \text{ W/m}^2 \cdot \text{K}^4)[4 \times (0.6 \text{ m})(1 \text{ m})](318.6 \text{ K})^4 - (260 \text{ K})^4]$$

$$= 1291 + 702 = \mathbf{1993 \text{ W}}$$

We could also determine the heat transfer by finding the average temperature of the inner wall, which is $(272.6 + 256.1)/2 = 264.4°\text{C}$, and applying Newton's law of cooling at that surface:

$$\dot{Q}_{\text{chimney}} = h_i A_i (T_i - T_{\text{wall, in}})$$

$$= (70 \text{ W/m}^2 \cdot \text{K})[4 \times (0.2 \text{ m})(1 \text{ m})](300 - 264.4)°\text{C} = 1994 \text{ W}$$

The difference between the two results is due to the approximate nature of the numerical analysis.

Discussion We used a relatively crude numerical model to solve this problem to keep the complexities at a manageable level. The accuracy of the solution obtained can be improved by using a finer mesh and thus a greater number of nodes. Also, when radiation is involved, it is more accurate (but more laborious) to determine the heat losses for each node and add them up instead of using the average temperature.

5-6 ■ TRANSIENT HEAT CONDUCTION

So far in this chapter we have applied the finite difference method to *steady* heat transfer problems. In this section we extend the method to solve *transient* problems.

We applied the finite difference method to steady problems by *discretizing* the problem in the space variables and solving for temperatures at *discrete* points called the nodes. The solution obtained is valid for any time since under steady conditions the temperatures do not change with time. In transient problems, however, the temperatures change with time as well as position, and thus the finite difference solution of transient problems requires *discretization in time* in addition to discretization in space, as shown in Figure 5-40. This is done by selecting a suitable time step Δt and solving for the unknown nodal temperatures repeatedly for each Δt until the solution at the desired time is obtained. For example, consider a hot metal object that is taken out of the oven at an initial temperature of T_i at time $t = 0$ and is allowed to cool in ambient air. If a time step of $\Delta t = 5$ min is chosen, the determination of the temperature distribution in the metal piece after 3 h requires the determination of the temperatures $3 \times 60/5 = 36$ times, or in 36 time steps. Therefore, the computation time of this problem will be 36 times that of a steady problem. Choosing a smaller Δt will increase the accuracy of the solution, but it will also increase the computation time.

In transient problems, the *superscript i* is used as the *index* or *counter* of time steps, with $i = 0$ corresponding to the specified initial condition. In the

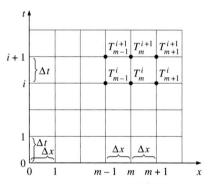

FIGURE 5-40

Finite difference formulation of time-dependent problems involves discrete points in time as well as space.

case of the hot metal piece discussed above, $i = 1$ corresponds to $t = 1 \times \Delta t = 5$ min, $i = 2$ corresponds to $t = 2 \times \Delta t = 10$ min, and a general time step i corresponds to $t_i = i\Delta t$. The notation T_m^i is used to represent the temperature at the node m at time step i.

The formulation of transient heat conduction problems differs from that of steady ones in that the transient problems involve an *additional term* representing the *change in the energy content* of the medium with time. This additional term appears as a first derivative of temperature with respect to time in the differential equation, and as a change in the internal energy content during Δt in the energy balance formulation. The nodes and the volume elements in transient problems are selected as they are in the steady case, and, again assuming all heat transfer is *into* the element for convenience, the energy balance on a volume element during a time interval Δt can be expressed as

$$
\begin{pmatrix}
\text{Heat transferred into} \\
\text{the volume element} \\
\text{from all of its surfaces} \\
\text{during } \Delta t
\end{pmatrix}
+
\begin{pmatrix}
\text{Heat generated} \\
\text{within the} \\
\text{volume element} \\
\text{during } \Delta t
\end{pmatrix}
=
\begin{pmatrix}
\text{The change in the} \\
\text{energy content of} \\
\text{the volume element} \\
\text{during } \Delta t
\end{pmatrix}
$$

or

$$
\Delta t \times \sum_{\text{All sides}} \dot{Q} + \Delta t \times \dot{G}_{\text{element}} = \Delta E_{\text{element}} \tag{5-37}
$$

where the rate of heat transfer $\dot{Q}$ normally consists of conduction terms for interior nodes, but may involve convection, heat flux, and radiation for boundary nodes.

Noting that $\Delta E_{\text{element}} = mC\Delta T = \rho V_{\text{element}} C\Delta T$, where ρ is density and C is the specific heat of the element, dividing the relation above by Δt gives

$$
\sum_{\text{All sides}} \dot{Q} + \dot{G}_{\text{element}} = \frac{\Delta E_{\text{element}}}{\Delta t} = \rho V_{\text{element}} C \frac{\Delta T}{\Delta t} \tag{5-38}
$$

or, for any node m in the medium and its volume element,

$$
\sum_{\text{All sides}} \dot{Q} + \dot{G}_{\text{element}} = \rho V_{\text{element}} C \frac{T_m^{i+1} - T_m^i}{\Delta t} \tag{5-39}
$$

where T_m^i and T_m^{i+1} are the temperatures of node m at times $t_i = i\Delta t$ and $t_{i+1} = (i+1)\Delta t$, respectively, and $T_m^{i+1} - T_m^i$ represents the temperature change of the node during the time interval Δt between the time steps i and $i + 1$ (Fig. 5-41).

Note that the ratio $(T_m^{i+1} - T_m^i)/\Delta t$ is simply the finite difference approximation of the partial derivative $\partial T/\partial t$ that appears in the differential equations of transient problems. Therefore, we would obtain the same result for the finite difference formulation if we followed a strict mathematical approach instead of the energy balance approach used above. Also note that

FIGURE 5-41

The change in the energy content of the volume element of a node during a time interval Δt.

Volume element
(can be any shape)

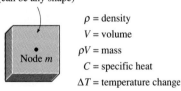

ρ = density
V = volume
ρV = mass
C = specific heat
ΔT = temperature change

$\Delta U = \rho V C \Delta T = \rho V C (T_m^{i+1} - T_m^i)$

Node m

the finite difference formulations of steady and transient problems differ by the single term on the right side of the equal sign, and the format of that term remains the same in all coordinate systems regardless of whether heat transfer is one-, two-, or three-dimensional. For the special case of $T_m^{i+1} = T_m^i$ (i.e., no change in temperature with time), the formulation reduces to that of steady case, as expected.

The nodal temperatures in transient problems normally change during each time step, and you may be wondering whether to use temperatures at the *previous* time step i or the *new* time step $i + 1$ for the terms on the left side of Equation 5-39. Well, both are reasonable approaches and both are used in practice. The finite difference approach is called the **explicit method** in the first case and the **implicit method** in the second case, and they are expressed in the general form as (Fig. 5-42)

$$\text{Explicit method:} \quad \sum_{\text{All sides}} \dot{Q}^i + \dot{G}_{\text{element}}^i = \rho V_{\text{element}} C \frac{T_m^{i+1} - T_m^i}{\Delta t} \quad (5\text{-}40)$$

$$\text{Implicit method:} \quad \sum_{\text{All sides}} \dot{Q}^{i+1} + \dot{G}_{\text{element}}^{i+1} = \rho V_{\text{element}} C \frac{T_m^{i+1} - T_m^i}{\Delta t} \quad (5\text{-}41)$$

It appears that the time derivative is expressed in *forward difference* form in the explicit case and *backward difference* form in the implicit case. Of course, it is also possible to mix the two fundamental formulations above and come up with more elaborate formulations, but such formulations offer little insight and are beyond the scope of this text. Note that both formulations are simply expressions between the nodal temperatures before and after a time interval and are based on determining the new temperatures T_m^{i+1} using the *previous* temperatures T_m^i. *The explicit and implicit formulations given above are quite general and can be used in any coordinate system regardless of the dimension of heat transfer.* The volume elements in multidimensional cases simply have more surfaces and thus involve more terms in the summation.

The explicit and implicit methods have their advantages and disadvantages, and one method is not necessarily better than the other one. Below you will see that the *explicit method* is easy to implement but imposes a limit on the allowable time step to avoid instabilities in the solution, and the *implicit method* requires the nodal temperatures to be solved simultaneously for each time step but imposes no limit on the magnitude of the time step. We will limit the discussion to one- and two-dimensional cases to keep the complexities at a manageable level, but the analysis can readily be extended to three-dimensional cases and other coordinate systems.

Transient Heat Conduction in a Plane Wall

Consider transient one-dimensional heat conduction in a plane wall of thickness L with heat generation $\dot{g}(x, t)$ that may vary with time and position and

If expressed at $i + 1$: Implicit method

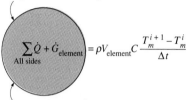

$$\sum_{\text{All sides}} \dot{Q} + \dot{G}_{\text{element}} = \rho V_{\text{element}} C \frac{T_m^{i+1} - T_m^i}{\Delta t}$$

If expressed at i: Explicit method

FIGURE 5-42

The formulation of explicit and implicit methods differs at the time step (previous or new) at which the heat transfer and heat generation terms are expressed.

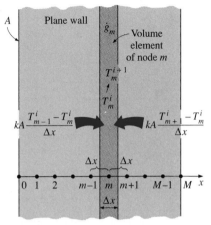

FIGURE 5-43

The nodal points and volume elements for the transient finite difference formulation of one-dimensional conduction in a plane wall.

constant conductivity k with a mesh size of $\Delta x = L/M$ and nodes $0, 1, 2, \ldots$, M in the x-direction, as shown in Figure 5-43. Noting that the volume element of a general interior node m involves heat conduction from two sides and the volume of the element is $V_{\text{element}} = A\Delta x$, the transient finite difference formulation for an interior node can be expressed on the basis of Equation 5-39 as

$$kA \frac{T_{m-1} - T_m}{\Delta x} + kA \frac{T_{m+1} - T_m}{\Delta x} + \dot{g}_m A\Delta x = \rho A\Delta x C \frac{T_m^{i+1} - T_m^i}{\Delta t} \quad (5\text{-}42)$$

Canceling the surface area A and multiplying by $\Delta x/k$, it simplifies to

$$T_{m-1} - 2T_m + T_{m+1} + \frac{\dot{g}_m \Delta x^2}{k} = \frac{\Delta x^2}{\alpha \Delta t}(T_m^{i+1} - T_m^i) \quad (5\text{-}43)$$

where $\alpha = k/\rho C$ is the *thermal diffusivity* of the wall material. We now define a dimensionless **mesh Fourier number** as

$$\tau = \frac{\alpha \Delta t}{\Delta x^2} \quad (5\text{-}44)$$

Then Equation 5-43 reduces to

$$T_{m-1} - 2T_m + T_{m+1} + \frac{\dot{g}_m \Delta x^2}{k} = \frac{T_m^{i+1} - T_m^i}{\tau} \quad (5\text{-}45)$$

Note that the left side of this equation is simply the finite difference formulation of the problem for the steady case. This is not surprising since the formulation must reduce to the steady case for $T_m^{i+1} = T_m^i$. Also, we are still not committed to explicit or implicit formulation since we did not indicate the time step on the left side of the equation. We now obtain the *explicit* finite difference formulation by expressing the left side at time step i as

$$T_{m-1}^i - 2T_m^i + T_{m+1}^i + \frac{\dot{g}_m^i \Delta x^2}{k} = \frac{T_m^{i+1} - T_m^i}{\tau} \quad \text{(explicit)} \quad (5\text{-}46)$$

This equation can be solved *explicitly* for the new temperature T_m^{i+1} (and thus the name *explicit* method) to give

$$T_m^{i+1} = \tau(T_{m-1}^i + T_{m+1}^i) + (1 - 2\tau)T_m^i + \tau \frac{\dot{g}_m^i \Delta x^2}{k} \quad (5\text{-}47)$$

for all interior nodes $m = 1, 2, 3, \ldots, M - 1$ in a plane wall. Expressing the left side of Equation 5-45 at time step $i + 1$ instead of i would give the *implicit* finite difference formulation as

$$T_{m-1}^{i+1} - 2T_m^{i+1} + T_{m+1}^{i+1} + \frac{\dot{g}_m^{i+1} \Delta x^2}{k} = \frac{T_m^{i+1} - T_m^i}{\tau} \quad \text{(implicit)} \quad (5\text{-}48)$$

which can be rearranged as

$$\tau T_{m-1}^{i+1} - (1 + 2\tau) T_m^{i+1} + \tau T_{m+1}^{i+1} + \tau \frac{\dot{g}_m^{i+1} \Delta x^2}{k} + T_m^i = 0 \quad (5\text{-}49)$$

The application of either explicit or implicit formulation above to each of the $M - 1$ interior nodes gives $M - 1$ equations. The remaining two equations are obtained by applying the same method to the two boundary nodes unless, of course, the boundary temperatures are specified as constants (invariant with time). For example, the formulation of the convection boundary condition at the left boundary (node 0) for the explicit case can be expressed as (Fig. 5-44)

$$hA(T_\infty - T_0^i) + kA \frac{T_1^i - T_0^i}{\Delta x} + \dot{g}_0^i A \frac{\Delta x}{2} = \rho A \frac{\Delta x}{2} C \frac{T_0^{i+1} - T_0^i}{\Delta t} \quad (5\text{-}50)$$

which simplifies to

$$T_0^{i+1} = \left(1 - 2\tau - 2\tau \frac{h\Delta x}{k}\right) T_0^i + 2\tau T_1^i + 2\tau \frac{h\Delta x}{k} T_\infty + \tau \frac{\dot{g}_0^i \Delta x}{k} \quad (5\text{-}51)$$

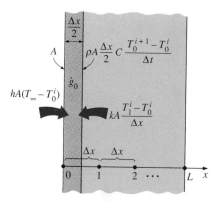

FIGURE 5-44

Schematic for the explicit finite difference formulation of the convection condition at the left boundary of a plane wall.

Note that in the case of no heat generation and $\tau = 0.5$, the explicit finite difference formulation for a general interior node reduces to $T_m^{i+1} = (T_{m-1}^i + T_{m-1}^i)/2$, which has the interesting interpretation that *the temperature of an interior node at the new time step is simply the average of the temperatures of its neighboring nodes at the previous time step.*

Once the formulation (explicit or implicit) is complete and the initial condition is specified, the solution of a transient problem is obtained by *marching* in time using a step size of Δt as follows: select a suitable time step Δt and determine the nodal temperatures from the initial condition. Taking the initial temperatures as the *previous* solution T_m^i at $t = 0$, obtain the new solution T_m^{i+1} at all nodes at time $t = \Delta t$ using the transient finite difference relations. Now using the solution just obtained at $t = \Delta t$ as the *previous* solution T_m^i, obtain the new solution T_m^{i+1} at $t = 2\Delta t$ using the same relations. Repeat the process until the solution at the desired time is obtained.

Stability Criterion for Explicit Method: Limitation on Δt

The explicit method is easy to use, but it suffers from an undesirable feature that severely restricts its utility: the explicit method is not unconditionally stable, and the largest permissible value of the time step Δt is limited by the stability criterion. If the time step Δt is not sufficiently small, the solutions obtained by the explicit method may oscillate wildly and diverge from the actual solution. To avoid such divergent oscillations in nodal temperatures, the value of Δt must be maintained below a certain upper limit established by the **stability criterion.** It can be shown mathematically or by a physical argument based on the second law of thermodynamics that *the stability criterion is satisfied if the coefficients of all T_m^i in the T_m^{i+1} expressions (called the **primary coefficients**) are greater than or equal to zero for all nodes m*

(Fig. 5-45). Of course, all the terms involving T_m^i for a particular node must be grouped together before this criterion is applied.

Different equations for different nodes may result in different restrictions on the size of the time step Δt, and the criterion that is most restrictive should be used in the solution of the problem. A practical approach is to identify the equation with the *smallest primary coefficient* since it is the most restrictive and to determine the allowable values of Δt by applying the stability criterion to that equation only. A Δt value obtained this way will also satisfy the stability criterion for all other equations in the system.

For example, in the case of transient one-dimensional heat conduction in a plane wall with specified surface temperatures, the explicit finite difference equations for all the nodes (which are *interior nodes*) are obtained from Equation 5-47. The coefficient of T_m^i in the T_m^{i+1} expression is $1 - 2\tau$, which is independent of the node number m, and thus the stability criterion for all nodes in this case is $1 - 2\tau \geq 0$ or

$$\tau = \frac{\alpha \Delta t}{\Delta x^2} \leq \frac{1}{2} \qquad \binom{\text{interior nodes, one-dimensional heat}}{\text{transfer in rectangular coordinates}} \qquad (5\text{-}52)$$

When the material of the medium and thus its thermal diffusivity α is known and the value of the mesh size Δx is specified, the largest allowable value of the time step Δt can be determined from the relation above. For example, in the case of a brick wall ($\alpha = 0.45 \times 10^{-6}$ m²/s) with a mesh size of $\Delta x = 0.01$ m, the upper limit of the time step is

$$\Delta t \leq \frac{1}{2} \frac{\Delta x^2}{\alpha} = \frac{(0.01 \text{ m})^2}{2(0.45 \times 10^{-6} \text{ m}^2/\text{s})} = 111\text{s} = 1.85 \text{ min}$$

The boundary nodes involving convection and/or radiation are more restrictive than the interior nodes and thus require smaller time steps. Therefore, the most restrictive boundary node should be used in the determination of the maximum allowable time step Δt when a transient problem is solved with the explicit method.

To gain a better understanding of the stability criterion, consider the explicit finite difference formulation for an interior node of a plane wall (Eq. 5-47) for the case of no heat generation,

$$T_m^{i+1} = \tau(T_{m-1}^i + T_{m+1}^i) + (1 - 2\tau)T_m^i$$

Assume that at some time step i the temperatures T_{m-1}^i and T_{m+1}^i are equal but less than T_m^i (say, $T_{m-1}^i = T_{m+1}^i = 50°C$ and $T_m^i = 80°C$). At the next time step, we expect the temperature of node m to be between the two values (say, 70°C). However, if the value of τ exceeds 0.5 (say, $\tau = 1$), the temperature of node m at the next time step will be less than the temperature of the neighboring nodes (it will be 20°C), which is physically impossible and violates the second law of thermodynamics (Fig. 5-46). Requiring the new temperature of node m to remain above the temperature of the neighboring nodes is equivalent to requiring the value of τ to remain below 0.5.

Explicit formulation:

$$T_0^{i+1} = a_0 T_0^i + \cdots$$
$$T_1^{i+1} = a_1 T_1^i + \cdots$$
$$\vdots$$
$$T_m^{i+1} = a_m T_m^i + \cdots$$
$$\vdots$$
$$T_M^{i+1} = a_M T_M^i + \cdots$$

Stability criterion:

$$a_m \geq 0, \quad m = 0, 1, 2, \ldots m, \ldots M$$

FIGURE 5-45

The stability criterion of the explicit method requires all primary coefficients to be positive or zero.

FIGURE 5-46

The violation of the stability criterion in the explicit method may result in the violation of the second law of thermodynamics and thus divergence of solution.

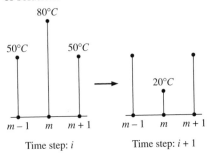

The implicit method is *unconditionally stable,* and thus we can use any time step we please with that method (of course, the smaller the time step, the better the accuracy of the solution). The disadvantage of the implicit method is that it results in a set of equations that must be solved *simultaneously* for each time step. Both methods are used in practice.

EXAMPLE 5-6 Transient Heat Conduction in a Large Uranium Plate

Consider a large uranium plate of thickness L = 4 cm, thermal conductivity k = 28 W/m · °C, and thermal diffusivity α = 12.5 × 10^{-6} m²/s that is initially at a uniform temperature of 200°C. Heat is generated uniformly in the plate at a constant rate of $\dot{g}$ = 5 × 10^6 W/m³. At time t = 0, one side of the plate is brought into contact with iced water and is maintained at 0°C at all times, while the other side is subjected to convection to an environment at T_∞ = 30°C with a heat transfer coefficient of h = 45 W/m² · °C, as shown in Figure 5-47. Considering a total of three equally spaced nodes in the medium, two at the boundaries and one at the middle, estimate the exposed surface temperature of the plate 2.5 min after the start of cooling using (a) the explicit method and (b) the implicit method.

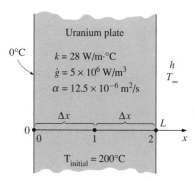

FIGURE 5-47
Schematic for Example 5-6.

Solution We have solved this problem in Example 5-1 for the steady case, and here we repeat it for the transient case to demonstrate the application of the transient finite difference methods. Again we assume one-dimensional heat transfer in rectangular coordinates and constant thermal conductivity. The number of nodes is specified to be M = 3, and they are chosen to be at the two surfaces of the plate and at the middle, as shown in the figure. Then the nodal spacing Δx becomes

$$\Delta x = \frac{L}{M-1} = \frac{0.04 \text{ m}}{3-1} = 0.02 \text{ m}$$

We number the nodes as 0, 1, and 2. The temperature at node 0 is given to be T_0 = 0°C at all times, and the temperatures at nodes 1 and 2 are to be determined. This problem involves only two unknown nodal temperatures, and thus we need to have only two equations to determine them uniquely. These equations are obtained by applying the finite difference method to nodes 1 and 2.

(a) Node 1 is an interior node, and the *explicit* finite difference formulation at that node is obtained directly from Equation 5-47 by setting m = 1:

$$T_1^{i+1} = \tau(T_0 + T_2^i) + (1 - 2\tau) T_1^i + \tau \frac{\dot{g}_1 \Delta x^2}{k} \qquad (1)$$

FIGURE 5-48
Schematic for the explicit finite difference formulation of the convection condition at the right boundary of a plane wall.

Node 2 is a boundary node subjected to convection, and the finite difference formulation at that node is obtained by writing an energy balance on the volume element of thickness $\Delta x/2$ at that boundary by assuming heat transfer to be into the medium at all sides (Fig. 5-48):

$$hA(T_\infty - T_2^i) + kA\frac{T_1^i - T_2^i}{\Delta x} + \dot{g}_2 A\frac{\Delta x}{2} = \rho A\frac{\Delta x}{2} C\frac{T_2^{i+1} - T_2^i}{\Delta t}$$

Dividing by $kA/2\Delta x$ and using the definitions of thermal diffusivity α = $k/\rho C$ and the dimensionless mesh Fourier number τ = $\alpha \Delta t/(\Delta x)^2$ give

$$\frac{2h\Delta x}{k}(T_\infty - T_2^i) + 2(T_1^i - T_2^i) + \frac{\dot{g}_2 \Delta x^2}{k} = \frac{T_2^{i+1} - T_2^i}{\tau}$$

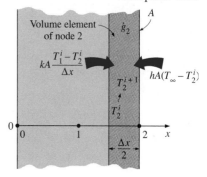

which can be solved for T_2^{i+1} to give

$$T_2^{i+1} = \left(1 - 2\tau - 2\tau \frac{h\Delta x}{k}\right) T_2^i + \tau\left(2T_1^i + 2\frac{h\Delta x}{k} T_\infty + \frac{\dot{g}_2 \Delta x^2}{k}\right) \quad (2)$$

Note that we did not use the superscript i for quantities that do not change with time. Next we need to determine the upper limit of the time step Δt from the stability criterion, which requires the coefficient of T_1^i in Equation 1 and the coefficient of T_2^i in the second equation to be greater than or equal to zero. The coefficient of T_2^i is smaller in this case, and thus the stability criterion for this problem can be expressed as

$$1 - 2\tau - 2\tau \frac{h\Delta x}{k} \geq 0 \quad \rightarrow \quad \tau \leq \frac{1}{2(1 + h\Delta x/k)} \quad \rightarrow \quad \Delta t \leq \frac{\Delta x^2}{2\alpha(1 + h\Delta x/k)}$$

since $\tau = \alpha \Delta t/(\Delta x)^2$. Substituting the given quantities, the maximum allowable value of the time step is determined to be

$$\Delta t \leq \frac{(0.02\ \text{m})^2}{2(12.5 \times 10^{-6}\ \text{m}^2/\text{s})[1 + (45\ \text{W/m}^2 \cdot °\text{C})(0.02\ \text{m})/28\ \text{W/m} \cdot °\text{C}]} = 15.5\ \text{s}$$

Therefore, any time step less than 15.5 s can be used to solve this problem. For convenience, let us choose the time step to be $\Delta t = 15$ s. Then the mesh Fourier number becomes

$$\tau = \frac{\alpha \Delta t}{(\Delta x)^2} = \frac{(12.5 \times 10^{-6}\ \text{m}^2/\text{s})(15\ \text{s})}{(0.02\ \text{m})^2} = 0.46875 \quad (\text{for } \Delta t = 15\ \text{s})$$

Substituting this value of τ and other given quantities, the explicit finite difference equations (1) and (2) developed above reduce to

$$T_1^{i+1} = 0.0625 T_1^i + 0.46875 T_2^i + 33.482$$
$$T_2^{i+1} = 0.9375 T_1^i + 0.032366 T_2^i + 34.386$$

The initial temperature of the medium at $t = 0$ and $i = 0$ is given to be 200°C throughout, and thus $T_1^0 = T_2^0 = 200$°C. Then the nodal temperatures at T_1^1 and T_2^1 at $t = \Delta t = 15$ s are determined from the equations above to be

$$T_1^1 = 0.0625 T_1^0 + 0.46875 T_2^0 + 33.482$$
$$= 0.0625 \times 200 + 0.46875 \times 200 + 33.482 = 139.7°\text{C}$$
$$T_2^1 = 0.9375 T_1^0 + 0.032366 T_2^0 + 34.386$$
$$= 0.9375 \times 200 + 0.032366 \times 200 + 34.386 = 228.4°\text{C}$$

Similarly, the nodal temperatures T_1^2 and T_2^2 at $t = 2\Delta t = 2 \times 15 = 30$ s are determined to be

$$T_1^2 = 0.0625 T_1^1 + 0.46875 T_2^1 + 33.482$$
$$= 0.0625 \times 139.7 + 0.46875 \times 228.4 + 33.482 = 149.3°\text{C}$$
$$T_2^2 = 0.9375 T_1^1 + 0.032366 T_2^1 + 34.386$$
$$= 0.9375 \times 139.7 + 0.032366 \times 228.4 + 34.386 = 172.8°\text{C}$$

Continuing in the same manner and using the program in Figure 5-49, the temperatures at nodes 1 and 2 are determined for $i = 1, 2, 3, 4, 5, \ldots, 50$ and are given in Table 5-3. Therefore, the temperature at the exposed boundary surface 2.5 min after the start of cooling is

$$T_L^{2.5\ \text{min}} = T_2^{10} = 139.0°\text{C}$$

```
PROGRAM EXPLICIT
DIMENSION TNEW(2),TOLD(2)
N=2
DO 5 I=1,N
TOLD (I)=200
5 CONTINUE
DO 20 K=1,50
TNEW(1)=0.0625*TOLD(1)+
+0.46875*TOLD(2)+33.482
TNEW(2)=0.032366*TOLD(2)+
+0.9375*TOLD(1)+34.386
WRITE(9,10)K,(TNEW(L),L=1,N)
10 FORMAT (' ',I3,9(F8.1))
DO 25 J=1,N
TOLD (J)=TNEW (J)
25 CONTINUE
20 CONTINUE
END
```

FIGURE 5-49

The complete FORTRAN program used to solve the transient problem in Example 5-6 using the explicit method.

(b) Node 1 is an interior node, and the *implicit* finite difference formulation at that node is obtained directly from Equation 5-49 by setting $m = 1$:

$$\tau T_0 - (1 + 2\tau)\, T_1^{i+1} + \tau T_2^{i+1} + \tau \frac{\dot{g}_0\, \Delta x^2}{k} + T_1^i = 0 \qquad (3)$$

Node 2 is a boundary node subjected to convection, and the implicit finite difference formulation at that node can be obtained from the explicit formulation above by expressing the left side of the equation at time step $i + 1$ instead of i as

$$\frac{2h\Delta x}{k}(T_\infty - T_2^{i+1}) + 2(T_1^{i+1} - T_2^{i+1}) + \frac{\dot{g}_2\, \Delta x^2}{k} = \frac{T_2^{i+1} - T_2^i}{\tau}$$

which can be rearranged as

$$2\tau T_1^{i+1} - \left(1 + 2\tau + 2\tau \frac{h\Delta x}{k}\right) T_2^{i+1} + 2\tau \frac{h\Delta x}{k} T_\infty + \tau \frac{\dot{g}_2\, \Delta x^2}{k} + T_2^i = 0 \qquad (4)$$

Again we did not use the superscript i or $i + 1$ for quantities that do not change with time. The implicit method imposes no limit on the time step, and thus we can choose any value we want. However, we will again choose $\Delta t = 15$ s, and thus $\tau = 0.46875$, to make a comparison with part (a) possible. Substituting this value of τ and other given quantities, the two implicit finite difference equations developed above reduce to

$$-1.9375 T_1^{i+1} + 0.46875 T_2^{i+1} + T_1^i + 33.482 = 0$$
$$0.9375 T_1^{i+1} - 1.9676 T_2^{i+1} + T_2^i + 34.386 = 0$$

Again $T_1^0 = T_2^0 = 200°C$ at $t = 0$ and $i = 0$ because of the initial condition, and for $i = 0$, the two equations above reduce to

$$-1.9375 T_1^1 + 0.46875 T_2^1 + 200 + 33.482 = 0$$
$$0.9375 T_1^1 - 1.9676 T_2^1 + 200 + 34.386 = 0$$

The unknown nodal temperatures T_1^1 and T_2^1 at $t = \Delta t = 15$ s are determined by solving the two equations above simultaneously to be

$$T_1^1 = 168.8°C \qquad \text{and} \qquad T_2^1 = 199.6°C$$

Similarly, for $i = 1$, the equations above reduce to

$$-1.9375 T_1^2 + 0.46875 T_2^2 + 168.8 + 33.482 = 0$$
$$0.9375 T_1^2 - 1.9676 T_2^2 + 199.6 + 34.386 = 0$$

The unknown nodal temperatures T_1^2 and T_2^2 at $t = \Delta t = 2 \times 15 = 30$ s are determined by solving the two equations above simultaneously to be

$$T_1^2 = 150.5°C \qquad \text{and} \qquad T_2^2 = 190.6°C$$

Continuing in this manner, the temperatures at nodes 1 and 2 are determined for $i = 2, 3, 4, 5, \ldots, 40$ and are listed in Table 5-4, and the temperature at the exposed boundary surface (node 2) 2.5 min after the start of cooling is obtained to be

$$T_L^{2.5\ min} = T_2^{10} = 143.9°C$$

which is close to the result obtained by the explicit method. Note that either method could be used to obtain satisfactory results to transient problems, except, perhaps, for the first few time steps. The implicit method is preferred when it is

TABLE 5-3

The variation of the nodal temperatures in Example 5-6 with time obtained by the *explicit* method

Time step, i	Time, s	Node temperature, °C	
		T_1^i	T_2^i
0	0	200.0	200.0
1	15	139.7	228.4
2	30	149.3	172.8
3	45	123.8	179.9
4	60	125.6	156.3
5	75	114.6	157.1
6	90	114.3	146.9
7	105	109.5	146.3
8	120	108.9	141.8
9	135	106.7	141.1
10	150	106.3	139.0
20	300	103.8	136.1
30	450	103.7	136.0
40	600	103.7	136.0

TABLE 5-4

The variation of the nodal temperatures in Example 5-6 with time obtained by the *implicit* method

Time step, i	Time, s	Node temperature, °C	
		T_1^i	T_2^i
0	0	200.0	200.0
1	15	168.8	199.6
2	30	150.5	190.6
3	45	138.6	180.4
4	60	130.3	171.2
5	75	124.1	163.6
6	90	119.5	157.6
7	105	115.9	152.8
8	120	113.2	149.0
9	135	111.0	146.1
10	150	109.4	143.9
20	300	104.2	136.7
30	450	103.8	136.1
40	600	103.8	136.1

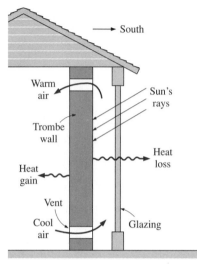

FIGURE 5-50

Schematic of a Trombe wall
(Example 5-7).

TABLE 5-5

**The hourly variation of monthly
average ambient temperature and
solar heat flux incident on a vertical
surface for January in Reno, Nevada**

Time of day	Ambient temper- ature, °F	Solar ra- diation, Btu/h · ft²
7 AM–10 AM	33	114
10 AM–1 PM	43	242
1 PM–4 PM	45	178
4 PM–7 PM	37	0
7 PM–10 PM	32	0
10 PM–1 AM	27	0
1 AM–4 AM	26	0
4 AM–7 AM	25	0

desirable to use large time steps, and the explicit method is preferred when one wishes to avoid the simultaneous solution of a system of algebraic equations.

EXAMPLE 5-7 Solar Energy Storage in Trombe Walls

Dark painted thick masonry walls called Trombe walls are commonly used on south sides of passive solar homes to absorb solar energy, store it during the day, and release it to the house during the night (Fig. 5-50). The idea was proposed by E. L. Morse of Massachusetts in 1881 and is named after Professor Felix Trombe of France, who used it extensively in his designs in the 1970s. Usually a single or double layer of glazing is placed outside the wall and transmits most of the solar energy while blocking heat losses from the exposed surface of the wall to the outside. Also, air vents are commonly installed at the bottom and top of the Trombe walls so that the house air enters the parallel flow channel between the Trombe wall and the glazing, rises as it is heated, and enters the room through the top vent.

Consider a house in Reno, Nevada, whose south wall consists of a 1-ft-thick Trombe wall whose thermal conductivity is $k = 0.40$ Btu/h · ft · °F and whose thermal diffusivity is $\alpha = 4.78 \times 10^{-6}$ ft²/s. The variation of the ambient temperature T_{out} and the solar heat flux $\dot{q}_{solar}$ incident on a south-facing vertical surface throughout the day for a typical day in January is given in Table 5-5 in 3-h intervals. The Trombe wall has single glazing with an absorptivity-transmissivity product of $\kappa = 0.77$ (that is, 77 percent of the solar energy incident is absorbed by the exposed surface of the Trombe wall), and the average combined heat transfer coefficient for heat loss from the Trombe wall to the ambient is determined to be $h_{out} = 0.7$ Btu/h · ft² · °F. The interior of the house is maintained at $T_{in} = 70$°F at all times, and the heat transfer coefficient at the interior surface of the Trombe wall is $h_{in} = 1.8$ Btu/h · ft² · °F. Also, the vents on the Trombe wall are kept closed, and thus the only heat transfer between the air in the house and the Trombe wall is through the interior surface of the wall. Assuming the temperature of the Trombe wall to vary linearly between 70°F at the interior surface and 30°F at the exterior surface at 7 AM and using the explicit finite difference method with a uniform nodal spacing of $\Delta x = 0.2$ ft, determine the temperature distribution along the thickness of the Trombe wall after 12, 24, 36, and 48 hours. Also, determine the net amount of heat transferred to the house from the Trombe wall during the first day and the second day. Assume the wall is 10 ft high and 25 ft long.

Solution The passive solar heating of a house through a Trombe wall is considered. The temperature distribution in the wall in 12 h intervals and the amount of heat transfer during the first and second days are to be determined.

Assumptions 1 Heat transfer is one-dimensional since the exposed surface of the wall is large relative to its thickness. 2 Thermal conductivity is constant. 3 The heat transfer coefficients are constant.

Properties The wall properties are given to be $k = 0.40$ Btu/h · ft · °F, $\alpha = 4.78 \times 10^{-6}$ ft²/s, and $\kappa = 0.77$.

Analysis The nodal spacing is given to be $\Delta x = 0.2$ ft, and thus the total number of nodes along the Trombe wall is

$$M = \frac{L}{\Delta x} + 1 = \frac{1 \text{ ft}}{0.2 \text{ ft}} + 1 = 6$$

We number the nodes as 0, 1, 2, 3, 4, and 5, with node 0 on the interior surface of the Trombe wall and node 5 on the exterior surface, as shown in Figure 5-51. Nodes 1 through 4 are interior nodes, and the explicit finite difference formulations of these nodes are obtained directly from Equation 5-47 to be

Node 1 ($m = 1$): $T_1^{i+1} = \tau(T_0^i + T_2^i) + (1 - 2\tau)T_1^i$ (1)

Node 2 ($m = 2$): $T_2^{i+1} = \tau(T_1^i + T_3^i) + (1 - 2\tau)T_2^i$ (2)

Node 3 ($m = 3$): $T_3^{i+1} = \tau(T_2^i + T_4^i) + (1 - 2\tau)T_3^i$ (3)

Node 4 ($m = 4$): $T_4^{i+1} = \tau(T_3^i + T_5^i) + (1 - 2\tau)T_4^i$ (4)

The interior surface is subjected to convection, and thus the explicit formulation of node 0 can be obtained directly from Equation 5-51 to be

$$T_0^{i+1} = \left(1 - 2\tau - 2\tau \frac{h_{in} \Delta x}{k}\right)T_0^i + 2\tau T_1^i + 2\tau \frac{h_{in} \Delta x}{k} T_{in}$$

Substituting the quantities h_{in}, Δx, k, and T_{in}, which do not change with time, into the equation above gives

$$T_0^{i+1} = (1 - 3.80\tau)\, T_0^i + \tau(2T_1^i + 126.0)$$ (5)

The exterior surface of the Trombe wall is subjected to convection as well as to heat flux. The explicit finite difference formulation at that boundary is obtained by writing an energy balance on the volume element represented by node 5,

$$h_{out} A(T_{out}^i - T_5^i) + \kappa A \dot{q}_{solar}^i + kA \frac{T_4^i - T_5^i}{\Delta x} = \rho A \frac{\Delta x}{2} C \frac{T_5^{i+1} - T_5^i}{\Delta t}$$ (5-53)

which simplifies to

$$T_5^{i+1} = \left(1 - 2\tau - 2\tau \frac{h_{out} \Delta x}{k}\right) T_5^i + 2\tau T_4^i + 2\,\tau \frac{h_{out} \Delta x}{k} T_{out}^i + 2\tau \frac{\kappa \dot{q}_{solar}^i \Delta x}{k}$$ (5-54)

where $\tau = \alpha \Delta t / \Delta x^2$ is the dimensionless mesh Fourier number. Note that we kept the superscript i for quantities that vary with time. Substituting the quantities h_{out}, Δx, k, and κ, which do not change with time, into the equation above gives

$$T_5^{i+1} = (1 - 2.70\tau)\, T_5^i + \tau(2T_4^i + 0.70T_{out}^i + 0.770\dot{q}_{solar}^i)$$ (6)

where the unit of $\dot{q}_{solar}^i$ is Btu/h · ft^2.

Next we need to determine the upper limit of the time step Δt from the stability criterion since we are using the explicit method. This requires the identification of the smallest primary coefficient in the system. We know that the boundary nodes are more restrictive than the interior nodes, and thus we examine the formulations of the boundary nodes 0 and 5 only. The smallest and thus the most restrictive primary coefficient in this case is the coefficient of T_0^i in the formulation of node 0 since $1 - 3.8\tau < 1 - 2.7\tau$, and thus the stability criterion for this problem can be expressed as

$$1 - 3.80\tau \geq 0 \quad \rightarrow \quad \tau = \frac{\alpha \Delta t}{\Delta x^2} \leq \frac{1}{3.80}$$

Substituting the given quantities, the maximum allowable value of the time step is determined to be

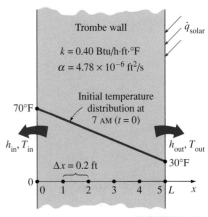

FIGURE 5-51

The nodal network for the Trombe wall discussed in Example 5-7.

Trombe wall

$\dot{q}_{solar}$

$k = 0.40$ Btu/h·ft·°F
$\alpha = 4.78 \times 10^{-6}$ ft^2/s

Initial temperature distribution at 7 AM ($t = 0$)

70°F

h_{in}, T_{in}

h_{out}, T_{out}

30°F

$\Delta x = 0.2$ ft

0 1 2 3 4 5 L x

$$\Delta t \le \frac{\Delta x^2}{3.80\alpha} = \frac{(0.2 \text{ ft})^2}{3.80 \times (4.78 \times 10^{-6} \text{ ft}^2/\text{s})} = 2202 \text{ s}$$

Therefore, any time step less than 2202 s can be used to solve this problem. For convenience, let us choose the time step to be $\Delta t = 900$ s $= 15$ min. Then the mesh Fourier number becomes

$$\tau = \frac{\alpha \Delta t}{(\Delta x)^2} = \frac{(4.78 \times 10^{-6} \text{ ft}^2/\text{s})(900 \text{ s})}{(0.2 \text{ ft})^2} = 0.10755 \qquad \text{(for } \Delta t = 15 \text{ min)}$$

Initially (at 7 AM or $t = 0$), the temperature of the wall is said to vary linearly between 70°F at node 0 and 30°F at node 5. Noting that there are five nodal spacings of equal length, the temperature change between two neighboring nodes is $(70 - 30)°\text{F}/5 = 8°\text{F}$. Therefore, the initial nodal temperatures are

$$T_0^0 = 70°\text{F}, \qquad T_1^0 = 62°\text{F}, \qquad T_2^0 = 54°\text{F},$$
$$T_3^0 = 46°\text{F}, \qquad T_4^0 = 38°\text{F}, \qquad T_5^0 = 30°\text{F}$$

Then the nodal temperatures at $t = \Delta t = 15$ min (at 7:15 AM) are determined from the equations above to be

$$T_0^1 = (1 - 3.80\tau) T_0^0 + \tau(2T_1^0 + 126.0)$$
$$= (1 - 3.80 \times 0.10755) 70 + 0.10755(2 \times 62 + 126.0) = 68.3° \text{ F}$$
$$T_1^1 = \tau(T_0^0 + T_2^0) + (1 - 2\tau) T_1^0$$
$$= 0.10755(70 + 54) + (1 - 2 \times 0.10755)62 = 62°\text{F}$$
$$T_2^1 = \tau(T_1^0 + T_3^0) + (1 - 2\tau) T_2^0$$
$$= 0.10755(62 + 46) + (1 - 2 \times 0.10755)54 = 54°\text{F}$$
$$T_3^1 = \tau(T_2^0 + T_4^0) + (1 - 2\tau) T_3^0$$
$$= 0.10755(54 + 38) + (1 - 2 \times 0.10755)46 = 46°\text{F}$$
$$T_4^1 = \tau(T_3^0 + T_5^0) + (1 - 2\tau) T_4^0$$
$$= 0.10755(46 + 30) + (1 - 2 \times 0.10755)38 = 38°\text{F}$$
$$T_5^1 = (1 - 2.70\tau) T_5^0 + \tau(2T_4^0 + 0.70T_{\text{out}}^0 + 0.770\dot{q}_{\text{solar}}^0)$$
$$= (1 - 2.70 \times 0.10755)30 + 0.10755(2 \times 38 + 0.70 \times 33 + 0.770 \times 114)$$
$$= 41.4°\text{F}$$

FIGURE 5-52

The variation of temperatures in the Trombe wall discussed in Example 5-7.

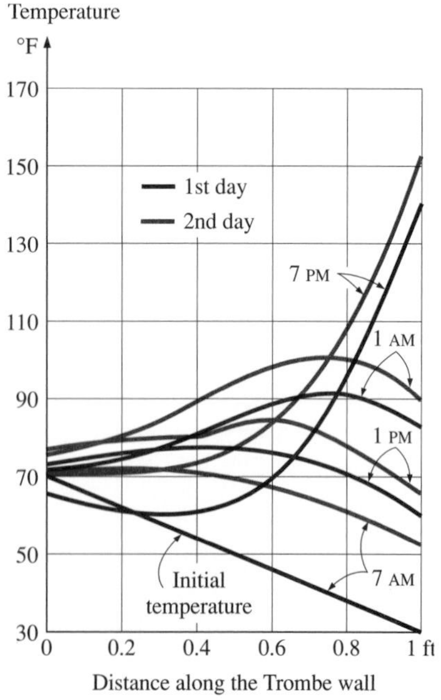

Temperature °F, with legend "1st day", "2nd day"; y-axis from 30 to 170; x-axis "Distance along the Trombe wall" from 0 to 1 ft; labels 7 PM, 1 AM, 1 PM, 7 AM, Initial temperature.

Note that the inner surface temperature of the Trombe wall dropped by 1.7°F and the outer surface temperature rose by 11.4°F during the first time step while the temperatures at the interior nodes remained the same. This is typical of transient problems in mediums that involve no heat generation. The nodal temperatures at the following time steps are determined similarly with the help of a computer. Note that the data for ambient temperature and the incident solar radiation change every 3 hours, which corresponds to 12 time steps, and this must be reflected in the computer program. For example, the value of $\dot{q}_{\text{solar}}^i$ must be taken to be $\dot{q}_{\text{solar}}^i = 75$ for $i = 1\text{--}12$, $\dot{q}_{\text{solar}}^i = 242$ for $i = 13\text{--}24$, $\dot{q}_{\text{solar}}^i = 178$ for $i = 25\text{--}36$, and $\dot{q}_{\text{solar}}^i = 0$ for $i = 37\text{--}96$.

The results after 6, 12, 18, 24, 30, 36, 42, and 48 h are given in Table 5-6 and are plotted in Figure 5-52 for the first day. Note that the interior temperature of the Trombe wall drops in early morning hours, but then rises as the solar energy absorbed by the exterior surface diffuses through the wall. The exterior surface temperature of the Trombe wall rises from 30 to 142°F in just 6 h because of the solar energy absorbed, but then drops to 53°F by next morning as a result of heat

TABLE 5-6

The temperatures at the nodes of a Trombe wall at various times

Time	Time step, i	Nodal temperatures, °F					
		T_0	T_1	T_2	T_3	T_4	T_5
0 h (7 AM)	0	70.0	62.0	54.0	46.0	38.0	30.0
6 h (1 PM)	24	65.3	61.7	61.5	69.7	94.1	142.0
12 h (7 PM)	48	71.6	74.2	80.4	88.4	91.7	82.4
18 h (1 AM)	72	73.3	75.9	77.4	76.3	71.2	61.2
24 h (7 AM)	96	71.2	71.9	70.9	67.7	61.7	53.0
30 h (1 PM)	120	70.3	71.1	74.3	84.2	108.3	153.2
36 h (7 PM)	144	75.4	81.1	89.4	98.2	101.0	89.7
42 h (1 AM)	168	75.8	80.7	83.5	83.0	77.4	66.2
48 h (7 AM)	192	73.0	75.1	72.2	66.0	66.0	56.3

loss at night. Therefore, it may be worthwhile to cover the outer surface at night to minimize the heat losses.

The rate of heat transfer from the Trombe wall to the interior of the house during each time step is determined from Newton's law using the average temperature at the inner surface of the wall (node 0) as

$$Q_{\text{Trombe wall}}^i = \dot{Q}_{\text{Trombe wall}}^i \Delta t = h_{\text{in}} A(T_0^i - T_{\text{in}}) \Delta t = h_{\text{in}} A[(T_0^i + T_0^{i-1})/2 - T_{\text{in}}]\Delta t$$

Therefore, the amount of heat transfer during the first time step ($i = 1$) or during the first 15-min period is

$$
\begin{aligned}
Q_{\text{Trombe wall}}^1 &= h_{\text{in}} A[(T_0^1 + T_0^0)/2 - T_{\text{in}}] \Delta t \\
&= (1.8 \text{ Btu/h} \cdot \text{ft}^2 \cdot °\text{F})(10 \times 25 \text{ ft}^2)[(68.3 + 70)/2 - 70°\text{F}](0.25 \text{ h}) \\
&= -95.6 \text{ Btu}
\end{aligned}
$$

The negative sign indicates that heat is transferred to the Trombe wall from the air in the house, which represents a heat loss. Then the total heat transfer during a specified time period is determined by adding the heat transfer amounts for each time step as

$$Q_{\text{Trombe wall}} = \sum_{i=1}^{I} \dot{Q}_{\text{Trombe wall}}^i = \sum_{i=1}^{I} h_{\text{in}} A[(T_0^i + T_0^{i-1})/2 - T_{\text{in}}] \Delta t \quad (5\text{-}55)$$

where I is the total number of time intervals in the specified time period. In this case $I = 48$ for 12 h, 96 for 24 h, and so on. Following the approach described above using a computer, the amount of heat transfer between the Trombe wall and the interior of the house is determined to be

$Q_{\text{Trombe wall}} = -17,048$ Btu after 12 h   ($-17,078$ Btu during the first 12 h)

$Q_{\text{Trombe wall}} = -2483$ Btu after 24 h   (14,565 Btu during the second 12 h)

$Q_{\text{Trombe wall}} = 5610$ Btu after 36 h   (8093 Btu during the third 12 h)

$Q_{\text{Trombe wall}} = 34,400$ Btu after 48 h   (28,790 Btu during the fourth 12 h)

Therefore, the house loses 2483 Btu through the Trombe wall the first day as a result of the low start-up temperature but delivers a total of 36,883 Btu of heat to the house the second day. It can be shown that the Trombe wall will deliver even more heat to the house during the third day since it will start the day at a higher average temperature.

Two-Dimensional Transient Heat Conduction

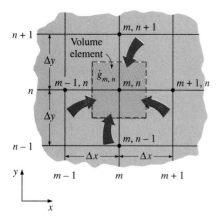

FIGURE 5-53
The volume element of a general interior node (m, n) for two-dimensional transient conduction in rectangular coordinates.

Consider a rectangular region in which heat conduction is significant in the x- and y-directions, and consider a unit depth of $\Delta z = 1$ in the z-direction. Heat may be generated in the medium at a rate of $\dot{g}(x, y, t)$, which may vary with time and position, with the thermal conductivity k of the medium assumed to be constant. Now divide the x-y-plane of the region into a *rectangular mesh* of nodal points spaced Δx and Δy apart in the x- and y-directions, respectively, and consider a general interior node (m, n) whose coordinates are $x = m\Delta x$ and $y = n\Delta y$, as shown in Figure 5-53. Noting that the volume element centered about the general interior node (m, n) involves heat conduction from four sides (right, left, top, and bottom) and the volume of the element is $V_{\text{element}} = \Delta x \times \Delta y \times 1 = \Delta x \Delta y$, the transient finite difference formulation for a general interior node can be expressed on the basis of Equation 5-39 as

$$k\Delta y \frac{T_{m-1,n} - T_{m,n}}{\Delta x} + k\Delta x \frac{T_{m,n+1} - T_{m,n}}{\Delta y} + k\Delta y \frac{T_{m+1,n} - T_{m,n}}{\Delta x}$$
$$+ k\Delta x \frac{T_{m,n-1} - T_{m,n}}{\Delta y} + \dot{g}_{m,n} \Delta x \Delta y = \rho \Delta x \Delta y\, C\, \frac{T_m^{i+1} - T_m^i}{\Delta t} \quad (5\text{-}56)$$

Taking a square mesh ($\Delta x = \Delta y = l$) and dividing each term by k gives after simplifying,

$$T_{m-1,n} + T_{m+1,n} + T_{m,n+1} + T_{m,n-1} - 4T_{m,n} + \frac{\dot{g}_{m,n} l^2}{k} = \frac{T_m^{i+1} - T_m^i}{\tau} \quad (5\text{-}57)$$

where again $\alpha = k/\rho C$ is the thermal diffusivity of the material and $\tau = \alpha \Delta t/l^2$ is the dimensionless mesh Fourier number. It can also be expressed in terms of the temperatures at the neighboring nodes in the following easy-to-remember form:

$$T_{\text{left}} + T_{\text{top}} + T_{\text{right}} + T_{\text{bottom}} - 4T_{\text{node}} + \frac{\dot{g}_{\text{node}} l^2}{k} = \frac{T_{\text{node}}^{i+1} - T_{\text{node}}^i}{\tau} \quad (5\text{-}58)$$

Again the left side of this equation is simply the finite difference formulation of the problem for the *steady case*, as expected. Also, we are still not committed to explicit or implicit formulation since we did not indicate the time step on the left side of the equation. We now obtain the *explicit* finite difference formulation by expressing the left side at time step i as

$$T^i_{\text{left}} + T^i_{\text{top}} + T^i_{\text{right}} + T^i_{\text{bottom}} - 4T^i_{\text{node}} + \frac{\dot{g}^i_{\text{node}} l^2}{k} = \frac{T^{i+1}_{\text{node}} - T^i_{\text{node}}}{\tau} \quad (5\text{-}59)$$

Expressing the left side at time step $i + 1$ instead of i would give the implicit formulation. The equation above can be solved *explicitly* for the new temperature T^{i+1}_{node} to give

$$T^{i+1}_{\text{node}} = \tau(T^i_{\text{left}} + T^i_{\text{top}} + T^i_{\text{right}} + T^i_{\text{bottom}}) + (1 - 4\tau) T^i_{\text{node}} + \tau \frac{\dot{g}^i_{\text{node}} l^2}{k}$$

$$(5\text{-}60)$$

for all interior nodes (m, n) where $m = 1, 2, 3, \ldots, M - 1$ and $n = 1, 2, 3, \ldots, N - 1$ in the medium. In the case of no heat generation and $\tau = \frac{1}{4}$, the explicit finite difference formulation for a general interior node reduces to $T^{i+1}_{\text{node}} = (T^i_{\text{left}} + T^i_{\text{top}} + T^i_{\text{right}} + T^i_{\text{bottom}})/4$, which has the interpretation that *the temperature of an interior node at the new time step is simply the average of the temperatures of its neighboring nodes at the previous time step* (Fig. 5-54).

The stability criterion that requires the coefficient of T^i_m in the T^{i+1}_m expression to be greater than or equal to zero for all nodes is equally valid for two- or three-dimensional cases and severely limits the size of the time step Δt that can be used with the explicit method. In the case of transient two-dimensional heat transfer in rectangular coordinates, the coefficient of T^i_m in the T^{i+1}_m expression is $1 - 4\tau$, and thus the stability criterion for all interior nodes in this case is $1 - 4\tau > 0$, or

$$\tau = \frac{\alpha \Delta t}{l^2} \leq \frac{1}{4} \qquad \begin{array}{l}\text{(interior nodes, two-dimensional heat} \\ \text{transfer in rectangular coordinates)}\end{array} \quad (5\text{-}61)$$

where $\Delta x = \Delta y = l$. When the material of the medium and thus its thermal diffusivity α are known and the value of the mesh size l is specified, the largest allowable value of the time step Δt can be determined from the relation above. Again the boundary nodes involving convection and/or radiation are more restrictive than the interior nodes and thus require smaller time steps. Therefore, the most restrictive boundary node should be used in the determination of the maximum allowable time step Δt when a transient problem is solved with the explicit method.

The application of Equation 5-60 to each of the $(M - 1) \times (N - 1)$ interior nodes gives $(M - 1) \times (N - 1)$ equations. The remaining equations are obtained by applying the method to the boundary nodes unless, of course, the boundary temperatures are specified as being constant. The development of the transient finite difference formulation of boundary nodes in two- (or three-) dimensional problems is similar to the development in the one-dimensional case discussed earlier. Again the region is partitioned between the nodes by forming volume elements around the nodes, and an energy balance is written for each boundary node on the basis of Equation 5-39. This is illustrated below with an example.

Time step i:

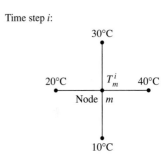

Time step $i + 1$:

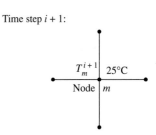

FIGURE 5-54

In the case of no heat generation and $\tau = \frac{1}{4}$, the temperature of an interior node at the new time step is the average of the temperatures of its neighboring nodes at the previous time step.

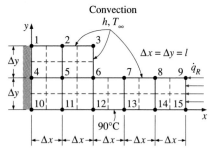

FIGURE 5-55

Schematic and nodal network for
Example 5-8.

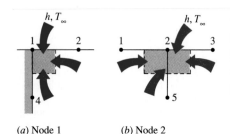

(a) Node 1   (b) Node 2

FIGURE 5-56

Schematics for energy balances on the
volume elements of nodes 1 and 2.

EXAMPLE 5-8   Transient Two-Dimensional Heat Conduction in L- Bars

Consider two-dimensional transient heat transfer in an L-shaped solid body
that is initially at a uniform temperature of 90°C and whose cross-section is
given in Figure 5-55. The thermal conductivity and diffusivity of the body are
$k = 15$ W/m·°C and $\alpha = 3.2 \times 10^{-6}$ m²/s, respectively, and heat is generated in
the body at a rate of $\dot{g} = 2 \times 10^6$ W/m³. The left surface of the body is insulated,
and the bottom surface is maintained at a uniform temperature of 90°C at all
times. At time $t = 0$, the entire top surface is subjected to convection to ambient
air at $T_\infty = 25$°C with a convection coefficient of h = 80 W/m²·°C, and the
right surface is subjected to heat flux at a uniform rate of $\dot{q}_R = 5000$ W/m². The
nodal network of the problem consists of 15 equally spaced nodes with
$\Delta x = \Delta y = 1.2$ cm, as shown in the figure. Five of the nodes are at the bottom
surface, and thus their temperatures are known. Using the explicit method, determine the temperature at the top corner (node 3) of the body after 1, 3, 5, 10, and
60 min.

Solution   This is a transient two-dimensional heat transfer problem in rectangular coordinates, and it was solved in Example 5-4 for the steady case. Therefore, the solution of this transient problem should approach the solution for the
steady case when the time is sufficiently large. The thermal conductivity and heat
generation rate are given to be constants. We observe that all nodes are boundary nodes except node 5, which is an interior node. Therefore, we will have to rely
on energy balances to obtain the finite difference equations. The region is partitioned among the nodes equitably as shown in the figure, and the explicit finite
difference equations are determined on the basis of the energy balance for the
transient case expressed as

$$\sum_{\text{All sides}} \dot{Q}^i + \dot{G}^i_{\text{element}} = \rho V_{\text{element}} C \frac{T_m^{i+1} - T_m^i}{\Delta t}$$

The quantities h, T_∞, $\dot{g}$, and $\dot{q}_R$ do not change with time, and thus we do not need
to use the superscript i for them. Also, the energy balance expressions are simplified using the definitions of thermal diffusivity $\alpha = k/\rho C$ and the dimensionless
mesh Fourier number $\tau = \alpha \Delta t / l^2$, where $\Delta x = \Delta y = l$.

(a) Node 1. (Boundary node subjected to convection and insulation, Fig. 5-56a)

$$h\frac{\Delta x}{2}(T_\infty - T_1^i) + k\frac{\Delta y}{2}\frac{T_2^i - T_1^i}{\Delta x} + k\frac{\Delta x}{2}\frac{T_4^i - T_1^i}{\Delta y} + \dot{g}_1\frac{\Delta x}{2}\frac{\Delta y}{2} = \rho\frac{\Delta x}{2}\frac{\Delta y}{2}C\frac{T_1^{i+1} - T_1^i}{\Delta t}$$

Dividing by $k/4$ and simplifying,

$$\frac{2hl}{k}(T_\infty - T_1^i) + 2(T_2^i - T_1^i) + 2(T_4^i - T_1^i) + \frac{\dot{g}_1 l^2}{k} = \frac{T_1^{i+1} - T_1^i}{\tau}$$

which can be solved for T_1^{i+1} to give

$$T_1^{i+1} = \left(1 - 4\tau - 2\tau\frac{hl}{k}\right)T_1^i + 2\tau\left(T_2^i + T_4^i + \frac{hl}{k}T_\infty + \frac{\dot{g}_1 l^2}{2k}\right)$$

(b) Node 2. (Boundary node subjected to convection, Fig. 5-56b)

$$h\Delta x(T_\infty - T_2^i) + k\frac{\Delta y}{2}\frac{T_3^i - T_2^i}{\Delta x} + k\Delta x\frac{T_5^i - T_2^i}{\Delta y}$$

$$+ k\frac{\Delta y}{2}\frac{T_1^i - T_2^i}{\Delta x} + \dot{g}_2\Delta x\frac{\Delta y}{2} = \rho\Delta x\frac{\Delta y}{2}C\frac{T_2^{i+1} - T_2^i}{\Delta t}$$

Dividing by $k/2$, simplifying, and solving for T_2^{i+1} give

$$T_2^{i+1} = \left(1 - 4\tau - 2\tau \frac{hl}{k}\right) T_2^i + \tau \left(T_1^i + T_3^i + 2T_5^i + \frac{2hl}{k} T_\infty + \frac{\dot{g}_2 l^2}{k}\right)$$

(c) *Node 3.* (Boundary node subjected to convection on two sides, Fig. 5-57a)

$$h\left(\frac{\Delta x}{2} + \frac{\Delta y}{2}\right)(T_\infty - T_3^i) + k\frac{\Delta x}{2}\frac{T_6^i - T_3^i}{\Delta y}$$

$$+ k\frac{\Delta y}{2}\frac{T_2^i - T_3^i}{\Delta x} + \dot{g}_3 \frac{\Delta x}{2}\frac{\Delta y}{2} = \rho \frac{\Delta x}{2}\frac{\Delta y}{2} C\frac{T_3^{i+1} - T_3^i}{\Delta t}$$

Dividing by $k/4$, simplifying, and solving for T_3^{i+1} give

$$T_3^{i+1} = \left(1 - 4\tau - 4\tau \frac{hl}{k}\right) T_3^i + 2\tau \left(T_4^i + T_6^i + 2\frac{hl}{k} T_\infty + \frac{\dot{g}_3 l^2}{2k}\right)$$

(d) *Node 4.* (On the insulated boundary, and can be treated as an interior node, Fig. 5-57b). Noting that $T_{10} = 90°C$, Equation 5-60 gives

$$T_4^{i+1} = (1 - 4\tau) T_4^i + \tau \left(T_1^i + 2T_5^i + 90 + \frac{\dot{g}_4 l^2}{k}\right)$$

(e) *Node 5.* (Interior node, Fig. 5-58a). Noting that $T_{11} = 90°C$, Equation 5-60 gives

$$T_5^{i+1} = (1 - 4\tau) T_5^i + \tau \left(T_2^i + T_4^i + T_6^i + 90 + \frac{\dot{g}_5 l^2}{k}\right)$$

(f) *Node 6.* (Boundary node subjected to convection on two sides, Fig. 5-58b)

$$h\left(\frac{\Delta x}{2} + \frac{\Delta y}{2}\right)(T_\infty - T_6^i) + k\frac{\Delta y}{2}\frac{T_7^i - T_6^i}{\Delta x} + k\Delta x\frac{T_{12}^i - T_6^i}{\Delta y} + k\Delta y\frac{T_5^i - T_6^i}{\Delta x}$$

$$+ k\frac{\Delta x}{2}\frac{T_3^i - T_6^i}{\Delta y} + \dot{g}_6 \frac{3\Delta x\Delta y}{4} = \rho \frac{3\Delta x\Delta y}{4} C\frac{T_6^{i+1} - T_6^i}{\Delta t}$$

Dividing by $3k/4$, simplifying, and solving for T_6^{i+1} give

$$T_6^{i+1} = \left(1 - 4\tau - 4\tau \frac{hl}{3k}\right) T_3^i + \frac{\tau}{3}\left[2T_3^i + 4T_5^i + 2T_7^i + 4 \times 90 + 4\frac{hl}{k} T_\infty + 3\frac{\dot{g}_6 l^2}{k}\right]$$

(g) *Node 7.* (Boundary node subjected to convection, Fig. 5-59a)

$$h\Delta x(T_\infty - T_7^i) + k\frac{\Delta y}{2}\frac{T_8^i - T_7^i}{\Delta x} + k\Delta x\frac{T_{13}^i - T_7^i}{\Delta y}$$

$$+ k\frac{\Delta y}{2}\frac{T_6^i - T_7^i}{\Delta x} + \dot{g}_7 \Delta x\frac{\Delta y}{2} = \rho \Delta x\frac{\Delta y}{2} C\frac{T_7^{i+1} - T_7^i}{\Delta t}$$

Dividing by $k/2$, simplifying, and solving for T_7^{i+1} give

$$T_7^{i+1} = \left(1 - 4\tau - 2\tau \frac{hl}{k}\right) T_7^i + \tau \left[T_6^i + T_8^i + 2 \times 90 + \frac{2hl}{k} T_\infty + \frac{\dot{g}_7 l^2}{k}\right]$$

(h) *Node 8.* This node is identical to node 7, and the finite difference formulation of this node can be obtained from that of node 7 by shifting the node numbers by 1 (i.e., replacing subscript m by subscript $m + 1$). It gives

$$T_8^{i+1} = \left(1 - 4\tau - 2\tau \frac{hl}{k}\right) T_8^i + \tau \left[T_7^i + T_9^i + 2 \times 90 + \frac{2hl}{k} T_\infty + \frac{\dot{g}_8 l^2}{k}\right]$$

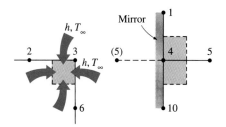

(a) Node 3   (b) Node 4

FIGURE 5-57

Schematics for energy balances on the volume elements of nodes 3 and 4.

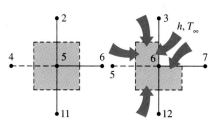

(a) Node 5   (b) Node 6

FIGURE 5-58

Schematics for energy balances on the volume elements of nodes 5 and 6.

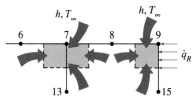

FIGURE 5-59

Schematics for energy balances on the volume elements of nodes 7 and 9.

(*i*) *Node 9.* (Boundary node subjected to convection on two sides, Fig. 5-59*b*)

$$h\frac{\Delta x}{2}(T_\infty - T_9^j) + \dot{q}_M\frac{\Delta y}{2} + k\frac{\Delta x}{2}\frac{T_{15}^j - T_9^j}{\Delta y}$$
$$+ \frac{k\Delta y}{2}\frac{T_8^j - T_9^j}{\Delta x} + \dot{g}_9\frac{\Delta x}{2}\frac{\Delta y}{2} = \rho\frac{\Delta x}{2}\frac{\Delta y}{2}C\frac{T_9^{j+1} - T_9^j}{\Delta t}$$

Dividing by $k/4$, simplifying, and solving for T_9^{j+1} give

$$T_9^{j+1} = \left(1 - 4\tau - 2\tau\frac{hl}{k}\right)T_9^j + 2\tau\left(T_8^j + 90 + \frac{\dot{q}_M l}{k} + \frac{hl}{k}T_\infty + \frac{\dot{g}_9 l^2}{2k}\right)$$

This completes the finite difference formulation of the problem. Next we need to determine the upper limit of the time step Δt from the stability criterion, which requires the coefficient of T_m^j in the T_m^{j+1} expression (the primary coefficient) to be greater than or equal to zero for all nodes. The smallest primary coefficient in the nine equations above is the coefficient of T_3^j in the expression, and thus the stability criterion for this problem can be expressed as

$$1 - 4\tau - 4\tau\frac{hl}{k} \geq 0 \quad \rightarrow \quad \tau \leq \frac{1}{4(1 + hl/k)} \quad \rightarrow \quad \Delta t \leq \frac{l^2}{4\alpha(1 + hl/k)}$$

since $\tau = \alpha\Delta t/l^2$. Substituting the given quantities, the maximum allowable value of the time step is determined to be

$$\Delta t \leq \frac{(0.012\text{ m})^2}{4(3.2 \times 10^{-6}\text{ m}^2/\text{s})[1 + (80\text{ W/m}^2 \cdot °\text{C})(0.012\text{ m})/(15\text{ W/m} \cdot °\text{C})]} = 10.6\text{ s}$$

Therefore, any time step less than 10.6 s can be used to solve this problem. For convenience, let us choose the time step to be $\Delta t = 10$ s. Then the mesh Fourier number becomes

$$\tau = \frac{\alpha\Delta t}{l^2} = \frac{(3.2 \times 10^{-6}\text{ m}^2/\text{s})(10\text{ s})}{(0.012\text{ m})^2} = 0.222 \quad (\text{for } \Delta t = 10\text{ s})$$

Substituting this value of τ and other given quantities, the transient finite difference equations developed above simplify to

$$T_1^{j+1} = 0.0836T_1^j + 0.444(T_2^j + T_4^j + 11.2)$$
$$T_2^{j+1} = 0.0836T_2^j + 0.222(T_1^j + T_3^j + 2T_5^j + 22.4)$$
$$T_3^{j+1} = 0.0552T_3^j + 0.444(T_2^j + T_6^j + 12.8)$$
$$T_4^{j+1} = 0.112T_4^j + 0.222(T_1^j + 2T_5^j + 109.2)$$
$$T_5^{j+1} = 0.112T_5^j + 0.222(T_2^j + T_4^j + T_6^j + 109.2)$$
$$T_6^{j+1} = 0.0931T_6^j + 0.074(2T_3^j + 4T_5^j + 2T_7^j + 424)$$
$$T_7^{j+1} = 0.0836T_7^j + 0.222(T_6^j + T_8^j + 202.4)$$
$$T_8^{j+1} = 0.0836T_8^j + 0.222(T_7^j + T_9^j + 202.4)$$
$$T_9^{j+1} = 0.0836T_9^j + 0.444(T_8^j + 105.2)$$

Using the specified initial condition as the solution at time $t = 0$ (for $i = 0$), sweeping through the nine equations above will give the solution at intervals of 10 s. Using a computer, the solution at the upper corner node (node 3) is determined to be 100.2, 105.9, 106.5, 106.6, and 106.6°C at 1, 3, 5, 10, and 60 min, respectively. Note that the last three solutions are practically identical to the solution for the steady case obtained in Example 5-4. This indicates that steady conditions are reached in the medium after about 5 min.

5-7 ■ CONTROLLING THE NUMERICAL ERROR

A comparison of the numerical results with the exact results for temperature distribution in a cylinder would show that the results obtained by a numerical method are approximate, and they may or may not be sufficiently close to the exact (true) solution values. The difference between a numerical solution and the exact solution is the **error** involved in the numerical solution, and it is primarily due to two sources:

- The **discretization error** (also called the *truncation* or *formulation* error), which is caused by the approximations used in the formulation of the numerical method.
- The **round-off error,** which is caused by the computer's use of a limited number of significant digits and continuously rounding (or chopping) off the digits it cannot retain.

Below we discuss both types of errors.

Discretization Error

The discretization error involved in numerical methods is due to replacing the *derivatives* by *differences* in each step, or the actual temperature distribution between two adjacent nodes by a straight line segment.

Consider the variation of the solution of a transient heat transfer problem with time at a specified nodal point. Both the numerical and actual (exact) solutions coincide at the beginning of the first time step, as expected, but the numerical solution deviates from the exact solution as the time t increases. The difference between the two solutions at $t = \Delta t$ is due to the approximation at the first time step only and is called the *local discretization error.* One would expect the situation to get worse with each step since the second step uses the erroneous result of the first step as its starting point and adds a second local discretization error on top of it, as shown in Figure 5-60. The accumulation of the local discretization errors continues with the increasing number of time steps, and the total discretization error at any step is called the *global* or *accumulated discretization error.* Note that the local and global discretization errors are identical for the first time step. The global discretization error usually increases with the increasing number of steps, but the opposite may occur when the solution function changes direction frequently, giving rise to local discretization errors of opposite signs, which tend to cancel each other.

To have an idea about the magnitude of the local discretization error, consider the Taylor series expansion of the temperature at a specified nodal point m about time t_i,

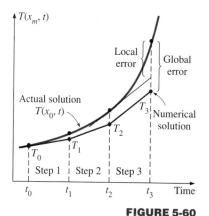

FIGURE 5-60

The local and global discretization errors of the finite difference method at the third time step at a specified nodal point.

$$T(x_m, t_i + \Delta t) = T(x_m, t_i) + \Delta t \frac{\partial T(x_m, t_i)}{\partial t} + \frac{1}{2} \Delta t^2 \frac{\partial^2 T(x_m, t_i)}{\partial t^2} + \cdots$$

$$(5\text{-}62)$$

The finite difference formulation of the time derivative at the same nodal point is expressed as

$$\frac{\partial T(x_m, t_i)}{\partial t} \cong \frac{T(x_m, t_i + \Delta t) - T(x_m, t_i)}{\Delta t} = \frac{T_m^{i+1} - T_m^i}{\Delta t} \qquad (5\text{-}63)$$

or

$$T(x_m, t_i + \Delta t) \cong T(x_m, t_i) + \Delta t \frac{\partial T(x_m, t_i)}{\partial t} \qquad (5\text{-}64)$$

which resembles the *Taylor series expansion* terminated after the first two terms. Therefore, the third and following terms in the Taylor series expansion represent the error involved in the finite difference approximation. For a sufficiently small time step, these terms decay rapidly as the order of derivative increases, and their contributions become smaller and smaller. The first term neglected in the Taylor series expansion is proportional to Δt^2, and thus the local discretization error of this approximation, which is the error involved in each step, is also proportional to Δt^2.

The local discretization error is the formulation error associated with a single step and gives an idea about the accuracy of the method used. However, the solution results obtained at every step except the first one involve the *accumulated error* up to that point, and the local error alone does not have much significance. What we really need to know is the global discretization error. At the worst case, the accumulated discretization error after I time steps during a time period t_0 is $i(\Delta t)^2 = (t_0/\Delta t)(\Delta t)^2 = t_0\Delta t$, which is proportional to Δt. Thus, we conclude that the local discretization error is proportional to the square of the step size Δt^2 while the global discretization error is proportional to the step size Δt itself. Therefore, the smaller the mesh size (or the size of the time step in transient problems), the smaller the error, and thus the more accurate is the approximation. For example, halving the step size will reduce the global discretization error by half. It should be clear from the discussions above that the discretization error can be minimized by decreasing the step size in space or time as much as possible. The discretization error approaches zero as the difference quantities such as Δx and Δt approach the differential quantities such as dx and dt.

Round-off Error

If we had a computer that could retain an infinite number of digits for all numbers, the difference between the exact solution and the approximate (numerical) solution at any point would entirely be due to discretization error. But we know that every computer (or calculator) represents numbers using a finite number of significant digits. The default value of the number of significant digits for many computers is 7, which is referred to as *single precision*. But the user may perform the calculations using 15 significant digits for the numbers, if he or she wishes, which is referred to as *double precision*. Of course, performing calculations in double precision will require more computer memory and a longer execution time.

In single precision mode with 7 significant digits, a computer will register the number 44444.666666 as 44444.67 or 44444.66, depending on the method of rounding the computer uses. In the first case, the excess digits are said to be *rounded* to the closest integer, whereas in the second case they are said to be *chopped off*. Therefore, the numbers $a = 44444.12345$ and $b = 44444.12032$ are equivalent for a computer that performs calculations using 7 significant digits. Such a computer would give $a - b = 0$ instead of the true value 0.00313.

The error due to retaining a limited number of digits during calculations is called the **round-off error.** This error is random in nature and there is no easy and systematic way of predicting it. It depends on the number of calculations, the method of rounding off, the type of computer, and even the sequence of calculations.

In algebra you learned that $a + b + c = a + c + b$, which seems quite reasonable. But this is not necessarily true for calculations performed with a computer, as demonstrated in Figure 5-61. Note that changing the sequence of calculations results in an error of 30.8 percent in just two operations. Considering that any significant problem involves thousands or even millions of such operations performed in sequence, we realize that the accumulated round-off error has the potential to cause serious error without giving any warning signs. Experienced programmers are very much aware of this danger, and they structure their programs to prevent any buildup of the round-off error. For example, it is much safer to multiply a number by 10 than to add it 10 times. Also, it is much safer to start any addition process with the smallest numbers and continue with larger numbers. This rule is particularly important when evaluating series with a large number of terms with alternating signs.

The round-off error is proportional to the *number of computations* performed during the solution. In the finite difference method, the number of calculations increases as the mesh size or the time step size decreases. Halving the mesh or time step size, for example, will double the number of calculations and thus the accumulated round-off error.

Controlling the Error in Numerical Methods

The total error in any result obtained by a numerical method is the sum of the *discretization error,* which decreases with decreasing step size, and the *round-off error,* which increases with decreasing step size, as shown in Figure 5-62. Therefore, decreasing the step size too much in order to get more accurate results may actually backfire and give less accurate results because of a faster increase in the round-off error. We should be careful not to let round-off error get out of control by avoiding a large number of computations with very small numbers.

Given:

$$a = 7777777$$
$$b = -7777776$$
$$c = 0.4444432$$

Find: $D = a + b + c$
$$E = a + c + b$$

Solution:

$$D = 7777777 - 7777776$$
$$+ 0.4444432$$
$$= 1 + 0.4444432$$
$$= 1.444443 \text{ (Correct result)}$$

$$E = 7777777 + 0.4444432$$
$$- 7777776$$
$$= 7777777 - 7777776$$
$$= 1.000000 \text{ (In error by 30.8\%)}$$

FIGURE 5-61

A simple arithmetic operation performed with a computer in single precision using 7 significant digits, which results in 30.8% error when the order of operation is reversed.

FIGURE 5-62

As the mesh or time step size decreases, the discretization error decreases but the round-off error increases.

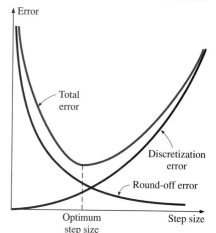

In practice, we will not know the exact solution of the problem, and thus we will not be able to determine the magnitude of the error involved in the numerical method. Knowing that the global discretization error is proportional to the step size is not much help either since there is no easy way of determining the value of the proportionality constant. Besides, the global discretization error alone is meaningless without a true estimate of the round-off error. Therefore, we recommend the following practical procedures to assess the accuracy of the results obtained by a numerical method.

- Start the calculations with a reasonable mesh size Δx (and time step size Δt for transient problems) based on experience. Then repeat the calculations using a mesh size of $\Delta x/2$. If the results obtained by halving the mesh size do not differ significantly from the results obtained with the full mesh size, we conclude that the discretization error is at an acceptable level. But if the difference is larger than we can accept, then we have to repeat the calculations using a mesh size $\Delta x/4$ or even a smaller one at regions of high temperature gradients. We continue in this manner until halving the mesh size does not cause any significant change in the results, which indicates that the discretization error is reduced to an acceptable level.
- Repeat the calculations using double precision holding the mesh size (and the size of the time step in transient problems) constant. If the changes are not significant, we conclude that the round-off error is not a problem. But if the changes are too large to accept, then we may try reducing the total number of calculations by increasing the mesh size or changing the order of computations. But if the increased mesh size gives unacceptable discretization errors, then we may have to find a reasonable compromise.

It should always be kept in mind that the results obtained by any numerical method may not reflect any trouble spots in certain problems that require special consideration such as hot spots or areas of high temperature gradients. The results that seem quite reasonable overall may be in considerable error at certain locations. This is another reason for always repeating the calculations at least twice with different mesh sizes before accepting them as the solution of the problem. Most commercial software packages have built-in routines that vary the mesh size as necessary to obtain highly accurate solutions. But it is a good engineering practice to be aware of any potential pitfalls of numerical methods and to examine the results obtained with a critical eye.

5-8 ■ SUMMARY

Analytical solution methods are limited to highly simplified problems in simple geometries, and it is often necessary to use a numerical method to solve real world problems with complicated geometries or nonuniform thermal conditions. The numerical *finite difference method* is based on replacing derivatives by differences, and the finite difference formulation of a heat transfer problem is obtained by selecting a sufficient number of points in the region, called the *nodal points* or *nodes,* and writing *energy balances* on the volume elements centered about the nodes.

For *steady* heat transfer, the *energy balance* on a volume element can be expressed in general as

$$\sum_{\text{all sides}} \dot{Q} + \dot{g} V_{\text{element}} = 0$$

whether the problem is one-, two-, or three-dimensional. For convenience in formulation, we always assume all heat transfer to be *into* the volume element from all surfaces toward the node under consideration, except for specified heat flux whose direction is already specified. The finite difference formulations for a general interior node under *steady* conditions are expressed for some geometries as follows:

One-dimensional steady conduction in a plane wall:

$$\frac{T_{m-1} - 2T_m + T_{m+1}}{(\Delta x)^2} + \frac{\dot{g}_m}{k} = 0$$

Two-dimensional steady conduction in rectangular coordinates:

$$T_{\text{left}} + T_{\text{top}} + T_{\text{right}} + T_{\text{bottom}} - 4T_{\text{node}} + \frac{\dot{g}_{\text{node}} l^2}{k} = 0$$

where Δx is the nodal spacing for the plane wall and $\Delta x = \Delta y = l$ is the nodal spacing for the two-dimensional case. Insulated boundaries can be viewed as mirrors in formulation, and thus the nodes on insulated boundaries can be treated as interior nodes by using mirror images.

The finite difference formulation at node 0 at the left boundary of a plane wall for steady one-dimensional heat conduction can be expressed as

$$\dot{Q}_{\text{left surface}} + kA \frac{T_1 - T_0}{\Delta x} + \dot{g}_0 (A \Delta x/2) = 0$$

where $A \Delta x / 2$ is the volume of the volume, $\dot{g}_0$ is the rate of heat generation per unit volume at $x = 0$, and A is the heat transfer area. The form of the first term depends on the boundary condition at $x = 0$ (convection, radiation, specified heat flux, etc.).

The finite difference formulation of heat conduction problems usually results in a system of N algebraic equations in N unknown nodal temperatures that need to be solved simultaneously. There are numerous systematic approaches available in the literature, and they are broadly classified as *direct* and *iterative* methods. The direct methods such as the *Gaussian elimination method* are based on a fixed number of well-defined steps that result in the solution in a systematic manner. The iterative methods such as the *Gauss-Seidel iteration method,* on the other hand, are based on an initial guess for the solution that is refined by iteration until a specified convergence criterion is satisfied. Several widely available *equation solvers* can also be used to solve a system of equations simultaneously at the press of a button.

The finite difference formulation of *transient* heat conduction problems is based on an energy balance that also accounts for the variation of the energy content of the volume element during a time interval Δt. The heat transfer and heat generation terms are expressed at the previous time step i in the *explicit method,* and at the new time step $i + 1$ in the *implicit method.* For a general node m, both methods are expressed as

Explicit method:
$$\sum_{\text{All sides}} \dot{Q}^i + \dot{G}^i_{\text{element}} = \rho V_{\text{element}} \, C \frac{T_m^{i+1} - T_m^i}{\Delta t}$$

Implicit method:
$$\sum_{\text{All sides}} \dot{Q}^{i+1} + \dot{G}^{i+1}_{\text{element}} = \rho V_{\text{element}} \, C \frac{T_m^{i+1} - T_m^i}{\Delta t}$$

where T_m^i and T_m^{i+1} are the temperatures of node m at times $t_i = i\Delta t$ and $t_{i+1} = (i + 1)\Delta t$, respectively, and $T_m^{i+1} - T_m^i$ represents the temperature change of the node during the time interval Δt between the time steps i and $i + 1$. The explicit and implicit formulations given above are quite general and can be used in any coordinate system regardless of the dimension of heat transfer.

The explicit formulation of a general interior node for one- and two-dimensional heat transfer in rectangular coordinates can be expressed as

One-dimensional case: $T_m^{i+1} = \tau(T_{m-1}^i + T_{m+1}^i) + (1 - 2\tau) T_m^i + \tau \dfrac{\dot{g}_m^i \, \Delta x^2}{k}$

Two-dimensional case: $T_{\text{node}}^{i+1} = \tau(T_{\text{left}}^i + T_{\text{top}}^i + T_{\text{right}}^i + T_{\text{bottom}}^i)$
$$+ (1 - 4\tau) T_{\text{node}}^i + \tau \frac{\dot{g}_{\text{node}}^i \, l^2}{k}$$

where

$$\tau = \frac{\alpha \Delta t}{\Delta x^2}$$

is the dimensionless *mesh Fourier number* and $\alpha = k/\rho C$ is the *thermal diffusivity* of the medium.

The implicit method is inherently stable, and any value of Δt can be used with that method as the time step. The largest value of the time step Δt in the explicit method is limited by the *stability criterion,* expressed as follows: *the coefficients of all T_m^i in the T_m^{i+1} expressions (called the primary coefficients) must be greater than or equal to zero for all nodes m.* The maximum value of Δt is determined by applying the stability criterion to the equation with the smallest primary coefficient since it is the most restrictive. For problems with specified temperatures or heat fluxes at all the boundaries, the stability criterion can be expressed as $\tau \leq \frac{1}{2}$ for one-dimensional problems and $\tau \leq \frac{1}{4}$ for the two-dimensional problems in rectangular coordinates.

The difference between a numerical solution and the exact solution represents the *error* in the numerical solution. The numerical error is primarily due to the *discretization error,* which is caused by the approximations used in the formulation of the numerical method (such as replacing the derivatives by differences and assuming the temperature distribution between two adjacent nodes to be linear), and the *round-off error,* which is caused by computers' using a limited number of significant digits and continuously rounding (or chopping) off the digits it cannot retain. The discretization error can be minimized by using sufficiently small values of nodal spacing and time steps, and the round-off error can be minimized by performing the calculations on a computer that represents the numbers with more significant digits.

REFERENCES AND SUGGESTED READING

1. D. A. Anderson, J. C. Tannehill, and R. H. Pletcher. *Computational Fluid Mechanics and Heat Transfer.* New York: Hemisphere, 1984.

2. C. A. Brebbia. *The Boundary Element Method for Engineers.* New York: Halsted Press, 1978.

3. G. E. Forsythe and W. R. Wasow. *Finite Difference Methods for Partial Differential Equations.* New York: John Wiley & Sons, 1960.

4. B. Gebhart. *Heat Conduction and Mass Diffusion.* New York: McGraw-Hill, 1993.

5. K. H. Huebner and E. A. Thornton. *The Finite Element Method for Engineers.* 2nd ed. New York: John Wiley & Sons, 1982.

6. Y. Jaluria and K. E. Torrance. *Computational Heat Transfer.* New York: Hemisphere, 1986.

7. W. J. Minkowycz, E. M. Sparrow, G. E. Schneider, and R. H. Pletcher. *Handbook of Numerical Heat Transfer.* New York: John Wiley & Sons, 1988.

8. G. E. Myers. *Analytical Methods in Conduction Heat Transfer.* New York: McGraw-Hill, 1971.

9. D. H. Norrie and G. DeVries. *An Introduction to Finite Element Analysis.* New York: Academic Press, 1978.

10. M. N. Özişik. *Finite Difference Methods in Heat Transfer.* Boca Raton, FL: CRC Press, 1994.

11. S. V. Patankhar. *Numerical Heat Transfer and Fluid Flow.* New York: Hemisphere, 1980.

12. T. M. Shih. *Numerical Heat Transfer.* New York: Hemisphere, 1984.

PROBLEMS

Why Numerical Methods?

5-1C   What are the limitations of the analytical solution methods?

5-2C   How do numerical solution methods differ from analytical ones? What are the advantages and disadvantages of numerical and analytical methods?

5-3C   What is the basis of the energy balance method? How does it differ from the formal finite difference method? For a specified nodal network, will these two methods result in the same or a different set of equations?

5-4C   Consider a heat conduction problem that can be solved both analytically, by solving the governing differential equation and applying the boundary conditions, and numerically, by a software package available on your computer. Which approach would you use to solve this problem? Explain your reasoning.

5-5C   Two engineers are to solve an actual heat transfer problem in a manufacturing facility. Engineer A makes the necessary simplifying assumptions and solves the problem analytically, while engineer B solves it numerically using a powerful software package. Engineer A claims he solved the problem exactly and thus his results are better, while engineer B claims that he used a more realistic model and thus his results are better. To resolve the dispute, you are asked to solve the problem experimentally in a lab. Which engineer do you think the experiments will prove right? Explain.

Finite Difference Formulation of Differential Equations

5-6C   Define the following terms used in the finite difference formulation: node, nodal network, volume element, nodal spacing, and difference equation.

5-7   Consider three consecutive nodes $n - 1$, n, and $n + 1$ in a plane wall. Using the finite difference form of the first derivative at the midpoints, show that the finite difference form of the second derivative can be expressed as

$$\frac{T_{n-1} - 2T_n + T_{n+1}}{\Delta x^2} = 0$$

5-8   The finite difference formulation of steady two-dimensional heat conduction in a medium with heat generation and constant thermal conductivity is given by

$$\frac{T_{m-1,n} - 2T_{m,n} + T_{m+1,n}}{\Delta x^2} + \frac{T_{m,n-1} - 2T_{m,n} + T_{m,n+1}}{\Delta y^2} + \frac{\dot{g}_{m,n}}{k} = 0$$

in rectangular coordinates. Modify this relation for the three-dimensional case.

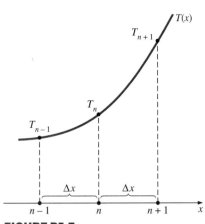

FIGURE P5-7

5-9 Consider steady one-dimensional heat conduction in a plane wall with variable heat generation and constant thermal conductivity. The nodal network of the medium consists of nodes 0, 1, 2, 3, and 4 with a uniform nodal spacing of Δx. Using the finite difference form of the first derivative (*not* the energy balance approach), obtain the finite difference formulation of the boundary nodes for the case of uniform heat flux $\dot{q}_0$ at the left boundary (node 0) and convection at the right boundary (node 4) with a convection coefficient of h and an ambient temperature of T_∞.

5-10 Consider steady one-dimensional heat conduction in a plane wall with variable heat generation and constant thermal conductivity. The nodal network of the medium consists of nodes 0, 1, 2, 3, 4, and 5 with a uniform nodal spacing of Δx. Using the finite difference form of the first derivative (*not* the energy balance approach), obtain the finite difference formulation of the boundary nodes for the case of insulation at the left boundary (node 0) and radiation at the right boundary (node 5) with an emissivity of ε and surrounding temperature of T_{surr}.

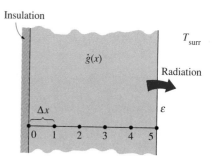

FIGURE P5-10

One-Dimensional Steady Heat Conduction

5-11C Explain how the finite difference form of a heat conduction problem is obtained by the energy balance method.

5-12C In the energy balance formulation of the finite difference method, it is recommended that all heat transfer at the boundaries of the volume element be assumed to be into the volume element even for steady heat conduction. Is this a valid recommendation even though it seems to violate the conservation of energy principle?

5-13C How is an insulated boundary handled in the finite difference formulation of a problem? How does a symmetry line differ from an insulated boundary in the finite difference formulation?

5-14C How can a node on an insulated boundary be treated as an interior node in the finite difference formulation of a plane wall? Explain.

5-15C Consider a medium in which the finite difference formulation of a general interior node is given in its simplest form as

$$\frac{T_{m-1} - 2T_m + T_{m+1}}{\Delta x^2} + \frac{\dot{g}_m}{k} = 0$$

(*a*) Is heat transfer in this medium steady or transient?
(*b*) Is heat transfer one-, two-, or three-dimensional?
(*c*) Is there heat generation in the medium?
(*d*) Is the nodal spacing constant or variable?
(*e*) Is the thermal conductivity of the medium constant or variable?

5-16 Consider steady heat conduction in a plane wall whose left surface (node 0) is maintained at 30°C while the right surface (node 8) is subjected to

FIGURE P5-16

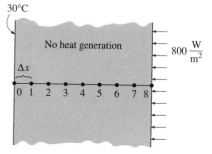

a heat flux of 800 W/m². Express the finite difference formulation of the boundary nodes 0 and 8 for the case of no heat generation. Also obtain the finite difference formulation for the rate of heat transfer at the left boundary.

5-17 Consider steady heat conduction in a plane wall with variable heat generation and constant thermal conductivity. The nodal network of the medium consists of nodes 0, 1, 2, 3, and 4 with a uniform nodal spacing of Δx. Using the energy balance approach, obtain the finite difference formulation of the boundary nodes for the case of uniform heat flux $\dot{q}_0$ at the left boundary (node 0) and convection at the right boundary (node 4) with a convection coefficient of h and an ambient temperature of T_∞.

5-18 Consider steady one-dimensional heat conduction in a plane wall with variable heat generation and constant thermal conductivity. The nodal network of the medium consists of nodes 0, 1, 2, 3, 4, and 5 with a uniform nodal spacing of Δx. Using the energy balance approach, obtain the finite difference formulation of the boundary nodes for the case of insulation at the left boundary (node 0) and radiation at the right boundary (node 5) with an emissivity of ε and surrounding temperature of T_{surr}.

5-19 Consider steady one-dimensional heat conduction in a plane wall with variable heat generation and constant thermal conductivity. The nodal network of the medium consists of nodes 0, 1, 2, 3, 4, and 5 with a uniform nodal spacing of Δx. The temperature at the right boundary (node 5) is specified. Using the energy balance approach, obtain the finite difference formulation of the boundary node 0 on the left boundary for the case of combined convection, radiation, and heat flux at the left boundary with an emissivity of ε, convection coefficient of h, ambient temperature of T_∞, surrounding temperature of T_{surr}, and uniform heat flux of $\dot{q}_0$. Also, obtain the finite difference formulation for the rate of heat transfer at the right boundary.

5-20 Consider steady one-dimensional heat conduction in a composite plane wall consisting of two layers A and B in perfect contact at the interface. The wall involves no heat generation. The nodal network of the medium consists of nodes 0, 1 (at the interface), and 2 with a uniform nodal spacing of Δx. Using the energy balance approach, obtain the finite difference formulation of this problem for the case of insulation at the left boundary (node 0) and radiation at the right boundary (node 2) with an emissivity of ε and surrounding temperature of T_{surr}.

5-21 Consider steady one-dimensional heat conduction in a plane wall with variable heat generation and variable thermal conductivity. The nodal network of the medium consists of nodes 0, 1, and 2 with a uniform nodal spacing of Δx. Using the energy balance approach, obtain the finite difference formulation of this problem for the case of specified heat flux $\dot{q}_0$ to the wall and convection at the left boundary (node 0) with a convection coefficient of h and ambient temperature of T_∞, and radiation at the right boundary (node 2) with an emissivity of ε and surrounding surface temperature of T_{surr}.

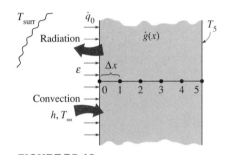

FIGURE P5-19

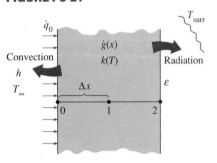

FIGURE P5-21

5-22 Consider steady one-dimensional heat conduction in a pin fin of constant diameter D with constant thermal conductivity. The fin is losing heat by convection to the ambient air at T_∞ with a heat transfer coefficient of h. The nodal network of the fin consists of nodes 0 (at the base), 1 (in the middle), and 2 (at the fin tip) with a uniform nodal spacing of Δx. Using the energy balance approach, obtain the finite difference formulation of this problem to determine T_1 and T_2 for the case of specified temperature at the fin base and negligible heat transfer at the fin tip. All temperatures are in °C.

5-23 Consider steady one-dimensional heat conduction in a pin fin of constant diameter D with constant thermal conductivity. The fin is losing heat by convection to the ambient air at T_∞ with a convection coefficient of h, and by radiation to the surrounding surfaces at an average temperature of T_{surr}. The nodal network of the fin consists of nodes 0 (at the base), 1 (in the middle), and 2 (at the fin tip) with a uniform nodal spacing of Δx. Using the energy balance approach, obtain the finite difference formulation of this problem to determine T_1 and T_2 for the case of specified temperature at the fin base and negligible heat transfer at the fin tip. All temperatures are in °C.

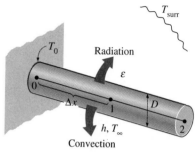

FIGURE P5-23

5-24 Consider a large uranium plate of thickness 5 cm and thermal conductivity $k = 28$ W/m · °C in which heat is generated uniformly at a constant rate of $\dot{g} = 8 \times 10^6$ W/m³. One side of the plate is insulated while the other side is subjected to convection to an environment at 30°C with a heat transfer coefficient of $h = 45$ W/m² · °C. Considering six equally spaced nodes with a nodal spacing of 1 cm, (a) obtain the finite difference formulation of this problem and (b) determine the nodal temperatures under steady conditions by solving those equations.

5-25 Consider an aluminum alloy fin ($k = 180$ W/m · °C) of triangular cross-section whose length is $L = 5$ cm, base thickness is $b = 1$ cm, and width w in the direction normal to the plane of paper is very large. The base of the fin is maintained at a temperature of $T_0 = 180$°C. The fin is losing heat by convection to the ambient air at $T_\infty = 25$°C with a heat transfer coefficient of $h = 25$ W/m² · °C and by radiation to the surrounding surfaces at an average temperature of $T_{surr} = 290$ K. Using the finite difference method with six equally spaced nodes along the fin in the x-direction, determine (a) the temperatures at the nodes and (b) the rate of heat transfer from the fin for $w = 1$ m. Take the emissivity of the fin surface to be 0.9 and assume steady one-dimensional heat transfer in the fin.

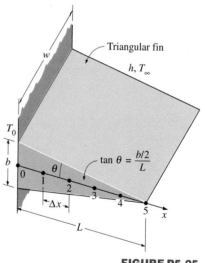

FIGURE P5-25

5-26 Consider a large plane wall of thickness $L = 0.4$ m, thermal conductivity $k = 2.3$ W/m · °C, and surface area $A = 20$ m². The left side of the wall is maintained at a constant temperature of 80°C, while the right side loses heat by convection to the surrounding air at $T_\infty = 15$°C with a heat transfer coefficient of $h = 24$ W/m² · °C. Assuming steady one-dimensional heat transfer and taking the nodal spacing to be 10 cm, (a) obtain the finite difference formulation for all nodes, (b) determine the nodal temperatures by solving those equations, and (c) evaluate the rate of heat transfer through the wall.

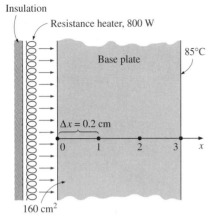

Insulation

Resistance heater, 800 W

Base plate

85°C

$\Delta x = 0.2$ cm

0   1   2   3   x

160 cm²

FIGURE P5-27

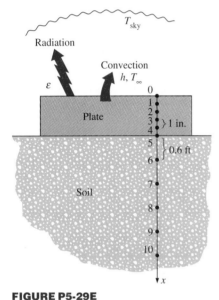

T_{sky}

Radiation

ε

Convection
h, T_∞

0
1
2
3 } 1 in.
4

Plate

5
} 0.6 ft
6

7

Soil

8

9

10

x

FIGURE P5-29E

FIGURE P5-31

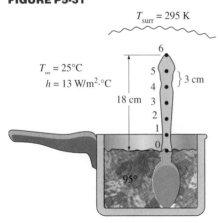

$T_{surr} = 295$ K

6

$T_\infty = 25°C$
$h = 13$ W/m²·°C

5
} 3 cm
4

18 cm 3

2

1

0

95°

5-27 Consider the base plate of a 800-W household iron having a thickness of $L = 0.6$ cm, base area of $A = 160$ cm², and thermal conductivity of $k = 20$ W/m·°C. The inner surface of the base plate is subjected to uniform heat flux generated by the resistance heaters inside. When steady operating conditions are reached, the outer surface temperature of the plate is measured to be 85°C. Disregarding any heat loss through the upper part of the iron and taking the nodal spacing to be 0.2 cm, (a) obtain the finite difference formulation for the nodes and (b) determine the inner surface temperature of the plate by solving those equations.   *Answer:* (b) 100°C

5-28 Consider a large plane wall of thickness $L = 0.3$ m, thermal conductivity $k = 2.5$ W/m·°C, and surface area $A = 12$ m². The left side of the wall is subjected to a heat flux of $\dot{q}_0 = 700$ W/m² while the temperature at that surface is measured to be $T_0 = 80°C$. Assuming steady one-dimensional heat transfer and taking the nodal spacing to be 6 cm, (a) obtain the finite difference formulation for the six nodes and (b) determine the temperature of the other surface of the wall by solving those equations.   *Answer:* (b) −4°C

5-29E A large steel plate having a thickness of $L = 5$ in., thermal conductivity of $k = 7.2$ Btu/h·ft·°F, and an emissivity of $\varepsilon = 0.6$ is lying on the ground. The exposed surface of the plate exchanges heat by convection with the ambient air at $T_\infty = 80°F$ with an average heat transfer coefficient of $h = 3.5$ Btu/h·ft²·°F as well as by radiation with the open sky at an equivalent sky temperature of $T_{sky} = 510$ R. The ground temperature below a certain depth (say, 3 ft) is not affected by the weather conditions outside and remains fairly constant at 50°F at that location. The thermal conductivity of the soil can be taken to be $k_{soil} = 0.49$ Btu/h·ft·°F, and the steel plate can be assumed to be in perfect contact with the ground. Assuming steady one-dimensional heat transfer and taking the nodal spacings to be 1 in. in the plate and 0.6 ft in the ground, (a) obtain the finite difference formulation for all 11 nodes shown in Figure P5-29E and (b) determine the top and bottom surface temperatures of the plate by solving those equations.

5-30 Repeat Problem 5-29E by disregarding radiation heat transfer from the upper surface.   *Answers:* (b) 78.7°F, 78.4°F

5-31 Consider a stainless steel spoon ($k = 15.1$ W/m·C, $\varepsilon = 0.6$) that is partially immersed in boiling water at 95°C in a kitchen at 25°C. The handle of the spoon has a cross-section of about 0.2 cm × 1 cm and extends 18 cm in the air from the free surface of the water. The spoon loses heat by convection to the ambient air with an average heat transfer coefficient of $h = 13$ W/m²·°C as well as by radiation to the surrounding surfaces at an average temperature of $T_{surr} = 295$ K. Assuming steady one-dimensional heat transfer along the spoon and taking the nodal spacing to be 3 cm, (a) obtain the finite difference formulation for all nodes, (b) determine the temperature of the tip of the spoon by solving those equations, and (c) determine the rate of heat transfer from the exposed surfaces of the spoon.

5-32 Repeat Problem 5-31 using a nodal spacing of 1.5 cm.

5-33 One side of a 2-m-high and 3-m-wide vertical plate at 120°C is to be cooled by attaching aluminum fins ($k = 237$ W/m · °C) of rectangular profile in an environment at 25°C. The fins are 2 cm long, 0.3 cm thick, and 0.4 cm apart. The heat transfer coefficient between the fins and the surrounding air for combined convection and radiation is estimated to be 30 W/m² · °C. Assuming steady one-dimensional heat transfer along the fin and taking the nodal spacing to be 0.5 cm, determine (a) the finite difference formulation of this problem, (b) the nodal temperatures along the fin by solving these equations, (c) the rate of heat transfer from a single fin, and (d) the rate of heat transfer from the entire finned surface of the plate.

5-34 A hot surface at 100°C is to be cooled by attaching 3-cm-long, 0.25-cm-diameter aluminum pin fins ($k = 237$ W/m · °C) with a center-to-center distance of 0.6 cm. The temperature of the surrounding medium is 30°C, and the combined heat transfer coefficient on the surfaces is 35 W/m² · °C. Assuming steady one-dimensional heat transfer along the fin and taking the nodal spacing to be 0.5 cm, determine (a) the finite difference formulation of this problem, (b) the nodal temperatures along the fin by solving these equations, (c) the rate of heat transfer from a single fin, and (d) the rate of heat transfer from a 1-m × 1-m section of the plate.

5-35 Repeat Problem 5-34 using copper fins ($k = 386$ W/m · °C) instead of aluminum ones.
  Answers: (b) 98.6°C, 97.5°C, 96.7°C, 96.0°C, 95.7°C, 95.5°C

5-36 Two 3-m-long and 0.5-cm-thick cast iron ($k = 52$ W/m · °C, $\varepsilon = 0.8$) steam pipes of outer diameter 10 cm are connected to each other through two 1-cm-thick flanges of outer diameter 20 cm, as shown in the figure. The steam flows inside the pipe at an average temperature of 200°C with a heat transfer coefficient of 180 W/m² · °C. The outer surface of the pipe is exposed to convection with ambient air at 8°C with a heat transfer coefficient of 25 W/m² · °C as well as radiation with the surrounding surfaces at an average temperature of $T_{surr} = 290$ K. Assuming steady one-dimensional heat conduction along the flanges and taking the nodal spacing to be 1 cm along the flange (a) obtain the finite difference formulation for all nodes, (b) determine the temperature at the tip of the flange by solving those equations, and (c) determine the rate of heat transfer from the exposed surfaces of the flange.

Solution Methods for Systems of Algebraic Equations

5-37C What are the two general categories of solution methods for a system of algebraic equations? Name a specific method for each category.

5-38C What is the Gaussian elimination method based on? Can this method be used to solve a system of nonlinear algebraic equations?

5-39C What is the Gauss-Seidel method based on? Can this method be used to solve a system of nonlinear algebraic equations?

5-40C Why do we try to avoid variable thermal conductivity or the radiation boundary conditions even in the numerical analysis of heat transfer problems?

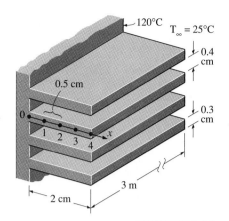

FIGURE P5-33

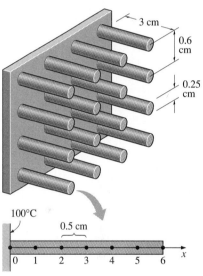

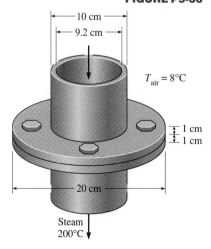

FIGURE P5-34

FIGURE P5-36

5-41 Using the Gauss-Seidel iteration method, solve the following systems of algebraic equations with an error less than 0.01 in each unknown.

(a) $3x_1 - x_2 + 3x_3 = 0$

$-x_1 + 2x_2 + x_3 = 3$

$2x_1 - x_2 - x_3 = 2$

(b) $4x_1 - 2x_2^2 + 0.5x_3 = -2$

$x_1^3 - x_2 + x_3^{1.4} = 11.964$

$x_1 + x_2 + x_3 = 3$

Answers: (a) $x_1 = 2$, $x_2 = 3$, $x_3 = -1$, (b) $x_1 = 2.53$, $x_2 = 2.36$, $x_3 = -1.90$

5-42 Using the Gauss-Seidel iteration method, solve the following systems of algebraic equations with an error less than 0.01 in each unknown.

(a) $3x_1 + 2x_2 - x_3 + x_4 = 6$

$x_1 + 2x_2 - x_4 = -3$

$-2x_1 + x_2 + 3x_3 + x_4 = 2$

$3x_2 + x_3 - 4x_4 = -6$

(b) $3x_1 + x_2^2 + 2x_3 = 8$

$-x_1^2 + 3x_2 + 2x_3^{1.5} = -6.293$

$2x_1 - x_2^4 + 4x_3 = -12$

5-43 Using the Gauss-Seidel iteration method, solve the following systems of algebraic equations with an error less than 0.01 in each unknown.

(a) $4x_1 - x_2 + 2x_3 + x_4 = -6$

$x_1 + 3x_2 - x_3 + 4x_4 = -1$

$-x_1 + 2x_2 + 5x_4 = 5$

$2x_2 - 4x_3 - 3x_4 = -5$

(b) $2x_1 + x_2^4 - 2x_3 + x_4 = 1$

$x_1^2 + 4x_2 + 2x_3^2 - 2x_4 = -3$

$-x_1 + x_2^4 + 5x_3 = 10$

$3x_1 - x_3^2 + 8x_4 = 15$

Two-Dimensional Steady Heat Conduction

5-44C Consider a medium in which the finite difference formulation of a general interior node is given in its simplest form as

$$T_{\text{left}} + T_{\text{top}} + T_{\text{right}} + T_{\text{bottom}} - 4T_{\text{node}} + \frac{\dot{g}_{\text{node}} l^2}{k} = 0$$

(a) Is heat transfer in this medium steady or transient?
(b) Is heat transfer one-, two-, or three-dimensional?
(c) Is there heat generation in the medium?
(d) Is the nodal spacing constant or variable?
(e) Is the thermal conductivity of the medium constant or variable?

5-45C Consider a medium in which the finite difference formulation of a general interior node is given in its simplest form as

$$T_{\text{node}} = (T_{\text{left}} + T_{\text{top}} + T_{\text{right}} + T_{\text{bottom}})/4$$

(a) Is heat transfer in this medium steady or transient?
(b) Is heat transfer one-, two-, or three-dimensional?
(c) Is there heat generation in the medium?
(d) Is the nodal spacing constant or variable?
(e) Is the thermal conductivity of the medium constant or variable?

5-46C What is an irregular boundary? What is a practical way of handling irregular boundary surfaces with the finite difference method?

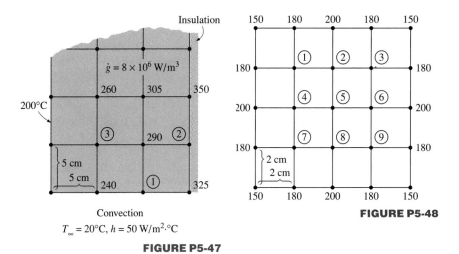

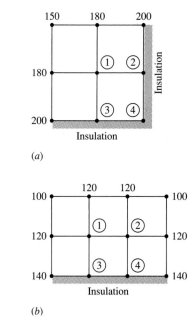

FIGURE P5-47

FIGURE P5-48

(a)

(b)

FIGURE P5-49

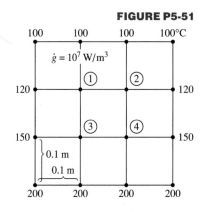

5-47 Consider steady two-dimensional heat transfer in a long solid body whose cross-section is given in the figure. The temperatures at the selected nodes and the thermal conditions at the boundaries are as shown. The thermal conductivity of the body is $k = 217$ W/m · °C, and heat is generated in the body uniformly at a rate of $\dot{g} = 8 \times 10^6$ W/m³. Using the finite difference method with a mesh size of $\Delta x = \Delta y = 5.0$ cm, determine (a) the temperatures at nodes 1, 2, and 3 and (b) the rate of heat loss from the bottom surface through a 1-m-long section of the body.

5-48 Consider steady two-dimensional heat transfer in a long solid body whose cross-section is given in the figure. The measured temperatures at selected points of the outer surfaces are as shown. The thermal conductivity of the body is $k = 45$ W/m · °C, and there is no heat generation. Using the finite difference method with a mesh size of $\Delta x = \Delta y = 2.0$ cm, determine the temperatures at the indicated points in the medium. *Hint:* Take advantage of symmetry.

5-49 Consider steady two-dimensional heat transfer in a long solid bar whose cross-section is given in the figure. The measured temperatures at selected points of the outer surfaces are as shown. The thermal conductivity of the body is $k = 20$ W/m · °C, and there is no heat generation. Using the finite difference method with a mesh size of $\Delta x = \Delta y = 1.0$ cm, determine the temperatures at the indicated points in the medium.
  Answers: $T_1 = 185$°C, $T_2 = T_3 = T_4 = 190$°C

5-50 Starting with an energy balance on a volume element, obtain the steady two-dimensional finite difference equation for a general interior node in rectangular coordinates for $T(x, y)$ for the case of variable thermal conductivity and uniform heat generation.

5-51 Consider steady two-dimensional heat transfer in a long solid body whose cross-section is given in the figure. The temperatures at the selected nodes and the thermal conditions on the boundaries are as shown. The ther-

FIGURE P5-51

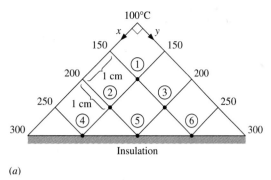

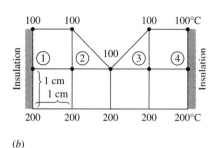

(a)

(b)

FIGURE P5-52

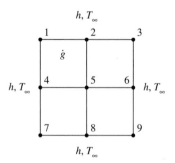

FIGURE P5-53

FIGURE P5-54E

mal conductivity of the body is $k = 180$ W/m · °C, and heat is generated in the body uniformly at a rate of $\dot{g} = 10^7$ W/m³. Using the finite difference method with a mesh size of $\Delta x = \Delta y = 10$ cm, determine (a) the temperatures at nodes 1, 2, 3, and 4 and (b) the rate of heat loss from the top surface through a 1-m-long section of the body.

5-52 Consider steady two-dimensional heat transfer in a long solid bar whose cross-section is given in the figure. The measured temperatures at selected points on the outer surfaces are as shown. The thermal conductivity of the body is $k = 20$ W/m · °C, and there is no heat generation. Using the finite difference method with a mesh size of $\Delta x = \Delta y = 1.0$ cm, determine the temperatures at the indicated points in the medium. *Hint:* Take advantage of symmetry.   *Answers:* (b) $T_1 = T_4 = 143$°C, $T_2 = T_3 = 136$°C

5-53 Consider steady two-dimensional heat transfer in an L-shaped solid body whose cross-section is given in the figure. The thermal conductivity of the body is $k = 45$ W/m · °C, and heat is generated in the body at a rate of $\dot{g} = 5 \times 10^6$ W/m³. The right surface of the body is insulated, and the bottom surface is maintained at a uniform temperature of 110°C. The entire top surface is subjected to convection with ambient air at $T_\infty = 20$°C with a heat transfer coefficient of $h = 55$ W/m² · °C, and the left surface is subjected to heat flux at a uniform rate of $\dot{q}_L = 8000$ W/m². The nodal network of the problem consists of 13 equally spaced nodes with $\Delta x = \Delta y = 1.5$ cm. Five of the nodes are at the bottom surface and thus their temperatures are known. (a) Obtain the finite difference equations at the remaining eight nodes and (b) determine the nodal temperatures by solving those equations.

5-54E Consider steady two-dimensional heat transfer in a long solid bar of square cross-section in which heat is generated uniformly at a rate of $\dot{g} = 0.19 \times 10^5$ Btu/h · ft³. The cross-section of the bar is 0.4 ft × 0.4 ft in size, and its thermal conductivity is $k = 16$ Btu/h · ft · °F. All four sides of the bar are subjected to convection with the ambient air at $T_\infty = 70$°F with a heat transfer coefficient of $h = 7.9$ Btu/h · ft² · °F. Using the finite difference method with a mesh size of $\Delta x = \Delta y = 0.2$ ft, determine (a) the temperatures at the nine nodes and (b) the rate of heat loss from the bar through a 1-ft-long section.   *Answer:* (b) 304.0 Btu/h

5-55 Hot combustion gases of a furnace are flowing through a concrete chimney ($k = 1.4$ W/m · °C) of rectangular cross-section. The flow section of the chimney is 20 cm × 40 cm, and the thickness of the wall is 10 cm. The average temperature of the hot gases in the chimney is $T_i = 280$°C, and the average convection heat transfer coefficient inside the chimney is $h_i = 75$ W/m² · °C. The chimney is losing heat from its outer surface to the ambient air at $T_o = 15$°C by convection with a heat transfer coefficient of $h_o = 18$ W/m² · °C and to the sky by radiation. The emissivity of the outer surface of the wall is $\varepsilon = 0.9$, and the effective sky temperature is estimated to be 250 K. Using the finite difference method with $\Delta x = \Delta y = 10$ cm and taking full advantage of symmetry, (a) obtain the finite difference formulation of this problem for steady two-dimensional heat transfer, (b) determine the temperatures at the nodal points of a cross-section, and (c) evaluate the rate of heat loss for a 1-m-long section of the chimney.

5-56 Repeat Problem 5-55 by disregarding radiation heat transfer from the outer surfaces of the chimney.

5-57 Consider a long concrete dam ($k = 0.6$ W/m · °C, $\alpha_s = 0.7$) of triangular cross-section whose exposed surface is subjected to solar heat flux of $\dot{q}_s = 800$ W/m² and to convection and radiation to the environment at 25°C with a combined heat transfer coefficient of 30 W/m² · °C. The 2-m-high vertical section of the dam is subjected to convection by water at 15°C with a heat transfer coefficient of 150 W/m² · °C, and heat transfer through the 2-m-long base is considered to be negligible. Using the finite difference method with a mesh size of $\Delta x = \Delta y = 1$ m and assuming steady two-dimensional heat transfer, determine the temperature of the top, middle, and bottom of the exposed surface of the dam.
Answers: 21.3°C, 43.2°C, 43.6°C

5-58E Consider steady two-dimensional heat transfer in a V-grooved solid body whose cross-section is given in the figure. The top surfaces of the groove are maintained at 32°F while the bottom surface is maintained at 212°F. The side surfaces of the groove are insulated. Using the finite difference method with a mesh size of $\Delta x = \Delta y = 1$ ft and taking advantage of symmetry, determine the temperatures at the middle of the insulated surfaces.

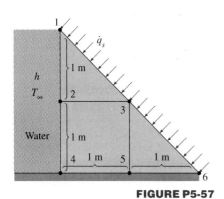

FIGURE P5-55

FIGURE P5-57

FIGURE P5-58E

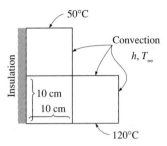

FIGURE P5-59

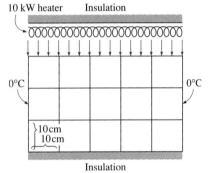

FIGURE P5-60

5-59 Consider a long solid bar whose thermal conductivity is $k = 12$ W/m · °C and whose cross-section is given in the figure. The top surface of the bar is maintained at 50°C while the bottom surface is maintained at 120°C. The left surface is insulated and the remaining three surfaces are subjected to convection with ambient air at $T_\infty = 25$°C with a heat transfer coefficient of $h = 30$ W/m² · °C. Using the finite difference method with a mesh size of $\Delta x = \Delta y = 10$ cm, (a) obtain the finite difference formulation of this problem for steady two-dimensional heat transfer and (b) determine the unknown nodal temperatures by solving those equations.
Answers: (b) 85.7°C, 86.4°C, 87.6°C

5-60 Consider a 5-m-long constantan block ($k = 23$ W/m · °C) 30 cm high and 50 cm wide. The block is completely submerged in iced water at 0°C that is well stirred, and the heat transfer coefficient is so high that the temperatures on both sides of the block can be taken to be 0°C. The bottom surface of the bar is covered with a low-conductivity material so that heat transfer through the bottom surface is negligible. The top surface of the block is heated uniformly by a 10-kW resistance heater. Using the finite difference method with a mesh size of $\Delta x = \Delta y = 10$ cm and taking advantage of symmetry, (a) obtain the finite difference formulation of this problem for steady two-dimensional heat transfer, (b) determine the unknown nodal temperatures by solving those equations, and (c) determine the rate of heat transfer from the block to the iced water.

Transient Heat Conduction

5-61C How does the finite difference formulation of a transient heat conduction problem differ from that of a steady heat conduction problem? What does the term $\rho A \Delta x C (T_m^{i+1} - T_m^i)/\Delta t$ represent in the transient finite difference formulation?

5-62C What are the two basic methods of solution of transient problems based on finite differencing? How do heat transfer terms in the energy balance formulation differ in the two methods?

5-63C The explicit finite difference formulation of a general interior node for transient heat conduction in a plane wall is given by

$$T_{m-1}^i - 2T_m^i + T_{m+1}^i + \frac{\dot{g}_m^i \, \Delta x^2}{k} = \frac{T_m^{i+1} - T_m^i}{\tau}$$

Obtain the finite difference formulation for the steady case by simplifying the relation above.

5-64C The explicit finite difference formulation of a general interior node for transient two-dimensional heat conduction is given by

$$T_{node}^{i+1} = \tau(T_{left}^i + T_{top}^i + T_{right}^i + T_{bottom}^i) + (1 - 4\tau)T_{node}^i + \tau \frac{\dot{g}_{node}^i l^2}{k}$$

Obtain the finite difference formulation for the steady case by simplifying the relation above.

5-65C Is there any limitation on the size of the time step Δt in the solution of transient heat conduction problems using (*a*) the explicit method and (*b*) the implicit method?

5-66C Express the general stability criterion for the explicit method of solution of transient heat conduction problems.

5-67C Consider transient one-dimensional heat conduction in a plane wall that is to be solved by the explicit method. If both sides of the wall are at specified temperatures, express the stability criterion for this problem in its simplest form.

5-68C Consider transient one-dimensional heat conduction in a plane wall that is to be solved by the explicit method. If both sides of the wall are subjected to specified heat flux, express the stability criterion for this problem in its simplest form.

5-69C Consider transient two-dimensional heat conduction in a rectangular region that is to be solved by the explicit method. If all boundaries of the region are either insulated or at specified temperatures, express the stability criterion for this problem in its simplest form.

5-70C The implicit method is unconditionally stable and thus any value of time step Δt can be used in the solution of transient heat conduction problems. To minimize the computation time, someone suggests using a very large value of Δt since there is no danger of instability. Do you agree with this suggestion? Explain.

5-71 Consider transient heat conduction in a plane wall whose left surface (node 0) is maintained at 50°C while the right surface (node 6) is subjected to a solar heat flux of 600 W/m². The wall is initially at a uniform temperature of 50°C. Express the explicit finite difference formulation of the boundary nodes 0 and 6 for the case of no heat generation. Also, obtain the finite difference formulation for the total amount of heat transfer at the left boundary during the first three time steps.

5-72 Consider transient heat conduction in a plane wall with variable heat generation and constant thermal conductivity. The nodal network of the medium consists of nodes 0, 1, 2, 3, and 4 with a uniform nodal spacing of Δx. The wall is initially at a specified temperature. Using the energy balance approach, obtain the explicit finite difference formulation of the boundary nodes for the case of uniform heat flux $\dot{q}_0$ at the left boundary (node 0) and convection at the right boundary (node 4) with a convection coefficient of h and an ambient temperature of T_∞. Do not simplify.

5-73 Repeat Problem 5-72 for the case of implicit formulation.

5-74 Consider transient heat conduction in a plane wall with variable heat generation and constant thermal conductivity. The nodal network of the medium consists of nodes 0, 1, 2, 3, 4, and 5 with a uniform nodal spacing of

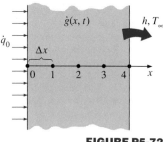

FIGURE P5-72

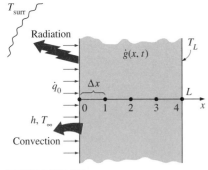

FIGURE P5-75

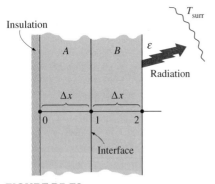

FIGURE P5-79

Δx. The wall is initially at a specified temperature. Using the energy balance approach, obtain the explicit finite difference formulation of the boundary nodes for the case of insulation at the left boundary (node 0) and radiation at the right boundary (node 5) with an emissivity of ε and surrounding temperature of T_{surr}.

5-75 Consider transient heat conduction in a plane wall with variable heat generation and constant thermal conductivity. The nodal network of the medium consists of nodes 0, 1, 2, 3, and 4 with a uniform nodal spacing of Δx. The wall is initially at a specified temperature. The temperature at the right boundary (node 4) is specified. Using the energy balance approach, obtain the explicit finite difference formulation of the boundary node 0 for the case of combined convection, radiation, and heat flux at the left boundary with an emissivity of ε, convection coefficient of h, ambient temperature of T_∞, surrounding temperature of T_{surr}, and uniform heat flux of $\dot{q}_0$ toward the wall. Also, obtain the finite difference formulation for the total amount of heat transfer at the right boundary for the first 20 time steps.

5-76 Starting with an energy balance on a volume element, obtain the two-dimensional transient explicit finite difference equation for a general interior node in rectangular coordinates for $T(x, y, t)$ for the case of constant thermal conductivity and no heat generation.

5-77 Starting with an energy balance on a volume element, obtain the two-dimensional transient implicit finite difference equation for a general interior node in rectangular coordinates for $T(x, y, t)$ for the case of constant thermal conductivity and no heat generation.

5-78 Starting with an energy balance on a disk volume element, derive the one-dimensional transient explicit finite difference equation for a general interior node for $T(z, t)$ in a cylinder whose side surface is insulated for the case of constant thermal conductivity with uniform heat generation.

5-79 Consider one-dimensional transient heat conduction in a composite plane wall that consists of two layers A and B with perfect contact at the interface. The wall involves no heat generation and initially is at a specified temperature. The nodal network of the medium consists of nodes 0, 1 (at the interface), and 2 with a uniform nodal spacing of Δx. Using the energy balance approach, obtain the explicit finite difference formulation of this problem for the case of insulation at the left boundary (node 0) and radiation at the right boundary (node 2) with an emissivity of ε and surrounding temperature of T_{surr}.

5-80 Consider transient one-dimensional heat conduction in a pin fin of constant diameter D with constant thermal conductivity. The fin is losing heat by convection to the ambient air at T_∞ with a heat transfer coefficient of h and by radiation to the surrounding surfaces at an average temperature of T_{surr}. The nodal network of the fin consists of nodes 0 (at the base), 1 (in the middle), and 2 (at the fin tip) with a uniform nodal spacing of Δx. Using the energy balance approach, obtain the explicit finite difference formulation of this problem for the case of a specified temperature at the fin base and negligible heat transfer at the fin tip.

5-81 Repeat Problem 5-80 for the case of implicit formulation.

5-82 Consider a large uranium plate of thickness $L = 8$ cm, thermal conductivity $k = 28$ W/m · °C, and thermal diffusivity $\alpha = 12.5 \times 10^{-6}$ m²/s that is initially at a uniform temperature of 100°C. Heat is generated uniformly in the plate at a constant rate of $\dot{g} = 10^6$ W/m³. At time $t = 0$, the left side of the plate is insulated while the other side is subjected to convection with an environment at $T_\infty = 20$°C with a heat transfer coefficient of $h = 35$ W/m² · °C. Using the explicit finite difference approach with a uniform nodal spacing of $\Delta x = 2$ cm, determine (*a*) the temperature distribution in the plate after 5 min and (*b*) how long it will take for steady conditions to be reached in the plate.

5-83 Consider a house whose south wall consists of a 30-cm-thick Trombe wall whose thermal conductivity is $k = 0.70$ W/m · °C and whose thermal diffusivity is $\alpha = 0.44 \times 10^{-6}$ m²/s. The variations of the ambient temperature T_{out} and the solar heat flux $\dot{q}_{\text{solar}}$ incident on a south-facing vertical surface throughout the day for a typical day in February are given in the table in 3-h intervals. The Trombe wall has single glazing with an absorptivity-transmissivity product of $\kappa = 0.76$ (that is, 76 percent of the solar energy incident is absorbed by the exposed surface of the Trombe wall), and the average combined heat transfer coefficient for heat loss from the Trombe wall to the ambient is determined to be $h_{\text{out}} = 3.4$ W/m² · °C. The interior of the house is maintained at $T_{\text{in}} = 20$°C at all times, and the heat transfer coefficient at the interior surface of the Trombe wall is $h_{\text{in}} = 9.1$ W/m² · °C. Also, the vents on the Trombe wall are kept closed, and thus the only heat transfer between the air in the house and the Trombe wall is through the interior surface of the wall. Assuming the temperature of the Trombe wall to

TABLE P5-83

The hourly variations of the monthly average ambient temperature and solar heat flux incident on a vertical surface

Time of day	Ambient temperature, °C	Solar insolation, W/m²
7 AM–10 AM	0	375
10 AM–1 PM	4	750
1 PM–4 PM	6	580
4 PM–7 PM	1	95
7 PM–10 PM	−2	0
10 PM–1 AM	−3	0
1 AM–4 AM	−4	0
4 AM–7 AM	4	0

FIGURE P5-83

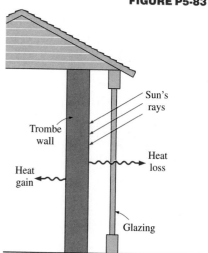

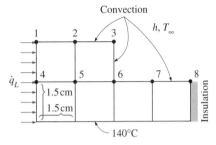

FIGURE P5-84

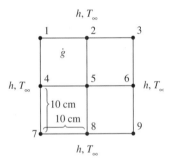

FIGURE P5-85

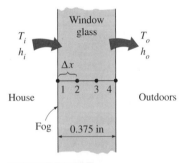

FIGURE P5-86E

vary linearly between 20°C at the interior surface and 0°C at the exterior surface at 7 AM and using the explicit finite difference method with a uniform nodal spacing of $\Delta x = 5$ cm, determine the temperature distribution along the thickness of the Trombe wall after 6, 12, 18, 24, 30, 36, 42, and 48 hours and plot the results. Also, determine the net amount of heat transferred to the house from the Trombe wall during the first day if the wall is 2.8 m high and 7 m long.

5-84 Consider two-dimensional transient heat transfer in an L-shaped solid bar that is initially at a uniform temperature of 140°C and whose cross-section is given in the figure. The thermal conductivity and diffusivity of the body are $k = 15$ W/m · °C and $\alpha = 3.2 \times 10^{-6}$ m²/s, respectively, and heat is generated in the body at a rate of $\dot{g} = 2 \times 10^7$ W/m³. The right surface of the body is insulated, and the bottom surface is maintained at a uniform temperature of 140°C at all times. At time $t = 0$, the entire top surface is subjected to convection with ambient air at $T_\infty = 25$°C with a heat transfer coefficient of $h = 80$ W/m² · °C, and the left surface is subjected to uniform heat flux at a rate of $\dot{q}_L = 8000$ W/m². The nodal network of the problem consists of 13 equally spaced nodes with $\Delta x = \Delta y = 1.5$ cm. Using the explicit method, determine the temperature at the top corner (node 3) of the body after 2, 5, and 30 min.

5-85 Consider a long solid bar ($k = 28$ W/m · °C and $\alpha = 12 \times 10^{-6}$ m²/s) of square cross-section that is initially at a uniform temperature of 20°C. The cross-section of the bar is 20 cm × 20 cm in size, and heat is generated in it uniformly at a rate of $\dot{g} = 2 \times 10^5$ W/m³. All four sides of the bar are subjected to convection to the ambient air at $T_\infty = 20$°C with a heat transfer coefficient of $h = 45$ W/m² · °C. Using the explicit finite difference method with a mesh size of $\Delta x = \Delta y = 10$ cm, determine the centerline temperature of the bar (a) after 10 min and (b) after steady conditions are established.

5-86E Consider a house whose windows are made of 0.375-in.-thick glass ($k = 0.48$ Btu/h · ft · °F and $\alpha = 4.2 \times 10^{-6}$ ft²/s). Initially, the entire house, including the walls and the windows, is at the outdoor temperature of $T_o = 35$°F. It is observed that the windows are fogged because the indoor temperature is below the dew-point temperature of 54°F. Now the heater is turned on and the air temperature in the house is raised to $T_i = 72$°F at a rate of 2°F rise per minute. The heat transfer coefficients at the inner and outer surfaces of the wall can be taken to be $h_i = 1.2$ and $h_o = 2.6$ Btu/h · ft² · °F, respectively, and the outdoor temperature can be assumed to remain constant. Using the explicit finite difference method with a mesh size of $\Delta x = 0.125$ in., determine how long it will take for the fog on the windows to clear up (i.e., for the inner surface temperature of the window glass to reach 54°F).

5-87 A common annoyance in cars in winter months is the formation of fog on the glass surfaces that blocks the view. A practical way of solving this problem is to blow hot air or to attach electric resistance heaters to the inner surfaces. Consider the rear window of a car that consists of a 0.4-cm-thick glass ($k = 0.84$ W/m · °C and $\alpha = 0.39 \times 10^{-6}$ m²/s). Strip heater wires of negligible thickness are attached to the inner surface of the glass, 4 cm apart. Each wire generates heat at a rate of 10 W/m length. Initially the entire car, including its windows, is at the outdoor temperature of $T_o = -3°C$. The heat transfer coefficients at the inner and outer surfaces of the glass can be taken to be $h_i = 6$ and $h_o = 20$ W/m² · °C, respectively. Using the explicit finite difference method with a mesh size of $\Delta x = 0.2$ cm along the thickness and $\Delta y = 1$ cm in the direction normal to the heater wires, determine the temperature distribution throughout the glass 15 min after the strip heaters are turned on. Also, determine the temperature distribution when steady conditions are reached.

5-88 Repeat Problem 5-87 using the implicit method with a time step of 1 min.

5-89 The roof of a house consists of a 15-cm-thick concrete slab ($k = 1.4$ W/m · °C and $\alpha = 0.69 \times 10^{-6}$ m²/s) that is 15 m wide and 20 m long. One evening at 6 PM, the slab is observed to be at a uniform temperature of 18°C. The average ambient air and the night sky temperatures for the entire night are predicted to be 6°C and 260 K, respectively. The convection heat transfer coefficients at the inner and outer surfaces of the roof can be taken to be $h_i = 5$ and $h_o = 12$ W/m² · °C, respectively. The house and the interior surfaces of the walls and the floor are maintained at a constant temperature of 20°C during the night, and the emissivity of both surfaces of the concrete roof is 0.9. Considering both radiation and convection heat transfers and using the explicit finite difference method with a time step of $\Delta t = 5$ min and a mesh size of $\Delta x = 3$ cm, determine the temperatures of the inner and outer surfaces of the roof at 6 AM. Also, determine the average rate of heat transfer through the roof during that night.

5-90 Consider a refrigerator whose outer dimensions are 1.80 m × 0.8 m × 0.7 m. The walls of the refrigerator are constructed of 3-cm-thick urethane insulation ($k = 0.026$ W/m · °C and $\alpha = 0.36 \times 10^{-6}$ m²/s) sandwiched between two layers of sheet metal with negligible thickness. The refrigerated space is maintained at 3°C and the average heat transfer coefficients at the inner and outer surfaces of the wall are 6 W/m² · °C and 9 W/m² · °C, respectively. Heat transfer through the bottom surface of the refrigerator is negligible. The kitchen temperature remains constant at about 25°C. Initially, the refrigerator contains 15 kg of food items at an average specific heat of 3.6 kJ/kg · °C. Now a malfunction occurs and the refrigerator stops running for 6 h as a result. Assuming the temperature of the contents of the refrigerator, including the air inside, rises uniformly during this period, predict the temperature inside the refrigerator after 6 h when the repairman arrives. Use the explicit finite difference method with a time step of $\Delta t = 1$ min and a mesh size of $\Delta x = 1$ cm and disregard corner effects (i.e., assume one-dimensional heat transfer in the walls).

Controlling the Numerical Error

5-91C Why do the results obtained using a numerical method differ from the exact results obtained analytically? What are the causes of this difference?

5-92C What is the cause of the discretization error? How does the global discretization error differ from the local discretization error?

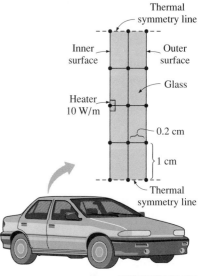

FIGURE P5-87

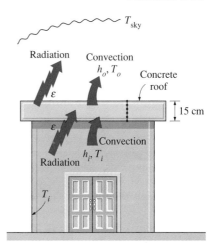

FIGURE P5-89

FIGURE P5-90

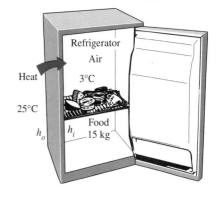

5-93C Can the global (accumulated) discretization error be less than the local error during a step? Explain.

5-94C How is the finite difference formulation for the first derivative related to the Taylor series expansion of the solution function?

5-95C Explain why the local discretization error of the finite difference method is proportional to the square of the step size. Also explain why the global discretization error is proportional to the step size itself.

5-96C What causes the round-off error? What kind of calculations are most susceptible to round-off error?

5-97C What happens to the discretization and the round-off errors as the step size is decreased?

5-98C Suggest some practical ways of reducing the round-off error.

5-99C What is a practical way of checking if the round-off error has been significant in calculations?

5-100C What is a practical way of checking if the discretization error has been significant in calculations?

Review Problems

5-101 Starting with an energy balance on the volume element, obtain the steady three-dimensional finite difference equation for a general interior node in rectangular coordinates for $T(x, y, z)$ for the case of constant thermal conductivity and uniform heat generation.

5-102 Starting with an energy balance on the volume element, obtain the three-dimensional transient explicit finite difference equation for a general interior node in rectangular coordinates for $T(x, y, z, t)$ for the case of constant thermal conductivity and no heat generation.

5-103 Consider steady one-dimensional heat conduction in a plane wall with variable heat generation and constant thermal conductivity. The nodal network of the medium consists of nodes 0, 1, 2, and 3 with a uniform nodal spacing of Δx. The temperature at the left boundary (node 0) is specified. Using the energy balance approach, obtain the finite difference formulation of boundary node 3 at the right boundary for the case of combined convection and radiation with an emissivity of ε, convection coefficient of h, ambient temperature of T_∞, and surrounding temperature of T_{surr}. Also, obtain the finite difference formulation for the rate of heat transfer at the left boundary.

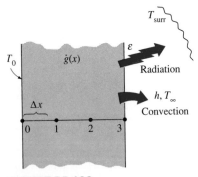

FIGURE P5-103

5-104 Consider one-dimensional transient heat conduction in a plane wall with variable heat generation and variable thermal conductivity. The nodal network of the medium consists of nodes 0, 1, and 2 with a uniform nodal spacing of Δx. Using the energy balance approach, obtain the explicit finite difference formulation of this problem for the case of specified heat flux $\dot{q}_0$

and convection at the left boundary (node 0) with a convection coefficient of h and ambient temperature of T_∞, and radiation at the right boundary (node 2) with an emissivity of ε and surrounding temperature of T_{surr}.

5-105 Repeat Problem 5-104 for the case of implicit formulation.

5-106 Consider steady one-dimensional heat conduction in a pin fin of constant diameter D with constant thermal conductivity. The fin is losing heat by convection with the ambient air at T_∞ (in °C) with a convection coefficient of h, and by radiation to the surrounding surfaces at an average temperature of T_{surr} (in K). The nodal network of the fin consists of nodes 0 (at the base), 1 (in the middle), and 2 (at the fin tip) with a uniform nodal spacing of Δx. Using the energy balance approach, obtain the finite difference formulation of this problem for the case of a specified temperature at the fin base and convection and radiation heat transfer at the fin tip.

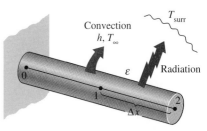

FIGURE P5-106

5-107 Starting with an energy balance on the volume element, obtain the two-dimensional transient explicit finite difference equation for a general interior node in rectangular coordinates for $T(x, y, t)$ for the case of constant thermal conductivity and uniform heat generation.

5-108 Starting with an energy balance on a disk volume element, derive the one-dimensional transient implicit finite difference equation for a general interior node for $T(z, t)$ in a cylinder whose side surface is subjected to convection with a convection coefficient of h and an ambient temperature of T_∞ for the case of constant thermal conductivity with uniform heat generation.

5-109E The roof of a house consists of a 5-in.-thick concrete slab ($k = 0.81$ Btu/h · ft · °F and $\alpha = 7.4 \times 10^{-6}$ ft²/s) that is 45 ft wide and 60 ft long. One evening at 6 PM, the slab is observed to be at a uniform temperature of 65°F. The ambient air temperature is predicted to be at about 50°F from 6 PM to 10 PM, 42°F from 10 PM to 2 AM, and 38°F from 2 AM to 6 AM, while the night sky temperature is expected to be about 445 R for the entire night. The convection heat transfer coefficients at the inner and outer surfaces of the roof can be taken to be $h_i = 0.9$ and $h_o = 2.1$ Btu/h · ft² · °F, respectively. The house and the interior surfaces of the walls and the floor are maintained at a constant temperature of 70°F during the night, and the emissivity of both surfaces of the concrete roof is 0.9. Considering both radiation and convection heat transfers and using the explicit finite difference method with a mesh size of $\Delta x = 1$ in. and a time step of $\Delta t = 5$ min, determine the temperatures of the inner and outer surfaces of the roof at 6 AM. Also, determine the average rate of heat transfer through the roof during that night.

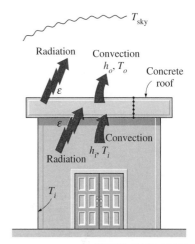

FIGURE P5-109E

FIGURE P5-110

5-110 Solar radiation incident on a large body of clean water ($k = 0.61$ W/m · °C and $\alpha = 0.15 \times 10^{-6}$ m²/s) such as a lake, a river, or a pond is mostly absorbed by water, and the amount of absorption varies with depth. For solar radiation incident at a 45° angle on a 1-m-deep large pond whose bottom surface is black (zero reflectivity), for example, 2.8 percent of the solar energy is reflected back to the atmosphere, 37.9 percent is absorbed by the bottom surface, and the remaining 59.3 percent is absorbed by the

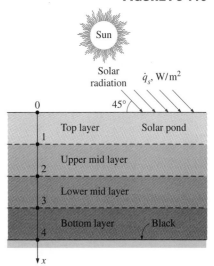

water body. If the pond is considered to be four layers of equal thickness (0.25 m in this case), it can be shown that 47.3 percent of the incident solar energy is absorbed by the top layer, 6.1 percent by the upper mid layer, 3.6 percent by the lower mid layer, and 2.4 percent by the bottom layer [for more information see Çengel and Özişik, *Solar Energy,* 33, no. 6 (1984), pp. 581–591]. The radiation absorbed by the water can be treated conveniently as heat generation in the heat transfer analysis of the pond.

Consider a large 1-m-deep pond that is initially at a uniform temperature of 15°C throughout. Solar energy is incident on the pond surface at 45° at an average rate of 500 W/m^2 for a period of 4 h. Assuming no convection currents in the water and using the explicit finite difference method with a mesh size of $\Delta x = 0.25$ m and a time step of $\Delta t = 15$ min, determine the temperature distribution in the pond under the most favorable conditions (i.e., no heat losses from the top or bottom surfaces of the pond). The solar energy absorbed by the bottom surface of the pond can be treated as a heat flux to the water at that surface in this case.

5-111 Reconsider Problem 5-110. The absorption of solar radiation in that case can be expressed more accurately as a fourth-degree polynomial as

$$\dot{g}(x) = \dot{q}_s(0.859 - 3.415x + 6.704x^2 - 6.339x^3 + 2.278x^4), \text{ W/m}^3$$

where $\dot{q}_s$ is the solar flux incident on the surface of the pond in W/m^2 and x is the distance from the free surface of the pond in m. Solve Problem 5-110 using the relation above for the absorption of solar radiation.

5-112 A hot surface at 100°C is to be cooled by attaching 8 cm long, 0.8 cm in diameter aluminum pin fins ($k = 237$ W/m · °C and $\alpha = 97.1 \times 10^{-6}$ m^2/s) to it with a center-to-center distance of 1.6 cm. The temperature of the surrounding medium is 30°C, and the heat transfer coefficient on the surfaces is 35 W/m^2 · °C. Initially, the fins are at a uniform temperature of 30°C, and at time $t = 0$, the temperature of the hot surface is raised to 100°C. Assuming one-dimensional heat conduction along the fin and taking the nodal spacing to be $\Delta x = 2$ cm and a time step to be $\Delta t = 0.5$ s, determine the nodal temperatures after 5 min by using the explicit finite difference method. Also, determine how long it will take for steady conditions to be reached.

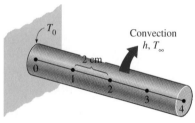

FIGURE P5-112

5-113E Consider a large plane wall of thickness $L = 0.3$ ft and thermal conductivity $k = 1.2$ Btu/h · ft · °F in space. The wall is covered with a material having an emissivity of $\varepsilon = 0.80$ and a solar absorptivity of $\alpha_s = 0.45$. The inner surface of the wall is maintained at 520 R at all times, while the outer surface is exposed to solar radiation that is incident at a rate of $\dot{q}_s = 300$ Btu/h · ft^2. The outer surface is also losing heat by radiation to deep space at 0 R. Using a uniform nodal spacing of $\Delta x = 0.1$ ft, (*a*) obtain the finite difference formulation for steady one-dimensional heat conduction and (*b*) determine the nodal temperatures by solving those equations.

Answers: (*b*) 467 R, 416 R, 361 R

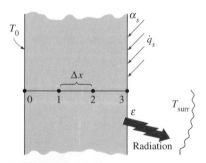

FIGURE P5-113E

5-114 Frozen food items can be defrosted by simply leaving them on the counter, but it takes too long. The process can be speeded up considerably for flat items such as steaks by placing them on a large piece of highly conducting metal, called the defrosting plate, which serves as a fin. The increased surface area enhances heat transfer and thus reduces the defrosting time.

Consider two 1.5-cm-thick frozen steaks at $-18°C$ that resemble a 15-cm-diameter circular object when placed next to each other. The steaks are now placed on a 1-cm-thick black-anodized circular aluminum defrosting plate ($k = 237$ W/m·°C, $\alpha = 97.1 \times 10^{-6}$ m²/s, and $\varepsilon = 0.90$) whose outer diameter is 30 cm. The properties of the frozen steaks are $\rho = 970$ kg/m³, $C_p = 1.55$ kJ/kg·°C, $k = 1.40$ W/m·°C, $\alpha = 0.93 \times 10^{-6}$ m²/s, and $\varepsilon = 0.95$, and the heat of fusion is $h_{if} = 187$ kJ/kg. The steaks can be considered to be defrosted when their average temperature is 0°C and all of the ice in the steaks is melted. Initially, the defrosting plate is at the room temperature of 20°C, and the wooden countertop it is placed on can be treated as insulation. Also, the surrounding surfaces can be taken to be at the same temperature as the ambient air, and the convection heat transfer coefficient for all exposed surfaces can be taken to be 12 W/m²·°C. Heat transfer from the lateral surfaces of the steaks and the defrosting plate can be neglected. Assuming one-dimensional heat conduction in both the steaks and the defrosting plate and using the explicit finite difference method, determine how long it will take to defrost the steaks. Use four nodes with a nodal spacing of $\Delta x = 0.5$ cm for the steaks, and three nodes with a nodal spacing of $\Delta r = 3.75$ cm for the exposed portion of the defrosting plate. Also, use a time step of $\Delta t = 5$ s. *Hint:* First, determine the total amount of heat transfer needed to defrost the steaks, and then determine how long it will take to transfer that much heat.

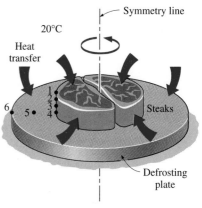

FIGURE P5-114

5-115 Repeat Problem 5-114 for a copper defrosting plate using a time step of $\Delta t = 3$ s.

Computer, Design, and Essay Problems

5-116 Write a two-page essay on the finite element method, and explain why it is used in most commercial engineering software packages. Also explain how it compares to the finite difference method.

5-117 Numerous professional software packages are available in the market for performing heat transfer analysis, and they are widely advertised in professional magazines such as the *Mechanical Engineering* magazine published by the American Society of Mechanical Engineers (ASME). Your company decides to purchase such a software package and asks you to prepare a report on the available packages, their costs, capabilities, ease of use, and compatibility with the available hardware, and other software as well as the reputation of the software company, their history, financial health, customer support, training, and future prospects, among other things. After a preliminary investigation, select the top three packages and prepare a full report on them.

5-118 Design a defrosting plate to speed up defrosting of flat food items such as frozen steaks and packaged vegetables and evaluate its performance using the finite difference method (see Prob. 5-114). Compare your design to the defrosting plates currently available on the market. The plate must perform well, and it must be suitable for purchase and use as a household utensil, durable, easy to clean, easy to manufacture, and affordable. The frozen food is expected to be at an initial temperature of $-18°C$ at the beginning of the thawing process and $0°C$ at the end with all the ice melted. Specify the material, shape, size, and thickness of the proposed plate. Justify your recommendations by calculations. Take the ambient and surrounding surface temperatures to be $20°C$ and the convection heat transfer coefficient to be $15 \ W/m^2 \cdot °C$ in your analysis. For a typical case, determine the defrosting time with and without the plate.

5-119 Design a fire-resistant safety box whose outer dimensions are $0.5 \ m \times 0.5 \ m \times 0.5 \ m$ that will protect its combustible contents from fire which may last up to 2 h. Assume the box will be exposed to an environment at an average temperature of $700°C$ with a combined heat transfer coefficient of $70 \ W/m^2 \cdot °C$ and the temperature inside the box must be below $150°C$ at the end of 2 h. The cavity of the box must be as large as possible while meeting the design constraints, and the insulation material selected must withstand the high temperatures to which it will be exposed. Cost, durability, and strength are also important considerations in the selection of insulation materials.

Forced Convection

CHAPTER **6**

So far, we have considered *conduction*, which is the mechanism of heat transfer through a solid or fluid in the absence of any fluid motion. We now consider *convection*, which is the mechanism of heat transfer through a fluid in the presence of bulk fluid motion.

Convection is classified as *natural* (or *free*) or *forced convection*, depending on how the fluid motion is initiated. In forced convection, the fluid is forced to flow over a surface or in a tube by external means such as a pump or a fan. In natural convection, any fluid motion is caused by natural means such as the buoyancy effect, which manifests itself as the rise of warmer fluid and the fall of the cooler fluid. Convection is also classified as *external* and *internal,* depending on whether the fluid is forced to flow over a surface or in a channel. Convection in external and internal flows exhibits very different characteristics.

In this chapter, we consider both *external* and *internal forced convection*. In Chapter 7 we will consider *natural convection*, which is associated with naturally caused fluid motions.

We start this chapter with a general physical description of the *convection* mechanism and the *velocity* and *thermal boundary layers*. We continue with the discussion of the dimensionless *Reynolds, Prandtl,* and *Nusselt numbers,* and their physical significance. We then present empirical relations for *friction* and *heat transfer coefficients* for flow over various geometries such as a flat plate, cylinder, and sphere, for both laminar and turbulent flow conditions. Finally, we discuss the characteristics of flow inside tubes and present the pressure drop and heat transfer correlations associated with it.

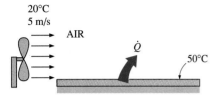

(a) Forced convection

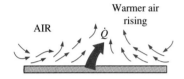

(b) Free convection

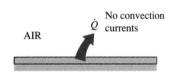

(c) Conduction

FIGURE 6-1

Heat transfer from a hot surface to the surrounding fluid by convection and conduction.

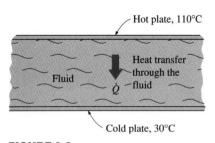

FIGURE 6-2

Heat transfer through a fluid sandwiched between two parallel plates.

6-1 ■ PHYSICAL MECHANISM OF FORCED CONVECTION

We mentioned earlier that there are three basic mechanisms of heat transfer: conduction, convection, and radiation. Conduction and convection are similar in that both mechanisms require the presence of a material medium. But they are different in that convection requires the presence of *fluid motion*.

Heat transfer through a *solid* is always by *conduction*, since the molecules of a solid remain at relatively fixed positions. Heat transfer through a *liquid* or *gas,* however, can be by *conduction* or *convection,* depending on the presence of any bulk fluid motion. Heat transfer through a fluid is by *convection* in the presence of bulk fluid motion and by *conduction* in the absence of it. Therefore, conduction in a fluid can be viewed as the *limiting case* of convection, corresponding to the case of quiescent fluid (Fig. 6-1).

Convection heat transfer is complicated by the fact that it involves *fluid motion* as well as *heat conduction*. The fluid motion *enhances* heat transfer, since it brings hotter and cooler chunks of fluid into contact, initiating higher rates of conduction at a greater number of sites in a fluid. Therefore, the rate of heat transfer through a fluid is much higher by convection than it is by conduction. In fact, the higher the *fluid velocity,* the higher the rate of *heat transfer.*

To clarify this point further, consider steady heat transfer through a fluid contained between two *parallel plates* maintained at different temperatures, as shown in Figure 6-2. The temperatures of the fluid and the plate will be the same at the points of contact because of the *continuity of temperature.* Assuming no fluid motion, the energy of the hotter fluid molecules near the hot plate will be transferred to the adjacent cooler fluid molecules. This energy will then be transferred to the next layer of the cooler fluid molecules. This energy will then be transferred to the next layer of the cooler fluid, and so on, until it is finally transferred to the other plate. This is what happens during *conduction* through a fluid. Now let us use a syringe to draw some fluid near the hot plate and inject it near the cold plate repeatedly. You can imagine that this will speed up the heat transfer process considerably, since some energy is *carried* to the other side as a result of fluid motion.

Consider the cooling of a *hot iron block* with a fan blowing air over its top surface, as shown in Figure 6-3. We know that heat will be transferred from the hot block to the surrounding cooler air, and the block will eventually cool. We also know that the block will cool faster if the fan is switched to a higher speed. Replacing air by water will enhance the convection heat transfer even more.

Experience shows that convection heat transfer strongly depends on the fluid properties *dynamic viscosity* μ, *thermal conductivity* k, *density* ρ, and *specific heat* C_p, as well as the *fluid velocity* $\mathcal{V}$. It also depends on the *geometry* and *roughness* of the solid surface, in addition to the *type of fluid flow* (such as being streamlined or turbulent). Thus, we expect the convection heat transfer relations to be rather complex because of the dependence of convection on so many variables. This is not surprising, since convection is the most complex mechanism of heat transfer.

Despite the complexity of convection, the rate of *convection heat transfer* is observed to be proportional to the temperature difference and is conveniently expressed by **Newton's law of cooling** as

$$\dot{q}_{conv} = h(T_s - T_\infty) \quad (W/m^2) \qquad (6\text{-}1)$$

or

$$\dot{Q}_{conv} = hA(T_s - T_\infty) \quad (W) \qquad (6\text{-}2)$$

where

h = convection heat transfer coefficient, W/m² · °C
A = heat transfer surface area, m²
T_s = temperature of the surface, °C
T_∞ = temperature of the fluid sufficiently far from the surface, °C

Judging from its units, the **convection heat transfer coefficient** can be defined as *the rate of heat transfer between a solid surface and a fluid per unit surface area per unit temperature difference.*

You should not be deceived by the simple appearance of this relation, because the convection heat transfer coefficient h depends on the several variables mentioned above, and thus is difficult to determine.

When a fluid is forced to flow over a solid surface, it is observed that the fluid layer in contact with the solid surface "sticks" to the surface. That is, a very thin layer of fluid assumes *zero velocity* at the wall. In fluid flow, this phenomenon is known as the *no-slip* condition. An implication of the no-slip condition is that heat transfer from the solid surface to the fluid layer adjacent to the surface is by *pure conduction*, since the fluid layer is motionless, and can be expressed as

$$\dot{q}_{conv} = \dot{q}_{cond} = -k_{fluid} \left.\frac{\partial T}{\partial y}\right|_{y=0} \quad (W/m^2) \qquad (6\text{-}3)$$

where T represents the temperature distribution in the fluid and $(\partial T/\partial y)_{y=0}$ is the *temperature gradient* at the surface. This heat is then *convected away* from the surface as a result of fluid motion. Note that convection heat transfer from a solid surface to a fluid is merely the conduction heat transfer from the solid surface to the fluid layer adjacent to the surface. Therefore, we can equate the expressions 6-1 and 6-3 for the heat flux to obtain

$$h = \frac{-k_{fluid}(\partial T/\partial y)_{y=0}}{T_s - T_\infty} \quad (W/m^2 \cdot °C) \qquad (6\text{-}4)$$

for the determination of the *convection heat transfer coefficient* when the temperature distribution within the fluid is known.

The convection heat transfer coefficient, in general, varies along the flow (or *x*-) direction. The *average* or *mean* convection heat transfer coefficient for a surface in such cases is determined by properly averaging the *local* convection heat transfer coefficients over the entire surface.

In convection studies, it is common practice to nondimensionalize the governing equations and combine the variables, which group together into

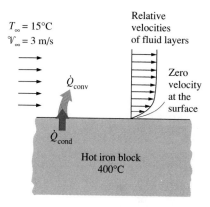

$T_\infty = 15°C$
$V_\infty = 3$ m/s

Relative velocities of fluid layers

Zero velocity at the surface

$\dot{Q}_{conv}$

$\dot{Q}_{cond}$

Hot iron block
400°C

FIGURE 6-3

The cooling of a hot block by forced convection.

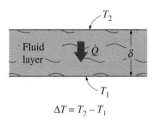

FIGURE 6-4

Heat transfer through a fluid layer of thickness δ and temperature difference ΔT.

Blowing on food

FIGURE 6-5

We resort to forced convection whenever we need to increase the rate of heat transfer.

dimensionless numbers in order to reduce the number of total variables. It is also common practice to *nondimensionalize* the heat transfer coefficient h with the **Nusselt number,** defined as

$$\text{Nu} = \frac{h\delta}{k} \tag{6-5}$$

where k is the thermal conductivity of the fluid and δ is the *characteristic length*. The Nusselt number is named after Wilhelm Nusselt, who made significant contributions to convective heat transfer in the first half of the 20th century, and it is viewed as the *dimensionless convection heat transfer coefficient*.

To understand the physical significance of the Nusselt number, consider a fluid layer of thickness δ and temperature difference $\Delta T = T_2 - T_1$, as shown in Figure 6-4. Heat transfer through the fluid layer will be by *convection* when the fluid involves some motion and by *conduction* when the fluid layer is motionless. Heat flux (the rate of heat transfer per unit time per unit surface area) in either case will be

$$\dot{q}_{\text{conv}} = h\Delta T$$

and

$$\dot{q}_{\text{cond}} = k\frac{\Delta T}{\delta}$$

Taking their ratio gives

$$\frac{\dot{q}_{\text{conv}}}{\dot{q}_{\text{cond}}} = \frac{h\Delta T}{k\Delta T/\delta} = \frac{h\delta}{k} = \text{Nu}$$

which is the *Nusselt number*. Therefore, the Nusselt number represents the enhancement of heat transfer through a fluid layer as a result of convection relative to conduction across the same fluid layer. The larger the Nusselt number, the more effective the convection. A Nusselt number Nu = 1 for a fluid layer represents heat transfer by pure conduction.

We use forced convection in daily life more often than you might think (Fig. 6-5). We resort to forced convection whenever we want to increase the rate of heat transfer from a hot object. For example, we turn on the *fan* on hot summer days to help our body cool more effectively. The higher the fan speed, the better we feel. We *stir* our soup and *blow* on a hot slice of pizza to make them cool faster. The air on *windy* winter days feels much colder than it actually is. The simplest solution to heating problems in electronics packaging is to use a large enough fan.

6-2 ■ VELOCITY BOUNDARY LAYER

Consider the flow of a fluid over a *flat plate*, as shown in Figure 6-6. The x-coordinate is measured along the plate surface from the *leading edge* of the plate in the direction of the flow, and y is measured from the surface in the normal direction. The fluid approaches the plate in the x-direction with a uniform velocity $\mathcal{V}_\infty$. For the sake of discussion, we can consider the fluid

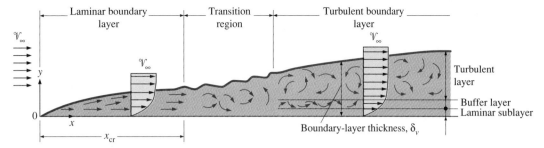

FIGURE 6-6

The development of the boundary layer for flow over a flat plate, and the different flow regimes.

to consist of *adjacent layers* piled on top of each other. The velocity of the particles in the first fluid layer adjacent to the plate becomes *zero* because of the no-slip condition. This motionless layer *slows down* the particles of the neighboring fluid layer as a result of friction between the particles of these two adjoining fluid layers at different velocities. This fluid layer then slows down the molecules of the next layer, and so on. Thus, the presence of the plate is felt up to some distance δ_v from the plate beyond which the fluid velocity $\mathcal{V}_\infty$ remains essentially unchanged. As a result, the fluid velocity at any x-location will vary from 0 at $y = 0$ to nearly $\mathcal{V}_\infty$ at $y = \delta_v$.

The region of the flow above the plate bounded by δ_v in which the effects of the viscous shearing forces caused by fluid viscosity are felt is called the **velocity boundary layer** or just the **boundary layer.** The *thickness* of the boundary layer, δ_v, is arbitrarily defined as the distance from the surface at which $\mathcal{V} = 0.99\mathcal{V}_\infty$.

The hypothetical line of $\mathcal{V} = 0.99\mathcal{V}_\infty$ divides the flow over a plate into two regions: the **boundary layer region,** in which the viscous effects and the velocity changes are significant, and the **inviscid flow region,** in which the frictional effects are negligible and the velocity remains essentially constant.

Consider two adjacent fluid layers in the boundary layer. The faster layer will try to drag along the slower one because of the friction between the two layers, exerting a *drag force* (or *friction force*) on it. The drag force per unit area is called the **shear stress** and is denoted by τ. Experimental studies indicate that the shear stress for most fluids is proportional to the *velocity gradient,* and the shear stress at the wall surface is expressed as

$$\tau_s = \mu \left. \frac{\partial \mathcal{V}}{\partial y} \right|_{y=0} \qquad (\text{N/m}^2) \qquad (6\text{-}6)$$

where the constant of proportionality μ is the *dynamic viscosity* of the fluid, whose unit is kg/m · s, or equivalently N · s/m².

The viscosity of a fluid is a measure of its *resistance to flow,* and it is a strong function of temperature. The viscosities of liquids *decrease* with temperature, whereas the viscosities of gases *increase* with temperature. The viscosities of some fluids at 20°C are listed in Table 6-1. Note that the viscosities of the different fluids differ by several orders of magnitude. Also note that it is more difficult to move an object in a higher-viscosity fluid such as engine oil than it is in a lower-viscosity fluid such as water.

TABLE 6-1

The dynamic viscosities of some liquids and air at 20°C

Fluid	μ, kg/m · s
Glycerin	1.49
Engine oil	0.800
Ethyl alcohol	0.00120
Water	0.00106
Freon-12	0.000262
Air	0.0000182

The determination of the surface shear stress τ_s from Equation 6-6 is not practical since it requires a knowledge of the flow velocity profile. A more practical approach in external flow is to relate τ_s to the free-stream velocity $\mathcal{V}_\infty$ as

$$\tau_s = C_f \frac{\rho \mathcal{V}_\infty^2}{2} \qquad (N/m^2) \qquad (6\text{-}7)$$

where C_f is the **friction coefficient** or the **drag coefficient,** whose value in most cases is determined experimentally, and ρ is the density of the fluid. Note that the friction coefficient, in general, will vary with location along the surface. Once the average friction coefficient over a given surface is available, the drag or friction force over the entire surface is determined from

$$F_D = C_f A \frac{\rho \mathcal{V}_\infty^2}{2} \qquad (N) \qquad (6\text{-}8)$$

where A is the surface area.

The friction coefficient is an important parameter in heat transfer studies since it is directly related to the heat transfer coefficient and the power requirements of the pump or fan.

Laminar and Turbulent Flows

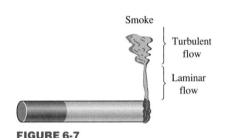

FIGURE 6-7

Laminar and turbulent flow regimes of cigarette smoke.

If you have been around smokers, you probably noticed that the cigarette smoke rises in a smooth plume for the first few centimeters and then starts fluctuating randomly in all directions as it continues its journey toward the lungs of nonsmokers (Fig. 6-7). Likewise, a careful inspection of flow over a flat plate reveals that the fluid flow in the boundary layer starts out as flat and streamlined but turns chaotic after some distance from the leading edge, as shown in Figure 6-6. The flow regime in the first case is said to be **laminar**, characterized by *smooth streamlines* and *highly ordered motion*, and **turbulent** in the second case, where it is characterized by *velocity fluctuations* and *highly disordered motion*. The **transition** from laminar to turbulent flow does not occur suddenly; rather, it occurs over some region in which the flow hesitates between laminar and turbulent flows before it becomes fully turbulent.

We can verify the existence of these laminar, transition, and turbulent flow regimes by injecting some dye into the flow stream. We will observe that the dye streak will form a *smooth line* when the flow is laminar, will have *bursts of fluctuations* in the transition regime, and will *zigzag rapidly and randomly* when the flow becomes fully turbulent.

Typical velocity profiles in laminar and turbulent flow are also given in Figure 6-6. Note that the velocity profile is approximately parabolic in laminar flow and becomes flatter in turbulent flow, with a sharp drop near the surface. The turbulent boundary layer can be considered to consist of three layers. The very thin layer next to the wall where the viscous effects are dominant is the **laminar sublayer.** The velocity profile in this layer is nearly linear, and the flow is streamlined. Next to the laminar sublayer is the **buffer**

layer, in which the turbulent effects are significant but not dominant of the diffusion effects, and next to it is the **turbulent layer,** in which the turbulent effects dominate.

The *intense mixing* of the fluid in turbulent flow as a result of rapid fluctuations enhances heat and momentum transfer between fluid particles, which increases the friction force on the surface and the convection heat transfer rate (Fig. 6-8). It also causes the boundary layer to enlarge. Both the friction and heat transfer coefficients reach maximum values when the flow becomes *fully turbulent.* So it will come as no surprise that a special effort is made in the design of heat transfer coefficients associated with turbulent flow. The enhancement in heat transfer in turbulent flow does not come for free, however. It may be necessary to use a larger pump or fan in turbulent flow to overcome the larger friction forces accompanying the higher heat transfer rate.

(*a*) Before
turbulence (°C)

(*b*) After
turbulence (°C)

FIGURE 6-8

The intense mixing in turbulent flow brings fluid particles at different temperatures into close contact, and thus enhances heat transfer.

Reynolds Number

The transition from laminar to turbulent flow depends on the *surface geometry, surface roughness, free-stream velocity, surface temperature,* and *type of fluid,* among other things. After exhaustive experiments in the 1880s, Osborn Reynolds discovered that the flow regime depends mainly on the ratio of the *inertia forces* to *viscous forces* in the fluid. This ratio is called the **Reynolds number** and is expressed for external flow as (Fig. 6-9)

$$\text{Re} = \frac{\text{Inertia forces}}{\text{Viscous forces}} = \frac{\mathcal{V}_\infty \delta}{\nu} \tag{6-9}$$

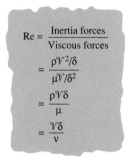

FIGURE 6-9

The Reynolds number can be viewed as the ratio of the inertia forces to viscous forces acting on a fluid volume element.

where
$\mathcal{V}_\infty$ = free-stream velocity, m/s
δ = characteristic length of the geometry, m
$\nu = \mu/\rho$ = kinematic viscosity of the fluid, m²/s

Note that the Reynolds number is a *dimensionless* quantity. Also note that *kinematic viscosity ν* differs from dynamic viscosity μ by the factor ρ. Kinematic viscosity has the unit m²/s, which is identical to the unit of thermal diffusivity, and can be viewed as *viscous diffusivity.* The characteristic length is the distance from the leading edge x in the flow direction for a flat plate and the diameter D for a circular cylinder or sphere.

At *large* Reynolds numbers, the inertia forces, which are proportional to the density and the velocity of the fluid, are large relative to the viscous forces, and thus the viscous forces cannot prevent the random and rapid fluctuations of the fluid. At *small* Reynolds numbers, however, the viscous forces are large enough to overcome the inertia forces and to keep the fluid "in line." Thus the flow is *turbulent* in the first case and *laminar* in the second.

The Reynolds number at which the flow becomes turbulent is called the **critical Reynolds number.** The value of the critical Reynolds number is different for different geometries. For flow over a *flat plate,* transition from laminar to turbulent occurs at the critical Reynolds number of

$$\text{Re}_{\text{critical, flat plate}} \approx 5 \times 10^5$$

This generally accepted value of the critical Reynolds number for a flat plate may vary somewhat depending on the surface roughness, the turbulence level, and the variation of pressure along the surface.

6-3 ■ THERMAL BOUNDARY LAYER

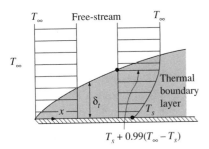

FIGURE 6-10

Thermal boundary layer on a flat plate (the fluid is hotter than the plate surface).

We have seen that a velocity boundary layer develops when a fluid flows over a surface as a result of the fluid layer adjacent to the surface assuming the surface velocity (i.e., zero velocity relative to the surface). Also, we defined the velocity boundary layer as the region in which the fluid velocity varies from zero to $0.99\mathcal{V}_\infty$. Likewise, a *thermal boundary layer* develops when a fluid at a specified temperature flows over a surface that is at a different temperature, as shown in Figure 6-10.

Consider the flow of a fluid at a uniform temperature of T_∞ over an isothermal flat plate at a temperature T_s. The fluid particles in the layer adjacent to the surface will reach thermal equilibrium with the plate and assume the surface temperature T_s. These fluid particles will then exchange energy with the particles in the adjoining fluid layer, and so on. As a result, a temperature profile will develop in the flow field that ranges from T_s at the surface to T_∞ sufficiently far from the surface. The flow region over the surface in which the temperature variation in the direction normal to the surface is significant is the **thermal boundary layer.** The *thickness* of the thermal boundary layer δ_t at any location along the surface is defined as *the distance from the surface at which the temperature difference $T - T_s$ equals* $0.99(T_\infty - T_s)$. Note that for the special case of $T_s = 0$, we have $T = 0.99T_\infty$ at the outer edge of the thermal boundary layer, which is analogous to $\mathcal{V} = 0.99\mathcal{V}_\infty$ for the velocity boundary layer.

The thickness of the thermal boundary layer increases in the flow direction, since the effects of heat transfer are felt at greater distances from the surface further down stream.

The convection heat transfer rate anywhere along the surface is directly related to the *temperature gradient* at that location. Therefore, the shape of the temperature profile in the thermal boundary layer dictates the convection heat transfer between a solid surface and the fluid flowing over it. In flow over a heated (or cooled) surface, both velocity and thermal boundary layers will develop simultaneously. Noting that the fluid velocity will have a strong influence on the temperature profile, the development of the velocity boundary layer relative to the thermal boundary layer will have a strong effect on the convection heat transfer.

The relative thickness of the velocity and the thermal boundary layers is best described by the *dimensionless* parameter **Prandtl number,** defined as

$$\text{Pr} = \frac{\text{Molecular diffusivity of momentum}}{\text{Molecular diffusivity of heat}} = \frac{\nu}{\alpha} = \frac{\mu C_p}{k} \quad (6\text{-}10)$$

It is named after Ludwig Prandtl, who introduced the concept of boundary layer in 1904 and made significant contributions to boundary layer theory.

The Prandtl numbers of fluids range from less than 0.01 for liquid metals to more than 100,000 for heavy oils (Table 6-2). Note that the Prandtl number is in the order of 10 for water.

The Prandtl numbers of gases are about 1, which indicates that both momentum and heat dissipate through the fluid at about the same rate. Heat diffuses very quickly in liquid metals ($Pr \ll 1$) and very slowly in oils ($Pr \gg 1$) relative to momentum. Consequently the thermal boundary layer is much thicker for liquid metals and much thinner for oils relative to the velocity boundary layer (Fig. 6-11).

6-4 ■ FLOW OVER FLAT PLATES

So far we have discussed the physical aspects of forced convection over surfaces. In this section, we will discuss the determination of the *heat transfer rate* to or from a flat plate, as well as the *drag force* exerted on the plate by the fluid for both laminar and turbulent flow cases. Surfaces that are slightly contoured such as turbine blades can also be approximated as flat plates with reasonable accuracy.

The friction and the heat transfer coefficients for a flat plate can be determined theoretically by solving the conservation of mass, momentum, and energy equations approximately or numerically. They can also be determined experimentally and expressed by empirical correlations. In either approach, it is found that the *average* Nusselt number can be expressed in terms of the Reynolds and Prandtl numbers in the form

$$\mathrm{Nu} = \frac{hL}{k} = C\,\mathrm{Re}_L^m\,\mathrm{Pr}^n \qquad (6\text{-}11)$$

where C, m, and n are constants and L is the *length* of the plate in the flow direction. The *local* Nusselt number at any point on the plate will depend on the distance of that point from the leading edge.

The fluid temperature in the thermal boundary layer varies from T_s at the surface to about T_∞ at the outer edge of the boundary. The fluid properties also vary with temperature, and thus with position across the boundary layer. In order to account for the variation of the properties with temperature properly, the fluid properties are usually evaluated at the so-called **film temperature,** defined as

$$T_f = \frac{T_s + T_\infty}{2} \qquad (6\text{-}12)$$

which is the *arithmetic average* of the surface and the free-stream temperatures. The fluid properties are then assumed to remain constant at those values during the entire flow.

The local friction and heat transfer coefficients *vary* along the surface of the flat plate as a result of the changes in the velocity and thermal boundary layers in the flow direction. We are usually interested in the heat transfer and drag force on the *entire* surface, which can be determined using the *average* heat transfer and friction coefficients. But sometimes we are also interested

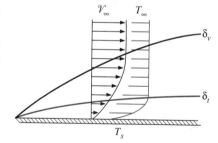
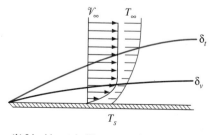

TABLE 6-2

Typical ranges of Prandtl numbers for common fluids

Fluid	Pr
Liquid metals	0.004–0.030
Gases	0.7–1.0
Water	1.7–13.7
Light organic fluids	5–50
Oils	50–100,000
Glycerin	2000–100,000

FIGURE 6-11

The relative thicknesses of the velocity and thermal boundary layers for liquid metals and oils.

(*a*) Oils

(*b*) Liquid metals (like mercury)

in the heat flux and the drag force at a certain location. In such cases, we need to know the *local* values of the heat transfer and friction coefficients. With this in mind, below we present correlations for both local and average friction and heat transfer coefficients. The *local* quantities are identified with the subscript x.

Recall that the *average* friction and heat transfer coefficients for the entire plate can be determined from the corresponding *local* values by integration from

$$C_f = \frac{1}{L} \int_0^L C_{f,x} \, dx \qquad (6\text{-}13)$$

and

$$h = \frac{1}{L} \int_0^L h_x \, dx \qquad (6\text{-}14)$$

Once C_f and h are available, the drag force and the heat transfer rate can be determined from Equations 6-2 and 6-8. Next we discuss the local and average friction and heat transfer coefficients over a flat plate for *laminar, turbulent*, and *combined laminar and turbulent* flow conditions.

1 Laminar Flow

The *local* friction coefficient and the Nusselt number at location x for laminar flow over a flat plate are given by

$$C_{f,x} = \frac{0.664}{\text{Re}_x^{1/2}} \qquad (6\text{-}15)$$

and

$$\text{Nu}_x = \frac{h_x x}{k} = 0.332 \, \text{Re}_x^{1/2} \, \text{Pr}^{1/3} \qquad (\text{Pr} \geq 0.6) \qquad (6\text{-}16)$$

where x is the distance from the leading edge of the plate and $\text{Re}_x = \mathcal{V}_\infty x/\nu$ is the Reynolds number at location x. Note that $C_{f,x}$ is proportional to $1/\text{Re}_x^{1/2}$ and thus to $x^{-1/2}$. Likewise, $\text{Nu}_x = h_x x/k$ is proportional to $x^{1/2}$ and thus h_x is proportional to $x^{-1/2}$. Therefore, both $C_{f,x}$ and h_x are supposedly *infinite* at the leading edge ($x = 0$) and decrease by a factor of $x^{-1/2}$ in the flow direction. The variation of the boundary layer thickness δ, the friction coefficient C_f, and the convection heat transfer coefficient h along an isothermal flat plate is shown in Figure 6-12.

The *average* friction coefficient and the Nusselt number over the entire plate are determined by substituting the relations above into Equations 6-13 and 6-14 and performing the simple integrations (Fig. 6-13). We get

$$C_f = \frac{1.328}{\text{Re}_L^{1/2}} \qquad (6\text{-}17)$$

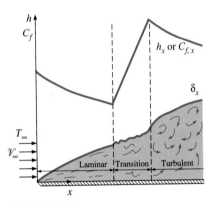

FIGURE 6-12

The variation of the local friction and heat transfer coefficients for flow over a flat plate.

FIGURE 6-13

The average friction coefficient over a surface is determined by integrating the local friction coefficient over the entire surface.

$$
\begin{aligned}
C_f &= \frac{1}{L} \int_0^L C_{f,x} \, dx \\
&= \frac{1}{L} \int_0^L \frac{0.664}{\text{Re}_x^{1/2}} \, dx \\
&= \frac{0.664}{L} \int_0^L \left(\frac{\mathcal{V}_m x}{\nu} \right)^{-1/2} dx \\
&= \frac{0.664}{L} \left(\frac{\mathcal{V}_m}{\nu} \right)^{-1/2} \frac{x^{1/2}}{\frac{1}{2}} \Big|_0^L \\
&= \frac{2 \times 0.664}{L} \left(\frac{\mathcal{V}_m L}{\nu} \right)^{-1/2} \\
&= \frac{1.328}{\text{Re}_L^{1/2}}
\end{aligned}
$$

and

$$\text{Nu} = \frac{hL}{k} = 0.664 \, \text{Re}_L^{1/2} \, \text{Pr}^{1/3} \qquad (\text{Pr} \ge 0.6) \qquad (6\text{-}18)$$

The relations above give the average friction and heat transfer coefficients for the entire plate when the flow is *laminar* over the *entire* plate.

Taking the critical Reynolds number to be $\text{Re}_{cr} = 5 \times 10^5$, the length of the plate x_{cr} over which the flow is laminar can be determined from

$$\text{Re}_{cr} = 5 \times 10^5 = \frac{\mathcal{V}_\infty x_{cr}}{\nu} \qquad (6\text{-}19)$$

Thus, the relations above can be used for $x \le x_{cr}$.

2 Turbulent Flow

The *local* friction coefficient and the Nusselt number at location x for turbulent flow over a flat plate are given by

$$C_{f,x} = \frac{0.0592}{\text{Re}_x^{1/5}} \qquad (5 \times 10^5 \le \text{Re}_x \le 10^7) \qquad (6\text{-}20)$$

and

$$\text{Nu}_x = \frac{h_x x}{k} = 0.0296 \, \text{Re}_x^{4/5} \, \text{Pr}^{1/3} \qquad \left(\begin{array}{l} 0.6 \le \text{Pr} \le 60 \\ 5 \times 10^5 \le \text{Re}_x \le 10^7 \end{array}\right) \qquad (6\text{-}21)$$

where again x is the distance from the leading edge of the plate and $\text{Re}_x = \mathcal{V}_\infty x/\nu$ is the Reynolds number at location x. The local friction and heat transfer coefficients are higher in turbulent flow than they are in laminar flow because of the intense mixing that occurs in the turbulent boundary layer. Note that both $C_{f,x}$ and h_x reach their highest values when the flow becomes fully turbulent, and then decrease by a factor of $x^{-1/5}$ in the flow direction, as shown in Figure 6-12.

The *average* friction coefficient and the Nusselt number over the entire plate in turbulent flow are determined by substituting the relations above into Equations 6-13 and 6-14 and performing the simple integrations. We get

$$C_f = \frac{0.074}{\text{Re}_L^{1/5}} \qquad (5 \times 10^5 \le \text{Re}_L \le 10^7) \qquad (6\text{-}22)$$

and

$$\text{Nu} = \frac{hL}{k} = 0.037 \, \text{Re}_L^{4/5} \, \text{Pr}^{1/3} \qquad \left(\begin{array}{l} 0.6 \le \text{Pr} \le 60 \\ 5 \times 10^5 \le \text{Re}_L \le 10^7 \end{array}\right) \qquad (6\text{-}23)$$

The two relations above give the average friction and heat transfer coefficients for the entire plate only when the flow is *turbulent* over the *entire*

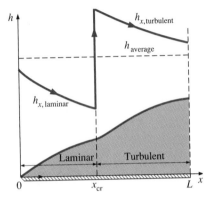

FIGURE 6-14

Graphical representation of the average heat transfer coefficient for a flat plate with combined laminar and turbulent flow.

plate, or when the laminar flow region of the plate is too small relative to the turbulent flow region (that is, $x_{cr} \ll L$).

3 Combined Laminar and Turbulent Flow

In some cases, a flat plate is sufficiently long for the flow to become turbulent, but not long enough to disregard the laminar flow region. In such cases, the *average* friction coefficient and the Nusselt number over the entire plate are determined by performing the integrations in Equations 6-13 and 6-14 over two parts: the laminar region $0 \le x \le x_{cr}$ and the turbulent region $x_{cr} < x \le L$ as

$$C_f = \frac{1}{L}\left(\int_0^{x_{cr}} C_{f,x,\text{laminar}}\, dx + \int_{x_{cr}}^{L} C_{f,x,\text{turbulent}}\, dx \right) \tag{6-24}$$

and

$$h = \frac{1}{L}\left(\int_0^{x_{cr}} h_{x,\text{laminar}}\, dx + \int_{x_{cr}}^{L} h_{x,\text{turbulent}}\, dx \right) \tag{6-25}$$

Note that we included the transition region with the turbulent region. Again taking the critical Reynolds number to be $\text{Re}_{cr} = 5 \times 10^5$ and performing the integrations above after substituting the indicated expressions, the *average* friction coefficient and the Nusselt number over the *entire* plate are determined to be (Fig. 6-14)

$$C_f = \frac{0.074}{\text{Re}_L^{1/5}} - \frac{1742}{\text{Re}_L} \qquad (5 \times 10^5 \le \text{Re}_L \le 10^7) \tag{6-26}$$

and

$$\text{Nu} = \frac{hL}{k} = (0.037\,\text{Re}_L^{4/5} - 871)\text{Pr}^{1/3} \quad \left(\begin{array}{l} 0.6 \le \text{Pr} \le 60 \\ 5 \times 10^5 \le \text{Re}_L \le 10^7 \end{array}\right) \tag{6-27}$$

The constants in the two relations above will be different for different critical Reynolds numbers.

When solving a forced convection problem over a flat plate, the first thing we do is determine Re_L, which is the Reynolds number at the rear end of the plate, using the fluid properties at the film temperature. If $\text{Re}_L < 5 \times 10^5$, the flow over the entire plate is laminar, and we use the laminar flow relations. If $\text{Re}_L > 5 \times 10^5$, then we use the turbulent flow or the combined laminar and turbulent flow relations, as appropriate. Once the average friction and heat transfer coefficients are calculated, we can determine the heat transfer rate and the drag force from

$$\dot{Q} = hA(T_s - T_\infty) \qquad \text{(W)} \tag{6-28}$$

and

$$F_D = C_f A \frac{\rho \mathcal{V}_\infty^2}{2} \qquad \text{(N)} \tag{6-29}$$

where $A = wL$ is the surface area of a flat plate of length L and width w.

These relations have been obtained for the case of *isothermal* surfaces but could also be used approximately for the case of nonisothermal surfaces by assuming the surface temperature to be constant at some average value. Also, the surfaces are assumed to be *smooth,* and the free stream to be *turbulent free.*

When a flat plate is subjected to *uniform heat flux* instead of uniform temperature, the local Nusselt number is given by

$$\text{Nu}_x = 0.453\ \text{Re}_x^{0.5}\ \text{Pr}^{1/3} \qquad (6\text{-}30)$$

for *laminar* flow and

$$\text{Nu}_x = 0.0308\ \text{Re}_x^{0.8}\ \text{Pr}^{1/3} \qquad (6\text{-}31)$$

for *turbulent* flow. The relations above give values that are 36 percent higher for laminar flow and 4 percent higher for turbulent flow relative to the isothermal plate case.

EXAMPLE 6-1 Flow of Hot Oil over a Flat Plate

Engine oil at 60°C flows over a 5-m-long flat plate whose temperature is 20°C with a velocity of 2 m/s (Fig. 6-15). Determine the total drag force and the rate of heat transfer per unit width of the entire plate.

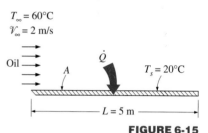

FIGURE 6-15
Schematic for Example 6-1.

Solution Hot engine oil flows over a flat plate. The total drag force and the rate of heat transfer per unit width of the plate are to be determined.

Assumptions **1** Steady operating conditions exist. **2** The critical Reynolds number is $\text{Re}_{cr} = 5 \times 10^5$. **3** Radiation effects are negligible.

Properties The properties of engine oil at the film temperature of $T_f = (T_s + T_\infty)/2 = (20 + 60)/2 = 40°C$ are (Table A-10).

$$\rho = 876\ \text{kg/m}^3 \qquad \text{Pr} = 2870$$
$$k = 0.144\ \text{W/m}\cdot°\text{C} \qquad \nu = 242 \times 10^{-6}\ \text{m}^2/\text{s}$$

Analysis Noting that $L = 5$ m, the Reynolds number at the end of the plate is

$$\text{Re}_L = \frac{\mathcal{V}_\infty L}{\nu} = \frac{(2\ \text{m/s})(5\ \text{m})}{242 \times 10^{-6}\ \text{m}^2/\text{s}} = 4.13 \times 10^4$$

which is less than the critical Reynolds number, Thus we have *laminar flow* over the entire plate, and the average friction coefficient is determined from

$$C_f = 1.328\ \text{Re}_L^{-0.5} = 1.328 \times (4.13 \times 10^4)^{-0.5} = 0.00653$$

Then the drag force acting on the plate per unit width becomes

$$F_D = C_f A \frac{\rho \mathcal{V}_\infty^2}{2} = 0.00653 \times (5 \times 1\ \text{m}^2) \frac{(876\ \text{kg/m}^3)(2\ \text{m/s})^2}{2} = \mathbf{57.2\ N}$$

This force corresponds to the weight of a mass of about 6 kg. Therefore, a person who applies an equal and opposite force to the plate to keep it from moving will feel like he or she is spending as much power as is necessary to hold a 6 kg mass from dropping.

Similarly, the Nusselt number is determined using the laminar flow relations for a flat plate,

$$\text{Nu} = \frac{hL}{k} = 0.664 \, \text{Re}_L^{0.5} \, \text{Pr}^{1/3} = 0.664 \times (4.13 \times 10^4)^{0.5} \times 2870^{1/3} = 1918$$

Then,

$$h = \frac{k}{L} \text{Nu} = \frac{0.144 \, \text{W/m} \cdot °\text{C}}{5 \, \text{m}} (1918) = 55.2 \, \text{W/m}^2 \cdot °\text{C}$$

and

$$\dot{Q} = hA(T_\infty - T_s) = (55.2 \, \text{W/m}^2 \cdot °\text{C})(5 \times 1 \, \text{m}^2)(60 - 20)°\text{C} = \textbf{11,040 W}$$

Discussion Note that heat transfer is always from the higher-temperature medium to the lower-temperature one. In this case, it is from the oil to the plate. Both the drag force and the heat transfer rate are per m width of the plate. The total quantities for the entire plate can be obtained by multiplying these quantities by the actual width of the plate.

EXAMPLE 6-2   Cooling of a Hot Block by Forced Air at High Elevation

The local atmospheric pressure in Denver, Colorado (elevation 1610m), is 83.4 kPa. Air at this pressure and 20°C flows with a velocity of 8 m/s over a 1.5 m × 6 m flat plate whose temperature is 134°C (Fig. 6-16). Determine the rate of heat transfer from the plate if the air flows parallel to the (a) 6-m-long side and (b) the 1.5-m side.

Solution The top surface of a hot block is to be cooled by forced air. The rate of heat transfer is to be determined for two cases.

Assumptions **1** Steady operating conditions exist. **2** The critical Reynolds number is $\text{Re}_{cr} = 5 \times 10^5$. **3** Radiation effects are negligible. **4** Air is an ideal gas.

Properties The properties k, μ, C_p, and Pr of ideal gases are independent of pressure, while the properties v and α are inversely proportional to density and thus pressure. The properties of air at the film temperature of $T_f = (T_s + T_\infty)/2 = (134 + 20)/2 = 77°\text{C} = 350 \, \text{K}$ and 1 atm pressure are (Table A-11)

$$k = 0.0297 \, \text{W/m} \cdot °\text{C} \qquad \text{Pr} = 0.706$$
$$v_{@ \, 1 \, atm} = 2.06 \times 10^{-5} \, \text{m}^2/\text{s}$$

The atmospheric pressure in Denver is $P = (83.4 \, \text{kPa})/(101.325 \, \text{kPa/atm}) = 0.823$ atm. Then the kinematic viscosity of air in Denver becomes

$$v = v_{@ \, 1 \, atm}/P(\text{atm}) = (2.06 \times 10^{-5} \, \text{m}^2/\text{s})/0.823 = 2.50 \times 10^{-5} \, \text{m}^2/\text{s}$$

Analysis (a) When air flow is parallel to the long side, we have $L = 6$ m, and the Reynolds number at the end of the plate becomes

$$\text{Re}_L = \frac{V_\infty L}{v} = \frac{(8 \, \text{m/s})(6 \, \text{m})}{2.50 \times 10^{-5} \, \text{m}^2/\text{s}} = 1.92 \times 10^6$$

which is greater than the critical Reynolds number. Thus, we have combined laminar and turbulent flow, and the average Nusselt number for the entire plate is determined from

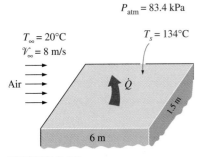

$P_{atm} = 83.4 \, \text{kPa}$

$T_\infty = 20°\text{C}$

$V_\infty = 8 \, \text{m/s}$

$T_s = 134°\text{C}$

Air

$\dot{Q}$

1.5 m

6 m

FIGURE 6-16

Schematic for Example 6-2.

$$\text{Nu} = \frac{hL}{k} = (0.037 \, \text{Re}_L^{0.8} - 871)\text{Pr}^{1/3}$$

$$= [0.037(1.92 \times 10^6)^{0.8} - 871]0.706^{1/3}$$

$$= 2727$$

Then

$$h = \frac{k}{L}\text{Nu} = \frac{0.0297 \, \text{W/m} \cdot {}^\circ\text{C}}{6 \, \text{m}}(2727) = 13.5 \, \text{W/m}^2 \cdot {}^\circ\text{C}$$

$$A = wL = (1.5 \, \text{m})(6 \, \text{m}) = 9 \, \text{m}^2$$

and

$$\dot{Q} = hA(T_s - T_\infty) = (13.5 \, \text{W/m}^2 \cdot {}^\circ\text{C})(9 \, \text{m}^2)(134 - 20){}^\circ\text{C} = \textbf{13,850 W}$$

Note that if we disregarded the laminar region and assumed turbulent flow over the plate, we would get Nu = 3520 from Eq. 6-23, which is 28 percent higher than the value we calculated above. Therefore, that assumption would cause a 28-percent error in heat transfer calculation in this case.

(b) When air flow is parallel to the short side, we have L = 1.5 m, and the Reynolds number at the end of the plate becomes

$$\text{Re}_L = \frac{\mathcal{V}_\infty L}{\nu} = \frac{(8 \, \text{m/s})(1.5 \, \text{m})}{2.50 \times 10^{-5} \, \text{m}^2/\text{s}} = 4.80 \times 10^5$$

which is less than the critical Reynolds number. Thus we have laminar flow over the entire plate, and the average Nusselt number is determined from

$$\text{Nu} = \frac{hL}{k} = 0.664 \, \text{Re}_L^{0.5} \, \text{Pr}^{1/3} = 0.664 \times (4.8 \times 10^5)^{0.5} \times 0.706^{1/3} = 410$$

Then

$$h = \frac{k}{L}\text{Nu} = \frac{0.0297 \, \text{W/m} \cdot {}^\circ\text{C}}{1.5 \, \text{m}}(410) = 8.12 \, \text{W/m}^2 \cdot {}^\circ\text{C}$$

and

$$\dot{Q} = hA(T_s - T_\infty) = (8.12 \, \text{W/m}^2 \cdot {}^\circ\text{C})(9 \, \text{m}^2)(134 - 20){}^\circ\text{C} = \textbf{8330 W}$$

which is considerably less than the heat transfer rate determined in case (a).

Discussion Note that the *direction* of fluid flow can have a significant effect on convection heat transfer to or from a surface (Fig. 6-17). In this case, we can increase the heat transfer rate by 67 percent by simply blowing the air parallel to the long side of the rectangular plate instead of the short side.

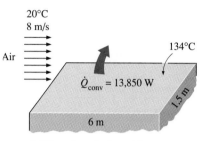

(a) Flow along the long side

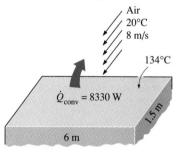

(b) Flow along the short side

FIGURE 6-17

The direction of fluid flow can have a significant effect on convection heat transfer.

6-5 ■ FLOW ACROSS CYLINDERS AND SPHERES

In the preceding section, we considered fluid flow over *flat* surfaces. In this section, we consider flow over cylinders and spheres, which is frequently encountered in practice. For example, the tubes in a tube-and-shell heat exchanger involve both *internal flow* through the tubes and *external flow* over the tubes, and both flows must be considered in the analysis of heat transfer between the two fluids. Below we consider external flow only.

The characteristic length for a circular cylinder or sphere is taken to be the *external diameter D*. Thus, the Reynolds number is defined as

$$\mathrm{Re} = \frac{\mathcal{V}_\infty D}{\nu}$$

where $\mathcal{V}_\infty$ is the uniform velocity of the fluid as it approaches the cylinder or sphere. The critical Reynolds number for flow across a circular cylinder or sphere is $\mathrm{Re_{cr}} \approx 2 \times 10^5$. That is, the boundary layer remains laminar for $\mathrm{Re} < 2 \times 10^5$ and becomes turbulent for $\mathrm{Re} > 2 \times 10^5$.

Cross flow over a cylinder exhibits complex flow patterns, as shown in Figure 6-18. The fluid approaching the cylinder will branch out and encircle the cylinder, forming a boundary layer that wraps around the cylinder. The fluid particles on the midplane will strike the cylinder at the stagnation point, bringing the fluid to a complete stop and thus raising the pressure at that point. The pressure decreases in the flow direction while the fluid velocity increases.

At very low free-stream velocities ($\mathrm{Re} < 4$), the fluid completely wraps around the cylinder and the two arms of the fluid meet on the rear side of the cylinder in an orderly manner. Thus the fluid follows the curvature of the cylinder. At higher velocities, the fluid still hugs the cylinder on the frontal side, but it is too fast to remain attached to the surface as it approaches the top of the cylinder. As a result, the boundary layer detaches from the surface, forming a wake behind the cylinder. This point is called the **separation point.** Flow in the wake region is characterized by random vortex formation and pressures much lower than the stagnation point pressure.

The flow separation phenomenon is analogous to fast vehicles jumping off on hills. At low velocities, the wheels of the vehicle always remain in contact with the road surface. But at high velocities, the vehicle is too fast to follow the curvature of the road and takes off at the hill, losing contact with the road.

Flow separation occurs at about $\theta \approx 80°$ when the boundary layer is *laminar* and at about $\theta \approx 140°$ when it is *turbulent* (Fig. 6-19). The delay of separation in turbulent flow is caused by the rapid fluctuations of the fluid in the transverse direction.

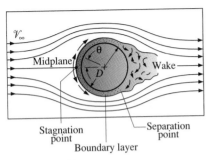

FIGURE 6-18

Typical flow patterns in cross flow over a cylinder.

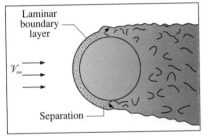

(a) Laminar flow ($\mathrm{Re} < 2 \times 10^5$)

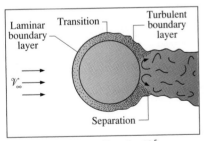

(b) Turbulence occurs ($\mathrm{Re} > 2 \times 10^5$)

FIGURE 6-19

Turbulence delays flow separation.

The Drag Coefficient

The nature of the flow across a cylinder or sphere strongly affects the drag coefficient C_D and the heat transfer coefficient h. The *drag force* acting on a body in cross flow is caused by two effects: the *friction drag,* which is due to the shear stress at the surface, and the *pressure drag,* which is due to pressure differential between the front and the rear sides of the body when a wake is formed in the rear. The high pressure in the vicinity of the stagnation point and the low pressure at the opposite side in the wake produce a net force on the body in the direction of flow. The drag force is primarily due to friction drag at low Reynolds numbers ($\mathrm{Re} < 4$) and to pressure drag at high

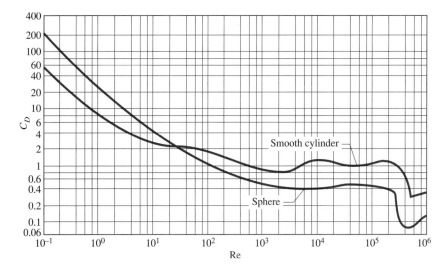

FIGURE 6-20

Average drag coefficient for cross flow over a smooth circular cylinder and a smooth sphere (from Schlichting, Ref. 13).

Reynolds numbers (Re > 5000). Both effects are significant at intermediate Reynolds numbers.

The average drag coefficients C_D for cross flow over a single circular cylinder and a sphere are given in Figure 6-20. The large reduction in C_D for Re > 2×10^5 is caused by the transition to *turbulent* flow, which moves the separation point further on the rear of the body, reducing the size of the wake and thus the magnitude of the pressure drag. Once the drag coefficient is available, the drag force acting on a body in cross flow can be determined from

$$F_D = C_D A_N \frac{\rho \mathcal{V}_\infty^2}{2} \qquad \text{(N)}$$

where A_N is the *frontal area* (area normal to the direction of flow). It is equal to LD for a cylinder of length L and $\frac{1}{4}\pi D^2$ for a sphere.

The term *drag coefficient* is also used commonly in daily life. Car manufacturers try to attract consumers by pointing out the *low drag coefficients* of their cars (Fig. 6-21). The surfaces of golf balls are intentionally roughened to induce *turbulence* at a lower Reynolds number to take advantage of the sharp *drop* in the drag coefficient at the onset of turbulence in the boundary layer. *Airplanes* are built to resemble birds, and *submarines* to resemble fish in order to minimize the drag coefficients, and thus the fuel consumption. The airplanes retract their wheels, just like the birds retracting their feet, in order to reduce the drag coefficients.

FIGURE 6-21

Modern vehicles are shaped so as to minimize the drag coefficient and thus to maximize the fuel efficiency.

The Heat Transfer Coefficient

Flows across cylinders and spheres, in general, involve *flow separation*, which is difficult to handle analytically. Therefore, such flows must be studied experimentally. Indeed, flow across cylinders and spheres has been studied experimentally by numerous investigators, and several empirical correlations are developed for the heat transfer coefficient.

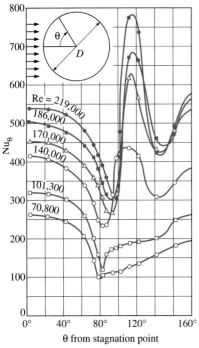

FIGURE 6-22

Variation of the local heat transfer coefficient along the circumference of a circular cylinder in cross flow of air (from Geidt, Ref. 5).

The complicated flow pattern across a cylinder discussed earlier greatly influences heat transfer. The variation of the local Nusselt number Nu_θ around the periphery of a cylinder subjected to cross flow of air is given in Figure 6-22. Note that, for all cases, the value of Nu_θ starts out relatively high at the stagnation point ($\theta = 0°$) but decreases with increasing θ as a result of the thickening of the laminar boundary layer. On the two curves at the bottom corresponding to Re = 70,800 and 101,300, Nu_θ reaches a minimum at $\theta \approx 80°$, which is the separation point in laminar flow. Then Nu_θ increases with increasing θ as a result of the intense mixing in the separated flow region (the wake). The curves at the top corresponding to Re = 140,000–219,000 differ from the first two curves in that they have *two* minima for Nu_θ. The sharp increase in Nu_θ at about $\theta \approx 90°$ is due to transition from laminar to turbulent flow. The later decrease in Nu_θ is again due to the thickening of the boundary layer. Nu_θ reaches its second minimum at about $\theta \approx 140°$, which is the flow separation point in turbulent flow, and increases with θ as a result of the intense mixing in the turbulent wake region.

The discussions above on the local heat transfer coefficients are insightful; however, they are of little value in heat transfer calculations since the calculation of heat transfer requires the *average* heat transfer coefficient over the entire surface. Of the several such relations available in the literature for the average Nusselt number for cross flow over a *cylinder,* we present the one proposed by Churchill and Bernstein:

$$Nu_{cyl} = \frac{hD}{k} = 0.3 + \frac{0.62\, Re^{1/2}\, Pr^{1/3}}{[1 + (0.4/Pr)^{2/3}]^{1/4}} \left[1 + \left(\frac{Re}{282,000} \right)^{5/8} \right]^{4/5} \quad (6\text{-}32)$$

This relation is quite comprehensive in that it correlates all available data well for Re Pr > 0.2. The fluid properties are evaluated at the *film temperature* $T_f = \frac{1}{2}(T_\infty + T_s)$, which is the average of the free-stream and surface temperatures.

For flow over a *sphere,* Whitaker recommends the following comprehensive correlation:

$$Nu_{sph} = \frac{hD}{k} = 2 + [0.4\, Re^{1/2} + 0.06\, Re^{2/3}]\, Pr^{0.4} \left(\frac{\mu_\infty}{\mu_s} \right)^{1/4} \quad (6\text{-}33)$$

which is valid for $3.5 \le Re \le 80,000$ and $0.7 \le Pr \le 380$. The fluid properties in this case are evaluated at the free-stream temperature T_∞, except for μ_s, which is evaluated at the surface temperature T_s. Although the two relations above are considered to be quite accurate, the results obtained from them can be off by as much as 30 percent.

The average Nusselt number for flow across cylinders can be expressed compactly as

$$Nu_{cyl} = \frac{hD}{k} = C\, Re^m\, Pr^n \quad (6\text{-}34)$$

TABLE 6-3

Empirical correlations for the average Nusselt number for forced convection over circular and noncircular cylinders in cross flow (from Zhukauskas, Ref. 18, and Jakob, Ref. 8)

Cross-section of the cylinder	Fluid	Range of Re	Nusselt number
Circle	Gas or liquid	0.4–4 4–40 40–4000 4000–40,000 40,000–400,000	$Nu = 0.989Re^{0.330} Pr^{1/3}$ $Nu = 0.911Re^{0.385} Pr^{1/3}$ $Nu = 0.683Re^{0.466} Pr^{1/3}$ $Nu = 0.193Re^{0.618} Pr^{1/3}$ $Nu = 0.027Re^{0.805} Pr^{1/3}$
Square	Gas	5000–100,000	$Nu = 0.102Re^{0.675} Pr^{1/3}$
Square (tilted 45°)	Gas	5000–100,000	$Nu = 0.246Re^{0.588} Pr^{1/3}$
Hexagon	Gas	5000–100,000	$Nu = 0.153Re^{0.638} Pr^{1/3}$
Hexagon (tilted 45°)	Gas	5000–19,500 19,500–100,000	$Nu = 0.160Re^{0.638} Pr^{1/3}$ $Nu = 0.0385Re^{0.782} Pr^{1/3}$
Vertical plate	Gas	4000–15,000	$Nu = 0.228Re^{0.731} Pr^{1/3}$
Ellipse	Gas	2500–15,000	$Nu = 0.248Re^{0.612} Pr^{1/3}$

where $n = \frac{1}{3}$ and the experimentally determined constants C and m are given in Table 6-3 for circular as well as various noncircular cylinders. The characteristic length D for use in the calculation of the Reynolds and the Nusselt numbers for different geometries is as indicated on the figure. All fluid properties are evaluated at the *film temperature* $T_f = \frac{1}{2}(T_\infty + T_s)$.

The relations for cylinders above are for *single* cylinders or cylinders oriented such that the flow over them is not affected by the presence of others. Also, they are applicable to *smooth* surfaces. *Surface roughness* and the *free-*

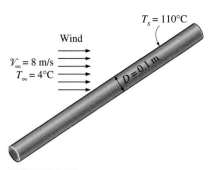

$T_s = 110°C$

Wind

$\mathcal{V}_\infty = 8$ m/s
$T_\infty = 4°C$

$D = 0.1$ m

FIGURE 6-23
Schematic for Example 6-3.

stream turbulence may affect the drag and heat transfer coefficients significantly. Drag and heat transfer coefficients for flow over *tube bundles* can be obtained from some of the standard heat transfer texts listed at the end of this chapter. Equation 6-34 provides a simpler alternative to Equation 6-32 for flow over cylinders. However, Equation 6-32 is more accurate, and thus should be preferred in calculations whenever possible.

EXAMPLE 6-3   Heat Loss from a Steam Pipe in Windy Air

A long 10-cm-diameter steam pipe whose external surface temperature is 110°C passes through some open area that is not protected against the winds (Fig. 6-23). Determine the rate of heat loss from the pipe per unit of its length when the air is at 1 atm pressure and 4°C and the wind is blowing across the pipe at a velocity of 8 m/s.

Solution   A steam pipe is exposed to windy air. The rate of heat loss from the steam is to be determined.

Assumptions   **1** Steady operating conditions exist. **2** Radiation effects are negligible. **3** Air is an ideal gas.

Properties   The properties of air at the film temperature of $T_f = (T_s + T_\infty)/2 = (110 + 4)/2 = 57°C = 330$ K and 1 atm pressure are (Table A-11)

$$k = 0.0283 \text{ W/m} \cdot °C \qquad Pr = 0.708$$
$$v = 1.86 \times 10^{-5} \text{ m}^2\text{/s}$$

Analysis   This is an *external flow* problem, since we are interested in the heat transfer from the pipe to the air that is flowing outside the pipe. The Reynolds number of the flow is

$$Re = \frac{\mathcal{V}_\infty D}{v} = \frac{(8 \text{ m/s})(0.1 \text{ m})}{1.86 \times 10^{-5} \text{ m}^2\text{/s}} = 43{,}011$$

Then the Nusselt number in this case can be determined from

$$Nu = \frac{hD}{k} = 0.3 + \frac{0.62 \, Re^{1/2} \, Pr^{1/3}}{[1 + (0.4/Pr)^{2/3}]^{1/4}} \left[1 + \left(\frac{Re}{282{,}000}\right)^{5/8}\right]^{4/5}$$

$$= 0.3 + \frac{0.62(43{,}011)^{1/2} \, (0.708)^{1/3}}{[1 + (0.4/0.708)^{2/3}]^{1/4}} \left[1 + \left(\frac{43{,}011}{282{,}000}\right)^{5/8}\right]^{4/5}$$

$$= 125.1$$

and

$$h = \frac{k}{D} Nu = \frac{0.0283 \text{ W/m} \cdot °C}{0.1 \text{ m}}(125.1) = 35.4 \text{ W/m}^2 \cdot °C$$

Then the rate of heat transfer from the pipe per unit of its length becomes

$$A = pL = \pi DL = \pi(0.1 \text{ m})(1 \text{ m}) = 0.314 \text{ m}^2$$
$$\dot{Q} = hA(T_s - T_\infty) = (35.4 \text{ W/m}^2 \cdot C)(0.314 \text{ m}^2)(110 - 4)°C = \textbf{1178W}$$

The rate of heat loss from the entire pipe can be obtained by multiplying the value above by the length of the pipe in m.

Discussion The simpler Nusselt number relation in Table 6-3 in this case would give Nu = 129, which is 3 percent higher than the value obtained above using Equation 6-32.

EXAMPLE 6-4   Cooling of a Steel Ball by Forced Air

A 25-cm-diameter stainless steel ball (ρ = 8055 kg/m³, C_p = 480 J/kg · °C) is removed from the oven at a uniform temperature of 300°C (Fig. 6-24). The ball is then subjected to the flow of air at 1 atm pressure and 27°C with a velocity of 3 m/s. The surface temperature of the ball eventually drops to 200°C. Determine the average convection heat transfer coefficient during this cooling process and estimate how long this cooling process will take.

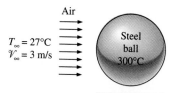

Air

$T_\infty = 27°C$
$\mathcal{V}_\infty = 3$ m/s

Steel
ball
300°C

FIGURE 6-24
Schematic for Example 6-4.

Solution   A hot stainless steel ball is cooled by forced air. The average convection heat transfer coefficient and the cooling time are to be determined.

Assumptions   **1** Steady operating conditions exist. **2** Radiation effects are negligible. **3** Air is an ideal gas. **4** The outer surface temperature of the ball is uniform at all times.

The surface temperature of the ball during cooling is changing. Therefore, the convection heat transfer coefficient between the ball and the air will also change. To avoid this complexity, we take the surface temperature of the ball to be constant at the average temperature of (300 + 200)/2 = 250°C in the evaluation of the heat transfer coefficient and use the value obtained for the entire cooling process.

Properties   The dynamic viscosity of air at the surface temperature is $\mu_s = \mu_{@\ 250°C}$ = 2.76 × 10⁻⁵ kg/m · s. The properties of air at the free-stream temperature of 27°C and 1 atm are (Table A-11)

$$k = 0.0261 \text{ W/m} \cdot °C \qquad \nu = 1.57 \times 10^{-5} \text{ m}^2/\text{s}$$
$$\mu = 1.85 \times 10^{-5} \text{ kg/m} \cdot \text{s} \qquad Pr = 0.712$$

Analysis   This is an *external flow* problem since the air flows outside the ball. The Reynolds number of the flow is determined from

$$Re = \frac{\mathcal{V}_\infty D}{\nu} = \frac{(3 \text{ m/s})(0.25 \text{ m})}{1.57 \times 10^{-5} \text{ m}^2/\text{s}} = 47{,}800$$

Then the Nusselt number can be determined from

$$Nu = \frac{hD}{k} = 2 + [0.4\ Re^{1/2} + 0.06\ Re^{2/3}]\ Pr^{0.4} \left(\frac{\mu_\infty}{\mu_s}\right)^{1/4}$$

$$= 2 + [0.4(47{,}800)^{1/2} + 0.06(47{,}800)^{2/3}](0.712)^{0.4} \left(\frac{1.85 \times 10^{-5}}{2.76 \times 10^{-5}}\right)^{1/4}$$

$$= 139$$

Then the average convection heat transfer coefficient becomes

$$h = \frac{k}{D} Nu = \frac{0.0261 \text{ W/m} \cdot °C}{0.25 \text{ m}} (139) = \textbf{14.5 W/m}^2 \cdot \textbf{°C}$$

In order to estimate the time of cooling of the ball from 300°C to 200°C, we determine the *average* rate of heat transfer from Newton's law of cooling by using the *average* surface temperature. That is,

$$A = \pi D^2 = \pi(0.25 \text{ m})^2 = 0.196 \text{ m}^2$$

$$\dot{Q}_{ave} = hA(T_{s,ave} - T_\infty) = (14.5 \text{ W/m}^2 \cdot {}^{\circ}\text{C})(0.196 \text{ m}^2)(250 - 27){}^{\circ}\text{C} = 634 \text{ W}$$

Next we determine the *total* heat transferred from the ball, which is simply the change in the energy of the ball as it cools from 300°C to 200°C:

$$m = \rho V = \rho \tfrac{1}{6}\pi D^3 = (8055 \text{ kg/m}^3)\tfrac{1}{6}\pi(0.25 \text{ m})^3 = 65.9 \text{ kg}$$

$$Q_{total} = mC_p(T_2 - T_1) = (65.9 \text{ kg})(480 \text{ J/kg} \cdot {}^{\circ}\text{C})(300 - 200){}^{\circ}\text{C} = 3,163,200 \text{ J}$$

In the above calculation, we assumed that the *entire ball* is at 200°C, which is not necessarily true. The inner region of the ball will probably be at a higher temperature than its surface. With this assumption, the time of cooling is determined to be

$$\Delta t \approx \frac{Q}{\dot{Q}_{ave}} = \frac{3,163,200 \text{ J}}{594 \text{ J/s}} = 5325 \text{ s} = \textbf{1 h 29 min}$$

Discussion   The time of cooling could also be determined more accurately using the transient temperature charts or relations introduced in Chapter 4. But the simplifying assumptions we made above can be justified if all we need is a ballpark value. It will be naive to expect the time of cooling to be exactly 1 h 29 min, but, using our engineering judgment, it is realistic to expect the time of cooling to be somewhere between one and two hours.

6-6 ■ FLOW IN TUBES

Liquid or gas flow through or *pipes* or *ducts* is commonly used in practice in heating and cooling applications. The fluid in such applications is forced to flow by a fan or pump through a tube that is sufficiently long to accomplish the desired heat transfer. In this section, we will discuss the *friction* and *heat transfer coefficients* that are directly related to the *pressure drop* and *heat flux* for flow through tubes. These quantities are then used to determine the pumping power requirement and the length of the tube.

There is a *fundamental* difference between external and internal flows. In *external flow*, which we have considered so far, the fluid had a free surface, and thus the boundary layer over the surface was free to grow indefinitely. In *internal flow*, however, the fluid is completely confined by the inner surfaces of the tube, and thus there is a limit on how much the boundary layer can grow.

General Considerations

The fluid velocity in a tube changes from *zero* at the surface to a *maximum* at the tube center. In fluid flow, it is convenient to work with an *average* or *mean* velocity $\mathcal{V}_m$, which remains constant in incompressible flow when the cross-sectional area of the tube is constant. The mean velocity in actual heating and cooling applications may change somewhat because of the changes in density with temperature. But, in practice, we evaluate the fluid properties at some average temperature and treat them as constants. The convenience in working with constant properties usually more than justifies the slight loss in accuracy.

The value of the mean velocity $\mathcal{V}_m$ is determined from the requirement that the *conservation of mass* principle be satisfied (Fig. 6-25). That is, the mass flow rate through the tube evaluated using the mean velocity $\mathcal{V}_m$ from

$$\dot{m} = \rho \mathcal{V}_m A_c \qquad (\text{kg/s})$$

will be equal to the actual mass flow rate. Here ρ is the density of the fluid and A_c is the cross-sectional area, which is equal to $A_c = \frac{1}{4}\pi D^2$ for a circular tube.

When a fluid is heated or cooled as it flows through a tube, the temperature of a fluid at any cross-section changes from T_s at the surface of the wall at that cross-section to some maximum (or minimum in the case of heating) at the tube center. In fluid flow it is convenient to work with an *average* or *mean* temperature T_m that remains constant at a cross-section. The mean temperature T_m *will change* in the flow direction, however, whenever the fluid is heated or cooled.

The value of the mean temperature T_m is determined from the requirement that the *conservation of energy* principle be satisfied. That is, the energy transported by the fluid through a cross-section in actual flow will be equal to the energy that would be transported through the same cross-section if the fluid were at a constant temperature T_m. This can be expressed mathematically as (Fig. 6-26)

$$\dot{E}_{\text{fluid}} = \dot{m} C_p T_m = \int_{\dot{m}} C_p T \, \delta \dot{m} = \int_{A_c} C_p T(\rho \mathcal{V} \, dA_c) \qquad (\text{kJ/s}) \quad (6\text{-}35)$$

where C_p is the specific heat of the fluid and $\dot{m}$ is the mass flow rate. Note that the product $\dot{m} C_p T_m$ at any cross-section along the tube represents the *energy flow* with the fluid at that cross-section. You will recall that in the absence of any work interactions (such as electric resistance heating), the conservation of energy equation for the steady flow of a fluid in a tube can be expressed as (Fig. 6-27)

$$\dot{Q} = \dot{m} C_p (T_e - T_i) \qquad (\text{kJ/s}) \qquad (6\text{-}36)$$

where T_i and T_e are the mean fluid temperatures at the inlet and exit of the tube, respectively, and $\dot{Q}$ is the rate of heat transfer to or from the fluid. Note that the temperature of a fluid flowing in a tube remains constant in the absence of any energy interactions through the wall of the tube.

Perhaps we should mention that the friction between the fluid layers in a tube does cause a slight rise in fluid temperature as a result of the mechanical energy being converted to sensible heat energy. But this *frictional heating* is too small to warrant any consideration in calculations, and thus is disregarded. For example, in the absence of any heat transfer, no noticeable difference will be detected between the inlet and exit temperatures of a fluid

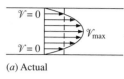

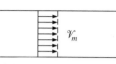

(a) Actual

(b) Idealized

FIGURE 6-25

Actual and idealized velocity profiles for flow in a tube (the mass flow rate of the fluid is the same for both cases).

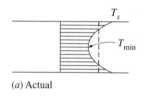

(a) Actual

(b) Idealized

FIGURE 6-26

Actual and idealized temperature profiles for flow in a tube (the rate at which energy is transported with the fluid is the same for both cases).

FIGURE 6-27

The heat transfer to a fluid flowing in a tube is equal to the increase in the energy of the fluid.

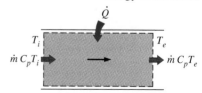

Energy balance:
$$\dot{Q} = \dot{m} C_p (T_e - T_i)$$

flowing in a tube. Thus, it is reasonable to assume that any temperature change in the fluid is due to heat transfer.

The thermal conditions at the surface of a tube can usually be approximated with reasonable accuracy to be *constant surface temperature* (T_s = constant) or *constant surface heat flux* ($\dot{q}_s$ = constant). For example, the constant surface temperature condition is realized when a phase change process such as boiling or condensation occurs at the outer surface of a tube. The constant surface heat flux condition is realized when the tube is subjected to radiation or electric resistance heating uniformly from all directions.

The convection heat flux at any location on the tube can be expressed as

$$\dot{q} = h(T_s - T_m) \qquad (\text{W/m}^2) \qquad (6\text{-}37)$$

where h is the *local* heat transfer coefficient and T_s and T_m are the surface and the mean fluid temperatures at that location. Note that the mean fluid temperature T_m of a fluid flowing in a tube must change during heating or cooling. Therefore, when h = constant, the surface temperature T_s must change when $\dot{q}_s$ = constant, and the surface heat flux $\dot{q}_s$ must change when T_s = constant. Thus we may have either T_s = constant or $\dot{q}_s$ = constant at the surface of a tube, but not both. Below we consider convection heat transfer for these two common cases.

Constant Surface Heat Flux ($\dot{q}_s$ = constant)

In the case of $\dot{q}_s$ = constant, the rate of heat transfer can also be expressed as

$$\dot{Q} = \dot{q}_s A = \dot{m} C_p (T_e - T_i) \qquad (\text{W}) \qquad (6\text{-}38)$$

Then the mean fluid temperature at the tube exit becomes

$$T_e = T_i + \frac{\dot{q}_s A}{\dot{m} C_p} \qquad (6\text{-}39)$$

Note that the mean fluid temperature increases *linearly* in the flow direction in the case of constant surface heat flux, since the surface area increases linearly in the flow direction (A is equal to the perimeter, which is constant, times the tube length).

The surface temperature in this case can be determined from $\dot{q} = h(T_s - T_m)$. Note that when h is constant, $T_s - T_m$ = constant, and thus the surface temperature will also increase *linearly* in the flow direction (Fig. 6-28). Of course, this is true when the variation of the specific heat C_p with T is disregarded and C_p is assumed to remain constant.

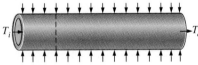

FIGURE 6-28

Variation of the *tube surface* and the *mean fluid* temperatures along the tube for the case of constant surface heat flux.

Constant Surface Temperature (T_s = constant)

From Newton's law of cooling, the rate of heat transfer to or from a fluid flowing in a tube can be expressed as

$$\dot{Q} = hA\Delta T_{ave} = hA(T_s - T_m)_{ave} \qquad (6\text{-}40)$$

where h is the average convection heat transfer coefficient, A is the heat transfer surface area (it is equal to πDL for a circular pipe of length L), and ΔT_{ave} is some appropriate *average* temperature difference between the fluid and the surface. Below we discuss two suitable ways of expressing ΔT_{ave}.

In the constant surface temperature (T_s = constant) case, ΔT_{ave} can be expressed *approximately* by the **arithmetic mean temperature difference** ΔT_{am} as

$$\Delta T_{ave} \approx \Delta T_{am} = \frac{\Delta T_i + \Delta T_e}{2} = \frac{(T_s - T_i) + (T_s - T_e)}{2}$$

$$= T_s - \frac{T_i + T_e}{2} = T_s - T_b \qquad (6\text{-}41)$$

where $T_b = \frac{1}{2}(T_i + T_e)$ is the *bulk mean fluid temperature,* which is the *arithmetic average* of the mean fluid temperatures at the inlet and the exit of the tube.

Note that the *arithmetic mean temperature difference* ΔT_{am} is simply the *average* of the *temperature differences* between the surface and the fluid at the inlet and the exit of the tube. Inherent in this definition is the assumption that the mean fluid temperature varies linearly along the tube, which is hardly ever the case when T_s = constant. This simple approximation often gives acceptable results, but not always. Therefore, we need a better way to evaluate ΔT_{ave}.

Consider the heating of a fluid in a tube of constant cross-section whose inner surface is maintained at a constant temperature of T_s. We know that the mean temperature of the fluid T_m will increase in the flow direction as a result of heat transfer. The energy balance on a differential control volume shown in Figure 6-29 gives

$$\dot{m}C_p dT_m = h(T_s - T_m)dA \qquad (6\text{-}42)$$

That is, the increase in the energy of the fluid (represented by an increase in its mean temperature by dT_m) is equal to the heat transferred to the fluid from the tube surface by convection. Noting that the differential surface area is $dA = pdx$ where p is the perimeter of the tube, and that $dT_m = -d(T_s - T_m)$, since T_s is constant, the relation above can be rearranged as

$$\frac{d(T_s - T_m)}{T_s - T_m} = -\frac{hp}{\dot{m}C_p}dx \qquad (6\text{-}43)$$

Integrating from $x = 0$ (tube inlet where $T_m = T_i$) to $x = L$ (tube exit where $T_m = T_e$) gives

$$\ln\frac{T_s - T_e}{T_s - T_i} = -\frac{hA}{\dot{m}C_p} \qquad (6\text{-}44)$$

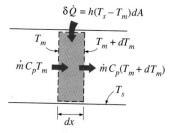

$$\delta\dot{Q} = h(T_s - T_m)dA$$

FIGURE 6-29

Energy interactions for a differential control volume in a tube.

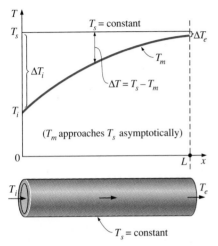

FIGURE 6-30

The variation of the *mean fluid* temperature along the tube for the case of constant surface temperature.

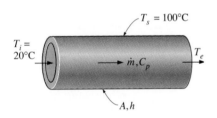

NTU = $hA / \dot{m}C_p$	T_e, °C
0.01	20.8
0.05	23.9
0.10	27.6
0.50	51.5
1.00	70.6
5.00	99.5
10.00	100.0

FIGURE 6-31

An NTU greater than 5 indicates that the fluid flowing in a tube will reach the surface temperature at the exit regardless of the inlet temperature.

where $A = pL$ is the surface area of the tube and h is the constant *average* convection heat transfer coefficient. Taking the exponential of both sides and solving for T_e gives the following very useful relation for the determination of the *mean fluid temperature at the tube exit*:

$$T_e = T_s - (T_s - T_i)e^{-hA/\dot{m}C_p} \qquad (6\text{-}45)$$

This relation can also be used to determine the mean fluid temperature $T_m(x)$ at any x by replacing $A = pL$ by px.

Note that the temperature difference between the fluid and the surface *decays exponentially* in the flow direction, and the rate of decay depends on the magnitude of the exponent $hA/\dot{m}C_p$, as shown in Figure 6-30. This dimensionless parameter is called the *number of transfer units,* denoted by NTU, and is a measure of the effectiveness of the heat transfer systems. For NTU > 5, the exit temperature of the fluid becomes almost equal to the surface temperature, $T_e \approx T_s$ (Fig. 6-31). Noting that the fluid temperature can approach the surface temperature but cannot cross it, an NTU of about 5 indicates that the limit is reached for heat transfer, and the heat transfer will not increase no matter how much we extend the length of the tube. A small value of NTU, on the other hand, indicates more opportunities for heat transfer, and the heat transfer will continue increasing as the tube length is increased. A large NTU and thus a large heat transfer surface area (which means a large tube) may be desirable from a heat transfer point of view, but it may be unacceptable from an economic point of view. The selection of heat transfer equipment usually reflects a compromise between heat transfer performance and cost.

Solving Equation 6-44 for $\dot{m}C_p$ gives

$$\dot{m}C_p = \frac{hA}{\ln \dfrac{T_s - T_e}{T_s - T_i}} \qquad (6\text{-}46)$$

Substituting this into Equation 6-38, we obtain

$$\dot{Q} = hA \, \Delta T_{\ln} \qquad (6\text{-}47)$$

where

$$\Delta T_{\ln} = \frac{T_e - T_i}{\ln \dfrac{T_s - T_e}{T_s - T_i}} = \frac{\Delta T_e - \Delta T_i}{\ln (\Delta T_e / \Delta T_i)} \qquad (6\text{-}48)$$

is the **logarithmic mean temperature difference.** Note that $\Delta T_i = T_s - T_i$ and $\Delta T_e = T_s - T_e$ are the temperature differences between the surface and the fluid at the inlet and the exit of the tube, respectively. The $\Delta T_{\ln}$ relation above appears to be prone to misuse, but it is practically failsafe, since using T_i in place of T_e and vice versa in the numerator and/or the denominator will, at most, affect the sign, not the magnitude. Also, it can be used for both heating ($T_s > T_i$ and T_e) and cooling ($T_s < T_i$ and T_e) of a fluid in a tube.

The logarithmic mean temperature difference ΔT_{ln} is obtained by tracing the actual temperature profile of the fluid along the tube, and is an *exact* representation of the *average temperature difference* between the fluid and the surface. It truly reflects the exponential decay of the local temperature difference. When ΔT_e differs from ΔT_i by no more that 40 percent, the error in using the arithmetic mean temperature difference is less that 1 percent. But the error increases to undesirable levels when ΔT_e differs from ΔT_i by greater amounts. Therefore, we should always use the logarithmic mean temperature difference when determining the convection heat transfer in a tube whose surface is maintained at a constant temperature T_s.

Pressure Drop

A quantity of interest in the analysis of tube flow is the *pressure drop* ΔP along the flow, since this quantity is directly related to the *power requirements* of the fan or pump to maintain the flow. The pressure drop during flow in a tube of length L is expressed as (Fig. 6-32)

$$\Delta P = f \frac{L}{D} \frac{\rho V_m^2}{2} \qquad (\text{N/m}^2) \qquad (6\text{-}49)$$

where f is the **friction factor.** The required **pumping power** to overcome a specified pressure drop ΔP is determined from

$$\dot{W}_{\text{pump}} = \dot{V}\Delta P = \frac{\dot{m}\Delta P}{\rho} \qquad (\text{W}) \qquad (6\text{-}50)$$

where $\dot{V} = V_m A_c = \dot{m}/\rho$ is the *volume flow rate* of the fluid through the tube.

Flow Regimes in a Tube

The flow in a tube can be laminar or turbulent, depending on the flow conditions. The type of flow in a tube can be verified experimentally by injecting a small amount of colored fluid into the main flow, as Reynolds did in the 1880s. The streak of colored fluid is *straight and smooth* in laminar flow, as shown in Figure 6-33, but *fluctuates rapidly and randomly* in turbulent flow.

The *Reynolds number* for flow in a circular tube of diameter D is defined as

$$\text{Re} = \frac{V_m D}{\nu} \qquad (6\text{-}51)$$

where V_m is the mean fluid velocity and ν is the kinematic viscosity of the fluid. The Reynolds number again provides a convenient criterion for determining the flow regime in a tube, although the *roughness* of the tube surface and the *fluctuations* in the flow have considerable influence. The critical Reynolds number for flow in a tube is generally accepted to be 2300. Therefore,

50 kPa — $\dot{m}$ — 35 kPa

(a) Flow resistance

50°C — $\dot{Q}$ — 35°C

(b) Thermal resistance

50 V — T — 35 V

(c) Electric resistance

FIGURE 6-32

Resistance to *fluid flow, heat flow,* and *current flow* causes, respectively, the *pressure,* the *temperature,* and the *voltage* to drop in the flow direction.

(a) Laminar flow

(b) Turbulent flow

FIGURE 6-33

The behavior of colored fluid injected into the flow in laminar and turbulent flows in a tube.

Re < 2300	laminar flow
2300 ≤ Re ≤ 4000	transition to turbulence
Re > 4000	turbulent flow

The transition from laminar to turbulent flow in a tube is quite different than it is over a flat plate, where the Reynolds number is zero at the leading edge and increases linearly in the flow direction. Consequently, the flow starts out laminar and becomes turbulent when a critical Reynolds number is reached. Therefore, the flow over a flat plate is partly laminar and partly turbulent. In flow in a tube, however, the Reynolds number is constant. Therefore, the flow is either laminar or turbulent over practically the entire length of the tube.

Hydrodynamic and Thermal Entry Lengths

Consider a fluid entering a circular tube at a uniform velocity. As in external flow, the fluid particles in the layer in contact with the surface of the tube will come to a complete stop. This layer will also cause the fluid particles in the adjacent layers to slow down gradually as a result of friction. To make up for this velocity reduction, the velocity of the fluid at the midsection of the tube will have to increase to keep the mass flow rate through the tube constant. As a result, a *velocity boundary layer* develops along the tube. The thickness of this boundary layer increases in the flow direction until the boundary layer reaches the tube center and thus fills the entire tube, as shown in Figure 6-34. The region from the tube inlet to the point at which the boundary layer merges at the centerline is called the **hydrodynamic entry region,** and the length of this region is called the **hydrodynamic entry length** L_h. The region beyond the hydrodynamic entry region in which the velocity profile is fully developed and remains unchanged is called the **hydrodynamically developed region.** The velocity profile in the hydrodynamically developed region is *parabolic* in laminar flow and somewhat *flatter* in turbulent flow.

Now consider a fluid at a uniform temperature entering a circular tube that is at a different temperature. This time, the fluid particles in the layer in contact with the surface of the tube will assume the surface temperature. This will initiate convection heat transfer in the tube and the development of a *thermal boundary layer* along the tube. The thickness of this boundary layer also increases in the flow direction until the boundary layer reaches the tube

FIGURE 6-34

The development of the velocity boundary layer in a tube. (The developed velocity profile will be parabolic in laminar flow, as shown, but somewhat blunt in turbulent flow.)

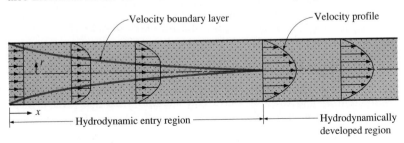

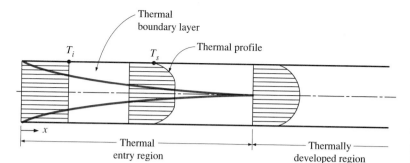

FIGURE 6-35

The development of the thermal boundary layer in a tube. (The fluid in the tube is being cooled.)

center and thus fills the entire tube, as shown in Figure 6-35. The region of flow over which the thermal boundary layer develops and reaches the tube center is called the **thermal entry region,** and the length of this region is called the **thermal entry length** L_t. The region beyond the thermal entry region in which the dimensionless temperature profile expressed as $(T - T_s)/(T_m - T_s)$ remains unchanged is called the **thermally developed region.** The region in which the flow is both hydrodynamically and thermally developed is called the **fully developed flow.**

Note that the *temperature profile* in the thermally developed region may *vary* with x in the flow direction. That is, unlike the velocity profile, the temperature profile can be different at different cross-sections of the tube in the developed region, and it usually is. However, it can be shown that the dimensionless temperature profile defined above remains unchanged in the thermally developed region when the temperature or heat flux at the tube surface remains constant.

In laminar flow in a tube, the magnitude of the dimensionless Prandtl number Pr is a measure of the relative growth of the velocity and thermal boundary layers. For fluids with Pr $\approx$ 1, such as gases, the two boundary layers essentially coincide with each other. For fluids with Pr $\gg$ 1, such as oils, the velocity boundary layer outgrows the thermal boundary layer. As a result, the hydrodynamic entry length is smaller than the thermal entry length. The opposite is true for fluids with Pr $\ll$ 1 such as liquid metals.

The hydrodynamic and thermal entry lengths in *laminar flow* are given approximately as

$$
\begin{aligned}
L_{h,\text{laminar}} &\approx 0.05 \, \text{Re} \, D \\
L_{t,\text{laminar}} &\approx 0.05 \, \text{Re} \, \text{Pr} \, D
\end{aligned}
\qquad (6\text{-}52)
$$

In *turbulent flow,* the hydrodynamic and thermal entry lengths are known to be independent of Re or Pr and are generally taken to be

$$
L_{h,\text{turbulent}} \approx L_{t,\text{turbulent}} \approx 10D \qquad (6\text{-}53)
$$

The friction coefficient is related to the shear stress at the surface, which is related to the slope of the velocity profile at the surface. Noting that the velocity profile remains unchanged in the hydrodynamically developed

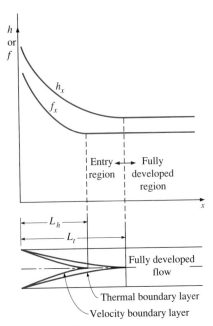

FIGURE 6-36

Variation of the friction factor and the convection heat transfer coefficient in the flow direction for flow in a tube (Pr > 1).

region, the friction coefficient also remains constant in that region. A similar argument can be given for the heat transfer coefficient in the thermally developed region. Thus, we conclude that *the friction and the heat transfer coefficients in the fully developed flow region remain constant.*

Consider a fluid that is being heated (or cooled) in a tube as it flows through it. The friction factor and the heat transfer coefficient are *highest* at the tube inlet where the thickness of the boundary layers is zero, and decrease gradually to the fully developed values, as shown in Figure 6-36. Therefore, the pressure drop and heat flux are *higher* in the entry regions of a tube, and the effect of the entry region is always to *enhance* the average friction and heat transfer coefficients for the entire tube. This enhancement can be significant for short tubes but negligible for long ones.

Precise correlations for the friction and heat transfer coefficients for the entry regions are available in the literature. However, the tubes used in practice in forced convection are usually many times the length of either entry region, and thus the flow through the tubes is assumed to be fully developed for the entire length of the tube. This approach, which we will also use for simplicity, gives *reasonable* results for long tubes and *conservative* results for short ones.

Laminar Flow in Tubes

We mentioned earlier that flow in smooth tubes is laminar for Re < 2300. The theory for laminar flow is well developed, and both the friction and heat transfer coefficients for fully developed laminar flow in smooth circular tubes can be determined analytically by solving the governing differential equations. Combining the conservation of *mass* and *momentum* equations in the axial direction for a tube and solving them subject to the no-slip condition at the boundary and the condition that the velocity profile is symmetric about the tube center give the following *parabolic* velocity profile for the hydrodynamically developed laminar flow:

$$\mathcal{V}(r) = 2\mathcal{V}_m \left(1 - \frac{r^2}{R^2}\right) \tag{6-54}$$

where $\mathcal{V}_m$ is the *mean fluid velocity* and R is the *radius* of the tube. Note that the maximum velocity occurs at the tube center ($r = 0$), and it is $\mathcal{V}_{max} = 2\mathcal{V}_m$. Knowing the velocity profile, the shear stress at the wall becomes

$$\tau_s = -\mu \frac{d\mathcal{V}}{dr}\bigg|_{r=R} = -2\mu\mathcal{V}_m \left(-\frac{2r}{R^2}\right)_{r=R} = \frac{8\mu\mathcal{V}_m}{D} \tag{6-55}$$

But we also have the following practical definition of shear stress:

$$\tau_s = C_f \frac{\rho\mathcal{V}_m^2}{2} \tag{6-56}$$

where C_f is the friction coefficient. Combining Equations 6-55 and 6-56 and solving for C_f gives

$$C_f = \frac{8\mu\mathcal{V}_m}{D}\frac{2}{\rho\mathcal{V}_m^2} = \frac{16\mu}{\rho\mathcal{V}_mD} = \frac{16}{\mathrm{Re}} \qquad (6\text{-}57)$$

The *friction factor f*, which is the parameter of interest in the pressure drop calculations, is related to the friction coefficient C_f by $f = 4C_f$. Therefore,

$$f = \frac{64}{\mathrm{Re}} \qquad \text{(laminar flow)} \qquad (6\text{-}58)$$

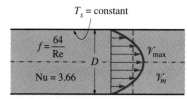

Note that the friction factor f is related to the *pressure drop* in the fluid, whereas the friction coefficient C_f is related to the *drag force* on the surface directly. Of course, these two coefficients are simply a constant multiple of each other.

The Nusselt number in the fully developed laminar flow region in a circular tube is determined in a similar manner from the conservation of energy equation to be (Fig. 6-37)

$$\mathrm{Nu} = 3.66 \qquad \text{for } T_s = \text{constant} \qquad \text{(laminar flow)}$$

$$\mathrm{Nu} = 4.36 \qquad \text{for } \dot{q}_s = \text{constant} \qquad \text{(laminar flow)}$$

FIGURE 6-37

In laminar flow in a tube with constant surface temperature, both the *friction factor* and the *heat transfer coefficient* remain constant in the fully developed region.

A general relation for the *average* Nusselt number for the hydro-dynamically and/or thermally *developing laminar flow* in a *circular* tube is given by Sieder and Tate as

$$\mathrm{Nu} = 1.86\left(\frac{\mathrm{Re}\,\mathrm{Pr}\,D}{L}\right)^{1/3}\left(\frac{\mu_b}{\mu_s}\right)^{0.14} \qquad (\mathrm{Pr} > 0.5) \qquad (6\text{-}59)$$

All properties are evaluated at the bulk mean fluid temperature, except for μ_s, which is evaluated at the surface temperature.

The Nusselt number Nu and the friction factor f are given in Table 6-4 for *fully developed laminar flow* in tubes of various cross-sections. The Reynolds and Nusselt numbers for flow in these tubes are based on the **hydraulic diameter** D_h defined as

$$D_h = \frac{4A_c}{p} \qquad (6\text{-}60)$$

where A_c is the cross-sectional area of the tube and p is its perimeter. The hydraulic diameter is defined such that it reduces to ordinary diameter D for circular tubes since $A_c = \pi D^2/4$ and $p = \pi D$. Once the Nusselt number is available, the convection heat transfer coefficient is determined from $h = k\,\mathrm{Nu}/D_h$. It turns out that for a fixed surface area, the *circular tube* gives the most heat transfer for the least pressure drop, which explains the overwhelming popularity of circular tubes in heat transfer equipment.

The effect of *surface roughness* on the friction factor and the heat transfer coefficient in laminar flow is negligible.

TABLE 6-4

Nusselt number and friction factor for fully developed laminar flow in tubes of various cross-sections ($D_h = 4A_c/p$, $Re = \mathscr{V}_m D_h/\nu$, and $Nu = hD_h/k$)

Cross-section of tube	a/b or θ°	Nusselt number		Friction factor f
		T_s = const.	$\dot{q}_s$ = const.	
Circle	—	3.66	4.36	64.00/Re
Hexagon	—	3.35	4.00	60.20/Re
Square	—	2.98	3.61	56.92/Re
Rectangle	a/b			
	1	2.98	3.61	56.92/Re
	2	3.39	4.12	62.20/Re
	3	3.96	4.79	68.36/Re
	4	4.44	5.33	72.92/Re
	6	5.14	6.05	78.80/Re
	8	5.60	6.49	82.32/Re
	∞	7.54	8.24	96.00/Re
Ellipse	a/b			
	1	3.66	4.36	64.00/Re
	2	3.74	4.56	67.28/Re
	4	3.79	4.88	72.96/Re
	8	3.72	5.09	76.60/Re
	16	3.65	5.18	78.16/Re
Triangle	θ			
	10°	1.61	2.45	50.80/Re
	30°	2.26	2.91	52.28/Re
	60°	2.47	3.11	53.32/Re
	90°	2.34	2.98	52.60/Re
	120°	2.00	2.68	50.96/Re

Turbulent Flow in Tubes

We mentioned earlier that flow in smooth tubes is turbulent at $Re > 4000$. Turbulent flow is commonly utilized in practice because of the higher heat transfer coefficients associated with it. Most correlations for the friction and

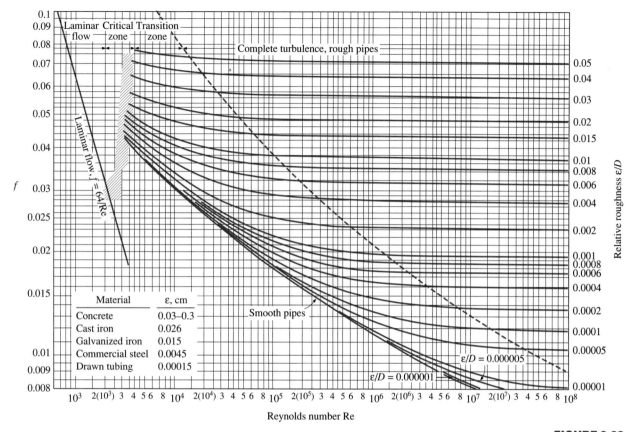

FIGURE 6-38

Friction factor for fully developed flow in circular tubes (the Moody chart).

heat transfer coefficients in turbulent flow are based on experimental studies because of the difficulty in dealing with turbulent flow theoretically.

For *smooth* tubes, the friction factor in fully developed turbulent flow can be determined from

$$f = 0.184 \, \mathrm{Re}^{-0.2} \qquad \text{(smooth tubes)} \qquad (6\text{-}61)$$

The friction factor for flow in tubes with *smooth* as well as *rough surfaces* over a wide range of Reynolds numbers is given in Figure 6-38, which is known as the **Moody diagram.** Note that the friction factor and thus the pressure drop for flow in a tube can vary several times as a result of surface roughness.

The Nusselt number in turbulent flow is related to the friction factor through the famous **Chilton–Colburn analogy** expressed as

$$\mathrm{Nu} = 0.125 f \, \mathrm{Re} \, \mathrm{Pr}^{1/3} \qquad \text{(turbulent flow)} \qquad (6\text{-}62)$$

Substituting the f relation from Equation 6-61 into Equation 6-62 gives the following relation for the Nusselt number for *fully developed turbulent flow in smooth tubes:*

$$\text{Nu} = 0.023 \, \text{Re}^{0.8} \, \text{Pr}^{1/3} \qquad \binom{0.7 \le \text{Pr} \le 160}{\text{Re} > 10{,}000} \qquad (6\text{-}63)$$

which is known as the **Colburn equation.** The accuracy of this equation can be improved by modifying it as

$$\text{Nu} = 0.023 \, \text{Re}^{0.8} \, \text{Pr}^{n} \qquad \binom{0.7 \le \text{Pr} \le 160}{\text{Re} > 10{,}000} \qquad (6\text{-}64)$$

where $n = 0.4$ for *heating* and 0.3 for *cooling* of the fluid flowing through the tube. This equation is known as the **Dittus–Boulter equation,** and it is preferred to the Colburn equation. The fluid properties are evaluated at the *bulk mean fluid temperature* $T_b = \frac{1}{2}(T_i + T_e)$, which is the arithmetic average of the mean fluid temperatures at the inlet and the exit of the tube.

The relations above are not very sensitive to the *thermal conditions* at the tube surfaces and can be used for both $T_s = $ constant and $\dot{q}_s = $ constant cases. Despite their simplicity, the correlations above give sufficiently accurate results for most engineering purposes. They can also be used to obtain rough estimates of the friction factor and the heat transfer coefficients in the transition region $2300 \le \text{Re} \le 4000$, especially when the Reynolds number is closer to 4000 than it is to 2300.

The Nusselt number for *rough surfaces* can also be determined from Equation 6-62 by substituting the friction factor f value from the Moody chart. Note that tubes with rough surfaces have much higher heat transfer coefficients than tubes with smooth surfaces. Therefore, tube surfaces are often intentionally *roughened, corrugated,* or *finned* in order to *enhance* the convection heat transfer coefficient and thus the convection heat transfer rate (Fig. 6-39). Heat transfer in turbulent flow in a tube has been increased by as much as 400 percent by roughening the surface. Roughening the surface, of course, also increases the friction factor and thus the power requirement for the pump or the fan.

The turbulent flow relations above can also be used for *noncircular tubes* with reasonable accuracy by replacing the diameter D in the evaluation of the Reynolds number by the hydraulic diameter $D_h = 4A_c/p$.

(*a*) Finned surface

(*b*) Roughened surface

FIGURE 6-39

Tube surfaces are often *roughened, corrugated,* or *finned* in order to *enhance* convection heat transfer.

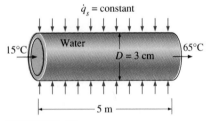

FIGURE 6-40

Schematic for Example 6-5.

EXAMPLE 6-5   Heating of Water by Resistance Heaters in a Tube

Water is to be heated from 15°C to 65°C as it flows through a 3-cm-internal-diameter 5-m-long tube (Fig. 6-40). The tube is equipped with an electric resistance heater that provides uniform heating throughout the surface of the tube. The outer surface of the heater is well insulated, so that in steady operation all the heat generated in the heater is transferred to the water in the tube. If the system is to provide hot water at a rate of 10 L/min, determine the power rating of the resistance heater. Also, estimate the inner surface temperature of the pipe at the exit.

Solution   Water is to be heated in a tube equipped with an electric resistance heater on its surface. The power rating of the heater and the inner surface temperature are to be determined.

Assumptions **1** Steady flow conditions exist. **2** The surface heat flux is uniform. **3** The inner surfaces of the tube are smooth.

Properties The properties of water at the bulk mean temperature of $T_b = (T_i + T_e)/2 = (15 + 65)/2 = 40°C$ are (Table A-9).

$$\rho = 992.1 \text{ kg/m}^3 \qquad\qquad C_p = 4179 \text{ J/kg} \cdot °C$$
$$k = 0.631 \text{ W/m} \cdot °C \qquad\qquad Pr = 4.32$$
$$\nu = \mu/\rho = 0.658 \times 10^{-6} \text{ m}^2/s$$

Analysis This is an *internal flow* problem since the water is flowing in a pipe. The cross-sectional and heat transfer surface areas are

$$A_c = \tfrac{1}{4}\pi D^2 = \tfrac{1}{4}\pi(0.03 \text{ m})^2 = 7.069 \times 10^{-4} \text{ m}^2$$
$$A = pL = \pi DL = \pi(0.03 \text{ m})(5 \text{ m}) = 0.471 \text{ m}^2$$

The volume flow rate of water is given as $\dot{V} = 10$ L/min $= 0.01$ m³/min. Then the mass flow rate of water becomes

$$\dot{m} = \rho\dot{V} = (992.1 \text{ kg/m}^3)(0.01 \text{ m}^3/\text{min}) = 9.921 \text{ kg/min} = 0.1654 \text{ kg/s}$$

To heat the water at this mass flow rate from 15°C to 65°C, heat must be supplied to the water at a rate of

$$\dot{Q} = \dot{m}C_p(T_e - T_i)$$
$$= (0.1654 \text{ kg/s})(4.179 \text{ kJ/kg} \cdot °C)(65 - 15)°C$$
$$= 34.6 \text{ kJ/s} = 34.6 \text{ kW}$$

All of this energy must come from the resistance heater. Therefore, the power rating of the heater must be **34.6 kW**

The surface temperature T_s of the tube at any location can be determined from

$$\dot{q}_s = h(T_s - T_m) \quad \rightarrow \quad T_s = T_m + \frac{\dot{q}_s}{h}$$

where h is the heat transfer coefficient and T_m is the mean temperature of the fluid at that location. The surface heat flux is constant in this case, and its value can be determined from

$$\dot{q}_s = \frac{\dot{Q}}{A} = \frac{34.6 \text{ kW}}{0.471 \text{ m}^2} = 73.46 \text{ kW/m}^2$$

To determine the heat transfer coefficient, we first need to find the mean velocity of water and the Reynolds number:

$$\mathcal{V}_m = \frac{\dot{V}}{A_c} = \frac{0.010 \text{ m}^3/\text{min}}{7.069 \times 10^{-4} \text{ m}^2} = 14.15 \text{ m/min} = 0.236 \text{ m/s}$$
$$Re = \frac{\mathcal{V}_m D}{\nu} = \frac{(0.236 \text{ m/s})(0.03 \text{ m})}{0.658 \times 10^{-6} \text{ m}^2/s} = 10,760$$

which is greater than 4000. Therefore, the flow is turbulent in this case and the entry lengths are roughly

$$L_h \approx L_t \approx 10D = 10 \times (0.03 \text{ m}) = 0.3 \text{ m}$$

which is much shorter than the total length of the pipe. Therefore, we can assume fully developed turbulent flow in the entire pipe and determine the Nusselt number from

$$Nu = \frac{hD}{k} = 0.023\ Re^{0.8}\ Pr^{0.4} = 0.023(10{,}760)^{0.8}\ (4.34)^{0.4} = 69.5$$

Then,

$$h = \frac{k}{D}Nu = \frac{0.631\ W/m \cdot {}^\circ C}{0.03\ m}(69.5) = 1462\ W/m^2 \cdot {}^\circ C$$

and the surface temperature of the pipe at the exit becomes

$$T_s = T_m + \frac{\dot{q}_s}{h} = 65{}^\circ C + \frac{73{,}460\ W/m^2}{1462\ W/m^2 \cdot {}^\circ C} = \mathbf{115{}^\circ C}$$

Discussion   Note that the inner surface temperature of the pipe will be 50°C higher than the mean water temperature at the pipe exit. This temperature difference of 50°C between the water and the surface will remain constant throughout the fully developed flow region.

EXAMPLE 6-6   Heat Loss from the Ducts of a Heating System in the Attic

Hot air at atmospheric pressure and 80°C enters an 8-m-long uninsulated square duct of cross-section 0.2 m × 0.2 m that passes through the attic of a house at a rate of 0.15 m³/s (Fig. 6-41). The duct is observed to be nearly isothermal at 60°C. Determine the exit temperature of the air and the rate of heat loss from the duct to the attic space.

Solution   Heat loss from uninsulated square ducts of a heating system in the attic is considered. The exit temperature and the rate of heat loss are to be determined.

Assumptions   **1** Steady operating conditions exist. **2** The inner surfaces of the duct are smooth. **3** Air is an ideal gas.

Properties   We do not know the exit temperature of the air in the duct, and thus we cannot determine the bulk mean temperature of air, which is the temperature at which the properties are to be determined. The mean temperature of air at the inlet is 80°C or 353 K, and we expect this temperature to drop somewhat as a result of heat loss through the duct whose surface is at a lower temperature. Thus it is reasonable to assume a bulk mean temperature of 350 K for air (we will check this assumption later) for the purpose of evaluating the properties of air. At this temperature and 1 atm we read (Table A-11)

$$\rho = 1.009\ kg/m^3 \qquad C_p = 1008\ J/kg \cdot {}^\circ C$$
$$k = 0.0297\ W/m \cdot {}^\circ C \qquad Pr = 0.706$$
$$\nu = 2.06 \times 10^{-5}\ m^2/s$$

Analysis   This is an *internal flow* problem since the air is flowing in a duct. The characteristic length (which is the hydraulic diameter), the mean velocity, and the Reynolds number in this case are

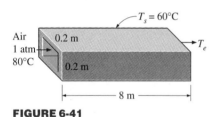

FIGURE 6-41

Schematic for Example 6-6.

$$D_h = \frac{4A_c}{p} = \frac{4a^2}{4a} = a = 0.2 \text{ m}$$

$$\mathcal{V}_m = \frac{\dot{V}}{A_c} = \frac{0.15 \text{ m}^3/\text{s}}{(0.2 \text{ m})^2} = 3.75 \text{ m/s}$$

$$\text{Re} = \frac{\mathcal{V}_m D_h}{\nu} = \frac{(3.75 \text{ m/s})(0.2 \text{ m})}{2.06 \times 10^{-5} \text{ m}^2/\text{s}} = 36{,}408$$

which is greater than 4000. Therefore, the flow is turbulent and the entry lengths in this case are roughly

$$L_h \approx L_t \approx 10 D_h = 10 \times (0.2 \text{ m}) = 2 \text{ m}$$

which is much shorter than the total length of the duct. Therefore, we can assume fully developed turbulent flow in the entire duct and determine the Nusselt number from

$$\text{Nu} = \frac{h D_h}{k} = 0.023 \, \text{Re}^{0.8} \, \text{Pr}^{0.3} = 0.023(36{,}408)^{0.8}(0.706)^{0.3} = 92.3$$

Then,

$$h = \frac{k}{D_h} \text{Nu} = \frac{0.0297 \text{ W/m} \cdot {}^\circ\text{C}}{0.2 \text{ m}} (92.3) = 13.7 \text{ W/m}^2 \cdot {}^\circ\text{C}$$

$$A = pL = 4aL = 4 \times (0.2 \text{ m})(8 \text{ m}) = 6.4 \text{ m}^2$$

$$\dot{m} = \rho \dot{V} = (1.009 \text{ kg/m}^3)(0.15 \text{ m}^3/\text{s}) = 0.151 \text{ kg/s}$$

Next, we determine the exit temperature of air from

$$T_e = T_s - (T_s - T_i)e^{-hA/\dot{m}C_p}$$

$$= 60^\circ\text{C} - [(60 - 80)^\circ\text{C}] \exp\left[-\frac{(13.7 \text{ W/m}^2 \cdot {}^\circ\text{C})(6.4 \text{ m}^2)}{(0.151 \text{ kg/s})(1008 \text{ J/kg} \cdot {}^\circ\text{C})}\right]$$

$$= \mathbf{71.2^\circ C}$$

Then the logarithmic mean temperature difference and the rate of heat loss from the air become

$$\Delta T_{\text{ln}} = \frac{T_e - T_i}{\ln \dfrac{T_s - T_e}{T_s - T_i}} = \frac{71.2 - 80}{\ln \dfrac{60 - 71.2}{60 - 80}} = 15.2^\circ\text{C}$$

$$\dot{Q} = hA \Delta T_{\text{ln}} = (13.7 \text{ W/m}^2 \cdot {}^\circ\text{C})(6.4 \text{ m}^2)(15.2^\circ\text{C}) = \mathbf{1368 \ W}$$

Therefore, the air will lose heat at a rate of 1368 W as it flows through the duct in the attic.

Discussion Having calculated the exit temperature of the air, we can now determine the actual bulk mean fluid temperature from

$$T_b = \frac{T_i + T_e}{2} = \frac{80 + 71.2}{2} = 75.6^\circ\text{C} = 348.6 \text{ K}$$

which is sufficiently close to the assumed value of 350 K at which we evaluated the properties of air. Therefore, it is not necessary to re-evaluate the properties at this T_b and to repeat the calculations.

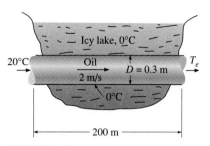

FIGURE 6-42
Schematic for Example 6-7.

EXAMPLE 6-7 Flow of Oil in a Pipeline through the Icy Waters of a Lake

Consider the flow of oil at 20°C in a 30-cm-diameter pipeline at an average velocity of 2 m/s (Fig. 6-42). A 200-m-long section of the pipeline passes through icy
waters of a lake at 0°C. Measurements indicate that the surface temperature of
the pipe is very nearly 0°C. Disregarding the thermal resistance of the pipe material, determine (a) the temperature of the oil when the pipe leaves the lake, (b)
the rate of heat transfer from the oil, and (c) the pumping power required to
overcome the pressure losses and to maintain the flow of the oil in the pipe.

Solution   Oil flows in a pipeline that passes through icy waters of a lake at 0°C.
The exit temperature of the oil, the rate of heat loss, and the pumping power
needed to overcome pressure losses are to be determined.

Assumptions   **1** Steady operating conditions exist. **2** The surface temperature of
the pipe is very nearly 0°C. **3** The thermal resistance of the pipe is negligible.
4 The inner surfaces of the pipeline are smooth. **5** The flow is hydrodynamically
developed when the pipeline reaches the lake.

Properties   We do not know the exit temperature of the oil, and thus we cannot
determine the bulk mean temperature, which is the temperature at which the
properties of oil are to be evaluated. The mean temperature of the oil at the inlet
is 20°C, and we expect this temperature to drop somewhat as a result of heat loss
to the icy waters of the lake. We evaluate the properties of the oil at the inlet
temperature, but we will repeat the calculations, if necessary, using properties at
the evaluated bulk mean temperature. At 20°C we read (Table A-10)

$$\rho = 888 \text{ kg/m}^3 \qquad \nu = 901 \times 10^{-6} \text{ m}^2/\text{s}$$
$$k = 0.145 \text{ W/m} \cdot °\text{C} \qquad C_p = 1880 \text{ J/kg} \cdot °\text{C}$$
$$\mu = 0.800 \text{ kg/m} \cdot \text{s} \qquad \text{Pr} = 10,400$$

Analysis   (a) This is an *internal flow* problem since the oil is flowing in a pipe.
The Reynolds number in this case is

$$\text{Re} = \frac{V_m D_h}{\nu} = \frac{(2 \text{ m/s})(0.3 \text{ m})}{901 \times 10^{-6} \text{ m}^2/\text{s}} = 666$$

which is less than the critical Reynolds number of 2300. Therefore, the flow is
laminar, and the thermal entry length in this case is roughly

$$L_t \approx 0.05 \text{ Re Pr } D = 0.05 \times 666 \times 10,400 \times (0.3 \text{ m}) \approx 104,000 \text{ m}$$

which is much greater than the total length of the pipe. This is typical of fluids with
high Prandtl numbers. Therefore, we assume thermally developing flow and determine the Nusselt number from

$$\text{Nu} = \frac{hD}{k} = 1.86 \left(\frac{\text{Re Pr } D}{L} \right)^{1/3} \left(\frac{\mu_b}{\mu_s} \right)^{0.14}$$

$$= 1.86 \left(\frac{666 \times 10,400 \times 0.3 \text{ m}}{200 \text{ m}} \right)^{1/3} \left(\frac{0.8}{3.85} \right)^{0.14} = 32.6$$

where the dynamic viscosity μ_s is determined at the surface temperature of 0°C.
Note that this Nusselt number is considerably higher than the fully developed
value of 3.66. Then,

$$h = \frac{k}{D} \text{Nu} = \frac{0.145 \text{ W/m} \cdot °\text{C}}{0.3 \text{ m}} (32.6) = 15.8 \text{ W/m}^2°\text{C}$$

Also,

$$A = pL = \pi DL = \pi(0.3 \text{ m})(200 \text{ m}) = 188.5 \text{ m}^2$$

$$\dot{m} = \rho A_c \mathcal{V}_m = (888 \text{ kg/m}^3)[\tfrac{1}{4}\pi(0.3 \text{ m})^2](2 \text{ m/s}) = 125.5 \text{ kg/s}$$

Next we determine the exit temperature of oil from

$$T_e = T_s - (T_s - T_i)e^{-hA/\dot{m}C_p}$$

$$= 0°C - [(0 - 20)°C] \exp\left[-\frac{(15.8 \text{ W/m}^2 \cdot °C)(188.5 \text{ m}^2)}{(125.5 \text{ kg/s})(1880 \text{ J/kg} \cdot °C)}\right]$$

$$= \mathbf{19.75°C}$$

Thus, the mean temperature of oil drops by a mere 0.25°C as it crosses the lake. This makes the bulk mean oil temperature 19.875°C, which is practically identical to the inlet mean temperature of 20°C. Therefore, we do not need to re-evaluate the properties at this bulk temperature and repeat the calculations.

(b) The logarithmic mean temperature difference and the rate of heat loss from the oil are

$$\Delta T_{\ln} = \frac{T_e - T_i}{\ln\dfrac{T_s - T_e}{T_s - T_i}} = \frac{19.75 - 20}{\ln\dfrac{0 - 19.75}{0 - 20}} = 19.875°C$$

$$\dot{Q} = hA \, \Delta T_{\ln} = (15.8 \text{ W/m}^2 \cdot °C)(188.5 \text{ m}^2)(19.875°C) = \mathbf{59{,}190 \text{ W}}$$

Therefore, the oil will lose heat at a rate of 59,190 W as it flows through the pipe in the icy waters of the lake. Note that $\Delta T_{\ln}$ is identical to the arithmetic mean temperature in this case, since $\Delta T_i \approx \Delta T_e$.

(c) The laminar flow of oil is hydrodynamically developed. Therefore, the friction factor can be determined from

$$f = \frac{64}{\text{Re}} = \frac{64}{666} = 0.0961$$

Then the pressure drop in the pipe and the required pumping power become

$$\Delta P = f\frac{L}{D}\frac{\rho\mathcal{V}_m^2}{2} = 0.0961\frac{200 \text{ m}}{0.3 \text{ m}}\frac{(888 \text{ kg/m}^3)(2 \text{ m/s})^2}{2} = 113{,}780 \text{ N/m}^2$$

$$\dot{W}_{\text{pump}} = \frac{\dot{m}\Delta P}{\rho} = \frac{(125.5 \text{ kg/s})(113{,}780 \text{ N/m}^2)}{888 \text{ kg/m}^3} = \mathbf{16.1 \text{ kW}}$$

Discussion   We will need a 16.1-kW pump just to overcome the friction in the pipe as the oil flows in the 200-m-long pipe through the lake.

We will close this chapter with the following external flow problem based on actual practical experience.

EXAMPLE 6–8 Cooling of Plastic Sheets by Forced Air

The forming section of a plastics plant puts out a continuous sheet of plastic that is 4 ft wide and 0.04 in. thick at a rate of 30 ft/min. The temperature of the plastic sheet is 200°F when it is exposed to the surrounding air, and a 2-ft-long section of the plastic sheet is subjected to air flow at 80°F at a velocity of 10 ft/s on both sides along its surfaces normal to the direction of motion of the sheet, as shown

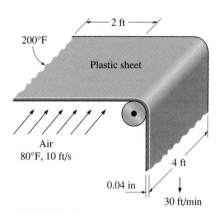

FIGURE 6-43
Schematic for Example 6-8.

in Figure 6-43. Determine (a) the rate of heat transfer from the plastic sheet to the air by forced convection and radiation and (b) the temperature of the plastic sheet at the end of the cooling section. Take the density, specific heat, and emissivity of the plastic sheet to be $\rho = 75$ lbm/ft^3, $C_p = 0.4$ Btu/lbm $\cdot$ °F, and $\varepsilon = 0.9$.

Solution Plastic sheets are cooled as they leave the forming section of a plastics plant. The rate of heat loss from the plastic sheet by convection and radiation and the exit temperature of the plastic sheet are to be determined.

Assumptions **1** Steady operating conditions exist. **2** The critical Reynolds number is $Re_{cr} = 5 \times 10^5$. **3** Air is an ideal gas. **4** The local atmospheric pressure is 1 atm.

Properties The properties of the plastic sheet are given in the problem statement. The properties of air at the film temperature of $T_f = (T_s + T_\infty)/2 = (200 + 80)/2 = 140$°F and 1 atm pressure are (Table A-11E)

$$k = 0.0162 \text{ Btu/h} \cdot \text{ft} \cdot \text{°F} \qquad Pr = 0.72$$
$$\nu = 0.204 \times 10^{-3} \text{ ft}^2/\text{s}$$

Analysis (a) We expect the temperature of the plastic sheet to drop somewhat as it flows through the 2-ft-long cooling section, but at this point we do not know the magnitude of that drop. Therefore, we assume the plastic sheet to be isothermal at 200°F to get started. We will repeat the calculations if necessary to account for the temperature drop of the plastic sheet.

Noting that $L = 4$ ft, the Reynolds number at the end of the air flow across the plastic sheet becomes

$$Re_L = \frac{\mathscr{V}_\infty L}{\nu} = \frac{(10 \text{ ft/s})(4 \text{ ft})}{0.204 \times 10^{-3} \text{ ft}^2/\text{s}} = 1.96 \times 10^5$$

which is less than the critical Reynolds number. Thus, we have *laminar flow* over the entire sheet, and the Nusselt number is determined, using the laminar flow relations for a flat plate, to be

$$Nu = \frac{hL}{k} = 0.664 \, Re_L^{0.5} \, Pr^{1/3} = 0.664 \times (1.96 \times 10^5)^{0.5} \times 0.72^{1/3} = 263.5$$

Then,

$$h = \frac{k}{L} Nu = \frac{0.0162 \text{ Btu/h} \cdot \text{ft} \cdot \text{°F}}{4 \text{ ft}} (263.5) = 1.07 \text{ Btu/h} \cdot \text{ft}^2 \cdot \text{°F}$$
$$A = (2 \text{ ft})(4 \text{ ft})(2 \text{ sides}) = 16 \text{ ft}^2$$

and

$$\dot{Q}_{conv} = hA(T_s - T_\infty)$$
$$= (1.07 \text{ Btu/h} \cdot \text{ft}^2 \cdot \text{°F})(16 \text{ ft}^2)(200 - 80)\text{°F}$$
$$= 2054 \text{ Btu/h}$$

$$\dot{Q}_{rad} = \varepsilon \sigma A(T_s^4 - T_{surr}^4)$$
$$= (0.9)(0.1714 \text{ Btu/h} \cdot \text{ft}^2 \cdot \text{R}^4)(16 \text{ ft}^2)[(660 \text{ R})^4 - (540 \text{ R})^4]$$
$$= 2584 \text{ Btu/h}$$

Therefore, the rate of cooling of the plastic sheet by combined convection and radiation is

$$\dot{Q}_{total} = \dot{Q}_{conv} + \dot{Q}_{rad} = 2054 + 2584 = \textbf{4638 Btu/h}$$

(b) To find the temperature of the plastic sheet at the end of the cooling section, we need to know the mass of the plastic rolling out per unit time (or the mass flow rate), which is determined from

$$\dot{m} = \rho A_c \mathcal{V} = (75 \text{ lbm/ft}^3)\left(\frac{4 \times 0.04}{12} \text{ ft}^2\right)\left(\frac{30}{60} \text{ ft/s}\right) = 0.5 \text{ lbm/s}$$

Then, an energy balance on the cooled section of the plastic sheet yields

$$\dot{Q} = \dot{m}C_p(T_2 - T_1) \quad \rightarrow \quad T_2 = T_1 + \frac{\dot{Q}}{\dot{m}C_p}$$

Noting that $\dot{Q}$ is a negative quantity (heat loss) for the plastic sheet and substituting, the temperature of the plastic sheet as it leaves the cooling section is determined to be

$$T_2 = 200°F + \frac{-4638 \text{ Btu/h}}{(0.5 \text{ lbm/s})(0.4 \text{ Btu/lbm} \cdot °F)}\left(\frac{1 \text{ h}}{3600 \text{ s}}\right) = \textbf{193.6°F}$$

Discussion The average temperature of the plastic sheet drops by about 6.4°F as it passes through the cooling section. The calculations now can be repeated by taking the average temperature of the plastic sheet to be 196.8°F instead of 200°F for better accuracy, but the change in the results will be insignificant because of the small change in temperature.

6-7 ■ SUMMARY

Convection is the mode of heat transfer that involves conduction as well as bulk fluid motion. The rate of convection heat transfer in external flow is expressed by *Newton's law of cooling* as

$$\dot{Q}_{conv} = hA(T_s - T_\infty) \qquad (W)$$

where T_s is the surface temperature and T_∞ is the free-stream temperature. The heat transfer coefficient h is usually expressed in the dimensionless form as the *Nusselt number*

$$\text{Nu} = \frac{h\delta}{k}$$

where δ is the *characteristic length*. The region of the flow in which the effects of the viscous shearing forces caused by fluid viscosity are felt is called the *velocity boundary layer* or just the *boundary layer*. The *drag force* acting over the entire surface in external flow is determined from

$$F_D = C_f A \frac{\rho \mathcal{V}_\infty^2}{2} \qquad (N)$$

where C_f is the average friction coefficient and A is the surface area for flow over a flat plate, but the frontal area for flow over a cylinder or sphere.

Fluid flow over a flat plate starts out as smooth and streamlined but turns chaotic after some distance from the leading edge. The flow regime is said in the first case to be *laminar,* characterized by smooth streamlines and highly ordered motion, and to be *turbulent* in the second case, where it is characterized by velocity fluctuations and highly disordered motion. The intense mixing in turbulent flow enhances both the drag force and the heat transfer. The flow regime depends mainly on the ratio of the inertia forces to viscous forces in the fluid. This ratio is called the *Reynolds number* and is expressed as

$$\text{Re} = \frac{\mathcal{V}\delta}{\nu}$$

where $\mathcal{V}$ is the free-stream velocity $\mathcal{V}_\infty$ for external flow and the mean fluid velocity $\mathcal{V}_m$ for internal flow. Also, δ is the characteristic length of the geometry, which is the distance from the leading edge for a flat plate, the outer diameter for flow over cylinders or spheres, and the inner diameter for flow inside circular tubes. The characteristic length for noncircular tubes is the *hydraulic diameter* D_h defined as

$$D_h = \frac{4A_c}{p}$$

where A_c is the cross-sectional area of the tube and p is its perimeter. The Reynolds number at which the flow becomes turbulent is called the *critical Reynolds number.* The value of the critical Reynolds number is about 5×10^5 for flow over a flat plate, 2×10^5 for flow over cylinders and spheres, and 2300 for flow inside tubes.

When the flow over the entire flat plate is laminar, the average friction coefficient and the Nusselt number can be determined from

$$C_f = \frac{1.328}{\text{Re}_L^{1/2}}$$

and

$$\text{Nu} = \frac{hL}{k} = 0.664\,\text{Re}_L^{1/2}\,\text{Pr}^{1/3} \qquad (\text{Pr} \geq 0.6)$$

After transition to turbulent flow at a critical Reynolds number $\text{Re}_{cr} = 5 \times 10^5$, the average friction coefficient and the Nusselt number over the entire plate become

$$C_f = \frac{0.074}{\text{Re}_L^{1/5}} - \frac{1742}{\text{Re}_L} \qquad (5 \times 10^5 \leq \text{Re}_L \leq 10^7)$$

and

$$\text{Nu} = \frac{hL}{k} = (0.037\,\text{Re}_L^{4/5} - 871)\,\text{Pr}^{1/3} \qquad \left(\begin{matrix} 0.6 \leq \text{Pr} \leq 60 \\ 5 \times 10^5 \leq \text{Re}_L \leq 10^7 \end{matrix}\right)$$

In order to properly account for the variation of the properties with temperature, the fluid properties are usually evaluated at the *film temperature,* defined as $T_f = \frac{1}{2}(T_s + T_\infty)$, which is the *arithmetic average* of the surface and the free-stream temperatures.

The average Nusselt numbers for cross flow over a *cylinder* and *sphere* can be determined from

$$\text{Nu}_{\text{cyl}} = \frac{hD}{k} = 0.3 + \frac{0.62\,\text{Re}^{1/2}\,\text{Pr}^{1/3}}{[1 + (0.4/\text{Pr})^{2/3}]^{1/4}}\left[1 + \left(\frac{\text{Re}}{28{,}200}\right)^{5/8}\right]^{4/5}$$

which is valid for Re Pr > 0.2, and

$$\text{Nu}_{\text{sph}} = \frac{hD}{k} = 2 + [0.4\,\text{Re}^{1/2} + 0.06\,\text{Re}^{2/3}]\,\text{Pr}^{0.4}\left(\frac{\mu_\infty}{\mu_s}\right)^{1/4}$$

which is valid for $3.5 \le \text{Re} \le 80{,}000$ and $0.7 \le \text{Pr} \le 380$. The fluid properties are evaluated at the film temperature $T_f = \frac{1}{2}(T_\infty + T_s)$ in the case of a cylinder, and at the free-stream temperature T_∞ (except for μ_s, which is evaluated at the surface temperature T_s) in the case of a sphere.

For flow in a tube, the mean velocity $\mathcal{V}_m$ is the average velocity of the fluid. The mean temperature T_m at a cross-section can be viewed as the average temperature at that cross-section. The mean velocity $\mathcal{V}_m$ remains constant, but the mean temperature T_m changes along the tube unless the fluid is not heated or cooled. The heat transfer to a fluid during steady flow in a tube can be expressed as

$$\dot{Q} = \dot{m}C_p(T_e - T_i) \qquad \text{(kJ/s)}$$

where T_i and T_e are the mean fluid temperatures at the inlet and exit of the tube.

The conditions at the surface of a tube can usually be approximated with reasonable accuracy to be *constant surface temperature* (T_s = constant) or *constant surface heat flux* ($\dot{q}_s$ = constant). In the case of $\dot{q}_s$ = constant, the rate of heat transfer can be expressed as

$$\dot{Q} = \dot{q}_s A = \dot{m}C_p(T_e - T_i) \qquad \text{(W)}$$

Then mean fluid temperature at the tube exit becomes

$$T_e = T_i + \frac{\dot{q}_s A}{\dot{m}C_p}$$

In the case of T_s = constant, the rate of heat transfer is expressed as

$$\dot{Q} = hA\,\Delta T_{\text{ln}}$$

where

$$\Delta T_{\text{ln}} = \frac{T_e - T_i}{\ln\dfrac{T_s - T_e}{T_s - T_i}} = \frac{\Delta T_e - \Delta T_i}{\ln(\Delta T_e/\Delta T_i)}$$

is the *logarithmic mean temperature difference*. Note that $\Delta T_i = T_s - T_i$ and $\Delta T_e = T_s - T_e$ are the temperature differences between the surface and the fluid at the inlet and the exit of the tube, respectively. Then the mean fluid temperature at the tube exit in this case can be determined from

$$T_e = T_s - (T_s - T_i)e^{-hA/\dot{m}C_p}$$

The pressure drop during flow in a tube of length L is expressed as

$$\Delta P = f\frac{L}{D}\frac{\rho V_m^2}{2} \qquad (\text{N/m}^2)$$

where f is the *friction factor*. The required *pumping power* to overcome a specified pressure drop ΔP is determined from

$$\dot{W}_{\text{pump}} = \dot{V}\,\Delta P = \frac{\dot{m}\Delta P}{\rho} \qquad (\text{W})$$

where $\dot{V} = V_m A_c = \dot{m}/\rho$ is the volume flow rate of the fluid through the tube.

The region from the tube inlet to the point at which the boundary layer merges at the centerline is called the *hydrodynamic entry region*, and the length of this region is called the *hydrodynamic entry length* L_h. The region beyond the hydrodynamic entry region in which the velocity profile is fully developed and remains unchanged is called the *hydrodynamically developed region*. The region of flow over which the thermal boundary layer develops and reaches the tube center is called the *thermal entry region*, and the length of this region is called the *thermal entry length* L_t. The region beyond the thermal entry region in which the dimensionless temperature profile expressed as $(T - T_s)/(T_m - T_s)$ remains unchanged is called the *thermally developed region*. The region in which the flow is both hydrodynamically and thermally developed is called the *fully developed flow*. The hydrodynamic and thermal entry lengths are given approximately as

$$L_{h,\text{laminar}} \approx 0.05\,\text{Re}\,D$$
$$L_{t,\text{laminar}} = 0.05\,\text{Re}\,\text{Pr}\,D$$
$$L_{h,\text{turbulent}} \approx L_{t,\text{turbulent}} \approx 10D$$

The friction and the heat transfer coefficients in the fully developed flow region remain constant.

In fully developed laminar flow, the friction factor is determined to be $f = 64/\text{Re}$. The Nusselt number is determined to be Nu = 3.66 for the case of T_s = constant and Nu = 4.36 for the case of $\dot{q}_s$ = constant. The average Nusselt number for the hydrodynamically and/or thermally developed laminar flow in a circular tube is given as

$$\text{Nu} = 1.86\left(\frac{\text{Re}\,\text{Pr}\,D}{L}\right)^{1/3}\left(\frac{\mu_b}{\mu_s}\right)^{0.14} \qquad (\text{Pr} > 0.5)$$

The recommended relations for the friction factor f and the Nusselt number for fully developed turbulent flow in smooth circular tubes are

$$f = 0.184 \, \text{Re}^{-0.2}$$

and

$$\text{Nu} = 0.023 \, \text{Re}^{0.8} \, \text{Pr}^n \qquad \left(\begin{array}{c} 0.7 \le \text{Pr} \le 160 \\ \text{Re} > 10{,}000 \end{array} \right)$$

where $n = 0.4$ for *heating* and 0.3 for *cooling* of the fluid flowing through the tube. The fluid properties are evaluated at the *bulk mean fluid temperature* $T_b = \frac{1}{2}(T_i + T_e)$, which is the arithmetic average of the mean fluid temperatures at the inlet and the exit of the tube.

REFERENCES AND SUGGESTED READING

1. Y. Bayazitoglu and M. N. Özişik. *Elements of Heat Transfer.* New York: McGraw-Hill, 1988.

2. S. W. Churchill and M. Bernstein. "A Correlating Equation for Forced Convection from Gases and Liquids to a Circular Cylinder in Cross Flow." *Journal of Heat Transfer* 99 (1977), pp. 300–6.

3. A. P. Colburn. *Transactions of the AIChE* 26 (1933), p. 174.

4. F. W. Dittus and L. M. K. Boelter. *University of California Publications on Engineering* 2 (1930), p. 433.

5. W. H. Giedt. "Investigation of Variation of Point Unit-Heat Transfer Coefficient around a Cylinder Normal to an Air Stream." *Transactions of the ASME* 71 (1949), pp. 375–81.

6. J. P. Holman. *Heat Transfer.* 7th ed. New York: McGraw-Hill, 1990.

7. F. P. Incropera and D. P. DeWitt. *Introduction to Heat Transfer.* 2nd ed. New York: John Wiley & Sons, 1990.

8. M. Jakob. *Heat Transfer.* Vol 1. New York: John Wiley & Sons, 1949.

9. F. Kreith and M. S. Bohn. *Principles of Heat Transfer.* 5th ed. St. Paul, MN: West Publishing, 1993.

10. L. F. Moody. "Friction Factor for Pipe Flow." *Transactions of the ASME* 66 (1944), pp. 671–84.

11. M. N. Özişik. *Heat Transfer—A Basic Approach.* New York: McGraw-Hill, 1985.

12. O. Reynolds. "On the Experimental Investigation of the Circumstances Which Determine Whether the Motion of Water Shall Be Direct or Sinuous, and the Law of Resistance in Parallel Channels." *Philosophical Transactions of the Royal Society of London* 174 (1883), pp. 935–82.

13. H. Schlichting. *Boundary Layer Theory.* 7th ed. New York: McGraw-Hill, 1979.

14. E. N. Sieder and G. E. Tate. "Heat Transfer and Pressure Drop of Liquids in Tubes." *Industrial Engineering Chemistry* 28 (1936), pp. 1429–35.

15. L. C. Thomas. *Heat Transfer.* Englewood Cliffs, NJ: Prentice Hall, 1992.

16. S. Whitaker. "Forced Convection Heat Transfer Correlations for Flow in Pipe, Past Flat Plates, Single Cylinders, and for Flow in Packed Beds and Tube Bundles." *AIChE Journal* 18 (1972), pp. 361–71.

17. F. M. White. *Heat and Mass Transfer.* Reading, MA: Addison-Wesley, 1988.

18. A. Zhukauskas. "Heat Transfer from Tubes in Cross Flow." In *Advances in Heat Transfer,* ed. J. P. Hartnett and T. F. Irvine, Jr. Vol. 8. New York: Academic Press, 1972.

PROBLEMS*

Physical Mechanism of Forced Convection

6-1C What is forced convection? How does it differ from natural convection? Is convection caused by winds forced or natural convection?

6-2C What is external forced convection? How does it differ from internal forced convection? Can a heat transfer system involve both internal and external convection at the same time? Give an example.

6-3C In which mode of heat transfer is the convection heat transfer coefficient usually higher, natural convection or forced convection? Why?

6-4C Consider a hot baked potato. Will the potato cool faster or slower when we blow the warm air coming from our lungs on it instead of letting it cool naturally in the cooler air in the room? Explain.

6-5C What is the physical significance of the Prandtl number? Does the value of the Prandtl number depend on the type of flow or the flow geometry? Does the Prandtl number of air change with pressure? Does it change with temperature?

6-6C What is the physical significance of the Reynolds number? How is it defined for (a) flow over a flat plate of length L, (b) flow over a cylinder of outer diameter D_o, (c) flow in a circular tube of inner diameter D_i, and (d) flow in a rectangular tube of cross-section $a \times b$?

6-7C What is the physical significance of the Nusselt number? How is it defined for (a) flow over a flat plate of length L, (b) flow over a cylinder of outer diameter D_o, (c) flow in a circular tube of inner diameter D_i, and (d) flow in a rectangular tube of cross-section $a \times b$?

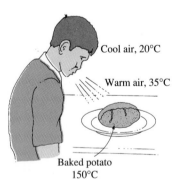

Cool air, 20°C

Warm air, 35°C

Baked potato
150°C

FIGURE P6-4C

*Students are encouraged to answer *all* the concept "C" questions.

6-8C When is heat transfer through a fluid conduction and when is it convection? For what case is the rate of heat transfer higher? How does the convection heat transfer coefficient differ from the thermal conductivity of a fluid?

6-9C How does turbulent flow differ from laminar flow? For which flow is (*a*) the friction coefficient and (*b*) the heat transfer coefficient higher?

6-10C Will a thermal boundary layer develop in flow over a surface even if both the fluid and the surface are at the same temperature?

Flow over Flat Plates

6-11C What fluid property is responsible for the development of the velocity boundary layer? For what kind of fluids will there be no velocity boundary layer on a flat plate?

6-12C Under what conditions can a curved surface be treated as a flat plate in convection calculations?

6-13C What is the no-slip condition on a surface?

6-14C Consider laminar forced convection from a horizontal flat plate. Will the heat flux be higher at the leading edge or at the tail of the plate? Why?

6-15C What does the friction coefficient represent in flow over a flat plate? How is it related to the drag force acting on the plate?

6-16C For flow over a flat plate, how does the flow in the thermal boundary layer differ from the flow outside the thermal boundary layer?

6-17C Consider laminar flow over a flat plate. Will the friction coefficient change with position? How about the heat transfer coefficient?

6-18C How are the average friction and heat transfer coefficients determined in flow over a flat plate?

6-19 Engine oil at 80°C flows over a 6-m-long flat plate whose temperature is 30°C with a velocity of 3 m/s. Determine the total drag force and the rate of heat transfer over the entire plate per unit width.

6-20 The local atmospheric pressure in Denver, Colorado (elevation 1610 m), is 83.4 kPa. Air at this pressure and at 30°C flows with a velocity of 6 m/s over a 2.5-m × 8-m flat plate whose temperature is 120°C. Determine the rate of heat transfer from the plate if the air flows parallel to the (*a*) 8-m-long side and (*b*) the 2.5-m side.

6-21 During a cold winter day, wind at 55 km/h is blowing parallel to a 4-m-high and 10-m-long wall of a house. If the air outside is at 5°C and the surface temperature of the wall is 12°C, determine the rate of heat loss from that wall by convection. What would your answer be if the wind velocity has doubled?   *Answers:* 9212 W, 16,408 W

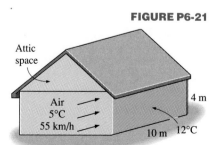

FIGURE P6-21

Attic space

Air
5°C
55 km/h

4 m

10 m 12°C

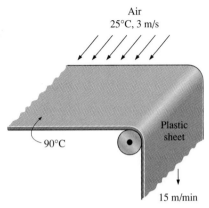

Air
25°C, 3 m/s

90°C

Plastic
sheet

15 m/min

FIGURE P6-24

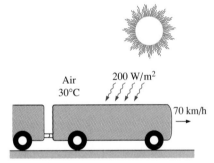

Air
30°C

200 W/m²

70 km/h

FIGURE P6-25

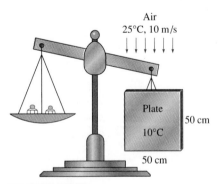

Air
25°C, 10 m/s

Plate

50 cm

10°C

50 cm

FIGURE P6-27

6-22E   Air at 65°F flows over a 10-ft-long flat plate at 7 ft/s. Determine the local friction and heat transfer coefficients at intervals of 1 ft, and plot the results against the distance from the leading edge.

6-23   Consider a hot automotive engine, which can be approximated as a 0.5-m-high, 0.40-m-wide, and 0.8-m-long rectangular block. The bottom surface of the block is at a temperature of 80°C and has an emissivity of 0.95. The ambient air is at 30°C, and the road surface is at 25°C. Determine the rate of heat transfer from the bottom surface of the engine block by convection and radiation as the car travels at a velocity of 80 km/h. Assume the flow to be turbulent over the entire surface because of the constant agitation of the engine block.

6-24   The forming section of a plastics plant puts out a continuous sheet of plastic that is 1.2 m wide and 2 mm thick at a rate of 15 m/min. The temperature of the plastic sheet is 90°C when it is exposed to the surrounding air, and the sheet is subjected to air flow at 25°C at a velocity of 3 m/s on both sides along its surfaces normal to the direction of motion of the sheet. The width of the air cooling section is such that a fixed point on the plastic sheet passes through that section in 2 s. Determine the rate of heat transfer from the plastic sheet to the air and the drag force the air exerts on the plastic sheet in the direction of air flow.

6-25   The top surface of the passenger car of a train moving at a velocity of 70 km/h is 2.8 m wide and 8 m long. The top surface is absorbing solar radiation at a rate of 200 W/m², and the temperature of the ambient air is 30°C. Assuming the roof of the car to be perfectly insulated and the radiation heat exchange with the surroundings to be small relative to convection, determine the equilibrium temperature of the top surface of the car.
   Answer: 35°C

6-26   A 15-cm × 15-cm circuit board dissipating 15 W of power uniformly is cooled by air, which approaches the circuit board at 50°C with a velocity of 5 m/s. Disregarding any heat transfer from the back surface of the board, determine the surface temperature of the electronic components (*a*) at the leading edge and (*b*) at the end of the board. Assume the flow to be turbulent since the electronic components are expected to act as turbulators.

6-27   The weight of a thin flat plate 50 cm × 50 cm in size is balanced by a counterweight that has a mass of 2 kg, as shown in Figure P6-27. Now a fan is turned on, and air at 25°C flows downward over both surfaces of the plate with a free-stream velocity of 10 m/s. Determine the mass of the counterweight that needs to be added in order to balance the plate in this case. Also determine the initial rate of heat transfer to the plate if the plate was initially at a uniform temperature of 10°C.

6-28   Consider laminar flow of a fluid over a flat plate maintained at a constant temperature. Now the free-stream velocity of the fluid is doubled. Determine the change in the drag force on the plate and the rate of heat transfer between the fluid and the plate. Assume the flow to remain laminar.

6-29E Consider a refrigeration truck traveling at 55 mph at a location where the air temperature is 80°F. The refrigerated compartment of the truck can be considered to be a 9-ft-wide, 8-ft-high, and 20-ft-long rectangular box. The refrigeration system of the truck can provide 3 tons of refrigeration (i.e., it can remove heat at a rate of 600 Btu/min). The outer surface of the truck is coated with a low-emissivity material, and thus radiation heat transfer is very small. Determine the average temperature of the outer surface of the refrigeration compartment of the truck if the refrigeration system is observed to be operating at half the capacity. Assume the air flow over the entire outer surface to be turbulent and the heat transfer coefficient at the front and rear surfaces to be equal to that on side surfaces.

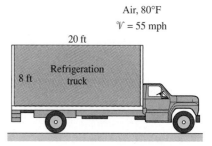

FIGURE P6-29E

6-30 Solar radiation is incident on the glass cover of a solar collector at a rate of 700 W/m². The glass transmits 88 percent of the incident radiation and has an emissivity of 0.90. The entire hot water needs of a family in summer can be met by two collectors 1.2 m high and 1 m wide. The two collectors are attached to each other on one side so that they appear like a single collector 1.2 m × 2 m in size. The temperature of the glass cover is measured to be 35°C on a day when the surrounding air temperature is 23°C and the wind is blowing at 30 km/h. The effective sky temperature for radiation exchange between the glass cover and the open sky is −40°C. Water enters the tubes attached to the absorber plate at a rate of 1 kg/min. Assuming the backsurface of the absorber plate to be heavily insulated and the only heat loss to occur through the glass cover, determine (a) the total rate of heat loss from the collector, (b) the collector efficiency, which is the ratio of the amount of heat transferred to the water to the solar energy incident on the collector, and (c) the temperature rise of water as it flows through the collector.
   Answers: (a) 1262 W, (b) 0.15, (c) 3.1°C

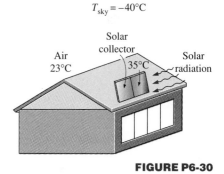

FIGURE P6-30

6-31 A transformer that is 10 cm long, 6.2 cm wide, and 5 cm high is to be cooled by attaching a 10 cm × 6.2 cm wide polished aluminum heat sink (emissivity = 0.03) to its top surface. The heat sink has seven fins, which are 5 mm high, 2 mm thick, and 10 cm long. A fan blows air at 25°C parallel to the passages between the fins. The heat sink is to dissipate 20 W of heat and the base temperature of the heat sink is not to exceed 60°C. Assuming the fins and the base plate to be nearly isothermal and the radiation heat transfer to be negligible, determine the minimum free-stream velocity the fan needs to supply to avoid overheating.

6-32 Repeat Problem 6-31 assuming the heat sink to be black-anodized and thus to have an effective emissivity of 0.90. Note that in radiation calculations the base area (10 cm × 6.2 cm) is to be used, not the total surface area.

6-33 An array of power transistors, dissipating 3 W of power each, are to be cooled by mounting them on a 25-cm × 25-cm square aluminum plate and blowing air at 35°C over the plate with a fan at a velocity of 4 m/s. The average temperature of the plate is not to exceed 65°C. Assuming the heat transfer from the back side of the plate to be negligible and disregarding radiation, determine the number of transistors that can be placed on this plate.

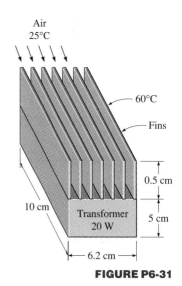

FIGURE P6-31

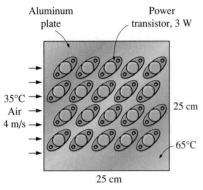

Aluminum plate

Power transistor, 3 W

35°C
Air
4 m/s

25 cm

65°C

25 cm

FIGURE P6-33

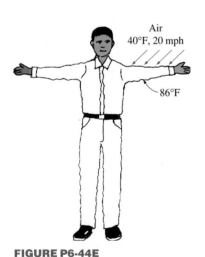

Air
40°F, 20 mph

86°F

FIGURE P6-44E

6-34 Repeat Problem 6-33 for a location at an elevation of 1610 m where the atmospheric pressure is 83.4 kPa.   *Answer:* 8

Flow across Cylinders and Spheres

6-35C How is the film temperature for flow over an isothermal cylinder defined?

6-36C In flow over cylinders, why does the drag coefficient suddenly drop when the flow becomes turbulent? Isn't turbulence supposed to increase the drag coefficient instead of decreasing it?

6-37C In flow over blunt bodies such as a cylinder, how does the pressure drag differ from the friction drag?

6-38C Why is flow separation in flow over cylinders delayed in turbulent flow?

6-39C Which bicyclist is more likely to go faster: the one who keeps his head and his body in the most upright position or the one who leans down and brings his body closer to his knees? Why?

6-40C Which car is more likely to be more fuel-efficient: the one with sharp corners or the one that is contoured to resemble an ellipse? Why?

6-41C Consider laminar flow of air across a hot circular cylinder. At what point on the cylinder will the heat transfer be highest? What would your answer be if the flow were turbulent?

6-42 A long 8-cm-diameter steam pipe whose external surface temperature is 90°C passes through some open area that is not protected against the winds. Determine the rate of heat loss from the pipe per unit of its length when the air is at 1 atm pressure and 7°C and the wind is blowing across the pipe at a velocity of 50 km/h.

6-43 A stainless steel ball ($\rho = 8055$ kg/m^3, $C_p = 480$ J/kg · °C) of diameter $D = 15$ cm is removed from the oven at a uniform temperature of 350°C. The ball is then subjected to the flow of air at 1 atm pressure and 30°C with a velocity of 6 m/s. The surface temperature of the ball eventually drops to 250°C. Determine the average convection heat transfer coefficient during this cooling process and estimate how long this process has taken.

6-44E A person extends his uncovered arms into the windy air outside at 40°F and 20 mph in order to feel nature closely. Initially, the skin temperature of the arm is 86°F. Treating the arm as a 2-ft-long and 3-in.-diameter cylinder, determine the rate of heat loss from the arm.

6-45 An average person generates heat at a rate of 84 W while resting. Assuming one-quarter of this heat is lost from the head and disregarding radiation, determine the average surface temperature of the head when it is not covered and is subjected to winds at 10°C and 35 km/h. The head can be approximated as a 30-cm-diameter sphere.   *Answer:* 12.7°C

6-46 Consider the flow of a fluid across a cylinder maintained at a constant temperature. Now the free-stream velocity of the fluid is doubled. Determine the change in the drag force on the cylinder and the rate of heat transfer between the fluid and the cylinder.

6-47 A 6-mm-diameter electrical transmission line carries an electric current of 50 A and has a resistance of 0.002 ohm per meter length. Determine the surface temperature of the wire during a windy day when the air temperature is 10°C and the wind is blowing across the transmission line at 40 km/h. Also determine the drag force exerted on the wire by the wind.

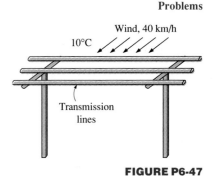

FIGURE P6-47

6-48 A heating system is to be designed to keep the wings of an aircraft cruising at a velocity of 900 km/h above freezing temperatures during flight at 12,200-m altitude where the standard atmospheric conditions are −55.4°C and 18.8 kPa. Approximating the wing as a cylinder of elliptical cross-section whose minor axis is 30 cm and disregarding radiation, determine the average convection heat transfer coefficient on the wing surface and the average rate of heat transfer per unit surface area.

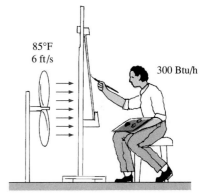

FIGURE P6-49

6-49 A long aluminum wire of diameter 3 mm is extruded at a temperature of 350°C. The wire is subjected to cross air flow at 35°C at a velocity of 6 m/s. Determine the rate of heat transfer from the wire to the air per meter length when it is first exposed to the air.

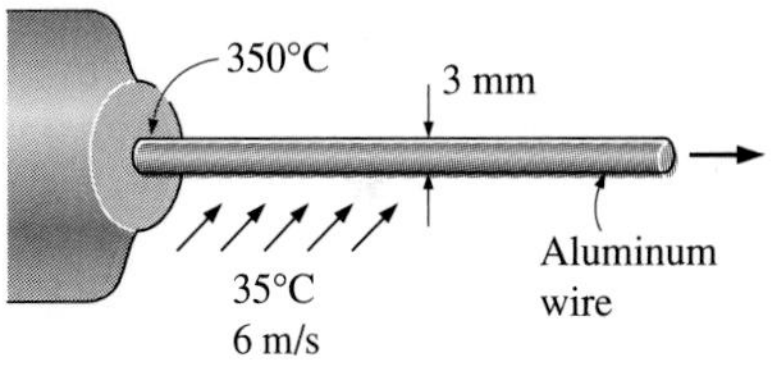

6-50E Consider a person who is trying to keep cool on a hot summer day by turning a fan on and exposing his entire body to air flow. The air temperature is 85°F and the fan is blowing air at a velocity of 6 ft/s. If the person is doing light work and generating sensible heat at a rate of 300 Btu/h, determine the average temperature of the outer surface (skin or clothing) of the person. The average human body can be treated as a 1-ft-diameter cylinder with an exposed surface area of 18 ft^2. Disregard any heat transfer by radiation. What would your answer be if the air velocity were doubled?

Answers: 91.7°F, 88.9°F

6-51 An incandescent light bulb is an inexpensive but highly inefficient device that converts electrical energy into light. It converts about 10 percent of the electrical energy it consumes into light while converting the remaining 90 percent into heat. (A fluorescent light bulb will give the same amount of light while consuming only one-fourth of the electrical energy, and it will last 10 times longer than an incandescent light bulb.) The glass bulb of the lamp heats up very quickly as a result of absorbing all that heat and dissipating it to the surroundings by convection and radiation.

Consider a 10-cm-diameter 100-W light bulb cooled by a fan that blows air at 25°C to the bulb at a velocity of 2 m/s. The surrounding surfaces are also at 25°C, and the emissivity of the glass is 0.9. Assuming 10 percent of the energy passes through the glass bulb as light with negligible absorption and the rest of the energy is absorbed and dissipated by the bulb itself, determine the equilibrium temperature of the glass bulb.

FIGURE P6-50E

FIGURE P6-51

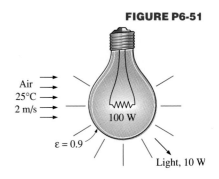

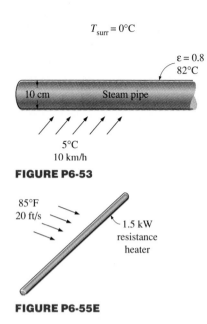

$T_{\text{surr}} = 0°C$

$\varepsilon = 0.8$
82°C

10 cm — Steam pipe

5°C
10 km/h

FIGURE P6-53

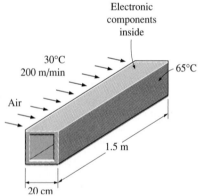

85°F
20 ft/s

1.5 kW
resistance
heater

FIGURE P6-55E

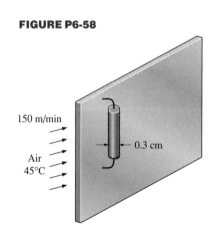

Electronic
components
inside

30°C
200 m/min

65°C

Air

1.5 m

20 cm

FIGURE P6-56

FIGURE P6-58

150 m/min

0.3 cm

Air
45°C

6-52 Consider a 3-mm-diameter raindrop at 8°C that is failing freely in atmospheric air at 20°C. Determine the terminal velocity of the raindrop, which is the velocity at which the drag force equals the weight of the drop, and the rate of heat transfer by convection as the raindrop descends at the terminal velocity. Assume there is no evaporation or condensation.

6-53 During a plant visit, it was noticed that a 12-m-long section of a 10-cm-diameter steam pipe is completely exposed to the ambient air. The temperature measurements indicate that the average temperature of the outer surface of the steam pipe is 82°C when the ambient temperature is 5°C. There are also light winds in the area at 10 km/h. The emissivity of the outer surface of the pipe is 0.8, and the average temperature of the surfaces surrounding the pipe, including the sky, is estimated to be 0°C. Determine the amount of heat lost from the steam during a 10-h-long work day.

Steam is supplied by a gas-fired steam generator that has an efficiency of 80 percent, and the plant pays $0.54/therm of natural gas (1 therm = 105,500 kJ). If the pipe is insulated and 90 percent of the heat loss is saved, determine the amount of money this facility will save a year as a result of insulating the steam pipes. Assume the plant operates every day of the year for 10 h. State your assumptions.

6-54 Reconsider Problem 6-53. There seems to be some uncertainty about the average temperature of the surfaces surrounding the pipe used in radiation calculations, and you are asked to determine if it makes any significant difference in overall heat transfer. Repeat the calculations in Problem 6-53 for average surrounding surface temperatures of −20°C and 25°C, and determine the change in the values obtained.

6-55E A 12-ft-long, 1.5-kW electrical resistance wire is made of 0.1-in.-diameter stainless steel ($k = 8.7$ Btu/h · ft · °F). The resistance wire operates in an environment at 85°F. Determine the surface temperature of the wire if it is cooled by a fan blowing air at a velocity of 20 ft/s.

6-56 The components of an electronic system are located in a 1.5-m-long horizontal duct whose cross-section is 20 cm × 20 cm. The components in the duct are not allowed to come into direct contact with cooling air, and thus are cooled by air at 30°C flowing over the duct with a velocity of 200 m/min. If the surface temperature of the duct is not to exceed 65°C, determine the total power rating of the electronic devices that can be mounted into the duct.
Answer: 643 W

6-57 Repeat Problem 6-56 for a location at 4000-m altitude where the atmospheric pressure is 61.66 kPa.

6-58 A 0.4-W cylindrical electronic component with diameter 0.3 cm and length 1.8 cm and mounted on a circuit board is cooled by air flowing across it at a velocity of 150 m/min. If the air temperature is 45°C, determine the surface temperature of the component.

6-59C   Show that the Reynolds number for flow in a circular tube of diameter D can be expressed as $Re = 4\dot{m}/\pi D \mu$.

6-60C   Which fluid requires a larger pump to move at a specified velocity in a specified tube: water or engine oil? Why?

6-61C   What are the generally accepted values of the critical Reynolds numbers for (a) flow over a flat plate, (b) flow over a circular cylinder, and (c) flow in a tube?

6-62C   In the fully developed region of flow in a circular tube, will the velocity profile change in the flow direction? How about the temperature profile?

6-63C   Consider the flow of oil in a tube. How will the hydrodynamic and thermal entry lengths compare if the flow is laminar? How would they compare if the flow were turbulent?

6-64C   Consider the flow of mercury (a liquid metal) in a tube. How will the hydrodynamic and thermal entry lengths compare if the flow is laminar? How would they compare if the flow were turbulent?

6-65C   What is the difference between the friction factor and the friction coefficient?

6-66C   What do the mean velocity $\mathcal{V}_m$ and the mean temperature T_m represent in flow through circular tubes of constant diameter?

6-67C   Consider fluid flow in a tube whose surface temperature remains constant. What is the appropriate temperature difference for use in Newton's law of cooling with an average heat transfer coefficient?

6-68C   What is the physical significance of the number of transfer units $NTU = hA/\dot{m}C_p$? What do a small and a large NTU tell the heat transfer engineer about a heat transfer system?

6-69C   How is the friction factor for flow in a tube related to the pressure drop? How is the pressure drop related to the pumping power requirement for a given mass flow rate?

6-70C   What does the logarithmic mean temperature difference represent for flow in a tube whose surface temperature is constant? Why do we use the logarithmic mean temperature instead of the arithmetic mean temperature?

6-71C   How is the hydrodynamic entry length defined for flow in a tube? How about the thermal entry length? In what region is the flow in a tube fully developed?

6-72C   Consider laminar forced convection in a circular tube. Will the friction factor be higher near the inlet of the tube or near the exit? Why? What would your response be if the flow were turbulent?

6-73C Consider laminar forced convection in a circular tube. Will the heat flux be higher near the inlet of the tube or near the exit? Why?

6-74C Consider turbulent forced convection in a circular tube. Will the heat flux be higher near the inlet of the tube or near the exit? Why?

6-75C How does surface roughness affect the pressure drop and the heat transfer in a tube if the fluid flow is turbulent? What would your response be if the flow in the tube were laminar?

6-76 Water is to be heated from 12°C to 70°C as it flows through a 2-cm-internal-diameter, 7-m-long tube. The tube is equipped with an electric resistance heater, which provides uniform heating throughout the surface of the tube. The outer surface of the heater is well insulated, so that in steady operation all the heat generated in the heater is transferred to the water in the tube. If the system is to provide hot water at a rate of 8 L/min, determine the power rating of the resistance heater. Also, estimate the inner surface temperature of the pipe at the exit.

6-77 Hot air at atmospheric pressure and 85°C enters a 10-m-long uninsulated square duct of cross-section 0.15 m × 0.15 m that passes through the attic of a house at a rate of 0.10 m³/s. The duct is observed to be nearly isothermal at 70°C. Determine the exit temperature of the air and the rate of heat loss from the duct to the air space in the attic.
Answers: 75.6°C, 946 W

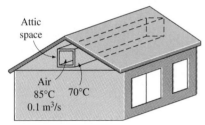

Attic space

Air
85°C   70°C
0.1 m³/s

FIGURE P6-77

6-78 Consider an air solar collector that is 1 m wide and 5 m long and has a constant spacing of 3 cm between the glass cover and the collector plate. Air enters the collector at 30°C at a rate of 0.15 m³/s through the 1-m-wide edge and flows along the 5-m-long passage way. If the average temperatures of the glass cover and the collector plate are 20°C and 60°C, respectively, determine (*a*) the net rate of heat transfer to the air in the collector and (*b*) the temperature rise of air as it flows through the collector.

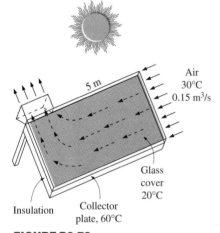

5 m

Air
30°C
0.15 m³/s

Glass cover
20°C

Insulation   Collector
plate, 60°C

FIGURE P6-78

6-79 Consider the flow of oil at 10°C in a 40-cm-diameter pipeline at an average velocity of 0.5 m/s. A 300-m-long section of the pipeline passes through icy waters of a lake at 0°C. Measurements indicate that the surface temperature of the pipe is very nearly 0°C. Disregarding the thermal resistance of the pipe material, determine (*a*) the temperature of the oil when the pipe leaves the lake, (*b*) the rate of heat transfer from the oil, and (*c*) the pumping power required to overcome the pressure losses and to maintain the flow of the oil in the pipe.

6-80 Consider laminar flow of a fluid through a square channel maintained at a constant temperature. Now the bulk fluid velocity of the fluid is doubled. Determine the change in the pressure drop of the fluid and the rate of heat transfer between the fluid and the walls of the channel. Assume the flow regime remains unchanged.   *Answers:* 2, 1.26

6-81 Repeat Problem 6-80 for turbulent flow.

6-82E The hot water needs of a household are to be met by heating water at 55°F to 200°F by a parabolic solar collector at a rate of 4 lbm/s. Water flows through a 1.25-in.-diameter thin aluminum tube whose outer surface is black-anodized in order to maximize its solar absorption ability. The centerline of the tube coincides with the focal line of the collector, and a glass sleeve is placed outside the tube to minimize the heat losses. If solar energy is transferred to water at a net rate of 350 Btu/h per ft length of the tube, determine the required length of the parabolic collector to meet the hot water requirements of this house. Also, determine the surface temperature of the tube at the exit.

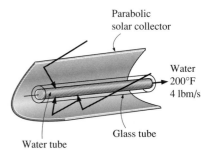

FIGURE P6-82E

6-83 A 15-cm × 20-cm printed circuit board whose components are not allowed to come into direct contact with air for reliability reasons is to be cooled by passing cool air through a 20-cm-long channel of rectangular cross-section 0.2 cm × 14 cm drilled into the board. The heat generated by the electronic components is conducted across the thin layer of the board to the channel, where it is removed by air that enters the channel at 20°C. The heat flux at the top surface of the channel can be considered to be uniform, and heat transfer through other surfaces is negligible. If the velocity of the air at the inlet of the channel is not to exceed 4 m/s and the surface temperature of the channel is to remain under 50°C, determine the maximum total power of the electronic components that can safely be mounted on this circuit board.

6-84 Repeat Problem 6-83 by replacing air with helium, which has six times the thermal conductivity of air.

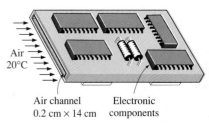

FIGURE P6-83

6-85 Air enters a 7-m-long section of a rectangular duct of cross-section 15 cm × 20 cm at 50°C at an average velocity of 7 m/s. If the walls of the duct are maintained at 10°C, determine (a) the outlet temperature of the air, (b) the rate of heat transfer from the air, and (c) the fan power needed to overcome the pressure losses in this section of the duct.

Answers: (a) 32.8°C, (b) 3674 W, (c) 4.22 W

6-86 Hot air at 60°C leaving the furnace of a house enters a 12-m-long section of a sheet metal duct of rectangular cross-section 20 cm × 20 cm at an average velocity of 4 m/s. The thermal resistance of the duct is negligible, and the outer surface of the duct, whose emissivity is 0.3, is exposed to the cold air at 10°C in the basement, with a convection heat transfer coefficient of 10 W/m² · °C. Taking the walls of the basement to be at 10°C also, determine (a) the temperature at which the hot air will leave the basement and (b) the rate of heat loss from the hot air in the duct to the basement.

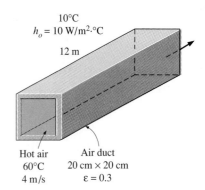

FIGURE P6-86

6-87 The components of an electronic system dissipating 90 W are located in a 1-m-long horizontal duct whose cross-section is 16 cm × 16 cm. The components in the duct are cooled by forced air, which enters at 32°C at a rate of 0.65 m³/min. Assuming 85 percent of the heat generated inside is transferred to air flowing through the duct and the remaining 15 percent is lost through the outer surfaces of the duct, determine (a) the exit temperature of air and (b) the highest component surface temperature in the duct.

6-88 Repeat Problem 6-87 for a circular horizontal duct of 15-cm diameter.

6-89 Consider a hollow-core printed circuit board 12 cm high and 18 cm long, dissipating a total of 20 W. The width of the air gap in the middle of the PCB is 0.25 cm. The cooling air enters the 12-cm-wide core at 32°C at a rate of 0.8 L/s. Assuming the heat generated to be uniformly distributed over the two side surfaces of the PCB, determine (*a*) the temperature at which the air leaves the hollow core and (*b*) the highest temperature on the inner surface of the core.   *Answers:* (*a*) 53.7°C, (*b*) 72.0°C

6-90 Repeat Problem 6-89 for a hollow-core PCB dissipating 35 W.

6-91E Water at 54°F is heated by passing it through 0.75-in.-internal-diameter thin-walled copper tubes. Heat is supplied to the water by steam that condenses outside the copper tubes at 250°F. If water is to be heated to 140°F at a rate of 0.7 lbm/s, determine (*a*) the length of the copper tube that needs to be used and (*b*) the pumping power required to maintain this flow at the specified rate. Assume the entire copper tube to be at the steam temperature of 250°F.

6-92 A computer cooled by a fan contains eight PCBs, each dissipating 10 W of power. The height of the PCBs is 12 cm and the length is 18 cm. The clearance between the tips of the components on the PCB and the back surface of the adjacent PCB is 0.3 cm. The cooling air is supplied by a 25-W fan mounted at the inlet. If the temperature rise of air as it flows through the case of the computer is not to exceed 10°C, determine (*a*) the flow rate of the air that the fan needs to deliver, (*b*) the fraction of the temperature rise of air that is due to the heat generated by the fan and its motor, and (*c*) the highest allowable inlet air temperature if the surface temperature of the components is not to exceed 70°C anywhere in the system.
    Answers: (*a*) 0.008 kg/s, (*b*) 31 percent, (*c*) 8.3°C

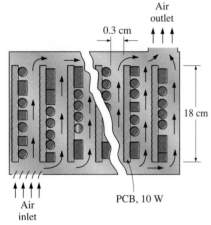

Air outlet

0.3 cm

18 cm

Air inlet

PCB, 10 W

FIGURE P6-92

Review Problems

6-93 Consider a house that is maintained at 22°C at all times. The walls of the house have *R*-3.38 insulation in SI units (i.e., they have an *L/k* value or a thermal resistance of 3.38 m$^2 \cdot$ °C/W). During a cold winter night, the outside air temperature is 4°C and wind at 50 km/h is blowing parallel to a 3-m-high and 8-m-long wall of the house. If the heat transfer coefficient on the interior surface of the wall is 8 W/m$^2 \cdot$ °C, determine the rate of heat loss from that wall of the house. Draw the thermal resistance network and disregard radiation heat transfer.   *Answer:* 122 W

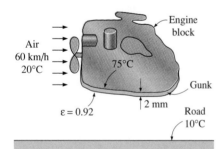

Engine block

Air
60 km/h
20°C

75°C

Gunk

ε = 0.92

2 mm

Road
10°C

FIGURE P6-94

6-94 An automotive engine can be approximated as a 0.4-m-high, 0.60-m-wide, and 0.7-m-long rectangular block. The bottom surface of the block is at a temperature of 75°C and has an emissivity of 0.92. The ambient air is at 20°C, and the road surface is at 10°C. Determine the rate of heat transfer from the bottom surface of the engine block by convection and radiation as the car travels at a velocity of 60 km/h. Assume the flow to be turbulent over the entire surface because of the constant agitation of the engine block. How will the heat transfer be affected when a 2-mm-thick gunk (*k* = 3 W/m · °C) has formed at the bottom surface as a result of the dirt and oil collected at that surface over time? Assume the metal temperature under the gunk still to be 75°C.

6-95 The thickness of the velocity boundary layer over a flat plate increases in the flow direction and is given by

$$\delta = \frac{5x}{\sqrt{Re_x}}$$

where $Re_x = \mathcal{V}_\infty x/v$ is the Reynolds number at a distance x from the leading edge of the plate. Calculate the thickness of the boundary layer during flow over a 3-m-long flat plate at intervals of 25 cm, and plot the boundary layer over the plate for the flow of (*a*) air, (*b*) water, and (*c*) engine oil at 20°C at a free-stream velocity of 3 m/s.

6-96E The passenger compartment of a minivan traveling at 60 mph can be modeled as 3.2-ft-high, 6-ft-wide, and 11-ft-long rectangular box whose walls have an insulating value of R-3 (i.e., a wall thickness–to–thermal conductivity ratio of 3 h · ft^2 · °F/Btu). The interior of a minivan is maintained at an average temperature of 70°F during a trip at night while the outside air temperature is 90°F. The average heat transfer coefficient on the interior surfaces of the van is 1.2 Btu/h · ft^2 · °F. The air flow over the exterior surfaces can be assumed to be turbulent because of the intense vibrations involved, and the heat transfer coefficient on the front and back surfaces can be taken to be equal to that on the top surface. Disregarding any heat gain or loss by radiation, determine the rate of heat transfer from the ambient air to the van.

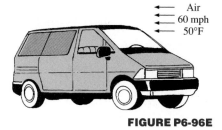

Air
60 mph
50°F

FIGURE P6-96E

6-97 Consider a house that is maintained at a constant temperature of 22°C. One of the walls of the house has three single-pane glass windows that are 1.5 m high and 1.2 m long. The glass ($k = 0.78$ W/m · °C) is 0.5 cm thick, and the heat transfer coefficient on the inner surface of the glass is 8 W/m^2 · C. Now winds at 60 km/h start to blow parallel to the surface of this wall. If the air temperature outside is −2°C, determine the rate of heat loss through the windows of this wall. Assume radiation heat transfer to be negligible.

6-98 The compressed air requirements of a manufacturing facility are met by a 150-hp compressor located in a room that is maintained at 25°C. In order to minimize the compressor work, the intake port of the compressor is connected to the outside through an 8-m-long, 20-cm-diameter duct made of thin aluminum sheet. The compressor takes in air at a rate of 0.27 m^3/s at the outdoor conditions of 10°C and 95 kPa. Disregarding the thermal resistance of the duct and taking the heat transfer coefficient on the outer surface of the duct to be 10 W/m^2 · °C, determine (*a*) the power used by the compressor to overcome the pressure drop in this duct, (*b*) the rate of heat transfer to the incoming cooler air, and (*c*) the temperature rise of air as it flows through the duct.

FIGURE P6-98

Air, 0.27 m^3/s
10°C, 95 kPa

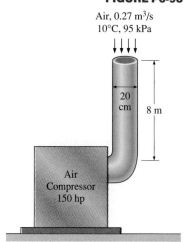

20 cm

8 m

Air Compressor 150 hp

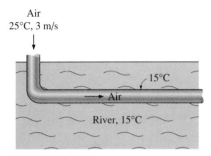

Air
25°C, 3 m/s

15°C

Air

River, 15°C

FIGURE P6-100

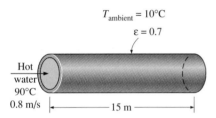

$T_{ambient} = 10°C$

$\varepsilon = 0.7$

Hot
water
90°C
0.8 m/s $\longmapsto$ —— 15 m ——$\longmapsto$

FIGURE P6-103

FIGURE P6-106

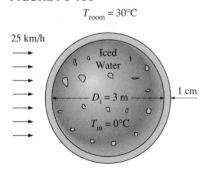

$T_{room} = 30°C$

25 km/h

Iced
Water

$D_i = 3$ m

1 cm

$T_{in} = 0°C$

6-99 Consider a person who is trying to keep cool on a hot summer day by turning a fan on and exposing his body to air flow. The air temperature is 32°C, and the fan is blowing air at a velocity of 5 m/s. The surrounding surfaces are at 40°C, and the emissivity of the person can be taken to be 0.9. If the person is doing light work and generating sensible heat at a rate of 90 W, determine the average temperature of the outer surface (skin or clothing) of the person. The average human body can be treated as a 30-cm-diameter cylinder with an exposed surface area of 1.7 m². *Answer:* 35°C

6-100 A house built on a riverside is to be cooled in summer by utilizing the cool water of the river, which flows at an average temperature of 15°C. A 15-m-long section of a circular duct of 20-cm diameter passes through the water. Air enters the underwater section of the duct at 25°C at a velocity of 3 m/s. Assuming the surface of the duct to be at the temperature of the water, determine the outlet temperature of air as it leaves the underwater portion of the duct. Also, determine the fan power needed to overcome the flow resistance in this section of the duct.

6-101 Repeat Problem 6-100 assuming that a 0.15-mm-thick layer of mineral deposit ($k = 3$ W/m · °C) formed on the inner surface of the pipe.

6-102E The exhaust gases of an automotive engine leave the combustion chamber and enter a 8-ft-long and 3.5-in.-diameter thin-walled steel exhaust pipe at 940°F and 16.1 psia at a rate of 0.2 lbm/s. The surrounding ambient air is at a temperature of 75°F, and the heat transfer coefficient on the outer surface of the exhaust pipe is 3 Btu/h · ft² · °F. Assuming the exhaust gases to have the properties of air, determine (*a*) the velocity of the exhaust gases at the inlet of the exhaust pipe and (*b*) the temperature at which the exhaust gases will leave the pipe and enter the air.

6-103 Hot water at 90°C enters a 15-m section of a cast iron pipe ($k = 52$ W/m · °C) whose inner and outer diameters are 4 and 4.6 cm, respectively, at an average velocity of 0.8 m/s. The outer surface of the pipe, whose emissivity is 0.7, is exposed to the cold air at 10°C in a basement, with a convection heat transfer coefficient of 15 W/m² · °C. Taking the walls of the basement to be at 10°C also, determine (*a*) the rate of heat loss from the water and (*b*) the temperature at which the water leaves the basement.

6-104 Repeat Problem 6-103 for a pipe made of copper ($k = 386$ W/m · °C) instead of cast iron.

6-105 Four power transistors, each dissipating 15 W, are mounted on a thin vertical aluminum plate ($k = 237$ W/m · °C) 22 cm × 22 cm in size. The heat generated by the transistors is to be dissipated by both surfaces of the plate to the surrounding air at 25°C, which is blown over the plate by a fan at a velocity of 250 m/min. The entire plate can be assumed to be nearly isothermal, and the exposed surface area of the transistor can be taken to be equal to its base area. Determine the temperature of the aluminum plate.

6-106 A 3-m-internal-diameter spherical tank made of 1-cm-thick stainless steel ($k = 15$ W/m · °C) is used to store iced water at 0°C. The tank is located outdoors at 30°C and is subjected to winds at 25 km/h. Assuming the entire steel tank to be at 0°C and thus its thermal resistance to be negligible, determine (*a*) the rate of heat transfer to the iced water in the tank and (*b*) the amount of ice at 0°C that melts during a 24-h period. The heat of fusion of water at atmospheric pressure is h_{if} = 333.7 kJ/kg. Disregard any heat transfer by radiation.

6-107 Repeat Problem 6-106, assuming the inner surface of the tank to be at 0°C but by taking the thermal resistance of the tank and heat transfer by radiation into consideration. Assume the average surrounding surface temperature for radiation exchange to be 15°C and the outer surface of the tank to have an emissivity of 0.9.   *Answers:* (*a*) 13,630 W, (*b*) 3529 kg

6-108 D. B. Tuckerman and R. F. Pease of Stanford University demonstrated in the early 1980s that integrated circuits can be cooled very effectively by fabricating a series of microscopic channels 0.3 mm high and 0.05 mm wide in the back of the substrate and covering them with a plate to confine the fluid flow within the channels. They were able to dissipate 790 W of power generated in a 1-cm² silicon chip at a junction-to-ambient temperature difference of 71°C using water as the coolant flowing at a rate of 0.01 L/s through 100 such channels under a 1-cm × 1-cm silicon chip. Heat is transferred primarily through the base area of the channel, and it was found that the increased surface area and thus the fin effect are of lesser importance. Disregarding the entrance effects and ignoring any heat transfer from the side and cover surfaces, determine (*a*) the temperature rise of water as it flows through the microchannels and (*b*) the average surface temperature of the base of the microchannels for a power dissipation of 50 W. Assume the water enters the channels at 20°C.

6-109E A transistor with a height of 0.25 in. and a diameter of 0.22 in. is mounted on a circuit board. The transistor is cooled by air flowing over it at a velocity of 500 ft/min. If the air temperature is 130°F and the transistor case temperature is not to exceed 165°F, determine the amount of power this transistor can dissipate safely.

6-110 Liquid-cooled systems have high heat transfer coefficients associated with them, but they have the inherent disadvantage that they present potential leakage problems. Therefore, air is proposed to be used as the microchannel coolant. Repeat Problem 6-108 using air as the cooling fluid instead of water, entering at a rate of 0.5 L/s.

6-111 A desktop computer is to be cooled by a fan. The electronic components of the computer consume 45 W of power under full-load conditions. The computer is to operate in environments at temperatures up to 50°C and at elevations up to 3000 m where the atmospheric pressure is 70.12 kPa. The exit temperature of air is not to exceed 60°C to meet the reliability requirements. Also, the average velocity of air is not to exceed 120 m/min at the exit of the computer case, where the fan is installed to keep the noise level down. Determine the flow rate of the fan that needs to be installed and the diameter of the casing of the fan.

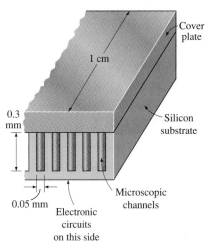

FIGURE P6-108

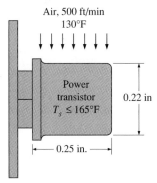

FIGURE P6-109E

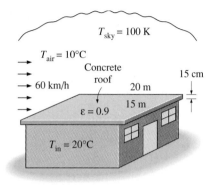

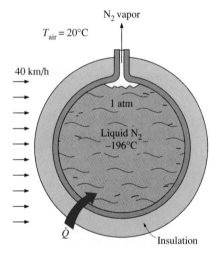

FIGURE P6-112

FIGURE P6-114

6-112 The roof of a house consists of a 15-cm-thick concrete slab ($k = 2$ W/m · °C) 15 m wide and 20 m long. The convection heat transfer coefficient on the inner surface of the roof is 5 W/m² · °C. On a clear winter night, the ambient air is reported to be at 10°C, while the night sky temperature is 100 K. The house and the interior surfaces of the wall are maintained at a constant temperature of 20°C. The emissivity of both surfaces of the concrete roof is 0.9. Considering both radiation and convection heat transfer, determine the rate of heat transfer through the roof when wind at 60 km/h is blowing over the roof.

If the house is heated by a furnace burning natural gas with an efficiency of 85 percent, and the price of natural gas is $0.60/therm (1 therm = 105,500 kJ of energy content), determine the money lost through the roof that night during a 14-h period.   *Answers:* 30.4 kW, $10.30

6-113 Steam at 250°C flows in a stainless steel pipe ($k = 15$ W/m · °C) whose inner and outer diameters are 4 cm and 4.6 cm, respectively. The pipe is covered with 3.5-cm-thick glass wool insulation ($k = 0.038$ W/m · °C) whose outer surface has an emissivity of 0.3. Heat is lost to the surrounding air and surfaces at 3°C by convection and radiation. Taking the heat transfer coefficient inside the pipe to be 80 W/m² · °C, determine the rate of heat loss from the steam per unit length of the pipe when air is flowing across the pipe at 4 m/s.

6-114 The boiling temperature of nitrogen at atmospheric pressure at sea level (1 atm pressure) is −196°C. Therefore, nitrogen is commonly used in low-temperature scientific studies, since the temperature of liquid nitrogen in a tank open to the atmosphere will remain constant at −196°C until it is depleted. Any heat transfer to the tank will result in the evaporation of some liquid nitrogen, which has a heat of vaporization of 198 kJ/kg and a density of 810 kg/m³ at 1 atm.

Consider a 4-m-diameter spherical tank that is initially filled with liquid nitrogen at 1 atm and −196°C. The tank is exposed to 20°C ambient air and 40 km/h winds. The temperature of the thin-shelled spherical tank is observed to be almost the same as the temperature of the nitrogen inside. Disregarding any radiation heat exchange, determine the rate of evaporation of the liquid nitrogen in the tank as a result of heat transfer from the ambient air if the tank is (*a*) not insulated, (*b*) insulated with 5-cm-thick fiberglass insulation ($k = 0.035$ W/m · °C), and (*c*) insulated with 2-cm-thick super-insulation that has an effective thermal conductivity of 0.00005 W/m · °C.

6-115 Repeat Problem 6-114 for liquid oxygen, which has a boiling temperature of −183°C, a heat of vaporization of 213 kJ/kg, and a density of 1140 kg/m³ at 1 atm pressure.

6-116 A 0.3-cm-thick, 12-cm-high, and 18-cm-long circuit board houses 80 closely spaced logic chips on one side, each dissipating 0.04 W. The board is impregnated with copper fillings and has an effective thermal conductivity of 16 W/m · °C. All the heat generated in the chips is conducted across the circuit board and is dissipated from the back side of the board to the ambient air at 40°C, which is forced to flow over the surface by a fan at a free-stream velocity of 400 m/min. Determine the temperatures on the two sides of the circuit board. *Answers:* 46.28°C, 46.31°C

6-117E It is well known that cold air feels much colder in windy weather than what the thermometer reading indicates because of the "chilling effect" of the wind. This effect is due to the increase in the convection heat transfer coefficient with increasing air velocities. The *equivalent windchill temperature* in °F is given by (1993 *ASHRAE Handbook of Fundamentals,* Atlanta, GA, p. 8.15)

$$T_{\text{equiv}} = 91.4 - (91.4 - T_{\text{ambient}})(0.475 - 0.0203V + 0.304\sqrt{V})$$

where V is the wind velocity in mph and T_{ambient} is the ambient air temperature in °F in calm air, which is taken to be air with light winds at speeds up to 4 mph. The constant 91.4°F in the above equation is the mean skin temperature of a resting person in a comfortable environment. Windy air at a temperature T_{ambient} and velocity V will feel as cold as calm air at a temperature T_{equiv}. The equation above is valid for winds up to 43 mph. Winds at higher velocities produce little additional chilling effect. Determine the equivalent wind chill temperature of an environment at 10°F at wind speeds of 10, 20, 30, and 40 mph. Exposed flesh can freeze within one minute at a temperature below −25°F in calm weather. Does a person need to be concerned about this possibility in any of the cases above?

FIGURE P6-117E

Computer, Design, and Essay Problems

6-118 Electronic boxes such as computers are commonly cooled by a fan. Write an essay on forced air cooling of electronic boxes and on the selection of the fan.

6-119 Obtain information on frostbite and the conditions under which it occurs. Using the relation in Problem 6-117E, prepare a table that shows how long people can stay in cold and windy weather for specified temperatures and wind speeds before the exposed flesh is in danger of experiencing frostbite.

6-120 Write an article on forced convection cooling with air, helium, water, and a dielectric liquid. Discuss the advantages and disadvantages of each fluid in heat transfer. Explain the circumstances under which a certain fluid will be most suitable for the cooling job.

Natural Convection

7

In the preceding chapter, we considered heat transfer by *forced convection* where a fluid was *forced* to move over a surface or in a tube by external means such as a pump or a fan. In this chapter, we consider *natural convection,* where any fluid motion occurs by natural means such as buoyancy. The fluid motion in forced convection is quite *noticeable,* since a fan or a pump can transfer enough momentum to the fluid to move it in a certain direction. The fluid motion in natural convection, however, is often not noticeable because of the low velocities involved.

Convection heat transfer coefficient is a strong function of *velocity*: the higher the velocity, the higher the convection heat transfer coefficient. The fluid velocities associated with natural convection are low, typically under 1 m/s. Therefore, the *heat transfer coefficients* encountered in natural convection are usually *much lower* than those encountered in forced convection. Yet several types of heat transfer equipment are designed to operate under natural convection conditions instead of forced convection, because natural convection does not require the use of a fluid mover.

We start this chapter with a discussion of the physical mechanism of *natural convection* and the *Grashof number.* We then present the correlations to evaluate heat transfer by natural convection for various geometries, including enclosures and finned surfaces. Finally, we discuss simultaneous forced and natural convection.

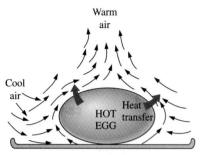

FIGURE 7-1

The cooling of a boiled egg in a cooler environment by natural convection.

FIGURE 7-2

The warming up of a cold drink in a warmer environment by natural convection.

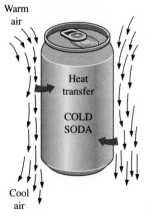

7-1 ■ PHYSICAL MECHANISM OF NATURAL CONVECTION

Many familiar heat transfer applications involve *natural convection* as the *primary* mechanism of heat transfer. Some examples are cooling of electronic equipment such as power transistors, TVs, and VCRs; heat transfer from electric baseboard heaters or steam radiators; heat transfer from the refrigeration coils and power transmission lines; and heat transfer from the bodies of animals and human beings. Natural convection in gases is usually accompanied by radiation of comparable magnitude except for low-emissivity surfaces.

We know that a hot *boiled egg* (or a hot *baked potato*) on a plate eventually cools to the surrounding air temperature (Fig. 7-1). The egg is cooled by transferring heat by *convection* to the air and by radiation to the surrounding surfaces. Disregarding heat transfer by radiation, the physical mechanism of cooling a hot egg (or any hot object) in a cooler environment can be explained as follows:

As soon as the hot egg is exposed to cooler air, the temperature of the outer surface of the egg shell will drop somewhat, and the temperature of the air adjacent to the shell will rise as a result of heat conduction from the shell to the air. Consequently, the egg will soon be surrounded by a thin layer of warmer air, and heat will then be transferred from this warmer layer to the outer layers of air. The cooling process in this case would be rather *slow* since the egg would always be *blanketed* by warm air, and it would have no direct contact with the cooler air farther away. We may not notice any *air motion* in the vicinity of the egg, but careful measurements indicate otherwise.

The temperature of the air adjacent to the egg is higher, and thus its density is lower, since at constant pressure the *density* of a gas is inversely proportional to its *temperature*. Thus, we have a situation in which some low-density or "light" gas is surrounded by a high-density or "heavy" gas, and the natural laws dictate that *the light gas rise*. This is no different than the oil in a vinegar-and-oil salad dressing rising to the top (note that $\rho_{oil} < \rho_{vinegar}$). This phenomenon is characterized incorrectly by the phrase "heat rises," which is understood to mean *heated air rises*. The space vacated by the warmer air in the vicinity of the egg is replaced by the cooler air nearby, and the presence of cooler air in the vicinity of the egg speeds up the cooling process. The rise of warmer air and the flow of cooler air into its place continues until the egg is cooled to the temperature of the surrounding air. The motion that results from the continual replacement of the heated air in the vicinity of the egg by the cooler air nearby is called a **natural convection current,** and the heat transfer that is enhanced as a result of this natural convection current is called **natural convection heat transfer.** Note that in the absence of natural convection currents, heat transfer from the egg to the air surrounding it would be by *conduction* only, and the rate of heat transfer from the egg would be much *lower*.

Natural convection is just as effective in the heating of cold surfaces in a warmer environment as it is in the cooling of hot surfaces in a cooler environment, as shown in Figure 7-2. Note that the direction of fluid motion is reversed in this case.

In a gravitational field, there seems to be a net force that pushes upward a light fluid placed in a heavier fluid. The upward force exerted by a fluid on a body completely or partially immersed in it is called the **buoyancy force.** The magnitude of the buoyancy force is equal to the weight of the *fluid displaced* by the body. That is,

$$F_{\text{buoyancy}} = \rho_{\text{fluid}}\, g V_{\text{body}} \qquad (7\text{-}1)$$

where ρ_{fluid} is the average density of the *fluid* (not the body), g is the gravitational acceleration, and V_{body} is the volume of the portion of the body immersed in the fluid (for bodies completely immersed in the fluid, it is the total volume of the body). In the absence of other forces, the net vertical force acting on a body is the difference between the weight of the body and the buoyancy force. That is,

$$
\begin{aligned}
F_{\text{net}} &= W - F_{\text{buoyancy}} \\
&= \rho_{\text{body}}\, g V_{\text{body}} - \rho_{\text{fluid}}\, g V_{\text{body}} \qquad (7\text{-}2)\\
&= (\rho_{\text{body}} - \rho_{\text{fluid}})\, g V_{\text{body}}
\end{aligned}
$$

Note that this force is *proportional* to the difference in the *densities* of the fluid and the body immersed in it. Thus, a body immersed in a fluid will experience a "weight loss" in an amount equal to the weight of the fluid it displaces. This is known as *Archimedes' principle*.

To have a better understanding of the buoyancy effect, consider an egg dropped into water. If the average density of the egg is greater than the density of water (a sign of freshness), the egg will settle at the bottom of the container. Otherwise, it will rise to the top. When the density of the egg equals the density of water, the egg will settle somewhere in the water while remaining completely immersed, acting like a "weightless object" in space. This occurs when the upward buoyancy force acting on the egg equals the weight of the egg, which acts downward.

The *buoyancy effect* has far-reaching implications in life. For one thing, without buoyancy, heat transfer between a hot (or cold) surface and the fluid surrounding it would be by *conduction* instead of by *natural convection*. The natural convection currents encountered in the oceans, lakes, and the atmosphere owe their existence to buoyancy. Also, light boats as well as heavy warships made of steel float on water because of buoyancy (Fig. 7-3). Ships are designed on the basis of the principle that the entire weight of a ship and its contents is equal to the weight of the water that the submerged volume of the ship can contain. (Note that a larger portion of the hull of a ship will sink in fresh water than it does in salty water.) The "chimney effect" that induces the upward flow of hot combustion gases through a chimney is also due to the buoyancy effect, and the upward force acting on the gases in the chimney is proportional to the difference between the densities of the hot gases in the chimney and the cooler air outside. Note that there is *no gravity* in space, and thus there will be no natural convection heat transfer in a spacecraft, even if the spacecraft is filled with atmospheric air.

In heat transfer studies, the primary variable is *temperature,* and it is desirable to express the net buoyancy force (Eq. 7-2) in terms of temperature

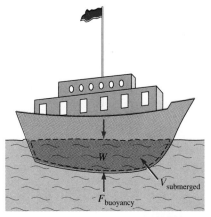

FIGURE 7-3

It is the buoyancy force that keeps the ships afloat in water ($W = F_{\text{Buoyancy}}$ for floating objects).

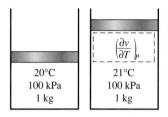

(*a*) A substance with a large β

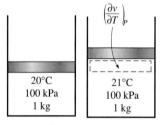

(*b*) A substance with a small β

FIGURE 7-4

The coefficient of volume expansion is a measure of the change in volume of a substance with temperature at constant pressure.

differences. But this requires expressing the density difference in terms of a temperature difference, which requires a knowledge of a property that represents the *variation of the density of a fluid with temperature at constant pressure. The property that provides that information is the* **volume expansion coefficient** β, defined as (Fig. 7-4)

$$\beta = \frac{1}{v}\left(\frac{\partial v}{\partial T}\right)_P = -\frac{1}{\rho}\left(\frac{\partial \rho}{\partial T}\right)_P \qquad (1/\text{K}) \qquad (7\text{-}3)$$

It can also be expressed approximately by replacing derivatives by differences as

$$\beta \approx -\frac{1}{\rho}\frac{\Delta\rho}{\Delta T} \quad \rightarrow \quad \Delta\rho \approx -\rho\beta\Delta T \qquad (\text{at constant } P)$$

We can show easily that the volume expansion coefficient β of an *ideal gas* $(P = \rho RT)$ at a temperature T is equivalent to the inverse of the temperature:

$$\beta_{\text{ideal gas}} = \frac{1}{T} \qquad (1/\text{K}) \qquad (7\text{-}4)$$

where T is the *absolute* temperature. Note that a large value of β for a fluid means a large change in density with temperature, and that the product $\beta\Delta T$ represents the fraction of volume change of a fluid that corresponds to a temperature change ΔT at constant pressure. Also note that the buoyancy force is proportional to the *density difference,* which is proportional to the *temperature difference* at constant pressure. Therefore, the larger the temperature difference between the fluid adjacent to a hot (or cold) surface and the fluid away from it, the *larger* the buoyancy force and the *stronger* the natural convection currents, and thus the *higher* the heat transfer rate.

The magnitude of the natural convection heat transfer between a surface and a fluid is directly related to the *mass flow rate* of the fluid. The higher the mass flow rate, the higher the heat transfer rate. In fact, it is the very high flow rates that increase the heat transfer coefficient by orders of magnitude when forced convection is used. In natural convection, no blowers are used, and therefore the flow rate cannot be controlled externally. The flow rate in this case is established by the dynamic balance of *buoyancy* and *friction*.

As we have discussed earlier, the *buoyancy force* is caused by the density difference between the heated (or cooled) fluid adjacent to the surface and the fluid surrounding it, and is proportional to this density difference and the volume occupied by the warmer fluid. It is also well known that whenever two bodies in contact (solid–solid, solid–fluid, or fluid–fluid) move relative to each other, a *friction force* develops at the contact surface in the direction opposite to that of the motion. This opposing force slows down the fluid and thus reduces the flow rate of the fluid. Under steady conditions, the air flow rate driven by buoyancy is established at the point where these two effects *balance* each other. The friction force increases as more and more solid surfaces are introduced, seriously disrupting the fluid flow and heat transfer.

For that reason, heat sinks with closely spaced fins are not suitable for natural convection cooling.

Most heat transfer correlations in natural convection are based on experimental measurements. The instrument used in natural convection experiments most often is the *Mach–Zehnder interferometer,* which gives a plot of isotherms in the fluid in the vicinity of a surface. The operation principle of interferometers is based on the fact that at low pressure, the lines of constant temperature for a gas correspond to the lines of constant density, and that the index of refraction of a gas is a function of its density. Therefore, the degree of refraction of light at some point in a gas is a measure of the temperature gradient at that point. An interferometer produces a map of interference fringes, which can be interpreted as lines of *constant temperature* as shown in Figure 7-5. The smooth and parallel lines in (*a*) indicate that the flow is *laminar,* whereas the eddies and irregularities in (*b*) indicate that the flow is *turbulent.* Note that the lines are closest near the surface, indicating a *higher temperature gradient.*

The Grashof Number

We mentioned in the preceding chapter that the flow regime in forced convection is governed by the dimensionless *Reynolds number,* which represents the ratio of inertial forces to viscous forces acting on the fluid. The flow regime in natural convection is governed by another dimensionless number, called the **Grashof number,** which represents the ratio of the *buoyancy force* to the *viscous force* acting on the fluid. That is,

$$Gr = \frac{\text{Buoyancy forces}}{\text{Viscous forces}} = \frac{g \Delta \rho V}{\rho v^2} = \frac{g \beta \Delta T V}{v^2}$$

Since $\Delta \rho \approx \rho \beta \Delta T$, it is formally expressed as (Fig. 7-6)

$$Gr = \frac{g \beta (T_s - T_\infty) \delta^3}{v^2} \qquad (7\text{-}5)$$

where
$\quad g$ = gravitational acceleration, m/s^2
$\quad \beta$ = coefficient of volume expansion, 1/K ($\beta = 1/T$ for ideal gases)
$\quad T_s$ = temperature of the surface, °C
$\quad T_\infty$ = temperature of the fluid sufficiently far from the surface, °C
$\quad \delta$ = characteristic length of the geometry, m
$\quad v$ = kinematic viscosity of the fluid, m^2/s

The role played by the *Reynolds number* in forced convection is played by the *Grashof number* in natural convection. As such, the Grashof number provides the main criterion in determining whether the fluid flow is laminar or turbulent in natural convection. For vertical plates, for example, the critical Grashof number is observed to be about 10^9. Therefore, the flow regime on a vertical plate becomes turbulent at Grashof numbers greater than 10^9.

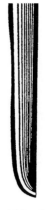

(*a*) Laminar flow (*b*) Turbulent flow

FIGURE 7-5

Isotherms in natural convection over a hot plate in air.

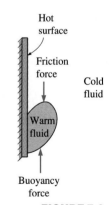

FIGURE 7-6

The Grashof number Gr is a measure of the relative magnitudes of the *buoyancy force* and the opposing *friction force* acting on the fluid.

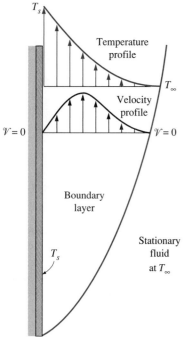

FIGURE 7-7

Typical velocity and temperature profiles for natural convection flow over a hot vertical plate at temperature T_s inserted in a fluid at temperature T_∞.

FIGURE 7-8

Natural convection heat transfer correlations are usually expressed in terms of the Rayleigh number raised to a constant n multiplied by another constant C, both of which are determined experimentally.

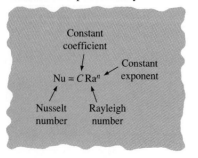

The heat transfer rate in natural convection from a solid surface to the surrounding fluid is expressed by Newton's law of cooling as

$$\dot{Q}_{conv} = hA(T_s - T_\infty) \qquad (W) \qquad (7\text{-}6)$$

where A is the heat transfer surface area and h is the average heat transfer coefficient on the surface.

7-2 ■ NATURAL CONVECTION OVER SURFACES

Natural convection heat transfer on a surface depends on the geometry of the surface as well as its orientation. It also depends on the variation of temperature on the surface and the thermophysical properties of the fluid involved.

The *velocity* and *temperature profiles* for natural convection over a vertical hot plate immersed in a quiescent fluid body are given in Figure 7-7. As in forced convection, the thickness of the boundary layer increases in the flow direction. Unlike forced convection, however, the fluid velocity is *zero* at the *outer edge* of the velocity boundary layer as well as at the surface of the plate. This is expected since the fluid beyond the boundary layer is stationary. Thus, the fluid velocity increases with distance from the surface, reaches a maximum, and gradually decreases to zero at a distance sufficiently far from the surface. The *temperature* of the fluid will equal the plate temperature at the surface and gradually decrease to the temperature of the surrounding fluid at a distance sufficiently far from the surface, as shown in the figure. In the case of *cold surfaces*, the shape of the velocity and temperature profiles remains the same but their direction is reversed.

Natural Convection Correlations

Although we understand the mechanism of natural convection well, the complexities of fluid motion make it very difficult to obtain simple analytical relations for heat transfer by solving the governing equations of motion and energy. Some analytical solutions exist for natural convection, but such solutions lack generality since they are obtained for simple geometries under some simplifying assumptions. Therefore, with the exception of some simple cases, heat transfer relations in natural convection are based on experimental studies. Of the numerous such correlations of varying complexity and claimed accuracy available in the literature for any given geometry, we present below the *simpler* ones for two reasons: first, the accuracy of simpler relations is usually within the range of uncertainty associated with a problem, and second, we would like to keep the emphasis on the physics of the problems instead of formula manipulation.

The simple empirical correlations for the average *Nusselt number* Nu in natural convection are of the form (Fig. 7-8)

$$Nu = \frac{h\delta}{k} = C(Gr\,Pr)^n = C\,Ra^n \qquad (7\text{-}7)$$

where Ra is the **Rayleigh number,** which is the product of the Grashof and Prandtl numbers:

$$Ra = Gr \, Pr = \frac{g\beta(T_s - T_\infty)\delta^3}{v^2} Pr \qquad (7\text{-}8)$$

The values of the constants C and n depend on the *geometry* of the surface and the *flow regime*, which is characterized by the range of the Rayleigh number. The value of n is usually $\frac{1}{4}$ for laminar flow and $\frac{1}{3}$ for turbulent flow. The value of the constant C is normally less than 1.

Simple relations for the average Nusselt number for various geometries are given in Table 7-1, together with sketches of the geometries. Also given in this table are the characteristic lengths of the geometries and the ranges of Rayleigh number in which the relation is applicable. All fluid properties are to be evaluated at the film temperature $T_f = \frac{1}{2}(T_s + T_\infty)$.

These relations have been obtained for the case of isothermal surfaces but could also be used approximately for the case of nonisothermal surfaces by assuming the surface temperature to be constant at some average value. The use of these relations is illustrated below with examples.

EXAMPLE 7-1   Heat Loss from Hot Water Pipes

A 6-m-long section of an 8-cm-diameter horizontal hot water pipe shown in Figure 7-9 passes through a large room whose temperature is 18°C. If the outer surface temperature of the pipe is 70°C, determine the rate of heat loss from the pipe by natural convection.

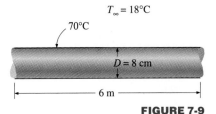

$T_\infty = 18°C$

70°C

$D = 8$ cm

6 m

FIGURE 7-9

Schematic for Example 7-1.

Solution   A horizontal hot water pipe passes through a large room. The rate of heat loss from the pipe by natural convection is to be determined.

Assumptions   **1** Steady operating conditions exist. **2** Air is an ideal gas. **3** The local atmospheric pressure is 1 atm.

Properties   The properties of air at the film temperature of $T_f = (T_s + T_\infty)/2 = (70 + 18)/2 = 44°C = 317$ K and 1 atm pressure are (Table A-11)

$$k = 0.0273 \text{ W/m} \cdot °C \qquad Pr = 0.710$$
$$v = 1.74 \times 10^{-5} \text{ m}^2/\text{s} \qquad \beta = \frac{1}{T_f} = \frac{1}{317 \text{ K}} = 0.00315 \text{ K}^{-1}$$

Analysis   The characteristic length in this case is the outer diameter of the pipe, $\delta = D = 0.08$ m. Then the Rayleigh number becomes

$$Ra = \frac{g\beta(T_s - T_\infty)\delta^3}{v^2} Pr$$

$$= \frac{(9.8 \text{ m/s}^2)(0.00315 \text{ K}^{-1})[(70 - 18) \text{ K}](0.08 \text{ m})^3}{(1.74 \times 10^{-5} \text{ m}^2/\text{s})^2}(0.710) = 1.930 \times 10^6$$

Then the natural convection Nusselt number in this case can be determined from Equation 7-16 to be

$$Nu = \left\{ 0.6 + \frac{0.387 \, Ra^{1/6}}{[1 + (0.559/Pr)^{9/16}]^{8/27}} \right\}^2$$

$$= \left\{ 0.6 + \frac{0.387(1.930 \times 10^6)^{1/6}}{[1 + (0.559/0.710)^{9/16}]^{8/27}} \right\}^2 = 17.2$$

$$.6 + \frac{4.3182}{1.2046}$$

TABLE 7-1

Empirical correlations for the average Nusselt number for natural convection over surfaces

Geometry	Characteristic length δ	Range of Ra	Nu	
Vertical plate T_s L	L	10^4–10^9 10^9–10^{13} Entire range	$\mathrm{Nu} = 0.59\mathrm{Ra}^{1/4}$ $\mathrm{Nu} = 0.1\mathrm{Ra}^{1/3}$ $\mathrm{Nu} = \left\{ 0.825 + \dfrac{0.387\mathrm{Ra}^{1/6}}{(1 + (0.492/\mathrm{Pr})^{9/16})^{8/27}} \right\}^2$ (complex but more accurate)	(7-9) (7-10) (7-11)
Inclined plate θ L	L		Use vertical plate equations as a first degree of approximation. Replace g by $g\cos\theta$ for Ra $< 10^9$	
Horizontal plate (Surface area A and perimeter p) (*a*) Upper surface of a hot plate (or lower surface of a cold plate) Hot surface — T_s (*b*) Lower surface of a hot plate (or upper surface of a cold plate) T_s Hot surface	A/p	10^4–10^7 10^7–10^{11} 10^5–10^{11}	$\mathrm{Nu} = 0.54\mathrm{Ra}^{1/4}$ $\mathrm{Nu} = 0.15\mathrm{Ra}^{1/3}$ $\mathrm{Nu} = 0.27\mathrm{Ra}^{1/4}$	(7-12) (7-13) (7-14)
vertical cylinder T_s L	L		A vertical cylinder can be treated as a vertical plate when $D \geq \dfrac{35L}{\mathrm{Gr}^{1/4}}$	(7-15)
Horizontal cylinder T_s D	D	10^{-5}–10^{12}	$\mathrm{Nu} = \left\{ 0.6 + \dfrac{0.387\mathrm{Ra}^{1/6}}{(1 + (0.559/\mathrm{Pr})^{9/16})^{8/27}} \right\}^2$	(7-16)
Sphere D	D	$\mathrm{Ra} \leq 10^{11}$ $(\mathrm{Pr} \geq 0.5)$	$\mathrm{Nu} = 2 + \dfrac{0.589\mathrm{Ra}^{1/4}}{(1 + (0.469/\mathrm{Pr})^{9/16})^{4/9}}$	(7-17)

Then

$$h = \frac{k}{D}\text{Nu} = \frac{0.0273 \text{ W/m} \cdot ^\circ\text{C}}{0.08 \text{ m}}(17.2) = 5.9 \text{ W/m}^2 \cdot ^\circ\text{C}$$

$$A = \pi DL = \pi(0.08 \text{ m})(6 \text{ m}) = 1.51 \text{ m}^2$$

and

$$\dot{Q} = hA(T_s - T_\infty) = (5.9 \text{ W/m}^2 \cdot ^\circ\text{C})(1.51 \text{ m}^2)(70 - 18)^\circ\text{C} = \textbf{463 W}$$

Therefore, the pipe will lose heat to the air in the room at a rate of 463 W by natural convection.

Discussion The pipe will lose heat to the surroundings by radiation as well as by natural convection. Assuming the outer surface of the pipe to be black (emissivity $\varepsilon = 1$) and the inner surfaces of the walls of the room to be at room temperature, the radiation heat transfer in this case is determined to be (Fig. 7-10)

$$\dot{Q}_{\text{rad}} = \varepsilon A \sigma (T_s^4 - T_\infty^4)$$
$$= (1)(1.51 \text{ m}^2)(5.67 \times 10^{-8} \text{ W/m}^2 \cdot \text{K}^4)[(70 + 273 \text{ K})^4 - (18 + 273 \text{ K})^4]$$
$$= 571 \text{ W}$$

which is as large as that for natural convection. The emissivity of a real surface is less than 1, and thus the radiation heat transfer for a real surface will be less. But radiation will still be significant for most systems cooled by natural convection. Therefore, a radiation analysis should normally accompany a natural convection analysis unless the emissivity of the surface is low.

EXAMPLE 7-2   Cooling of a Plate in Different Orientations

Consider a 0.6-m × 0.6-m thin square plate in a room at 30°C. One side of the plate is maintained at a temperature of 74°C, while the other side is insulated, as shown in Figure 7-11. Determine the rate of heat transfer from the plate by natural convection if the plate is (a) vertical, (b) horizontal with hot surface facing up, and (c) horizontal with hot surface facing down.

Solution A hot plate with an insulated back is considered. The rate of heat loss by natural convection is to be determined for different orientations.

Assumptions **1** Steady operating conditions exist. **2** Air is an ideal gas. **3** The local atmospheric pressure is 1 atm.

Properties The properties of air at the film temperature of $T_f = (T_s + T_\infty)/2 = (74 + 30)/2 = 52^\circ\text{C} = 325$ K and 1 atm pressure are (Table A-11)

$$k = 0.0279 \text{ W/m} \cdot ^\circ\text{C} \qquad \text{Pr} = 0.709$$
$$\nu = 1.815 \times 10^{-5} \text{ m}^2/\text{s}$$
$$\beta = \frac{1}{T_f} = \frac{1}{325 \text{ K}} = 0.00308 \text{ K}^{-1}$$

Analysis (a) *Vertical*. The characteristic length in this case is the height of the plate, which is $\delta = 0.6$ m. The Rayleigh number is

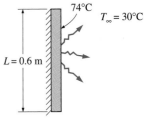

$T_\infty = 18^\circ\text{C}$

$\dot{Q}_{\text{nat conv}} = 463 \text{ W}$

$T_s = 70^\circ\text{C}$

$\dot{Q}_{\text{rad, max}} = 457 \text{ W}$

FIGURE 7-10
Radiation heat transfer is usually comparable to natural convection in magnitude and should be considered in heat transfer analysis.

FIGURE 7-11
Schematic for Example 7-2.

74°C

$T_\infty = 30^\circ\text{C}$

$L = 0.6 \text{ m}$

(a) Vertical

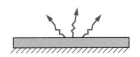

(b) Hot surface facing up

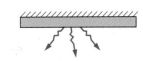

(c) Hot surface facing down

$$Ra = \frac{g\beta(T_s - T_\infty)\delta^3}{\nu^2} Pr$$

$$= \frac{(9.8 \text{ m/s}^2)(0.00308 \text{ K}^{-1})[(74 - 30) \text{ K}](0.6 \text{ m})^3}{(1.815 \times 10^{-5} \text{m}^2/\text{s})^2}(0.709) = 6.174 \times 10^8$$

Then the natural convection Nusselt number can be determined from Equation 7-9 to be

$$Nu = 0.59 \, Ra^{1/4} = 0.59(6.174 \times 10^8)^{1/4} = 93.0$$

Then

$$h = \frac{k}{\delta} Nu = \frac{0.0279 \text{ W/m} \cdot °C}{0.6 \text{ m}}(93.0) = 4.3 \text{ W/m}^2 \cdot °C$$

$$A = L^2 = (0.6 \text{ m})^2 = 0.36 \text{ m}^2$$

and

$$\dot{Q} = hA(T_s - T_\infty) = (4.3 \text{ W/m}^2 \cdot °C)(0.36 \text{ m}^2)(74 - 30)°C = \textbf{68.1 W}$$

(b) *Horizontal with hot surface facing up.* The characteristic length and the Rayleigh number in this case are

$$\delta = \frac{A}{p} = \frac{L^2}{4L} = \frac{L}{4} = \frac{0.6 \text{ m}}{4} = 0.15 \text{ m}$$

and

$$Ra = \frac{g\beta(T_s - T_\infty)\delta^3}{\nu^2} Pr$$

$$= \frac{(9.8 \text{ m/s}^2)(0.00308 \text{ K}^{-1})[(74 - 30) \text{ K}](0.15 \text{ m})^3}{(1.815 \times 10^{-5} \text{ m}^2/\text{s})^2}(0.708) = 9.43 \times 10^6$$

Then the natural convection Nusselt number can be determined from Equation 7-12 to be

$$Nu = 0.54 \, Ra^{1/4} = 0.54(9.43 \times 10^6)^{1/4} = 29.9$$

Therefore,

$$h = \frac{k}{\delta} Nu = \frac{0.0279 \text{ W/m} \cdot °C}{0.15 \text{ m}}(29.9) = 5.56 \text{ W/m}^2 \cdot °C$$

$$A = L^2 = (0.6 \text{ m})^2 = 0.36 \text{ m}^2$$

and

$$\dot{Q} = hA(T_s - T_\infty) = (5.56 \text{ W/m}^2 \cdot °C)(0.36 \text{ m}^2)(74 - 30)°C = \textbf{88.1 W}$$

(c) *Horizontal with hot surface facing down.* The characteristic length δ, the heat transfer surface area A, and the Rayleigh number in this case are the same as those determined in (b). But the natural convection Nusselt number is to be determined from Equation 7-14.

$$Nu = 0.27 \, Ra^{1/4} = 0.27(9.43 \times 10^6)^{1/4} = 15.0$$

Then,

$$h = \frac{k}{\delta} Nu = \frac{0.0279 \, \text{W/m} \cdot {}^\circ\text{C}}{0.15 \, \text{m}}(15.0) = 2.8 \, \text{W/m}^2 \cdot {}^\circ\text{C}$$

and

$$\dot{Q} = hA(T_s - T_\infty) = (2.8 \, \text{W/m}^2 \cdot {}^\circ\text{C})(0.36 \, \text{m}^2)(74 - 30){}^\circ\text{C} = \textbf{44.4 W}$$

Note that the natural convection heat transfer is the lowest in the case of the hot surface facing down. This is not surprising, since the hot air is "trapped" under the plate in this case and cannot get away from the plate easily. As a result, the cooler air in the vicinity of the plate will have difficulty reaching the plate, which results in a reduced rate of heat transfer.

Discussion The plate will lose heat to the surroundings by radiation as well as by natural convection. Assuming the surface of the plate to be black (emissivity $\varepsilon = 1$) and the inner surfaces of the walls of the room to be at room temperature, the radiation heat transfer in this case is determined to be

$$\begin{aligned}
\dot{Q}_{\text{rad}} &= \varepsilon A \sigma (T_s^4 - T_\infty^4) \\
&= (1)(0.36 \, \text{m}^2)(5.67 \times 10^{-8} \, \text{W/m}^2 \cdot \text{K}^4)[(74 + 273 \, \text{K})^4 - (30 + 273 \, \text{K})^4] \\
&= 124 \, \text{W}
\end{aligned}$$

which is larger than that for natural convection heat transfer for each case. The emissivity of a real surface is less than 1, and thus the radiation heat transfer for a real surface will be less. But radiation can still be significant and needs to be considered in surfaces cooled by natural convection.

7-3 ▪ NATURAL CONVECTION INSIDE ENCLOSURES

A considerable portion of heat loss from a typical residence occurs through the windows. We certainly would insulate the windows, if we could, in order to conserve energy. The problem is finding an insulating material that is transparent. An examination of the thermal conductivities of the insulting materials reveals that *air* is a *better insulator* than most common insulating materials. Besides, it is transparent. Therefore, it makes sense to insulate the windows with a layer of air. Of course, we need to use another sheet of glass to trap the air. The result is an *enclosure,* which is known as a *double-pane window* in this case. Other examples of enclosures include wall cavities, solar collectors, and cryogenic chambers involving concentric cylinders or spheres.

Enclosures are frequently encountered in practice, and heat transfer through them is of practical interest. Heat transfer in enclosed spaces is complicated by the fact that the fluid in the enclosure, in general, does not remain stationary. In a vertical enclosure, the fluid adjacent to the hotter surface rises and the fluid adjacent to the cooler one falls, setting off a rotationary motion within the enclosure that enhances heat transfer through the

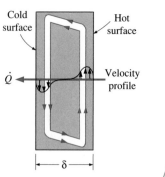

FIGURE 7-12

Convective currents in a vertical rectangular enclosure.

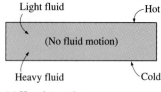

(a) Hot plate at the top

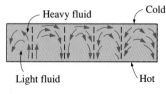

(b) Hot plate at the bottom

FIGURE 7-13

Convective currents in a horizontal enclosure with (a) hot plate at the top and (b) hot plate at the bottom.

enclosure. Typical flow patterns in vertical and horizontal rectangular enclosures are shown in Figures 7-12 and 7-13.

The characteristics of heat transfer through a horizontal enclosure depend on whether the hotter plate is at the top or at the bottom, as shown in Figure 7-13. When the *hotter plate* is at the *top*, no convection currents will develop in the enclosure, since the lighter fluid will always be on top of the heavier fluid. Heat transfer in this case will be by *pure conduction*, and we will have Nu = 1. When the *hotter plate* is at the *bottom*, the heavier fluid will be on top of the lighter fluid, and there will be a tendency for the lighter fluid to topple the heavier fluid and rise to the top, where it will come in contact with the cooler plate and cool down. Until that happens, however, the heat transfer is still by *pure conduction* and Nu = 1. When Ra > 1708, the buoyant force overcomes the fluid resistance and initiates natural convection currents, which are observed to be in the form of hexagonal cells called *Bénard cells.* For Ra > 3 × 10^5, the cells break down and the fluid motion becomes turbulent.

The Rayleigh number for an enclosure is determined from

$$Ra = \frac{g\beta(T_1 - T_2)\delta^3}{\nu^2} Pr \qquad (7\text{-}18)$$

where the characteristic length δ is the distance between the hot and cold surfaces, and T_1 and T_2 are the temperatures of the hot and cold surfaces, respectively. All fluid properties are to be evaluated at the average fluid temperature $T_{av} = \frac{1}{2}(T_1 + T_2)$.

Simple empirical correlations for the Nusselt number for various enclosures are given in Table 7-2. Once the Nusselt number is available, the heat transfer coefficient and the rate of heat transfer through the enclosure can be determined from

$$h = \frac{k}{\delta}Nu \qquad (7\text{-}19)$$

and

$$\dot{Q} = hA(T_1 - T_2) = k\,Nu\,A\,\frac{T_1 - T_2}{\delta} \qquad (7\text{-}20)$$

where

$$A = \begin{cases} HL & \text{rectangular enclosures} \\ \dfrac{\pi L(D_2 - D_1)}{\ln(D_2/D_1)} & \text{concentric cylinders} \\ \pi D_1 D_2 & \text{concentric spheres} \end{cases} \qquad (7\text{-}21)$$

For *inclined rectangular enclosures,* highly accurate but complex correlations are available in the literature. In the absence of such relations, the Nusselt number correlations for vertical enclosures can be used for inclined

TABLE 7-2

Empirical correlations for the average Nusselt number for natural convection in enclosures (the characteristic length δ is as indicated on the respective diagram)

Geometry	Fluid	H/δ	Range of Pr	Range of Ra	Nusselt number	
Vertical rectangular enclosure (or vertical cylindrical enclosure)	Gas or liquid	—	—	Ra < 2000	$Nu = 1$	(7-22)
	Gas	11–42	0.5–2	2×10^3–2×10^5	$Nu = 0.197 Ra^{1/4}\left(\dfrac{H}{\delta}\right)^{-1/9}$	(7-23)
		11–42	0.5–2	2×10^5–10^7	$Nu = 0.073 Ra^{1/3}\left(\dfrac{H}{\delta}\right)^{-1/9}$	(7-24)
	Liquid	10–40	1–20,000	10^4–10^7	$Nu = 0.42 Pr^{0.012} Ra^{1/4}\left(\dfrac{H}{\delta}\right)^{-0.3}$	(7-25)
		1–40	1–20	10^6–10^9	$Nu = 0.046 Ra^{1/3}$	(7-26)
Inclined rectangular enclosure					Use the correlations for vertical enclosures as a first-degree approximation for $\theta \le 20°$ by replacing g in the Ra relation by $g \cos \theta$	
Horizontal rectangular enclosure (hot surface at the top)	Gas or liquid	—	—	—	$Nu = 1$	(7-27)
Horizontal rectangular enclosure (hot surface at the bottom)	Gas or liquid	—	—	Ra < 1700	$Nu = 1$	(7-28)
	Gas	—	0.5–2	1.7×10^3–7×10^3	$Nu = 0.059 Ra^{0.4}$	(7-29)
		—	0.5–2	7×10^3–3.2×10^5	$Nu = 0.212 Ra^{1/4}$	(7-30)
		—	0.5–2	Ra > 3.2×10^5	$Nu = 0.061 Ra^{1/3}$	(7-31)
	Liquid	—	1–5000	1.7×10^3–6×10^3	$Nu = 0.012 Ra^{0.6}$	(7-32)
		—	1–5000	6×10^3–3.7×10^4	$Nu = 0.375 Ra^{0.2}$	(7-33)
		—	1–20	3.7×10^4–10^8	$Nu = 0.13 Ra^{0.3}$	(7-34)
		—	1–20	Ra > 10^8	$Nu = 0.057 Ra^{1/3}$	(7-35)
Concentric rectangular cylinders	Gas or liquid	—	1–5000	6.3×10^3–10^6	$Nu = 0.11 Ra^{0.29}$	(7-36)
		—	1–5000	10^6–10^8	$Nu = 0.40 Ra^{0.20}$	(7-37)
Concentric spheres	Gas or liquid	—	0.7–4000	10^2–10^9	$Nu = 0.228 Ra^{0.226}$	(7-38)

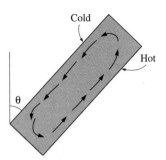

FIGURE 7-14

An inclined rectangular enclosure heated from below.

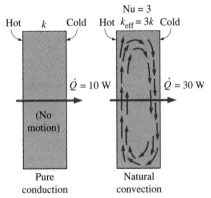

FIGURE 7-15

A Nusselt number of 3 for an enclosure indicates that heat transfer through the enclosure by *natural convection* is 3 times that by *pure conduction*.

enclosures heated from below for inclination angles up to about $\theta = 20°$ from the vertical by replacing g in the Ra relation by $g \cos \theta$ (Fig. 7-14).

Effective Thermal Conductivity

You will recall from Chapter 3 that the rate of steady heat conduction across a layer of thickness δ, surface area A, and thermal conductivity k is

$$\dot{Q}_{cond} = kA \frac{T_1 - T_2}{\delta} \tag{7-39}$$

where T_1 and T_2 are the temperatures on the two sides of the layer. A comparison of this relation with Equation 7-20 reveals that the convection heat transfer in an enclosure is analogous to heat conduction across the fluid layer in the enclosure provided that the thermal conductivity k is replaced by kNu. That is, the fluid in an enclosure behaves like a fluid whose thermal conductivity is kNu as a result of convection currents. Therefore, the quantity kNu is called the **effective thermal conductivity** of the enclosure. That is,

$$k_{eff} = k\text{Nu} \tag{7-40}$$

Note that for the special case of Nu $= 1$, the effective thermal conductivity of the enclosure becomes equal to the conductivity of the fluid. This is expected since this case corresponds to pure conduction (Fig. 7-15).

EXAMPLE 7-3   Heat Loss through a Double-Pane Window

The vertical 0.8-m-high, 2-m-wide double-pane window shown in Figure 7-16 consists of two sheets of glass separated by a 2-cm air gap at atmospheric pressure. If the glass surface temperatures across the air gap are measured to be 12°C and 2°C, determine the rate of heat transfer through the window.

Solution   Two glasses of a double-pane window are maintained at specified temperatures. The rate of heat transfer through the window is to be determined.

Assumptions   **1** Steady operating conditions exist. **2** Air is an ideal gas.

Properties   The properties of air at the average temperature of $T_{ave} = (T_1 + T_2)/2 = (12 + 2)/2 = 7°C = 280$ K and 1 atm pressure are (Table A-11)

$$k = 0.0246 \text{ W/m} \cdot °\text{C} \qquad \text{Pr} = 0.717$$
$$v = 1.40 \times 10^{-5} \text{ m}^2/\text{s} \qquad \beta = \frac{1}{T_f} = \frac{1}{280 \text{ K}} = 0.00357 \text{ K}^{-1}$$

Analysis   We have a rectangular enclosure filled with air. The characteristic length in this case is the distance between the two glasses, $\delta = 0.02$ m. Then the Rayleigh number becomes

$$\text{Ra} = \frac{g\beta(T_1 - T_2)\delta^3}{v^2} \text{Pr}$$

$$= \frac{(9.8 \text{ m/s}^2)(0.00357 \text{ K}^{-1})[(12 - 2) \text{ K}](0.02 \text{ m})^3}{(1.40 \times 10^{-5} \text{m}^2/\text{s})^2}(0.717) = 1.024 \times 10^4$$

Then the natural convection Nusselt number in this case can be determined from Equation 7-23 to be

$$Nu = 0.197\,Ra^{1/4}\left(\frac{H}{\delta}\right)^{-1/9} = 0.197(1.024 \times 10^4)^{1/4}\left(\frac{0.8\text{ m}}{0.02\text{ m}}\right)^{-1/9} = 1.32$$

Then

$$A = H \times L = (0.8\text{ m})(2\text{ m}) = 1.6\text{ m}^2$$

and

$$\dot{Q} = k\,Nu\,A\,\frac{T_1 - T_2}{\delta}$$

$$= (0.0246\text{ W/m}\cdot{}^\circ\text{C})(1.32)(1.6\text{ m}^2)\frac{(12-2)^\circ\text{C}}{0.02\text{ m}} = \mathbf{25.9\ W}$$

Therefore, heat will be lost through the window at a rate of 25.9 W.

Discussion Recall that a Nusselt number of Nu = 1 for an enclosure corresponds to pure conduction heat transfer through the enclosure. The air in the enclosure in this case remains still, and no natural convection currents occur in the enclosure. The Nusselt number in our case is 1.32, which indicates that heat transfer through the enclosure is 1.32 times that by pure conduction. The increase in heat transfer is due to the natural convection currents that develop in the enclosure.

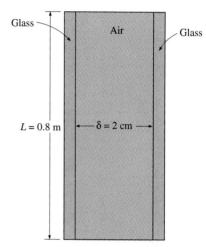

FIGURE 7-16

Schematic for Example 7-3.

EXAMPLE 7-4   Heat Transfer through a Spherical Enclosure

The two concentric spheres of diameters $D_1 = 20$ cm and $D_2 = 30$ cm shown in Figure 7-17 are separated by air at 1 atm pressure. The surface temperatures of the two spheres enclosing the air are $T_1 = 320$ K and $T_2 = 280$ K, respectively. Determine the rate of heat transfer from the inner sphere to the outer sphere by natural convection.

Solution Two surfaces of a spherical enclosure are maintained at specified temperatures. The rate of heat transfer through the enclosure is to be determined.

Assumptions **1** Steady operating conditions exist. **2** Air is an ideal gas.

Properties The properties of air at the average temperature of $T_{ave} = (T_1 + T_2)/2 = (320 + 280)/2 = 300$ K and 1 atm pressure are (Table A-11)

$$k = 0.0261\text{ W/m}\cdot{}^\circ\text{C} \qquad Pr = 0.712$$
$$\nu = 1.57 \times 10^{-5}\text{ m}^2\text{/s}$$
$$\beta = \frac{1}{T_f} = \frac{1}{300\text{ K}} = 0.00333\text{ K}^{-1}$$

Analysis We have a spherical enclosure filled with air. The characteristic length in this case is the distance between the two spheres, which is determined to be

$$\delta = \tfrac{1}{2}(D_2 - D_1) = \tfrac{1}{2}(0.3 - 0.2)\text{m} = 0.05\text{ m}$$

Then the Rayleigh number becomes

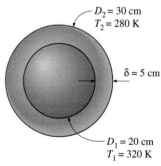

FIGURE 7-17

Schematic for Example 7-4.

$$Ra = \frac{g\beta(T_1 - T_2)\delta^3}{\nu^2}Pr$$

$$= \frac{(9.8 \text{ m/s}^2)(0.00333 \text{ K}^{-1})[(320 - 280) \text{ K}](0.05 \text{ m})^3}{(1.57 \times 10^{-5} \text{ m}^2/\text{s})^2}(0.712)$$

$$= 4.713 \times 10^5$$

Then the natural convection Nusselt number in this case can be determined from Equation 7-38 to be

$$Nu = 0.228 \, Ra^{0.226} = 0.228(4.713 \times 10^5)^{0.226} = 4.37$$

That is, the air in the spherical enclosure will act like a stationary fluid whose thermal conductivity is 4.37 times that of air as a result of natural convection currents. Then,

$$A = \pi D_1 D_2 = \pi(0.2 \text{ m})(0.3 \text{ m}) = 0.188 \text{ m}^2$$

and

$$\dot{Q} = k \, Nu \, A \frac{T_1 - T_2}{\delta}$$

$$= (0.0261 \text{ W/m} \cdot °\text{C})(4.37)(0.188 \text{ m}^2)\frac{(320 - 280) \text{ K}}{0.05 \text{ m}} = \mathbf{17.2 \ W}$$

Therefore, heat will be lost from the inner sphere to the outer one at a rate of 17.2 W.

Discussion Assuming the surfaces of the spheres to be black (emissivity $\varepsilon = 1$), the rate of heat transfer between the two spheres by radiation is

$$\dot{Q}_{rad} = \varepsilon A_1 \sigma(T_1^4 - T_2^4)$$

$$= (1)\pi(0.2 \text{ m})^2(5.67 \times 10^{-8} \text{ W/m}^2 \cdot \text{K}^4)[(320 \text{ K})^4 - (280 \text{ K})^4]$$

$$= 30.9 \text{ W}$$

Thus, the maximum heat transfer by radiation is greater than the heat transfer by natural convection in this case. The emissivity of a real surface is less than 1, and thus the radiation heat transfer for a real enclosure will be less. But radiation can still be significant and needs to be considered.

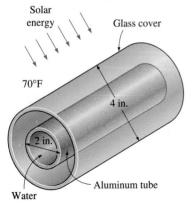

FIGURE 7-18

Schematic for Example 7-5.

EXAMPLE 7-5   Heating Water in a Tube of Solar Energy

A solar collector consists of a horizontal aluminum tube having an outer diameter of 2 in. enclosed in a concentric thin glass tube of 4-in.-diameter (Fig. 7-18). Water is heated as it flows through the tube, and the annular space between the aluminum and the glass tubes is filled with air at 1 atm pressure. The pump circulating the water fails during a clear day, and the water temperature in the tube starts rising. The aluminum tube absorbs solar radiation at a rate of 30 Btu/h per foot length, and the temperature of the ambient air outside is 70°F. Disregarding any heat loss by radiation, determine the temperature of the aluminum tube when steady operation is established (i.e., when the rate of heat loss from the tube equals the amount of solar energy gained by the tube).

Solution   The circulating pump of a solar collector that consists of a horizontal tube and its glass cover fails one day. The temperature of the tube is determined when steady conditions are reached.

Assumptions **1** Steady operating conditions exist. **2** The tube and its cover are isothermal. **3** Air is an ideal gas. **4** Heat loss by radiation is negligible.

Properties The properties of air should be evaluated at the average temperature. But we do not know the exit temperature of the air in the duct, and thus we cannot determine the bulk fluid and glass cover temperatures at this point, and thus we cannot evaluate the average temperatures. Therefore, we will use the properties at an anticipated average temperature of 100°F (Table A-11E),

$$k = 0.0154 \text{ Btu/h} \cdot \text{ft} \cdot °\text{F} \qquad \text{Pr} = 0.72$$
$$\nu = 0.18 \times 10^{-3} \text{ ft}^2/\text{s}$$
$$\beta = \frac{1}{T_f} = \frac{1}{(100 + 460) \text{ R}} = 0.001786 \text{ R}^{-1}$$

Analysis We have a horizontal cylindrical enclosure filled with air at 1 atm pressure. The problem involves heat transfer from the aluminum tube to the glass cover and from the outer surface of the glass cover to the surrounding ambient air. When steady operation is reached, these two heat transfer rates must equal the rate of heat gain. That is,

$$\dot{Q}_{\text{tube-glass}} = \dot{Q}_{\text{glass-ambient}} = \dot{Q}_{\text{solar gain}} = 30 \text{ Btu/h} \qquad \text{(per foot of tube)}$$

The heat transfer surface area of the glass cover is

$$A_2 = A_{\text{glass}} = (\pi DL)_{\text{glass}} = \pi\left(\frac{4}{12} \text{ ft}\right)(1 \text{ ft}) = 1.047 \text{ ft}^2 \qquad \text{(per foot of tube)}$$

To determine the Rayleigh number, we need to know the surface temperature of the glass, which is not available. Therefore, it is clear that the solution will require a trial-and-error approach. Assuming the glass cover temperature to be 100°F, the Rayleigh number, the Nusselt number, the convection heat transfer coefficient, and the rate of natural convection heat transfer from the glass cover to the ambient air are determined to be

$$\text{Ra} = \frac{g\beta(T_s - T_\infty)\delta^3}{\nu^2}\text{Pr}$$

$$= \frac{(32.2 \text{ ft/s}^2)(0.001786 \text{ R}^{-1})[(100 - 70) \text{ R}](\frac{4}{12} \text{ ft})^3}{(0.18 \times 10^{-3} \text{ ft}^2/\text{s})^2}(0.72) = 1.420 \times 10^6$$

$$\text{Nu} = \left\{0.6 + \frac{0.387 \text{ Ra}^{1/6}}{[1 + (0.559/\text{Pr})^{9/16}]^{8/27}}\right\}^2$$

$$= \left\{0.6 + \frac{0.387(1.420 \times 10^6)^{1/6}}{[1 + (0.559/0.72)^{9/16}]^{8/27}}\right\}^2 = 16.1$$

$$h = \frac{k}{D}\text{Nu} = \frac{0.0154 \text{ Btu/h} \cdot \text{ft} \cdot °\text{F}}{\frac{4}{12} \text{ ft}}(16.1) = 0.743 \text{ Btu/h} \cdot \text{ft}^2 \cdot °\text{F}$$

$$\dot{Q}_{\text{glass}} = hA_2(T_2 - T_\infty) = (0.743 \text{ Btu/h} \cdot \text{ft}^2 \cdot °\text{F})(1.047 \text{ ft}^2)(100 - 70)°\text{F}$$
$$= 23.3 \text{ Btu/h}$$

which is less than 30 Btu/h. Therefore, the assumed temperature of 100°F for the glass cover is low. Repeating the calculations for a temperature of 110°F gives 33.8 Btu/h, which is high. Then the glass cover temperature corresponding to 30 Btu/h is determined by interpolation to be 106.4°F.

The temperature of the aluminum tube is determined in a similar manner using the natural convection relations for two horizontal concentric cylinders. The characteristic length in this case is the distance between the two cylinders, which is determined to be

$$\delta = \tfrac{1}{2}(D_2 - D_1) = \tfrac{1}{2}(4 - 2) \text{ in.} = 1 \text{ in.}$$

Also,

$$A = \frac{\pi L(D_2 - D_1)}{\ln(D_2/D_1)} = \frac{\pi(1 \text{ ft})(\tfrac{4}{12} - \tfrac{2}{12}) \text{ ft}}{\ln\left(\tfrac{4}{2}\right)} = 0.755 \text{ ft}^2$$

We start the calculations by assuming the tube temperature to be 200°F. This gives

$$\text{Ra} = \frac{g\beta(T_1 - T_2)\delta^3}{\nu^2} \text{Pr}$$

$$= \frac{(32.2 \text{ ft/s}^2)(0.001786 \text{ R}^{-1})[(200 - 107) \text{ R}](\tfrac{1}{12} \text{ ft})^3}{(0.18 \times 10^{-3} \text{ ft}^2/\text{s})^2}(0.72) = 6.88 \times 10^4$$

$$\text{Nu} = 0.11 \, \text{Ra}^{0.29} = 0.11(6.88 \times 10^4)^{0.29} = 2.78$$

$$\dot{Q}_{\text{tube}} = k \, \text{Nu} \, A \frac{T_1 - T_2}{\delta}$$

$$= (0.0154 \text{ Btu/h} \cdot \text{ft} \cdot {}^\circ\text{F})(2.78)(0.755 \text{ ft}^2)\frac{(200 - 107)\text{R}}{\tfrac{1}{12} \text{ ft}} = 36.1 \text{ Btu/h}$$

which is more than 30 Btu/h. Therefore, the assumed temperature of 200°F for the tube is high. Repeating the calculations for a temperature of 180°F gives 26.4 Btu/h, which is low. Then the tube temperature corresponding to 30 Btu/h is determined by interpolation to be **188°F**. Therefore, the tube will reach an equilibrium temperature of 188°F when the pump fails.

This result above is obtained by using air properties at 100°F. It appears that this result can be improved by repeating the calculations above using air properties at the average temperature of 88.5°F for heat transfer from the glass cover to the ambient air and at 147.5°F for heat transfer from the tube to the glass cover. Also, we have not considered heat loss by radiation in the calculations, and thus the tube temperature determined above is probably too high. This problem is considered again in Chapter 9 by accounting for the effect of radiation heat transfer.

7-4 ■ NATURAL CONVECTION FROM FINNED SURFACES

Finned surfaces of various shapes, called *heat sinks*, are frequently used in the cooling of electronic devices. Energy dissipated by these devices is transferred to the heat sinks by *conduction* and from the heat sinks to the ambient air by *natural* or *forced convection*, depending on the power dissipation requirements. Natural convection is the preferred mode of heat transfer since it involves *no moving parts*, like the electronic components themselves. However, in the natural convection mode, the components are more likely to run at a *higher temperature* and thus undermine reliability. A properly selected heat sink may considerably *lower* the operation temperature of the components and thus reduce the risk of failure.

A question that often arises in the selection of a heat sink is whether to select one with *closely packed* fins or *widely spaced* fins for a given base area (Fig. 7-19). A heat sink with closely packed fins will have greater surface area for heat transfer but a smaller heat transfer coefficient because of the extra resistance the additional fins will introduce to fluid flow through the interfin passages. A heat sink with widely spaced fins, on the other hand, will have a higher heat transfer coefficient but a smaller surface area. Therefore, there must be an *optimum spacing* that maximizes the natural convection heat transfer from the heat sink for a given base area WL, where W and L are the width and height of the base of the heat sink, respectively, as shown in Figure 7-20. When the fins are essentially isothermal and the fin thickness t is small relative to the fin spacing S, the optimum fin spacing for a vertical heat sink is determined by Bar-Cohen and Rohsenow to be

$$S_{opt} = 2.714 \frac{L}{\mathrm{Ra}^{1/4}} \qquad (7\text{-}41)$$

where the fin length L in the vertical direction is taken to be the characteristic length in the evaluation of the Rayleigh number. The heat transfer coefficient for the optimum spacing case was determined to be

$$h = 1.31 \frac{k}{S_{opt}} \qquad (7\text{-}42)$$

Then the rate of heat transfer by natural convection from the fins can be determined from

$$\dot{Q} = h(2nLH)(T_s - T_\infty) \qquad (7\text{-}43)$$

where $n = W/(S + t) \approx W/S$ is the number of fins on the heat sink and T_s is the surface temperature of the fins.

As we mentioned earlier, the magnitude of the natural convection heat transfer is directly related to the *mass flow rate* of the fluid, which is established by the dynamic balance of two opposing effects: *buoyancy* and *friction*.

The fins of a heat sink introduce both effects: *inducing extra buoyancy* as a result of the elevated temperature of the fin surfaces and *slowing down the fluid* by acting as an added obstacle on the flow path. As a result, increasing the number of fins on a heat sink can either enhance or reduce natural convection, depending on which effect is dominant. The buoyancy-driven air flow rate is established at the point where these two effects balance each other. The friction force increases as more and more solid surfaces are introduced, seriously disrupting fluid flow and heat transfer. Under some conditions, the increase in friction may more than offset the increase in buoyancy. This in turn will tend to reduce the flow rate and thus the heat transfer. For that reason, heat sinks with closely spaced fins are not suitable for natural convection cooling.

When the heat sink involves closely spaced fins, the narrow channels formed tend to block or "suffocate" the fluid, especially when the heat sink

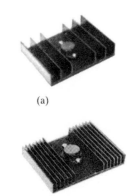

(a)

(b)

FIGURE 7-19

Heat sinks with (*a*) widely spaced and (*b*) closely packed fins (courtesy of Vemaline Products).

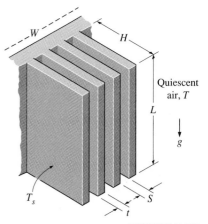

FIGURE 7-20

Various dimensions of a finned surface oriented vertically.

is long. As a result, the blocking action produced overwhelms the extra buoyancy and downgrades the heat transfer characteristics of the heat sink. Then, at a fixed power setting, the heat sink runs at a higher temperature relative to the no-shroud case. When the heat sink involves widely spaced fins, the shroud does not introduce a significant increase in resistance to flow, and the buoyancy effects dominate. As a result, heat transfer by natural convection may improve, and at a fixed power level the heat sink may run at a lower temperature.

When extended surfaces such as fins are used to enhance natural convection heat transfer between a solid and a fluid, the flow rate of the fluid in the vicinity of the solid adjusts itself to incorporate the changes in buoyancy and friction. It is obvious that this enhancement technique will work to advantage only when the increase in buoyancy is greater than the additional friction introduced. One does not need to be concerned with pressure drop or pumping power when studying natural convection since no pumps or blowers are used in this case. Therefore, an enhancement technique in natural convection is evaluated on heat transfer performance alone.

The failure rate of an electronic component increases almost exponentially with operating temperature. The cooler the electronic device operates, the more reliable it is. A rule of thumb is that semiconductor failure rate is halved for each 10°C reduction in junction operating temperature. The desire to lower the operating temperature without having to resort to forced convection has motivated researchers to investigate enhancement techniques for natural convection. Sparrow and Prakash have demonstrated that, under certain conditions, the use of discrete plates in lieu of continuous plates of the same surface area increases heat transfer considerably. In other experimental work, using transistors as the heat source, Çengel and Zing have demonstrated that temperature recorded on the transistor case dropped by as much as 30°C when a shroud was used, as opposed to the corresponding no-shroud case.

EXAMPLE 7-6 Optimum Fin Spacing of a Heat Sink

A 12-cm-wide and 18-cm-high vertical hot surface in 25°C air is to be cooled by a heat sink with equally spaced fins of rectangular profile (Fig. 7-21). The fins are 0. 1 cm thick and l8 cm long in the vertical direction and have a height of 2.4 cm from the base. Determine the optimum fin spacing and the rate of heat transfer by natural convection from the heat sink if the base temperature is 80°C.

Solution A heat sink with equally spaced rectangular fins is to be used to cool a hot surface. The optimum fin spacing and the rate of heat transfer from the heat sink are to be determined.

Assumptions **1** Steady operating conditions exist. **2** Air is an ideal gas. **3** The atmospheric pressure at that location is 1 atm. **4** The thickness t of the fins is very small relative to the fin spacing S so that Equations 7-41 and 7-42 for optimum fin spacing are applicable.

Properties The properties of air at the film temperature of $T_f = (T_s + T_\infty)/2 = (80 + 25)/2 = 52.5°C$ and 325.5 K and 1 atm pressure are (Table A-11)

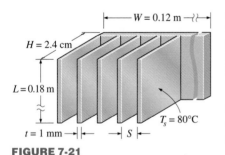

FIGURE 7-21

Schematic for Example 7-6.

$$k = 0.0279 \text{ W/m} \cdot °\text{C} \qquad \text{Pr} = 0.709$$

$$v = 1.82 \times 10^{-5} \text{ m}^2/\text{s} \qquad \beta = \frac{1}{T_f} = \frac{1}{325.5 \text{ K}} = 0.003072 \text{ K}^{-1}$$

Analysis   The characteristic length in this case is the length of the fins in the vertical direction, which is given to be $L = 0.18$ m. Then the Rayleigh number becomes

$$\text{Ra} = \frac{g\beta(T_s - T_\infty)\delta^3}{v^2} \text{Pr}$$

$$= \frac{(9.8 \text{ m/s}^2)(0.003072 \text{ K}^{-1})[(80 - 25) \text{ K}](0.18 \text{ m})^3}{(1.82 \times 10^{-5} \text{ m}^2/\text{s})^2}(0.709) = 2.067 \times 10^7$$

The optimum fin spacing is determined from Equation 7-41 to be

$$S_\text{opt} = 2.714 \frac{L}{\text{Ra}^{1/4}} = 2.714 \frac{0.18 \text{ m}}{(2.067 \times 10^7)^{1/4}} = 0.0072 \text{ m} = \textbf{7.2 mm}$$

which is about 7 times the thickness of the fins. Therefore, the assumption of negligible fin thickness in this case is acceptable for practical purposes. The number of fins and the heat transfer coefficient for this optimum fin spacing case are

$$n = \frac{W}{S + t} = \frac{0.12 \text{ m}}{(0.0072 + 0.001) \text{ m}} \approx 15 \text{ fins}$$

$$h = 1.31 \frac{k}{S_\text{opt}} = 1.31 \frac{0.0279 \text{ W/m} \cdot °\text{C}}{0.0072 \text{ m}} = 5.08 \text{ W/m}^2 \cdot °\text{C}$$

Then the rate of natural convection heat transfer becomes

$$\dot{Q} = h(2nLH)(T_s - T_\infty)$$

$$= (5.08 \text{ W/m}^2 \cdot °\text{C})[2 \times 15 \times (0.18 \text{ m})(0.024 \text{ m})](80 - 25)°\text{C} = 36.2 \text{ W}$$

Therefore, the heat sink can dissipate heat by natural convection at a rate of 181 W.

7-5 ■ COMBINED NATURAL AND FORCED CONVECTION

The presence of a temperature gradient in a fluid in a gravity field always gives rise to natural convection currents, and thus heat transfer by natural convection. Therefore, forced convection is always accompanied by natural convection.

We mentioned earlier that the convection heat transfer coefficient, natural or forced, is a strong function of the fluid velocity. Heat transfer coefficients encountered in forced convection are typically much higher than those encountered in natural convection because of the higher fluid velocities associated with forced convection. As a result, we tend to ignore natural convection in heat transfer analyses that involve forced convection, although we recognize that natural convection always accompanies forced convection. The error involved in ignoring natural convection is negligible at high velocities but may be considerable at low velocities associated with forced

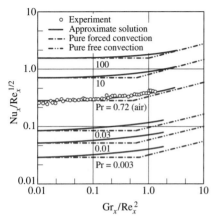

FIGURE 7-22

Variation of the local Nusselt number NU_x for combined natural and forced convection from a hot isothermal vertical plate (from Lloyd and Sparrow, Ref. 12).

convection. Therefore, it is desirable to have a criterion to assess the relative magnitude of natural convection in the presence of forced convection.

For a given fluid, it is observed that the parameter Gr/Re^2 represents the importance of natural convection relative to forced convection. This is not surprising since the convection heat transfer coefficient is a strong function of the Reynolds number Re in forced convection and the Grashof number Gr in natural convection.

A plot of the nondimensionalized heat transfer coefficient for combined natural and forced convection on a vertical plate is given in Figure 7-22 for different fluids. We note from this figure that natural convection is negligible when $Gr/Re^2 < 0.1$, forced convection is negligible when $Gr/Re^2 > 10$, and neither is negligible when $0.1 < Gr/Re^2 < 10$. Therefore, both natural and forced convection must be considered in heat transfer calculations when the Gr and Re^2 are of the same order of magnitude (one is within a factor of 10 times the other). Note that forced convection is small relative to natural convection only in the rare case of extremely low forced flow velocities.

Natural convection may *help* or *hurt* forced convection heat transfer, depending on the relative directions of *buoyancy-induced* and the *forced convection* motions (Fig. 7-23):

1. In *assisting flow,* the buoyant motion is in the *same* direction as the forced motion. Therefore, natural convection assists forced convection and *enhances* heat transfer. An example is upward forced flow over a hot surface.

2. In *opposing flow,* the buoyant motion is in the *opposite* direction to the forced motion. Therefore, natural convection resists forced convection and *decreases* heat transfer. An example is upward forced flow over a cold surface.

3. In *transverse flow,* the buoyant motion is *perpendicular* to the forced motion. Transverse flow enhances fluid mixing and thus *enhances* heat transfer. An example is horizontal forced flow over a hot or cold cylinder or sphere.

FIGURE 7-23

Natural convection can *enhance* or *inhibit* heat transfer, depending on the relative directions of *buoyancy-induced motion* and the *forced convection motion.*

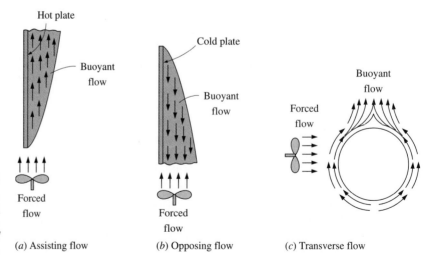

(a) Assisting flow   (b) Opposing flow   (c) Transverse flow

When determining heat transfer under combined natural and forced convection conditions, it is tempting to add the contributions of natural and forced convection in assisting flows and to subtract them in opposing flows. However, the evidence indicates differently. A review of experimental data suggests a correlation of the form

$$\text{Nu}_{\text{combined}} = (\text{Nu}_{\text{forced}}^n \pm \text{Nu}_{\text{natural}}^n)^{1/n} \qquad (7\text{-}44)$$

where $\text{Nu}_{\text{forced}}$ and $\text{Nu}_{\text{natural}}$ are determined from the correlations for *pure forced* and *pure natural convection*, respectively. The plus sign is for *assisting* and *transverse* flows and the minus sign is for *opposing* flows. The value of the exponent n varies between 3 and 4, depending on the geometry involved. It is observed that $n = 3$ correlates experimental data for vertical surfaces well. Larger values of n are better suited for horizontal surfaces.

A question that frequently arises in the cooling of heat-generating equipment such as electronic components is whether to use a fan (or a pump if the cooling medium is a liquid)—that is, whether to utilize *natural* or *forced* convection in the cooling of the equipment. The answer depends on the maximum allowable operating temperature. Recall that the convection heat transfer rate from a surface at temperature T_s in a medium at T_∞ is given by

$$\dot{Q}_{\text{conv}} = hA(T_s - T_\infty)$$

where h is the convection heat transfer coefficient and A is the surface area. Note that for a fixed value of power dissipation and surface area, h and T_s are *inversely proportional*. Therefore, the device will operate at a *higher* temperature when h is low (typical of natural convection) and at a *lower* temperature when h is high (typical of forced convection).

Natural convection is the preferred mode of heat transfer since no blowers or pumps are needed and thus all the problems associated with these, such as noise, vibration, power consumption, and malfunctioning, are avoided. Natural convection is adequate for cooling *low-power-output* devices, especially when they are attached to extended surfaces such as heat sinks. For *high-power-output* devices, however, we have no choice but to use a blower or a pump to keep the operating temperature below the maximum allowable level. For *very-high-power-output* devices, even forced convection may not be sufficient to keep the surface temperature at the desirable levels. In such cases, we may have to use *boiling* and *condensation* to take advantage of the very high heat transfer coefficients associated with phase change processes.

7-6 ■ SUMMARY

In this chapter, we have considered *natural convection* heat transfer where any fluid motion occurs by natural means such as buoyancy. The fluid velocities associated with natural convection are low. Therefore, the heat transfer coefficients encountered in natural convection are usually much lower than those encountered in forced convection.

The upward force exerted by a fluid on a body completely or partially immersed in it is called the *buoyancy force*, whose magnitude is equal to the weight of the fluid displaced by the body. The *volume expansion coefficient* β of a substance represents the variation of the density of that substance with temperature at constant pressure and is defined as

$$\beta = \frac{1}{v}\left(\frac{\partial v}{\partial T}\right)_P = -\frac{1}{\rho}\left(\frac{\partial \rho}{\partial T}\right)_P \qquad (1/\text{K})$$

For an ideal gas, it reduces to

$$\beta_{\text{ideal gas}} = \frac{1}{T} \qquad (1/\text{K})$$

where T is the absolute temperature. The instrument used in natural convection experiments most often is the *Mach–Zehnder interferometer,* which gives a plot of isotherms in the fluid in the vicinity of a surface.

The flow regime in natural convection is governed by a dimensionless number called the *Grashof number,* which represents the ratio of the buoyancy force to the viscous force acting on the fluid and is expressed as

$$\text{Gr} = \frac{g\beta(T_s - T_\infty)\delta^3}{v^2}$$

where

g = gravitational acceleration, m/s^2
β = coefficient of volume expansion, 1/K ($\beta = 1/T$ for ideal gases)
T_s = temperature of the surface, °C
T_∞ = temperature of the fluid sufficiently far from the surface, °C
δ = characteristic length of the geometry, m
v = kinematic viscosity of the fluid, m^2/s

The Grashof number provides the main criterion in determining whether the fluid flow is laminar or turbulent in natural convection. The heat transfer rate in natural convection from a solid surface to the surrounding fluid is expressed by Newton's law of cooling as

$$\dot{Q}_{\text{conv}} = hA(T_s - T_\infty) \qquad (\text{W})$$

where A is the heat transfer surface area and h is the average heat transfer coefficient on the surface.

Most heat transfer relations in natural convection are based on experimental studies, and the simple empirical correlations for the average *Nusselt number* Nu in natural convection are of the form

$$\text{Nu} = \frac{h\delta}{k} = C(\text{Gr Pr})^n = C\,\text{Ra}^n$$

where Ra is the *Rayleigh number,* which is the product of the Grashof and Prandtl numbers:

$$\text{Ra} = \text{Gr Pr} = \frac{g\beta(T_s - T_\infty)\delta^3}{v^2}\,\text{Pr}$$

The values of the constants C and n depend on the geometry of the surface and the flow regime, which is characterized by the range of the Rayleigh number. Simple relations for the average Nusselt number for various geometries are given in Table 7-1 together with a sketch of the geometry. All fluid properties are to be evaluated at the film temperature $T_f = \frac{1}{2}(T_s + T_\infty)$.

Simple empirical correlations for the Nusselt number for various enclosures are given in Table 7-2. Once the Nusselt number is available, the rate of heat transfer through the enclosure can be determined from

where
$$\dot{Q} = hA(T_1 - T_2) = k\,\text{Nu}\,A\,\frac{T_1 - T_2}{\delta}$$

$$A = \begin{cases} HL & \text{rectangular enclosures} \\ \dfrac{\pi L(D_2 - D_1)}{\ln(D_2/D_1)} & \text{concentric cylinders} \\ \pi D_1 D_2 & \text{concentric spheres} \end{cases}$$

The quantity $k\,\text{Nu}$ is called the *effective thermal conductivity* of the enclosure, since a fluid in an enclosure behaves like a quiescent fluid whose thermal conductivity is $k\,\text{Nu}$ as a result of convection currents.

For a given fluid, the parameter Gr/Re^2 represents the importance of natural convection relative to forced convection. Natural convection is negligible when $\text{Gr}/\text{Re}^2 < 0.1$, forced convection is negligible when $\text{Gr}/\text{Re}^2 > 10$, and neither is negligible when $0.1 < \text{Gr}/\text{Re}^2 < 10$.

REFERENCES AND SUGGESTED READING

1. A. Bar-Cohen. "Fin Thickness for an Optimized Natural Convection Array of Rectangular Fins." *Journal of Heat Transfer* 101 (1979), pp. 564–66.

2. A. Bar-Cohen and W. M. Rohsenow. "Thermally Optimum Spacing of Vertical Natural Convection Cooled Parallel Plates." *Journal of Heat Transfer* 106 (1984), pp. 116–23.

3. Y. Bayazitoglu and M. N. Özişik. *Elements of Heat Transfer.* New York: McGraw-Hill, 1988.

4. Y. A. Çengel and P. T. L. Zing. "Enhancement of Natural Convection Heat Transfer from Heat Sinks by Shrouding." *Proceedings of ASME/JSME Thermal Engineering Conference,* Honolulu, HA, March 22–27, 1987, Vol. 3, pp. 451–57.

5. S. W. Churchill. "Combined Free and Forced Convection around Immersed Bodies." *Heat Exchanger Design Handbook.* Section 2.5.9. New York: Hemisphere Publishing, 1986.

6. S. W. Churchill. "A Comprehensive Correlating Equation for Laminar Assisting Forced and Free Convection." *AIChE Journal* 23 (1977), pp. 10–16.

7. E. R. G. Eckerd and E. Soehngen. "Interferometric Studies on the Stability and Transition to Turbulence of a Free Convection Boundary Layer." *Proceedings of General Discussion, Heat Transfer ASME–IME,* London, 1951.

8. E. R. G. Eckerd and E. Soehngen. "Studies on Heat Transfer in Laminar Free Convection with Zehnder–Mach Interferometer." *USAF Technical Report 5747,* December 1948.

9. J. P. Holman. *Heat Transfer.* 7th ed. New York: McGraw-Hill, 1990.

10. F. P. Incropera and D. P. DeWitt. *Introduction to Heat Transfer.* 2nd ed. New York: John Wiley & Sons, 1990.

11. F. Kreith and M. S. Bohn. *Principles of Heat Transfer.* 5th ed. St. Paul, MN: West Publishing, 1993.

12. J. R. Lloyd and E. M. Sparrow. "Combined Forced and Free Convection Flow on Vertical Surfaces." *International Journal of Heat and Mass Transfer* 13 (1970), p. 434.

13. M. N. Özişik. *Heat Transfer—A Basic Approach.* New York: McGraw-Hill, 1985.

14. E. M. Sparrow and C. Prakash. "Enhancement of Natural Convection Heat Transfer by a Staggered Array of Vertical Plates." *Journal of Heat Transfer* 102 (1980), pp. 215–20.

15. E. M. Sparrow and S. B. Vemuri. "Natural Convection/Radiation Heat Transfer from Highly Populated Pin Fin Arrays." *Journal of Heat Transfer.* 107 (1985), pp. 190–97.

16. L. C. Thomas. *Heat Transfer.* Englewood Cliffs, NJ: Prentice Hall, 1992.

17. F. M. White. *Heat and Mass Transfer.* Reading, MA: Addison-Wesley, 1988.

PROBLEMS*

Physical Mechanism of Natural Convection

7-1C What is natural convection? How does it differ from forced convection? What force causes natural convection currents?

7-2C In which mode of heat transfer is the convection heat transfer coefficient usually higher, natural convection or forced convection? Why?

7-3C Consider a hot boiled egg in a spacecraft that is filled with air at atmospheric pressure and temperature at all times. Will the egg cool faster or slower when the spacecraft is in space instead of on the ground? Explain.

---

*Students are encouraged to answer *all* the concept "C" questions.

7-4C What is buoyancy force? Compare the relative magnitudes of the buoyancy force acting on a body immersed in the following mediums: (a) air, (b) water, (c) mercury, and (d) an evacuated chamber.

7-5C When will the hull of a ship sink in water deeper: when the ship is sailing in fresh water or in sea water? Why?

7-6C A person weighs himself on a waterproof spring scale placed at the bottom of a 1-m-deep swimming pool. Will the person weigh more or less in water? Why?

7-7C Consider two fluids, one with a large coefficient of volume expansion and the other with a small one. In what fluid will a hot surface initiate stronger natural convection currents? Why? Assume the viscosity of the fluids to be the same.

7-8C Consider a fluid whose volume does not change with temperature at constant pressure. What can you say abut natural convection heat transfer in this medium?

7-9C What do the lines on an interferometer photograph represent? What do closely packed lines on the same photograph represent?

7-10C Physically, what does the Grashof number represent? How does the Grashof number differ from the Reynolds number?

7-11 Show that the volume expansion coefficient of an ideal gas is $\beta = 1/T$, where T is the absolute temperature.

Natural Convection over Surfaces

7-12C How does the Rayleigh number differ from the Grashof number?

7-13C Under what conditions can the outer surface of a vertical cylinder be treated as a vertical plate in natural convection calculations?

7-14C Will a hot horizontal plate whose back side is insulated cool faster or slower when its hot surface is facing down instead of up?

7-15C Consider laminar natural convection from a vertical hot plate. Will the heat flux be higher at the top or at the bottom of the plate? Why?

7-16 An 8-m-long section of a 6-cm-diameter horizontal hot water pipe passes through a large room whose temperature is 22°C. If the temperature and the emissivity of the outer surface of the pipe are 65°C and 0.8, respectively, determine the rate of heat loss from the pipe by (a) natural convection and (b) radiation.

7-17 Consider a wall-mounted power transistor that dissipates 0.18 W of power in an environment at 35°C. The transistor is 0.45 cm long and has a diameter of 0.4 cm. The emissivity of the outer surface of the transistor is 0.1, and the average temperature of the surrounding surfaces is 25°C. Disregarding any heat transfer from the base surface, determine the surface temperature of the transistor.

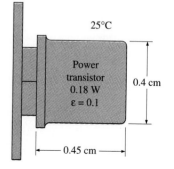

FIGURE P7-17

25°C

Power transistor
0.18 W
$\varepsilon = 0.1$

0.4 cm

0.45 cm

7-18E Consider a 2-ft × 2-ft thin square plate in a room at 75°F. One side of the plate is maintained at a temperature of 130°F, while the other side is insulated. Determine the rate of heat transfer from the plate by natural convection if the plate is (a) vertical, (b) horizontal with hot surface facing up, and (c) horizontal with hot surface facing down.

7-19 A 500-W cylindrical resistance heater is 1 m long and 0.5 cm in diameter. The resistance wire is placed horizontally in a fluid at 20°C. Determine the outer surface temperature of the resistance wire in steady operation if the fluid is (a) air and (b) water. Ignore any heat transfer by radiation. For water take $\beta = 0.000365 \text{ K}^{-1}$.

7-20 Water is boiling in a 12-cm-deep pan with an outer diameter of 25 cm that is placed on top of a stove. The ambient air and the surrounding surfaces are at a temperature of 25°C, and the emissivity of the outer surface of the pan is 0.95. Assuming the entire pan to be at an average temperature of 98°C, determine the rate of heat loss from the cylindrical side surface of the pan to the surroundings by (a) natural convection and (b) radiation. (c) If water is boiling at a rate of 2 kg/h at 100°C, determine the ratio of the heat lost from the side surfaces of the pan to that by the evaporation of water. The heat of vaporization of water at 100°C is 2257 kJ/kg.

Answers: 50 W, 56.1 W, 0.085

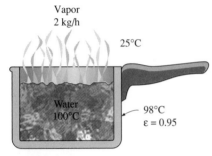

Vapor
2 kg/h

25°C

Water
100°C

98°C
$\varepsilon = 0.95$

FIGURE P7-20

7-21 Repeat Problem 7-20 for a pan whose outer surface is polished and has an emissivity of 0.1.

7-22 In a plant that manufactures canned aerosol paints, the cans are temperature-tested in water baths at 55°C before they are shipped to ensure that they will withstand temperatures up to 55°C during transportation and shelving. The cans, moving on a conveyor, enter the open hot water bath, which is 0.5 m deep, 1 m wide, and 3.5 m long, and move slowly in the hot water toward the other end. Some of the cans fail the test and explode in the water bath. The water container is made of sheet metal, and the entire container is at about the same temperature as the hot water. The emissivity of the outer surface of the container is 0.7. If the temperature of the surrounding air and surfaces is 20°C, determine the rate of heat loss from the four side surfaces of the container (disregard the top surface, which is open).

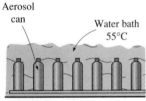

Aerosol
can

Water bath
55°C

FIGURE P7-22

The water is heated electrically by resistance heaters, and the cost of electricity is $0.085/kWh. If the plant operates 24 h a day 365 days a year and thus 8760 h a year, determine the annual cost of the heat losses from the container for this facility.

7-23 Reconsider Problem 7-22. In order to reduce the heating cost of the hot water, it is proposed to insulate the side and bottom surfaces of the container with 5-cm-thick fiberglass insulation ($k = 0.035 \text{ W/m} \cdot °C$) and to wrap the insulation with aluminum foil ($\varepsilon = 0.1$) in order to minimize the heat loss by radiation. An estimate is obtained from a local insulation contractor, who proposes to do the insulation job for $350, including materials and labor. Would you support this proposal? How long will it take for the insulation to pay for itself from the energy it saves?

7-24 Consider a 15-cm × 20-cm printed circuit board (PCB) that has electronic components on one side. The board is placed in a room at 20°C. The heat loss from the back surface of the board is negligible. If the circuit board is dissipating 8 W of power in steady operation, determine the average temperature of the hot surface of the board, assuming the board is (*a*) vertical, (*b*) horizontal with hot surface facing up, and (*c*) horizontal with hot surface facing down. Take the emissivity of the surface of the board to be 0.8 and assume the surrounding surfaces to be at the same temperature as the air in the room.   *Answers:* (*a*) 46°C, (*b*) 42°C, (*c*) 50°C

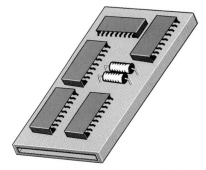

FIGURE P7-24

7-25 A manufacturer makes absorber plates that are 1.2 m × 0.8 m in size for use in solar collectors. The back side of the plate is heavily insulated, while its front surface is coated with black chrome, which has an absorptivity of 0.87 for solar radiation and an emissivity of 0.09. Consider such a plate placed horizontally outdoors in calm air at 25°C. Solar radiation is incident on the plate at a rate of 700 W/m². Taking the effective sky temperature to be 10°C, determine the equilibrium temperature of the absorber plate. What would your answer be if the absorber plate is made of ordinary aluminum plate that has a solar absorptivity of 0.28 and an emissivity of 0.07?

7-26 Repeat Problem 7-25 for an aluminum plate painted flat black (solar absorptivity 0.98 and emissivity 0.98) and also for a plate painted white (solar absorptivity 0.26 and emissivity 0.90).

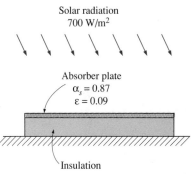

FIGURE P7-25

7-27 The following experiment is conducted to determine the natural convection heat transfer coefficient for a horizontal cylinder that is 50 cm long and 2 cm in diameter. A 50-cm-long resistance heater is placed along the centerline of the cylinder, and the surfaces of the cylinder are polished to minimize the radiation effect. The two circular side surfaces of the cylinder are well insulated. The resistance heater is turned on, and the power dissipation is maintained constant at 40 W. If the average surface temperature of the cylinder is measured to be 120°C in the 20°C room air when steady operation is reached, determine the natural convection heat transfer coefficient. If the emissivity of the outer surface of the cylinder is 0.1 and a 5 percent error is acceptable, do you think we need to do any correction for the radiation effect? Assume the surrounding surfaces to be at 20°C also.

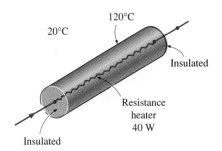

FIGURE P7-27

7-28 Thick fluids such as asphalt and waxes and the pipes in which they flow are often heated in order to reduce the viscosity of the fluids and thus to reduce the pumping costs. Consider the flow of such a fluid through a 100-m-long pipe of outer diameter 30 cm in calm ambient air at 0°C. The pipe is heated electrically, and a thermostat keeps the outer surface temperature of the pipe constant at 25°C. The emissivity of the outer surface of the pipe is 0.8, and the effective sky temperature is −30°C, Determine the power rating of the electric resistance heater, in kW, that needs to be used. Also, determine the cost of electricity associated with heating the pipe during a 10-h period under the above conditions if the price of electricity is $0.09/kWh.   *Answers:* 29.2 kW, $26.2

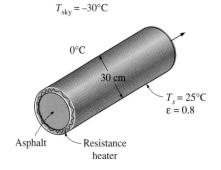

FIGURE P7-28

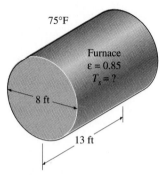

75°F

Furnace
$\varepsilon = 0.85$
$T_s = ?$

8 ft

13 ft

FIGURE P7-30E

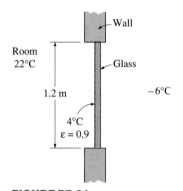

Wall

Room
22°C

Glass

1.2 m

$-6°C$

4°C
$\varepsilon = 0.9$

FIGURE P7-31

FIGURE P7-33

170°C
20°C $\varepsilon = 0.7$

6.03 cm

60 m

Steam

7-29   Reconsider Problem 7-28. To reduce the heating cost of the pipe, it is proposed to insulate it with sufficiently thick fiberglass insulation ($k = 0.035$ W/m · °C) wrapped with aluminum foil ($\varepsilon = 0.1$) to cut down the heat losses by 85 percent. Assuming the pipe temperature to remain constant at 25°C, determine the thickness of the insulation that needs to be used. How much money will the insulation save during this 10-h period?
   Answers: 1.3 cm, $22.3

7-30E   Consider an industrial furnace that resembles a 13-ft-long horizontal cylindrical enclosure 8 ft in diameter whose end surfaces are well insulated. The furnace burns natural gas at a rate of 48 therms/h (1 therm = 100,000 Btu). The combustion efficiency of the furnace is 82 percent (i.e., 18 percent of the chemical energy of the fuel is lost through the flue gases as a result of incomplete combustion and the flue gases leaving the furnace at high temperature). If the heat loss from the outer surfaces of the furnace by natural convection and radiation is not to exceed 1 percent of the heat generated inside, determine the highest allowable surface temperature of the furnace. Assume the air and wall surface temperature of the room to be 75°F, and take the emissivity of the outer surface of the furnace to be 0.85. If the cost of natural gas is $0.48/therm and the furnace operates 3000 h per year, determine the annual cost of this heat loss to the plant.

7-31   Consider a 1.2-m-high and 2-m-wide glass window with a thickness of 6 mm thermal conductivity $k = 0.78$ W/m · °C, and emissivity $\varepsilon = 0.9$. The room and the walls that face the window are maintained at 22°C, and the average temperature of the inner surface of the window is measured to be 4°C. If the temperature of the outdoors is -6°C, determine (*a*) the convection heat transfer coefficient on the inner surface of the window, (*b*) the rate of total heat transfer through the window, and (*c*) the combined natural convection and radiation heat transfer coefficient on the outer surface of the window. Is it reasonable to neglect the thermal resistance of the glass in this case?

7-32   A 3-mm-diameter and 12-m-long electric wire is tightly wrapped with a 1.5-mm-thick plastic cover whose thermal conductivity and emissivity are $k = 0. 15$ W/m · °C and $\varepsilon = 0.9$. Electrical measurements indicate that a current of 10 A passes through the wire and there is a voltage drop of 8 V along the wire. If the insulated wire is exposed to calm atmospheric air at $T_\infty = 30$°C, determine the temperature at the interface of the wire and the plastic cover in steady operation. Take the surrounding surfaces to be at about the same temperature as the air.   *Answer:* 57.9°C

7-33   During a visit to a plastic sheeting plant, it was observed that a 60-m-long section of a 2-in. nominal (6.03-cm outer-diameter) steam pipe extended from one end of the plant to the other with no insulation on it. The temperature measurements at several locations revealed that the average temperature of the exposed surfaces of the steam pipe was 170°C, while the temperature of the surrounding air was 20°C. The outer surface of the pipe appeared to be oxidized, and its emissivity can be taken to be 0.7. Taking the temperature of the surrounding surfaces to be 20°C also, determine the rate of heat loss from the steam pipe.

Steam is generated in a gas furnace that has an efficiency of 78 percent, and the plant pays $0.538 per therm (1 therm = 105,500 kJ) of natural gas. The plant operates 24 h a day 365 days a year, and thus 8760 h a year. Determine the annual cost of the heat losses from the steam pipe for this facility.

7-34 Reconsider Problem 7-33. In order to reduce heat losses, it is proposed to insulate the steam pipe with 5-cm-thick fiberglass insulation ($k = 0.038$ W/m · °C) and to wrap it with aluminum foil ($\varepsilon = 0.1$) in order to minimize the radiation losses. Also, an estimate is obtained from a local insulation contractor, who proposed to do the insulation job for $750, including materials and labor. Would you support this proposal? How long will it take for the insulation to pay for itself from the energy it saves? Assume the temperature of the steam pipe to remain constant at 170°C.

7-35 A 30-cm × 30-cm circuit board that contains 121 square chips on one side is to be cooled by combined natural convection and radiation by mounting it on a vertical surface in a room at 25°C. Each chip dissipates 0.05 W of power, and the emissivity of the chip surfaces is 0.7. Assuming the heat transfer from the back side of the circuit board to be negligible, and the temperature of the surrounding surfaces to be the same as the air temperature of the room, determine the surface temperature of the chips.
Answer: 33°C

7-36 Repeat Problem 7-35 assuming the circuit board to be positioned horizontally with (*a*) chips facing up and (*b*) chips facing down.

7-37 The side surfaces of a 2-m-high cubic industrial furnace burning natural gas are not insulated, and the temperature at the outer surface of this section is measured to be 110°C. The temperature of the furnace room, including its surfaces, is 30°C, and the emissivity of the outer surface of the furnace is 0.7. It is proposed that this section of the furnace wall be insulated with glass wool insulation ($k = 0.038$ W/m · °C) wrapped by a reflective sheet ($\varepsilon = 0.2$) in order to reduce the heat loss by 90 percent. Assuming the outer surface temperature of the metal section still remains at about 110°C, determine the thickness of the insulation that needs to be used.

The furnace operates continuously throughout the year and has an efficiency of 78 percent. The price of the natural gas is $0.55/therm (1 therm = 105,500 kJ of energy content). If the installation of the insulation will cost $550 for materials and labor, determine how long it will take for the insulation to pay for itself from the energy it saves.

7-38 A 1.5-m-diameter, 5-m-long cylindrical propane tank is initially filled with liquid propane, whose density is 581 kg/m³. The tank is exposed to the ambient air at 25°C in calm weather. The outer surface of the tank is polished so that the radiation heat transfer is negligible. Now a crack develops at the top of the tank, and the pressure inside drops to 1 atm while the temperature drops to −42°C, which is the boiling temperature of propane at 1 atm. The heat of vaporization of propane at 1 atm is 425 kJ/kg. The propane is slowly vaporized as a result of the heat transfer from the ambient air into the tank,

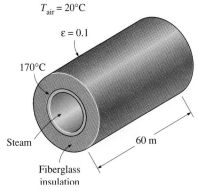

$T_{air} = 20$°C

$\varepsilon = 0.1$

170°C

Steam

Fiberglass insulation

60 m

FIGURE P7-34

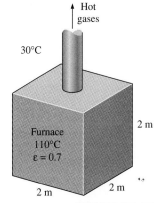

Hot gases

30°C

Furnace
110°C
$\varepsilon = 0.7$

2 m

2 m

2 m

FIGURE P7-37

FIGURE P7-38

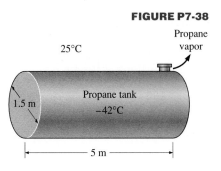

25°C

Propane vapor

1.5 m

Propane tank
−42°C

5 m

and the propane vapor escapes the tank at −42°C through the crack. Assuming the propane tank to be at about the same temperature as the propane inside at all times, determine how long it will take for the tank to empty if it is not insulated.

7-39E An average person generates heat at a rate of 287 Btu/h while resting in a room at 77°F. Assuming one-quarter of this heat is lost from the head and taking the emissivity of the skin to be 0.9, determine the average surface temperature of the head when it is not covered. The head can be approximated as a 12-in.-diameter sphere, and the interior surfaces of the room can be assumed to be at the room temperature.

7-40 An incandescent light bulb is an inexpensive but highly inefficient device that converts electrical energy into light. It converts about 10 percent of the electrical energy it consumes into light while converting the remaining 90 percent into heat. The glass bulb of the lamp heats up very quickly as a result of absorbing all that heat and dissipating it to the surroundings by convection and radiation. Consider an 8-cm-diameter 60-W light bulb in a room at 25°C. The emissivity of the glass is 0.9. Assuming that 10 percent of the energy passes through the glass bulb as light with negligible absorption and the rest of the energy is absorbed and dissipated by the bulb itself by natural convection and radiation, determine the equilibrium temperature of the glass bulb. Assume the interior surfaces of the room to be at room temperature.   *Answer:* 175°C

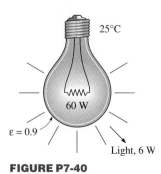

25°C

60 W

ε = 0.9

Light, 6 W

FIGURE P7-40

Natural Convection inside Enclosures

7-41C The upper and lower compartments of a well-insulated container are separated by two parallel sheets of glass with an air space between them. One of the compartments is to be filled with a hot fluid and the other with a cold fluid. If it is desired that heat transfer between the two compartments be minimal, would you recommend putting the hot fluid into the upper or the lower compartment of the container? Why?

7-42C Someone claims that the air space in a double-pane window enhances the heat transfer from a house because of the natural convection currents that occur in the air space and recommends that the double-pane window be replaced by a single sheet of glass whose thickness is equal to the sum of the thicknesses of the two glasses of the double-pane window to save energy. Do you agree with this claim?

7-43C Consider a double-pane window consisting of two glass sheets separated by a 1-cm-wide air space. Someone suggests inserting a thin vinyl sheet in the middle of the two glasses to form two 0.5-cm-wide compartments in the window in order to reduce natural convection heat transfer through the window. From a heat transfer point of view, would you be in favor of this idea to reduce heat losses through the window?

7-44C What does the effective conductivity of an enclosure represent? How is the ratio of the effective conductivity to thermal conductivity related to the Nusselt number?

7-45 Show that the thermal resistance of a rectangular enclosure can be expressed as $R = \delta/(Ak\ \text{Nu})$, where k is the thermal conductivity of the fluid in the enclosure.

7-46E A vertical 4-ft-high and 6-ft-wide double-pane window consists of two sheets of glass separated by a 1-in. air gap at atmospheric pressure. If the glass surface temperatures across the air gap are measured to be 65°F and 40°F, determine the rate of heat transfer through the window by (*a*) natural convection and (*b*) radiation. Also, determine the *R*-value of insulation of this window such that multiplying the inverse of the *R*-value by the surface area and the temperature difference gives the total rate of heat transfer through the window. The effective emissivity for use in radiation calculations between two large parallel glass plates can be taken to be 0.82.

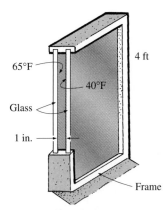

FIGURE P7-46E

7-47 Two concentric spheres of diameters 15 cm and 25 cm are separated by air at 1 atm pressure. The surface temperatures of the two spheres enclosing the air are $T_1 = 350$ K and $T_2 = 275$ K, respectively. Determine the rate of heat transfer from the inner sphere to the outer sphere by natural convection.

7-48 Flat-plate solar collectors are often tilted up toward the sun in order to intercept a greater amount of direct solar radiation. The tilt angle from the horizontal also affects the rate of heat loss from the collector. Consider a 2-m-high and 3-m-wide solar collector that is tilted at an angle θ from the horizontal. The back side of the absorber is heavily insulated. The absorber plate and the glass cover, which are spaced 2.5 cm from each other, are maintained at temperatures of 80°C and 32°C, respectively. Determine the rate of heat loss from the absorber plate by natural convection for $\theta = 0°$, 20°, and 90°.

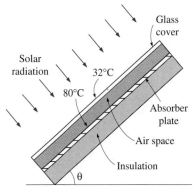

FIGURE P7-48

7-49 A simple solar collector is built by placing a 5-cm-diameter clear plastic tube around a garden hose whose outer diameter is 1.6 cm. The hose is painted black to maximize solar absorption, and some plastic rings are used to keep the spacing between the hose and the clear plastic cover constant. During a clear day, the temperature of the hose is measured to be 65°C, while the ambient air temperature is 26°C. Determine the rate of heat loss from the water in the hose per meter of its length by natural convection. Also, discuss how the performance of this solar collector can be improved.
  Answer: 6.5 W

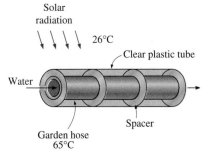

FIGURE P7-49

Natural Convection from Finned Surfaces

7-50C Why are finned surfaces frequently used in practice? Why are the finned surfaces referred to as heat sinks in the electronics industry?

7-51C Why are heat sinks with closely packed fins not suitable for natural convection heat transfer, although they increase the heat transfer surface area more?

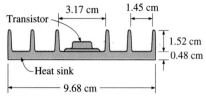

FIGURE P7-53

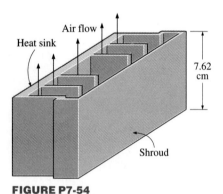

FIGURE P7-54

7-52C Consider a heat sink with optimum fin spacing. Explain how heat transfer from this heat sink will be affected by (a) removing some of the fins on the heat sink and (b) doubling the number of fins on the heat sink by reducing the fin spacing. The base area of the heat sink remains unchanged at all times.

7-53 Aluminum heat sinks of rectangular profile are commonly used to cool electronic components. Consider a 7.62-cm-long and 9.68-cm-wide commercially available heat sink whose cross-section and dimensions are as shown in Figure P7-53. The heat sink is oriented vertically and is used to cool a power transistor that can dissipate up to 125 W of power. The back surface of the heat sink is insulated. The surfaces of the heat sink are untreated, and thus they have a low emissivity (under 0.1). Therefore, radiation heat transfer from the heat sink can be neglected. During an experiment conducted in room air at 22°C, the base temperature of the heat sink was measured to be 120°C when the power dissipation of the transistor was 15 W. Assuming the entire heat sink to be at the base temperature, determine the average natural convection heat transfer coefficient for this case. *Answer:* 7.1 W/m² · °C

7-54 Reconsider the heat sink in Problem 7-53. In order to enhance heat transfer, a shroud (a thin rectangular metal plate) whose surface area is equal to the base area of the heat sink is placed very close to the tips of the fins such that the interfin spaces are converted into rectangular channels. The base temperature of the heat sink in this case was measured to be 108°C. Noting that the shroud loses heat to the ambient air from both sides, determine the average natural convection heat transfer coefficient in this shrouded case. (For complete details, see Çengel and Zing, Ref. 4).

7-55E A 6-in.-wide and 8-in.-high vertical hot surface in 78°F air is to be cooled by a heat sink with equally spaced fins of rectangular profile. The fins are 0.08 in. thick and 8 in. long in the vertical direction and have a height of 1.2 in. from the base. Determine the optimum fin spacing and the rate of heat transfer by natural convection from the heat sink if the base temperature is 180°F.

7-56 A 12.1-cm-wide and 18-cm-high vertical hot surface in 25°C air is to be cooled by an aluminum alloy heat sink ($k = 177$ W/m.°C) with 25 equally spaced fins of rectangular profile. The fins are 0.1 cm thick and 18 cm long in the vertical direction. Determine the optimum fin height and the rate of heat transfer by natural convection from the heat sink if the base temperature is 70°C.

Combined Natural and Forced Convection

7-57C When is natural convection negligible and when is it not negligible in forced convection heat transfer?

7-58C Under what conditions does natural convection enhance forced convection, and under what conditions does it hurt forced convection?

7-59C When neither natural nor forced convection is negligible, is it correct to calculate each independently and add them to determine the total convection heat transfer?

7-60 Consider a 5-m-long vertical plate at 85°C in air at 30°C. Determine the forced motion velocity above which natural convection heat transfer from this plate is negligible. *Answer:* 9.05 m/s

7-61 Consider a 3-m-long vertical plate at 60°C in water at 25°C. Determine the forced motion velocity above which natural convection heat transfer from this plate is negligible. Take $\beta = 0.0004$ K^{-1} for water.

7-62 In a production facility, thin square plates 2 m × 2 m in size coming out of the oven at 300°C are cooled by blowing ambient air at 30°C horizontally parallel to their surfaces. Determine the air velocity above which the natural convection effects on heat transfer are less than 10 percent and thus are negligible.

7-63 A 12-cm-high and 20-cm-wide circuit board houses 100 closely spaced logic chips on its surface, each dissipating 0.05 W. The board is cooled by a fan that blows air over the hot surface of the board at 35°C at a velocity of 0.5 m/s. The heat transfer from the back surface of the board is negligible. Determine the average temperature on the surface of the circuit board assuming the air flows vertically upwards along the 12-cm-long side by (*a*) ignoring natural convection and (*b*) considering the contribution of natural convection. Disregard any heat transfer by radiation.

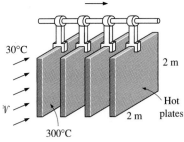

FIGURE P7-62

Review Problems

7-64E A 0.1-W small cylindrical resistor mounted on a lower part of a vertical circuit board is 0.3 in. long and has a diameter of 0.2 in. The view of the resistor is largely blocked by another circuit board facing it, and the heat transfer through the connecting wires is negligible. The air is free to flow through the large parallel flow passages between the boards as a result of natural convection currents. If the air temperature at the vicinity of the resistor is 120°F, determine the approximate surface temperature of the resistor.
Answer: 212°F

FIGURE P7-64E

7-65 An ice chest whose outer dimensions are 30 cm × 40 cm × 40 cm is made of 3-cm-thick styrofoam ($k = 0.033$ W/m · °C). Initially, the chest is filled with 40 kg of ice at 0°C, and the inner surface temperature of the ice chest can be taken to be 0°C at all times. The heat of fusion of water at 0°C is 333.7 kJ/kg. and the surrounding ambient air is at 20°C. Disregarding any heat transfer from the 40 cm × 40 cm base of the ice chest, determine how long it will take for the ice in the chest to melt completely if the ice chest is subjected to (*a*) calm air and (*b*) winds at 50 km/h. Assume the heat transfer coefficient on the front, back, and top surfaces to be the same as that on the side surfaces.

7-66 An electronic box that consumes 180 W of power is cooled by a fan blowing air into the box enclosure. The dimensions of the electronic box are 15 cm × 50 cm × 50 cm, and all surfaces of the box are exposed to the ambient except the base surface. Temperature measurements indicate that the box is at an average temperature of 32°C when the ambient temperature and the temperature of the surrounding walls are 25°C. If the emissivity of the

FIGURE P7-66

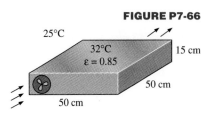

outer surface of the box is 0.85, determine the fraction of the heat lost from the outer surfaces of the electronic box.

7-67   A 6-m internal-diameter spherical tank made of 1.5-cm-thick stainless steel ($k = 15$ W/m · °C) is used to store iced water at 0°C. The walls of the room are also at 20°C. The outer surface of the tank is black (emissivity $\varepsilon = 1$), and heat transfer between the outer surface of the tank and the surroundings is by natural convection and radiation. Assuming the entire steel tank to be at 0°C and thus the thermal resistance of the tank to be negligible, determine (*a*) the rate of heat transfer to the iced water in the tank and (*b*) the amount of ice at 0°C that melts during a 24-h period.
Answers: (*a*) 15,044 W, (*b*) 3895 kg

7-68   Consider a 1.2-m-high and 2-m-wide double-pane window consisting of two 3-mm-thick layers of glass ($k = 0.78$ W/m · °C) separated by an 11-mm-wide air space. Determine the steady rate of heat transfer through this window and the temperature of its inner surface for a day during which the room is maintained at 24°C while the temperature of the outdoors is −5°C. Take the heat transfer coefficients on the inner and outer surfaces of the window to be $h_1 = 10$ W/m² · °C and $h_2 = 25$ W/m² · °C and disregard any heat transfer by radiation.

7-69   An electric resistance space heater is designed such that it resembles a rectangular box 50 cm high, 80 cm long, and 15 cm wide filled with 45 kg of oil. The heater is to be placed against a wall, and thus heat transfer from its back surface is negligible for safety considerations. The surface temperature of the heater is not to exceed 45°C in a room at 25°C. Disregarding heat transfer from the bottom and top surfaces of the heater in anticipation that the top surface will be used as a shelf, determine the power rating of the heater in W. Take the emissivity of the outer surface of the heater to be 0.8 and the average temperature of the ceiling and wall surfaces to be the same as the room air temperature.

   Also, determine how long it will take for the heater to reach steady operation when it is first turned on (i.e., for the oil temperature to rise from 25°C to 45°C). State your assumptions in the calculations.

7-70   Skylights or "roof windows" are commonly used in homes and manufacturing facilities since they let natural light in during day time and thus reduce the lighting costs. However, they offer little resistance to heat transfer, and large amounts of energy are lost through them in winter unless they are equipped with a motorized insulating cover that can be used in cold weather and at nights to reduce heat losses. Consider a 1-m-wide and 2.5-m-long horizontal skylight on the roof of a house that is kept at 20°C. The glazing of the skylight is made of a single layer of 0.5-cm-thick glass ($k = 0.78$ W/m · °C and $\varepsilon = 0.9$). Determine the rate of heat loss through the skylight when the air temperature outside is −8°C and the effective sky temperature is −30°C. Compare your result with the rate of heat loss through an equivalent surface area of the roof that has a common R-5.34 construction in SI units (i.e., a thickness-to-effective-thermal-conductivity ratio of 5.34 m² · °C/W).

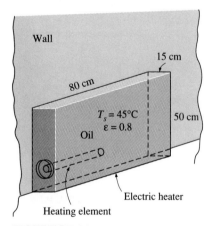

FIGURE P7-69

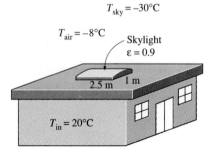

FIGURE P7-70

7-71 A solar collector consists of a horizontal copper tube of outer diameter 5 cm enclosed in a concentric thin glass tube of 9 cm diameter. Water is heated as it flows through the tube, and the annular space between the copper and glass tube is filled with air at 1 atm pressure. During a clear day, the temperatures of the tube surface and the glass cover are measured to be 60°C and 32°C, respectively. Determine the rate of heat loss from the collector by natural convection per meter length of the tube.   *Answer:* 19.6 W

7-72 A solar collector consists of a horizontal aluminum tube of outer diameter 4 cm enclosed in a concentric thin glass tube of 7 cm diameter. Water is heated as it flows through the aluminum tube, and the annular space between the aluminum and glass tubes is filled with air at 1 atm pressure. The pump circulating the water fails during a clear day, and the water temperature in the tube starts rising. The aluminum tube absorbs solar radiation at a rate of 20 W per meter length, and the temperature of the ambient air outside is 30°C. Approximating the surfaces of the tube and the glass cover as being black (emissivity $\varepsilon = 1$) in radiation calculations and taking the effective sky temperature to be 10°C, determine the temperature of the aluminum tube when equilibrium is established (i.e., when the net heat loss from the tube by convection and radiation equals the amount of solar energy absorbed by the tube).

7-73E The components of an electronic system dissipating 180 W are located in a 4-ft-long horizontal duct whose cross-section is 6 in. × 6 in. The components in the duct are cooled by forced air, which enters at 85°F at a rate of 22 cfm and leaves at 100°F. The surfaces of the sheet metal duct are not painted, and thus radiation heat transfer from the outer surfaces is negligible. If the ambient air temperature is 80°F, determine (*a*) the heat transfer from the outer surfaces of the duct to the ambient air by natural convection and (*b*) the average temperature of the duct.

7-74E Repeat Problem 7-73E for a circular horizontal duct of diameter 4 in.

7-75E Repeat Problem 7-73E assuming the fan fails and thus the entire heat generated inside the duct must be rejected to the ambient air by natural convection through the outer surfaces of the duct.

7-76 Consider a cold aluminum canned drink that is initially at a uniform temperature of 5°C. The can is 12.5 cm high and has a diameter of 6 cm. The emissivity of the outer surface of the can is 0.6. Disregarding any heat transfer from the bottom surface of the can, determine how long it will take for the average temperature of the drink to rise to 7°C if the surrounding air and surfaces are at 25°C.   *Answer:* 11.4 min

7-77 Consider a 2-m-high electric hot water heater that has a diameter of 40 cm and maintains the hot water at 55°C. The tank is located in a small room at 25°C whose walls and the ceiling are at about the same temperature. The tank is placed in a 46-cm-diameter sheet metal shell of negligible thickness, and the space between the tank and the shell is filled with foam insulation. The average temperature and emissivity of the outer surface of the shell are 40°C and 0.7, respectively. The price of electricity is $0.08/kWh. Hot

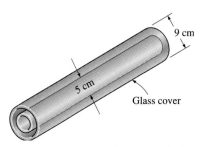

FIGURE P7-71

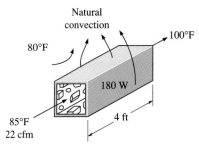

FIGURE P7-73E

FIGURE P7-77

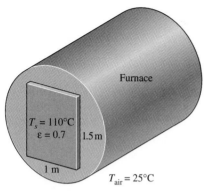

FIGURE P7-78

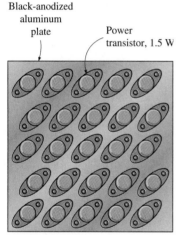

FIGURE P7-79

water tank insulation kits large enough to wrap the entire tank are available on the market for about $30. If such an insulation is installed on this water tank by the homeowner himself, how long will it take for this additional insulation to pay for itself? Disregard any heat loss from the top and bottom surfaces, and assume the insulation to reduce the heat losses by 80 percent.

7-78 During a plant visit, it was observed that a 1.5-m-high and 1-m-wide section of the vertical front section of a natural gas furnace wall was too hot to touch. The temperature measurements on the surface revealed that the average temperature of the exposed hot surface was 110°C, while the temperature of the surrounding air was 25°C. The surface appeared to be oxidized, and its emissivity can be taken to be 0.7. Taking the temperature of the surrounding surfaces to be 25°C also, determine the rate of heat loss from this furnace.

The furnace has an efficiency of 79 percent, and the plant pays $0.58 per therm (1 therm = 105,500 kJ) of natural gas. If the plant operates 10 h a day, 260 days a year, and thus 2600 h a year, determine the annual cost of the heat loss from this vertical hot surface on the front section of the furnace wall.

7-79 A group of 25 power transistors, dissipating 1.5 W each, are to be cooled by attaching them to a black-anodized square aluminum plate and mounting the plate on the wall of a room at 30°C. The emissivity of the transistor and the plate surfaces is 0.9. Assuming the heat transfer from the back side of the plate to be negligible and the temperature of the surrounding surfaces to be the same as the air temperature of the room, determine the size of the plate if the average surface temperature of the plate is not to exceed 50°C.

7-80 Repeat Problem 7-79 assuming the plate to be positioned horizontally with (a) transistors facing up and (b) transistors facing down.

7-81E Hot water is flowing at an average velocity of 4 ft/s through a cast iron pipe ($k = 30$ Btu/h · ft · °F) whose inner and outer diameters are 1.0 in. and 1.2 in., respectively. The pipe passes through a 50-ft-long section of a basement whose temperature is 60°F. The emissivity of the outer surface of the pipe is 0.5, and the walls of the basement are also at about 60°F. If the inlet temperature of the water is 150°F and the heat transfer coefficient on the inner surface of the pipe is 30 Btu/h · ft² · °F, determine the temperature drop of water as it passes through the basement.

7-82 Consider a flat-plate solar collector placed horizontally on the flat roof of a house. The collector is 1.5 m wide and 6 m long, and the average temperature of the exposed surface of the collector is 42°C. Determine the rate of heat loss from the collector by natural convection during a calm day when the ambient air temperature is 15°C. Also, determine the heat loss by radiation by taking the emissivity of the collector surface to be 0.9 and the effective sky temperature to be −30°C.   *Answers:* 1295 W, 2921 W

7-83 Solar radiation is incident on the glass cover of a solar collector at a rate of 650 W/m². The glass transmits 88 percent of the incident radiation and has an emissivity of 0.90. The hot water needs of a family in summer can be met completely by a collector 1.5 m high and 2 m wide, and tilted 40°C from the horizontal. The temperature of the glass cover is measured to be 35°C on a calm day when the surrounding air temperature is 23°C. The effective sky temperature for radiation exchange between the glass cover and the open sky is −40°C. Water enters the tubes attached to the absorber plate at a rate of 1 kg/min. Assuming the back surface of the absorber plate to be heavily insulated and the only heat loss occurs through the glass cover, determine (*a*) the total rate of heat loss from the collector, (*b*) the collector efficiency, which is the ratio of the amount of heat transferred to the water to the solar energy incident on the collector, and (*c*) the temperature rise of water as it flows through the collector.

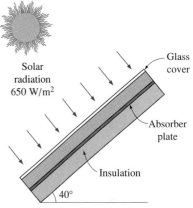

FIGURE P7-83

Computer, Design, and Essay Problems

7-84 Write a computer program to evaluate the variation of temperature with time of thin square metal plates that are removed from an oven at a specified temperature and placed vertically in a large room. The thickness, the size, the initial temperature, the emissivity, and the thermophysical properties of the plate as well as the room temperature are to be specified by the user. The program should evaluate the temperature of the plate at specified intervals and tabulate the results against time. The computer should list the assumptions made during calculations before printing the results.

For each step or time interval, assume the surface temperature to be constant and evaluate the heat loss during that time interval and the temperature drop of the plate as a result of this heat loss. This gives the temperature of the plate at the end of a time interval, which is to serve as the initial temperature of the plate for the beginning of the next time interval.

Try your program for 0.2-cm-thick vertical copper plates of 40 cm × 40 cm in size initially at 300°C cooled in a room at 25°C. Take the surface emissivity to be 0.9. Use a time interval of 1 s in calculations, but print the results at 10-s intervals for a total cooling period of 15 min.

7-85 Repeat Problem 7-84 for a vertical slender cylindrical metal object instead of a square plate. The height and the diameter of the cylinder are to be specified by the user.

7-86 Write a computer program to optimize the spacing between the two glasses of a double-pane window. Assume the spacing is filled with dry air at atmospheric pressure. The program should evaluate the recommended practical value of the spacing to minimize the heat losses and list it when the size of the window (the height and the width) and the temperatures of the two glasses are specified.

7-87 Contact a manufacturer of aluminum heat sinks and obtain their product catalog for cooling electronic components by natural convection and radiation. Write an essay on how to select a suitable heat sink for an electronic component when its maximum power dissipation and maximum allowable surface temperature are specified.

7-88 The top surfaces of practically all flat-plate solar collectors are covered with glass in order to reduce the heat losses from the absorber plate underneath. Although the glass cover reflects or absorbs about 15 percent of the incident solar radiation, it saves much more from the potential heat losses from the absorber plate, and thus it is considered to be an essential part of a well-designed solar collector. Inspired by the energy efficiency of double-pane windows, someone proposes to use double glazing on solar collectors instead of a single glass. Investigate if this is a good idea for the town in which you live. Use local weather data and base your conclusion on heat transfer analysis and economic considerations.

Boiling and Condensation

CHAPTER

8

We know from thermodynamics that when the temperature of a liquid at a specified pressure is raised to the saturation temperature T_{sat} at that pressure, *boiling* occurs. Likewise, when the temperature of a vapor is lowered to T_{sat}, *condensation* occurs. In this chapter we study the rates of heat transfer during such liquid-to-vapor and vapor-to-liquid phase transformations.

Although boiling and condensation exhibit some unique features, they are considered to be forms of *convection* heat transfer since they involve fluid motion (such as the rise of the bubbles to the top and the flow of condensate to the bottom). Boiling and condensation differ from other forms of convection in that they depend on the *latent heat of vaporization h_{fg}* of the fluid and the *surface tension σ* at the liquid–vapor interface, in addition to the properties of the fluid in each phase. Noting that under equilibrium conditions the temperature remains constant during a phase-change process at a fixed pressure, large amounts of heat (due to the large latent heat of vaporization released or absorbed) can be transferred during boiling and condensation essentially at constant temperature. In practice, however, it is necessary to maintain some difference between the surface temperature T_s and T_{sat} for effective heat transfer. Heat transfer coefficients h associated with boiling and condensation are typically several times higher than those encountered in other forms of convection processes that involve a single phase.

We start this chapter with a discussion of the *boiling curve* and the modes of pool boiling such as *free convection boiling, nucleate boiling,* and *film boiling.* We then discuss boiling in the presence of forced convection. In the second part of this chapter, we describe the physical mechanism of *film condensation* and discuss condensation heat transfer in several geometrical arrangements and orientations. Finally, we introduce *dropwise condensation* and discuss ways of maintaining it.

FIGURE 8-1

A liquid-to-vapor phase change process is called *evaporation* if it occurs at a liquid–vapor interface and *boiling* if it occurs at a solid–liquid interface.

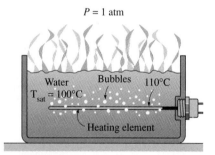

FIGURE 8-2

Boiling occurs when a liquid is brought into contact with a surface at a temperature above the saturation temperature of the liquid.

8-1 ■ BOILING HEAT TRANSFER

Many familiar engineering applications involve condensation and boiling heat transfer. In a household refrigerator, for example, the refrigerant absorbs heat from the refrigerated space by boiling in the *evaporator* section and rejects heat to the kitchen air by condensing in the *condenser* section (the long coils behind the refrigerator). Also, in steam power plants, heat is transferred to the steam in the *boiler* where water is vaporized, and the waste heat is rejected from the steam in the *condenser* where the steam is condensed. Some electronic components are cooled by boiling by immersing them in a fluid with an appropriate boiling temperature.

Boiling is a liquid-to-vapor phase change process just like evaporation, but there are significant differences between the two. **Evaporation** occurs at the *liquid–vapor interface* when the vapor pressure is less than the saturation pressure of the liquid at a given temperature. Water in a lake at 20°C, for example, will evaporate to air at 20°C and 60 percent relative humidity since the saturation pressure of water at 20°C is 2.3 kPa and the vapor pressure of air at 20°C and 60 percent relative humidity is 1.4 kPa (evaporation rates are determined in Chap. 11). Other examples of evaporation are the drying of clothes, fruits, and vegetables; the evaporation of sweat to cool the human body; and the rejection of waste heat in wet cooling towers. Note that evaporation involves no bubble formation or bubble motion (Fig. 8-1).

Boiling, on the other hand, occurs at the *solid–liquid interface* when a liquid is brought into contact with a surface maintained at a temperature T_s sufficiently above the saturation temperature T_{sat} of the liquid (Fig. 8-2). At 1 atm, for example, liquid water in contact with a solid surface at 110°C will boil since the saturation temperature of water at 1 atm is 100°C. The boiling process is characterized by the rapid formation of *vapor bubbles* at the solid–liquid interface that detach from the surface when they reach a certain size and attempt to rise to the free surface of the liquid. When cooking, we do not say water is boiling until we see the bubbles rising to the top. Boiling is a complicated phenomenon because of the large number of variables involved in the process and the complex fluid motion patterns caused by the bubble formation and growth.

As a form of convection heat transfer, the *boiling heat flux* (heat transfer per unit time per unit surface area) from a solid surface to the fluid is expressed from Newton's law of cooling as

$$\dot{q}_{\text{boiling}} = h(T_s - T_{\text{sat}}) = h\Delta T_{\text{excess}} \qquad (\text{W/m}^2) \qquad (8\text{-}1)$$

where $\Delta T_{\text{excess}} = T_s - T_{\text{sat}}$ is called the *excess temperature,* which represents the excess of the surface above the saturation temperature of the fluid.

In the preceding chapters we considered forced and free convection heat transfer involving a single phase of a fluid only. The analysis of such convection processes involves the thermophysical properties ρ, μ, k, and C_p of the fluid. The analysis of boiling heat transfer involves these properties of the liquid (indicated by the subscript l) or vapor (indicated by the subscript v) as well as the properties h_{fg} (the latent heat of vaporization) and σ (the surface tension). The h_{fg} represents the energy absorbed as a unit mass of liquid vaporizes at a specified temperature or pressure and is the primary quantity of energy transferred during boiling heat transfer. The h_{fg} values of water at various temperatures are given in Table A-9.

Bubbles owe their existence to the *surface-tension* σ at the liquid–vapor interface due to the attraction force on molecules at the interface toward the liquid phase. The surface tension decreases with increasing temperature and becomes zero at the critical temperature. This explains why no bubbles are formed during boiling at supercritical pressures and temperatures. Surface tension has the unit N/m.

The boiling processes in practice do not occur under *equilibrium* conditions, and normally the bubbles are not in thermodynamic equilibrium with the surrounding liquid. That is, the temperature and pressure of the vapor in a bubble are usually different than those of the liquid. The pressure difference between the liquid and the vapor is balanced by the surface tension at the interface. The temperature difference between the vapor in a bubble and the surrounding liquid is the driving force for heat transfer between the two phases. When the liquid is at a *lower temperature* than the bubble, heat will be transferred from the bubble into the liquid, causing some of the vapor inside the bubble to condense and the bubble eventually to collapse. When the liquid is at a *higher temperature* than the bubble, heat will be transferred from the liquid to the bubble, causing the bubble to grow and rise to the top under the influence of buoyancy.

Boiling is classified as *pool boiling* or *flow boiling,* depending on the presence of bulk fluid motion (Fig. 8-3). Boiling is called **pool boiling** in the absence of bulk fluid flow and **flow boiling** (or *forced convection boiling*) in the presence of it. In pool boiling, the fluid is stationary, and any motion of the fluid is due to natural convection currents and the motion of the bubbles under the influence of buoyancy. The boiling of water in a pan on top of a stove is an example of pool boiling. Pool boiling of a fluid can also be achieved by placing a heating coil in the fluid. In flow boiling, the fluid is forced to move in a heated pipe or surface by external means such as a pump. Therefore, flow boiling is always accompanied by other convection effects.

Pool and flow boiling are further classified as *subcooled boiling* or *saturated boiling,* depending on the bulk liquid temperature (Fig. 8-4). Boiling is said to be **subcooled** (or *local*) when the temperature of the main body of the liquid is below the saturation temperature T_{sat} (i.e., the bulk of the liquid is subcooled) and **saturated** (or *bulk*) when the temperature of the liquid is equal to T_{sat} (i.e., the bulk of the liquid is saturated). At the early stages of boiling, the bubbles are confined to a narrow region near the hot surface. This is because the liquid adjacent to the hot surface vaporizes as a result of being heated above its saturation temperature. But these bubbles disappear soon after they move away from the hot surface as a result of heat transfer from the bubbles to the cooler liquid surrounding them. This happens when the bulk of the liquid is at a lower temperature than the saturation temperature. The bubbles serve as "energy movers" from the hot surface into the liquid body by absorbing heat from the hot surface and releasing it into the liquid as they condense and collapse. Boiling in this case is confined to a region in

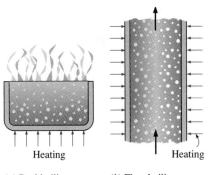

Heating   Heating

(*a*) Pool boiling   (*b*) Flow boiling

FIGURE 8-3

Classification of boiling on the basis of the presence of bulk fluid motion.

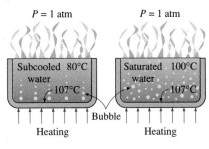

(*a*) Subcooled boiling   (*b*) Saturated boiling

FIGURE 8-4

Classification of boiling on the basis of the presence of bulk liquid temperature.

the locality of the hot surface and is appropriately called *local* or *subcooled* boiling. When the entire liquid body reaches the saturation temperature, the bubbles start rising to the top. We can see bubbles throughout the bulk of the liquid, and boiling in this case is called the *bulk* or *saturated* boiling. Next, we consider different boiling regimes in detail.

8-2 ■ POOL BOILING

Above we presented some general discussions on boiling. Now we turn our attention to the physical mechanisms involved in *pool boiling,* that is, the boiling of stationary fluids. In pool boiling, the fluid is not forced to flow by a mover such as a pump, and any motion of the fluid is due to natural convection currents and the motion of the bubbles under the influence of buoyancy.

As a familiar example of pool boiling, consider the boiling of tap water in a pan on top of a stove. The water will initially be at about 15°C, far below the saturation temperature of 100°C at standard atmospheric pressure. At the early stages of boiling, you will not notice anything significant except some bubbles that stick to the surface of the pan. These bubbles are caused by the release of air molecules dissolved in liquid water and should not be confused with vapor bubbles. As the water temperature rises, you will notice chunks of liquid water rolling up and down as a result of natural convection currents, followed by the first vapor bubbles forming at the bottom surface of the pan. These bubbles get smaller as they detach from the surface and start rising, and eventually collapse in the cooler water above. This is *subcooled boiling* since the bulk of the liquid water has not reached saturation temperature yet. The intensity of bubble formation increases as the water temperature rises further, and you will notice waves of vapor bubbles coming from the bottom and rising to the top when the water temperature reaches the saturation temperature (100°C at standard atmospheric conditions). This full scale boiling is the *saturated boiling.*

Boiling Regimes and the Boiling Curve

Boiling is probably the most familiar form of heat transfer, yet it remains to be the least understood form. After hundreds of papers written on the subject, we still do not fully understand the process of bubble formation and we must still rely on empirical or semi-empirical relations to predict the rate of boiling heat transfer.

The pioneering work on boiling was done in 1934 by S. Nukiyama, who used electrically heated nichrome and platinum wires immersed in liquids in his experiments. Nukiyama noticed that boiling takes different forms, depending on the value of the excess temperature ΔT_{excess}. Four different boiling regimes are observed: *natural convection boiling, nucleate boiling, transition boiling,* and *film boiling* (Fig. 8-5). These regimes are illustrated on the **boiling curve** in Figure 8-6, which is a plot of boiling heat flux versus the excess temperature. Although the boiling curve given in this figure is for water, the general shape of the boiling curve remains the same for different fluids. The specific shape of the curve depends on the fluid–heating surface

FIGURE 8-5

Different boiling regimes in pool boiling.

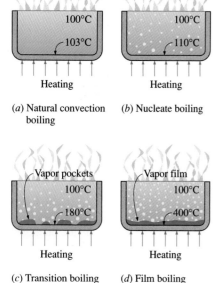

(*a*) Natural convection boiling

(*b*) Nucleate boiling

(*c*) Transition boiling

(*d*) Film boiling

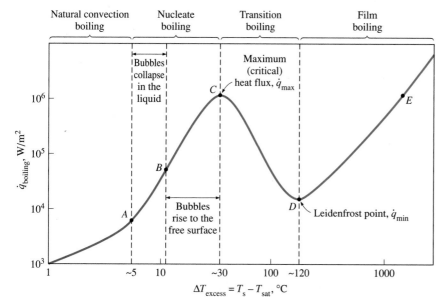

FIGURE 8-6

Typical boiling curve for water at 1 atm pressure.

material combination and the fluid pressure, but it is practically independent of the geometry of the heating surface. Below we describe each boiling regime in detail.

Natural Convection Boiling (to point *A* on the boiling curve)

We learned in thermodynamics that a pure substance at a specified pressure starts boiling when it reaches the saturation temperature at that pressure. But in practice we do not see any bubbles forming on the heating surface until the liquid is heated a few degrees above the saturation temperature (about 2 to 6°C for water). Therefore, the liquid is slightly *superheated* in this case (a *metastable* condition) and evaporates when it rises to the free surface. The fluid motion in this mode of boiling is governed by natural convection currents, and heat transfer from the heating surface to the fluid is by natural convection.

Nucleate Boiling (between points *A* and *C*)

The first bubbles start forming at point *A* of the boiling curve at various preferential sites on the heating surface. The bubbles form at an increasing rate at an increasing number of nucleation sites as we move along the boiling curve toward point *C*.

The nucleate boiling regime can be separated into two distinct regions. In region *A–B*, *isolated bubbles* are formed at various preferential nucleation sites on the heated surface. But these bubbles are dissipated in the liquid shortly after they separate from the surface. The space vacated by the rising bubbles is filled by the liquid in the vicinity of the heater surface, and the process is repeated. The stirring and agitation caused by the entrainment of the liquid to the heater surface is primarily responsible for the increased heat transfer coefficient and heat flux in this region of nucleate boiling.

In region B–C, the heater temperature is further increased, and bubbles form at such great rates at such a large number of nucleation sites that they form numerous *continuous columns of vapor* in the liquid. These bubbles move all the way up to the free surface, where they break up and release their vapor content. The large heat fluxes obtainable in this region are caused by the combined effect of liquid entrainment and evaporation.

At large values of ΔT_{excess}, the rate of evaporation at the heater surface reaches such high values that a large fraction of the heater surface is covered by bubbles, making it difficult for the liquid to reach the heater surface and wet it. Consequently, the heat flux increases at a lower rate with increasing ΔT_{excess}, and reaches a maximum at point C. The heat flux at this point is called the **critical** (or **maximum**) **heat flux,** $\dot{q}_{max}$. For water, the critical heat flux exceeds 1 MW/m^2.

Nucleate boiling is the most desirable boiling regime in practice because high heat transfer rates can be achieved in this regime with relatively small values of ΔT_{excess}, typically under 30°C for water. The photographs in Figure 8-7 show the nature of bubble formation and bubble motion associated with nucleate, transition, and film boiling.

Transition Boiling (between points C and D on the boiling curve)

As the heater temperature and thus the ΔT_{excess} is increased past point C, the heat flux decreases, as shown in Figure 8-6. This is because a larger fraction of the heater surface is covered by a vapor film, which acts as an insulation because of the low thermal conductivity of the vapor relative to that of the liquid. In the transition boiling regime, both nucleate and film boiling partially occur. Nucleate boiling at point C is completely replaced by film boiling at point D. Operation in the transition boiling regime, which is also called the *unstable film boiling regime,* is avoided in practice. For water, transition boiling occurs over the excess temperature range from about 30°C to about 120°C.

Film Boiling (beyond point D)

In this region the heater surface is completely covered by a continuous stable vapor film. Point D, where the heat flux reaches a minimum, is called the **Leidenfrost point,** in honor of J. C. Leidenfrost, who observed in 1756 that liquid droplets on a very hot surface jump around and slowly boil away. The presence of a vapor film between the heater surface and the liquid is responsible for the low heat transfer rates in the film boiling region. The heat transfer rate increases with increasing excess temperature as a result of heat transfer from the heated surface to the liquid through the vapor film by radiation, which becomes significant at high temperatures.

A typical boiling process will not follow the boiling curve beyond point C, as Nukiyama has observed during his experiments. Nukiyama noticed, with surprise, that when the power applied to the nichrome wire immersed in water exceeded $\dot{q}_{max}$ even slightly, the wire temperature increased suddenly

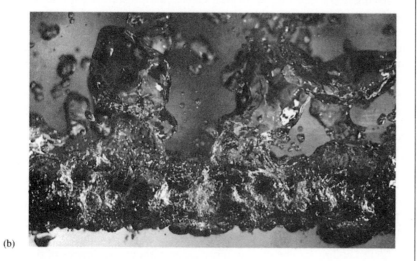

FIGURE 8-7

Various boiling regimes during boiling of methanol on a horizontal 1-cm-diameter steam-heated copper tube: (*a*) nucleate boiling, (*b*) transition boiling, and (*c*) film boiling (from J. W. Westwater and J. G. Santangelo, University of Illinois at Champaign-Urbana).

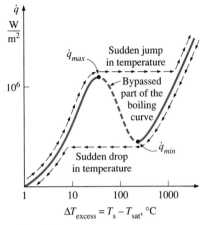

FIGURE 8-8
The actual boiling curve obtained with
heated platinum wire in water as the
heat flux is increased and then
decreased.

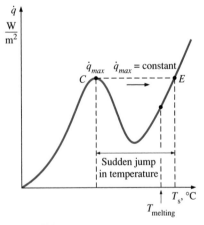

FIGURE 8-9
An attempt to increase the boiling heat
flux beyond the *critical* value often
causes the temperature of the heating
element to jump suddenly to a value that
is above the melting point, resulting in
burnout.

to the melting point of the wire and *burnout* occurred beyond his control.
When he repeated the experiments with platinum wire, which has a much
higher melting point, he was able to avoid burnout and maintain heat fluxes
higher than $\dot{q}_{max}$. When he gradually reduced power, he obtained the cooling
curve shown in Figure 8-8 with a sudden drop in excess temperature when
$\dot{q}_{min}$ is reached. Note that the boiling process cannot follow the transition
boiling part of the boiling curve past point C unless the power applied is
reduced suddenly.

The *burnout phenomenon* in boiling can be explained as follows: In order
to move beyond point C where $\dot{q}_{max}$ occurs, we must increase the heater
surface temperature T_s. To increase T_s, however, we must increase the heat
flux. But the fluid cannot receive this increased energy at an excess temperature just beyond point C. Therefore, the heater surface ends up absorbing the
increased energy, causing the heater surface temperature T_s to rise. But the
fluid can receive even less energy at this increased excess temperature, causing the heater surface temperature T_s to rise even further. This continues until
the surface temperature reaches a point at which it no longer rises and the
heat supplied can be transferred to the fluid steadily. This is point E on the
boiling curve, which corresponds to very high surface temperatures. Therefore, any attempt to increase the heat flux beyond $\dot{q}_{max}$ will cause the operation point on the boiling curve to jump suddenly from point C to point E.
However, surface temperature that corresponds to point E is beyond the melting point of most heater materials, and *burnout* occurs. Therefore, point C on
the boiling curve is also called the **burnout point,** and the heat flux at this
point the **burnout heat flux** (Fig. 8-9).

Most boiling heat transfer equipment in practice operate slightly below
$\dot{q}_{max}$ to avoid any disastrous burnout. However, in cryogenic applications
involving fluids with very low boiling points such as oxygen and nitrogen,
point E usually falls below the melting point of the heater materials, and
steady film boiling can be used in those cases without any danger of burnout.

Heat Transfer Correlations in Pool Boiling

Boiling regimes discussed above differ considerably in their character, and
thus different heat transfer relations need to be used for different boiling
regimes. In the *natural convection boiling* regime, boiling is governed by
natural convection currents, and heat transfer rates in this case can be determined accurately using natural convection relations presented in Chapter 7.

Nucleate Boiling

In the *nucleate boiling* regime, the rate of heat transfer strongly depends on
the nature of nucleation (the number of active nucleation sites on the surface,
the rate of bubble formation at each site, etc.), which is difficult to predict.
The type and the condition of the heated surface also affect the heat transfer.
These complications made it difficult to develop theoretical relations for heat
transfer in the nucleate boiling regime, and people had to rely on relations
based on experimental data. The most widely used correlation for the rate of

heat transfer in the nucleate boiling regime was proposed in 1952 by Rohsenow, and expressed as

$$\dot{q}_{nucleate} = \mu_l h_{fg} \left[\frac{g(\rho_l - \rho_v)}{\sigma} \right]^{1/2} \left[\frac{C_{pl}(T_s - T_{sat})}{C_{sf} h_{fg} Pr_l^n} \right]^3 \qquad (8\text{-}2)$$

where

$\dot{q}_{nucleate}$ = nucleate boiling heat flux, W/m²
μ_l = viscosity of the liquid, kg/(m · s)
h_{fg} = enthalpy of vaporization, J/kg
g = gravitational acceleration, m/s²
ρ_l = density of the liquid, kg/m³
ρ_v = density of the vapor, kg/m³
σ = surface tension of liquid–vapor interface, N/m
C_{pl} = specific heat of the liquid, J/(kg · °C)
T_s = surface temperature of the heater, °C
T_{sat} = saturation temperature of the fluid, °C
C_{sf} = experimental constant that depends on surface–fluid combination
Pr_l = Prandtl number of the liquid
n = experimental constant that depends on the fluid

It can be shown easily that using property values in the specified units in the Rohsenow equation produces the desired unit W/m² for the boiling heat flux, thus saving one from having to go through tedious unit manipulations (Fig. 8-10).

The surface tension at the vapor–liquid interface is given in Table 8-1 for water, and Table 8-2 for some other fluids. Experimentally determined values of the constant C_{sf} are given in Table 8-3 for various fluid–surface combinations. These values can be used for *any geometry* since it is found that the rate of heat transfer during nucleate boiling is essentially independent of the geometry and orientation of the heated surface. The fluid properties in Equation 8-2 are to be evaluated at the saturation temperature T_{sat}.

The *condition* of the heater surface greatly affects heat transfer, and the Rohsenow equation given above is applicable to *clean* and relatively *smooth* surfaces. The results obtained using the Rohsenow equation can be in error by ±100% for the heat transfer rate for a given excess temperature and by ±30% for the excess temperature for a given heat transfer rate. Therefore, care should be exercised in the interpretation of the results.

Recall from thermodynamics that the enthalpy of vaporization h_{fg} of a pure substance decreases with increasing pressure (or temperature) and reaches zero at the critical point. Noting that h_{fg} appears in the denominator of the Rohsenow equation, we should see a significant rise in the rate of heat transfer at *high pressures* during nucleate boiling.

Peak Heat Flux

In the design of boiling heat transfer equipment, it is extremely important for the designer to have a knowledge of the maximum heat flux in order to avoid the danger of burnout. The *maximum* (or *critical*) *heat flux* in nucleate pool

$$\dot{q} = \left(\frac{kg}{m \cdot s} \right) \left(\frac{J}{kg} \right)$$

$$\times \left(\frac{\dfrac{m \ kg}{s^2 \ m^3}}{\dfrac{N}{m}} \right)^{1/2} \left(\frac{\dfrac{J}{kg \cdot °C} °C}{\dfrac{J}{kg}} \right)^3$$

$$= \frac{W}{m} \left(\frac{1}{m^2} \right)^{1/2} (1)^3$$

$$= W/m^2$$

FIGURE 8-10

Equation 8-2 gives the boiling heat flux in W/m² when the quantities are expressed in the units specified in their descriptions.

TABLE 8-1

Surface tension of liquid–vapor interface for water

T, °C	σ, N/m*
0	0.0757
20	0.0727
40	0.0696
60	0.0662
80	0.0627
100	0.0589
120	0.0550
140	0.0509
160	0.0466
180	0.0422
200	0,0377
220	0.0331
240	0.0284
260	0.0237
280	0.0190
300	0.0144
320	0.0099
340	0.0056
360	0.0019
374	0.0

*Multiply by 0.06852 to convert to lbf/ft or by 2.2046 to convert to lbm/s².

TABLE 8-2

Surface tension of some fluids (from Suryanarayana, Ref. 23; originally based on data from Jasper, Ref. 12)

Substance and temp. range	Surface tension, σ, N/m* (T in °C)
Ammonia	
−75 to −40°C	$0.0264 + 0.000223T$
Benzene	
10 to 80°C	$0.0315 − 0.000129T$
Butane	
−70 to −20°C	$0.0149 − 0.000121T$
Carbon dioxide	
−30 to −20°C	$0.0043 − 0.000160T$
Ethyl alcohol	
10 to 70°C	$0.0241 − 0.000083T$
Mercury	
5 to 200°C	$0.4906 − 0.000205T$
Methyl alcohol	
10 to 60°C	$0.0240 − 0.000077T$
Pentane	
10 to 30°C	$0.0183 − 0.000110T$
Propane	
−90 to −10°C	$0.0092 − 0.000087T$

*Multiply by 0.06852 to convert to lbf/ft or by 2.2046 to convert to lbm/s².

TABLE 8-3

Values of the coefficient C_{sf} and n for various fluid–surface combinations

Fluid-heating surface combination	C_{sf}	n
Water–copper (polished)	0.0130	1.0
Water–copper (scored)	0.0068	1.0
Water–stainless steel (mechanically polished)	0.0130	1.0
Water–stainless steel (ground and polished)	0.0060	1.0
Water–stainless steel (teflon pitted)	0.0058	1.0
Water–stainless steel (chemically etched)	0.0130	1.0
Water–brass	0.0060	1.0
Water–nickel	0.0060	1.0
Water–platinum	0.0130	1.0
n-Pentane–copper (polished)	0.0154	1.7
n-Pentane–chromium	0.0150	1.7
Benzene–chromium	0.1010	1.7
Ethyl alcohol–chromium	0.0027	1.7
Carbon tetrachloride–copper	0.0130	1.7
Isopropanol–copper	0.0025	1.7

boiling was determined theoretically by S. S. Kutateladze in Russia in 1948 and N. Zuber in the United States in 1958 using quite different approaches, and is expressed as (Fig. 8-11)

$$\dot{q}_{max} = C_{cr} h_{fg} [\sigma g \rho_v^2 (\rho_l − \rho_v)]^{1/4} \tag{8-3}$$

where C_{cr} is a constant whose value depends on the heater geometry. Exhaustive experimental studies by Lienhard and his coworkers indicated that the value of C_{cr} is about 0.15. Specific values of C_{cr} for different heater geometries are listed in Table 8-4. Note that the heaters are classified as being large or small based on the value of the parameter L^*.

Equation 8-3 will give the maximum heat flux in W/m² if the properties are used *in the units specified* earlier in their descriptions following Equation 8-2. The maximum heat flux is independent of the fluid–heating surface combination, as well as the viscosity, thermal conductivity, and the specific heat of the liquid.

Note that ρ_v increases but σ and h_{fg} decrease with increasing pressure, and thus the change in $\dot{q}_{max}$ with pressure depends on which effect dominates. The experimental studies of Cichelli and Bonilla indicate that $\dot{q}_{max}$ increases with pressure up to about one-third of the critical pressure, and then starts to decrease and becomes zero at the critical pressure. Also note that $\dot{q}_{max}$ is proportional to h_{fg}, and large maximum heat fluxes can be obtained using fluids with a large enthalpy of vaporization, such as water.

Minimum Heat Flux

Minimum heat flux, which occurs at the Leidenfrost point, is of practical interest since it represents the lower limit for the heat flux in the film boiling

TABLE 8-4

461

Values of the coefficient C_{cr} for use in Equation 8-3 for maximum heat flux (dimensionless parameter $L^* = L[g(\rho_l - \rho_v)/\sigma]^{1/2}$)

Heater geometry	C_{cr}	Charac. dimension of heater, L	Range of L^*
Large horizontal flat heater	0.149	Width or diameter	$L^* > 27$
Small horizontal flat heater[1]	$18.9K_1$	Width or diameter	$9 < L^* < 20$
Large horizontal cylinder	0.12	Radius	$L^* > 1.2$
Small horizontal cylinder	$0.12L^{*-0.25}$	Radius	$0.15 < L^* < 1.2$
Large sphere	0.11	Radius	$L^* > 4.26$
Small sphere	$0.227L^{*-0.5}$	Radius	$0.15 < L^* < 4.26$

[1]$K_1 = \sigma/[g(\rho_l - \rho_v)A_{heater}]$

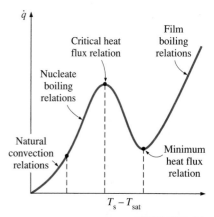

FIGURE 8-11

Different relations are used to determine the heat flux in different boiling regimes.

regime. Using the stability theory, Zuber derived the following expression for the minimum heat flux for a *large horizontal plate*,

$$\dot{q}_{min} = 0.09\rho_v \, h_{fg} \left[\frac{\sigma g(\rho_l - \rho_v)}{(\rho_l + \rho_v)^2} \right]^{1/4} \qquad (8-4)$$

where the constant 0.09 was determined by Berenson in 1961. He replaced the theoretically determined value of $\frac{\pi}{24}$ by 0.09 to match the experimental data better. Still, the relation above can be in error by 50 percent or more.

Film Boiling

Using an analysis similar to Nusselt's theory on filmwise condensation presented in the next section, Bromley developed a theory for the prediction of heat flux for stable *film boiling* on the outside of a horizontal cylinder. The heat flux for film boiling on a *horizontal cylinder* or *sphere* of diameter D is given by

$$\dot{q}_{film} = C_{film} \left[\frac{gk_v^3 \, \rho_v \, (\rho_l - \rho_v)[h_{fg} + 0.4C_{pv} \, (T_s - T_{sat})]}{\mu_v \, D(T_s - T_{sat})} \right]^{1/4} (T_s - T_{sat}) \quad (8-5)$$

where k_v is the thermal conductivity of the vapor in W/m · °C and

$$C_{film} = \begin{cases} 0.62 \text{ for horizontal cylinders} \\ 0.67 \text{ for spheres} \end{cases}$$

Other properties are as listed before in connection with Equation 8-2. We used a modified latent heat of vaporization in Equation 8-5 to account for the heat transfer associated with the superheating of the vapor.

The *vapor* properties are to be evaluated at the *film temperature*, given as $T_f = (T_s + T_{sat})/2$, which is the *average temperature* of the vapor film. The liquid properties and h_{fg} are to be evaluated at the saturation temperature at the specified pressure. Again, this relation will give the film boiling heat flux in W/m² if the properties are used *in the units specified* earlier in their descriptions following Equation 8-2.

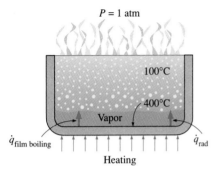

$P = 1$ atm

100°C

400°C

Vapor

$\dot{q}_{\text{film boiling}}$ $\dot{q}_{\text{rad}}$

Heating

FIGURE 8-12

At high heater surface temperatures,
radiation heat transfer becomes
significant during film boiling.

At high surface temperatures (typically above 300°C), heat transfer across the vapor film by *radiation* becomes significant and needs to be considered (Fig. 8-12). Treating the vapor film as a transparent medium sandwiched between two large parallel plates and approximating the liquid as a blackbody, *radiation heat transfer* can be determined from

$$\dot{q}_{\text{rad}} = \varepsilon\sigma \, (T_s^4 - T_{\text{sat}}^4) \tag{8-6}$$

where ε is the emissivity of the heating surface and $\sigma = 5.67 \times 10^{-8}$ W/m$^2 \cdot$ K^4 is the Stefan-Boltzman constant. Note that the temperature in this case *must* be expressed in K, not °C, and that surface tension and the Stefan-Boltzman constant share the same symbol.

You may be tempted to simply add the convection and radiation heat transfers to determine the total heat transfer during film boiling. However, these two mechanisms of heat transfer adversely affect each other, causing the total heat transfer to be less then their sum. For example, the radiation heat transfer from the surface to the liquid enhances the rate of evaporation, and thus the thickness of the vapor film, which impedes convection heat transfer. For $\dot{q}_{\text{rad}} < \dot{q}_{\text{film}}$, Bromley determined that the relation

$$\dot{q}_{\text{total}} = \dot{q}_{\text{film}} + \frac{3}{4} \, \dot{q}_{\text{rad}} \tag{8-7}$$

correlates experimental data well.

Operation in the *transition boiling* regime is normally avoided in the design of heat transfer equipment, and thus no major attempt has been made to develop general correlations for boiling heat transfer in this regime.

Note that the gravitational acceleration g, whose value is 9.8 m/s^2 at sea level, appears in all of the relations above for boiling heat transfer. The effects of low and high gravity (as encountered in aerospace applications and turbomachinery) are studied experimentally. The studies confirm that the critical heat flux and heat flux in film boiling are proportional to $g^{1/4}$. However, they indicate that heat flux in nucleate boiling is practically independent of gravity g, instead of being proportional to $g^{1/2}$, as dictated by Equation 8-2.

Enhancement of Heat Transfer in Pool Boiling

FIGURE 8-13

The cavities on a rough surface act as
nucleation sites and enhance boiling
heat transfer.

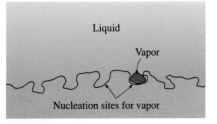

Liquid

Vapor

Nucleation sites for vapor

The pool boiling heat transfer relations given above apply to smooth surfaces. Below we will discuss some methods to enhance heat transfer in pool boiling.

We pointed out earlier that the rate of heat transfer in the *nucleate boiling* regime strongly depends on the number of active nucleation sites on the surface, and the rate of bubble formation at each site. Therefore, any modification that will enhance *nucleation* on the heating surface will also enhance *heat transfer* in nucleate boiling. It is observed that *irregularities* on the heating surface, including roughness and dirt, serve as additional nucleation sites during boiling, as shown in Figure 8-13. For example, the first bubbles in a pan filled with water are most likely to form at the *scratches* at the bottom surface. These scratches act like "nests" for the bubbles to form and thus

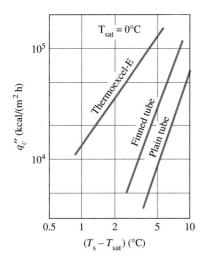

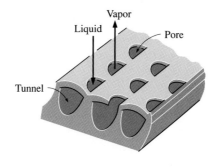

FIGURE 8-14

The enhancement of boiling heat transfer in Freon-12 by a mechanically roughened surface, thermoexcel-E.

increase the rate of bubble formation. Berensen has shown that heat flux in the nucleate boiling regime can be increased by a factor of 10 by *roughening* the heating surface. However, these high heat transfer rates cannot be sustained for long since the effect of surface roughness is observed to decay with time, and the heat flux eventually to drop to values for smooth surfaces. The effect of surface roughness is negligible on the critical heat flux and the heat flux in film boiling.

Surfaces that provide enhanced heat transfer in nucleate boiling *permanently* are being manufactured and are available in the market. Enhancement in nucleation and thus heat transfer in such special surfaces is achieved either by *coating* the surface with a thin layer (much less than 1 mm) of very porous material or by *forming cavities* on the surface mechanically to facilitate continuous vapor formation. Such surfaces are reported to enhance heat transfer in the nucleate boiling regime by a factor of up to 10, and the critical heat flux by a factor of 3. The enhancement provided by one such material prepared by machine roughening, the thermoexcel-E, is shown in Figure 8-14. The use of finned surfaces is also known to enhance nucleate boiling heat transfer and the critical heat flux.

Boiling heat transfer can also be enhanced by other techniques such as *mechanical agitation* and *surface vibration*. These techniques are not practical, however, because of the complications involved.

EXAMPLE 8-1 Nucleate Boiling of Water in a Pan

Water is to be boiled at atmospheric pressure in a mechanically polished stainless steel pan placed on top of a heating unit, as shown in Figure 8-15. The inner surface of the bottom of the pan is maintained at 108°C. If the diameter of the bottom of the pan is 30 cm, determine (*a*) the rate of heat transfer to the water and (*b*) the rate of evaporation of water.

Solution Water is boiled at 1 atm pressure and thus at $T_{sat} = 100°C$ on a stainless steel surface maintained at $T_s = 108°C$. The rate of heat transfer to the water and the rate of evaporation of water are to be determined.

FIGURE 8-15

Schematic for Example 8-1.

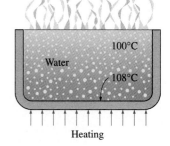

Assumptions **1** Steady operating conditions exist. **2** Heat losses from the heater and the pan are negligible.

Properties The properties of water at the saturation temperature of 100°C are $\sigma = 0.0589$ N/m (Table 8-1) and, from Table A-9,

$$\rho_l = 957.9 \text{ kg/m}^3 \qquad h_{fg} = 2257.0 \times 10^3 \text{ J/kg}$$
$$\rho_v = 0.6 \text{ kg/m}^3 \qquad \mu_l = 0.282 \times 10^{-3} \text{ kg} \cdot \text{m/s}$$
$$Pr_l = 1.75 \qquad C_{pl} = 4217 \text{ J/kg} \cdot °C$$

Also, $C_{sf} = 0.0130$ and $n = 1.0$ for the boiling of water on a mechanically polished stainless steel surface (Table 8-3). Note that we expressed the properties in units specified under Equation 8-2 in connection with their definitions in order to avoid unit manipulations.

Analysis (*a*) The excess temperature in this case is $\Delta T = T_s - T_{sat} = 108 - 100 = 8°C$ which is relatively low (less than 30°C). Therefore, nucleate boiling will occur. The heat flux in this case can be determined from the Rohsenow relation to be

$$\dot{q}_{nucleate} = \mu_l h_{fg} \left[\frac{g(\rho_l - \rho_v)}{\sigma} \right]^{1/2} \left[\frac{C_{pl}(T_s - T_{sat})}{C_{sf} h_{fg} Pr_l^n} \right]^3$$

$$= (0.282 \times 10^{-3})(2257 \times 10^3) \left[\frac{9.8(957.9 - 0.6)}{0.0589} \right]^{1/2}$$

$$\times \left(\frac{4217(108 - 100)}{0.0130(2257 \times 10^3)1.75} \right)^3$$

$$= 72,045 \text{ W/m}^2$$

The surface area of the bottom of the pan is

$$A = \pi D^2/4 = \pi(0.3 \text{ m})^2/4 = 0.07069 \text{ m}^2$$

Then the rate of heat transfer during nucleate boiling becomes

$$\dot{Q}_{boiling} = A\dot{q}_{nucleate} = (0.07069 \text{ m}^2)(72,045 \text{ W/m}^2) = \textbf{5093 W}$$

(*b*) The rate of evaporation of water is determined from

$$\dot{m}_{evaporation} = \frac{\dot{Q}_{boiling}}{h_{fg}} = \frac{5093 \text{ J/s}}{2257 \times 10^3 \text{ J/kg}} = \textbf{2.26} \times \textbf{10}^{-3} \textbf{ kg/s}$$

That is, water in the pan will boil at a rate of more than 2 grams per second.

EXAMPLE 8-2   Peak Heat Flux in Nucleate Boiling

Water in a tank is to be boiled at sea level by a 1-cm-diameter nickel plated heating element equipped with electrical resistance wires inside, as shown in Figure 8-16. Determine the maximum heat flux that can be attained in the nucleate boiling regime and the surface temperature of the heater surface in that case.

FIGURE 8-16

Schematic for Example 8-2.

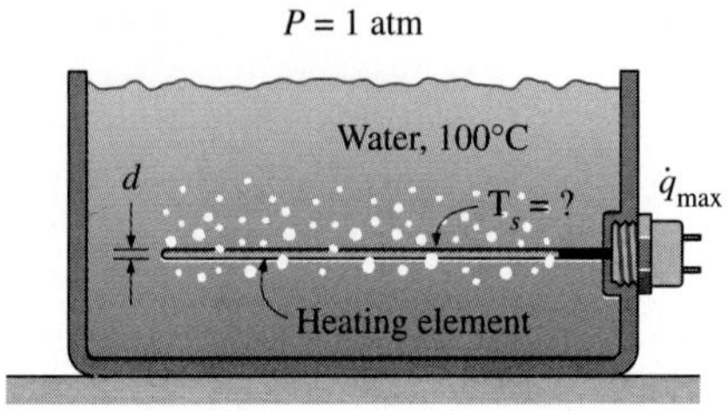

Solution Water is boiled at 1 atm pressure and thus at $T_{sat} = 100°C$ on a stainless steel surface. The maximum (critical) heat flux and the surface temperature for that case are to be determined.

Assumptions **1** Steady operating conditions exist. **2** Heat losses from the boiler are negligible.

Properties The properties of water at the saturation temperature of 100°C are $\sigma = 0.0589$ N/m (Table 8-1) and, from Table A-9,

$$\rho_l = 957.9 \text{ kg/m}^3 \qquad h_{fg} = 2257 \times 10^3 \text{ J/kg}$$
$$\rho_v = 0.6 \text{ kg/m}^3 \qquad \mu_l = 0.282 \times 10^{-3} \text{ kg} \cdot \text{m/s}$$
$$Pr_l = 1.75 \qquad C_{pl} = 4217 \text{ J/kg} \cdot \text{°C}$$

Also, $C_{sf} = 0.0060$ and $n = 1.0$ for the boiling of water on a nickel plated surface (Table 8-3). Note that we expressed the properties in units specified under Equations 8-2 and 8-3 in connection with their definitions in order to avoid unit manipulations.

Analysis The heating element in this case can be considered to be a short cylinder whose characteristic dimension is its radius. That is, $L = r = 0.005$ m. The dimensionless parameter L^* and the constant C_{cr} are determined from Table 8-4 to be

$$L^* = L\left(\frac{g(\rho_l - \rho_v)}{\sigma}\right)^{1/2} = (0.005)\left(\frac{(9.8)(957.9 - 0.6)}{0.0589}\right)^{1/2} = 2.00 > 1.2$$

which corresponds to $C_{cr} = 0.12$.

Then the maximum or critical heat flux is determined from Equation 8-3 to be

$$\dot{q}_{max} = C_{cr} h_{fg} [\sigma g \rho_v^2 (\rho_l - \rho_v)]^{1/4}$$
$$= 0.12(2257 \times 10^3)[0.0589 \times 9.8 \times (0.6)^2(957.9 - 0.6)]^{1/4}$$
$$= \textbf{1,017,200 W/m}^2$$

The Rohsenow relation, which gives the nucleate boiling heat flux for a specified surface temperature, can also be used to determine the surface temperature when the heat flux is given. Substituting the maximum heat flux into Equation 8-2 together with other properties gives

$$\dot{q}_{nucleate} = \mu_l h_{fg} \left[\frac{g(\rho_l - \rho_v)}{\sigma}\right]^{1/2} \left[\frac{C_{pl}(T_s - T_{sat})}{C_{sf} h_{fg} Pr_l^n}\right]^3$$

$$1,017,200 = (0.282 \times 10^{-3})(2257 \times 10^3)\left[\frac{9.8(957.9 - 0.6)}{0.0589}\right]^{1/2}$$
$$\left[\frac{4217(T_s - 100)}{0.0130(2257 \times 10^3)1.75}\right]^3$$

$$T_s = \textbf{119.3°C}$$

Discussion Note that heat fluxes on the order of 1 MW/m^2 can be obtained in nucleate boiling with a temperature difference of less than 20°C.

EXAMPLE 8-3   Film Boiling of Water on a Heating Element

Water is boiled at atmospheric pressure by a horizontal polished copper heating element of diameter $D = 5$ mm and emissivity $\varepsilon = 0.05$ immersed in water, as shown in Figure 8-17. If the surface temperature of the heating wire is 350°C, determine the rate of heat transfer from the wire to the water per unit length of the wire.

Solution Water is boiled at 1 atm and thus at $T_{sat} = 100$°C by a horizontal polished copper heating element at $T_s = 350$°C. The rate of heat transfer to the water per unit length of the heater is to be determined.

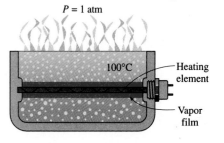

FIGURE 8-17

Schematic for Example 8-3.

Assumptions **1** Steady operating conditions exist. **2** Heat losses from the boiler are negligible.

Properties The properties of water at the saturation temperature of 100°C are $h_{fg} = 2257 \times 10^3$ J/kg and $\rho_l = 957.9$ kg/m^3 (Table A-9). The properties of vapor at the film temperature of $T_f = (T_{sat} + T_s)/2 = (100 + 350)/2 = 225$°C = 498 K (which is sufficiently close to 500 K) are, from Table A-11,

$$\rho_v = 0.441 \text{ kg/m}^3 \qquad\qquad C_{pv} = 1977 \text{ J/kg} \cdot °C$$
$$\mu_v = 1.73 \times 10^{-5} \text{ kg/m} \cdot s \qquad k_v = 0.0357 \text{ W/m} \cdot °C$$

Note that we expressed the properties in units that will cancel each other in boiling heat transfer relations. Also note that we used vapor properties at 1 atm pressure from Table A-11 instead of the properties of saturated vapor from Table A-9 at 250°C since the latter are at the saturation pressure of 4.0 MPa.

Analysis The excess temperature in this case is $\Delta T = T_s - T_{sat} = 350 - 100 = 250$°C, which is much larger than 30°C for water. Therefore, film boiling will occur. The film boiling heat flux in this case can be determined from Equation 8-5 to be

$$\dot{q}_{film} = 0.62 \left[\frac{gk_v^3 \rho_v (\rho_l - \rho_v)[h_{fg} + 0.4C_{pv}(T_s - T_{sat})]}{\mu_v D(T_s - T_{sat})} \right]^{1/4} (T_s - T_{sat})$$

$$= 0.62 \left[\frac{9.8(0.0357)^3 (0.441)(957.9 - 0.441)[(2257 \times 10^3) + 0.4 \times 1977(250)]}{(1.73 \times 10^{-5})(5 \times 10^{-3})(250)} \right]^{1/4}$$

$$\times 250$$

$$= 59{,}264 \text{ W/m}^2$$

The radiation heat flux is determined from Equation 8-6 to be

$$\dot{q}_{rad} = \varepsilon\sigma (T_s^4 - T_{sat}^4)$$
$$= (0.05)(5.67 \times 10^{-8} \text{ W/m}^2 \cdot \text{K}^4)[(250 + 273 \text{ K})^4 - (100 + 273 \text{ K})^4]$$
$$= 157 \text{ W/m}^2$$

Note that heat transfer by radiation is negligible in this case because of the low emissivity of the surface and the relatively low surface temperature of the heating element. Then the total heat flux becomes (Eq. 8-7)

$$\dot{q}_{total} = \dot{q}_{film} + \frac{3}{4}\dot{q}_{rad} = 59{,}264 + \frac{3}{4} \times 157 = 59{,}382 \text{ W/m}^2$$

Finally, the rate of heat transfer from the heating element to the water is determined by multiplying the heat flux by the heat transfer surface area,

$$\dot{Q}_{total} = A\dot{q}_{total} = (\pi DL)\dot{q}_{total}$$
$$= (\pi \times 0.005 \text{ m} \times 1 \text{ m})(59{,}382 \text{ W/m}^2)$$
$$= \mathbf{933W}$$

Discussion Note that the 5-mm-diameter copper heating wire will consume about 1 kW of electric power per unit length in steady operation in the film boiling regime. This energy is transferred to the water through the vapor film that forms around the wire.

The *pool boiling* we considered so far involves a pool of seemingly motion-less liquid, with vapor bubbles rising to the top as a result of buoyancy effects. In **flow boiling,** the fluid is forced to move by an external source such as a pump as it undergoes a phase-change process. The boiling in this case exhibits the combined effects of natural and/or forced convection and pool boiling. The flow boiling is also classified as either *external* and *internal flow boiling* depending on whether the fluid is forced to flow over a heated surface or inside a heated tube.

External flow boiling over a plate or cylinder is similar to pool boiling, but the added motion increases both the nucleate boiling heat flux and the critical heat flux considerably, as shown in Figure 8-18. Note that the higher the velocity, the higher the nucleate boiling heat flux and the critical heat flux. In experiments with water, critical heat flux values as high as 35 MW/m^2 have been obtained (compare this to the pool boiling value of 1.3 MW/m^2 at 1 atm pressure) by increasing the fluid velocity.

Internal flow boiling is much more complicated in nature because there is no free surface for the vapor to escape, and thus both the liquid and the vapor are forced to flow together. The two-phase flow in a tube exhibits different flow boiling regimes, depending on the relative amounts of the liquid and the vapor phases. This complicates the analysis even further.

The different stages encountered in flow boiling in a heated tube are illustrated in Figure 8-19 together with the variation of the heat transfer coefficient along the tube. Initially, the liquid is subcooled and heat transfer to the liquid is by *forced convection.* Then bubbles start forming on the inner surfaces of the tube, and the detached bubbles are drafted into the main-stream. This gives the fluid flow a bubbly appearance, and thus the name *bubbly flow regime.* As the fluid is heated further, the bubbles grow in size and eventually coalesce into slugs of vapor. Up to half of the volume in the tube in this *slug-flow regime* is occupied by vapor. After a while the core of the flow consists of vapor only, and the liquid is confined only in the annular space between the vapor core and the tube walls. This is the *annular-flow regime,* and very high heat transfer coefficients are realized in this regime. As the heating continues, the annular liquid layer gets thinner and thinner, and eventually dry spots start to appear on the inner surfaces of the tube. The appearance of dry spots is accompanied by a sharp decrease in the heat transfer coefficient. This *transition regime* continues until the inner surface of the tube is completely dry. Any liquid at this moment is in the form of droplets suspended in the vapor core, which resembles a mist, and we have a *mist-flow regime* until all the liquid droplets are vaporized. At the end of the mist-flow regime we have saturated vapor, which becomes superheated with any further heat transfer.

Note that the tube contains a liquid before the bubbly flow regime and a vapor after the mist-flow regime. Heat transfer in those two cases can be determined using the appropriate relations for single-phase convection heat transfer. Many correlations are proposed for the determination of heat transfer in the two-phase flow (bubbly flow, slug-flow, annular-flow, and

Flow Boiling

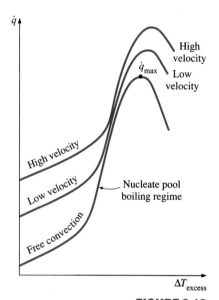

FIGURE 8-18

The effect of forced convection on external flow boiling for different flow velocities.

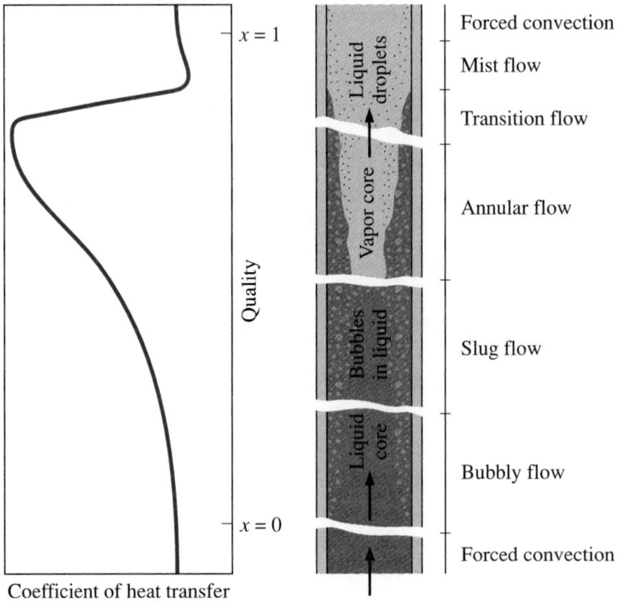

FIGURE 8-19

Different flow regimes encountered in flow boiling in a tube under forced convection.

mist-flow) cases, but they are beyond the scope of this introductory text. A crude estimate for heat flux in flow boiling can be obtained by simply adding the forced convection and pool boiling heat fluxes.

8-4 ■ CONDENSATION HEAT TRANSFER

Condensation occurs when the temperature of a vapor is reduced *below* its saturation temperature T_{sat}. This is usually done by bringing the vapor into contact with a solid surface whose temperature T_s is *below* the saturation temperature T_{sat} of the vapor. But condensation can also occur on the free surface of a liquid or even in a gas when the temperature of the liquid or the gas to which the vapor is exposed is below T_{sat}. In the latter case, the liquid droplets suspended in the gas form a fog. In this chapter, we will consider condensation on solid surfaces only.

Two distinct forms of condensation are observed: *film condensation* and *dropwise condensation*. In **film condensation,** the condensate wets the surface and forms a liquid film on the surface that slides down under the influence of gravity. The thickness of the liquid film increases in the flow direction as more vapor condenses on the film. This is how condensation normally occurs in practice. In **dropwise condensation,** the condensed vapor forms droplets on the surface instead of a continuous film, and the surface is covered by countless droplets of varying diameters (Fig. 8-20).

In film condensation, the surface is blanketed by a liquid film of increasing thickness, and this "liquid wall" between solid surface and the vapor serves as a *resistance* to heat transfer. The heat of vaporization h_{fg} released as the vapor condenses must pass through this resistance before it can reach the solid surface and be transferred to the medium on the other side. In dropwise

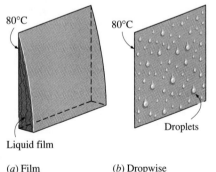

(a) Film condensation   (b) Dropwise condensation

FIGURE 8-20

When a vapor is exposed to a surface at a temperature below T_{sat}, condensation in the form of a liquid film or individual droplets occurs on the surface.

condensation, however, the droplets slide down when they reach a certain size, clearing the surface and exposing it to vapor. There is no liquid film in this case to resist heat transfer. As a result, heat transfer rates that are more than 10 times larger than those associated with film condensation can be achieved with dropwise condensation. Therefore, dropwise condensation is the preferred mode of condensation in heat transfer applications, and people have long tried to achieve sustained dropwise condensation by using various vapor additives and surface coatings. These attempts have not been very successful, however, since the dropwise condensation achieved did not last long and converted to film condensation after some time. Therefore, it is common practice to be conservative and assume film condensation in the design of heat transfer equipment.

8-5 ■ FILM CONDENSATION

We now consider film condensation on a vertical plate, as shown in Figure 8-21. The liquid film starts forming at the top of the plate and flows downward under the influence of gravity. The thickness of the film δ *increases* in the flow direction x because of continued condensation at the liquid–vapor interface. Heat in the amount h_{fg} (the latent heat of vaporization) is released during condensation and is *transferred* through the film to the plate surface at temperature T_s. Note that T_s must be below the saturation temperature T_{sat} of the vapor for condensation to occur.

Typical velocity and temperature profiles of the condensate are also given in Figure 8-21. Note that the *velocity* of the condensate at the wall is zero because of the "no-slip" condition and reaches a *maximum* at the liquid–vapor interface. The *temperature* of the condensate is T_{sat} at the interface and decreases gradually to T_s at the wall.

As was the case in forced convection involving a single phase, heat transfer in condensation also depends on whether the condensate flow is *laminar* or *turbulent*. Again the criterion for the flow regime is provided by the Reynolds number, which is defined as

$$\text{Re} = \frac{D_h \rho_l \mathcal{V}_l}{\mu_l} = \frac{4 A_c \rho_l \mathcal{V}_l}{p\mu_l} = \frac{4 \rho_l \mathcal{V}_l \delta}{\mu_l} = \frac{4\dot{m}}{p\mu_l} \tag{8-8}$$

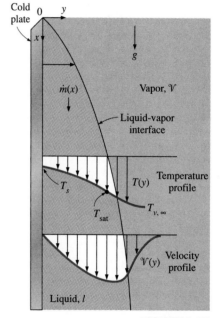

FIGURE 8-21

Film condensation on a vertical plate.

where

$D_h = 4A_c/p = 4\delta$ = hydraulic diameter of the condensate flow, m

p = wetted perimeter of the condensate, m

$A_c = p\delta$ = wetted perimeter × film thickness, m^2, cross-sectional area of the condensate flow at the lowest part of the flow

ρ_l = density of the liquid, kg/m^3

μ_l = viscosity of the liquid, kg/m · s

$\mathcal{V}$ = average velocity of the condensate at the lowest part of the flow, m/s

$\dot{m} = \rho_l A \mathcal{V}_l$ = mass flow rate of the condensate at the lowest part, kg/s

The evaluation of the hydraulic diameter D_h for some common geometries is illustrated in Figure 8-22. Note that the hydraulic diameter is again defined such that it reduces to the ordinary diameter for flow in a circular tube, as was done in Chapter 6 for internal flow, and it is equivalent to 4 times the thickness of the condensate film at the location where the hydraulic diameter is evaluated. That is, $D_h = 4\delta$.

The latent heat of vaporization h_{fg} is the heat released as a unit mass of vapor condenses, and it normally represents the heat transfer per unit mass of condensate formed during condensation. However, the condensate in an actual condensation process is cooled further to some average temperature between T_{sat} and T_s, releasing *more heat* in the process. Therefore, the actual heat transfer will be larger. Rohsenow showed in 1956 that the cooling of the liquid below the saturation temperature can be accounted for by replacing h_{fg} by the **modified latent heat of vaporization** h_{fg}^*, defined as

$$h_{fg}^* = h_{fg} + 0.68C_{pl}\,(T_{sat} - T_s) \qquad (8\text{-}9a)$$

where C_{pl} is the specific heat of the liquid at the average film temperature.

We can have a similar argument for vapor that enters the condenser as **superheated vapor** at a temperature T_v instead of as saturated vapor. In this case the vapor must be cooled first to T_{sat} before it can condense, and this heat must be transferred to the wall as well. The amount of heat released as a unit mass of superheated vapor at a temperature T_v is cooled to T_{sat} is simply $C_{pv}(T_v - T_{sat})$, where C_{pv} is the specific heat of the vapor at the average temperature of $(T_v + T_{sat})/2$. The modified latent heat of vaporization in this case becomes

$$h_{fg}^* = h_{fg} + 0.68C_{pl}\,(T_{sat} - T_s) + C_{pv}\,(T_v - T_{sat}) \qquad (8\text{-}9b)$$

With these considerations, the rate of heat transfer can be expressed as

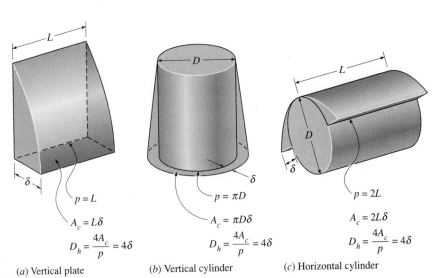

FIGURE 8-22

The wetted perimeter p, the condensate cross-sectional area A_c, and the hydraulic diameter D_h for some common geometries.

$p = L$

$A_c = L\delta$

$D_h = \dfrac{4A_c}{p} = 4\delta$

(a) Vertical plate

$p = \pi D$

$A_c = \pi D\delta$

$D_h = \dfrac{4A_c}{p} = 4\delta$

(b) Vertical cylinder

$p = 2L$

$A_c = 2L\delta$

$D_h = \dfrac{4A_c}{p} = 4\delta$

(c) Horizontal cylinder

$$\dot{Q}_{\text{conden}} = hA(T_{\text{sat}} - T_s) = \dot{m}h_{fg}^* \qquad (8\text{-}10)$$

where A is the heat transfer area (the surface area on which condensation occurs). Solving for $\dot{m}$ from the equation above and substituting it into Equation 8-8 gives yet another relation for the Reynolds number,

$$\text{Re} = \frac{4\dot{Q}_{\text{conden}}}{p\mu_l h_{fg}^*} = \frac{4Ah(T_{\text{sat}} - T_s)}{p\mu_l h_{fg}^*} \qquad (8\text{-}11)$$

This relation is convenient to use to determine the Reynolds number when the condensation heat transfer coefficient or the rate of heat transfer is known.

The temperature of the liquid film varies from T_{sat} on the liquid–vapor interface to T_s at the wall surface. Therefore, the properties of the liquid should be evaluated at the *film temperature* $T_f = (T_{\text{sat}} + T_s)/2$, which is approximately the *average* temperature of the liquid. The h_{fg}, however, should be evaluated at T_{sat} since it is not affected by the subcooling of the liquid.

Flow regimes

The Reynolds number for condensation on the outer surfaces of vertical tubes or plates increases in the flow direction due to the increase of the liquid film thickness δ. The flow of liquid film exhibits *different regimes*, depending on the value of the Reynolds number. It is observed that the outer surface of the liquid film remains *smooth* and *wave-free* for about Re $\leq$ 30, as shown in Figure 8-23, and thus the flow is clearly *laminar*. Ripples or waves appear on the free surface of the condensate flow as the Reynolds number increases, and the condensate flow becomes fully *turbulent* at about Re $\approx$ 1800. The condensate flow is called *wavy-laminar* in the range of $450 < \text{Re} < 1800$ and *turbulent* for Re > 1800. However, some disagreement exists about the value of Re at which the flow becomes wavy-laminar or turbulent.

Heat Transfer Correlations for Film Condensation

Below we discuss relations for the average heat transfer coefficient h for the case of *laminar* film condensation for various geometries.

1 Vertical Plates

Consider a vertical plate of height L and width b maintained at a constant temperature T_s that is exposed to vapor at the saturation temperature T_{sat}. The downward direction is taken as the positive x-direction with the origin placed at the top of the plate where condensation initiates, as shown in Figure 8-24. The surface temperature is below the saturation temperature ($T_s < T_{\text{sat}}$) and thus the vapor condenses on the surface. The liquid film flows downward under the influence of gravity. The film thickness δ and thus the mass flow

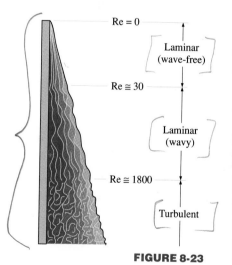

FIGURE 8-23

Flow regimes during film condensation on a vertical plate.

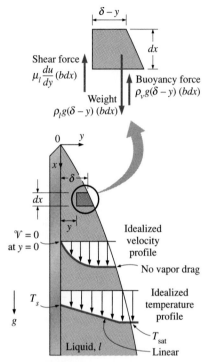

FIGURE 8-24

The volume element of condensate on a vertical plate considered in Nusselt's analysis.

rate of the condensate increases with x as a result of continued condensation on the existing film. Then heat transfer from the vapor to the plate must occur through the film, which offers resistance to heat transfer. Obviously the thicker the film, the larger its thermal resistance and thus the lower the rate of heat transfer.

The analytical relation for the heat transfer coefficient in film condensation on a vertical plate described above was first developed by Nusselt in 1916 under the following simplifying assumptions:

1. Both the plate and the vapor are maintained at *constant temperatures* of T_s and T_{sat}, respectively, and the temperature across the liquid film varies *linearly*.

2. Heat transfer across the liquid film is by pure *conduction* (no convection currents in the liquid film).

3. The velocity of the vapor is low (or zero) so that it exerts *no drag* on the condensate (no viscous shear on the liquid–vapor interface).

4. The flow of the condensate is *laminar* and the properties of the liquid are constant.

5. The acceleration of the condensate layer is negligible.

Then Newton's second law of motion for the volume element shown in Figure 8-24 in the vertical x-direction can be written as

$$\sum F_x = ma_x = 0$$

since the acceleration of the fluid is zero. Noting that the only force acting downward is the weight of the liquid element, and the forces acting upward are the viscous shear (or fluid friction) force at the left and the buoyancy force, the force balance on the volume element becomes

$$F_{\text{downward}}\downarrow = F_{\text{upward}}\uparrow$$
$$\text{Weight} = \text{Viscous shear force} + \text{Buoyancy force}$$

$$\rho_l g(\delta - y)(bdx) = \mu_l \frac{du}{dy}(bdx) + \rho_v g(\delta - y)(bdx)$$

Canceling the plate width b and solving for du/dy gives

$$\frac{du}{dy} = \frac{g(\rho_l - \rho_v)g(\delta - y)}{\mu_l}$$

Integrating from $y = 0$ where $u = 0$ (because of the no-slip boundary condition) to $y = y$ where $u = u(y)$ gives

$$u(y) = \frac{g(\rho_l - \rho_v)g}{\mu_l}\left(y\delta - \frac{y^2}{2}\right) \tag{8-12}$$

The mass flow rate of the condensate at a location x, where the boundary thickness is δ, is determined from

$$\dot{m}(x) = \int_A \rho_l u(y) dA = \int_{y=0}^{\delta} \rho_l u(y) b dy \qquad (8\text{-}13)$$

Substituting the $u(y)$ relation from Equation 8-12 into the equation above gives

$$\dot{m}(x) = \frac{g b \rho_l (\rho_l - \rho_v) \delta^3}{3 \mu_l} \qquad (8\text{-}14)$$

whose derivative with respect to x is

$$\frac{d\dot{m}}{dx} = \frac{g b \rho_l (\rho_l - \rho_v) \delta^2}{\mu_l} \frac{d\delta}{dx} \qquad (8\text{-}15)$$

which represents the rate of condensation of vapor over a vertical distance dx. The rate of heat transfer from the vapor to the plate through the liquid film is simply equal to the heat released as the vapor is condensed and is expressed as

$$d\dot{Q} = h_{fg} d\dot{m} = k_l (b dx) \frac{T_{sat} - T_s}{\delta} \quad \rightarrow \quad \frac{d\dot{m}}{dx} = \frac{k_l b}{h_{fg}} \frac{T_{sat} - T_s}{\delta} \qquad (8\text{-}16)$$

Equating Equations 8-15 and 8-16 for $d\dot{m}/dx$ to each other and separating the variables give

$$\delta^3 d\delta = \frac{\mu_l k_l (T_{sat} - T_s)}{g \rho_l (\rho_l - \rho_v) h_{fg}} dx \qquad (8\text{-}17)$$

Integrating from $x = 0$ where $\delta = 0$ (the top of the plate) to $x = x$ where $\delta = \delta(x)$, the liquid film thickness at any location x is determined to be

$$\delta(x) = \left[\frac{4 \mu_l k_l (T_{sat} - T_s) x}{g \rho_l (\rho_l - \rho_v) h_{fg}} \right]^{1/4} \qquad (8\text{-}18)$$

The heat transfer rate from the vapor to the plate at a location x can be expressed as

$$\dot{q}_x = h_x (T_{sat} - T_s) = k_l \frac{T_{sat} - T_s}{\delta} \quad \rightarrow \quad h_x = \frac{k_l}{\delta(x)} \qquad (8\text{-}19)$$

Substituting the $\delta(x)$ expression from Equation 8-18, the local heat transfer coefficient h_x is determined to be

$$h_x = \left[\frac{g \rho_l (\rho_l - \rho_v) h_{fg} k_l^3}{4 \mu_l (T_{sat} - T_s) x} \right]^{1/4} \qquad (8\text{-}20)$$

The average heat transfer coefficient over the entire plate is determined from its definition by substituting the h_x relation and performing the integration. It gives

$$h = h_{ave} = \frac{1}{L} \int_0^L h_x dx = \frac{4}{3} h_{x=L} = 0.943 \left[\frac{g \rho_l (\rho_l - \rho_v) h_{fg} k_l^3}{\mu_l (T_{sat} - T_s) L} \right]^{1/4} \qquad (8\text{-}21)$$

The relation above, which is obtained with the simplifying assumptions stated earlier, provides a lot of insight on the functional dependence of the condensation heat transfer coefficient. However, it is observed to underpredict heat transfer because it does not take into account the effects of the nonlinear temperature profile in the liquid film and the cooling of the liquid below the saturation temperature. Both of these effects can be accounted for by replacing h_{fg} by h_{fg}^* given by Equation 8-9. With this modification, the *average heat transfer coefficient* for laminar film condensation over a vertical flat plate of height L is determined to be

$$h_{vert} = 0.943 \left[\frac{g\rho_l (\rho_l - \rho_v)h_{fg}^* k_l^3}{\mu_l (T_{sat} - T_s)L} \right]^{1/4} \quad (W/m^2 \cdot {}^\circ C), \ 0 < Re < 30 \quad (8\text{-}22)$$

where
$$g = \text{gravitational acceleration, m/s}^2$$
$$\rho_l, \rho_v = \text{densities of the liquid and vapor, respectively, kg/m}^3$$
$$\mu_l = \text{viscosity of the liquid, kg/m} \cdot \text{s}$$
$$h_{fg}^* = h_{fg} + 0.68 C_{pl} (T_{sat} - T_s) = \text{modified latent heat of vaporization, J/kg}$$
$$k_l = \text{thermal conductivity of the liquid, W/m} \cdot {}^\circ C$$
$$L = \text{height of the vertical plate, m}$$
$$T_s = \text{surface temperature of the plate, } {}^\circ C$$
$$T_{sat} = \text{saturation temperature of the condensing fluid, } {}^\circ C$$

At a given temperature, $\rho_v \ll \rho_l$ and thus $\rho_l - \rho_v \approx \rho_l$ except near the critical point of the substance. Using this approximation and substituting Equations 8-14 and 8-18 at $x = L$ into Equation 8-8 by noting that $\delta_{x=L} = \dfrac{k_l}{h_{x=L}}$ and $h_{vert} = \dfrac{4}{3} h_{x=L}$ (Eqs. 8-19 and 8-21) give

$$Re \cong \frac{4g\rho_l (\rho_l - \rho_v)\delta^3}{3\mu_l^2} = \frac{4g\rho_l^2}{3\mu_l^2}\left(\frac{k_l}{h_{x=L}}\right)^3 = \frac{4g}{3v_l^2}\left(\frac{k_l}{3h_{vert}/4}\right)^3 \quad (8\text{-}23)$$

Then the heat transfer coefficient h_{vert} in terms of Re becomes

$$h_{vert} \cong 1.47 k_l \, Re^{-1/3}\left(\frac{g}{v_l^2}\right)^{1/3}, \quad \begin{array}{l} 0 < Re < 30 \\ \rho_v \ll \rho_l \end{array} \quad (8\text{-}24)$$

The results obtained from the theoretical relations above are in excellent agreement with the experimental results. It can be shown easily that using property values in Equations 8-22 and 8-24 in the *specified units* gives the condensation heat transfer coefficient in $W/m^2 \cdot {}^\circ C$, thus saving one from having to go through tedious unit manipulations each time (Fig. 8-25). This is also true for the equations below. All properties of the *liquid* are to be evaluated at the film temperature $T_f = (T_{sat} + T_s)/2$. The h_{fg} and ρ_v are to be evaluated at the saturation temperature T_{sat}.

$$h_{vert} = \left(\frac{\dfrac{m}{s^2}\dfrac{kg}{m^3}\dfrac{kg}{m^3}\dfrac{J}{kg}\left(\dfrac{W}{m \cdot {}^\circ C}\right)^3}{\dfrac{kg}{m \cdot s} \cdot {}^\circ C \cdot m} \right)^{1/4}$$

$$= \left[\frac{m}{s}\frac{1}{m^6}\frac{W^3}{m^3 \cdot {}^\circ C^3}\frac{J}{{}^\circ C} \right]$$

$$= \left(\frac{W^4}{m^8 \cdot {}^\circ C^4} \right)^{1/4}$$

$$= W/m^2 \cdot {}^\circ C$$

FIGURE 8-25

Equation 8-22 gives the condensation heat transfer coefficient in $W/m^2 \cdot {}^\circ C$ when the quantities are expressed in the units specified in their descriptions.

At Reynolds numbers greater than about 30, it is observed that waves form at the liquid–vapor interface although the flow in liquid film remains laminar. The flow in this case is said to be *wavy laminar*. The waves at the liquid–vapor interface tend to increase heat transfer. But the waves also complicate the analysis and make it very difficult to obtain analytical solutions. Therefore, we have to rely on experimental studies. The increase in heat transfer due to the wave effect is, on average, about 20 percent, but it can exceed 50 percent. The exact amount of enhancement depends on the Reynolds number. Based on his experimental studies, Kutateladze (1963) recommended the following relation for the average heat transfer coefficient in wavy laminar condensate flow for $\rho_v \ll \rho_l$ and $30 < Re < 1800$,

$$h_{\text{vert, wavy}} = \frac{Re\, k_l}{1.08\, Re^{1.22} - 5.2} \left(\frac{g}{v_l^2}\right)^{1/3}, \quad \begin{array}{c} 30 < Re < 1800 \\ \rho_v \ll \rho_l \end{array} \quad (8\text{-}25)$$

A simpler alternative to the relation above proposed by Kutateladze (1963) is

$$h_{\text{vert, wavy}} = 0.8\, Re^{0.11}\, h_{\text{vert (smooth)}} \qquad (8\text{-}26)$$

which relates the heat transfer coefficient in wavy laminar flow to that in wave-free laminar flow. McAdams (1954) went even further and suggested accounting for the increase in heat transfer in the wavy region by simply increasing the heat transfer coefficient determined from Equation 8-22 for the laminar case by 20 percent. Holman (1990) suggested using Equation 8-22 for the wavy region also, with the understanding that this is a conservative approach that provides a safety margin in thermal design. In this book we will use Equation 8-25.

A relation for the Reynolds number in the wavy laminar region can be determined by substituting the h relation in Equation 8-25 into the Re relation in Equation 8-11 and simplifying. It yields

$$Re_{\text{vert, wavy}} = \left[4.81 + \frac{3.70\, Lk_l(T_{\text{sat}} - T_s)}{\mu_l\, h_{fg}^*} \left(\frac{g}{v_l^2}\right)^{1/3} \right]^{0.820}, \quad \rho_v \ll \rho_l \quad (8\text{-}27)$$

Turbulent Flow on Vertical Plates

At a Reynolds number of about 1800, the condensate flow becomes turbulent. Several empirical relations of varying degrees of complexity are proposed for the heat transfer coefficient for turbulent flow. Again assuming $\rho_v \ll \rho_l$ for simplicity, Labuntsov (1957) proposed the following relation for the turbulent flow of condensate on *vertical plates*:

$$h_{\text{vert, turbulent}} = \frac{Re\, k_l}{8750 + 58\, Pr^{-0.5}(Re^{0.75} - 253)} \left(\frac{g}{v_l^2}\right)^{1/3}, \quad \begin{array}{c} Re > 1800 \\ \rho_v \ll \rho_l \end{array} \quad (8\text{-}28)$$

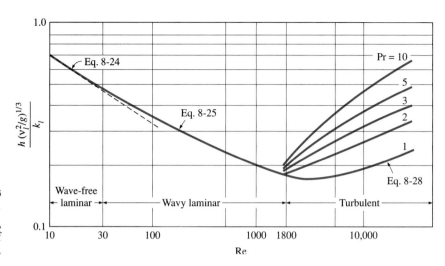

FIGURE 8-26

Nondimensionalized heat transfer coefficients for the wave-free laminar, wavy laminar, and turbulent flow of condensate on vertical plates.

The physical properties of the condensate are again to be evaluated at the film temperature $T_f = (T_{sat} + T_s)/2$. The Re relation in this case is obtained by substituting the h relation above into the Re relation in Eq. 8-11, which gives

$$\text{Re}_{\text{vert, turbulent}} = \left[\frac{0.0690 \, Lk_l \, \text{Pr}^{0.5} \, (T_{sat} - T_s)}{\mu_l \, h_{fg}^*} \left(\frac{g}{v_l^2} \right)^{1/3} - 151 \, \text{Pr}^{0.5} + 253 \right]^{4/3}$$

$$(8\text{-}29)$$

Nondimensionalized heat transfer coefficients for the wave-free laminar, wavy laminar, and turbulent flow of condensate on vertical plates are plotted in Figure 8-26.

2 Inclined Plates

Equation 8-12 was developed for vertical plates, but it can also be used for laminar film condensation on the upper surfaces of plates that are *inclined* by an angle θ from the *vertical*, by replacing g in that equation by $g \cos \theta$ (Fig. 8-27). This approximation gives satisfactory results especially for $\theta \leq 60°$. Note that the condensation heat transfer coefficients on vertical and inclined plates are related to each other by

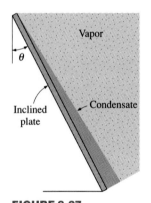

FIGURE 8-27

Film condensation on an inclined plate.

$$h_{\text{inclined}} = h_{\text{vert}} \, (\cos \theta)^{1/4} \qquad \text{(laminar)} \qquad (8\text{-}30)$$

The relation above is developed for laminar flow of condensate, but it can also be used for wavy laminar flows as an approximation.

3 Vertical Tubes

Equation 8-22 for vertical plates can also be used to calculate the average heat transfer coefficient for laminar film condensation on the outer surfaces of vertical tubes provided that the tube diameter is large relative to the thickness of the liquid film.

4 Horizontal Tubes and Spheres

Nusselt's analysis of film condensation on vertical plates can also be extended to horizontal tubes and spheres. The average heat transfer coefficient for film condensation on the outer surfaces of a *horizontal tube* is determined to be

$$h_{horiz} = 0.729 \left[\frac{g\rho_l(\rho_l - \rho_v) h_{fg}^* k_l^3}{\mu_l(T_{sat} - T_s)D} \right]^{1/4} \quad (W/m^2 \cdot {}^\circ C) \quad (8\text{-}31)$$

where D is the diameter of the horizontal tube. This relation can easily be modified for a *sphere* by replacing the constant 0.729 by 0.815.

A comparison of the heat transfer coefficient relations for a vertical tube of height L and a horizontal tube of diameter D yields

$$\frac{h_{vert}}{h_{horiz}} = 1.29 \left(\frac{D}{L} \right)^{1/4} \quad (8\text{-}32)$$

Setting $h_{vertical} = h_{horizontal}$ gives $L = 1.29^4 D = 2.77D$, which implies that for a tube whose length is 2.77 times its diameter, the average heat transfer coefficient for laminar film condensation will be the *same* whether the tube is positioned horizontally or vertically. For $L > 2.77D$, the heat transfer coefficient will be higher in the horizontal position. Considering that the length of a tube in any practical application is several times its diameter, it is common practice to place the tubes in a condenser *horizontally* to *maximize* the condensation heat transfer coefficient on the outer surfaces of the tubes.

5 Horizontal Tube Banks

Horizontal tubes stacked on top of each other as shown in Figure 8-28 are commonly used in condenser design. The average thickness of the liquid film at the lower tubes is much larger as a result of condensate falling on top of them from the tubes directly above. Therefore, the average heat transfer coefficient at the lower tubes in such arrangements is smaller. Assuming the condensate from the tubes above to the ones below drain smoothly, the average film condensation heat transfer coefficient for all tubes in a vertical tier can be expressed as

$$h_{horiz,\,N\,tubes} = 0.729 \left[\frac{g\rho_l(\rho_l - \rho_v) h_{fg}^* k_l^3}{\mu_l(T_{sat} - T_s)\,ND} \right]^{1/4} = \frac{1}{N^{1/4}} h_{horiz,\,1\,tube} \quad (8\text{-}33)$$

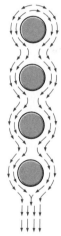

FIGURE 8-28

Film condensation on a vertical tier of horizontal tubes.

Note that the relation above can be obtained from the heat transfer coefficient relation for a horizontal tube by replacing D in that relation by ND. This relation does not account for the increase in heat transfer due to the ripple formation and turbulence caused during drainage, and thus generally yields conservative results.

Effect of Vapor Velocity

In the analysis above we assumed the vapor velocity to be small and thus the vapor drag exerted on the liquid film to be negligible, which is usually

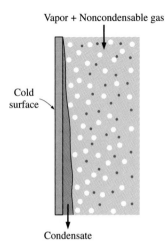

Vapor + Noncondensable gas

Cold
surface

Condensate

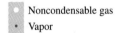

Noncondensable gas

Vapor

FIGURE 8-29

The presence of a noncondensable gas
in a vapor prevents the vapor molecules
from reaching the cold surface easily,
and thus impedes condensation heat
transfer.

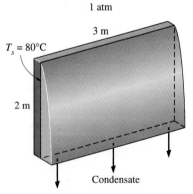

1 atm

3 m

$T_s = 80°C$

2 m

Condensate

FIGURE 8-30

Schematic for Example 8-4.

the case. However, when the vapor velocity is high, the vapor will "pull" the liquid at the interface along since the vapor velocity at the interface must drop to the value of the liquid velocity. If the vapor flows downward (i.e., in the same direction as the liquid), this additional force will increase the average velocity of the liquid and thus decrease the film thickness. This, in turn, will decrease the thermal resistance of the liquid film and thus increase heat transfer. Upward vapor flow has the opposite effects: the vapor exerts a force on the liquid in the opposite direction to flow, thickens the liquid film, and thus decreases heat transfer. Condensation in the presence of high vapor flow is studied (e.g., Shekriladze and Gomelauri, 1966) and heat transfer relations are obtained, but a detailed analysis of this topic is beyond the scope of this introductory text.

The Presence of Noncondensable Gases in Condensers

Most condensers used in steam power plants operate at pressures well below the atmospheric pressure (usually under 0.1 atm) to maximize cycle thermal efficiency, and operation at such low pressures raises the possibility of air (a noncondensable gas) leaking into the condensers. Experimental studies show that the presence of noncondensable gases in the vapor has a detrimental effect on condensation heat transfer. Even small amounts of a noncondensable gas in the vapor cause significant drops in heat transfer coefficient during condensation. For example, the presence of less than 1 percent (by mass) of air in steam can reduce the condensation heat transfer coefficient by more than half. Therefore, it is common practice to periodically vent out the noncondensable gases that accumulate in the condensers to ensure proper operation.

The drastic reduction in the condensation heat transfer coefficient in the presence of a noncondensable gas can be explained as follows: When the vapor mixed with a noncondensable gas condenses, only the noncondensable gas remains in the vicinity of the surface (Fig. 8-29). This gas layer acts as a *barrier* between the vapor and the surface, and makes it difficult for the vapor to reach the surface. The vapor now must diffuse through the noncondensable gas first before reaching the surface, and this reduces the effectiveness of the condensation process.

Experimental studies show that heat transfer in the presence of a noncondensable gas strongly depends on the nature of the vapor flow and the flow velocity. As you would expect, a *high flow velocity* is more likely to remove the stagnant noncondensable gas from the vicinity of the surface, and thus *improve* heat transfer.

EXAMPLE 8-4 Condensation of Steam on a Vertical Plate

Saturated steam at atmospheric pressure condenses on a 2-m-high and 3-m-wide vertical plate that is maintained at 80°C by circulating cooling water through the other side (Fig. 8-30). Determine (*a*) the rate of heat transfer by condensation to the plate and (*b*) the rate at which the condensate drips off the plate at the bottom.

Solution Saturated steam at 1 atm and thus at $T_{sat} = 100°C$ condenses on a vertical plate at 80°C. The rates of heat transfer and condensation are to be determined.

Assumptions **1** Steady operating conditions exist. **2** The plate is isothermal. **3** The condensate flow is wavy-laminar over the entire plate (this assumption will be verified). **4** The density of vapor is much smaller than the density of liquid, $\rho_v \ll \rho_l$.

Properties The properties of water at the saturation temperature of 100°C are $h_{fg} = 2257 \times 10^3$ J/kg and $\rho_v = 0.60$ kg/m³. The properties of liquid water at the film temperature of $T_f = (T_{sat} + T_s)/2 = (100 + 80)/2 = 90$°C are (Table A-9)

$$\rho_l = 965.3 \text{ kg/m}^3 \qquad\qquad C_{pl} = 4206 \text{ J/kg} \cdot °\text{C}$$
$$\mu_l = 0.315 \times 10^{-3} \text{ kg/m} \cdot \text{s} \qquad k_l = 0.675 \text{ W/m} \cdot °\text{C}$$
$$\nu_l = \mu_l/\rho_l = 0.326 \times 10^{-6} \text{ m}^2/\text{s}$$

Analysis (a) The modified latent heat of vaporization is

$$h_{fg}^* = h_{fg} + 0.68C_{pl}(T_{sat} - T_s)$$
$$= 2257 \times 10^3 \text{ J/kg} + 0.68 \times 4206 \text{ J/kg} \cdot °\text{C}(100 - 80)°\text{C}$$
$$= 2314 \times 10^3 \text{ J/kg}$$

For wavy-laminar flow, the Reynolds number is determined from Equation 8-27 to be

$$\text{Re} = \text{Re}_{\text{vertical, wavy}} = \left[4.81 + \frac{3.70Lk_l(T_{sat} - T_s)}{\mu_l\, h_{fg}^*}\left(\frac{g}{\nu_l^2}\right)^{1/3} \right]^{0.820}$$

$$= \left[4.81 + \frac{3.70(3 \text{ m})(0.675 \text{ W/m} \cdot °\text{C})(100 - 90)°\text{C}}{(0.315 \times 10^{-3} \text{ kg/m} \cdot \text{s})(2314 \times 10^3 \text{ J/kg})} \right.$$
$$\left. \times \left(\frac{9.8 \text{ m/s}^2}{(0.326 \times 10^{-6} \text{ m}^2/\text{s})^2}\right)^{1/3} \right]^{0.82}$$

$$= 1287$$

which is between 30 and 1800, and thus our assumption of wavy laminar flow is correct. Then the condensation heat transfer coefficient is determined from Equation 8-25 to be

$$h = h_{\text{vertical, wavy}} = \frac{\text{Re}\, k_l}{1.08\, \text{Re}^{1.22} - 5.2}\left(\frac{g}{\nu_l^2}\right)^{1/3}$$

$$= \frac{1287 \times (0.675 \text{ W/m} \cdot °\text{C})}{1.08(1287)^{1.22} - 5.2}\left(\frac{9.8 \text{ m/s}^2}{(0.326 \times 10^{-6} \text{ m}^2/\text{s})^2}\right)^{1/3} = 5848 \text{ W/m}^2 \cdot °\text{C}$$

The heat transfer surface area of the plate is $A = w \times L = (3 \text{ m})(2 \text{ m}) = 6 \text{ m}^2$. Then the rate of heat transfer during this condensation process becomes

$$\dot{Q} = hA(T_{sat} - T_s) = (5848 \text{ W/m}^2 \cdot °\text{C})(6 \text{ m}^2)(100 - 80)°\text{C} = \textbf{701,800 W}$$

(b) The rate of condensation of water is determined from

$$\dot{m}_{\text{condensation}} = \frac{\dot{Q}}{h_{fg}^*} = \frac{701,800 \text{ J/s}}{2314 \times 10^3 \text{ J/kg}} = \textbf{0.303 kg/s}$$

That is, steam will condense on the surface at a rate of 303 grams per second.

EXAMPLE 8-5 Condensation of Steam on a Tilted Plate

What would your answer be to the preceding example problem if the plate were tilted 30° from the vertical, as shown in Figure 8-31?

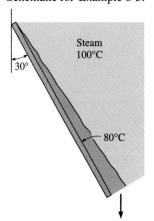

FIGURE 8-31
Schematic for Example 8-5.

Solution (a) The heat transfer coefficient in this case can be determined from the vertical plate relation by replacing g by $g\cos\theta$. But we will use Equation 8-30 instead since we already know the value for the vertical plate from the preceding example:

$$h = h_{\text{inclined}} = h_{\text{vert}}\,(\cos\theta)^{1/4} = (5848\ \text{W/m}^2\cdot°\text{C})(\cos 30°)^{1/4} = 5641\ \text{W/m}^2\cdot°\text{C}$$

The heat transfer surface area of the plate is still 6 m². Then the rate of condensation heat transfer in the tilted plate case becomes

$$\dot{Q} = hA(T_{\text{sat}} - T_s) = (5641\ \text{W/m}^2\cdot°\text{C})(6\ \text{m}^2)(100 - 80)°\text{C} = \textbf{677,000 W}$$

(b) The rate of condensation of steam is again determined from

$$\dot{m}_{\text{condensation}} = \frac{\dot{Q}}{h_{fg}^*} = \frac{677,000\ \text{J/s}}{2314 \times 10^3\ \text{J/kg}} = \textbf{0.293 kg/s}$$

Discussion Note that the rate of condensation decreased by about 3.6 percent when the plate is tilted.

Steam, 40°C   30°C
Cooling water

FIGURE 8-32
Schematic for Example 8-6.

EXAMPLE 8-6 Condensation of Steam on Horizontal Tubes

The condenser of a steam power plant operates at a pressure of 7.38 kPa. Steam at this pressure condenses on the outer surfaces of horizontal pipes through which cooling water circulates. The outer diameter of the pipes is 3 cm, and the outer surfaces of the pipes are maintained at 30°C (Fig. 8-32). Determine (a) the rate of heat transfer to the cooling water circulating in the pipes and (b) the rate of condensation of steam per unit length of a horizontal pipe.

Solution (a) Saturated steam at a pressure of 7.38 kPa and thus at a saturation temperature of $T_{\text{sat}} = 40°\text{C}$ (Table A-9) condenses on a horizontal tube at 30°C. The rates of heat transfer and condensation are to be determined.

Assumptions **1** Steady operating conditions exist. **2** The tube is isothermal.

Properties The properties of water at the saturation temperature of 40°C are $h_{fg} = 2407 \times 10^3$ J/kg and $\rho_v = 0.05$ kg/m³. The properties of liquid water at the film temperature of $T_f = (T_{\text{sat}} + T_s)/2 = (40 + 30)/2 = 35°\text{C}$ are (Table A-9)

$$\rho_l = 994\ \text{kg/m}^3 \qquad\qquad C_{pl} = 4178\ \text{J/kg}\cdot°\text{C}$$
$$\mu_l = 0.720 \times 10^{-3}\ \text{kg/m}\cdot\text{s} \qquad\qquad k_l = 0.623\ \text{W/m}\cdot°\text{C}$$

Analysis The modified latent heat of vaporization is

$$h_{fg}^* = h_{fg} + 0.68 C_{pl}(T_{\text{sat}} - T_s)$$
$$= 2407 \times 10^3\ \text{J/kg} + 0.68 \times 4178\ \text{J/kg}\cdot°\text{C}(40 - 30)°\text{C}$$
$$= 2435 \times 10^3\ \text{J/kg}$$

Noting that $\rho_v \ll \rho_l$ (since $0.05 \ll 994$), the heat transfer coefficient for condensation on a single horizontal tube is determined from Equation 8-31 to be

$$h = h_{\text{horizontal}} = 0.729\left[\frac{g\rho_l(\rho_l - \rho_v)\,h_{fg}^*\,k_l^3}{\mu_l(T_{\text{sat}} - T_s)D}\right]^{1/4} \cong 0.729\left[\frac{g\rho_l^2\,h_{fg}^*\,k_l^3}{\mu_1(T_{\text{sat}} - T_s)D}\right]^{1/4}$$

$$= 0.729\left[\frac{(9.8\ \text{m/s}^2)(994\ \text{kg/m}^3)^2\,(2435 \times 10^3\ \text{J/kg})(0.623\ \text{W/m}\cdot°\text{C})^3}{(0.720 \times 10^{-3}\ \text{kg/m}\cdot\text{s})(40 - 30)°\text{C}(0.03\ \text{m})}\right]^{1/4}$$

$$= 9292\ \text{W/m}^2\cdot°\text{C}$$

The heat transfer surface area of the pipe per unit of its length is $A = \pi DL = \pi(0.03 \text{ m})(1 \text{ m}) = 0.09425 \text{ m}^2$. Then the rate of heat transfer during this condensation process becomes

$$\dot{Q} = hA(T_{sat} - T_s) = (9292 \text{ W/m}^2 \cdot °C)(0.09425 \text{ m}^2)(40 - 30)°C = \textbf{8758 W}$$

(b) The rate of condensation of steam is determined from

$$\dot{m}_{condensation} = \frac{\dot{Q}}{h_{fg}^*} = \frac{8578 \text{ J/s}}{2435 \times 10^3 \text{ J/kg}} = \textbf{0.00360 kg/s}$$

Therefore, steam will condense on the horizontal tube at a rate of 3.6 g/s or 12.9 kg/h per meter of its length.

EXAMPLE 8-7 Condensation of Steam on Horizontal Tube Banks

Repeat the proceeding example problem for the case of 12 horizontal tubes arranged in a rectangular array of 3 tubes high and 4 tubes wide, as shown in Figure 8-33.

Solution (a) Condensation heat transfer on a tube is not influenced by the presence of other tubes in its neighborhood unless the condensate from other tubes drips on it. In our case, the horizontal tubes are arranged in four vertical tiers, each tier consisting of 3 tubes. The average heat transfer coefficient for a vertical tier of N horizontal tubes is related to the one for a single horizontal tube by Equation 8-33 and is determined to be

$$h_{horiz, N \, tubes} = \frac{1}{N^{1/4}} h_{horiz, 1 \, tube} = \frac{1}{3^{1/4}} (9292 \text{ W/m}^2 \cdot °C) = 7060 \text{ W/m}^2 \cdot °C$$

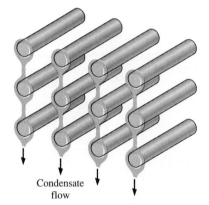

Condensate
flow

FIGURE 8-33
Schematic for Example 8-7.

Each vertical tier consists of 3 tubes, and thus the heat transfer coefficient determined above is valid for each of the four tiers. In other words, the value above can be taken to be the average heat transfer coefficient for all 12 tubes.

The surface area for all 12 tubes per unit length of the tubes is

$$A = N_{total} \, \pi DL = 12\pi(0.03 \text{ m})(1 \text{ m}) = 1.1310 \text{ m}^2$$

Then the rate of heat transfer during this condensation process becomes

$$\dot{Q} = hA(T_{sat} - T_s) = (7060 \text{ W/m}^2 \cdot °C)(1.131 \text{ m}^2)(40 - 30)°C = \textbf{79,850 W}$$

(b) The rate of condensation of steam is again determined from

$$\dot{m}_{condensation} = \frac{\dot{Q}}{h_{fg}^*} = \frac{79,850 \text{ J/s}}{2435 \times 10^3 \text{ J/kg}} = \textbf{0.0328 kg/s}$$

Therefore, steam will condense on the horizontal pipes at a rate of 32.8 grams second per meter length of the tubes.

■ FILM CONDENSATION INSIDE HORIZONTAL TUBES

far we have discussed film condensation on the *outer surfaces* of tubes other geometries, which is characterized by negligible vapor velocity the unrestricted flow of the condensate. Most condensation processes ountered in refrigeration and air-conditioning applications, however,

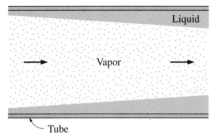

FIGURE 8-34

Condensate flow in a horizontal tube
with large vapor velocities.

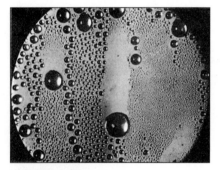

FIGURE 8-35

Dropwise condensation of steam on a
vertical surface (from Hampson and
Özişik, Ref. 9).

involve condensation on the *inner surfaces* of horizontal or vertical tubes. Heat transfer analysis of condensation inside tubes is complicated by the fact that it is strongly influenced by the vapor velocity and the rate of liquid accumulation on the walls of the tubes (Fig. 8-34).

For *low vapor velocities,* Chato recommends the following expression for condensation

$$h_{\text{internal}} = 0.555 \left[\frac{g\rho_l(\rho_l - \rho_v)k_l^3}{\mu_l(T_{\text{sat}} - T_s)} \left(h_{fg} + \frac{3}{8} C_{pl}(T_{\text{sat}} - T_s) \right) \right]^{1/4} \qquad (8\text{-}34)$$

for

$$\text{Re}_{\text{vapor}} = \left(\frac{\rho_v \mathcal{V}_v D}{\mu_v} \right)_{\text{inlet}} < 35{,}000 \qquad (8\text{-}35)$$

where the Reynolds number of the vapor is to be evaluated at the tube *inlet* conditions using the internal tube diameter as the characteristic length. Heat transfer coefficient correlations for higher vapor velocities are given by Rohsenow.

8-7 ■ DROPWISE CONDENSATION

Dropwise condensation, characterized by countless droplets of varying diameters on the condensing surface instead of a continuous liquid film, is one of the most effective mechanisms of heat transfer, and extremely large heat transfer coefficients can be achieved with this mechanism (Fig. 8-35).

In dropwise condensation, the small droplets that form at the nucleation sites on the surface grow as a result of continued condensation, coalesce into large droplets, and slide down when they reach a certain size, clearing the surface and exposing it to vapor. There is no liquid film in this case to resist heat transfer. As a result, with dropwise condensation, heat transfer coefficients can be achieved that are more than 10 times larger than those associated with film condensation. Large heat transfer coefficients enable designers to achieve a specified heat transfer rate with a smaller surface area, and thus a smaller (and less expensive) condenser. Therefore, dropwise condensation is the preferred mode of condensation in heat transfer applications.

The challenge in dropwise condensation is not to achieve it, but rather, to *sustain* it for prolonged periods of time. Dropwise condensation is achieved by *adding* a promoting chemical into the vapor, *treating* the surface with a promoter chemical, or *coating* the surface with a polymer such as teflon or a noble metal such as gold, silver, rhodium, palladium, or platinum. The *promoters* used include various waxes and fatty acids such as oleic, stearic, and linoic acids. They lose their effectiveness after a while, however, because of fouling, oxidation, and the removal of the promoter from the surface. It is possible to sustain dropwise condensation for over a year by the combined effects of surface coating and periodic injection of the promoter into the vapor. However, any gain in heat transfer must be weighed against the cost associated with sustaining dropwise condensation.

Dropwise condensation has been studied experimentally for a number of surface–fluid combinations. Of these, the studies on the condensation of steam on copper surfaces has attracted the most attention because of their widespread use in steam power plants. P. Griffith recommends the following simple correlations for dropwise condensation of *steam* on *copper surfaces*:

$$h_{\text{dropwise}} = \begin{cases} 51{,}104 + 2044T_{\text{sat}}, & 22°\text{C} < T_{\text{sat}} < 100°\text{C} \quad \text{(8-36)} \\ 255{,}310 & T_{\text{sat}} > 100°\text{C} \quad \text{(8-37)} \end{cases}$$

where T_{sat} is in °C and the heat transfer coefficient h_{dropwise} is in $\text{W/m}^2 \cdot °\text{C}$.

The very high heat transfer coefficients achievable with dropwise condensation are of little significance if the material of the condensing surface is not a good conductor like copper or if the thermal resistance on the other side of the surface is too large. In steady operation, heat transfer from one medium to another depends on the sum of the thermal resistances on the path of heat flow, and a large thermal resistance may overshadow all others and dominate the heat transfer process. In such cases, improving the accuracy of a small resistance (such as one due to condensation or boiling) makes hardly any difference in overall heat transfer calculations,

8-8 ■ SUMMARY

Boiling occurs when a liquid is in contact with a surface maintained at a temperature T_s sufficiently above the saturation temperature T_{sat} of the liquid. Boiling is classified as *pool boiling* or *flow boiling* depending on the presence of bulk fluid motion. Boiling is called *pool boiling* in the absence of bulk fluid flow and *flow boiling* (or *forced convection boiling*) in its presence. Pool and flow boiling are further classified as *subcooled boiling* and *saturated boiling* depending on the bulk liquid temperature. Boiling is said to be *subcooled* (or *local*) when the temperature of the main body of the liquid is below the saturation temperature T_{sat} and *saturated* (or *bulk*) when the temperature of the liquid is equal to T_{sat}. Boiling exhibits different regimes depending on the value of the excess temperature ΔT_{excess}. Four different boiling regimes are observed: natural convection boiling, nucleate boiling, transition boiling, and film boiling. These regimes are illustrated on the *boiling curve*. The rate of evaporation and the rate of heat transfer in nucleate boiling increase with increasing ΔT_{excess} and reach a maximum at some point. The heat flux at this point is called the *critical* (or *maximum*) *heat flux*, $\dot{q}_{\text{max}}$. The rate of heat transfer in nucleate pool boiling is determined from

$$\dot{q}_{\text{nucleate}} = \mu_l \, h_{fg} \left[\frac{g(\rho_l - \rho_v)}{\sigma} \right]^{1/2} \left[\frac{C_{pl}(T_s - T_{\text{sat}})}{C_{sf} \, h_{fg} \, \text{Pr}_l^n} \right]^3$$

The *maximum* (or *critical*) *heat flux* in nucleate pool boiling is determined from

$$\dot{q}_{\text{max}} = C_{\text{cr}} \, h_{fg} [\sigma g \rho_v^2 (\rho_l - \rho_v)]^{1/4}$$

where the value of the constant C_{cr} is about 0. 15. The minimum heat flux is given by

$$\dot{q}_{min} = 0.09\rho_v\,h_{fg}\left[\frac{\sigma g(\rho_l - \rho_v)}{(\rho_l + \rho_v)^2}\right]^{1/4}$$

The heat flux for stable *film boiling* on the outside of a *horizontal cylinder* or *sphere* of diameter D is given by

$$\dot{q}_{film} = C_{film}\left[\frac{gk_v^3\,\rho_v(\rho_l - \rho_v)[h_{fg} + 0.4C_{pv}\,(T_s - T_{sat})]}{\mu_v\,D(T_s - T_{sat})}\right]^{1/4}(T_s - T_{sat})$$

where the constant $C_{film} = 0.62$ for horizontal cylinders and 0.67 for spheres, and k_v is the thermal conductivity of the vapor. The *vapor* properties are to be evaluated at the *film temperature* $T_f = (T_{sat} + T_s)/2$, which is the *average temperature* of the vapor film. The liquid properties and h_{fg} are to be evaluated at the saturation temperature at the specified pressure.

Two distinct forms of condensation are observed in nature: film condensation and dropwise condensation. In *film condensation*, the condensate wets the surface and forms a liquid film on the surface that slides down under the influence of gravity. In *dropwise condensation*, the condensed vapor forms countless droplets of varying diameters on the surface instead of a continuous film.

The Reynolds number for the condensate flow is defined as

$$\text{Re} = \frac{D_h\,\rho_l\,\mathcal{V}_l}{\mu_l} = \frac{4\,A\rho_l\,\mathcal{V}_l}{p\mu_l} = \frac{4\dot{m}}{p\mu_l}$$

and

$$\text{Re} = \frac{4\dot{Q}_{conden}}{p\mu_l\,h_{fg}^*} = \frac{4\,Ah(T_{sat} - T_s)}{p\mu_l\,h_{fg}^*}$$

where h_{fg}^* is the *modified latent heat of vaporization*, defined as

$$h_{fg}^* = h_{fg} + 0.68C_{pl}\,(T_{sat} - T_s)$$

and represents heat transfer during condensation per unit mass of condensate. Here C_{pl} is the specific heat of the liquid in J/kg · °C.

Using some simplifying assumptions, the *average heat transfer coefficient* for film condensation on a vertical plate of height L is determined to be

$$h_{vert} = 0.943\left[\frac{g\rho_l\,(\rho_l - \rho_v)\,h_{fg}^*k_l^3}{\mu_l\,(T_{sat} - T_s)L}\right]^{1/4}\qquad(\text{W/m}^2 \cdot {}^\circ\text{C})$$

All properties of the *liquid* are to be evaluated at the film temperature $T_f = (T_{sat} + T_s)/2$. The h_{fg} and ρ_v are to be evaluated at T_{sat}. Condensate flow is *smooth* and *wave-free laminar* for about $\text{Re} \leq 30$, *wavy-laminar* in the range of $450 < \text{Re} < 1800$, and fully *turbulent* for $\text{Re} > 1800$. Heat transfer coefficients in the wavy-laminar and turbulent flow regions are determined from

$$h_{vert,wavy} = \frac{Re\, k_l}{1.08Re^{1.22} - 5.2}\left(\frac{g}{v_l^2}\right)^{1/3}, \qquad \begin{matrix} 30 < Re < 1800 \\ \rho_v \ll \rho_l \end{matrix}$$

$$h_{vert,turbulent} = \frac{Re\, k_l}{8750 + 58\, Pr^{-0.5}\,(Re^{0.75} - 253)}\left(\frac{g}{v_l^2}\right)^{1/3}, \qquad \begin{matrix} Re > 1800 \\ \rho_v \ll \rho_l \end{matrix}$$

Equations for vertical plates can also be used for laminar film condensation on the upper surfaces of the plates that are inclined by an angle θ from the vertical, by replacing g in that equation by $g\cos\theta$. Vertical plate equations can also be used to calculate the average heat transfer coefficient for laminar film condensation on the outer surfaces of vertical tubes provided that the tube diameter is large relative to the thickness of the liquid film.

The average heat transfer coefficient for film condensation on the outer surfaces of a *horizontal tube* is determined to be

$$h_{horiz} = 0.729\left[\frac{g\rho_l(\rho_l - \rho_v)h_{fg}^*k_l^3}{\mu_l(T_{sat} - T_s)D}\right]^{1/4} \qquad (W/m^2\cdot°C)$$

where D is the diameter of the horizontal tube. This relation can easily be modified for a *sphere* by replacing the constant 0.729 by 0.815. It can also be used for *N horizontal tubes* stacked on top of each other by replacing D in the denominator by ND.

For low vapor velocities, film condensation heat transfer *inside horizontal tubes* can be determined from

$$h_{internal} = 0.555\left[\frac{g\rho_l(\rho_l - \rho_v)k_l^3}{\mu_l(T_{sat} - T_s)}\left(h_{fg} + \frac{3}{8}C_{pl}(T_{sat} - T_s)\right)\right]^{1/4}$$

and

$$Re_{vapor} = \left(\frac{\rho_v\, \mathcal{V}_v\, D}{\mu_v}\right)_{inlet} < 35,000$$

where the Reynolds number of the vapor is to be evaluated at the tube inlet conditions using the internal tube diameter as the characteristic length. Finally, the heat transfer coefficient for *dropwise condensation* of steam on copper surfaces is given by

$$h_{dropwise} = \begin{cases} 51,104 + 2044T_{sat}, & 22°C < T_{sat} < 100°C \\ 255,310, & T_{sat} > 100°C \end{cases}$$

where T_{sat} is in °C and the heat transfer coefficient $h_{dropwise}$ is in $W/m^2\cdot°C$.

REFERENCES AND SUGGESTED READING

1. N. Arai, T. Fukushima, A. Arai, T. Nakajima, K. Fujie, and Y. Nakayama. "Heat Transfer Tubes Enhancing Boiling and Condensation in Heat Exchangers of a Refrigeration Machine." *ASHRAE Journal* 83 (1977), p. 58.

2. Y. Bayazitoglu and M. N. Özişik. *Elements of Heat Transfer.* New York: McGraw-Hill, 1988.

3. P. J. Berensen. "Film Boiling Heat Transfer for a Horizontal Surface." *Journal of Heat Transfer* 83 (1961), pp. 351–58.

4. P. J. Berensen. "Experiments in Pool Boiling Heat Transfer." *International Journal of Heat Mass Transfer* 5 (1962), pp. 985–99.

5. L. A. Bromley. "Heat Transfer in Stable Film Boiling." *Chemical Engineering Prog.* 46 (1950), pp. 221–27.

6. J. C. Chato. "Laminar Condensation inside Horizontal and Inclined Tubes." *ASHRAE Journal* 4 (1962), p. 52.

7. M. T. Cichelli and C. F. Bonilla. "Heat Transfer to Liquids Boiling under Pressure." *Transactions of AIChE* 41 (1945), pp. 755–87.

8. P. Griffith. "Dropwise Condensation." In *Heat Exchanger Design Handbook,* ed. E. U. Schlunder, Vol 2, Ch. 2.6.5. New York: Hemisphere, 1983.

9. H. Hampson and N. Özişik. "An Investigation into the Condensation of Steam." *Proceedings of the Institute of Mechanical Engineers,* London 1B (1952), pp. 282–94.

10. J. P. Holman. *Heat Transfer.* 7th ed. New York: McGraw-Hill, 1990.

11. F. P. Incropera and D. P. DeWitt. *Introduction to Heat Transfer.* 2nd ed. New York: John Wiley & Sons, 1990.

12. J. J. Jasper. "The Surface Tension of Pure Liquid Compounds." *Journal of Physical and Chemical Reference Data* 1, No. 4 (1972), pp. 841–1009.

13. S. S. Kutateladze. *Fundamentals of Heat Transfer.* New York: Academic Press, 1963.

14. S. S. Kutateladze. "On the Transition to Film Boiling under Natural Convection." *Kotloturbostroenie* 3 (1948), p. 48.

15. D. A. Labuntsov. "Heat Transfer in Film Condensation of Pure Steam on Vertical Surfaces and Horizontal Tubes." *Teploenergetika* 4 (1957), pp. 72–80.

16. J. H. Lienhard and V. K. Dhir. "Extended Hydrodynamic Theory of the Peak and Minimum Pool Boiling Heat Fluxes." *NASA Report,* NASA-CR-2270, July 1973.

17. J. H. Lienhard and V. K. Dhir. "Hydrodynamic Prediction of Peak Pool Boiling Heat Fluxes from Finite Bodies." *Journal of Heat Transfer* 95 (1973), pp. 152–58.

18. W. H. McAdams. *Heat Transmission.* 3rd ed. New York: McGraw-Hill, 1954.

19. A. F. Mills. *Basic Heat and Mass Transfer.* Burr Ridge, IL: Richard D. Irwin, 1995.

20. W. M. Rohsenow. "A Method of Correlating Heat Transfer Data for Surface Boiling of Liquids." *ASME Transactions* 74 (1952), pp. 969–75.

21. W. M. Rohsenow. "Film Condensation." In *Handbook of Heat Transfer,* ed. W. M. Rohsenow and J. P. Hartnett, Ch. 12A. New York: McGraw-Hill, 1973.

22. I. G. Shekriladze, I. G. Gomelauri, and V. I. Gomelauri. "Theoretical Study of Laminar Film Condensation of Flowing Vapor." *International Journal of Heat Mass Transfer* 9 (1966), pp. 591–92.

23. N. V. Suryanarayana. *Engineering Heat Transfer.* St. Paul, MN: West Publishing, 1995.

24. L. C. Thomas. *Heat Transfer.* Englewood Cliffs, NJ: Prentice Hall, 1992.

25. J. W. Westwater and J. G. Santangelo. *Ind. Eng. Chem.* 47 (1955), p. 1605.

26. F. W. White. *Heat and Mass Transfer.* Reading, MA: Addison-Wesley, 1988.

27. N. Zuber. "On the Stability of Boiling Heat Transfer. *ASME Transactions* 80 (1958), pp. 711–20.

PROBLEMS*

Boiling Heat Transfer

8-1C   What is boiling? What mechanisms are responsible for the very high heat transfer coefficients in nucleate boiling?

8-2C   Does the amount of heat absorbed as 1 kg of saturated liquid water boils at 100°C have to be equal to the amount of heat released as 1 kg of saturated water vapor condenses at 100°C?

8-3C   What is the difference between evaporation and boiling?

8-4C   What is the difference between pool boiling and flow boiling?

8-5C   What is the difference between subcooled and saturated boiling?

8-6C   Draw the boiling curve and identify the different boiling regimes. Also, explain the characteristics of each regime.

8-7C   How does film boiling differ from nucleate boiling? Is the boiling heat flux necessarily higher in the stable film boiling regime than it is in the nucleate boiling regime?

8-8C   Draw the boiling curve and identify the burnout point on the curve. Explain how burnout is caused. Why is the burnout point avoided in the design of boilers?

8-9C   Discuss some methods of enhancing pool boiling heat transfer permanently.

---

*Students are encouraged to answer *all* the concept "C" questions.

1 atm

110°C

FIGURE P8-11

1 atm

3 kW

FIGURE P8-15

FIGURE P8-17

1 atm

Coffee
maker

1 L

8-10C Name the different boiling regimes in the order they occur in a vertical tube during flow boiling.

8-11 Water is to be boiled at atmospheric pressure in a mechanically polished steel pan placed on top of a heating unit. The inner surface of the bottom of the pan is maintained at 110°C. If the diameter of the bottom of the pan is 35 cm, determine (*a*) the rate of heat transfer to the water and (*b*) the rate of evaporation.

8-12 Water is to be boiled at atmospheric pressure on a 3-cm diameter mechanically polished steel heater. Determine the maximum heat flux that can be attained in the nucleate boiling regime and the surface temperature of the heater surface in that case.

8-13E Water is boiled at atmospheric pressure by a horizontal polished copper heating element of diameter $D = 0.5$ in. and emissivity $\varepsilon = 0.08$ immersed in water. If the surface temperature of the heating element is 788°F, determine the rate of heat transfer to the water per unit length of the heating element. *Answer:* 2478 Btu/h

8-14E Repeat Problem 8-13E for a heating element temperature of 988°F.

8-15 Water is to be boiled at sea level in a 30-cm-diameter mechanically polished AISI 304 stainless steel pan placed on top of a 3-kW electric burner. If 60 percent of the heat generated by the burner is transferred to the water during boiling, determine the temperature of the inner surface of the bottom of the pan. Also, determine the temperature difference between the inner and outer surfaces of the bottom of the pan if it is 4 mm thick.

8-16 Repeat Problem 8-15 for a location at an elevation of 1500 m where the atmospheric pressure is 84.5 kPa and thus the boiling temperature of water is 95°C. *Answers:* 100.6°C, 6.83°C

8-17 Water is boiled at sea level in a coffee maker equipped with a 20-cm long 0.4-cm diameter immersion-type electric heating element made of mechanically polished stainless steel. The coffee maker initially contains 1 L of water at 18°C. Once boiling starts, it is observed that half of the water in the coffee maker evaporates in 25 min. Determine the power rating of the electric heating element immersed in water and the surface temperature of the heating element. Also determine how long it will take for this heater to raise the temperature of 1 L of cold water from 18°C to the boiling temperature.

8-18 Repeat Problem 8-17 for a copper heating element.

8-19 A 50-cm-long, 2-cm-diameter brass heating element is to be used to boil water at 120°C. If the surface temperature of the heating element is not to exceed 125°C, determine the highest rate of steam production in the boiler, in kg/h. *Answer:* 14.9 kg/h

8-20 To understand the burnout phenomenon, boiling experiments are conducted in water at atmospheric pressure using an electrically heated 30-cm-long, 3-mm-diameter nickel-plated horizontal wire. Determine (*a*) the critical heat flux and (b) the increase in the temperature of the wire as the operating point jumps from the nucleate boiling to the film boiling regime at the critical heat flux. Take the emissivity of the wire to be 0.5.

8-21 Water is boiled at 1 atm pressure in a 20-cm-internal-diameter teflon-pitted stainless steel pan on an electric range. If it is observed that the water level in the pan drops by 10 cm in 30 min, determine the inner surface temperature of the pan. *Answer:* 111.5°C

8-22 Repeat Problem 8-21 for a polished copper pan.

8-23 In a gas-fired boiler, water is boiled at 150°C by hot gases flowing through 50-m-long, 5-cm-outer-diameter mechanically polished stainless steel pipes submerged in water. If the outer surface temperature of the pipes is 165°C, determine (a) the rate of heat transfer from the hot gases to water, (b) the rate of evaporation, (c) the ratio of the critical heat flux to the present heat flux, and (d) the surface temperature of the pipe at which critical heat flux occurs. *Answers:* (a) 10,865 kW, (b) 5.139 kg/s, (c) 1.34, (d) 166.5°C

8-24 Repeat Problem 8-23 for a boiling temperature of 160°C.

8-25E Water is boiled at 250°F by a 2-ft-long and 0.5-in.-diameter nickel-plated electric heating element maintained at 280°F. Determine (a) the boiling heat transfer coefficient, (b) the electric power consumed by the heating element, and (c) the rate of evaporation of water.

8-26E Repeat Problem 8-25E for a platinum-plated heating element.

8-27 Cold water enters a steam generator at 20°C and leaves as saturated steam at 100°C. Determine the fraction of heat used to preheat the liquid water from 20°C to the saturation temperature of 100°C in the steam generator. *Answer:* 12.9 percent

8-28 Cold water enters a steam generator at 20°C and leaves as saturated steam at the boiler pressure. At what pressure will the amount of heat needed to preheat the water to saturation temperature be equal to the heat needed to vaporize the liquid at the boiler pressure?

8-29 A 50-cm-long, 2-mm-diameter electric resistance wire submerged in water is used to determine the boiling heat transfer coefficient in water at 1 atm experimentally. The wire temperature is measured to be 130°C when a wattmeter indicates the electric power consumed to be 4.1 kW. Using Newton's law of cooling, determine the boiling heat transfer coefficient.

Condensation Heat Transfer

8-30C What is condensation? How does it occur?

8-31C What is the difference between film and dropwise condensation? Which is a more effective mechanism of heat transfer?

8-32C In condensate flow, how is the wetted perimeter defined? How does wetted perimeter differ from ordinary perimeter?

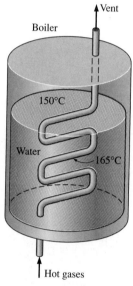

Vent

Boiler

150°C

Water

165°C

Hot gases

FIGURE P8-23

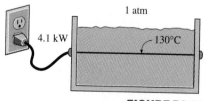

1 atm

4.1 kW

130°C

FIGURE P8-29

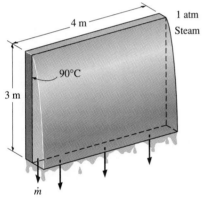

FIGURE P8-40

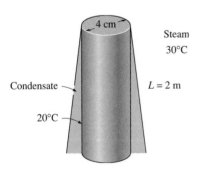

FIGURE P8-42

8-33C What is the modified latent heat of vaporization? For what is it used? How does it differ from the ordinary latent heat of vaporization?

8-34C Consider film condensation on a vertical plate. Will the heat flux be higher at the top or at the bottom of the plate? Why?

8-35C Consider film condensation on the outer surfaces of a tube whose length is 10 times its diameter. For which orientation of the tube will the heat transfer rate be the highest: horizontal or vertical? Explain. Disregard the base and top surfaces of the tube.

8-36C Consider film condensation on the outer surfaces of four long tubes. For which orientation of the tubes will the condensation heat transfer coefficient be the highest: (*a*) vertical, (*b*) horizontal side by side, (*c*) horizontal but in a vertical tier (directly on top of each other), or (*d*) a horizontal stack of two tubes high and two tubes wide?

8-37C How does the presence of a noncondensable gas in a vapor influence the condensation heat transfer?

8-38 The Reynolds number for condensate flow is defined as $Re = 4\dot{m}/p\mu_l$, where p is the wetted perimeter. Obtain simplified relations for the Reynolds number by expressing p and $\dot{m}$ by their equivalence for the following geometries: (*a*) a vertical plate of height L and width w, (*b*) a tilted plate of height L and width w inclined at an angle θ from the vertical, (*c*) a vertical cylinder of length L and diameter D, (*d*) a horizontal cylinder of length L and diameter D, and (*e*) a sphere of diameter D.

8-39 Consider film condensation on the outer surfaces of N horizontal tubes arranged in a vertical tier. For what value of N will the average heat transfer coefficient for the entire stack of tubes be equal to half of what it is for a single horizontal tube? *Answer:* 16

8-40 Saturated steam at 1 atm condenses on a 3-m-high and 4-m-wide vertical plate that is maintained at 90°C by circulating cooling water through the other side. Determine (*a*) the rate of heat transfer by condensation to the plate, and (*b*) the rate at which the condensate drips off the plate at the bottom. *Answers:* (*a*) 753,500 W, (*b*) 0.330 kg/s

8-41 Repeat Problem 8-40 for the case of the plate being tilted 60° from the vertical.

8-42 Saturated steam at 30°C condenses on the outside of a 4-cm-outer-diameter, 2-m-long vertical tube. The temperature of the tube is maintained at 20°C by the cooling water. Determine (*a*) the rate of heat transfer from the steam to the cooling water, (*b*) the rate of condensation of steam, and (*c*) the approximate thickness of the liquid film at the bottom of the tube.

8-43E Saturated steam at 95°F is condensed on the outer surfaces of an array of horizontal pipes through which cooling water circulates. The outer diameter of the pipes is 1 in. and the outer surfaces of the pipes are maintained at 85°F. Determine (*a*) the rate of heat transfer to the cooling water circulating in the pipes and (b) the rate of condensation of steam per unit length of a single horizontal pipe.

8-44E Repeat Problem 8-43E for the case of 20 horizontal pipes arranged in a rectangular array of 4 pipes high and 5 pipes wide.

8-45 Saturated steam at 55°C is to be condensed at a rate of 10 kg/h on the outside of a 3-cm-outer-diameter vertical tube whose surface is maintained at 45°C by the cooling water. Determine the tube length required.

8-46 Repeat Problem 8-45 for a horizontal tube.   *Answer:* 0.70 m

8-47 Saturated steam at 100°C condenses on a 2-m × 2-m plate that is tilted 40° from the vertical. The plate is maintained at 80°C by cooling it from the other side. Determine (a) the average heat transfer coefficient over the entire plate and (b) the rate at which the condensate drips off the plate at the bottom.

8-48 Saturated ammonia vapor at 10°C condenses on the outside of a 2-cm-outer-diameter, 5-m-long horizontal tube whose outer surface is maintained at -10°C. Determine (a) the rate of heat transfer from the ammonia and (b) the rate of condensation of ammonia. For ammonia, $h_{fg} = 1226$ kJ/kg and $\rho_v = 4.863$ kg/m³ at 10°C.

8-49 The condenser of a steam power plant operates at a pressure of 4.25 kPa. The condenser consists of 100 horizontal tubes arranged in a 10 × 10 square array. The tubes are 8 m long and have an outer diameter of 3 cm. If the tube surfaces are at 20°C, determine (a) the rate of heat transfer from the steam to the cooling water and (b) the rate of condensation of steam in the condenser.   *Answers:* (a) 1239 kW, (b) 503.9 kg/s

8-50 A large heat exchanger has several columns of tubes, with 20 tubes in each column. The outer diameter of the tubes is 1.5 cm. Saturated steam at 50°C condenses on the outer surfaces of the tubes, which are maintained at 20°C. Determine (a) the average heat transfer coefficient and (b) the rate of condensation of steam per m length of a column.

8-51 Saturated refrigerant-134a vapor at 30°C is to be condensed in a 5-m-long, 1-cm-diameter horizontal tube that is maintained at a temperature of 20°C. If the refrigerant enters the tube at a rate of 2.5 kg/min, determine the fraction of the refrigerant that will have condensed at the end of the tube. The properties of refrigerant-134a can be taken to be

$\rho_l = 1187$ kg/m³ $\mu_l = 0.201 \times 10^{-3}$ kg/m · s $k_l = 0.0796$ W/m · °C
$\rho_v = 37.5$ kg/m³ $C_{pl} = 1447$ J/kg · °C $h_{fg} = 173.3$ kJ/kg

8-52 Repeat Problem 8-51 for a tube length of 8 m.   *Answer:* 17 percent

Review Problems

8-53 Steam at 40°C condenses on the outside of a 3-cm diameter thin horizontal copper tube by cooling water that enters the tube at 25°C at an average velocity of 2 m/s and leaves at 35°C. Determine the rate of condensation of steam, the average overall heat transfer coefficient between the steam and the cooling water, and the tube length.

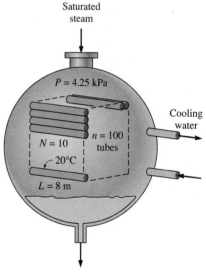

FIGURE P8-49

FIGURE P8-53

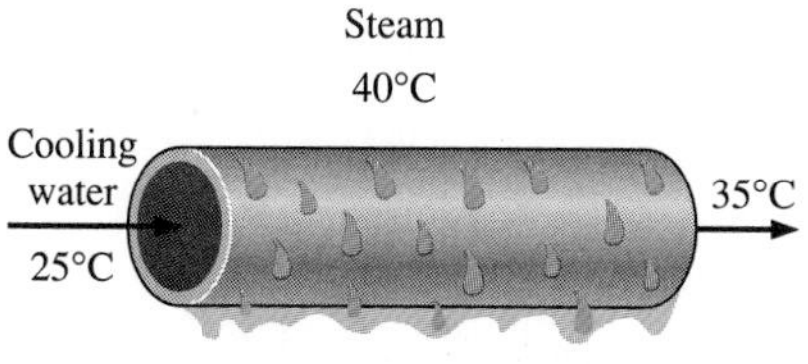

8-54 Saturated ammonia vapor at 25°C condenses on the outside of a 2-m-long, 2.5-cm-outer-diameter vertical tube maintained at 15°C. Determine (a) the average heat transfer coefficient, (b) the rate of heat transfer, and (c) the rate of condensation of ammonia. For ammonia, $h_{fg} = 1167$ kJ/kg and $\rho_v = 7.76$ kg/m^3 at 25°C.

8-55 Saturated isobutane vapor in a binary geothermal power plant is to be condensed outside an array of 8 horizontal tubes. Determine the ratio of the condensation rate for the cases of the tubes being arranged in a horizontal tier versus in a vertical tier of horizontal tubes.   *Answer:* 1.68

8-56E The condenser of a steam power plant operates at a pressure of 0.95 psia. The condenser consists of 144 horizontal tubes arranged in a 12 × 12 square array. The tubes are 15 ft long and have an outer diameter of 1.2 in. If the outer surfaces of the tubes are maintained at 80°F, determine (a) the rate of heat transfer from the steam to the cooling water and (b) the rate of condensation of steam in the condenser.

8-57E Repeat Problem 8-56E for a tube diameter of 2 in.

8-58 Water is boiled at 100°C electrically by a 50-cm-long, 2-mm-diameter horizontal resistance wire made of chemically etched stainless steel. Determine (a) the rate of heat transfer to the water and the rate of evaporation of water if the temperature of the wire is 115°C and (b) the maximum rate of evaporation in the nucleate boiling regime.
    Answers: (a) 1492 W, 2.38 kg/h, (b) 1280 kW/m^2

8-59E Saturated steam at 100°F is condensed on a 6-ft-high vertical plate that is maintained at 80°F. Determine the rate of heat transfer from the steam to the plate and the rate of condensation per foot width of the plate.

8-60 Saturated refrigerant-134a vapor at 30°C is to be condensed on the outer surface of a 7-m-long, 1.5-cm-diameter horizontal tube that is maintained at a temperature of 25°C. Determine the rate at which the refrigerant will condense, in kg/min. Use the following properties for refrigerant-134a:

$\rho_l = 1187$ kg/m^3 $\mu_l = 0.201 \times 10^{-3}$ kg/m·s $k_l = 0.0796$ W/m·°C
$\rho_v = 37.5$ kg/m^3 $C_{pl} = 1447$ J/kg·°C $h_{fg} = 173.3$ kJ/kg

8-61 Repeat Problem 8-60 for a tube diameter of 3 cm.

8-62 Saturated steam at 270.1 kPa condenses inside a horizontal, 4-m-long, 3-cm-internal-diameter pipe whose surface is maintained at 110°C. Assuming low vapor velocity, determine the average heat transfer coefficient and the rate of condensation of the steam inside the pipe.
    Answers: 3345 W/m^2·°C, 0.0116 kg/s

8-63 A 1.5-cm-diameter silver sphere initially at 30°C is suspended in a room filled with saturated steam at 100°C. Using the lumped system analysis, determine how long it will take for the temperature of the ball to rise to 50°C. Also, determine the amount of steam that condenses during this process and verify that the lumped system analysis is applicable.

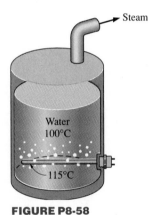

Steam

Water
100°C

115°C

FIGURE P8-58

8-64 Repeat Problem 8-63 for a 3-cm-diameter copper ball.

8-65 You have probably noticed that water vapor that condenses on a canned drink slides down, clearing the surface for further condensation. Therefore, condensation in this case can be considered to be dropwise. Determine the condensation heat transfer coefficient on a cold canned drink at 5°C that is placed in a large container filled with saturated steam at 95°C.

8-66 A resistance heater made of 2-mm-diameter nickel wire is used to heat water at 1 atm pressure. Determine the highest temperature at which this heater can operate safely without the danger of burning out.
Answer: 109.6°C

Steam
95°C

5°C

Soda

FIGURE P8-65

Computer, Design, and Essay Problems

8-67 Design the condenser of a steam power plant that has a thermal efficiency of 40 percent and generates 10 MW of net electric power. Steam enters the condenser as saturated vapor at 10 kPa, and it is to be condensed outside horizontal tubes through which cooling water from a nearby river flows. The temperature rise of the cooling water is limited to 8°C, and the velocity of the cooling water in the pipes is limited to 6 m/s to keep the pressure drop at an acceptable level. Specify the pipe diameter, total pipe length, and the arrangement of the pipes to minimize the condenser volume.

8-68 The refrigerant in a household refrigerator is condensed as it flows through the coil that is typically placed behind the refrigerator. Heat transfer from the outer surface of the coil to the surroundings is by natural convection and radiation. Obtaining information about the operating conditions of the refrigerator, including the pressures and temperatures of the refrigerant at the inlet and the exit of the coil, show that the coil is selected properly, and determine the safety margin in the selection.

8-69 Water-cooled steam condensers are commonly used in steam power plants. Obtain information about water-cooled steam condensers by doing a literature search on the topic and also by contacting some condenser manufacturers. In a report, describe the various types, the way they are designed, the limitation on each type, and the selection criteria.

8-70 Steam boilers have long been used to provide process heat as well as to generate power. Write an essay on the history of steam boilers and the evolution of modern supercritical steam power plants. What was the role of the American Society of Mechanical Engineers in this development?

8-71 The technology for power generation using geothermal energy is well established, and numerous geothermal power plants throughout the world are currently generating electricity economically. Binary geothermal plants utilize a volatile secondary fluid such as isobutane, n-pentane, and R-114 in a closed loop. Consider a binary geothermal plant with R-114 as the working fluid that is flowing at a rate of 600 kg/s. The R-114 is vaporized in a boiler at 115°C by the geothermal fluid that enters at 165°C and is condensed at 30°C outside the tubes by cooling water that enters the tubes at 18°C. Design

the condenser of this binary plant using the following data for R-114 at
condenser conditions:

$$\rho_l = 1445 \text{ kg/m}^3 \qquad \mu_l = 0.349 \times 10^{-3} \text{ kg/m} \cdot \text{s} \qquad k_l = 0.0643 \text{ W/m} \cdot {}^\circ\text{C}$$
$$\rho_v = 18.6 \text{ kg/m}^3 \qquad C_{pl} = 961 \text{ J/kg} \cdot {}^\circ\text{C} \qquad h_{fg} = 121.5 \text{ kJ/kg}$$

Specify (a) the length, diameter, and number of tubes and their arrangement
in the condenser, (b) the mass flow rate of cooling water, and (c) the flow rate
of make-up water needed if a cooling tower is used to reject the waste heat
from the cooling water. The liquid velocity is to remain under 6 m/s and the
length of the tubes is limited to 8 m.

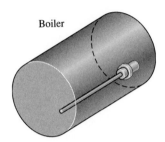

Boiler

FIGURE P8-72

8-72 A manufacturing facility requires saturated steam at 120°C at a rate of
1.2 kg/min. Design an electric steam boiler for this purpose under the following constraints:

- The boiler will be in cylindrical shape with a height-to-diameter ratio
  of 1.5. The boiler can be horizontal or vertical.
- The boiler will operate in the nucleate boiling regime, and the design
  heat flux will not exceed 60 percent of the critical heat flux to provide
  an adequate safety margin.
- A commercially available plug-in type electrical heating element made
  of mechanically polished stainless steel will be used. The diameter of
  the heater cannot be between 0.5 cm and 3 cm.
- Half of the volume of the boiler should be occupied by steam, and the
  boiler should be large enough to hold enough water for 2 h supply of
  steam. Also, the boiler will be well insulated.

You are to specify the following: (1) The height and inner diameter of the
tank, (2) the length, diameter, power rating, and surface temperature of the
electric heating element, (3) the maximum rate of steam production during
short periods of overload conditions, and how it can be accomplished.

8-73 Repeat Problem 8-72 for a boiler that produces steam at 150°C at a
rate of 2.5 kg/min.

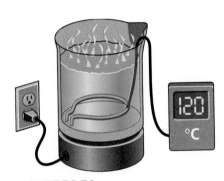

FIGURE P8-74

8-74 Conduct the following experiment to determine the boiling heat transfer coefficient. You will need a portable immersion-type electric heating
element, an indoor-outdoor thermometer, and metal glue (all can be purchased for about $15 in a hardware store). You will also need a piece of string
and a ruler to calculate the surface area of the heater. First, boil water in a
pan using the heating element and measure the temperature of the boiling
water away from the heating element. Based on your reading, estimate the
elevation of your location, and compare it to the actual value. Then glue the
tip of the thermocouple wire of the thermometer to the mid-section of the
heater surface. The temperature reading in this case will give the surface
temperature of the heater. Assuming the rated power of the heater to be the
actual power consumption during heating (you can check this by measuring
the electric current and voltage), calculate the heat transfer coefficients from
Newton's law of cooling.

Radiation Heat Transfer

9

So far, we have considered the conduction and convection modes of heat transfer, which are related to the nature of the materials involved and the presence of fluid motion, among other things. We now turn our attention to a third mechanism of heat transfer: *radiation*, which is characteristically different from the other two.

We start this chapter with a discussion of *electromagnetic waves* and the *electromagnetic spectrum*, with particular emphasis on *thermal radiation*. Then we introduce the idealized *blackbody, blackbody radiation*, and the *blackbody radiation function*, together with the *Stefan–Boltzmann law, Planck's distribution law*, and *Wien's displacement law*. This is followed by a discussion of radiation properties of materials such as *emissivity, absorptivity, reflectivity*, and *transmissivity* and their dependence on wavelength and temperature. The *greenhouse effect* is presented as an example of the consequences of the wavelength dependence of radiation properties. A separate section is devoted to the discussions of *atmospheric* and *solar radiation* because of their importance.

The second part of this chapter starts with a discussion of *view factors* and the rules associated with them. View factor *expressions* and *charts* for some common configurations are given, and the *crossed-strings method* is presented. We then discuss *radiation heat transfer*, first between black surfaces and then between nonblack surfaces using the *radiation network* approach. Finally, we consider *radiation shields* and discuss the *radiation effect* on temperature measurements and comfort.

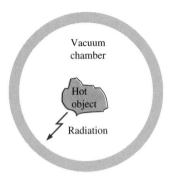

FIGURE 9-1

A hot object in a vacuum chamber loses heat by radiation only.

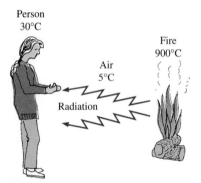

FIGURE 9-2

Unlike conduction and convection, heat transfer by radiation can occur between two bodies, even when they are separated by a medium colder than both of them.

9-1 ■ INTRODUCTION

Consider a hot object that is placed in an evacuated chamber whose walls are at room temperature (Fig. 9-1). Our experience tells us that the hot object will eventually cool down and reach thermal equilibrium with its surroundings. That is, it will lose heat until its temperature reaches the temperature of the walls of the chamber. Heat transfer between the object and the chamber could not have taken place by conduction or convection, because these two mechanisms cannot occur in a vacuum. Therefore, heat transfer must have occurred through another mechanism that involves the emission of the sensible internal energy of the object. This mechanism is *radiation*.

Radiation differs from the other two heat transfer mechanisms in that it does not require the presence of a material medium to take place. In fact, energy transfer by radiation is fastest (at the speed of light) and it suffers no attenuation in a *vacuum*. Also, radiation transfer occurs in solids as well as liquids and gases. In most practical applications, all three modes of heat transfer occur concurrently at varying degrees. But heat transfer through an evacuated space can occur only by radiation. For example, the energy of the sun reaches the earth by radiation.

You will recall that heat transfer by conduction or convection takes place in the direction of decreasing temperature; that is, from a high-temperature medium to a lower-temperature one. It is interesting that radiation heat transfer can occur between two bodies separated by a medium colder than both bodies (Fig. 9-2). For example, solar radiation reaches the surface of the earth after passing through extremely cold air layers at high altitudes. Also, the radiation-absorbing surfaces inside a greenhouse reach high temperatures even when its plastic or glass cover remains relatively cool.

The theoretical foundation of radiation was established in 1864 by physicist James Clerk Maxwell, who postulated that accelerated charges or changing electric currents give rise to electric and magnetic fields. These rapidly moving fields are called **electromagnetic waves** or **electromagnetic radiation,** and they represent the energy emitted by matter as a result of the changes in the electronic configurations of the atoms or molecules. In 1887, Heinrich Hertz experimentally demonstrated the existence of such waves. Electromagnetic waves transport energy just like other waves, and all electromagnetic waves travel at the *speed of light*. Electromagnetic waves are characterized by their *frequency* ν and *wavelength* λ. These two properties in a medium are related by

$$\lambda = \frac{c}{\nu} \tag{9-1}$$

where c is the speed of light in that medium. In a vacuum, $c = c_0 = 2.998 \times 10^8$ m/s. The speed of light in a medium is related to the speed of light in a vacuum by $c = c_0/n$, where n is the *index of refraction* of that medium. The index of refraction is essentially unity for air and most gases and about 1.5 for water and glass. The commonly used unit of wavelength is the *micrometer* (μm), where 1μm $= 10^{-6}$ m. Unlike the wavelength and the speed of propagation, the frequency of an electromagnetic wave depends only on the source and is independent of the medium through which the wave travels. The *frequency* (the number of oscillations per second) of an electromagnetic wave can range from a few cycles to millions of cycles and even higher per second, depending on the source. Note from Equation 9-1 that the wavelength and the frequency of electromagnetic radiation are inversely proportional.

In radiation studies, it has proven useful to view electromagnetic radiation as the propagation of a collection of discrete packets of energy called **photons** or **quanta,** as proposed by Max Planck in 1900 in conjunction with his *quantum theory.* In this view, each photon of frequency ν is considered to have an energy of

$$e = h\nu = \frac{hc}{\lambda} \tag{9-2}$$

where $h = 6.625 \times 10^{-34}$ J $\cdot$ s is *Planck's constant.* Note from the second part of Equation 9-2 that h and c are constants, and thus the energy of a photon is inversely proportional to its wavelength. Therefore, shorter-wavelength radiation possesses larger photon energies. It is no wonder that we try to avoid very-short-wavelength radiation such as gamma rays and X-rays since they are highly destructive.

9-2 ■ THERMAL RADIATION

Although all electromagnetic waves have the same general features, waves of different wavelength differ significantly in their behavior. The electromagnetic radiation encountered in practice covers a wide range of wavelengths, varying from less than 10^{-10} μm for cosmic rays to more than 10^{10} μm for electrical power waves. The **electromagnetic spectrum** also includes gamma rays, X-rays, ultraviolet radiation, visible light, infrared radiation, thermal radiation, microwaves, and radio waves, as shown in Figure 9-3.

Different types of electromagnetic radiation are produced differently through different mechanisms. For example, *gamma rays* are produced by nuclear reactions, *X-rays* by the bombardment of metals with high-energy electrons, *microwaves* by special types of electron tubes such as klystrons and magnetrons, and *radio waves* by the excitation of some crystals or by the flow of alternating current through electric conductors.

The short-wavelength gamma rays and X-rays are primarily of concern to nuclear engineers, while the long-wavelength microwaves and radio waves are of concern to electrical engineers. The type of electromagnetic radiation that is pertinent to heat transfer is the **thermal radiation** emitted as a result of vibrational and rotational motions of molecules, atoms, and electrons of a substance. Temperature is a measure of the strength of these activities at the microscopic level, as discussed in Chapter 1, and the rate of thermal radiation emission increases with increasing temperature. Thermal radiation is continuously emitted by all matter whose temperature is above absolute zero. That is, everything around us such as walls, furniture, and our friends constantly emits (and absorbs) radiation (Fig. 9-4). Thermal radiation is also defined as

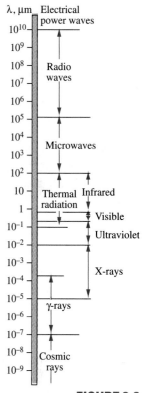

FIGURE 9-3

The electromagnetic wave spectrum.

FIGURE 9-4

Everything around us constantly emits thermal radiation.

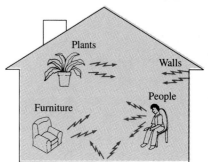

TABLE 9-1

The wavelength ranges of different colors

Color	Wavelength band
Violet	0.40-0.44 μm
Blue	0.44–0.49 μm
Green	0.49–0.54 μm
Yellow	0.54–0.60 μm
Orange	0.60–0.673 μm
Red	0.63–0.76 μm

the portion of the electromagnetic spectrum that extends from about 0.1 to 100 μm, since the radiation emitted by bodies because of their temperature falls almost entirely into this wavelength range. Thus, thermal radiation includes the entire visible and infrared (IR) radiation as well as a portion of the ultraviolet (UV) radiation.

What we call **light** is simply the *visible* portion of the electromagnetic spectrum that lies between 0.40 and 0.76 μm. Light is characteristically no different than other electromagnetic radiation, except that it happens to trigger the sensation of seeing in the human eye. Light or the visible spectrum consists of narrow bands of color from violet (0.40–0.44 μm) to red (0.63–0.76 μm), as shown in Table 9-1. The color of a surface depends on its ability to *reflect* certain wavelengths. For example, a surface that reflects radiation in the wavelength range 0.63–0.76 μm while absorbing the rest of the visible radiation appears red to the eye. A surface that reflects all of the light appears *white,* while a surface that absorbs all of the light incident on it appears *black*.

A body that emits some radiation in the visible range is called a light source. The sun is obviously our primary light source. The electromagnetic radiation emitted by the sun is known as **solar radiation,** and nearly all of it falls into the wavelength band 0.3–3 μm. Almost half of solar radiation is light (i.e., it falls into the visible range), with the remaining being ultraviolet and infrared.

The radiation emitted by bodies at room temperature falls into the **infrared** region of the spectrum, which extends from 0.76 to 100 μm. Bodies start emitting noticeable visible radiation at temperatures above 800 K. The tungsten filament of a light bulb must be heated to temperatures above 2000 K before it can emit any significant amount of radiation in the visible range.

The **ultraviolet** radiation occupies the low-wavelength end of the thermal radiation spectrum and lies between the wavelengths 0.01 and 0.40 μm. Ultraviolet rays are to be avoided since they can kill microorganisms and cause serious damage to humans and other living organisms. About 12 percent of solar radiation is in the ultraviolet range, and it would be devastating if it were to reach the surface of the earth. Fortunately, the ozone (O_3) layer in the atmosphere acts as a protective blanket and absorbs most of this ultraviolet radiation. The ultraviolet rays that remain in sunlight are still sufficient to cause serious sunburns in sun worshippers, and prolonged exposure to direct sunlight is the leading cause of skin cancer, which can be lethal. Recent discoveries of "holes" in the ozone layer have prompted the international community to ban the use of ozone-destroying chemicals such as the widely used refrigerant Freon-12 in order to save the earth. Ultraviolet radiation is also produced artificially in fluorescent lamps for use in medicine as a bacteria killer and in tanning parlors as an artificial tanner. The connection between skin cancer and ultraviolet rays has caused dermatologists to issue strong warnings against its use for tanning.

Microwave ovens utilize electromagnetic radiation in the **microwave** region of the spectrum generated by microwave tubes called *magnetrons*. Microwaves in the range of 10^2–10^5 μm are very suitable for use in cooking since they are *reflected* by metals, *transmitted* by glass and plastics, and *absorbed* by food (especially water) molecules. Thus, the electric energy converted to radiation in a microwave oven eventually becomes part of the internal energy of the food. The fast and efficient cooking of microwave ovens has made them some of the essential appliances in modern kitchens (Fig. 9-5).

Radars and cordless telephones also use electromagnetic radiation in the microwave region. The wavelength of the electromagnetic waves used in radio and TV broadcasting usually ranges between 1 and 1000 m in the **radio wave** region of the spectrum.

In heat transfer studies, we are interested in the energy emitted by bodies because of their temperature only. Therefore, we will limit our consideration to *thermal radiation,* which we will simply call *radiation.* The relations developed below are restricted to thermal radiation only and may not be applicable to other forms of electromagnetic radiation.

The electrons, atoms, and molecules of all solids, liquids, and gases above absolute zero temperature are constantly in motion, and thus radiation is constantly emitted, as well as being absorbed or transmitted throughout the entire volume of matter. That is, radiation is a **volumetric phenomenon.** However, for opaque (nontransparent) solids such as metals, wood, and rocks, radiation is considered to be a **surface phenomenon,** since the radiation emitted by the interior regions can never reach the surface, and the radiation incident on such bodies is usually absorbed within a few microns from the surface (Fig. 9-6). Note that the radiation characteristics of surfaces can be changed completely by applying thin layers of coatings on them.

9-3 ■ BLACKBODY RADIATION

A body at a temperature above absolute zero emits radiation in all directions over a wide range of wavelengths. The amount of radiation energy emitted from a surface at a given wavelength depends on the material of the body and the condition of its surface as well as the surface temperature. Therefore, different bodies may emit different amounts of radiation per unit surface area, even when they are at the same temperature. Thus, it is natural to be curious about the *maximum* amount of radiation that can be emitted by a surface at a given temperature. Satisfying this curiosity requires the definition of an idealized body, called a *blackbody,* to serve as a standard against which the radiative properties of real surfaces may be compared.

A **blackbody** is defined as *a perfect emitter and absorber of radiation.* At a specified temperature and wavelength, no surface can emit more energy than a blackbody. A blackbody absorbs *all* incident radiation, regardless of wavelength and direction. Also, a blackbody emits radiation energy uniformly in all directions (Fig. 9-7). That is, a blackbody is a *diffuse* emitter. The term *diffuse* means "independent of direction."

The radiation energy emitted by a blackbody per unit time and per unit surface area was determined experimentally by Joseph Stefan in 1879 and is expressed as

$$E_b = \sigma T^4 \qquad \text{(W/m}^2) \tag{9-3}$$

FIGURE 9-5
Food is heated or cooked in a microwave oven by absorbing the electromagnetic radiation energy generated by the magnetron of the oven.

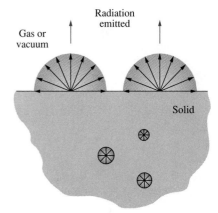

FIGURE 9-6
Radiation in opaque solids is considered a surface phenomenon since the radiation emitted only by the molecules at the surface can escape the solid.

FIGURE 9-7
A blackbody is said to be a *diffuse* emitter since it emits radiation energy uniformly in all directions.

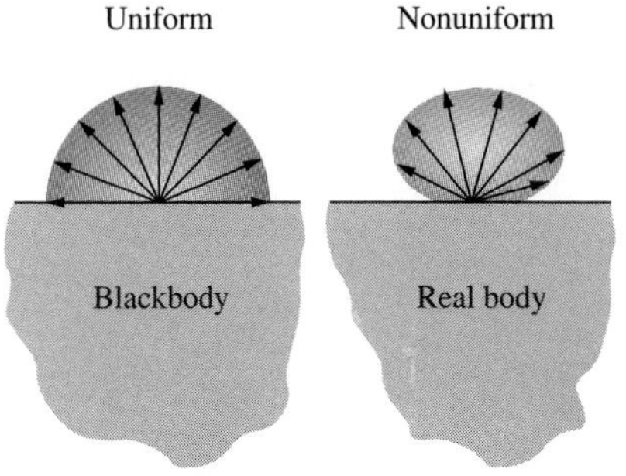

where $\sigma = 5.67 \times 10^{-8}$ W/m$^2 \cdot$ K^4 is the *Stefan–Boltzmann constant* and T is the absolute temperature of the surface in K. This relation was theoretically verified in 1884 by Ludwig Boltzmann. Equation 9-3 is known as the **Stefan–Boltzmann law** and E_b is called the **blackbody emissive power.** Note that the emission of thermal radiation is proportional to the *fourth power* of the absolute temperature.

Although a blackbody would appear *black* to the eye, a distinction should be made between the idealized blackbody and an ordinary black surface. Any surface that absorbs light (the visible portion of radiation) would appear black to the eye, and a surface that reflects it completely would appear white. Considering that visible radiation occupies a very narrow band of the spectrum from 0.4 to 0.76 μm, we cannot make any judgments about the blackness of a surface on the basis of visual observations. For example, snow and white paint reflect light and thus appear white. But they are essentially black for infrared radiation since they strongly absorb long-wavelength radiation. Surfaces coated with lampblack paint approach idealized blackbody behavior.

Another type of body that closely resembles a blackbody is a *large cavity with a small opening,* as shown in Figure 9-8. Radiation coming in through the opening of area A will undergo multiple reflections, and thus it will have several chances to be absorbed by the interior surfaces of the cavity before any part of it can possibly escape. Also, if the surface of the cavity is isothermal at temperature T, the radiation emitted by the interior surfaces will stream through the opening after undergoing multiple reflections, and thus it will have a diffuse nature. Therefore, the cavity will act as a perfect absorber and perfect emitter, and the opening will resemble a blackbody of surface area A at temperature T, regardless of the actual radiative properties of the cavity.

The Stefan–Boltzmann law in Equation 9-3 gives the *total* blackbody emissive power E_b, which is the sum of the radiation emitted over all wavelengths. Sometimes we need to know the **spectral blackbody emissive power,** which is *the amount of radiation energy emitted by a blackbody at an absolute temperature T per unit time, per unit surface area, and per unit wavelength about the wavelength λ.* For example, we are more interested in the amount of radiation an incandescent light bulb emits in the visible wavelength spectrum than we are in the total amount of radiation that the light bulb emits.

The relation for the spectral blackbody emissive power $E_{b\lambda}$ was developed by Max Planck in 1901 in conjunction with his famous quantum theory. This relation is known as **Planck's distribution law** and is expressed as

$$E_{b\lambda}(T) = \frac{C_1}{\lambda^5[\exp(C_2/\lambda T) - 1]} \qquad (\text{W/m}^2 \cdot \mu\text{m}) \qquad (9\text{-}4)$$

where
$$C_1 = 2\pi h c_0^2 = 3.742 \times 10^8 \text{ W} \cdot \mu\text{m}^4/\text{m}^2$$
$$C_2 = h c_0/k = 1.439 \times 10^4 \ \mu\text{m} \times \text{K}$$

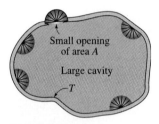

FIGURE 9-8

A large isothermal cavity at temperature T with a small opening of area A closely resembles a blackbody of surface area A at the same temperature.

Small opening
of area A

Large cavity

T

Also, T is the absolute temperature of the surface, λ is the wavelength of the radiation emitted, and $k = 1.3805 \times 10^{-23}$ J/K is *Boltzmann's constant*. This relation is valid for a surface in a *vacuum* or a *gas*. For other mediums, it needs to be modified by replacing C_1 by C_1/n^2, where n is the index of refraction of the medium. Note that the term *spectral* indicates dependence on wavelength.

The variation of the blackbody emissive power with wavelength is plotted in Figure 9-9 for selected temperatures. Several observations can be made from this figure:

1. The emitted radiation is a continuous function of *wavelength*. At any specified temperature, it increases with wavelength, reaches a peak, and then decreases with increasing wavelength.

2. At any wavelength, the amount of emitted radiation *increases* with increasing temperature.

3. As temperature increases, the curves get steeper and shift to the left to the shorter-wavelength region. Consequently, a larger fraction of the radiation is emitted at *shorter wavelengths* at higher temperatures.

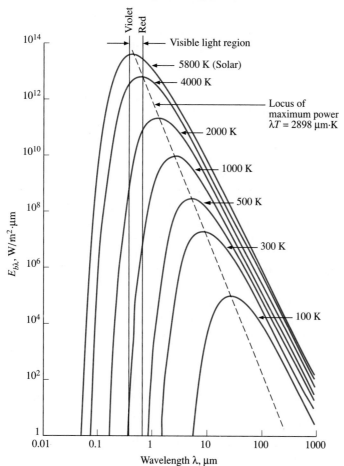

FIGURE 9-9

The variation of the blackbody emissive power with wavelength for several temperatures.

4. The radiation emitted by the *sun*, which is considered to be a blackbody at 5762 K (or roughly at 5800 K), reaches its peak in the visible region of the spectrum. Therefore, the sun is in tune with our eyes. On the other hand, surfaces at $T \leq 800$ K emit almost entirely in the infrared region and thus are not visible to the eye unless they reflect light coming from other sources.

As the temperature increases, the peak of the curve in Figure 9-9 shifts toward shorter wavelengths. The wavelength at which the peak occurs for a specified temperature is given by **Wien's displacement law** as

$$(\lambda T)_{\text{max power}} = 2897.8 \ \mu\text{m} \cdot \text{K} \qquad (9\text{-}5)$$

This relation was originally developed by Willy Wien in 1894 using classical thermodynamics, but it can also be obtained by differentiating Equation 9-4 with respect to λ while holding T constant and setting the result equal to zero. A plot of Wien's displacement law, which is the locus of the peaks of the radiation emission curves, is also given in Figure 9-9.

The peak of the solar radiation, for example, occurs at $\lambda = 2897.8/5762 = 0.50 \ \mu\text{m}$, which is near the middle of the visible range. The peak of the radiation emitted by a surface at room temperature ($T = 298$ K) occurs at 9.72 μm, which is well into the infrared region of the spectrum.

An electrical resistance heater starts radiating heat soon after it is plugged in, and we can feel the emitted radiation energy by holding our hands against the heater. But this radiation is entirely in the infrared region and thus cannot be sensed by our eyes. The heater would appear dull red when its temperature reaches about 1000 K, since it will start emitting a detectable amount (about 1 W/m$^2 \cdot \mu$m) of visible red radiation at that temperature. As the temperature rises even more, the heater appears bright red and is said to be *red hot*. When the temperature reaches about 1500 K, the heater emits enough radiation in the entire visible range of the spectrum to appear almost *white* to the eye, and it is called *white hot*.

Although it cannot be sensed directly by the human eye, infrared radiation can be detected by infrared cameras, which transmit the information to microprocessors to display visual images of objects at night. *Rattlesnakes* can sense the infrared radiation or the "body heat" coming off warm-blooded animals, and thus they can see at night without using any instruments. Similarly, honeybees are sensitive to ultraviolet radiation.

It should be clear from the discussion above that the color of an object is not due to emission, which is primarily in the infrared region, unless the surface temperature of the object exceeds about 1000 K. Instead, the color of a surface depends on the absorption and reflection characteristics of the surface and is due to selective absorption and reflection of the incident visible radiation coming from a light source such as the sun or an incandescent light bulb. A piece of clothing containing a pigment that reflects red while absorbing the remaining parts of the incident light appears "red" to the eye (Fig. 9-10). Leaves appear "green" because their cells contain the pigment chlorophyll, which strongly reflects green while absorbing other colors.

FIGURE 9-10

A surface that reflects red while absorbing the remaining parts of the incident light appears red to the eye.

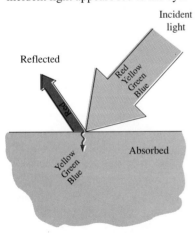

It is left as an exercise to show that integration of the *spectral* blackbody emissive power $E_{b\lambda}$ over the entire wavelength spectrum gives the *total* blackbody emissive power E_b:

$$E_b(T) = \int_0^\infty E_{b\lambda}(T)\, d\lambda = \sigma T^4 \qquad (\text{W/m}^2) \qquad (9\text{-}6)$$

Thus, we obtained the Stefan–Boltzmann law (Eq. 9-3) by integrating Planck's distribution law (Eq. 9-4) over all wavelengths. Note that on an $E_{b\lambda}$–λ chart, $E_{b\lambda}$ corresponds to any value on the curve, whereas E_b corresponds to the area under the entire curve for a specified temperature (Fig. 9-11). Also, the term *total* means "integrated over all wavelengths."

EXAMPLE 9-1   Radiation Emission from a Black Ball

Consider a 20-cm-diameter spherical ball at 800K suspended in the air as shown in Figure 9-12. Assuming that the ball closely approximates a blackbody, determine (a) the total blackbody emissive power, (b) the total amount of radiation emitted by the ball in 5 min, and (c) the spectral blackbody emissive power at a wavelength of 3 μm.

Solution An isothermal sphere is suspended in the air. The total blackbody emissive power, the total radiation emitted in 5 minutes, and the spectral blackbody emissive power at 3 μm are to be determined.

Assumptions The ball behaves as a blackbody.

Analysis (a) The total blackbody emissive power is determined from the Stefan–Boltzmann law to be

$$E_b = \sigma T^4 = (5.67 \times 10^{-8}\ \text{W/m}^2 \cdot \text{K}^4)(800\ \text{K})^4 = \textbf{23,224 W/m}^2$$

That is, the ball emits 23,224 J of energy in the form of electromagnetic radiation per second per m² of the surface area of the ball.

(b) The total amount of radiation energy emitted from the entire ball in 5 min is determined by multiplying the blackbody emissive power obtained above by the total surface area of the ball and the given time interval:

$$A = \pi D^2 = \pi(0.2\ \text{m})^2 = 0.1257\ \text{m}^2$$

$$\Delta t = (5\ \text{min})\left(\frac{60\ \text{s}}{1\ \text{min}}\right) = 300\ \text{s}$$

$$Q_{\text{rad}} = E_b A\, \Delta t = (23,224\ \text{W/m}^2)(0.1257\ \text{m}^2)(300\ \text{s})\left(\frac{1\ \text{kJ}}{1000\ \text{W} \cdot \text{s}}\right)$$

$$= \textbf{875.8 kJ}$$

That is, the ball loses 875.8 kJ of its internal energy in the form of electromagnetic waves to the surroundings in 5 min, which is enough energy to raise the temperature of 1 kg of water by 50°C. Note that the surface temperature of the ball cannot remain constant at 800 K unless there is an equal amount of energy flow to the surface from the surroundings or from the interior regions of the ball through some mechanisms such as chemical or nuclear reactions.

(c) The spectral blackbody emissive power at a wavelength of 3 μm is determined from Planck's distribution law to be

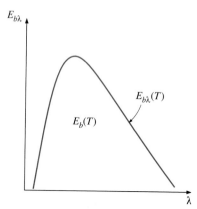

$E_{b\lambda}$

$E_{b\lambda}(T)$

$E_b(T)$

λ

FIGURE 9-11

On an $E_{b\lambda}$ − λ chart, the area under a curve for a given temperature represents the total radiation energy emitted by a blackbody at that temperature.

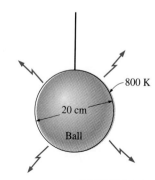

800 K

20 cm

Ball

FIGURE 9-12

The spherical ball considered in Example 9-1.

$$E_{b\lambda} = \frac{C}{\lambda^5\left[\exp\left(\dfrac{C_2}{\lambda T}\right) - 1\right]} = \frac{3.743 \times 10^8 \ \text{W} \cdot \mu\text{m}^4/\text{m}^2}{(3 \ \mu\text{m})^5\left[\exp\left(\dfrac{1.4387 \times 10^4 \ \mu\text{m} \cdot \text{K}}{(3\mu\text{m})(800 \ \text{K})}\right) - 1\right]}$$

$$= 3848.4 \ \text{W/m}^2 \cdot \mu\text{m}$$

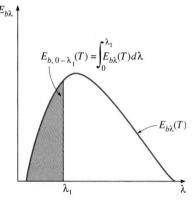

FIGURE 9-13

On an $E_{b\lambda} - \lambda$ chart, the area under the curve to the left of the $\lambda = \lambda_1$ line represents the radiation energy emitted by a blackbody in the wavelength range $0–\lambda_1$ for the given temperature.

The Stefan–Boltzmann law $E_b(T) = \sigma T^4$ gives the *total* radiation emitted by a blackbody at all wavelengths from $\lambda = 0$ to $\lambda = \infty$. But we are often interested in the amount of radiation emitted over *some wavelength band*. For example, an incandescent light bulb is judged on the basis of the radiation it emits in the visible range rather than the radiation it emits at all wavelengths.

The radiation energy emitted by a blackbody per unit area over a wavelength band from $\lambda = 0$ to λ is determined from (Fig. 9-13)

$$E_{b,\,0-\lambda}(T) = \int_0^\lambda E_{b\lambda}(T)\,d\lambda \qquad (\text{W/m}^2) \qquad (9\text{-}7)$$

It looks like we can determine $E_{b,\,0-\lambda}$ by substituting the $E_{b\lambda}$ relation from Equation 9-4 and performing this integration. But it turns out that this integration does not have a simple closed-form solution, and performing a numerical integration each time we need a value of $E_{b,\,0-\lambda}$ is not practical. Therefore, we define a dimensionless quantity f_λ called the **blackbody radiation function** as

$$f_\lambda(T) = \frac{\displaystyle\int_0^\lambda E_{b\lambda}(T)\,d\lambda}{\sigma T^4} \qquad (9\text{-}8)$$

The function f_λ represents *the fraction of radiation emitted from a blackbody at temperature* T *in the wavelength band from* $\lambda = 0$ *to* λ. The values of f_λ are listed in Table 9-2 as a function of λT, where λ is in μm and T is in K.

The fraction of radiation energy emitted by a blackbody at temperature T over a finite wavelength band from $\lambda = \lambda_1$ to $\lambda = \lambda_2$ is determined from (Fig. 9-14)

$$f_{\lambda_1-\lambda_2}(T) = f_{\lambda_2}(T) - f_{\lambda_1}(T) \qquad (9\text{-}9)$$

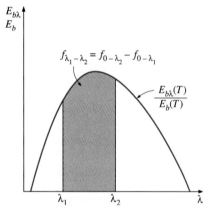

FIGURE 9-14

Graphical representation of the fraction of radiation emitted in the wavelength band from λ_1 to λ_2.

where $f_{\lambda_1}(T)$ and $f_{\lambda_2}(T)$ are blackbody radiation functions corresponding to $\lambda_1 T$ and $\lambda_2 T$, respectively.

EXAMPLE 9-2   Emission of Radiation from an Incandescent Light Bulb

The temperature of the filament of an incandescent light bulb is 2500 K. Assuming the filament to be a blackbody, determine the fraction of the radiant energy emitted by the filament that falls in the visible range. Also, determine the wavelength at which the emission of radiation from the filament peaks.

TABLE 9-2

Blackbody radiation functions f_λ

λT, $\mu m \cdot K$	f_λ	λT, $\mu m \cdot K$	f_λ
200	0.000000	6200	0.754140
400	0.000000	6400	0.769234
600	0.000000	6600	0.783199
800	0.000016	6800	0.796129
1000	0.000321	7000	0.808109
1200	0.002134	7200	0.819217
1400	0.007790	7400	0.829527
1600	0.019718	7600	0.839102
1800	0.039341	7800	0.848005
2000	0.066728	8000	0.856288
2200	0.100888	8500	0.874608
2400	0.140256	9000	0.890029
2600	0.183120	9500	0.903085
2800	0.227897	10,000	0.914199
3000	0.273232	10,500	0.923710
3200	0.318102	11,000	0.931890
3400	0.361735	11,500	0.939959
3600	0.403607	12,000	0.945098
3800	0.443382	13,000	0.955139
4000	0.480877	14,000	0.962898
4200	0.516014	15,000	0.969981
4400	0.548796	16,000	0.973814
4600	0.579280	18,000	0.980860
4800	0.607559	20,000	0.985602
5000	0.633747	25,000	0.992215
5200	0.658970	30,000	0.995340
5400	0.680360	40,000	0.997967
5600	0.701046	50,000	0.998953
5800	0.720158	75,000	0.999713
6000	0.737818	100,000	0.999905

Solution The temperature of the filament of an incandescent light bulb is given. The fraction of visible radiation emitted by the filament and the wavelength at which the emission peaks are to be determined.

Assumptions The filament behaves as black body.

Analysis The visible range of the electromagnetic spectrum extends from $\lambda_1 = 0.4$ μm to $\lambda_2 = 0.76$ μm. Noting that $T = 2500$ K, the blackbody radiation functions corresponding to $\lambda_1 T$ and $\lambda_2 T$ are determined from Table 9-2 to be

$$\lambda_1 T = (0.40\ \mu m)(2500\ K) = 1000\ \mu m \cdot K \longrightarrow f_{\lambda_1} = 0.000321$$

$$\lambda_2 T = (0.76\ \mu m)(2500\ K) = 1900\ \mu m \cdot K \longrightarrow f_{\lambda_2} = 0.053035$$

That is, 0.03 percent of the radiation is emitted at wavelengths less than 0.4 μm and 5.3 percent at wavelengths less than 0.76 μm. Then the fraction of radiation emitted between these two wavelengths is (Fig. 9-15)

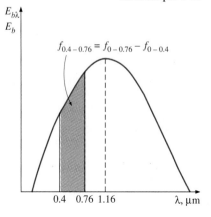

FIGURE 9-15

Graphical representation of the fraction of radiation emitted in the visible range in Example 9-2.

$$f_{\lambda_1 - \lambda_2} = f_{\lambda_2} - f_{\lambda_1} = 0.053035 - 0.000321 = \mathbf{0.0527135}$$

That is, only about 5 percent of the radiation emitted by the filament of the light bulb falls in the visible range. The remaining 95 percent of the radiation appears in the infrared region in the form of radiant heat or "invisible light," as it used to be called. This is certainly a very inefficient way of converting electrical energy to light and explains why fluorescent tubes are a wiser choice for lighting.

The wavelength at which the emission of radiation from the filament peaks is easily determined from Wien's displacement law (Eq. 9-5) to be

$$(\lambda T)_{\text{max power}} = 2897.8 \ \mu\text{m} \cdot \text{K} \quad \rightarrow \quad \lambda_{\text{max power}} = \frac{2897.8 \ \mu\text{m} \cdot \text{K}}{2500 \ \text{K}} = \mathbf{1.16 \ \mu\text{m}}$$

Discussion Note that the radiation emitted from the filament peaks in the infrared region.

9-4 ■ RADIATION PROPERTIES

Most materials encountered in practice, such as metals, wood, and bricks, are *opaque* to thermal radiation, and radiation is considered to be a *surface phenomenon* for such materials. That is, thermal radiation is emitted or absorbed within the first few microns of the surface, and thus we speak of radiation properties of *surfaces* for opaque materials.

Some other materials, such as glass and water, allow visible radiation to penetrate to considerable depths before any significant absorption takes place. Radiation through such *semitransparent* materials obviously cannot be considered to be a surface phenomenon since the entire volume of the material interacts with radiation. On the other hand, both glass and water are practically opaque to infrared radiation. Therefore, materials can exhibit different behavior at different wavelengths, and the dependence on wavelength is an important consideration in the study of radiation properties such as emissivity, absorptivity, reflectivity, and transmissivity of materials.

In the preceding section, we defined a *blackbody* as a perfect emitter and absorber of radiation and said that no body can emit more radiation than a blackbody at the same temperature. Therefore, a blackbody can serve as a convenient *reference* in describing the emission and absorption characteristics of real surfaces.

Emissivity

The **emissivity** of a surface is defined as *the ratio of the radiation emitted by the surface to the radiation emitted by a blackbody at the same temperature.* The emissivity of a surface is denoted by ε, and it varies between zero and one, $0 \leq \varepsilon \leq 1$. Emissivity is a measure of how closely a surface approximates a blackbody, for which $\varepsilon = 1$.

The emissivity of a real surface is not a constant. Rather, it varies with the *temperature* of the surface as well as the *wavelength* and the *direction* of the emitted radiation. Therefore, different emissivities can be defined for a surface, depending on the effects considered. For example, the emissivity of a surface at a specified wavelength is called *spectral emissivity* and is denoted by ε_λ. Likewise, the emissivity in a specified direction is called *directional emissivity*, denoted by ε_θ, where θ is the angle between the direction of radiation and the normal of the surface. The emissivity of a surface averaged over all directions is called the *hemispherical emissivity*, and the emissivity averaged over all wavelengths is called the *total emissivity*. Thus, the *total hemispherical emissivity* ε of a surface is simply the average emissivity over all directions and wavelengths and can be expressed as

$$\varepsilon(T) = \frac{E(T)}{E_b(T)} = \frac{E(T)}{\sigma T^4} \qquad (9\text{-}10)$$

where $E(T)$ is the total emissive power of the real surface. Equation 9-10 can be rearranged as

$$E(T) = \varepsilon(T)\sigma T^4 \qquad (\text{W/m}^2) \qquad (9\text{-}11)$$

Thus, the radiation emitted by the unit area of a real surface at temperature T is obtained by multiplying the radiation emitted by a blackbody at the same temperature by the emissivity of the surface.

Spectral emissivity is defined in a similar manner as

$$\varepsilon_\lambda(T) = \frac{E_\lambda(T)}{E_{b\lambda}(T)} \qquad (9\text{-}12)$$

where $E_\lambda(T)$ is the spectral emissive power of the real surface.

Radiation is a complex phenomenon as it is, and the consideration of wavelength and direction dependence of properties, assuming sufficient data exist, makes it even more complicated. Therefore, the *gray* and *diffuse* approximations are commonly utilized in radiation calculations. A surface is said to be *diffuse* if its properties are *independent of direction* and *gray* if its properties are *independent of wavelength*. Therefore, the emissivity of a gray, diffuse surface is simply the total hemispherical emissivity of that surface because of independence of direction and wavelength (Fig. 9-16).

A few comments about the validity of the diffuse approximation are in order. Although real surfaces do not emit radiation in a perfectly diffuse manner as a blackbody does, they usually come close. The variation of emissivity with direction for both conductors and nonconductors is given in Figure 9-17. Here θ is the angle measured from the normal of the surface, and thus $\theta = 0$ for radiation emitted in a direction normal to the surface. Note that ε_θ remains nearly constant for about $\theta < 40°$ for conductors such as metals and for $\theta < 70°$ for nonconductors such as plastics. Therefore, the directional emissivity of a surface in the normal direction is representative of the hemispherical emissivity of the surface. In radiation analysis, it is common practice to assume the surfaces to be diffuse emitters with an emissivity equal to the value in the normal ($\theta = 0$) direction.

The effect of the gray approximation on emissivity and emissive power of a real surface is illustrated in Figure 9-18. Note that the radiation emission

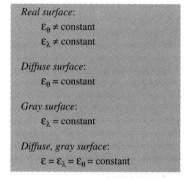

Real surface:
  $\varepsilon_\theta \neq$ constant
  $\varepsilon_\lambda \neq$ constant

Diffuse surface:
  $\varepsilon_\theta =$ constant

Gray surface:
  $\varepsilon_\lambda =$ constant

Diffuse, gray surface:
  $\varepsilon = \varepsilon_\lambda = \varepsilon_\theta =$ constant

FIGURE 9-16

The effect of diffuse and gray approximations on the emissivity of a surface.

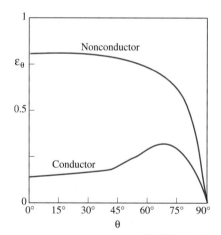

FIGURE 9-17

Typical variations of emissivity with direction for electrical conductors and nonconductors.

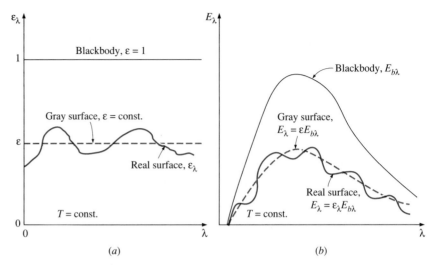

FIGURE 9-18

Comparison of the emissivity (a) and
emissive power (b) of a real surface
with those of a gray surface and a
blackbody at the same temperature.

from a real surface, in general, differs from the Planck distribution, and the
emission curve may have several peaks and valleys.

A gray surface should emit as much radiation as the real surface it represents at the same temperature. Therefore, the areas under the emission
curves of the real and gray surfaces must be equal. That is, $\varepsilon(T)\sigma T^4 = \int_0^\infty \varepsilon_\lambda(T)E_{b\lambda}(T)\,d\lambda$. This requirement yields the following expression for the
average emissivity:

$$\varepsilon(T) = \frac{\int_0^\infty \varepsilon_\lambda(T)E_{b\lambda}(T)d\lambda}{\sigma T^4} \tag{9-13}$$

To perform this integration, we need to know the variation of spectral emissivity with wavelength at the specified temperature. The integrand is usually
a complicated function, and the integration has to be performed numerically.
However, the integration can be performed quite easily by dividing the spectrum into a sufficient number of *wavelength bands* and assuming the emissivity to remain constant over each band; that is, by expressing the function
$\varepsilon(T)$ as a step function. This simplification offers great convenience for little
sacrifice of accuracy, since it allows us to transform the integration into a
summation in terms of blackbody emission functions.

As an example, consider the emissivity function plotted in Figure 9-19.
It seems like this function can be approximated reasonably well by a step
function of the form

$$\varepsilon_\lambda = \begin{cases} \varepsilon_1 = \text{constant}, & 0 \le \lambda < \lambda_1 \\ \varepsilon_2 = \text{constant}, & \lambda_1 \le \lambda < \lambda_2 \\ \varepsilon_3 = \text{constant}, & \lambda_2 \le \lambda < \infty \end{cases} \tag{9-14}$$

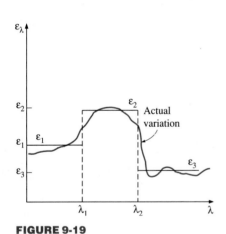

FIGURE 9-19

Approximating the actual variation of
emissivity with wavelength
by a step function.

FIGURE 9-20

The variation of normal emissivity with (*a*) wavelength and (*b*) temperature for various materials.

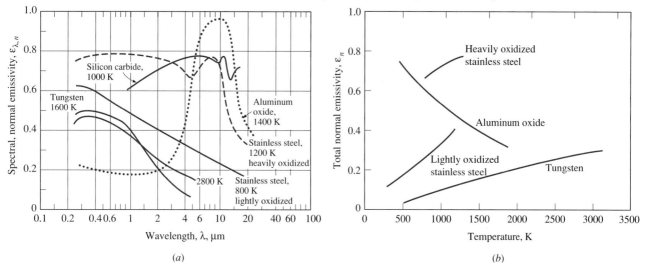

(*a*)

(*b*)

Then the average emissivity can be determined from Equation 9-13 by breaking the integral into three parts and utilizing the definition of the blackbody radiation function as

$$
\varepsilon(T) = \frac{\varepsilon_1 \int_0^{\lambda_1} E_{b\lambda}(T)\, d\lambda}{\sigma T^4} + \frac{\varepsilon_2 \int_{\lambda_1}^{\lambda_2} E_{b\lambda}(T)\, d\lambda}{\sigma T^4} + \frac{\varepsilon_3 \int_{\lambda_2}^{\infty} E_{b\lambda}(T)\, d\lambda}{\sigma T^4}
$$

$$
= \varepsilon_1 f_{0-\lambda_1}(T) + \varepsilon_2 f_{\lambda_1-\lambda_2}(T) + \varepsilon_3 f_{\lambda_2-\infty}(T)
$$

(9-15)

The emissivities of common materials are listed in Table A-14 in the appendix, and the variation of emissivity with wavelength and temperature is illustrated in Figure 9-20. Typical ranges of emissivity of various materials are given in Figure 9-21. Note that metals generally have low emissivities, as low as 0.02 for polished surfaces, and nonmetals such as ceramics and organic materials have high ones. The emissivity of metals increases with temperature. Also, oxidation causes significant increases in the emissivity of metals. Heavily oxidized metals can have emissivities comparable to those of nonmetals.

Care should be exercised in the use and interpretation of radiation property data reported in the literature, since the properties strongly depend on the surface conditions such as oxidation, roughness, type of finish, and cleanliness. Consequently, there is considerable discrepancy and uncertainty in the reported values. This uncertainty is largely due to the difficulty in characterizing and describing the surface conditions precisely.

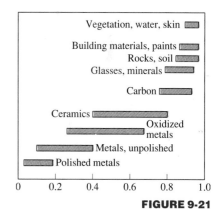

FIGURE 9-21

Typical ranges of emissivity for various materials.

EXAMPLE 9-3 Average Emissivity of a Surface and Emissive Power

The spectral emissivity function of an opaque surface at 800 K is approximated as (Fig. 9-22)

$$\varepsilon_\lambda = \begin{cases} \varepsilon_1 = 0.3, & 0 \le \lambda < 3 \ \mu m \\ \varepsilon_2 = 0.8, & 3 \ \mu m \le \lambda < 7 \ \mu m \\ \varepsilon_3 = 0.1, & 7 \ \mu m \le \lambda < \infty \end{cases}$$

Determine the average emissivity of the surface and its emissive power.

Solution The variation of emissivity of a surface at a specified temperature with wavelength is given. The average emissivity of the surface and its emissive power are to be determined.

Analysis The variation of the emissivity of the surface with wavelength is given as a step function. Therefore, the average emissivity of the surface can be determined from Equation 9-13 by breaking the integral into three parts, as shown in Equation 9-15:

$$\varepsilon(T) = \frac{\varepsilon_1 \int_0^{\lambda_1} E_{b\lambda}(T)\,d\lambda}{\sigma T^4} + \frac{\varepsilon_2 \int_{\lambda_1}^{\lambda_2} E_{b\lambda}(T)\,d\lambda}{\sigma T^4} + \frac{\varepsilon_3 \int_{\lambda_2}^{\infty} E_{b\lambda}(T)\,d\lambda}{\sigma T^4}$$

$$= \varepsilon_1 f_{0-\lambda_1}(T) + \varepsilon_2 f_{\lambda_1-\lambda_2}(T) + \varepsilon_3 f_{\lambda_2-\infty}(T)$$

$$= \varepsilon_1 f_{\lambda_1} + \varepsilon_2 (f_{\lambda_2} - f_{\lambda_1}) + \varepsilon_3 (1 - f_{\lambda_2})$$

where f_{λ_1} and f_{λ_2} are blackbody radiation functions corresponding to $\lambda_1 T$ and $\lambda_2 T$. These functions are determined from Table 9-2 to be

$$\lambda_1 T = (3 \ \mu m)(800 \ K) = 2400 \ \mu m \cdot K \quad \rightarrow \quad f_{\lambda_1} = 0.140256$$

$$\lambda_2 T = (7 \ \mu m)(800 \ K) = 5600 \ \mu m \cdot K \quad \rightarrow \quad f_{\lambda_2} = 0.701046$$

Note that $f_{0-\lambda_1} = f_{\lambda_1} - f_0 = f_{\lambda_1}$, since $f_0 = 0$, and $f_{\lambda_2-\infty} = f_\infty - f_{\lambda_2} = 1 - f_{\lambda_2}$, since $f_\infty = 1$. Substituting,

$$\varepsilon = 0.3 \times 0.140256 + 0.8(0.701046 - 0.140256) + 0.1(1 - 0.701046)$$

$$= \mathbf{0.521}$$

That is, the surface will emit as much radiation energy at 800 K as a gray surface having a constant emissivity $\varepsilon = 0.521$. The emissive power of the surface is

$$E = \varepsilon \sigma T^4 = 0.521(5.67 \times 10^{-8} \ W/m^2 \cdot K^4)(800 \ K)^4 = \mathbf{12,100 \ W/m^2}$$

Discussion Note that the surface emits 12,100 J of radiation energy per second per m^2 area of the surface.

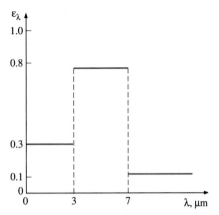

FIGURE 9-22

The spectral emissivity of the surface
considered in Example 9-3.

Absorptivity, Reflectivity, and Transmissivity

Everything around us constantly emits radiation, and the emissivity represents the emission characteristics of those bodies. This means that every body, including our own, is constantly bombarded by radiation coming from all directions over a range of wavelengths. *The radiation energy incident on a surface per unit surface area per unit time* is called **irradiation** and is denoted by *G*.

When radiation strikes a surface, part of it is absorbed, part of it is reflected, and the remaining part, if any, is transmitted, as illustrated in

Figure 9-23. *The fraction of irradiation absorbed by the surface* is called the **absorptivity** α, *the fraction reflected by the surface* is called the **reflectivity** ρ, and *the fraction transmitted* is called the **transmissivity** τ. That is,

$$\text{Absorptivity:} \quad \alpha = \frac{\text{Absorbed radiation}}{\text{Incident radiation}} = \frac{G_{\text{abs}}}{G} \qquad 0 \le \alpha \le 1 \qquad (9\text{-}16a)$$

$$\text{Reflectivity:} \quad \rho = \frac{\text{Reflected ratiation}}{\text{Incident radiation}} = \frac{G_{\text{ref}}}{G} \qquad 0 \le \rho \le 1 \qquad (9\text{-}16b)$$

$$\text{Transmissivity:} \quad \tau = \frac{\text{Transmitted radiation}}{\text{Incident radiation}} = \frac{G_{\text{tr}}}{G}, \qquad 0 \le \tau \le 1 \qquad (9\text{-}16c)$$

where G is the radiation energy incident on the surface, and G_{abs}, G_{ref}, and G_{tr} are the absorbed, reflected, and transmitted portions of it, respectively. The first law of thermodynamics requires that the sum of the absorbed, reflected, and transmitted radiation energy be equal to the incident radiation. That is,

$$G_{\text{abs}} + G_{\text{ref}} + G_{\text{tr}} = G$$

Dividing each term of this relation by G yields

$$\alpha + \rho + \tau = 1 \qquad (9\text{-}17)$$

For opaque surfaces, $\tau = 0$, and thus

$$\alpha + \rho = 1 \qquad (9\text{-}18)$$

This is an important property relation since it allows us to determine both the absorptivity and reflectivity of an opaque surface by measuring either of these properties.

The definitions above are for *total hemispherical* properties, since G represents the radiation energy incident on the surface from all directions over the hemispherical space and over all wavelengths. Thus, α, ρ, and τ are the *average* properties of a medium for all directions and all wavelengths. However, like emissivity, these properties can also be defined for a specific wavelength or direction. For example, the *spectral* absorptivity, reflectivity, and transmissivity of a surface are defined in a similar manner as

$$\alpha_\lambda = \frac{G_{\lambda,\text{abs}}}{G_\lambda}, \qquad \rho_\lambda = \frac{G_{\lambda,\text{ref}}}{G_\lambda}, \qquad \tau_\lambda = \frac{G_{\lambda,\text{tr}}}{G_\lambda} \qquad (9\text{-}19)$$

where G_λ is the radiation energy incident at the wavelength λ and $G_{\lambda,\text{abs}}$, $G_{\lambda,\text{ref}}$, and $G_{\lambda,\text{tr}}$ are the absorbed, reflected, and transmitted portions of it, respectively. Similar definitions can be given for *directional* properties in direction θ by replacing all occurrences of the subscripts λ in Equation 9-19 by θ.

The average absorptivity, reflectivity, and transmissivity of a surface can also be defined in terms of their spectral counterparts as

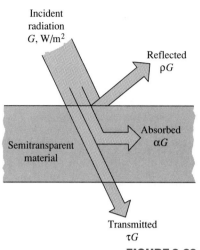

Incident
radiation
G, W/m^2

Reflected
ρG

Absorbed
αG

Semitransparent
material

Transmitted
τG

FIGURE 9-23

The absorption, reflection, and transmission of incident radiation by a semitransparent material.

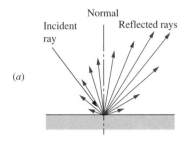

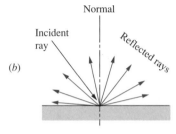

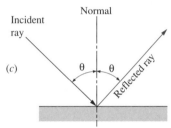

FIGURE 9-24

Different types of reflection from a surface: (*a*) actual or irregular, (*b*) diffuse, and (*c*) specular or mirrorlike.

FIGURE 9-25

Variation of absorptivity with the temperature of the source of irradiation for various common materials at room temperature.

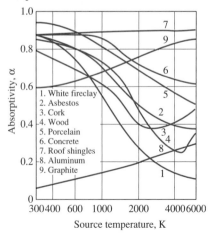

$$
\alpha = \frac{\int_0^\infty \alpha_\lambda\, G_\lambda\, d\lambda}{\int_0^\infty G_\lambda\, d\lambda}, \qquad
\rho = \frac{\int_0^\infty \rho_\lambda\, G_\lambda\, d\lambda}{\int_0^\infty G_\lambda\, d\lambda}, \qquad
\tau = \frac{\int_0^\infty \tau_\lambda\, G_\lambda\, d\lambda}{\int_0^\infty G_\lambda\, d\lambda} \qquad (9\text{-}20)
$$

The reflectivity differs somewhat from the other properties in that it is *bidirectional* in nature. That is, the value of the reflectivity of a surface depends not only on the direction of the incident radiation but also the direction of reflection. Therefore, the reflected rays of a radiation beam incident on a real surface in a specified direction will form an irregular shape, as shown in Figure 9-24. Such detailed reflectivity data do not exist for most surfaces, and even if they did, they would be of little value in radiation calculations since this would usually add more complication to the analysis than it is worth.

In practice, for simplicity, surfaces are assumed to reflect in a perfectly *specular* or *diffuse* manner. In **specular** (or *mirrorlike*) **reflection,** *the angle or reflection equals the angle of incidence of the radiation beam*. In **diffuse reflection,** *radiation is reflected equally in all directions,* as shown in Figure 9-24. Reflection from smooth and polished surfaces approximates specular reflection, whereas reflection from rough surfaces approximates diffuse reflection. In radiation analysis, smoothness is defined relative to wavelength. A surface is said to be *smooth* if the height of the surface roughness is much smaller than the wavelength of the incident radiation.

Unlike emissivity, the absorptivity of a material is practically independent of surface temperature. However, the absorptivity depends strongly on the temperature of the source at which the incident radiation is originating. This is also evident from Figure 9-25, which shows the absorptivities of various materials at room temperature as functions of the temperature of the radiation source. For example, the absorptivity of the concrete roof of a house is about 0.6 for solar radiation (source temperature 5762 K) and 0.9 for radiation originating from the surrounding trees and buildings (source temperature 300 K), as illustrated in Figure 9-26.

Notice that the absorptivity of aluminum increases with temperature, a characteristic for metals, and the absorptivity of electric nonconductors, in general, decreases with temperature. This decrease is most pronounced for surfaces that appear white to the eye. For example, the absorptivity of a white painted surface is low for solar radiation, although it is rather high for infrared radiation.

Kirchhoff's Law

Consider a small body of surface area A, emissivity ε, and absorptivity α at temperature T contained in a large isothermal enclosure at the same temperature, as shown in Figure 9-27. Recall that a large isothermal enclosure forms a blackbody cavity regardless of the radiative properties of the enclosure surface, and the body in the enclosure is too small to interfere with the blackbody nature of the cavity. Therefore, the radiation incident on any part

of the surface of the small body is equal to the radiation emitted by a blackbody at temperature T. That is, $G = E_b(T) = \sigma T^4$, and the radiation absorbed
by the small body per unit of its surface area is

$$G_{abs} = \alpha G = \alpha \sigma T^4$$

The radiation emitted by the small body is (Eq. 9-3)

$$E_{emit} = \varepsilon \sigma T^4$$

Considering that the small body is in thermal equilibrium with the enclosure,
the net rate of heat transfer to the body must be zero. Therefore, the radiation
emitted by the body must be equal to the radiation absorbed by it:

$$A \varepsilon \sigma T^4 = A \alpha \sigma T^4$$

Thus, we conclude that

$$\varepsilon(T) = \alpha(T) \qquad (9\text{-}21)$$

That is, *the total hemispherical emissivity of a surface at temperature T is
equal to its total hemispherical absorptivity for radiation coming from a
blackbody at the same temperature.* This relation, which greatly simplifies
the radiation analysis, was first developed by Gustav Kirchhoff in 1860 and
is now called **Kirchhoff's law.** Note that this relation is derived under the
condition that the surface temperature is equal to the temperature of the
source of irradiation, and the reader is cautioned against using it when considerable difference (more than a few hundred degrees) exists between the
surface temperature and the temperature of the source of irradiation.

The derivation above can also be repeated for radiation at a specified
wavelength to obtain the *spectral* form of Kirchhoff's law:

$$\varepsilon_\lambda(T) = \alpha_\lambda(T) \qquad (9\text{-}22)$$

This relation is valid when the irradiation or the emitted radiation is independent of direction. The form of Kirchhoff's law that involves no restrictions
is the *spectral directional* form expressed as $\varepsilon_{\lambda,\theta}(T) = \alpha_{\lambda,\theta}(T)$. That is, the
emissivity of a surface at a specified wavelength, direction, and temperature
is always equal to its absorptivity at the same wavelength, direction, and
temperature.

It is very tempting to use Kirchhoff's law in radiation analysis since the
relation $\varepsilon = \alpha$ together with $\rho = 1 - \alpha$ enables us to determine all three
properties of an opaque surface from a knowledge of only *one* property.
Although using Kirchhoff's law gives acceptable results in most cases, in
practice, care should be exercised when there is considerable difference between the surface temperature and the temperature of the source of incident
radiation.

The Greenhouse Effect

You have probably noticed that when you leave your car under direct sunlight
on a sunny day, the interior of the car gets much warmer than the air outside,

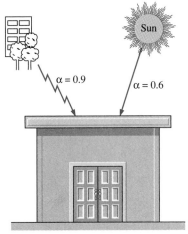

FIGURE 9-26

The absorptivity of a material may be
quite different for radiation originating
from sources at different temperatures.

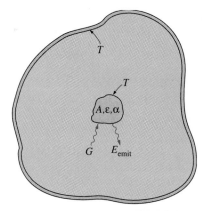

FIGURE 9-27

The small body contained in a large
isothermal enclosure used in the
development of Kirchhoff's law.

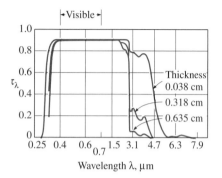

FIGURE 9-28

The spectral transmissivity of low-iron glass at room temperature for different thicknesses.

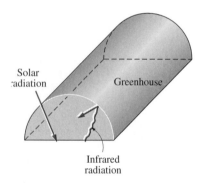

FIGURE 9-29

A greenhouse traps energy by allowing the solar radiation to come in but not allowing the infrared radiation to go out.

and you may have wondered why the car acts like a *heat trap*. The answer lies in the spectral transmissivity curve of the *glass,* which resembles an inverted U, as shown in Figure 9-28. We observe from this figure that glass at thicknesses encountered in practice transmits over 90 percent of radiation in the visible range and is practically opaque (nontransparent) to radiation in the longer-wavelength infrared regions of the electromagnetic spectrum (roughly $\lambda > 3$ μm). Therefore, glass has a transparent window in the wavelength range 0.3 μm $< \lambda <$ 3 μm in which over 90 percent of solar radiation is emitted. On the other hand, the entire radiation emitted by surfaces at room temperature falls in the infrared region. Consequently, glass allows the solar radiation to enter but does not allow the infrared radiation from the interior surfaces to leave. This causes a rise in the interior temperature as a result of the energy build-up in the car. This heating effect, which is due to the non-gray characteristic of glass (or clear plastics), is known as the **greenhouse effect,** since it is utilized primarily in greenhouses (Fig. 9-29).

The greenhouse effect is also experienced on a larger scale on earth. The surface of the earth, which warms up during the day as a result of the absorption of solar energy, cools down at night by radiating its energy into deep space as infrared radiation. The combustion gases such as CO_2 and water vapor in the atmosphere transmit the bulk of the solar radiation but absorb the infrared radiation emitted by the surface of the earth. Thus, there is concern that the energy trapped on earth will eventually cause global warming and thus drastic changes in weather patterns.

In *humid* places such as coastal areas, there is not a drastic change between the daytime and nighttime temperatures, because the humidity acts as a barrier on the path of the infrared radiation coming from the earth, and thus slows down the cooling process at night. In areas with clear skies such as deserts, there is a large swing between the daytime and nighttime temperatures because of the absence of such barriers for infrared radiation.

9-5 ■ ATMOSPHERIC AND SOLAR RADIATION

The sun is our primary source of energy. The energy coming off the sun, called *solar energy,* reaches us in the form of electromagnetic waves after experiencing considerable interactions with the atmosphere. The radiation energy emitted or reflected by the constituents of the atmosphere form the *atmospheric radiation.* Below we give an overview of the solar and atmospheric radiation because of their importance and relevance to daily life. Also, our familiarity with solar energy makes it an effective tool in developing a better understanding for some of the new concepts introduced earlier. Detailed treatment of this exciting subject can be found in numerous books devoted to this topic.

The *sun* is a nearly spherical body that has a diameter of $D \approx 1.39 \times 10^9$ m and a mass of $m \approx 2 \times 10^{30}$ kg and is located at a mean distance of $L = 1.50 \times 10^{11}$ m from the earth. It emits radiation energy continuously at a rate of $E_{sun} \approx 3.8 \times 10^{26}$ W. Less than a billionth of this energy (about 1.7×10^{17} W) strikes the earth, which is sufficient to keep the earth warm and to maintain life through the photosynthesis process. The energy of the sun is due to the continuous *fusion* reaction during which two hydrogen atoms fuse to form one atom of helium. Therefore, the sun is essentially a *nuclear reactor,* with temperatures as high as 40,000,000 K in its core region. The temperature drops to about 6000 K in the outer region of the sun, called the convective zone, as a result of the dissipation of this energy by radiation.

The solar energy reaching the earth's atmosphere is determined by a series of measurements taken in the late 1960s by using high-altitude aircraft, balloons, and spacecraft to be 1353 W/m². This quantity is called the *solar constant* G_s:

$$G_s = 1353 \text{ W/m}^2 \qquad (9\text{-}23)$$

The **solar constant** represents *the rate at which solar energy is incident on a surface normal to the sun's rays at the outer edge of the atmosphere when the earth is at its mean distance from the sun* (Fig. 9-30). Owing to the ellipticity of the earth's orbit, the distance between the sun and the earth, and thus the actual value of the solar constant, changes throughout the year. It varies from a maximum of 1399 W/m² on December 21 to a minimum of 1310 W/m² on June 21. (Note that the earth is farthest away from the sun in summer in the northern hemisphere.) However, this variation, which remains within ±3.4 percent of the mean value, is considered negligible for most practical purposes, and G_s is taken to be a *constant* at its mean value of 1353 W/m².

The measured value of the solar constant can be used to estimate the effective surface temperature of the sun from the requirement that

$$(4\pi L^2)G_s = (4\pi r^2)\,\sigma T_{\text{sun}}^4 \qquad (9\text{-}24)$$

where L is the mean distance between the sun's center and the earth and r is the radius of the sun. The left-hand side of this equation represents the total solar energy passing through a spherical surface whose radius is the mean earth–sun distance, and the right-hand side represents the total energy that leaves the sun's outer surface. The conservation of energy principle requires that these two quantities be equal to each other, since the solar energy experiences no attenuation (or enhancement) on its way through the vacuum (Fig. 9-31). The **effective surface temperature** of the sun is determined from Equation 9-24 to be $T_{\text{sun}} = 5762$ K. That is, the sun can be treated as a blackbody at a temperature of 5762 K. This is also confirmed by the measurements of the spectral distribution of the solar radiation just outside the atmosphere plotted in Figure 9-32, which shows only small deviations from the idealized blackbody behavior.

The spectral distribution of solar radiation on the ground plotted in Figure 9-32 shows that the solar radiation undergoes considerable *attenuation* as it passes through the atmosphere as a result of *absorption* and *scattering.* About 99 percent of the atmosphere is contained within a distance of 30 km from the earth's surface. The several dips on the spectral distribution of radiation on the earth's surface are due to *absorption* by the gases O_2, O_3 (ozone), H_2O, and CO_2. Absorption by *oxygen* occurs in a narrow band about $\lambda = 0.76$ μm. The *ozone* absorbs *ultraviolet* radiation at wavelengths below 0.3 μm almost completely, and radiation in the range 0.3–0.4 μm considerably. Thus, the ozone layer in the upper regions of the atmosphere protects biological systems on earth from harmful ultraviolet radiation. In turn, we

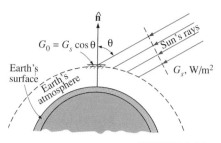

FIGURE 9-30

Solar radiation reaching the earth's atmosphere and the solar constant.

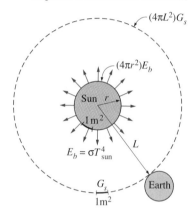

FIGURE 9-31

The total solar energy passing through concentric spheres remains constant, but the energy falling per unit area decreases with increasing radius.

FIGURE 9-32

Spectral distribution of solar radiation just outside the atmosphere, at the surface of the earth on a typical day, and comparison with blackbody radiation at 5762 K.

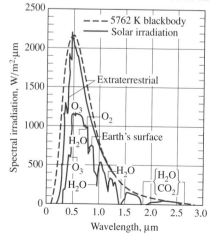

must protect the ozone layer from the destructive chemicals commonly used as refrigerants, cleaning agents, and propellants in aerosol cans. The use of these chemicals is now banned in many countries. The ozone gas also absorbs some radiation in the visible range. Absorption in the infrared region is dominated by *water vapor* and *carbon dioxide*. The dust particles and other pollutants in the atmosphere also absorb radiation at various wavelengths.

As a result of these absorptions, the solar energy reaching the *earth's surface* is weakened considerably, to about 950 W/m² on a clear day and much less on cloudy or smoggy days. Also, practically all of the solar radiation reaching the earth's surface falls in the wavelength band from 0.3 to 2.5 μm.

Another mechanism that attenuates solar radiation as it passes through the atmosphere is *scattering* or *reflection* by air molecules and the many other kinds of particles such as dust, smog, and water droplets suspended in the atmosphere. Scattering is mainly governed by the size of the particle relative to the wavelength of radiation. The oxygen and nitrogen molecules primarily scatter radiation at very short wavelengths, comparable to the size of the molecules themselves. Therefore, radiation at wavelengths corresponding to violet and blue colors is scattered the most. This molecular scattering in all directions is what gives the sky its bluish color. The same phenomenon is responsible for red sunrises and sunsets. Early in the morning and late in the afternoon, the sun's rays pass through a greater thickness of the atmosphere than they do at midday, when the sun is at the top. Therefore, the violet and blue colors of the light encounter a greater number of molecules by the time they reach the earth's surface, and thus a greater fraction of them are scattered (Fig. 9-33). Consequently, the light that reaches the earth's surface consists primarily of colors corresponding to longer wavelengths such as red, orange, and yellow. The clouds appear in reddish-orange color during sunrise and sunset because the light they reflect is reddish-orange at those times. For the same reason, a red traffic light is visible from a longer distance than is a green light under the same circumstances.

The solar energy incident on a surface on earth is considered to consist of *direct* and *diffuse* parts. The part of solar radiation that reaches the earth's surface without being scattered or absorbed by the atmosphere is called **direct solar radiation** G_D. The scattered radiation is assumed to reach the earth's surface uniformly from all directions and is called **diffuse solar radiation** G_d. Then the *total solar energy* incident on the unit area of a *horizontal surface* on the ground is (Fig. 9-34)

$$G_{\text{solar}} = G_D \cos \theta + G_d \qquad (\text{W/m}^2) \qquad (9\text{-}25)$$

where θ is the angle of incidence of direct solar radiation (the angle that the sun's rays make with the normal of the surface). The diffuse radiation varies from about 10 percent of the total radiation on a clear day to nearly 100 percent on a totally cloudy day.

The gas molecules and the suspended particles in the atmosphere *emit radiation* as well as absorb it. The atmospheric emission is primarily due to the CO_2 and H_2O molecules and is concentrated in the regions from 5 to 8 μm

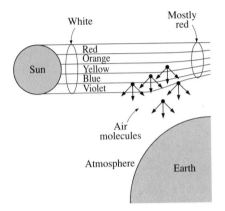

FIGURE 9-33

Air molecules scatter blue light much more than they do red light. At sunset, the light travels through a thicker layer of atmosphere, which removes much of the blue from the natural light, letting the red dominate.

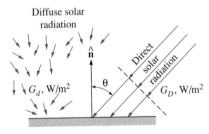

FIGURE 9-34

The direct and diffuse radiation incident on a horizontal surface at the earth's surface.

and above 13 μm. Although this emission is far from resembling the distribution of radiation from a blackbody, it is found convenient in radiation calculations to treat the atmosphere as a blackbody at some lower fictitious temperature that emits an equivalent amount of radiation energy. This fictitious temperature is called the **effective sky temperature** T_{sky}. Then the radiation emission from the atmosphere to the earth's surface is expressed as

$$G_{sky} = \sigma T_{sky}^4 \qquad (W/m^2) \qquad (9\text{-}26)$$

The value of T_{sky} depends on the atmospheric conditions. It ranges from about 230 K for cold, clear-sky conditions to about 285 K for warm, cloudy-sky conditions.

Note that the effective sky temperature does not deviate much from the room temperature. Thus, in the light of Kirchhoff's law, we can take the absorptivity of a surface to be equal to its emissivity at room temperature, $\alpha = \varepsilon$. Then the sky radiation absorbed by a surface can be expressed as

$$E_{sky,\,absorbed} = \alpha G_{sky} = \alpha \sigma T_{sky}^4 = \varepsilon \sigma T_{sky}^4 \qquad (W/m^2) \qquad (9\text{-}27)$$

The net rate of radiation heat transfer to a surface exposed to solar and atmospheric radiation is determined from an energy balance (Fig. 9-35):

$$\begin{aligned}
\dot{q}_{net,\,rad} &= \sum E_{absorbed} - \sum E_{emitted} \\
&= E_{solar,\,absorbed} + E_{sky,\,absorbed} - E_{emitted} \qquad (9\text{-}28) \\
&= \alpha_s G_{solar} + \varepsilon \sigma T_{sky}^4 - \varepsilon \sigma T_s^4 \\
&= \alpha_s G_{solar} + \varepsilon \sigma (T_{sky}^4 - T_s^4) \qquad (W/m^2)
\end{aligned}$$

where T_s is the temperature of the surface in K and ε is its emissivity at room temperature. A positive result for $\dot{q}_{net,\,rad}$ indicates a radiation heat gain by the surface and a negative result indicates a heat loss.

The absorption and emission of radiation by the *elementary gases* such as H_2, O_2, and N_2 at moderate temperatures are negligible, and a medium filled with these gases can be treated as a *vacuum* in radiation analysis. The absorption and emission of gases with *larger molecules* such as H_2O and CO_2, however, can be *significant* and may need to be considered when considerable amounts of such gases are present in a medium. For example, a 1-m-thick layer of water vapor at 1 atm pressure and 100°C emits more than 50 percent of the energy that a blackbody would emit at the same temperature.

In solar energy applications, the spectral distribution of incident solar radiation is very different than the spectral distribution of emitted radiation by the surfaces, since the former is concentrated in the short-wavelength region and the latter in the infrared region. Therefore, the radiation properties of surfaces will be quite different for the incident and emitted radiation, and the surfaces cannot be assumed to be gray. Instead, the surfaces are assumed to have two sets of properties: one for solar radiation and another for infrared radiation at room temperature. Table 9-3 lists the *emissivity* ε and the *solar absorptivity* α_s of the surfaces of some common materials. Surfaces that are intended to *collect solar energy,* such as the absorber surfaces of solar collectors, are desired to have high α_s but low ε values to

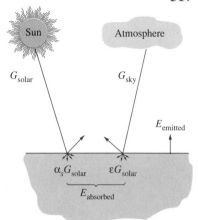

FIGURE 9-35

Radiation interactions of a surface exposed to solar and atmospheric radiation.

TABLE 9-3

Comparison of the solar absorptivity α_s of some surfaces with their emissivity ε at room temperature

Surface	α_s	ε
Aluminum		
Polished	0.09	0.03
Anodized	0.14	0.84
Foil	0.15	0.05
Copper		
Polished	0.18	0.03
Tarnished	0.65	0.75
Stainless steel		
Polished	0.37	0.60
Dull	0.50	0.21
Plated metals		
Black nickel oxide	0.92	0.08
Black chrome	0.87	0.09
Concrete	0.60	0.88
White marble	0.46	0.95
Red brick	0.63	0.93
Asphalt	0.90	0.90
Black paint	0.97	0.97
White paint	0.14	0.93
Snow	0.28	0.97
Human skin (caucasian)	0.62	0.97

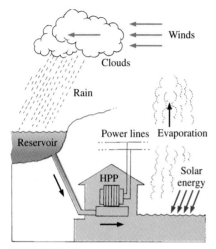

FIGURE 9-36

The cycle that water undergoes in a
hydroelectric power plant.

maximize the absorption of solar radiation and to minimize the emission of radiation. Surfaces that are intended to *remain cool* under the sun, such as the outer surfaces of fuel tanks and refrigerator trucks, are desired to have just the opposite properties. Surfaces are often given the desired properties by coating them with thin layers of *selective* materials. A surface can be kept cool, for example, by simply painting it white.

We close this section by pointing out that what we call *renewable energy* is usually nothing more than the manifestation of solar energy in different forms. Such energy sources include wind energy, hydroelectric power, ocean thermal energy, ocean wave energy, and wood. For example, no hydroelectric power plant can generate electricity year after year unless the water evaporates by absorbing solar energy and comes back as a rainfall to replenish the water source (Fig. 9-36). Although solar energy is sufficient to meet the entire energy needs of the world, currently it is not economical to do so because of the low concentration of solar energy on earth and the high capital cost of harnessing it.

EXAMPLE 9-4   Selective Absorber and Reflective Surfaces

Consider a surface exposed to solar radiation. At some time, the direct and diffuse components of solar radiation are $G_D = 400$ and $G_d = 300$ W/m², and the direct radiation makes a 20° angle with the normal of the surface. The surface temperature is observed to be 320 K at that time. Assuming an effective sky temperature of 260 K, determine the net rate of radiation heat transfer for the following cases (Fig. 9-37):

(a) $\alpha_s = 0.9$ and $\varepsilon = 0.9$ (gray absorber surface)
(b) $\alpha_s = 0.1$ and $\varepsilon = 0.1$ (gray reflector surface)
(c) $\alpha_s = 0.9$ and $\varepsilon = 0.1$ (selective absorber surface)
(d) $\alpha_s = 0.1$ and $\varepsilon = 0.9$ (selective reflector surface)

Solution   A surface is exposed to solar and sky radiation. The net rate of radiation heat transfer is to be determined for four different combinations of emissivities and solar absorptivities.

Analysis   The total solar energy incident on the surface is determined from Equation 9-25 to be

$$G_{solar} = G_D \cos \theta + G_d$$
$$= (400 \text{ W/m}^2) \cos 20° + (300 \text{ W/m}^2)$$
$$= 675.9 \text{ W/m}^2$$

Then the net rate of radiation heat transfer for each of the four cases is determined from Equation 9-28:

$$\dot{q}_{net, rad} = \alpha_s G_{solar} + \varepsilon\sigma(T_{sky}^4 - T_s^4)$$

(a) $\alpha_s = 0.9$ and $\varepsilon = 0.9$ (gray absorber surface):

$$\dot{q}_{net, rad} = 0.9(675.9 \text{ W/m}^2) + 0.9(5.67 \times 10^{-8} \text{ W/m}^2 \cdot \text{K}^4)[(260 \text{ K})^4 - (320 \text{ K})^4]$$
$$= \textbf{306.5 W/m}^2$$

(b) $\alpha_s = 0.1$ and $\varepsilon = 0.1$ (gray reflector surface):

$$\dot{q}_{net, rad} = 0.1(675.9 \text{ W/m}^2) + 0.1(5.67 \times 10^{-8} \text{ W/m}^2 \cdot \text{K}^4)[(260 \text{ K})^4 - (320 \text{ K})^4]$$
$$= \textbf{34.1 W/m}^2$$

(c) $\alpha_s = 0.9$ and $\varepsilon = 0.1$ (selective absorber surface):

$$\dot{q}_{net,\,rad} = 0.9(675.9 \text{ W/m}^2) + 0.1(5.67 \times 10^{-8} \text{ W/m}^2 \cdot \text{K}^4)[(260 \text{ K})^4 - (320 \text{ K})^4]$$
$$= \textbf{574.8 W/m}^2$$

(d) $\alpha_s = 0.1$ and $\varepsilon = 0.9$ (selective reflector surface):

$$\dot{q}_{net,\,rad} = 0.1(675.9 \text{ W/m}^2) + 0.9(5.67 \times 10^{-8} \text{ W/m}^2 \cdot \text{K}^4)[(260 \text{ K})^4 - (320 \text{ K})^4]$$
$$= \textbf{−234.3 W/m}^2$$

Discussion   Note that the surface of an ordinary gray material of high absorptivity gains heat at a rate of 306.5 W/m². The amount of heat gain increases to 574.8 W/m² when the surface is coated with a selective material that has the same absorptivity for solar radiation but a low emissivity for infrared radiation. Also note that the surface of an ordinary gray material of high reflectivity still gains heat at a rate of 34.1 W/m². When the surface is coated with a selective material that has the same reflectivity for solar radiation but a high emissivity for infrared radiation, the surface loses 234.3 W/m² instead. Therefore, the temperature of the surface will decrease when a selective reflector surface is used.

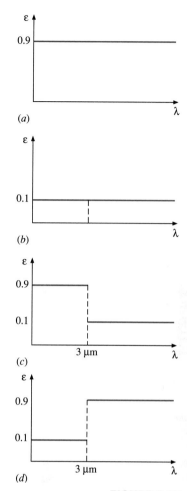

FIGURE 9-37

Graphical representation of the spectral emissivities of the four surfaces considered in Example 9-4.

9-6 ■ THE VIEW FACTOR

So far, we have considered the radiation properties of surfaces and the radiation interactions of a single surface. We are now in a position to consider *radiation heat transfer* between two or more surfaces, which is the primary quantity of interest in practice. Radiation heat transfer between surfaces depends on the *orientation* of the surfaces relative to each other as well as their radiation properties and temperatures, as illustrated in Figure 9-38. For example, a camper will make the most use of a campfire on a cold night by standing as close to the fire as possible and by blocking as much of the radiation coming from the fire by turning her front to the fire instead her side. Likewise, a person will maximize the amount of solar radiation incident on him by lying down on his back instead of standing up on his feet.

To account for the effects of orientation on radiation heat transfer between two surfaces, we define a new parameter called the *view factor,* which is a purely geometric quantity and is independent of the surface properties and temperature. It is also called the *shape factor, configuration factor,* and *angle factor.* The view factor based on the assumption that the surfaces are diffuse emitters and diffuse reflectors is called the *diffuse view factor,* and the view factor based on the assumption that the surfaces are diffuse emitters but specular reflectors is called the *specular view factor.* In this book, we will consider radiation exchange between diffuse surfaces only, and thus the term *view factor* will simply imply *diffuse view factor.*

The **view factor** from a surface i to a surface j is denoted by $F_{i \rightarrow j}$ and is defined as

$F_{i \rightarrow j} = $ *the fraction of the radiation leaving surface i that strikes surface j directly*

Therefore, the view factor $F_{1 \rightarrow 2}$ represents the fraction of the radiation leaving surface 1 that strikes surface 2 and $F_{2 \rightarrow 1}$ represents the fraction of the

FIGURE 9-38

Radiation heat exchange between surfaces depends on the *orientation* of the surfaces relative to each other, and this dependence on orientation is accounted for by the *view factor.*

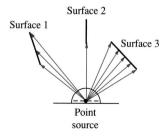

(a) Plane surface

(b) Convex surface

(c) Concave surface

FIGURE 9-39

The view factor from a surface to itself is *zero* for *plane* or *convex* surfaces and *nonzero* for *concave* surfaces.

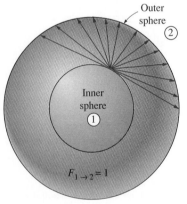

FIGURE 9-40

In a geometry that consists of two concentric spheres, the view factor $F_{1 \to 2} = 1$, since the entire radiation leaving the surface of the smaller sphere will be intercepted by the larger sphere.

radiation leaving surface 2 that strikes surface 1 directly. Note that the radiation that strikes a surface does not need to be absorbed by that surface. Also, radiation that strikes a surface after being reflected by other surfaces is not considered in the evaluation of the view factors. For the special case of $j = i$, we have

$F_{i \to i}$ = *the fraction of radiation leaving surface i that strikes itself directly*

Noting that in the absence of strong electromagnetic fields radiation beams travel in straight paths, the view factor from a surface to itself will be zero unless the surface "sees" itself. Therefore, $F_{i \to i} = 0$ for *plane* or *convex* surfaces and $F_{i \to i} \neq 0$ for *concave* surfaces, as illustrated in Figure 9-39.

The value of the view factor ranges between *zero* and *one*. The limiting case $F_{i \to j} = 0$ indicates that the two surfaces do not have a direct view of each other, and thus radiation leaving surface *i* cannot strike surface *j* directly. The other limiting case $F_{i \to j} = 1$ indicates that surface *j* completely surrounds surface *i*, so that the entire radiation leaving surface *i* is intercepted by surface *j*. For example, in a geometry consisting of two concentric spheres, the entire radiation leaving the surface of the smaller sphere (surface 1) will strike the larger sphere (surface 2), and thus $F_{1 \to 2} = 1$, as illustrated in Figure 9-40.

The view factor has proven to be very useful in radiation analysis because it allows us to express the *fraction of radiation* leaving a surface that strikes another surface in terms of the orientation of these two surfaces relative to each other. The underlying assumption in this process is that the radiation a surface receives from a source is directly proportional to the angle the surface subtends when viewed from the source. This would be the case only if the radiation coming off the source is *uniform* in all directions throughout its surface and the medium between the surfaces does not *absorb*, *emit*, or *scatter* radiation. That is, it will be the case when the surfaces are *isothermal* and *diffuse* emitters and reflectors and the surfaces are separated by a *nonparticipating* medium such as a vacuum or air.

The view factor $F_{1 \to 2}$ between two surfaces A_1 and A_2 can be determined in a systematic manner first by expressing the view factor between two differential areas dA_1 and dA_2 in terms of the spatial variables and then by performing the necessary integrations. However, this approach is not practical, since, even for simple geometries, the resulting integrations are usually very complex and difficult to perform.

View factors for hundreds of common geometries are evaluated and the results are given in analytical, graphical, and tabular form in several publications. View factors for selected geometries are given in Tables 9-4 and 9-5 in *analytical* form and in Figs. 9-41 to 9-44 in *graphical* form. The view factors in Table 9-4 are for three-dimensional geometries. The view factors in Table 9-5, on the other hand, are for geometries that are *infinitely long* in the direction perpendicular to the plane of the paper and are therefore two-dimensional.

TABLE 9-4

521

The View Factor

View factor expressions for some common geometries of finite size (3D)

Geometry	Relation
Aligned parallel rectangles	$\overline{X} = X/L,\ \overline{Y} = Y/L$ $$F_{i \to j} = \frac{2}{\pi \overline{X} \overline{Y}} \left\{ \ln \left[\frac{(1 + \overline{X}^2)(1 + \overline{Y}^2)}{1 + \overline{X}^2 + \overline{Y}^2} \right]^{1/2} \right.$$ $$+ \overline{X}(1 + \overline{Y}^2)^{1/2} \tan^{-1} \frac{\overline{X}}{(1 + \overline{Y}^2)^{1/2}}$$ $$+ \overline{Y}(1 + \overline{X}^2)^{1/2} \tan^{-1} \frac{\overline{Y}}{(1 + \overline{X}^2)^{1/2}}$$ $$\left. - \overline{X} \tan^{-1} \overline{X} - \overline{Y} \tan^{-1} \overline{Y} \right\}$$
Coaxial parallel disks	$R_i = r_i/L,\ R_j = r_j/L$ $$S = 1 + \frac{1 + R_j^2}{R_i^2}$$ $$F_{i \to j} = \frac{1}{2} \left\{ S - \left[S^2 - 4 \left(\frac{R_j}{R_i} \right)^2 \right]^{1/2} \right\}$$
Perpendicular rectangles with a common edge	$H = Z/X,\ W = Y/X$ $$F_{i \to j} = \frac{1}{\pi W} \left(W \tan^{-1} \frac{1}{W} + H \tan^{-1} \frac{1}{H} \right.$$ $$- (H^2 + W^2)^{1/2} \tan^{-1} \frac{1}{(H^2 + W^2)^{1/2}}$$ $$+ \frac{1}{4} \ln \left\{ \frac{(1 + W^2)(1 + H^2)}{1 + W^2 + H^2} \right.$$ $$\times \left[\frac{W^2(1 + W^2 + H^2)}{(1 + W^2)(W^2 + H^2)} \right]^{W^2}$$ $$\left. \left. \times \left[\frac{H^2(1 + H^2 + W^2)}{(1 + H^2)(H^2 + W^2)} \right]^{H^2} \right\} \right)$$

View Factor Relations

Radiation analysis on an enclosure consisting of N surfaces requires the evaluation of N^2 view factors, and this evaluation process is probably the most time-consuming part of a radiation analysis. However, it is neither practical nor necessary to evaluate all of the view factors directly. Once a sufficient number of view factors are available, the rest of them can be determined by utilizing some fundamental relations for view factors, as discussed below.

TABLE 9-5

View factor expressions for some infinitely long (2D) geometries

Geometry	Relation
Parallel plates with midlines connected by perpendicular	$W_i = w_i/L,\ W_j = w_j/L$ $F_{i \to j} = \dfrac{[(W_i + W_j)^2 + 4]^{1/2} - (W_j - W_i)^2 + 4]^{1/2}}{2W_i}$
Inclined plates of equal width and with a common edge	$F_{i \to j} = 1 - \sin \dfrac{1}{2}\alpha$
Perpendicular plates with a common edge	$F_{i \to j} = \dfrac{1}{2}\left\{ 1 + \dfrac{w_j}{w_i} - \left[1 + \left(\dfrac{w_j}{w_i}\right)^2 \right]^{1/2} \right\}$
Three-sided enclosure	$F_{i \to j} = \dfrac{w_i + w_j - w_k}{2w_i}$
Infinite plane and row of cylinders	$F_{i \to j} = 1 - \left[1 - \left(\dfrac{D}{s}\right)^2 \right]^{1/2}$ $+ \dfrac{D}{s} \tan^{-1}\left(\dfrac{s^2 - D^2}{D^2}\right)^{1/2}$

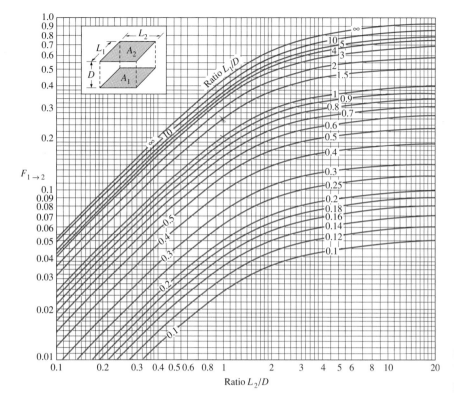

FIGURE 9-41

View factor between two aligned parallel rectangles of equal size.

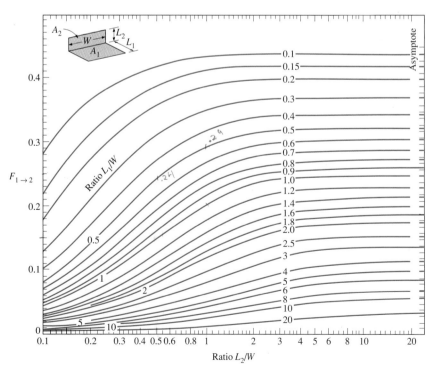

FIGURE 9-42

View factor between two perpendicular rectangles with a common edge.

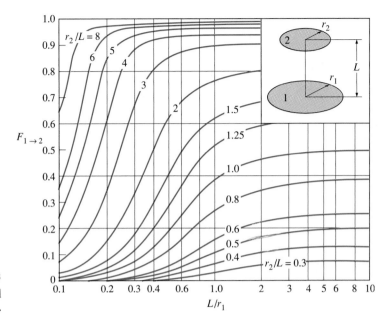

FIGURE 9-43

View factor between two coaxial
parallel disks.

FIGURE 9-44

View factors for two concentric cylinders of finite length: (*a*) outer cylinder to inner cylinder; (*b*) outer cylinder to itself.

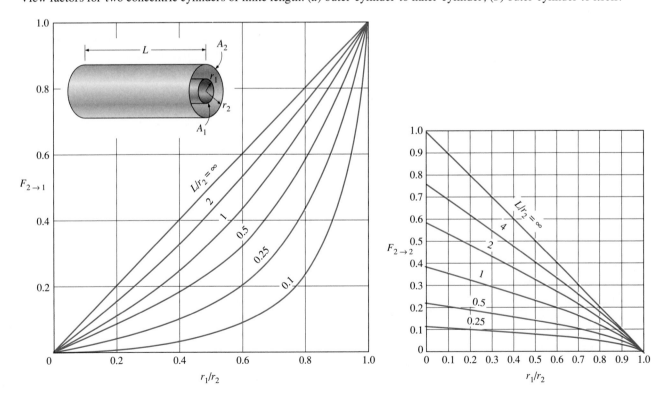

1 The Reciprocity Rule

The view factors $F_{i \to j}$ and $F_{j \to i}$ are *not* equal to each other unless the areas of the two surfaces are equal. That is,

$$F_{j \to i} = F_{i \to j} \quad \text{when} \quad A_i = A_j$$
$$F_{j \to i} \neq F_{i \to j} \quad \text{when} \quad A_i \neq A_j$$

Using the radiation intensity concept and going through some manipulations, it can be shown that the pair of view factors $F_{i \to j}$ and $F_{j \to i}$ are related to each other by

$$A_i F_{i \to j} = A_j F_{j \to i} \tag{9-29}$$

This relation is known as the **reciprocity rule,** and it enables us to determine the counterpart of a view factor from a knowledge of the view factor itself and the areas of the two surfaces. When determining the pair of view factors $F_{i \to j}$ and $F_{j \to i}$, it makes sense to evaluate first the easier one directly and then the harder one by applying the reciprocity rule.

2 The Summation Rule

The radiation analysis of a surface normally requires the consideration of the radiation coming in or going out in all directions. Therefore, most radiation problems encountered in practice involve enclosed spaces. When formulating a radiation problem, we usually form an *enclosure* consisting of the surfaces interacting radiatively. Even openings are treated as imaginary surfaces with radiation properties equivalent to those of the opening.

The conservation of energy principle requires that the entire radiation leaving any surface i of an enclosure be intercepted by the surfaces of the enclosure. Therefore, *the sum of the view factors from surface i of an enclosure to all surfaces of the enclosure, including to itself, must equal unity.* This is known as the **summation rule** for an enclosure and is expressed as (Fig. 9-45)

$$\sum_{j=1}^{N} F_{i \to j} = 1 \tag{9-30}$$

where N is the number of surfaces of the enclosure. For example, applying the summation rule to surface 1 of a three-surface enclosure yields

$$\sum_{j=1}^{3} F_{1 \to j} = F_{1 \to 1} + F_{1 \to 2} + F_{1 \to 3} = 1$$

The notation $F_{i \to j}$ is *instructive* for beginners, since it emphasizes that the view factor is for radiation that travels from surface i to surface j. However, this notation becomes rather awkward when it has to be used many times in a problem. In such cases, it is convenient to replace it by its *shorthand* version F_{ij}.

The summation rule can be applied to each surface of an enclosure by varying i from 1 to N. Therefore, the summation rule applied to each of the N surfaces of an enclosure gives N relations for the determination of the view

FIGURE 9-45

Radiation leaving any surface i of an enclosure must be intercepted completely by the surfaces of the enclosure. Therefore, the sum of the view factors from surface i to each one of the surfaces of the enclosure must be unity.

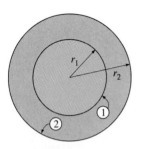

FIGURE 9-46

The geometry considered in
Example 9-5.

factors. Also, the reciprocity rule gives $\frac{1}{2}N(N-1)$ additional relations. Then the total number of view factors that need to be evaluated directly for an N-surface enclosure becomes

$$N^2 - [N + \tfrac{1}{2}N(N-1)] = \tfrac{1}{2}N(N-1)$$

For example, for a six-surface enclosure, we need to determine only $\frac{1}{2} \times 6(6-1) = 15$ of the $6^2 = 36$ view factors directly. The remaining 21 view factors can be determined from the 21 equations that are obtained by applying the reciprocity and the summation rules.

EXAMPLE 9-5   View Factors Associated with Two Concentric Spheres

Determine the view factors associated with an enclosure formed by two spheres, shown in Figure 9-46.

Solution   The view factors associated with two concentric spheres are to be determined.

Assumptions   The surfaces are diffuse emitters and reflectors.

Analysis   The outer surface of the smaller sphere (surface 1) and inner surface of the larger sphere (surface 2) form a two-surface enclosure. Therefore, $N = 2$ and this enclosure involves $N^2 = 2^2 = 4$ view factors, which are F_{11}, F_{12}, F_{21}, and F_{22}. In this two-surface enclosure, we need to determine only

$$\tfrac{1}{2}N(N-1) = \tfrac{1}{2} \times 2(2-1) = 1$$

view factor directly. The remaining three view factors can be determined by the application of the summation and reciprocity rules. But it turns out that we can determine not only one but *two* view factors directly in this case by a simple *inspection:*

$$F_{11} = 0, \quad \text{since no radiation leaving surface 1 strikes itself}$$
$$F_{12} = 1, \quad \text{since all radiation leaving surface 1 strikes surface 2}$$

Actually it would be sufficient to determine only one of these view factors by inspection, since we could always determine the other one from the summation rule applied to surface 1 as $F_{11} + F_{12} = 1$.

The view factor F_{21} is determined by applying the reciprocity rule to surfaces 1 and 2:

$$A_1 F_{12} = A_2 F_{21}$$

which yields

$$F_{21} = \frac{A_1}{A_2} F_{12} = \frac{4\pi r_1^2}{4\pi r_2^2} \times 1 = \left(\frac{r_1}{r_2}\right)^2$$

Finally, the view factor F_{22} is determined by applying the summation rule to surface 2:

$$F_{21} + F_{22} = 1$$

and thus

$$F_{22} = 1 - F_{21} = 1 - \left(\frac{r_1}{r_2}\right)^2$$

Discussion Note that when the outer sphere is much larger than the inner sphere ($r_2 \gg r_1$), F_{22} approaches one. This is expected, since the fraction of radiation leaving the outer sphere that is intercepted by the inner sphere will be negligible in that case. Also note that the two spheres considered above do not need to be concentric. However, the radiation analysis will be most accurate for the case of concentric spheres, since the radiation is most likely to be uniform on the surfaces in that case.

3   The Superposition Rule

Sometimes the view factor associated with a given geometry is not available in standard tables and charts. In such cases, it is desirable to express the given geometry as the sum or difference of some geometries with known view factors, and then to apply the **superposition rule,** which can be expressed as follows: *the view factor from a surface i to a surface j is equal to the sum of the view factors from surface i to the parts of surface j.* Note that the reverse of this is not true. That is, the view factor from a surface *j* to a surface *i* is *not* equal to the sum of the view factors from the parts of surface *j* to surface *i*.

Consider the geometry in Figure 9-47, which is infinitely long in the direction perpendicular to the plane of the paper. The radiation that leaves surface 1 and strikes the combined surfaces 2 and 3 is equal to the sum of the radiation that strikes surfaces 2 and 3. Therefore, the view factor from surface 1 to the combined surfaces of 2 and 3 is

$$F_{1 \to (2,3)} = F_{1 \to 2} + F_{1 \to 3} \qquad (9\text{-}31)$$

Suppose we need to find the view factor $F_{1 \to 3}$. A quick check of the view factor expressions and charts in this section will reveal that such a view factor cannot be evaluated directly. However, the view factor $F_{1 \to 3}$ can be determined from Equation 9-31 after determining both $F_{1 \to 2}$ and $F_{1 \to (2,3)}$ from the chart in Figure 9-42. Therefore, it may be possible to determine some difficult view factors with relative ease by expressing one or both of the areas as the sum or differences of ares and then applying the superposition rule.

To obtain a relation for the view factor $F_{(2,3) \to 1}$, we multiply Equation 9-31 by A_1,

$$A_1 F_{1 \to (2,3)} = A_1 F_{1 \to 2} + A_1 F_{1 \to 3}$$

and apply the reciprocity rule to each term to get

$$(A_2 + A_3)F_{(2,3) \to 1} = A_2 F_{2 \to 1} + A_3 F_{3 \to 1}$$

or

$$F_{(2,3) \to 1} = \frac{A_2 F_{2 \to 1} + A_3 F_{3 \to 1}}{A_2 + A_3} \qquad (9\text{-}32)$$

Areas that are expressed as the sum of more than two parts can be handled in a similar manner.

EXAMPLE 9-6   Fraction of Radiation Leaving through an Opening

Determine the fraction of the radiation leaving the base of the cylindrical enclosure shown in Figure 9-48 that escapes through a coaxial ring opening at its top

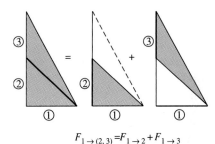

$$F_{1 \to (2,3)} = F_{1 \to 2} + F_{1 \to 3}$$

FIGURE 9-47

The view factor from a surface to a composite surface is equal to the sum of the view factors from the surface to the parts of the composite surface.

FIGURE 9-48

The cylindrical enclosure considered in Example 9-6.

surface. The radius and the length of the enclosure are $r_1 = 10$ cm and $L = 10$ cm, while the inner and outer radii of the ring are $r_2 = 5$ cm and $r_3 = 8$ cm, respectively.

Solution The fraction of radiation leaving the base of a cylindrical enclosure through a coaxial ring opening at its top surface is to be determined.

Assumptions The base surface is a diffuse emitter and reflector.

Analysis We are asked to determine the fraction of the radiation leaving the base of the enclosure that escapes through an opening at the top surface. Actually, what we are asked to determine is simply the *view factor* $F_{1 \rightarrow ring}$ from the base of the enclosure to the ring-shaped surface at the top.

We do not have an analytical expression or chart for view factors between a circular area and a coaxial ring, and so we cannot determine $F_{1 \rightarrow ring}$ directly. However, we do have a chart for view factors between two coaxial parallel disks, and we can always express a ring in terms of disks.

Let the base surface of radius $r_1 = 10$ cm be surface 1, the circular area of $r_2 = 5$ cm at the top be surface 2, and the circular area of $r_3 = 8$ cm be surface 3. Using the superposition rule, the view factor from surface 1 to surface 3 can be expressed as

$$F_{1 \rightarrow 3} = F_{1 \rightarrow 2} + F_{1 \rightarrow ring}$$

since surface 3 is the sum of surface 2 and the ring area. The view factors $F_{1 \rightarrow 2}$ and $F_{1 \rightarrow 3}$ are determined from the chart in Figure 9-43 as follows:

$$\frac{L}{r_1} = \frac{10 \text{ cm}}{10 \text{ cm}} = 1 \quad \text{and} \quad \frac{r_2}{L} = \frac{5 \text{ cm}}{10 \text{ cm}} = 0.5 \xrightarrow{\text{(Fig. 9-43)}} F_{1 \rightarrow 2} = 0.11$$

$$\frac{L}{r_1} = \frac{10 \text{ cm}}{10 \text{ cm}} = 1 \quad \text{and} \quad \frac{r_3}{L} = \frac{8 \text{ cm}}{10 \text{ cm}} = 0.8 \xrightarrow{\text{(Fig. 9-43)}} F_{1 \rightarrow 3} = 0.28$$

Therefore,

$$F_{1 \rightarrow ring} = F_{1 \rightarrow 3} - F_{1 \rightarrow 2} = 0.28 - 0.11 = \mathbf{0.17}$$

which is the desired result. Note that $F_{1 \rightarrow 2}$ and $F_{1 \rightarrow 3}$ represent the fractions of radiation leaving the base that strike the circular surfaces 2 and 3, respectively, and their difference gives the fraction that strikes the ring area.

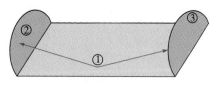

$F_{1 \rightarrow 2} = F_{1 \rightarrow 3}$

(Also, $F_{2 \rightarrow 1} = F_{3 \rightarrow 1}$)

FIGURE 9-49

Two surfaces that are symmetric about a third surface will have the same view factor from the third surface.

4 The Symmetry Rule

The determination of the view factors in a problem can be simplified further if the geometry involved possesses some sort of symmetry. Therefore, it is good practice to check for the presence of any *symmetry* in a problem before attempting to determine the view factors directly. The presence of symmetry can be determined *by inspection*, keeping the definition of the view factor in mind. Identical surfaces that are oriented in an identical manner with respect to another surface will intercept identical amounts of radiation leaving that surface. Therefore, the **symmetry rule** can be expressed as follows: *two (or more) surfaces that possess symmetry about a third surface will have identical view factors from that surface* (Fig. 9-49).

The symmetry rule can also be expressed as follows: *if the surfaces j and k are symmetric about the surface i then $F_{i \rightarrow j} = F_{i \rightarrow k}$*. Using the reciprocity rule, we can show that the relation $F_{j \to i} = F_{k \to i}$ is also true in this case.

EXAMPLE 9-7 View Factors Associated with a Tetragon

Determine the view factors from the base of the pyramid shown in Figure 9-50 to each of its four side surfaces. The base of the pyramid is a square, and its side surfaces are isosceles triangles.

Solution The view factors from the base of a tetragon to each of its four side surfaces for the case of a square base are to be determined.

Assumptions The surfaces are diffuse emitters and reflectors.

Analysis The base of the pyramid (surface 1) and its four side surfaces (surfaces 2, 3, 4, and 5) form a five-surface enclosure. The first thing we notice about this enclosure is its symmetry. The four side surfaces are symmetric about the base surface. Then, from the *symmetry rule,* we have

$$F_{12} = F_{13} = F_{14} = F_{15}$$

Also, the *summation rule* applied to surface 1 yields

$$\sum_{j=1}^{5} F_{1j} = F_{11} + F_{12} + F_{13} + F_{14} + F_{15} = 1$$

However, $F_{11} = 0$, since the base is a *flat* surface. Then the two relations above yield

$$F_{12} = F_{13} = F_{14} = F_{15} = \textbf{0.25}$$

Discussion Note that each of the four side surfaces of the pyramid receive one-fourth of the entire radiation leaving the base surface, as expected. Also note that the presence of symmetry greatly simplified the determination of the view factors.

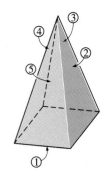

FIGURE 9-50
The pyramid considered in Example 9-7.

EXAMPLE 9-8 View Factors Associated with a Long Triangular Duct

Determine the view factor from any one side to any other side of the infinitely long triangular duct whose cross-section is given in Figure 9-51.

Solution The view factors associated with an infinitely long triangular duct are to be determined.

Assumptions The surfaces are diffuse emitters and reflectors.

Analysis The widths of the sides of the triangular cross-section of the duct are L_1, L_2, and L_3, and the surface areas corresponding to them are A_1, A_2, and A_3, respectively. Since the duct is infinitely long, the fraction of radiation leaving any surface that escapes through the ends of the duct is negligible. Therefore, the infinitely long duct can be considered to be a three-surface enclosure, $N = 3$.

This enclosure involves $N^2 = 3^2 = 9$ view factors, and we need to determine

$$\tfrac{1}{2}N(N-1) = \tfrac{1}{2} \times 3(3-1) = 3$$

of these view factors directly. Fortunately, we can determine all three of them by inspection to be

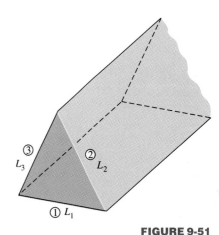

FIGURE 9-51
The infinitely long triangular duct considered in Example 9-8.

$$F_{11} = F_{22} = F_{33} = 0$$

since all three surfaces are flat. The remaining six view factors can be determined by the application of the summation and reciprocity rules.

Applying the summation rule to each of the three surfaces gives

$$F_{11} + F_{12} + F_{13} = 1$$
$$F_{21} + F_{22} + F_{23} = 1$$
$$F_{31} + F_{32} + F_{33} = 1$$

Noting that $F_{11} = F_{22} = F_{33} = 0$ and multiplying the first equation by A_1, the second by A_2, and the third by A_3 gives

$$A_1F_{12} + A_1F_{13} = A_1$$
$$A_2F_{21} + A_2F_{23} = A_2$$
$$A_3F_{31} + A_3F_{32} = A_3$$

Finally, applying the three reciprocity rules $A_1F_{12} = A_2F_{21}$, $A_1F_{13} = A_3F_{31}$, and $A_2F_{23} = A_3F_{32}$ gives

$$A_1F_{12} + A_1F_{13} = A_1$$
$$A_1F_{12} + A_2F_{23} = A_2$$
$$A_1F_{13} + A_2F_{23} = A_3$$

This is a set of three algebraic equations with three unknowns, which can be solved to obtain

$$F_{12} = \frac{A_1 + A_2 - A_3}{2A_1} = \frac{L_1 + L_2 - L_3}{2L_1}$$

$$F_{13} = \frac{A_1 + A_3 - A_2}{2A_1} = \frac{L_1 + L_3 - L_2}{2L_1} \qquad (9\text{-}33)$$

$$F_{23} = \frac{A_2 + A_3 - A_1}{2A_2} = \frac{L_2 + L_3 - L_1}{2L_2}$$

Discussion   Note that we have replaced the areas of the side surfaces by their corresponding widths for simplicity, since $A = Ls$ and the length s can be factored out and canceled. We can generalize this result as follows: *the view factor from a surface of a very long triangular duct to another surface is equal to the sum of the widths of these two surfaces minus the width of the third surface, divided by twice the width of the first surface.*

View Factors between Infinitely Long Surfaces: The Crossed-Strings Method

Many problems encountered in practice involve geometries of constant cross-section such as channels and ducts that are *very long* in one direction relative to the other directions. Such geometries can conveniently be considered to be *two-dimensional*, since any radiation interaction through their end surfaces will be negligible. Then they can be modeled as being *infinitely long*, and the view factor between their surfaces can be determined by the amazingly simple *crossed-strings method* developed by H. C. Hottel in the 1950s. The surfaces of the geometry do not need to be flat; they can be convex, concave, or any irregular shape.

To demonstrate the method, consider the geometry shown in Figure 9-52, and let us try to find the view factor $F_{1 \rightarrow 2}$ between surfaces 1 and 2. The first thing we do is identify the endpoints of the surfaces (the points A, B, C, and D) and connect them to each other with tightly stretched strings, which are indicated by dashed lines. Hottel has shown that the view factor $F_{1 \rightarrow 2}$ can be expressed in terms of the lengths of these stretched strings, which are straight lines, as

$$F_{1 \rightarrow 2} = \frac{(L_5 + L_6) - (L_3 + L_4)}{2L_1} \tag{9-34}$$

Note that $L_5 + L_6$ is the sum of the lengths of the *crossed strings*, and $L_3 + L_4$ is the sum of the lengths of the *uncrossed strings* attached to the endpoints. Therefore, Hottel's crossed-string method can be expressed verbally as

$$F_{i \rightarrow j} = \frac{\Sigma \,(\text{Crossed strings}) - \Sigma \,(\text{Uncrossed strings})}{2 \times (\text{String on surface } i)} \tag{9-35}$$

The crossed-strings method is applicable even when the two surfaces considered share a common edge, as in a triangle. In such cases, the common edge can be treated as an imaginary string of zero length. The method can also be applied to surfaces that are partially blocked by other surfaces by allowing the strings to bend around the blocking surfaces.

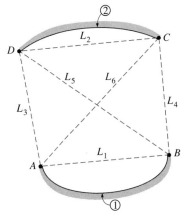

FIGURE 9-52

Determination of the view factor $F_{1 \rightarrow 2}$ by the application of the crossed-strings method.

EXAMPLE 9-9 The Crossed-Strings Method for View Factors

Two infinitely long parallel plates of widths $a = 12$ cm and $b = 5$ cm are located a distance $c = 6$ cm apart, as shown in Figure 9-53. (a) Determine the view factor $F_{1 \rightarrow 2}$ from surface 1 to surface 2 by using the crossed-strings method. (b) Derive the crossed-strings formula by forming triangles on the given geometry and using Equation 9-33 for view factors between the sides of triangles.

Solution The view factors between two infinitely long parallel plates are to be determined using the crossed-strings method, and the formula for the view factor is to be derived.

Assumptions The surfaces are diffuse emitters and reflectors.

Analysis (a) First we label the endpoints of both surfaces and draw straight dashed lines between the endpoints, as shown in Figure 9-53. Then we identify the crossed and uncrossed strings and apply the crossed-strings method (Eq. 9-35) to determine the view factor $F_{1 \rightarrow 2}$:

$$F_{1 \rightarrow 2} = \frac{\Sigma \,(\text{Crossed strings}) - \Sigma \,(\text{Uncrossed strings})}{2 \times (\text{String on surface 1})} = \frac{(L_5 + L_6) - (L_3 + L_4)}{2L_1}$$

where

$$L_1 = a = 12 \text{ cm} \qquad L_4 = \sqrt{7^2 + 6^2} = 9.22 \text{ cm}$$
$$L_2 = b = 5 \text{ cm} \qquad L_5 = \sqrt{5^2 + 6^2} = 7.81 \text{ cm}$$
$$L_3 = c = 6 \text{ cm} \qquad L_6 = \sqrt{12^2 + 6^2} = 13.42 \text{ cm}$$

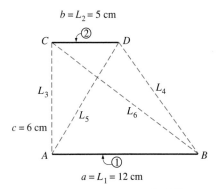

FIGURE 9-53

The two infinitely long parallel plates considered in Example 9-9.

Substituting,

$$F_{1 \to 2} = \frac{[(7.81 + 13.42) - (6 + 9.22)] \text{ cm}}{2 \times 12 \text{ cm}} = \textbf{0.250}$$

(b) The geometry is infinitely long in the direction perpendicular to the plane of the paper, and thus the two plates (surfaces 1 and 2) and the two openings (imaginary surfaces 3 and 4) form a four-surface enclosure. Then applying the summation rule to surface 1 yields

$$F_{11} + F_{12} + F_{13} + F_{14} = 1$$

But $F_{11} = 0$ since it is a flat surface. Therefore,

$$F_{12} = 1 - F_{13} - F_{14}$$

where the view factors F_{13} and F_{14} can be determined by considering the triangles ABC and ABD, respectively, and applying Equation 9-33 for view factors between the sides of triangles. We obtain

$$F_{13} = \frac{L_1 + L_3 - L_6}{2L_1}, \qquad F_{14} = \frac{L_1 + L_4 - L_5}{2L_1}$$

Substituting,

$$F_{12} = 1 - \frac{L_1 + L_3 - L_6}{2L_1} - \frac{L_1 + L_4 - L_5}{2L_1}$$

$$= \frac{(L_5 + L_6) - (L_3 + L_4)}{2L_1}$$

which is the desired result. This is also a miniproof of the crossed-strings method for the case of two infinitely long plain parallel surfaces.

9-7 ■ RADIATION HEAT TRANSFER: BLACK SURFACES

So far, we have considered the nature of radiation, the radiation properties of materials, and the view factors, and we are now in a position to consider the rate of heat transfer between surfaces by radiation. The analysis of radiation exchange between surfaces, in general, is complicated because of reflection: a radiation beam leaving a surface may be reflected several times, with partial reflection occurring at each surface, before it is completely absorbed. The analysis is simplified greatly when the surfaces involved can be approximated as blackbodies because of the absence of reflection. In this section, we consider radiation exchange between *black surfaces* only; we will extend the analysis to reflecting surfaces in the next section.

Consider two black surfaces of arbitrary shape maintained at uniform temperatures T_1 and T_2, as shown in Figure 9-54. Recognizing that radiation leaves a black surface at a rate of $E_b = \sigma T^4$ per unit surface area and that the view factor $F_{1 \to 2}$ represents the fraction of radiation leaving surface 1 that strikes surface 2, the *net* rate of radiation heat transfer from surface 1 to surface 2 can be expressed as

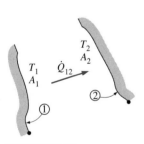

FIGURE 9-54

Two general black surfaces maintained at uniform temperatures T_1 and T_2.

$$\dot{Q}_{1 \to 2} = \begin{pmatrix} \text{Radiation leaving} \\ \text{the entire surface 1} \\ \text{that strikes surface 2} \end{pmatrix} - \begin{pmatrix} \text{Radiation leaving} \\ \text{the entire surface 2} \\ \text{that strikes surface 1} \end{pmatrix} \quad \text{(9-36)}$$

$$= A_1 E_{b1} F_{1 \to 2} - A_2 E_{b2} F_{2 \to 1} \quad \text{(W)}$$

Applying the reciprocity rule $A_1 F_{1 \to 2} = A_2 F_{2 \to 1}$ yields

$$\dot{Q}_{1 \to 2} = A_1 F_{1 \to 2} \sigma(T_1^4 - T_2^4) \quad \text{(W)} \quad \text{(9-37)}$$

which is the desired relation. A negative value for $\dot{Q}_{1 \to 2}$ indicates that net radiation heat transfer is from surface 2 to surface 1.

Now consider an *enclosure* consisting of *N black* surfaces maintained at specified temperatures. The *net* radiation heat transfer *from* any surface *i* of this enclosure is determined by adding up the net radiation heat transfers from surface *i* to each of the surfaces of the enclosure:

$$\dot{Q}_i = \sum_{j=1}^{N} \dot{Q}_{i \to j} = \sum_{j=1}^{N} A_i F_{i \to j} \sigma (T_i^4 - T_j^4) \quad \text{(W)} \quad \text{(9-38)}$$

Again a negative value for $\dot{Q}$ indicates that net radiation heat transfer is *to* surface *i* (i.e., surface *i gains* radiation energy instead of losing). Also, the net heat transfer from a surface to itself is zero, regardless of the shape of the surface.

EXAMPLE 9-10   Radiation Heat Transfer in a Black Cubical Furnace

Consider the 5-m × 5-m × 5-m cubical furnace shown in Figure 9-55, whose surfaces closely approximate black surfaces. The base, top, and side surfaces of the furnace are maintained at uniform temperatures of 800 K, 1500 K, and 500 K, respectively. Determine (*a*) the net rate of radiation heat transfer between the base and the side surfaces, (*b*) the net rate of radiation heat transfer between the base and the top surface, and (*c*) the net radiation heat transfer from the base surface.

Solution   The surfaces of a cubical furnace are black and are maintained at uniform temperatures. The net rate of radiation heat transfer between the base and side surfaces, between the base and the top surface, from the base surface are to be determined.

Assumptions   The surfaces are black and isothermal.

Analysis   (*a*) Considering that the geometry involves six surfaces, we may be tempted at first to treat the furnace as a six-furnace enclosure. However, the four side surfaces possess the same properties, and thus we can treat them as a single side surface in radiation analysis. We consider the base surface to be surface 1, the top surface to be surface 2, and the side surfaces to be surface 3. Then the problem reduces to determining $\dot{Q}_{1 \to 3}$, $\dot{Q}_{1 \to 2}$, and $\dot{Q}_1$.

The net rate of radiation heat transfer $\dot{Q}_{1 \to 3}$ from surface 1 to surface 3 can be determined from Equation 9-37, since both surfaces involved are black, by replacing the subscript 2 by 3:

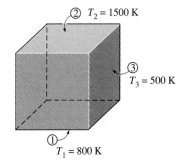

FIGURE 9-55

The cubical furnace of black surfaces considered in Example 9-10.

$$\dot{Q}_{1 \rightarrow 3} = A_1 F_{1 \rightarrow 3} \sigma (T_1^4 - T_3^4)$$

But first we need to evaluate the view factor $F_{1 \rightarrow 3}$. After checking the view factor charts and tables, we realize that we cannot determine this view factor directly. However, we can determine the view factor $F_{1 \rightarrow 2}$ directly from Figure 9-41 to be $F_{1 \rightarrow 2} = 0.2$, and we know that $F_{1 \rightarrow 1} = 0$, since surface 1 is a plane. Then applying the summation rule to surface 1 yields

$$F_{1 \rightarrow 1} + F_{1 \rightarrow 2} + F_{1 \rightarrow 3} = 1$$

or

$$F_{1 \rightarrow 3} = 1 - F_{1 \rightarrow 1} - F_{1 \rightarrow 2} = 1 - 0 - 0.2 = 0.8$$

Substituting,

$$\dot{Q}_{1 \rightarrow 3} = (25 \text{ m}^2)(0.8)(5.67 \times 10^{-8} \text{ W/m}^2 \cdot \text{K}^4)[(800 \text{ K})^4 - (500 \text{ K})^4]$$
$$= \textbf{393,611 W}$$

(b) The net rate of radiation heat transfer $\dot{Q}_{1 \rightarrow 2}$ from surface 1 to surface 2 is determined in a similar manner from Equation 9-37 to be

$$\dot{Q}_{1 \rightarrow 2} = A_1 F_{1 \rightarrow 2} \sigma (T_1^4 - T_2^4)$$
$$= (25 \text{ m}^2)(0.2)(5.67 \times 10^{-8} \text{ W/m}^2 \cdot \text{K}^4)[(800 \text{ K})^4 - (1500 \text{ K})^4]$$
$$= \textbf{-1,319,097 W}$$

The negative sign indicates that net radiation heat transfer is from surface 2 to surface 1.

(c) The net radiation heat transfer from the base surface $\dot{Q}_1$ is determined from Equation 9-38 by replacing the subscript i by 1 and taking N = 3:

$$\dot{Q}_1 = \sum_{j=1}^{3} \dot{Q}_{1 \rightarrow j} = \dot{Q}_{1 \rightarrow 1} + \dot{Q}_{1 \rightarrow 2} + \dot{Q}_{1 \rightarrow 3}$$
$$= 0 + (-1,319,097 \text{ W}) + (393,611 \text{ W})$$
$$= \textbf{-925,486 W}$$

Again the negative sign indicates that net radiation heat transfer is to surface 1. That is, the base of the furnace is gaining net radiation at a rate of about 925 kW.

9-8 ■ RADIATION HEAT TRANSFER: DIFFUSE, GRAY SURFACES

The analysis of radiation transfer in enclosures consisting of black surfaces is relatively easy, as we have seen above, but most enclosures encountered in practice involve nonblack surfaces, which allow multiple reflections to occur. Radiation analysis of such enclosures becomes very complicated unless some simplifying assumptions are made.

To make a simple radiation analysis possible, it is common to assume the surfaces of an enclosure to be *opaque, diffuse,* and *gray.* That is, the surfaces are nontransparent, they are diffuse emitters and diffuse reflectors, and their radiation properties are independent of wavelength. Also, each surface of the enclosure is *isothermal,* and both the incoming and outgoing radiation are *uniform* over each surface. But first we introduce the concept of radiosity.

Radiosity

Surfaces emit radiation as well as reflect it, and thus the radiation leaving a surface consists of emitted and reflected parts. The calculation of radiation heat transfer between surfaces involves the *total* radiation energy streaming away from a surface, with no regard for its origin. Thus, we need to define a new quantity to represent the *total radiation energy leaving a surface per unit time and per unit area.* This quantity is called the **radiosity** and is denoted by J (Fig. 9-56).

For a surface i that is *gray* and *opaque* ($\varepsilon_i = \alpha_i$ and $\alpha_i + \rho_i = 1$), the radiosity can be expressed as

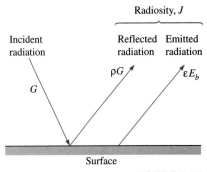

Radiosity, J

Incident radiation

Reflected radiation   Emitted radiation

ρG εE_b

G

Surface

FIGURE 9-56

Radiosity represents the sum of the radiation energy emitted and reflected by a surface.

$$J_i = \left(\begin{array}{c}\text{Radiation emitted} \\ \text{by surface } i\end{array}\right) + \left(\begin{array}{c}\text{Radiation reflected} \\ \text{by surface } i\end{array}\right)$$
$$= \varepsilon_i E_{bi} + \rho_i G_i \qquad (9\text{-}39)$$
$$= \varepsilon_i E_{bi} + (1 - \varepsilon_i)G_i \qquad (\text{W/m}^2)$$

where $E_{bi} = \sigma T_i^4$ is the blackbody emissive power of surface i and G_i is irradiation (i.e., the radiation energy incident on surface i per unit time per unit area).

For a surface that can be approximated as a *blackbody* ($\varepsilon_i = 1$), the radiosity relation reduces to

$$J_i = E_{bi} = \sigma T_i^4 \qquad (\text{blackbody}) \qquad (9\text{-}40)$$

That is, *the radiosity of a blackbody is equal to its emissive power.* That is expected, since a blackbody does not reflect any radiation, and thus radiation coming from a blackbody is due to emission only.

Net Radiation Heat Transfer to (or from) a Surface

During a radiation interaction, a surface *loses* energy by emitting radiation and *gains* energy by absorbing radiation emitted by other surfaces. A surface experiences a net gain or a net loss of energy, depending on which quantity is larger. The *net* rate of radiation heat transfer from a surface i of surface area A_i is denoted by $\dot{Q}_i$ and is expressed as

$$\dot{Q}_i = \left(\begin{array}{c}\text{Radiation leaving} \\ \text{entire surface } i\end{array}\right) - \left(\begin{array}{c}\text{Radiation incident} \\ \text{on entire surface } i\end{array}\right)$$
$$= A_i(J_i - G_i) \qquad (\text{W}) \qquad (9\text{-}41)$$

Solving for G_i from Equation 9-39 and substituting into Equation 9-41 yields

$$\dot{Q}_i = A_i\left(J_i - \frac{J_i - \varepsilon_i E_{bi}}{1 - \varepsilon_i}\right) = \frac{A_i \varepsilon_i}{1 - \varepsilon_i}(E_{bi} - J_i) \qquad (\text{W}) \qquad (9\text{-}42)$$

In an electrical analogy to Ohm's law, this equation can be rearranged as

$$\dot{Q}_i = \frac{E_{bi} - J_i}{R_i} \qquad \text{(W)} \qquad (9\text{-}43)$$

where

$$R_i = \frac{1 - \varepsilon_i}{A_i \varepsilon_i} \qquad (9\text{-}44)$$

is the **surface resistance** to radiation. The quantity $E_{bi} - J_i$ corresponds to a *potential difference* and the net rate of heat transfer corresponds to *current* in the electrical analogy, as illustrated in Figure 9-57.

The direction of the net radiation heat transfer depends on the relative magnitudes of J_i (the radiosity) and E_{bi} (the emissive power of a blackbody at the temperature of the surface). It will be *from* the surface if $E_{bi} > J_i$ and *to* the surface if $J_i > E_{bi}$. A negative value for $\dot{Q}_i$ indicates that heat transfer is *to* the surface. All of this radiation energy gained must be removed from the other side of the surface through some mechanism if the surface temperature is to remain constant.

The surface resistance to radiation for a *blackbody* is *zero* since $\varepsilon_i = 1$ and $J_i = E_{bi}$. The net rate of radiation heat transfer in this case is determined directly from Equation 9-41.

Some surfaces encountered in numerous practical heat transfer applications are modeled as being *adiabatic* since their back sides are well insulated and the net heat transfer through them is zero. When the convection effects on the front (heat transfer) side of such a surface is negligible and steady-state conditions are reached, the surface must lose as much radiation energy as it gains, and thus $\dot{Q}_i = 0$. In such cases, the surface is said to *reradiate* all the radiation energy it receives, and such a surface is called a **reradiating surface.** Setting $\dot{Q}_i = 0$ in Equation 9-43 yields

$$J_i = E_{bi} = \sigma T_i^4 \qquad \text{(W/m}^2\text{)} \qquad (9\text{-}45)$$

Therefore, the *temperature* of a reradiating surface under steady conditions can easily be determined from the equation above once its radiosity is known. Note that the temperature of a reradiating surface is *independent of its emissivity.* In radiation analysis, the surface resistance of a reradiating surface is disregarded since there is no net heat transfer through it. (This is like the fact that there is no need to consider a resistance in an electrical network if no current is flowing through it.)

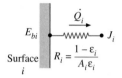

FIGURE 9-57

Electrical analogy of surface resistance to radiation.

Net Radiation Heat Transfer between Any Two Surfaces

Consider two diffuse, gray, and opaque surfaces of arbitrary shape maintained at uniform temperatures, as shown in Figure 9-58. Recognizing that the radiosity J represents the rate of radiation leaving a surface per unit surface area and that the view factor $F_{i \to j}$ represents the fraction of radiation leaving surface i that strikes surface j, the *net* rate of radiation heat transfer from surface i to surface j can be expressed as

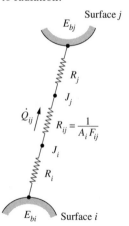

FIGURE 9-58

Electrical analogy of space resistance to radiation.

$$\dot{Q}_{i \to j} = \begin{pmatrix} \text{Radiation leaving} \\ \text{the entire surface } i \\ \text{that strikes surface } j \end{pmatrix} - \begin{pmatrix} \text{Radiation leaving} \\ \text{the entire surface } j \\ \text{that strikes surface } i \end{pmatrix} \quad \text{(9-46)}$$

$$= A_i J_i F_{i \to j} - A_j J_j F_{j \to i} \qquad \text{(W)}$$

Applying the reciprocity relation $A_i F_{i \to j} = A_j F_{j \to i}$ yields

$$\dot{Q}_{i \to j} = A_i F_{i \to j} (J_i - J_j) \qquad \text{(W)} \qquad \text{(9-47)}$$

Again in analogy to Ohm's law, this equation can be rearranged as

$$\dot{Q}_{i \to j} = \frac{J_i - J_j}{R_{i \to j}} \qquad \text{(W)} \qquad \text{(9-48)}$$

where

$$R_{i \to j} = \frac{1}{A_i F_{i \to j}} \qquad \text{(9-49)}$$

is the **space resistance** to radiation. Again the quantity $J_i - J_j$ corresponds to a *potential difference,* and the net rate of heat transfer between two surfaces corresponds to *current* in the electrical analogy, as illustrated in Figure 9-58.

The direction of the net radiation heat transfer between two surfaces depends on the relative magnitudes of J_i and J_j. A positive value for $\dot{Q}_{i \to j}$ indicates that net heat transfer is *from* surface i *to* surface j. A negative value indicates the opposite.

In an N-surface enclosure, the conservation of energy principle requires that the net heat transfer from surface i be equal to the sum of the net heat transfers from surface i to each of the N surfaces of the enclosure. That is,

$$\dot{Q}_i = \sum_{j=1}^{N} \dot{Q}_{i \to j} = \sum_{j=1}^{N} A_i F_{i \to j} (J_i - J_j) = \sum_{j=1}^{N} \frac{J_i - J_j}{R_{i \to j}} \qquad \text{(W)} \quad \text{(9-50)}$$

The network representation of net radiation heat transfer from surface i to the remaining surfaces of an N-surface enclosure is given in Figure 9-59. Note that $\dot{Q}_{i \to i}$ (the net rate of heat transfer from a surface to itself) is zero regardless of the shape of the surface. Combining Equations 9-43 and 9-50 gives

$$\frac{E_{bi} - J_i}{R_i} = \sum_{j=1}^{N} \frac{J_i - J_j}{R_{i \to j}} \qquad \text{(W)} \qquad \text{(9-51)}$$

which has the electrical analogy interpretation that *the net radiation flow from a surface through its surface resistance is equal to the sum of the radiation flows from that surface to all other surfaces through the corresponding space resistances.*

Methods of Solving Radiation Problems

In the radiation analysis of an enclosure, either the temperature or the net rate of heat transfer must be given for each of the surfaces to obtain a unique solution for the unknown surface temperatures and heat transfer rates. There

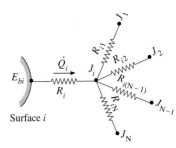

FIGURE 9-59

Network representation of net radiation heat transfer from surface i to the remaining surfaces of an N-surface enclosure.

are two methods commonly used to solve radiation problems. In the first method, Equations 9-50 (for surfaces with specified heat transfer rates) and 9-51 (for surfaces with specified temperatures) are simplified and rearranged as

$$\text{Surfaces with specified net heat transfer rate } \dot{Q}_i \qquad \dot{Q}_i = A_i \sum_{j=1}^{N} F_{i \rightarrow j}(J_i - J_j) \qquad (9\text{-}52a)$$

$$\text{Surfaces with specified temperature } T_i \qquad \sigma T_i^4 = J_i + \frac{1 - \varepsilon_i}{\varepsilon_i} \sum_{j=1}^{N} F_{i \rightarrow j}(J_i - J_j) \qquad (9\text{-}52b)$$

Note that $\dot{Q}_i = 0$ for insulated (or reradiating) surfaces, and $\sigma T_i^4 = J_i$ for black surfaces since $\varepsilon_i = 1$ in that case. Also, the term corresponding to $j = i$ will drop out from either relation since $J_i - J_j = J_i - J_i = 0$ in that case.

The equations above give N linear algebraic equations for the determination of the N unknown radiosities for an N-surface enclosure. Once the radiosities $J_1, J_2, \ldots, J_N$ are available, the unknown heat transfer rates can be determined from Equation 9-52a while the unknown surface temperatures can be determined from Equation 9-52b. The temperatures of insulated or reradiating surfaces can be determined from $\sigma T_i^4 = J_i$. A positive value for $\dot{Q}_i$ indicates net radiation heat transfer *from* surface i to other surfaces in the enclosure while a negative value indicates net radiation heat transfer *to* the surface.

The systematic approach described above for solving radiation heat transfer problems is very suitable for use with today's popular equation solvers such as EES, Mathcad, and Matlab, especially when there are a large number of surfaces, and is known as the **direct method** (formerly, the *matrix method,* since it resulted in matrices and the solution required a knowledge of linear algebra). The second method described below, called the **network method,** is based on the electrical network analogy.

The network method was first introduced by A. K. Oppenheim in the 1950s and found widespread acceptance because of its simplicity and emphasis on the physics of the problem. The application of the method is straightforward: draw a surface resistance associated with each surface of an enclosure and connect them with space resistances. Then solve the radiation problem by treating it as an electrical network problem where the radiation heat transfer replaces the current and radiosity replaces the potential.

The network method is not practical for enclosures with more than three of four surfaces, however, because of the increased complexity of the network. Below we apply the method to solve radiation problems in two- and three-surface enclosures.

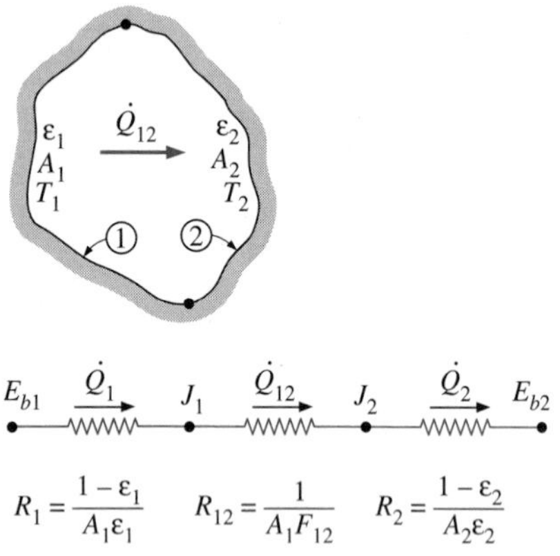

FIGURE 9-60

Schematic of a two-surface enclosure and the radiation network associated with it.

Radiation Heat Transfer in Two-Surface Enclosures

Consider an enclosure consisting of two opaque surfaces at specified temperatures T_1 and T_2, as shown in Figure 9-60, and try to determine the net rate of radiation heat transfer between the two surfaces with the network method. Surfaces 1 and 2 have emissivities ε_1 and ε_2 and surface areas A_1 and A_2 and are maintained at uniform temperatures T_1 and T_2, respectively. There are only two surfaces in the enclosure, and thus we can write

$$\dot{Q}_{12} = \dot{Q}_1 = -\dot{Q}_2$$

That is, the net rate of radiation transfer from surface 1 to surface 2 must equal the net rate of radiation transfer *from* surface 1 and the net rate of radiation transfer *to* surface 2.

The radiation network of this two-surface enclosure consists of two surface resistances and one space resistance, as shown in Figure 9-60. In an electrical network, the electric current flowing through these resistances connected in series would be determined by dividing the potential difference between points A and B by the total resistance between the same two points. The net rate of radiation transfer is determined in the same manner and is expressed as

$$\dot{Q}_{12} = \frac{E_{b1} - E_{b2}}{R_1 + R_{12} + R_2} = \dot{Q}_1 = -\dot{Q}_2$$

or

$$\dot{Q}_{12} = \frac{\sigma(T_1^4 - T_2^4)}{\dfrac{1 - \varepsilon_1}{A_1 \varepsilon_1} + \dfrac{1}{A_1 F_{12}} + \dfrac{1 - \varepsilon_2}{A_2 \varepsilon_2}} \qquad \text{(W)} \qquad (9\text{-}53)$$

This important result is applicable to any two gray, diffuse, opaque surfaces that form an enclosure. The view factor F_{12} depends on the geometry and must be determined in advance. Simplified forms of Equation 9-53 for some familiar arrangements that form a two-surface enclosure are given in Table 9-6. Note that $F_{12} = 1$ for all of these special cases.

EXAMPLE 9-11 Radiation Heat Transfer between Large Parallel Plates

Two very large parallel plates are maintained at uniform temperatures $T_1 = 800$ K and $T_2 = 500$ K and have emissivities $\varepsilon_1 = 0.2$ and $\varepsilon_2 = 0.7$, respectively, as shown in Figure 9-61. Determine the net rate of radiation heat transfer between the two surfaces per unit surface area of the plates.

Solution Two large parallel plates are maintained at uniform temperatures. The net rate of radiation heat transfer between the plates is to be determined.

Assumptions Both surfaces are opaque, diffuse, and gray.

Analysis The net rate of radiation heat transfer between the two plates per unit area is readily determined from Equation 9-55 to be

$$\dot{q}_{12} = \frac{\dot{Q}_{12}}{A} = \frac{\sigma(T_1^4 - T_2^4)}{\dfrac{1}{\varepsilon_1} + \dfrac{1}{\varepsilon_2} - 1} = \frac{(5.67 \times 10^{-8}\ \text{W/m}^2 \cdot \text{K}^4)[(800\ \text{K})^4 - (500\ \text{K})^4]}{\dfrac{1}{0.2} + \dfrac{1}{0.7} - 1}$$

$$= 3625\ \text{W/m}^2$$

Discussion Note that heat at a net rate of 3625 W is transferred from plate 1 to plate 2 by radiation per unit surface area of either plate.

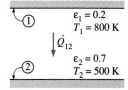

FIGURE 9-61
The two parallel plates considered in
Example 9-11.

TABLE 9-6

Small object in a large cavity A_1, T_1, ε_1 A_2, T_2, ε_2	$\dfrac{A_1}{A_2} \approx 0$ $F_{12} = 1$	$\dot{Q}_{12} = A_1 \sigma \varepsilon_1 (T_1^4 - T_2^4)$ (9–54)
Infinitely large parallel plates $\underline{\quad A_1, T_1, \varepsilon_1 \quad}$ $\underline{\quad A_2, T_2, \varepsilon_2 \quad}$	$A_1 = A_2 = A$ $F_{12} = 1$	$\dot{Q}_{12} = \dfrac{A\sigma(T_1^4 - T_2^4)}{\dfrac{1}{\varepsilon_1} + \dfrac{1}{\varepsilon_2} - 1}$ (9–55)
Infinitely long concentric cylinders 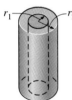 $r_1 \qquad r_2$	$\dfrac{A_1}{A_2} = \dfrac{r_1}{r_2}$ $F_{12} = 1$	$\dot{Q}_{12} = \dfrac{A_1\sigma(T_1^4 - T_2^4)}{\dfrac{1}{\varepsilon_1} + \dfrac{1 - \varepsilon_2}{\varepsilon_2}\left(\dfrac{r_1}{r_2}\right)}$ (9–56)
Concentric spheres $r_1 \qquad r_2$	$\dfrac{A_1}{A_2} = \left(\dfrac{r_1}{r_2}\right)^2$ $F_{12} = 1$	$\dot{Q}_{12} = \dfrac{A_1\sigma(T_1^4 - T_2^4)}{\dfrac{1}{\varepsilon_1} + \dfrac{1 - \varepsilon_2}{\varepsilon_2}\left(\dfrac{r_1}{r_2}\right)^2}$ (9–57)

Radiation Heat Transfer in Three-Surface Enclosures

We now consider an enclosure consisting of three opaque, diffuse, gray surfaces, as shown in Figure 9-62. Surfaces 1, 2, and 3 have surface areas A_1, A_2, and A_3; emissivities ε_1, ε_2, and ε_3; and uniform temperatures T_1, T_2, and T_3, respectively. The radiation network of this geometry is constructed by following the standard procedure: draw a surface resistance associated with each of the three surfaces and connect these surface resistances with space resistances, as shown in the figure. Relations for the surface and space resistances

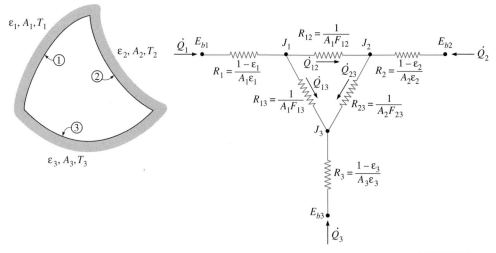

FIGURE 9-62

Schematic of a three-surface enclosure and the radiation network associated with it.

are given by Equations 9-44 and 9-49. The three endpoint potentials E_{b1}, E_{b2}, and E_{b3} are considered known, since the surface temperatures are specified. Then all we need to find are the radiosities J_1, J_2, and J_3. The three equations for the determination of these three unknowns are obtained from the requirement that *the algebraic sum of the currents (net radiation heat transfer) at each node must equal zero.* That is,

$$\frac{E_{b1} - J_1}{R_1} + \frac{J_2 - J_1}{R_{12}} + \frac{J_3 - J_1}{R_{13}} = 0$$

$$\frac{J_1 - J_2}{R_{12}} + \frac{E_{b2} - J_2}{R_2} + \frac{J_3 - J_2}{R_{23}} = 0 \qquad (9\text{-}58)$$

$$\frac{J_1 - J_3}{R_{13}} + \frac{J_2 - J_3}{R_{23}} + \frac{E_{b3} - J_3}{R_3} = 0$$

Once the radiosities J_1, J_2, and J_3 are available, the net rate of radiation heat transfers at each surface can be determined from Equation 9-50.

The set of equations above simplify further if one or more surfaces are "special" in some way. For example, $J_i = E_{bi} = \sigma T_i^4$ for a *black* or *reradiating* surface. Also, $\dot{Q}_i = 0$ for a reradiating surface. Finally, when the net rate of radiation heat transfer $\dot{Q}_i$ is specified at surface i instead of the temperature, the term $(E_{bi} - J_i)/R_i$ should be replaced by the specified $\dot{Q}_i$.

EXAMPLE 9-12   Radiation Heat Transfer in a Cylindrical Furnace

Consider a cylindrical furnace with $r_0 = H = 1$ m, as shown in Figure 9-63. The top (surface 1) and the base (surface 2) of the furnace has emissivities $\varepsilon_1 = 0.8$ and $\varepsilon_2 = 0.4$, respectively, and are maintained at uniform temperatures $T_1 = 700$ K and $T_2 = 500$ K. The side surface closely approximates a blackbody and is maintained at a temperature of $T_3 = 400$ K. Determine the net rate of radiation heat transfer at each surface during steady-state operation and explain how these surfaces can be maintained at specified temperatures.

FIGURE 9-63

The cylindrical furnace considered in Example 9-12.

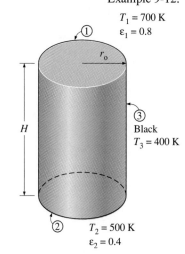

Solution The surfaces of a cylindrical furnace are maintained at uniform temperatures. The net rate of radiation heat transfer at each surface during steady operation is to be determined.

Assumptions **1** Steady operating conditions exist. **2** The surfaces are opaque, diffuse, and gray. **3** Convection heat transfer is not considered.

Analysis We will solve this problem systematically using the direct method to demonstrate its use. The cylindrical furnace can be considered to be a three-surface enclosure with surface areas of

$$A_1 = A_2 = \pi r_o^2 = \pi(1 \text{ m})^2 = 3.14 \text{ m}^2$$
$$A_3 = 2\pi r_o H = 2\pi(1 \text{ m})(1 \text{ m}) = 6.28 \text{ m}^2$$

The view factor from the base to the top surface is determined from Figure 9-43 to be $F_{12} = 0.38$. Then the view factor from the base to the side surface is determined by applying the summation rule to be

$$F_{11} + F_{12} + F_{13} = 1 \quad \rightarrow \quad F_{13} = 1 - F_{11} - F_{12} = 1 - 0 - 0.38 = 0.62$$

since the base surface is flat and thus $F_{11} = 0$. Noting that the top and bottom surfaces are symmetric about the side surface, $F_{21} = F_{12} = 0.38$ and $F_{23} = F_{13} = 0.62$. The view factor F_{31} is determined from the reciprocity rule,

$$A_1 F_{13} = A_3 F_{31} \quad \rightarrow \quad F_{31} = F_{13}(A_1/A_3) = (0.62)(0.314/0.628) = 0.31$$

Also, $F_{32} = F_{31} = 0.31$ because of symmetry. Now that all the view factors are available, we apply Equation 9-52b to each surface to determine the radiosities:

Top surface ($i = 1$): $\qquad \sigma T_1^4 = J_1 + \dfrac{1 - \varepsilon_1}{\varepsilon_1}[F_{1 \rightarrow 2}(J_1 - J_2) + F_{1 \rightarrow 3}(J_1 - J_3)]$

Bottom surface ($i = 2$): $\qquad \sigma T_2^4 = J_2 + \dfrac{1 - \varepsilon_2}{\varepsilon_2}[F_{2 \rightarrow 1}(J_2 - J_1) + F_{2 \rightarrow 3}(J_2 - J_3)]$

Side surface ($i = 3$): $\qquad \sigma T_3^4 = J_3 + 0$ (since surface 3 is black and thus $\varepsilon_3 = 1$)

Substituting the known quantities,

$$(5.67 \times 10^{-8} \text{ W/m}^2 \cdot \text{K}^4)(700 \text{ K})^4 = J_1 + \frac{1 - 0.8}{0.8}[0.38(J_1 - J_2) + 0.68(J_1 - J_3)]$$

$$(5.67 \times 10^{-8} \text{ W/m}^2 \cdot \text{K}^4)(500 \text{ K})^4 = J_2 + \frac{1 - 0.4}{0.4}[0.28(J_2 - J_1) + 0.68(J_2 - J_3)]$$

$$(5.67 \times 10^{-8} \text{ W/m}^2 \cdot \text{K}^4)(400 \text{ K})^4 = J_3$$

Solving the equations above for J_1, J_2, and J_3 gives

$$J_1 = 11{,}418 \text{ W/m}^2, \, J_2 = 4562 \text{ W/m}^2, \quad \text{and} \quad J_3 = 1452 \text{ W/m}^2$$

Then the net rates of radiation heat transfer at the three surfaces are determined from Equation 9-52a to be

$$\dot{Q}_1 = A_1[F_{1 \rightarrow 2}(J_1 - J_2) + F_{1 \rightarrow 3}(J_1 - J_3)]$$
$$= (3.14 \text{ m}^2)[0.38(11{,}418 - 4562) + 0.62(11{,}418 - 1452)]\text{W/m}^2 = \textbf{27,582 W}$$

$$\dot{Q}_2 = A_2[F_{2 \rightarrow 1}(J_2 - J_1) + F_{2 \rightarrow 3}(J_2 - J_3)]$$
$$= (3.12 \text{ m}^2)[0.38(4562 - 11{,}418) + 0.62(4562 - 1452)]\text{W/m}^2 = \textbf{-2126 W}$$

$$\dot{Q}_3 = A_3[F_{3 \rightarrow 1}(J_3 - J_1) + F_{3 \rightarrow 2}(J_3 - J_2)]$$
$$= (6.28 \text{ m}^2)[0.31(1452 - 11{,}418) + 0.31(1452 - 4562)]\text{W/m}^2 = \textbf{-25,456 W}$$

Note that the direction of net radiation heat transfer is *from* the top surface *to* the base and side surfaces, and the algebraic sum of these three quantities must be equal to zero. That is,

$$\dot{Q}_1 + \dot{Q}_2 + \dot{Q}_3 = 27{,}582 + (-2126) + (-25{,}456) = 0$$

Discussion   To maintain the surfaces at the specified temperatures, we must supply heat to the top surface continuously at a rate of 27,582 W while removing 2126 W from the base and 25,456 W from the side surfaces.

The direct method presented above is straightforward, and it does not require the evaluation of radiation resistances. Also, it can be applied to enclosures with any number of surfaces in the same manner.

EXAMPLE 9-13   Radiation Heat Transfer in a Triangular Furnace

A furnace is shaped like a long equilateral triangular duct, as shown in Figure 9-64. The width of each side is 1 m. The base surface has an emissivity of 0.7 and is maintained at a uniform temperature of 600 K. The heated left-side surface closely approximates a blackbody at 1000 K. The right-side surface is well insulated. Determine the rate at which energy must be supplied to the heated side externally per unit length of the duct in order to maintain these operating conditions.

Solution   Two of the surfaces of a long equilateral triangular furnace are maintained at uniform temperatures while the third surface is insulated. The external rate of heat transfer to the heated side per unit length of the duct during steady operation is to be determined.

Assumptions   **1** Steady operating conditions exist. **2** The surfaces are opaque, diffuse, and gray. **3** Convection heat transfer is not considered.

Analysis   The furnace can be considered to be a three-surface enclosure with a radiation network as shown in the figure, since the duct is very long and thus the end effects are negligible. We observe that the view factor from any surface to any other surface in the enclosure is 0.5 because of symmetry. Surface 3 is a reradiating surface since the net rate of heat transfer at that surface is zero. Then we must have $\dot{Q}_1 = -\dot{Q}_2$, since the entire heat lost by surface 1 must be gained by surface 2. The radiation network in this case is a simple series–parallel connection, and we can determine $\dot{Q}_1$ directly from

$$\dot{Q}_1 = \frac{E_{b1} - E_{b2}}{R_1 + \left(\dfrac{1}{R_{12}} + \dfrac{1}{R_{13} + R_{23}}\right)^{-1}} = \frac{E_{b1} - E_{b2}}{\dfrac{1 - \varepsilon_1}{A_1 \varepsilon_1} + \left(A_1 F_{12} + \dfrac{1}{1/A_1 F_{13} + 1/A_2 F_{23}}\right)^{-1}}$$

where

$$A_1 = A_2 = A_3 = wL = 1\ \text{m} \times 1\ \text{m} = 1\ \text{m}^2 \qquad \text{(per unit length of the duct)}$$
$$F_{12} = F_{13} = F_{23} = 0.5 \qquad \text{(symmetry)}$$
$$E_{b1} = \sigma T_1^4 = (5.67 \times 10^{-8}\ \text{W/m}^2 \cdot \text{K}^4)(600\ \text{K})^4 = 7348\ \text{W/m}^2$$
$$E_{b2} = \sigma T_2^4 = (5.67 \times 10^{-8}\ \text{W/m}^2 \cdot \text{K}^4)(1000\ \text{K})^4 = 56{,}700\ \text{W/m}^2$$

Substituting,

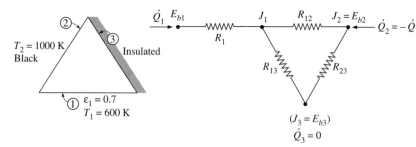

FIGURE 9-64

The triangular furnace considered in
Example 9-13.

$$\dot{Q}_1 = \frac{(56{,}700 - 7348)\ \text{W/m}^2}{\dfrac{1-0.7}{0.7 \times 1\ \text{m}^2} + \left[(0.5 \times 1\ \text{m}^2) + \dfrac{1}{1/(0.5 \times 1\ \text{m}^2) + 1/(0.5 \times 1\ \text{m}^2)} \right]^{-1}}$$

$$= \textbf{28,009 W}$$

Therefore, energy at a rate of 28,009 W must be supplied to the heated surface
per unit length of the duct to maintain steady operation in the furnace.

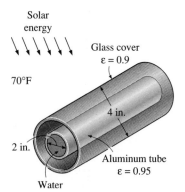

FIGURE 9-65

Schematic for Example 9-14.

EXAMPLE 9-14   Heat Transfer through Tubular Solar Collector

A solar collector consists of a horizontal aluminum tube having an outer diameter
of 2 in. enclosed in a concentric thin glass tube of 4-in. diameter, as shown in
Figure 9-65. Water is heated as it flows through the tube, and the annual space
between the aluminum and the glass tubes is filled with air at 1 atm pressure.
The pump circulating the water fails during a clear day, and the water temperature in the tube starts rising. The aluminum tube absorbs solar radiation at a rate
of 30 Btu/h per foot length, and the temperature of the ambient air outside is 70°F.
The emissivities of the tube and the glass cover are 0.95 and 0.9, respectively.
Taking the effective sky temperature to be 50°F, determine the temperature of
the aluminum tube when steady operating conditions are established (i.e, when
the rate of heat loss from the tube equals the amount of solar energy gained
by the tube).

Solution   The circulating pump of a solar collector that consists of a horizontal
tube and its glass cover fails one day. The temperature of the tube is to be
determined when steady conditions are reached.

Assumptions   **1** Steady operating conditions exist. **2** The tube and its cover are
isothermal. **3** Air is an ideal gas. **4** The surfaces are opaque, diffuse, and gray for
infrared radiation. The outer surface is transparent to solar radiation.

Properties   The properties of air should be evaluated at the average temperature. But we do not know the exit temperature of the air in the duct, and thus we
cannot determine the bulk tube and glass cover temperatures at this point, and
thus we cannot evaluate the average temperatures. Therefore, we will use the
properties at an anticipated average temperature of 100°F (Table A-11),

$$k = 0.0154\ \text{Btu/h} \cdot \text{ft} \cdot °\text{F} \qquad \text{Pr} = 0.72$$

$$\nu = 0.18 \times 10^{-3}\ \text{ft}^2/\text{s}$$

$$\beta = \frac{1}{T_f} = \frac{1}{(100+460)\ \text{R}} = 0.001786\ \text{R}^{-1}$$

The results obtained can then be refined for better accuracy, if necessary, using the evaluated surface temperatures.

Analysis This problem was solved in Chapter 7 by disregarding radiation heat transfer. Now we will repeat the solution by considering natural convection and radiation occurring simultaneously. We have a horizontal cylindrical enclosure filled with air at 1 atm pressure. The problem involves heat transfer from the aluminum tube to the glass cover and from the outer surface of the glass cover to the surrounding ambient air. When thermal equilibrium is established and steady operation is reached, these two heat transfer rates must equal the rate of heat gain. That is,

$$\dot{Q}_{\text{tube-glass}} = \dot{Q}_{\text{glass-ambient}} = \dot{Q}_{\text{solar gain}} = 30 \text{ Btu/h} \qquad \text{(per foot of tube)}$$

The heat transfer surface area of the glass cover is

$$A_2 = A_{\text{glass}} = (\pi DL)_{\text{glass}} = \pi(\tfrac{4}{12} \text{ ft})(1 \text{ ft}) = 1.047 \text{ ft}^2 \qquad \text{(per foot length of tube)}$$

To determine the Rayleigh number, we need to know the surface temperature of the glass, which is not available. Therefore, it is clear that the solution will require a trial-and-error approach. Assuming the glass cover temperature to be 80°F, the Raleigh number, the Nusselt number, the convection heat transfer coefficient, and the rate of natural convection heat transfer from the glass cover to the ambient air are determined to be

$$\text{Ra} = \frac{g\beta(T_s - T_\infty)\delta^3}{\nu^2} \text{Pr}$$

$$= \frac{(32.2 \text{ ft/s}^2)(0.001786 \text{ R}^{-1})[(80 - 70)\text{K}](\tfrac{4}{12} \text{ ft})^3}{(0.18 \times 10^{-3} \text{ ft}^2/\text{s})^2}(0.72) = 4.733 \times 10^5$$

$$\text{Nu} = \left\{ 0.6 + \frac{0.387 \text{ Ra}^{1/6}}{[1 + (0.559/\text{Pr})^{2/16}]^{8/27}} \right\}$$

$$= \left\{ 0.6 + \frac{0.387(4.733 \times 10^5)^{1/6}}{[1 + (0.559/0.72)^{9/16}]^{8/27}} \right\}^2 = 11.8$$

$$h = \frac{k}{D}\text{Nu} = \frac{0.0154 \text{ Btu/ft} \cdot °F}{\tfrac{4}{12} \text{ ft}}(11.8) = 0.546 \text{ Btu/h} \cdot \text{ft}^2 \cdot °F$$

$$\dot{Q}_{2,\text{conv}} = hA_2(T_2 - T_\infty) = (0.546 \text{ Btu/h} \cdot \text{ft}^2 \cdot °F)(1.047 \text{ ft}^2)(80 - 70)°F$$
$$= 5.7 \text{ Btu/h}$$

Also,

$$\dot{Q}_{2,\text{rad}} = \varepsilon\sigma A_2(T_2^4 - T_{\text{surr}}^4)$$
$$= (0.9)(0.1714 \times 10^{-8} \text{ Btu/h} \cdot \text{ft}^2 \cdot \text{R}^4)(1.047 \text{ ft}^2)[(540 \text{ R})^4 - (510 \text{ R})^4]$$
$$= 28.1 \text{ Btu/h}$$

and

$$\dot{Q}_{2,\text{total}} = \dot{Q}_{2,\text{conv}} + \dot{Q}_{2,\text{rad}} = 5.7 + 28.1 = 33.8 \text{ Btu/h}$$

which is more than 30 Btu/h. Therefore, the assumed temperature of 80°F for the glass cover is high. Repeating the calculations for a temperature of 75°F gives 25.5 Btu/h, which is low. Then the glass cover temperature corresponding to 30 Btu/h is determined by interpolation to be 78°F.

The temperature of the aluminum tube is determined in a similar manner using the natural convection radiation relations for two horizontal concentric cylinders. The characteristic length in this case is the distance between the two cylinders, which is determined to be

$$\delta = \tfrac{1}{2}(D_2 - D_1) = \tfrac{1}{2}(4 - 2) \text{ in.} = 1 \text{ in.}$$

Also,

$$A_1 = A_{\text{tube}} = (\pi D L)_{\text{tube}} = \pi(\tfrac{2}{12} \text{ ft})(1 \text{ ft}) = 0.524 \text{ ft}^2 \qquad \text{(per foot length of tube)}$$

$$A = \frac{\pi L(D_2 - D_1)}{\ln (D_2/D_1)} = \frac{\pi(1 \text{ ft})(\tfrac{4}{12} - \tfrac{2}{12}) \text{ ft}}{\ln (\tfrac{4}{2})} = 0.755 \text{ ft}^2$$

We start the calculations by assuming the tube temperature to be 120°F. This gives

$$\text{Ra} = \frac{g\beta(T_1 - T_2)\delta^3}{\nu^2} \text{Pr}$$

$$= \frac{(32.2 \text{ ft/s}^2)(0.001786 \text{ R}^{-1})[(120 - 78)\text{R}](\tfrac{1}{12} \text{ ft})^3}{(0.18 \times 10^{-3} \text{ ft}^2/\text{s})^2}(0.72) = 3.11 \times 10^4$$

$$\text{Nu} = 0.11 \text{ Ra}^{0.29} = 0.11(3.11 \times 10^4)^{0.29} = 2.21$$

$$\dot{Q}_{1,\text{conv}} = k\text{Nu}A \frac{T_1 - T_2}{\delta}$$

$$= (0.0154 \text{ Btu/h} \cdot \text{ft} \cdot °\text{F})(2.21)(0.755 \text{ ft}^2)\frac{(120 - 78)\text{R}}{\tfrac{1}{12} \text{ ft}} = 13.0 \text{ Btu/h}$$

Also,

$$\dot{Q}_{1,\text{rad}} = \frac{\sigma A_1(T_1^4 - T_2^4)}{\dfrac{1}{\varepsilon_1} + \dfrac{1 - \varepsilon_2}{\varepsilon_2}\left(\dfrac{D_1}{D_2}\right)}$$

$$= \frac{(0.1714 \times 10^{-8} \text{ Btu/h} \cdot \text{ft}^2 \cdot \text{R}^4)(0.524 \text{ ft}^2)](580 \text{ R})^4 - (538 \text{ R})^4]}{\dfrac{1}{0.95} + \dfrac{1 - 0.9}{0.9}\left(\dfrac{2 \text{ in.}}{4 \text{ in.}}\right)}$$

$$= 23.8 \text{ Btu/h}$$

and

$$\dot{Q}_{1,\text{total}} = \dot{Q}_{1,\text{conv}} + \dot{Q}_{1,\text{rad}} = 13.0 + 23.8 = 36.8 \text{ But/h}$$

which is more than 30 Btu/h. Therefore, the assumed temperature of 120°F for the tube is high. Repeating the calculations for a temperature of 110°F gives 26.7 Btu/h, which is low. Then the tube temperature corresponding to 30 Btu/h is determined by interpolation to be **113°F**. Therefore, the tube will reach an equilibrium temperature of 113°F when the pump fails. Recall that disregarding radiation in the previous chapter gave a result of 188°F, which seems to be quite unrealistic.

Discussion   Radiation should always be considered in systems that are heated or cooled by natural convection, unless the surfaces involved are polished and thus have emissivities close to zero.

9-9 ■ RADIATION SHIELDS AND THE RADIATION EFFECT

Radiation heat transfer between two surfaces can be reduced greatly by inserting a thin, high-reflectivity (low-emissivity) sheet of material between the two surfaces. Such highly reflective thin plates or shells are called **radiation shields.** Multilayer radiation shields constructed of about 20 sheets per cm thickness separated by evacuated space are commonly used in cryogenic and space applications. Radiation shields are also used in temperature measurements of fluids to reduce the error caused by the radiation effect when the temperature sensor is exposed to surfaces that are much hotter or colder than the fluid itself. The role of the radiation shield is to reduce the rate of radiation heat transfer by placing additional resistances in the path of radiation heat flow. The lower the emissivity of the shield, the higher the resistance.

Radiation heat transfer between two large parallel plates of emissivities ε_1 and ε_2 maintained at uniform temperatures T_1 and T_2 is given by Equation 9-55:

$$\dot{Q}_{12,\text{ no shield}} = \frac{A\sigma(T_1^4 - T_2^4)}{\dfrac{1}{\varepsilon_1} + \dfrac{1}{\varepsilon_2} - 1}$$

Now consider a radiation shield placed between these two plates, as shown in Figure 9-66. Let the emissivities of the shield facing plates 1 and 2 be $\varepsilon_{3,1}$ and $\varepsilon_{3,2}$, respectively. Note that the emissivity of different surfaces of the shield may be different. The radiation network of this geometry is constructed, as usual, by drawing a surface resistance associated with each surface and connecting these surface resistances with space resistances, as shown in the figure. The resistances are connected in series, and thus the rate of radiation heat transfer is

$$\dot{Q}_{12,\text{ one shield}} = \frac{E_{b1} - E_{b2}}{\dfrac{1-\varepsilon_1}{A_1\varepsilon_1} + \dfrac{1}{A_1 F_{12}} + \dfrac{1-\varepsilon_{3,1}}{A_3\varepsilon_{3,1}} + \dfrac{1-\varepsilon_{3,2}}{A_3\varepsilon_{3,2}} + \dfrac{1}{A_3 F_{32}} + \dfrac{1-\varepsilon_2}{A_2\varepsilon_2}}$$

$$(9\text{-}59)$$

Noting the $F_{13} = F_{23} = 1$ and $A_1 = A_2 = A_3 = A$ for parallel plates, Equation 9-59 simplifies to

$$\dot{Q}_{12,\text{ one shield}} = \frac{A\sigma(T_1^4 - T_2^4)}{\left(\dfrac{1}{\varepsilon_1} + \dfrac{1}{\varepsilon_2} - 1\right) + \left(\dfrac{1}{\varepsilon_{3,1}} + \dfrac{1}{\varepsilon_{3,2}} - 1\right)} \qquad (9\text{-}60)$$

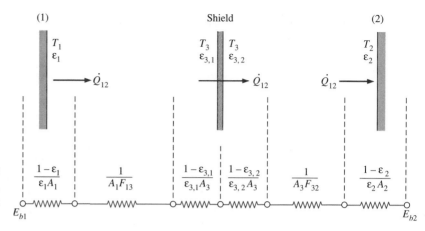

FIGURE 9-66

The radiation shield placed between two parallel plates and the radiation network associated with it.

where the terms in the second set of parentheses in the denominator represent the additional resistance to radiation introduced by the shield. The appearance of the equation above suggests that parallel plates involving multiple radiation shields can be handled by adding a group of terms like those in the second set of parentheses to the denominator for each radiation shield. Then the radiation heat transfer through large parallel plates separated by N radiation shields becomes

$$\dot{Q}_{12,\,N\text{ shields}} = \frac{A\sigma(T_1^4 - T_2^4)}{\left(\dfrac{1}{\varepsilon_1} + \dfrac{1}{\varepsilon_2} - 1\right) + \left(\dfrac{1}{\varepsilon_{3,1}} + \dfrac{1}{\varepsilon_{3,2}} - 1\right) + \cdots + \left(\dfrac{1}{\varepsilon_{N,1}} + \dfrac{1}{\varepsilon_{N,2}} - 1\right)}$$

(9-61)

If the emissivities of all surfaces are equal, Equation 9-61 reduces to

$$\dot{Q}_{12,\,N\text{ shields}} = \frac{A\sigma(T_1^4 - T_2^4)}{(N+1)\left(\dfrac{1}{\varepsilon} + \dfrac{1}{\varepsilon} - 1\right)} = \frac{1}{N+1}\dot{Q}_{12,\text{ no shields}}$$

(9-62)

Therefore, when all emissivities are equal, 1 shield reduces the rate of radiation heat transfer to one-half, 9 shields reduce it to one-tenth, and 19 shields reduce it to one-twentieth (or 5 percent) of what it was when there were no shields.

The equilibrium temperature of the radiation shield T_3 in Figure 9-66 can be determined by expressing Equation 9-55 for $\dot{Q}_{13}$ or $\dot{Q}_{23}$ (which involves T_3) after evaluating $\dot{Q}_{12}$ from Equation 9-60 and noting that $\dot{Q}_{12} = \dot{Q}_{13} = \dot{Q}_{23}$ when steady-state conditions are reached.

Radiation shields used to retard the rate of radiation heat transfer between concentric cylinders and spheres can be handled in a similar manner. In case of one shield, Equation 9-59 can be used by taking $F_{13} = F_{23} = 1$ for both cases and by replacing the A's by the proper area relations.

Radiation Effect on Temperature Measurements

A temperature measuring device indicates the temperature of its *sensor,* which is supposed to be, but is not necessarily, the temperature of the medium that the sensor is in. When a thermometer (or any other temperature measuring device such as a thermocouple) is placed in a medium, heat transfer takes place between the sensor of the thermometer and the medium by convection until the sensor reaches the temperature of the medium. But when the sensor is surrounded by surfaces that are at a different temperature than the fluid, radiation exchange will take place between the sensor and the surrounding surfaces. When the heat transfers by convection and radiation balance each other, the sensor will indicate a temperature that falls between the fluid and surface temperatures. Below we develop a procedure to account for the radiation effect and to determine the actual fluid temperature.

Consider a thermometer that is used to measure the temperature of a fluid flowing through a large channel whose walls are at a lower temperature than the fluid (Fig. 9-67). Equilibrium will be established and the reading of the thermometer will stabilize when heat gain by convection, as measured by the sensor, equals heat loss by radiation (or vice versa). That is, on a unit-area basis,

$$\dot{q}_{\text{conv, to sensor}} = \dot{q}_{\text{rad, from sensor}}$$
$$h(T_f - T_{\text{th}}) = \varepsilon_{\text{th}}\sigma(T_{\text{th}}^4 - T_w^4)$$

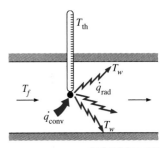

FIGURE 9-67

A thermometer used to measure the temperature of a fluid in a channel.

or

$$T_f = T_{\text{th}} + \frac{\varepsilon_{\text{th}}\sigma(T_{\text{th}}^4 - T_w^4)}{h} \qquad \text{(K)} \qquad (9\text{-}63)$$

where
T_f = actual temperature of the fluid, K
T_{th} = temperature value measured by the thermometer, K
T_w = temperature of the surrounding surfaces, K
h = convection heat transfer coefficient, W/m² · K
ε = emissivity of the sensor of the thermometer

The last term in Equation 9-63 is due to the *radiation effect* and represents the radiation correction. Note that the radiation correction term is most significant when the convection heat transfer coefficient is small and the emissivity of the surface of the sensor is large. Therefore, the sensor should be coated with a material of high reflectivity (low emissivity) to reduce the radiation effect.

Placing the sensor in a radiation shield without interfering with the fluid flow also reduces the radiation effect. The sensors of temperature measurement devices used outdoors must be protected from direct sunlight since the radiation effect in that case is sure to reach unacceptable levels.

The radiation effect is also a significant factor in *human comfort* in heating and air-conditioning applications. A person who feels fine in a room at a specified temperature may feel chilly in another room at the same

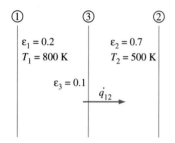

$\varepsilon_1 = 0.2$
$T_1 = 800$ K

$\varepsilon_2 = 0.7$
$T_2 = 500$ K

$\varepsilon_3 = 0.1$
$\dot{q}_{12}$

FIGURE 9-68

Schematic for Example 9-15.

temperature as a result of the radiation effect if the walls of the second room are at a considerably lower temperature. For example, most people will feel comfortable in a room at 22°C if the walls of the room are also roughly at that temperature. When the wall temperature drops to 5°C for some reason, the interior temperature of the room must be raised to at least 27°C to maintain the same level of comfort. Therefore, well-insulated buildings conserve energy not only by reducing the heat loss or heat gain, but also by allowing the thermostats to be set at a lower temperature in winter and at a higher temperature in summer without compromising the comfort level.

EXAMPLE 9-15 Radiation Shields

A thin aluminum sheet with an emissivity of 0.1 on both sides is placed between two very large parallel plates that are maintained at uniform temperatures $T_1 = 800$ K and $T_2 = 500$ K and have emissivities $\varepsilon_1 = 0.2$ and $\varepsilon_2 = 0.7$, respectively, as shown in Figure 9-68. Determine the net rate of radiation heat transfer between the two plates per unit surface area of the plates and compare the result to that without the shield.

Solution A thin aluminum sheet is placed between two large parallel plates maintained at uniform temperatures. The net rates of radiation heat transfer between the two plates with and without the radiation shield are to be determined.

Assumptions The surfaces are opaque, diffuse, and gray.

Analysis The net rate of radiation heat transfer between these two plates without the shield was determined in Example 9-11 to be 3625 W/m². Heat transfer in the presence of one shield is determined from Equation 9-60 to be

$$
\dot{q}_{12,\,\text{one shield}} = \frac{\dot{Q}_{12,\,\text{one shield}}}{A} = \frac{\sigma(T_1^4 - T_2^4)}{\left(\dfrac{1}{\varepsilon_1} + \dfrac{1}{\varepsilon_2} - 1\right) + \left(\dfrac{1}{\varepsilon_{3,1}} + \dfrac{1}{\varepsilon_{3,2}} - 1\right)}
$$

$$
= \frac{(5.67 \times 10^{-8}\ \text{W/m}^2 \cdot \text{K}^4)[(800\ \text{K})^4 - (500\ \text{K})^4]}{\left(\dfrac{1}{0.2} + \dfrac{1}{0.7} - 1\right) + \left(\dfrac{1}{0.1} + \dfrac{1}{0.1} - 1\right)}
$$

$$
= 805.6\ \text{W/m}^2
$$

Discussion Note that the rate of radiation heat transfer reduces to about one-fourth of what it was as a result of placing a radiation shield between the two parallel plates.

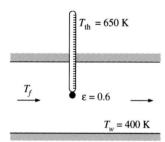

$T_{\text{th}} = 650$ K

T_f

$\varepsilon = 0.6$

$T_w = 400$ K

FIGURE 9-69

Schematic for Example 9-16.

EXAMPLE 9-16 Radiation Effect on Temperature Measurements

A thermocouple used to measure the temperature of hot air flowing in a duct whose walls are maintained at $T_w = 400$ K shows a temperature reading of $T_{\text{th}} = 650$ K (Fig. 9-69). Assuming the emissivity of the thermocouple junction to be $\varepsilon = 0.6$ and the convection heat transfer coefficient to be $h = 80$ W/m² · °C, determine the actual temperature of the air.

Solution The temperature of air in a duct is measured by a thermocouple that also exchanges heat by radiation with the surrounding surfaces. The radiation effect on the temperature measurement is to be quantified, and the actual air temperature is to be determined.

Assumptions   The surfaces are opaque, diffuse, and gray.

Analysis   The walls of the duct are at a considerably lower temperature than the air in it, and thus we expect the thermocouple to show a reading lower than the actual air temperature as a result of radiation effect. The actual air temperature is determined from Equation 9-63 to be

$$T_f = T_{th} + \frac{\varepsilon_{th}\sigma(T_{th}^4 - T_w^4)}{h}$$

$$= (650 \text{ K}) + \frac{0.6 \times (5.67 \times 10^{-8} \text{ W/m}^2 \cdot \text{K}^4)[(650 \text{ K})^4 - (400 \text{ K})^4]}{80 \text{ W/m}^2 \cdot ^\circ\text{C}}$$

$$= 715.0 \text{ K}$$

Note that the radiation effect causes a difference of 65°C (or 65 K since °C ≡ K for temperature differences) in temperature reading in this case.

9-10 ■ SUMMARY

Radiation propagates in the form of electromagnetic waves. The *frequency* v and *wavelength* λ of electromagnetic waves in a medium are related by $\lambda = c/v$, where c is the speed of light in that medium. All matter whose temperature is above absolute zero continuously emits *thermal radiation* as a result of vibrational and rotational motions of molecules, atoms, and electrons of a substance. Temperature is a measure of the strength of these activities at the microscopic level.

A *blackbody* is defined as a *perfect emitter and absorber of radiation*. At a specified temperature and wavelength, no surface can emit more energy than a blackbody. A blackbody absorbs *all* incident radiation, regardless of wavelength and direction. The radiation energy emitted by a blackbody per unit time and per unit surface area is called the *blackbody emissive power* and is expressed by the *Stefan–Boltzmann law* as

$$E_b = \sigma T^4 \qquad (\text{W/m}^2)$$

where $\sigma = 5.67 \times 10^{-8}$ W/m² · K⁴ is the *Stefan–Boltzmann constant, E_b* is the blackbody emissive power, and T is the absolute temperature of the surface in K. At any specified temperature, the spectral blackbody emissive power $E_{b\lambda}$ increases with wavelength, reaches a peak, and then decreases with increasing wavelength. The wavelength at which the peak occurs for a specified temperature is given by *Wien's displacement law* as

$$(\lambda T)_{\text{max power}} = 2897.8 \text{ μm} \cdot \text{K}$$

The *blackbody radiation function* f_λ represents the fraction of radiation emitted by a blackbody at temperature T in the wavelength band from $\lambda = 0$ to λ. The fraction of radiation energy emitted by a blackbody at temperature T over a finite wavelength band from $\lambda = \lambda_1$ to $\lambda = \lambda_2$ is determined from

$$f_{\lambda_1-\lambda_2}(T) = f_{\lambda_2}(T) - f_{\lambda_1}(T)$$

where $f_{\lambda_1}(T)$ and $f_{\lambda_2}(T)$ are the blackbody radiation functions corresponding to $\lambda_1 T$ and $\lambda_2 T$, respectively.

The *emissivity* of a surface is defined as the ratio of the radiation emitted by the surface to the radiation emitted by a blackbody at the same temperature. The *total hemispherical emissivity* ε of a surface is simply the average emissivity over all directions and wavelengths and is expressed as

$$\varepsilon(T) = \frac{E(T)}{E_b(T)} = \frac{E(T)}{\sigma T^4}$$

where $E(T)$ is the total emissive power of the surface.

The consideration of wavelength and direction dependence of properties makes radiation calculations very complicated. Therefore, the *gray* and *diffuse* approximations are commonly utilized in radiation calculations. A surface is said to be *diffuse* if its properties are *independent of direction* and *gray* if its properties are *independent of wavelength*.

The radiation energy incident on a surface per unit surface area per unit time is called *irradiation* and is denoted by G. When irradiation strikes a surface, part of it is absorbed, part of it is reflected, and the remaining part, if any, is transmitted. The fraction of irradiation absorbed by the surface is called the *absorptivity* α. The fraction of irradiation reflected by the surface is called the *reflectivity* ρ, and the fraction transmitted is called the *transmissivity* τ. The sum of the absorbed, reflected, and transmitted fractions of radiation energy must be equal to unity,

$$\alpha + \rho + \tau = 1$$

For *opaque* surfaces, $\tau = 0$, and thus

$$\alpha + \rho = 1$$

Surfaces are usually assumed to reflect in a perfectly *specular* or *diffuse* manner for simplicity. In *specular* (or *mirrorlike*) *reflection,* the angle of reflection equals the angle of incidence of the radiation beam. In *diffuse reflection,* radiation is reflected equally in all directions. Reflection from smooth and polished surfaces approximates specular reflection, whereas reflection from rough surfaces approximates diffuse reflection.

Kirchhoff's law of radiation is expressed as

$$\varepsilon(T) = \alpha(T)$$

That is, the total hemispherical emissivity of a surface at temperature T is equal to its total hemispherical absorptivity for radiation coming from a blackbody at the same temperature.

The *view factor* from a surface i to a surface j is denoted by $F_{i \rightarrow j}$ and is defined as the fraction of the radiation leaving surface i that strikes surface j directly. The view factor $F_{i \rightarrow i}$ represents the fraction of the radiation leaving surface i that strikes itself directly. $F_{i \rightarrow i} = 0$ for *plane* or *convex* surfaces and $F_{i \to i} \neq 0$ for *concave* surfaces. For view factors, the *reciprocity rule* is expressed as

$$A_i F_{i \to j} = A_j F_{j \to i}$$

The sum of the view factors from surface i of an enclosure to all surfaces of the enclosure, including to itself, must equal unity. This is known as the *summation rule* for an enclosure. The *superposition rule* is expressed as follows: the view factor from a surface i to a surface j is equal to the sum of the view factors from surface i to the parts of surface j. The symmetry rule is expressed as follows: if the surfaces j and k are symmetric about the surface i then $F_{i \to j} = F_{i \to k}$.

The rate of net radiation heat transfer between two *black* surfaces is determined from

$$\dot{Q}_{1 \to 2} = A_1 F_{1 \to 2} \sigma (T_1^4 - T_2^4) \qquad \text{(W)}$$

The *net* radiation heat transfer from any surface i of a *black* enclosure is determined by adding up the net radiation heat transfers from surface i to each of the surfaces of the enclosure:

$$\dot{Q}_i = \sum_{j=1}^{N} \dot{Q}_{i \to j} = \sum_{j=1}^{N} A_i F_{i \to j} \sigma (T_i^4 - T_j^4) \qquad \text{(W)}$$

The total radiation energy leaving a surface per unit time and per unit area is called the *radiosity* and is denoted by J. The *net* rate of radiation heat transfer from a surface i of surface area A_i is expressed as

$$\dot{Q}_i = \frac{E_{bi} - J_i}{R_i} \qquad \text{(W)}$$

where

$$R_i = \frac{1 - \varepsilon_i}{A_i \varepsilon_i}$$

is the *surface resistance* to radiation. The *net* rate of radiation heat transfer from surface i to surface j can be expressed as

$$\dot{Q}_{i \to j} = \frac{J_i - J_j}{R_{i \to j}} \qquad \text{(W)}$$

where

$$R_{i \to j} = \frac{1}{A_i F_{i \to j}}$$

is the *space resistance* to radiation. The *network method* is applied to radiation enclosure problems by drawing a surface resistance associated with each surface of an enclosure and connecting them with space resistances. Then the problem is solved by treating it as an electrical network problem where the radiation heat transfer replaces the current and the radiosity replaces the potential. The *direct method* is based on the following two equations:

Surfaces with specified
net heat transfer rate $\dot{Q}_i$
$$\dot{Q}_i = A_i \sum_{j=1}^{N} F_{i \to j} (J_i - J_j)$$

Surfaces with specified
temperature T_i
$$\sigma T_i^4 = J_i + \frac{1 - \varepsilon_i}{\varepsilon_i} \sum_{j=1}^{N} F_{i \to j} (J_i - J_j)$$

The first group (for surfaces with specified heat transfer rates) and the second group (for surfaces with specified temperatures) of equations give N linear algebraic equations for the determination of the N unknown radiosities for an N-surface enclosure. Once the radiosities $J_1, J_2, \ldots, J_N$ are available, the unknown surface temperatures and heat transfer rates can be determined from the equations above.

The net rate of radiation transfer between any two gray, diffuse, opaque surfaces that form an enclosure is given by

$$\dot{Q}_{12} = \frac{\sigma(T_1^4 - T_2^4)}{\dfrac{1 - \varepsilon_1}{A_1 \varepsilon_1} + \dfrac{1}{A_1 F_{12}} + \dfrac{1 - \varepsilon_2}{A_2 \varepsilon_2}} \qquad \text{(W)}$$

Radiation heat transfer between two surfaces can be reduced greatly by inserting between the two surfaces thin, high-reflectivity (low-emissivity) sheets of material called *radiation shields*. Radiation heat transfer between two large parallel plates separated by N radiation shields is

$$\dot{Q}_{12, N \text{ shields}} = \frac{A\sigma(T_1^4 - T_2^4)}{\left(\dfrac{1}{\varepsilon_1} + \dfrac{1}{\varepsilon_2} - 1\right) + \left(\dfrac{1}{\varepsilon_{3,1}} + \dfrac{1}{\varepsilon_{3,2}} - 1\right) + \cdots + \left(\dfrac{1}{\varepsilon_{N,1}} + \dfrac{1}{\varepsilon_{N,2}} - 1\right)}$$

The radiation effect in temperature measurements can be properly accounted for by the relation

$$T_f = T_{\text{th}} + \frac{\varepsilon_{\text{th}} \sigma(T_{\text{th}}^4 - T_w^4)}{h} \qquad \text{(K)}$$

where T_f is the actual temperature of the fluid, T_{th} is the temperature value measured by the thermometer, and T_w is the temperature of the surrounding walls, all in K.

REFERENCES AND SUGGESTED READING

1. A. G. H. Dietz. "Diathermanous Materials and Properties of Surfaces." In *Space Heating with Solar Energy,* R. W. Hamilton. Cambridge, MA: MIT Press, 1954.

2. J. A. Duffy and W. A. Backman. *Solar Energy Thermal Process.* New York: John Wiley & Sons, 1974.

3. D. C. Hamilton and W. R. Morgan. "Radiation Interchange Configuration Factors." National Advisory Committee for Aeronautics, Technical Note 2836, 1952.

4. J. P. Holman. *Heat Transfer.* 7th ed. New York: McGraw-Hill, 1990.

5. H. C. Hottel. "Radiant Heat Transmission." In *Heat Transmission,* 3d ed., ed. W. H. McAdams. New York: McGraw-Hill, 1954.

6. J. R. Howell. *A Catalog of Radiation Configuration Factors.* New York: McGraw-Hill, 1982.

7. F. P. Incropera and D. P. DeWitt. *Introduction to Heat Transfer.* 2d ed. New York: John Wiley & Sons, 1990.

8. A. K. Oppenheim. "Radiation Analysis by the Network Method." *Transactions of the ASME* 78 (1956) pp. 725–35.

9. M. N. Özişik. *Heat Transfer—A Basic Approach.* New York: McGraw-Hill, 1985.

10. W. Sieber. *Zeitschrift für Technische Physics* 22 (1941), pp. 130–35.

11. L. C. Thomas. *Heat Transfer.* Englewood Cliffs, NJ: Prentice Hall, 1992.

12. Y. S. Touloukain and D. P. DeWitt. "Nonmetallic Solids." In *Thermal Radiative Properties.* Vol. 8. New York: IFI/Plenum, 1970.

13. Y. S. Touloukian and D. P. DeWitt. "Metallic Elements and Alloys." In *Thermal Radiative Properties,* Vol. 7. New York: IFI/Plenum, 1970.

PROBLEMS*

Electromagnetic and Thermal Radiation

9-1C What is an electromagnetic wave? How does it differ from a sound wave?

9-2C By what properties is an electromagnetic wave characterized? How are these properties related to each other?

9-3C What is visible light? How does it differ from the other forms of electromagnetic radiation?

9-4C How do ultraviolet and infrared radiation differ? Do you think your body emits any radiation in the ultraviolet range?

9-5C What is thermal radiation? How does it differ from the other forms of electromagnetic radiation?

9-6C What is the cause of color? Why do some objects appear blue to the eye while others appear red? Is the color of a surface at room temperature related to the radiation it emits?

*Students are encouraged to answer *all* the concept "C" questions.

9-7C Why is radiation usually treated as a surface phenomenon?

9-8C Why do skiers get sunburned so easily?

9-9C How does microwave cooking differ from conventional cooking?

9-10 Electricity is generated and transmitted in power lines at a frequency of 60 Hz (1 Hz = 1 cycle per second). Determine the wavelength of the electromagnetic waves generated by the passage of electricity in power lines.

9-11 A microwave oven is designed to operate at a frequency of 2.8×10^9 Hz. Determine the wavelength of these microwaves and the energy of each microwave.

9-12 A radio station is broadcasting radiowaves at a wavelength of 300 m. Determine the frequency of these waves.   *Answer:* 1.0×10^6 Hz

9-13 A cordless telephone is designed to operate at a frequency of 8.5×10^8 Hz. Determine the wavelength of these telephone waves.

Blackbody Radiation

9-14C What is a blackbody? Does a blackbody actually exist?

9-15C Define the total and spectral blackbody emissive powers. How are they related to each other? How do they differ?

9-16C Why did we define the blackbody radiation function? What does it represent? For what is it used?

9-17C Consider two identical bodies, one at 1000 K and the other at 1500 K. Which body emits more radiation in the shorter-wavelength region? Which body emits more radiation at a wavelength of 20 μm?

9-18 Consider a 20-cm $\times$ 20-cm $\times$ 20-cm cubical body at 1000 K suspended in the air. Assuming the body closely approximates a blackbody, determine (a) the rate at which the cube emits radiation energy, in W, and (b) the spectral blackbody emissive power at a wavelength of 4 μm.

9-19E The sun can be treated as a blackbody at an effective surface temperature of 10,372 R. Determine the rate at which infrared radiation energy ($\lambda = 0.76$–100 μm) is emitted by the sun, in Btu/h · ft^2.

9-20 The temperature of the filament of an incandescent light bulb is 3200 K. Treating the filament as a blackbody, determine the fraction of the radiant energy emitted by the filament that falls in the visible range. Also, determine the wavelength at which the emission of radiation from the filament peaks.

9-21 An incandescent light bulb is desired to emit at least 20 percent of its energy at wavelengths shorter than 1 μm. Determine the minimum temperature to which the filament of the light bulb must be heated.

9-22 It is desired that the radiation energy emitted by a light source reach a maximum in the blue range ($\lambda = 0.47$ μm). Determine the temperature of this light source and the fraction of radiation it emits in the visible range ($\lambda = 0.40$–0.76 μm).

9-23 A 3-mm-thick glass window transmits 90 percent of the radiation between $\lambda = 0.3$ and 3.0 μm and is essentially opaque for radiation at other wavelengths. Determine the rate of radiation transmitted through a 2-m $\times$ 2-m glass window from blackbody sources at (*a*) 5800 K and (*b*) 1000 K.
Answers: (*a*) 218,400 kW, (*b*) 55.8 kW

Radiation Properties

9-24C Define the properties emissivity and absorptivity. When are these two properties equal to each other?

9-25C Define the properties reflectivity and transmissivity and discuss the different forms of reflection.

9-26C What is a graybody? How does it differ from a blackbody? What is a diffuse gray surface?

9-27C What is the greenhouse effect? Why is it a matter of great concern among atmospheric scientists?

9-28C We can see the inside of a microwave oven during operation through its glass door, which indicates that visible radiation is escaping the oven. Do you think that the harmful microwave radiation might also be escaping?

9-29 The spectral emissivity function of an opaque surface at 1000 K is approximated as

$$\varepsilon_\lambda = \begin{cases} \varepsilon_1 = 0.4, & 0 \leq \lambda < 2 \ \mu\text{m} \\ \varepsilon_2 = 0.7, & 2 \ \mu\text{m} \leq \lambda < 6 \ \mu\text{m} \\ \varepsilon_3 = 0.3, & 6 \ \mu\text{m} \leq \lambda < \infty \end{cases}$$

Determine the average emissivity of the surface and the rate of radiation emission from the surface, in W/m^2. *Answers:* 0.575, 32.6 kW/m^2

9-30 The reflectivity of aluminum coated with lead sulfate is 0.35 for radiation at wavelengths less than 3 μm and 0.95 for radiation greater than 3 μm. Determine the average reflectivity of this surface for solar radiation ($T \approx 5800$ K) and radiation coming from surfaces at room temperature ($T \approx 300$ K). Also, determine the emissivity and absorptivity of this surface at both temperatures. Do you think this material is suitable for use in solar collectors?

9-31 A furnace that has a 20-cm $\times$ 20-cm glass window can be considered to be a blackbody at 1200 K. If the transmissivity of the glass is 0.8 for radiation at wavelengths less than 3 μm and zero for radiation at greater than 3 μm, determine the fraction and the rate of radiation coming from the furnace and transmitted through the window.

9-32 The emissivity of a tungsten filament can be approximated to be 0.5 for radiation at wavelengths less than 1 μm and 0.15 for radiation at greater

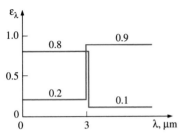

FIGURE P9-33

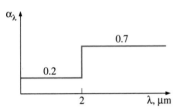

FIGURE P9-35

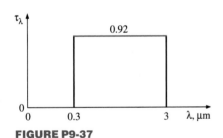

FIGURE P9-37

than 1 μm. Determine the average emissivity of the filament at (*a*) 1500 K and (*b*) 3000 K. Also, determine the absorptivity and reflectivity of the filament at both temperatures.

9-33   The variations of the spectral emissivity of two surfaces are as given in Figure P9-33. Determine the average emissivity of each surface at $T = 3000$ K. Also, determine the average absorptivity and reflectivity of each surface for radiation coming from a source at 3000 K. Which surface is more suitable to serve as a solar absorber?

9-34   The emissivity of a surface coated with aluminum oxide can be approximated to be 0.2 for radiation at wavelengths less than 5 μm and 0.9 for radiation at wavelengths greater than 5 μm. Determine the average emissivity of this surface at (*a*) 5800 K and (*b*) 300 K. What can you say about the absorptivity of this surface for radiation coming from sources at 5800 K and 300 K?   *Answers:* (*a*) 0.203, (*b*) 0.89

9-35   The variation of the spectral absorptivity of a surface is as given in Figure P9-35. Determine the average absorptivity and reflectivity of the surface for radiation that originates from a source at $T = 2500$ K. Also, determine the average emissivity of this surface at 3000 K.

9-36E   A 5-in.-diameter spherical ball is known to emit radiation at a rate of 120 Btu/h when its surface temperature is 800 R. Determine the average emissivity of the ball at this temperature.

9-37   The variation of the spectral transmissivity of a 0.6-cm-thick glass pane is as given in Figure P9-37. Determine the average transmissivity of this pane for solar radiation ($T \approx 5800$ K) and radiation coming from surfaces at room temperature ($T \approx 300$ K). Also, determine the amount of solar radiation transmitted through the pane for incident solar radiation of 650 W/m².
     Answers: 0.848, 0.00015, 551.1 W/m²

Atmospheric and Solar Radiation

9-38C   What is the solar constant? How is it used to determine the effective surface temperature of the sun? How would the value of the solar constant change if the distance between the earth and the sun doubled?

9-39C   What changes would you notice if the sun emitted radiation at an effective temperature of 2000 K instead of 5762 K?

9-40C   Explain why the sky is blue and the sunset is yellow-orange.

9-41C   When the earth is closest to the sun, we have winter in the northern hemisphere. Explain why. Also explain why we have summer in the northern hemisphere when the earth is farthest away from the sun.

9-42C   What is the effective sky temperature?

9-43C   You have probably noticed warning signs on the highways stating that bridges may be icy even when the roads are not. Explain how this can happen.

9-44C Unless you live in a warm southern state, you have probably had to scrape ice from the windshield and windows of your car many mornings. You may have noticed, with frustration, that the thickest layer of ice always forms on the windshield instead of the side windows. Explain why this is the case.

9-45C Explain why surfaces usually have quite different absorptivities for solar radiation and for radiation originating from the surrounding bodies.

9-46 A surface has an absorptivity of $\alpha_s = 0.85$ for solar radiation and an emissivity of $\varepsilon = 0.5$ at room temperature. The surface temperature is observed to be 350 K when the direct and the diffuse components of solar radiation are $G_D = 350$ and $G_d = 400$ W/m², respectively, and the direct radiation makes a 30°angle with the normal of the surface. Taking the effective sky temperature to be 280 K, determine the net rate of radiation to the surface at that time.

9-47E Solar radiation is incident on the outer surface of a spaceship at a rate of 400 Btu/h · ft². The surface has an absorptivity of $\alpha_s = 0.10$ for solar radiation and an emissivity of $\varepsilon = 0.8$ at room temperature. The outer surface radiates heat into space at 0 R. If there is no net heat transfer into the spaceship, determine the equilibrium temperature of the surface.
    Answer: 413.3 R

9-48 The air temperature on a clear night is observed to remain at about 4°C. Yet water is reported to have frozen that night. Taking the convection heat transfer coefficient to be 10 W/m² · °C, determine the value of the effective sky temperature that night.

9-49 The absorber surface of a solar collector is made of aluminum coated with black chrome ($\alpha_s = 0.87$ and $\varepsilon = 0.09$). Solar radiation is incident on the surface at a rate of 600 W/m². The air and the effective sky temperatures are 25°C and 15°C, respectively, and the convection heat transfer coefficient is 10 W/m² · °C. For an absorber surface temperature of 70°C, determine the net rate of solar energy delivered by the absorber plate to the water circulating behind it.

9-50 Determine the equilibrium temperature of the absorber surface in Problem 9-49 if the back side of the absorber is insulated.

The View Factor

9-51C What does the view factor represent? When is the view factor from a surface to itself not zero?

9-52C How can you determine the view factor F_{12} when the view factor F_{21} is available?

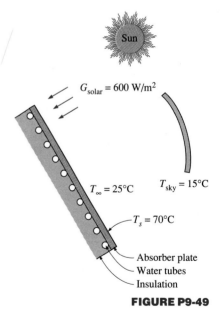

Sun

$G_{solar} = 600$ W/m²

$T_\infty = 25°C$ $T_{sky} = 15°C$

$T_s = 70°C$

Absorber plate
Water tubes
Insulation

FIGURE P9-49

9-53C What are the summation rule and the superposition rule for view factors?

9-54C What is the crossed-strings method? For what kind of geometries is the crossed-strings method applicable?

9-55 Consider an enclosure consisting of seven surfaces. How many view factors does this geometry involve? How many of these view factors can be determined by the application of the reciprocity and the summation rules?

9-56 Consider an enclosure consisting of five surfaces. How many view factors does this geometry involve? How many of these view factors can be determined by the application of the reciprocity and summation rules?

9-57 Consider an enclosure consisting of 12 surfaces. How many view factors does this geometry involve? How many of these view factors can be determined by the application of the reciprocity and the summation rules?
Answers: 144, 78

9-58 Determine the view factors F_{13} and F_{23} between the rectangular surfaces shown in Figure P9-58.

9-59 Consider a cylindrical enclosure whose height is twice the diameter of its base. Determine the view factor from the side surface of this cylindrical enclosure to its base surface.

9-60 Consider a hemispherical furnace with a flat circular base of diameter D. Determine the view factor from the dome of this furnace to its base. *Answer:* 0.5

9-61 Determine the view factors F_{12} and F_{21} for the very long ducts shown in Figure P9-61 without using any view factor tables or charts. Neglect end effects.

9-62 Determine the view factors from the very long grooves shown in Figure P9-62 to the surroundings without using any view factor tables or charts. Neglect end effects.

9-63 Determine the view factors from the base of a cube to each of the other five surfaces. *Answer:* 0.2

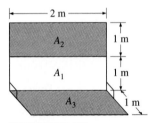

FIGURE P9-58

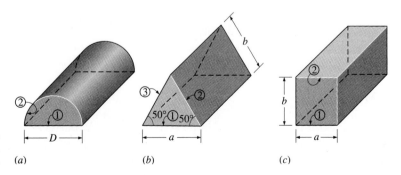

FIGURE P9-61
(a) Semicylindrical duct.
(b) Triangular duct.
(c) Rectangular duct.

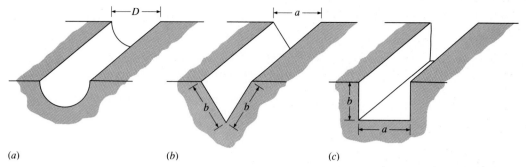

(a)   (b)   (c)

FIGURE P9-62

(a) Semicylindrical groove.
(b) Triangular groove.
(c) Rectangular groove.

9-64 Consider a conical enclosure of height h and base diameter D. Determine the view factor from the conical side surface to a hole of diameter d located at the center of the base.

9-65 Determine the four view factors associated with an enclosure formed by two very long concentric cylinders of radii r_1 and r_2. Neglect the end effects.

9-66 Determine the view factor F_{12} between the rectangular surfaces shown in Figure P9-66.

9-67 Two infinitely long parallel cylinders of diameter D are located a distance s apart from each other. Determine the view factor F_{12} between these two cylinders.

9-68 Three infinitely long parallel cylinders of diameter D are located a distance s apart from each other. Determine the view factor between the cylinder in the middle and the surroundings.

FIGURE P9-64

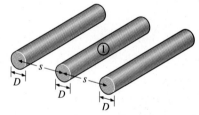

FIGURE P9-68

Radiation Heat Transfer between Surfaces

9-69C Why is the radiation analysis of enclosures that consist of black surfaces relatively easy? How is the rate of radiation heat transfer between two surfaces expressed in this case?

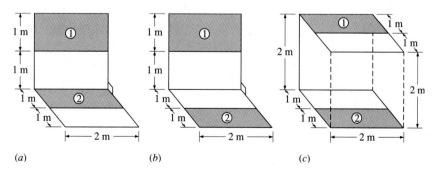

(a)   (b)   (c)   **FIGURE P9-66**

9-70C How does radiosity for a surface differ from the emitted energy? For what kind of surfaces are these two quantities identical?

9-71C What are the radiation surface and space resistances? How are they expressed? For what kind of surfaces is the radiation surface resistance zero?

9-72C What are the two methods used in radiation analysis? How do these two methods differ?

9-73C What is a reradiating surface? What simplifications does a reradiating surface offer in the radiation analysis?

9-74E Consider a 10-ft × 10-ft × 10-ft cubical furnace whose top and side surfaces closely approximate black surfaces and whose base surface has an emissivity $\varepsilon = 0.7$. The base, top, and side surfaces of the furnace are maintained at uniform temperatures of 800 R, 1600 R, and 2400 R, respectively. Determine the net rate of radiation heat transfer between (a) the base and side surfaces and (b) the base and top surfaces. Also, determine the net rate of radiation heat transfer to the base surface.

9-75 Two very large parallel plates are maintained at uniform temperatures of $T_1 = 600$ K and $T_2 = 400$ K and have emissivities $\varepsilon_1 = 0.5$ and $\varepsilon_2 = 0.9$, respectively. Determine the net rate of radiation heat transfer between the two surfaces per unit area of the plates.

9-76 A furnace is of cylindrical shape with $r = h = 2$ m. The base, top, and side surfaces of the furnace are all black and are maintained at uniform temperatures of 500, 700, and 800 K, respectively. Determine the net rate of radiation heat transfer to or from the top surface during steady operation.

9-77 Consider a hemispherical furnace of diameter $D = 5$ m with a flat base. The dome of the furnace is black, and the base has an emissivity of 0.7. The base and the dome of the furnace are maintained at uniform temperatures of 400 and 1000 K, respectively. Determine the net rate of radiation heat transfer from the dome to the base surface during steady operation.
    Answer: 759 kW

9-78 Two very long concentric cylinders of diameters $D_1 = 0.2$ m and $D_2 = 0.5$ m are maintained at uniform temperatures of $T_1 = 800$ K and $T_2 = 500$ K and have emissivities $\varepsilon_1 = 1$ and $\varepsilon_2 = 0.7$, respectively. Determine the net rate of radiation heat transfer between the two cylinders per unit length of the cylinders.

9-79 The following experiment is conducted to determine the emissivity of a certain material. A long cylindrical rod of diameter $D_1 = 0.01$ m is coated with this new material and is placed in an evacuated long cylindrical enclosure of diameter $D_2 = 0.1$ m and emissivity $\varepsilon_2 = 0.95$, which is cooled externally and maintained at a temperature of 200 K at all times. The rod is heated by passing electric current through it. When steady operating conditions are reached, it is observed that the rod is dissipating electric power at a rate of 8 W per unit of its length and its surface temperature is 500 K. Based on these measurements, determine the emissivity of the coating on the rod.

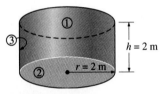

FIGURE P9-76

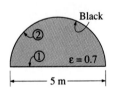

FIGURE P9-77

9-80E A furnace is shaped like a long semicylindrical duct of diameter $D = 15$ ft. The base and the dome of the furnace have emissivities of 0.5 and 0.9 and are maintained at uniform temperatures of 700 and 1800 R, respectively. Determine the net rate of radiation heat transfer from the dome to the base surface per unit length during steady operation.

9-81 Two parallel disks of diameter $D = 0.6$ m separated by $L = 0.4$ m are located directly on top of each other. Both disks are black and are maintained at a temperature of 700 K. The back sides of the disks are insulated, and the environment that the disks are in can be considered to be a blackbody at $T_\infty = 300$ K. Determine the net rate of radiation heat transfer from the disks to the environment. *Answer:* 5505 W

9-82 A furnace is shaped like a long equilateral-triangular duct where the width of each side is 2 m. Heat is supplied from the base surface, whose emissivity is $\varepsilon_1 = 0.8$, at a rate of 800 W/m^2 while the side surfaces, whose emissivities are 0.5, are maintained at 500 K. Neglecting the end effects, determine the temperature of the base surface. Can you treat this geometry as a two-surface enclosure?

9-83 Consider a 4-m × 4-m × 4-m cubical furnace whose floor and ceiling are black and whose side surfaces are reradiating. The floor and the ceiling of the furnace are maintained at temperatures of 550 K and 1100 K, respectively. Determine the net rate of radiation heat transfer between the floor and the ceiling of the furnace.

9-84 Two concentric spheres of diameters $D_1 = 0.3$ m and $D_2 = 0.8$ m are maintained at uniform temperatures $T_1 = 700$ K and $T_2 = 400$ K and have emissivities $\varepsilon_1 = 0.5$ and $\varepsilon_2 = 0.7$, respectively. Determine the net rate of radiation heat transfer between the two spheres. Also, determine the convection heat transfer coefficient at the outer surface if both the surrounding medium and the surrounding surfaces are at 25°C. Assume the emissivity of the outer surface is 0.2.

9-85 A spherical tank of diameter $D = 2$ m that is filled with liquid nitrogen at 100 K is kept in an evacuated cubic enclosure whose sides are 3 m long. The emissivities of the spherical tank and the enclosure are $\varepsilon_1 = 0.1$ and $\varepsilon_2 = 0.8$, respectively. If the temperature of the cubic enclosure is measured to be 240 K, determine the net rate of radiation heat transfer to the liquid nitrogen. *Answer:* 228 W

9-86 Repeat Problem 9-85 by replacing the cubic enclosure by a spherical enclosure whose diameter is 3 m.

9-87 Consider a circular grill whose diameter is 0.3 m. The bottom of the grill is covered with hot coal bricks at 1100 K, while the wire mesh on top of the grill is covered with steaks initially at 5°C. The distance between the coal bricks and the steaks is 0.20 m. Treating both the steaks and the coal bricks as blackbodies, determine the initial rate of radiation heat transfer from the coal bricks to the steaks. Also, determine the initial rate of radiation heat transfer to the steaks if the side opening of the grill is covered by aluminum

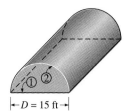

$\leftarrow D = 15$ ft$\rightarrow$

FIGURE P9-80

FIGURE P9-85

FIGURE P9-87

Steaks

Coal bricks

0.20 m

foil, which can be approximated as a reradiating surface.
    Answers: 1671 W, 3755 W

9-88E A 9-ft-high room with a base area of 12 ft × 12 ft is to be heated by electric resistance heaters placed on the ceiling, which is maintained at a uniform temperature of 90°F at all times. The floor of the room is at 65°F and has an emissivity of 0.8. The side surfaces are well insulated. Treating the ceiling as a blackbody, determine the rate of heat loss from the room through the floor.

Radiation Shields and the Radiation Effect

9-89C What is a radiation shield? Why is it used?

9-90C What is the radiation effect? How does it influence the temperature measurements?

9-91C Give examples of radiation effects that affect human comfort.

9-92 Consider a person whose exposed surface area is 1.7 m², emissivity is 0.7, and surface temperature is 32°C. Determine the rate of heat loss from that person by radiation in a large room whose walls are at a temperature of (*a*) 300 K and (*b*) 280 K.   *Answers:* (*a*) 37.4 W, (*b*) 169.2 W

9-93 A thin aluminum sheet with an emissivity of 0.15 on both sides is placed between two very large parallel plates, which are maintained at uniform temperatures $T_1 = 900$ K and $T_2 = 650$ K and have emissivities $\varepsilon_1 = 0.5$ and $\varepsilon_2 = 0.8$, respectively. Determine the net rate of radiation heat transfer between the two plates per unit surface area of the plates and compare the result with that without the shield.

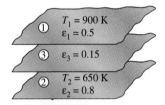

$T_1 = 900$ K
$\varepsilon_1 = 0.5$

$\varepsilon_3 = 0.15$

$T_2 = 650$ K
$\varepsilon_2 = 0.8$

FIGURE P9-93

9-94 Two very large parallel plates are maintained at uniform temperatures of $T_1 = 1000$ K and $T_2 = 500$ K and have emissivities of $\varepsilon_1 = \varepsilon_2 = 0.2$, respectively. It is desired to reduce the net rate of radiation heat transfer between the two plates to one-fifth by placing thin aluminum sheets with an emissivity of 0.2 on both sides between the plates. Determine the number of sheets that need to be inserted.

9-95 Five identical thin aluminum sheets with emissivities of 0.1 on both sides are placed between two very large parallel plates, which are maintained at uniform temperatures of $T_1 = 800$ K and $T_2 = 450$ K and have emissivities of $\varepsilon_1 = \varepsilon_2 = 0.1$, respectively. Determine the net rate of radiation heat transfer between the two plates per unit surface area of the plates and compare the result to that without the shield.

9-96E Two parallel disks of diameter $D = 3$ ft separated by $L = 2$ ft are located directly on top of each other. The disks are separated by a radiation shield whose emissivity is 0.15. Both disks are black and are maintained at temperatures of 1200 R and 700 R, respectively. The environment that the disks are in can be considered to be a blackbody at $T_\infty = 540$ R. Determine the net rate of radiation heat transfer through the shield under steady conditions.   *Answer:* 799 Btu/h

FIGURE P9-96E

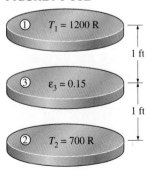

$T_1 = 1200$ R

1 ft

$\varepsilon_3 = 0.15$

1 ft

$T_2 = 700$ R

9-97 A radiation shield that has the same emissivity ε_3 on both sides is placed between two large parallel plates, which are maintained at uniform temperatures of $T_1 = 650$ K and $T_2 = 400$ K and have emissivities of $\varepsilon_1 = 0.6$ and $\varepsilon_2 = 0.9$, respectively. Determine the emissivity of the radiation shield if the radiation heat transfer between the plates is to be reduced to 15 percent of that without the radiation shield.

9-98 Two coaxial cylinders of diameters $D_1 = 0.10$ m and $D_2 = 0.30$ m and emissivities $\varepsilon_1 = 0.7$ and $\varepsilon_2 = 0.4$ are maintained at uniform temperatures of $T_1 = 750$ K and $T_2 = 500$ K, respectively. Now a coaxial radiation shield of diameter $D_3 = 0.20$ m and emissivity $\varepsilon_3 = 0.2$ is placed between the two cylinders. Determine the net rate of radiation heat transfer between the two cylinders per unit length of the cylinders and compare the result with that without the shield.

Review Problems

9-99 A thermocouple used to measure the temperature of hot air flowing in a duct whose walls are maintained at $T_w = 500$ K shows a temperature reading of $T_{th} = 850$ K. Assuming the emissivity of the thermocouple junction to be $\varepsilon = 0.6$ and the convection heat transfer coefficient to be $h = 60$ W/m² · °C, determine the actual temperature of the air.
  Answer: 1111 K

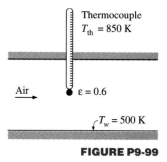

FIGURE P9-99

9-100 A thermocouple shielded by aluminum foil of emissivity 0.1 is used to measure the temperature of hot gases flowing in a duct whose walls are maintained at $T_w = 380$ K. The thermometer shows a temperature reading of $T_{th} = 530$ K. Assuming the emissivity of the thermocouple junction to be $\varepsilon = 0.8$ and the convection heat transfer coefficient to be $h = 120$ W/m² · °C, determine the actual temperature of the gas. What would the thermometer reading be if no radiation shield was used?

9-101 The spectral emissivity function of an opaque surface at 1200 K is approximated as

$$\varepsilon_\lambda = \begin{cases} \varepsilon_1 = 0.0, & 0 \le \lambda < 2 \ \mu m \\ \varepsilon_2 = 0.8, & 2 \ \mu m \le \lambda < 6 \ \mu m \\ \varepsilon_3 = 0.6, & 6 \ \mu m \le \lambda < \infty \end{cases}$$

Determine the average emissivity of the surface and the rate of radiation emission from the surface, in W/m².

9-102E Consider a sealed 8-in.-high electronic box whose base dimensions are 12 in. × 12 in. placed in a vacuum chamber. The emissivity of the outer surface of the box is 0.95. If the electronic components in the box dissipate a total of 100 W of power and the outer surface temperature of the box is not to exceed 130°F, determine the highest temperature at which the surrounding surfaces must be kept if this box is to be cooled by radiation alone. Assume the heat transfer from the bottom surface of the box to the stand to be negligible. *Answer:* 43°F

FIGURE P9-102E

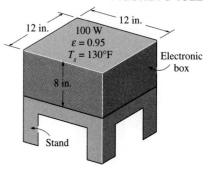

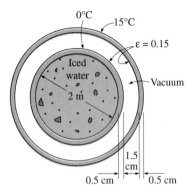

FIGURE P9-103

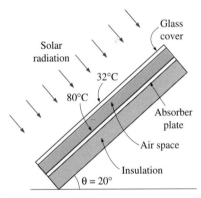

FIGURE P9-105

FIGURE P9-107

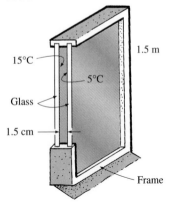

9-103 A 2-m-internal-diameter double-walled spherical tank is used to store iced water at 0°C. Each wall is 0.5 cm thick, and the 1.5-cm-thick air space between the two walls of the tank is evacuated in order to minimize heat transfer. The surfaces surrounding the evacuated space are polished so that each surface has an emissivity of 0.15. The temperature of the outer wall of the tank is measured to be 15°C. Assuming the inner wall of the steel tank to be at 0°C, determine (a) the rate of heat transfer to the iced water in the tank and (b) the amount of ice at 0°C that melts during a 24-h period.

9-104 Two concentric spheres of diameters $D_1 = 15$ cm and $D_2 = 25$ cm are separated by air at 1 atm pressure. The surface temperatures of the two spheres enclosing the air are $T_1 = 350$ K and $T_2 = 275$ K, respectively, and their emissivities are 0.5. Determine the rate of heat transfer from the inner sphere to the outer sphere by (a) natural convection and (b) radiation.

9-105 Consider a 1.5-m-high and 3-m-wide solar collector that is tilted at an angle 20° from the horizontal. The distance between the glass cover and the absorber plate is 3 cm, and the back side of the absorber is heavily insulated. The absorber plate and the glass cover are maintained at temperatures of 80°C and 32°C, respectively. The emissivity of the glass surface is 0.9 and that of the absorber plate is 0.8. Determine the rate of heat loss from the absorber plate by natural convection and radiation.

Answers: 713 W, 1289 W

9-106E A solar collector consists of a horizontal aluminum tube having an outer diameter of 2.5 in. enclosed in a concentric thin glass tube of diameter 5 in. Water is heated as it flows through the tube, and the annular space between the aluminum and the glass tube is filled with air at 0.5 atm pressure. The pump circulating the water fails during a clear day, and the water temperature in the tube starts rising. The aluminum tube absorbs solar radiation at a rate of 30 Btu/h per foot length, and the temperature of the ambient air outside is 75°F. The emissivities of the tube and the glass cover are 0.9. Taking the effective sky temperature to be 60°F, determine the temperature of the aluminum tube when thermal equilibrium is established (i.e, when the rate of heat loss from the tube equals the amount of solar energy gained by the tube).

9-107 A vertical 1.5-m-high and 3-m-wide double-pane window consists of two sheets of glass separated by a 1.5-cm-thick air gap. In order to reduce heat transfer through the window, the air space between the two glasses is partially evacuated to 0.3 atm pressure. The emissivities of the glass surfaces are 0.9. Taking the glass surface temperatures across the air gap to be 15°C and 5°C, determine the rate of heat transfer through the window by natural convection and radiation.

9-108 A simple solar collector is built by placing a 6-cm-diameter clear plastic tube around a garden hose whose outer diameter is 2 cm. The hose is painted black to maximize solar absorption, and some plastic rings are used to keep the spacing between the hose and the clear plastic cover constant. The emissivities of the hose surface and the glass cover are 0.9, and the

effective sky temperature is estimated to be 15°C. The temperature of the plastic tube is measured to be 40°C, while the ambient air temperature is 25°C. Determine the rate of heat loss from the water in the hose by natural convection and radiation per meter of its length under steady conditions.
    Answers: 5.9 W, 26.2 W

9-109 A solar collector consists of a horizontal copper tube of outer diameter 5 cm enclosed in a concentric thin glass tube of diameter 9 cm. Water is heated as it flows through the tube, and the annular space between the copper and the glass tubes is filled with air at 1 atm pressure. The emissivities of the tube surface and the glass cover are 0.85 and 0.9, respectively. During a clear day, the temperatures of the tube surface and the glass cover are measured to be 60°C and 32°C, respectively. Determine the rate of heat loss from the collector by natural convection and radiation per meter length of the tube.

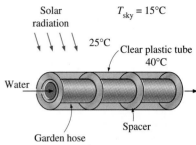

FIGURE P9-108

Computer, Design, and Essay Problems

9-110 Consider an enclosure consisting of N diffuse and gray surfaces. The emissivity and temperature of each surface as well as all the view factors between the surfaces are specified. Using the matrix method and obtaining a subroutine for solving a system of N linear equations simultaneously, write a program to determine the net rate of radiation heat transfer for each surface.

9-111 Radiation shields are commonly used in the design of superinsulations for use in space and cryogenic applications. Write an essay on superinsulations and how they are used in different applications.

9-112 Thermal comfort in a house is strongly affected by the so-called radiation effect, which is due to radiation heat transfer between the person and surrounding surfaces. A person feels much colder in the morning, for example, because of the lower surface temperature of the walls at that time, although the thermostat setting of the house is fixed. Write an essay on the radiation effect, how it affects human comfort, and how it is accounted for in heating and air-conditioning applications.

Heat Exchangers

10

Heat exchangers are devices that facilitate the *exchange of heat* between *two fluids* that are at different temperatures while keeping them from mixing with each other. Heat exchangers are commonly used in practice in a wide range of applications, from heating and air-conditioning systems in a household, to chemical processing and power production in large plants. Heat exchangers differ from mixing chambers in that they do not allow the two fluids involved to mix. In a car radiator, for example, heat is transferred from the hot water flowing through the radiator tubes to the air flowing through the closely spaced thin plates outside attached to the tubes.

Heat transfer in a heat exchanger usually involves *convection* in each fluid and *conduction* through the wall separating the two fluids. In the analysis of heat exchangers, it is convenient to work with an *overall heat transfer coefficient U* that accounts for the contribution of all these effects on heat transfer. The rate of heat transfer between the two fluids at a location in a heat exchanger depends on the magnitude of the temperature difference at that location, which varies along the heat exchanger. In the analysis of heat exchangers, it is usually convenient to work with the *logarithmic mean temperature difference LMTD,* which is an equivalent mean temperature difference between the two fluids for the entire heat exchanger.

Heat exchangers are manufactured in a variety of types, and thus we start this chapter with the *classification* of heat exchangers. We then discuss the determination of the overall heat transfer coefficient in heat exchangers, and the logarithmic mean temperature difference LMTD for some configurations. We then introduce the *correction factor F* to account for the deviation of the mean temperature difference from the LMTD in complex configurations. Next we discuss the effectiveness-NTU method, which enables us to analyze heat exchangers when the outlet temperatures of the fluids are not known. Finally, we discuss the selection of heat exchangers.

10-1 ■ TYPES OF HEAT EXCHANGERS

Different heat transfer applications require different types of hardware and different configurations of heat transfer equipment. The attempt to match the heat transfer hardware to the heat transfer requirements within the specified constraints has resulted in numerous types of innovative heat exchanger designs.

The simplest type of heat exchanger consists of two concentric pipes of different diameters, as shown in Figure 10-1, called the **double-pipe** heat exchanger. One fluid in a double-pipe heat exchanger flows through the smaller pipe while the other fluid flows through the annular space between the two pipes. Two types of flow arrangement are possible in a double-pipe heat exchanger: in **parallel flow,** both the hot and cold fluids enter the heat exchanger at the same end and move in the *same* direction. In **counter flow,** on the other hand, the hot and cold fluids enter the heat exchanger at opposite ends and flow in *opposite* directions.

Another type of heat exchanger, which is specifically designed to realize a large heat transfer surface area per unit volume, is the **compact** heat exchanger. The ratio of the heat transfer surface area of a heat exchanger to its

FIGURE 10-1

Different flow regimes and associated temperature profiles in a double-pipe heat exchanger.

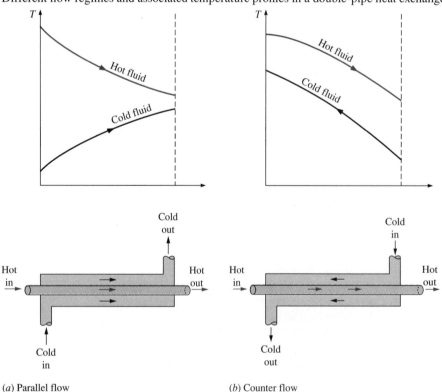

(*a*) Parallel flow   (*b*) Counter flow

volume is called the *area density* β. A heat exchanger with $\beta > 700$ m²/m³ (or 200 ft²/ft³) is classified as being compact. Examples of compact heat exchangers are car radiators ($\beta \approx 1000$ m²/m³), glass ceramic gas turbine heat exchangers ($\beta \approx 6000$ m²/m³), the regenerator of a Stirling engine ($\beta \approx 15,000$ m²/m³), and the human lung ($\beta \approx 20,000$ m²/m³). Compact heat exchangers enable us to achieve high heat transfer rates between two fluids in a small volume, and they are commonly used in applications with strict limitations on the weight and volume of heat exchangers (Fig. 10-2).

The large surface area in compact heat exchangers is obtained by attaching closely spaced *thin plate* or *corrugated fins* to the walls separating the two fluids. Compact heat exchangers are commonly used in gas-to-gas and gas-to-liquid (or liquid-to-gas) heat exchangers to counteract the low heat transfer coefficient associated with gas flow with increased surface area. In a car radiator, which is a water-to-air compact heat exchanger, for example, it is no surprise that fins are attached to the air side of the tube surface.

In compact heat exchangers, the two fluids usually move *perpendicular* to each other, and such flow configuration is called **cross-flow.** The cross-flow is further classified as *unmixed* and *mixed flow,* depending on the flow configuration, as shown in Figure 10-3. In (*a*) the cross-flow is said to be *unmixed* since the plate fins force the fluid to flow through a particular interfin spacing and prevent it from moving in the transverse direction (i.e., parallel to the tubes). The cross-flow in (*b*) is said to be *mixed* since the fluid now is free to move in the transverse direction. Both fluids are unmixed in a car radiator. The presence of mixing in the fluid can have a significant effect on the heat transfer characteristics of the heat exchanger.

Perhaps the most common type of heat exchanger in industrial applications in the **shell-and-tube** heat exchanger, shown in Figure 10-4. Shell-and-

FIGURE 10-2

A gas-to-liquid compact heat exchanger for a residential air-conditioning system.

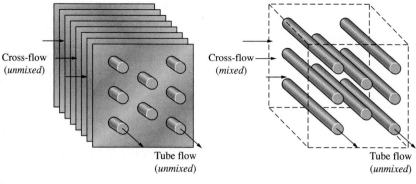

FIGURE 10-3

Different flow configurations in cross-flow heat exchangers.

(*a*) Both fluids unmixed

(*b*) One fluid mixed, one fluid unmixed

tube heat exchangers contain a large number of tubes (sometimes several hundred) packed in a shell with their axes parallel to that of the shell. Heat transfer takes place as one fluid flows inside the tubes while the other fluid flows outside the tubes through the shell. *Baffles* are commonly placed in the shell to force the shell-side fluid to flow across the shell to enhance heat transfer and to maintain uniform spacing between the tubes. Despite their widespread use, shell-and-tube heat exchangers are not suitable for use in automotive, aircraft, and marine applications because of their relatively large size and weight. Note that the tubes in a shell-and-tube heat exchanger open to some large flow areas called *headers* at both ends of the shell, where the tube-side fluid accumulates before entering the tubes and after leaving them.

Shell-and-tube heat exchangers are further classified according to the number of shell and tube passes involved. Heat exchangers in which all the tubes make one U-turn in the shell, for example, are called *one-shell pass and two-tube pass* heat exchangers. Likewise, a heat exchanger that involves two passes in the shell and four passes in the tubes is called a *two-shell pass and four-tube pass* heat exchanger (Fig. 10-5).

An innovative type of heat exchanger that has found widespread use is the **plate and frame** (or just plate) heat exchanger, which consists of a series

FIGURE 10-4

The schematic of a shell-and-tube heat exchanger (one-shell pass and one-tube pass).

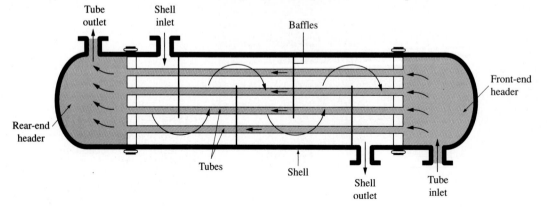

of plates with corrugated flat flow passages (Fig. 10-6). The hot and cold fluids flow in alternate passages, and thus each cold fluid stream is surrounded by two hot fluid streams, resulting in very effective heat transfer. Also, plate heat exchangers can grow with increasing demand for heat transfer by simply mounting more plates. They are well suited for liquid-to-liquid heat exchange applications, provided that the hot and cold fluid streams are at about the same pressure.

Another type of heat exchanger that involves the alternate passage of the hot and cold fluid streams through the same flow area is the **regenerative** heat exchanger. The *static*-type regenerative heat exchanger is basically a porous mass that has a large heat storage capacity, such as a ceramic wire mesh. Hot and cold fluids flow through this porous mass alternatively. Heat is transferred from the hot fluid to the matrix of the regenerator during the flow of the hot fluid, and from the matrix to the cold fluid during the flow of the cold fluid. Thus, the matrix serves as a temporary heat storage medium.

The *dynamic*-type regenerator involves a rotating drum and continuous flow of the hot and cold fluid through different portions of the drum so that

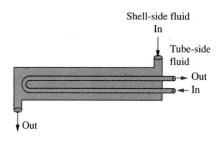

(a) One-shell pass and two-tube passes

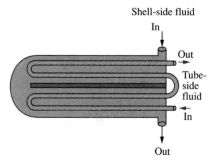

(b) Two-shell passes and four-tube passes

FIGURE 10-5

Multipass flow arrangements in shell-and-tube heat exchangers.

FIGURE 10-6

A plate-and-frame liquid-to-liquid heat exchanger (courtesy of Trante Corp.).

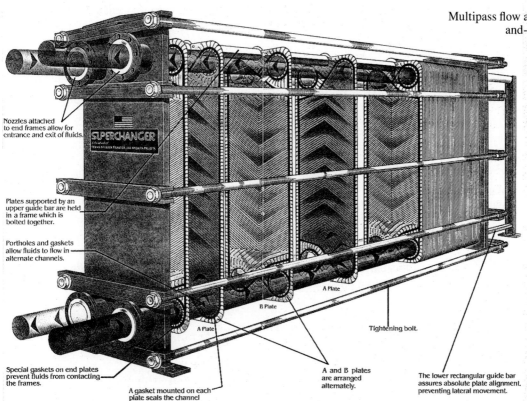

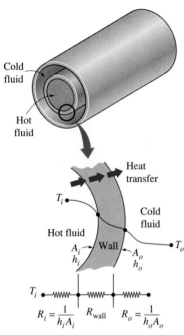

FIGURE 10-7

Thermal resistance network associated with heat transfer in a double-pipe heat exchanger.

FIGURE 10-8

The two heat transfer surface areas associated with a double-pipe heat exchanger (for thin tubes, $D_i \approx D_o$ and thus $A_i \approx A_o$).

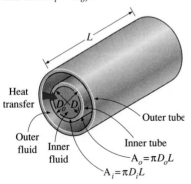

any portion of the drum passes periodically through the hot stream, storing heat, and then through the cold stream, *rejecting* this *stored* heat. Again the drum serves as the medium to transport the heat from the hot to the cold fluid stream.

Heat exchangers are often given specific names to reflect the specific application for which they are used. For example, a *condenser* is a heat exchanger in which one of the fluids gives up heat and condenses as it flows through the heat exchanger. A *boiler* is another heat exchanger in which one of the fluids absorbs heat and vaporizes. A *space radiator* is a heat exchanger that transfers heat from the hot fluid to the surrounding space by radiation.

10-2 ■ THE OVERALL HEAT TRANSFER COEFFICIENT

A heat exchanger typically involves two flowing fluids separated by a solid wall. Heat is first transferred from the hot fluid to the wall by *convection*, through the wall by *conduction*, and from the wall to the cold fluid again by *convection*. Any radiation effects are usually included in the convection heat transfer coefficients.

The thermal resistance network associated with this heat transfer process involves two convection and one conduction resistances, as shown in Figure 10-7. Here the subscripts i and o represent the inner and outer surfaces of the inner tube. For a double-pipe heat exchanger, we have $A_i = \pi D_i L$ and $A_o = \pi D_o L$, and the *thermal resistance* of the tube wall in this case is

$$R_{wall} = \frac{\ln (D_o/D_i)}{2\pi k L} \quad (10\text{-}1)$$

where k is the thermal conductivity of the wall material and L is the length of the tube. Then the *total thermal resistance* becomes

$$R = R_{total} = R_i + R_{wall} + R_o = \frac{1}{h_i A_i} + \frac{\ln (D_o/D_i)}{2\pi k L} + \frac{1}{h_o A_o} \quad (10\text{-}2)$$

The A_i is the area of the *inner surface* of the wall that separates the two fluids, and A_o is the area of the *outer surface* of the wall. In other words, A_i and A_o are surface areas of the separating wall wetted by the inner and the outer fluids, respectively. When one fluid flows inside a circular tube and the other outside of it, we have $A_i = \pi D_i L$ and $A_o = \pi D_o L$ (Fig. 10-8).

In the analysis of heat exchangers, it is convenient to combine all the thermal resistances in the path of heat flow from the hot fluid to the cold one into a single resistance R, as we discussed in Chapter 3, and to express the rate of heat transfer between the two fluids as

$$\dot{Q} = \frac{\Delta T}{R} = UA\Delta T = U_i A_i \Delta T = U_o A_o \Delta T \quad (10\text{-}3)$$

where U is the **overall heat transfer coefficient**, whose unit is W/m² · °C, which is identical to the unit of the ordinary convection coefficient h. Canceling ΔT, Equation 10-3 reduces to

$$\frac{1}{UA} = \frac{1}{U_i A_i} = \frac{1}{U_o A_o} = R = \frac{1}{h_i A_i} + R_{\text{wall}} + \frac{1}{h_o A_o} \qquad (10\text{-}4)$$

Perhaps you are wondering why we have *two* overall heat transfer coefficients U_i and U_o for a heat exchanger. The reason is that every heat exchanger has two heat transfer surface areas A_i and A_o, which, in general, are not equal to each other.

Note that $U_i A_i = U_o A_o$, but $U_i \neq U_o$ unless $A_i = A_o$. Therefore, the overall heat transfer coefficient U of a heat exchanger is meaningless unless the area on which it is based is specified. This is especially the case when one side of the tube wall is finned and the other side is not, since the surface area of the finned side is several times that of the unfinned side.

When the wall thickness of the tube is small and the thermal conductivity of the tube material is high, as is usually the case, the thermal resistance of the tube is negligible ($R_{\text{wall}} \approx 0$) and the inner and outer surfaces of the tube are almost identical ($A_i \approx A_o \approx A$). Then Equation 10-4 for the overall heat transfer coefficient simplifies to

$$\frac{1}{U} \approx \frac{1}{h_i} + \frac{1}{h_o} \qquad (10\text{-}5)$$

where $U \approx U_i \approx U_o$. The individual convection heat transfer coefficients inside and outside the tube, h_i and h_o, are determined using the convection relations discussed in earlier chapters.

The overall heat transfer coefficient U in Equation 10-5 is dominated by the *smaller* convection coefficient, since the inverse of a large number is small. When one of the convection coefficients is *much smaller* than the other (say, $h_i \ll h_o$), we have $1/h_i \gg 1/h_o$, and thus $U \approx h_i$. Therefore, the smaller heat transfer coefficient creates a *bottleneck* on the path of heat flow and seriously impedes heat transfer. This situation arises frequently when one of the fluids is a gas and the other is a liquid. In such cases, fins are commonly used on the gas side to enhance the product UA and thus the heat transfer on that side.

Representative values of the overall heat transfer coefficient U are given in Table 10-1. Note that the overall heat transfer coefficient ranges from about 10 W/m$^2 \cdot$ °C for gas-to-gas heat exchangers to about 10,000 W/m$^2 \cdot$ °C for heat exchangers that involve phase changes. This is not surprising, since gases have very low thermal conductivities, and phase-change processes involve very high heat transfer coefficients.

When the tube is *finned* on one side to enhance heat transfer, the total heat transfer surface area on the fined side becomes

$$A = A_{\text{total}} = A_{\text{fin}} + A_{\text{unfinned}} \qquad (10\text{-}6)$$

where A_{fin} is the surface area of the fins and A_{unfinned} is the area of unfinned portion of the tube surface. For short fins of high thermal conductivity, we can use this total area in the convection resistance relation $R_{\text{conv}} = 1/hA$ since the fins in this case will be very nearly isothermal. Otherwise, we should determine the effective surface are A from

TABLE 10-1

Representative values of the overall heat transfer coefficients in heat exchangers

Type of heat exchanger	U, W/m² · °C*
Water-to-water	850–1700
Water-to-oil	100–350
Water-to-gasoline or kerosene	300–1000
Feedwater heaters	1000–8500
Steam-to-light fuel oil	200–400
Steam-to-heavy fuel oil	50–200
Steam condenser	1000–6000
Freon condenser (water cooled)	300–1000
Ammonia condenser (water cooled)	800–1400
Alcohol condensers (water cooled)	250–700
Gas-to-gas	10–40
Water-to-air in finned tubes (water in tubes)	30–60[†]
	400–850[†]
Steam-to-air in finned tubes (steam in tubes)	30–300[†]
	400–4000[‡]

*Multiply the listed values by 0.176 to convert them to Btu/h · ft² · °F.

[†]Based on air-side surface area.

[‡]Based on water- or steam-side surface area.

$$A = A_{\text{unfinned}} + \eta_{\text{fin}} A_{\text{fin}} \qquad (10\text{-}7)$$

where η_{fin} is the fin efficiency. This way, the temperature drop along the fins is accounted for. Note that $\eta_{\text{fin}} = 1$ for isothermal fins, and thus Equation 10-7 reduces to Equation 10-6 in that case.

Fouling Factor

The performance of heat exchangers usually deteriorates with time as a result of accumulation of *deposits* on heat transfer surfaces. The layer of deposits represents *additional resistance* to heat transfer and causes the rate of heat transfer in a heat exchanger to decrease. The net effect of these accumulations on heat transfer is represented by a **fouling factor** R_f, which is a measure of the *thermal resistance* introduced by fouling.

The most common type of fouling is the *precipitation* of solid deposits in a fluid on the heat transfer surfaces. You can observe this type of fouling even in your house. If you check the inner surfaces of your teapot after prolonged use, you will probably notice a layer of calcium-based deposits on the surfaces at which boiling occurs. This is especially the case in areas where the water is hard. The scales of such deposits come off by scratching, and the surfaces can be cleaned of such deposits by chemical treatment. Now imagine

FIGURE 10-9

Precipitation fouling of ash particles on superheater tubes (from *Steam, Its Generation, and Use,* Babcock and Wilcox Co., 1978).

those mineral deposits forming on the inner surfaces of fine tubes in a heat exchanger (Fig. 10-9) and the detrimental effect it may have on the flow passage area and the heat transfer. To avoid this potential problem, water in power and process plants is extensively treated and its solid contents are removed before it is allowed to circulate through the system. The solid ash particles in the flue gases accumulating on the surfaces of air preheaters create similar problems.

Another form of fouling, which is common in the chemical process industry, is *corrosion* and other *chemical fouling*. In this case, the surfaces are fouled by the accumulation of the products of chemical reactions on the surfaces. This form of fouling can be avoided by coating metal pipes with glass or using plastic pipes instead of metal ones. Heat exchangers may also be fouled by the growth of algae in warm fluids. This type of fouling is called *biological fouling* and can be prevented by chemical treatment.

In applications where it is likely to occur, fouling should be considered in the design and selection of heat exchangers. In such applications, it may be necessary to select a larger and thus more expensive heat exchanger to ensure that it meets the design heat transfer requirements even after fouling occurs. The periodic cleaning of heat exchangers and the resulting down time are additional penalties associated with fouling.

The fouling factor is obviously zero for a new heat exchanger and increases with time as the solid deposits build up on the heat exchanger surface. The fouling factor depends on the *operating temperature* and the *velocity* of the fluids, as well as the length of service. Fouling increases with *increasing temperature* and *decreasing velocity*.

The overall heat transfer coefficient relation given above is valid for clean surfaces and needs to be modified to account for the effects of fouling on both the inner and the outer surfaces of the tube. For an unfinned shell-and-tube heat exchanger, it can be expressed as

$$\frac{1}{UA} = \frac{1}{U_iA_i} = \frac{1}{U_oA_o} = R = \frac{1}{h_iA_i} + \frac{R_{f,i}}{A_i} + \frac{\ln(D_o/D_i)}{2\pi kL} + \frac{R_{f,o}}{A_o} + \frac{1}{h_oA_o} \quad (10\text{-}8)$$

where $A_i = \pi D_i L$ and $A_o = \pi D_o L$ are the areas of inner and outer surfaces, and R_{fi} and R_{fo} are the fouling factors at those surfaces.

Representative values of fouling factors are given in Table 10-2. More comprehensive tables of fouling factors are available in handbooks. As you would expect, considerable uncertainty exists in these values, and they should be used as a guide in the selection and evaluation of heat exchangers to account for the effects of anticipated fouling on heat transfer. Note that most fouling factors in the table are of the order of 10^{-4} $m^2 \cdot °C/W$, which is equivalent to the thermal resistance of a 0.2-mm-thick limestone layer ($k = 2.9$ W/m $\cdot$ °C) per unit surface area. Therefore, in the absence of specific data, we can assume the surfaces to be coated with 0.2 mm of limestone as a starting point to account for the effects of fouling.

TABLE 10-2

Representative fouling factors (thermal resistance due to fouling for a unit surface area)

Fluid	R_f, $m^2 \cdot °C/W$
Distilled water, sea water, river water, boiler feedwater:	
Below 50°C	0.0001
Above 50°C	0.0002
Fuel oil	0.0009
Steam (oil-free)	0.0001
Refrigerants (liquid)	0.0002
Refrigerants (vapor)	0.0004
Alcohol vapors	0.0001
Air	0.0004

Source: Tubular Exchange Manufacturers Association.

EXAMPLE 10-1   Overall Heat Transfer Coefficient of a Heat Exchanger

Hot oil is to be cooled in a double-tube counter-flow heat exchanger. The copper inner tube has a diameter of 2 cm and negligible thickness. The inner diameter of the outer tube (the shell) is 3 cm. Water flows through the tube at a rate of 0.5 kg/s, and the oil through the shell at a rate of 0.8 kg/s. Taking the average temperatures of the water and the oil to be 45°C and 80°C, respectively, determine the overall heat transfer coefficient of this heat exchanger.

Solution   Hot oil is cooled by water in a double-tube counterflow heat exchanger. The overall heat transfer coefficient is to be determined.

Assumptions   **1** The thermal resistance of the inner tube is negligible since the tube material is highly conductive and its thickness is negligible. **2** Both the oil and water flow are fully developed. **3** Properties of the oil and water are constant.

Properties   The properties of water at 45°C are (Table A-9)

$$\rho = 990 \text{ kg/m}^3 \qquad Pr = 3.91$$
$$k = 0.637 \text{ W/m} \cdot °C \quad \nu = \mu/\rho = 0.602 \times 10^{-6} \text{ m}^2/s$$

The properties of oil at 80°C are (Table A-10).

$$\rho = 852 \text{ kg/m}^3 \qquad Pr = 490$$
$$k = 0.138 \text{ W/m} \cdot °C \quad \nu = 37.5 \times 10^{-6} \text{ m}^2/s$$

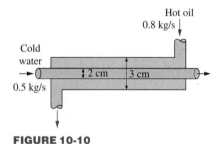

FIGURE 10-10
Schematic for Example 10-1.

Analysis   The schematic of the heat exchanger is given in Figure 10-10. The overall heat transfer coefficient U can be determined from Equation 10-5:

$$\frac{1}{U} \approx \frac{1}{h_i} + \frac{1}{h_o}$$

where h_i and h_o are the convection heat transfer coefficients inside and outside the tube, respectively, which are to be determined using the forced convection relations discussed in Chapter 6.

The hydraulic diameter for a circular tube is the diameter of the tube itself, $D_h = D = 0.02$ m. The average velocity of water in the tube and the Reynolds number are

$$\mathcal{V}_m = \frac{\dot{m}}{\rho A_c} = \frac{\dot{m}}{\rho(\frac{1}{4}\pi D^2)} = \frac{0.5 \text{ kg/s}}{(990 \text{ kg/m}^3)[\frac{1}{4}\pi(0.02 \text{ m})^2]} = 1.61 \text{ m/s}$$

and

$$\text{Re} = \frac{\mathcal{V}_m D_h}{\nu} = \frac{(1.61 \text{ m/s})(0.02 \text{ m})}{0.602 \times 10^{-6} \text{ m}^2/\text{s}} = 53{,}490$$

which is greater than 4000. Therefore, the flow of water is turbulent. Assuming the flow to be fully developed, the Nusselt number can be determined from

$$\text{Nu} = \frac{hD_h}{k} = 0.023 \text{ Re}^{0.8}\text{Pr}^{0.4} = 0.023(53{,}490)^{0.8}(3.91)^{0.4} = 240.6$$

Then,

$$h = \frac{k}{D_h}\text{Nu} = \frac{0.637 \text{ W/m} \cdot {}^\circ\text{C}}{0.02 \text{ m}}(240.6) = 7663 \text{ W/m}^2 \cdot {}^\circ\text{C}$$

Now we repeat the analysis above for oil. The properties of oil at 80°C are

$$\rho = 852 \text{ kg/m}^3 \qquad \nu = 37.5 \times 10^{-6} \text{ m}^2/\text{s}$$
$$k = 0.138 \text{ W/m} \cdot {}^\circ\text{C} \qquad \text{Pr} = 490$$

The hydraulic diameter for the annular space is

$$D_h = D_o - D_i = 0.03 - 0.02 = 0.01\text{m}$$

The average velocity and the Reynolds number in this case are

$$\mathcal{V}_m = \frac{\dot{m}}{\rho A_c} = \frac{\dot{m}}{\rho[\frac{1}{4}\pi(D_o^2 - D_i^2)]} = \frac{0.8 \text{ kg/s}}{(852 \text{ kg/m}^3)[\frac{1}{4}\pi(0.03^2 - 0.02^2)] \text{ m}^2} = 2.39 \text{ m/s}$$

and

$$\text{Re} = \frac{\mathcal{V}_m D_h}{\nu} = \frac{(2.39 \text{ m/s})(0.01 \text{ m})}{37.5 \times 10^{-6} \text{ m}^2/\text{s}} = 637$$

which is less than 4000. Therefore, the flow of oil is laminar. Assuming fully developed flow, the Nusselt number on the tube side of the annular space Nu_i corresponding to $D_i/D_o = 0.02/0.03 = 0.667$ can be determined from Table 10-3 by interpolation to be

$$\text{Nu} = 5.45$$

and

$$h_o = \frac{k}{D_h}\text{Nu} = \frac{0.138 \text{ W/m} \cdot {}^\circ\text{C}}{0.01 \text{ m}}(5.45) = 75.2 \text{ W/m}^2 \cdot {}^\circ\text{C}$$

Then the overall heat transfer coefficient for this heat exchanger becomes

$$U = \frac{1}{\dfrac{1}{h_i} + \dfrac{1}{h_o}} = \frac{1}{\dfrac{1}{7663 \text{ W/m}^2 \cdot {}^\circ\text{C}} + \dfrac{1}{75.2 \text{ W/m}^2 \cdot {}^\circ\text{C}}} = \textbf{74.5 W/m}^2 \cdot {}^\circ\textbf{C}$$

Discussion Note that $U \approx h_o$ in this case, since $h_i \gg h_o$. This confirms our earlier statement that the overall heat transfer coefficient in a heat exchanger is

TABLE 10-3

Nusselt number for fully developed laminar flow in a circular annulus with one surface insulated and the other isothermal

D_i/D_o	Nu_i	Nu_o
0.00	—	3.66
0.05	17.46	4.06
0.10	11.56	4.11
0.25	7.37	4.23
0.50	5.74	4.43
1.00	4.86	4.86

Source: Kays and Perkins, Ref. 8.

dominated by the smaller heat transfer coefficient when the difference between the two values is large.

To improve the overall heat transfer coefficient and thus the heat transfer in this heat exchanger, we must use some enhancement techniques on the oil side, such as a finned surface.

EXAMPLE 10-2   Effect of Fouling on the Overall Heat Transfer Coefficient

A double-pipe (shell-and-tube) heat exchanger is constructed of a stainless steel ($k = 15.1$ W/m $\cdot$ °C) inner tube of inner diameter $D_i = 1.5$ cm and outer diameter $D_o = 1.9$ cm and an outer shell of inner diameter 3.2 cm. The convection heat transfer coefficient is given to be $h_i = 800$ W/m$^2 \cdot$ °C on the inner surface of the tube and $h_o = 1200$ W/m$^2 \cdot$ °C on the outer surface. For a fouling factor of $R_{f,i} = 0.0004$ m$^2 \cdot$ °C/W on the tube side and $R_{f,o} = 0.0001$ m$^2 \cdot$ °C/W on the shell side, determine (a) the thermal resistance of the heat exchanger per unit length and (b) the overall heat transfer coefficients, U_i and U_o based on the inner and outer surface areas of the tube, respectively.

Solution   The heat transfer coefficients and the fouling factors on tube and shell side of a heat exchanger are given. The thermal resistance and the overall heat transfer coefficients based on the inner and outer areas are to be determined.

Assumptions   The heat transfer coefficients and the fouling factors are constant and uniform.

Analysis   (a) The schematic of the heat exchanger is given in Figure 10-11. The thermal resistance for an unfinned shell-and-tube heat exchanger with fouling on both heat transfer surfaces is given by Equation 10-8 as

$$R = \frac{1}{UA} = \frac{1}{U_i A_i} = \frac{1}{U_o A_o} = \frac{1}{h_i A_i} + \frac{R_{f,i}}{A_i} + \frac{\ln (D_o/D_i)}{2\pi kL} + \frac{R_{f,o}}{A_o} + \frac{1}{h_o A_o}$$

where

$$A_i = \pi D_i L = \pi(0.015 \text{ m})(1 \text{ m}) = 0.0471 \text{ m}^2$$
$$A_o = \pi D_o L = \pi(0.019 \text{ m})(1 \text{ m}) = 0.0597 \text{ m}^2$$

Substituting, the total thermal resistance is determined to be

$$R = \frac{1}{(800 \text{ W/m}^2 \cdot \text{°C})(0.0471 \text{ m}^2)} + \frac{0.0004 \text{ m}^2 \cdot \text{°C/W}}{0.0471 \text{ m}^2}$$

$$+ \frac{\ln (0.019/0.015)}{2\pi(15.1 \text{ W/m} \cdot \text{°C})(1 \text{ m})}$$

$$+ \frac{0.0001 \text{ m}^2 \cdot \text{°C/W}}{0.0597 \text{ m}^2} + \frac{1}{(1200 \text{ W/m}^2 \cdot \text{°C})(0.0597 \text{ m}^2)}$$

$$= (0.02654 + 0.00849 + 0.0025 + 0.00168 + 0.01396)\text{°C/W}$$

$$= \mathbf{0.0532°C/W}$$

Discussion   Note that about 19 percent of the total thermal resistance in this case is due to fouling and about 5 percent of it is due to the steel tube separating the two fluids. The rest (76 percent) is due to the convection resistances on the two sides of the inner tube.

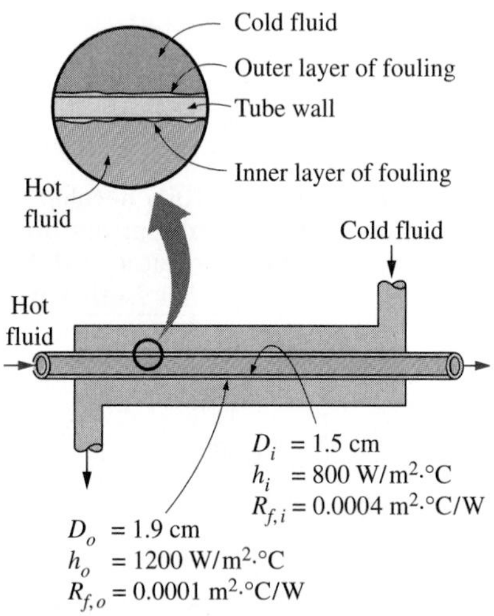

Cold fluid
Outer layer of fouling
Tube wall
Inner layer of fouling

Hot fluid
Cold fluid

Hot fluid

$D_i = 1.5$ cm
$h_i = 800$ W/m$^2 \cdot$ °C
$R_{f,i} = 0.0004$ m$^2 \cdot$ °C/W

$D_o = 1.9$ cm
$h_o = 1200$ W/m$^2 \cdot$ °C
$R_{f,o} = 0.0001$ m$^2 \cdot$ °C/W

FIGURE 10-11

Schematic for Example 10-2.

(b) Knowing the total thermal resistance and the heat transfer surface areas, the overall heat transfer coefficient based on the inner and outer surfaces of the tube are determined again from Equation 10-8 to be

$$U_i = \frac{1}{RA_i} = \frac{1}{(0.0532 \text{ °C/W})(0.0471 \text{ m}^2)} = \textbf{399.1 W/m}^2 \cdot \textbf{°C}$$

and

$$U_o = \frac{1}{RA_o} = \frac{1}{(0.0532 \text{ °C/W})(0.0597 \text{ m}^2)} = \textbf{314.9 W/m}^2 \cdot \textbf{°C}$$

Discussion Note that the two overall heat transfer coefficients differ significantly (by 27 percent) in this case because of the considerable difference between the heat transfer surface areas on the inner and the outer sides of the tube. For tubes of negligible thickness, the difference between the two overall heat transfer coefficients would be negligible.

10-3 ■ ANALYSIS OF HEAT EXCHANGERS

Heat exchangers are commonly used in practice, and an engineer often finds himself or herself in a position to *select a heat exchanger* that will achieve a *specified temperature change* in a fluid stream of known mass flow rate, or to *predict the outlet temperatures* of the hot and cold fluid streams in a *specified heat exchanger.*

In the following sections, we will discuss the two methods used in the analysis of heat exchangers. Of these, the *log mean temperature difference* (or LMTD) method is best suited for the first task and the *effectiveness-NTU* method for the second task stated above. But first we present some general considerations.

Heat exchangers usually operate for long periods of time with no change in their operating conditions. Therefore, they can be modeled as *steady-flow* devices. As such, the mass flow rate of each fluid remains constant, and the fluid properties such as temperature and velocity at any inlet or outlet remain the same. Also, the fluid streams experience little or no change in their velocities and elevations, and thus the *kinetic* and *potential energy changes* are negligible. The *specific heat* of a fluid, in general, changes with temperature. But, in a specified temperature range, it can be treated as a constant at some average value with little loss in accuracy. *Axial heat conduction* along the tube is usually insignificant and can be considered negligible. Finally, the outer surface of the heat exchanger is assumed to be *perfectly insulated,* so that there is no heat loss to the surrounding medium, and any heat transfer occurs between the two fluids only.

The idealizations stated above are closely approximated in practice, and they greatly simplify the analysis of a heat exchanger with little sacrifice of accuracy. Therefore, they are commonly used. Under these assumptions, the *first law of thermodynamics* requires that the rate of heat transfer from the hot fluid be equal to the rate of heat transfer to the cold one. That is,

$$\dot{Q} = \dot{m}_c C_{pc}(T_{c,\text{out}} - T_{c,\text{in}}) \tag{10-9}$$

and

$$\dot{Q} = \dot{m}_h C_{ph}(T_{h,\text{in}} - T_{h,\text{out}}) \tag{10-10}$$

where the subscripts c and h stand for *cold* and *hot* fluids, respectively, and

$$\dot{m}_c, \dot{m}_h = \text{mass flow rates}$$
$$C_{pc}, C_{ph} = \text{specific heats}$$
$$T_{c,\text{out}}, T_{h,\text{out}} = \text{outlet temperatures}$$
$$T_{c,\text{in}}, T_{h,\text{in}} = \text{inlet temperatures}$$

Note that the heat transfer rate $\dot{Q}$ is taken to be a positive quantity, and its direction is understood to be from the hot fluid to the cold one in accordance with the second law of thermodynamics.

In heat exchanger analysis, it is often convenient to combine the product of the *mass flow rate* and the *specific heat* of a fluid into a single quantity. This quantity is called the **heat capacity rate** and is defined as

$$C = \dot{m} C_p \tag{10-11}$$

The heat capacity rate of a fluid stream represents the rate of heat transfer needed to change the temperature of the fluid stream by 1°C as it flows through a heat exchanger. Note that in a heat exchanger, the fluid with a *large* heat capacity rate will experience a *small* temperature change, and the fluid with a *small* heat capacity rate will experience a *large* temperature change. Therefore, *doubling* the mass flow rate of a fluid while leaving everything else unchanged will *halve* the temperature change of that fluid.

With the definition of the heat capacity rate above, Equations 10-9 and 10-10 can also be expressed as

$$\dot{Q} = C_c(T_{c,\text{out}} - T_{c,\text{in}}) \tag{10-12}$$

and

$$\dot{Q} = C_h(T_{h,\text{in}} - T_{h,\text{out}}) \tag{10-13}$$

That is, the heat transfer rate in a heat exchanger is equal to the heat capacity rate of either fluid multiplied by the temperature change of that fluid. Note that *the only time the temperature rise of a cold fluid is equal to the temperature drop of the hot fluid is when the heat capacity rates of the two fluids are equal to each other* (Fig. 10-12).

Two special types of heat exchangers commonly used in practice are *condensers* and *boilers*. One of the fluids in a condenser or a boiler undergoes a phase-change process, and the rate of heat transfer is expressed as

FIGURE 10-12

Two fluids that have the same mass flow rate and the same specific heat experience the same temperature change in a well-insulated heat exchanger.

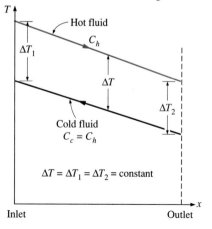

$$\dot{Q} = \dot{m}h_{fg} \qquad (10\text{-}14)$$

where $\dot{m}$ is the rate of evaporation or condensation of the fluid and h_{fg} is the enthalpy of vaporization of the fluid at the specified temperature or pressure.

An ordinary fluid absorbs or releases a large amount of heat essentially at constant temperature during a phase-change process, as shown in Figure 10-13. The heat capacity rate of a fluid during a phase-change process must approach infinity since the temperature change is practically zero. That is, $C = \dot{m}C_p \rightarrow \infty$ when $\Delta T \rightarrow 0$, so that the heat transfer rate $\dot{Q} = \dot{m}C_p\Delta T$ is a finite quantity. Therefore, in heat exchanger analysis, a condensing or boiling fluid is conveniently modeled as a fluid whose heat capacity rate is *infinity*.

The rate of heat transfer in a heat exchanger can also be expressed in an analogous manner to Newton's law of cooling as

$$\dot{Q} = UA\,\Delta T_m \qquad (10\text{-}15)$$

where U is the overall heat transfer coefficient, A is the heat transfer area, and ΔT_m is an appropriate average temperature difference between the two fluids. Here the surface area A can be determined precisely using the dimensions of the heat exchanger. However, the overall heat transfer coefficient U and the temperature difference ΔT between the hot and cold fluids, in general, are not constant and vary along the heat exchanger.

The average value of the overall heat transfer coefficient can be determined as described in the preceding section by using the average convection coefficients for each fluid. It turns out that the appropriate form of the mean temperature difference between the two fluids is *logarithmic* in nature, and its determination is presented in the next section.

10-4 ■ THE LOG MEAN TEMPERATURE DIFFERENCE METHOD

Earlier, we mentioned that the temperature difference between the hot and cold fluids varies along the heat exchanger, and it is convenient to have a *mean temperature difference* ΔT_m for use in the relation

$$\dot{Q} = UA\Delta T_m$$

In order to develop a relation for the equivalent average temperature difference between the two fluids, consider the *parallel-flow double-pipe* heat exchanger shown in Figure 10-14. Note that the temperature difference ΔT between the hot and cold fluids is large at the inlet of the heat exchanger but decreases exponentially toward the outlet. As you would expect, the temperature of the hot fluid decreases and the temperature of the cold fluid increases along the heat exchanger, but the temperature of the cold fluid can never exceed that of the hot fluid no matter how long the heat exchanger is.

Assuming the outer surface of the heat exchanger to be well insulated so that any heat transfer occurs between the two fluids, and disregarding any changes in kinetic and potential energy, an energy balance on each fluid in a

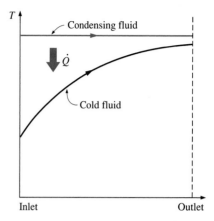

(a) Condenser ($C_h \rightarrow \infty$)

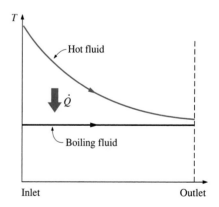

(b) Boiler ($C_c \rightarrow \infty$)

FIGURE 10-13

Variation of fluid temperatures in a heat exchanger when one of the fluids condenses or boils.

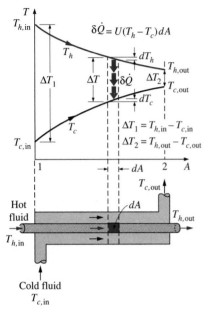

FIGURE 10-14

Variation of the fluid temperatures in a parallel-flow double-pipe heat exchanger.

differential section of the heat exchanger can be expressed as

$$\delta\dot{Q} = -\dot{m}_h C_{ph} dT_h \tag{10-16}$$

and

$$\delta\dot{Q} = \dot{m}_c C_{pc} dT_c \tag{10-17}$$

That is, the rate of heat loss from the hot fluid at any section of a heat exchanger is equal to the rate of heat gain by the cold fluid in that section. The temperature change of the hot fluid is a *negative* quantity, and so a *negative sign* is added to Equation 10-16 to make the heat transfer rate $\dot{Q}$ a positive quantity. Solving the equations above for dT_h and dT_c gives

$$dT_h = -\frac{\delta\dot{Q}}{\dot{m}_h C_{ph}} \tag{10-18}$$

and

$$dT_c = \frac{\delta\dot{Q}}{\dot{m}_c C_{pc}} \tag{10-19}$$

Taking their difference, we get

$$dT_h - dT_c = d(T_h - T_c) = -\delta\dot{Q}\left(\frac{1}{\dot{m}_h C_{ph}} + \frac{1}{\dot{m}_c C_{pc}}\right) \tag{10-20}$$

The rate of heat transfer in the differential section of the heat exchanger can also be expressed as

$$\delta\dot{Q} = U(T_h - T_c)dA \tag{10-21}$$

Substituting this equation into Equation 10-20 and rearranging gives

$$\frac{d(T_h - T_c)}{T_h - T_c} = -U\,dA\left(\frac{1}{\dot{m}_h C_{ph}} + \frac{1}{\dot{m}_c C_{pc}}\right) \tag{10-22}$$

Integrating from the inlet of the heat exchanger to its outlet, we obtain

$$\ln\frac{T_{h,\text{out}} - T_{c,\text{out}}}{T_{h,\text{in}} - T_{c,\text{in}}} = -UA\left(\frac{1}{\dot{m}_h C_{ph}} + \frac{1}{\dot{m}_c C_{pc}}\right) \tag{10-23}$$

Finally, solving Equations 10-9 and 10-10 for $\dot{m}_c C_{pc}$ and $\dot{m}_h C_{ph}$ and substituting into Equation 10-23 gives, after some rearrangement,

$$\dot{Q} = UA\,\Delta T_{\text{lm}} \tag{10-24}$$

where

$$\Delta T_{\text{lm}} = \frac{\Delta T_1 - \Delta T_2}{\ln(\Delta T_1/\Delta T_2)} \tag{10-25}$$

is the **log mean temperature difference,** which is the suitable form of the average temperature difference for use in the analysis of heat exchangers. Here ΔT_1 and ΔT_2 represent the temperature difference between the two fluids at the two ends (inlet and outlet) of the heat exchanger. It makes no difference which end of the heat exchanger is designated as the *inlet* or the *outlet* (Fig. 10-15).

The temperature difference between the two fluids decreases from ΔT_1 at the inlet to ΔT_2 at the outlet. Thus, it is tempting to use the arithmetic mean temperature $\Delta T_{am} = \frac{1}{2}(\Delta T_1 + \Delta T_2)$ as the average temperature difference. The logarithmic mean temperature difference ΔT_{lm} is obtained by tracing the actual temperature profile of the fluids along the heat exchanger and is an *exact* representation of the *average temperature difference* between the hot and cold fluids. It truly reflects the exponential decay of the local temperature difference.

Note that ΔT_{lm} is always less than ΔT_{am}. Therefore, using ΔT_{am} in calculations instead of ΔT_{lm} will overestimate the rate of heat transfer in a heat exchanger between the two fluids. When ΔT_1 differs from ΔT_2 by no more than 40 percent, the error in using the arithmetic mean temperature difference is less than 1 percent. But the error increases to undesirable levels when ΔT_1 differs from ΔT_2 by greater amounts. Therefore, we should always use the *logarithmic mean temperature difference* when determining the rate of heat transfer in a heat exchanger.

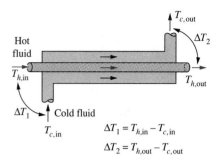

(a) Parallel-flow heat exchangers

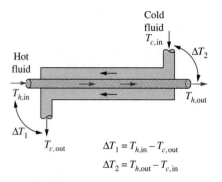

(b) Counter-flow heat exchangers

FIGURE 10-15

The ΔT_1 and ΔT_2 expressions in parallel-flow and counter-flow heat exchangers.

Counter-Flow Heat Exchangers

The variation of temperatures of hot and cold fluids in a counter-flow heat exchanger is given in Figure 10-16. Note that the hot and cold fluids enter the heat exchanger from opposite ends, and the outlet temperature of the *cold fluid* in this case may exceed the outlet temperature of the *hot fluid*. In the limiting case, the cold fluid will be heated to the inlet temperature of the hot fluid. However, the outlet temperature of the cold fluid can *never* exceed the inlet temperature of the hot fluid, since this would be a violation of the second law of thermodynamics.

The relation above for the log mean temperature difference is developed using a parallel-flow heat exchanger, but we can show by repeating the analysis above for a counter-flow heat exchanger that is also applicable to counter-flow heat exchangers. But this time, ΔT_1 and ΔT_2 are expressed as shown in Figure 10-15.

For specified inlet and outlet temperatures, the log mean temperature difference for a *counter-flow* heat exchanger is always *greater* than that for a parallel-flow heat exchanger. That is, $\Delta T_{lm, CF} > \Delta T_{lm, PF}$, and thus a smaller surface area (and thus a smaller heat exchanger) is needed to achieve a specified heat transfer rate in a counter-flow heat exchanger. Therefore, it is common practice to use counter-flow arrangements in heat exchangers.

In a counter-flow heat exchanger, the temperature difference between the hot and the cold fluids will remain constant along the heat exchanger when the *heat capacity rates* of the two fluids are *equal* (that is, $\Delta T = $ constant when $C_h = C_c$ or $\dot{m}_h C_{ph} = \dot{m}_c C_{pc}$). Then we have $\Delta T_1 = \Delta T_2$, and the log mean temperature difference relation above gives $\Delta T_{lm} = \frac{0}{0}$, which is indeterminate. It can be shown by the application of l'Hôpital's rule that in this case we have $\Delta T_{lm} = \Delta T_1 = \Delta T_2$, as expected.

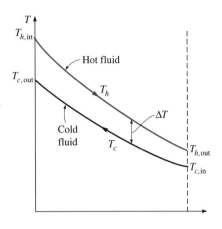

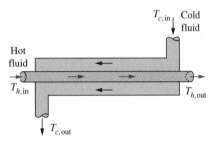

FIGURE 10-16

The variation of the fluid temperatures in a counter-flow double-pipe heat exchanger.

FIGURE 10-17

The determination of the heat transfer rate for cross-flow and multipass shell-and-tube heat exchangers using the correction factor.

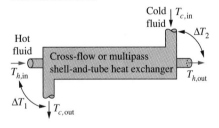

Heat transfer rate:

$$\dot{Q} = UAF\,\Delta T_{lm,CF}$$

where

$$\Delta T_{lm,CF} = \frac{\Delta T_1 - \Delta T_2}{\ln(\Delta T_1/\Delta T_2)}$$

$$\Delta T_1 = T_{h,in} - T_{c,out}$$

$$\Delta T_2 = T_{h,out} - T_{c,in}$$

and $\qquad F = \ldots$ (Fig. 10–18)

A *condenser* or a *boiler* can be considered to be either a parallel- or counter-flow heat exchanger since both approaches give the same result.

Multipass and Cross-Flow Heat Exchangers: Use of a Correction Factor

The log mean temperature difference ΔT_{lm} relation developed earlier is limited to parallel-flow and counter-flow heat exchangers only. Similar relations are also developed for *cross-flow* and *multipass shell-and-tube* heat exchangers, but the resulting expressions are too complicated because of the complex flow conditions.

In such cases, it is convenient to relate the equivalent temperature difference to the log mean temperature difference relation for the counter-flow case as

$$\Delta T_{lm} = F\Delta T_{lm,CF} \tag{10-26}$$

where F is the **correction factor,** which depends on the *geometry* of the heat exchanger and the inlet and outlet temperatures of the hot and cold fluid streams. The $\Delta T_{lm,CF}$ is the log mean temperature difference for the case of a *counter-flow* heat exchanger with the same inlet and outlet temperatures and is determined from Equation 10-25 by taking $\Delta T_1 = T_{h,in} - T_{c,out}$ and $\Delta T_2 = T_{h,out} - T_{c,in}$ (Fig. 10-17).

The correction factor is less than unity for a cross-flow and multipass shell-and-tube heat exchanger. That is, $F \leq 1$. The limiting value of $F = 1$ corresponds to the counter-flow heat exchanger. Thus, the correction factor F for a heat exchanger is *a measure of deviation of the ΔT_{lm} from the corresponding values for the counter-flow case.*

The correction factor F for common cross-flow and shell-and-tube heat exchanger configurations is given in Figure 10-18 versus two temperature ratios P and R defined as

$$P = \frac{t_2 - t_1}{T_1 - t_1} \tag{10-27}$$

and

$$R = \frac{T_1 - T_2}{t_2 - t_1} = \frac{(\dot{m}C_p)_{tube\,side}}{(\dot{m}C_p)_{shell\,side}} \tag{10-28}$$

where the subscripts 1 and 2 represent the *inlet* and *outlet,* respectively. Note that for a shell-and-tube heat exchanger, T and t represent the *shell-* and *tube-side* temperatures, respectively, as shown in the correction factor charts. It makes no difference whether the hot or the cold fluid flows through the shell or the tube. The determination of the correction factor F requires the availability of the *inlet* and the *outlet* temperatures for both the cold and hot fluids.

Note that the value of P ranges from 0 to 1. The value of R, on the other hand, ranges from 0 to infinity, with $R = 0$ corresponding to the phase-change (condensation or boiling) on the shell-side and $R \to \infty$ to phase-change on

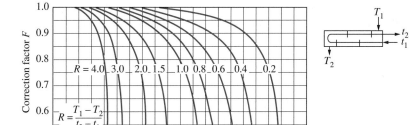

$R = \dfrac{T_1 - T_2}{t_2 - t_1}$ $P = \dfrac{t_2 - t_1}{T_1 - t_1}$

(a) One-shell pass and 2, 4, 6, etc. (any multiple of 2), tube passes

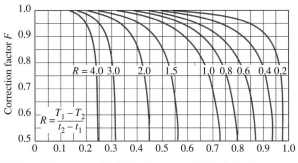
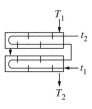

$R = \dfrac{T_1 - T_2}{t_2 - t_1}$ $P = \dfrac{t_2 - t_1}{T_1 - t_1}$

(b) Two-shell passes and 4, 8, 12, etc. (any multiple of 4), tube passes

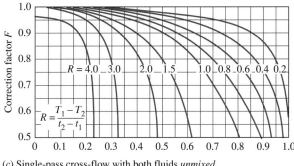
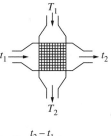

$R = \dfrac{T_1 - T_2}{t_2 - t_1}$ $P = \dfrac{t_2 - t_1}{T_1 - t_1}$

(c) Single-pass cross-flow with both fluids *unmixed*

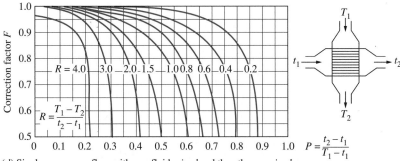

$R = \dfrac{T_1 - T_2}{t_2 - t_1}$ $P = \dfrac{t_2 - t_1}{T_1 - t_1}$

(d) Single-pass cross-flow with one fluid *mixed* and the other *unmixed*

FIGURE 10-18

Correction factor F charts for common
shell-and-tube and cross-flow heat
exchangers (from Bowman, Mueller,
and Nagle, Ref. 2).

the tube side. The correction factor is $F = 1$ for both of these limiting cases. Therefore, the correction factor for a *condenser* or *boiler* is $F = 1$, regardless of the configuration of the heat exchanger.

EXAMPLE 10-3   The Condensation of Steam in a Condenser

Steam in the condenser of a steam power plant is to be condensed at a temperature of 30°C with cooling water from a nearby lake, which enters the tubes of the condenser at 14°C and leaves at 22°C. The surface area of the tubes is 45 m², and the overall heat transfer coefficient is 2100 W/m² · °C. Determine the mass flow rate of the cooling water needed and the rate of condensation of the steam in the condenser.

Solution   Steam is condensed by cooling water in the condenser of a power plant. The mass flow rate of the cooling water and the rate of condensation are to be determined.

Assumptions   **1** Steady operating conditions exist. **2** The heat exchager is well insulated so that heat loss to the surroundings is negligible and thus heat transfer from the hot fluid is equal to the heat transfer to the cold fluid. **3** Changes in the kinetic and potential energies of fluid streams are negligible. **4** There is no fouling. **5** Fluid properties are constant.

Properties   The heat of vaporization of water at 30°C is $h_{fg} = 2431$ kJ/kg and the specific heat of cold water at the average temperature of 18°C is $C_p = 4184$ J/kg.°C (Table A-9).

Analysis   The schematic of the condenser is given in Figure 10-19. The condenser can be treated as a counter-flow heat exchanger since the temperature of one of the fluids (the steam) remains constant.

The temperature difference between the steam and the cooling water at the two ends of the condenser is

$$\Delta T_1 = T_{h,\,in} - T_{c,\,out} = (30 - 22)°C = 8°C$$
$$\Delta T_2 = T_{h,\,out} - T_{c,\,in} = (30 - 14)°C = 16°C$$

That is, the temperature difference between the two fluids varies from 8°C at one end to 16°C at the other. The proper average temperature difference between the two fluids is the *logarithmic mean temperature difference* (not the arithmetic), which is determined from

$$\Delta T_{lm} = \frac{\Delta T_1 - \Delta T_2}{\ln (\Delta T_1/\Delta T_2)} = \frac{8 - 16}{\ln (8/16)} = 11.5°C$$

This is a little less than the arithmetic mean temperature difference of $\frac{1}{2}(8 + 16) = 12°C$. Then the heat transfer rate in the condenser is determined from

$$\dot{Q} = UA\Delta T_{lm} = (2100 \text{ W/m}^2 \cdot °C)(45 \text{ m}^2)(11.5°C) = \mathbf{1,086,750 \ W}$$

Therefore, the steam will lose heat at a rate of 1,086.75 kW as it flows through the condenser, and the cooling water will gain practically all of it, since the condenser is well insulated.

The mass flow rate of the cooling water and the rate of the condensation of the steam are determined from $\dot{Q} = [\dot{m}C(T_{out} - T_{in})]_{\text{cooling water}} = (\dot{m}h_{fg})_{\text{steam}}$ to be

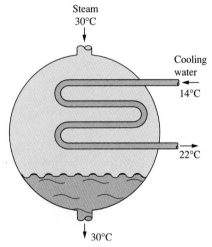

Steam
30°C

Cooling
water
14°C

22°C

30°C

FIGURE 10-19

Schematic for Example 10-3.

$$\dot{m}_{\text{cooling water}} = \frac{\dot{Q}}{C_p(T_{\text{out}} - T_{\text{in}})}$$

$$= \frac{1{,}086.75 \text{ kJ/s}}{(4.184 \text{ kJ/kg} \cdot \text{°C})(22 - 14)\text{°C}} = \textbf{32.5 kg/s}$$

and

$$\dot{m}_{\text{steam}} = \frac{\dot{Q}}{h_{fg}} = \frac{1{,}086.75 \text{ kJ/s}}{2431 \text{ kJ/kg}} = \textbf{0.45 kg/s}$$

Therefore, we need to circulate about 72 kg of cooling water for each 1 kg of steam condensing to remove the heat released during the condensation process.

EXAMPLE 10-4 Heating Water in a Counter-Flow Heat Exchanger

A counter-flow double-pipe heat exchanger is to heat water from 20°C to 80°C at a rate of 1.2 kg/s. The heating is to be accomplished by geothermal water available at 160°C at a mass flow rate of 2 kg/s. The inner tube is thin-walled and has a diameter of 1.5 cm. If the overall heat transfer coefficient of the heat exchanger is 640 W/m^2 · °C, determine the length of the heat exchanger required to achieve the desired heating.

Solution Water is heated in a counter-flow double-pipe heat exchanger by geothermal water. The required length of the heat exchanger is to be determined.

Assumptions **1** Steady operating conditions exist. **2** The heat exchanger is well insulated so that heat loss to the surroundings is negligible and thus heat transfer from the hot fluid is equal to the heat transfer to the cold fluid. **3** Changes in the kinetic and potential energies of fluid streams are negligible. **4** There is no fouling. **5** Fluid properties are constant.

Properties We take the specific heats of water and geothermal fluid to be 4.18 and 4.31 kJ/kg.°C, respectively.

Analysis The schematic of the heat exchanger is given in Figure 10-20. The rate of heat transfer in the heat exchanger can be determined from

$$\dot{Q} = [\dot{m}C_p(T_{\text{out}} - T_{\text{in}})]_{\text{water}} = (1.2 \text{ kg/s})(4.18 \text{ kJ/kg} \cdot \text{°C})(80 - 20)\text{°C} = 301.0 \text{ kW}$$

Noting that all of this heat is supplied by the geothermal water, the outlet temperature of the geothermal water is determined to be

$$\dot{Q} = [\dot{m}C_p(T_{\text{in}} - T_{\text{out}})]_{\text{geothermal}} \longrightarrow T_{\text{out}} = T_{\text{in}} - \frac{\dot{Q}}{\dot{m}C_p}$$

$$= 160\text{°C} - \frac{301.0 \text{ kW}}{(2 \text{ kg/s})(4.31 \text{ kJ/kg} \cdot \text{°C})}$$

$$= 125.1\text{°C}$$

Knowing the inlet and outlet temperatures of both fluids, the logarithmic mean temperature difference for this counter-flow heat exchanger becomes

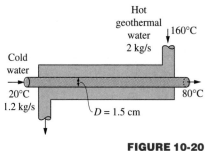

FIGURE 10-20
Schematic for Example 10-4.

$$\Delta T_1 = T_{h,\,in} - T_{c,\,out} = (160 - 80)°C = 80°C$$

$$\Delta T_2 = T_{h,\,out} - T_{c,\,in} = (125.1 - 20)°C = 105.1°C$$

and

$$\Delta T_{lm} = \frac{\Delta T_1 - \Delta T_2}{\ln (\Delta T_1/\Delta T_2)} = \frac{80 - 105.1}{\ln (80/105.1)} = 92.0°C$$

Then the surface area of the heat exchanger is determined to be

$$\dot{Q} = UA\,\Delta T_{lm} \quad \longrightarrow \quad A = \frac{\dot{Q}}{U\Delta T_{lm}} = \frac{301,000\ \text{W}}{(640\ \text{W/m}^2 \cdot °C)(92.0°C)} = 5.11\ \text{m}^2$$

To provide this much heat transfer surface area, the length of the tube must be

$$A = \pi DL \quad \longrightarrow \quad L = \frac{A}{\pi D} = \frac{5.11\ \text{m}^2}{\pi(0.015\ \text{m})} = \textbf{108.4 m}$$

Discussion The inner tube of this counter-flow heat exchanger (and thus the heat exchanger itself) needs to be over 100 m long to achieve the desired heat transfer, which is impractical. In cases like this, we need to use a plate heat exchanger or a multipass shell-and-tube heat exchanger with multiple passes of tube bundles.

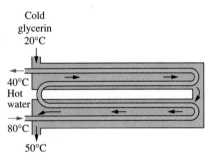

Cold
glycerin
20°C

40°C
Hot
water

80°C

50°C

FIGURE 10-21
Schematic for Example 10-5.

EXAMPLE 10-5   Heating of Glycerin in a Multipass Heat Exchanger

A 2-shell passes and 4-tube passes heat exchanger is used to heat glycerin from 20°C to 50°C by hot water, which enters the thin-walled 2-cm-diameter tubes at 80°C and leaves at 40°C (Fig. 10-21). The total length of the tubes in the heat exchanger is 60 m. The convection heat transfer coefficient is 25 W/m² · °C on the glycerin (shell) side and 160 W/m² · °C on the water (tube) side. Determine the rate of heat transfer in the heat exchanger (*a*) before any fouling occurs and (*b*) after fouling with a fouling factor of 0.0006 m² · °C/W occurs on the outer surfaces of the tubes.

Solution   Glycerin is heated in a 2-shell passes and 4-tube passes heat exchanger by hot water. The rate of heat transfer for the cases of fouling and no fouling are to be determined.

Assumptions **1** Steady operating conditions exist. **2** The heat exchanger is well insulated so that heat loss to the surroundings is negligible and thus heat transfer from the hot fluid is equal to heat transfer to the cold fluid. **3** Changes in the kinetic and potential energies of fluid streams are negligible. **4** Heat transfer coefficients and fouling factors are constant and uniform. **5** The thermal resistance of the inner tube is negligible since the tube is thin-walled and highly conductive.

Analysis The tubes are said to be thin-walled, and thus it is reasonable to assume the inner surface area of the tubes to be equal to the outer surface area. Then the heat transfer surface area of this heat exchanger becomes

$$A = \pi DL = \pi(0.02\ \text{m})(60\ \text{m}) = 3.77\ \text{m}^2$$

The rate of heat transfer in this heat exchanger can be determined from

$$\dot{Q} = UAF\,\Delta T_{\text{lm, }CF}$$

where F is the correction factor and $\Delta T_{\text{lm, }CF}$ is the log mean temperature difference for the counter-flow arrangement. These two quantities are determined from

$$\Delta T_1 = T_{h,\,\text{in}} - T_{c,\,\text{out}} = (80 - 50)\text{°C} = 30\text{°C}$$

$$\Delta T_2 = T_{h,\,\text{out}} - T_{c,\,\text{in}} = (40 - 20)\text{°C} = 20\text{°C}$$

$$\Delta T_{\text{lm, }CF} = \frac{\Delta T_1 - \Delta T_2}{\ln(\Delta T_1/\Delta T_2)} = \frac{30 - 20}{\ln(30/20)} = 24.7\text{°C}$$

and

$$\left.\begin{array}{l} P = \dfrac{t_2 - t_1}{T_1 - t_1} = \dfrac{40 - 80}{20 - 80} = 0.67 \\[3mm] R = \dfrac{T_1 - T_2}{t_2 - t_1} = \dfrac{20 - 50}{40 - 80} = 0.75 \end{array}\right\} \quad F = 0.91 \qquad \text{(Fig. 10-18}b\text{)}$$

(a) In the case of no fouling, the overall heat transfer coefficient U is determined from

$$U = \frac{1}{\dfrac{1}{h_i} + \dfrac{1}{h_o}} = \frac{1}{\dfrac{1}{160\ \text{W/m}^2 \cdot \text{°C}} + \dfrac{1}{25\ \text{W/m}^2 \cdot \text{°C}}} = 21.6\ \text{W/m}^2 \cdot \text{°C}$$

Then the rate of heat transfer becomes

$$\dot{Q} = UAF\,\Delta T_{\text{lm, }CF} = (21.6\ \text{W/m}^2 \cdot \text{°C})(3.77\text{m}^2)(0.91)(24.7\text{°C}) = \mathbf{1730\ W}$$

(b) When there is fouling on one of the surfaces, the overall heat transfer coefficient U is determined from

$$U = \frac{1}{\dfrac{1}{h_i} + \dfrac{1}{h_o} + R_f} = \frac{1}{\dfrac{1}{160\ \text{W/m}^2 \cdot \text{°C}} + \dfrac{1}{25\ \text{W/m}^2 \cdot \text{°C}} + 0.0006\ \text{m}^2 \cdot \text{°C/W}}$$

$$= 21.3\ \text{W/m}^2 \cdot \text{°C}$$

The rate of heat transfer in this case becomes

$$\dot{Q} = UAF\Delta T_{\text{lm, }CF} = (21.3\ \text{W/m}^2 \cdot \text{°C})(3.77\text{m}^2)(0.91)(24.7\text{°C}) = \mathbf{1805\ W}$$

Discussion Note that the rate of heat transfer decreases as a result of fouling, as expected. The decrease is not dramatic, however, because of the relatively low convection heat transfer coefficients involved.

EXAMPLE 10-6 Cooling of an Automotive Radiator

A test is conducted to determine the overall heat transfer coefficient in an automotive radiator that is a compact cross-flow water-to-air heat exchanger with both fluids (air and water) unmixed (Fig. 10-22). The radiator has 40 tubes of internal diameter 0.5 cm and length 65 cm in a closely spaced plate-finned matrix. Hot water enters the tubes at 90°C at a rate of 0.6 kg/s and leaves at 65°C. Air flows across the radiator through the interfin spaces and is heated from 20°C to 40°C. Determine the overall heat transfer coefficient U_i of this radiator based on the inner surface area of the tubes.

FIGURE 10-22
Schematic for Example 10-6.

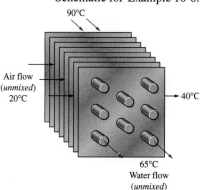

Solution During an experiment involving an automotive radiator, the inlet and exit temperatures of water and air and the mass flow rate of water are measured. The overall heat transfer coefficient based on the inner surface area is to be determined.

Assumptions **1** Steady operating conditions exist. **2** Changes in the kinetic and potential energies of fluid streams are negligible. **3** Fluid properties are constant.

Properties The specific heat of water at the average temperature of $(90 + 65)/2 = 77.5°C$ is 4.195 kJ/kg · °C.

Analysis The rate of heat transfer in this radiator from the hot water to the air is determined from an energy balance on water flow,

$$\dot{Q} = [\dot{m}C_p(T_{in} - T_{out})]_{water} = (0.6 \text{ kg/s})(4.195 \text{ kJ/kg} \cdot °C)(90 - 65)°C = 62.93 \text{ kW}$$

The tube-side heat transfer area is the total surface area of the tubes and is determined from

$$A_i = n\pi D_i L = (40)\pi(0.005 \text{ m})(0.65 \text{ m}) = 0.408 \text{ m}^2$$

Knowing the rate of heat transfer and the surface area, the overall heat transfer coefficient in this heat exchanger can be determined from

$$\dot{Q} = U_i A_i F \Delta T_{lm, CF} \longrightarrow U_i = \frac{\dot{Q}}{A_i F \Delta T_{lm, CF}}$$

where F is the correction factor and $\Delta T_{lm, CF}$ is the log mean temperature difference for the counter-flow arrangement. These two quantities are found to be

$$\Delta T_1 = T_{h, in} - T_{c, out} = (90 - 40)°C = 50°C$$
$$\Delta T_2 = T_{h, out} - T_{c, in} = (65 - 20)°C = 45°C$$

$$\Delta T_{lm, CF} = \frac{\Delta T_1 - \Delta T_2}{\ln(\Delta T_1/\Delta T_2)} = \frac{50 - 45}{\ln(50/45)} = 47.6°C$$

and

$$\left.\begin{array}{l} P = \dfrac{t_2 - t_1}{T_1 - t_1} = \dfrac{65 - 90}{20 - 90} = 0.36 \\[2mm] R = \dfrac{T_1 - T_2}{t_2 - t_1} = \dfrac{20 - 40}{65 - 90} = 0.80 \end{array}\right\} \quad F = 0.97 \qquad \text{(Fig. 10-18c)}$$

Substituting, the overall heat transfer coefficient U_i is determined to be

$$U_i = \frac{\dot{Q}}{A_i F \Delta T_{lm, CF}} = \frac{62,930 \text{ W}}{(0.408 \text{ m}^2)(0.97)(47.6°C)} = \textbf{3341 W/m}^2 \cdot \textbf{°C}$$

Note that the overall heat transfer coefficient on the air side would be much lower because of the large surface area involved on that side.

10-5 ■ THE EFFECTIVENESS–NTU METHOD

The log mean temperature difference (LMTD) method discussed in the previous section is easy to use in heat exchanger analysis when the inlet and the outlet temperatures of the hot and cold fluids are known or can be determined from an energy balance. Once ΔT_{lm}, the mass flow rates, and the overall heat transfer coefficient are available, the heat transfer surface area of the heat exchanger can be determined from

$$\dot{Q} = UA\,\Delta T_{lm}$$

Therefore, the LMTD method is very suitable for determining the *size* of a heat exchanger to realize prescribed outlet temperatures when the mass flow rates and the inlet and outlet temperatures of the hot and cold fluids are specified.

With the LMTD method, the task is to *select* a heat exchanger that will meet the prescribed heat transfer requirements. The procedure to be followed by the selection process is as follows:

1   Select the type of heat exchanger suitable for the application.

2   Determine any unknown inlet or outlet temperature and the heat transfer rate using an energy balance.

3   Calculate the log mean temperature difference ΔT_{lm} and the correction factor F, if necessary.

4   Obtain (select or calculate) the value of the overall heat transfer coefficient U.

5   Calculate the heat transfer surface area A.

The task is completed by selecting a heat exchanger that has a heat transfer surface area equal to or larger than A.

A second kind of problem encountered in heat exchanger analysis is the determination of the *heat transfer rate* and the *outlet temperatures* of the hot and cold fluids for prescribed fluid mass flow rates and inlet temperatures when the *type* and *size* of the heat exchanger are specified. The heat transfer surface area A of the heat exchanger in this case is known, but the *outlet temperatures* are not. Here the task is to determine the heat transfer performance of a specified heat exchanger or to determine if a heat exchanger available in storage will do the job.

The LMTD method could still be used for this alternative problem, but the procedure would require tedious iterations, and thus it is not practical. In an attempt to eliminate the iterations from the solution of such problems, Kays and London came up with a new method in 1955 called the **effectiveness-NTU method,** which greatly simplified heat exchanger analysis.

This new method is based on a dimensionless parameter called the **heat transfer effectiveness** ε, defined as

$$\varepsilon = \frac{\dot{Q}}{\dot{Q}_{max}} = \frac{\text{Actual heat transfer rate}}{\text{Maximum possible heat transfer rate}} \qquad \text{(10-29)}$$

The *actual* heat transfer rate in a heat exchanger can be determined from an energy balance on the hot or cold fluids and can be expressed as

$$\dot{Q} = C_c(T_{c,\,out} - T_{c,\,in}) = C_h(T_{h,\,in} - T_{h,\,out}) \qquad (10\text{-}30)$$

where $C_c = \dot{m}_c C_{pc}$ and $C_h = \dot{m}_c C_{ph}$ are the heat capacity rates of the cold and the hot fluids, respectively.

To determine the maximum possible heat transfer rate in a heat exchanger, we first recognize that the *maximum temperature difference* in a heat exchanger is the difference between the *inlet* temperatures of the hot and cold fluids. That is,

$$\Delta T_{max} = T_{h,\,in} - T_{c,\,in} \qquad (10\text{-}31)$$

The heat transfer in a heat exchanger will reach its maximum value when (1) the cold fluid is heated to the inlet temperature of the hot fluid or (2) the hot fluid is cooled to the inlet temperature of the cold fluid. These two limiting conditions will not be reached simultaneously unless the heat capacity rates of the hot and cold fluids are identical (i.e., $C_c = C_h$). When $C_c \neq C_h$, which is usually the case, the fluid with the *smaller* heat capacity rate will experience a larger temperature change, and thus it will be the first to experience the maximum temperature, at which point the heat transfer will come to a halt. Therefore, the maximum possible heat transfer rate in a heat exchanger is (Fig. 10-23)

$$\dot{Q}_{max} = C_{min}(T_{h,\,in} - T_{c,\,in}) \qquad (10\text{-}32)$$

where C_{min} is the smaller of $C_h = \dot{m}_h C_{ph}$ and $C_c = \dot{m}_c C_{pc}$. This is further clarified by the following example.

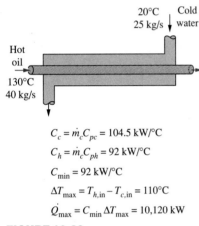

$$C_c = \dot{m}_c C_{pc} = 104.5 \text{ kW/°C}$$
$$C_h = \dot{m}_c C_{ph} = 92 \text{ kW/°C}$$
$$C_{min} = 92 \text{ kW/°C}$$
$$\Delta T_{max} = T_{h,\,in} - T_{c,\,in} = 110\text{°C}$$
$$\dot{Q}_{max} = C_{min} \Delta T_{max} = 10{,}120 \text{ kW}$$

FIGURE 10-23

The determination of the maximum rate of heat transfer in a heat exchanger.

EXAMPLE 10-7   Upper Limit for Heat Transfer in a Heat Exchanger

Cold water enters a counter-flow heat exchanger at 10°C at a rate of 8 kg/s, where it is heated by a hot water stream that enters the heat exchanger at 70°C at a rate of 2 kg/s. Assuming the specific heat of water to remain constant at $C_p = 4.18$ kJ/ kg · °C, determine the maximum heat transfer rate and the outlet temperatures of the cold and the hot water streams for this limiting case.

Solution   Cold and hot water streams enter a heat exchanger at specified temperatures and flow rates. The maximum rate of heat transfer in the heat exchanger is to be determined.

Assumptions   **1** Steady operating conditions exist. **2** The heat exchanger is well insulated so that heat loss to the surroundings is negligible and thus heat transfer from the hot fluid is equal to heat transfer to the cold fluid. **3** Changes in the kinetic and potential energies of fluid streams are negligible. **4** Heat transfer coefficients and fouling factors are constant and uniform. **5** The thermal resistance of the inner tube is negligible since the tube is thin-walled and highly conductive.

Properties   The specific heat of water is given to be $C_p = 4.18$ kJ/kg · °C.

Analysis A schematic of the heat exchanger is given in Figure 10-24. The heat capacity rates of the hot and cold fluids are determined from

$$C_h = \dot{m}_h C_{ph} = (2 \text{ kg/s})(4.18 \text{ kJ/kg} \cdot °C) = 8.36 \text{ kW/°C}$$

and

$$C_c = \dot{m}_c C_{pc} = (8 \text{ kg/s})(4.18 \text{ kJ/kg} \cdot °C) = 33.44 \text{ kW/°C}$$

Therefore

$$C_{min} = C_h = 8.36 \text{ kW/°C}$$

which is the smaller of the two heat capacity rates. Then the maximum heat transfer rate is determined from Equation 10-32 to be

$$\begin{aligned}
\dot{Q}_{max} &= C_{min}(T_{h, in} - T_{c, in}) \\
&= (8.36 \text{ kW/°C})(70 - 10)°C \\
&= \textbf{501.6 kW}
\end{aligned}$$

That is, the maximum possible heat transfer rate in this heat exchanger is 501.6 kW. This value would be approached in a counter-flow heat exchanger with a *very large* heat transfer surface area.

The maximum temperature difference in this heat exchanger is $\Delta T_{max} = T_{h, in} - T_{c, in} = (70 - 10)°C = 60°C$. Therefore, the hot water cannot be cooled by more than 60°C (to 10°C) in this heat exchanger, and the cold water cannot be heated by more than 60°C (to 70°C), no matter what we do. The outlet temperatures of the cold and the hot streams in this limiting case are determined to be

$$\dot{Q} = C_c(T_{c, out} - T_{c, in}) \quad \longrightarrow \quad T_{c, out} = T_{c, in} + \frac{\dot{Q}}{C_c} = 10°C + \frac{501.6 \text{ kW}}{33.44 \text{ kW/°C}} = \textbf{25°C}$$

$$\dot{Q} = C_h(T_{h, in} - T_{h, out}) \quad \longrightarrow \quad T_{h, out} = T_{h, in} - \frac{\dot{Q}}{C_h} = 70°C - \frac{501.6 \text{ kW}}{8.38 \text{ kW/°C}} = \textbf{10°C}$$

Discussion Note that the hot water is cooled to the limit of 10°C (the inlet temperature of the cold water stream), but the cold water is heated to 25°C only when maximum heat transfer occurs in the heat exchanger. This is not surprising, since the mass flow rate of the hot water is only one-fourth that of the cold water, and, as a result, the temperature of the cold water increases by 0.25°C for each 1°C drop in the temperature of the hot water.

You may be tempted to think that the cold water should be heated to 70°C in the limiting case of maximum heat transfer. But this will require the temperature of the hot water to drop to −170°C (below 10°C), which is impossible. Therefore, heat transfer in a heat exchanger reaches its maximum value when the fluid with the smaller heat capacity rate (or the smaller mass flow rate when both fluids have the same specific heat value) experiences the maximum temperature change. This example explains why we use C_{min} in the evaluation of $\dot{Q}_{max}$ instead of C_{max}.

We can show that the hot water will leave at the inlet temperature of the cold water and vice versa in the limiting case of maximum heat transfer when the mass flow rates of the hot and cold water streams are identical (Fig. 10-25). We can also show that the outlet temperature of the cold water will reach the 70°C limit when the mass flow rate of the hot water is greater than that of the cold water.

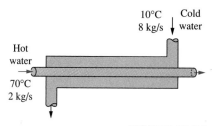

FIGURE 10-24

Schematic for Example 10-7.

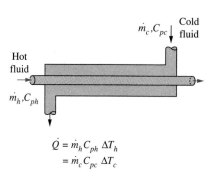

FIGURE 10-25

The temperature rise of the cold fluid in a heat exchanger will be equal to the temperature drop of the hot fluid when the mass flow rates and the specific heats of the hot and cold fluids are identical.

The determination of $\dot{Q}_{max}$ requires the availability of the *inlet temperature* of the hot and cold fluids and their *mass flow rates,* which are usually specified. Then, once the effectiveness of the heat exchanger is known, the actual heat transfer rate $\dot{Q}$ can be determined from

$$\dot{Q} = \varepsilon \dot{Q}_{max} = \varepsilon C_{min}(T_{h, in} - T_{c, in}) \qquad (10\text{-}33)$$

Therefore, the effectiveness of a heat exchanger enables us to determine the heat transfer rate without knowing the *outlet temperatures* of the fluids.

The effectiveness of a heat exchanger depends on the *geometry* of the heat exchanger as well as the *flow arrangement.* Therefore, different types of heat exchangers have different effectiveness relations. Below we illustrate the development of the effectiveness ε relation for the double-pipe *parallel-flow* heat exchanger.

Equation 10-23 developed in the previous section for a parallel-flow heat exchanger can be rearranged as

$$\ln \frac{T_{h, out} - T_{c, out}}{T_{h, in} - T_{c, in}} = -\frac{UA}{C_c}\left(1 + \frac{C_c}{C_h}\right) \qquad (10\text{-}34)$$

Also, solving Equation 10-30 for $T_{h, out}$ gives

$$T_{h, out} = T_{h, in} - \frac{C_c}{C_h}(T_{c, out} - T_{c, in}) \qquad (10\text{-}35)$$

Substituting this relation into Equation 10-34 after adding and subtracting $T_{c, in}$ gives

$$\ln \frac{T_{h, in} - T_{c, in} + T_{c, in} - T_{c, out} - \dfrac{C_c}{C_h}(T_{c, out} - T_{c, in})}{T_{h, in} - T_{c, in}} = -\frac{UA}{C_c}\left(1 + \frac{C_c}{C_h}\right)$$

which simplifies to

$$\ln \left[1 - \left(1 + \frac{C_c}{C_h}\right)\frac{T_{c, out} - T_{c, in}}{T_{h, in} - T_{c, in}} \right] = -\frac{UA}{C_c}\left(1 + \frac{C_c}{C_h}\right) \qquad (10\text{-}36)$$

We now manipulate the definition of effectiveness to obtain

$$\varepsilon = \frac{\dot{Q}}{\dot{Q}_{max}} = \frac{C_c(T_{c, out} - T_{c, in})}{C_{min}(T_{h, in} - T_{c, in})} \longrightarrow \frac{T_{c, out} - T_{c, in}}{T_{h, in} - T_{c, in}} = \varepsilon\frac{C_{min}}{C_c}$$

Substituting this result into Equation 10-36 and solving for ε gives the following relation for the effectiveness of a *parallel-flow* heat exchanger:

$$\varepsilon_{\text{parallel flow}} = \frac{1 - \exp\left[-\dfrac{UA}{C_c}\left(1 + \dfrac{C_c}{C_h}\right)\right]}{\left(1 + \dfrac{C_c}{C_h}\right)\dfrac{C_{min}}{C_c}} \qquad (10\text{-}37)$$

Taking either C_c or C_h to be C_{min} (both approaches give the same result), the relation above can be expressed more conveniently as

$$\varepsilon_{\text{parallel flow}} = \frac{1 - \exp\left[-\frac{UA}{C_{min}}\left(1 + \frac{C_{min}}{C_{max}}\right)\right]}{1 + \frac{C_{min}}{C_{max}}} \qquad (10\text{-}38)$$

Again C_{min} is the *smaller* heat capacity ratio and C_{max} is the larger one, and it makes no difference whether C_{min} belongs to the hot or cold fluid.

Effectiveness relations of the heat exchangers typically involve the *dimensionless* group UA/C_{min}. This quantity is called the **number of transfer units NTU** and is expressed as

$$\text{NTU} = \frac{UA}{C_{min}} = \frac{UA}{(\dot{m}C_p)_{min}} \qquad (10\text{-}39)$$

where U is the overall heat transfer coefficient and A is the heat transfer surface area of the heat exchanger. Note that NTU is proportional to A. Therefore, for specified values of U and C_{min}, the value of NTU *is a measure of the heat transfer surface area A*. Thus, the larger the NTU, the larger the heat exchanger.

In heat exchanger analysis, it is also convenient to define another dimensionless quantity called **capacity ratio C** as

$$C = \frac{C_{min}}{C_{max}} \qquad (10\text{-}40)$$

It can be shown that the effectiveness of a heat exchanger is a function of the number of transfer units NTU and the capacity ratio C. That is,

$$\varepsilon = \text{function}\,(UA/C_{min}, C_{min}/C_{max}) = \text{function}\,(\text{NTU}, C)$$

Effectiveness relations have been developed for a large number of heat exchangers, and the results are given in Table 10-4. The effectivenesses of some common types of heat exchangers are also plotted in Figure 10-26. More extensive effectiveness charts and relations are available in the literature. The dashed lines in Figure 10-26*f* are for the case of C_{min} unmixed and C_{max} mixed, and the solid lines are for the opposite case. The analytic relations for the effectiveness give more accurate results than the charts, since reading errors in charts are unavoidable, and the relations are very suitable for computerized design analysis of heat exchangers.

We make the following observations from the effectiveness relations and charts given above:

1 The value of the effectiveness ranges from 0 to 1. It increases rapidly with NTU for small values (up to about NTU = 1.5) but rather slowly for larger values. Therefore, the use of a heat exchanger with a large NTU

FIGURE 10-26

Effectiveness for heat exchangers (from Kays and London, Ref. 7).

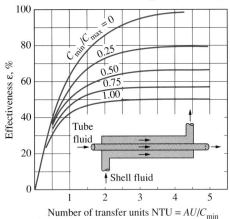

(a) Parallel-flow

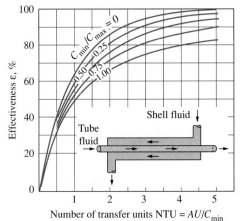

(b) Counter-flow

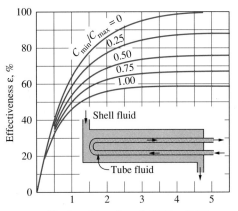

(c) One-shell pass and 2, 4, 6, tube passes

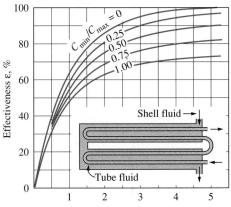

(d) Two-shell passes and 4, 8, 12, tube passes

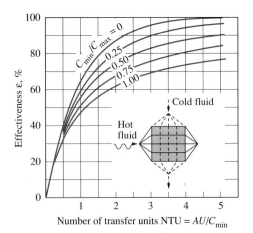

(e) Cross-flow with both fluids unmixed

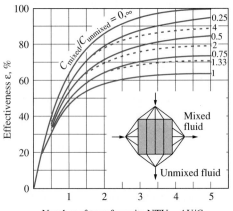

(f) Cross-flow with one fluid mixed and the other unmixed

TABLE 10-4

Effectiveness relations for heat exchangers: $NTU = UA/C_{min}$
and $C = C_{min}/C_{max} = (\dot{m}C_p)_{min}/(\dot{m}C_p)_{max}$

Heat exchanger type	Effectiveness relation
1 *Double pipe:*	
Parallel-flow	$\varepsilon = \dfrac{1 - \exp\left[-NTU(1 + C)\right]}{1 + C}$
Counter-flow	$\varepsilon = \dfrac{1 - \exp\left[-NTU(1 - C)\right]}{1 - C\exp\left[-NTU(1 - C)\right]}$
2 *Shell and tube:* One-shell pass 2, 4, . . . tube passes	$\varepsilon = 2\left\{1 + C + \sqrt{1 + C^2}\,\dfrac{1 + \exp\left[-NTU\sqrt{1 + C^2}\right]}{1 - \exp\left[-NTU\sqrt{1 + C^2}\right]}\right\}^{-1}$
3 *Cross-flow (single-pass)*	
Both fluids unmixed	$\varepsilon = 1 - \exp\left\{\dfrac{NTU^{0.22}}{C}\left[\exp\left(-C\,NTU^{0.78}\right) - 1\right]\right\}$
C_{max} mixed, C_{min} unmixed	$\varepsilon = \dfrac{1}{C}(1 - \exp\{1 - C[1 - \exp(-NTU)]\})$
C_{min} mixed, C_{max} unmixed	$\varepsilon = 1 - \exp\left\{-\dfrac{1}{C}[1 - \exp(-C\,NTU)]\right\}$
4 *All heat exchangers with C = 0*	$\varepsilon = 1 - \exp(-NTU)$

Source: Kays and London, Ref. 7.

(usually larger than 3) and thus a large size cannot be justified economically, since a large increase in NTU in this case corresponds to a small increase in effectiveness. Thus, a heat exchanger with a very high effectiveness may be highly desirable from a heat transfer point of view but rather undesirable from an economical point of view

2 For a given NTU and capacity ratio $C = C_{min}/C_{max}$, the *counter-flow* heat exchanger has the *highest* effectiveness, followed closely by the cross-flow heat exchangers with both fluids unmixed. As you might expect, the lowest effectiveness values are encountered in parallel-flow heat exchangers (Fig. 10-27).

3 The effectiveness of a heat exchanger is independent of the capacity ratio C for NTU values of less than about 0.3.

4 The value of the capacity ratio C ranges between 0 and 1. For a given NTU, the effectiveness becomes a *maximum* for $C = 0$ and a *minimum* for $C = 1$. The case $C = C_{min}/C_{max} \to 0$ corresponds to $C_{max} \to \infty$, which is

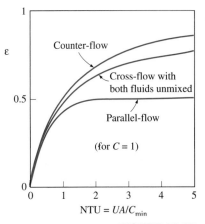

FIGURE 10-27

For a specified NTU and capacity ratio C, the counter-flow heat exchanger has the highest effectiveness and the parallel-flow the lowest.

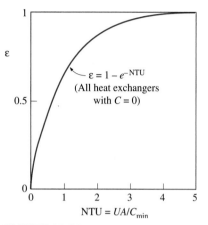

FIGURE 10-28

The effectiveness relation reduces to $\varepsilon = \varepsilon_{max} = 1 - \exp(-NTU)$ for all heat exchangers when the capacity ratio $C = 0$.

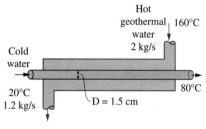

FIGURE 10-29

Schematic for Example 10-8.

realized during a phase-change process in a *condenser* or *boiler*. All effectiveness relations in this case reduce to

$$\varepsilon = \varepsilon_{max} = 1 - \exp(-NTU) \qquad (10\text{-}41)$$

regardless of the type of heat exchanger (Fig. 10-28). Note that the temperature of the condensing or boiling fluid remains constant in this case. The effectiveness is the *lowest* in the other limiting case of $C = C_{min}/C_{max} = 1$, which is realized when the heat capacity rates of the two fluids are equal.

Once the quantities $C = C_{min}/C_{max}$ and $NTU = UA/C_{min}$ have been evaluated, the effectiveness ε can be determined from either the charts or (preferably) the effectiveness relation for the specified type of heat exchanger. Then the rate of heat transfer $\dot{Q}$ and the outlet temperatures $T_{h,out}$ and $T_{c,out}$ can be determined from Equations 10-33 and 10-30, respectively. Note that the analysis of heat exchangers with unknown outlet temperatures is a straightforward matter with the effectiveness-NTU method but requires rather tedious iterations with the LMTD method.

We mentioned earlier that when all the inlet and outlet temperatures are specified, the *size* of the heat exchanger can easily be determined using the LMTD method. Alternatively, it can also be determined from the effectiveness-NTU method by first evaluating the effectiveness ε from its definition (Eq. 10-29) and then the NTU from the appropriate NTU relation in Table 10-5.

Note that the relations in Table 10-5 are equivalent to those in Table 10-4. Both sets of relations are given for convenience. The relations in Table 10-4 give the effectiveness directly when NTU is known, and the relations in Table 10-5 give the NTU directly when the effectiveness ε is known.

EXAMPLE 10-8   Using the Effectiveness-NTU Method

Repeat Example 10-4, which was solved with the LMTD method, using the effectiveness-NTU method.

Solution   The schematic of the heat exchanger is redrawn in Figure 10-29, and the same assumptions are utilized.

Analysis   In the effectiveness-NTU method, we first determine the heat capacity rates of the hot and cold fluids and identify the smaller one:

$$C_h = \dot{m}_h C_{ph} = (2 \text{ kg/s})(4.31 \text{ kJ/kg} \cdot {}^\circ\text{C}) = 8.62 \text{ kW/}^\circ\text{C}$$
$$C_c = \dot{m}_c C_{pc} = (1.2 \text{ kg/s})(4.18 \text{ kJ/kg} \cdot {}^\circ\text{C}) = 5.02 \text{ kW/}^\circ\text{C}$$

Therefore,

$$C_{min} = C_c = 5.02 \text{ kW/}^\circ\text{C}$$

and

$$C = C_{min}/C_{max} = 5.02/8.62 = 0.583$$

Then the maximum heat transfer rate is determined from Equation 10-32 to be

$$\begin{aligned}\dot{Q}_{max} &= C_{min}(T_{h,in} - T_{c,in})\\ &= (5.02 \text{ kW/}^\circ\text{C})(160 - 20)^\circ\text{C}\\ &= 702.8 \text{ kW}\end{aligned}$$

TABLE 10-5

NTU relations for heat exchangers NTU $= UA/C_{min}$
and $C = C_{min}/C_{max} = (\dot{m}C_p)_{min}/(\dot{m}C_p)_{max}$

Heat exchanger type	NTU relation
1 *Double-pipe:*	
Parallel-flow	$NTU = -\dfrac{\ln[1 - \varepsilon(1 + C)]}{1 + C}$
Counter-flow	$NTU = \dfrac{1}{C - 1}\ln\left(\dfrac{\varepsilon - 1}{\varepsilon C - 1}\right)$
2 *Shell and tube:*	
One-shell pass 2, 4, ... tube passes	$NTU = -\dfrac{1}{\sqrt{1 + C^2}}\ln\left(\dfrac{2/\varepsilon - 1 - C - \sqrt{1 + C^2}}{2/\varepsilon - 1 - C + \sqrt{1 + C^2}}\right)$
3 *Cross-flow (single-pass)*	
C_{max} mixed, C_{min} unmixed	$NTU = -\ln\left[1 + \dfrac{\ln(1 - \varepsilon C)}{C}\right]$
C_{min} mixed, C_{max} unmixed	$NTU = -\dfrac{\ln[C\ln(1 - \varepsilon) + 1]}{C}$
4 *All heat exchangers with C = 0*	$NTU = -\ln(1 - \varepsilon)$

Source: Kays and London, Ref. 7.

That is, the maximum possible heat transfer rate in this heat exchanger is 702.8 kW. The actual rate of heat transfer in the heat exchanger is

$$\dot{Q} = [\dot{m}C_p(T_{out} - T_{in})]_{water} = (1.2 \text{ kg/s})(4.18 \text{ kJ/kg} \cdot °\text{C})(80 - 20)°\text{C} = 301.0 \text{ kW}$$

Thus, the effectiveness of the heat exchanger is

$$\varepsilon = \frac{\dot{Q}}{\dot{Q}_{max}} = \frac{301.0 \text{ kW}}{702.8 \text{ kW}} = 0.428$$

Knowing the effectiveness, the NTU of this counter-flow heat exchanger can be determined from Figure 10-26*b* or the appropriate relation from Table 10-5. We choose the latter approach for greater accuracy:

$$NTU = \frac{1}{C - 1}\ln\left(\frac{\varepsilon - 1}{\varepsilon C - 1}\right) = \frac{1}{0.583 - 1}\ln\left(\frac{0.428 - 1}{0.428 \times 0.583 - 1}\right) = 0.651$$

Then the heat transfer surface area becomes

$$NTU = \frac{UA}{C_{min}} \longrightarrow A = \frac{NTU\,C_{min}}{U} = \frac{(0.651)(5020 \text{ W/°C})}{640 \text{ W/m}^2 \cdot °\text{C}} = 5.11 \text{ m}^2$$

To provide this much heat transfer surface area, the length of the tube must be

$$A = \pi DL \longrightarrow L = \frac{A}{\pi D} = \frac{5.11 \text{ m}^2}{\pi(0.015 \text{ m})} = \textbf{108.4 m}$$

Discussion   Note that we obtained the same result with the effectiveness-NTU method in a systematic and straightforward manner.

EXAMPLE 10-9   Cooling Hot Oil by Water in a Multipass Heat Exchanger

Hot oil is to be cooled by water in a 1-shell-pass and 8-tube-passes heat exchanger. The tubes are thin-walled and are made of copper with an internal diameter of 1.4 cm. The length of each tube pass in the heat exchanger is 5 m, and the overall heat transfer coefficient is 310 W/m² · °C. Water flows through the tubes at a rate of 0.2 kg/s, and the oil through the shell at a rate of 0.3 kg/s. The water and the oil enter at temperatures of 20°C and 150°C, respectively. Determine the rate of heat transfer in the heat exchanger and the outlet temperatures of the water and the oil.

Solution   Hot oil is to be cooled by water in a heat exchanger. The mass flow rates and the inlet temperatures are given. The rate of heat transfer and the outlet temperatures are to be determined.

Assumptions   **1** Steady operating conditions exist. **2** The heat exchanger is well insulated so that heat loss to the surroundings is negligible and thus heat transfer from the hot fluid is equal to the heat transfer to the cold fluid. **3** The thickness of the tube is negligible since it is thin-walled. **4** Changes in the kinetic and potential energies of fluid streams are negligible. **5** The overall heat transfer coefficient is constant and uniform.

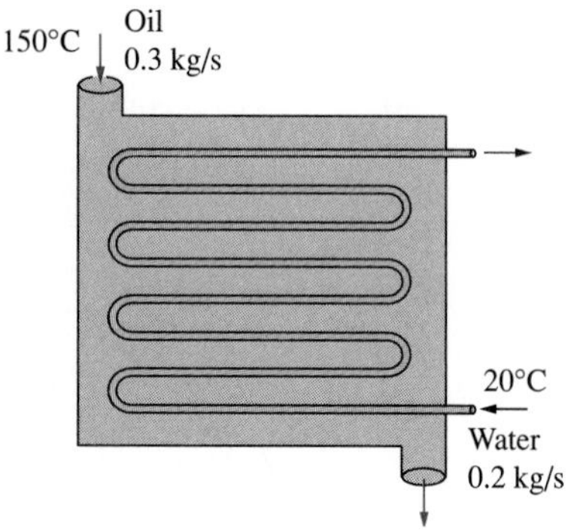

Analysis   The schematic of the heat exchanger is given in Figure 10-30. The outlet temperatures are not specified, and they cannot be determined from an energy balance. The use of the LMTD method in this case will involve tedious iterations, and thus the ε-NTU method is indicated. The first step in the ε-NTU method is to determine the heat capacity rates of the hot and cold fluids and identify the smaller one:

$$C_h = \dot{m}_h C_{ph} = (0.3 \text{ kg/s})(2.13 \text{ kJ/kg} \cdot °C) = 0.639 \text{ kW/°C}$$
$$C_c = \dot{m}_c C_{pc} = (0.2 \text{ kg/s})(4.18 \text{ kJ/kg} \cdot °C) = 0.836 \text{ kW/°C}$$

Therefore,

$$C_{min} = C_h = 0.639 \text{ kW/°C}$$

and

$$C = \frac{C_{min}}{C_{max}} = \frac{0.639}{0.836} = 0.764$$

Then the maximum heat transfer rate is determined from Equation 10-32 to be

$$\dot{Q}_{max} = C_{min}(T_{h, in} - T_{c, in})$$
$$= (0.639 \text{ kW/°C})(150 - 20)°C = 83.1 \text{ kW}$$

That is, the maximum possible heat transfer rate in this heat exchanger is 83.1 kW. The heat transfer surface area is

$$A = n(\pi DL) = 8\pi(0.014 \text{ m})(5 \text{ m}) = 1.76 \text{ m}^2$$

Then the NTU of this heat exchanger becomes

FIGURE 10-30
Schematic for Example 10-9.

Oil
150°C   0.3 kg/s

20°C
Water
0.2 kg/s

$$\text{NTU} = \frac{UA}{C_{min}} = \frac{(310 \text{ W/m}^2 \cdot {}^\circ\text{C})(1.76 \text{ m}^2)}{639 \text{ W/}{}^\circ\text{C}} = 0.853$$

The effectiveness of this heat exchanger corresponding to $C = 0.764$ and NTU = 0.853 is determined from Figure 10-26c to be

$$\varepsilon = 0.47$$

We could also determine the effectiveness from the third relation in Table 10-4 more accurately but with more labor. Then the actual rate of heat transfer becomes

$$\dot{Q} = \varepsilon \dot{Q}_{max} = (0.47)(83.1 \text{ kW}) = \textbf{39.1 kW}$$

Finally, the outlet temperatures of the cold and the hot fluid streams are determined to be

$$\dot{Q} = C_c(T_{c,\,out} - T_{c,\,in}) \quad \longrightarrow \quad T_{c,\,out} = T_{c,\,in} + \frac{\dot{Q}}{C_c}$$
$$= 20{}^\circ\text{C} + \frac{39.1 \text{ kW}}{0.836 \text{ kW/}{}^\circ\text{C}} = \textbf{66.8}{}^\circ\textbf{C}$$

$$\dot{Q} = C_h(T_{h,\,in} - T_{h,\,out}) \quad \longrightarrow \quad T_{h,\,out} = T_{h,\,in} - \frac{\dot{Q}}{C_h}$$
$$= 150{}^\circ\text{C} - \frac{39.1 \text{ kW}}{0.639 \text{ kW/}{}^\circ\text{C}} = \textbf{88.8}{}^\circ\textbf{C}$$

Therefore, the temperature of the cooling water will rise from 20°C to 66.8°C as it cools the hot oil from 150°C to 88.8°C in this heat exchanger.

10-6 ▪ THE SELECTION OF HEAT EXCHANGERS

Heat exchangers are complicated devices, and the results obtained with the simplified approaches presented above should be used with care. For example, we assumed that the overall heat transfer coefficient U is constant throughout the heat exchanger and that the convection heat transfer coefficients can be predicted using the convection correlations. However, it should be kept in mind that the uncertainty in the predicted value of U can even exceed 30 percent. Thus, it is natural to tend to overdesign the heat exchangers in order to avoid unpleasant surprises.

Heat transfer enhancement in heat exchangers is usually accompanied by *increased pressure drop*, and thus *higher pumping power.* Therefore, any gain from the enhancement in heat transfer should be weighed against the cost of the accompanying pressure drop. Also, some thought should be given to which fluid should pass through the tube side and which through the shell side. Usually, the more viscous fluid is more suitable for the shell side (larger passage area and thus lower pressure drop) and the fluid with the higher pressure for the tube side.

Engineers in industry often find themselves in a position to select heat exchangers to accomplish certain heat transfer tasks. Usually, the goal is to heat or cool a certain fluid at a known mass flow rate and temperature to a desired temperature. Thus, the rate of heat transfer in the prospective heat exchanger is

$$\dot{Q}_{\max} = \dot{m}C_p(T_{\text{in}} - T_{\text{out}})$$

which gives the heat transfer requirement of the heat exchanger before having any idea about the heat exchanger itself.

An engineer going through catalogs of heat exchanger manufacturers will be overwhelmed by the type and number of readily available off-the-shelf heat exchangers. The proper selection depends on several factors.

Heat Transfer Rate

This is the most important quantity in the selection of a heat exchanger. A heat exchanger should be capable of transferring heat at the specified rate in order to achieve the desired temperature change of the fluid at the specified mass flow rate.

Cost

Budgetary limitations usually play the most important role in the selection of heat exchangers, except for some specialized cases where "money is no object." An off-the-shelf heat exchanger has a definite cost advantage over those made to order. However, in some cases, none of the existing heat exchangers will do, and it may be necessary to undertake the expensive and time-consuming task of designing and manufacturing a heat exchanger from scratch to suit the needs. This is often the case when the heat exchanger is an integral part of the overall device to be manufactured.

The operation and maintenance costs of the heat exchanger are also important considerations in assessing the overall cost.

Pumping Power

In a heat exchanger, both fluids are usually forced to flow by pumps or fans that consume electrical power. The annual cost of electricity associated with the operation of the pumps and fans can be determined from

> Operating cost = (Pumping power, kW) × (Hours of operation, h)
> × (Price of electricity, $/kWh)

where the pumping power is the total electrical power consumed by the motors of the pumps and fans. For example, a heat exchanger that involves a 1-hp pump and a $\frac{1}{3}$-hp fan (1 hp = 0.746 kW) operating 8 h a day and 5 days a week will consume 2017 kWh of electricity per year, which will cost $161.4 at an electricity cost of 8 cents/kWh.

Minimizing the pressure drop and the mass flow rate of the fluids will *minimize* the operating cost of the heat exchanger, but it will *maximize* the size of the heat exchanger and thus the initial cost. As a rule of thumb, doubling the mass flow rate will reduce the initial cost by *half* but will increase the pumping power requirements by a factor of roughly *eight*.

Typically, fluid velocities encountered in heat exchangers range between 0.7 and 7 m/s for liquids and between 3 and 30 m/s for gases. Low velocities are helpful in avoiding erosion, tube vibrations, and noise as well as pressure drop.

Size and Weight

Normally, the *smaller* and the *lighter* the heat exchanger, the better it is. This is especially the case in the *automotive* and *aerospace* industries, where size and weight requirements are most stringent. Also, a larger heat exchanger normally carries a higher price tag. The space available for the heat exchanger in some cases limits the length of the tubes that can be used.

Type

The type of heat exchanger to be selected depends primarily on the type of *fluids* involved, the *size* and *weight* limitations, and the presence of any *phase-change* processes. For example, a heat exchanger is suitable to cool a liquid by a gas if the surface area on the gas side is many times that on the liquid side. On the other hand, a plate or shell-and-tube heat exchanger is very suitable for cooling a liquid by another liquid.

Materials

The materials used in the construction of the heat exchanger may be an important consideration in the selection of heat exchangers. For example, the thermal and structural *stress effects* need not be considered at pressures below 15 atm or temperatures below 150°C. But these effects are major considerations above 70 atm or 550°C and seriously limit the acceptable materials of the heat exchanger.

A temperature difference of 50°C or more between the tubes and the shell will probably pose *differential thermal expansion* problems and needs to be considered. In the case of corrosive fluids, we may have to select expensive *corrosion-resistant* materials such as stainless steel or even titanium if we are not willing to replace low-cost heat exchangers frequently.

Other Considerations

There are other considerations in the selection of heat exchangers that may or may not be important, depending on the application. For example, being *leak-tight* is an important consideration when *toxic* or *expensive* fluids are involved. Ease of servicing, low maintenance cost, and safety and reliability are some other important considerations in the selection process. Quietness is one of the primary considerations in the selection of liquid-to-air heat exchangers used in heating and air-conditioning applications.

EXAMPLE 10-10   Installing a Heat Exchanger to Save Energy and Money

In a dairy plant, milk is pasteurized by hot water supplied by a natural gas furnace. The hot water is then discharged to an open floor drain at 80°C at a rate of 15 kg/min. The plant operates 24 h a day and 365 days a year. The furnace has an efficiency of 80 percent, and the cost of the natural gas is $0.40 per therm (1 therm = 105,500 kJ). The average temperature of the cold water entering the furnace throughout the year is 15°C. The drained hot water cannot be returned to the furnace and recirculated, because it is contaminated during the process.

In order to save energy, installation of a water-to-water heat exchanger to preheat the incoming cold water by the drained hot water is proposed. Assuming that the heat exchanger will recover 75 percent of the available heat in the hot water, determine the heat transfer rating of the heat exchanger that needs to be purchased and suggest a suitable type. Also, determine the amount of money this heat exchanger will save the company per year from natural gas savings.

Solution A water-to-water heat exchanger is to be installed to transfer energy from drained hot water to the incoming cold water to preheat it. The rate of heat transfer in the heat exchanger and the amount of energy and money saved per year are to be determined.

Assumptions **1** Steady operating conditions exist. **2** The effectiveness of the heat exchanger remains constant.

Properties We use the specific heat of water at room temperature, $C_p = 4.18$ kJ/ kg · °C (Table A-9), and treat it as a constant.

Analysis A schematic of the prospective heat exchanger is given in Figure 10-31. The heat recovery from the hot water will be a maximum when it leaves the heat exchanger at the inlet temperature of the cold water. Therefore,

$$\dot{Q}_{max} = \dot{m}_h C_p (T_{h,\,in} - T_{c,\,in})$$

$$= \left(\frac{15}{60}\text{ kg/s}\right)(4.18\text{ kJ/kg} \cdot °\text{C})(80 - 15)°\text{C}$$

$$= 67.9\text{ kJ/s}$$

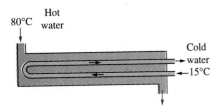

80°C Hot water

Cold water
15°C

FIGURE 10-31
Schematic for Example 10-10.

That is, the existing hot water stream has the potential to supply heat at a rate of 67.9 kJ/s to the incoming cold water. This value would be approached in a counter-flow heat exchanger with a *very large* heat transfer surface area. A heat exchanger of reasonable size and cost can capture 75 percent of this heat transfer potential. Thus, the heat transfer rating of the prospective heat exchanger must be

$$\dot{Q} = \varepsilon \dot{Q}_{max} = (0.75)(67.9\text{ kJ/s}) = \textbf{50.9 kJ/s}$$

That is, the heat exchanger should be able to deliver heat at a rate of 50.9 kJ/s from the hot to the cold water. An ordinary plate or *shell-and-tube* heat exchanger should be adequate for this purpose, since both sides of the heat exchanger involve the same fluid at comparable flow rates and thus comparable heat transfer coefficients. (Note that if we were heating air with hot water, we would have to specify a heat exchanger that has a large surface area on the air side.)

The heat exchanger will operate 24 h a day and 365 days a year. Therefore, the annual operating hours are

$$\text{Operating hours} = (24\text{ h/day})(365\text{ days/year}) = 8760\text{ h/year}$$

Noting that this heat exchanger saves 50.9 kJ of energy per second, the energy saved during an entire year will be

$$\text{Energy saved} = (\text{Heat transfer rate})(\text{Operation time})$$
$$= (50.9\text{ kJ/s})(8760\text{ h/year})(3600\text{ s/h})$$
$$= 1.605 \times 10^9\text{ kJ/year}$$

The furnace is said to be 80-percent efficient. That is, for each 80 units of heat supplied by the furnace, natural gas with an energy content of 100 units must be supplied to the furnace. Therefore, the energy savings determined above result in fuel savings in the amount of

$$\text{Fuel saved} = \frac{\text{Energy saved}}{\text{Furnace efficiency}} = \frac{1.605 \times 10^9 \text{ kJ/year}}{0.80}\left(\frac{1 \text{ therm}}{105{,}500 \text{ kJ}}\right)$$

$$= 19{,}020 \text{ therms/year}$$

since 1 therm = 105,500 kJ. Noting that the price of natural gas is $0.40 per therm, the amount of money saved becomes

$$\text{Money saved} = (\text{Fuel saved}) \times (\text{Price of fuel})$$

$$= (19{,}020 \text{ therms/year})(\$0.40/\text{therm})$$

$$= \mathbf{\$7607/year}$$

Therefore, the installation of the proposed heat exchanger will save the company $7607 a year, and the installation cost of the heat exchanger will probably be paid from the fuel savings in a short time.

10-7 ■ SUMMARY

Heat exchangers are devices that allow the exchange of heat between two fluids without allowing them to mix with each other. Heat exchangers are manufactured in a variety of types, the simplest being the *double-pipe* heat exchanger. In a *parallel-flow* type, both the hot and cold fluids enter the heat exchanger at the same end and move in the same direction, whereas in a *counter-flow* type, the hot and cold fluids enter the heat exchanger at opposite ends and flow in opposite directions. In *compact* heat exchangers, the two fluids move perpendicular to each other, and such a flow configuration is called *cross-flow*. Other common types of heat exchangers in industrial applications are the *plate* and the *shell-and-tube* heat exchangers.

Heat transfer in a heat exchanger usually involves convection in each fluid and conduction through the wall separating the two fluids. In the analysis of heat exchangers, it is convenient to work with an *overall heat transfer coefficient U* or a *total thermal resistance R*, expressed as

$$\frac{1}{UA} = \frac{1}{U_i A_i} = \frac{1}{U_o A_o} = R = \frac{1}{h_i A_i} + R_{\text{wall}} + \frac{1}{h_o A_o}$$

where the subscripts i and o stand for the inner and outer surfaces of the wall that separates the two fluids, respectively. When the wall thickness of the tube is small and the thermal conductivity of the tube material is high, the above relation simplifies to

$$\frac{1}{U} \approx \frac{1}{h_i} + \frac{1}{h_o}$$

where $U \approx U_i \approx U_o$. The effects of fouling on both the inner and the outer surfaces of the tubes of a heat exchanger can be accounted for by

$$\frac{1}{UA} = \frac{1}{U_i A_i} = \frac{1}{U_o A_o} = R = \frac{1}{h_i A_i} + \frac{R_{f,i}}{A_i} + \frac{\ln(D_o/D_i)}{2\pi k L} + \frac{R_{f,o}}{A_o} + \frac{1}{h_o A_o}$$

where $A_i = \pi D_i L$ and $A_o = \pi D_o L$ are the areas of the inner and outer surfaces and $R_{f,i}$ and $R_{f,o}$ are the fouling factors at those surfaces.

In a well-insulated heat exchanger, the rate of heat transfer from the hot fluid is equal to the rate of heat transfer to the cold one. That is,

$$\dot{Q} = \dot{m}_c C_{pc}(T_{c,\text{out}} - T_{c,\text{in}}) = C_c(T_{c,\text{out}} - T_{c,\text{in}})$$

and

$$\dot{Q} = \dot{m}_h C_{ph}(T_{h,\text{in}} - T_{h,\text{out}}) = C_h(T_{h,\text{in}} - T_{h,\text{out}})$$

where the subscripts c and h stand for the cold and hot fluids, respectively, and the product of the mass flow rate and the specific heat of a fluid $C = \dot{m} C_p$ is called the *heat capacity rate*.

Of the two methods used in the analysis of heat exchangers, the *log mean temperature difference* (or LMTD) method is best suited for determining the size of a heat exchanger when all the inlet and the outlet temperatures are known. The *effectiveness-NTU* method is best suited to predict the outlet temperatures of the hot and cold fluid streams in a specified heat exchanger. In the LMTD method, the rate of heat transfer is determined from

$$\dot{Q} = UA\Delta T_{\text{lm}}$$

where

$$\Delta T_{\text{lm}} = \frac{\Delta T_1 - \Delta T_2}{\ln(\Delta T_1/\Delta T_2)}$$

is the *log mean temperature difference*, which is the suitable form of the average temperature difference for use in the analysis of heat exchangers. Here ΔT_1 and ΔT_2 represent the temperature differences between the two fluids at the two ends (inlet and outlet) of the heat exchanger. For cross-flow and multipass shell-and-tube heat exchangers, the logarithmic mean temperature difference is related to the counter-flow one $\Delta T_{\text{lm},CF}$ as

$$\Delta T_{\text{lm}} = F\Delta T_{\text{lm},CF}$$

where F is the *correction factor*, which depends on the geometry of the heat exchanger and the inlet and outlet temperatures of the hot and cold fluid streams.

The *effectiveness* of a heat exchanger is defined as

$$\varepsilon = \frac{\dot{Q}}{\dot{Q}_{\text{max}}} = \frac{\text{Actual heat transfer rate}}{\text{Maximum possible heat transfer rate}}$$

where

$$\dot{Q}_{\text{max}} = C_{\text{min}}(T_{h,\text{in}} - T_{c,\text{in}})$$

and C_{min} is the smaller of $C_h = \dot{m}_h C_{ph}$ and $C_c = \dot{m}_c C_{pc}$. The effectiveness of heat exchangers can be determined from effectiveness relations or charts.

The selection or design of a heat exchanger depends on several factors such as the heat transfer rate, cost, pressure drop, size, weight, construction type, materials, and operating environment.

REFERENCES AND SUGGESTED READING

1 N. Afgan and E. U. Schlunder. *Heat Exchanger: Design and Theory Sourcebook*. Washington DC: McGraw-Hill/Scripta, 1974.

2 R. A. Bowman, A. C. Mueller, and W. M. Nagle. "Mean Temperature Difference in Design." *Transactions of the ASME* 62 (1940), p. 283.

3 A. P. Fraas. *Heat Exchanger Design*. 2d ed. New York: John Wiley & Sons, 1989.

4 K. A. Gardner. "Variable Heat Transfer Rate Correction in Multipass Exchangers, Shell Side Film Controlling." *Transactions of the ASME* 67 (1945), pp. 31–8.

5 J. P. Holman. *Heat Transfer*. 7th ed. New York: McGraw-Hill, 1990.

6 F. P. Incropera and D. P. DeWitt. *Introduction to Heat Transfer*. 2nd ed. New York: John Wiley & Sons, 1990.

7 W. M. Kays and A. L. London. *Compact Heat Exchangers*. 3rd ed. New York: McGraw-Hill, 1984.

8 W. M. Kays and H. C. Perkins. In *Handbook of Heat Transfer*, ed. W. M. Rohsenow and J. P. Hartnett. New York: McGraw-Hill, 1972, Chap. 7.

9 A. C. Mueller. "Heat Exchangers." In *Handbook of Heat Transfer*, ed. W. M. Rohsenow and J. P. Hartnett. New York: McGraw-Hill, 1972, Chap. 18.

10 M. N. Özişik. *Heat Transfer—A Basic Approach*. New York: McGraw-Hill, 1985.

11 E. U. Schlunder. *Heat Exchanger Design Handbook*. Washington, DC: Hemisphere, 1982.

12 *Standards of Tubular Exchanger Manufacturers Association*. New York: Tubular Exchanger Manufacturers Association, latest ed.

13 R. A. Stevens, J. Fernandes, and J. R. Woolf. "Mean Temperature Difference in One, Two, and Three Pass Crossflow Heat Exchangers." *Transactions of the ASME* 79 (1957), pp. 287–97.

14 J. Taborek, G. F. Hewitt, and N. Afgan. *Heat Exchangers: Theory and Practice*. New York: Hemisphere, 1983.

15 L. C. Thomas. *Heat Transfer*. Englewood Cliffs, NJ: Prentice Hall, 1992.

16 G. Walker. *Industrial Heat Exchangers*. Washington, DC: Hemisphere, 1982.

17 F. M. White. *Heat and Mass Transfer*. Reading, MA: Addison-Wesley, 1988.

PROBLEMS*

Types of Heat Exchangers

10-1C   Classify heat exchangers according to flow type and explain the characteristics of each type.

10-2C   Classify heat exchangers according to construction type and explain the characteristics of each type.

10-3C   When is a heat exchanger classified as being compact? Do you think a double-pipe heat exchanger can be classified as a compact heat exchanger?

10-4C   How does a cross-flow heat exchanger differ from a counter-flow one? What is the difference between mixed and unmixed fluids in cross-flow?

10-5C   What is the role of the baffles in a shell-and-tube heat exchanger? How does the presence of baffles affect the heat transfer and the pumping power requirements? Explain.

10-6C   Draw a 1-shell-pass and 6-tube-passes shell-and-tube heat exchanger. What are the advantages and disadvantages of using 6 tube passes instead of just 2 of the same diameter?

10-7C   Draw a 2-shell-passes and 8-tube-passes shell-and-tube heat exchanger. What is the primary reason for using so many tube passes?

10-8C   What is a regenerative heat exchanger? How does a static type of regenerative heat exchanger differ from a dynamic type?

The Overall Heat Transfer Coefficient

10-9C   What are the heat transfer mechanisms involved during heat transfer from the hot to the cold fluid?

10-10C   Under what conditions is the thermal resistance of the tube in a heat exchanger negligible?

10-11C   Consider a double-pipe parallel-flow heat exchanger of length L. The inner and outer diameters of the inner tube are D_1 and D_2, respectively, and the inner diameter of the outer tube is D_3. Explain how you would determine the two heat transfer surface areas A_i and A_o. When is it reasonable to assume $A_i \approx A_o \approx A$?

10-12C   Is the approximation $h_i \approx h_o \approx h$ for the convection heat transfer coefficient in a heat exchanger a reasonable one when the thickness of the tube wall is negligible?

10-13C   Under what conditions can the overall heat transfer coefficient of a heat exchanger be determined from $U = (1/h_i + 1/h_o)^{-1}$?

---

*Students are encouraged to answer *all* the concept "C" questions.

10-14C What are the restrictions on the relation $UA = U_i A_i = U_o A_o$ for a heat exchanger? Here A is the heat transfer surface area and U is the overall heat transfer coefficient.

10-15C In a thin-walled double-pipe heat exchanger, when is the approximation $U = h_i$ a reasonable one? Here U is the overall heat transfer coefficient and h_i is the convection heat transfer coefficient inside the tube.

10-16C What are the common causes of fouling in a heat exchanger? How does fouling affect heat transfer and pressure drop?

10-17C How is the thermal resistance due to fouling in a heat exchanger accounted for? How do the fluid velocity and temperature affect fouling?

10-18 A double-pipe heat exchanger is constructed of a copper ($k = 380$ W/m · °C) inner tube of internal diameter $D_i = 1.2$ cm and external diameter $D_o = 1.6$ cm and an outer tube of diameter 3.0 cm. The convection heat transfer coefficient is reported to be $h_i = 700$ W/m² · °C on the inner surface of the tube and $h_o = 1400$ W/m² · °C on its outer surface. For a fouling factor $R_{f,i} = 0.0005$ m² · °C/W on the tube side and $R_{f,o} = 0.0002$ m² · °C/W on the shell side, determine (*a*) the thermal resistance of the heat exchanger per unit length and (*b*) the overall heat transfer coefficients U_i and U_o based on the inner and outer surface areas of the tube, respectively.

10-19 Water at an average temperature of 107°C and an average velocity of 3.5 m/s flows through a 5-m-long stainless steel tube ($k = 14.2$ W/m · °C) in a boiler. The inner and outer diameters of the tube are $D_i = 1.0$ cm and $D_o = 1.4$ cm, respectively. If the convection heat transfer coefficient at the outer surface of the tube where boiling is taking place is $h_o = 8400$ W/m² · °C, determine the overall heat transfer coefficient U_i of this boiler based on the inner surface area of the tube.

10-20 Repeat Problem 10-19, assuming a fouling factor $R_{f,i} = 0.0005$ m² · °C/W on the inner surface of the tube.

10-21 A long thin-walled double-pipe heat exchanger with tube and shell diameters of 1.0 cm and 2.5 cm, respectively, is used to condense refrigerant 134a at 20°C by water. The refrigerant flows through the tube, with a convection heat transfer coefficient of $h_i = 5000$ W/m² · °C. Water flows through the shell at a rate of 0.3 kg/s. Determine the overall heat transfer coefficient of this heat exchanger. *Answer:* 2100 W/m² · °C

10-22 Repeat Problem 10-21 by assuming a 2-mm-thick layer of limestone ($k = 1.3$ W/m · °C) forms on the outer surface of the inner tube.

10-23E Water at an average temperature of 140°F and an average velocity of 8 ft/s flows through a thin-walled $\frac{3}{4}$-in.-diameter tube. The water is cooled by air that flows across the tube with a velocity of $U_\infty = 20$ ft/s at an average temperature of 80°F. Determine the overall heat transfer coefficient.

Analysis of Heat Exchangers

10-24C   What are the common approximations made in the analysis of heat exchangers?

10-25C   Under what conditions is the heat transfer relation

$$\dot{Q} = \dot{m}_c C_{pc}(T_{c,\,\text{out}} - T_{c,\,\text{in}}) = \dot{m}_h C_{ph}\,(T_{h,\,\text{in}} - T_{h,\,\text{out}})$$

valid for a heat exchanger?

10-26C   What is the heat capacity rate? What can you say about the temperature changes of the hot and cold fluids in a heat exchanger if both fluids have the same capacity rate? What does a heat capacity of infinity for a fluid in a heat exchanger mean?

10-27C   Consider a condenser in which steam at a specified temperature is condensed by rejecting heat to the cooling water. If the heat transfer rate in the condenser and the temperature rise of the cooling water is known, explain how the rate of condensation of the steam and the mass flow rate of the cooling water can be determined. Also, explain how the total thermal resistance R of this condenser can be evaluated in this case.

10-28C   Under what conditions will the temperature rise of the cold fluid in a heat exchanger be equal to the temperature drop of the hot fluid?

The Log Mean Temperature Difference Method

10-29C   In the heat transfer relation $\dot{Q} = UA\Delta T_{\text{lm}}$ for a heat exchanger, what is ΔT_{lm} called? How is it determined for a parallel-flow and counter-flow heat exchanger?

10-30C   How does the log mean temperature difference for a heat exchanger differ from the arithmetic mean temperature difference (AMTD)? For specified inlet and outlet temperatures, which one of these two quantities is larger?

10-31C   The temperature difference between the hot and cold fluids in a heat exchanger is given to be ΔT_1 at one end and ΔT_2 at the other end. Can the logarithmic temperature difference ΔT_{lm} of this heat exchanger be greater than both ΔT_1 and ΔT_2? Explain.

10-32C   Can the logarithmic mean temperature difference ΔT_{lm} of a heat exchanger be a negative quantity? Explain.

10-33C   Can the outlet temperature of the cold fluid in a heat exchanger be higher than the outlet temperature of the hot fluid in a parallel-flow heat exchanger? How about in a counter-flow heat exchanger? Explain.

10-34C   For specified inlet and outlet temperatures, for what kind of heat exchanger will the ΔT_{lm} be greatest: double-pipe parallel-flow, double-pipe counter-flow, cross-flow, or multipass shell-and-tube heat exchanger?

10-35C   In the heat transfer relation $\dot{Q} = UAF\Delta T_{\text{lm}}$ for a heat exchanger, what is the quantity F called? What does it represent? Can F be greater than one?

10-36C When the outlet temperatures of the fluids in a heat exchanger are not known, is it still practical to use the LMTD method? Why?

10-37C Explain how the LMTD method can be used to determine the heat transfer surface area of a multipass shell-and-tube heat exchanger when all the necessary information, including the outlet temperatures, is given.

10-38 Steam in the condenser of a steam power plant is to be condensed at a temperature of 50°C (h_{fg} = 2305 kJ/kg) with cooling water (C_p = 4180 J/kg · °C) from a nearby lake, which enters the tubes of the condenser at 18°C and leaves at 27°C. The surface area of the tubes is 58 m², and the overall heat transfer coefficient is 2400 W/m² · °C. Determine the mass flow rate of the cooling water needed and the rate of condensation of the steam in the condenser.   *Answers:* 101 kg/s, 1.65 kg/s

10-39 A double-pipe parallel-flow heat exchanger is to heat water (C_p = 4180 J/kg · °C) from 25 °C to 60°C at a rate of 0.2 kg/s. The heating is to be accomplished by geothermal water (C_p = 4310 J/kg · °C) available at 140°C at a mass flow rate of 0.3 kg/s. The inner tube is thin-walled and has a diameter of 0.8 cm. If the overall heat transfer coefficient of the heat exchanger is 550 W/m² · °C, determine the length of the heat exchanger required to achieve the desired heating.

10-40E A 1-shell-pass and 8-tube-passes heat exchanger is used to heat glycerin (C_p = 0.60 Btu/lbm · °F) from 75°F to 140°F by hot water (C_p = 1.0 Btu/lbm · °F) that enters the thin-walled 0.5-in.-diameter tubes at 190°F and leaves at 120°F. The total length of the tubes in the heat exchanger is 500 ft. The convection heat transfer coefficient is 4 Btu/h · ft² · °F on the glycerin (shell) side and 50 Btu/h · ft² · °F on the water (tube) side. Determine the rate of heat transfer in the heat exchanger (*a*) before any fouling occurs and (*b*) after fouling with a fouling factor of 0.002 h · ft² · °F/Btu occurs on the outer surfaces of the tubes.

10-41 A test is conducted to determine the overall heat transfer coefficient in a shell-and-tube oil-to-water heat exchanger that has 24 tubes of internal diameter 1.2 cm and length 2 m in a single shell. Cold water (C_p = 4180 J/kg · °C) enters the tubes at 20°C at a rate of 5 kg/s and leaves at 55°C. Oil (C_p = 2150 J/kg · °C) flows through the shell and is cooled from 120°C to 45°C. Determine the overall heat transfer coefficient U_i of this heat exchanger based on the inner surface area of the tubes.   *Answer:* 13.9 kW/m² · °C

10-42 A double-pipe counter-flow heat exchanger is to cool ethylene glycol (C_p = 2560 J/kg · °C) flowing at a rate of 2 kg/s from 80°C to 40°C by water (C_p = 4180 J/kg · °C) that enters at 20°C and leaves at 55°C. The overall heat transfer coefficient based on the inner surface area of the tube is 250 W/m² · °C. Determine (*a*) the rate of heat transfer, (*b*) the mass flow rate of water, and (*c*) the heat transfer surface area on the inner side of the tube.

10-43 Water (C_p = 4180 J/kg · °C) enters the 2.5-cm-internal-diameter tube of a double-pipe counter-flow heat exchanger at 17°C at a rate of 3 kg/s. It is heated by steam condensing at 120°C (h_{fg} = 2203 kJ/kg) in the shell. If the overall heat transfer coefficient of the heat exchanger is 1500 W/m² · °C, determine the length of the tube required in order to heat the water to 80°C.

10-44 A thin-walled double-pipe counter-flow heat exchanger is to be used to cool oil (C_p = 2200 J/kg · °C) from 150°C to 40°C at a rate of 2 kg/s by water (C_p = 4180 J/kg · °C) that enters at 22°C at a rate of 1.5 kg/s. The diameter of the tube is 2.5 cm, and its length is 6 m. Determine the overall heat transfer coefficient of this heat exchanger.

10-45 Consider a water-to-water double-pipe heat exchanger whose flow arrangement is not known. The temperature measurements indicate that the cold water enters at 20°C and leaves at 50°C, while the hot water enters at 80°C and leaves at 45°C. Do you think this is a parallel-flow or counter-flow heat exchanger? Explain.

10-46 Cold water (C_p = 4180 J/kg · °C) leading to a shower enters a thin-walled double-pipe counter-flow heat exchanger at 15°C at a rate of 0.25 kg/s and is heated to 45°C by hot water (C_p = 4190 J/kg · °C) that enters at 100°C at a rate of 3 kg/s. If the overall heat transfer coefficient is 950 W/m² · °C, determine the rate of heat transfer and the heat transfer surface area of the heat exchanger.

10-47 Engine oil (C_p = 2100 J/kg · °C) is to be heated from 20°C to 60°C at a rate of 0.3 kg/s in a 2-cm-diameter thin-walled copper tube by condensing steam outside at a temperature of 130°C (h_{fg} = 2174 kJ/kg). For an overall heat transfer coefficient of 650 W/m² · °C, determine the rate of heat transfer and the length of the tube required to achieve it.   *Answers:* 25.2 kW, 7.0 m

10-48E Geothermal water (C_p = 1.03 Btu/lbm · °F) is to be used as the heat source to supply heat to the hydronic heating system of a house at a rate of 30 Btu/s in a double-pipe counter-flow heat exchanger. Water (C_p = 1.0 Btu/lbm · °F) is heated from 140°F to 200°F in the heat exchanger as the geothermal water is cooled from 250°F to 180°F. Determine the mass flow rate of each fluid and the total thermal resistance of this heat exchanger.

10-49 Glycerin (C_p = 2400 J/kg · °C) at 20°C and 0.3 kg/s is to be heated by ethylene glycol (C_p = 2500 J/kg · °C) at 60°C in a thin-walled double-pipe parallel-flow heat exchanger. The temperature difference between the two fluids is 15°C at the outlet of the heat exchanger. If the overall heat transfer coefficient is 240 W/m² · °C and the heat transfer surface area is 7.6 m², determine (a) the rate of heat transfer, (b) the outlet temperature of the glycerin, and (c) the mass flow rate of the ethylene glycol.

10-50 Air (C_p = 1005 J/kg · °C) is to be preheated by hot exhaust gases in a cross-flow heat exchanger before it enters the furnace. Air enters the heat exchanger at 95 kPa and 20°C at a rate of 0.8 m³/s. The combustion gases (C_p = 1100 J/kg · °C) enter at 180°C at a rate of 1.1 kg/s and leave at 95°C. The product of the overall heat transfer coefficient and the heat transfer surface area is AU = 1200 W/°C. Assuming both fluids to be unmixed, determine the rate of heat transfer and the outlet temperature of the air.

10-51 A shell-and-tube heat exchanger with 2-shell passes and 12-tube passes is used to heat water (C_p = 4180 J/kg · °C) in the tubes from 20°C to

70°C at a rate of 4.5 kg/s. Heat is supplied by hot oil (C_p = 2300 J/kg · °C) that enters the shell side at 170°C at a rate of 10 kg/s. For a tube-side overall heat transfer coefficient of 600 W/m² · °C, determine the heat transfer surface area on the tube side.   *Answer:* 15 m²

10-52   Repeat Problem 10-51 for a mass flow rate of 2 kg/s for water.

10-53   A shell-and-tube heat exchanger with 2-shell passes and 8-tube passes is used to heat ethyl alcohol (C_p = 2670 J/kg · °C) in the tubes from 25°C to 70°C at a rate of 2.1 kg/s. The heating is to be done by water (C_p = 4190 J/kg · °C) that enters the shell side at 95°C and leaves at 45°C. If the overall heat transfer coefficient is 800 W/m² · °C, determine the heat transfer surface area of the heat exchanger.

10-54   A shell-and-tube heat exchanger with 2-shell passes and 12-tube passes is used to heat water (C_p = 4180 J/kg · °C) with ethylene glycol (C_p = 2680 J/kg · °C). Water enters the tubes at 22°C at a rate of 0.8 kg/s and leaves at 70°C. Ethylene glycol enters the shell at 110°C and leaves at 60°C. If the overall heat transfer coefficient based on the tube side is 280 W/m² · °C, determine the rate of heat transfer and the heat transfer surface area on the tube side.

10-55E   Steam is to be condensed on the shell side of a 1-shell-pass and 8-tube-passes condenser, with 50 tubes in each pass at 90°F (h_{fg} = 1043 Btu/lbm) at a rate of 20 lbm/s. Cooling water (C_p = 1.0 Btu/lbm · °F) enters the tubes at 60°F and leaves at 73°F. The tubes are thin-walled and have a diameter of $\frac{3}{4}$ in. and length of 5 ft per pass. If the overall heat transfer coefficient is 600 Btu/h · ft² · °F, determine (*a*) the rate of heat transfer, (*b*) the rate of condensation of steam, and (*c*) the mass flow rate of cold water.

10-56   A shell-and-tube heat exchanger with 1-shell pass and 10-tube passes is used to heat glycerin (C_p = 2480 J/kg · °C) in the shell, with hot water in the tubes. The tubes are thin-walled and have a diameter of 1.5 cm and length of 2 m per pass. The water enters the tubes at 100°C at a rate of 5 kg/s and leaves at 55°C. The glycerin enters the shell at 15°C and leaves at 55°C. Determine the mass flow rate of the glycerin and the overall heat transfer coefficient of the heat exchanger.

The Effectiveness-NTU Method

10-57C   Under what conditions is the effectiveness-NTU method definitely preferred over the LMTD method in heat exchanger analysis?

10-58C   What does the effectiveness of a heat exchanger represent? Can effectiveness be greater than one? On what factors does the effectiveness of a heat exchanger depend?

10-59C   For a specified fluid pair, inlet temperatures, and mass flow rates, what kind of heat exchanger will have the highest effectiveness: double-pipe parallel-flow, double-pipe counter-flow, cross-flow, or multipass shell-and-tube heat exchanger?

10-60C   Explain how you can evaluate the outlet temperatures of the cold and hot fluids in a heat exchanger after its effectiveness is determined.

10-61C   Can the temperature of the hot fluid drop below the inlet temperature of the cold fluid at any location in a heat exchanger? Explain.

10-62C   Can the temperature of the cold fluid rise above the inlet temperature of the hot fluid at any location in a heat exchanger? Explain.

10-63C   Consider a heat exchanger in which both fluids have the same specific heats but different mass flow rates. Which fluid will experience a larger temperature change: the one with the lower or higher mass flow rate?

10-64C   Explain how the maximum possible heat transfer rate $\dot{Q}_{max}$ in a heat exchanger can be determined when the mass flow rates, specific heats, and the inlet temperatures of the two fluids are specified. Does the value of $\dot{Q}_{max}$ depend on the type of the heat exchanger?

10-65C   Consider two double-pipe counter-flow heat exchangers that are identical except that one is twice as long as the other one. Which heat exchanger is more likely to have a higher effectiveness?

10-66C   Consider a double-pipe counter-flow heat exchanger. In order to enhance heat transfer, the length of the heat exchanger is now doubled. Do you think, its effectiveness will also double?

10-67C   Consider a shell-and-tube water-to-water heat exchanger with identical mass flow rates for both the hot and cold water streams. Now the mass flow rate of the cold water is reduced by half. Will the effectiveness of this heat exchanger increase, decrease, or remain the same as a result of this modification? Explain. Assume the overall heat transfer coefficient and the inlet temperatures remain the same.

10-68C   Under what conditions can a counter-flow heat exchanger have an effectiveness of one? What would your answer be for a parallel-flow heat exchanger?

10-69C   How is the NTU of a heat exchanger defined? What does it represent? Is a heat exchanger with a very large NTU (say, 10) necessarily a good one to buy?

10-70C   Consider a heat exchanger that has an NTU of 4. Someone proposes to double the size of the heat exchanger and thus double the NTU to 8 in order to increase the effectiveness of the heat exchanger and thus save energy. Would you support this proposal?

10-71C   Consider a heat exchanger that has an NTU of 0.1. Someone proposes to triple the size of the heat exchanger and thus triple the NTU to 0.3 in order to increase the effectiveness of the heat exchanger and thus save energy. Would you support this proposal?

10-72   Air ($C_p = 1005$ J/kg $\cdot$ °C) enters a cross-flow heat exchanger at 12°C at a rate of 3 kg/s, where it is heated by a hot water stream ($C_p = 4190$ J/kg $\cdot$ °C) that enters the heat exchanger at 90°C at a rate of 1 kg/s. Determine the maximum heat transfer rate and the outlet temperatures of the cold and the hot water streams for that case.

10-73 Hot oil (C_p = 2200 J/kg · °C) is to be cooled by water (C_p = 4180 J/kg · °C) in a 2-shell-passes and 12-tube-passes heat exchanger. The tubes are thin-walled and are made of copper with an internal diameter of 1.8 cm. The length of each tube pass in the heat exchanger is 3 m, and the overall heat transfer coefficient is 340 W/m² · °C. Water flows through the tubes at a rate of 0.1 kg/s, and the oil through the shell at a rate of 0.2 kg/s. The water and the oil enter at temperatures 18°C and 160°C, respectively. Determine the rate of heat transfer in the heat exchanger and the outlet temperatures of the water and the oil.   *Answers:* 36.2 kW, 104.6°C, 77.7°C

10-74 Consider an oil-to-oil double-pipe heat exchanger whose flow arrangement is not known. The temperature measurements indicate that the cold oil enters at 20°C and leaves at 55°C, while the hot oil enters at 80°C and leaves at 45°C. Do you think this is a parallel-flow or counter-flow heat exchanger? Why? Assuming the mass flow rates of both fluids to be identical, determine the effectiveness of this heat exchanger.

10-75E Hot water enters a double-pipe counter-flow water-to-oil heat exchanger at 200°F and leaves at 100°F. Oil enters at 70°F and leaves at 130°F. Determine which fluid has the smaller heat capacity rate and calculate the effectiveness of this heat exchanger.

10-76 A thin-walled double-pipe parallel-flow heat exchanger is used to heat a chemical whose specific heat is 1800 J/kg · °C with hot water (C_p = 4180 J/kg · °C). The chemical enters at 20°C at a rate of 3 kg/s, while the water enters at 110°C at a rate of 2 kg/s. The heat transfer surface area of the heat exchanger is 7 m² and the overall heat transfer coefficient is 1200 W/m² · °C. Determine the outlet temperatures of the chemical and the water.

10-77 A cross-flow air-to-water heat exchanger with an effectiveness of 0.65 is used to heat water (C_p = 4180 J/kg · °C) with hot air (C_p = 1010 J/kg · °C). Water enters the heat exchanger at 20°C at a rate of 4 kg/s, while air enters at 100°C at a rate of 9 kg/s. If the overall heat transfer coefficient based on the water side is 260 W/m² · °C, determine the heat transfer surface area of the heat exchanger on the water side. Assume both fluids are unmixed.   *Answer:* 52.4 m²

10-78 Water (C_p = 4180 J/kg · °C) enters the 2.5-cm-internal-diameter tube of a double-pipe counter-flow heat exchanger at 17°C at a rate of 3 kg/s. Water is heated by steam condensing at 120°C (h_{fg} = 2203 kJ/kg) in the shell. If the overall heat transfer coefficient of the heat exchanger is 900 W/m² · °C, determine the length of the tube required in order to heat the water to 80°C using (*a*) the LMTD method and (*b*) the ε-NTU method.

10-79 Ethanol is vaporized at 78°C (h_{fg} = 846 kJ/kg) in a double-pipe parallel-flow heat exchanger at a rate of 0.03 kg/s by hot oil (C_p = 2200 J/kg · °C) that enters at 120°C. If the heat transfer surface area and the overall heat transfer coefficients are 7.8 m² and 210 W/m² · °C, respectively, determine the outlet temperature and the mass flow rate of the oil using (a) the LMTD method and (b) the ε-NTU method.

10-80 Water (C_p = 4180 J/kg · °C) is to be heated by solar-heated hot air (C_p = 1010 J/kg · °C) in a double-pipe counter-flow heat exchanger. Air enters the heat exchanger at 90°C at a rate of 0.3 kg/s, while water enters at 22°C at a rate of 0.1 kg/s. The overall heat transfer coefficient based on the inner side of the tube is given to be 80 W/m² · °C. The length of the tube is 12 m and the internal diameter of the tube is 1.2 cm. Determine the outlet temperatures of the water and the air.

10-81E A thin-walled double-pipe heat exchanger is to be used to cool oil (C_p = 0.525 Btu/lbm · °F) from 300°F to 105°F at a rate of 5 lbm/s by water (C_p = 1.0 Btu/lbm · °F) that enters at 70°F at a rate of 3 lbm/s. The diameter of the tube is 1 in. and its length is 20 ft. Determine the overall heat transfer coefficient of this heat exchanger using (a) the LMTD method and (b) the ε-NTU method.

10-82 Cold water (C_p = 4180 J/kg · °C) leading to a shower enters a thin-walled double-pipe counter-flow heat exchanger at 15°C at a rate of 0.25 kg/s and is heated to 45°C by hot water (C_p = 4190 J/kg · °C) that enters at 100°C at a rate of 3 kg/s. If the overall heat transfer coefficient is 950 W/m² · °C, determine the rate of heat transfer and the heat transfer surface area of the heat exchanger using the ε-NTU method.
    Answers: 31.35 kW, 0.482 m²

10-83 Glycerin (C_p = 2400 J/kg · °C) at 20°C and 0.3 kg/s is to be heated by ethylene glycol (C_p = 2500 J/kg · °C) at 60°C and the same mass flow rate in a thin-walled double-pipe parallel-flow heat exchanger. If the overall heat transfer coefficient is 240 W/m² · °C and the heat transfer surface area is 7.6 m², determine (a) the rate of heat transfer and (b) the outlet temperatures of the glycerin and the glycol.

10-84 A cross-flow heat exchanger consists of 40 thin-walled tubes of 1-cm diameter located in a duct of 1 m × 1 m cross-section. There are no fins attached to the tubes. Cold water (C_p = 4180 J/kg · °C) enters the tubes at 18°C with an average velocity of 3 m/s, while hot air (C_p = 1010 J/kg · °C) enters the channel at 130°C and 105 kPa at an average velocity of 12 m/s. If the overall heat transfer coefficient is 80 W/m² · °C, determine the outlet temperatures of both fluids and the rate of heat transfer.

10-85 A shell-and-tube heat exchanger with 2-shell passes and 8-tube passes is used to heat ethyl alcohol (C_p = 2670 J/kg · °C) in the tubes from 25°C to 70°C at a rate of 2.1 kg/s. The heating is to be done by water (C_p = 4190 J/kg · °C) that enters the shell at 95°C and leaves at 45°C. If the overall heat transfer coefficient is 800 W/m² · °C, determine the heat transfer surface area of the heat exchanger using (a) the LMTD method and (b) the ε-NTU method.   *Answer:* 17.4 m²

10-86 Steam is to be condensed on the shell side of a 1-shell-pass and 8-tube-passes condenser, with 50 tubes in each pass, at 30°C (h_{fg} = 2430 kJ/kg). Cooling water (C_p = 4180 J/kg · °C) enters the tubes at 15°C at a rate of 1800 kg/h. The tubes are thin-walled, and have a diameter of 1.5 cm and length of 2 m per pass. If the overall heat transfer coefficient is 3000 W/m² · °C, determine (*a*) the rate of heat transfer and (*b*) the rate of condensation of steam.

10-87 Cold water (C_p = 4180 J/kg · °C) enters the tubes of a heat exchanger with 2-shell-passes and 20-tube-passes at 20°C at a rate of 3 kg/s, while hot oil (C_p = 2200 J/kg · °C) enters the shell at 130°C at the same mass flow rate. The overall heat transfer coefficient based on the outer surface of the tube is 300 W/m² · °C and the heat transfer surface area on that side is 20 m². Determine the rate of heat transfer using (*a*) the LMTD method and (*b*) the ε-NTU method.

Selection of Heat Exchangers

10-88C A heat exchanger is to be selected to cool a hot liquid chemical at a specified rate to a specified temperature. Explain the steps involved in the selection process.

10-89C There are two heat exchangers that can meet the heat transfer requirements of a facility. One is smaller and cheaper but requires a larger pump, while the other is larger and more expensive but has a smaller pressure drop and thus requires a smaller pump. Both heat exchangers have the same life expectancy and meet all other requirements. Explain which heat exchanger you would choose under what conditions.

10-90C There are two heat exchangers that can meet the heat transfer requirements of a facility. Both have the same pumping power requirements, the same useful life, and the same price tag. But one is heavier and larger in size. Under what conditions would you choose the smaller one?

10-91 A heat exchanger is to cool oil (C_p = 2200 J/kg · °C) at a rate of 20 kg/s from 120°C to 50°C by air. Determine the heat transfer rating of the heat exchanger and propose a suitable type.

10-92 A shell-and-tube process heater is to be selected to heat water (C_p = 4190 J/kg · °C) from 20°C to 90°C by steam flowing on the shell side. The heat transfer load of the heater is 600 kW. If the inner diameter of the tubes is 1 cm and the velocity of water is not to exceed 3 m/s, determine how many tube passes need to be used in the heat exchanger.

10-93 The condenser of a large power plant is to remove 500 MW of heat from steam condensing at 30°C (h_{fg} = 2430 kJ/kg). The cooling is to be accomplished by cooling water (C_p = 4180 J/kg · °C) from a nearby river, which enters the tubes at 18°C and leaves at 26°C. The tubes of the heat exchanger have an internal diameter of 2 cm, and the overall heat transfer

coefficient is 3500 W/m$^2 \cdot$°C. Determine the total length of the tubes required in the condenser. What type of heat exchanger is suitable for this task?
 Answer: 312.3 km

10-94 Repeat Problem 10-93 for a heat transfer load of 300 MW.

Review Problems

10-95 Hot oil is to be cooled in a multipass shell-and-tube heat exchanger by water. The oil flows through the shell, with a heat transfer coefficient of $h_o = 35$ W/m$^2 \cdot$°C, and the water flows through the tube with an average velocity of 3 m/s. The tube is made of brass ($k = 110$ W/m$\cdot$°C) with internal and external diameters of 1.3 cm and 1.5 cm, respectively. Using water properties at 25°C, determine the overall heat transfer coefficient of this heat exchanger based on the inner surface.

10-96 Repeat Problem 10-95 by assuming a fouling factor $R_{f,o} = 0.0004$ m$^2 \cdot$°C/W on the outer surface of the tube.

10-97 Cold water ($C_p = 4180$ J/kg$\cdot$°C) enters the tubes of a heat exchanger with 2-shell passes and 20-tube passes at 20°C at a rate of 3 kg/s, while hot oil ($C_p = 2200$ J/kg$\cdot$°C) enters the shell at 130°C at the same mass flow rate and leaves at 60°C. If the overall heat transfer coefficient based on the outer surface of the tube is 300 W/m$^2 \cdot$°C, determine (*a*) the rate of heat transfer and (*b*) the heat transfer surface area on the outer side of the tube.
 Answers: (*a*) 462 kW, (*b*) 29.2 m^2

10-98E Water ($C_p = 1.0$ Btu/lbm$\cdot$°F) is to be heated by solar-heated hot air ($C_p = 0.24$ Btu/lbm$\cdot$°F) in a double-pipe counter-flow heat exchanger. Air enters the heat exchanger at 190°F at a rate of 0.7 lbm/s and leaves at 120°F. Water enters at 70°F at a rate of 0.2 lbm/s. The overall heat transfer coefficient based on the inner side of the tube is given to be 20 Btu/h$\cdot$ft$^2 \cdot$°F. Determine the length of the tube required for a tube internal diameter of 0.5 in.

10-99 By taking the limit as $\Delta T_2 \longrightarrow \Delta T_1$, show that when $\Delta T_1 = \Delta T_2$ for a heat exchanger, the ΔT_{lm} relation reduces to $\Delta T_{lm} = \Delta T_1 = \Delta T_2$.

10-100 The condenser of a room air conditioner is designed to reject heat at a rate of 15,000 kJ/h from Refrigerant-134a as the refrigerant is condensed at a temperature of 40°C. Air ($C_p = 1005$ J/kg$\cdot$°C) flows across the finned condenser coils, entering at 25°C and leaving at 35°C. If the overall heat transfer coefficient based on the refrigerant side is 150 W/m$^2 \cdot$°C, determine the heat transfer area on the refrigerant side.   *Answer:* 3.05 m^2

10-101 Air ($C_p = 1005$ J/kg$\cdot$°C) is to be preheated by hot exhaust gases in a cross-flow heat exchanger before it enters the furnace. Air enters the heat exchanger at 95 kPa and 20°C at a rate of 0.8 m^3/s. The combustion gases ($C_p = 1100$ J/kg$\cdot$°C) enter at 180°C at a rate of 1.1 kg/s and leave at 95°C. The product of the overall heat transfer coefficient and the heat transfer surface area is $AU = 1200$ W/°C. Assuming both fluids to be unmixed, determine the rate of heat transfer.

10-102 In a chemical plant, a certain chemical is heated by hot water supplied by a natural gas furnace. The hot water ($C_p = 4180$ J/kg · °C) is then discharged at 60°C at a rate of 8 kg/min. The plant operates 8 h a day, 5 days a week, 52 weeks a year. The furnace has an efficiency of 78 percent, and the cost of the natural gas is $0.54 per therm (1 therm = 100,000 Btu = 105,500 kJ). The average temperature of the cold water entering the furnace throughout the year is 14°C. In order to save energy, it is proposed to install a water-to-water heat exchanger to preheat the incoming cold water by the drained hot water. Assuming that the heat exchanger will recover 72 percent of the available heat in the hot water, determine the heat transfer rating of the heat exchanger that needs to be purchased and suggest a suitable type. Also, determine the amount of money this heat exchanger will save the company per year from natural gas savings.

Computer, Design, and Essay Problems

10-103 Write an interactive computer program that will give the effectiveness of a heat exchanger and the outlet temperatures of both the hot and cold fluids when the type of fluids, the inlet temperatures, the mass flow rates, the heat transfer surface area, the overall heat transfer coefficient, and the type of heat exchanger are specified. The program should allow the user to select from the fluids water, engine oil, glycerin, ethyl alcohol, and ammonia. Assume constant specific heats at about room temperature.

10-104 Repeat the problem above, accounting for the variation of specific heats with temperature.

10-105 Water flows through a shower head steadily at a rate of 8 kg/min. The water is heated in an electric water heater from 15°C to 45°C. In an attempt to conserve energy, it is proposed to pass the drained warm water at a temperature of 38°C through a heat exchanger to preheat the incoming cold water. Design a heat exchanger that is suitable for this task, and discuss the potential savings in energy and money for your area.

10-106 Open the engine compartment of your car and search for heat exchangers. How many do you have? What type are they? Why do you think those specific types are selected? If you were redesigning the car, would you use different kinds? Explain.

10-107 Write an essay on the static and dynamic types of regenerative heat exchangers and compile information about the manufacturers of such heat exchangers. Choose a few models by different manufacturers and compare their costs and performance.

Mass Transfer

To this point we have restricted our attention to heat transfer problems that did not involve any mass transfer. However, many significant heat transfer problems encountered in practice involve mass transfer. For example, about one-third of the heat loss from a resting person is due to the evaporation. It turns out that mass transfer is analogous to heat transfer in many respects, and there is close resemblance between the heat and mass transfer relations. In this chapter we discuss the mass transfer mechanisms and develop relations for the mass transfer rate for some situations commonly encountered in practice.

Distinction should be made between *mass transfer* and the *bulk fluid motion* (or *fluid flow*) that occurs on a macroscopic level as a fluid is transported from one location to another. Mass transfer requires the presence of two regions at different chemical compositions, and mass transfer refers to the movement of a chemical species from a high concentration region toward a lower concentration one relative to the other chemical species present in the medium. The primary driving force for fluid flow is the *pressure difference*, whereas for mass transfer it is the *concentration difference*. Therefore, we do not speak of mass transfer in a homogeneous medium.

We begin this chapter by pointing out numerous analogies between heat and mass transfer and draw several parallels between them. We then discuss boundary conditions associated with mass transfer and one-dimensional steady and transient mass diffusion. Following a discussion of mass transfer in a moving medium, we close this chapter by considering convection mass transfer and simultaneous heat and mass transfer.

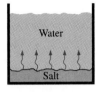 

(a) Before   *(b)* After

FIGURE 11-1

Whenever there is concentration difference of a physical quantity in a medium, nature tends to equalize things by forcing a flow from the high to the low concentration region.

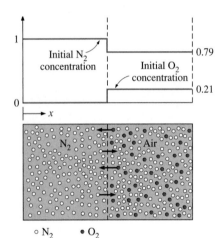

FIGURE 11-2

A tank that contains N_2 and air in its two compartments, and the diffusion of N_2 into the air when the partition is removed.

11-1 ■ INTRODUCTION

It is a common observation that whenever there is an imbalance of a commodity in a medium, the nature tends to redistribute it until a "balance" or "equality" is established. This tendency is often referred to as the *driving force,* which is the mechanism behind many naturally occurring transport phenomena.

If we define the amount of a commodity per unit volume as the **concentration** of that commodity, we can say that the flow of a commodity is always in the direction of decreasing concentration; that is, from the region of high concentration to the region of low concentration (Fig. 11-1). The commodity simply creeps away during redistribution, and thus the flow is a *diffusion process.* The rate of flow of the commodity is proportional to the *concentration gradient dC/dx,* which is the change in the concentration C per unit length in the direction of flow x, and the area A normal to the direction of flow and is expressed as

$$\text{Flow rate} \propto (\text{Normal area})(\text{Concentration gradient})$$

or

$$\dot{Q} = -k_{\text{diff}} A \frac{dC}{dx} \qquad (11\text{-}1)$$

Here the proportionality constant k_{diff} is the *diffusion coefficient* of the medium, which is a measure of how fast a commodity diffuses in the medium, and the negative sign is added to make the flow in the positive direction a positive quantity (note that dC/dx is a negative quantity since concentration decreases in the flow direction). You may recall that *Fourier's law of heat conduction, Ohm's law of electrical conduction,* and *Newton's law of viscosity* are all in the form of Equation 11-1.

To understand the diffusion process better, consider a tank that is divided into two equal parts by a partition. Initially, the left half of the tank contains nitrogen N_2 gas while the right half contains air (about 21 percent O_2 and 79 percent N_2) at the same temperature and pressure. The O_2 and N_2 molecules are indicated by dark and light circles, respectively. When the partition is removed, we know that the N_2 molecules will start diffusing into the air while the O_2 molecules diffuse into the N_2, as shown in Figure 11-2. If we wait long enough, we will have a homogeneous mixture of N_2 and O_2 in the tank. This mass diffusion process can be explained by considering an imaginary plane indicated by the dashed line in the figure as follows: Gas molecules move randomly, and thus the probability of a molecule moving to the right or to the left is the same. Consequently, half of the molecules on one side of the dashed line at any given moment will move to the other side. Since the concentration of N_2 is greater on the left side than it is on the right side, more N_2 molecules will move toward the right than toward the left, resulting in a net flow of N_2 toward the right. As a result, N_2 is said to be *transferred* to the right. A similar argument can be given for O_2 being transferred to the left. The process continues until uniform concentrations of N_2 and O_2 are established throughout the tank so that the number of N_2 (or O_2) molecules moving to the right equals the number moving to the left, resulting in zero net transfer of N_2 or O_2 across an imaginary plane.

The molecules in a gas mixture continually collide with each other, and the diffusion process is strongly influenced by this collision process. The collision of like molecules is of little consequence since both molecules are identical and it makes no difference which molecule crosses a certain plane. The collisions of unlike molecules, however, influence the rate of diffusion since unlike molecules may have different masses and thus different momentums, and thus the diffusion process will be dominated by the heavier molecules. The diffusion coefficients and thus diffusion rates of gases depend strongly on *temperature* since the temperature is a measure of the average velocity of gas molecules. Therefore, the diffusion rates will be higher at higher temperatures.

Mass transfer can also occur in liquids and solids as well as in gases. For example, a cup of water left in a room will eventually evaporate as a result of water molecules diffusing into the air *(liquid-to-gas mass transfer)*. A piece of solid CO_2 (dry ice) will also get smaller and smaller in time as the CO_2 molecules diffuse into the air *(solid-to-gas mass transfer)*. A spoon of sugar in a cup of coffee will eventually move up and sweeten the coffee although the sugar molecules are much heavier than the water molecules, and the molecules of a colored pencil inserted into a glass of water will diffuse into the water as evidenced by the gradual spread of color in the water *(solid-to-liquid mass transfer)*. Of course, the mass transfer can also occur from a gas to a liquid or solid if the concentration of the species is higher in the gas phase. For example, a small fraction of O_2 in the air diffuses into the water and meets the oxygen needs of marine animals. The diffusion of carbon into iron during case-hardening, doping of semiconductors for transistors, and the migration of doped molecules in semiconductors at high temperature are examples of solid-to-solid diffusion processes (Fig. 11-3).

Another factor that influences the diffusion process is the *molecular spacing*. The larger the spacing, in general, the higher the diffusion rate. Therefore, the diffusion rates are typically much higher in gases than they are in liquids and much higher in liquids than in solids. Diffusion coefficients in gas mixtures are a few orders of magnitude larger than these of liquid or solid solutions.

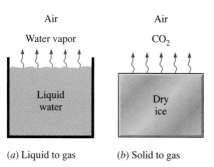

(a) Liquid to gas   *(b)* Solid to gas

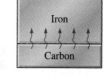

(c) Solid to liquid   *(d)* Solid to solid

FIGURE 11-3

Some examples of mass transfer that involve a liquid and/or a solid.

11-2 ■ ANALOGY BETWEEN HEAT AND MASS TRANSFER

We have spent a considerable amount of time studying heat transfer, and we could spend just as much time (perhaps more) studying mass transfer. However, the mechanisms of heat and mass transfer are analogous to each other, and thus we can develop an understanding of mass transfer in a short time with little effort by simply drawing *parallels* between heat and mass transfer. Establishing those "bridges" between the two seemingly unrelated areas will make it possible to use our heat transfer knowledge to solve mass transfer problems. Alternately, gaining a working knowledge of mass transfer will help us to better understand the heat transfer processes by thinking of heat as a massless substance as they did in the 19th century. The short-lived caloric

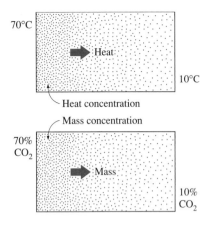

Heat

70°C

Heat concentration

10°C

Mass concentration

70%
CO₂

Mass

10%
CO₂

FIGURE 11-4

Analogy between heat and mass transfer.

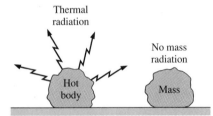

Thermal radiation

Hot body

No mass radiation

Mass

FIGURE 11-5

Unlike heat radiation, there is no such thing as mass radiation.

FIGURE 11-6

Analogy between heat conduction and mass diffusion.

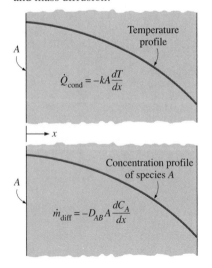

Temperature profile

A

$\dot{Q}_{cond} = -kA\dfrac{dT}{dx}$

x

Concentration profile of species A

A

$\dot{m}_{diff} = -D_{AB}A\dfrac{dC_A}{dx}$

theory of heat is the origin of most heat transfer terminology used today and served its purpose well until it was replaced by the kinetic theory. Mass is, in essence, energy since mass and energy can be converted to each other according to Einstein's formula $E = mc^2$ where c is the speed of light. Therefore, we can look at mass and heat as two different forms of energy and exploit this to advantage without going overboard.

Temperature

The driving force for heat transfer is the *temperature difference*. In contrast, the driving force for mass transfer is the *concentration difference*. We can view temperature as a measure of "heat concentration," and thus a high temperature region as one that has a high heat concentration (Fig. 11-4). Therefore, both heat and mass are transferred from the more concentrated regions to the less concentrated ones. If there is no temperature difference between two regions, then there is no heat transfer. Likewise, if there is no difference between the concentrations of a species at different parts of a medium, there will be no mass transfer.

Conduction

You will recall that heat is transferred by conduction, convection, and radiation. Mass, however, is transferred by *conduction* (called diffusion) and *convection* only, and there is no such thing as "mass radiation" (unless there is something Scotty knows that we don't when he "beams" people to anywhere in space at the speed of light) (Fig. 11-5). The rate of heat conduction in a direction x is proportional to the temperature gradient dT/dx in that direction and is expressed by **Fourier's law of heat conduction** as

$$\dot{Q}_{cond} = -kA\frac{dT}{dx} \qquad (11\text{-}2)$$

where k is the thermal conductivity of the medium and A is the area normal to the direction of heat transfer. Likewise, the rate of mass diffusion $\dot{m}_{diff}$ of a chemical species A in a stationary medium in the direction x is proportional to the concentration gradient dC/dx in that direction and is expressed by **Fick's law of diffusion** by (Fig. 11-6)

$$\dot{m}_{diff} = -D_{AB}A\frac{dC_A}{dx} \qquad (11\text{-}3)$$

where D_{AB} is the **diffusion coefficient** (or *mass diffusivity*) of the species in the mixture and C_A is the concentration of the species in the mixture at that location.

It can be shown that the differential equations for both heat conduction and mass diffusion are of the same form. Therefore, the solutions of mass diffusion equations can be obtained from the solutions of corresponding heat conduction equations for the same type of boundary conditions by simply switching the corresponding coefficients and variables.

Heat Generation

Heat generation refers to the conversion of some form of energy such as electrical, chemical, or nuclear energy into *sensible heat* energy in the medium. Heat generation occurs throughout the medium and exhibits itself as a rise in temperature. Similarly, some mass transfer problems involve chemical reactions that occur within the medium and result in the *generation of a species* throughout. Therefore, species generation is a *volumetric phenomenon,* and the rate of generation may vary from point to point in the medium. Such reactions that occur within the medium are called **homogeneous reactions** and are analogous to internal heat generation. In contrast, some chemical reactions result in the generation of a species *at the surface* as a result of chemical reactions occurring at the surface due to contact between the medium and the surroundings. This is a *surface phenomenon,* and as such it needs to be treated as a boundary condition. In mass transfer studies, such reactions are called **heterogeneous reactions** and are analogous to *specified surface heat flux.*

Convection

You will recall that *heat convection* is the heat transfer mechanism that involves both *heat conduction* (molecular diffusion) and *bulk fluid motion.* Fluid motion enhances heat transfer considerably by removing the heated fluid near the surface and replacing it by the cooler fluid further away. In the limiting case of no bulk fluid motion, convection reduces to conduction. Likewise, **mass convection** (or *convective mass transfer*) is the mass transfer mechanism between a surface and a moving fluid that involves both *mass diffusion* and *bulk fluid motion.* Fluid motion also enhances mass transfer considerably by removing the high concentration fluid near the surface and replacing it by the lower concentration fluid further away. In mass convection, we define a *concentration boundary layer* in an analogous manner to the thermal boundary layer and define new dimensionless numbers that are counterparts of the Nusselt and Prandtl numbers.

The rate of heat convection for external flow was expressed conveniently by *Newton's law of cooling* as

$$\dot{Q}_{\text{conv}} = h_{\text{conv}} A (T_s - T_\infty) \tag{11-4}$$

where h_{conv} is the heat transfer coefficient, A is the surface area, and $T_s - T_\infty$ is the temperature difference across the thermal boundary layer. Likewise, the rate of mass convection can be expressed as (Fig. 11-7)

$$\dot{m}_{\text{conv}} = h_{\text{mass}} A (C_s - C_\infty) \tag{11-5}$$

where h_{mass} is the *mass transfer coefficient,* A is the surface area, and $C_s - C_\infty$ is a suitable concentration difference across the concentration boundary layer.

Various aspects of the analogy between heat and mass convection are explored in Section 11-9. The analogy is valid for *low mass transfer rate* cases in which the flow rate of species undergoing mass flow is low (under 10 percent) relative to the total flow rate of the liquid or gas mixture.

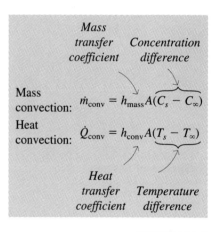

FIGURE 11-7

Analogy between convection heat transfer and convection mass transfer.

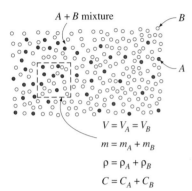

$A + B$ mixture — B

— A

$$V = V_A = V_B$$
$$m = m_A + m_B$$
$$\rho = \rho_A + \rho_B$$
$$C = C_A + C_B$$

Mass basis:

$$\rho_A = \frac{m_A}{V}, \ \rho = \frac{m}{V}, \ w_A = \frac{\rho_A}{\rho}$$

Mole basis:

$$C_A = \frac{N_A}{V}, \ C = \frac{N}{V}, \ y_A = \frac{C_A}{C}$$

Relation between them:

$$C_A = \frac{\rho_A}{M_A}, \ w_A = y_A \frac{M_A}{M}$$

FIGURE 11-8

Different ways of expressing the concentration of species A of a binary mixture A and B.

11-3 ■ MASS DIFFUSION

Fick's law of diffusion, proposed in 1855, states that the rate of diffusion of a chemical species at a location in a gas mixture (or liquid or solid solution) is proportional to the *concentration gradient* of that species at that location. Although a higher concentration for a species means more molecules of that species per unit volume, the concentration of a species can be expressed in several ways. Below we describe two common ways.

1   Mass Basis

On a *mass basis,* concentration is expressed in terms of **density** (or *mass concentration*), which is mass per unit volume. Considering a small volume V at a location within the mixture, the densities of a species (subscript i) and of the mixture (no subscript) at that location are given by (Fig. 11-8)

Partial density of species i: $\quad \rho_i = m_i/V$	(kg/m^3)
Total density of mixture: $\quad \rho = m/V = \sum m_i/V = \sum \rho_i$	

Therefore, the *density of a mixture* at a location is equal to the sum of the *densities of its constituents* at that location. Mass concentration can also be expressed in dimensionless form in terms of **mass fraction** w as

$$\textit{Mass fraction of species i:} \quad w_i = \frac{m_i}{m} = \frac{m_i/V}{m/V} = \frac{\rho_i}{\rho} \qquad (11\text{-}6)$$

Note that the mass fraction of a species ranges between 0 and 1, and the conservation of mass requires that the sum of the mass fractions of the constituents of a mixture be equal to 1. That is, $\sum w_i = 1$. Also note that the density and mass fraction of a constituent in a mixture, in general, vary with location unless the concentration gradients are zero.

2   Mole Basis

On a *mole basis,* concentration is expressed in terms of **molar concentration** (or *molar density*), which is the amount of matter in kmol per unit volume. Again considering a small volume V at a location within the mixture, the molar concentrations of a species (subscript i) and of the mixture (no subscript) at that location are given by

Partial molar concentration of species i: $C_i = N_i/V$	(kmol/m^3)
Total molar concentration of mixture: $\quad C = N/V = \sum N_i/V = \sum C_i$	

Therefore, the molar concentration of a mixture at a location is equal to the sum of the molar concentrations of its constituents at that location. Molar concentration can also be expressed in dimensionless form in terms of **mole fraction** y as

$$\textit{Mole fraction of species i:} \quad y_i = \frac{N_i}{N} = \frac{N_i/V}{N/V} = \frac{C_i}{C} \qquad (11\text{-}7)$$

Again the mole fraction of a species ranges between 0 and 1, and the sum of the mole fractions of the constituents of a mixture is unity, $\sum y_i = 1$.

The mass m and mole number N of a substance are related to each other by $m = NM$ (or, for a unit volume, $\rho = CM$) where M is the *molar mass* (also called the *molecular weight*) of the substance. This is expected since the mass of 1 kmol of the substance is M kg, and thus the mass of N kmol is NM kg. Therefore, the mass and molar concentrations are related to each other by

$$C_i = \frac{\rho_i}{M_i} \quad \text{(for species } i\text{)} \quad \text{and} \quad C = \frac{\rho}{M} \quad \text{(for the mixture)} \quad \text{(11-8)}$$

where M is the molar mass of the mixture which can be determined from

$$M = \frac{m}{N} = \frac{\sum N_i M_i}{N} = \sum \frac{N_i}{N} M_i = \sum y_i M_i \quad \text{(11-9)}$$

The mass and mole fractions of species i of a mixture are related to each other by

$$w_i = \frac{\rho_i}{\rho} = \frac{C_i M_i}{CM} = y_i \frac{M_i}{M} \quad \text{(11-10)}$$

Two different approaches are presented above for the description of concentration at a location, and you may be wondering which approach is better to use. Well, the answer depends on the situation on hand. Both approaches are equivalent, and the better approach for a given problem is the one that yields the desired solution more easily.

Special Case: Ideal Gas Mixtures

At low pressures, a gas or gas mixture can conveniently be approximated as an ideal gas with negligible error. For example, a mixture of dry air and water vapor at atmospheric conditions can be treated as an ideal gas with an error much less than 1 percent. The total pressure of a gas mixture P is equal to the sum of the partial pressures P_i of the individual gases in the mixture and is expressed as $P = \sum P_i$. Here P_i is called the **partial pressure** of species i, which is the pressure species i would exert if it existed alone at the mixture temperature and volume. This is known as **Dalton's law of additive pressures.** Then using the ideal gas relation $PV = NR_uT$ where R_u is the universal gas constant for both the species i and the mixture, the **pressure fraction** of species i can be expressed as (Fig. 11-9)

$$\frac{P_i}{P} = \frac{N_i R_u T/V}{N R_u T/V} = \frac{N_i}{N} = y_i \quad \text{(11-11)}$$

Therefore, the *pressure fraction* of species i of an ideal gas mixture is equivalent to the *mole fraction* of that species and can be used in place of it in mass transfer analysis.

| 2 mol A |
| 6 mol B |
| $P = 120$ kPa |

A mixture of two ideal gases A and B

$$y_A = \frac{N_A}{N} = \frac{2}{2+6} = 0.25$$

$$P_A = y_A P = 0.25 \times 120 = 30 \text{ kPa}$$

FIGURE 11-9

For ideal gas mixtures, pressure fraction of a gas is equal to its mole fraction.

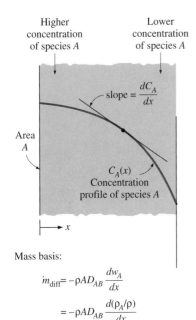

Higher
concentration
of species A

Lower
concentration
of species A

slope $= \dfrac{dC_A}{dx}$

Area
A

$C_A(x)$
Concentration
profile of species A

$\longrightarrow x$

Mass basis:

$$\dot{m}_{\text{diff}} = -\rho A D_{AB} \frac{dw_A}{dx}$$

$$= -\rho A D_{AB} \frac{d(\rho_A/\rho)}{dx}$$

$$= -A D_{AB} \frac{d\rho_A}{dx} \text{ (if } \rho = \text{constant)}$$

Mole basis:

$$\dot{N}_{\text{diff},\,A} = -C A D_{AB} \frac{dy_A}{dx}$$

$$= -C A D_{AB} \frac{d(C_A/C)}{dx}$$

$$= -A D_{AB} \frac{dC_A}{dx} \text{ (if } C = \text{constant)}$$

FIGURE 11-10

Various expressions of Fick's law of
diffusion for a binary mixture.

Fick's Law of Diffusion: Stationary Medium Consisting of Two Species

We mentioned earlier that the rate of mass diffusion of a chemical species in a stagnant medium in a specified direction is proportional to the local concentration gradient in that direction. This linear relationship between the rate of diffusion and the concentration gradient proposed by Fick in 1855 is known as **Fick's law of diffusion** and can be expressed as

Mass flux = Constant of proportionality × Concentration gradient

But the concentration of a species in a gas mixture or liquid or solid solution can be defined in several ways such as density, mass fraction, molar concentration, and mole fraction, as discussed above, and thus Fick's law can be expressed mathematically in many ways. It turns out that it is best to express the concentration gradient in terms of the mass or mole fraction, and the most appropriate formulation of *Fick's law* for the diffusion of a species A in a stationary binary mixture of species A and B in a specified direction x is given by (Fig. 11-10)

$$\textit{Mass basis: } j_{\text{diff},\,A} = \frac{\dot{m}_{\text{diff},\,A}}{A} = -\rho D_{AB} \frac{d(\rho_A/\rho)}{dx} = -\rho D_{AB} \frac{dw_A}{dx} \text{ (kg/s} \cdot \text{m}^2)$$

$$\textit{Mole basis: } \bar{j}_{\text{diff},\,A} = \frac{\dot{N}_{\text{diff},\,A}}{A} = -C D_{AB} \frac{d(C_A/C)}{dx} = -C D_{AB} \frac{dy_A}{dx} \text{ (mol/s} \cdot \text{m}^2)$$

$$(11\text{-}12)$$

Here j_A is the **(diffusive) mass flux** of species A (mass transfer by diffusion per unit time and per unit area normal to the direction of mass transfer, in kg/s · m²) and $\bar{j}_A$ is the **(diffusive) molar flux** (in kmol/s · m²). The mass flux of a species at a location is proportional to the density of the mixture at that location. Note that $\rho = \rho_A + \rho_B$ is the density and $C = C_A + C_B$ is the molar concentration of the binary mixture, and in general, they may vary throughout the mixture. Therefore, $\rho d(\rho_A/\rho) \neq d\rho_A$ or $C d(C_A/C) \neq dC_A$. But in the special case of constant mixture density ρ or constant molar concentration C, the relations above simplify to

Mass basis ($\rho = $ constant):	$j_{\text{diff},\,A} = -D_{AB} \dfrac{d\rho_A}{dx}$	(kg/s · m²)
Mole basis ($C = $ constant):	$\bar{j}_{\text{diff},\,A} = -D_{AB} \dfrac{dC_A}{dx}$	(kmol/s · m²)

$$(11\text{-}13)$$

The constant density or constant molar concentration assumption is usually appropriate for *solid* and *dilute liquid* solutions, but often this is not the case for gas mixtures or concentrated liquid solutions. Therefore, Equations 11-12 should be used in the latter case. In this introductory treatment we will limit our consideration to one-dimensional mass diffusion. For two- or three-dimensional cases, Fick's law can conveniently be expressed in vector form by simply replacing the derivatives in the above relations by the corresponding gradients (such as $\mathbf{j}_A = -\rho D_{AB} \nabla w_A$).

Remember that the constant of proportionality in Fourier's law was defined as the transport property *thermal conductivity*. Similarly, the constant of proportionality in Fick's law is defined as another transport property called the **binary diffusion coefficient** or **mass diffusivity,** D_{AB}. The unit of mass diffusivity is m²/s, which is the same as the units of *thermal diffusivity* or *momentum diffusivity* (also called *kinematic viscosity*) (Fig. 11-11).

Because of the complex nature of mass diffusion, the diffusion coefficients are usually determined experimentally. The kinetic theory of gases indicates that the diffusion coefficient for dilute gases at ordinary pressures is essentially independent of mixture composition and tends to increase with temperature while decreasing with pressure as

$$D_{AB} \propto \frac{T^{3/2}}{P} \quad \text{or} \quad \frac{D_{AB,1}}{D_{AB,2}} = \frac{P_2}{P_1}\left(\frac{T_1}{T_2}\right)^{3/2} \quad (11\text{-}14)$$

This relation is useful in determining the diffusion coefficient for gases at different temperatures and pressures from a knowledge of the diffusion coefficient at a specified temperature and pressure. More general but complicated relations that account for the effects of molecular collisions are also available. The diffusion coefficients of some gases in air at 1 atm pressure are given in Table 11-1 at various temperatures.

The diffusion coefficients of solids and liquids also tend to increase with temperature while exhibiting a strong dependence on the composition. The diffusion process in solids and liquids is a great deal more complicated than that in gases, and the diffusion coefficients in this case are almost exclusively determined experimentally.

The binary diffusion coefficient for several binary gas mixtures and solid and liquid solutions are given in Tables 11-2 and 11-3. We make the following two observations from these tables:

1. The diffusion coefficients, in general, are *highest in gases* and *lowest in solids*. The diffusion coefficients of gases are several orders of magnitude greater than those of liquids.

2. Diffusion coefficients *increase with temperature*. The diffusion coefficient (and thus the mass diffusion rate) of carbon through iron during a hardening process, for example, increases by 6000 times as the temperature is raised from 500°C to 1000°C.

Due to its practical importance, the diffusion of *water vapor* in *air* has been the topic of several studies, and some empirical formulas have been developed for the diffusion coefficient D_{H_2O-air}. Recently Marrero and Mason proposed the following popular formula (Table 11-4):

$$D_{H_2O-Air} = 1.87 \times 10^{-10}\,\frac{T^{2.072}}{P} \quad (\text{m}^2/\text{s}), \quad 280\text{ K} < T < 450\text{ K} \quad (11\text{-}15)$$

where P is total pressure in atm and T is the temperature in K.

The primary driving mechanism of mass diffusion is the concentration gradient, and mass diffusion due to a concentration gradient is known as the

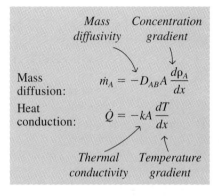

FIGURE 11-11

Analogy between Fourier's law of heat conduction and Fick's law of mass diffusion.

TABLE 11-1

Binary diffusion coefficients of some gases in air at 1 atm pressure (from Mills, Ref. 13, Table A.17a, p. 869.

$T,$ K	Binary diffusion coefficient,* m²/s × 10⁵			
	O_2	CO_2	H_2	NO
200	0.95	0.74	3.75	0.88
300	1.88	1.57	7.77	1.80
400	5.25	2.63	12.5	3.03
500	4.75	3.85	17.1	4.43
600	6.46	5.37	24.4	6.03
700	8.38	6.84	31.7	7.82
800	10.5	8.57	39.3	9.78
900	12.6	10.5	47.7	11.8
1000	15.2	12.4	56.9	14.1
1200	20.6	16.9	77.7	19.2
1400	26.6	21.7	99.0	24.5
1600	33.2	27.5	125	30.4
1800	40.3	32.8	152	37.0
2000	48.0	39.4	180	44.8

*Multiply by 10.76 to convert to ft²/s.

TABLE 11-2

Binary diffusion coefficients of dilute gas mixtures at 1 atm
(from Barrer, Ref. 2; Geankoplis, Ref. 5; Perry, Ref. 14; and Reid et al., Ref. 15)

Substance A	Substance B	T, K	D_{AB} or D_{BA}, m^2/s	Substance A	Substance B	T, K	D_{AB} or D_{BA}, m^2/s
Air	Acetone	273	1.1×10^{-5}	Argon, Ar	Nitrogen, N_2	293	1.9×10^{-5}
Air	Ammonia, NH_3	298	2.6×10^{-5}	Carbon dioxide, CO_2	Benzene	318	0.72×10^{-5}
Air	Benzene	298	0.88×10^{-5}	Carbon dioxide, CO_2	Hydrogen, H_2	273	5.5×10^{-5}
Air	Carbon dioxide	298	1.6×10^{-5}	Carbon dioxide, CO_2	Nitrogen, N_2	293	1.6×10^{-5}
Air	Chlorine	273	1.2×10^{-5}	Carbon dioxide, CO_2	Oxygen, O_2	273	1.4×10^{-5}
Air	Ethyl alcohol	298	1.2×10^{-5}	Carbon dioxide, CO_2	Water vapor	298	1.6×10^{-5}
Air	Ethyl ether	298	0.93×10^{-5}	Hydrogen, H_2	Nitrogen, N_2	273	6.8×10^{-5}
Air	Helium, He	298	7.2×10^{-5}	Hydrogen, H_2	Oxygen, O_2	273	7.0×10^{-5}
Air	Hydrogen, H_2	298	7.2×10^{-5}	Oxygen, O_2	Ammonia	293	2.5×10^{-5}
Air	Iodine, I_2	298	0.83×10^{-5}	Oxygen, O_2	Benzene	296	0.39×10^{-5}
Air	Methanol	298	1.6×10^{-5}	Oxygen, O_2	Nitrogen, N_2	273	1.8×10^{-5}
Air	Mercury	614	4.7×10^{-5}	Oxygen, O_2	Water vapor	298	2.5×10^{-5}
Air	Napthalene	300	0.62×10^{-5}	Water vapor	Argon, Ar	298	2.4×10^{-5}
Air	Oxygen, O_2.	298	2.1×10^{-5}	Water vapor	Helium, He	298	9.2×10^{-5}
Air	Water vapor	298	2.5×10^{-5}	Water vapor	Nitrogen, N_2	298	2.5×10^{-5}

Note: The effect of pressure and temperature on D_{AB} can be accounted for through $D_{AB} \sim T^{3/2}/P$. Also, multiply D_{AB} values by 10.76 to convert them to ft^2/s.

TABLE 11-3

Binary diffusion coefficients of dilute liquid solutions and solid solutions at 1 atm
(from Barrer, Ref. 2; Reid et al., Ref. 15; Thomas, Ref. 19; and van Black, Ref. 20)

(a) Diffusion through liquids				(b) Diffusion through solids			
Substance A (solute)	Substance B (solvent)	T, K	D_{AB}, m^2/s	Substance A (solute)	Substance B (solvent)	T, K	D_{AB}, m^2/s
Ammonia	Water	285	1.6×10^{-9}	Carbon dioxide	Natural rubber	298	1.1×10^{-10}
Benzene	Water	293	1.0×10^{-9}	Nitrogen	Natural rubber	298	1.5×10^{-10}
Carbon dioxide	Water	298	2.0×10^{-9}	Oxygen	Natural rubber	298	2.1×10^{-10}
Chlorine	Water	285	1.4×10^{-9}	Helium	Pyrex	773	2.0×10^{-12}
Ethanol	Water	283	0.84×10^{-9}	Helium	Pyrex	293	4.5×10^{-15}
Ethanol	Water	288	1.0×10^{-9}	Helium	Silicon dioxide	298	4.0×10^{-14}
Ethanol	Water	298	1.2×10^{-9}	Hydrogen	Iron	298	2.6×10^{-13}
Glucose	Water	298	0.69×10^{-9}	Hydrogen	Nickel	358	1.2×10^{-12}
Hydrogen	Water	298	6.3×10^{-9}	Hydrogen	Nickel	438	1.0×10^{-11}
Methane	Water	275	0.85×10^{-9}	Cadmium	Copper	293	2.7×10^{-19}
Methane	Water	293	1.5×10^{-9}	Zinc	Copper	773	4.0×10^{-18}
Methane	Water	333	3.6×10^{-9}	Zinc	Copper	1273	5.0×10^{-13}
Methanol	Water	288	1.3×10^{-9}	Antimony	Silver	293	3.5×10^{-25}
Nitrogen	Water	298	2.6×10^{-9}	Bismuth	Lead	293	1.1×10^{-20}
Oxygen	Water	298	2.4×10^{-9}	Mercury	Lead	293	2.5×10^{-19}
Water	Ethanol	298	1.2×10^{-9}	Copper	Aluminum	773	4.0×10^{-14}
Water	Ethylene glycol	298	0.18×10^{-9}	Copper	Aluminum	1273	1.0×10^{-10}
Water	Methanol	298	1.8×10^{-9}	Carbon	Iron (fcc)	773	5.0×10^{-15}
Chloroform	Methanol	288	2.1×10^{-9}	Carbon	Iron (fcc)	1273	3.0×10^{-11}

ordinary diffusion. However, diffusion may also be caused by other effects. Temperature gradients in a medium can cause **thermal diffusion** (also called the **soret effect**), and pressure gradients may result in **pressure diffusion.** Both of these effects are usually negligible, however, unless the gradients are very large. In centrifuges, the pressure gradient generated by the centrifugal effect is used to separate liquid solutions and gaseous isotopes. An external force field such as an electric or magnetic field applied on a mixture or solution can be used successfully to separate electrically charged or magnetized molecules (as in an electrolyte or ionized gas) from the mixture. This is called **forced diffusion.** Also, when the pores of a porous solid such as silica-gel are smaller than the mean free path of the gas molecules, the molecular collisions may be negligible and a free molecule flow may be initiated. This is known as **knodsen diffusion.** When the size of the gas molecules is comparable to the pore size, adsorbed molecules move along the pore walls. This is known as **surface diffusion.** Finally, particles whose diameter is under $0.1 \ \mu m$ such as mist and soot particles act like large molecules, and the diffusion process of such particles due to the concentration gradient is called **Brownian motion.** Large particles (those whose diameter is greater than $1 \ \mu m$) are not affected by diffusion as the motion of such particles is governed by Newton's laws. In our elementary treatment of mass diffusion, we will assume these additional effects to be nonexistent or negligible, as is usually the case, and refer the interested reader to advanced books on these topics.

TABLE 11-4

In a binary ideal gas mixture of species A and B, the diffusion coefficient of A in B is equal to the diffusion coefficient of B in A, and both increase with temperature

T, °C	D_{H_2O-Air} or D_{Air-H_2O} at 1 atm, in m^2/s (from Eq. 11-15)
0	2.09×10^{-5}
5	2.17×10^{-5}
10	2.25×10^{-5}
15	2.33×10^{-5}
20	2.42×10^{-5}
25	2.50×10^{-5}
30	2.59×10^{-5}
35	2.68×10^{-5}
40	2.77×10^{-5}
50	2.96×10^{-5}
100	3.99×10^{-5}
150	5.18×10^{-5}

EXAMPLE 11-1 Determining Mass Fractions from Mole Fractions

The composition of dry standard atmosphere is given on a molar basis to be 78.1% N_2, 20.9% O_2, and 1.0% Ar and other constituents (Fig. 11-12). Treating other constituents as Ar, determine the mass fractions of the constituents of air.

Solution The molar fractions of the constituents of air are given. The mass fractions are to be determined.

Assumptions The small amounts of other gases in air are treated as argon.

Properties The molar masses of N_2, O_2, and Ar are 28.0, 32.0, and 39.9 kg/kmol, respectively (Table A-1).

Analysis The molar mass of air is determined to be

$$M = \sum y_i M_i = 0.781 \times 28.0 + 0.209 \times 32.0 + 0.01 \times 39.9 = 29.0 \text{ kg/kmol}$$

Then the mass fractions of constituent gases are determined from Equation 11-10 to be

N_2:
$$w_{N_2} = y_{N_2} \frac{M_{N_2}}{M} = (0.781)\frac{28.0}{29.0} = \mathbf{0.754}$$

O_2:
$$w_{O_2} = y_{O_2} \frac{M_{O_2}}{M} = (0.209)\frac{32.0}{29.0} = \mathbf{0.231}$$

Ar:
$$w_{Ar} = y_{Ar} \frac{M_{Ar}}{M} = (0.01)\frac{39.9}{29.0} = \mathbf{0.014}$$

Therefore, the mass fractions of N_2, O_2, and Ar in dry standard atmosphere are 75.4%, 23.1%, and 1.4%, respectively.

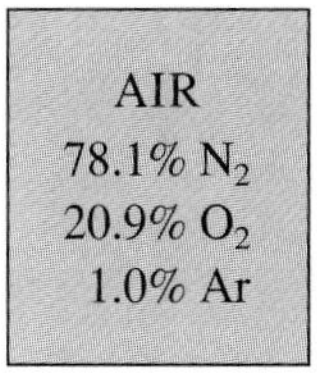

AIR

78.1% N_2
20.9% O_2
1.0% Ar

FIGURE 11-12

Schematic for Example 11-1.

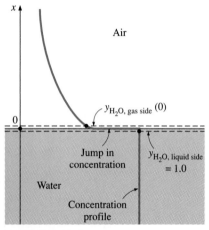

x

Air

$y_{H_2O, \text{gas side}}(0)$

0

Jump in
concentration

$y_{H_2O, \text{liquid side}} = 1.0$

Water

Concentration
profile

FIGURE 11-13

Unlike temperature, the concentration
of species on the two sides of a liquid–
gas (or solid–gas or solid–liquid)
interface are usually not the same.

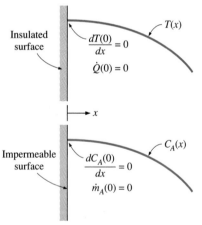

Insulated
surface

$\dfrac{dT(0)}{dx} = 0$

$\dot{Q}(0) = 0$

$T(x)$

x

Impermeable
surface

$\dfrac{dC_A(0)}{dx} = 0$

$\dot{m}_A(0) = 0$

$C_A(x)$

FIGURE 11-14

An impermeable surface in mass
transfer is analogous to an insulated
surface in heat transfer.

11-4 ■ BOUNDARY CONDITIONS

We mentioned earlier that the mass diffusion equation is analogous to the heat diffusion (conduction) equation, and thus we need comparable boundary conditions to determine the species concentration distribution in a medium. Two common types of boundary conditions are the (1) *specified species concentration,* which corresponds to specified temperature, and (2) *specified species flux,* which corresponds to specified heat flux.

Despite their apparent similarity, an important difference exists between temperature and concentration: temperature is necessarily a *continuous* function, but concentration, in general, is not. The wall and air temperatures at a wall surface, for example, are always the same. The concentrations of air on the two sides of a water–air interface, however, are obviously very different (in fact, the concentration of air in water is close to zero). Likewise, the concentrations of water on the two sides of a water–air interface are also different even when air is saturated (Fig. 11-13). Therefore, when specifying a boundary condition, specifying the location is not enough. We also need to specify the side of the boundary. To do this, we consider two imaginary surfaces on the two sides of the interface that are infinitesimally close to the interface. Whenever there is a doubt, we indicate the desired side of the interface by specifying its phase as a subscript. For example, the water (liquid or vapor) concentration at the liquid and gas sides of a water–air interface at $x = 0$ can be expressed on a molar basis is

$$y_{H_2O, \text{liquid side}}(0) = y_1 \qquad \text{and} \qquad y_{H_2O, \text{gas side}}(0) = y_2 \qquad (11\text{-}16)$$

Using Fick's law, the constant species flux boundary condition for a diffusing species A at a boundary at $x = 0$ is expressed, in the absence of any blowing or suction, as

$$-CD_{AB} \left. \frac{dy_A}{dx} \right|_{x=0} = \bar{j}_{A,0} \qquad \text{or} \qquad -\rho D_{AB} \left. \frac{dw_A}{dx} \right|_{x=0} = j_{A,0} \qquad (11\text{-}17)$$

where $\bar{j}_{A,0}$ and $j_{A,0}$ are the specified mole and mass fluxes of species A at the boundary, respectively. The special case of zero mass flux ($\bar{j}_{A,0} = j_{A,0} = 0$) corresponds to an **impermeable surface** for which $dy_A(0)/dx = dw_A(0)/dx = 0$ (Fig. 11-14).

To apply the *specified concentration* boundary condition, we must know the concentration of a species at the boundary. This information is usually obtained from the requirement that *thermodynamic equilibrium* must exist at the interface of two phases of a species. In the case of air–water interface, the concentration values of water vapor in the air are easily determined from saturation data, as shown in the example below.

EXAMPLE 11-2   Mole Fraction of Water Vapor at the Surface of a Lake

Determine the mole fraction of the water vapor at the surface of a lake whose temperature is 15°C and compare it to the mole fraction of water in the lake (Fig. 11-15). Take the atmospheric pressure at lake level to be 92 kPa.

Solution   The mole fraction of the water vapor at the surface of a lake and the mole fraction of water in the lake are to be determined and compared.

Assumptions **1** Both the air and water vapor are ideal gases. **2** The mole fraction of dissolved air in water is negligible.

Properties The saturation pressure of water at 15°C is 1.705 kPa (Table A-9).

Analysis The air at the water surface will be saturated. Therefore, the partial pressure of water vapor in the air at the lake surface will simply be the saturation pressure of water at 15°C,

$$P_{vapor} = P_{sat @ 15°C} = 1.705 \text{ kPa}$$

Assuming both the air and vapor to be ideal gases, the mole fraction of water vapor in the air at the surface of the lake is determined from Equation 11-11 to be

$$y_{vapor} = \frac{P_{vapor}}{P} = \frac{1.705 \text{ kPa}}{92 \text{ kPa}} = \textbf{0.0185} \qquad \text{(or 1.85 percent)}$$

Water contains some dissolved air, but the amount is negligible. Therefore, we can assume the entire lake to be liquid water. Then its mole fraction becomes

$$y_{water, \text{ liquid side}} \cong 1.0 \qquad \text{(or 100 percent)}$$

Discussion Note that the concentration of water on a molar basis is 100 percent just beneath the air–water interface and 1.85 percent just above it, even though the air is assumed to be saturated (so this is the highest value at 15°C). Therefore, huge discontinuities can occur in the concentrations of a species across phase boundaries.

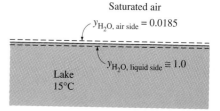

FIGURE 11-15

Schematic for Example 11-2.

The situation is similar at *solid–liquid* interfaces. Again, at a given temperature, only a certain amount of solid can be dissolved in a liquid, and the solubility of the solid in the liquid is determined from the requirement that thermodynamic equilibrium exists between the solid and the solution at the interface. The **solubility** represents *the maximum amount of solid that can be dissolved in a liquid at a specified temperature* and is widely available in chemistry handbooks. In Table 11-5 we present sample solubility data for sodium chloride (NaCl) and calcium bicarbonate $[Ca(HCO_3)_2]$ at various temperatures. For example, the solubility of salt (NaCl) in water at 310 K is 36.5 kg per 100 kg of water. Therefore, the mass fraction of salt in the brine at the interface is simply

$$w_{salt, \text{ liquid side}} = \frac{m_{salt}}{m} = \frac{36.5 \text{ kg}}{(100 + 36.5) \text{ kg}} = 0.267 \qquad \text{(or 26.7 percent)}$$

whereas the mass fraction of salt in the pure solid salt is $w = 1.0$. Note that water becomes *saturated* with salt when 36.5 kg of salt are dissolved in 100 kg of water at 310 K.

Many processes involve the absorption of a gas into a liquid. Most gases are weakly soluble in liquids (such as air in water), and for such dilute solutions the mole fractions of a species i in the gas and liquid phases at the interface are observed to be proportional to each other. That is, $y_{i, \text{ gas side}} \propto y_{i, \text{ liquid side}}$ or $P_{i, \text{ gas side}} \propto P \, y_{i, \text{ liquid side}}$ since $y_{i, \text{ gas side}} = P_{i, \text{ gas side}}/P$ for ideal gas mixtures. This is known as **Henry's law** and is expressed as

TABLE 11-5

Solubility of two inorganic compounds in water at various temperatures, in kg, in 100 kg of water [from *Handbook of Chemistry* (New York: McGraw-Hill, 1961)]

Tempera-ture, K	Salt, NaCl	Calcium bicarbonate, Ca(HCO$_3$)$_2$
	Salute	
273.15	35.7	16.15
280	35.8	16.30
290	35.9	16.53
300	36.2	16.75
310	36.5	16.98
320	36.9	17.20
330	37.2	17.43
340	37.6	17.65
350	38.2	17.88
360	38.8	18.10
370	39.5	18.33
373.15	39.8	18.40

$$y_{i,\text{ liquid side}} = \frac{P_{i,\text{ gas side}}}{H} \text{ (at interface)} \qquad (11\text{-}18)$$

where H is **Henry's constant,** which is the product of the total pressure of the gas mixture and the proportionality constant. For a given species, it is a function of temperature only and is practically independent of pressure for pressures under about 5 atm. Values of Henry's constant for a number of aqueous solutions are given in Table 11-6 for various temperatures. From this table and the equation above we make the following observations:

1. The concentration of a gas dissolved in a liquid is inversely proportional to Henry's constant. Therefore, the larger Henry's constant, the smaller the concentration of dissolved gases in the liquid.

2. Henry's constant increases (and thus the fraction of a dissolved gas in the liquid decreases) with increasing temperature. Therefore, the dissolved gases in a liquid can be driven off by heating the liquid (Fig. 11-16).

3. The concentration of a gas dissolved in a liquid is proportional to the partial pressure of the gas. Therefore, the amount of gas dissolved in a liquid can be increased by increasing the pressure of the gas. This can be used to advantage in the carbonation of soft drinks with CO_2 gas.

Strictly speaking, the result obtained from Equation 11-18 for the mole fraction of dissolved gas is valid for the liquid layer just beneath the interface and not necessarily the entire liquid. The latter will be the case only when thermodynamic phase equilibrium is established throughout the entire liquid body.

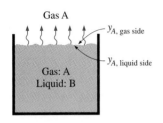

$y_{A,\text{ gas side}} \propto y_{A,\text{ liquid side}}$

or

$\dfrac{P_{A,\text{ gas side}}}{P} \propto y_{A,\text{ liquid side}}$

or

$P_{A,\text{ gas side}} = H y_{A,\text{ liquid side}}$

FIGURE 11-16

Dissolved gases in a liquid can be driven off by heating the liquid.

EXAMPLE 11-3   Mole Fraction of Dissolved Air in Water

Determine the mole fraction of air dissolved in water at the surface of a lake whose temperature is 17°C (Fig. 11-17). Take the atmospheric pressure at lake level to be 92 kPa.

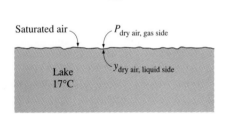

FIGURE 11-17

Schematic for Example 11-3.

TABLE 11-6

Henry's constant H (in bars) for selected gases in water at low to moderate pressures (for gas i, $H = P_{i,\text{ gas side}}/y_{i,\text{ water side}}$) (from Mills, Ref. 13, Table A.21, p. 874)

Solute	290 K	300 K	310 K	320 K	330 K	340 K
H_2S	440	560	700	830	980	1140
CO_2	1280	1710	2170	2720	3220	—
O_2	38,000	45,000	52,000	57,000	61,000	65,000
H_2	67,000	72,000	75,000	76,000	77,000	76,000
CO	51,000	60,000	67,000	74,000	80,000	84,000
Air	62,000	74,000	84,000	92,000	99,000	104,000
N_2	76,000	89,000	101,000	110,000	118,000	124,000

Solution The mole fraction of air dissolved in water at the surface of a lake is to be determined.

Assumptions **1** Both the air and water vapor are ideal gases. **2** Air is weakly soluble in water so that Henry's law is applicable.

Properties The saturation pressure of water at 17°C is 1.92 kPa (Table A-9). Henry's constant for air dissolved in water at 290 K is $H = 62{,}000$ bar (Table 11-6).

Analysis This example is similar to the previous example. Again the air at the water surface will be saturated, and thus the partial pressure of water vapor in the air at the lake surface will be the saturation pressure of water at 17°C,

$$P_{vapor} = P_{sat \, @ \, 17°C} = 1.92 \text{ kPa}$$

Assuming both the air and vapor to be ideal gases, the partial pressure of dry air is determined to be

$$P_{dry \, air} = P - P_{vapor} = 92 - 1.92 = 90.08 \text{ kPa} = 0.9008 \text{ bar}$$

Note that with little loss in accuracy (an error of about 2 percent), we could have ignored the vapor pressure since the amount of vapor in air is so small. Then the mole fraction of air in the water becomes

$$y_{dry \, air, \, liquid \, state} = \frac{P_{dry \, air, \, gas \, side}}{H} = \frac{0.9008 \text{ bar}}{62{,}000 \text{ bar}} = \textbf{1.45} \times \textbf{10}^{-5}$$

which is very small, as expected. Therefore, the concentration of air in water just below the air–water interface is 1.45 moles per 100,000 moles. But obviously this is enough oxygen for fish and other creatures in the lake. Note that the amount of air dissolved in water will decrease with increasing depth.

We mentioned earlier that the use of Henry's law is limited to dilute gas–liquid solutions; that is, a liquid with a small amount of gas dissolved in it. Then the question that arises naturally is, what do we do when the gas is highly soluble in the liquid (or solid), such as ammonia in water? In this case the linear relationship of Henry's law does not apply, and the mole fraction of a gas dissolved in the liquid (or solid) is usually expressed as a function of the partial pressure of the gas in the gas phase and the temperature. An approximate relation in this case for the *mole fractions* of a species on the *liquid* and *gas sides* of the interface is given by **Raoult's law** as

$$P_{i, \, gas \, side} = y_{i, \, gas \, side} P = y_{i, \, liquid \, side} P_{i, \, sat}(T) \qquad (11\text{-}19)$$

where $P_{i, \, sat}(T)$ is the *saturation pressure* of the species i at the interface temperature and P is the *total pressure* on the gas phase side. Tabular data are available in chemical handbooks for common solutions such as the ammonia–water solution that is widely used in absorption-refrigeration systems.

Gases may also dissolve in *solids,* but the diffusion process in this case can be very complicated. The dissolution of a gas may be independent of the structure of the solid, or it may depend strongly on its porosity. Some dissolution processes (such as the dissolution of hydrogen in titanium, similar to the dissolution of CO_2 in water) are *reversible,* and thus maintaining the gas content in the solid requires constant contact of the solid with a reservoir of that gas. Some other dissolution processes are *irreversible.* For example, oxygen gas dissolving in titanium forms TiO_2 on the surface, and the process does not reverse itself.

The concentration of the gas species i in the solid at the interface $C_{i,\text{ solid side}}$ is proportional to the *partial pressure* of the species i in the gas $P_{i,\text{ gas side}}$ on the gas side of the interface and is expressed as

$$C_{i,\text{ solid side}} = \mathcal{S} \times P_{i,\text{ gas side}} \qquad (\text{kmol/m}^3) \qquad (11\text{-}20)$$

where $\mathcal{S}$ is the **solubility.** Expressing the pressure in bars and noting that the unit of molar concentration is kmol of species i per m^3, the unit of solubility is $\text{kmol/m}^3 \cdot \text{bar}$. Solubility data for selected gas–solid combinations are given in Table 11-7. The product of the *solubility* of a gas and the *diffusion coefficient* of the gas in a solid is referred to as the **permeability** $\mathcal{P}$, which is a measure of the ability of the gas to penetrate a solid. That is, $\mathcal{P} = \mathcal{S}D_{AB}$ where D_{AB} is the diffusivity of the gas in the solid. Permeability is inversely proportional to thickness and has the unit $\text{kmol/s} \cdot \text{bar}$.

Finally, if a process involves the *sublimation* of a pure solid (such as ice or solid CO_2) or the *evaporation* of a pure liquid (such as water) in a different medium such as air, the mole (or mass) fraction of the substance in the liquid or solid phase is simply taken to be 1.0, and the partial pressure and thus the mole fraction of the substance in the gas phase can readily be determined from the saturation data of the substance at the specified temperature. Also, the assumption of thermodynamic equilibrium at the interface is very reasonable for pure solids, pure liquids, and solutions, except when chemical reactions are occurring at the interface.

TABLE 11-7

Solubility of selected gases and solids (for gas i, $\mathcal{S} = C_{i,\text{ solid side}}/P_{i,\text{ gas side}}$) (from Barrer, Ref. 2)

Gas	Solid	T, K	$\mathcal{S}$ $\text{kmol/m}^3 \cdot \text{bar}$
O_2	Rubber	298	0.00312
N_2	Rubber	298	0.00156
CO_2	Rubber	298	0.04015
He	SiO_2	293	0.00045
H_2	Ni	358	0.00901

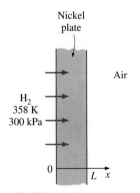

FIGURE 11-18
Schematic for Example 11-4.

EXAMPLE 11-4   Diffusion of Hydrogen Gas into a Nickel Plate

Consider a nickel plate that is in contact with hydrogen gas at 358 K and 300 kPa. Determine the molar and mass density of hydrogen in the nickel at the interface (Fig. 11-18).

Solution   A nickel plate is exposed to hydrogen. The molar and mass density of hydrogen in the nickel at the interface is to be determined.

Assumptions   Nickel and hydrogen are in thermodynamic equilibrium at the interface.

Properties   The molar mass of hydrogen is $M = 2$ kg/kmol (Table A-1). The solubility of hydrogen in nickel at 358 K is 0.00901 $\text{kmol/m}^3 \cdot \text{bar}$ (Table 11-7).

Analysis   Noting that 300 kPa = 3 bar, the molar density of hydrogen in the nickel at the interface is determined from Equation 11-20 to be

$$C_{H_2,\text{ solid side}} = \mathcal{S} \times P_{H_2,\text{ gas side}}$$
$$= (0.00901 \text{ kmol/m}^3 \cdot \text{bar})(3 \text{ bar}) = \mathbf{0.027 \text{ kmol/m}^3}$$

It corresponds to a mass density of

$$\rho_{H_2,\text{ solid side}} = C_{H_2,\text{ solid side}}M_{H_2}$$
$$= (0.027 \text{ kmol/m}^3)(2) = \textbf{0.054 kg/m}^3$$

That is, there will be 0.027 kmol (or 0.054 kg) of H_2 gas in each m^3 volume of nickel adjacent to the interface.

11-5 ■ STEADY MASS DIFFUSION THROUGH A WALL

Many practical mass transfer problems involve the diffusion of a species through a plane-parallel medium that does not involve any homogeneous chemical reactions under *one-dimensional steady* conditions. Such mass transfer problems are analogous to the steady one-dimensional heat conduction problems in a plane wall with no heat generation and can be analyzed similarly. In fact, many of the relations developed in Chapter 3 can be used for mass transfer by replacing temperature by mass (or molar) fraction, thermal conductivity by ρD_{AB} (or CD_{AB}), and heat flux by mass (or molar) flux (Table 11-8).

Consider a *solid plane wall* (medium B) of area A, thickness L, and density ρ. The wall is subjected on both sides to different concentrations of a species A to which it is permeable. The boundary surfaces at $x = 0$ and $x = L$ are located within the solid adjacent to the interfaces, and the mass fractions of A at those surfaces are maintained at $w_{A,1}$ and $w_{A,2}$, respectively, at all times (Fig. 11-19). The mass fraction of species A in the wall will vary in the x-direction only and can be expressed as $w_A(x)$. Therefore, mass transfer through the wall in this case can be modeled as *steady* and *one-dimensional*. Below we determine the rate of mass diffusion of species A through the wall using a similar approach to that used in Chapter 3 for heat conduction.

The concentration of species A at any point will not change with time since operation is steady, and there will be no production or destruction of species A since no chemical reactions are occurring in the medium. Then the **conservation of mass** principle for species A can be expressed as *the mass flow rate of species A through the wall at any cross-section is the same.* That is

$$\dot{m}_{\text{diff},A} = j_A A = \text{constant} \qquad \text{(kg/s)}$$

Then Fick's law of diffusion becomes

$$j_A = \frac{\dot{m}_{\text{diff},A}}{A} = -\rho D_{AB}\frac{dw_A}{dx} = \text{constant}$$

Separating the variables in the above equation and integrating across the wall from $x = 0$ where $w(0) = w_{A,1}$ to $x = L$ where $w(L) = w_{A,2}$, we get

$$\frac{\dot{m}_{\text{diff},A}}{A}\int_0^L dx = -\int_{w_{A,1}}^{w_{A,2}} \rho D_{AB}\, dw_A \qquad (11\text{-}21)$$

where the mass transfer rate $\dot{m}_{\text{diff},A}$ and the wall area A are taken out of the integral sign since both are constants. If the density ρ and the mass diffusion

TABLE 11-8

Analogy between heat conduction and mass diffusion in a stationary medium

Heat conduction	Mass diffusion	
	Mass basis	**Molar basis**
T	w_i	y_i
k	ρD_{AB}	CD_{AB}
$\dot{q}$	j_i	$\bar{j}_i$
α	D_{AB}	D_{AB}
L	L	L

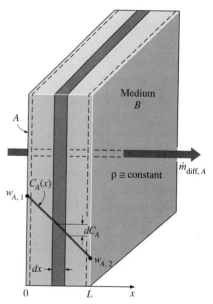

FIGURE 11-19

Schematic for steady one-dimensional mass diffusion of species A through a plane wall.

coefficient D_{AB} vary little along the wall, they can be assumed to be constant. The integration can be performed in that case to yield

$$\dot{m}_{\text{diff},A,\text{wall}} = \rho D_{AB} A \frac{w_{A,1} - w_{A,2}}{L} = D_{AB} A \frac{\rho_{A,1} - \rho_{A,2}}{L} \qquad \text{(kg/s)} \quad (11\text{-}22)$$

This relation can be rearranged as

$$\dot{m}_{\text{diff},A,\text{wall}} = \frac{w_{A,1} - w_{A,2}}{L/\rho D_{AB} A} = \frac{w_{A,1} - w_{A,2}}{R_{\text{diff},\text{wall}}} \qquad (11\text{-}23)$$

where

$$R_{\text{diff},\text{wall}} = \frac{L}{\rho D_{AB} A}$$

is the **diffusion resistance** of the wall, in s/kg, which is analogous to the electrical or conduction resistance of a plane wall of thickness L and area A (Fig. 11-20). Thus, we conclude that *the rate of mass diffusion through a plane wall is proportional to the average density, the wall area, and the concentration difference across the wall, but is inversely proportional to the wall thickness.* Also, once the rate of mass diffusion is determined, the mass fraction $w_A(x)$ at any location x can be determined by replacing $w_{A,2}$ in Equation 11-22 by $w_A(x)$ and L by x.

The preceding analysis can be repeated on a molar basis with the following result,

$$\dot{N}_{\text{diff},A,\text{wall}} = C D_{AB} A \frac{y_{A,1} - y_{A,2}}{L} = D_{AB} A \frac{C_{A,1} - C_{A,2}}{L} = \frac{y_{A,1} - y_{A,2}}{\overline{R}_{\text{diff},\text{wall}}}$$
$$(11\text{-}24)$$

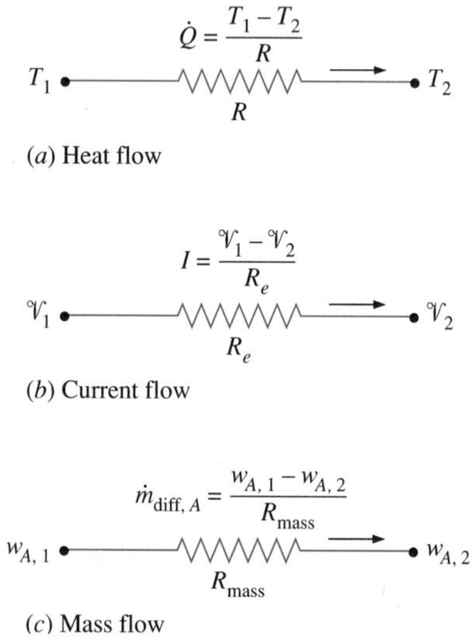

where $\overline{R}_{\text{diff},\text{wall}} = L/C D_{AB} A$ is the **molar diffusion resistance** of the wall in s/kmol. Note that mole fractions are accompanied by molar concentrations and mass fractions are accompanied by density. Either relation can be used to determine the diffusion rate of species A across the wall, depending on whether the mass or molar fractions of species A are known at the boundaries. Also, the concentration gradients on both sides of an interface are different, and thus diffusion resistance networks cannot be constructed in an analogous manner to thermal resistance networks.

In developing the relations above, we assumed the density and the diffusion coefficient of the wall to be nearly constant. This assumption is reasonable when a small amount of species A diffuses through the wall and thus *the concentration of A is small.* The species A can be a gas, a liquid, or a solid. Also, the wall can be a plane layer of a liquid or gas provided that it is *stationary.*

The analogy between heat and mass transfer also applies to *cylindrical* and *spherical* geometries. Repeating the approach outlined in Chapter 3 for heat conduction, we obtain the following analogous relations for steady one-

$$\dot{Q} = \frac{T_1 - T_2}{R}$$

(a) Heat flow

$$I = \frac{\mathcal{V}_1 - \mathcal{V}_2}{R_e}$$

(b) Current flow

$$\dot{m}_{\text{diff},A} = \frac{w_{A,1} - w_{A,2}}{R_{\text{mass}}}$$

(c) Mass flow

FIGURE 11-20

Analogy between thermal, electrical, and mass diffusion resistance concepts.

dimensional mass transfer through nonreacting cylindrical and spherical layers (Fig. 11-21)

$$\dot{m}_{\text{diff},A,\text{cyl}} = 2\pi L\rho D_{AB}\frac{w_{A,1}-w_{A,2}}{\ln(r_2/r_1)} = 2\pi LD_{AB}\frac{\rho_{A,1}-\rho_{A,2}}{\ln(r_2/r_1)} \qquad (11\text{-}25)$$

$$\dot{m}_{\text{diff},A,\text{sph}} = 4\pi r_1 r_2\rho D_{AB}\frac{w_{A,1}-w_{A,2}}{r_2-r_1} = 4\pi r_1 r_2 D_{AB}\frac{\rho_{A,1}-\rho_{A,2}}{r_2-r_1} \qquad (11\text{-}26)$$

or, on a molar basis,

$$\dot{N}_{\text{diff},A,\text{cyl}} = 2\pi LCD_{AB}\frac{y_{A,1}-y_{A,2}}{\ln(r_2/r_1)} = 2\pi LD_{AB}\frac{C_{A,1}-C_{A,2}}{\ln(r_2/r_1)} \qquad (11\text{-}27)$$

$$\dot{N}_{\text{diff},A,\text{sph}} = 4\pi r_1 r_2 CD_{Ab}\frac{y_{A,1}-y_{A,2}}{r_2-r_1} = 4\pi r_1 r_2 D_{AB}\frac{C_{A,1}-C_{A,2}}{r_2-r_1} \qquad (11\text{-}28)$$

Here, L is the length of the cylinder, r_1 is the inner radius, and r_2 is the outer radius for the cylinder or the sphere. Again, the boundary surfaces at $r = r_1$ and $r = r_2$ are located within the solid adjacent to the interfaces, and the mass fractions of A at those surfaces are maintained at $w_{A,1}$ and $w_{A,2}$, respectively, at all times. (We could make similar statements for the density, molar concentration, and mole fraction of species A at the boundaries.)

We mentioned earlier that the concentration of the gas species in a solid at the interface is proportional to the partial pressure of the adjacent gas and was expressed as $C_{A,\text{solid side}} = \mathcal{S}_{AB}P_{A,\text{gas side}}$ where $\mathcal{S}_{AB}$ is the *solubility* (in kmol/m^3 bar) of the gas A in the solid B. We also mentioned that the product of solubility and the diffusion coefficient is called the *permeability*, $\mathcal{P}_{Ab} = \mathcal{S}_{AB}D_{AB}$ (in kmol/m $\cdot$ s $\cdot$ bar). Then the molar flow rate of a gas through a solid under steady one-dimensional conditions can be expressed in terms of the partial pressures of the adjacent gas on the two sides of the solid by replacing C_A in the relations above by $\mathcal{S}_{AB}P_A$ or $\mathcal{P}_{AB}P_A/D_{AB}$. In the case of a *plane wall*, for example, it gives (Fig. 11-22)

$$\dot{N}_{\text{diff},A,\text{wall}} = D_{AB}\mathcal{S}_{AB}A\frac{P_{A,1}-P_{A,2}}{L} = \mathcal{P}_{AB}A\frac{P_{A,1}-P_{A,2}}{L} \qquad \text{(kmol/s)} \quad (11\text{-}29)$$

where $P_{A,1}$ and $P_{A,2}$ are the *partial pressures* of gas A on the two sides of the wall. Similar relations can be obtained for cylindrical and spherical walls by following the same procedure. Also, if the permeability is given on a mass basis (in kg/m $\cdot$ s $\cdot$ bar), then the relation above will give the diffusion mass flow rate.

Noting that 1 kmol of an ideal gas at the standard conditions of 0°C and 1 atm occupies a volume of 22.414 m^3, the volume flow rate of the gas through the wall by diffusion can be determined from

$$\dot{V}_{\text{diff},A} = 22.414\dot{N}_{\text{diff},A} \qquad \text{(standard m}^3\text{/s, at 0°C and 1 atm)}$$

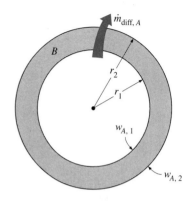

FIGURE 11-21

One-dimensional mass diffusion through a cylindrical or spherical shell.

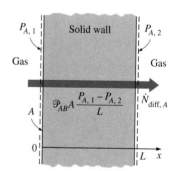

FIGURE 11-22

The diffusion rate of a gas species through a solid can be determined from a knowledge of the partial pressures of the gas on both sides and the permeability of the solid to that gas.

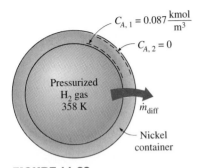

$C_{A,1} = 0.087 \dfrac{kmol}{m^3}$

$C_{A,2} = 0$

Pressurized
H₂ gas
358 K

$\dot{m}_{diff}$

Nickel
container

FIGURE 11-23
Schematic for Example 11-5.

The volume flow rate at other conditions can be determined from the ideal gas relation $P_A \dot{V} = \dot{N}_A R_u T$.

EXAMPLE 11-5 Diffusion of Hydrogen through a Spherical Container

Pressurized hydrogen gas is stored at 358 K in a 4.8-m-outer-diameter spherical container made of nickel (Fig. 11-23). The shell of the container is 6 cm thick. The molar concentration of hydrogen in the nickel at the inner surface is determined to be 0.087 kmol/m³. The concentration of hydrogen in the nickel at the outer surface is negligible. Determine the mass flow rate of hydrogen by diffusion through the nickel container.

Solution Pressurized hydrogen gas is stored in a spherical container. The diffusion rate of hydrogen through the container is to be determined.

Assumptions **1** Mass diffusion is *steady* and *one-dimensional* since the hydrogen concentration in the tank and thus at the inner surface of the container is practically constant, and the hydrogen concentration in the atmosphere and thus at the outer surface is practically zero. Also, there is thermal symmetry about the center. **2** There are no chemical reactions in the nickel shell that result in the generation or depletion of hydrogen.

Properties The binary diffusion coefficient for hydrogen in the nickel at the specified temperature is 1.2×10^{-12} m²/s (Table 11-3b).

Analysis We can consider the total molar concentration to be constant ($C = C_A + C_B \cong C_B = $ constant), and the container to be a *stationary* medium since there is no diffusion of nickel molecules ($\dot{N}_B = 0$) and the concentration of the hydrogen in the container is extremely low ($C_A \ll 1$). Then the molar flow rate of hydrogen through this spherical shell by diffusion can readily be determined from Equation 11-28 to be

$$\dot{N}_{diff} = 4\pi r_1 r_2 D_{AB} \frac{C_{A,1} - C_{A,2}}{r_2 - r_1}$$

$$= 4\pi(2.34 \text{ m})(2.40 \text{ m})(1.2 \times 10^{-12} \text{ m}^2/\text{s}) \frac{(0.087 - 0) \text{ kmol/m}^3}{2.40 - 2.34}$$

$$= 1.228 \times 10^{-10} \text{ kmol/s}$$

The mass flow rate is determined by multiplying the molar flow rate by the molar mass of hydrogen, which is $M = 2$ kg/kmol,

$$\dot{m}_{diff} = M\dot{N}_{diff} = (2 \text{ kg/kmol})(1.228 \times 10^{-10} \text{ kmol/s}) = \mathbf{2.46 \times 10^{-10} \text{ kg/s}}$$

Therefore, hydrogen will leak out through the shell of the container by diffusion at a rate of 2.46×10^{-10} kg/s or 7.8 g/year. Note that the concentration of hydrogen in the nickel at the inner surface depends on the temperature and pressure of the hydrogen in the tank and can be determined as explained in the previous example. Also, the assumption of zero hydrogen concentration in nickel at the outer surface is reasonable since there is only a trace amount of hydrogen in the atmosphere (0.5 part per million by mole numbers).

Moisture greatly influences the *performance* and *durability* of building materials, and thus moisture transmission is an important consideration in the construction and maintenance of buildings.

The *dimensions* of wood and other hygroscopic substances change with moisture content. For example, a variation of 4.5 percent in moisture content causes the volume of white oak wood to change by 2.5 percent. Such cyclic changes of dimensions weaken the joints and can jeopardize the structural integrity of building components, causing "squeaking" at the minimum. Excess moisture can also cause changes in the *appearance* and *physical properties* of materials: *corrosion* and *rusting* in metals, *rotting* in woods, and *peeling of paint* on the interior and exterior wall surfaces. Soaked wood with a water content of 24 to 31 percent is observed to decay rapidly at temperatures 10 to 38°C. Also, *molds* grow on wood surfaces at relative humidities above 85 percent. The expansion of water during freezing may damage the cell structure of porous materials.

Moisture content also affects the *effective conductivity* of porous mediums such as soils, building materials, and insulations, and thus heat transfer through them. Several studies have indicated that heat transfer increases almost linearly with moisture content, at a rate of 3 to 5 percent for each percent increase in moisture content by volume. Insulation with 5 percent moisture content by volume, for example, increases heat transfer by 15 to 25 percent relative to dry insulation (ASHRAE *Handbook of Fundamentals,* Chap. 20) (Fig. 11-24). Moisture migration may also serve as a transfer mechanism for latent heat by alternate evaporation and condensation. During a hot and humid day, for example, water vapor may migrate through a wall and condense on the inner side, releasing the heat of vaporization, with the process reversing during a cool night. Moisture content also affects the *specific heat* and thus the heat storage characteristics of building materials.

Moisture migration in the walls, floors, or ceilings of buildings and in other applications is controlled by either **vapor barriers** or **vapor retarders.** *Vapor barriers* are materials that are impermeable to moisture, such as sheet metals, heavy metal foils, and thick plastic layers, and they effectively bar the vapor from migrating. *Vapor retarders,* on the other hand, *retard* or *slow down* the flow of moisture through the structures but do not totally eliminate it. Vapor retarders are available as solid, flexible, or coating materials, but they usually consist of a thin sheet or coating. Common forms of vapor retarders are *reinforced plastics or metals, thin foils, plastic films, treated papers, coated felts,* and *polymeric or asphaltic paint coatings.* In applications such as the building of walls where vapor penetration is unavoidable because of numerous openings such as electrical boxes, telephone lines, and plumbing passages, vapor retarders are used instead of vapor barriers to allow the vapor that somehow leaks in to exit to the outside instead of trapping it in. Vapor retarders with a permeance of 57.4×10^{-9} kg/s · m^2 are commonly used in residential buildings.

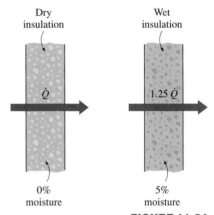

FIGURE 11-24

A 5 percent moisture content can increase heat transfer through wall insulation by 25 percent.

TABLE 11-9

Saturation pressure of water at various temperatures

Temperature, °C	Saturation pressure, Pa
−40	13
−36	20
−32	31
−28	47
−24	70
−20	104
−16	151
−12	218
−8	310
−4	438
0	611
5	872
10	1,228
15	1,705
20	2,339
25	3,169
30	4,246
35	5,628
40	7,384
50	12,349
100	101,330
200	1.55×10^6
300	8.58×10^6

The insulation on *chilled water lines* and other impermeable surfaces that are always cold must be wrapped with a *vapor barrier jacket,* or such cold surfaces must be insulated with a material that is impermeable to moisture. This is because moisture that migrates through the insulation to the cold surface will condense and remain there indefinitely with no possibility of vaporizing and moving back to the outside. The accumulation of moisture in such cases may render the insulation useless, resulting in excessive energy consumption.

Atmospheric air can be viewed as a mixture of dry air and water vapor, and the atmospheric pressure is the sum of the pressure of dry air and the pressure of water vapor, which is called the **vapor pressure** P_v. Air can hold a certain amount of moisture only, and the ratio of the actual amount of moisture in the air at a given temperature to the maximum amount air can hold at that temperature is called the **relative humidity** ϕ. The relative humidity ranges from 0 for dry air to 100 percent for *saturated* air (air that cannot hold any more moisture). The partial pressure of water vapor in saturated air is called the **saturation pressure** P_{sat}. Table 11-9 lists the saturation pressure at various temperatures.

The amount of moisture in the air is completely specified by the temperature and the relative humidity, and the vapor pressure is related to relative humidity ϕ by

$$P_v = \phi P_{sat} \qquad (11\text{-}30)$$

where P_{sat} is the saturation (or boiling) pressure of water at the specified temperature. Then the mass flow rate of moisture through a plain layer of thickness L and normal area A can be expressed as

$$\dot{m}_v = \mathscr{P} A \frac{P_{v,1} - P_{v,2}}{L} = \mathscr{P} A \frac{\phi_1 P_{sat,1} - \phi_2 P_{sat,2}}{L} \qquad \text{(kg/s)} \quad (11\text{-}31)$$

where $\mathscr{P}$ is the vapor permeability of the material, which is usually expressed on a mass basis in the unit ng/s $\cdot$ m $\cdot$ Pa where ng = 10^{-12} kg and 1 Pa = 10^{-5} bar. Note that vapor migrates or diffuses from a region of higher vapor pressure toward a region of lower vapor pressure.

The permeability of most construction materials is usually expressed for a given thickness instead of per unit thickness. It is called the **permeance** $\mathscr{M}$, which is the ratio of the permeability of the material to its thickness. That is,

$$\text{Permeance} = \frac{\text{Permeability}}{\text{Thickness}}$$

$$\mathscr{M} = \frac{\mathscr{P}}{L} \qquad (\text{kg/s} \cdot \text{m}^2 \cdot \text{Pa}) \qquad (11\text{-}32)$$

The reciprocal of permeance is called (unit) **vapor resistance** and is expressed as

$$\text{Vapor resistance} = \frac{1}{\text{Permeance}}$$

$$R_v = \frac{1}{\mathcal{M}} = \frac{L}{\mathcal{P}} \qquad (\text{s} \cdot \text{m}^2 \cdot \text{Pa/kg}) \qquad (11\text{-}33)$$

Note that vapor resistance represents the resistance of a material to water vapor transmission.

It should be pointed out that the amount of moisture that enters or leaves a building by *diffusion* is usually negligible compared to the amount that enters with *infiltrating air* or leaves with *exfiltrating air*. The primary cause of interest in the moisture diffusion is its impact on the performance and longevity of building materials.

The overall vapor resistance of a *composite* building structure that consists of several layers in series is the sum of the resistances of the individual layers and is expressed as

$$R_{v,\text{total}} = R_{v,1} + R_{v,2} + \cdots + R_{v,n} = \sum R_{v,i} \qquad (11\text{-}34)$$

Then the rate of vapor transmission through a composite structure can be determined in an analogous manner to heat transfer from

$$\dot{m}_v = A \frac{\Delta P_v}{R_{v,\text{total}}} \qquad (\text{kg/s}) \qquad (11\text{-}35)$$

Vapor permeance of common building materials is given in Table 11-10.

EXAMPLE 11-6   Condensation and Freezing of Moisture in the Walls

The condensation and even freezing of moisture in walls without effective vapor retarders is a real concern in cold climates, and it undermines the effectiveness of insulations. Consider a wood frame wall that is built around 38 mm × 90 mm (2 × 4 nominal) wood studs. The 90-mm-wide cavity between the studs is filled with glass fiber insulation. The inside is finished with 13-mm gypsum wallboard and the outside with 13-mm wood fiberboard and 13-mm × 200-mm wood bevel lapped siding. Using manufacturer's data, the thermal and vapor resistances of various components for a unit wall area are determined to be as follows:

Construction	R-value, $\text{m}^2 \cdot {}^\circ\text{C/W}$	R_v-value, $\text{s} \cdot \text{m}^2 \cdot \text{Pa/ng}$
1. Outside surface, 24 km/h wind	0.030	—
2. Painted wood bevel lapped siding	0.14	0.019
3. Wood fiberboard sheeting, 13 mm	0.23	0.0138
4. Glass fiber insulation, 90 mm	2.45	0.0004
5. Painted gypsum wallboard, 13 mm	0.079	0.012
6. Inside surface, still air	0.12	—
TOTAL	3.05	0.0452

The indoor conditions are 20°C and 60 percent relative humidity while the outside conditions are −16°C and 70 percent relative humidity. Determine if condensation or freezing of moisture will occur in the insulation.

TABLE 11-10

Typical vapor permeance of common building materials (from ASHRAE, Ref. 1, Chap. 22, Table 9)*

Materials and its thickness	Permeance $\text{ng/s} \cdot \text{m}^2 \cdot \text{Pa}$
Concrete (1:2:4 mix, 1 m)	4.7
Brick, masonry, 100 mm	46
Plaster on metal lath, 19 mm	860
Plaster on wood lath, 19mm	630
Gypsum wall board, 9.5 mm	2860
Plywood, 6.4 mm	40–109
Still air, 1 m	174
Mineral wool insulation (unprotected), 1 m	245
Expanded polyurethane insulation board, 1 m	0.58–2.3
Aluminum foil, 0.025 mm	0.0
Aluminum foil, 0.009 mm	2.9
Polyethylene, 0.051 mm	9.1
Polyethylene, 0.2 mm	2.3
Polyester, 0.19 mm	4.6
Vapor retarder latex paint, 0.070 mm	26
Exterior acrylic house and trim paint, 0.040 mm	313
Building paper, unit mass of 0.16–0.68 kg/m²	0.1–2400

*Data vary greatly. Consult manufacturer for more accurate data. Multiply by 1.41×10^{-6} to convert to $\text{lbm/s} \cdot \text{ft}^2 \cdot \text{psi}$. Also, 1 ng = 10^{-12} kg.

Solution The thermal and vapor resistances of different layers of a wall are given. The possibility of condensation or freezing of moisture in the wall is to be investigated.

Assumptions **1** Steady operating conditions exist. **2** Heat transfer through the wall is one-dimensional. **3** Thermal and vapor resistances of different layers of the wall and the heat transfer coefficients are constant.

Properties The thermal and vapor resistances are as given in the problem statement. The saturation pressures of water at 20°C and −16°C are 2339 Pa and 151 Pa, respectively (Table 11-9).

Analysis The schematic of the wall as well as the different elements used in its construction are shown in Figure 11-25. Condensation is most likely to occur at the coldest part of insulation, which is the part adjacent to the exterior sheathing. Noting that the total thermal resistance of the wall is 3.05 m² · °C/W, the rate of heat transfer through a unit area $A = 1$ m² of the wall is

$$\dot{Q}_{wall} = A\frac{T_i - T_o}{R_{total}} = (1 \text{ m}^2)\frac{[20 - (-16)°C]}{3.05 \text{ m}^2 \cdot °C/W} = 11.8 \text{ W}$$

The thermal resistance of the exterior part of the wall beyond the insulation is $0.03 + 0.14 + 0.23 = 0.40$ m² · °C/W. Then the temperature of the insulation–outer sheathing interface is

$$T_I = T_o + \dot{Q}_{wall}R_{ext} = -16°C + (11.8 \text{ W})(0.40°C/W) = -11.3°C$$

The saturation pressure of water at −11.3°C is 234 Pa, as shown in Table 11-9, and if there is condensation or freezing, the vapor pressure at the insulation–outer sheathing interface will have to be this value. The vapor pressure at the indoors and the outdoors is

$$P_{v,1} = \phi_1 P_{sat,1} = 0.60 \times (2339 \text{ Pa}) = 1404 \text{ Pa}$$
$$P_{v,2} = \phi_2 P_{sat,2} = 0.70 \times (151 \text{ Pa}) = 106 \text{ Pa}$$

Then the rate of moisture flow through the interior and exterior parts of the wall becomes

$$\dot{m}_{v,interior} = A\left(\frac{\Delta P}{R_v}\right)_{interior} = A\frac{P_{v,1} - P_{v,I}}{R_{v,interior}}$$
$$= (1 \text{ m}^2)\frac{(1404 - 234) \text{ Pa}}{(0.012 + 0.0004) \text{ Pa} \cdot \text{m}^2 \cdot \text{s/ng}} = 94{,}355 \text{ ng/s} = 94.4 \text{ μg/s}$$

$$\dot{m}_{v,exterior} = A\left(\frac{\Delta P}{R_v}\right)_{exterior} = A\frac{P_{v,I} - P_{v,2}}{R_{v,exterior}}$$
$$= (1 \text{ m}^2)\frac{(234 - 106) \text{ Pa}}{(0.019 + 0.0138)\text{Pa} \cdot \text{m}^2 \cdot \text{s/ng}} = 3902 \text{ ng/s} = 3.9 \text{ μg/s}$$

That is, moisture is flowing toward the interface at a rate of 94.4 μg/s but flowing from the interface to the outdoors at a rate of only 3.9 μg/s. Noting that the interface pressure cannot exceed 234 Pa, these results indicate that moisture is freezing in the insulation at a rate of

$$\dot{m}_{v,freeing} = \dot{m}_{v,interior} - \dot{m}_{v,exterior} = 94.4 - 3.9 = \textbf{90.5 μg/s}$$

This corresponds to 7.82 g during a 24-h period, which can be absorbed by the insulation or sheathing, and then flows out when the conditions improve. How-

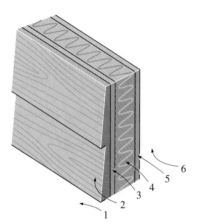

FIGURE 11-25
Schematic for Example 11-6.

ever, excessive condensation (or frosting at temperatures below 0°C) of moisture in the walls during long cold spells can cause serious problems. This problem can be avoided or minimized by installing vapor barriers on the interior side of the wall, which will limit the moisture flow rate to 3.9 µg/s. Note that if there were no condensation or freezing, the flow rate of moisture through a 1 m² section of the wall would be 28.7 µg/s (can you verify this?).

11-7 ■ TRANSIENT MASS DIFFUSION

The steady analysis discussed earlier is useful when determining the leakage rate of a species through a stationary layer. But sometimes we are interested in the diffusion of a species into a body during a limited time before steady operating conditions are established. Such problems are studied using **transient analysis.** For example, the surface of a mild steel component is commonly hardened by packing the component in a carbonaceous material in a furnace at high temperature. During the short time period in the furnace, the carbon molecules diffuse through the surface of the steel component, but they penetrate to a depth of only a few millimeters. The carbon concentration decreases exponentially from the surface to the inner parts, and the result is a steel component with a very hard surface and a relatively soft core region (Fig. 11-26).

The same process is used in the gem industry to color clear stones. For example, a clear sapphire is given a brilliant blue color by packing it in titanium and iron oxide powders and baking it in an oven at about 2000°C for about a month. The titanium and iron molecules penetrate less than 0.5 mm in the sapphire during this process. Diffusion in solids is usually done at high temperatures to take advantage of the high diffusion coefficients at high temperatures and thus to keep the diffusion time at a reasonable level. Such diffusion or "doping" is also commonly practiced in the production of n- or p-type semiconductor materials used in the manufacture of electronic components. Drying processes such as the drying of coal, timber, food, and textiles constitute another major application area of transient mass diffusion.

Transient mass diffusion in a stationary medium is analogous to transient heat transfer provided that the solution is dilute and thus the density of the medium ρ is constant. In Chapter 4 we presented analytical and graphical solutions for one-dimensional transient heat conduction problems in solids with constant properties, no heat generation, and uniform initial temperature. The analogous one-dimensional transient mass diffusion problems satisfy the following:

1. The *diffusion coefficient is constant.* This is valid for an isothermal medium since D_{AB} varies with temperature (corresponds to constant thermal diffusivity).

2. There are *no homogeneous reactions* in the medium that generate or deplete the diffusing species A (corresponds to no heat generation).

3. Initially ($t = 0$) the concentration of species A is *constant* throughout the medium (corresponds to uniform initial temperature).

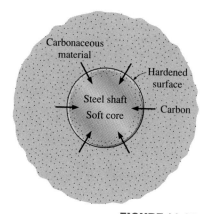

FIGURE 11-26

The surface hardening of a mild steel component by the diffusion of carbon molecules is a transient mass diffusion process.

Heat conduction	Mass diffusion
T	C, y, ρ or w
α	D_{AB}
$\theta = \dfrac{T(x,t) - T_\infty}{T_i - T_\infty}$	$\theta_{mass} = \dfrac{w_A(x,t) - w_{A,\infty}}{w_{A,i} - w_{A,\infty}}$ *
$\dfrac{T(x,t) - T_s}{T_i - T_s}$	$\dfrac{w_A(x,t) - w_A}{w_{A,i} - w_A}$
$\xi = \dfrac{x}{2\sqrt{\alpha t}}$	$\xi_{mass} = \dfrac{x}{2\sqrt{D_{AB}t}}$
$Bi = \dfrac{h_{conv}L}{k}$	$Bi_{mass} = \dfrac{h_{mass}L}{D_{AB}}$
$\tau = \dfrac{\alpha t}{L^2}$	$\tau = \dfrac{D_{AB}t}{L^2}$

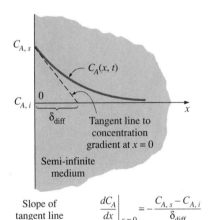

FIGURE 11-27

The concentration profile of species A in a semi-infinite medium during transient mass diffusion and the penetration depth.

Then the solution of a mass diffusion problem can be obtained directly from the analytical or graphical solution of the corresponding heat conduction problem given in Chapter 4. The analogous quantities between heat and mass transfer are summarized in Table 11-11 for easy reference. For the case of a semi-infinite medium with constant surface concentration, for example, the solution can be expressed in an analogous manner to Equation 4-24 as

$$\frac{C_A(x,t) - C_{A,i}}{C_{A,s} - C_{A,i}} = \text{erfc}\left(\frac{x}{2\sqrt{D_{AB}t}}\right) \qquad (11\text{-}36)$$

where $C_{A,i}$ is the initial concentration of species A at time $t = 0$ and $C_{A,s}$ is the concentration at the inner side of the exposed surface of the medium. By using the definitions of molar fraction, mass fraction, and density, it can be shown that for dilute solutions

$$\frac{C_A(x,t) - C_{A,i}}{C_{A,s} - C_{A,i}} = \frac{\rho_A(x,t) - \rho_{A,i}}{\rho_{A,s} - \rho_{A,i}} = \frac{w_A(x,t) - w_{A,i}}{w_{A,s} - w_{A,i}} = \frac{y_A(x,t) - y_{A,i}}{y_{A,s} - y_{A,i}} \qquad (11\text{-}37)$$

since the total density or total molar concentration of dilute solutions is usually constant ($\rho = $ constant or $C = $ constant). Therefore, other measures of concentration can be used in Equation 11-36.

A quantity of interest in mass diffusion processes is the depth of diffusion at a given time. This is usually characterized by the **penetration depth** defined as *the location x where the tangent to the concentration profile at the surface ($x = 0$) intercepts the $C_A = C_{A,i}$ line*, as shown in Figure 11-27. Obtaining the concentration gradient at $x = 0$ by differentiating Equation 11-36, the penetration depth is determined to be

$$\delta_{diff} = \frac{C_{A,s} - C_{A,i}}{-(dC_A/dx)_{x=0}} = \frac{C_{A,s} - C_{A,i}}{(C_{A,s} - C_{A,i})/\sqrt{\pi D_{AB}t}} = \sqrt{\pi D_{AB}t} \qquad (11\text{-}38)$$

Therefore, the penetration depth is proportional to the square root of both the diffusion coefficient and time. The diffusion coefficient of zinc in copper at 1000°C, for example, is 5.0×10^{-12} m²/s. Then the penetration depth of zinc in copper in 10 h is

$$\delta_{diff} = \sqrt{\pi D_{AB}t} = \sqrt{\pi(5.0 \times 10^{-12}\ \text{m}^2/\text{s})(10 \times 3600\ \text{s})}$$
$$= 0.00038\ \text{m} = 0.38\ \text{mm}$$

That is, zinc will penetrate to a depth of about 0.38 mm in an appreciable amount in 10 h, and there will hardly be any zinc in the copper block beyond a depth of 0.38 mm.

The diffusion coefficients in solids are typically very low (on the order of 10^{-9} to 10^{-15} m²/s), and thus the diffusion process usually affects a thin layer at the surface. A solid can conveniently be treated as a semi-infinite medium during transient mass diffusion regardless of its size and shape when the penetration depth is small relative to the thickness of the solid. When this is not the case, solutions for one-dimensional transient mass diffusion through a plane wall, cylinder, and sphere can be obtained from the solutions of analogous heat conduction problems using the Heisler charts or one-term solutions presented in Chapter 4.

EXAMPLE 11-7   Hardening of Steel by the Diffusion of Carbon

The surface of a mild steel component is commonly hardened by packing the component in a carbonaceous material in a furnace at a high temperature for a predetermined time. Consider such a component with a uniform initial carbon concentration of 0.15 percent by mass. The component is now packed in a carbonaceous material and is placed in a high temperature furnace. The diffusion coefficient of carbon in steel at the furnace temperature is given to be 4.8×10^{-10} m^2/s, and the equilibrium concentration of carbon in the iron at the interface is determined from equilibrium data to be 1.2 percent by mass. Determine how long the component should be kept in the furnace for the mass concentration of carbon 0.5 mm below the surface to reach 1 percent (Fig. 11-28).

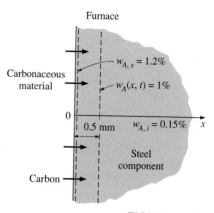

FIGURE 11-28
Schematic for Example 11-7.

Solution   A steel component is to be surface hardened by packing it in a carbonaceous material in a furnace. The length of time the component should be kept in the furnace is to be determined.

Assumptions   Carbon penetrates into a very thin layer beneath the surface of the component, and thus the component can be modeled as a semi-infinite medium regardless of its thickness or shape.

Properties   The relevant properties are given in the problem statement.

Analysis   This problem is analogous to the one-dimensional transient heat conduction problem in a semi-infinite medium with specified surface temperature, and thus can be solved accordingly. Using mass fraction for concentration since the data are given in that form, the solution can be expressed as

$$\frac{w_A(x, t) - w_{A,i}}{w_{A,s} - w_{A,i}} = \text{erfc}\left(\frac{x}{2\sqrt{D_{AB}t}}\right)$$

Substituting the specified quantities gives

$$\frac{0.01 - 0.0015}{0.012 - 0.0015} = 0.81 = \text{erfc}\left(\frac{x}{2\sqrt{D_{AB}t}}\right)$$

The argument whose complementary error function is 0.81 is determined from Table 4-3 to be 0.17. That is,

$$\frac{x}{2\sqrt{D_{AB}t}} = 0.17$$

Then solving for the time t gives

$$t = \frac{x^2}{4D_{AB}(0.17)^2} = \frac{(0.0005 \text{ m})^2}{4 \times (4.8 \times 10^{-10} \text{ m}^2/\text{s})(0.17)^2} = 4505 \text{ s} = \textbf{1 h 15 min}$$

Discussion   The steel component in this case must be held in the furnace for 1 h and 15 min to achieve the desired level of hardening. The diffusion coefficient of carbon in steel increases exponentially with temperature, and thus this process is commonly done at high temperatures to keep the diffusion time at a reasonable level.

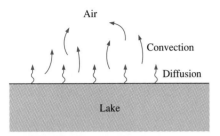

Air

Convection

Diffusion

Lake

FIGURE 11-29

In a moving medium, mass transfer is
due to both diffusion and convection.

11-8 ■ DIFFUSION IN A MOVING MEDIUM

To this point we have limited our consideration to mass diffusion in a *stationary medium,* and thus the only motion involved was the creeping motion of molecules in the direction of decreasing concentration, and there was no motion of the mixture as a whole. Many practical problems, such as the evaporation of water from a lake under the influence of the wind or the mixing of two fluids as they flow in a pipe, involve diffusion in a **moving medium** where the bulk motion is caused by an external force. Mass diffusion in such cases is complicated by the fact that chemical species are transported both by *diffusion* and by the bulk motion of the medium (i.e., *convection*). The velocities and mass flow rates of species in a moving medium consist of two components: one due to *molecular diffusion* and one due to *convection* (Fig. 11-29).

Diffusion in a moving medium, in general, is difficult to analyze since various species can move at different velocities in different directions. Turbulence will complicate the things even more. To gain a firm understanding of the physical mechanism while keeping the mathematical complexities to a minimum, we limit our consideration to systems that involve only *two components* (species A and B) in *one-dimensional flow* (velocity and other properties change in one direction only, say the x-direction). We also assume the total density (or molar concentration) of the medium remains constant. That is, $\rho = \rho_A + \rho_B = $ constant (or $C = C_A + C_B = $ constant) but the densities of species A and B may vary in the x-direction.

Several possibilities are summarized in Figure 11-30. In the trivial case (case *a*) of a *stationary homogeneous mixture,* there will be no mass transfer by molecular diffusion or convection since there is no concentration gradient or bulk motion. The next case (case *b*) corresponds to the *flow of a well-mixed fluid mixture* through a pipe. Note that there is no concentration gradients and thus molecular diffusion in this case, and all species move at the bulk flow velocity of $\mathcal{V}$. The mixture in the third case (case *c*) is *stationary* ($\mathcal{V} = 0$) and thus it corresponds to ordinary molecular diffusion in stationary mediums, which we discussed before. Note that the velocity of a species at a location in this case is simply the **diffusion velocity,** which is the average velocity of a group of molecules at that location moving under the influence of concentration gradient. Finally, the last case (case *d*) involves both *molecular diffusion* and *convection,* and the velocity of a species in this case is equal to the sum of the bulk flow velocity and the diffusion velocity. Note that the flow and the diffusion velocities can be in the same or opposite directions, depending on the direction of the concentration gradient. The diffusion velocity of a species is *negative* when the bulk flow is in the positive x-direction and the concentration gradient is positive (i.e., the concentration of the species increases in the x-direction).

Noting that the mass flow rate at any flow section is expressed as $\dot{m} = \rho \mathcal{V} A$ where ρ is the density, $\mathcal{V}$ is the velocity, and A is the cross-sectional area, the conservation of mass relation for the flow of a mixture that involves two species A and B can be expressed as

	Species	Density	Velocity	Mass flow rate
(a) Homogeneous mixture without bulk motion (no concentration gradients and thus no diffusion) $\mathcal{V} = 0$	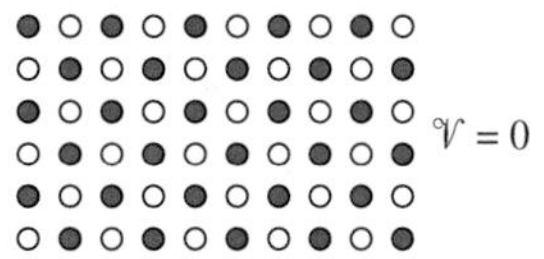 Species A	$\rho_A = $ constant	$\mathcal{V}_A = 0$	$\dot{m}_A = 0$
	Species B	$\rho_B = $ constant	$\mathcal{V}_B = 0$	$\dot{m}_B = 0$
	Mixture of A and B	$\rho = \rho_A + \rho_B$ = constant	$\mathcal{V} = 0$	$\dot{m} = 0$
(b) Homogeneous mixture with bulk motion (no concentration gradients and thus no diffusion) $\longrightarrow \mathcal{V}$	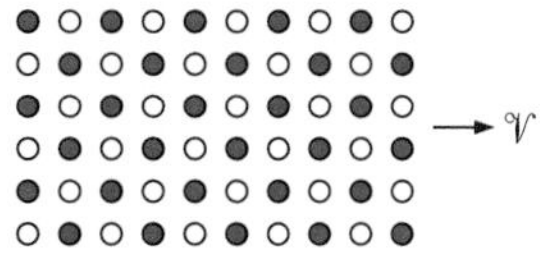 Species A	$\rho_A = $ constant	$\mathcal{V}_A = \mathcal{V}$	$\dot{m}_A = \rho_A \mathcal{V}_A A$
	Species B	$\rho_B = $ constant	$\mathcal{V}_B = \mathcal{V}$	$\dot{m}_B = \rho_B \mathcal{V}_B A$
	Mixture of A and B	$\rho = \rho_A + \rho_B$ = constant	$\mathcal{V} = \mathcal{V}$	$\dot{m} = \rho \mathcal{V} A$ $= \dot{m}_A + \dot{m}_B$
(c) Nonhomogeneous mixture without bulk motion (stationary medium with concentration gradients) $\mathcal{V} = 0$ $\mathcal{V}_{\text{diff}, A} \longrightarrow \quad \longleftarrow \mathcal{V}_{\text{diff}, B}$	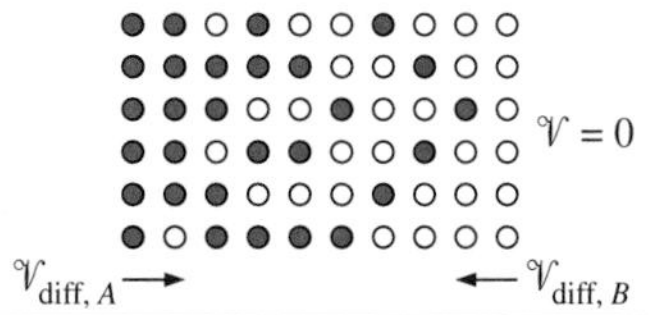 Species A	$\rho_A \neq $ constant	$\mathcal{V}_A = \mathcal{V}_{\text{diff}, A}$	$\dot{m}_A = \rho_A \mathcal{V}_{\text{diff}, A} A$
	Species B	$\rho_B \neq $ constant	$\mathcal{V}_B = \mathcal{V}_{\text{diff}, B}$	$\dot{m}_B = \rho_B \mathcal{V}_{\text{diff}, B} A$
	Mixture of A and B	$\rho = \rho_A + \rho_B$ = constant	$\mathcal{V} = 0$	$\dot{m} = \rho \mathcal{V} A = 0$ (thus $\dot{m}_A = -\dot{m}_B$)
(d) Nonhomogeneous mixture with bulk motion (moving medium with concentration gradients) $\longrightarrow \mathcal{V}$ $\mathcal{V}_{\text{diff}, A} \longrightarrow \quad \longleftarrow \mathcal{V}_{\text{diff}, B}$	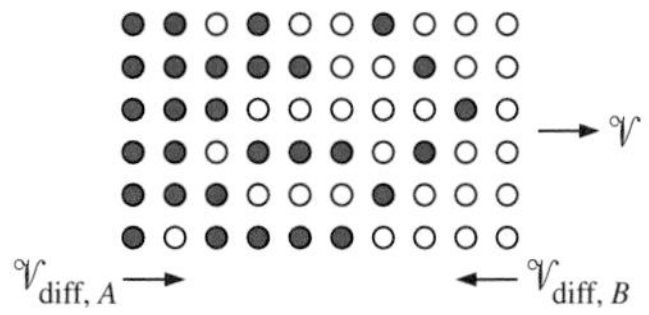 Species A	$\rho_A \neq $ constant	$\mathcal{V}_A = \mathcal{V} + \mathcal{V}_{\text{diff}, A}$	$\dot{m}_A = \rho_A \mathcal{V}_{\text{diff}, A} A$
	Species B	$\rho_B \neq $ constant	$\mathcal{V}_B = \mathcal{V} - \mathcal{V}_{\text{diff}, B}$	$\dot{m}_B = \rho_B \mathcal{V}_{\text{diff}, B} A$
	Mixture of A and B	$\rho = \rho_A + \rho_B$ = constant	$\mathcal{V} = \mathcal{V}$	$\dot{m} = \rho \mathcal{V} A$ $= \dot{m}_A + \dot{m}_B$

FIGURE 11-30

Various quantities associated with a mixture of two species A and B at a location x under one-dimensional flow or no-flow conditions. (The density of the mixture $\rho = \rho_A + \rho_B$ is assumed to remain constant.)

$$\dot{m} = \dot{m}_A + \dot{m}_B$$

or

$$\rho \mathcal{V} A = \rho_A \mathcal{V}_A A + \rho_B \mathcal{V}_B A$$

Canceling A and solving for $\mathcal{V}$ gives

$$\mathcal{V} = \frac{\rho_A \mathcal{V}_A + \rho_B \mathcal{V}_B}{\rho} = \frac{\rho_A}{\rho} \mathcal{V}_A + \frac{\rho_B}{\rho} \mathcal{V}_B = w_A \mathcal{V}_A + w_B \mathcal{V}_B \quad (11\text{-}39)$$

where $\mathcal{V}$ is called the **mass-average velocity** of the flow, which is the velocity that would be measured by a velocity sensor such as a pitot tube, a turbine device, or a hot wire anemometer inserted into the flow.

The special case $\mathcal{V} = 0$ corresponds to a **stationary medium,** which can now be defined more precisely as *a medium whose mass-average velocity is zero.* Therefore, mass transport in a stationary medium is by diffusion only, and zero mass-average velocity indicates that there is no bulk fluid motion.

When there is *no concentration gradient* (and thus no molecular mass diffusion) in the fluid, the velocity of all species will be equal to the *mass-average velocity of the flow.* That is, $\mathcal{V} = \mathcal{V}_A = \mathcal{V}_B$. But when there is a concentration gradient, there will also be a simultaneous flow of species in the direction of decreasing concentration at a diffusion velocity of $\mathcal{V}_{\text{diff}}$. Then

$\mathcal{V}_{\text{diff},A} = 0$
$\mathcal{V}_A = \mathcal{V}$

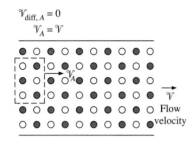

(a) No concentration gradient

$\mathcal{V}_{\text{diff},A} \neq 0$
$\mathcal{V}_A = \mathcal{V} + \mathcal{V}_{\text{diff},A}$

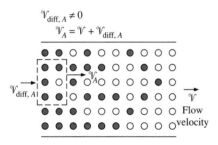

(b) Mass concentration gradient and thus
mass diffusion

FIGURE 11-31

The velocity of a species at a point is
equal to the sum of the bulk flow
velocity and the diffusion velocity
of that species at that point.

the average velocity of the species A and B can be determined by superimposing the average flow velocity and the diffusion velocity as (Fig. 11-31)

$$\mathcal{V}_A = \mathcal{V} + \mathcal{V}_{\text{diff},A}$$
$$\mathcal{V}_B = \mathcal{V} + \mathcal{V}_{\text{diff},B} \tag{11-40}$$

Similarly, we apply the superposition principle to the species mass flow rates to get

$$\dot{m}_A = \rho_A \mathcal{V}_A A = \rho_A(\mathcal{V} + \mathcal{V}_{\text{diff},A})A = \rho_A \mathcal{V} A + \rho_A \mathcal{V}_{\text{diff},A} A = \dot{m}_{\text{conv},A} + \dot{m}_{\text{diff},A}$$
$$\dot{m}_B = \rho_B \mathcal{V}_B A = \rho_B(\mathcal{V} + \mathcal{V}_{\text{diff},B})A = \rho_B \mathcal{V} A + \rho_B \mathcal{V}_{\text{diff},B} A = \dot{m}_{\text{conv},B} + \dot{m}_{\text{diff},B}$$

$$\tag{11-41}$$

Using Fick's law of diffusion, the total mass fluxes $j = \dot{m}/A$ can be expressed as

$$j_A = \rho_A \mathcal{V} + \rho_A \mathcal{V}_{\text{diff},A} = \frac{\rho_A}{\rho} \rho \mathcal{V} - \rho D_{AB} \frac{dw_A}{dx} = w_A(j_A + j_B) - \rho D_{AB} \frac{dw_A}{dx}$$

$$j_B = \rho_B \mathcal{V} + \rho_B \mathcal{V}_{\text{diff},B} = \frac{\rho_B}{\rho} \rho \mathcal{V} - \rho D_{BA} \frac{dw_B}{dx} = w_B(j_A + j_B) - \rho D_{BA} \frac{dw_B}{dx}$$

$$\tag{11-42}$$

Note that the diffusion velocity of a species is negative when the molecular diffusion occurs in the negative x-direction (opposite to flow direction). The mass diffusion rates of the species A and B at a specified location x can be expressed as

$$\dot{m}_{\text{diff},A} = \rho_A \mathcal{V}_{\text{diff},A} A = \rho_A(\mathcal{V}_A - \mathcal{V})A$$
$$\dot{m}_{\text{diff},B} = \rho_B \mathcal{V}_{\text{diff},B} A = \rho_B(\mathcal{V}_B - \mathcal{V})A \tag{11-43}$$

By substituting the $\mathcal{V}$ relation from Equation 11-39 into the equations above, it can be shown that at any cross-section

$$\dot{m}_{\text{diff},A} + \dot{m}_{\text{diff},B} = 0 \quad \rightarrow \quad \dot{m}_{\text{diff},A} = -\dot{m}_{\text{diff},B} \quad \rightarrow \quad -\rho A D_{AB} \frac{dw_A}{dx} = \rho A D_{BA} \frac{dw_B}{dx} \tag{11-44}$$

which indicates that the rates of diffusion of species A and B must be equal in magnitude but opposite in sign. This is a consequence of the assumption $\rho = \rho_A + \rho_B = $ constant, and it indicates that *anytime the species A diffuses in one direction, an equal amount of species B must diffuse in the opposite direction to maintain the density (or the molar concentration) constant.* This behavior is closely approximated by dilute gas mixtures and dilute liquid or solid solutions. For example, when a small amount of gas diffuses into a liquid, it is reasonable to assume the density of the liquid to remain constant.

Note that for a binary mixture, $w_A + w_B = 1$ at any location x. Taking the derivative with respect to x gives

$$\frac{dw_A}{dx} = -\frac{dw_B}{dx} \tag{11-45}$$

Thus we conclude from Equation 11-44 that (Fig. 11-32)

$$D_{AB} = D_{BA} \qquad (11\text{-}46)$$

That is, in the case of constant total concentration, the diffusion coefficient of species A into B is equal to the diffusion coefficient of species B into A.

We now repeat the analysis presented above with molar concentration C and the molar flow rate $\dot{N}$. The conservation of matter in this case is expressed as

$$\dot{N} = \dot{N}_A + \dot{N}_B$$

or

$$\rho \overline{\mathcal{V}} A = \rho_A \overline{\mathcal{V}}_A A + \rho_B \overline{\mathcal{V}}_B A \qquad (11\text{-}47)$$

Canceling A and solving for $\overline{\mathcal{V}}$ gives

$$\overline{\mathcal{V}} = \frac{C_A \overline{\mathcal{V}}_A + C_B \overline{\mathcal{V}}_B}{C} = \frac{C_A}{C} \overline{\mathcal{V}}_A + \frac{C_B}{C} \overline{\mathcal{V}}_B = y_A \overline{\mathcal{V}}_A + y_B \overline{\mathcal{V}}_B \qquad (11\text{-}48)$$

where $\overline{\mathcal{V}}$ is called the **molar-average velocity** of the flow. Note that $\overline{\mathcal{V}} \neq \mathcal{V}$ unless the mass and molar fractions are the same. The molar flow rates of species are determined similarly to be

$$\dot{N}_A = C_A \mathcal{V}_A A = C_A (\overline{\mathcal{V}} + \overline{\mathcal{V}}_{\text{diff},A}) A = C_A \overline{\mathcal{V}} A + C_A \overline{\mathcal{V}}_{\text{diff},A} A = \dot{N}_{\text{conv},A} + \dot{N}_{\text{diff},A}$$
$$\dot{N}_B = C_B \mathcal{V}_B A = C_B (\overline{\mathcal{V}} + \overline{\mathcal{V}}_{\text{diff},B}) A = C_B \overline{\mathcal{V}} A + C_B \overline{\mathcal{V}}_{\text{diff},B} A = \dot{N}_{\text{conv},B} + \dot{N}_{\text{diff},B} \qquad (11\text{-}49)$$

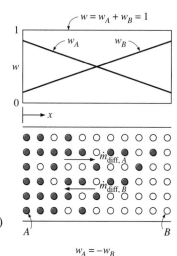

$$w_A = -w_B$$
$$\frac{dw_A}{dx} = -\frac{dw_B}{dx}$$
$$\dot{m}_{\text{diff},A} = -\dot{m}_{\text{diff},B}$$
$$D_{AB} = D_{BA}$$

FIGURE 11-32

In a binary mixture of species A and B with $\rho = \rho_A + \rho_B = $ constant, the rates of mass diffusion of species A and B are equal magnitude and opposite in direction.

Using Fick's law of diffusion, the total molar fluxes $\bar{j} = \dot{N}/A$ and diffusion molar flow rates $\dot{N}_{\text{diff}}$ can be expressed as

$$\bar{j}_A = C_A \overline{\mathcal{V}} + C_A \overline{\mathcal{V}}_{\text{diff},A} = \frac{C_A}{C} C \overline{\mathcal{V}} - CD_{AB} \frac{dy_A}{dx} = y_A(\bar{j}_A + \bar{j}_B) - CD_{AB} \frac{dy_A}{dx}$$

$$\bar{j}_B = C_B \overline{\mathcal{V}} + C_B \overline{\mathcal{V}}_{\text{diff},B} = \frac{C_B}{C} C \overline{\mathcal{V}} - CD_{BA} \frac{dy_B}{dx} = y_B(\bar{j}_A + \bar{j}_B) - CD_{BA} \frac{dy_B}{dx} \qquad (11\text{-}50)$$

and

$$\dot{N}_{\text{diff},A} = C_A \overline{\mathcal{V}}_{\text{diff},A} A = C_A(\mathcal{V}_A - \overline{\mathcal{V}}) A \qquad (11\text{-}51)$$
$$\dot{N}_{\text{diff},B} = C_B \overline{\mathcal{V}}_{\text{diff},B} A = C_B(\mathcal{V}_B - \overline{\mathcal{V}}) A$$

By substituting the $\overline{\mathcal{V}}$ relation from Equation 11-48 into the two equations above, it can be shown that

$$\dot{N}_{\text{diff},A} + \dot{N}_{\text{diff},B} = 0 \quad \rightarrow \quad \dot{N}_{\text{diff},A} = -\dot{N}_{\text{diff},B} \qquad (11\text{-}52)$$

which again indicates that the rates of diffusion of species A and B must be equal in magnitude but opposite in sign.

It is important to note that when working with molar units, a medium is said to be stationary when the *molar-average velocity* is zero. The average velocity of the molecules will be zero in this case, but the apparent velocity of the mixture as measured by a velocimeter placed in the flow will not necessarily be zero because of the different masses of different molecules. In a *mass-based stationary medium,* for each unit mass of species A moving in one direction, a unit mass of species B moves in the opposite direction. In a *mole-based stationary medium,* however, for each mole of species A moving in one direction, one mole of species B moves in the opposite direction. But this may result in a net mass flow rate in one direction that can be measured by a velocimeter since the masses of different molecules are different.

You may be wondering whether to use the mass analysis or molar analysis in a problem. The two approaches are equivalent, and either approach can be used in mass transfer analysis. But sometimes it may be easier to use one of the approaches, depending on what is given. When mass-average velocity is known or can easily be obtained, obviously it is more convenient to use the mass-based formulation. When the *total pressure* and *temperature* of a mixture are constant, however, it is more convenient to use the molar formulation, as explained below.

Special Case: Gas Mixtures at Constant Pressure and Temperature

Consider a gas mixture whose total pressure and temperature are constant throughout. When the mixture is homogeneous, the mass density ρ, the molar density (or concentration) C, the gas constant R, and the molar mass M of the mixture are the same throughout the mixture. But when the concentration of one or more gases in the mixture is not constant, setting the stage for mass diffusion, then the mole fractions y_i of the species will vary throughout the mixture. As a result, the gas constant R, the molar mass M, and the mass density ρ of the mixture will also vary since, assuming ideal gas behavior,

$$M = \sum y_i M_i, \qquad R = \frac{R_u}{M}, \qquad \text{and} \qquad \rho = \frac{P}{RT}$$

where $R_u = 8.314$ kJ/kmol $\cdot$ K is the universal gas constant. Therefore, the assumption of *constant mixture density* (ρ = constant) in such cases will not be accurate unless the gas or gases with variable concentrations constitute a very small fraction of the mixture. However, the *molar density C of a mixture remains constant* when the mixture pressure P and temperature T are constant since

$$P = \rho RT = \rho \frac{R_u}{M} T = CR_u T \tag{11-53}$$

The condition C = constant offers considerable simplification in mass transfer analysis, and thus it is more convenient to use the molar formulation when dealing with gas mixtures at constant total pressure and temperature (Fig. 11-33).

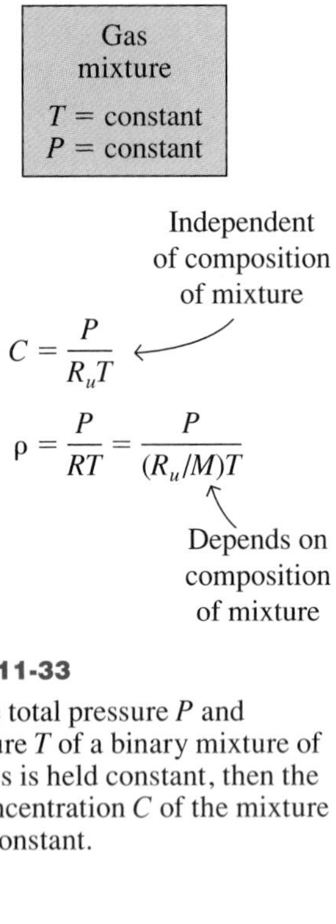

Gas
mixture

T = constant
P = constant

Independent
of composition
of mixture

$$C = \frac{P}{R_u T} \longleftarrow$$

$$\rho = \frac{P}{RT} = \frac{P}{(R_u/M)T}$$

Depends on
composition
of mixture

FIGURE 11-33

When the total pressure P and temperature T of a binary mixture of ideal gases is held constant, then the molar concentration C of the mixture remains constant.

Diffusion of Vapor through a Stationary Gas: Stefan Flow

Many engineering applications such as heat pipes, cooling ponds, and the familiar perspiration involve condensation, evaporation, and transpiration in the presence of a noncondensable gas, and thus the *diffusion* of a vapor through a stationary (or stagnant) gas. To understand and analyze such processes, consider a liquid layer of species A in a tank surrounded by a gas of species B, such as a layer of liquid water in a tank open to the atmospheric air (Fig. 11-34), at constant pressure P and temperature T. Equilibrium exists between the liquid and vapor phases at the interface ($x = 0$), and thus the vapor pressure at the interface must equal the saturation pressure of species A at the specified temperature. We assume the gas to be insoluble in the liquid, and both the gas and the vapor to behave as ideal gases.

If the surrounding gas at the top of the tank ($x = L$) is not saturated, the vapor pressure at the interface will be greater than the vapor pressure at the top of the tank ($P_{A,0} > P_{A,L}$ and thus $y_{A,0} > y_{A,L}$ since $y_A = P_A/P$), and this pressure (or concentration) difference will drive the vapor upward from the air–water interface into the stagnant gas. The upward flow of vapor will be sustained by the evaporation of water at the interface. Under steady conditions, the molar (or mass) flow rate of vapor throughout the stagnant gas column remains constant. That is,

$$\bar{j}_A = \dot{N}_A/A = \text{constant} \qquad (\text{or } j_A = \dot{m}_A/A = \text{constant})$$

The pressure and temperature of the gas–vapor mixture are said to be constant, and thus the molar density of the mixture must be constant throughout the mixture, as shown earlier. That is, $C = C_A + C_B = \text{constant}$, and it is more convenient to work with mole fractions or molar concentrations in this case instead of mass fractions or densities since $\rho \neq \text{constant}$.

Noting that $y_A + y_B = 1$ and that $y_{A,0} > y_{A,L}$, we must have $y_{B,0} < y_{B,L}$. That is, the mole fraction of the gas must be decreasing downward by the same amount that the mole fraction of the vapor is increasing. Therefore, gas must be diffusing from the top of the column toward the liquid interface. However, the gas is said to be *insoluble* in the liquid, and thus there can be no net mass flow of the gas downward. Then under steady conditions, there must be an *upward bulk fluid motion* with an average velocity $\mathcal{V}$ that is just large enough to balance the diffusion of air downward so that the net molar (or mass) flow rate of the gas at any point is zero. In other words, the upward bulk motion offsets the downward diffusion, and for each air molecule that moves downward, there is another air molecule that moves upward. As a result, the air appears to be *stagnant* (it does not move). That is,

$$\bar{j}_B = \dot{N}_B/A = 0 \qquad (\text{or } j_B = \dot{m}_B/A = 0)$$

The diffusion medium is no longer stationary because of the bulk motion. The implication of the bulk motion of the gas is that it transports vapor as well as the gas upward with a velocity of $\mathcal{V}$, which results in *additional* mass flow of vapor upward. Therefore, the molar flux of the vapor can be expressed as

$$\bar{j}_A = \dot{N}_A/A = \bar{j}_{A,\text{conv}} + \bar{j}_{A,\text{diff}} = y_A(\bar{j}_A + \bar{j}_B) - CD_{AB}\frac{dy_A}{dx} \qquad (11\text{-}54)$$

Noting that $\bar{j}_B = 0$, it simplifies to

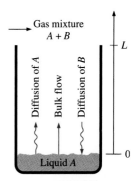

Gas mixture
$A + B$

Diffusion of A

Bulk flow

Diffusion of B

L

0

Liquid A

FIGURE 11-34

Diffusion of a vapor A through a stagnant gas B.

$$\bar{j}_A = y_A \bar{j}_A - CD_{AB} \frac{dy_A}{dx} \tag{11-55}$$

Solving for $\bar{j}_A$ gives

$$\bar{j}_A = -\frac{CD_{AB}}{1 - y_A} \frac{dy_A}{dx} \quad \rightarrow \quad -\frac{1}{1 - y_A} \frac{dy_A}{dx} = \frac{\bar{j}_A}{CD_{AB}} = \text{constant} \tag{11-56}$$

since $\bar{j}_A = \text{constant}$, $C = \text{constant}$, and $D_{AB} = \text{constant}$. Separating the variables and integrating from $x = 0$ where $y_A(0) = y_{A,0}$ to $x = L$ where $y_A(L) = y_{A,L}$ gives

$$-\int_{A,0}^{y_{A,L}} \frac{dy_A}{1 - y_A} = \int_0^L \frac{\bar{j}_A}{CD_{AB}} dx \tag{11-57}$$

Performing the integrations,

$$\ln \frac{1 - y_{A,L}}{1 - y_{A,0}} = \frac{\bar{j}_A}{CD_{AB}} L \tag{11-58}$$

Then the molar flux of vapor A, which is the *evaporation rate of species A per unit interface area*, becomes

$$\bar{j}_A = \dot{N}_A/A = \frac{CD_{AB}}{L} \ln \frac{1 - y_{A,L}}{1 - y_{A,0}} \qquad (\text{kmol/s} \cdot \text{m}^2) \tag{11-59}$$

This relation is known as **Stefan's law,** and the *induced convective flow* described above that enhances mass diffusion is called the **Stefan flow.**

An expression for the variation of the mole fraction of A with x can be determined by performing the integration in Equation 11-57 to the upper limit of x where $y_A(x) = y_A$ (instead of to L where $y_A(L) = y_{A,L}$). It yields

$$\ln \frac{1 - y_A}{1 - y_{A,0}} = \frac{\bar{j}_A}{CD_{AB}} x \tag{11-60}$$

Substituting the $\bar{j}_A$ expression from Equation 11-59 into the above relation and rearranging gives

$$\frac{1 - y_A}{1 - y_{A,0}} = \left(\frac{1 - y_{A,L}}{1 - y_{A,0}}\right)^{x/L} \quad \text{and} \quad \frac{y_B}{y_{B,0}} = \left(\frac{y_{B,L}}{y_{B,0}}\right)^{x/L} \tag{11-61}$$

The second relation for the variation of the mole fraction of the stationary gas B is obtained from the first one by substituting $1 - y_A = y_B$ since $y_A + y_B = 1$.

To maintain isothermal conditions in the tank during evaporation, heat must be supplied to the tank at a rate of

$$\dot{Q} = \dot{m}_A h_{fg,A} = \bar{j}_A A h_{fg,A} = (\bar{j}_A M_A) A h_{fg,A} \qquad (\text{kJ/s}) \tag{11-62}$$

where A is the surface area of the liquid–vapor interface, h_{fg} is the latent heat of vaporization, and M is the molar mass of species A.

Equimolar Counterdiffusion

Consider two large reservoirs connected by a channel of length L, as shown in Figure 11-35. The entire system contains a binary mixture of gases A and B at a uniform temperature T and pressure P throughout. The concentrations of species are maintained constant in each of the reservoirs such that $y_{A,0} > y_{A,L}$ and $y_{B,0} < y_{B,L}$. The resulting concentration gradients will cause the species A to diffuse in the positive x-direction and the species B in the opposite direction. Assuming the gases to behave as ideal gases and thus $P = CR_uT$, the total molar concentration of the mixture C will remain constant throughout the mixture since P and T are constant. That is,

$$C = C_A + C_B = \text{constant} \qquad (\text{kmol/m}^3)$$

This requires that for each molecule of A that moves to the right, a molecule of B moves to the left, and thus the molar flow rates of species A and B must be equal in magnitude and opposite in sign. That is,

$$\dot{N}_A = -\dot{N}_B \qquad \text{or} \qquad \dot{N}_A + \dot{N}_B = 0 \qquad (\text{kmol/s})$$

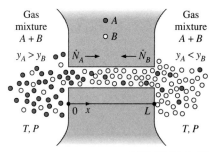

FIGURE 11-35

Equimolar isothermal counterdiffusion of two gases A and B.

This process is called **equimolar counterdiffusion** for obvious reasons. The net molar flow rate of the mixture for such a process, and thus the molar-average velocity, is zero since

$$\dot{N} = \dot{N}_A + \dot{N}_B = 0 \quad \rightarrow \quad CA\mathcal{V} = 0 \quad \rightarrow \quad \mathcal{V} = 0$$

Therefore, the mixture is *stationary* on a molar basis and thus mass transfer is by diffusion only (there is no mass transfer by convection) so that

$$\bar{j}_A = \dot{N}_A/A = -CD_{AB}\frac{dy_A}{dx} \qquad \text{and} \qquad \bar{j}_B = \dot{N}_B/A = -CD_{BA}\frac{dy_B}{dx} \qquad (11\text{-}63)$$

Under steady conditions, the molar flow rates of species A and B can be determined directly from Equation 11-24 developed earlier for one-dimensional steady diffusion in a stationary medium, noting that $P = CR_uT$ and thus $C = P/R_uT$ for each constituent gas and the mixture. For one-dimensional flow through a channel of uniform cross-sectional area A with no homogeneous chemical reactions, they are expressed as

$$\dot{N}_{\text{diff},A} = CD_{AB}A\frac{y_{A,1} - y_{A,2}}{L} = D_{AB}A\frac{C_{A,1} - C_{A,2}}{L} = \frac{D_{AB}}{R_uT}A\frac{P_{A,0} - P_{A,L}}{L}$$

$$\dot{N}_{\text{diff},B} = CD_{BA}A\frac{y_{B,1} - y_{B,2}}{L} = D_{BA}A\frac{C_{B,1} - C_{B,2}}{L} = \frac{D_{BA}}{R_uT}A\frac{P_{B,0} - P_{B,L}}{L} \qquad (11\text{-}64)$$

The relations above imply that the mole fraction, molar concentration, and the partial pressure of either gas vary linearly during equimolar counterdiffusion.

It is interesting to note that the mixture is *stationary* on a molar basis, but it is not stationary on a mass basis unless the molar masses of A and B are equal. Although the net molar flow rate through the channel is zero, the net mass flow rate of the mixture through the channel is not zero and can be determined from

$$\dot{m} = \dot{m}_A + \dot{m}_B = \dot{N}_A M_A + \dot{N}_B M_B = \dot{N}_A(M_A - M_B) \qquad (11\text{-}65)$$

since $\dot{N}_B = -\dot{N}_A$. Note that the direction of net mass flow rate is the flow direction of the gas with the larger molar mass. A velocity measurement device such as an anemometer placed in the channel will indicate a velocity of $\mathcal{V} = \dot{m}/\rho A$ where ρ is the total density of the mixture at the site of measurement.

EXAMPLE 11-8 Venting of Helium into the Atmosphere by Diffusion

The pressure in a pipeline that transports helium gas at a rate of 2 kg/s is maintained at 1 atm by venting helium to the atmosphere through a 5-mm-internal-diameter tube that extends 15 m into the air, as shown in Figure 11-36. Assuming both the helium and the atmospheric air to be at 25°C, determine (*a*) the mass flow rate of helium lost to the atmosphere through an individual tube, (*b*) the mass flow rate of air that infiltrates into the pipeline, and (*c*) the flow velocity at the bottom of the tube where it is attached to the pipeline that will be measured by an anemometer in steady operation.

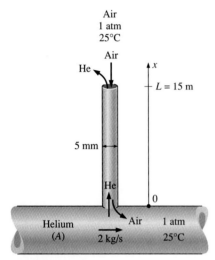

FIGURE 11-36

Schematic for Example 11-8

Solution   The pressure in a helium pipeline is maintained constant by venting to the atmosphere through a long tube. The mass flow rates of helium and air through the tube and the net flow velocity at the bottom are to be determined.

Assumptions   **1** Steady operating conditions exist. **2** Helium and atmospheric air are ideal gases. **3** No chemical reactions occur in the tube. **4** Air concentration in the pipeline and helium concentration in the atmosphere are negligible so that the mole fraction of the helium is 1 in the pipeline and 0 in the atmosphere (we will check this assumption later).

Properties   The diffusion coefficient of helium in air (or air in helium) at normal atmospheric conditions is $D_{AB} = 7.20 \times 10^{-5}$ m²/s (Table 11-2). The molar masses of air and helium are 29 and 4 kg/kmol, respectively (Table A-1).

Analysis   This is a typical equimolar counterdiffusion process since the problem involves two large reservoirs of ideal gas mixtures connected to each other by a channel, and the concentrations of species in each reservoir (the pipeline and the atmosphere) remain constant.

(*a*)   The flow area, which is the cross-sectional area of the tube, is

$$A = \pi D^2/4 = \pi(0.005 \text{ m})^2/4 = 1.963 \times 10^{-5} \text{ m}^2$$

Noting that the pressure of helium is 1 atm at the bottom of the tube ($x = 0$) and 0 at the top ($x = L$), its molar flow rate is determined from Equation 11-64 to be

$$\dot{N}_{\text{helium}} = \dot{N}_{\text{diff}, A} = \frac{D_{AB} A}{R_u T} \frac{P_{A,0} - P_{A,L}}{L}$$

$$= \frac{(7.20 \times 10^{-5} \text{ m}^2/\text{s})(1.963 \times 10^{-5} \text{ m}^2)}{(8.314 \text{ kPa} \cdot \text{m}^3/\text{kmol} \cdot \text{K})(298 \text{ K})} \left(\frac{1 \text{ atm} - 0}{15 \text{ m}}\right)\left(\frac{101.3 \text{ kPa}}{1 \text{ atm}}\right)$$

$$= 3.85 \times 10^{-12} \text{ kmol/s}$$

Therefore,

$$\dot{m}_{helium} = (\dot{N}M)_{helium} = (3.85 \times 10^{-12} \text{ kmol/s})(4 \text{ kg/kmol}) = \mathbf{1.54 \times 10^{-11} \text{ kg/s}}$$

which corresponds to about 0.5 g per year.

(b) Noting that $\dot{N}_B = -\dot{N}_A$ during an equimolar counterdiffusion process, the molar flow rate of air into the helium pipeline is equal to the molar flow rate of helium. The mass flow rate of air into the pipeline is

$$\dot{m}_{air} = (\dot{N}M)_{air} = (-3.85 \times 10^{-12} \text{ kmol/s})(29 \text{ kg/kmol}) = \mathbf{-112 \times 10^{-12} \text{ kg/s}}$$

The mass fraction of air in the helium pipeline is

$$w_{air} = \frac{|\dot{m}_{air}|}{\dot{m}_{total}} = \frac{112 \times 10^{-12} \text{ kg/s}}{(2 + 112 \times 10^{-12} - 1.54 \times 10^{-11}) \text{ kg/s}} = 5.6 \times 10^{-11} \approx 0$$

which validates our original assumption of negligible air in the pipeline.

(c) The net mass flow rate through the tube is

$$\dot{m}_{net} = \dot{m}_{helium} + \dot{m}_{air} = 1.54 \times 10^{-11} - 112 \times 10^{-12} = \mathbf{-9.66 \times 10^{-11} \text{ kg/s}}$$

The mass fraction of air at the bottom of the tube is very small, as shown above, and thus the density of the mixture at $x = 0$ can simply be taken to be the density of helium, which is

$$\rho \cong \rho_{helium} = \frac{P}{RT} = \frac{101.325 \text{ kPa}}{(2.0769 \text{ kPa} \cdot \text{m}^3/\text{kg} \cdot \text{K})(298 \text{ K})} = 0.01637 \text{ kg/m}^3$$

Then the average flow velocity at the bottom part of the tube becomes

$$\mathcal{V} = \frac{\dot{m}}{\rho A} = \frac{-9.66 \times 10^{-11} \text{ kg/s}}{(0.01637 \text{ kg/m}^3)(1.963 \times 10^{-5} \text{ m}^2)} = \mathbf{-3.01 \times 10^{-5} \text{ m/s}}$$

which is difficult to measure by even the most sensitive velocity measurement devices. The negative sign indicates flow in the negative x-direction (toward the pipeline).

EXAMPLE 11-9 Measuring the Diffusion Coefficient by the Stefan Tube

A 3-cm-diameter Stefan tube is used to measure the binary diffusion coefficient of water vapor in air at 20°C at an elevation of 1600 m where the atmospheric pressure is 83.5 kPa. The tube is partially filled with water, and the distance from the water surface to the open end of the tube is 40 cm (Fig. 11-37). Dry air is blown over the open end of the tube so that water vapor rising to the top is removed immediately and the concentration of vapor at the top of the tube is zero. In 15 days of continuous operation at constant pressure and temperature, the amount of water that has evaporated is measured to be 1.23 g. Determine the diffusion coefficient of water vapor in air at 20°C and 83.5 kPa.

Solution The amount of water that evaporates from a Stefan tube at a specified temperature and pressure over a specified time period is measured. The diffusion coefficient of water vapor in air is to be determined.

Assumptions **1** Water vapor and atmospheric air are ideal gases. **2** The amount of air dissolved in liquid water is negligible. **3** Heat is transferred to the water from

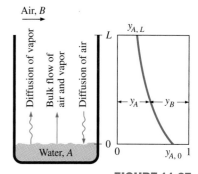

FIGURE 11-37

Schematic for Example 11-9.

the surroundings to make up for the latent heat of vaporization so that the temperature of water remains constant at 20°C.

Properties The saturation pressure of water at 20°C is 2.34 kPa (Table A-9).

Analysis The vapor pressure at the air–water interface is the saturation pressure of water at 20°C, and the mole fraction of water vapor (species A) at the interface is determined from

$$y_{vapor,0} = y_{A,0} = \frac{P_{vapor,0}}{P} = \frac{2.34 \text{ kPa}}{83.5 \text{ kPa}} = 0.0280$$

Dry air is blown on top of the tube and, thus, $y_{vapor,L} = y_{A,L} = 0$. Also, the total molar density throughout the tube remains constant because of the constant temperature and pressure conditions and is determined to be

$$C = \frac{P}{R_u T} = \frac{83.5 \text{ kPa}}{(8.314 \text{ kPa} \cdot \text{m}^3/\text{kmol} \cdot \text{K})(293 \text{ K})} = 0.0343 \text{ kmol/m}^3$$

The cross-sectional area of the tube is

$$A = \pi D^2/4 = \pi(0.03 \text{ m})^2/4 = 7.069 \times 10^{-4} \text{ m}^2$$

The evaporation rate is given to be 1.23 g per 15 days. Then the molar flow rate of vapor is determined to be

$$\dot{N}_A = \dot{N}_{vapor} = \frac{\dot{m}_{vapor}}{M_{vapor}} = \frac{1.23 \times 10^{-3} \text{ kg}}{(15 \times 24 \times 3600 \text{ s})(18 \text{ kg/kmol})} = 5.27 \times 10^{-11} \text{ kmol/s}$$

Finally, substituting the information above into Equation 11-59 we get

$$\frac{5.27 \times 10^{-11} \text{ kmol/s}}{7.069 \times 10^{-4} \text{ m}^2} = \frac{(0.0343 \text{ kmol/m}^3)D_{AB}}{0.4 \text{ m}} \ln \frac{1-0}{1-0.028}$$

which gives

$$D_{AB} = 3.06 \times 10^{-5} \text{ m}^2/\text{s}$$

for the binary diffusion coefficient of water vapor in air at 20°C and 83.5 kPa.

11-9 ■ MASS CONVECTION

So far we have considered *mass diffusion*, which is the transfer of mass due to a concentration gradient. Now we consider *mass convection* (or *convective mass transfer*), which is the transfer of mass between a surface and a moving fluid due to both *mass diffusion* and *bulk fluid motion*. We mentioned earlier that fluid motion enhances heat transfer considerably by removing the heated fluid near the surface and replacing it by the cooler fluid further away. Likewise, fluid motion enhances mass transfer considerably by removing the high concentration fluid near the surface and replacing it by the lower concentration fluid further away. In the limiting case of no bulk fluid motion, mass convection reduces to mass diffusion, just as convection reduces to conduction. The analogy between heat and mass convection holds for both *forced* and *natural* convection, *laminar* and *turbulent* flow, and *internal* and *external* flow.

Like heat convection, mass convection is also complicated because of the complications associated with fluid flow such as the *surface geometry, flow regime, flow velocity,* and the *variation of the fluid properties* and *composition.* Therefore, we will have to rely on experimental relations to determine mass transfer. Also, mass convection is usually analyzed on a *mass basis* rather than on a molar basis. Therefore, we will present formulations in terms of mass concentration (density ρ or mass fraction w) instead of molar concentration (molar density C or mole fraction y). But the formulations on a molar basis can be obtained using the relation $C = \rho/M$ where M is the molar mass. Also, for simplicity, we will restrict our attention to convection in fluids that are (or can be treated as) *binary mixtures.*

Consider the flow of air over the free surface of a water body such as a lake under isothermal conditions. If the air is not saturated, the concentration of water vapor will vary from a maximum at the water surface where the air is always saturated to the free steam value far from the surface. In heat convection, we defined the region in which temperature gradients exist as the *thermal boundary layer.* Similarly, in mass convection, we define the region of the fluid in which concentration gradients exist as the **concentration boundary layer,** as shown in Figure 11-38. In **external flow,** the thickness of the concentration boundary layer δ_c for a species A at a specified location on the surface is defined as the normal distance y from the surface at which

$$\frac{\rho_{A,s} - \rho_A}{\rho_{A,s} - \rho_{A,\infty}} = 0.99$$

where $\rho_{A,s}$ and $\rho_{A,\infty}$ are the densities of species A at the surface (on the fluid side) and the free stream, respectively.

In **internal flow,** we have a **concentration entrance region** where the concentration profile develops, in addition to the hydrodynamic and thermal entry regions (Fig. 11-39). The concentration boundary layer continues to develop in the flow direction until its thickness reaches the tube center and the boundary layers merge. The distance from the tube inlet to the location where this merging occurs is called the **concentration entry length** L_c, and the region beyond that point is called the **fully developed region,** which is characterized by

$$\frac{\partial}{\partial x}\left(\frac{\rho_{A,s} - \rho_A}{\rho_{A,s} - \rho_{A,b}}\right) = 0 \qquad (11\text{-}66)$$

where $\rho_{A,b}$ is the *bulk mean density* of species A defined as

$$\rho_{A,b} = \frac{1}{A\mathcal{V}_{ave}}\int_A \rho_A \mathcal{V}dA \qquad (11\text{-}67)$$

Therefore, the nondimensionalized concentration difference profile as well as the mass transfer coefficient remain constant in the fully developed region. This is analogous to the friction and heat transfer coefficients remaining constant in the fully developed region.

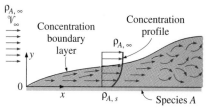

FIGURE 11-38

The development of a concentration boundary layer for species A during external flow on a flat surface.

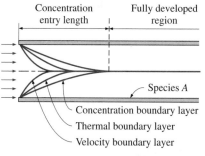

FIGURE 11-39

The development of the velocity, thermal, and concentration boundary layers in internal flow.

$$\text{Heat transfer:} \quad \text{Pr} = \frac{\nu}{\alpha}$$

$$\text{Mass transfer:} \quad \text{Sc} = \frac{\nu}{D_{AB}}$$

FIGURE 11-40
In mass transfer, the Schmidt number plays the role of the Prandtl number in heat transfer.

$$\text{Le} = \frac{\text{Sc}}{\text{Pr}} \quad \frac{\alpha}{D_{AB}}$$

Thermal diffusivity

Mass diffusivity

FIGURE 11-41
Lewis number is a measure of heat diffusion relative to mass diffusion.

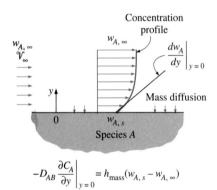

$$-D_{AB} \frac{\partial C_A}{\partial y}\bigg|_{y=0} = h_{\text{mass}}(w_{A,\,s} - w_{A,\,\infty})$$

FIGURE 11-42
Mass transfer at a surface occurs by diffusion because of the no-slip boundary condition, just like heat transfer occurring by conduction.

In heat convection, the relative magnitudes of momentum and heat diffusion in the velocity and thermal boundary layers are expressed by the dimensionless *Prandtl number,* defined as (Fig. 11-40)

$$\textit{Prandtl number:} \qquad \text{Pr} = \frac{\nu}{\alpha} = \frac{\text{Momentum diffusivity}}{\text{Thermal diffusivity}} \qquad (11\text{-}68)$$

The corresponding quantity in mass convection is the dimensionless **Schmidt number,** defined as

$$\textit{Schmidt number:} \qquad \text{Sc} = \frac{\nu}{D_{AB}} = \frac{\text{Momentum diffusivity}}{\text{Mass diffusivity}} \qquad (11\text{-}69)$$

which represents the relative magnitudes of molecular momentum and mass diffusion in the velocity and concentration boundary layers, respectively.

The relative growth of the velocity and thermal boundary layers in laminar flow is governed by the Prandtl number, whereas the relative growth of the velocity and concentration boundary layers is governed by the Schmidt number. A Prandtl number of near unity ($\text{Pr} \approx 1$) indicates that momentum and heat transfer by diffusion are comparable, and velocity and thermal boundary layers almost coincide with each other. *A Schmidt number of near unity ($\text{Sc} \approx 1$) indicates that momentum and mass transfer by diffusion are comparable, and velocity and concentration boundary layers almost coincide with each other.*

It seems like we need one more dimensionless number to represent the relative magnitudes of heat and mass diffusion in the thermal and concentration boundary layers. That is the **Lewis number,** defined as (Fig. 11-41)

$$\textit{Lewis number:} \qquad \text{Le} = \frac{\text{Sc}}{\text{Pr}} = \frac{\alpha}{D_{AB}} = \frac{\text{Thermal diffusivity}}{\text{Mass diffusivity}} \qquad (11\text{-}70)$$

The relative thicknesses of velocity, thermal, and concentration boundary layers in laminar flow are expressed as

$$\frac{\delta_{\text{velocity}}}{\delta_{\text{thermal}}} = \text{Pr}^n, \qquad \frac{\delta_{\text{velocity}}}{\delta_{\text{concentration}}} = \text{Sc}^n, \quad \text{and} \quad \frac{\delta_{\text{thermal}}}{\delta_{\text{concentration}}} = \text{Le}^n \qquad (11\text{-}71)$$

where $n = \frac{1}{3}$ for most applications in all three relations. These relations, in general, are not applicable to turbulent boundary layers since turbulent mixing in this case may dominate the diffusion processes.

Note that species transfer at the surface ($y = 0$) is by diffusion only because of the *no-slip boundary condition,* and mass flux of species A at the surface can be expressed by Fick's law as (Fig. 11-42)

$$j_A = \dot{m}_A/A = -\rho D_{AB} \frac{\partial w_A}{\partial y}\bigg|_{y=0} \qquad (\text{kg/s} \cdot \text{m}^2) \qquad (11\text{-}72)$$

This is analogous to heat transfer at the surface being by conduction only and expressing it by Fourier's law.

The rate of heat convection for external flow was expressed conveniently by *Newton's law of cooling* as

$$\dot{Q}_{\text{conv}} = h_{\text{conv}}A(T_s - T_\infty)$$

where h_{conv} is the average heat transfer coefficient, A is the surface area, and $T_s - T_\infty$ is the temperature difference across the thermal boundary layer. Likewise, the rate of mass convection can be expressed as

$$\dot{m}_{\text{conv}} = h_{\text{mass}}A(\rho_{A,s} - \rho_{A,\infty}) = h_{\text{mass}}\rho A(w_{A,s} - w_{A,\infty}) \quad \text{(kg/s)} \quad (11\text{-}73)$$

where h_{mass} is the average **mass transfer coefficient,** in m/s; A is the surface area; $\rho_{A,s} - \rho_{A,\infty}$ is the mass concentration difference of species A across the concentration boundary layer; and ρ is the average density of the fluid in the boundary layer. The product $h_{\text{mass}}\rho$, whose unit is kg/m$^2 \cdot$ s, is called the *mass transfer conductance*. If the local mass transfer coefficient varies in the flow direction, the *average mass transfer coefficient* can be determined from

$$h_{\text{mass, ave}} = \frac{1}{A}\int_A h_{\text{mass}}\, dA \quad (11\text{-}74)$$

In heat convection analysis, it is often convenient to express the heat transfer coefficient in a nondimensionalized form in terms of the dimensionless *Nusselt number,* defined as

Nusselt number: $$\text{Nu} = \frac{h_{\text{conv}}L}{k} \quad (11\text{-}75)$$

where L is the characteristic length and k is the thermal conductivity of the fluid. The corresponding quantity in mass convection is the dimensionless **Sherwood number,** defined as (Fig. 11-43)

Sherwood number: $$\text{Sh} = \frac{h_{\text{mass}}L}{D_{AB}} \quad (11\text{-}76)$$

where h_{mass} is the mass transfer coefficient and D_{AB} is the mass diffusivity. The Nusselt and Sherwood numbers represent the effectiveness of heat and mass convection at the surface, respectively.

Sometimes it is more convenient to express the heat and mass transfer coefficients in terms of the dimensionless **Stanton number** as

Heat transfer Stanton number: $$\text{St} = \frac{h_{\text{conv}}}{\rho \mathcal{V} C_p} = \text{Nu}\frac{1}{\text{Re Pr}} \quad (11\text{-}77)$$

and

Mass transfer Stanton number: $$\text{St}_{\text{mass}} = \frac{h_{\text{mass}}}{\mathcal{V}} = \text{Sh}\frac{1}{\text{Re Sc}} \quad (11\text{-}78)$$

Heat transfer: $\quad \text{Nu} = \dfrac{h_{\text{conv}}L}{k}$

Mass transfer: $\quad \text{Sh} = \dfrac{h_{\text{mass}}L}{D_{AB}}$

FIGURE 11-43

In mass transfer, the Sherwood number plays the role the Nusselt number plays in heat transfer.

Heat convection	Mass convection
T	C, y, ρ, or w
h_{conv}	h_{mass}
$\delta_{thermal}$	$\delta_{concentration}$
$Re = \dfrac{\mathcal{V}L}{\nu}$	$Re = \dfrac{\mathcal{V}L}{\nu}$
$Gr = \dfrac{g\beta(T_s - T_\infty)L^3}{\nu^2}$	$Gr = \dfrac{g(\rho_\infty - \rho_s)L^3}{\rho\nu^2}$
$Pr = \dfrac{\nu}{\alpha}$	$Sc = \dfrac{\nu}{D_{AB}}$
$St = \dfrac{h_{conv}}{\rho\mathcal{V}C_p}$	$St_{mass} = \dfrac{h_{mass}}{\mathcal{V}}$
$Nu = \dfrac{h_{conv}L}{k}$	$Sh = \dfrac{h_{mass}L}{D_{AB}}$
$Nu = f(Re, Pr)$	$Sh = f(Re, Sc)$
$Nu = f(Gr, Pr)$	$Sh = f(Gr, Sc)$

20°C	Fresh water
No convection currents	SOLAR POND
70°C $\rho_{brine} > \rho_{water}$	Brine
Salt	

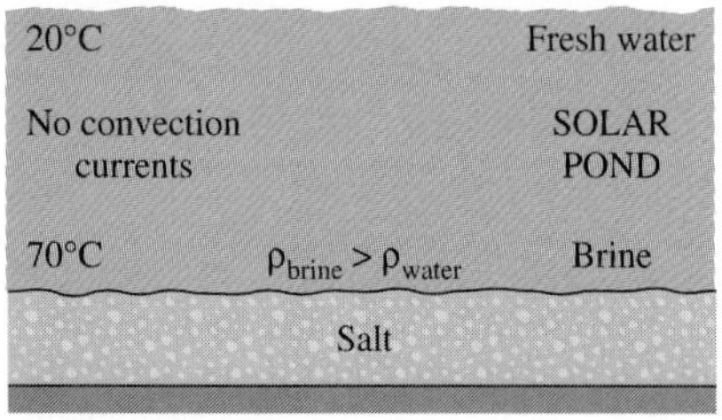

FIGURE 11-44

A hot fluid at the bottom will rise and initiate natural convection currents only if its density is lower.

where $\mathcal{V}$ is the free steam velocity in external flow and the bulk mean fluid velocity in internal flow.

For a given geometry, the average Nusselt number in forced convection depends on the Reynolds and Prandtl numbers, whereas the average Sherwood number depends on the Reynolds and Schmidt numbers. That is,

Nusselt number:	$Nu = f(Re, Pr)$
Sherwood number:	$Sh = f(Re, Sc)$

where the functional form of f is the same for both the Nusselt and Sherwood numbers in a given geometry, provided that the thermal and concentration boundary conditions are of the same type. Therefore, *the Sherwood number can be obtained from the Nusselt number expression by simply replacing the Prandtl number by the Schmidt number.* This shows what a powerful tool analogy can be in the study of natural phenomena (Table 11-12).

In *natural convection mass transfer,* the analogy between the Nusselt and Sherwood numbers still holds, and thus $Sh = f(Gr, Sc)$. But the Grashof number in this case should be determined directly from

$$Gr = \frac{g(\rho_\infty - \rho_s)L^3}{\rho\nu^2} = \frac{g(\Delta\rho/\rho)L^3}{\nu^2} \qquad (11\text{-}79)$$

which is applicable to both temperature- and/or concentration-driven natural convection flows. Note that in *homogeneous* fluids (i.e., fluids with no concentration gradients), density differences are due to temperature differences only, and thus we can replace $\Delta\rho/\rho$ by $\beta\Delta T$ for convenience, as we did in natural convection heat transfer. However, in *nonhomogeneous* fluids, density differences are due to the *combined effects* of *temperature* and *concentration differences,* and $\Delta\rho/\rho$ cannot be replaced by $\beta\Delta T$ in such cases even when all we care about is heat transfer and we have no interest in mass transfer. For example, hot water at the bottom of a pond rises to the top. But when salt is placed at the bottom, as it is done in solar ponds, the salty water (brine) at the bottom will not rise because it is now heavier than the fresh water at the top (Fig. 11-44).

Concentration-driven natural convection flows are based on the densities of different species in a mixture being different. Therefore, at isothermal conditions, there will be no natural convection in a gas mixture that is composed of gases with identical molar masses. Also, the case of a hot surface facing up corresponds to diffusing fluid having a lower density than the mixture (and thus rising under the influence of buoyancy), and the case of a hot surface facing down corresponds to the diffusing fluid having a higher density. For example, the evaporation of water into air corresponds to a hot surface facing up since water vapor is lighter than the air and it will tend to rise. But this will not be the case for gasoline unless the temperature of the gasoline–air mixture at the gasoline surface is so high that thermal expansion overwhelms the density differential due to higher gasoline concentration near the surface.

Analogy between Friction, Heat Transfer, and Mass Transfer Coefficients

Consider the flow of a fluid over a flat plate of length L with free steam conditions of T_∞, $\mathcal{V}_\infty$, and $w_{A,\infty}$ (Fig. 11-45). Noting that convection at the surface ($y = 0$) is equal to diffusion because of the no-slip condition, the friction, heat transfer, and mass transfer conditions at the surface can be expressed as

Wall friction:
$$\tau_s = \mu \left. \frac{\partial \mathcal{V}}{\partial y} \right|_{y=0} = \frac{f}{2} \rho \mathcal{V}_\infty^2 \qquad (11\text{-}80)$$

Heat transfer:
$$\dot{q}_s = -k \left. \frac{\partial T}{\partial y} \right|_{y=0} = h_{\text{heat}}(T_s - T_\infty) \qquad (11\text{-}81)$$

Mass transfer:
$$j_{A,s} = -D_{AB} \left. \frac{\partial w_A}{\partial y} \right|_{y=0} = h_{\text{mass}}(w_{A,s} - w_{A,\infty}) \qquad (11\text{-}82)$$

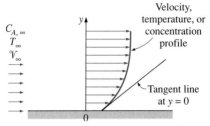

FIGURE 11-45

The friction, heat, and mass transfer coefficients for flow over a surface are proportional to the slope of the tangent line of the velocity, temperature, and concentration profiles, respectively, at the surface.

The relations above can be rewritten for internal flow by using *bulk mean properties* instead of free stream properties. After some simple mathematical manipulations, the three relations above can be rearranged as

Wall friction:
$$\left. \frac{d(\mathcal{V}/\mathcal{V}_\infty)}{d(y/L)} \right|_{y=0} = \frac{f}{2} \frac{\rho \mathcal{V}_\infty L}{\mu} = \frac{f}{2} \text{Re} \qquad (11\text{-}83)$$

Heat transfer:
$$\left. \frac{d[(T - T_s)/(T_\infty - T_s)]}{d(y/L)} \right|_{y=0} = \frac{h_{\text{heat}} L}{k} = \text{Nu} \qquad (11\text{-}84)$$

Mass transfer:
$$\left. \frac{d[(w_A - w_{A,s})/(w_{A,\infty} - w_{A,s})]}{d(y/L)} \right|_{y=0} = \frac{h_{\text{mass}} L}{D_{AB}} = \text{Sh} \qquad (11\text{-}85)$$

The left sides of the three relations above are the slopes of the normalized velocity, temperature, and concentration profiles at the surface, and the right sides are the dimensionless numbers discussed earlier.

Special Case: Pr ≈ Sc ≈ 1 (Reynolds Analogy)

Now consider the hypothetical case in which the molecular diffusivities of momentum, heat, and mass are identical. That is, $\nu = \alpha = D_{AB}$ and thus Pr = Sc = Le = 1. In this case the normalized velocity, temperature, and concentration profiles will coincide, and thus the slope of these three curves at the surface (the left sides of Eqs. 11-83 through 11-85) will be identical (Fig. 11-46). Then we can set the right sides of those three equations equal to each other and obtain

$$\frac{f}{2} \text{Re} = \text{Nu} = \text{Sh} \qquad \text{or} \qquad \frac{f}{2} \frac{\mathcal{V}_\infty L}{\nu} = \frac{h_{\text{heat}} L}{k} = \frac{h_{\text{mass}} L}{D_{AB}} \qquad (11\text{-}86)$$

Noting that Pr = Sc = 1, we can also write the equations above as

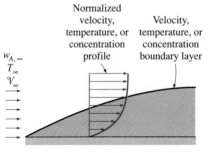

FIGURE 11-46

When the molecular diffusivities of momentum, heat, and mass are equal to each other, the velocity, temperature, and concentration boundary layers coincide.

Reynolds analogy
$\nu = \alpha = D_{AB}$
(or Pr = Sc = Le)

$$\frac{f}{2} = \frac{Nu}{Re\ Pr} = \frac{Sh}{Re\ Sc} \quad \text{or} \quad \frac{f}{2} = St = St_{mass} \quad (11\text{-}87)$$

The relation above is known as the **Reynolds analogy,** and it enables us to determine the seemingly unrelated friction, heat transfer, and mass transfer coefficients when only one of them is known or measured. (Actually the original Reynolds analogy proposed by O. Reynolds is 1874 is $St = f/2$, which is then extended to include mass transfer.) However, it should always be remembered that the analogy is restricted to situations for which $Pr \approx Sc \approx 1$. Of course the first part of the analogy between friction and heat transfer coefficients can always be used for gases since their Prandtl number is close to unity.

General Case: $Pr \neq Sc \neq 1$ (Chilton–Colburn Analogy)

The Reynolds analogy is a very useful relation, and it is certainly desirable to extend it to a wider range of Pr and Sc numbers. Several attempts have been made in this regard, but the simplest and the best known is the one suggested by Chilton and Colburn in 1934 as

$$\frac{f}{2} = St\ Pr^{2/3} = St_{mass}\ Sc^{2/3} \quad (11\text{-}88)$$

for $0.6 < Pr < 60$ and $0.6 < Sc < 3000$. The equation above is known as the **Chilton–Colburn analogy.** Using the definition of heat and mass Stanton numbers, the analogy between heat and mass transfer can be expressed more conveniently as (Fig. 11-47)

$$\frac{St}{St_{mass}} = \left(\frac{Sc}{Pr}\right)^{2/3}$$

$$\text{or} \quad \frac{h_{heat}}{h_{mass}} = \rho C_p \left(\frac{Sc}{Pr}\right)^{2/3} = \rho C_p \left(\frac{\alpha}{D_{AB}}\right)^{2/3} = \rho C_p Le^{2/3} \quad (11\text{-}89)$$

For air–water vapor mixtures at 298 K, the mass and thermal diffusivities are $D_{AB} = 2.5 \times 10^{-5}$ m²/s and $\alpha = 2.18 \times 10^{-5}$ m²/s and thus the Lewis number is $Le = \alpha/D_{AB} = 0.872$. (We simply use the α value of dry air instead of the moist air since the fraction of vapor in the air at atmospheric conditions is low.) Then $(\alpha/D_{AB})^{2/3} = 0.872^{2/3} = 0.913$, which is close to unity. Also, the Lewis number is relatively insensitive to variations in temperature. Therefore, for air–water vapor mixtures, the relation between heat and mass transfer coefficients can be expressed with a good accuracy as

$$h_{heat} \cong \rho C_p h_{mass} \quad \text{(air–water vapor mixtures)} \quad (11\text{-}90)$$

where ρ and C_p are the density and specific heat of air at mean conditions (or ρC_p is the specific heat of air per unit volume). Equation 11-90 is known as the **Lewis relation** and is commonly used in air-conditioning applications.

Chilton-Colburn Analogy

General:

$$h_{mass} = \frac{h_{heat}}{\rho C_p} \left(\frac{D_{AB}}{\alpha}\right)^{2/3}$$

$$= \frac{1}{2} f \mathcal{V} \left(\frac{D_{AB}}{\nu}\right)^{2/3}$$

Special case: $\nu = \alpha = D_{AB}$

$$h_{mass} = \frac{h_{heat}}{\rho C_p} = \frac{1}{2} f \mathcal{V}$$

FIGURE 11-47

When the friction or heat trasnfer coefficient is known, the mass transfer coefficient can be determined directly from the Chilton–Colburn analogy.

Another important consequence of Le $\cong$ 1 is that the *wet-bulb* and *adiabatic saturation temperatures* of moist air are nearly identical. In turbulent flow, the Lewis relation can be used even when the Lewis number is not 1 since eddy mixing in turbulent flow overwhelms any molecular diffusion, and heat and mass are transported at the same rate.

The Chilton–Colburn analogy has been observed to hold quite well in laminar or turbulent flow over plane surfaces. But this is not always the case for internal flow and flow over irregular geometries, and in such cases specific relations developed should be used. When dealing with flow over blunt bodies, it is important to note that f in the relations above is the *skin friction coefficient,* not the total drag coefficient, which also includes the pressure drag.

Limitation on the Heat–Mass Convection Analogy

Caution should be exercised when using the analogy in Equation 11-88 since there are a few factors that put some shadow on the accuracy of that relation. For one thing, the Nusselt numbers are usually evaluated for smooth surfaces, but many mass transfer problems involve wavy or roughened surfaces. Also, many Nusselt relations are obtained for constant surface temperature situations, but the concentration may not be constant over the entire surface because of the possible surface dryout. The blowing or suction at the surface during mass transfer may also cause some deviation, especially during high speed blowing or suction.

Finally, the heat–mass convection analogy is valid for **low mass flux** cases in which the flow rate of species undergoing mass flow is low relative to the total flow rate of the liquid or gas mixture so that the mass transfer between the fluid and the surface does not affect the *flow velocity.* (Note that convection relations are based on *zero* fluid velocity at the surface, which is true only when there is no net mass transfer at the surface.) Therefore, the heat–mass convection analogy is not applicable when the rate of mass transfer of a species is high relative to the flow rate of that species.

Consider, for example, the evaporation and transfer of water vapor into air in an air washer, an evaporative cooler, a wet cooling tower, or just at the free surface of a lake or river (Fig. 11-48). Even at a temperature of 40°C, the vapor pressure at the water surface is the saturation pressure of 7.4 kPa, which corresponds to a mole fraction of 0.074 or a mass fraction of $w_{A,s} = 0.047$ for the vapor. Then the mass fraction difference across the boundary layer will be, at most, $\Delta w = w_{A,s} - w_{A,\infty} = 0.047 - 0 = 0.047$. For the evaporation of water into air, the error involved in the low mass flux approximation is roughly $\Delta w/2$, which is 2.5 percent in the worst case considered above. Therefore, in processes that involve the evaporation of water into air, we can use the heat–mass convection analogy with confidence. However, the mass fraction of vapor approaches 1 as the water temperature approaches the saturation temperature, and thus the low mass flux approximation is not applicable to mass transfer in boilers, condensers, and the evaporation of fuel droplets in combustion chambers. In this chapter we limit our consideration to low mass flux applications.

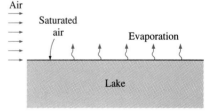

FIGURE 11-48

Evaporation from the free surface of water into air.

TABLE 11-13

Sherwood number relations in mass convection for specified concentration at the surface corresponding to the Nusselt number relations in heat convection for specified surface temperature

Convective heat transfer	Convective mass transfer
1. *Forced Convection over a Flat Plate*	
(*a*) Laminar flow (Re < 5 × 10⁵)	
$Nu = 0.664\, Re_L^{0.5}\, Pr^{1/3}$, $Pr > 0.6$	$Sh = 0.664\, Re_L^{0.5}\, Sc^{1/3}$, $Sc > 0.5$
(*b*) Turbulent flow (5 × 10⁵ < Re < 10⁷)	
$Nu = 0.037\, Re_L^{0.8}\, Pr^{1/3}$, $Pr > 0.6$	$Sh = 0.037\, Re_L^{0.8}\, Sc^{1/3}$, $Sc > 0.5$
2. *Fully Developed Flow in Smooth Circular Pipes*	
(*a*) Laminar flow (Re < 2300)	
$Nu = 3.66$	$Sh = 3.66$
(*b*) Turbulent flow (Re > 10,000)	
$Nu = 0.023\, Re^{0.8}\, Pr^{0.4}$, $0.7 < Pr\ 160$	$Sh = 0.023\, Re^{0.8}\, Sc^{0.4}$, $0.7 < Sc\ 160$
3. *Natural Convection over Surfaces*	
(*a*) Vertical plate	
$Nu = 0.59(Gr\ Pr)^{1/4}$, $10^5 < Gr\ Pr < 10^9$	$Sh = 0.59(Gr\ Sc)^{1/4}$, $10^5 < Gr\ Sc < 10^9$
$Nu = 0.1(Gr\ Pr)^{1/3}$, $10^9 < Gr\ Pr < 10^{13}$	$Sh = 0.1(Gr\ Sc)^{1/3}$, $10^9 < Gr\ Sc < 10^{13}$
(*b*) Upper surface of a horizontal plate	
Surface is hot ($T_s > T_\infty$)	Fluid near the surface is light ($\rho_s < \rho_\infty$)
$Nu = 0.54(Gr\ Pr)^{1/4}$, $10^4 < Gr\ Pr < 10^7$	$Sh = 0.54(Gr\ Sc)^{1/4}$, $10^4 < Gr\ Sc < 10^7$
$Nu = 0.15(Gr\ Pr)^{1/3}$, $10^7 < Gr\ Pr < 10^{11}$	$Sh = 0.15(Gr\ Sc)^{1/3}$, $10^7 < Gr\ Sc < 10^{11}$
(*c*) Lower surface of a horizontal plate	
Surface is hot ($T_s > T_\infty$)	Fluid near the surface is light ($\rho_s < \rho_\infty$)
$Nu = 0.27(Gr\ Pr)^{1/4}$, $10^5 < Gr\ Pr < 10^{11}$	$Sh = 0.27(Gr\ Sc)^{1/4}$, $10^5 < Gr\ Sc < 10^{11}$

Mass Convection Relations

Under low mass flux conditions, the mass convection coefficients can be determined by either (1) determining the friction or heat transfer coefficient and then using the Chilton–Colburn analogy or (2) picking the appropriate Nusselt number relation for the given geometry and analogous boundary conditions, replacing the Nusselt number by the Sherwood number and the Prandtl number by the Schmidt number, as shown in Table 11-13 for some representative cases. The first approach is obviously more convenient when the friction or heat transfer coefficient is already known. Otherwise, the second approach should be preferred since it is generally more accurate, and the Chilton–Colburn analogy offers no significant advantage in this case. Relations for convection mass transfer in other geometries can be written similarly using the corresponding heat transfer relation in Chapters 6 and 7.

FIGURE 11-49

Schematic for Example 11-10.

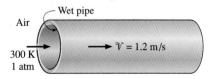

EXAMPLE 11-10   Mass Convection inside a Circular Pipe

Consider a circular pipe of inner diameter $D = 0.015$ m whose inner surface is covered with a layer of liquid water as a result of condensation (Fig. 11-49). In order to dry the pipe, air at 300 K and 1 atm is forced to flow through it with an average velocity of 1.2 m/s. Using the analogy between heat and mass transfer, determine the mass transfer coefficient inside the pipe for fully developed flow.

Solution The liquid layer on the inner surface of a circular pipe is dried by blowing air through it. The mass transfer coefficient is to be determined.

Assumptions **1** The low mass flux model and thus the analogy between heat and mass transfer is applicable since the mass fraction of vapor in the air is low (about 2 percent for saturated air at 300 K). **2** The flow is fully developed.

Properties Because of low mass flux conditions, we can use dry air properties for the mixture at the specified temperature of 300 K and 1 atm, for which $\nu = 1.57 \times 10^{-5}$ m²/s (Table A-11). The mass diffusivity of water vapor in the air at 300 K is determined from Equation 11-15 to be

$$D_{AB} = D_{H_2O\text{-air}} = 1.87 \times 10^{-10}\frac{T^{2.072}}{P} = 1.87 \times 10^{-10}\frac{300^{2.072}}{1} = 2.54 \times 10^{-5}\text{ m}^2/\text{s}$$

Analysis The Reynolds number for this internal flow is

$$\text{Re} = \frac{\mathcal{V}D}{\nu} = \frac{(1.2\text{ m/s})(0.015\text{ m})}{1.57 \times 10^{-5}\text{ m}^2/\text{s}} = 1146$$

which is less than 2300 and thus the flow is laminar. Therefore, based on the analogy between heat and mass transfer, the Nusselt and the Sherwood numbers in this case are Nu = Sh = 3.66. Using the definition of Sherwood number, the mass transfer coefficient is determined to be

$$h_{\text{mass}} = \frac{\text{Sh}D_{AB}}{D} = \frac{(3.66)(2.54 \times 10^{-5}\text{ m}^2/\text{s})}{0.015\text{ m}} = \mathbf{0.00620\text{ m/s}}$$

The mass transfer rate (or the evaporation rate) in this case can be determined by defining the logarithmic mean concentration difference in an analogous manner to the logarithmic mean temperature difference.

EXAMPLE 11-11   Analogy between Heat and Mass Transfer

Heat transfer coefficients in complex geometries with complicated boundary conditions can be determined by mass transfer measurements on similar geometries under similar flow conditions using volatile solids such as naphthalene and dichlorobenzene and utilizing the Chilton–Colburn analogy between heat and mass transfer at low mass flux conditions. The amount of mass transfer during a specified time period is determined by weighing the model or measuring the surface recession.

During a certain experiment involving the flow of dry air at 25°C and 1 atm at a free stream velocity of 2 m/s over a body covered with a layer of naphthalene, it is observed that 12 g of naphthalene has sublimated in 15 min (Fig. 11-50). The surface area of the body is 0.3 m². Both the body and the air were kept at 25°C during the study. The vapor pressure of naphthalene at 25°C is 11 Pa and the mass diffusivity of naphthalene in air at 25°C is $D_{AB} = 0.61 \times 10^{-5}$ m²/s. Determine the heat transfer coefficient under the same flow conditions over the same geometry.

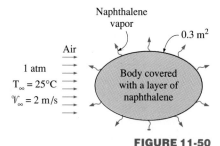

FIGURE 11-50

Schematic for Example 11-11.

Solution Air is blown over a body covered with a layer of naphthalene, and the rate of sublimation is measured. The heat transfer coefficient under the same flow conditions over the same geometry is to be determined.

Assumptions **1** The low mass flux conditions exist so that the Chilton–Colburn analogy between heat and mass transfer is applicable (will be verified). **2** Both air and naphthalene vapor are ideal gases.

Properties The molar mass of naphthalene is 128.2 kg/kmol. Because of low mass flux conditions, we can use dry air properties for the mixture at the specified temperature of 298 K and 1 atm, at which $\rho = 1.19$ kg/m³, $C_P = 1005$ J/kg·K, and $\alpha = 2.18 \times 10^{-5}$ m²/s (Table A-11).

Analysis The incoming air is free of naphthalene, and thus the mass fraction of naphthalene at free stream conditions is zero, $w_{A,\infty} = 0$. Noting that the vapor pressure of naphthalene at the surface is 11 Pa, its mass fraction at the surface is determined to be

$$w_{A,s} = \frac{P_{A,s}}{P}\left(\frac{M_A}{M_{air}}\right) = \frac{11 \text{ Pa}}{101,325 \text{ Pa}}\left(\frac{128.2 \text{ kg/kmol}}{29 \text{ kg/kmol}}\right) = 4.8 \times 10^{-4}$$

which confirms that the low mass flux approximation is valid. The rate of evaporation of naphthalene in this case is

$$\dot{m}_{evap} = \frac{m}{\Delta t} = \frac{0.012 \text{ kg}}{(15 \times 60 \text{ s})} = 1.33 \times 10^{-5} \text{ kg/s}$$

Then the mass convection coefficient becomes

$$h_{mass} = \frac{\dot{m}}{\rho A(w_{A,s} - w_{A,\infty})} = \frac{1.33 \times 10^{-5} \text{ kg/s}}{(1.19 \text{ kg/m}^3)(0.3 \text{ m}^2)(4.8 \times 10^{-4} - 0)} = 0.0776 \text{ m/s}$$

Using the analogy between heat and mass transfer, the average heat transfer coefficient is determined from Equation 11-89 to be

$$h_{heat} = \rho C_p h_{mass}\left(\frac{\alpha}{D_{AB}}\right)^{2/3}$$

$$= (1.19 \text{ kg/m}^3)(1005 \text{ J/kg·°C})(0.0776 \text{ m/s})\left(\frac{2.18 \times 10^{-5} \text{ m}^2/\text{s}}{0.61 \times 10^{-5} \text{ m}^2/\text{s}}\right)^{2/3}$$

$$= \mathbf{217 \text{ W/m}^2 \cdot {}^\circ\text{C}}$$

Discussion Because of the convenience it offers, naphthalene has been used in numerous heat transfer studies to determine convection heat transfer coefficients.

▬▬▬▬▬▬▬▬▬▬▬▬▬▬▬▬▬▬▬▬▬▬▬▬▬▬▬▬▬▬▬

11-10 ■ SIMULTANEOUS HEAT AND MASS TRANSFER

Many mass transfer processes encountered in practice occur isothermally, and thus they do not involve any heat transfer. But some engineering applications involve the vaporization of a liquid and the diffusion of this vapor into the surrounding gas. Such processes require the transfer of the latent heat of vaporization h_{fg} to the liquid in order to vaporize it, and thus such problems involve simultaneous heat and mass transfer. To generalize, any mass transfer problem involving *phase change* (evaporation, sublimation, condensation, melting, etc.) must also involve *heat transfer,* and the solution of such problems needs to be analyzed by considering *simultaneous heat and mass transfer.* Some examples of simultaneous heat and mass transfer problems are drying, evaporative cooling, transpiration (or sweat) cooling, cooling by dry ice, combustion of fuel droplets, and ablation cooling of space vehicles during reentry, and even ordinary events like rain, snow, and hail. In warmer locations, for example, the snow melts and the rain evaporates before reaching the ground (Fig. 11-51).

To understand the mechanism of simultaneous heat and mass transfer, consider the *evaporation of water* from a swimming pool into air. Let us assume that the water and the air are initially at the same temperature. If the air is saturated (a relative humidity of $\phi = 100$ percent), there will be no heat or mass transfer as long as the isothermal conditions remain. But if the air is not saturated ($\phi < 100$ percent), there will be a difference between the concentration of water vapor at the water–air interface (which is always saturated) and some distance above the interface (the concentration boundary layer). Concentration difference is the driving force for mass transfer, and thus this concentration difference will drive the water into the air. But the water must vaporize first, and it must absorb the latent heat of vaporization in order to vaporize. Initially, the entire heat of vaporization will come from the water near the interface since there is no temperature difference between the water and the surroundings and thus there cannot be any heat transfer. The temperature of water near the surface must drop as a result of the sensible heat loss, which also drops the saturation pressure and thus vapor concentration at the interface.

This temperature drop creates temperature differences within the water at the top as well as between the water and the surrounding air. These temperature differences drive heat transfer toward the water surface from both the air and the deeper parts of the water, as shown in Figure 11-52. If the evaporation rate is high and thus the demand for the heat of vaporization is higher than the amount of heat that can be supplied from the lower parts of the water body and the surroundings, the deficit will be made up from the sensible heat of the water at the surface, and thus the temperature of water at the surface will drop further. The process will continue until the latent heat of vaporization is equal to the rate of heat transfer to the water at the surface. Once the steady operation conditions are reached and the interface temperature stabilizes, the energy balance on a thin layer of liquid at the surface can be expressed as

$$\dot{Q}_{\text{sensible, transferred}} = \dot{Q}_{\text{latent, absorbed}} \quad \text{or} \quad \dot{Q} = \dot{m}_v h_{fg} \quad (11\text{-}91)$$

where $\dot{m}_v$ is the rate of evaporation and h_{fg} is the latent heat of vaporization of water at the surface temperature. Various expressions for $\dot{m}_v$ under various approximations are given in Table 11-14. The mixture properties such as the specific heat C_p and molar mass M should normally be evaluated at the *mean film composition* and *mean film temperature.* However, when dealing with air–water vapor mixtures at atmospheric conditions or other low mass flux

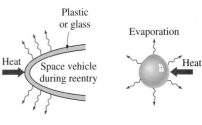

(a) Ablation

(b) Evaporation of rain droplet

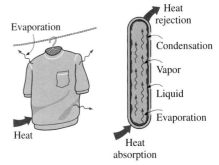

(c) Drying of clothes   (d) Heat pipes

FIGURE 11-51

Many problems encountered in practice involve simultaneous heat and mass transfer.

FIGURE 11-52

Various mechanisms of heat transfer involved during the evaporation of water from the surface of a lake.

TABLE 11-14

Various expressions for evaporation rate of a liquid into a gas through an interface area A under various approximations (subscript v stands for vapor, s for liquid–gas interface, and ∞ away from surface)

Assumption	Evaporation rate
General	$\dot{m}_v = h_{mass} A(\rho_{v,s} - \rho_{v,\infty})$
Assuming vapor to be an ideal gas, $P_v = \rho_v R_v T$	$\dot{m}_v = \dfrac{h_{mass} A}{R_v}\left(\dfrac{P_{v,s}}{T_s} - \dfrac{P_{v,\infty}}{T_\infty}\right)$
Using Chilton–Colburn analogy, $h_{heat} = \rho C_p h_{mass} Le^{2/3}$	$\dot{m}_v = \dfrac{h_{heat} A}{\rho C_p Le^{2/3} R_v}\left(\dfrac{P_{v,s}}{T_s} - \dfrac{P_{v,\infty}}{T_\infty}\right)$
Using $\dfrac{1}{T_s} - \dfrac{1}{T_\infty} \approx \dfrac{1}{T}$, where $T = \dfrac{T_s + T_\infty}{2}$ and $P = \rho R T = \rho(R_u/M)T$	$\dot{m}_v = \dfrac{h_{heat} A}{C_p Le^{2/3}}\dfrac{M_v}{M}\dfrac{P_{v,s} - P_{v,\infty}}{P}$

situations, we can simply use the properties of the gas with reasonable accuracy.

The $\dot{Q}$ in Equation 11-91 represents all forms of heat from all sources transferred to the surface, including convection and radiation from the surroundings and conduction from the deeper parts of the water due to the sensible energy of the water itself or due to heating the water body by a resistance heater, heating coil, or even chemical reactions in the water. If heat transfer from the water body to the surface as well as radiation from the surroundings is negligible, which is often the case, then the heat loss by evaporation must equal heat gain by convection. That is,

$$\dot{Q}_{conv} = \dot{m}_v h_{fg} \qquad \text{or} \qquad h_{conv} A(T_\infty - T_s) = \frac{h_{conv} A h_{fg}}{C_p Le^{2/3}}\frac{M_v}{M}\frac{P_{v,s} - P_{v,\infty}}{P}$$

Canceling $h_{conv} A$ from both sides of the second equation above gives

$$T_s = T_\infty - \frac{h_{fg}}{C_p Le^{2/3}}\frac{M_v}{M}\frac{P_{v,s} - P_{v,\infty}}{P} \qquad (11\text{-}92)$$

which is a relation for the temperature difference between the liquid and the surrounding gas under steady conditions.

EXAMPLE 11-12   Evaporative Cooling of a Canned Drink

During a hot summer day, a canned drink is to be cooled by wrapping it in a cloth that is kept wet continually, and blowing air to it with a fan (Fig. 11-53). If the environment conditions are 1 atm, 30°C, and 40 percent relative humidity, determine the temperature of the drink when steady conditions are reached.

Solution   Air is blown over a canned drink wrapped in a wet cloth to cool it by simultaneous heat and mass transfer. The temperature of the drink when steady conditions are reached is to be determined.

FIGURE 11-53

Schematic for Example 11-12.

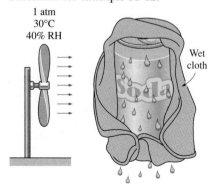

1 atm
30°C
40% RH

Wet
cloth

Assumptions **1** The low mass flux conditions exist so that the Chilton–Colburn analogy between heat and mass transfer is applicable since the mass fraction of vapor in the air is low (about 2 percent for saturated air at 300 K). **2** Both air and water vapor at specified conditions are ideal gases (the error involved in this assumption is less than 1 percent). **3** Radiation effects are negligible.

Properties Because of low mass flux conditions, we can use dry air properties for the mixture at the average temperature of $(T_\infty + T_s)/2$ which cannot be determined at this point because of the unknown surface temperature T_s. We know that $T_s < T_\infty$ and, for the purpose of property evaluation, we take T_s to be 20°C. Then the properties of water at 20°C and the properties of dry air at the average temperature of 25°C and 1 atm are (Tables A-9 and A-11)

Water: h_{fg} = 2454 kJ/kg, P_v = 2.34 kPa; also, P_v = 4.25 kPa at 30°C

Dry air: C_P = 1.005 kJ/kg · °C, α = 2.18 × 10^{-5} m²/s

The molar masses of water and air are 18 and 29 kg/kmol, respectively (Table A-1). Also, the mass diffusivity of water vapor in air at 25°C is $D_{H_2O\text{-air}}$ = 2.50 × 10^{-5} m²/s (Table 11-4).

Analysis Utilizing the Chilton–Colburn analogy, the surface temperature of the drink can be determined from Equation 11-92,

$$T_s = T_\infty - \frac{h_{fg}}{C_p \text{Le}^{2/3}} \frac{M_v}{M} \frac{P_{v,s} - P_{v,\infty}}{P}$$

where the Lewis number is

$$\text{Le} = \frac{\alpha}{D_{AB}} = \frac{2.18 \times 10^{-5} \text{ m}^2/\text{s}}{2.5 \times 10^{-5} \text{ m}^2/\text{s}} = 0.872$$

Note that we could take the Lewis number to be 1 for simplicity, but we chose to incorporate it for better accuracy.

The air at the surface is saturated, and thus the vapor pressure at the surface is simply the saturation pressure of water at the surface temperature (2.34 kPa). The vapor pressure of air far from the surface is determined from

$$P_{v,\infty} = \phi P_{\text{sat @ } T_\infty} = (0.40)P_{\text{sat @ 30°C}} = (0.40)(4.25 \text{ kPa}) = 1.70 \text{ kPa}$$

Noting that the atmospheric pressure is 1 atm = 101.3 kPa, substituting gives

$$T_s = 30°C - \frac{2454 \text{ kJ/kg}}{(1.005 \text{ KJ/kg} \cdot °C)(0.872)^{2/3}} \frac{18 \text{ kg/kmol}}{29 \text{ kg/kmol}} \frac{(2.34 - 1.70) \text{ kPa}}{101.3 \text{ kPa}} = \textbf{19.5°C}$$

Therefore, the temperature of the drink can be lowered to 19.5°C by this process.

EXAMPLE 11-13 Heat Loss from Uncovered Hot Water Baths

Hot water baths with open tops are commonly used in manufacturing facilities for various reasons. In a plant that manufactures spray paints, the pressurized paint cans are temperature tested by submerging them in hot water at 50°C in a 40-cm-deep rectangular bath and keeping them there until the cans are heated to 50°C to ensure that the cans can withstand temperatures up to 50°C during transportation and storage (Fig. 11-54). The water bath is 1 m wide and 3.5 m long, and its top surface is open to ambient air to facilitate easy observation for the workers. If the average conditions in the plant are 92 kPa, 25°C, and

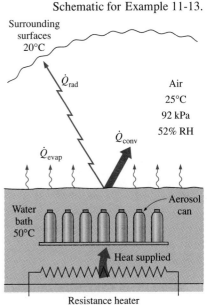

FIGURE 11-54
Schematic for Example 11-13.

52 percent relative humidity, determine the rate of heat loss from the top surface of the water bath by (a) radiation, (b) natural convection, and (c) evaporation. Assume the water is well agitated and maintained at a uniform temperature of 50°C at all times by a heater, and take the average temperature of the surrounding surfaces to be 20°C.

Solution Spray paint cans are temperature tested by submerging them in an uncovered hot water bath. The rates of heat loss from the top surface of the bath by radiation, natural convection, and evaporation are to be determined.

Assumptions **1** The low mass flux conditions exist so that the Chilton–Colburn analogy between heat and mass transfer is applicable since the mass fraction of vapor in the air is low (about 2 percent for saturated air at 300 K). **2** Both air and water vapor at specified conditions are ideal gases (the error involved in this assumption is less than 1 percent). **3** Water is maintained at a uniform temperature of 50°C.

Properties Relevant properties for each mode of heat transfer are determined below in respective sections.

Analysis (a) The emissivity of liquid water is given in Table A-14 to be 0.95. Then the radiation heat loss from the water to the surrounding surfaces becomes

$$\dot{Q}_{rad} = \varepsilon A \sigma (T_s^4 - T_{surr}^4)$$
$$= (0.95)(3.5 \text{ m}^2)(5.67 \times 10^{-8} \text{ W/m}^2 \times \text{K}^4)[(323 \text{ K})^4 - (293 \text{K})^4]$$
$$= \textbf{663 W}$$

(b) The air–water vapor mixture is dilute and thus we can use dry air properties for the mixture at the average temperature of $(T_\infty + T_S)/2 = (25 + 50)/2 = 37.5°C = 310.5$ K. Noting that the total atmospheric pressure is $92/101.3 = 0.908$ atm, the properties of dry air at 310.5 K and 0.908 atm are

$$k = 0.0269 \text{ W/m} \cdot °\text{C}, \qquad Pr = 0.711 \text{ (independent of pressure)}$$
$$\alpha = (2.36 \times 10^{-5} \text{ m}^2/\text{s})/0.908 = 2.60 \times 10^{-5} \text{ m}^2/\text{s}$$
$$\nu = (1.67 \times 10^{-5} \text{ m}^2/\text{s})/0.908 = 1.84 \times 10^{-5} \text{ m}^2/\text{s}$$

The properties of water at 50°C are

$$h_{fg} = 2383 \text{ kJ/kg} \qquad \text{and} \qquad P_v = 12.35 \text{ kPa}$$

The air at the surface is saturated, and thus the vapor pressure at the surface is simply the saturation pressure of water at the surface temperature (12.35 kPa). The vapor pressure of air far from the water surface is determined from

$$P_{v,\infty} = \phi P_{sat \, @ \, T_\infty} = (0.52)P_{sat \, @ \, 25°C} = (0.52)(3.17 \text{ kPa}) = 1.65 \text{ kPa}$$

Treating the water vapor and the air as ideal gases and noting that the total atmospheric pressure is the sum of the vapor and dry air pressures, the densities of the water vapor, dry air, and their mixture at the water–air interface and far from the surface are determined to be

At the surface:

$$\rho_{v,s} = \frac{P_{v,s}}{R_v T_S} = \frac{12.35 \text{ kPa}}{(0.4615 \text{ kPa} \cdot \text{m}^3/\text{kg} \cdot \text{K})(323 \text{ K})} = 0.0828 \text{ kg/m}^3$$

$$\rho_{a,s} = \frac{P_{a,s}}{R_a T_S} = \frac{(92 - 12.35) \text{ kPa}}{(0.287 \text{ kPa} \cdot \text{m}^3/\text{kg} \cdot \text{K})(323 \text{ K})} = 0.8592 \text{ kg/m}^3$$

$$\rho_S = \rho_{v,s} + \rho_{a,s} = 0.0828 + 0.8592 = 0.9420 \text{ kg/m}^3$$

and

Away from the surface:

$$\rho_{v,\infty} = \frac{P_{v,\infty}}{R_v T_\infty} = \frac{1.65 \text{ kPa}}{(0.4615 \text{ kPa} \cdot \text{m}^3/\text{kg} \cdot \text{K})(298 \text{ K})} = 0.0120 \text{ kg/m}^3$$

$$\rho_{a,\infty} = \frac{P_{a,\infty}}{R_a T_\infty} = \frac{(92 - 1.65) \text{ kPa}}{(0.287 \text{ kPa} \cdot \text{m}^3/\text{kg} \cdot \text{K})(298 \text{ K})} = 1.0564 \text{ kg/m}^3$$

$$\rho_\infty = \rho_{v,\infty} + \rho_{a,\infty} = 0.0120 + 1.0564 = 1.0684 \text{ kg/m}^3$$

The area of the top surface of the water bath is $A = (3.5 \text{ m})(1 \text{ m}) = 3.5 \text{ m}^2$ and its perimeter is $p = 2(3.5 + 1) = 9$ m. Therefore, the characteristic length is

$$L = \frac{A}{p} = \frac{3.5 \text{ m}^2}{9 \text{ m}} = 0.389 \text{ m}$$

Then using densities (instead of temperatures) since the mixture is not homogeneous, the Grashoff number is determined to be

$$Gr = \frac{g(\rho_\infty - \rho_S)L^3}{\rho \nu^2}$$

$$= \frac{(9.81 \text{ m/s}^2)(1.0684 - 0.9420 \text{ kg/m}^3)(0.389 \text{ m})^3}{[(0.9420 + 1.0684)/2 \text{ kg/m}^3](1.84 \times 10^{-5} \text{ m}^2/\text{s})^2}$$

$$= 2.14 \times 10^8$$

Recognizing that this is a natural convection problem with hot horizontal surface facing up, the Nusselt number and the convection heat transfer coefficients are determined to be

$$Nu = 0.15(Gr \, Pr)^{1/3} = 0.15(2.14 \times 10^8 \times 0.711)^{1/3} = 80.1$$

and

$$h_{conv} = \frac{Nu k}{L} = \frac{(80.1)(0.0269 \text{ W/m} \cdot °\text{C})}{0.389 \text{ m}} = 5.54 \text{ W/m}^2 \cdot °\text{C}$$

Then the natural convection heat transfer rate becomes

$$\dot{Q}_{conv} = h_{conv} A(T_s - T_\infty)$$

$$= (5.54 \text{ W/m}^2 \cdot °\text{C})(3.5 \text{ m}^2)(50 - 25)°\text{C} = \textbf{485 W}$$

Note that the magnitude of natural convection heat transfer is comparable to that of radiation, as expected.

(c) Utilizing the analogy between heat and mass convection, the mass transfer coefficient is determined the same way by replacing Pr by Sc. The mass diffusivity of water vapor in air at the average temperature of 310.5 K is determined from Equation 11-15 to be

$$D_{AB} = D_{H_2O-air} = 1.87 \times 10^{-10} \frac{T^{2.072}}{P} = 1.87 \times 10^{-10} \frac{310.5^{2.072}}{0.908} = 3.00 \times 10^{-5} \text{ m}^2/\text{s}$$

The Schmidt number is determined from its definition to be

$$Sc = \frac{\nu}{D_{AB}} = \frac{1.84 \times 10^{-5} \text{ m}^2/\text{s}}{3.00 \times 10^{-5} \text{ m}^2/\text{s}} = 0.613$$

The Sherwood number and the mass transfer coefficients are determined to be

$$Sh = 0.15(Gr\, Sc)^{1/3} = 0.15(2.14 \times 10^8 \times 0.613)^{1/3} = 76.2$$

and

$$h_{mass} = \frac{Sh D_{AB}}{L} = \frac{(76.2)(3.00 \times 10^{-5} \text{ m}^2/\text{s})}{0.389 \text{ m}} = 0.00588 \text{ m/s}$$

Then the evaporation rate and the rate of heat transfer by evaporation become

$$\dot{m}_v = h_{mass} A(\rho_{v,s} - \rho_{v,\infty})$$
$$= (0.00588 \text{ m/s})(3.5 \text{ m}^2)(0.0828 - 0.0120) \text{kg/m}^3$$
$$= 0.00146 \text{ kg/s} = 5.25 \text{ kg/h}$$

and

$$\dot{Q}_{evap} = \dot{m}_v h_{fg} = (0.00146 \text{ kg/s})(2383 \text{ kJ/kg}) = 3.479 \text{ kW} = \textbf{3479 W}$$

which is more than 7 times the rate of heat transfer by natural convection.

Finally, noting that the direction of heat transfer is always from high to low temperature, all forms of heat transfer determined above are in the same direction, and the total rate of heat loss from the water to the surrounding air and surfaces is

$$\dot{Q}_{total} = \dot{Q}_{rad} + \dot{Q}_{conv} + \dot{Q}_{evap} = 663 + 485 + 3479 = \textbf{4627 W}$$

Discussion   Note that if the water bath is heated electrically, a 4.6 kW resistance heater will be needed just to make up for the heat loss from the top surface. The total heater size will have to be larger to account for the heat losses from the side and bottom surfaces of the bath as well as the heat absorbed by the spray paint cans as they are heated to 50°C. Also note that water needs to be supplied to the bath at a rate of 5.25 kg/h to make up for the water loss by evaporation. Also, in reality, the surface temperature will probably be a little lower than the bulk water temperature, and thus the heat transfer rates will be somewhat lower than indicated above.

11-11 ■ SUMMARY

Mass transfer is the movement of a chemical species from a high concentration region toward a lower concentration one relative to the other chemical species present in the medium. Heat and mass transfer are analogous to each other, and several parallels can be drawn between them. The driving forces are the *temperature difference* in heat transfer and the *concentration difference* in mass transfer. Fick's law of mass diffusion is of the same form as Fourier's law of heat conduction. The species generation in a medium due to *homogeneous reactions* is analogous to heat generation. Also, mass convection due to bulk fluid motion is analogous to heat convection. Constant surface temperature corresponds to constant concentration at the surface, and an adiabatic wall corresponds to an impermeable wall. However, concentration is usually not a continuous function at a phase interface.

The concentration of a species A can be expressed in terms of density ρ_A or molar concentration C_A. It can also be expressed in dimensionless form in terms of *mass* or *molar fraction* as

Mass fraction of species A: $\quad w_A = \dfrac{m_A}{m} = \dfrac{m_A/V}{m/V} = \dfrac{\rho_A}{\rho}$

Mole fraction of species A: $\quad y_A = \dfrac{N_A}{N} = \dfrac{N_A/V}{N/V} = \dfrac{C_A}{C}$

In the case of an ideal gas mixture, the mole fraction of a gas is equal to its pressure fraction. *Fick's law* for the diffusion of a species A in a stationary binary mixture of species A and B in a specified direction x is expressed as

Mass basis: $\quad j_{\text{diff},A} = \dfrac{\dot{m}_{\text{diff},A}}{A} = -\rho D_{AB}\dfrac{d(\rho_A/\rho)}{dx} = -\rho D_{AB}\dfrac{dw_A}{dx} \quad (\text{kg/s} \cdot \text{m}^2)$

Mole basis: $\quad \bar{j}_{\text{diff},A} = \dfrac{\dot{N}_{\text{diff},A}}{A} = -CD_{AB}\dfrac{d(C_A/c)}{dx} = -CD_{AB}\dfrac{dy_A}{dx} \quad (\text{mol/s} \cdot \text{m}^2)$

where D_{AB} is the *diffusion coefficient* (or *mass diffusivity*) of the species in the mixture, $j_{\text{diff},A}$ is the diffusive *mass flux* of species A, and $\bar{j}_{\text{diff},A}$ is the *molar flux*.

The mole fractions of a species i in the gas and liquid phases at the interface of a dilute mixture are proportional to each other and are expressed by *Henry's law* as

$$y_{i,\text{ liquid side}} = \dfrac{P_{i,\text{ gas side}}}{H}$$

where H is *Henry's constant*. When the mixture is not dilute, an approximate relation for the mole fractions of a species on the liquid and gas sides of the interface are expressed approximately by *Raoult's law* as

$$P_{i,\text{ gas side}} = y_{i,\text{ gas side}}P = y_{i,\text{ liquid side}}P_{i,\text{ sat}}(T)$$

where $P_{i,\text{ sat}}(T)$ is the saturation pressure of the species i at the interface temperature and P is the *total pressure* on the gas phase side.

The concentration of the gas species i in the solid at the interface $C_{i,\text{ solid side}}$ is proportional to the *partial pressure* of the species i in the gas $P_{i,\text{ gas side}}$ on the gas side of the interface and is expressed as

$$C_{i,\text{ solid side}} = \mathscr{S} \times P_{i,\text{ gas side}} \quad (\text{kmol/m}^3)$$

where $\mathscr{S}$ is the *solubility*. The product of the *solubility* of a gas and the *diffusion coefficient* of the gas in a solid is referred to as the *permeability* $\mathscr{P}$, which is a measure of the ability of the gas to penetrate a solid.

In the absence of any chemical reactions, the mass transfer rates $\dot{m}_{\text{diff},A}$ through a plane wall of area A and thickness L and cylindrical and spherical shells of inner and outer radii r_1 and r_2 under one-dimensional steady conditions are expressed as

$$\dot{m}_{\text{diff},A,\text{wall}} = \rho D_{AB} A \frac{w_{A,1} - w_{A,2}}{L} = D_{AB} A \frac{\rho_{A,1} - \rho_{A,2}}{L}$$

$$\dot{m}_{\text{diff},A,\text{cyl}} = 2\pi L \rho D_{AB} \frac{w_{A,1} - w_{A,2}}{\ln(r_2/r_1)} = 2\pi L D_{AB} \frac{\rho_{A,1} - \rho_{A,2}}{\ln(r_2/r_1)}$$

$$\dot{m}_{\text{diff},A,\text{sph}} = 4\pi r_1 r_2 \rho D_{AB} \frac{w_{A,1} - w_{A,2}}{r_2 - r_1} = 4\pi r_1 r_2 D_{AB} \frac{\rho_{A,1} - \rho_{A,2}}{r_2 - r_1}$$

The mass flow rate of a gas through a solid plane wall under steady one-dimensional conditions can also be expressed in terms of the partial pressures of the adjacent gas on the two sides of the solid as

$$\dot{m}_{\text{diff},A,\text{wall}} = D_{AB} \mathcal{S}_{AB} A \frac{P_{A,1} - P_{A,2}}{L} = \mathcal{P}_{AB} A \frac{P_{A,1} - P_{A,2}}{L} \qquad \text{(kg/s)}$$

where $P_{A,1}$ and $P_{A,2}$ are the partial pressures of gas A on the two sides of the wall.

During mass transfer in a *moving medium*, chemical species are transported both by molecular diffusion and by the bulk fluid motion, and the velocities of the species are expressed as

$$\mathcal{V}_A = \mathcal{V} + \mathcal{V}_{\text{diff},A}$$
$$\mathcal{V}_B = \mathcal{V} + \mathcal{V}_{\text{diff},B}$$

where $\mathcal{V}$ is the *mass-average velocity* of the flow. It is the velocity that would be measured by a velocity sensor and is expressed as

$$\mathcal{V} = w_A \mathcal{V}_A + w_B \mathcal{V}_B$$

The special case $\mathcal{V} = 0$ corresponds to a *stationary medium*. Using Fick's law of diffusion, the total mass fluxes $j = \dot{m}/A$ in a moving medium are expressed as

$$j_A = \rho_A \mathcal{V} + \rho_A \mathcal{V}_{\text{diff},A} = w_A(j_A + j_B) - \rho D_{AB} \frac{dw_A}{dx}$$

$$j_B = \rho_B \mathcal{V} + \rho_B \mathcal{V}_{\text{diff},B} = w_B(j_A + j_B) - \rho D_{BA} \frac{dw_B}{dx}$$

The *rate of mass convection* of species A in a binary mixture is expressed in an analogous manner to Newton's law of cooling as

$$\dot{m}_{\text{conv}} = h_{\text{mass}} A(\rho_{A,s} - \rho_{A,\infty}) = h_{\text{mass}} \rho A(w_{A,s} - w_{A,\infty}) \qquad \text{(kg/s)}$$

where h_{mass} is the average *mass transfer coefficient*, in m/s.

The counterparts of the Prandtl and Nusselt numbers in mass convection are the *Schmidt number* Sc and the *Sherwood number* Sh, defined as

$$\text{Sc} = \frac{\nu}{D_{AB}} = \frac{\text{Momentum diffusivity}}{\text{Mass diffusivity}} \qquad \text{and} \qquad \text{Sh} = \frac{h_{\text{mass}} L}{D_{AB}}$$

The relative magnitudes of heat and mass diffusion in the thermal and concentration boundary layers are represented by the *Lewis number,* defined as

$$Le = \frac{Sc}{Pr} = \frac{\alpha}{D_{AB}} = \frac{\text{Thermal diffusivity}}{\text{Mass diffusivity}}$$

Heat and mass transfer coefficients are sometimes expressed in terms of the dimensionless *Stanton number,* defined as

$$St_{heat} = \frac{h_{conv}}{\rho \mathcal{V} C_p} = Nu \frac{1}{Re \, Pr} \quad \text{and} \quad St_{mass} = \frac{h_{mass}}{\mathcal{V}} = Sh \frac{1}{Re \, Sc}$$

where $\mathcal{V}$ is the free stream velocity in external flow and the bulk mean fluid velocity in internal flow. For a given geometry and boundary conditions, the Sherwood number in natural or forced convection can be determined from the corresponding Nusselt number expression by simply replacing the Prandtl number by the Schmidt number. But in natural convection, the Grashoff number should be expressed in terms of density difference instead of temperature difference.

When the molecular diffusivities of momentum, heat, and mass are identical, we have $v = \alpha = D_{AB}$, and thus $Pr = Sc = Le = 1$. The similarity between momentum, heat, and mass transfer in this case is given by the *Reynolds analogy,* expressed as

$$\frac{f}{2} Re = Nu = Sh \quad \text{or} \quad \frac{f}{2} \frac{\mathcal{V}_\infty L}{v} = \frac{h_{heat} L}{k} = \frac{h_{mass} L}{D_{AB}} \quad \text{or} \quad \frac{f}{2} = St = St_{mass}$$

For the general case of $Pr \neq Sc \neq 1$, it is modified as

$$\frac{f}{2} = St \, Pr^{2/3} = St_{mass} \, Sc^{2/3}$$

which is known as the *Chilton–Colburn analogy.* The analogy between heat and mass transfer is expressed more conveniently as

$$h_{heat} = \rho C_p Le^{2/3} h_{mass} = \rho C_p (\alpha/D_{AB})^{2/3} h_{mass}$$

For air–water vapor mixtures, $Le \cong 1$, and thus the relation above simplifies further. The heat–mass convection analogy is limited to *low mass flux* cases in which the flow rate of species undergoing mass flow is low relative to the total flow rate of the liquid or gas mixture. The mass transfer problems that involve phase change (evaporation, sublimation, condensation, melting, etc.) also involve heat transfer, and such problems are analyzed by considering heat and mass transfer simultaneously.

REFERENCES AND SUGGESTED READING

1. American Society of Heating, Refrigeration, and Air Conditioning Engineers. *Handbook of Fundamentals.* Atlanta: ASHRAE, 1993.

2. R. M. Barrer. *Diffusion in and through Solids*. New York: Macmillan, 1941.

3. R. B. Bird. "Theory of Diffusion." *Advances in Chemical Engineering* 1 (1956), p. 170.

4. R. B. Bird, W. E. Stewart, and E. N. Lightfoot. *Transport Phenomena*. New York: John Wiley & Sons, 1960.

5. C. J. Geankoplis. *Mass Transport Phenomena*. New York: Holt, Rinehart, and Winston, 1972.

6. *Handbook of Chemistry and Physics* 56th ed. Cleveland, OH: Chemical Rubber Publishing Co., 1976.

7. J. O. Hirshfelder, F. Curtis, and R. B. Bird. *Molecular Theory of Gases and Liquids*. New York: John Wiley & Sons, 1954.

8. J. P. Holman. *Heat Transfer*. 7th ed. New York: McGraw-Hill, 1990.

9. F. P. Incropera and D. P. De Witt. *Fundamentals of Heat and Mass Transfer*. 2nd ed. New York: John Wiley & Sons, 1985.

10. *International Critical Tables*. Vol. 3. New York: McGraw-Hill, 1928.

11. W. M. Kays and M. E. Crawford. *Convective Heat and Mass Transfer*. 2nd ed. New York: McGraw-Hill, 1980.

12. T. R. Marrero and E. A. Mason. "Gaseous Diffusion Coefficients." *Journal of Phys. Chem. Ref. Data* 1 (1972), pp. 3–118.

13. A. F. Mills. *Basic Heat and Mass Transfer*. Burr Ridge, IL: Richard D. Irwin, 1995.

14. J. H. Perry, ed. *Chemical Engineer's Handbook*. 4th ed. New York: McGraw-Hill, 1963.

15. R. D. Reid, J. M. Prausnitz, and T. K. Sherwood. *The Properties of Gases and Liquids*. 3rd ed. New York: McGraw-Hill, 1977.

16. A. H. P. Skelland. *Diffusional Mass Transfer*. New York: John Wiley & Sons, 1974.

17. D. B. Spalding. *Convective Mass Transfer*. New York: McGraw-Hill, 1963.

18. W. F. Stoecker and J. W. Jones. *Refrigeration and Air Conditioning*. New York: McGraw-Hill, 1982.

19. L. C. Thomas. *Mass Transfer Supplement—Heat Transfer*. Englewood Cliffs, NJ: Prentice Hall, 1991.

20. L. Van Black. *Elements of Material Science and Engineering*. Reading, MA: Addison-Wesley, 1980.

Analogy between Heat and Mass Transfer

11-1C How does mass transfer differ from bulk fluid flow? Can mass transfer occur in a homogeneous medium?

11-2C How is the concentration of a commodity defined? How is the concentration gradient defined? How is the diffusion rate of a commodity related to the concentration gradient?

11-3C Give examples for (*a*) liquid-to-gas, (*b*) solid-to-liquid, (*c*) solid-to-gas, and (*d*) gas-to-liquid mass transfer.

11-4C Someone suggests that thermal (or heat) radiation can also be viewed as mass radiation since, according to Einstein's formula, an energy transfer in the amount of E corresponds to a mass transfer in the amount of $m = E/c^2$. What do you think?

11-5C What is the driving force for (*a*) heat transfer, (*b*) electric current flow, (*c*) fluid flow, and (*d*) mass transfer?

11-6C What do (*a*) homogeneous reactions and (*b*) heterogeneous reactions represent in mass transfer? To what do they correspond in heat transfer?

Mass Diffusion

11-7C Both Fourier's law of heat conduction and Fick's law of mass diffusion can be expressed as $\dot{Q} = -kA(dT/dx)$. What do the quantities $\dot{Q}$, k, A, and T represent in (*a*) heat conduction and (*b*) mass diffusion?

11-8C Mark the following statements as being True or False for a binary mixture of substances A and B.

_____ (*a*) The density of a mixture is always equal to the sum of the densities of its constituents.

_____ (*b*) The ratio of the density of component A to the density of component B is equal to the mass fraction of component A.

_____ (*c*) If the mass fraction of component A is greater than 0.5, then at least half of the moles of the mixture are component A.

_____ (*d*) If the molar masses of A and B are equal to each other, then the mass fraction of A will be equal to the mole fraction of A.

_____ (*e*) If the mass fractions of A and B are both 0.5, then the molar mass of the mixture is simply the arithmetic average of the molar masses of A and B.

11-9C Mark the following statements as being True or False for a binary mixture of substances A and B.

_____ (*a*) The molar concentration of a mixture is always equal to the sum of the molar concentrations of its constituents.

_____

*Students are encouraged to answer *all* the concept "C" questions.

_____ (b) The ratio of the molar concentration of A to the molar concentration of B is equal to the mole fraction of component A.

_____ (c) If the mole fraction of component A is greater than 0.5, then at least half of the mass of the mixture is component A.

_____(d) If both A and B are ideal gases, then the pressure fraction of A is equal to its mole fraction.

_____ (e) If the mole fractions of A and B are both 0.5, then the molar mass of the mixture is simply the arithmetic average of the molar masses of A and B.

11-10C Fick's law of diffusion is expressed on the mass and mole basis as $\dot{m}_{\text{diff},A} = -\rho A D_{AB}(dw_A/dx)$ and $\dot{N}_{\text{diff},A} = -CAD_{AB}(dy_A/dx)$, respectively. Are the diffusion coefficients D_{AB} in the two relations the same or different?

11-11C How does the mass diffusivity of a gas mixture change with (a) temperature and (b) pressure?

11-12C At a given temperature and pressure, do you think the mass diffusivity of air in water vapor will be equal to the mass diffusivity of water vapor in air? Explain.

11-13C At a given temperature and pressure, do you think the mass diffusivity of copper in aluminum will be equal to the mass diffusivity of aluminum in copper? Explain.

11-14C In a mass production facility, steel components are to be hardened by carbon diffusion. Would you carry out the hardening process at room temperature or in a furnace at a high temperature, say 900°C? Why?

11-15C Someone claims that the mass and the mole fractions for a mixture of CO_2 and N_2O gases are identical. Do you agree? Explain.

11-16 The composition of moist air is given on a molar basis is to be 78 percent N_2, 20 percent O_2, and 2 percent water vapor. Determine the mass fractions of the constituents of air.
     Answers: 76.4% N_2, 22.4% O_2, 1.2% H_2O

11-17E A gas mixture consists of 5 lbm of O_2, 8 lbm of N_2, and 10 lbm of CO_2. Determine (a) the mass fraction of each component, (b) the mole fraction of each component, and (c) the average molar mass of the mixture.

11-18 A gas mixture consists of 8 kmol of H_2 and 2 kmol of N_2. Determine the mass of each gas and the apparent gas constant of the mixture.

11-19 The molar analysis of a gas mixture at 290 K and 250 kPa is 65 percent N_2, 20 percent O_2, and 15 percent CO_2. Determine the mass fraction and partial pressure of each gas.

11-20 Determine the binary diffusion coefficient of CO_2 in air at (a) 200 K and 1 atm, (b) 400 K and 0.8 atm, and (c) 600 K and 3 atm.

11-21 Repeat Problem 11-20 for O_2 in N_2.

11-22E The relative humidity of air at 80°F and 14.7 psia is increased from 30 percent to 90 percent during a humidification process at constant temperature and pressure. Determine the percent error involved in assuming the density of air to have remained constant.   *Answer:* 2.1%

11-23 The diffusion coefficient of hydrogen in steel is given as a function of temperature as

$$D_{AB} = 1.65 \times 10^{-6} \exp(-4630/T) \quad (m^2/s)$$

where T is in K. Determine the diffusion coefficients from 200 K to 1200 K in 200 K increments and plot the results.

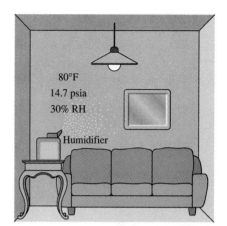

80°F
14.7 psia
30% RH

Humidifier

FIGURE P11-22E

Boundary Conditions

11-24C Write three boundary conditions for mass transfer (on a mass basis) for species A at $x = 0$ that correspond to specified temperature, specified heat flux, and convection boundary conditions in heat transfer.

11-25C What is an impermeable surface in mass transfer? How is it expressed mathematically (on a mass basis)? To what does it correspond in heat transfer?

11-26C Consider the free surface of a lake exposed to the atmosphere. If the air at the lake surface is saturated, will the mole fraction of water vapor in air at the lake surface be the same as the mole fraction of water in the lake (which is nearly 1)?

11-27C When prescribing a boundary condition for mass transfer at a solid–gas interface, why do we need to specify the side of the surface (whether the solid or the gas side)? Why did we not do it in heat transfer?

11-28C Using properties of saturated water, explain how you would determine the mole fraction of water vapor at the surface of a lake when the temperature of the lake surface and the atmospheric pressure are specified.

11-29C Using solubility data of a solid in a specified liquid, explain how you would determine the mass fraction of the solid in the liquid at the interface at a specified temperature.

11-30C Using solubility data of a gas in a solid, explain how you would determine the molar concentration of the gas in the solid at the solid–gas interface at a specified temperature.

11-31C Using Henry's constant data for a gas dissolved in a liquid, explain how you would determine the mole fraction of the gas dissolved in the liquid at the interface at a specified temperature.

11-32C What is permeability? How is the permeability of a gas in a solid related to the solubility of the gas in that solid?

11-33E Determine the mole fraction of the water vapor at the surface of a lake whose temperature at the surface is 60°F, and compare it to the mole fraction of water in the lake. Take the atmospheric pressure at lake level to be 13.8 psia.

11-34 Determine the mole fraction of dry air at the surface of a lake whose temperature is 15°C. Take the atmospheric pressure at lake level to be 100 kPa.   *Answer:* 98.3%

11-35 Consider a rubber plate that is in contact with nitrogen gas at 298 K and 250 kPa. Determine the molar and mass densities of nitrogen in the rubber at the interface.   *Answers:* 0.0039 kmol/m³, 0.1092 kg/m³

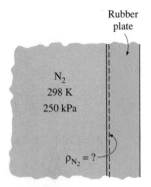

Rubber plate

N_2
298 K
250 kPa

$\rho_{N_2} = ?$

FIGURE P11-35

11-36 A wall made of natural rubber separates O_2 and N_2 gases at 25°C and 500 kPa. Determine the molar concentrations of O_2 and N_2 in the wall.

11-37 Consider a glass of water in a room at 20°C and 97 kPa. If the relative humidity in the room is 100 percent and the water and the air are in thermal and phase equilibrium, determine (*a*) the mole fraction of the water vapor in the air and (*b*) the mole fraction of air in the water.

11-38E Water is sprayed into air at 80°F and 14.3 psia, and the falling water droplets are collected in a container on the floor. Determine the mass and mole fractions of air dissolved in the water.

11-39 Consider a carbonated drink in a bottle at 27°C and 130 kPa. Assuming the gas space above the liquid consists of a saturated mixture of CO_2 and water vapor and treating the drink as water, determine (*a*) the mole fraction of the water vapor in the CO_2 gas and (*b*) the mass of dissolved CO_2 in a 200-ml drink.   *Answers:* (*a*) 2.77%, (*b*) 0.36 g

CO_2
H_2O

27°C
130 kPa

FIGURE P11-39

Steady Mass Diffusion through a Wall

11-40C Write down the relations for steady one-dimensional heat conduction and mass diffusion through a plane wall, and identify the quantities in the two equations that correspond to each other.

11-41C Consider steady one-dimensional mass diffusion through a wall. Mark the following statements as being True or False.

____ (*a*) Other things being equal, the higher the density of the wall, the higher the rate of mass transfer.

____ (*b*) Other things being equal, doubling the thickness of the wall will double the rate of mass transfer.

____ (*c*) Other things being equal, the higher the temperature, the higher the rate of mass transfer.

____ (*d*) Other things being equal, doubling the mass fraction of the diffusing species at the high concentration side will double the rate of mass transfer.

11-42C Consider one-dimensional mass diffusion of species *A* through a plane wall of thickness *L*. Under what conditions will the concentration profile of species *A* in the wall be a straight line?

11-43C Consider one-dimensional mass diffusion of species A through a plane wall. Does the species A content of the wall change during steady mass diffusion? How about during transient mass diffusion?

11-44 Helium gas is stored at 293 K in a 3-m-outer-diameter spherical container made of 5-cm-thick Pyrex. The molar concentration of helium in the Pyrex is 0.00073 kmol/m³ at the inner surface and negligible at the outer surface. Determine the mass flow rate of helium by diffusion through the Pyrex container. *Answer:* 7.2×10^{-15} kg/s

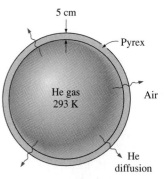

5 cm

Pyrex

He gas
293 K

Air

He
diffusion

FIGURE P11-44

11-45 A thin plastic membrane separates hydrogen from air. The molar concentrations of hydrogen in the membrane at the inner and outer surfaces are determined to be 0.065 and 0.003 kmol/m³, respectively. The binary diffusion coefficient of hydrogen in plastic at the operation temperature is 5.3×10^{-10} m²/s. Determine the mass flow rate of hydrogen by diffusion through the membrane under steady conditions if the thickness of the membrane is (*a*) 2 mm and (*b*) 0.5 mm.

11-46 The solubility of hydrogen gas in steel in terms of its mass fraction is given as $w_{H_2} = 2.09 \times 10^{-4} \exp(-3950/T)P_{H_2}^{0.5}$ where P_{H_2} is the partial pressure of hydrogen in bars and T is the temperature in K. If natural gas is transported in a 1-cm-thick, 3-m-internal-diameter steel pipe at 500 kPa pressure and the mole fraction of hydrogen in the natural gas is 8 percent, determine the highest rate of hydrogen loss through a 100-m-long section of the pipe at steady conditions at a temperature of 293 K if the pipe is exposed to air. Take the diffusivity of hydrogen in steel to be 2.9×10^{-13} m²/s.
Answer: 2.035×10^{-17} kg/s

11-47 Helium gas is stored at 293 K and 500 kPa in a 1-cm-thick, 2-m-inner-diameter spherical tank made of fused silica (SiO_2). The area where the container is located is well ventilated. Determine (*a*) the mass flow rate of helium by diffusion through the tank and (*b*) the pressure drop in the tank in one week as a result of the loss of helium gas.

11-48 You probably have noticed that balloons inflated with helium gas rise in the air the first day during a party but they fall down the next day and act like ordinary balloons filled with air. This is because the helium in the balloon slowly leaks out through the wall while air leaks in by diffusion.

Consider a balloon that is made of 0.1-mm-thick soft rubber and has a diameter of 15 cm when inflated. The pressure and temperature inside the balloon are initially 110 kPa and 25°C. The permeability of rubber to helium, oxygen, and nitrogen at 25°C are 9.4×10^{-13}, 7.05×10^{-13}, and 2.6×10^{-13} kmol/m · s · bars, respectively. Determine the initial rates of diffusion of helium, oxygen, and nitrogen through the balloon wall and the mass fraction of helium that escapes the balloon during the first 5 h assuming the helium pressure inside the balloon remains nearly constant. Assume air to be 21 percent oxygen and 79 percent nitrogen by mole numbers and take the room conditions to be 100 kPa and 25°C.

110 kPa
25°C
He

Air

FIGURE P11-48

11-49 Reconsider the balloon discussed in Problem 11-48. Assuming the volume to remain constant and disregarding the diffusion of air into the

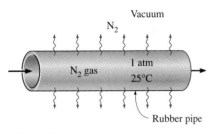

FIGURE P11-50

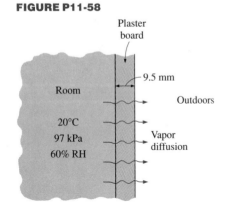

balloon, obtain a relation for the variation of pressure in the balloon with time. Using the results obtained and the numerical values given in the problem, determine how long it will take for the pressure inside the balloon to drop to 100 kPa.

11-50 Pure N_2 gas at 1 atm and 25°C is flowing through a 10-m-long, 3-cm-inner diameter pipe made of 1-mm-thick rubber. Determine the rate at which N_2 leaks out of the pipe if the medium surrounding the pipe is (*a*) a vacuum and (*b*) atmospheric air at 1 atm and 25°C with 21 percent O_2 and 79 percent N_2.   *Answers: (a)* 4.48×10^{-10} kmol/s, *(b)* 9.4×10^{-11} kmol/s

Water Vapor Migration in Buildings

11-51C Consider a tank that contains moist air at 3 atm and whose walls are permeable to water vapor. The surrounding air at 1 atm pressure also contains some moisture. Is it possible for the water vapor to flow into the tank from surroundings? Explain.

11-52C Express the mass flow rate of water vapor through a wall of thickness L in terms of the partial pressure of water vapor on both sides of the wall and the permeability of the wall to the water vapor.

11-53C How does the condensation or freezing of water vapor in the wall affect the effectiveness of the insulation in the wall? How does the moisture content affect the effective thermal conductivity of soil?

11-54C Moisture migration in the walls, floors, and ceilings of buildings is controlled by vapor barriers or vapor retarders. Explain the difference between the two, and discuss which is more suitable for use in the walls of residential buildings.

11-55C What are the adverse effects of excess moisture on the wood and metal components of a house and the paint on the walls?

11-56C Why are the insulations on the chilled water lines always wrapped with vapor barrier jackets?

11-57C Explain how vapor pressure of the ambient air is determined when the temperature, total pressure, and relative humidity of the air are given.

11-58 The diffusion of water vapor through plaster boards and its condensation in the wall insulation in cold weather are of concern since they reduce the effectiveness of insulation. Consider a house that is maintained at 20°C and 60 percent relative humidity at a location where the atmospheric pressure is 97 kPa. The inside of the walls is finished with 9.5-mm thick gypsum wallboard. Taking the vapor pressure at the outer side of the wallboard to be zero, determine the maximum amount of water vapor that will diffuse through a 3-m × 8-m section of a wall during a 24-h period. The permeance of the 9.5-mm-thick gypsum wallboard to water vapor is 2.86×10^{-9} kg/s · m² · Pa.

11-59 Reconsider Problem 11-58. In order to reduce the migration of water vapor through the wall, it is proposed to use a 0.2-mm-thick polyethylene film whose permeance is 2.3×10^{-12} kg/s $\cdot$ m^2 $\cdot$ Pa. Determine the amount of water vapor that will diffuse through the wall in this case during a 24-h period.   *Answer:* 6.7 g

11-60   The roof of a house is 15 m $\times$ 8 m and is made of a 20-cm-thick concrete layer. The interior of the house is maintained at 25°C and 50 percent relative humidity and the local atmospheric pressure is 100 kPa. Determine the amount of water vapor that will migrate through the roof in 24 h if the average outside conditions during that period are 3°C and 30 percent relative humidity. The permeability of concrete to water vapor is 24.7×10^{-12} kg/s $\cdot$ m $\cdot$ Pa.

11-61   Reconsider Problem 11-60. In order to reduce the migration of water vapor, the inner surface of the wall is painted with vapor retarder latex paint whose permeance is 26×10^{-12} kg/s $\cdot$ m^2 $\cdot$ Pa. Determine the amount of water vapor that will diffuse through the roof in this case during a 24-h period.

11-62   A glass of milk left on top of a counter in the kitchen at 25°C, 88 kPa, and 50 percent relative humidity is tightly sealed by a sheet of 0.009-mm-thick aluminum foil whose permeance is 2.9×10^{-12} kg/s $\cdot$ m^2 $\cdot$ Pa. The inner diameter of the glass is 12 cm. Assuming the air in the glass to be saturated at all times, determine how much the level of the milk in the glass will recede in 12 h.   *Answer:* 0.00079 mm

25°C
88 kPa
50% RH   Moisture migration

Aluminum foil

Milk
25°C

FIGURE P11-62

Transient Mass Diffusion

11-63C   In transient mass diffusion analysis, can we treat the diffusion of a solid into another solid of finite thickness (such as the diffusion of carbon into an ordinary steel component) as a diffusion process in a semi-infinite medium? Explain.

11-64C   Define the penetration depth for mass transfer, and explain how it can be determined at a specified time when the diffusion coefficient is known.

11-65C   When the density of a species A in a semi-infinite medium is known at the beginning and at the surface, explain how you would determine the concentration of the species A at a specified location and time.

11-66   A steel part whose initial carbon content is 0.12 percent by mass is to be case-hardened in a furnace at 1150 K by exposing it to a carburizing gas. The diffusion coefficient of carbon in steel is strongly temperature dependent, and at the furnace temperature it is given to be $D_{AB} = 7.2 \times 10^{-12}$ m^2/s. Also, the mass fraction of carbon at the exposed surface of the steel part is maintained at 0.011 by the carbon-rich environment in the furnace. If the hardening process is to continue until the mass fraction of carbon at a depth of 0.7 mm is raised to 0.32 percent, determine how long the part should be held in the furnace.   *Answer:* 9 h

11-67   Repeat Problem 11-66 for a furnace temperature of 500 K at which the diffusion coefficient of carbon in steel is $D_{AB} = 2.1 \times 10^{20}$ m^2/s.

FIGURE P11-66

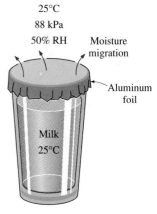

1150 K

Carbon

Steel part

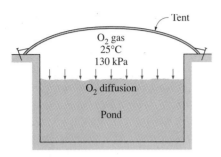

FIGURE P11-68

11-68 A pond whose initial oxygen content is zero is to be oxygenated by forming a tent over the water surface and filling the tent with oxygen gas at 25°C and 130 kPa. Determine the mole fraction of oxygen at a depth of 2 cm from the surface after 12 h.

11-69 A long nickel bar whose diameter is 5 cm has been stored in a hydrogen-rich environment at 358 K and 300 kPa for a long time, and thus it contains hydrogen gas throughout uniformly. Now the bar is taken into a well-ventilated area so that the hydrogen concentration at the outer surface remains at almost zero at all times. Determine how long it will take for the hydrogen concentration at the center of the bar to drop by half. The diffusion coefficient of hydrogen in the nickel bar at the room temperature of 298 K can be taken to be $D_{AB} = 1.2 \times 10^{-12}$ m²/s.   *Answer:* 3.3 years

Diffusion in a Moving Medium

11-70C Define the following terms: mass-average velocity, diffusion velocity, stationary medium, and moving medium.

11-71C What is diffusion velocity? How does it affect the mass-average velocity? Can the velocity of a species in a moving medium relative to a fixed reference point be zero in a moving medium? Explain.

11-72C What is the difference between mass-average velocity and mole-average velocity during mass transfer in a moving medium? If one of these velocities is zero, will the other also necessarily be zero? Under what conditions will these two velocities be the same for a binary mixture?

11-73C Consider one-dimensional mass transfer in a moving medium that consists of species A and B with $\rho = \rho_A + \rho_B =$ constant. Mark the following statements as being True or False.

_____ (a) The rates of mass diffusion of species A and B are equal in magnitude and opposite in direction.
_____ (b) $D_{AB} = D_{BA}$.
_____ (c) During equimolar counterdiffusion through a tube, equal numbers of moles of A and B move in opposite directions, and thus a velocity measurement device placed in the tube will read zero.
_____ (d) The lid of a tank containing propane gas (which is heavier than air) is left open. If the surrounding air and the propane in the tank are at the same temperature and pressure, no propane will escape the tank and no air will enter.

11-74C What is Stefan flow? Write the expression for Stefan's law and indicate what each variable represents.

11-75E The pressure in a pipeline that transports helium gas at a rate of 5 lbm/s is maintained at 14.5 psia by venting helium to the atmosphere through a $\frac{1}{4}$-in. internal diameter tube that extends 30 ft into the air. Assuming both the helium and the atmospheric air to be at 80°F, determine (a) the mass flow rate of helium lost to the atmosphere through an individual tube, (b) the

FIGURE P11-75E

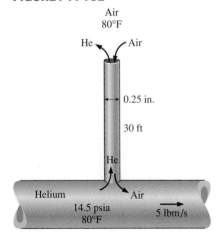

mass flow rate of air that infiltrates into the pipeline, and (c) the flow velocity at the bottom of the tube where it is attached to the pipeline that will be measured by an anemometer in steady operation.

11-76E Repeat Problem 11-75E for a pipeline that transports carbon dioxide instead of helium.

11-77 A tank whose shell is 2-cm thick contains hydrogen gas at the atmospheric conditions of 25°C and 90 kPa. The charging valve of the tank has an internal diameter of 3-cm and extends 8 cm above the tank. If the lid of the tank is left open so that hydrogen and air can undergo equimolar counter-diffusion through the 10-cm-long passageway, determine the mass flow rate of hydrogen lost to the atmosphere through the valve at the initial stages of the process.   *Answer:* 1.85×10^{-8} kg/s

11-78E A 1-in.-diameter Stefan tube is used to measure the binary diffusion coefficient of water vapor in air at 70°F and 13.8 psia. The tube is partially filled with water with a distance from the water surface to the open end of the tube of 10 in. Dry air is blown over the open end of the tube so that water vapor rising to the top is removed immediately and the concentration of vapor at the top of the tube is zero. During 10 days of continuous operation at constant pressure and temperature, the amount of water that has evaporated is measured to be 0.0015 lbm. Determine the diffusion coefficient of water vapor in air at 70°F and 13.8 psia.

11-79 An 8-cm-internal-diameter, 30-cm-high pitcher half filled with water is left in a dry room at 15°C and 87 kPa with its top open. If the water is maintained at 15°C at all times also, determine how long it will take for the water to evaporate completely.   *Answer:* 750 days

11-80 A large tank containing ammonia at 1 atm and 25°C is vented to the atmosphere through a 3-m-long tube whose internal diameter is 1 cm. Determine the rate of loss of ammonia and the rate of infiltration of air into the tank.

Mass Convection

11-81C Heat convection is expressed by Newton's law of cooling as $\dot{Q} = hA(T_s - T_\infty)$. Express mass convection in an analogous manner on a mass basis, and identify all the quantities in the expression and state their units.

11-82C What is a concentration boundary layer? How is it defined for flow over a plate?

11-83C What is the physical significance of the Schmidt number? How is it defined? To what dimensionless number does it correspond in heat transfer? What does a Schmidt number of 1 indicate?

11-84C What is the physical significance of the Sherwood number? How is it defined? To what dimensionless number does it correspond in heat transfer? What does a Sherwood number of 1 indicate for a plain fluid layer?

Room
15°C
87 kPa

Water
vapor

Water
15°C

FIGURE P11-79

11-85C What is the physical significance of the Lewis number? How is it defined? What does a Lewis number of 1 indicate?

11-86C In natural convection mass transfer, the Grashof number is evaluated using density difference instead of temperature difference. Can the Grashof number evaluated this way be used in heat transfer calculations also?

11-87C Using the analogy between heat and mass transfer, explain how the mass transfer coefficient can be determined from the relations for the heat transfer coefficient.

11-88C It is well known that warm air in a cooler environment rises. Now consider a warm mixture of air and gasoline (C_8H_{18}) on top of an open gasoline can. Do you think this gas mixture will rise in a cooler environment?

11-89C Consider two identical cups of coffee, one with no sugar and the other with plenty of sugar at the bottom. Initially, both cups are at the same temperature. If left unattended, which cup of coffee will cool faster?

11-90C Under what conditions will the normalized velocity, thermal, and concentration boundary layers coincide during flow over a flat plate?

11-91C What is the relation $(f/2)$ Re = Nu = Sh known as? Under what conditions is it valid? What is the practical importance of it?

11-92C What is the name of the relation $f/2 = \text{St Pr}^{2/3} = \text{St}_{mass}\text{Sc}^{2/3}$ and what are the names of the variables in it? Under what conditions is it valid? What is the importance of it in engineering?

11-93C What is the relation $h_{heat} = \rho C_p h_{mass}$ known as? For what kind of mixtures is it valid? What is the practical importance of it?

11-94C What is the low mass flux approximation in mass transfer analysis? Can the evaporation of water from a lake be treated as a low mass flux process?

11-95E Consider a circular pipe of inner diameter $D = 0.5$ in. whose inner surface is covered with a thin layer of liquid water as a result of condensation. In order to dry the pipe, air at 540 R and 1 atm is forced to flow through it with an average velocity of 4 ft/s. Using the analogy between heat and mass transfer, determine the mass transfer coefficient inside the pipe for fully developed flow.   *Answer:* 0.024 ft/s

11-96 The average heat transfer coefficient for air flow over an odd-shaped body is to be determined by mass transfer measurements and using the Chilton–Colburn analogy between heat and mass transfer. The experiment is conducted by blowing dry air at 1 atm at a free stream velocity of 2 m/s over a body covered with a layer of naphthalene. The surface area of the body is 0.75 m², and it is observed that 100 g of naphthalene has sublimated in 45 min. During the experiment, both the body and the air were kept at 25°C, at which the vapor pressure and mass diffusivity of naphthalene are 11 Pa and $D_{AB} = 0.61 \times 10^{-5}$ m²/s, respectively. Determine the heat transfer coefficient under the same flow conditions over the same geometry.

FIGURE P11-96

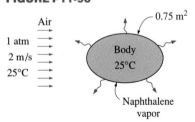

Air
1 atm
2 m/s
25°C
Body
25°C
0.75 m²
Naphthalene vapor

11-97 Consider a 15-cm-internal-diameter, 10-m-long circular duct whose interior surface is wet. The duct is to be dried by forcing dry air at 1 atm and 15°C through it at an average velocity of 3 m/s. The duct passes through a chilled room, and it remains at an average temperature of 15°C at all times. Determine the mass transfer coefficient in the duct.

11-98 Dry air at 15°C and 92 kPa flows over a 2-m-long wet surface with a free stream velocity of 4 m/s. Determine the average mass transfer coefficient. *Answer:* 0.00515 m/s

11-99 Consider a 5-m × 5-m wet concrete patio with an average water film thickness of 0.3 mm. Now wind at 50 km/h is blowing over the surface. If the air is at 1 atm, 15°C, and 35 percent relative humidity, determine how long it will take for the patio to dry completely. *Answer:* 18.4 min

11-100E A 2-in.-diameter spherical naphthalene ball is suspended in a room at 1 atm and 80°F. Determine the average mass transfer coefficient between the naphthalene and the air if air is forced to flow over naphthalene with a free stream velocity of 15 ft/s. The Schmidt number of naphthalene in air at room temperature is 2.35. *Answer:* 0.0525 ft/s

11-101 Consider a 3-mm-diameter raindrop that is falling freely in atmospheric air at 25°C. Taking the temperature of the raindrop to be 9°C, determine the terminal velocity of the raindrop at which the drag force equals the weight of the drop and the average mass transfer coefficient at that time.

11-102 In a manufacturing facility, wet brass plates coming out of a water bath are to be dried by passing them through a section where dry air at 1 atm and 25°C is blown parallel to their surfaces. If the plates are at 20°C and there are no dry spots, determine the rate of evaporation from both sides of a plate.

11-103E Air at 80°F, 1 atm, and 30 percent relative humidity is blown over the surface of a 15-in. × 15-in. square pan filled with water at a free stream velocity of 10 ft/s. If the water is maintained at a uniform temperature of 80°F, determine the rate of evaporation of water and the amount of heat that needs to be supplied to the water to maintain its temperature constant.

11-104E Repeat Problem 11-103E for temperature of 60°F for both the air and water.

Simultaneous Heat and Mass Transfer

11-105C Does a mass transfer process have to involve heat transfer? Describe a process that involves both heat and mass transfer.

11-106C Consider a shallow body of water. Is it possible for this water to freeze during a cold and dry night even when the ambient air and surrounding surface temperatures never drop to 0°C? Explain.

11-107C During evaporation from a water body to air, under what conditions will the latent heat of vaporization be equal to convection heat transfer from the air?

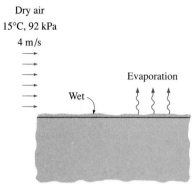

Dry air
15°C, 92 kPa
4 m/s

Evaporation

Wet

FIGURE P11-98

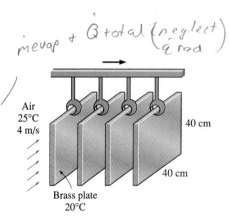

$\dot{m}_{evap} + \dot{Q}_{total} \left(\begin{array}{c} \text{neglect} \\ \dot{q}_{rad} \end{array} \right)$

Air
25°C
4 m/s

40 cm

40 cm

Brass plate
20°C

FIGURE P11-102

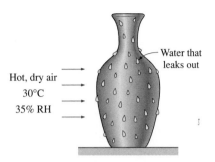

Hot, dry air
30°C
35% RH

Water that leaks out

FIGURE P11-108

11-108 Jugs made of porous clay were commonly used to cool water in the past. A small amount of water that leaks out keeps the outer surface of the jug wet at all times, and hot and relatively dry air flowing over the jug causes this water to evaporate. Part of the latent heat of evaporation comes from the water in the jug, and the water is cooled as a result. If the environment conditions are 1 atm, 30°C, and 35 percent relative humidity, determine the temperature of the water when steady conditions are reached.

11-109E During a hot summer day, a 2-L bottle drink is to be cooled by wrapping it in a cloth kept wet continually and blowing air to it with a fan. If the environment conditions are 1 atm, 80°F, and 30 percent relative humidity, determine the temperature of the drink when steady conditions are reached.

11-110 A glass bottle washing facility uses a well-agitated hot water bath at 55°C with open top that is placed on the ground. The bathtub is 1-m high, 2-m wide, and 4-m long and is made of sheet metal so that the outer side surfaces are also at about 55°C. The bottles enter at a rate of 800 per minute at ambient temperature and leave at the water temperature. Each bottle has a mass of 150 g and removes 0.6 g of water as it leaves the bath wet. Make-up water is supplied at 15°C. If the average conditions in the plant are 1 atm, 25°C, and 50 percent relative humidity, and the average temperature of the surrounding surfaces is 15°C, determine (*a*) the amount of heat and water removed by the bottles themselves per second; (*b*) the rate of heat loss from the top surface of the water bath by radiation, natural convection, and evaporation; (*c*) the rate of heat loss from the side surfaces by natural convection and radiation; and (*d*) the rate at which heat and water must be supplied to maintain steady operating conditions. Disregard heat loss through the bottom surface of the bath and take the emissivities of sheet metal and water to be 0.61 and 0.95, respectively.
Answers: (*a*) 61,337 W, (*b*) 1495 W, (*c*) 3780 W, (*d*) 79,884 W, 44.9 kg/h

11-111 Repeat Problem 11-110 for a water bath temperature of 50°C.

11-112 One way of increasing heat transfer from the head on a hot summer day is to wet it. This is especially effective in windy weather, as you may have noticed. Approximating the head as a 30-cm-diameter sphere at 30°C with an emissivity of 0.95, determine the total rate of heat loss from the head at ambient air conditions of 1 atm, 25°C, 40 percent relative humidity, and 25 km/h winds if the head is (*a*) dry and (*b*) wet.
Answers: (*a*) 40.8 W, (*b*) 351.8 W

11-113 A 2-m-deep 20-m × 20-m heated swimming pool is maintained at a constant temperature of 30°C at a location where the atmospheric pressure is 1 atm. If the ambient air is at 20°C and 60 percent relative humidity and the effective sky temperature is 0°C, determine the rate of heat loss from the top surface of the pool by (*a*) radiation, (*b*) natural convection, and (*c*) evaporation. (*d*) Assuming the heat losses to the ground to be negligible, determine the size of the heater.

11-114 Repeat Problem 11-113 for a pool temperature of 25°C.

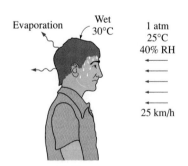

Evaporation

Wet
30°C

1 atm
25°C
40% RH

25 km/h

FIGURE P11-112

11-115C Mark the following statements as being True or False.

____ (a) The units of mass diffusivity, heat diffusivity, and momentum diffusivity are all the same.

____ (b) If the molar concentration (or molar density) C of a mixture is constant, then its density ρ must also be constant.

____ (c) If the mass-average velocity of a binary mixture is zero, then the mole-average velocity of the mixture must also be zero.

____ (d) If the mole fractions of A and B of a mixture are both 0.5, then the molar mass of the mixture is simply the arithmetic average of the molar masses of A and B.

11-116 Using Henry's law, show that the dissolved gases in a liquid can be driven off by heating the liquid.

11-117 Show that for an ideal gas mixture maintained at a constant temperature and pressure, the molar concentration C of the mixture remains constant but this is not necessarily the case for the density ρ of the mixture.

11-118E A gas mixture in a tank at 600 R and 20 psia consists of 1 lbm of CO_2 and 3 lbm of CH_4. Determine the volume of the tank and the partial pressure of each gas.

11-119 Dry air whose molar analysis is 78.1 percent N_2, 20.9 percent O_2, and 1 percent Ar flows over a water body until it is saturated. If the pressure and temperature of air remain constant at 1 atm and 25°C during the process, determine (a) the molar analysis of the saturated air and (b) the density of air before and after the process. What do you conclude from your results?

11-120 Consider a glass of water in a room at 25°C and 100 kPa. If the relative humidity in the room is 70 percent and the water and the air are at the same temperature, determine (a) the mole fraction of the water vapor in the room air, (b) the mole fraction of the water vapor in the air adjacent to the water surface, and (c) the mole fraction of air in the water near the surface.   *Answers:* (a) 2.22%, (b) 3.17%, (c) 1.34×10^{-5}

11-121 The diffusion coefficient of carbon in steel is given as

$$D_{AB} = 2.67 \times 10^{-5} \exp(-17{,}400/T) \qquad m^2/s$$

where T is in K. Determine the diffusion coefficient from 300 K to 1500 K in 100 K increments and plot the results.

11-122 A carbonated drink is fully charged with CO_2 gas at 17°C and 600 kPa such that the entire bulk of the drink is in thermodynamic equilibrium with the CO_2–water vapor mixture. Now consider a 2-L soda bottle. If the CO_2 gas in that bottle were to be released and stored in a container at 25°C and 100 kPa, determine the volume of the container.   *Answer:* 12.7 L

11-123 Consider a brick house that is maintained at 20°C and 60 percent relative humidity at a location where the atmospheric pressure is 85 kPa. The walls of the house are made of 20-cm thick brick whose permeance is 23 ×

25°C
100 kPa
70% RH

Air-water interface

Water
25°C

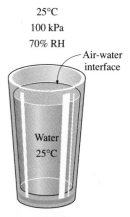

FIGURE P11-120

FIGURE P11-122

CO_2
Water

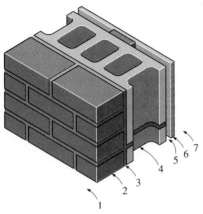

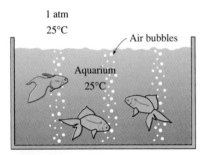

1 atm
25°C

Air bubbles

Aquarium
25°C

10^{-9} kg/s · m² · Pa. Taking the vapor pressure at the outer side of the wallboard to be zero, determine the maximum amount of water vapor that will diffuse through a 4-m × 7-m section of a wall during a 24-h period.

11-124E Consider a masonry cavity wall that is built around 6-in.-thick concrete blocks. The outside is finished with 4-in. face brick with $\frac{1}{2}$-in. cement mortar between the bricks and concrete blocks. The inside finish consists of $\frac{1}{2}$-in. gypsum wallboard separated from the concrete block by $\frac{3}{4}$-in.-thick air space. The thermal and vapor resistances of various components for a unit wall area are as follows:

Construction	R-value, h · ft² · °F/Btu	R_v-value, s · ft² · psi/lbm
1. Outside surface, 15 mph wind	0.17	—
2. Face brick, 4 in.	0.43	15,000
3. Cement mortar, 0.5 in.	0.10	1930
4. Concrete block, 6-in.	4.20	23,000
5. Air space, $\frac{3}{4}$-in.	1.02	77.6
6. Gypsum wallboard, 0.5 in.	0.45	332
7. Inside surface, still air	0.68	—

The indoor conditions are 70°F and 65 percent relative humidity while the outdoor conditions are 32°F and 40 percent relative humidity. Determine the rates of heat and water vapor transfer through a 9-ft × 25-ft section of the wall.   *Answers:* 1436 Btu/h, 4.03 lbm/h

11-125 The oxygen needs of fish in aquariums are usually met by forcing air to the bottom of the aquarium by a compressor. The air bubbles provide a large contact area between the water and the air, and as the bubbles rise, oxygen and nitrogen gases in the air dissolve in water while some water evaporates into the bubbles. Consider an aquarium that is maintained at room temperature of 25°C at all times. The air bubbles are observed to rise to the free surface of water in 2 s. If the air entering the aquarium is completely dry and the diameter of the air bubbles is 4 mm, determine the mole fraction of water vapor at the center of the bubble when it leaves the aquarium. Assume no fluid motion in the bubble so that water vapor propagates in the bubble by diffusion only.   *Answer:* 3.13%

11-126 Oxygen gas is forced into an aquarium at 1 atm and 25°C, and the oxygen bubbles are observed to rise to the free surface in 2 s. Determine the penetration depth of oxygen into water from a bubble during this time period.

11-127 Consider a 30-cm-diameter pan filled with water at 15°C in a room at 20°C, 1 atm, and 30 percent relative humidity. Determine (*a*) the rate of heat transfer by convection, (*b*) the rate of evaporation of water, and (*c*) the rate of heat transfer to the water needed to maintain its temperature at 15°C. Disregard any radiation effects.

11-128 Repeat Problem 11-127 assuming a fan blows air over the water surface at a velocity of 3 m/s. Take the radius of the pan to be the characteristic length.

11-129 Naphthalene is commonly used as a repellent against moths to protect clothing during storage. Consider a 1-cm-diameter spherical naphthalene ball hanging in a closet at 25°C and 1 atm. Considering the variation of diameter with time, determine how long it will take for the naphthalene to sublimate completely. The density and vapor pressure of naphthalene at 25°C are 0.11 Pa and 1100 kg/m³ and 11 Pa, respectively, and the mass diffusivity of naphthalene in air at 25°C is $D_{AB} = 0.61 \times 10^{-5}$ m²/s.
Answer: 45.7 days

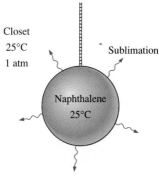

FIGURE P11-129

11-130E A swimmer extends his wet arms into the windy air outside at 1 atm, 40°F, 50 percent relative humidity, and 20 mph. If the average skin temperature is 80°F, determine the rate at which water evaporates from both arms and the corresponding rate of heat transfer by evaporation. The arm can be modeled as a 2-ft-long and 3-in.-diameter cylinder with adiabatic ends.

11-131 A thick part made of nickel is put into a room filled with hydrogen at 3 atm and 85°C. Determine the hydrogen concentration at a depth of 2-mm from the surface after 24 h. *Answer:* 4.1×10^{-7} kmol/m³

11-132 A membrane made of 0.1-mm-thick soft rubber separates pure O_2 at 1 atm and 25°C from air at 1.2 atm pressure. Determine the mass flow rate of O_2 through the membrane per unit area and the direction of flow.

11-133E The top section of an 8-ft-deep 100-ft × 100-ft heated solar pond is maintained at a constant temperature of 80°F at a location where the atmospheric pressure is 1 atm. If the ambient air is at 70°F and 100 percent relative humidity and wind is blowing at an average velocity of 40 mph, determine the rate of heat loss from the top surface of the pond by (*a*) forced convection, (*b*) radiation, and (*c*) evaporation. Take the average temperature of the surrounding surfaces to be 60°F.

11-134E Repeat Problem 11-133E for a solar pond surface temperature of 90°F.
Answers: (*a*) 299,400 Btu/h, (*b*) 1,066,000 Btu/h, (*c*) 3,396,000 Btu/h

Computer, Design, and Essay Problems

11-135 Write an essay on diffusion caused by effects other than the concentration gradient such as thermal diffusion, pressure diffusion, forced diffusion, knodsen diffusion, and surface diffusion.

11-136 Write a computer program that will convert the mole fractions of a gas mixture to mass fractions when the molar masses of the components of the mixture are specified.

11-137 One way of generating electricity from solar energy involves the collection and storage of solar energy in large artificial lakes of a few meters deep, called solar ponds. Solar energy is stored at the bottom part of the pond at temperatures close to boiling, and the rise of hot water to the top is prevented by planting salt to the bottom of the pond. Write an essay on the operation of solar pond power plants, and find out how much salt is used per

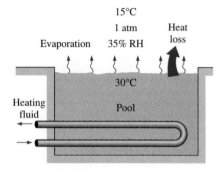

15°C
1 atm
Evaporation   35% RH   Heat loss

30°C

Heating fluid   Pool

FIGURE P11-139

year per m². If the cost is not a factor, can sugar be used instead of salt to maintain the concentration gradient? Explain.

11-138   The condensation and even freezing of moisture in building walls without effective vapor retarders is a real concern in cold climates as it undermines the effectiveness of the insulation. Investigate how the builders in your area are coping with this problem, whether they are using vapor retarders or vapor barriers in the walls, and where they are located in the walls. Prepare a report on your findings and explain the reasoning for the current practice.

11-139   You are asked to design a heating system for a swimming pool that is 2-m deep, 25-m long, and 25-m wide. Your client desires that the heating system be large enough to raise the water temperature from 20°C to 30°C in 3 h. The heater must also be able to maintain the pool at 30°C at the outdoor design conditions of 15°C, 1 atm, 35 percent relative humidity, 40 mph winds, and effective sky temperature of 10°C. Heat losses to the ground are expected to be small and can be disregarded. The heater considered is a natural gas furnace whose efficiency is 80 percent. What heater size (in Btu/h input) would you recommend that your client buy?

Applications

PART

2

Heating and Cooling of Buildings

The houses in the past were built to keep the rain, snow, and thieves out with hardly any attention given to heat losses and *energy conservation*. Houses had little or no insulation, and the structures had numerous cracks through which air leaked. We have seen dramatic changes in the construction of residential and commercial buildings in the 20th century as a result of increased awareness of limited energy resources together with the escalating energy prices and the demand for a higher level of thermal comfort. Today, most local codes specify the minimum level of insulation to be used in the walls and the roof of new houses, and often require the use of double-pane windows. As a result, today's houses are well insulated, weatherproofed, and nearly air tight, and provide better thermal comfort.

The failures and successes of the past often shed light to the future, and thus we start this chapter with a *brief history* of heating and cooling to put things into historical perspective. Then we discuss the criteria for *thermal comfort,* which is the primary reason for installing heating and cooling systems. In the remainder of the chapter, we present calculation procedures for the *heating and cooling loads* of buildings using the most recent information and design criteria established by the American Society of Heating, Refrigerating and Air-Conditioning Engineers, Inc. (ASHRAE), which publishes and periodically revises the most authoritative handbooks in the field. This chapter is intended to introduce the readers to an exciting application area of heat transfer, and to help them develop a deeper understanding of the fundamentals of heat transfer using this familiar setup. The reader is referred to ASHRAE handbooks for more information.

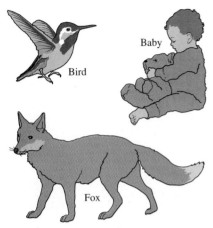

FIGURE 12-1

Most animals come into this world with
built-in insulation, but human beings
come with a delicate skin.

FIGURE 12-2

In 1775, ice was made by evacuating
the air in a water tank.

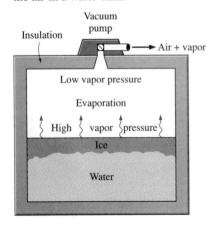

12-1 ■ A BRIEF HISTORY

Unlike animals such as a fox or a bear that are born with built-in furs, human beings come into this world with little protection against the harsh environmental conditions (Fig. 12-1). Therefore, we can claim that the search for thermal comfort dates back to the beginning of human history. It is believed that early human beings lived in caves that provided shelter as well as protection from extreme thermal conditions. Probably the first form of heating system used was *open fire,* followed by fire in dwellings through the use of a *chimney* to vent out the combustion gases. The concept of *central heating* dates back to the times of the Romans, who heated homes by utilizing double-floor construction techniques and passing the fire's fumes through the opening between the two floor layers. The Romans were also the first to use *transparent windows* made of mica or glass to keep the wind and rain out while letting the light in. Wood and coal were the primary energy sources for heating, and oil and candles were used for lighting. The ruins of south-facing houses indicate that the value of *solar heating* was recognized early in the history.

The development of the first *steam heating* system by James Watt dates back to 1770. When the American Society of Heating and Ventilating Engineers was established in New York in 1894, central heating systems using cast iron warm air furnaces and boilers were in common use. *Fans* were added in 1899 to move the air mechanically, and later *automatic firing* replaced the manual firing. The steam heating systems gained widespread acceptance in the early 1900s by the introduction of fluid-operated *thermostatic traps* to improve the fluid circulation. Gravity-driven hot water heating systems were developed in parallel with steam systems. Suspended and floor-type unit heaters, unit ventilators, and panel heaters were developed in the 1920s. *Unit heaters* and *panel heaters* usually used steam, hot water, or electricity as the heat source. It became common practice to conceal the radiators in the 1930s, and the *baseboard radiator* was developed in 1944. Today, air heating systems with a duct distribution network dominate the residential and commercial buildings.

The development of *cooling systems* took the back seat in the history of thermal comfort since there was no quick way of creating "coolness." Therefore, early attempts at cooling were passive measures such as blocking off direct sunlight and using thick stone walls to store coolness at night. A more sophisticated approach was to take advantage of *evaporative cooling* by running water through the structure, as done in the Alhambra castle. Of course, natural *ice* and *snow* served as "cold storage" mediums and provided some cooling.

In 1775, Dr. William Cullen made *ice* in Scotland by evacuating the air in a water tank (Fig. 12-2). It was also known at those times that some chemicals lowered temperatures. For example, the temperature of snow can be dropped to $-33°C$ ($-27°F$) by mixing it with calcium chloride. This process was commonly used to make ice cream. In 1851, Ferdinand Carre designed the first *ammonia absorption refrigeration system,* while Dr. John Gorrie received a patent for an *open air refrigeration cycle* to produce ice and refrigerated air. In 1853, Alexander Twining of Connecticut produced 1600 pounds (726 kg) of ice a day using sulfuric ether as the refrigerant. In 1872, David Boyle developed an ammonia compression machine that produced ice. Mechanical refrigeration at those times was used primarily to make ice and preserve perishable commodities such as meat and fish (Sauer and Howell, Ref. 7).

Comfort cooling was obtained by ice or by chillers that used ice. *Air cooling systems* for thermal comfort were built in the 1890s, but they did not find widespread use until the development of mechanical refrigeration in the early 1900s. In 1905, 200 Btu/min (or 12,000 Btu/h) was established as **1 ton of refrigeration,** and in 1902 a 400-ton air-conditioning system was installed in the New York Stock Exchange. The system operated reliably for 20 years. A modern air-conditioning system was installed in the Boston Floating Hospital in 1908, which was a first for a hospital. In a monumental paper presented in 1911, Willis Carrier (1876–1950), known as the "Father of Air Conditioning," laid out the formulas related to the dry-bulb, wet-bulb, and dew-point temperatures of air and the sensible, latent, and total heat loads. By 1922, the centrifugal refrigeration machine developed by Carrier made water chilling for medium and large commercial and industrial facilities practical and economical. In 1928 the Milan Building in San Antonio, Texas, was the first commercial building designed with and built for comfort air-conditioning specifications (Sauer and Howell, Ref. 7).

Frigidaire introduced the first *room air conditioner* in the late 1920s (Fig. 12-3). The halocarbon refrigerants such as Freon-12 were developed in 1930. The concept of a *heat pump* was described by Sadi Carnot in 1824, and the operation of such a device called the "heat multiplier" was first described by William Thomson (Lord Kelvin) in 1852. T. G. N. Haldane built an experimental heat pump in 1930, and a heat pump was marketed by De La Vargne in 1933. General Electric introduced the heat pump in the mid 1930s, and heat pumps were being mass produced in 1952. *Central air-conditioning systems* were being installed routinely in the 1960s. The oil crises of the 1970s sent shock waves among the consumers and the producers of energy-consuming equipment, which had taken energy for granted, and brought about a renewed interest in the development of energy-efficient systems and more effective insulation materials. Today most residential and commercial buildings are equipped with modern air-conditioning systems that can heat, cool, humidify, dehumidify, clean, and even deodorize the air—in other words, *condition* the air to people's desires.

12-2 ■ HUMAN BODY AND THERMAL COMFORT

The term **air-conditioning** is usually used in a restricted sense to imply cooling, but in its broad sense it means *to condition* the air to the desired level by heating, cooling, humidifying, dehumidifying, cleaning, and deodorizing. The purpose of the air-conditioning system of a building is to provide *complete thermal comfort* for its occupants. Therefore, we need to understand the thermal aspects of the *human body* in order to design an effective air-conditioning system.

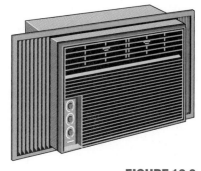

FIGURE 12-3

The first room air conditioner was introduced by Frigidaire in the late 1920s.

FIGURE 12-4

Two fast-dancing people supply more heat to a room than a 1-kW resistance heater.

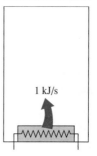

1.2 kJ/s

1 kJ/s

TABLE 12-1

Metabolic rates during various activities (from ASHRAE Handbook of Fundamentals, Ref. 1, Chap. 8, Table 4).

Activity	Metabolic rate* W/m^2
Resting:	
Sleeping	40
Reclining	45
Seated, quiet	60
Standing, relaxed	70
Walking (on the level):	
2 mph (0.89 m/s)	115
3 mph (1.34 m/s)	150
4 mph (1.79 m/s)	220
Office Activities:	
Reading, seated	55
Writing	60
Typing	65
Filing, seated	70
Filing, standing	80
Walking about	100
Lifting/packing	120
Driving/Flying:	
Car	60–115
Aircraft, routine	70
Heavy vehicle	185
Miscellaneous Occupational Activities:	
Cooking	95–115
Cleaning house	115–140
Machine work:	
Light	115–140
Heavy	235
Handling 50-kg bags	235
Pick and shovel work	235–280
Miscellaneous Leisure Activities:	
Dancing, social	140–255
Calisthenics/exercise	175–235
Tennis, singles	210–270
Basketball	290–440
Wrestling, competitive	410–505

*Multiply by 1.8 m^2 to obtain metabolic rates for an average man. Multiply by 0.3171 to convert to Btu/h · ft^2.

The building blocks of living organisms are *cells,* which resemble miniature factories performing various functions necessary for the survival of organisms. The human body contains about 100 trillion cells with an average diameter of 0.01 mm. In a typical cell, thousands of chemical reactions occur every second during which some molecules are broken down and energy is released and some new molecules are formed. The high level of chemical activity in the cells that maintain the human body temperature at a temperature of 37.0°C (98.6°F) while performing the necessary bodily functions is called the **metabolism.** In simple terms, metabolism refers to the burning of foods such as carbohydrates, fat, and protein. The metabolizable energy content of foods is usually expressed by nutritionists in terms of the capitalized Calorie. One Calorie is equivalent to 1 Cal = 1 kcal = 4.1868 kJ.

The rate of metabolism at the resting state is called the *basal metabolic rate,* which is the rate of metabolism required to keep a body performing the necessary bodily functions such as breathing and blood circulation at zero external activity level. The metabolic rate can also be interpreted as the energy consumption rate for a body. For an *average man* (30 years old, 70 kg, 1.73 m high, 1.8 m^2 surface area), the basal metabolic rate is 84 W. That is, the body is converting chemical energy of the food (or of the body fat if the person had not eaten) into heat at a rate of 84 J/s, which is then dissipated to the surroundings. The metabolic rate increases with the *level of activity,* and it may exceed 10 times the basal metabolic rate when someone is doing strenuous exercise. That is, two people doing heavy exercising in a room may be supplying more energy to the room than a 1-kW resistance heater (Fig. 12-4). An average man generates heat at a rate of 108 W while reading, writing, typing, or listening to a lecture in a classroom in a seated position. The maximum metabolic rate of an average man is 1250 W at age 20 and 730 at age 70. The corresponding rates for women are about 30 percent lower. Maximum metabolic rates of trained athletes can exceed 2000 W.

Metabolic rates during various activities are given in Table 12-1 per unit body surface area. The **surface area** of a nude body was given by D. DuBois in 1916 as

$$A = 0.202 m^{0.425} h^{0.725} \quad (m^2) \qquad (12\text{-}1)$$

where m is the mass of the body in kg and h is the height in m. *Clothing* increases the exposed surface area of a person by up to about 50 percent. The metabolic rates given in the table are sufficiently accurate for most purposes, but there is considerable uncertainty at high activity levels. More accurate values can be determined by measuring the rate of respiratory *oxygen consumption,* which ranges from about 0.25 L/min for an average resting man to more than 2 L/min during extremely heavy work. The entire energy released during metabolism can be assumed to be released as *heat* (in sensible or latent forms) since the external mechanical work done by the muscles is very small. Besides, the work done during most activities such as walking or riding an exercise bicycle is eventually converted to heat through friction.

The comfort of the human body depends primarily on three environmental factors: the temperature, relative humidity, and air motion. The temperature of the environment is the single most important index of comfort.

Extensive research is done on human subjects to determine the "**thermal comfort zone**" and to identify the conditions under which the body feels comfortable in an environment. It has been observed that most normally clothed people resting or doing light work feel comfortable in the *operative temperature* (roughly, the average temperature of air and surrounding surfaces) range of 23 to 27°C or 73 to 80°F (Fig. 12-5). For unclothed people, this range is 29 to 31°C. Relative humidity also has a considerable effect on comfort since it is a measure of air's ability to absorb moisture and thus it affects the amount of heat a body can dissipate by evaporation. High relative humidity slows down heat rejection by evaporation, especially at high temperatures, and low relative humidity speeds it up. The desirable level of *relative humidity* is the broad range of 30 to 70 percent, with 50 percent being the most desirable level. Most people at these conditions feel neither hot nor cold, and the body does not need to activate any of the defense mechanisms to maintain the normal body temperature (Fig. 12-6).

Another factor that has a major effect on thermal comfort is **excessive air motion** or **draft,** which causes undesired local cooling of the human body. Draft is identified by many as a most annoying factor in work places, automobiles, and airplanes. Experiencing discomfort by draft is most common among people wearing indoor clothing and doing light sedentary work, and least common among people with high activity levels. The air velocity should be kept below 9 m/min (30 ft/min) in winter and 15 m/min (50 ft/min) in summer to minimize discomfort by draft, especially when the air is cool. A low level of air motion is desirable as it removes the warm, moist air that builds around the body and replaces it with fresh air. Therefore, air motion should be strong enough to remove heat and moisture from the vicinity of the body, but gentle enough to be unnoticed. High speed air motion causes discomfort outdoors as well. For example, an environment at 10°C (50°F) with 48 km/h winds feels as cold as an environment at −7°C (20°F) with 3 km/h winds because of the chilling effect of the air motion (the wind-chill factor).

A comfort system should provide *uniform conditions* throughout the living space to avoid discomfort caused by nonuniformities such as *drafts, asymmetric thermal radiation, hot or cold floors,* and *vertical temperature stratification.* **Asymmetric thermal radiation** is caused by the *cold surfaces* of large windows, uninsulated walls, or cold products and the *warm surfaces* of gas or electric radiant heating panels on the walls or ceiling, solar-heated masonry walls or ceilings, and warm machinery. Asymmetric radiation causes discomfort by exposing different sides of the body to surfaces at different temperatures and thus to different heat loss or gain by radiation. A person whose left side is exposed to a cold window, for example, will feel like heat is being drained from that side of his or her body (Fig. 12-7). For thermal comfort, the radiant temperature asymmetry should not exceed 5°C in the vertical direction and 10°C in the horizontal direction. The unpleasant effect of radiation asymmetry can be minimized by properly sizing and installing heating panels, using double-pane windows, and providing generous insulation at the walls and the roof.

Direct contact with **cold** or **hot floor surfaces** also causes localized discomfort in the feet. The temperature of the floor depends on the way it is

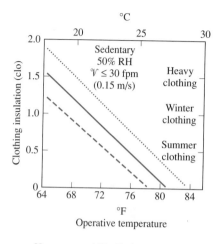

.... Upper acceptability limit
——— Optimum
- - - Lower acceptability limit

FIGURE 12-5

The effect of clothing on the environment temperature that feels comfortable (1 clo = 0.155 m² · °C/W = 0.880 ft² · °F · h/Btu) (from ASHRAE Standard 55-1981).

FIGURE 12-6

A thermally comfortable environment.

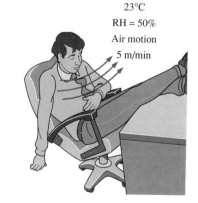

23°C
RH = 50%
Air motion
5 m/min

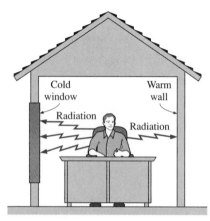

Cold window

Warm wall

Radiation

Radiation

FIGURE 12-7

Cold surfaces cause excessive heat loss
from the body by radiation, and thus
discomfort on that side of the body.

constructed (being directly on the ground or on top of a heated room, being made of wood or concrete, the using insulation, etc.) as well as the *floor covering used* such as pads, carpets, rugs, and linoleum. A floor temperature of 23 to 25°C is found to be comfortable to most people. The floor asymmetry loses its significance for people with footwear. An effective and economical way of raising the floor temperature is to use radiant heating panels instead of turning the thermostat up. Another nonuniform condition that causes discomfort is **temperature stratification** in a room that exposes the head and the feet to different temperatures. For thermal comfort, the temperature difference between the head and foot levels should not exceed 3°C. This effect can be minimized by using destratification fans.

It should be noted that no thermal environment will please everyone. No matter what we do, some people will express some discomfort. The thermal comfort zone is based on a 90 percent acceptance rate. That is, an environment is deemed comfortable if only 10 percent of the people are dissatisfied with it. Metabolism decreases somewhat with *age*, but it has no effect on the comfort zone. Research indicates that there is no appreciable difference between the environments preferred by old and young people. Experiments also show that *men* and *women* prefer almost the same environment. The metabolism rate of women is somewhat lower, but this is compensated by their slightly lower skin temperature and evaporative loss. Also, there is no significant variation in the comfort zone from one part of the world to another and from winter to summer. Therefore, the same thermal comfort conditions can be used *throughout the world* in any season. Also, people cannot *acclimatize* themselves to prefer different comfort conditions.

In a **cold environment,** the rate of heat loss from the body may exceed the rate of metabolic heat generation. Average specific heat of the human body is 3.49 kJ/kg · °C, and thus each 1°C drop in body temperature corresponds to a deficit of 244 kJ in body heat content for an average 70 kg man. A drop of 0.5°C in mean body temperature causes noticeable but acceptable discomfort. A drop of 2.6°C causes extreme discomfort. A sleeping person will wake up when his or her mean body temperature drops by 1.3°C (which normally shows up as a 0.5°C drop in the deep body and 3°C in the skin area). The drop of deep body temperature below 35°C may damage the body temperature regulation mechanism, while a drop below 28°C may be fatal. Sedentary people reported to feel *comfortable* at a *mean skin temperature* of 33.3°C, uncomfortably cold at 31°C, *shivering cold* at 30°C, and *extremely cold* at 29°C. People doing heavy work reported to feel comfortable at much lower temperatures, which shows that the activity level affects human performance and comfort. The extremities of the body such as hands and feet are most easily affected by cold weather, and their temperature is a better indication of comfort and performance. A hand-skin temperature of 20°C is perceived to be uncomfortably cold, 15°C to be extremely cold, and 5°C to be painfully cold. Useful work can be performed by hands without difficulty as long as the skin temperature of fingers remains above 16°C (ASHRAE *Handbook of Fundamentals*, Ref. 1, Chap. 8).

The first line of defense of the body against excessive heat loss in a cold environment is *to reduce the skin temperature* and thus the rate of heat loss from the skin by constricting the veins and decreasing the blood flow to the skin. This measure decreases the temperature of the tissues subjacent to the skin, but maintains the inner body temperature. The next preventive measure is increasing the rate of *metabolic heat generation* in the body by *shivering*, unless the person does it voluntarily by increasing his or her level of activity or puts on additional clothing. Shivering begins slowly in small muscle groups and may double the rate of metabolic heat production of the body at its initial stages. In the extreme case of total body shivering, the rate of heat production may reach 6 times the resting levels (Fig. 12-8). If this measure also proves inadequate, the deep body temperature starts *falling*. Body parts furthest away from the core such as the hands and feet are at greatest danger for tissue damage.

In **hot environments**, the rate of heat loss from the body may drop below the metabolic heat generation rate. This time the body activates the opposite mechanisms. First the body increases the *blood flow* and thus heat transport to the skin, causing the temperature of the skin and the subjacent tissues to rise and approach the deep body temperature. Under extreme heat conditions, the *heart rate* may reach 180 beats per minute in order to maintain adequate blood supply to the brain and the skin. At higher heart rates, the *volumetric efficiency* of the heart drops because of the very short time between the beats to fill the heart with blood, and the blood supply to the skin and more importantly to the brain drops. This causes the person to faint as a result of *heat exhaustion*. Dehydration makes the problem worse. A similar thing happens when a person working very hard for a long time stops suddenly. The blood that has flooded the skin has difficulty returning to the heart in this case since the relaxed muscles no longer force the blood back to the heart, and thus there is less blood available for pumping to the brain.

The next line of defense is releasing water from sweat glands and resorting to *evaporative cooling*, unless the person removes some clothing and reduces the activity level (Fig. 12-9). The body can maintain its core temperature at 37°C in this evaporative cooling mode indefinitely, even in environments at higher temperatures (as high as 200°C during military endurance tests), if the person drinks plenty of liquids to replenish his or her water reserves and the ambient air is sufficiently dry to allow the sweat to evaporate instead of rolling down the skin. If this measure proves inadequate, the body will have to start absorbing the metabolic heat and the deep body temperature will rise. A person can tolerate a temperature rise of 1.4°C without major discomfort but may *collapse* when the temperature rise reaches 2.8°C. People feel sluggish and their efficiency drops considerably when the core body temperature rises above 39°C. A core temperature above 41°C may damage hypothalamic proteins, resulting in cessation of sweating, increased heat production by shivering, and a *heat stroke* with an irreversible and life-threatening damage. Death can occur above 43°C.

A surface temperature of 46°C causes pain on the skin. Therefore, direct contact with a metal block at this temperature or above is painful. However, a person can stay in a room at 100°C for up to 30 min without any damage or pain on the skin because of the convective resistance at the skin surface and

FIGURE 12-8

The rate of metabolic heat generation may go up by 6 times the resting level during total body shivering in cold weather.

FIGURE 12-9

In hot environments, a body can dissipate a large amount of metabolic heat by sweating since the sweat absorbs the body heat and evaporates.

TABLE 12-2

Minimum fresh air requirements in buildings (from ASHRAE Standard 62-1989)

Appli- cation	Requirement	
	L/s per person	ft³/min per person
Classrooms, laundries, libraries, supermarkets	8	15
Dining rooms, conference rooms, offices	10	20
Hospital rooms	13	25
Hotel rooms	15 (per room)	30 (per room)
Smoking lounges	30	60
Retail stores	1.0–1.5 (per m²)	0.2–0.3 (per ft²)
Residential buildings	0.35 air change per hour, but not less than 7.5 L/s (or 15 ft³/min) per person	

evaporative cooling. We can even put our hands into at oven at 200°C for a short time without getting burned.

Another factor that affects thermal comfort, health, and productivity is **ventilation.** Fresh outdoor air can be provided to a building *naturally* by doing nothing, or *forcefully* by a mechanical ventilation system. In the first case, which is the norm in residential buildings, the necessary ventilation is provided by *infiltration through cracks and leaks* in the living space and by the opening of the windows and doors. The additional ventilation needed in the bathrooms and kitchens is provided by *air vents with dampers* or *exhaust fans.* With this kind of uncontrolled ventilation, however, the fresh air supply will be either too high, wasting energy, or too low, causing poor indoor air quality. But the current practice is not likely to change for residential buildings since there is not a public outcry for energy waste or air quality, and thus it is difficult to justify the cost and complexity of mechanical ventilation systems.

Mechanical ventilation systems are part of any heating and air conditioning system in *commercial buildings*, providing the necessary amount of fresh outdoor air and distributing it uniformly throughout the building. This is not surprising since many rooms in large commercial buildings have no windows and thus rely on mechanical ventilation. Even the rooms with windows are in the same situation since the windows are tightly sealed and cannot be opened in most buildings. It is not a good idea to oversize the ventilation system just to be on the "safe side" since exhausting the heated or cooled indoor air wastes energy. On the other hand, reducing the ventilation rates below the required minimum to conserve energy should also be avoided so that the indoor air quality can be maintained at the required levels. The minimum fresh air ventilation requirements are listed in Table 12-2. The values are based on controlling the CO_2 and other contaminants with an adequate margin of safety, which requires each person be supplied with at least 7.5 L/s (15 ft³/min) of fresh air.

Another function of the mechanical ventilation system is to **clean** the air by filtering it as it enters the building. Various types of filters are available for this purpose, depending on the cleanliness requirements and the allowable pressure drop.

12-3 ■ HEAT TRANSFER FROM THE HUMAN BODY

The metabolic heat generated in the body is dissipated to the environment through the skin and the lungs by convection and radiation as *sensible heat* and by evaporation as *latent heat* (Fig. 12-10). Latent heat represents the heat of vaporization of water as it evaporates in the lungs and on the skin by absorbing body heat, and latent heat is released as the moisture condenses on cold surfaces. The warming of the inhaled air represents sensible heat transfer in the lungs and is proportional to the temperature rise of inhaled air. The total rate of heat loss from the body can be expressed as

$$\dot{Q}_{body,\,total} = \dot{Q}_{skin} + \dot{Q}_{lungs}$$

$$= (\dot{Q}_{sensible} + \dot{Q}_{latent})_{skin} + (\dot{Q}_{sensible} + \dot{Q}_{latent})_{lungs} \qquad (12\text{-}2)$$

$$= (\dot{Q}_{convection} + \dot{Q}_{radiation} + \dot{Q}_{latent})_{skin} + (\dot{Q}_{convection} + \dot{Q}_{latent})_{lungs}$$

Therefore, the determination of heat transfer from the body by analysis alone is difficult. Clothing further complicates the heat transfer from the body, and thus we must rely on experimental data. Under steady conditions, the total rate of heat transfer from the body is equal to the rate of metabolic heat generation in the body, which varies from about 100 W for light office work to roughly 1000 W during heavy physical work.

Sensible heat loss from the skin depends on the temperatures of the skin, the environment, and the surrounding surfaces as well as the air motion. The *latent heat loss,* on the other hand, depends on the skin wettedness and the relative humidity of the environment as well. *Clothing* serves as insulation and reduces both the sensible and latent forms of heat loss. The heat transfer from the lungs through respiration obviously depends on the frequency of breathing and the volume of the lungs as well as the environmental factors that affect heat transfer from the skin.

Sensible heat from the clothed skin is first transferred to the clothing and then from the clothing to the environment. The convection and radiation heat losses from the outer surface of a clothed body can be expressed as

$$\dot{Q}_{conv} = h_{conv} A_{clothing} (T_{clothing} - T_{ambient}) \qquad \text{(W)} \qquad (12\text{-}3)$$

$$\dot{Q}_{rad} = h_{rad} A_{clothing} (T_{clothing} - T_{surr}) \qquad\qquad\qquad\qquad (12\text{-}4)$$

where

h_{conv} = convection heat transfer coefficient, as given in Table 12-3
h_{rad} = radiation heat transfer coefficient, 4.7 W/m² · °C for typical indoor conditions; the emissivity is assumed to be 0.95, which is typical
$A_{clothing}$ = outer surface area of a clothed person
$T_{clothing}$ = average temperature of exposed skin and clothing
$T_{ambient}$ = ambient air temperature
T_{surr} = average temperature of the surrounding surfaces

The convection heat transfer coefficients at 1 atm pressure are given in Table 12-3. Convection coefficients at pressures P other than 1 atm are obtained by multiplying the values at atmospheric pressure by $P^{0.55}$ where P is in atm. Also, it is recognized that the temperatures of different surfaces surrounding a person are probably different, and T_{surr} represents the **mean radiation temperature,** which is the temperature of an imaginary isothermal enclosure in which radiation heat exchange with the human body equals the radiation heat exchange with the actual enclosure. Noting that most clothing and building materials are essentially black, the *mean radiation temperature* of an enclosure that consists of N surfaces at different temperatures can be

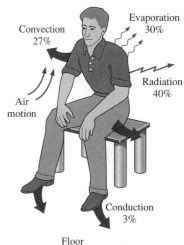

FIGURE 12-10

Mechanisms of heat loss from the human body and relative magnitudes for a resting person.

TABLE 12-3

Convection heat transfer coefficients for a clothed body at 1 atm ($\mathcal{V}$ is in m/s) (compiled from various sources)

Activity	h_{conv},* W/m² · °C
Seated in air moving at	
$0 < \mathcal{V} < 0.2$ m/s	3.1
$0.2 < \mathcal{V} < 4$ m/s	$8.3\mathcal{V}^{0.6}$
Walking in still air at	
$0.5 < \mathcal{V} < 2$ m/s	$8.6\mathcal{V}^{0.53}$
Walking on treadmill in still air at $0.5 < \mathcal{V} < 2$ m/s	$6.5\mathcal{V}^{0.39}$
Standing in moving air at	
$0 < \mathcal{V} < 0.15$ m/s	4.0
$0.15 < \mathcal{V} < 1.5$ m/s	$14.8\mathcal{V}^{0.69}$

*At pressures other than 1 atm, multiply by $P^{0.55}$, where P is in atm.

determined from

$$T_{surr} \cong F_{person\text{-}1} T_1 + F_{person\text{-}2} T_2 + \cdots + F_{person\text{-}N} T_N \qquad (12\text{-}5)$$

where T_i is the *temperature of the surface i* and $F_{person\text{-}i}$ is the *view factor* between the person and surface i.

Total sensible heat loss can also be expressed conveniently by combining the convection and radiation heat losses as

$$\dot{Q}_{conv+rad} = h_{combined} A_{clothing} (T_{clothing} - T_{operative}) \qquad (12\text{-}6)$$
$$= (h_{conv} + h_{rad}) A_{clothing} (T_{clothing} - T_{operative}) \qquad (W) \qquad (12\text{-}7)$$

where the **operative temperature** $T_{operative}$ is the average of the mean radiant and ambient temperatures weighed by their respective convection and radiation heat transfer coefficients and is expressed as (Fig. 12-11)

$$T_{operative} = \frac{h_{conv} T_{ambient} + h_{rad} T_{surr}}{h_{conv} + h_{rad}} \cong \frac{T_{ambient} + T_{surr}}{2} \qquad (12\text{-}8)$$

Note that the operative temperature will be the arithmetic average of the ambient and surrounding surface temperatures when the convection and radiation heat transfer coefficients are equal to each other. Another environmental index used in thermal comfort analysis is the **effective temperature,** which combines the effects of temperature and humidity. Two environments with the same effective temperature will evoke the same thermal response in people even though they are at different temperatures and humidities.

Heat transfer through the *clothing* can be expressed as

$$\dot{Q}_{conv+rad} = \frac{A_{clothing} (T_{skin} - T_{clothing})}{R_{clothing}} \qquad (12\text{-}9)$$

where $R_{clothing}$ is the **unit thermal resistance of clothing** in $m^2 \cdot °C/W$, which involves the combined effects of conduction, convection, and radiation between the skin and the outer surface of clothing. The thermal resistance of clothing is usually expressed in the unit **clo** where 1 clo = 0.155 $m^2 \cdot °C/W$ = 0.880 $ft^2 \cdot °F \cdot h/Btu$. The thermal resistance of trousers, long-sleeve shirt, long-sleeve sweater, and T-shirt is 1.0 clo, or 0.155 $m^2 \cdot °C/W$. Summer clothing such as light slacks and short-sleeved shirt has an insulation value of 0.5 clo, whereas winter clothing such as heavy slacks, long-sleeve shirt, and a sweater or jacket has an insulation value of 0.9 clo.

Then the total sensible heat loss can be expressed in terms of the skin temperature instead of the inconvenient clothing temperature as (Fig. 12-12)

$$\dot{Q}_{conv+rad} = \frac{A_{clothing} (T_{skin} - T_{operative})}{R_{clothing} + \dfrac{1}{h_{combined}}} \qquad (12\text{-}10)$$

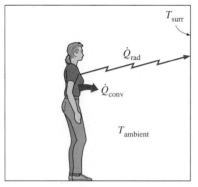

T_{surr}

$\dot{Q}_{rad}$

$\dot{Q}_{conv}$

$T_{ambient}$

(*a*) Convection and radiation, separate

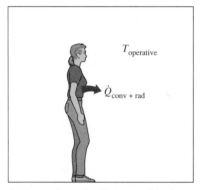

$T_{operative}$

$\dot{Q}_{conv+rad}$

(*b*) Convection and radiation, combined

FIGURE 12-11

Heat loss by convection and radiation from the body can be combined into a single term by defining an equivalent operative temperature.

At a state of thermal comfort, the average skin temperature of the body is observed to be 33°C (91.5°F). No discomfort is experienced as the skin temperature fluctuates by ±1.5°C (2.5°F). This is the case whether the body is clothed or unclothed.

Evaporative or **latent heat loss** from the skin is proportional to the difference between the water vapor pressure at the skin and the ambient air, and the skin wettedness, which is a measure of the amount of moisture on the skin. It is due to the combined effects of the *evaporation of sweat* and the *diffusion* of water through the skin and can be expressed as

$$\dot{Q}_{latent} = \dot{m}_{vapor}h_{fg} \tag{12-11}$$

where

$\dot{m}_{vapor}$ = the rate of evaporation from the body, kg/s

h_{fg} = the enthalpy of vaporization of water = 2430 kJ/kg at 30°C

Heat loss by evaporation is maximum when the skin is completely wetted. Also, clothing offers resistance to evaporation, and the rate of evaporation in clothed bodies depends on the moisture permeability of the clothes. The maximum evaporation rate for an average man is about 1 L/h (0.3 g/s), which represents an upper limit of 730 W for the evaporative cooling rate. A person can lose as much as 2 kg of water per hour during a workout on a hot day, but any excess sweat slides off the skin surface without evaporating (Fig. 12-13).

During *respiration*, the inhaled air enters at ambient conditions and exhaled air leaves nearly saturated at a temperature close to the deep body temperature (Fig. 12-14). Therefore, the body loses both sensible heat by convection and latent heat by evaporation from the lungs, and these can be expressed as

$$\dot{Q}_{conv,\,lungs} = \dot{m}_{air,\,lungs}\,C_{p,air}(T_{exhale} - T_{ambient}) \tag{12-12}$$

$$\dot{Q}_{latent,\,lungs} = \dot{m}_{vapor,\,lungs}\,h_{fg} = \dot{m}_{air,\,lungs}\,(w_{exhale} - w_{ambient})h_{fg} \tag{12-13}$$

where

$\dot{m}_{air,\,lungs}$ = rate of air intake to the lungs, kg/s

$C_{p,\,air}$ = specific heat of air = 1.0 kJ/kg · °C

T_{exhale} = temperature of exhaled air

w = humidity ratio (the mass of moisture per unit mass of dry air)

The rate of air intake to the lungs is directly proportional to the metabolic rate $\dot{Q}_{met}$. The rate of total heat loss from the lungs through respiration can be expressed approximately as

$$\dot{Q}_{conv\,+\,latent,\,lungs} = 0.0014\dot{Q}_{met}\,(34 - T_{ambient}) + 0.0173\dot{Q}_{met}\,(5.87 - P_{v,\,ambient}) \tag{12-14}$$

where $P_{v,\,ambient}$ is the vapor pressure of ambient air in kPa.

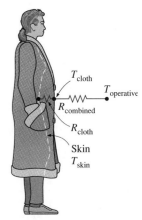

FIGURE 12-12

Simplified thermal resistance network for heat transfer from a clothed person.

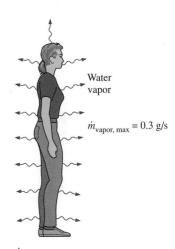

Water vapor

$\dot{m}_{vapor,\,max} = 0.3$ g/s

$\dot{Q}_{latent,\,max} = \dot{m}_{latent,\,max}\,h_{fg}$ at 30°C

$= (0.3$ g/s$)(2430$ kJ/kg$)$

$= 730$ W

FIGURE 12-13

An average person can lose heat at a rate of up to 730 W by evaporation.

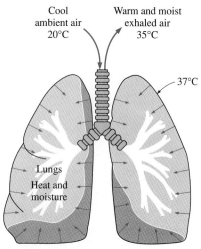

FIGURE 12-14
Part of the metabolic heat generated in the body is rejected to the air from the lungs during respiration.

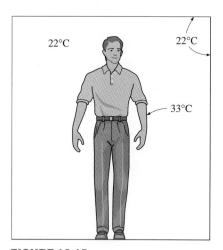

FIGURE 12-15
Schematic for Example 12-1.

The fraction of sensible heat varies from about 40 percent in the case of heavy work to about 70 percent during light work. The rest of the energy is rejected from the body by perspiration in the form of latent heat.

EXAMPLE 12-1 Effect of Clothing on Thermal Comfort

It is well established that a clothed or unclothed person feels comfortable when the skin temperature is about 33°C. Consider an average man wearing summer clothes whose thermal resistance is 0.6 clo. The man feels very comfortable while standing in a room maintained at 22°C. The air motion in the room is negligible, and the interior surface temperature of the room is about the same as the air temperature. If this man were to stand in that room unclothed, determine the temperature at which the room must be maintained for him to feel thermally comfortable.

Solution A man wearing summer clothes feels comfortable in a room at 22°C. The room temperature at which this man would feel thermally comfortable when unclothed is to be determined.

Assumptions **1** Steady conditions exist. **2** The latent heat loss from the person remains the same. **3** The heat transfer coefficients remain the same.

Analysis The body loses heat in sensible and latent forms, and the sensible heat consists of convection and radiation heat transfer. At low air velocities, the convection heat transfer coefficient for a standing man is given in Table 12-3 to be 4.0 W/m² · °C. The radiation heat transfer coefficient at typical indoor conditions is 4.7 W/m² · °C. Therefore, the surface heat transfer coefficient for a standing person for combined convection and radiation is

$$h_{combined} = h_{conv} + h_{rad} = 4.0 + 4.7 = 8.7 \text{ W/m}^2 \cdot °C$$

The thermal resistance of the clothing is given to be

$$R_{clothing} = 0.6 \text{ clo} = 0.6 \times 0.155 \text{ m}^2 \cdot °C/W = 0.093 \text{ m}^2 \cdot °C/W$$

Noting that the surface area of an average man is 1.8 m², the sensible heat loss from this person when clothed is determined to be (Fig. 12-15)

$$\dot{Q}_{sensible, clothed} = \frac{A(T_{skin} - T_{ambient})}{R_{clothing} + \dfrac{1}{h_{combined}}} = \frac{(1.8 \text{ m}^2)(33 - 22)°C}{0.093 \text{ m}^2 \cdot °C/W + \dfrac{1}{8.7 \text{ W/m}^2 \cdot °C}} = 95.2 \text{ W}$$

From a heat transfer point of view, taking the clothes off is equivalent to removing the clothing insulation or setting $R_{cloth} = 0$. The heat transfer in this case can be expressed as

$$\dot{Q}_{sensible, unclothed} = \frac{A(T_{skin} - T_{ambient})}{\dfrac{1}{h_{combined}}} = \frac{(1.8 \text{ m}^2)(33 - T_{ambient})°C}{\dfrac{1}{8.7 \text{ W/m}^2 \cdot °C}}$$

To maintain thermal comfort after taking the clothes off, the skin temperature of the person and the rate of heat transfer from him must remain the same. Then setting the equation above equal to 95.2 W gives

$$T_{ambient} = \textbf{26.9°C}$$

Therefore, the air temperature needs to be raised from 22 to 26.9°C to ensure that the person will feel comfortable in the room after he takes his clothes off

(Fig. 12-16). Note that the effect of clothing on latent heat is assumed to be negligible in the solution above. We also assumed the surface area of the clothed and unclothed person to be the same for simplicity, and these two effects should counteract each other.

 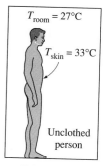

FIGURE 12-16

Clothing serves as insulation, and the room temperature needs to be raised when a person is unclothed to maintain the same comfort level.

12-4 ■ DESIGN CONDITIONS FOR HEATING AND COOLING

The size of a heating or cooling system for a building is determined on the basis of the desired indoor conditions that must be maintained based on the outdoor conditions that exist at that location. The desirable ranges of temperatures, humidities, and ventilation rates (the thermal comfort zone) discussed earlier constitute the typical **indoor design conditions,** and they remain fairly constant. For example, the recommended indoor temperature for general comfort heating is 22°C (or 72°F). The outdoor conditions at a location, on the other hand, vary greatly from year to year, month to month, and even hour to hour. The set of extreme outdoor conditions under which a heating or cooling system must be able to maintain a building at the indoor design conditions is called the **outdoor design conditions** (Fig. 12-17).

When designing a heating, ventilating, and air-conditioning (HVAC) system, perhaps the first thought that comes to mind is to select a system that is large enough to keep the indoors at the desired conditions at all times even under the *worst weather conditions.* But sizing an HVAC system on the basis of the most extreme weather on record is not practical since such an oversized system will have a higher initial cost, will occupy more space, and will probably have a higher operating cost because the equipment in this case will run at partial load most of time and thus at a lower efficiency. Most people would not mind experiencing an occasional slight discomfort under extreme weather conditions if it means a significant reduction in the initial and operating costs of the heating or cooling system. The question that arises naturally is *what is a good compromise between economics and comfort?*

To answer this question, we need to know what the weather will be like in the future. But even the best weather forecasters cannot help us with that. Therefore, we turn to the past instead of the future and bet that the past weather data averaged over several years will be representative of a typical year in the future. The weather data in Tables 12-4 and 12-5 are based on the records of numerous weather stations in the United States that recorded various weather data in hourly intervals. For ordinary buildings, it turns out that the economics and comfort meet at the 97.5 percent level in winter. That is, the heating system will provide thermal comfort 97.5 percent of the time but may fail to do so during 2.5 percent of the time (Fig. 12-18). For example, the 97.5 percent winter design temperature for Denver, Colorado, is −17°C, and thus the temperatures in Denver may fall below −17°C about 2.5 percent of the time during winter months in a typical year. Critical applications such as health care facilities and certain process industries may require the more stringent 99 percent level.

Table 12-4 lists the outdoor design conditions for both cases as well as summer comfort levels. The winter percentages are based on the weather

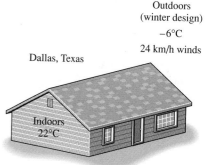

FIGURE 12-17

The size of a heating system is determined on the basis of heat loss during indoor and outdoor design conditions.

SALT LAKE CITY, UTAH

$97\frac{1}{2}\%$ Winter design temp = −13°C
No. of hours during winter (Dec., Jan., and Feb.) = 90 × 24 = 2160 hours

Therefore,

$$T_{outdoor} \begin{cases} > -13°C \text{ for } 2106 \text{ h } (97.5\%) \\ < -13°C \text{ for } 54 \text{ h } (2.5\%) \end{cases}$$

FIGURE 12-18

The 97.5 percent winter design temperature represents the outdoor temperature that will be exceeded during 97.5 percent of the time in winter.

TABLE 12-4

Weather data for selected cities in the United States (from ASHRAE *Handbook of Fundamentals*, Ref. 1, Chap. 24, Table 1)

State and station	Elevation		Winter				Summer					
			99%		$97\frac{1}{2}$%		Dry bulb, $2\frac{1}{2}$%		Wet bulb, $2\frac{1}{2}$%		Daily range	
	ft	m	°F	°C	°F	°C	°F	°C	°F	°C	°F	°C
Alabama, Birmingham AP	610	186	17	−8	21	.−6	94	34	75	24	21	12
Alaska, Anchorage AP	90	27	−23	−31	−18	−28	68	20	58	14	15	8
Arizona, Tucson AP	2584	788	28	−2	32	0	102	39	66	19	26	14
Arkansas, Little Rock AP	257	78.3	15	−9	20	−7	96	36	77	25	22	12
California, San Francisco AP	8	2.4	35	2	38	3	77	25	63	17	20	11
Colorado, Denver AP	5283	1610	−5	−21	1	−17	91	33	59	15	28	16
Connecticut, Bridgeport AP	7	2.1	−6	−21	9	−13	84	29	71	22	18	10
Delaware, Wilmington AP	78	24	10	−12	14	−10	89	32	74	23	20	11
Florida, Tallahassee AP	58	18	27	−3	30	−1	92	33	76	24	19	11
Georgia, Atlanta AP	1005	306	17	−8	22	−6	92	33	74	23	19	11
Hawaii, Honolulu AP	7	2.1	62	17	63	17	86	30	73	23	12	7
Idaho, Boise AP	2842	866	3	−16	10	−12	94	34	64	18	31	17
Illinois, Chicago O'Hare AP	658	201	−8	−22	−4	−20	89	32	74	23	20	11
Indiana, Indianapolis AP	793	242	−2	−19	2	−17	90	32	74	23	22	12
Iowa, Sioux City AP	1095	334	−11	−24	−7	−22	92	33	74	23	24	13
Kansas, Wichita AP	1321	403	3	−16	7	−14	98	37	73	23	23	13
Kentucky, Louisville AP	474	144	5	−15	10	−12	93	34	75	24	23	13
Louisiana, Shreveport AP	252	76.8	20	−7	25	−4	96	36	76	24	20	11
Maryland, Baltimore AP	146	44.5	10	−12	13	−11	91	33	75	24	21	12
Massachusetts, Boston AP	15	4.6	−6	−14	9	−13	88	31	71	22	16	9
Michigan, Lansing AP	852	260	−3	−19	1	−17	87	31	72	22	24	13
Minnesota, Minneapolis/St. Paul	822	251	−16	−27	−12	−24	89	32	73	23	22	12
Mississippi, Jackson AP	330	101	21	−6	25	−4	95	35	76	24	21	12
Missouri, Kansas City AP	742	226	2	−17	6	−14	96	36	74	23	20	11
Montana, Billings AP	3567	1087	−15	−26	−10	−23	91	33	64	18	31	17
Nebraska, Lincoln CO	1150	351	−5	−21	−2	−19	95	35	74	23	24	13
Nevada, Las Vegas AP	2162	659	25	−4	28	−12	106	41	65	18	30	17
New Mexico, Albuquerque AP	5310	1618	12	−11	16	−9	94	34	61	16	30	17
New York, Syracuse AP	424	129	−3	−19	2	−17	87	31	71	22	20	11
North Carolina, Charlotte AP	735	224	18	−8	22	−6	93	34	74	23	20	11
Ohio, Cleveland AP	777	237	1	−17	5	−15	88	31	72	22	22	12
Oklahoma, Stillwater	884	269	8	−13	13	−11	96	36	74	23	24	13
Oregon, Pendleton AP	1492	455	−2	−19	5	−15	93	34	64	18	29	16
Pennsylvania, Pittsburgh AP	1137	347	1	−17	5	−15	86	30	71	22	22	12
South Carolina, Charleston AFB	41	12	24	−4	27	−3	91	33	78	26	18	10
Tennessee, Memphis AP	263	80.2	13	−11	18	−8	95	35	76	24	21	12
Texas, Dallas AP	481	147	18	−8	22	−6	100	38	75	24	20	11
Utah, Salt Lake City	4220	1286	3	−16	8	−13	95	35	62	17	32	18
Virginia, Norfolk AP	26	7.9	20	−7	22	−6	91	33	76	24	18	10
Washington, Spokane AP	2357	718	−6	−21	2	−17	90	32	63	17	28	16

TABLE 12-5

Average winter temperatures and number of degree-days for selected cities in the United States (from ASHRAE *Handbook of Systems*, 1980)

State and station	Average winter temp. °F	°C	Degree-days,* °F-day July	Aug.	Sep.	Oct.	Nov.	Dec.	Jan.	Feb.	March	April	May	June	Yearly total
Alabama, Birmingham	54.2	12.7	0	0	6	93	363	555	592	462	363	108	9	0	2551
Alaska, Anchorage	23.0	5.0	245	291	516	930	1284	1572	1631	1316	1293	879	592	315	10,864
Arizona, Tucson	58.1	14.8	0	0	0	25	231	406	471	344	242	75	6	0	1800
California, San Francisco	53.4	12.2	82	78	60	143	306	462	508	395	363	279	214	126	3015
Colorado, Denver	37.6	3.44	6	9	117	428	819	1035	1132	938	887	558	288	66	6283
Florida, Tallahasse	60.1	15.9	0	0	0	28	198	360	375	286	202	86	0	0	1485
Georgia, Atlanta	51.7	11.28	0	0	18	124	417	648	636	518	428	147	25	0	2961
Hawaii, Honolulu	74.2	23.8	0	0	0	0	0	0	0	0	0	0	0	0	0
Idaho, Boise	39.7	4.61	0	0	132	415	792	1017	1113	854	722	438	245	81	5809
Illinois, Chicago	35.8	2.44	0	12	117	381	807	1166	1265	1086	939	534	260	72	6639
Indiana, Indianapolis	39.6	4.56	0	0	90	316	723	1051	1113	949	809	432	177	39	5699
Iowa, Sioux City	43.0	1.10	0	9	108	369	867	1240	1435	1198	989	483	214	39	6951
Kansas, Wichita	44.2	7.11	0	0	33	229	618	905	1023	804	645	270	87	6	4620
Kentucky, Louisville	44.0	6.70	0	0	54	248	609	890	930	818	682	315	105	9	4660
Louisiana, Shreveport	56.2	13.8	0	0	0	47	297	477	552	426	304	81	0	0	2184
Maryland, Baltimore	43.7	6.83	0	0	48	264	585	905	936	820	679	327	90	0	4654
Massachusetts, Boston	40.0	4.40	0	9	60	316	603	983	1088	972	846	513	208	36	5634
Michigan, Lansing	34.8	1.89	6	22	138	431	813	1163	1262	1142	1011	579	273	69	6909
Minnesota, Minneapolis	28.3	−1.72	22	31	189	505	1014	1454	1631	1380	1166	621	288	81	8382
Montana, Billings	34.5	1.72	6	15	186	487	897	1135	1296	1100	970	570	285	102	7049
Nebraska, Lincoln	38.8	4.11	0	6	75	301	726	1066	1237	1016	834	402	171	30	5864
Nevada, Las Vegas	53.5	12.28	0	0	0	78	387	617	688	487	335	111	6	0	2709
New York, Syracuse	35.2	2.11	6	28	132	415	744	1153	1271	1140	1004	570	248	45	6756
North Carolina, Charlotte	50.4	10.56	0	0	6	124	438	691	691	582	481	156	22	0	3191
Ohio, Cleveland	37.2	3.22	9	25	105	384	738	1088	1159	1047	918	552	260	66	6351
Oklahoma, Stillwater	48.3	9.39	0	0	15	164	498	766	868	664	527	189	34	0	3725
Pennsylvania, Pittsburgh	38.4	3.89	0	9	105	375	726	1063	1119	1002	874	480	195	39	5987
Tennessee, Memphis	50.5	10.6	0	0	18	130	447	698	729	585	456	147	22	0	3232
Texas, Dallas	55.3	13.3	0	0	0	62	321	524	601	440	319	90	6	0	2363
Utah, Salt Lake City	38.4	3.89	0	0	81	419	849	1082	1172	910	763	459	233	84	6052
Virginia, Norfolk	49.2	9.89	0	0	0	136	408	698	738	655	533	216	37	0	3421
Washington, Spokane	36.5	2.83	9	25	168	493	879	1082	1231	980	834	531	288	135	6655

*Based on degrees F; quantities may be converted to degree days based on degrees C by dividing by 1.8. This assumes 18°C corresponds to 65°F.

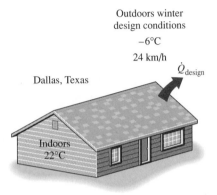

Outdoors winter
design conditions

−6°C

24 km/h

$\dot{Q}_{design}$

Dallas, Texas

Indoors
22°C

FIGURE 12-19

The design heat load of a building
represents the heat loss of a building
during design conditions at the indoors
and the outdoors.

FIGURE 12-20

Recommended winter design values for
heat transfer coefficients for combined
convection and radiation on the outer
and inner surfaces of a building.

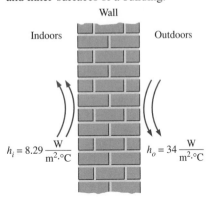

Wall

Indoors   Outdoors

$h_i = 8.29 \dfrac{W}{m^2 \cdot °C}$ $h_o = 34 \dfrac{W}{m^2 \cdot °C}$

data for the months of December, January, and February while the summer
percentages are based on the four months June through September. The three
winter months have a total of $31 + 31 + 28 = 90$ days and thus 2160 hours.
Therefore, the conditions of a house whose heating system is based on the
97.5 percent level may fall below the comfort level for $2160 \times 2.5\% = 54$ hours during the heating season of a typical year. However, most people
will not even notice it because everything in the house will start giving off
heat as soon as the temperature drops below the thermostat setting. This is
especially the case in buildings with large thermal masses. The minimum
temperatures usually occur between 6:00 AM and 8:00 AM solar time, and
thus commercial buildings that open late (such as shopping centers) may even
use less stringent outdoor design conditions (such as the 95 percent level) for
their heating systems. This is also the case with the cooling systems of residences that are unoccupied during the maximum temperatures, which occur
between 2:00 PM and 4:00 PM solar time in the summer.

The **heating** or **cooling loads** of a building represent the heat that must
be supplied to or removed from the interior of a building to maintain it at the
desired conditions. A distinction should be made between the *design load*
and the *actual load* of heating or cooling systems. The *design* (or *peak*)
heating load is usually determined with a steady-state analysis using the design conditions for the indoors and the outdoors for the purpose of *sizing* the
heating system (Fig. 12-19). This ensures that the system has the required
capacity to perform adequately at the anticipated worst conditions. But the
energy use of a building during a heating or cooling season is determined on
the basis of the *actual* heating or cooling load, which varies throughout the
day.

The **internal heat load** (the heat given off by people, lights, and appliances in a building) is usually not considered in the determination of the
design heating load but is considered in the determination of the design cooling load. This is to ensure that the heating system selected can heat the
building even when there is no contribution from people or appliances, and
the cooling system is capable of cooling it even when the heat given off by
people and appliances is at its highest level.

Wind increases heat transfer to or from the walls, roof, and windows
of a building by increasing the convection heat transfer coefficient and also
increasing the infiltration. Therefore, **wind speed** is another consideration
when determining the heating and cooling loads. The recommended values
of wind speed to be considered are 15 mph (6.7 m/s) for winter and 7.5 mph
(3.4 m/s) for summer. The corresponding design values recommended by
ASHRAE for heat transfer coefficients for combined convection and radiation on the *outer surface* of a building are

$$h_{o,\,winter} = 34.0 \text{ W/m}^2 \cdot °C = 6.0 \text{ Btu/h} \cdot ft^2 \cdot °F$$
$$h_{o,\,summer} = 22.7 \text{ W/m}^2 \cdot °C = 4.0 \text{ Btu/h} \cdot ft^2 \cdot °F$$

The recommended heat transfer coefficient value for the *interior surfaces* of
a building for both summer and winter is (Fig. 12-20)

$$h_i = 8.29 \text{ W/m}^2 \cdot °\text{C} = 1.46 \text{ Btu/h} \cdot \text{ft}^2 \cdot °\text{F}$$

For well-insulated buildings, the surface heat transfer coefficients constitute a small part of the overall heat transfer coefficients, and thus the effect of possible deviations from the above values is usually insignificant.

In summer, the **moisture level** of the outdoor air is much higher than that of indoor air. Therefore, the excess moisture that enters a house from the outside with infiltrating air needs to be condensed and removed by the cooling system. But this requires the removal of the latent heat from the moisture, and the cooling system must be large enough to handle this excess cooling load. To size the cooling system properly, we need to know the moisture level of the outdoor air at design conditions. This is usually done by specifying the wet-bulb temperature, which is a good indicator of the amount of moisture in the air. The moisture level of the cold outside air is very low in winter, and thus normally it does not affect the heating load of a building.

Solar radiation plays a major role on the heating and cooling of buildings, and you may think that it should be an important consideration in the evaluation of the design heating and cooling loads. Well, it turns out that peak heating loads usually occur early in the mornings just before sunrise. Therefore, solar radiation does not affect the *peak* or *design heating load* and thus the size of a heating system. However, it has a major effect on the actual heating load, and solar radiation can reduce the annual heating energy consumption of a building considerably.

EXAMPLE 12-2 Summer and Winter Design Conditions for Atlanta

Determine the outdoor design conditions for Atlanta, Georgia, for summer for the 2.5 percent level and for winter for the 97.5 percent and 99 percent levels.

Solution The climatic conditions for major cities in the United States are listed in Table 12-4, and for the indicated design levels we read

Winter:	$T_{outdoor} = -6°C$	(97.5 percent level)
Winter:	$T_{outdoor} = -8°C$	(99 percent level)
Summer:	$T_{outdoor} = 33°C$	
	$T_{wet-bulb} = 23°C$	(2.5 percent level)

Therefore, the heating and cooling systems in Atlanta for common applications should be sized for these outdoor conditions. Note that when the wet-bulb and ambient temperatures are available, the relative humidity and the humidity ratio of air can be determined from the psychrometric chart (Fig. 12-21).

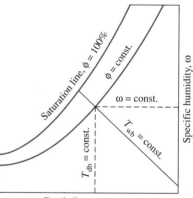

Dry-bulb temperature

FIGURE 12-21

Determination of the relative humidity and the humidity ratio of air from the psychrometric chart when the wet-bulb and ambient temperatures are given.

Sol-Air Temperature

The sun is the main heat source of the earth, and without the sun, the environment temperature would not be much higher than the deep space temperature of $-270°C$. The solar energy stored in the atmospheric air, the ground,

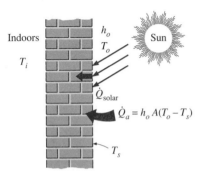

(a) Actual case

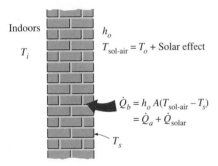

(b) Idealized case (no sun)

FIGURE 12-22
The sol-air temperature represents the equivalent outdoor air temperature that gives the same rate of heat flow to a surface as would the combination of incident solar radiation and convection/radiation with the environment.

and the structures such as buildings during the day is slowly released at night, and thus the variation of the outdoor temperature is governed by the *incident solar radiation* and the *thermal inertia* of the earth. Heat gain from the sun is the primary reason for installing cooling systems, and thus solar radiation has a major effect on the *peak* or *design cooling load* of a building, which usually occurs early in the afternoon as a result of the solar radiation entering through the glazing directly and the radiation absorbed by the walls and the roof that is released later in the day.

The effect of solar radiation for glazing such as windows is expressed in terms of the *solar heat gain factor* (SHGF), discussed later in this chapter. For opaque surfaces such as the walls and the roof, on the other hand, the effect of solar radiation is conveniently accounted for by considering the outside temperature to be higher by an amount equivalent to the effect of solar radiation. This is done by replacing the ambient temperature in the heat transfer relation through the walls and the roof by the **sol-air temperature,** which is defined as *the equivalent outdoor air temperature that gives the same rate of heat flow to a surface as would the combination of incident solar radiation, convection with the ambient air, and radiation exchange with the sky and the surrounding surfaces* (Fig. 12-22).

Heat flow into an exterior surface of a building subjected to solar radiation can be expressed as

$$\dot{Q}_{surface} = \dot{Q}_{conv + rad} + \dot{Q}_{solar} - \dot{Q}_{radiation\,correction}$$
$$= h_o A(T_{ambient} - T_{surface}) + \alpha_s A\dot{q}_{solar} - \varepsilon A\sigma(T_{ambient}^4 - T_{surr}^4) \quad (12\text{-}15)$$
$$= h_o A(T_{sol\text{-}air} - T_{surface})$$

where α_s is the **solar absorptivity** and ε is the **emissivity** of the surface, h_o is the combined convection and radiation heat transfer coefficient, $\dot{q}_{solar}$ is the solar radiation incident on the surface (in W/m^2 or Btu/h · ft^2) and

$$T_{sol\text{-}air} = T_{ambient} + \frac{\alpha_s \dot{q}_{solar}}{h_o} - \frac{\varepsilon\sigma(T_{ambient}^4 - T_{surr}^4)}{h_o} \quad (12\text{-}16)$$

is the sol-air temperature. The first term in Equation 12-15 represents the convection and radiation heat transfer to the surface when the average surrounding surface and sky temperature is equal to the ambient air temperature, $T_{surr} = T_{ambient}$, and the last term represents the correction for the radiation heat transfer when $T_{surr} \neq T_{ambient}$. The last term in the sol-air temperature relation represents the equivalent change in the ambient temperature corresponding to this radiation correction effect and ranges from about zero for vertical wall surfaces to 4°C (or 7°F) for horizontal or inclined roof surfaces facing the sky. This difference is due to the low effective sky temperature.

The sol-air temperature for a surface obviously depends on the absorptivity of the surface for solar radiation, which is listed in Table 12-6 for common exterior surfaces. Being conservative and taking $h_o = 17$ W/m^2 · °C = 3.0 Btu/h · ft^2 · °F, the summer design values of the ratio α_s/h_o for light- and dark-colored surfaces are determined to be (Fig. 12-23)

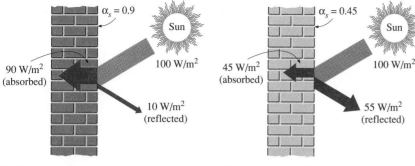

FIGURE 12-23

Dark-colored buildings absorb most of the incident solar radiation whereas light-colored ones reflect most of it.

$$\left(\frac{\alpha_s}{h_o}\right)_{\text{light}} = \frac{0.45}{17 \text{ W/m}^2 \cdot {}^\circ\text{C}} = 0.026 \text{ m}^2 \cdot {}^\circ\text{C/W} = 0.15 \text{ h} \cdot \text{ft}^2 \cdot {}^\circ\text{F/Btu}$$

$$\left(\frac{\alpha_s}{h_o}\right)_{\text{dark}} = \frac{0.90}{17 \text{ W/m}^2 \cdot {}^\circ\text{C}} = 0.052 \text{ m}^2 \cdot {}^\circ\text{C/W} = 0.30 \text{ h} \cdot \text{ft}^2 \cdot {}^\circ\text{F/Btu}$$

where we have assumed conservative values of 0.45 and 0.90 for the solar absorptivities of light- and dark-colored surfaces, respectively. The sol-air temperatures for light- and dark-colored surfaces are listed in Table 12-7 for July 21 at 40° N latitude versus solar time. Sol-air temperatures for other dates and latitudes can be determined from Equation 12-16 by using appropriate temperature and incident solar radiation data.

Once the sol-air temperature is available, heat transfer through a wall (or similarly a roof) can be expressed as

$$\dot{Q}_{\text{wall}} = UA(T_{\text{sol-air}} - T_{\text{inside}}) \qquad (12\text{-}17)$$

where A is the wall area and U is the overall heat transfer coefficient of the wall. Therefore, the rate of heat transfer through the wall will go up by UA for each degree rise in equivalent outdoor temperature due to solar radiation. Noting that the temperature rise due to solar radiation is

$$\Delta T_{\text{solar}} = \frac{\alpha_s \dot{q}_{\text{solar}}}{h_o} \qquad (12\text{-}18)$$

the rate of additional heat gain through the wall becomes

$$\dot{Q}_{\text{wall, solar}} = UA\Delta T_{\text{solar}} = UA\frac{\alpha_s \dot{q}_{\text{solar}}}{h_o} \qquad (12\text{-}19)$$

The total solar radiation incident on the entire wall is $\dot{Q}_{\text{solar}} = A\dot{q}_{\text{solar}}$. Therefore, the fraction of incident solar heat transferred to the interior of the house is

$$\text{Solar fraction transferred} = \frac{\dot{Q}_{\text{wall, solar}}}{\dot{Q}_{\text{solar}}} = \frac{\dot{Q}_{\text{wall, solar}}}{A\dot{q}_{\text{solar}}} = U\frac{\alpha_s}{h_o} \qquad (12\text{-}20)$$

TABLE 12-6

The reflectivity ρ_s and absorptivity α_s of common exterior surfaces for solar radiation (from Kreider and Rabl, Ref. 3, Table 6.1)

Surface	ρ_s	α_s
Natural Surfaces		
Fresh snow	0.75	0.25
Soils (clay, loam, etc.)	0.14	0.86
Water	0.07	0.93
Artificial Surfaces		
Bituminous and gravel roof	0.13	0.87
Blacktop, old	0.10	0.90
Dark building surfaces (red brick, dark paints, etc.)	0.27	0.73
Light building surfaces (light brick, light paints, etc.)	0.60	0.40
New concrete	0.35	0.65
Old concrete	0.25	0.75
Crushed rock surface	0.20	0.80
Earth roads	0.04	0.96
Vegetation		
Coniferous forest (winter)	0.07	0.93
Dead leaves	030	0.70
Forests in autumn, ripe field crops, plants, green grass	0.26	0.74
Dry grass	0.20	0.80

TABLE 12-7

Sol-air temperatures for July 21 at 40° latitude (from ASHRAE *Handbook of Fundamentals*, Ref. 1, Chap. 26, Table 1)

(*a*) SI units

Solar time	Air temp., °C	Light-colored surface, $\alpha/h_o = 0.026\ m^2 \cdot °C/W$									Solar time	Air temp., °C	Dark-colored surface, $\alpha/h_o = 0.052\ m^2 \cdot °C/W$								
		N	NE	E	SE	S	SW	W	NW	Horiz.			N	NE	E	SE	S	SW	W	NW	Horiz.
5	24.0	24.1	24.2	24.2	24.1	24.0	24.0	24.0	24.0	20.1	5	24.0	24.2	24.4	24.3	24.1	24.0	24.0	24.0	24.0	20.2
6	24.2	27.2	34.5	35.5	29.8	25.1	25.1	25.1	25.1	22.9	6	24.2	30.2	44.7	46.7	35.4	26.0	26.0	26.0	26.0	25.5
7	24.8	27.3	38.1	41.5	35.2	26.5	26.4	26.4	26.4	28.1	7	24.8	29.7	51.5	58.2	45.6	28.2	28.0	28.0	28.0	35.4
8	25.8	28.1	38.0	43.5	38.9	28.2	28.0	28.0	28.0	33.8	8	25.8	30.5	50.1	61.2	52.1	30.7	30.1	30.1	30.1	45.8
9	27.2	29.9	35.9	43.1	41.2	31.5	29.8	29.8	29.8	39.2	9	27.2	32.5	44.5	58.9	55.1	35.8	32.3	32.3	32.3	55.1
10	28.8	31.7	33.4	40.8	41.8	35.4	31.8	31.7	31.7	43.9	10	28.8	34.5	38.0	52.8	54.9	42.0	34.7	34.5	34.:	62.8
11	30.7	33.7	34.0	37.4	41.1	39.0	34.2	33.7	33.7	47.7	11	30.7	36.8	37.2	44.0	51.5	47.4	37.7	36.8	36.i	68.5
12	32.5	35.6	35.6	35.9	39.1	41.4	39.1	35.9	35.6	50.1	12	32.5	38.7	38.7	39.3	45.7	50.4	45.7	39.3	38.:	71.6
13	33.8	36.8	36.8	36.8	37.3	42.1	44.2	40.5	37.1	50.8	13	33.8	39.9	39.9	39.9	40.8	50.5	54.6	47.1	40.:	71.6
14	34.7	37.6	37.6	37.6	37.7	41.3	47.7	46.7	39.3	49.8	14	34.7	40.4	40.4	40.4	40.6	47.9	60.8	58.7	43.9	68.7
15	35.0	37.7	37.6	37.6	37.6	39.3	49.0	50.9	43.7	47.0	15	35.0	40.3	40.1	40.1	40.1	43.6	62.9	66.7	52.:	62.9
16	34.7	37.0	36.9	36.9	36.9	37.1	47.8	52.4	46.9	42.7	16	34.7	39.4	39.0	39.0	39.0	39.6	61.0	70.1	59.0	54.7
17	33.9	36.4	35.5	35.5	35.5	35.6	44.3	50.6	47.2	37.2	17	33.9	38.8	37.1	37.1	37.1	37.3	54.7	67.3	60.6	44.5
18	32.7	35.7	33.6	33.6	33.6	33.6	38.3	44.0	43.0	31.4	18	32.7	38.7	34.5	34.5	34.5	34.5	43.9	55.2	53.2	34.0
19	31.3	31.4	31.3	31.3	31.3	31.3	31.4	31.5	31.5	27.4	19	31.3	31.5	31.3	31.3	31.3	31.3	31.4	31.6	31.7	27.5
20	29.8	29.8	29.8	29.8	29.8	29.8	29.8	29.8	29.8	25.9	20	29.8	29.8	29.8	29.8	29.8	29.8	29.8	29.8	29.8	25.9
Avg.	29.0	30.0	32.0	33.0	32.0	31.0	32.0	33.0	32.0	32.0	Avg.	29.0	32.0	35.0	37.0	37.0	34.0	37.0	37.0	35.0	40.0

(*b*) English units

Solar time	Air temp., °F	Light-colored surface, $\alpha/h_o = 0.026\ m^2 \cdot °C/W$									Solar time	Air temp., °F	Dark-colored surface, $\alpha/h_o = 0.052\ m^2 \cdot °C/W$								
		N	NE	E	SE	S	SW	W	NW	Horiz.			N	NE	E	SE	S	SW	W	NW	Horiz.
5	74	74	74	74	74	74	74	74	74	67	5	74	74	75	75	74	74	74	74	74	67
6	74	80	93	95	84	76	76	76	76	72	6	74	85	112	115	94	77	77	77	77	77
7	75	80	99	106	94	78	78	78	78	81	7	75	84	124	136	113	81	81	81	81	94
8	77	81	99	109	101	82	81	81	81	92	8	77	85	121	142	125	86	85	85	85	114
9	80	85	96	109	106	88	85	85	85	102	9	80	90	112	138	131	96	89	89	89	131
10	83	88	91	105	107	95	88	88	88	111	10	83	94	100	127	131	107	94	94	94	145
11	87	93	93	99	106	102	93	93	93	118	11	87	98	99	111	125	118	100	98	98	156
12	90	96	96	96	102	106	102	96	96	122	12	90	101	101	102	114	123	114	102	101	162
13	93	99	99	99	99	108	112	105	99	124	13	93	104	104	104	106	124	131	117	105	162
14	94	99	99	99	99	106	118	116	102	122	14	94	105	105	105	105	118	142	138	111	156
15	95	100	100	100	100	103	121	124	111	117	15	95	105	104	104	104	111	146	153	127	146
16	94	98	98	98	98	99	118	126	116	109	16	94	102	102	102	102	103	142	159	138	131
17	93	98	96	96	96	96	112	124	117	99	17	93	102	99	99	99	99	131	154	142	112
18	91	97	93	93	93	93	101	112	110	89	18	91	102	94	94	94	94	111	132	129	94
19	87	87	87	87	87	87	87	87	87	80	19	87	87	87	87	87	87	87	88	88	80
20	85	85	85	85	85	85	85	85	85	78	20	85	85	85	85	85	85	85	85	85	78
Avg.	83	86	88	90	90	87	90	90	88	90	Avg.	83	89	94	99	97	93	97	99	94	104

Note: Sol-air temperatures are calculated based on a radiation correction of 7°F (3.9°C) for horizontal surfaces and 0°F (0°C) for vertical surfaces.

EXAMPLE 12-3   Effect of Solar Heated Walls on Design Heat Load

The west masonry wall of a house is made of 4-in. thick red face brick, 4-in.-thick common brick, $\frac{3}{4}$-in.-thick air space, and $\frac{1}{2}$-in. thick gypsum board, and its overall heat transfer coefficient is 0.29 Btu/h · ft² · °F, which includes the effects of convection on both the interior and exterior surfaces (Fig. 12-24). The house is located at 40° N latitude, and its cooling system is to be sized on the basis of the heat gain at 15:00 hour (3 PM) solar time on July 21. The interior of the house is to be maintained at 75°F, and the exposed surface area of the wall is 210 ft². If the design ambient air temperature at that time at that location is 90°F, determine (a) the design heat gain through the wall, (b) the fraction of this heat gain due to solar heating, and (c) the fraction of incident solar radiation transferred into the house through the wall.

Solution   The west wall of a house is subjected to solar radiation at summer design conditions. The design heat gain, the fraction of heat gain due to solar heating, and the fraction of solar radiation that is transferred to the house are to be determined.

Assumptions   **1** Steady conditions exist. **2** Thermal properties of the wall and the heat transfer coefficients are constant.

Properties   The overall heat transfer coefficient of the wall is given to be 0.29 Btu/h · ft² · °F.

Analysis   (a) The house is located at 40° N latitude, and thus we can use the sol-air temperature data directly from Table 12-7. At 15:00 the tabulated air temperature is 95°F, which is 5°F higher than the air temperature given in the problem. But we can still use the data in that table provided that we subtract 5°F from all temperatures. Therefore, the sol-air temperature on the west wall in this case is $124 - 5 = 119$°F, and the heat gain through the wall is determined to be

$$\dot{Q}_{wall} = UA(T_{sol\text{-}air} - T_{inside})$$
$$= (0.29 \text{ Btu/h} \cdot \text{ft}^2 \cdot °F)(210 \text{ ft}^2)(119 - 75)°F = \textbf{2680 Btu/h}$$

(b) Heat transfer is proportional to the temperature difference, and the overall temperature difference in this case is $119 - 75 = 44$°F. Also, the difference between the sol-air temperature and the ambient air temperature is (Fig. 12-25)

$$\Delta T_{solar} = T_{sol\text{-}air} - T_{ambient} = (119 - 90)°F = 29°F$$

which is the equivalent temperature rise of the ambient air due to solar heating. The fraction of heat gain due to solar heating is equal to the ratio of the solar temperature difference to the overall temperature difference, and is determined to be

$$\text{Solar fraction} = \frac{\dot{Q}_{wall,\,solar}}{\dot{Q}_{wall,\,total}} = \frac{UA\Delta T_{solar}}{UA\Delta T_{total}} = \frac{\Delta T_{solar}}{\Delta T_{total}} = \frac{29°F}{44°F} = \textbf{0.66} \text{ (or 66\%)}$$

Therefore, almost two-thirds of the heat gain through the west wall in this case is due to solar heating of the wall.

(c) The outer layer of the wall is made of red brick, which is dark colored. Therefore, the value of α_s/h_o is 0.30 h · ft² · °F/Btu. Then the fraction of incident solar energy transferred to the interior of the house is determined directly from Equation 12-20 to be

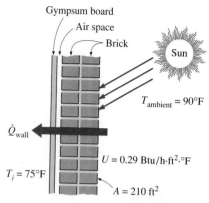

FIGURE 12-24

Schematic for Example 12-3.

FIGURE 12-25

The difference between the sol-air temperature and the ambient air temperature represents the equivalent temperature rise of the ambient air due to solar heating.

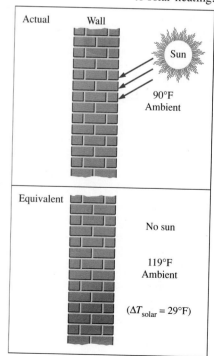

Solar fraction transferred $= U\dfrac{\alpha_s}{h_o} = (029\ \text{Btu/h}\cdot\text{ft}^2\cdot°\text{F})(0.30\ \text{h}\cdot\text{ft}^2\cdot°\text{F/Btu}) = \mathbf{0.087}$

Therefore, less than 10 percent of the solar energy incident on the surface will be transferred to the house. Note that a glass wall would transmit about 10 times more energy into the house.

12-5 ■ HEAT GAIN FROM PEOPLE, LIGHTS, AND APPLIANCES

The conversion of chemical or electrical energy to heat in a building constitutes the **internal heat gain** or **internal load** of a building. The primary sources of internal heat gain are people, lights, appliances, and miscellaneous equipment such as computers, printers, and copiers (Fig. 12-26). Internal heat gain is usually ignored in *design heating load* calculations to ensure that the heating system can do the job even when there is no heat gain, but it is always considered in *design cooling load* calculations since the internal heat gain usually constitutes a significant fraction of design cooling load.

FIGURE 12-26

The heat given off by people, lights, and equipment represents the internal heat gain of a building.

People

The average amount of heat given off by a person depends on the level of activity, and can range from about 100 W for a resting person to more than 500 W for a physically very active person. Typical rates of heat dissipation by people are given in Table 12-8 for various activities in various application areas. Note that *latent heat* constitutes about one-third of the total heat dissipated during resting, but rises to almost two-thirds the level during heavy physical work. Also, about 30 percent of the sensible heat is lost by convection and the remaining 70 percent by radiation. The *latent* and *convective* sensible heat losses represent the "instant" cooling load for people since they need to be removed immediately. The *radiative sensible heat,* on the other hand, is first absorbed by the surrounding surfaces and then released gradually with some delay.

It is interesting to note that an average person dissipates latent heat at a minimum rate of 30 W while resting. Noting that the enthalpy of vaporization of water at 33°C is 2423 kJ/kg, the amount of water an average person loses a day by evaporation at the skin and the lungs is (Fig. 12-27)

FIGURE 12-27

If the moisture leaving an average resting person's body in one day were collected and condensed it would fill a 1-L container.

$$\text{Daily water loss} = \frac{\text{Latent heat loss per day}}{\text{Heat of vaporization}}$$
$$= \frac{(0.030\ \text{kJ/s})(24 \times 3600\ \text{s/day})}{2423\ \text{kJ/kg}} = 1.07\ \text{kg/day}$$

which justifies the sound advice that a person must drink at least 1 L of water every day. Therefore, a family of four will supply 4 L of water a day to the air in the house while just resting. This amount will be much higher during heavy work.

TABLE 12-8

Heat gain from people in conditioned spaces (from ASHRAE *Handbook of Fundamentals*, Ref. 1, Chap. 26, Table 3).

Degree of activity	Typical application	Total heat, W*		Sensible heat, W*	Latent heat, W*
		Adult male	Adjusted M/F/C[1]		
Seated at theater	Theater—matinee	115	95	65	30
Seated at theater, night	Theater—evening	115	105	70	35
Seated, very light work	Offices, hotels, apartments	130	115	70	45
Moderately active office work	Offices, hotels, apartments	140	130	75	55
Standing, light work; walking	Department or retail store	160	130	75	55
Walking, standing	Drug store, bank	160	145	75	70
Sedentary work	Restaurant[2]	145	160	80	80
Light bench work	Factory	235	220	80	80
Moderate dancing	Dance hall	265	250	90	90
Walking 4.8 km/h (3 mph); light machine work	Factory	295	295	110	110
Bowling[3]	Bowling alley	440	425	170	255
Heavy work	Factory	440	425	170	255
Heavy machine work; lifting	Factory	470	470	185	285
Athletics	Gymnasium	585	525	210	315

Note: Tabulated values are based on a room temperature of 24°C (75°F). For a room temperature of 27°C (80°F), the total heat gain remains the same but the sensible heat values should be decreased by about 20 percent, and the latent heat values should be increased accordingly. All values are rounded to nearest 5 W. The fraction of sensible heat that is radiant ranges from 54 to 60 percent in calm air ($\mathcal{V} < 0.2$ m/s) and from 19 to 38 percent in moving air ($0.2 < \mathcal{V} < 4$ m/s).

*Multiply by 3.412 to convert to Btu/h.

[1]Adjusted heat gain is based on normal percentage of men, women, and children for the application listed, with the postulate that the gain from an adult female is 85 percent of that for an adult male and that the gain from a child is 75 percent of that for an adult male.

[2]Adjusted heat gain includes 18 W (60 Btu/h) for food per individual (9 W sensible and 9 W latent).

[3]Figure one person per alley actually bowling, and all others are sitting (117 W) or standing or walking slowly (231 W).

Heat given off by people usually constitutes a significant fraction of the sensible and latent heat gain of a building, and may dominate the cooling load in high occupancy buildings such as theaters and concert halls. The rate of heat gain from people given in Table 12-8 is quite accurate, but there is considerable uncertainty in the internal load due to people because of the difficulty in predicting the number of occupants in a building at any given time. The design cooling load of a building should be determined assuming full occupancy. In the absence of better data, the number of occupants can be estimated on the basis of one occupant per 1 m² in auditoriums, 2.5 m² in schools, 3–5 m² in retail stores, and 10–15 m² in offices.

Lights

Lighting constitutes about 7 percent of the total energy use in residential buildings and 25 percent in commercial buildings. Therefore, lighting can have a significant impact on the heating and cooling loads of a building. Not

TABLE 12-9

Comparison of different lighting systems

Type of lighting	Efficacy, lumens/W	Life, h	Comments
Combustion Candle	0.2	10	Very inefficient. Best for emergencies.
Incandescent Ordinary Halogen	5–20 15–25	1000 2000	Low initial cost; low efficiency. Better efficiency; excellent color rendition.
Fluorescent Ordinary High output Compact Metal halide	40–60 70–90 50–80 55–125	10,000 10,000 10,000	Being replaced by high-output types. Commonly in offices and plants. Fits into the sockets of incandescent lights. High efficiency; good color rendition.
Gaseous Discharge Mercury vapor High-pressure sodium Low-pressure sodium	50–60 100–150 up to 200	10,000 15,000	Both indoor and outdoor use. Good color rendition. Indoor and outdoor use. Distinct yellow light. Best for outdoor use.

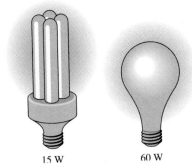

15 W   60 W

FIGURE 12-28

A 15-W compact fluorescent lamp provides as much light as a 60-W incandescent lamp.

counting the candle light used for emergencies and romantic settings, and the kerosene lamps used during camping, all modern lighting equipment is powered by electricity. The basic types of electric lighting devices are incandescent, fluorescent, and gaseous discharge lamps.

The amount of heat given off per lux of lighting varies greatly with the type of lighting, and thus we need to know the type of lighting installed in order to predict the lighting internal heat load accurately. The lighting efficacy of common types of lighting is given in Table 12-9. Note that incandescent lights are the least efficient lighting sources, and thus they will impose the greatest load on cooling systems (Fig. 12-28). So it is no surprise that practically all office buildings use high-efficiency fluorescent lights despite their higher initial cost. Note that incandescent lights waste energy by (1) consuming more electricity for the same amount of lighting and (2) making the cooling system work harder and longer to remove the heat given off. Office spaces are usually well lit, and the lighting energy consumption in office buildings is about 20 to 30 W/m² (2 to 3 W/ft²) of floor space.

The energy consumed by the lights is dissipated by convection and radiation. The convection component of the heat constitutes about 40 percent for fluorescent lamps, and it represents the instantaneous part of the cooling load due to lighting. The remaining part is in the form of radiation that is absorbed and reradiated by the walls, floors, ceiling, and the furniture, and thus they affect the cooling load with time delay. Therefore, lighting may continue contributing to the cooling load by reradiation even after the lights have been turned off. Sometimes it may be necessary to consider time lag effects when determining the design cooling load.

The ratio of the lighting wattage in use to the total wattage installed is called the **usage factor**, and it must be considered when determining the heat gain due to lighting at a given time since installed lighting does not give off heat unless it is on. For commercial applications such as supermarkets and shopping centers, the usage factor is taken to be unity.

Equipment and Appliances

Most equipment and appliances are driven by electric motors, and thus the heat given off by an appliance in steady operation is simply the power consumed by its motor. For a fan, for example, part of the power consumed by the motor is transmitted to the fan to drive it, while the rest is converted to heat because of the inefficiency of the motor. The fan transmits the energy to the air molecules and increases their kinetic energy. But this energy is also converted to heat as the fast-moving molecules are slowed down by other molecules and stopped as a result of friction. Therefore, we can say that the entire energy consumed by the motor of the fan in a room is eventually converted to heat in that room. Of course, if the motor is in one room (say, room A) and the fan is in another (say, room B), then the heat gain of room B will be equal to the power transmitted to the fan only, while the heat gain of room A will be the heat generated by the motor due to its inefficiency (Fig. 12-29).

The power rating $\dot{W}_{motor}$ on the label of a motor represents the power that the motor will supply under full load conditions. But a motor usually operates at part load, sometimes at as low as 30 to 40 percent, and thus it consumes and delivers much less power than the label indicates. This is characterized by the **load factor** f_{load} of the motor during operation, which is $f_{load} = 1.0$ for full load. Also, there is an inefficiency associated with the conversion of electrical energy to rotational mechanical energy. This is characterized by the **motor efficiency** η_{motor}, which decreases with decreasing load factor. Therefore, it is not a good idea to oversize the motor since oversized motors operate at a low load factor and thus at a lower efficiency. Another factor that affects the amount of heat generated by a motor is how long a motor actually operates. This is characterized by the **usage factor** f_{usage}, with $f_{usage} = 1.0$ for continuous operation. Motors with very low usage factors such as the motors of dock doors can be ignored in calculations. Then the heat gain due to a motor inside a conditioned space can be expressed as

$$\dot{Q}_{motor,total} = \dot{W}_{motor} \times f_{load} \times f_{usage}/\eta_{motor} \quad \text{(W)} \quad \text{(12-21)}$$

Heat generated in conditioned spaces by electric, gas, and steam appliances such as a range, refrigerator, freezer, TV, dishwasher, clothes washer, drier, computers, printers, and copiers can be significant, and thus must be considered when determining the peak cooling load of a building. There is considerable uncertainty in the estimated heat gain from appliances owing to the variations in appliances and the varying usage schedules. The exhaust hoods in the kitchen complicate things further. Also, some office equipment such as printers and copiers consume considerable power in the standby mode. A 350-W laser printer, for example, may consume 175 W and a 600-W computer may consume 530 W when in standby mode.

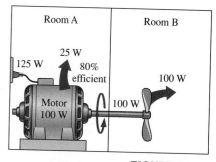

FIGURE 12-29

An 80 percent efficient motor that drives a 100-W fan contributes 25 W and 100 W to the heat loads of the motor and equipment rooms, respectively.

The heat gain from office equipment in a typical office with computer terminals on most desks can be up to 47 W/m². This value can be 10 times as large for computer rooms that house mainframe computers. When the equipment inventory of a building is known, the equipment heat gain can be determined more accurately using the data given in the ASHRAE *Handbook of Fundamentals* (Ref. 1).

The presence of thermostatic controls and typical usage practices make it highly unlikely for all the appliances in a conditioned space to operate at full load. A more realistic approach is to take 50 percent of the total nameplate ratings of the appliances to represent the maximum use. Therefore, the peak heat gain from appliances is taken to be

$$\dot{Q}_{\text{unhooded appliance}} = 0.5\dot{Q}_{\text{appliance, input}} \quad \text{(W)} \quad (12\text{-}22)$$

regardless of the type of energy or fuel used. For cooling load estimate, about 34 percent of heat gain can be assumed to be latent heat, with the remaining 66 percent to be sensible in this case.

In hooded appliances, the air heated by convection and the moisture generated are removed by the hood. Therefore, the only heat gain from hooded appliances is radiation, which constitutes up to 32 percent of the energy consumed by the appliance (Fig. 12-30). Therefore, the design value of heat gain from hooded electric or steam appliances is simply half of this 32 percent.

EXAMPLE 12-4 Energy Consumption of Electric and Gas Burners

The efficiency of cooking equipment affects the internal heat gain from them since an inefficient appliance consumes a greater amount of energy for the same task, and the excess energy consumed shows up as heat in the living space. The efficiency of open burners is determined to be 73 percent for electric units and 38 percent for gas units (Fig. 12-31). Consider a 2-kW unhooded electric open burner in an area where the unit costs of electricity and natural gas are $0.09/kWh and $0.55/therm, respectively. Determine the amount of electrical energy used directly for cooking, the cost of energy per "utilized" kWh, and the contribution of this burner to the design cooling load. Repeat the calculations for the gas burner.

Solution The efficiency of the electric heater is given to be 73 percent. Therefore, a burner that consumes 2 kW of electrical energy will supply

$$\dot{Q}_{\text{utilized}} = (\text{Energy input}) \times (\text{Efficiency}) = (2 \text{ kW})(0.73) = \textbf{1.46 kW}$$

of useful energy. The unit cost of utilized energy is inversely proportional to the efficiency and is determined from

$$\text{Cost of utilized energy} = \frac{\text{Cost of energy input}}{\text{Efficiency}} = \frac{\$0.09/\text{kWh}}{0.73} = \textbf{\$0.123/kWh}$$

The design heat gain from an unhooded appliance is taken to be half of its rated energy consumption and is determined to be

$$\dot{Q}_{\text{unhooded appliance}} = 0.5\dot{Q}_{\text{appliance, input}}$$
$$= 0.5 \times (2 \text{ kW}) = \textbf{1 kW} \quad \text{(electric burner)}$$

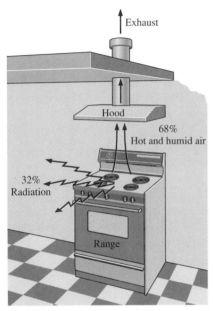

FIGURE 12-30

In hooded appliances, about 68 percent of the generated heat is vented out with the heated and humidified air.

FIGURE 12-31

Schematic of the 73 percent efficient electric heating unit and 38 percent efficient gas burner discussed in Example 12-4.

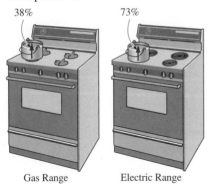

Gas Range   Electric Range

Noting that the efficiency of a gas burner is 38 percent, the energy input to a gas burner that supplies utilized energy at the same rate (1.46 kW) is

$$\dot{Q}_{input,\,gas} = \frac{\dot{Q}_{utilized}}{Efficiency} = \frac{1.46\ kW}{0.38} = \textbf{3.84 kW} \ (=13,100\ Btu/h)$$

since 1 kW = 3412 Btu/h. Therefore, a gas burner should have a rating of at least 13,100 Btu/h to perform as well as the electric unit.

Noting that 1 therm = 29.3 kWh, the unit cost of utilized energy in the case of a gas burner is determined similarly to be

$$Cost\ of\ utilized\ energy = \frac{Cost\ of\ energy\ input}{Efficiency} = \frac{\$0.55/(29.3\ kWh)}{0.38} = \textbf{\$0.049/kWh}$$

which is about one-quarter of the unit cost of utilized electricity. Therefore, despite its higher efficiency, cooking with an electric burner will cost four times as much compared to a gas burner in this case. This explains why cost-conscious consumers always ask for gas appliances, and it is not wise to use electricity for heating purposes.

Finally, the design heat gain from this unhooded gas burner is determined to be

$$\dot{Q}_{unhooded\,appliance} = 0.5\dot{Q}_{appliance,\,input}$$
$$= 0.5 \times (3.84\ kW) = \textbf{1.92 kW} \qquad (gas\ burner)$$

which is 92 percent larger than that of the electric burner. Therefore, an unhooded gas appliance will contribute more to the heat gain than a comparable electric appliance.

EXAMPLE 12-5   Heat Gain of an Exercise Room

An exercise room has 10 weight-lifting machines that have no motors and 7 treadmills each equipped with a 2-hp motor (Fig. 12-32). The motors operate at an average load factor of 0.6, at which their efficiency is 0.75. During peak evening hours, 17 pieces of exercising equipment are used continuously, and there are also four people doing light exercises while waiting in line for one piece of the equipment. Determine the rate of heat gain of the exercise room from people and the equipment at peak load conditions. How much of this heat gain is in the latent form?

Solution   The 10 weight-lifting machines do not have any motors, and thus they do not contribute to the internal heat gain directly. The usage factors of the motors of the treadmills are taken to be unity since they are used constantly during heat periods. Noting that 1 hp = 746 W, the total heat generated by the motors is

$$\dot{Q}_{motors} = (No.\ of\ motors) \times \dot{W}_{motor} \times f_{load} \times f_{usage}/\eta_{motor}$$
$$= 7 \times (2 \times 746\ W) \times 0.60 \times 1.0/0.75 = 8355\ W$$

The average rate of heat dissipated by people in an exercise room is given in Table 12-8 to be 525 W, of which 315 W is in latent form. Therefore, the heat gain from 21 people is

$$\dot{Q}_{people} = (No.\ of\ people) \times \dot{Q}_{person} = 21 \times (525\ W) = 11,025\ W$$

FIGURE 12-32

Schematic for Example 12-5.

Then the total rate of heat gain (or the internal heat load) of the exercise room during peak period becomes

$$\dot{Q}_{total} = \dot{Q}_{motors} + \dot{Q}_{people} = 8355 + 11{,}025 = \textbf{19{,}380 W}$$

The entire heat given off by the motors is in sensible form. Therefore, the latent heat gain is due to people only, which is determined to be

$$\dot{Q}_{latent} = (\text{No. of people}) \times \dot{Q}_{latent,\,per\,person} = 21 \times (315\ \text{W}) = \textbf{6615 W}$$

The remaining 12,765 W of heat gain is in the sensible form.

12-6 ■ HEAT TRANSFER THROUGH WALLS AND ROOFS

Under steady conditions, the rate of heat transfer through any section of a building wall or roof can be determined from

$$\dot{Q} = UA(T_i - T_o) = \frac{A(T_i - T_o)}{R} \qquad (12\text{-}23)$$

TABLE 12-10

Unit thermal resistance (the R-value) of common components used in buildings

Component	R-value $m^2 \cdot °C/W$	$ft^2 \cdot h \cdot °F/Btu$	Component	R-value $m^2 \cdot °C/W$	$ft^2 \cdot h \cdot °F/Btu$
Outside surface (winter)	0.030	0.17	Wood stud, nominal 2 in. ×		
Outside surface (summer)	0.044	0.25	6 in. (5.5 in. or 140 mm wide)	0.98	5.56
Inside surface, still air	0.12	0.68	Clay tile, 100 mm (4 in.)	0.18	1.01
Plane air space, vertical,			Acoustic tile	0.32	1.79
ordinary surfaces			Asphalt shingle roofing	0.077	0.44
($\varepsilon_{eff} = 0.82$):			Building paper	0.011	0.06
13 mm ($\frac{1}{2}$ in.)	0.16	0.90	Concrete block, 100 mm		
20 mm ($\frac{3}{4}$ in.)	0.17	0.94	(4 in.):		
40 mm (1.5 in.)	0.16	0.90	Lightweight	0.27	1.51
90 mm (3.5 in.)	0.16	0.91	Heavyweight	0.13	0.71
Insulation, 25 mm (1 in.)			Plaster or gypsum board,		
Glass fiber	0.70	4.00	13 mm ($\frac{1}{2}$ in.)	0.079	0.45
Mineral fiber batt	0.66	3.73	Wood fiberboard, 13 mm ($\frac{1}{2}$ in.)	0.23	1.31
Urethane			Plywood, 13 mm ($\frac{1}{2}$ in.)	0.11	0.62
rigid foam	0.98	5.56	Concrete, 200 mm (8 in.):		
Stucco, 25 mm (1 in.)	0.037	0.21	Lightweight	1.17	6.67
Face brick, 100 mm (4 in.)	0.075	0.43	Heavyweight	0.12	0.67
Common brick, 100 mm (4 in.)	0.12	0.79	Cement mortar, 13 mm ($\frac{1}{2}$ in.)	0.018	0.10
Steel siding	0.00	0.00	Wood bevel lapped siding,		
Slag, 13 mm ($\frac{1}{2}$ in.)	0.067	0.38	13 mm × 200 mm		
Wood, 25 mm (1 in.)	0.22	1.25	($\frac{1}{2}$ in. × 8 in.)	0.14	0.81
Wood stud, nominal 2 in. ×					
4 in. (3.5 in. or 90 mm wide)	0.63	3.58			

where T_i and T_o are the indoor and outdoor air temperatures, A is the heat transfer area, U is the overall heat transfer coefficient (the U-factor), and $R = 1/U$ is the overall unit thermal resistance (the R-value). Walls and roofs of buildings consist of various layers of materials, and the structure and operating conditions of the walls and the roofs may differ significantly from one building to another. Therefore, it is not practical to list the R-values (or U-factors) of different kinds of walls or roofs under different conditions. Instead, the overall R-value is determined from the thermal resistances of the individual components using the thermal resistance network. The overall thermal resistance of a structure can be determined most accurately in a lab by actually assembling the unit and testing it as a whole, but this approach is usually very time consuming and expensive. The analytical approach described here is fast and straightforward, and the results are usually in good agreement with the experimental values.

The unit thermal resistance of a plane layer of thickness L and thermal conductivity k can be determined from $R = L/k$. The thermal conductivity and other properties of common building materials are given in the appendix. The unit thermal resistances of various components used in building structures are listed in Table 12-10 for convenience.

Heat transfer through a wall or roof section is also affected by the convection and radiation heat transfer coefficients at the exposed surfaces. The effects of convection and radiation on the inner and outer surfaces of walls and roofs are usually combined into the *combined convection and radiation heat transfer coefficients* (also called *surface conductances*) h_i and h_o, respectively, whose values are given in Table 12-11 for ordinary surfaces ($\varepsilon = 0.9$) and reflective surfaces ($\varepsilon = 0.2$ or 0.05). Note that surfaces having a low emittance also have a low surface conductance due to the reduction in radiation heat transfer. The values in the table are based on a surface temperature of 21°C (72°F) and a surface–air temperature difference of 5.5°C (10°F). Also, the equivalent surface temperature of the environment is assumed to be equal to the ambient air temperature. Despite the convenience it offers, this assumption is not quite accurate because of the additional radiation heat loss from the surface to the clear sky. The effect of sky radiation can be accounted for approximately by taking the outside temperature to be the average of the outdoor air and sky temperatures.

The inner surface heat transfer coefficient h_i remains fairly constant throughout the year, but the value of h_o varies considerable because of its dependence on the orientation and wind speed, which can vary from less than 1 km/h in calm weather to over 40 km/h during storms. The commonly used values of h_i and h_o for peak load calculations are

$$h_i = 8.29 \text{ W/m}^2 \cdot °\text{C} = 1.46 \text{ Btu/h} \cdot \text{ft}^2 \cdot °\text{F} \quad \text{(winter and summer)}$$

$$h_o = \begin{cases} 34.0 \text{ W/m}^2 \cdot °\text{C} = 6.0 \text{ Btu/h} \cdot \text{ft}^2 \cdot °\text{F} & \text{(winter)} \\ 22.7 \text{ W/m}^2 \cdot °\text{C} = 4.0 \text{ Btu/h} \cdot \text{ft}^2 \cdot °\text{F} & \text{(summer)} \end{cases}$$

which correspond to design wind conditions of 24 km/h (15 mph) for winter and 12 km/h (7.5 mph) for summer. The corresponding surface thermal

TABLE 12-11

Combined convection and radiation heat transfer coefficients at window, wall, or roof surfaces (from ASHRAE *Handbook of Fundamentals*, Ref. 1, Chap. 22, Table 1).

Posi-tion	Direc-tion of heat flow	h, W/m² · °C*		
		Surface emittance, ε		
		0.90	0.20	0.05
Still Air (both indoors and outdoors)				
Horiz.	Up ↑	9.26	5.17	4.32
Horiz.	Down ↓	6.13	2.10	1.25
45° slope	Up ↑	9.09	5.00	4.15
45° slope	Down ↓	7.50	3.41	2.56
Vertical	Horiz. →	8.29	4.20	3.35

Moving Air (any position, any direction)

Winter condition (winds at 15 mph or 24 km/h)		34.0	—	—
Summer condition (winds at 7.5 mph or 12 km/h)		22.7	—	—

*Multiply by 0.176 to convert to Btu/h · ft² · °F. Surface resistance can be obtained from $R = 1/h$.

TABLE 12-12

Emissivities ε of various surfaces
and the effective emissivity of
air spaces (from ASHRAE
Handbook of Fundamentals,
Ref. 1, Chap. 22, Table 3).

Surface	ε	Effective emissivity of air space	
		$\varepsilon_1 = \varepsilon$ $\varepsilon_2 = 0.9$	$\varepsilon_1 = \varepsilon$ $\varepsilon_2 = \varepsilon$
Aluminum foil, bright	0.05*	0.05	0.03
Aluminum sheet	0.12	0.12	0.06
Aluminum-coated paper, polished	0.20	0.20	0.11
Steel, galvanized, bright	0.25	0.24	0.15
Aluminum paint	0.50	0.47	0.35
Building materials: Wood, paper, masonry, nonmetallic paints	0.90	0.82	0.82
Ordinary glass	0.84	0.77	0.72

*Surface emissivity of aluminum foil
increases to 0.30 with barely visible
condensation, and to 0.70 with clearly
visible condensation.

resistances (R-values) are determined from $R_i = 1/h_i$ and $R_o = 1/h_o$. The surface conductance values under still air conditions can be used for interior surfaces as well as exterior surfaces in calm weather.

Building components often involve *trapped air spaces* between various layers. Thermal resistances of such air spaces depend on the thickness of the layer, the temperature difference across the layer, the mean air temperature, the emissivity of each surface, the orientation of the air layer, and the direction of heat transfer. The emissivities of surfaces commonly encountered in buildings are given in Table 12-12. The **effective emissivity** of a plane-parallel air space is given by

$$\frac{1}{\varepsilon_{\text{effective}}} = \frac{1}{\varepsilon_1} + \frac{1}{\varepsilon_2} - 1 \tag{12-24}$$

where ε_1 and ε_2 are the emissivities of the surfaces of the air space. Table 12-12 also lists the effective emissivities of air spaces for the cases where (1) the emissivity of one surface of the air space is ε while the emissivity of the other surface is 0.9 (a building material) and (2) the emissivity of both surfaces is ε. Note that the effective emissivity of an air space between building materials is $0.82/0.03 = 27$ times that of an air space between surfaces covered with aluminum foil. For specified surface temperatures, radiation heat transfer through an air space is proportional to effective emissivity, and thus the rate of radiation heat transfer in the ordinary surface case is 27 times that of the reflective surface case.

Table 12-13 lists the thermal resistances of 20-mm-, 40-mm-, and 90-mm- (0.75-in., 1.5-in., and 3.5-in.) thick air spaces under various conditions. The thermal resistance values in the table are applicable to air spaces of uniform thickness bounded by plane, smooth, parallel surfaces with no air leakage. Thermal resistances for other temperatures, emissivities, and air spaces can be obtained by interpolation and moderate extrapolation. Note that the presence of a low-emissivity surface reduces radiation heat transfer across an air space and thus significantly increases the thermal resistance. The thermal effectiveness of a low-emissivity surface will decline, however, if the condition of the surface changes as a result of some effects such as condensation, surface oxidation, and dust accumulation.

The R-value of a wall or roof structure that involves layers of uniform thickness is determined easily by simply adding up the unit thermal resistances of the layers that are in series. But when a structure involves components such as wood studs and metal connectors, then the thermal resistance network involves parallel connections and possible two-dimensional effects. The overall R-value in this case can be determined by assuming (1) parallel heat flow paths through areas of different construction or (2) isothermal planes normal to the direction of heat transfer. The first approach usually overpredicts the overall thermal resistance, whereas the second approach usually underpredicts it. The parallel heat flow path approach is more suitable for wood frame walls and roofs, whereas the isothermal planes approach is more suitable for masonry or metal frame walls.

The thermal contact resistance between different components of building structures ranges between 0.01 and 0.1 $m^2 \cdot °C/W$, which is negligible in most

TABLE 12-13

Unit thermal resistances (*R*-values) of well-sealed plane air spaces
(from ASHRAE *Handbook of Fundamentals*, Ref. 1, Chap. 22, Table 2)
(*a*) SI units (in m² · °C/W)

Position of air space	Direction of heat flow	Mean temp., °C	Temp. diff., °C	20-mm air space Effective emissivity, ε_{eff}				40-mm air space Effective emissivity, ε_{eff}				90-mm air space Effective emissivity, ε_{eff}			
				0.03	0.05	0.5	0.82	0.03	0.05	0.5	0.82	0.03	0.05	0.5	0.82
Horizontal	Up ↑	32.2	5.6	0.41	0.39	0.18	0.13	0.45	0.42	0.19	0.14	0.50	0.47	0.20	0.14
		10.0	16.7	0.30	0.29	0.17	0.14	0.33	0.32	0.18	0.14	0.27	0.35	0.19	0.15
		10.0	5.6	0.40	0.39	0.20	0.15	0.44	0.42	0.21	0.16	0.49	0.47	0.23	0.16
		−17.8	11.1	0.32	0.32	0.20	0.16	0.35	0.34	0.22	0.17	0.40	0.38	0.23	0.18
45° slope	Up ↑	32.2	5.6	0.52	0.49	0.20	0.14	0.51	0.48	0.20	0.14	0.56	0.52	0.21	0.14
		10.0	16.7	0.35	0.34	0.19	0.14	0.38	0.36	0.20	0.15	0.40	0.38	0.20	0.15
		10.0	5.6	0.51	0.48	0.23	0.17	0.51	0.48	0.23	0.17	0.55	0.52	0.24	0.17
		−17.8	11.1	0.37	0.36	0.23	0.18	0.40	0.39	0.24	0.18	0.43	0.41	0.24	0.19
Vertical	Horizontal →	32.2	5.6	0.62	0.57	0.21	0.15	0.70	0.64	0.22	0.15	0.65	0.60	0.22	0.15
		10.0	16.7	0.51	0.49	0.23	0.17	0.45	0.43	0.22	0.16	0.47	0.45	0.22	0.16
		10.0	5.6	0.65	0.61	0.25	0.18	0.67	0.62	0.26	0.18	0.64	0.60	0.25	0.18
		−17.8	11.1	0.55	0.53	0.28	0.21	0.49	0.47	0.26	0.20	0.51	0.49	0.27	0.20
45° slope	Down ↓	32.2	5.6	0.62	0.58	0.21	0.15	0.89	0.80	0.24	0.16	0.85	0.76	0.24	0.16
		10.0	16.7	0.60	0.57	0.24	0.17	0.63	0.59	0.25	0.18	0.62	0.58	0.25	0.18
		10.0	5.6	0.67	0.63	0.26	0.18	0.90	0.82	0.28	0.19	0.83	0.77	0.28	0.19
		−17.8	11.1	0.66	0.63	0.30	0.22	0.68	0.64	0.31	0.22	0.67	0.64	0.31	0.22
Horizontal	Down ↓	32.2	5.6	0.62	0.58	0.21	0.15	1.07	0.94	0.25	0.17	1.77	1.44	0.28	0.18
		10.0	16.7	0.66	0.62	0.25	0.18	1.10	0.99	0.30	0.20	1.69	1.44	0.33	0.21
		10.0	5.6	0.68	0.63	0.26	0.18	1.16	1.04	0.30	0.20	1.96	1.63	0.34	0.22
		−17.8	11.1	0.74	0.70	0.32	0.23	1.24	1.13	0.39	0.26	1.92	1.68	0.43	0.29

(*b*) English units (in h · ft² · °F/Btu)

Position of air space	Direction of heat flow	Mean temp., °F	Temp. diff., °F	0.75-in. air space Effective emissivity, ε_{eff}				1.5-in. air space Effective emissivity, ε_{eff}				3.5-in. air space Effective emissivity, ε_{eff}			
				0.03	0.05	0.5	0.82	0.03	0.05	0.5	0.82	0.03	0.05	0.5	0.82
Horizontal	Up ↑	90	10	2.34	2.22	1.04	0.75	2.55	2.41	1.08	0.77	2.84	2.66	1.13	0.80
		50	30	1.71	1.66	0.99	0.77	1.87	1.81	1.04	0.80	2.09	2.01	1.10	0.84
		50	10	2.30	2.21	1.16	0.87	2.50	2.40	1.21	0.89	2.80	2.66	1.28	0.93
		0	20	1.83	1.79	1.16	0.93	2.01	1.95	1.23	0.97	2.25	2.18	1.32	1.03
45° slope	Up ↑	90	10	2.96	2.78	1.15	0.81	2.92	2.73	1.14	0.80	3.18	2.96	1.18	0.82
		50	30	1.99	1.92	1.08	0.82	2.14	2.06	1.12	0.84	2.26	2.17	1.15	0.86
		50	10	2.90	2.75	1.29	0.94	2.88	2.74	1.29	0.94	3.12	2.95	1.34	0.96
		0	20	2.13	2.07	1.28	1.00	2.30	2.23	1.34	1.04	2.42	2.35	1.38	1.06
Vertical	Horizontal →	90	10	3.50	3.24	1.22	0.84	3.99	3.66	1.27	0.87	3.69	3.40	1.24	0.85
		50	30	2.91	2.77	1.30	0.94	2.58	2.46	1.23	0.90	2.67	2.55	1.25	0.91
		50	10	3.70	3.46	1.43	1.01	3.79	3.55	1.45	1.02	3.63	3.40	1.42	1.01
		0	20	3.14	3.02	1.58	1.18	2.76	2.66	1.48	1.12	2.88	2.78	1.51	1.14
45° slope	Down ↓	90	10	3.53	3.27	1.22	0.84	5.07	4.55	1.36	0.91	4.81	4.33	1.34	0.90
		50	30	3.43	3.23	1.39	0.99	3.58	3.36	1.42	1.00	3.51	3.30	1.40	1.00
		50	10	3.81	3.57	1.45	1.02	5.10	4.66	1.60	1.09	4.74	4.36	1.57	1.08
		0	20	3.75	3.57	1.72	1.26	3.85	3.66	1.74	1.27	3.81	3.63	1.74	1.27
Horizontal	Down ↓	90	10	3.55	3.29	1.22	0.85	6.09	5.35	1.43	0.94	10.07	8.19	1.57	1.00
		50	30	3.77	3.52	1.44	1.02	6.27	5.63	1.70	1.14	9.60	8.17	1.88	1.22
		50	10	3.84	3.59	1.45	1.02	6.61	5.90	1.73	1.15	11.15	9.27	1.93	1.24
		0	20	4.18	3.96	1.81	1.30	7.03	6.43	2.19	1.49	10.90	9.52	2.47	1.62

cases. However, it may be significant for metal building components such as steel framing members.

EXAMPLE 12-6   The R-Value of a Wood Frame Wall

Determine the overall unit thermal resistance (the R-value) and the overall heat transfer coefficient (the U-factor) of a wood frame wall that is built around 38-mm × 90-mm (2 × 4 nominal) wood studs with a center-to-center distance of 400 mm. The 90-mm-wide cavity between the studs is filled with glass fiber insulation. The inside is finished with 13-mm gypsum wallboard and the outside with 13-mm wood fiberboard and 13-mm × 200-mm wood bevel lapped siding. The insulated cavity constitutes 75 percent of the heat transmission area while the studs, plates, and sills constitute 21 percent. The headers constitute 4 percent of the area, and they can be treated as studs.

Also, determine the rate of heat loss through the walls of a house whose perimeter is 50 m and wall height is 2.5 m in Las Vegas, Nevada, whose winter design temperature is −2°C. Take the indoor design temperature to be 22°C and assume 20 percent of the wall area is occupied by glazing.

Solution   The R-value and the U-factor of a wood frame wall as well as the rate of heat loss through such a wall in Las Vegas are to be determined.

Assumptions   **1** Steady operating conditions exist. **2** Heat transfer through the wall is one-dimensional. **3** Thermal properties of the wall and the heat transfer coefficients are constant.

Properties   The R-values of different materials are given in Table 12-10.

Analysis   The schematic of the wall as well as the different elements used in its construction are shown below. Heat transfer through the insulation and through the studs will meet different resistances, and thus we need to analyze the thermal resistance for each path separately. Once the unit thermal resistances and the U-factors for the insulation and stud sections are available, the overall average thermal resistance for the entire wall can be determined from

$$R_{overall} = 1/U_{overall}$$

where

$$U_{overall} = (U \times f_{area})_{insulation} + (U \times f_{area})_{stud}$$

and the value of the area fraction f_{area} is 0.75 for the insulation section and 0.25 for the stud section since the headers that constitute a small part of the wall are to be treated as studs. Using the available R-values from Table 12-10 and calculating others, the total R-values for each section can be determined in a systematic manner in the table on next page.

We conclude that the overall unit thermal resistance of the wall is 2.23 m² · °C/W, and this value accounts for the effects of the studs and headers. It corresponds to an R-value of 2.23 × 5.68 = 12.7 (or nearly R-13) in English units. Note that if there were no wood studs and headers in the wall, the overall thermal resistance would be 3.05 m² · °C/W, which is 37 percent greater than 2.23 m² ·

°C/W. Therefore, the wood studs and headers in this case serve as thermal bridges in wood frame walls, and their effect must be considered in the thermal analysis of buildings.

Schematic	Construction	R-value, m²·°C/W	
		Between studs	At studs
	1. Outside surface, 24 km/h wind	0.030	0.030
	2. Wood bevel lapped siding	0.14	0.14
	3. Wood fiberboard sheeting, 13 mm	0.23	0.23
	4a. Glass fiber insulation, 90 mm	2.45	—
	4b. Wood stud, 38 mm × 90 mm	—	0.63
	5. Gypsum wallboard, 13 mm	0.079	0.079
	6. Inside surface, still air	0.12	0.12

	Between studs	At studs
Total unit thermal resistance of each section, R (in m²·°C/W)	3.05	1.23
The U-factor of each section, $U = 1/R$, in W/m²·°C	0.328	0.813
Area fraction of each section, f_{area}	0.75	0.25

Overall U-factor: $U = \Sigma f_{area,i}\, U_i = 0.75 \times 0.328 + 0.25 \times 0.813 = $ **0.449 W/m²·°C**

Overall unit thermal resistance: $R = 1/U = $ **2.23 m²·°C/W**

The perimeter of the building is 50 m and the height of the walls is 2.5 m. Noting that glazing constitutes 20 percent of the walls, the total wall area is

$$A_{wall} = 0.80(\text{Perimeter})(\text{Height}) = 0.80(50 \text{ m})(2.5 \text{ m}) = 100 \text{ m}^2$$

Then the rate of heat loss through the walls under design conditions becomes

$$\dot{Q}_{wall} = (UA)_{wall}\,(T_i - T_o)$$
$$= (0.449 \text{ W/m}^2 \cdot °\text{C})(100 \text{ m}^2)[22 - (-2)°\text{C}]$$
$$= \textbf{1078 W}$$

Discussion Note that a 1-kW resistance heater in this house will make up almost all the heat lost through the walls, except through the doors and windows, when the outdoor air temperature drops to −2°C.

EXAMPLE 12-7   The *R*-Value of a Wall with Rigid Foam

The 13-mm-thick wood fiberboard sheathing of the wood stud wall discussed in the previous example is replaced by a 25-mm-thick rigid foam insulation. Determine the percent increase in the *R*-value of the wall as a result.

Solution The overall R-value of the existing wall was determined in Example 12-6 to be 2.23 $m^2 \cdot$ °C/W. Noting that the R-values of the fiberboard and the foam insulation are 0.23 $m^2 \cdot$ °C/W and 0.98 $m^2 \cdot$ °C/W, respectively, and the added and removed thermal resistances are in series, the overall R-value of the wall after modification becomes

$$R_{new} = R_{old} - R_{removed} + R_{added}$$
$$= 2.23 - 0.23 + 0.98$$
$$= 2.98 \ m^2 \cdot °C/W$$

This represents an increase of $(2.98 - 2.23)/2.23 = 0.34$ or **34 percent** in the R-value of the wall. This example demonstrated how to evaluate the new R-value of a structure when some structural members are added or removed.

EXAMPLE 12-8 The *R*-Value of a Masonry Wall

Determine the overall unit thermal resistance (the R-value) and the overall heat transfer coefficient (the U-factor) of a masonry cavity wall that is built around 6-in.-thick concrete blocks made of lightweight aggregate with 3 cores filled with perlite ($R = 4.2$ h $\cdot$ ft$^2 \cdot$ °F/Btu). The outside is finished with 4-in. face brick with $\frac{1}{2}$-in. cement mortar between the bricks and concrete blocks. The inside finish consists of $\frac{1}{2}$ in. gypsum wallboard separated from the concrete block by $\frac{3}{4}$-in.-thick (1-in. × 3-in. nominal) vertical furring ($R = 4.2$ h $\cdot$ ft$^2 \cdot$ °F/Btu) whose center-to-center distance is 16 in. Both sides of the $\frac{3}{4}$-in.-thick air space between the concrete block and the gypsum board are coated with reflective aluminum foil ($\varepsilon = 0.05$) so that the effective emissivity of the air space is 0.03. For a mean temperature of 50°F and a temperature difference of 30°F, the R-value of the air space is 2.91 h $\cdot$ ft$^2 \cdot$ °F/Btu. The reflective air space constitutes 80 percent of the heat transmission area, while the vertical furring constitutes 20 percent.

Solution The R-value and the U-factor of a masonry cavity wall are to be determined.

Assumptions **1** Steady operating conditions exist. **2** Heat transfer through the wall is one-dimensional. **3** Thermal properties of the wall and the heat transfer coefficients are constant.

Properties The R-values of different materials are given in Table 12-10.

Analysis The schematic of the wall as well as the different elements used in its construction are shown below. Following the approach described above and using the available R-values from Table 12-10, the overall R-value of the wall is determined in the table below.

		R-value, h·ft²·°F/Btu	
	Construction	Between furring	At furring
	1. Outside surface, 15 mph wind	0.17	0.17
	2. Face brick, 4 in.	0.43	0.43
	3. Cement mortar, 0.5 in.	0.10	0.10
	4. Concrete block, 6 in.	4.20	4.20
	5a. Reflective air space, $\frac{3}{4}$ in.	2.91	—
	5b. Nominal 1 × 3 vertical furring	—	0.94
	6. Gypsum wallboard, 0.5 in.	0.45	0.45
	7. Inside surface, still air	0.68	0.68

Total unit thermal resistance of each section, R 8.94   6.97

The U-factor of each section, $U = 1/R$, in Btu/h·ft²·°F   0.112   0.143

Area fraction of each section, f_{area} 0.80   0.20

Overall U-factor: $U = \sum f_{area,i} U_i = 0.80 \times 0.112 + 0.20 \times 0.143 = \mathbf{0.118\ Btu/h \cdot ft^2 \cdot °F}$

Overall unit thermal resistance: $R = 1/U = \mathbf{8.46\ h \cdot ft^2 \cdot °F/Btu}$

Therefore, the overall unit thermal resistance of the wall is 8.46 h·ft²·°F/Btu and the overall U-factor is 0.118 Btu/h·ft²·°F. These values account for the effects of the vertical furring.

EXAMPLE 12-9   The *R*-Value of a Pitched Roof

Determine the overall unit thermal resistance (the R-value) and the overall heat transfer coefficient (the U-factor) of a 45° pitched roof built around nominal 2-in. × 4-in. wood studs with a center-to-center distance of 16 in. The 3.5-in.-wide air space between the studs does not have any reflective surface and thus its effective emissivity is 0.84. For a mean temperature of 90°F and a temperature difference of 30°F, the R-value of the air space is 0.86 h·ft²·°F/Btu. The lower part of the roof is finished with $\frac{1}{2}$-in. gypsum wallboard and the upper part with $\frac{5}{8}$-in. plywood, building paper, and asphalt shingle roofing. The air space constitutes 75 percent of the heat transmission area, while the studs and headers constitute 25 percent.

Solution   The R-value and the U-factor of a 45° pitched roof are to be determined.

Assumptions   **1** Steady operating conditions exist. **2** Heat transfer through the roof is one-dimensional. **3** Thermal properties of the roof and the heat transfer coefficients are constant.

Properties The *R*-values of different materials are given in Table 12-10.

Analysis The schematic of the pitched roof as well as the different elements used in its construction are shown below. Following the approach described above and using the available *R*-values from Table 12-10, the overall *R*-value of the roof can be determined in the table below.

Schematic		R-value, h · ft² · °F/Btu	
	Construction	Between studs	At studs
	1. Outside surface, 15 mph wind	0.17	0.17
	2. Asphalt shingle roofing	0.44	0.44
	3. Building paper	0.10	0.10
	4. Plywood deck, $\frac{5}{8}$ in.	0.78	0.78
	5a. Nonreflective air space, 3.5 in.	0.86	—
	5b. Wood stud, 2 in. by 4 in.	—	3.58
	6. Gypsum wallboard, 0.5 in.	0.45	0.45
	7. Inside surface, 45° slope, still air	0.63	0.63

	Between studs	At studs
Total unit thermal resistance of each section, *R*	3.43	6.15
The *U*-factor of each section, $U = 1/R$, in Btu/h · ft² · °F	0.292	0.163
Area fraction of each section, f_{area}	0.75	0.25

Overall *U*-factor: $U = \sum f_{area,\,i} U_i = 0.75 \times 0.292 + 0.25 \times 0.163 = \textbf{0.260 Btu/h} \cdot \textbf{ft}^2 \cdot \textbf{°F}$
Overall unit thermal resistance: $R = 1/U = \textbf{3.85 h} \cdot \textbf{ft}^2 \cdot \textbf{°F/Btu}$

Therefore, the overall unit thermal resistance of this pitched roof is 3.85 h · ft² · °F/Btu and the overall *U*-factor is 0.260 Btu/h · ft² · °F. Note that the wood studs offer much larger thermal resistance to heat flow than the air space between the studs.

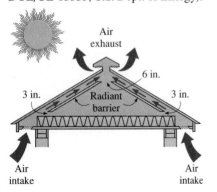

FIGURE 12-33

Ventilation paths for a naturally ventilated attic and the appropriate size of the flow areas around the radiant barrier for proper air circulation (from DOE/CE-0335P, U.S. Dept. of Energy).

The construction of wood frame flat ceilings typically involve 2-in. × 6-in. joists on 400-mm (16-in.) or 600-mm (24-in.) centers. The fraction of framing is usually taken to be 0.10 for joists on 400-mm centers and 0.07 for joists on 600-mm centers.

Most buildings have a combination of a ceiling and a roof with an attic space in between, and the determination of the *R*-value of the roof–attic–ceiling combination depends on whether the attic is vented or not. For adequately ventilated attics, the attic air temperature is practically the same as the outdoor air temperature, and thus heat transfer through the roof is governed by the *R*-value of the ceiling only. However, heat is also transferred between the roof and the ceiling by radiation, and it needs to be considered (Fig. 12-33). The major function of the roof in this case is to serve as a radiation shield by blocking off solar radiation. Effectively ventilating the attic in summer should not lead one to believe that heat gain to the building through the attic is greatly reduced. This is because most of the heat transfer through the attic is by radiation.

Radiation heat transfer between the ceiling and the roof can be minimized by covering at least one side of the attic (the roof or the ceiling side)

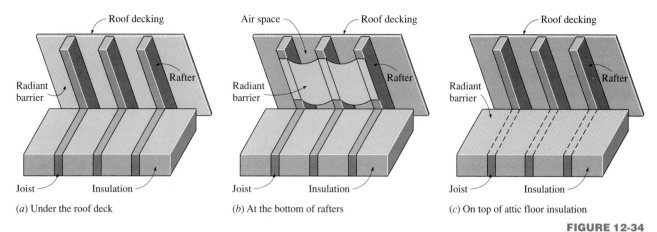

(a) Under the roof deck (b) At the bottom of rafters (c) On top of attic floor insulation

FIGURE 12-34

Three possible locations for an attic radiant barrier (from DOE/CE-0335P, U.S. Dept. of Energy).

by a reflective material, called *radiant barrier,* such as aluminum foil or aluminum-coated paper. Tests on houses with R-19 attic floor insulation have shown that radiant barriers can reduce summer ceiling heat gains by 16 to 42 percent compared to an attic with the same insulation level and no radiant barrier. Considering that the ceiling heat gain represents about 15 to 25 percent of the total cooling load of a house, radiant barriers will reduce the air conditioning costs by 2 to 10 percent. Radiant barriers also reduce the heat loss in winter through the ceiling, but tests have shown that the percentage reduction in heat losses is less. As a result, the percentage reduction in heating costs will be less than the reduction in the air-conditioning costs. Also, the values given are for new and undusted radiant barrier installations, and percentages will be lower for aged or dusty radiant barriers.

Some possible locations for attic radiant barriers are given in Figure 12-34. In whole house tests on houses with R-19 attic floor insulation, radiant barriers have reduced the ceiling heat gain by an average of 35 percent when the radiant barrier is installed on the attic floor, and by 24 percent when it is attached to the bottom of roof rafters. Test cell tests also demonstrated that the best location for radiant barriers is the attic floor, provided that the attic is not used as a storage area and is kept clean.

For unvented attics, any heat transfer must occur through (1) the ceiling, (2) the attic space, and (3) the roof (Fig. 12-35). Therefore, the overall R-value of the roof–ceiling combination with an unvented attic depends on the combined effects of the R-value of the ceiling and the R-value of the roof as well as the thermal resistance of the attic space. The attic space can be treated as an air layer in the analysis. But a more practical way of accounting for its effect is to consider surface resistances on the roof and ceiling surfaces facing each other. In this case, the R-values of the ceiling and the roof are first determined separately (by using convection resistances for the still-air case for the attic surfaces). Then it can be shown that the overall R-value of the ceiling–roof combination per unit area of the ceiling can be expressed as

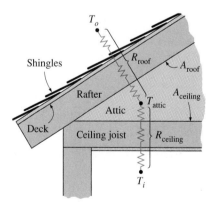

FIGURE 12-35

Thermal resistance network for a pitched roof–attic–ceiling combination for the case of an unvented attic.

$$R = R_{\text{ceiling}} + R_{\text{roof}} \left(\frac{A_{\text{ceiling}}}{A_{\text{roof}}} \right) \qquad (12\text{-}25)$$

where A_{ceiling} and A_{roof} are the ceiling and roof areas, respectively. The area ratio is equal to 1 for flat roofs and is less than 1 for pitched roofs. For a 45° pitched roof, the area ratio is $A_{\text{ceiling}}/A_{\text{roof}} = 1/\sqrt{2} = 0.707$. Note that the pitched roof has a greater area for heat transfer than the flat ceiling, and the area ratio accounts for the reduction in the unit R-value of the roof when expressed per unit area of the ceiling. Also, the direction of heat flow is up in winter (heat loss through the roof) and down in summer (heat gain through the roof).

The R-value of a structure determined by analysis assumes that the materials used and the quality of workmanship meet the standards. Poor workmanship and substandard materials used during construction may result in R-values that deviate from predicted values. Therefore, some engineers use a safety factor in their designs based on experience in critical applications.

12-7 ■ HEAT LOSS FROM BASEMENT WALLS AND FLOORS

The floors and the underground portion of the walls of a basement are in direct contact with the ground, which is usually at a different temperature than the basement, and thus there is heat transfer between the basement and the ground. This is *conduction heat transfer* because of the direct contact between the walls and the floor, and it depends on the temperature difference between the basement and the ground, the construction of the walls and the floor, and the thermal conductivity of the surrounding earth. There is considerable uncertainty in the ground heat loss calculations, and they probably constitute the least accurate part of heat load estimates of a building because of the large thermal mass of the ground and the large variation of the thermal conductivity of the soil [it varies between 0.5 and 2.5 W/m · °C (or 0.3 to 1.4 Btu/h · ft · °F), depending on the composition and moisture content]. However, ground heat losses are a small fraction of total heat load of a large building, and thus it has little effect on the overall heat load.

Temperature measurements of uninsulated basements indicate that heat conduction through the ground is not one-dimensional, and thus it cannot be estimated by a simple one-dimensional heat conduction analysis. Instead, heat conduction is observed to be two-dimensional with nearly circular concentric heat flow lines centered at the intersection of the wall and the earth (Fig. 12-36). When partial insulation is applied to the walls, the heat flow lines tend to be straight lines instead of circular. Also, a basement wall whose top portion is exposed to ambient air may act as a thermal bridge, conducting heat upward and dissipating it to the ambient from its top part. This vertical heat flow may be significant in some cases.

Despite its complexity, heat loss through the below-grade section of **basement walls** can be determined easily from

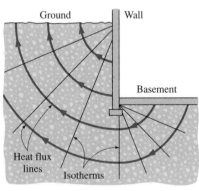

FIGURE 12-36

Radial isotherms and circular heat flow lines during heat flow from uninsulated basement.

$$\dot{Q}_{\text{basement walls}} = U_{\text{wall, ave}} A_{\text{wall}} (T_{\text{basement}} - T_{\text{ground surface}}) \qquad \text{(W)} \qquad (12\text{-}26)$$

where

$U_{wall,ave}$ = Average overall heat transfer coefficient between the basement wall and the surface of the ground

$A_{wall,ave}$ = Wall surface area of the basement (underground portion)

$T_{basement}$ = Interior air temperature of the basement

$T_{ground\,surface}$ = Mean ground surface temperature in winter

The overall heat transfer coefficients at different depths are given in Table 12-14*a* for depth increments of 0.3 m (or 1 ft) for uninsulated and insulated concrete walls. These values are based on a soil thermal conductivity of 1.38 W/m · °C (0.8 Btu/h · ft · °F). Note that the heat transfer coefficient values decrease with increasing depth since the heat at a lower section must pass through a longer path to reach the ground surface. For a specified wall, $U_{wall,ave}$ is simply the arithmetic average of the U_{wall} values corresponding to the different sections of the wall. Also note that heat loss through a depth increment is equal to the U_{wall} value of the increment multiplied by the perimeter of the building, the depth increment, and the temperature difference.

The interior air temperature of the basement can vary considerably, depending on whether it is being heated or not. In the absence of reliable data, the basement temperature can be taken to be 10°C since the heating system, water heater, and heating ducts are often located in the basement. Also, the ground surface temperature fluctuates about the mean winter ambient temperature by an amplitude A that varies with geographic location and the condition of the surface, as shown in Figure 12-37. Therefore, a reasonable value for the design temperature of ground surface can be obtained by subtracting A for the specified location from the mean winter air temperature. That is,

$$T_{ground\,surface} = T_{winter,mean} - A \qquad (12\text{-}27)$$

Heat loss through the **basement floor** is much smaller since the heat flow path to the ground surface is much longer in this case. It is calculated in a similar manner from

$$\dot{Q}_{basement\,floor} = U_{floor} A_{floor} (T_{basement} - T_{ground\,surface}) \qquad (W) \quad (12\text{-}28)$$

where U_{floor} is the overall heat transfer coefficient at the basement floor whose values are listed in Table 12-14*b*, A_{floor} is the floor area, and the temperature difference is the same as the one used for the basement wall.

The temperature of an unheated below-grade basement is between the temperatures of the rooms above and the ground temperature. Heat losses from the water heater and the space heater located in the basement usually keep the air near the basement ceiling sufficiently warm. Heat losses from the rooms above to the basement can be neglected in such cases. This will not be the case, however, if the basement has windows.

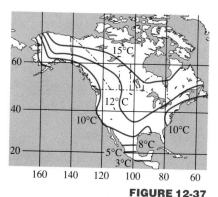

FIGURE 12-37

Lines of constant amplitude of annual soil temperature swings (from ASHRAE *Handbook of Fundamentals*, Ref. 1, Chap. 25, Fig. 6). Multiply by 1.8 to get values in °F.

EXAMPLE 12-10 Heat Loss from a Below-Grade Basement

Consider a basement in Chicago, where the mean winter temperature is 2.4°C. The basement is 8.5 m wide and 12 m long, and the basement floor is 2.1 m below grade (the ground level). The top 0.9-m section of the wall below the grade is insulated with R-2.20 m² · °C/W insulation. Assuming the interior temperature

TABLE 12-14

Heat transfer coefficients for heat loss through the basement walls, basement floors, and concrete floors on grade in both SI and English units (from ASHRAE *Handbook of Fundamentals*, Ref. 1, Chap. 25, Tables 14, 15, 16)

(*a*) Heat loss through below-grade basement walls

SI Units					English Units				
	U_{wall}, W/m$^2 \cdot$°C					U_{wall}, Btu/h $\cdot$ ft$^2 \cdot$ °F			
	Insulation level, m$^2 \cdot$ °C/W					Insulation level, h $\cdot$ ft$^2 \cdot$ °F/Btu			
Depth, m	No insulation	R-0.73	R-1.47	R-2.2	Depth, ft	No insulation	R-4.17	R-8.34	R-12.5
0.0–0.3	7.77	2.87	1.77	1.27	0–1	0.410	0.152	0.093	0.067
0.3–0.6	4.20	2.20	1.50	1.20	1–2	0.222	0.116	0.079	0.059
0.6–0.9	2.93	1.77	1.27	1.00	2–3	0.155	0.094	0.068	0.053
0.9–1.2	2.23	1.50	1.13	0.90	3–4	0.119	0.079	0.060	0.048
1.2–1.5	1.80	1.30	1.00	0.83	4–5	0.096	0.069	0.053	0.044
1.5–1.8	1.50	1.13	0.90	0.77	5–6	0.079	0.060	0.048	0.040
1.8–2.1	1.30	1.00	0.83	0.70	6–7	0.069	0.054	0.044	0.037

(*b*) Heat loss through below-grade basement floors

	U_{floor}, W/m$^2 \cdot$ °C					U_{floor}, Btu/h $\cdot$ ft$^2 \cdot$ °F			
Depth of wall below grade, m	Shortest width of building, m				Depth of wall below grade, ft	Shortest width of building, ft			
	6.0	7.3	8.5	9.7		20	24	28	32
1.5	0.18	0.16	0.15	0.13	5	0.032	0.029	0.026	0.023
1.8	0.17	0.15	0.14	0.12	6	0.030	0.027	0.025	0.022
2.1	0.16	0.15	0.13	0.12	7	0.029	0.026	0.023	0.021

(*c*) Heat loss through on-grade concrete basement floors

	U_{grade}, W/m $\cdot$ °C (per unit length of perimeter)					U_{grade}, Btu/h $\cdot$ ft $\cdot$ °F (per unit length of perimeter)			
Wall construction	Insulation (from edge to footer)	Weather conditions			Wall construction	Insulation (from edge to footer)	Weather conditions		
		Mild	Moderate	Severe			Mild	Moderate	Severe
8-in. (20-cm) block wall with brick	None	1.24	1.17	1.07	8-in. (20-cm) block wall with brick	None	0.62	0.68	0.72
	R-0.95	0.97	0.86	0.83		R-5.4	0.48	0.50	0.56
4-in. (10-cm) block wall with brick	None	1.61	1.45	1.38	4-in. (10-cm) block wall with brick	None	0.80	0.84	0.93
	R-0.95	0.93	0.85	0.81		R-5.4	0.47	0.49	0.54
Metal–stud wall with stucco	None	2.32	2.07	1.99	Metal–stud wall with stucco	None	1.15	1.20	1.34
	R-0.95	1.00	0.92	0.88		R-5.4	0.51	0.53	0.38
Poured concrete wall with heating pipes or ducts near perimeter	None	4.72	3.67	3.18	Poured concrete wall with heating pipes or ducts near perimeter	None	1.84	2.12	2.73
	R-0.95	1.56	1.24	1.11		R-5.4	0.64	0.72	0.90

of the basement is 22°C, determine the peak heat loss from the basement to the ground through its walls and floor.

Solution The peak heat loss from a below-grade basement in Chicago to the ground through its walls and the floor is to be determined.

Assumption Steady operating conditions exist.

Properties The heat transfer coefficients are given in Table 12-14.

Analysis The schematic of the basement is given in Figure 12-38. The floor and wall areas of the basement are

$$A_{wall} = \text{Height} \times \text{Perimeter} = 2 \times (2.1 \text{ m})(8.5 + 12)\text{m} = 86.1 \text{ m}^2$$
$$A_{floor} = \text{Length} \times \text{Width} = (8.5 \text{ m})(12 \text{ m}) = 102 \text{ m}^2$$

The amplitude of the annual soil temperature is determined from Figure 12-37 to be 12°C. Then the ground surface temperature for the design heat loss becomes

$$T_{ground\,surface} = T_{winter,\,mean} - A = 2.4 - 12 = -9.6°C$$

The top 0.9-m section of the wall below the grade is insulated with R-2.2, and the heat transfer coefficients through that section are given in Table 12-14a to be 1.27, 1.20, and 1.00 W/m² · °C through the first, second, and third 0.3-m-wide depth increments, respectively. The heat transfer coefficients through the un-insulated section of the wall that extends from the 0.9-m to the 2.1-m level are determined from the same table to be 2.23, 1.80, 1.50, and 1.30 W/m² · °C for each of the remaining 0.3-m-wide depth increments. The average overall heat transfer coefficient is

$$U_{wall,\,ave} = \frac{\sum U_{wall}}{\text{No. of increments}} = \frac{1.27 + 1.2 + 1.0 + 2.23 + 1.8 + 1.5 + 1.3}{7}$$
$$= 1.47 \text{ W/m}^2 \cdot °C$$

Then the heat loss through the basement wall becomes

$$\dot{Q}_{basement\,walls} = U_{wall,\,ave}\, A_{wall}\, (T_{basement} - T_{ground\,surface})$$
$$= (1.47 \text{ W/m}^2 \cdot °C)(86.1 \text{ m}^2)[22 - (-9.6)°C] = 4000 \text{ W}$$

The shortest width of the house is 8.5 m, and the depth of the foundation below grade is 2.1 m. The floor heat transfer coefficient is given in Table 12-14b to be 0.13 W/m² · °C. Then the heat loss through the floor of the basement becomes

$$\dot{Q}_{basement\,floor} = U_{floor}\, A_{floor}\, (T_{basement} - T_{ground\,surface})$$
$$= (0.13 \text{ W/m}^2 \cdot °C)(102 \text{ m}^2)[(22 - (-9.6)]°C = 419 \text{ W}$$

which is considerably less than the heat loss through the wall. The total heat loss from the basement is then determined to be

$$\dot{Q}_{basement} = \dot{Q}_{basement\,walls} + \dot{Q}_{basement\,floor} = 4000 + 419 = \textbf{4419 W}$$

This is the *design* or *peak* rate of heat transfer from the below-grade section of the basement, and this is the value to be used when sizing the heating system. The actual heat loss from the basement will be much less than that most of the time.

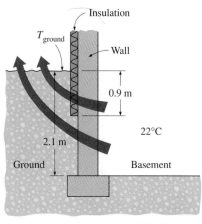

FIGURE 12-38
Schematic for Example 12-10.

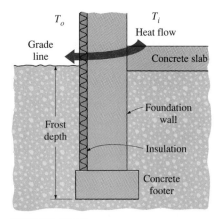

FIGURE 12-39
An on-grade concrete floor with
insulated foundation wall.

Concrete Floors on Grade (at Ground Level)

Many residential and commercial buildings do not have a basement, and the floor sits directly on the ground at or slightly above the ground level. Research indicates that heat loss from such floors is mostly through the perimeter to the outside air rather than through the floor into the ground, as shown in Figure 12-39. Therefore, total heat loss from a concrete slab floor is proportional to the perimeter of the slab instead of the area of the floor and is expressed as

$$\dot{Q}_{\text{floor on grade}} = U_{\text{grade}}\, p_{\text{floor}}\, (T_{\text{indoor}} - T_{\text{outdoor}}) \qquad \text{(W)} \quad (12\text{-}29)$$

where U_{grade} represents the rate of heat transfer from the slab per unit temperature difference between the indoor temperature T_{indoor} and the outdoor temperature T_{outdoor} and per unit length of the perimeter p_{floor} of the building.

Typical values of U_{grade} are listed in Table 12-14c for four common types of slab-on-grade construction for mild, moderate, and severe weather conditions. The ground temperature is not involved in the formulation since the slab is located above the ground level and heat loss to the ground is negligible. Note from the table that perimeter insulation of slab-on-grade reduces heat losses considerably, and thus it saves energy while enhancing comfort. Insulation is a must for radiating floors that contain heated pipes or ducts through which hot water or air is circulated since heat loss in the uninsulated case is about three times that of the insulated case. This is also the case when baseboard heaters are used on the floor near the exterior walls. Heat transfer through the floors and the basement is usually ignored in cooling load calculations.

Heat Loss from Crawl Spaces

A crawl space can be considered to be a small basement except that it may be vented year round to prevent the accumulation of moisture and radioactive gases such as radon. Venting the crawl space during the heating season creates a low temperature region underneath the house and causes considerable heat loss through the floor. The ceiling of the crawl space (i.e., the floor of the building) in such cases must be insulated. If the vents are closed during the heating season, then the walls of the crawl space can be insulated instead.

The temperature of the crawl space will be very close to the ambient air temperature when it is well ventilated. The heating ducts and hot water pipes passing through the crawl space must be adequately insulated in this case. In severe climates, it may even be necessary to insulate the cold water pipes to prevent freezing. The temperature of the crawl space will approach the indoor temperature when the vents are closed for the heating season. The air infiltration in this case is estimated to be 0.67 air change per hour.

When the crawl space temperature is known, heat loss through the **floor of the building** is determined from

$$\dot{Q}_{\text{building floor}} = U_{\text{building floor}}\, A_{\text{floor}}\, (T_{\text{indoor}} - T_{\text{crawl}}) \qquad \text{(W)} \quad (12\text{-}30)$$

where $U_{\text{building floor}}$ is the overall heat transfer coefficient for the floor, A_{floor} is the floor area, and T_{indoor} and T_{crawl} are the indoor and crawl space temperatures, respectively.

Overall heat transfer coefficients associated with the walls, floors, and ceilings of typical crawl spaces are given in Table 12-15. Note that heat loss through the uninsulated floor to the crawl space is three times that of the insulated floor. The ground temperature can be taken to be 10°C when calculating heat loss from the crawl space to the ground. Also, the infiltration heat loss from the crawl space can be determined from

$$\dot{Q}_{\text{infiltration, crawl}} = (\rho C_p \dot{V})_{\text{air}} (T_{\text{crawl}} - T_{\text{ambient}})$$
$$= \rho C_p (\text{ACH})(V_{\text{crawl}})(T_{\text{crawl}} - T_{\text{ambient}}) \qquad \text{(J/h)} \qquad \text{(12-31)}$$

where ACH is the air changes per hour, V_{crawl} is the volume of the crawl space, and T_{crawl} and T_{ambient} are the crawl space and ambient temperatures, respectively.

In the case of *closed vents*, the steady state temperature of the crawl space will be between the indoors and outdoors temperatures and can be determined from the energy balance expressed as

$$\dot{Q}_{\text{floor}} + \dot{Q}_{\text{infiltration}} + \dot{Q}_{\text{ground}} + \dot{Q}_{\text{wall}} = 0 \qquad \text{(W)} \qquad \text{(12-32)}$$

and assuming all heat transfer to be toward the crawl space for convenience in formulation.

EXAMPLE 12-11   Heat Loss to the Crawl Space through Floors

Consider a crawl space that is 8-m wide, 12-m long, and 0.70-m high, as shown in Figure 12-40, whose vent is kept open. The interior of the house is maintained at 22°C and the ambient temperature is −5°C. Determine the rate of heat loss through the floor of the house to the crawl space for the cases of (*a*) an insulated and (*b*) an uninsulated floor.

Solution   The vent of the crawl space is kept open. The rate of heat loss to the crawl space through insulated and uninsulated floors is to be determined.

Assumption   Steady operating conditions exist.

Properties   The overall heat transfer coefficient for the insulated floor is given in Table 12-15 to be 0.432 W/m² · °C.

Analysis   (*a*) The floor area of the house (or the ceiling area of the crawl space) is

$$A_{\text{floor}} = \text{Length} \times \text{Width} = (8\text{ m})(12\text{ m}) = 96\text{ m}^2$$

Then the heat loss from the house to the crawl space becomes

$$\dot{Q}_{\text{insulated floor}} = U_{\text{insulated floor}} A_{\text{floor}} (T_{\text{indoor}} - T_{\text{crawl}})$$
$$= (0.432\text{ W/m}^2 \cdot °\text{C})(96\text{ m}^2)[(22 - (-5)]°\text{C} = \mathbf{1120\ W}$$

(*b*) The heat loss for the uninsulated case is determined similarly to be

TABLE 12-15

Estimated *U*-values for insulated and uninsulated crawl spaces (from ASHRAE *Handbook of Fundamentals*, Ref. 1, Chap. 25, Table 13)

Application	*U*, **W/m² · °C**[1]	
	Un-insulated	In-sulated[2]
Floor above crawl space	1.42	0.432
Ground of crawl space	0.437	0.437
Wall of crawl space	2.77	1.07

[1]Multiply given values by 0.176 to convert them to Btu/h · ft² · °F.

[2]An insulation *R*-value of 1.94 m² · °C/W is used on the floor, and 0.95 m² · °C/W on the walls.

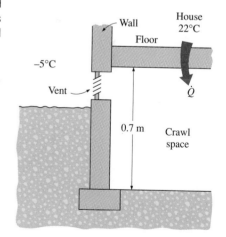

FIGURE 12-40

Schematic for Example 12-11.

$$\dot{Q}_{\text{uninsulated floor}} = U_{\text{uninsulated floor}} A_{\text{floor}} (T_{\text{indoor}} - T_{\text{crawl}})$$
$$= (1.42 \text{ W/m}^2 \cdot °C)(96 \text{ m}^2)[(22 - (-5)]°C = \textbf{3681 W}$$

which is more than three times the heat loss through the insulated floor. Therefore, it is a good practice to insulate floors when the crawl space is ventilated to conserve energy and enhance comfort.

12-8 ■ HEAT TRANSFER THROUGH WINDOWS

Windows are *glazed apertures* in the building envelope that typically consist of single or multiple *glazing* (glass or plastic), *framing,* and *shading.* In a building envelope, windows offer the *least resistance* to heat flow. In a typical house, about *one-third* of the total heat loss in winter occurs through the windows. Also, most air infiltration occurs at the edges of the windows. The solar heat gain through the windows is responsible for much of the cooling load in summer. The net effect of a window on the heat balance of a building depends on the characteristics and orientation of the window as well as the solar and weather data. Workmanship is very important in the construction and installation of windows to provide effective sealing around the edges while allowing them to be opened and closed easily.

Despite being so undesirable from an energy conservation point of view, windows are an essential part of any building envelope since they enhance the appearance of the building, allow *daylight* and *solar heat* to come in, and allow people to view and observe outside without leaving their home. For low-rise buildings, windows also provide easy exit areas during emergencies such as fire. Important considerations in the selection of windows are *thermal comfort* and *energy conservation.* A window should have a good light transmittance while providing effective resistance to heat flow. The lighting requirements of a building can be minimized by maximizing the use of natural daylight. Heat loss in winter through the windows can be minimized by using airtight double- or triple-pane windows with spectrally selective films or coatings, and letting in as much solar radiation as possible. Heat gain and thus cooling load in summer can be minimized by using effective internal or external shading on the windows.

Even in the absence of solar radiation and air infiltration, heat transfer through the windows is more complicated than it appears to be. This is because the structure and properties of the frame are quite different than the glazing. As a result, heat transfer through the frame and the edge section of the glazing adjacent to the frame is two-dimensional. Therefore, it is customary to consider the window in three regions when analyzing heat transfer through it: (1) the *center-of-glass,* (2) the *edge-of-glass,* and (3) the *frame* regions, as shown in Figure 12-41. Then the total rate of heat transfer through the window is determined by adding the heat transfer through each region as

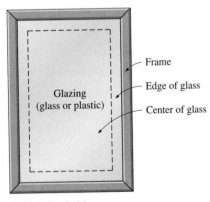

FIGURE 12-41

The three regions of a window considered in heat transfer analysis.

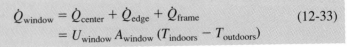

$$\dot{Q}_{\text{window}} = \dot{Q}_{\text{center}} + \dot{Q}_{\text{edge}} + \dot{Q}_{\text{frame}} \qquad (12\text{-}33)$$
$$= U_{\text{window}} A_{\text{window}} (T_{\text{indoors}} - T_{\text{outdoors}})$$

where

$$U_{\text{window}} = (U_{\text{center}} A_{\text{center}} + U_{\text{edge}} A_{\text{edge}} + U_{\text{frame}} A_{\text{frame}})/A_{\text{window}} \quad \text{(12-34)}$$

is the **U-factor** or the **overall heat transfer coefficient** of the window; A_{window} is the window area; A_{center}, A_{edge}, and A_{frame} are the areas of the center, edge, and frame sections of the window, respectively; and U_{center}, U_{edge}, and U_{frame} are the heat transfer coefficients for the center, edge, and frame sections of the window. Note that $A_{\text{window}} = A_{\text{center}} + A_{\text{edge}} + A_{\text{frame}}$, and the overall U-factor of the window is determined from the area-weighed U-factors of each region of the window. Also, the inverse of the U-factor is the R-value, which is the unit thermal resistance of the window (thermal resistance for a unit area).

Consider steady one-dimensional heat transfer through a single-pane glass of thickness L and thermal conductivity k. The thermal resistance network of this problem consists of surface resistances on the inner and outer surfaces and the conduction resistance of the glass in series, as shown in Figure 12-42, and the total resistance on a unit area basis can be expressed as

$$R_{\text{total}} = R_{\text{inside}} + R_{\text{glass}} + R_{\text{outside}} = \frac{1}{h_i} + \frac{L_{\text{glass}}}{k_{\text{glass}}} + \frac{1}{h_o} \quad \text{(12-35)}$$

Using common values of 3 mm for the thickness and 0.92 W/m · °C for the thermal conductivity of the glass and the winter design values of 8.29 and 34.0 W/m² · °C for the inner and outer surface heat transfer coefficients, the thermal resistance of the glass is determined to be

$$R_{\text{total}} = \frac{1}{8.29 \text{ W/m}^2 \cdot \text{°C}} + \frac{0.003 \text{ m}}{0.92 \text{ W/m} \cdot \text{°C}} + \frac{1}{34.0 \text{ W/m}^2 \cdot \text{°C}}$$
$$= 0.121 + 0.003 + 0.029 = 0.153 \text{ m}^2 \cdot \text{°C/W}$$

Note that the ratio of the glass resistance to the total resistance is

$$\frac{R_{\text{glass}}}{R_{\text{total}}} = \frac{0.003 \text{ m}^2 \cdot \text{°C/W}}{0.153 \text{ m}^2 \cdot \text{°C/W}} = 0.020 = 2.0\%$$

That is, the glass layer itself contributes about 2 percent of the total thermal resistance of the window, which is negligible. The situation would not be much different if we used acrylic, whose thermal conductivity is 0.19 W/m · °C, instead of glass. Therefore, we cannot reduce the heat transfer through the window effectively by simply increasing the thickness of the glass. But we can reduce it by trapping still air between two layers of glass. The result is a **double-pane window,** which has become the norm in window construction.

The thermal conductivity of air at room temperature is $k_{\text{air}} = 0.025$ W/m · °C, which is one-thirtieth that of glass. Therefore, the thermal resistance of 1-cm-thick still air is equivalent to the thermal resistance of a 30-cm-thick glass layer. Disregarding the thermal resistances of glass layers, the thermal resistance and U-factor of a double-pane window can be expressed as (Fig. 12-43)

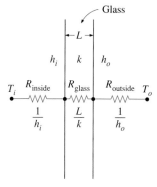

FIGURE 12-42

The thermal resistance network for heat transfer through a single glass.

FIGURE 12-43

The thermal resistance network for heat transfer through the center section of a double-pane window (the resistances of the glasses are neglected).

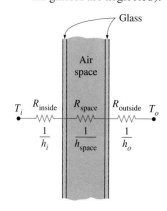

$$\frac{1}{U_{\text{double-pane (center region)}}} \cong \frac{1}{h_i} + \frac{1}{h_{\text{space}}} + \frac{1}{h_o} \qquad (12\text{-}36)$$

where $h_{\text{space}} = h_{\text{rad, space}} + h_{\text{conv, space}}$ is the combined radiation and convection heat transfer coefficient of the space trapped between the two glass layers.

Roughly half of the heat transfer through the air space of a double-pane window is by radiation and the other half is by conduction (or convection, if there is any air motion). Therefore, there are two ways to minimize h_{space} and thus the rate of heat transfer through a double-pane window:

1. *Minimize radiation heat transfer through the air space.* This can be done by reducing the emissivity of glass surfaces by coating them with low-emissivity (or "low-e" for short) material. Recall that the *effective emissivity* of two parallel plates of emissivities ε_1 and ε_2 is given by

$$\varepsilon_{\text{effective}} = \frac{1}{1/\varepsilon_1 + 1/\varepsilon_2 - 1}$$

The emissivity of an ordinary glass surface is 0.84. Therefore, the effective emissivity of two parallel glass surfaces facing each other is 0.72. But when the glass surfaces are coated with a film that has an emissivity of 0.1, the effective emissivity reduces to 0.05, which is one-fourteenth of 0.72. Then for the same surface temperatures, radiation heat transfer will also go down by a factor of 14. Even if only one of the surfaces is coated, the overall emissivity reduces to 0.1, which is the emissivity of the coating. Thus it is no surprise that about one-fourth of all windows sold for residences have a low-e coating. The heat transfer coefficient h_{space} for the air space trapped between the two vertical parallel glass layers is given in Table 12-16 for 13-mm- ($\frac{1}{2}$-in.) and 6-mm- ($\frac{1}{4}$-in.) thick air spaces for various effective emissivities and temperature differences.

It can be shown that coating just one of the two parallel surfaces facing each other by a material of emissivity ε reduces the effective emissivity nearly to ε. Therefore, it is usually more economical to coat only one of the facing surfaces. Note from Figure 12-44 that coating one of the interior surfaces of a double-pane window with a material having an emissivity of 0.1 reduces by half the rate of heat transfer through the center section of the window.

2. *Minimize conduction heat transfer through air space.* This can be done by *increasing* the distance d between the two glasses. However, this cannot be done indefinitely since increasing the spacing beyond a critical value initiates convection currents in the enclosed air space, which increases the heat transfer coefficient and thus defeats the purpose. Besides, increasing the spacing also increases the thickness of the necessary framing and the cost of the window. Experimental studies have shown that when the spacing d is less than about 13 mm, there is no convection, and heat transfer through the air is by conduction. But as the spacing is increased further, convection currents appear in the air space, and the increase in heat transfer coefficient offsets

TABLE 12–16

The heat transfer coefficient h_{space} for the air space trapped between the two vertical parallel glass layers for 13-mm- and 6-mm-thick air spaces (from Building Materials and Structures, Report 151, U.S. Dept. of Commerce).

(*a*) Air space thickness = 13 mm

T_{ave}, °C	ΔT, °C	h_{space}, W/m² · °C*			
		$\varepsilon_{effective}$			
		0.72	0.4	0.2	0.1
0	5	5.3	3.8	2.9	2.4
0	15	5.3	3.8	2.9	2.4
0	30	5.5	4.0	3.1	2.6
10	5	5.7	4.1	3.0	2.5
10	15	5.7	4.1	3.1	2.5
10	30	6.0	4.3	3.3	2.7
30	5	5.7	4.6	3.4	2.7
30	15	5.7	4.7	3.4	2.8
30	30	6.0	4.9	3.6	3.0

(*b*) Air space thickness = 6 mm

T_{ave}, °C	ΔT, °C	h_{space}, W/m² · °C*			
		$\varepsilon_{effective}$			
		0.72	0.4	0.2	0.1
0	5	7.2	5.7	4.8	4.3
0	50	7.2	5.7	4.8	4.3
10	5	7.7	6.0	5.0	4.5
10	50	7.7	6.1	5.0	4.5
30	5	8.8	6.8	5.5	4.9
30	50	8.8	6.8	5.5	4.9
50	5	10.0	7.5	6.0	5.2
50	50	10.0	7.5	6.0	5.2

*Multiply by 0.176 to convert to Btu/h · ft² · °F.

FIGURE 12–44

The variation of the *U*-factor for the center section of double- and triple-pane windows with spacing between the panes (from ASHRAE *Handbook of Fundamentals,* Ref. 1, Chap. 27, Fig. 1).

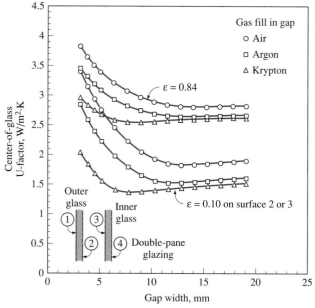

(*a*) Double-pane window

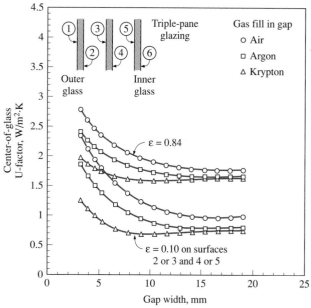

(*b*) Triple-pane window

any benefit obtained by the thicker air layer. As a result, the heat transfer coefficient remains nearly constant, as shown in Figure 12-44. Therefore, it makes no sense to use an air space thicker than 13 mm in a double-pane window unless a thin polyester film is used to divide the air space into two halves to suppress convection currents. The film provides added insulation without adding much to the weight or cost of the double-pane window. The thermal resistance of the window can be increased further by using triple- or quadruple-pane windows whenever it is economical to do so. Note that using a triple-pane window instead of a double-pane reduces the rate of heat transfer through the center section of the window by about one-third.

Another way of reducing conduction heat transfer through a double-pane window is to use a *less-conducting fluid* such as argon or krypton to fill the gap between the glasses instead of air. The gap in this case needs to be well sealed to prevent the gas from leaking outside. Of course, another alternative is to evacuate the gap between the glasses completely, but it is not practical to do so.

Edge-of-Glass *U*-Factor of a Window

The glasses in double- and triple-pane windows are kept apart from each other at a uniform distance by **spacers** made of metals or insulators like aluminum, fiberglass, wood, and butyl. Continuous spacer strips are placed around the glass perimeter to provide edge seal as well as uniform spacing. However, the spacers also serve as undesirable "thermal bridges" between the glasses, which are at different temperatures, and this short-circuiting may increase heat transfer through the window considerably. Heat transfer in the edge region of a window is two-dimensional, and lab measurements indicate that the edge effects are limited to a 6.5-cm-wide band around the perimeter of the glass.

The *U*-factor for the edge region of a window is given in Figure 12-45 relative to the *U*-factor for the center region of the window. The curve would be a straight diagonal line if the two *U*-values were equal to each other. Note that this is almost the case for insulating spacers such as wood and fiberglass. But the *U*-factor for the edge region can be twice that of the center region for conducting spacers such as those made of aluminum. Values for steel spacers fall between the two curves for metallic and insulating spacers. The edge effect is not applicable to single-pane windows.

Frame *U*-Factor

The framing of a window consists of the entire window except the glazing. Heat transfer through the framing is difficult to determine because of the different window configurations, different sizes, different constructions, and different combination of materials used in the frame construction. The type of glazing such as single pane, double pane, and triple pane affects the thickness of the framing and thus heat transfer through the frame. Most frames are made of *wood, aluminum, vinyl,* or *fiberglass.* However, using a combination of these materials (such as aluminum-clad wood and vinyl-clad aluminum) is also common to improve appearance and durability.

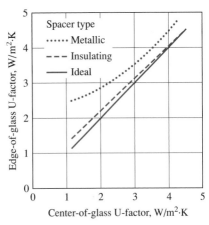

FIGURE 12-45

The edge-of-glass *U*-factor relative to the center-of-glass *U*-factor for windows with various spacers (from ASHRAE *Handbook of Fundamentals,* Ref. 1, Chap. 27, Fig. 2).

Aluminum is a popular framing material because it is inexpensive, durable, and easy to manufacture, and does not rot or absorb water like wood. However, from a heat transfer point of view, it is the least desirable framing material because of its high thermal conductivity. It will come as no surprise that the U-factor of solid aluminum frames is the highest, and thus a window with aluminum framing will lose much more heat than a comparable window with wood or vinyl framing. Heat transfer through the aluminum framing members can be reduced by using plastic inserts between components to serve as thermal barriers. The thickness of these inserts greatly affects heat transfer through the frame. For aluminum frames without the plastic strips, the primary resistance to heat transfer is due to the interior surface heat transfer coefficient. The U-factors for various frames are listed in Table 12-17 as a function of spacer materials and the glazing unit thicknesses. Note that the U-factor of metal framing and thus the rate of heat transfer through a metal window frame is more than three times that of a wood or vinyl window frame.

TABLE 12-17

Representative frame U-factors for fixed vertical windows (from ASHRAE Handbook of Fundamentals, Ref. 1, Chap. 27, Table 2)

Frame material	U-factor, W/m² · °C*
Aluminum:	
Single glazing (3 mm)	10.1
Double glazing (18 mm)	10.1
Triple glazing (33 mm)	10.1
Wood or vinyl:	
Single glazing (3 mm)	2.9
Double glazing (18 mm)	2.8
Triple glazing (33 mm)	2.7

*Multiply by 0.176 to convert to Btu/h · ft² · °F

Interior and Exterior Surface Heat Transfer Coefficients

Heat transfer through a window is also affected by the convection and radiation heat transfer coefficients between the glass surfaces and surroundings. The effects of convection and radiation on the inner and outer surfaces of glazings are usually combined into the combined convection and radiation heat transfer coefficients h_i and h_o, respectively. Under still air conditions, the combined heat transfer coefficient at the inner surface of a vertical window can be determined from

$$h_i = h_{conv} + h_{rad} = 1.77(T_g - T_i)^{0.25} + \frac{\varepsilon_g \sigma (T_g^4 - T_i^4)}{T_g - T_i} \quad \text{(W/m}^2 \cdot °\text{C)}$$

$$(12\text{-}37)$$

where T_g = glass temperature in K, T_i = indoor air temperature in K, ε_g = emissivity of the inner surface of the glass exposed to the room (taken to be 0.84 for uncoated glass), and $\sigma = 5.67 \times 10^{-8}$ W/m² · K⁴ is the Stefan–Boltzmann constant. Here the temperature of the interior surfaces facing the window is assumed to be equal to the indoor air temperature. This assumption is reasonable when the window faces mostly interior walls, but it becomes questionable when the window is exposed to heated or cooled surfaces or to other windows. The commonly used value of h_i for peak load calculation is

$$h_i = 8.29 \text{ W/m}^2 \cdot °\text{C} = 1.46 \text{ Btu/h} \cdot \text{ft}^2 \cdot °\text{F} \quad \text{(winter and summer)}$$

which corresponds to the winter design conditions of $T_i = 22°C$ and $T_g = -7°C$ for uncoated glass with $\varepsilon_g = 0.84$. But the same value of h_i can also be used for summer design conditions as it corresponds to summer conditions of $T_i = 24°C$ and $T_g = 32°C$. The values of h_i for various temperatures and glass emissivities are given in Table 12-18. The commonly used

TABLE 12-18

Combined convection and radiation heat transfer coefficient h_i at the inner surface of a vertical glass under still air conditions (in W/m² · °C)*

T_i, °C	T_g, °C	Glass emissivity, ε_g		
		0.05	0.20	0.84
20	17	2.6	3.5	7.1
20	15	2.9	3.8	7.3
20	10	3.4	4.2	7.7
20	5	3.7	4.5	7.9
20	0	4.0	4.8	8.1
20	-5	4.2	5.0	8.2
20	-10	4.4	5.1	8.3

*Multiply by 0.176 to convert to Btu/h · ft² · °F.

values of h_o for peak load calculations are the same as those used for outer wall surfaces (34.0 W/m² · °C for winter and 22.7 W/m² · °C for summer).

Overall *U*-Factor of Windows

The overall *U*-factors for various kinds of windows and skylights are evaluated using computer simulations and laboratory testing for winter design conditions; representative values are given in Table 12-19. Test data may provide more accurate information for specific products and should be preferred when available. However, the values listed in the table can be used to obtain satisfactory results under various conditions in the absence of product-specific data. The *U*-factor of a fenestration product that differs considerably from the ones in the table can be determined by (1) determining the fractions of the area that are frame, center-of-glass, and edge-of-glass (assuming a 65-mm-wide band around the perimeter of each glazing), (2) determining the *U*-factors for each section (the center-of-glass and edge-of-glass *U*-factors can be taken from the first two columns of Table 12-19 and the frame *U*-factor can be taken from Table 12-18 or other sources), and (3) multiplying the area fractions and the *U*-factors for each section and adding them up (or from Eq. 12-34 for U_{window}).

Glazed wall systems can be treated as fixed windows. Also, the data for double-door windows can be used for single-glass doors. Several observations can be made from the data in the table:

1. Skylight *U*-factors are considerably greater than those of vertical windows. This is because the skylight area, including the curb, can be 13 to 240 percent greater than the rough opening area. The slope of the skylight also has some effect.

2. The *U*-factor of multiple-glazed units can be reduced considerably by filling cavities with argon gas instead of dry air. The performance of CO_2-filled units is similar to those filled with argon. The *U*-factor can be reduced even further by filling the glazing cavities with krypton gas.

3. Coating the glazing surfaces with low-e (low-emissivity) films reduces the *U*-factor significantly. For multiple-glazed units, it is adequate to coat one of the two surfaces facing each other.

4. The thicker the air space in multiple-glazed units, the lower the *U*-factor, for a thickness of up to 13 mm ($\frac{1}{2}$ in.) of air space. For a specified number of glazings, the window with thicker air layers will have a lower *U*-factor. For a specified overall thickness of glazing, the higher the number of glazings, the lower the *U*-factor. Therefore, a triple-pane window with air spaces of 6.4 mm (two such air spaces) will have a lower *U*-value than a double-pane window with an air space of 12.7 mm.

5. Wood or vinyl frame windows have a considerably lower *U*-value than comparable metal-frame windows. Therefore, wood or vinyl frame windows are called for in energy-efficient designs.

TABLE 12-19　　　　　　　　　　　　　　　　　　　　　　　　　　　　749

Overall U-factors (heat transfer coefficients) for various windows and skylights in $W/m^2 \cdot {}^\circ C$
(from ASHRAE *Handbook of Fundamentals,* Ref. 1, Chap. 27, Table 5)

Type →	Glass section (glazing) only			Aluminum frame (without thermal break)			Wood or vinyl frame					
	Center-of-glass	Edge-of-glass		Fixed	Double door	Sloped skylight	Fixed		Double door		Sloped skylight	
Frame width →	(Not applicable)			32 mm (1 1/4 in.)	53 mm (2 in.)	19 mm (3/4 in.)	41 mm (1 5/8 in.)		88 mm (3 7/18 in.)		23 mm (7/8 in.)	
Spacer type →	—	Metal	Insul.	All	All	All	Metal	Insul.	Metal	Insul.	Metal	Insul.
Glazing Type												
Single Glazing												
3 mm (1/8 in.) glass	6.30	6.30	—	6.63	7.16	9.88	5.93	—	5.57	—	7.57	—
6.4 mm (1/4 in.) acrylic	5.28	5.28	—	5.69	6.27	8.86	5.02	—	4.77	—	6.57	—
3 mm (1/8 in.) acrylic	5.79	5.79	—	6.16	6.71	9.94	5.48	—	5.17	—	7.63	—
Double Glazing (no coating)												
6.4 mm air space	3.24	3.71	3.34	3.90	4.55	6.70	3.26	3.16	3.20	3.09	4.37	4.22
12.7 mm air space	2.78	3.40	2.91	3.51	4.18	6.65	2.88	2.76	2.86	2.74	4.32	4.17
6.4 mm argon space	2.95	3.52	3.07	3.66	4.32	6.47	3.03	2.91	2.98	2.87	4.14	3.97
12.7 mm argon space	2.61	3.28	2.76	3.36	4.04	6.47	2.74	2.61	2.73	2.60	4.14	3.97
Double Glazing [ε = 0.1, coating on one of the surfaces of air space (surface 2 or 3, counting from the outside toward inside)]												
6.4 mm air space	2.44	3.16	2.60	3.21	3.89	6.04	2.59	2.46	2.60	2.47	3.73	3.53
12.7 mm air space	1.82	2.71	2.06	2.67	3.37	6.04	2.06	1.92	2.13	1.99	3.73	3.53
6.4 mm argon space	1.99	2.83	2.21	2.82	3.52	5.62	2.21	2.07	2.26	2.12	3.32	3.09
12.7 mm argon space	1.53	2.49	1.83	2.42	3.14	5.71	1.82	1.67	1.91	1.78	3.41	3.19
Triple Glazing (no coating)												
6.4 mm air space	2.16	2.96	2.35	2.97	3.66	5.81	2.34	2.18	2.36	2.21	3.48	3.24
12.7 mm air space	1.76	2.67	2.02	2.62	3.33	5.67	2.01	1.84	2.07	1.91	3.34	3.09
6.4 mm argon space	1.93	2.79	2.16	2.77	3.47	5.57	2.15	1.99	2.19	2.04	3.25	3.00
12.7 mm argon space	1.65	2.58	1.92	2.52	3.23	5.53	1.91	1.74	1.98	1.82	3.20	2.95
Triple Glazing [ε = 0.1, coating on one of the surfaces of air spaces (surfaces 3 and 5, counting from the outside toward inside)]												
6.4 mm air space	1.53	2.49	1.83	2.42	3.14	5.24	1.81	1.64	1.89	1.73	2.92	2.66
12.7 mm air space	0.97	2.05	1.38	1.92	2.66	5.10	1.33	1.15	1.46	1.30	2.78	2.52
6.4 mm argon space	1.19	2.23	1.56	2.12	2.85	4.90	1.52	1.35	1.64	1.47	2.59	2.33
12.7 mm argon space	0.80	1.92	1.25	1.77	2.51	4.86	1.18	1.01	1.33	1.17	2.55	2.28

Notes:

(1) Multiply by 0.176 to obtain U-factors in $Btu/h \cdot ft^2 \cdot {}^\circ F$.

(2) The U-factors in this table include the effects of surface heat transfer coefficients and are based on winter conditions of $-18{}^\circ C$ outdoor air and $21{}^\circ C$ indoor air temperature, with 24 km/h (15 mph) winds outdoors and zero solar flux. Small changes in indoor and outdoor temperatures will not affect the overall U-factors much. Windows are assumed to be vertical, and the skylights are tilted 20° from the horizontal with upward heat flow. Insulation spacers are wood, fiberglass, or butyl. Edge-of-glass effects are assumed to extend the 65-mm band around perimeter of each glazing. The product sizes are 1.2 m × 1.8 m for fixed windows, 1.8 m × 2.0 m for double-door windows, and 1.2 m × 0.6 m for the skylights, but the values given can also be used for products of similar sizes. All data are based on 3-mm (1/8-in.) glass unless noted otherwise.

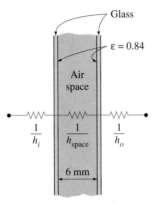

FIGURE 12-46

Schematic of Example 12-12.

EXAMPLE 12-12   U-Factor for Center-of-Glass Section of Windows

Determine the U-factor for the center-of-glass section of a double-pane window with a 6-mm air space for winter design conditions (Fig. 12-46). The glazings are made of clear glass that has an emissivity of 0.84. Take the average air space temperature at design conditions to be 0°C.

Solution   The U-factor for the center-of-glass section of a double-pane window is to be determined.

Assumptions   **1** Steady operating conditions exist. **2** Heat transfer through the window is one-dimensional. **3** The thermal resistance of glass sheets is negligible.

Properties   The emissivity of clear glass is 0.84.

Analysis   Disregarding the thermal resistance of glass sheets, which are small, the U-factor for the center region of a double-pane window is determined from

$$\frac{1}{U_{center}} \cong \frac{1}{h_i} + \frac{1}{h_{space}} + \frac{1}{h_o}$$

where h_i, h_{space}, and h_o are the heat transfer coefficients at the inner surface of the window, the air space between the glass layers, and the outer surface of the window, respectively. The values of h_i and h_o for winter design conditions were given earlier to be $h_i = 8.29$ W/m² · °C and $h_o = 34.0$ W/m² · °C. The effective emissivity of the air space of the double-pane window is

$$\varepsilon_{effective} = \frac{1}{1/\varepsilon_1 + 1/\varepsilon_2 - 1} = \frac{1}{1/0.84 + 1/0.84 - 1} = 0.72$$

For this value of emissivity and an average air space temperature of 0°C, we read $h_{space} = 7.2$ W/m² · °C from Table 12-16 for 6-mm-thick air space. Therefore,

$$\frac{1}{U_{center}} = \frac{1}{8.29} + \frac{1}{7.2} + \frac{1}{34.0} \quad \rightarrow \quad U_{center} = \textbf{3.46 W/m}^2 \textbf{· °C}$$

Discussion   The center-of-glass U-factor value of 3.24 W/m² · °C in Table 12-19 (fourth row and second column) is obtained by using a standard value of $h_o = 29$ W/m² · °C (instead of 34.0 W/m² · °C) and $h_{space} = 6.5$ W/m² · °C at an average air space temperature of −15°C.

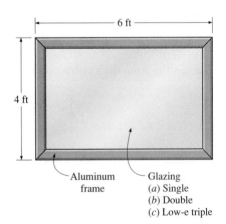

FIGURE 12-47

Schematic for Example 12-13.

EXAMPLE 12-13   Heat Loss through Aluminum Framed Windows

A fixed aluminum-framed window with glass glazing is being considered for an opening that is 4 ft high and 6 ft wide in the wall of a house that is maintained at 72°F (Fig. 12-47). Determine the rate of heat loss through the window and the inner surface temperature of the window glass facing the room when the outdoor air temperature is 15°F if the window is selected to be (a) $\frac{1}{8}$-in. single glazing, (b) double glazing with an air space of $\frac{1}{2}$ in., and (c) low-e-coated triple glazing with an air space of $\frac{1}{2}$ in.

Solution   The rate of heat loss through an aluminum framed window and the inner surface temperature are to be determined from the cases of single-pane, double-pane, and low-e triple-pane windows.

Assumptions   **1** Steady operating conditions exist. **2** Heat transfer through the window is one-dimensional. **3** Thermal properties of the windows and the heat transfer coefficients are constant.

Properties   The U-factors of the windows are given in Table 12-19.

Analysis   The rate of heat transfer through the window can be determined from

$$\dot{Q}_{window} = U_{overall} A_{window}(T_i - T_o)$$

where T_i and T_o are the indoor and outdoor air temperatures, respectively; $U_{overall}$ is the U-factor (the overall heat transfer coefficient) of the window; and A_{window} is the window area, which is determined to be

$$A_{window} = \text{Height} \times \text{Width} = (4\ \text{ft})(6\ \text{ft}) = 24\ \text{ft}^2$$

The U-factors for the three cases can be determined directly from Table 12-19 to be 6.63, 3.51, and 1.92 W/m² · °C, respectively, to be multiplied by the factor 0.176 to convert them to Btu/h · ft² · °F. Also, the inner surface temperature of the window glass can be determined from Newton's law

$$\dot{Q}_{window} = h_i A_{window}(T_i - T_{glass}) \quad \rightarrow \quad T_{glass} = T_i - \frac{\dot{Q}_{window}}{h_i A_{window}}$$

where h_i is the heat transfer coefficient on the inner surface of the window, which is determined from Table 12-18 to be $h_i = 8.3$ W/m² · °C = 1.46 Btu/h · ft² · °F. Then the rate of heat loss and the interior glass temperature for each case are determined as follows:

(*a*) Single glazing:

$$\dot{Q}_{window} = (6.63 \times 0.176\ \text{Btu/h} \cdot \text{ft}^2 \cdot °\text{F})(24\ \text{ft}^2)(72 - 15)°\text{F} = \mathbf{1596\ Btu/h}$$

$$T_{glass} = T_i - \frac{\dot{Q}_{window}}{h_i A_{window}} = 72°\text{F} - \frac{1596\ \text{Btu/h}}{(1.46\ \text{Btu/h} \cdot \text{ft}^2 \cdot °\text{F})(24\ \text{ft}^2)} = \mathbf{26.5°F}$$

(*b*) Double glazing ($\frac{1}{2}$ in. air space):

$$\dot{Q}_{window} = (3.51 \times 0.176\ \text{Btu/h} \cdot \text{ft}^2 \cdot °\text{F})(24\ \text{ft}^2)(72 - 15)°\text{F} = \mathbf{845\ Btu/h}$$

$$T_{glass} = T_i - \frac{\dot{Q}_{window}}{h_i A_{window}} = 72°\text{F} - \frac{845\ \text{Btu/h}}{(1.46\ \text{Btu/h} \cdot \text{ft}^2 \cdot °\text{F})(24\ \text{ft}^2)} = \mathbf{47.9°F}$$

(*c*) Triple glazing ($\frac{1}{2}$ in. air space, low-e coated):

$$\dot{Q}_{window} = (1.92 \times 0.176\ \text{Btu/h} \cdot \text{ft}^2 \cdot °\text{F})(24\ \text{ft}^2)(72 - 15)°\text{F} = \mathbf{462\ Btu/h}$$

$$T_{glass} = T_i - \frac{\dot{Q}_{window}}{h_i A_{window}} = 72°\text{F} - \frac{462\ \text{Btu/h}}{(1.46\ \text{Btu/h} \cdot \text{ft}^2 \cdot °\text{F})(24\ \text{ft}^2)} = \mathbf{58.8°F}$$

Therefore, heat loss through the window will be reduced by 47 percent in the case of double glazing and by 71 percent in the case of triple glazing relative to the single-glazing case. Also, in the case of single glazing, the low inner-glass surface temperature will cause considerable discomfort in the occupants because of the excessive heat loss from the body by radiation. It is raised from 26.5°F, which is below freezing, to 47.9°F in the case of double glazing and to 58.8°F in the case of triple glazing.

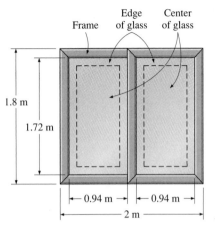

FIGURE 12-48
Schematic for Example 12-14.

EXAMPLE 12-14   U-Factor of a Double-Door Window

Determine the overall U-factor for a double-door-type, wood-framed double-pane window with metal spacers, and compare your result to the value listed in Table 12-19. The overall dimensions of the window are 1.80 m × 2.00 m, and the dimensions of each glazing are 1.72 m × 0.94 m (Fig. 12-48).

Solution   The overall U-factor for a double-door type window is to be determined, and the result is to be compared to the tabulated value.

Assumptions   **1** Steady operating conditions exist. **2** Heat transfer through the window is one-dimensional.

Properties   The U-factors for the various sections of windows are given in Tables 12-17 and 12-19.

Analysis   The areas of the window, the glazing, and the frame are

$$A_{window} = \text{Height} \times \text{Width} = (1.8 \text{ m})(2.0 \text{ m}) = 3.60 \text{ m}^2$$
$$A_{glazing} = 2 \times (\text{Height} \times \text{Width}) = 2(1.72 \text{ m})(0.94 \text{ m}) = 3.23 \text{ m}^2$$
$$A_{frame} = A_{window} - A_{glazing} = 3.60 - 3.23 = 0.37 \text{ m}^2$$

The edge-of-glass region consists of a 6.5-cm-wide band around the perimeter of the glazings, and the areas of the center and edge sections of the glazing are determined to be

$$A_{center} = 2 \times (\text{Height} \times \text{Width}) = 2(1.72 - 0.13 \text{ m})(0.94 - 0.13 \text{ m}) = 2.58 \text{ m}^2$$
$$A_{edge} = A_{glazing} - A_{center} = 3.23 - 2.58 = 0.65 \text{ m}^2$$

The U-factor for the frame section is determined from Table 12-17 to be $U_{frame} = 2.8 \text{ W/m}^2 \cdot °C$. The U-factors for the center and edge sections are determined from Table 12-19 (fifth row, second and third columns) to be $U_{center} = 3.24$ W/m$^2 \cdot °C$ and $U_{edge} = 3.71$ W/m$^2 \cdot °C$. Then the overall U-factor of the entire window becomes

$$U_{window} = (U_{center} A_{center} + U_{edge} A_{edge} + U_{frame} A_{frame})/A_{window}$$
$$= (3.24 \times 2.58 + 3.71 \times 0.65 + 2.8 \times 0.37)/3.60$$
$$= \textbf{3.28 W/m}^2 \cdot °\textbf{C}$$

The overall U-factor listed in Table 12-19 for the specified type of window is 3.20 W/m$^2 \cdot °C$, which is sufficiently close to the value obtained above.

12-9 ■ SOLAR HEAT GAIN THROUGH WINDOWS

The sun is the primary heat source of the earth, and the solar irradiance on a surface normal to the sun's rays beyond the earth's atmosphere at the mean earth–sun distance of 149.5 million km is called the **solar constant.** The accepted value of the solar constant is 1373 W/m^2 (435.4 Btu/h · ft^2), but its value changes by 3.5 percent from a maximum of 1418 W/m^2 on January 3 when the earth is closest to the sun, to a minimum of 1325 W/m^2 on July 4 when the earth is farthest away from the sun. The spectral distribution of solar radiation beyond the earth's atmosphere resembles the energy emitted by a blackbody at 5782°C, with about 9 percent of the energy contained in the ultraviolet region (at wavelengths between 0.29 to 0.4 μm), 39 percent in the visible region (0.4 to 0.7 μm), and the remaining 52 percent in the near-infrared region (0.7 to 3.5 μm). The peak radiation occurs at a wavelength of about 0.48 μm, which corresponds to the green color portion of the visible spectrum. Obviously a glazing material that transmits the visible part of the spectrum while absorbing the infrared portion is ideally suited for an application that calls for maximum daylight and minimum solar heat gain. Surprisingly, the ordinary window glass approximates this behavior remarkably well (Fig. 12-49).

Part of the solar radiation entering the earth's atmosphere is scattered and absorbed by air and water vapor molecules, dust particles, and water droplets in the clouds, and thus the solar radiation incident on earth's surface is less than the solar constant. The extent of the attenuation of solar radiation depends on the length of the path of the rays through the atmosphere as well as the composition of the atmosphere (the clouds, dust, humidity, and smog) along the path. Most ultraviolet radiation is absorbed by the ozone in the upper atmosphere, and the scattering of short wavelength radiation in the blue range by the air molecules is responsible for the blue color of the clear skies. At a solar altitude of 41.8°, the total energy of direct solar radiation incident at sea level on a clear day consists of about 3 percent ultraviolet, 38 percent visible, and 59 percent infrared radiation.

The part of solar radiation that reaches the earth's surface without being scattered or absorbed is called **direct radiation.** Solar radiation that is scattered or reemitted by the constituents of the atmosphere is called **diffuse radiation.** Direct radiation comes directly from the sun following a straight path, whereas diffuse radiation comes from all directions in the sky. The entire radiation reaching the ground on an overcast day is diffuse radiation. The radiation reaching a surface, in general, consists of three components: direct radiation, diffuse radiation, and radiation reflected onto the surface from surrounding surfaces (Fig. 12-50). Common surfaces such as grass, trees, rocks, and concrete reflect about 20 percent of the radiation while absorbing the rest. Snow-covered surfaces, however, reflect 70 percent of the incident radiation. Radiation incident on a surface that does not have a direct view of the sun consists of diffuse and reflected radiation. Therefore, at solar noon, solar radiations incident on the east, west, and north surfaces of a south-facing house are identical since they all consist of diffuse and reflected components. The difference between the radiations incident on the south and north walls in this case gives the magnitude of direct radiation incident on the south wall.

When solar radiation strikes a glass surface, part of it (about 8 percent for uncoated clear glass) is reflected back to outdoors, part of it (5 to 50 percent, depending on composition and thickness) is absorbed within the glass, and the remainder is transmitted indoors, as shown in Figure 12-51. The conservation of energy principle requires that the sum of the transmitted, reflected, and absorbed solar radiations be equal to the incident solar radiation. That is,

$$\tau_s + \rho_s + \alpha_s = 1$$

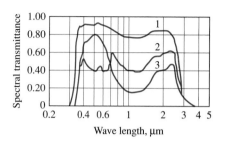

1. 3 mm regular sheet
2. 6 mm gray heat-absorbing plate/float
3. 6 mm green heat-absorbing plate/float

FIGURE 12-49

The variation of the transmittance of typical architectural glass with wavelength (from ASHRAE *Handbook of Fundamentals,* Ref. 1, Chap. 27, Fig. 11).

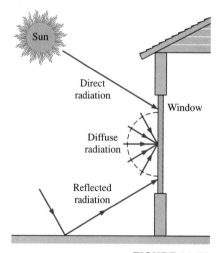

FIGURE 12-50

Direct, diffuse, and reflected components of solar radiation incident on a window.

TABLE 12-20

Hourly variation of solar radiation incident on various surfaces and the daily totals throughout the year at 40° latitude (from ASHRAE *Handbook of Fundamentals*, Ref. 1, Chap. 27, Table 15)

Date	Direction of surface	Solar radiation incident on the surface,* W/m² Solar time														Daily total
		5	6	7	8	9	10	11	12 noon	13	14	15	16	17	18	19	
Jan.	N	0	0	0	20	43	66	68	71	68	66	43	20	0	0	0	446
	NE	0	0	0	63	47	66	68	71	68	59	43	20	0	0	0	489
	E	0	0	0	402	557	448	222	76	68	59	43	20	0	0	0	1863
	SE	0	0	0	483	811	875	803	647	428	185	48	20	0	0	0	4266
	S	0	0	0	271	579	771	884	922	884	771	579	271	0	0	0	5897
	SW	0	0	0	20	48	185	428	647	803	875	811	483	0	0	0	4266
	W	0	0	0	20	43	59	68	76	222	448	557	402	0	0	0	1863
	NW	0	0	0	20	43	59	68	71	68	66	47	63	0	0	0	489
	Horizontal	0	0	0	51	198	348	448	482	448	348	198	51	0	0	0	2568
	Direct	0	0	0	446	753	865	912	926	912	865	753	446	0	0	0	—
Apr.	N	0	41	57	79	97	110	120	122	120	110	97	79	57	41	0	1117
	NE	0	262	508	462	291	134	123	122	120	110	97	77	52	17	0	2347
	E	0	321	728	810	732	552	293	131	120	110	97	77	52	17	0	4006
	SE	0	189	518	682	736	699	582	392	187	116	97	77	52	17	0	4323
	S	0	18	59	149	333	437	528	559	528	437	333	149	59	18	0	3536
	SW	0	17	52	77	97	116	187	392	582	699	736	682	518	189	0	4323
	W	0	17	52	77	97	110	120	392	293	552	732	810	728	321	0	4006
	NW	0	17	52	77	97	110	120	122	123	134	291	462	508	262	0	2347
	Horizontal	0	39	222	447	640	786	880	911	880	786	640	447	222	39	0	6938
	Direct	0	282	651	794	864	901	919	925	919	901	864	794	651	282	0	—
July	N	3	133	109	103	117	126	134	138	134	126	117	103	109	133	3	1621
	NE	8	454	590	540	383	203	144	138	134	126	114	95	71	39	0	3068
	E	7	498	739	782	701	531	294	149	134	126	114	95	71	39	0	4313
	SE	2	248	460	580	617	576	460	291	155	131	114	95	71	39	0	3849
	S	0	39	76	108	190	292	369	395	369	292	190	108	76	39	0	2552
	SW	0	39	71	95	114	131	155	291	460	576	617	580	460	248	2	3849
	W	0	39	71	95	114	126	134	149	294	531	701	782	739	498	7	4313
	NW	0	39	71	95	114	126	134	138	144	203	383	540	590	454	8	3068
	Horizontal	1	115	320	528	702	838	922	949	922	838	702	528	320	115	1	3902
	Direct	7	434	656	762	818	850	866	871	866	850	818	762	656	434	7	—
Oct.	N	0	0	7	40	62	77	87	90	87	77	62	40	7	0	0	453
	NE	0	0	74	178	84	80	87	90	87	87	62	40	7	0	0	869
	E	0	0	163	626	652	505	256	97	87	87	62	40	7	0	0	2578
	SE	0	0	152	680	853	864	770	599	364	137	66	40	7	0	0	4543
	S	0	0	44	321	547	711	813	847	813	711	547	321	44	0	0	5731
	SW	0	0	7	40	66	137	364	599	770	864	853	680	152	0	0	4543
	W	0	0	7	40	62	87	87	97	256	505	652	626	163	0	0	2578
	NW	0	0	7	40	62	87	87	90	87	80	84	178	74	0	0	869
	Horizontal	0	0	14	156	351	509	608	640	608	509	351	156	14	0	0	3917
	Direct	0	0	152	643	811	884	917	927	917	884	811	643	152	0	0	—

*Multiply by 0.3171 to convert to Btu/h · ft².

where τ_s is the transmissivity, ρ_s is the reflectivity, and α_s is the absorptivity of the glass for solar energy, which are the fractions of incident solar radiation transmitted, reflected, and absorbed, respectively. The standard 3-mm- ($\frac{1}{8}$-in.) thick single-pane double-strength clear window glass transmits 86 percent, reflects 8 percent, and absorbs 6 percent of the solar energy incident on it. The radiation properties of materials are usually given for normal incidence, but can also be used for radiation incident at other angles since the transmissivity, reflectivity, and absorptivity of the glazing materials remain essentially constant for incidence angles up to about 60° from the normal.

The hourly variation of solar radiation incident on the walls and windows of a house is given in Table 12-20. Solar radiation that is transmitted indoors is partially absorbed and partially reflected each time it strikes a surface, but all of it is eventually absorbed as sensible heat by the furniture, walls, people, and so forth. Therefore, the solar energy transmitted inside a building represents a heat gain for the building. Also, the solar radiation absorbed by the glass is subsequently transferred to the indoors and outdoors by convection and radiation. The sum of the *transmitted* solar radiation and the portion of the *absorbed* radiation that flows indoors constitutes the **solar heat gain** of the building.

The fraction of incident solar radiation that enters through the glazing is called the **solar heat gain coefficient** SHGC and is expressed as

$$SHGC = \frac{\text{Solar heat gain through the window}}{\text{Solar radiation incident on the window}} \quad (12\text{-}38)$$

$$= \frac{\dot{q}_{\text{solar, gain}}}{\dot{q}_{\text{solar, incident}}} = \tau_s + f_i \alpha_s$$

where α_s is the solar absorptivity of the glass and f_i is the inward flowing fraction of the solar radiation absorbed by the glass. Therefore, the dimensionless quantity SHGC is the sum of the fractions of the directly transmitted (τ_s) and the absorbed and reemitted ($f_i \alpha_s$) portions of solar radiation incident on the window. The value of SHGC ranges from 0 to 1, with 1 corresponding to an opening in the wall (or the ceiling) with no glazing. When the SHGC of a window is known, the total solar heat gain through that window is determined from

$$\dot{Q}_{\text{solar, gain}} = SHGC \times A_{\text{glazing}} \times \dot{q}_{\text{solar, incident}} \quad (W) \quad (12\text{-}39)$$

where A_{glazing} is the glazing area of the window and $\dot{q}_{\text{solar, incident}}$ is the solar heat flux incident on the outer surface of the window, in W/m².

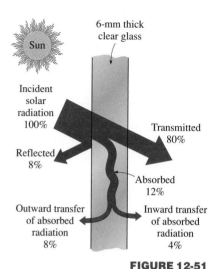

FIGURE 12-51

Distribution of solar radiation incident on a clear glass.

Notes on Table 12-20: Values given are for the 21st of the month for average days with no clouds. The values can be up to 15 percent higher at high elevations under very clear skies and up to 30 percent lower at very humid locations with very dusty industrial atmospheres. Daily totals are obtained using Simpson's rule for integration with 10-minute time intervals. Solar reflectance of the ground is assumed to be 0.2, which is valid for old concrete, crushed rock, and bright green grass. For a specified location, use solar radiation data obtained for that location. The direction of a surface indicates the direction a vertical surface is facing. For example, W represent the solar radiation incident on a west-facing wall per unit area of the wall.

Solar time may deviate from the local time. Solar noon at a location is the time when the sun is at the highest location (and thus when the shadows are shortest). Solar radiation data are symmetric about the solar noon: the value on a west wall two hours before the solar noon is equal to the value on an east wall two hours after the solar noon.

TABLE 12-21

Shading coefficient SC and solar transmissivity τ_{solar} for some common glass types for summer design conditions (from ASHRAE *Handbook of Fundamentals,* Ref. 1, Chap. 27, Table 11).

Type of glazing	Nominal thickness		τ_{solar}	SC*
	mm	in.		
(a) Single Glazing				
Clear	3	$\frac{1}{8}$	0.86	1.0
	6	$\frac{1}{4}$	0.78	0.95
	10	$\frac{3}{8}$	0.72	0.92
	13	$\frac{1}{2}$	0.67	0.88
Heat absorbing	3	$\frac{1}{8}$	0.64	0.85
	6	$\frac{1}{4}$	0.46	0.73
	10	$\frac{3}{8}$	0.33	0.64
	13	$\frac{1}{2}$	0.24	0.58
(b) Double Glazing				
Clear in,	3[a]	$\frac{1}{8}$	0.71[b]	0.88
clear out	6	$\frac{1}{4}$	0.61	0.82
Clear in, heat absorbing out[c]	6	$\frac{1}{4}$	0.36	0.58

*Multiply by 0.87 to obtain SHGC.

[a]The thickness of each pane of glass.

[b]Combined transmittance for assembled unit.

[c]Refers to gray-, bronze-, and green-tinted heat-absorbing float glass.

Another way of characterizing the solar transmission characteristics of different kinds of glazing and shading devices is to compare them to a well-known glazing material that can serve as a base case. This is done by taking the standard 3-mm- ($\frac{1}{8}$-in.) thick double-strength clear window glass sheet whose SHGC is 0.87 as the *reference glazing* and defining a **shading coefficient** SC as

$$SC = \frac{\text{Solar heat gain of product}}{\text{Solar heat gain of reference glazing}}$$
$$= \frac{\text{SHGC}}{\text{SHGC}_{ref}} = \frac{\text{SHGC}}{0.87} = 1.15 \times \text{SHGC} \qquad (12\text{-}40)$$

Therefore, the shading coefficient of a single-pane clear glass window is SC = 1.0. The shading coefficients of other commonly used fenestration products are given in Table 12-21 for summer design conditions. The values for winter design conditions may be slightly lower because of the higher heat transfer coefficients on the outer surface due to high winds and thus higher rate of outward flow of solar heat absorbed by the glazing, but the difference is small.

Note that the larger the shading coefficient, the smaller the shading effect, and thus the larger the amount of solar heat gain. A glazing material with a large shading coefficient will allow a large fraction of solar radiation to come in.

Shading devices are classified as *internal shading* and *external shading,* depending on whether the shading device is placed *inside* or *outside.* External shading devices are more effective in reducing the solar heat gain since they intercept the sun's rays before they reach the glazing. The solar heat gain through a window can be reduced by as much as 80 percent by exterior shading. Roof overhangs have long been used for exterior shading of windows. The sun is high in the horizon in summer and low in winter. A properly sized roof overhang or a horizontal projection blocks off the sun's rays completely in summer while letting in most of them in winter, as shown in Figure 12-52. Such shading structures can reduce the solar heat gain on the south, southeast, and southwest windows in the northern hemisphere considerably. A window can also be shaded from outside by vertical or horizontal architectural projections, insect or shading screens, and sun screens. To be effective, air must be able to move freely around the exterior device to carry away the heat absorbed by the shading and the glazing materials.

Some type of internal shading is used in most windows to provide privacy and aesthetic effects as well as some control over solar heat gain. Internal shading devices reduce solar heat gain by reflecting transmitted solar radiation back through the glazing before it can be absorbed and converted into heat in the building.

Draperies reduce the annual heating and cooling loads of a building by 5 to 20 percent, depending on the type and the user habits. In summer, they reduce heat gain primarily by reflecting back direct solar radiation (Fig. 12-53). The semiclosed air space formed by the draperies serves as an additional barrier against heat transfer, resulting in a lower *U*-factor for the window and thus a lower rate of heat transfer in summer and winter. The solar optical properties of draperies can be measured accurately, or they can be obtained directly from the manufacturers. The shading coefficient of draperies depends on the openness factor, which is the ratio of the open area between the fibers that permits the sun's rays to pass freely, to the total area of the fabric. Tightly woven fabrics allow little direct radiation to pass through, and thus they have a small openness factor. The *reflectance* of the surface of the drapery facing the glazing has a major effect on the amount of solar heat gain. *Light-colored* draperies made of closed or tightly woven fabrics maximize the back reflection and thus minimize the solar gain. *Dark-colored* draperies made of open or semi-open woven fabrics, on the other hand, minimize the back reflection and thus maximize the solar gain.

The shading coefficients of drapes also depend on the way they are hung. Usually, the width of drapery used is twice the width of the draped area to allow folding of the drapes and to give them their characteristic "full" or "wavy" appearance. A flat drape behaves like an ordinary window shade. A flat drape has a higher reflectance and thus a lower shading coefficient than a full drape.

External shading devices such as overhangs and tinted glazings do not require operation, and provide reliable service over a long time without significant degradation during their service life. Their operation does not depend on a person or an automated system, and these passive shading devices are considered fully effective when determining the peak cooling load and the annual energy use. The effectiveness of manually operated shading devices, on the other hand, varies greatly depending on the user habits, and this variation should be considered when evaluating performance.

The primary function of an indoor shading device is providing *thermal comfort* for the occupants. An unshaded window glass allows most of the incident solar radiation in, and also dissipates part of the solar energy it absorbs by emitting infrared radiation to the room. The emitted radiation and the transmitted direct sunlight may bother the occupants near the window. In winter, the temperature of the glass is lower than the room air temperature, causing excessive heat loss by radiation from the occupants. A shading device allows the control of direct solar and infrared radiation while providing various degrees of privacy and outward vision. The shading device is also at a higher temperature than the glass in winter, and thus reduces radiation loss from occupants. *Glare* from draperies can be minimized by using off-white colors. Indoor shading devices, especially draperies made of a closed-weave fabric, are effective in reducing *sounds* that originate in the room, but they are not as effective against the sounds coming from outside.

The type of climate in an area usually dictates the type of windows to be used in buildings. In *cold climates* where the heating load is much larger than the cooling load, the windows should have the highest transmissivity for the entire solar spectrum, and a high reflectivity (or low emissivity) for the far infrared radiation emitted by the walls and furnishings of the room. Low-e windows are well suited for such heating-dominated buildings. Properly designed and operated windows allow more heat into the building over a heating

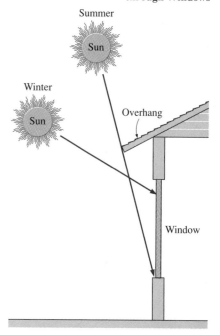

FIGURE 12-52
A properly sized overhang blocks off the sun's rays completely in summer while letting them in in winter.

FIGURE 12-53
Draperies reduce heat gain in summer by reflecting back solar radiation and reduce heat loss in winter by forming an air space before the window.

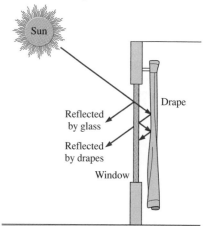

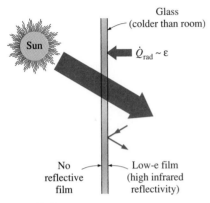

(a) Cold climates

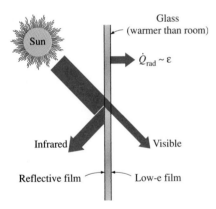

(b) Warm climates

FIGURE 12-54

Radiation heat transfer between a room and its windows is proportional to the emissivity of the glass surface, and low-e coatings on the inner surface of the windows reduce heat loss in winter and heat gain in summer.

season than it loses, making them energy contributors rather then energy losers. In *warm climates* where the cooling load is much larger than the heating load, the windows should allow the visible solar radiation (light) in, but should block off the infrared solar radiation. Such windows can reduce the solar heat gain by 60 percent with no appreciable loss in daylighting. This behavior is approximated by window glazings that are coated with a heat-absorbing film outside and a low-e film inside (Fig. 12-54). Properly selected windows can reduce the cooling load by 15 to 30 percent compared to windows with clear glass.

Note that radiation heat transfer between a room and its windows is proportional to the emissivity of the glass surface facing the room, $\varepsilon_{\text{glass}}$, and can be expressed as

$$\dot{Q}_{\text{rad, room-window}} = \varepsilon_{\text{glass}} A_{\text{glass}} \sigma (T_{\text{room}}^4 - T_{\text{glass}}^4)$$

Therefore, a low-e interior glass will reduce the heat loss by radiation in winter ($T_{\text{glass}} < T_{\text{room}}$) and heat gain by radiation in summer ($T_{\text{glass}} > T_{\text{room}}$).

Tinted glass and glass coated with reflective films reduce solar heat gain in summer and heat loss in winter. The conductive heat gains or losses can be minimized by using multiple-pane windows. Double-pane windows are usually called for in climates where the winter design temperature is less than 7°C (45°F). Double-pane windows with tinted or reflective films are commonly used in buildings with large window areas. Clear glass is preferred for showrooms since it affords maximum visibility from outside, but bronze-, gray-, and green-colored glass are preferred in office buildings since they provide considerable privacy while reducing glare.

EXAMPLE 12-15   Installing Reflective Films on Windows

A manufacturing facility located at 40° N latitude has a glazing area of 40 m² that consists of double-pane windows made of clear glass (SHGC = 0.766). To reduce the solar heat gain in summer, a reflective film that will reduce the SHGC to 0.261 is considered. The cooling season consists of June, July, August, and September, and the heating season October through April. The average daily solar heat fluxes incident on the west side at this latitude are 1.86, 2.66, 3.43, 4.00, 4.36, 5.13, 4.31, 3.93, 3.28, 2.80, 1.84, and 1.54 kWh/day · m² for January through December, respectively. Also, the unit cost of the electricity and natural gas are $0.08/kWh and $0.50/therm, respectively. If the coefficient of performance of the cooling system is 2.5 and efficiency of the furnace is 0.8, determine the net annual cost savings due to installing reflective coating on the windows. Also, determine the simple payback period if the installation cost of reflective film is $20/m² (Fig. 12-55).

Solution   The net annual cost savings due to installing reflective film on the west windows of a building and the simple payback period are to be determined.

Assumptions   **1** The calculations given below are for an average year. **2** The unit costs of electricity and natural gas remain constant.

Analysis   Using the daily averages for each month and noting the number of days of each month, the total solar heat flux incident on the glazing during summer and winter months are determined to be

$$Q_{solar, summer} = 5.13 \times 30 + 4.31 \times 31 + 3.93 \times 31 + 3.28 \times 30 = 508 \text{ kWh/year}$$

$$\begin{aligned}Q_{solar, winter} &= 2.80 \times 31 + 1.84 \times 30 + 1.54 \times 31 + 1.86 \times 31 \\ &\quad + 2.66 \times 28 + 3.43 \times 31 + 4.00 \times 30 \\ &= 548 \text{ kWh/year}\end{aligned}$$

Then the decrease in the annual cooling load and the increase in the annual heating load due to the reflective film become

$$\begin{aligned}\text{Cooling load decrease} &= Q_{solar, summer} A_{glazing} (\text{SHGC}_{without\,film} - \text{SHGC}_{with\,film}) \\ &= (508 \text{ kWh/year})(40 \text{ m}^2)(0.766 - 0.261) \\ &= 10{,}262 \text{ kWh/year}\end{aligned}$$

$$\begin{aligned}\text{Heating load increase} &= Q_{solar, winter} A_{glazing} (\text{SHGC}_{without\,film} - \text{SHGC}_{with\,film}) \\ &= (548 \text{ kWh/year})(40 \text{ m}^2)(0.766 - 0.261) \\ &= 11{,}070 \text{ kWh/year} = 377.7 \text{ therms/year}\end{aligned}$$

since 1 therm = 29.31 kWh. The corresponding decrease in cooling costs and the increase in heating costs are

$$\begin{aligned}\text{Decrease in cooling costs} &= (\text{Cooling load decrease})(\text{Unit cost of electricity})/\text{COP} \\ &= (10{,}262 \text{ kWh/year})(\$0.08/\text{kWh})/2.5 = \$328/\text{year}\end{aligned}$$

$$\begin{aligned}\text{Increase in heating costs} &= (\text{Heating load increase})(\text{Unit cost of fuel})/\text{Efficiency} \\ &= (377.7 \text{ therms/year})(\$0.50/\text{therm})/0.80 = \$236/\text{year}\end{aligned}$$

Then the net annual cost savings due to the reflective film become

$$\begin{aligned}\text{Cost savings} &= \text{Decrease in cooling costs} - \text{Increase in heating costs} \\ &= \$328 - \$236 = \textbf{\$92/year}\end{aligned}$$

The implementation cost of installing films is

$$\text{Implementation cost} = (\$20/\text{m}^2)(40 \text{ m}^2) = \$800$$

This gives a simple payback period of

$$\text{Simple payback period} = \frac{\text{Implementation cost}}{\text{Annual cost savings}} = \frac{\$800}{\$92/\text{year}} = \textbf{8.7 years}$$

Discussion The reflective film will pay for itself in this case in about nine years. This may be unacceptable to most manufacturers since they are not usually interested in any energy conservation measure that does not pay for itself within three years.

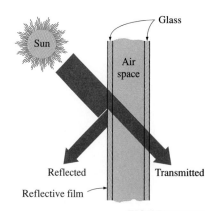

FIGURE 12-55
Schematic for Example 12-15.

12-10 ■ INFILTRATION HEAT LOAD AND WEATHERIZING

Most older homes and some poorly constructed new ones have numerous cracks, holes, and openings through which cold outdoor air exchanges with the warm air inside a building in winter, and vice versa in summer. This uncontrolled entry of outside air into a building through unintentional openings is called **infiltration,** and it wastes a significant amount of energy since

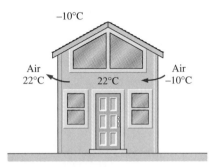

$$\text{Energy loss} = m_{air}C_{p,\,air}[22 - (-10)°C]$$

FIGURE 12-56

When cold outdoor air enters a building in winter, an equivalent amount of warm air must leave the house. This represents energy loss by infiltration.

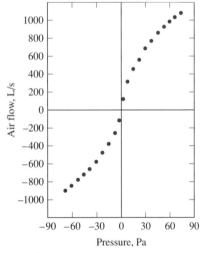

FIGURE 12-57

Typical data from a whole house pressurization test for the variation of air flow rate with pressure difference (from ASHRAE *Handbook of Fundamentals,* Ref. 1, Chap. 23, Fig. 8).

the air entering must be heated in winter and cooled in summer (Fig. 12-56). The warm air leaving the house represents *energy loss.* This is also the case for cool air leaving in summer since some electricity is used to cool that air. In homes that have not been properly weatherized, the air leaks account for about 30–40 percent of the total heat lost from the house in winter. That is, about one-third of the heating bill of such a house is due to the air leaks.

The rate of infiltration depends on the *wind velocity* and the *temperature difference* between the inside and the outside, and thus it varies throughout the year. The infiltration rates are much higher in winter than they are in summer because of the higher winds and larger temperature differences in winter. Therefore, distinction should be made between the *design infiltration rate* at design conditions, which is used to size heating or cooling equipment, and the *seasonal average infiltration rate,* which is used to properly estimate the seasonal energy consumption for heating or cooling. Infiltration appears to be providing "fresh outdoor air" to a building, but it is not a reliable ventilation mechanism since it depends on the weather conditions and the size and location of the cracks.

The air infiltration rate of a building can be determined by direct measurements by (1) *injecting a tracer gas* into a building and observing the decline of its concentration with time or (2) *pressurizing the building* to 10 to 75 Pa gage pressure by a large fan mounted on a door or window, and measuring the air flow required to maintain a specified indoor–outdoor pressure difference. The larger the air flow to maintain a pressure difference, the more the building may leak. Sulfur hexafluoride (SF_6) is commonly used as a tracer gas because it is inert, nontoxic, and easily detectable at concentrations as low as 1 part per billion. Pressurization testing is easier to conduct, and thus preferable to tracer gas testing. Pressurization test results for a whole house are given in Figure 12-57.

Despite their accuracy, direct measurement techniques are inconvenient, expensive, and time consuming. A practical alternative is to *predict* the air infiltration rate on the basis of extensive data available on existing buildings. One way of predicting the air infiltration rate is by determining the type and size of all the cracks at all possible locations (around doors and windows, lighting fixtures, wall–floor joints, etc., as shown in Fig. 12-58), as well as the pressure differential across the cracks at specified conditions, and calculating the air flow rates. This is known as the **crack method.**

A simpler and more practical approach is to "estimate" how many times the entire air in a building is replaced by the outside air per hour on the basis of experience with similar buildings under similar conditions. This is called the **air-change method,** and the infiltration rate in this case is expressed in terms of **air changes per hour** (ACH), defined as

$$\text{ACH} = \frac{\text{Flow rate of outdoor air into the building (per hour)}}{\text{Internal volume of the building}} = \frac{\dot{V}(m^3/h)}{V(m^3)} \tag{12-41}$$

The mass of air corresponding to 1 ACH is determined from $m = \rho V$ where ρ is the density of air whose value is determined at the *outdoor* temperature

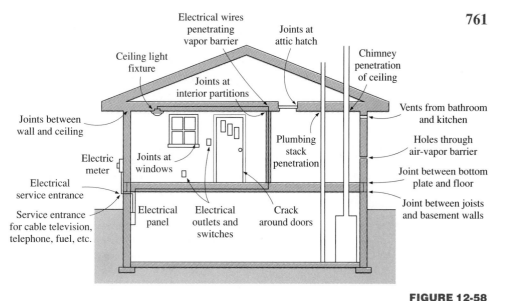

FIGURE 12-58
Typical air-leakage sites of a house (from U.S. Department of Energy pamphlet, FS 203, 1992).

and pressure. Therefore, the quantity ACH represents the number of building volumes of outdoor air that infiltrates (and eventually exfiltrates) per hour. At sea-level standard conditions of 1 atm (101.3 kPa or 14.7 psia) and 20°C (68°F), the density of air is

$$\rho_{\text{air, standard}} = 1.20 \text{ kg/m}^3 = 0.075 \text{ lbm/ft}^3$$

However, the atmospheric pressure and thus the density of air will drop by about 20 percent at 1500 m (5000 ft) elevation at 20°C, and by about 10 percent when the temperature rises to 50°C at 1 atm pressure. Therefore, local air density should be used in calculations to avoid such errors.

Infiltration rate values for hundreds of buildings throughout the United States have been measured during the last two decades, and the seasonal average infiltration rates have been observed to vary from about 0.2 ACH for newer energy-efficient tight buildings to about 2.0 ACH for older buildings. Therefore, infiltration rates can easily vary by a factor of 10 from one building to another. Seasonal average infiltration rates as low as 0.02 have been recorded. A study that involved 312 mostly new homes determined the average infiltration rate to be about 0.5 ACH. Another study that involved 266 mostly older homes determined the average infiltration rate to be about 0.9 ACH. The infiltration rates of some new office buildings with no outdoor air intake are measured to be between 0.1 and 0.6 ACH. Occupancy is estimated to add 0.1 to 0.15 ACH to unoccupied infiltration rate values. Also, the infiltration rate of a building can vary by a factor of 5, depending on the weather.

A minimum of 0.35 ACH is required to meet the *fresh air requirements* of residential buildings and to maintain indoor air quality, provided that at least 7.5 L/s (15 ft³/min) of fresh air is supplied per occupant to keep the

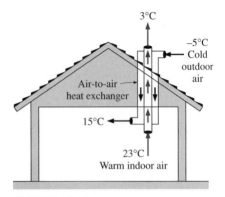

3°C

−5°C
Cold
outdoor
air

Air-to-air
heat exchanger

15°C

23°C
Warm indoor air

FIGURE 12-59

An air-to-air heat exchanger
recuperates some of the energy of warm
indoor air vented out of a building.

indoor CO_2 concentration level below 1000 parts per million (0.1 percent). Usually the infiltration rates of houses are above 0.35 ACH, and thus we do not need to be concerned about *mechanical ventilation*. However, the infiltration rates of some of today's energy-efficient buildings are below the required minimum, and additional fresh air must be supplied to such buildings by mechanical ventilation. It may be necessary to install a central ventilating system in addition to the bathroom and kitchen fans to bring the air quality to desired levels.

Venting the cold outside air directly into the house will obviously increase the heating load in winter. But part of the energy in the warm air vented out can be recovered by installing an *air-to-air heat exchanger* (also called an "economizer" or "heat recuperator") that transfers the heat from the exhausted stale air to the incoming fresh air without any mixing (Fig. 12-59). Such heat exchangers are commonly used in superinsulated houses, but the benefits of such heat exchangers must be weighed against the cost and complexity of their installation. The effectiveness of such heat exchangers is typically low (about 40 percent) because of the small temperature differences involved.

The primary cause of excessive infiltration is *poor workmanship,* but it may also be the settling and aging of the house. Infiltration is likely to develop where two surfaces meet such as the wall–foundation joint. Large differences between indoor and outdoor humidity and temperatures may aggravate the problem. Winds exert a dynamic pressure on the house, which forces the outside air through the cracks inside the house.

Infiltration should not be confused with **ventilation,** which is the *intentional* and *controlled* mechanism of air flow into or out of a building. Ventilation can be *natural* or *forced* (or *mechanical*), depending on how it is achieved. Ventilation accomplished by the opening of windows or doors is natural ventilation, whereas ventilation accomplished by an air mover such as a fan is forced ventilation. Forced ventilation gives the designer the greatest control over the magnitude and distribution of air flow throughout a building. The airtightness or air exchange rate of a building at any given time usually includes the effects of natural and forced ventilation as well as infiltration.

Air exchange, or the supply of fresh air, has a significant role on health, air quality, thermal comfort, and energy consumption. The *supply of fresh air* is a double-edged sword: too little of it will cause health and comfort problems such as the sick-building syndrome that was experienced in super-airtight buildings, and too much of it will waste energy. Therefore, the rate of fresh air supply should be just enough to maintain the indoor air quality at an acceptable level. The infiltration rate of older buildings is several times the required minimum flow rate of fresh air, and thus there is a high energy penalty associated with it.

Infiltration increases the *energy consumption* of a building in two ways: First, the incoming outdoor air must be heated (or cooled in summer) to the indoor air temperature. This represents the **sensible heat load** of infiltration and is expressed as

$$\dot{Q}_{\text{infiltration, sensible}} = \rho_o C_p \dot{V}(T_i - T_o) \qquad (12\text{-}42)$$
$$= \rho_o C_p (ACH)(V_{\text{building}})(T_i - T_o)$$

where ρ_o is the density of outdoor air; C_p is the specific heat of air (about 1 kJ/kg · °C or 0.24 Btu/lbm · °F); $\dot{V} = (ACH)(V_{\text{building}})$ is the volumetric flow rate of air, which is the number of air changes per hour times the volume of the building; and $T_i - T_o$ is the temperature difference between the indoor and outdoor air. Second, the moisture content of outdoor air, in general, is different than that of indoor air, and thus the incoming air may need to be humidified or dehumidified. This represents the **latent heat load** of infiltration and is expressed as (Fig. 12-60)

$$\dot{Q}_{\text{infiltration, latent}} = \rho_o h_{fg} \dot{V}(w_i - w_o) \qquad (12\text{-}43)$$
$$= \rho_o h_{fg}(ACH)(V_{\text{building}})(w_i - w_o)$$

where h_{fg} is the latent heat of vaporization at the indoor temperature (about 2340 kJ/kg or 1000 Btu/lbm) and $w_i - w_o$ is the humidity ratio difference between the indoor and outdoor air, which can be determined from the psychrometric charts. The latent heat load is particularly significant in summer months in hot and humid regions such as Florida and coastal Texas. In winter, the humidity ratio of outdoor air is usually much lower than that of indoor air, and the latent infiltration load in this case represents the energy needed to vaporize the required amount of water to raise the humidity of indoor air to the desired level.

Preventing Infiltration

Infiltration accounts for a significant part of the total heat loss, and sealing the sites of air leaks by *caulking* or *weather-stripping* should be the first step to reduce energy waste and heating and cooling costs. Weatherizing requires some work, of course, but it is relatively easy and inexpensive to do.

Caulking can be applied with a caulking gun inside and outside where two stationary surfaces such as a wall and a window frame meet. It is easy to apply and is very effective in fixing air leaks. Potential sites of air leaks that can be fixed by caulking are entrance points of electrical wires, plumbing, and telephone lines; the sill plates where walls meet the foundation; joints between exterior window frames and siding; joints between door frames and walls; and around exhaust fans.

Weather-stripping is a narrow piece of metal, vinyl, rubber, felt, or foam that seals the contact area between the fixed and movable sections of a joint. Weather-stripping is best suited for sites that involve moving parts such as doors and windows. It minimizes air leakage by closing off the gaps between the moving parts and their fixed frames when they are closed. All exterior doors and windows should be weatherized. There are various kinds of weather-stripping, and some kinds are more suitable for particular kinds of gaps. Some common types of weather-stripping are shown in Figure 12-61.

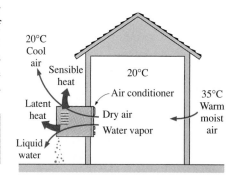

FIGURE 12-60

The energy removed while water vapor in the air is condensed constitutes the latent heat load of an air-conditioning system.

FIGURE 12-61

Some common types of weather-stripping.

Rolled vinyl with rigid metal backing

Nonreinforced, self-adhesive

Door sweep
(vinyl lip with metal, wood, or plastic retainer)

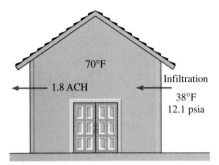

FIGURE 12-62
Schematic for Example 12-16.

EXAMPLE 12-16 Reducing Infiltration Losses by Winterizing

The average atmospheric pressure in Denver, Colorado (elevation = 5300 ft), is 12.1 psia, and the average winter temperature is 38°F. The pressurization test of a 9-ft-high, 2500 ft^2 older home revealed that the seasonal average infiltration rate of the house is 1.8 ACH (Fig. 12-62). It is suggested that the infiltration rate of the house can be reduced by one-third to 1.2 ACH by winterizing the doors and the windows. If the house is heated by natural gas whose unit cost is $0.58/therm and the heating season can be taken to be six months, determine how much the home owner will save from the heating costs per year by this winterization project. Assume the house is maintained at 70°F at all times and the efficiency of the furnace is 0.75. Also, the latent heat load during the heating season in Denver is negligible.

Solution A winterizing project will reduce the infiltration rate of a house from 1.8 ACH to 1.2 ACH. The resulting cost savings are to be determined.

Assumptions **1** The house is maintained at 70°F at all times. **2** The latent heat load during the heating season is negligible. **3** The infiltrating air is heated to 70°F before it exfiltrates.

Properties The specific heat of air at room temperature is 0.24 Btu/lbm · °F (Table A-11E). The molar mass of air is 0.3704 psia · ft^3/lbm · R (Table A-1E).

Analysis The density of air at the outdoor conditions is

$$\rho_o = \frac{P_o}{RT_o} = \frac{12.1 \text{ psia}}{(0.3704 \text{ psia} \cdot \text{ft}^3/\text{lbm} \cdot \text{R})(498 \text{ R})} = 0.0656 \text{ lbm/ft}^3$$

The volume of the building is

$$V_{\text{building}} = (\text{Floor area})(\text{Height}) = (2500 \text{ ft}^2)(9 \text{ ft}) = 22,500 \text{ ft}^3$$

The sensible infiltration heat load corresponding to the reduction in the infiltration rate of 0.6 ACH is

$$\begin{aligned}
\dot{Q}_{\text{infiltration, saved}} &= \rho_o C_p (ACH_{\text{saved}})(V_{\text{building}})(T_i - T_o) \\
&= (0.0656 \text{ lbm/ft}^3)(0.24 \text{ Btu/lbm} \cdot °\text{F})(0.6/\text{h})(22,500 \text{ ft}^3)(70 - 38)°\text{F} \\
&= 6800 \text{ Btu/h} = 0.068 \text{ therm/h}
\end{aligned}$$

since 1 therm = 100,000 Btu. The number of hours during a six-month period is $6 \times 30 \times 24 = 4320$ h. Noting that the furnace efficiency is 0.75 and the unit cost of natural gas is $0.58/therm, the energy and money saved during the six-month period are

$$\begin{aligned}
\text{Energy savings} &= (\dot{Q}_{\text{infiltration, saved}})(\text{No. of hours per year})/\text{Efficiency} \\
&= (0.068 \text{ therm/h})(4320 \text{ h/year})/0.75 \\
&= 392 \text{ therms/year}
\end{aligned}$$

$$\begin{aligned}
\text{Cost savings} &= (\text{Energy savings})(\text{Unit cost of energy}) \\
&= (392 \text{ therms/year})(\$0.58/\text{therm}) \\
&= \textbf{\$227/year}
\end{aligned}$$

Therefore, reducing the infiltration rate by one-third will reduce the heating costs of this home owner by $227 per year.

In the thermal analysis of buildings, two quantities of major interest are (1) the *size* or *capacity* of the heating and the cooling system and (2) the *annual energy consumption*. The size of a heating or cooling system is based on the *most demanding* situations under the anticipated *worst weather* conditions, whereas the average annual energy consumption is based on *average usage* situations under *average weather* conditions. Therefore, the calculation procedure of annual energy usage is quite different than that of design heating or cooling loads.

An analysis of annual energy consumption and cost usually accompanies the design heat load calculations and plays an important role in the selection of a heating or cooling system. Often a choice must be made among several systems that have the same capacity but different efficiencies and initial costs. More efficient systems usually consume less energy and money per year, but they cost more to purchase and install. The purchase of a more efficient but more expensive heating or cooling system can be economically justified only if the system saves more in the long run from energy costs than its initial cost differential.

The impact on the *environment* may also be an important consideration on the selection process: A system that consumes less fuel pollutes the environment less, and thus reduces all the adverse effects associated with environmental pollution. But it is difficult to quantify the environmental impact in an economic analysis unless a price is put on it.

One way of reducing the initial and operating costs of a heating or cooling system is to compromise the *thermal comfort* of occupants. This option should be avoided, however, since a small loss in employee productivity due to thermal discomfort can easily offset any potential gains from reduced energy use. The U.S. Department of Energy periodically conducts comprehensive energy surveys to determine the energy usage in residential as well as nonresidential buildings and the industrial sector. Two 1983 reports (DOE/EIA-0246 and DOE/EIA-0318) indicate that the national average natural gas usage of all commercial buildings in the United States is 70,000 Btu/ft² · year, which is worth about $0.50/ft² or $5/m² per year. The reports also indicate that the average annual electricity consumption of commercial buildings due to air-conditioning is about 12 kWh/ft² · year, which is worth about $1/ft² or $10/m² per year. Therefore, the average cost of heating and cooling of commercial buildings is about $15/m² per year. This corresponds to $300/year for a 20 m² floor space, which is large enough for an average office worker. But noting that the average salary and benefits of a worker are no less than $30,000 a year, it appears that the heating and cooling cost of a commercial building constitutes about 1 percent of the total cost (Fig. 12-63). Therefore, even a 1 percent loss in productivity due to thermal discomfort may cost the business owner more than the entire cost of energy. Likewise, the loss of business in retail stores due to unpleasant thermal conditions will cost the store owner many times what he or she is saving from energy. Thus, the message to the HVAC engineer is clear: *in the design of heating and cooling systems of commercial buildings, treat the thermal comfort conditions as design constraints rather than as variables. The cost of energy is a*

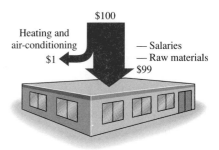

FIGURE 12-63

The heating and cooling cost of a commercial building constitutes about 1 percent of the total cost. Therefore, thermal comfort and thus productivity should not be risked to conserve energy.

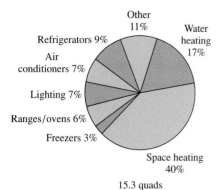

Other 11%

Refrigerators 9%

Air conditioners 7%

Lighting 7%

Ranges/ovens 6%

Freezers 3%

Water heating 17%

Space heating 40%

15.3 quads

(*a*) Residential buildings

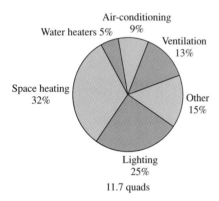

Water heaters 5%

Air-conditioning 9%

Ventilation 13%

Space heating 32%

Other 15%

Lighting 25%

11.7 quads

(*b*) Commercial buildings

FIGURE 12-64

Breakdown of energy consumption in residential and commercial buildings in 1986 (from U.S. Department of Energy).

FIGURE 12-65

The various quantities involved in the evaluation of the annual energy consumption of a building.

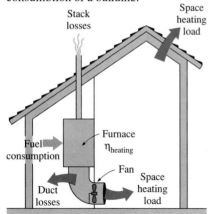

Stack losses

Space heating load

Fuel consumption

Furnace $\eta_{heating}$

Fan

Duct losses

Space heating load

very small fraction of the goods and services produced, and thus, do not incorporate any energy conservation measures that may result in a loss of productivity or loss of revenues.

When trying to minimize annual energy consumption, it is helpful to have a general idea about where most energy is used. A *breakdown* of energy usage in residential and commercial buildings is given in Figure 12-64. Note that space heating accounts for most energy usage in all buildings, followed by water heating in residential buildings and lighting in commercial buildings. Therefore, any conservation measure dealing with them will have the greatest impact.

For *existing buildings,* the amount and cost of energy (fuel or electricity) used for heating and cooling of a building can be determined by simply analyzing the *utility bills* for a typical year. For example, if a house uses natural gas for space and water heating, the natural gas consumption for space heating can be determined by estimating the average monthly usage for water heating from summer bills, multiplying it by 12 to estimate the yearly usage, and subtracting it from the total annual natural gas usage. Likewise, the annual electricity usage and cost for air-conditioning can be determined by simply evaluating the excess electricity usage during the cooling months and adding them up. If the bills examined are not for a typical year, corrections can be made by comparing the weather data for that year to the average weather data.

For buildings that are at the design or construction stage, the evaluation of annual energy consumption involves the determination of (1) the *space load* for heating or cooling due to heat transfer through the building envelope and infiltration, (2) the *efficiency* of the furnace where the fuel is burned or the COP of cooling or heat pump systems, and (3) the *parasitic energy* consumed by the distribution system (pumps or fans) and the energy lost or gained from the pipes or ducts (Fig. 12-65). The determination of the space load is similar to the determination of the peak load, except the average conditions are used for the weather instead of design conditions. The space heat load is usually based on the average temperature difference between the indoors and the outdoors, but internal heat gains and solar effects must also be considered for better accuracy. Very accurate results can be obtained by using hourly data for a whole year and by making a computer simulation using one of the commercial building energy analysis software packages.

The simplest and most intuitive way of estimating the annual energy consumption of a building is the **degree-day** (or **degree-hour**) **method,** which is a *steady-state* approach. It is based on constant indoor conditions during the heating or cooling season and assumes the efficiency of the heating or cooling equipment is not affected by the variation of outdoor temperature. These conditions will be closely approximated if all the thermostats in a building are set at the same temperature at the beginning of a heating or cooling season and are never changed, and a seasonal average efficiency is used (rather than the full-load or design efficiency) for the furnaces or coolers.

You may think that anytime the outdoor temperature T_o drops below the indoor temperature T_i at which the thermostat is set, the heater will turn on to make up for the heat losses to the outside. However, the internal heat generated by people, lights, and appliances in occupied buildings as well as the heat gain from the sun during the day, $\dot{Q}_{gain}$, will be sufficient to compensate for the heat losses from the building until the outdoor temperature drops below a certain value. The *outdoor temperature above which no heating is required* is called the **balance point temperature** $T_{balance}$ (or the *base temperature*) and is determined from (Fig. 12-66)

$$K_{overall}(T_i - T_{balance}) = \dot{Q}_{gain} \quad \rightarrow \quad T_{balance} = T_i - \frac{\dot{Q}_{gain}}{K_{overall}} \quad (12\text{-}44)$$

where $K_{overall}$ is the *overall heat transfer coefficient* of the building in W/°C or Btu/h · °F. There is considerable uncertainty associated with the determination of the balance point temperature, but based on the observations of typical buildings, it is usually taken to be 18°C in Europe and 65°F (18.3°C) in the United States for convenience. The rate of energy consumption of the heating system is

$$\dot{Q}_{heating} = \frac{K_{overall}}{\eta_{heating}}(T_{balance} - T_o)^+ \quad (12\text{-}45)$$

where $\eta_{heating}$ is the efficiency of the heating system, which is equal to 1.0 for electric resistance heating systems, COP for the heat pumps, and combustion efficiency (about 0.6 to 0.95) for furnaces. If $K_{overall}$, $T_{balance}$, and $\eta_{heating}$ are taken to be constants, the annual energy consumption for heating can be determined by integration (or by summation over daily or hourly averages) as

$$Q_{heating,\,year} = \frac{K_{overall}}{\eta_{heating}} \int [T_{balance} - T_o(t)]^+ \, dt \cong \frac{K_{overall}}{\eta_{heating}} DD_{heating} \quad (12\text{-}46)$$

where $DD_{heating}$ is the **heating degree-days.** The + sign above the parenthesis indicates that only positive values are to be counted, and the temperature difference is to be taken to be zero when $T_o > T_{balance}$. The number of degree-days for a heating season is determined from

$$DD_{heating} = (1 \text{ day}) \sum_{days} (T_{balance} - T_{o,\,ave,\,day})^+ \quad (°\text{C-day}) \quad (12\text{-}47)$$

where $T_{o,\,ave,\,day}$ is the *average* outdoor temperature for each day (without considering temperatures above $T_{balance}$), and the summation is performed daily (Fig. 12-67). Similarly, we can also define *heating degree-hours* by using hourly average outdoor temperatures and performing the summation hourly. Note that the number of degree-hours is equal to 24 times the number of degree-days. Heating degree-days for each month and the yearly total for a balance point temperature of 65°F are given in Table 12-5 for several cities. *Cooling degree-days* are defined in the same manner to evaluate the annual energy consumption for cooling, using the same balance point temperature.

Expressing the design energy consumption of a building for heating as $\dot{Q}_{design} = K_{overall}(T_i - T_o)_{design}/\eta_{heating}$ and comparing it to the annual energy consumption gives the following relation between energy consumption at designed conditions and the annual energy consumption (Table 12-22),

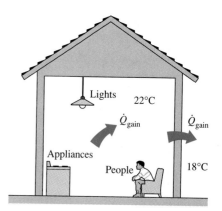

FIGURE 12-66

The heater of a building will not turn on as long as the internal heat gain makes up for the heat loss from a building (the *balance point* outdoor temperature).

For a given day:

Highest outdoor temperature: 50°F

Lowest outdoor temperature: 30°F

Average outdoor temperature: 40°F

Degree-days for that day for a balance-point temperature of 65°F:

$$DD = (1 \text{ day})(65 - 40)°\text{F}$$
$$= 25°\text{F-day}$$
$$= 600°\text{F-hour}$$

FIGURE 12-67

The outdoor temperatures for a day during which the heating degree-day is 25°F-day.

TABLE 12-22

The ratio of annual energy
consumption to the hourly
energy consumption at design
conditions at several locations for
$T_i = 70°F$ (from Eq. 12-48).

City	$T_{o,\text{design}}$	**°F-days**	**Ratio**
Tucson	32°F	1800	1137
Las Vegas	28°F	2709	1548
Charleston	11°F	4476	1821
Cleveland	5°F	6351	2345
Minneapolis	−12°F	8382	2453
Anchorage	−18°F	10,864	2963

TABLE 12-23

Approximate percent savings from
thermostat setback from 65°F for
14 hours per night and the entire
weekends (from National Frozen Food
Association/U.S. Department of
Energy, "Reducing Energy Costs
Means a Better Bottom Line").

°F-days	Amount of setback, °F			
	5°F	**10°F**	**15°F**	**20°F**
1000	13%	25%	38%	50%
2000	12	24	36	48
3000	11	22	33	44
4000	10	20	30	40
5000	9	19	28	38
6000	8	16	24	32
7000	7	15	22	30
8000	7	13	19	26
9000	6	11	16	22
10,000	5	9	14	18

$$\frac{Q_{\text{heating, year}}}{\dot{Q}_{\text{design}}} = \frac{DD_{\text{heating}}}{(T_i - T_o)_{\text{design}}} \qquad (12\text{-}48)$$

where $(T_i - T_o)_{\text{design}}$ is the design indoor–outdoor temperature difference.

Despite its simplicity, remarkably accurate results can be obtained with the *degree-day method* for most houses and single-zone buildings using a hand calculator. Besides, the degree-days characterize the *severity* of the weather at a location accurately, and the degree-day method serves as a valuable tool for gaining an *intuitive understanding* of annual energy consumption. But when the efficiency of the HVAC equipment changes considerably with the outdoor temperature, or the balance-point temperature varies significantly with time, it may be necessary to consider several bands (or "bins") of outdoor temperatures and to determine the energy consumption for each band using the equipment efficiency for those outdoor temperatures and the number of hours those temperatures are in effect. Then the annual energy consumption is obtained by adding the results of all bands. This modified degree-day approach is known as the **bin method,** and the calculations can still be performed using a hand calculator.

The steady-state methods become too crude and unreliable for buildings that experience large daily fluctuations, such as a typical, well-lit, crowded office building that is open Monday through Friday from 8 AM to 5 PM. This is especially the case when the building is equipped with programmable thermostats that utilize night setback to conserve energy. Also, the efficiency of a heat pump varies considerably with the outdoor temperatures, and the efficiencies of boilers and chillers are lower at part load. Further, the internal heat gain and necessary ventilation rate of commercial buildings vary greatly with occupancy. In such cases, it may be necessary to use a *dynamic method* such as the **transfer function method** to predict the annual energy consumption accurately. Such dynamic methods are based on performing *hourly calculations* for the entire year and adding the results. Obviously they require the use of a computer with a well-developed and hopefully user-friendly program. Very accurate results can be obtained with dynamic methods since they consider the hourly variation of indoor and outdoor conditions as well as the solar radiation, the thermal inertia of the building, the variation of the heat loss coefficient of the building, and the variation of equipment efficiency with outdoor temperatures. Even when a dynamic method is used to determine the annual energy consumption, the simple degree-day method can still be used as a check to ensure that the results obtained are in the proper range.

Some simple practices can result in significant *energy savings* in residential buildings while causing minimal discomfort. The annual energy consumption can be reduced by up to 50 percent by setting the thermostat back in winter and up in summer, and setting it back further at nights (Table 12-23). Reducing the thermostat setting in winter by 4°F (2.2°C) alone can save 12 to 18 percent; setting the thermostat back by 10°F (5.6°C) alone for 8 h on winter nights can save 7 to 13 percent. Setting the thermostat up in summer by 4°F (2.2°C) can reduce the energy consumption of residential cooling units by 18 to 32 percent. Cooling energy consumption can be reduced by up to 25 percent by sunscreening and by up to 9 percent by attic ventilation (ASHRAE *Handbook of Fundamentals,* Ref. 1, p. 28.14).

EXAMPLE 12-17 Energy and Money Savings by Winterization

You probably noticed that the heating bills are highest in December and January because the temperatures are the lowest in those months. Imagine that you have moved to Cleveland, Ohio, and your roommate offered to pay the remaining heating bills if you pay the December and January bills only. Should you accept this offer?

Solution It makes sense to accept this offer if the cost of heating in December and January is less than half of the heating bill for the entire winter. The energy consumption of a building for heating is proportional to the heating degree-days. For Cleveland, they are listed in Table 12-5 to be 1088°F-day for December, 1159°F-day for January, and 6371°F-day for the entire year (Table 12-24). The ratio of December–January degree-days to the annual degree-days is

$$\frac{DD_{\text{heating, Dec-Jan}}}{DD_{\text{heating, annual}}} = \frac{(1088 + 1159)°\text{F-day}}{6351°\text{F-day}} = 0.354$$

which is less than half. Therefore, this is a good offer and **should be accepted.**

TABLE 12-24

Monthly heating degree-days for Cleveland, Ohio, and the yearly total (Example 12-17).

| Month | Degree-days | |
	°F-days	°C-days
July	9	5
August	25	14
September	105	58
October	384	213
November	738	410
December	1088	604
January	1159	644
February	1047	582
March	918	510
April	552	307
May	260	144
June	66	37
Yearly total	6351	3528

EXAMPLE 12-18 Annual Heating Cost of a House

Using indoor and outdoor winter design temperatures of 70°F and 8°F, respectively, the design heat load of a 3000-ft² house in Salt Lake City, Utah, is determined to be 72,000 Btu/h (Fig. 12-68). The house is to be heated by natural gas that is to be burned in an 80 percent efficient furnace. If the unit cost of natural gas is $0.55/therm, estimate the annual gas consumption of this house and its cost.

Solution The annual gas consumption and its cost for a house in Salt Lake City with a design heat load of 72,000 Btu/h are to be determined.

Assumption The house is maintained at 70°F at all times during the heating season.

Analysis The gas consumption of the house for heating at design conditions is

$$\dot{Q}_{\text{design}} = \dot{Q}_{\text{design, load}}/\eta_{\text{heating}}$$
$$= (70,000 \text{ Btu/h})/0.80 = 87,500 \text{ Btu/h} = 0.875 \text{ therm/h}$$

The annual heating degree-days of Salt Lake City are listed in Table 12-5 to be 6052°F-day. Then the annual natural gas usage of the house can be determined from Equation 12-48 to be

$$Q_{\text{heating, year}} = \frac{DD_{\text{heating}}}{(T_i - T_o)_{\text{design}}} \dot{Q}_{\text{design}}$$
$$= \frac{6052°\text{F-day}}{(70 - 8)°\text{F}} \left(\frac{24 \text{ h}}{1 \text{ day}}\right)(0.875 \text{ therm/h}) = \textbf{2050 therms/year}$$

whose cost is

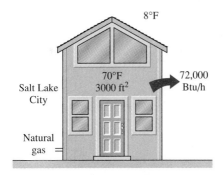

FIGURE 12-68

Schematic for Example 12-18.

FIGURE 12-69
Schematic for Example 12-19.

Annual heating cost = (Annual energy consumption)(Unit cost of energy)

= (2050 therms/year)($0.55/therm) = **$1128/year**

Therefore, it will cost $1128 per year to heat this house.

EXAMPLE 12-19   Choosing the Most Economical Air Conditioner

Consider a house whose annual air-conditioning load is estimated to be 40,000 kWh in an area where the unit cost of electricity is $0.09/kWh. Two air conditioners are considered for the house. Air conditioner A has a seasonal average COP of 2.5 and costs $2500 to purchase and install. Air conditioner B has a seasonal average COP of 5.0 and costs $4000 to purchase and install. If all else is equal, determine which air conditioner is a better buy (Fig. 12-69).

Solution   A decision is to be made between a cheaper but inefficient and an expensive but efficient air conditioner for a house.

Assumption   The two air conditioners are comparable in all aspects other than the initial cost and the efficiency.

Analysis   The unit that will cost less during its lifetime is a better buy. The total cost of a system during its lifetime (the initial, operation, maintenance, etc.) can be determined by performing a life cycle cost analysis. A simpler alternative is to determine the simple payback period. The energy and cost savings of the more efficient air conditioner in this case are

Energy savings = (Annual energy usage of A) − (Annual energy usage of B)

= (Annual cooling load)$(1/COP_A − 1/COP_B)$

= (40,000 kWh/year)(1/2.5 − 1/5.0)

= 8000 kWh/year

Cost savings = (Energy savings)(Unit cost of energy)

= (8000 kWh/year)($0.09/kWh) = **$720/year**

Therefore, the more efficient air conditioner will pay for the $1500 cost differential in this case in about two years. A cost-conscious consumer will have no difficulty in deciding that the more expensive but more efficient air conditioner B is clearly a better buy in this case since air conditioners last at least 15 years. But the decision would not be so easy if the unit cost of electricity at that location was $0.03/kWh instead of $0.09/kWh, or if the annual air-conditioning load of the house was just 10,000 kWh instead of 40,000 kWh.

12-12 ■ SUMMARY

In a broad sense, *air-conditioning* means to condition the air to the desired level by heating, cooling, humidifying, dehumidifying, cleaning, and deodorizing. The purpose of the air-conditioning system of a building is to provide complete thermal comfort for its occupants. The metabolic heat generated in the body is dissipated to the environment through the skin and lungs by convection and radiation as *sensible heat* and by evaporation as *latent heat*. The total *sensible heat loss* can be expressed by combining convection and radiation heat losses as

$$\dot{Q}_{conv+rad} = (h_{conv} + h_{rad})\, A_{clothing}(T_{clothing} - T_{operative}) = \frac{A_{clothing}(T_{skin} - T_{clothing})}{R_{clothing}}$$

where $R_{clothing}$ is the *unit thermal resistance of clothing*, which involves the combined effects of conduction, convection, and radiation between the skin and the outer surface of clothing. The *operative temperature* $T_{operative}$ is approximately the arithmetic average of the ambient and surrounding surface temperatures. Another environmental index used in thermal comfort analysis is the *effective temperature*, which combines the effects of temperature and humidity.

The desirable ranges of temperatures, humidities, and ventilation rates for indoors constitute the typical *indoor design conditions*. The set of extreme outdoor conditions under which a heating or cooling system must be able to maintain a building at the indoor design conditions is called the *outdoor design conditions*. The *heating* or *cooling loads* of a building represent the heat that must be supplied to or removed from the interior of a building to maintain it at the desired conditions. The effect of solar heating on opaque surfaces is accounted for by replacing the ambient temperature in the heat transfer relation through the walls and the roof by the *sol-air temperature*, which is defined as the equivalent outdoor air temperature that gives the same rate of heat flow to a surface as would the combination of incident solar radiation, convection with the ambient air, and radiation exchange with the sky and the surrounding surfaces.

Heat flow into an exterior surface of a building subjected to solar radiation can be expressed as

$$\dot{Q}_{surface} = h_o A(T_{sol\text{-}air} - T_{surface})$$

where

$$T_{sol\text{-}air} = T_{ambient} + \frac{\alpha_s \dot{q}_{solar}}{h_o} - \frac{\varepsilon\sigma(T_{ambient}^4 - T_{surr}^4)}{h_o}$$

and α_s is the *solar absorptivity* and ε is the *emissivity* of the surface, h_o is the combined convection and radiation heat transfer coefficient, and $\dot{q}_{solar}$ is the solar radiation incident on the surface.

The conversion of chemical or electrical energy to heat in a building constitutes the *internal heat gain* of a building. The primary sources of internal heat gain are people, lights, appliances, and miscellaneous equipment such as computers, printers, and copiers. The average amount of heat given off by a person depends on the level of activity and can range from about 100 W for a resting person to more than 500 W for a physically very active person. The heat gain due to a motor inside a conditioned space can be expressed as

$$\dot{Q}_{motor,\,total} = \dot{W}_{motor} \times f_{load} \times f_{usage}/\eta_{motor}$$

where $\dot{W}_{motor}$ is the power rating of the motor, f_{load} is the *load factor* of the motor during operation, f_{usage} is the *usage factor*, and η_{motor} is the *motor efficiency*.

Under steady conditions, the rate of heat transfer through any section of a *building wall* or *roof* can be determined from

$$\dot{Q} = UA(T_i - T_o) = \frac{A(T_i - T_o)}{R}$$

where T_i and T_o are the indoor and outdoor air temperatures, A is the heat transfer area, U is the overall heat transfer coefficient (the U-factor), and $R = 1/U$ is the overall unit thermal resistance (the R-value). The overall R-value of a wall or roof can be determined from the thermal resistances of the individual components using the thermal resistance network. The *effective emissivity* of a plane-parallel air space is given by

$$\frac{1}{\varepsilon_{\text{effective}}} = \frac{1}{\varepsilon_1} + \frac{1}{\varepsilon_2} - 1$$

where ε_1 and ε_2 are the emissivities of the surfaces of the air space.

Heat losses through the below-grade section of a *basement wall* and through the *basement floor* are given as

$$\dot{Q}_{\text{basement walls}} = U_{\text{wall, ave}} A_{\text{wall}}(T_{\text{basement}} - T_{\text{ground surface}})$$

$$\dot{Q}_{\text{basement floor}} = U_{\text{floor}} A_{\text{floor}}(T_{\text{basement}} - T_{\text{ground surface}})$$

where $U_{\text{wall, ave}}$ is the average overall heat transfer coefficient between the basement wall and the surface of the ground and U_{floor} is the overall heat transfer coefficient at the basement floor. Heat loss from floors that sit directly on the ground at or slightly above the ground level is mostly through the perimeter to the outside air rather than through the floor into the ground and is expressed as

$$\dot{Q}_{\text{floor on grade}} = U_{\text{grade}} p_{\text{floor}}(T_{\text{indoor}} - T_{\text{outdoor}})$$

where U_{grade} represents the rate of heat transfer from the slab per unit temperature difference between the indoor temperature T_{indoor} and the outdoor temperature T_{outdoor} and per unit length of the perimeter p_{floor} of the building. When the crawl space temperature is known, heat loss through the *floor of the building* is determined from

$$\dot{Q}_{\text{building floor}} = U_{\text{building floor}} A_{\text{floor}}(T_{\text{indoor}} - T_{\text{crawl}})$$

where $U_{\text{building floor}}$ is the overall heat transfer coefficient for the floor.

Windows are considered in three regions when analyzing heat transfer through them: (1) the *center-of-glass*, (2) the *edge-of-glass*, and (3) the *frame* regions. Total rate of heat transfer through the window is determined by adding the heat transfer through each region as

$$\dot{Q}_{\text{window}} = \dot{Q}_{\text{center}} + \dot{Q}_{\text{edge}} + \dot{Q}_{\text{frame}} = U_{\text{window}} A_{\text{window}}(T_{\text{indoors}} - T_{\text{outdoors}})$$

where

$$U_{\text{window}} = (U_{\text{center}} A_{\text{center}} + U_{\text{edge}} A_{\text{edge}} + U_{\text{frame}} A_{\text{frame}})/A_{\text{window}}$$

is the *U-factor* or the *overall heat transfer coefficient* of the window; A_{window} is the window area; A_{center}, A_{edge}, and A_{frame} are the areas of the center, edge, and frame sections of the window, respectively; and U_{center}, U_{edge}, and U_{frame}

are the heat transfer coefficients for the center, edge, and frame sections of the window.

The sum of the transmitted solar radiation and the portion of the absorbed radiation that flows indoors constitutes the *solar heat gain* of the building. The fraction of incident solar radiation that enters through the glazing is called the *solar heat gain coefficient* SHGC, and the total solar heat gain through that window is determined from

$$\dot{Q}_{\text{solar, gain}} = \text{SHGC} \times A_{\text{glazing}} \times \dot{q}_{\text{solar, incident}}$$

where A_{glazing} is the glazing area of the window and $\dot{q}_{\text{solar, incident}}$ is the solar heat flux incident on the outer surface of the window. Using the standard 3-mm-thick double-strength clear window glass sheet whose SHGC is 0.87 as the reference glazing, the *shading coefficient* SC is defined as

$$\text{SC} = \frac{\text{Solar heat gain of product}}{\text{Solar heat gain of reference glazing}}$$

$$= \frac{\text{SHGC}}{\text{SHGC}_{\text{ref}}} = \frac{\text{SHGC}}{0.87} = 1.15 \times \text{SHGC}$$

Shading devices are classified as *internal shading* and *external shading*, depending on whether the shading device is placed inside or outside.

The uncontrolled entry of outside air into a building through unintentional openings is called *infiltration*, and it wastes a significant amount of energy since the air entering must be heated in winter and cooled in summer. The *sensible* and *latent heat load* of infiltration are expressed as

$$\dot{Q}_{\text{infiltration, sensible}} = \rho_o C_p \dot{V}(T_i - T_o) = \rho_o C_p(\text{ACH})(V_{\text{building}})(T_i - T_o)$$

$$\dot{Q}_{\text{infiltration, latent}} = \rho_o h_{fg} \dot{V}(w_i - w_o) = \rho_o h_{fg}(\text{ACH})(V_{\text{building}})(w_i - w_o)$$

where ρ_o is the density of outdoor air; C_p is the specific heat of air (about 1 kJ/kg $\cdot$ °C or 0.24 Btu/lbm $\cdot$ °F); $\dot{V} = (\text{ACH})(V_{\text{building}})$ is the volumetric flow rate of air, which is the number of air changes per hour times the volume of the building; and $T_i - T_o$ is the temperature difference between the indoor and outdoor air. Also, h_{fg} is the latent heat of vaporization at indoor temperature (about 2340 kJ/kg or 1000 Btu/lbm) and $w_i - w_o$ is the humidity ratio difference between the indoor and outdoor air.

The *annual energy consumption* of a building depends on the space load for heating or cooling, the efficiency of the heating or cooling equipment, and the parasitic energy consumed by the pumps or fans and the energy lost or gained from the pipes or ducts. The annual energy consumption of a building can be estimated using the *degree-day method* as

$$Q_{\text{heating, year}} = \frac{K_{\text{overall}}}{\eta_{\text{heating}}} DD_{\text{heating}}$$

where DD_{heating} is the *heating degree-days*, K_{overall} is the overall heat transfer coefficient of the building in W/°C or Btu/h $\cdot$ °F, and η_{heating} is the efficiency of the heating system, which is equal to 1.0 for electric resistance heating systems, COP for the heat pumps, and combustion efficiency (about 0.6 to 0.95) for furnaces.

REFERENCES AND SUGGESTED READING

1. American Society of Heating, Refrigeration, and Air-Conditioning Engineers. *Handbook of Fundamentals.* Atlanta: ASHRAE, 1993.

2. *How to Reduce Your Energy Costs.* 2nd ed. Boston: Center for Information Sharing, Inc. (through Sierra Pacific Power Company), 1991.

3. J. F. Kreider and A. Rabl. *Heating and Cooling of Buildings.* New York: McGraw-Hill, 1994.

4. F. C. McQuiston and J. D. Parker. *Heating, Ventilating, and Air Conditioning.* 4th ed. New York: Wiley, 1994.

5. *Radiant Barrier Attic Fact Sheet.* DOE/CE-0335P. Washington, DC: U.S. Department of Energy, 1991.

6. "Replacement Windows." *Consumer Reports,* Yonkers, NY: Consumer Union. October 1993, p. 664.

7. H. J. Sauer, Jr., and R. H. Howell. *Principles of Heating, Ventilating, and Air Conditioning.* Atlanta: ASHRAE, 1994.

8. W. F. Stoecker and J. W. Jones. *Refrigeration and Air Conditioning.* New York: McGraw-Hill, 1982.

PROBLEMS*

A Brief History

12-1C In 1775, Dr. William Cullen made ice in Scotland by evacuating the air in a water tank. Explain how that device works, and discuss how the process can be made more efficient.

12-2C When was the first ammonia absorption refrigeration system developed? Who developed the formulas related to dry-bulb, wet-bulb, and dew-point temperatures, and when?

12-3C When was the concept of heat pump conceived and by whom? When was the first heat pump built, and when were the heat pumps mass produced?

Human Body and Thermal Comfort

12-4C What is metabolism? What is the range of metabolic rate for an average man? Why are we interested in metabolic rate of the occupants of a building when we deal with heating and air conditioning?

12-5C Why is the metabolic rate of women, in general, lower than that of men? What is the effect of clothing on the environmental temperature that feels comfortable?

---

*Students are encouraged to answer *all* the concept "C" questions.

12-6C What is asymmetric thermal radiation? How does it cause thermal discomfort in the occupants of a room?

12-7C How do (a) draft and (b) cold floor surfaces cause discomfort for a room's occupants?

12-8C What is stratification? Is it likely to occur at places with low or high ceilings? How does it cause thermal discomfort for a room's occupants? How can stratification be prevented?

12-9C Why is it necessary to ventilate buildings? What is the effect of ventilation on energy consumption for heating in winter and for cooling in summer? Is it a good idea to keep the bathroom fans on all the time? Explain.

Heat Transfer from the Human Body

12-10C Consider a person who is resting or doing light work. Is it fair to say that roughly one-third of the metabolic heat generated in the body is dissipated to the environment by convection, one-third by evaporation, and the remaining one-third by radiation?

12-11C What is sensible heat? How is the sensible heat loss from a human body affected by (a) skin temperature, (b) environment temperature, and (c) air motion?

12-12C What is latent heat? How is the latent heat loss from the human body affected by (a) skin wettedness and (b) relative humidity of the environment? How is the rate of evaporation from the body related to the rate of latent heat loss?

12-13C How is the insulating effect of clothing expressed? How does clothing affect heat loss from the body by convection, radiation, and evaporation? How does clothing affect heat gain from the sun?

12-14C Explain all the different mechanisms of heat transfer from the human body (a) through the skin and (b) through the lungs.

12-15C What is operative temperature? How is it related to the mean ambient and radiant temperatures? How does it differ from effective temperature?

12-16 The convection heat transfer coefficient for a clothed person while walking in still air at a velocity of 0.5 to 2 m/s is given by $h = 8.6 \mathcal{V}^{0.53}$ where $\mathcal{V}$ is in m/s and h is in $W/m^2 \cdot °C$. Plot the convection coefficient against the walking velocity, and compare the convection coefficients in that range to the average radiation coefficient of about 5 $W/m^2 \cdot °C$.

12-17 A clothed or unclothed person feels comfortable when the skin temperature is about 33°C. Consider an average man wearing summer clothes whose thermal resistance is 0.7 clo. The man feels very comfortable while standing in a room maintained at 20°C. If this man were to stand in that room unclothed, determine the temperature at which the room must be maintained for him to feel thermally comfortable. Assume the latent heat loss from the person to remain the same.   *Answer:* 26.4°C

Moisture
0.50 lbm

FIGURE P12-18E

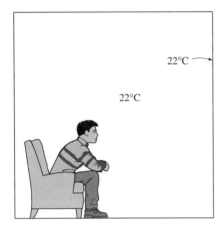

22°C

22°C

FIGURE P12-22

Radiant
heater

FIGURE P12-23

12-18E An average person produces 0.50 lbm of moisture while taking a shower and 0.12 lbm while bathing in a tub. Consider a family of four who shower once a day in a bathroom that is not ventilated. Taking the heat of vaporization of water to be 1050 Btu/lbm, determine the contribution of showers to the latent heat load of the air conditioner in summer per day.

12-19 An average (1.82 kg or 4.0 lbm) chicken has a basal metabolic rate of 5.47 W and an average metabolic rate of 10.2 W (3.78 W sensible and 6.42 W latent) during normal activity. If there are 100 chickens in a breeding room, determine the rate of total heat generation and the rate of moisture production in the room. Take the heat of vaporization of water to be 2430 kJ/kg.

12-20 Consider a large classroom with 150 students on a hot summer day. All the lights with 4.0 kW of rated power are kept on. The room has no external walls, and thus heat gain through the walls and the roof is negligible. Chilled air is available at 15°C, and the temperature of the return air is not to exceed 25°C. Determine the required flow rate of air, in kg/s, that needs to be supplied to the room.   *Answer:* 1.45 kg/s

12-21 A smoking lounge is to accommodate 15 heavy smokers. Determine the minimum required flow rate of fresh air that needs to be supplied to the lounge.   *Answer:* 0.45 m³/s

12-22 A person feels very comfortable in his house in light clothing when the thermostat is set at 22°C and the mean radiation temperature (the average temperature of the surrounding surfaces) is also 22°C. During a cold day, the average mean radiation temperature drops to 18°C. To what value must the indoor air temperature be raised to maintain the same level of comfort in the same clothing?

12-23 A car mechanic is working in a shop whose interior space is not heated. Comfort for the mechanic is provided by two radiant heaters that radiate heat at a total rate of 10 kJ/s. About 5 percent of this heat strikes the mechanic directly. The shop and its surfaces can be assumed to be at the ambient temperature, and the emissivity and absorptivity of the mechanic can be taken to be 0.95 and the surface area to be 1.8 m². The mechanic is generating heat at a rate of 350 W, half of which is latent, and is wearing medium clothing with a thermal resistance of 0.7 clo. Determine the lowest ambient temperature in which the mechanic can work comfortably.

Design Conditions for Heating and Cooling

12-24C What is winter outdoor design temperature? How does it differ from average winter outdoor temperature? How do 97.5 percent and 99 percent winter outdoor design temperatures differ from each other?

12-25C Weather data for two different cities A and B are compared. Is it possible for city A to have a lower winter design temperature but a higher average winter temperature? Explain.

12-26C What is the effect of solar radiation on (*a*) the design heating load in winter and (*b*) the annual energy consumption for heating? Answer the same question for the heat generated by people, lights, and appliances.

12-27C What is the effect of solar radiation on (*a*) the design cooling load in summer and (*b*) the annual energy consumption for cooling? Answer the same question for the heat generated by people, lights, and appliances.

12-28C Does the moisture level of the outdoor air affect the cooling load in summer, or can it just be ignored? Explain. Answer the same question for the heating load in winter.

12-29C The recommended design heat transfer coefficients for combined convection and radiation on the outer surface of a building are 34 W/m² · °C for winter and 22.7 W/m² · °C for summer. What is the reason for different values?

12-30C What is sol-air temperature? What is it used for? What is the effect of solar absorptivity of the outer surface of a wall on the sol-air temperature?

12-31C Consider the solar energy absorbed by the walls of a brick house. Will most of this energy absorbed by the wall be transferred to indoors or lost to the outdoors? Why?

12-32 Determine the outdoor design conditions for Lincoln, Nebraska, for summer for the 2.5 percent level and for winter for the 97.5 percent level.

12-33 Specify the indoor and outdoor design conditions for a hospital in Wichita, Kansas.

12-34 The south masonry wall of a house is made of 10-cm-thick red face brick, 10-cm-thick common brick, 19-mm-thick air space, and 13-mm-thick gypsum board, and its overall heat transfer coefficient is 1.6 W/m² · °C, which includes the effects of convection on both sides. The house is located at 40° N latitude, and its cooling system is to be sized on the basis of the heat gain at 3 PM solar time on July 21. The interior of the house is to be maintained at 22°C, and the area of the wall is 20 m². If the design ambient air temperature at that time is 35°C, determine (*a*) the design heat gain through the wall, (*b*) the fraction of this heat gain due to solar heating, and (*c*) the fraction of incident solar radiation transferred into the house through the wall.

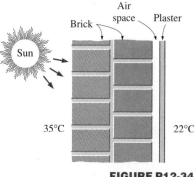

FIGURE P12-34

12-35E The west wall of a shopping center in Pittsburgh, Pennsylvania (40° latitude), is 150 ft long and 12 ft high. The 2.5 percent summer design temperature of Pittsburgh is 86°F, and the interior of the building is maintained at 72°F. The wall is made of face brick and concrete block with cement mortar in between, reflective air space, and a gypsum wallboard. The overall heat transfer coefficient of the wall is 0.14 Btu/h · ft² · °F, which includes the effects of convection on both sides. The cooling system is to be sized on the basis of the heat gain at 16:00 hour (4 PM) solar time in late July. Determine (*a*) the design heat gain through the wall and (*b*) the fraction of this heat gain due to solar heating.

FIGURE P12-36

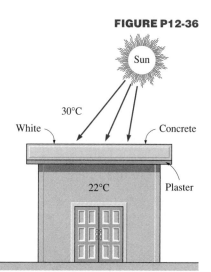

12-36 The roof of a building in Istanbul, Turkey (40° latitude), is made of a 30-cm-thick horizontal layer of concrete, painted white to minimize solar gain, with a layer of plaster inside. The overall heat transfer coefficient of the roof is 1.8 W/m² · °C, which includes the effects of convection on both the interior and exterior surfaces. The cooling system is to be sized on the basis of the heat gain at 16:00 hour solar time in late July. The interior of the building is to be maintained at 22°C, and the summer design temperature of Istanbul is 30°C. If the exposed surface area of the roof is 150 m², determine (*a*) the design heat gain through the roof and (*b*) the fraction of this heat gain due to solar heating.   *Answers:* (*a*) 4320 W, (*b*) 50 percent

Heat Gain from People, Lights, and Appliances

12-37C Is the heat given off by people in a concert hall an important consideration in the sizing of the air-conditioning system for that building or can it be ignored? Explain.

12-38C During a lighting retrofitting project, all the incandescent lamps of a building are replaced by high-efficiency fluorescent lamps. Explain how this retrofit will affect the (*a*) design cooling load, (*b*) annual energy consumption for cooling, and (*c*) annual energy consumption for heating for the building.

12-39C Give two good reasons why it is usually a good idea to replace incandescent light bulbs by compact fluorescent bulbs that may cost 40 times as much to purchase.

12-40C Explain how the motors and appliances in a building affect the (*a*) design cooling load, (*b*) annual energy consumption for cooling, and (*c*) annual energy consumption for heating of the building.

12-41C Define motor efficiency, and explain how it affects the design cooling load of a building and the annual energy consumption for cooling.

12-42C Consider a hooded range in a kitchen with a powerful fan that exhausts all the air heated and humidified by the range. Does the heat generated by this range still need to be considered in the determination of the cooling load of the kitchen or can it just be ignored since all the heated air is exhausted?

12-43 Consider a 3-kW hooded electric open burner in an area where the unit costs of electricity and natural gas are $0.09/kWh and $0.55/therm, respectively. The efficiency of open burners can be taken to be 73 percent for electric burners and 38 percent for gas burners. Determine the amount of the electrical energy used directly for cooking, the cost of energy per "utilized" kWh, and the contribution of this burner to the design cooling load. Repeat the calculations for the gas burner.

Hood

Open
burner

FIGURE P12-43

12-44 An exercise room has eight weight-lifting machines that have no motors and four treadmills each equipped with a 2.5-hp motor. The motors operate at an average load factor of 0.7, at which their efficiency is 0.77. During peak evening hours, all 12 pieces of exercising equipment are used continuously, and there are also two people doing light exercises while waiting in line for one piece of the equipment. Determine the rate of heat gain of the exercise room from people and the equipment at peak load conditions. How much of this heat gain is in the latent form?

12-45 A 75-hp motor with an efficiency of 91.0 percent is worn out and is replaced by a high-efficiency motor having an efficiency of 95.4 percent.

Determine the reduction in the rate of internal heat gain due to higher efficiency under full-load conditions.   *Answer:* 2836 W

12-46   The efficiencies of commercial hot plates are 48 and 23 percent for electric and gas models, respectively. For the same amount of "utilized" energy, determine the ratio of internal heat generated by gas hot plates to that by electric ones.

12-47   Consider a classroom for 40 students and one instructor. Lighting is provided by 18 fluorescent light bulbs, 40 W each, and the ballasts consume an additional 10 percent. Determine the rate of internal heat generation in this classroom when it is fully occupied.   *Answer:* 5507 W

12-48   A 60-hp electric car is powered by an electric motor mounted in the engine compartment. If the motor has an average efficiency of 88 percent, determine the rate of heat supply by the motor to the engine compartment at full load.

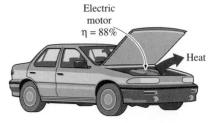

FIGURE P12-48

12-49   A room is cooled by circulating chilled water through a heat exchanger located in a room. The air is circulated through the heat exchanger by a 0.25-hp fan. Typical efficiency of small electric motors driving 0.25-hp equipment is 54 percent. Determine the contribution of the fan-motor assembly to the cooling load of the room.   *Answer:* 345 W

12-50   Consider an office room that is being cooled adequately by a 12,000 Btu/h window air conditioner. Now it is decided to convert this room into a computer room by installing several computers, terminals, and printers with a total rated power of 3.5 kW. The facility has several 4000 Btu/h air conditioners in storage that can be installed to meet the additional cooling requirements. Assuming a usage factor of 0.4 (i.e., only 40 percent of the rated power will be consumed at any given time) and additional occupancy of four people, determine how many of these air conditioners need to be installed in the room.

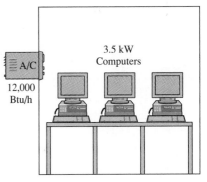

FIGURE P12-50

12-51   A restaurant purchases a new 8-kW electric range for its kitchen. Determine the increase in the design cooling load of the kitchen if the range is (*a*) hooded and (*b*) unhooded.   *Answers:* (*a*) 4.0 kW, (*b*) 1.28 kW

12-52   A department store expects to have 80 customers and 15 employees at peak times in summer. Determine the contribution of people to the sensible, latent, and total cooling load of the store.

12-53E   In a movie theater in winter, 500 people are watching a movie. The heat losses through the walls, windows, and the roof are estimated to be 150,000 Btu/h. Determine if the theater needs to be heated or cooled.

Heat Transfer through the Walls and Roofs

12-54C   What is the *R*-value of a wall? How does it differ from the unit thermal resistance of the wall? How is it related to the *U*-factor?

12-55C   What is effective emissivity for a plane-parallel air space? How is it determined? How is radiation heat transfer through the air space determined when the effective emissivity is known?

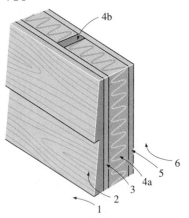

FIGURE P12-59

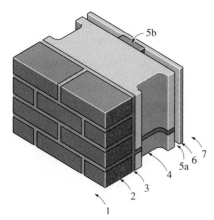

FIGURE P12-61E

FIGURE P12-62

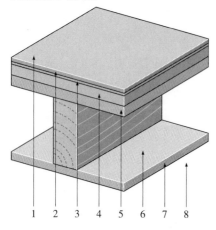

12-56C The unit thermal resistances (*R*-values) of both 40-mm and 90-mm vertical air spaces are given in Table 12-13 to be 0.22 $m^2 \cdot °C/W$, which implies that more than doubling the thickness of air space in a wall has no effect on heat transfer through the wall. Do you think this is a typing error? Explain.

12-57C What is a radiant barrier? What kind of materials are suitable for use as radiant barriers? Is it worthwhile to use radiant barriers in the attics of homes?

12-58C Consider a house whose attic space is ventilated effectively so that the air temperature in the attic is the same as the ambient air temperature at all times. Will the roof still have any effect on heat transfer through the ceiling? Explain.

12-59 Determine the summer *R*-value and the *U*-factor of a wood frame wall that is built around 38-mm × 140-mm wood studs with a center-to-center distance of 400 mm. The 140-mm-wide cavity between the studs is filled with mineral fiber batt insulation. The inside is finished with 13-mm gypsum wallboard and the outside with 13-mm wood fiberboard and 13-mm × 200-mm wood bevel lapped siding. The insulated cavity constitutes 80 percent of the heat transmission area, while the studs, headers, plates, and sills constitute 20 percent. *Answers:* 3.213 $m^2 \cdot °C/W$, 0.311 $W/m^2 \cdot °C$

12-60 The 13-mm-thick wood fiberboard sheathing of the wood stud wall in Problem 12-59 is replaced by a 25-mm-thick rigid foam insulation. Determine the percent increase in the *R*-value of the wall as a result.

12-61E Determine the winter *R*-value and the *U*-factor of a masonry cavity wall that is built around 4-in.-thick concrete blocks made of lightweight aggregate. The outside is finished with 4-in. face brick with $\frac{1}{2}$-in. cement mortar between the bricks and concrete blocks. The inside finish consists of $\frac{1}{2}$-in. gypsum wallboard separated from the concrete block by $\frac{3}{4}$-in.-thick (1-in. by 3-in. nominal) vertical furring whose center-to-center distance is 16 in. Neither side of the $\frac{3}{4}$-in.-thick air space between the concrete block and the gypsum board is coated with any reflective film. When determining the *R*-value of the air space, the temperature difference across it can be taken to be 30°F with a mean air temperature of 50°F. The air space constitutes 80 percent of the heat transmission area, while the vertical furring and similar structures constitute 20 percent.

12-62 Consider a flat ceiling that is built around 38-mm × 90-mm wood studs with a center-to-center distance of 400 mm. The lower part of the ceiling is finished with 13-mm gypsum wallboard, while the upper part consists of a wood subfloor (*R* = 0.166 $m^2 \cdot °C/W$), a 16-mm plywood, a layer of felt (*R* = 0.011 $m^2 \cdot °C/W$), and linoleum (*R* = 0.009 $m^2 \cdot °C/W$). Both sides of the ceiling are exposed to still air. The air space constitutes 82 percent of the heat transmission area, while the studs and headers constitute 18 percent. Determine the winter *R*-value and the *U*-factor of the ceiling assuming the 90-mm-wide air space between the studs (*a*) does not have any reflective surface, (*b*) has a reflective surface with $\varepsilon = 0.05$ on one side, and (*c*) has reflective surfaces with $\varepsilon = 0.05$ on both sides. Assume a mean temperature of 10°C and a temperature difference of 5.6°C for the air space.

12-63 Determine the winter *R*-value and the *U*-factor of a masonry cavity wall that consists of 100-mm common bricks, a 90-mm air space, 100-mm concrete blocks made of lightweight aggregate, 20-mm air space, and 13-mm gypsum wallboard separated from the concrete block by 20-mm-thick (1-in. × 3-in. nominal) vertical furring whose center-to-center distance is 400 mm. Neither side of the two air spaces is coated with any reflective films. When determining the *R*-value of the air spaces, the temperature difference across them can be taken to be 16.7°C with a mean air temperature of 10°C. The air space constitutes 84 percent of the heat transmission area, while the vertical furring and similar structures constitute 16 percent.
Answers: 1.02 m² · °C/W, 0.978 W/m² · °C

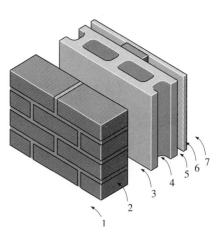

FIGURE P12-63

12-64 Repeat Problem 12-63 assuming one side of both air spaces is coated with a reflective film of $\varepsilon = 0.05$.

12-65 Determine the winter *R*-value and the *U*-factor of a masonry wall that consists of the following layers: 100-mm face bricks, 100-mm common bricks, 25-mm urethane rigid foam insulation, and 13-mm gypsum wallboard. *Answers:* 1.404 m² · °C/W, 0.712 W/m² · °C

12-66 The overall heat transfer coefficient (the *U*-value) of a wall under winter design conditions is $U = 1.55$ W/m² · °C. Determine the *U*-value of the wall under summer design conditions.

12-67 The overall heat transfer coefficient (the *U*-value) of a wall under winter design conditions is $U = 2.25$ W/m² · °C. Now a layer of 100-mm face brick is added to the outside, leaving a 20-mm air space between the wall and the bricks. Determine the new *U*-value of the wall. Also, determine the rate of heat transfer through a 3-m-high, 7-m-long section of the wall after modification when the indoor and outdoor temperatures are 22°C and −5°C, respectively.

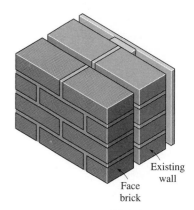

Existing wall

Face brick

FIGURE P12-67

12-68 Determine the summer and winter *R*-values, in m² · °C/W, of a masonry wall that consists of 100-mm face bricks, 13-mm of cement mortar, 100-mm lightweight concrete block, 40-mm air space, and 20-mm plasterboard. *Answers:* 0.809 and 0.795 m² · °C/W

12-69E The overall heat transfer coefficient of a wall is determined to be $U = 0.09$ Btu/h · ft² · °F under the conditions of still air inside and winds of 7.5 mph outside. What will the *U*-factor be when the wind velocity outside is doubled? *Answer:* 0.0907 Btu/h · ft² · °F

12-70 Two homes are identical, except that the walls of one house consist of 200-mm lightweight concrete blocks, 20-mm air space, and 20-mm plasterboard, while the walls of the other house involve the standard *R*-2.4 m² · °C/W frame wall construction. Which house do you think is more energy efficient?

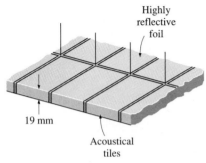

FIGURE P12-71

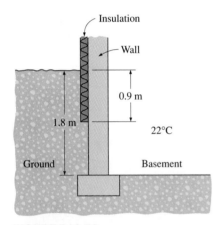

FIGURE P12-78

FIGURE P12-81E

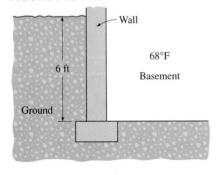

12-71 Determine the *R*-value of a ceiling that consists of a layer of 19-mm acoustical tiles whose top surface is covered with a highly reflective aluminum foil for winter conditions. Assume still air below and above the tiles.

Heat Loss from Basement Walls and Floors

12-72C What is the mechanism of heat transfer from the basement walls and floors to the ground? What is the effect of the composition and moisture content of the soil on this heat transfer?

12-73C Consider a basement wall that is completely below grade. Will heat loss through the upper half of the wall be greater or smaller than the heat loss through the lower half? Why?

12-74C Does a building lose more heat to the ground through the floor of a basement or through the below-grade section of the basement wall per unit surface area? Explain.

12-75C Is heat transfer from a floor on grade at ground level proportional to the surface area or perimeter of the floor?

12-76C Crawl spaces are often vented to prevent moisture accumulation and associated problems. How does venting the crawl space affect heat loss through the floor?

12-77C Consider a house with an unheated crawl space in winter. If the vents of the crawl space are tightly closed, do you think the cold water pipes should still be insulated to avoid the danger of freezing?

12-78 Consider a basement in Anchorage, Alaska, where the mean winter temperature is −5.0°C. The basement is 7 m wide and 10 m long, and the basement floor is 1.8 m below the ground level. The top 0.9-m section of the wall below the grade is insulated with R-2.20 m² · °C/W insulation. Assuming the interior temperature of the basement to be 20°C, determine the peak heat loss from the basement to the ground through its walls and floor.

12-79 Consider a crawl space that is 10 m wide, 18 m long, and 0.80 m high whose vent is kept open. The interior of the house is maintained at 21°C and the ambient temperature is 2.5°C. Determine the rate of heat loss through the floor of the house to the crawl space for the cases of (*a*) insulated and (*b*) uninsulated floor.   *Answers:* (*a*) 1439 W, (*b*) 4729 W

12-80 The dimensions of a house with basement in Boise, Idaho, are 18 m × 12 m. The 1.8-m-high portion of the basement wall is below the ground level and is not insulated. If the basement is maintained at 18°C, determine the rate of design heat loss from the basement through its walls and floor.

12-81E The dimensions of a house with basement in Boise, Idaho, are 60 ft × 32 ft. The 6-ft-high portion of the basement wall is below the ground level and is not insulated. If the basement is maintained at 68°F, determine the rate of design heat loss from the basement through its walls and floor.

12-82 A house in Baltimore, Maryland, has a concrete slab floor that sits directly on the ground at grade level. The house is 18 m long and 15 m wide, and the weather in Baltimore can be considered to be moderate. The walls of the house are made of 20-cm block wall with brick and are insulated from the edge to the footer with R-0.95 m² · °C/W insulation. If the house is maintained at 22°C, determine the heat loss from the floor at winter design conditions.   *Answer:* 1873 W

12-83 Repeat Problem 12-82 assuming the below-grade section of the wall is not insulated.

12-84 The crawl space of a house is 12 m wide, 20 m long, and 0.70 m high. The vents of the crawl space are kept closed, but air still infiltrates at a rate of 1.2 ACH. The indoor and outdoor design temperatures are 22°C and −5°C, respectively, and the deep-down ground temperature is 10°C. Determine the heat loss from the house to the crawl space, and the crawl space temperature assuming the walls, floor, and ceiling of the crawl space are (a) insulated and (b) uninsulated.   *Answers:* (a) 1.3°C, (b) 11.7°C

12-85 Repeat Problem 12-84 assuming the vents are tightly sealed for winter and thus air infiltration is negligible.

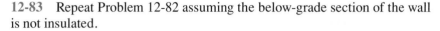

FIGURE P12-84

Heat Transfer through Windows

12-86C Why are the windows considered in three regions when analyzing heat transfer through them? Name those regions and explain how the overall U-value of the window is determined when the heat transfer coefficients for all three regions are known.

12-87C Consider three similar double-pane windows with air gap widths of 5, 10, and 20 mm. For which case will the heat transfer through the window will be a minimum?

12-88C In an ordinary double-pane window, about half of the heat transfer is by radiation. Describe a practical way of reducing the radiation component of heat transfer.

12-89C Consider a double-pane window whose air space width is 20 mm. Now a thin polyester film is used to divide the air space into two 10-mm-wide layers. How will the film affect (a) convection and (b) radiation heat transfer through the window?

12-90C Consider a double-pane window whose air space is flashed and filled with argon gas. How will replacing the air in the gap by argon affect (a) convection and (b) radiation heat transfer through the window?

12-91C Is the heat transfer rate through the glazing of a double-pane window higher at the center or edge section of the glass area? Explain.

12-92C How do the relative magnitudes of U-factors of windows with aluminum, wood, and vinyl frames compare? Assume the windows are identical except for the frames.

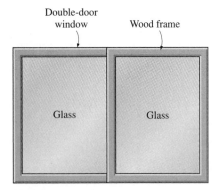

FIGURE P12-94

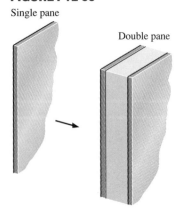

12-93 Determine the U-factor for the center-of-glass section of a double-pane window with a 13-mm air space for winter design conditions. The glazings are made of clear glass having an emissivity of 0.84. Take the average air space temperature at design conditions to be 10°C and the temperature difference across the air space to be 15°C.

12-94 A double-door wood-framed window with glass glazing and metal spacers is being considered for an opening that is 1.2 m high and 1.8 m wide in the wall of a house maintained at 20°C. Determine the rate of heat loss through the window and the inner surface temperature of the window glass facing the room when the outdoor air temperature is −8°C if the window is selected to be (*a*) 3-mm single glazing, (*b*) double glazing with an air space of 13 mm, and (*c*) low-e-coated triple glazing with an air space of 13 mm.

12-95 Determine the overall U-factor for a double-door-type wood-framed double-pane window with 13-mm air space and metal spacers, and compare your result to the value listed in Table 12-19. The overall dimensions of the window are 2.00 m × 2.40 m, and the dimensions of each glazing are 1.92 m × 1.14 m.

12-96 Consider a house in Atlanta, Georgia, that is maintained at 22°C and has a total of 20 m² of window area. The windows are double-door-type with wood frames and metal spacers. The glazing consists of two layers of glass with 12.7 mm of air space with one of the inner surfaces coated with reflective film. The winter average temperature of Atlanta is 11.3°C. Determine the average rate of heat loss through the windows in winter.   *Answer:* 456 W

12-97E Consider an ordinary house with R-13 walls (walls that have an R-value of 13 h · ft² · °F/Btu). Compare this to the R-value of the common double-door windows that are double pane with $\frac{1}{4}$ in. of air space and have aluminum frames. If the windows occupy only 20 percent of the wall area, determine if more heat is lost through the windows or through the remaining 80 percent of the wall area. Disregard infiltration losses.

12-98 The overall U-factor of a fixed wood-framed window with double glazing is given by the manufacturer to be $U = 2.76$ W/m² · °C under the conditions of still air inside and winds of 12 km/h outside. What will the U-factor be when the wind velocity outside is doubled?
    Answer: 2.88 W/m² · °C

12-99 The owner of an older house in Wichita, Kansas, is considering replacing the existing double-door type wood-framed single-pane windows with vinyl-framed double-pane windows with an air space of 6.4 mm. The new windows are of double-door type with metal spacers. The house is maintained at 22°C at all times, but heating is needed only when the outdoor temperature drops below 18°C because of the internal heat gain from people, lights, appliances, and the sun. The average winter temperature of Wichita is 7.1°C, and the house is heated by electric resistance heaters. If the unit cost of electricity is $0.07/kWh and the total window area of the house is 12 m², determine how much money the new windows will save the home owner per month in winter.

12-100C   What fraction of the solar energy is in the visible range (*a*) outside the earth's atmosphere and (*b*) at sea level on earth? Answer the same question for infrared radiation.

12-101C   Describe the solar radiation properties of a window that is ideally suited for minimizing the air-conditioning load.

12-102C   Define the SHGC (solar heat gain coefficient), and explain how it differs from the SC (shading coefficient). What are the values of the SHGC and SC of a single-pane clear-glass window?

12-103C   What does the SC (shading coefficient) of a device represent? How do the SCs of clear glass and heat-absorbing glass compare?

12-104C   What is a shading device? Is an internal or external shading device more effective in reducing the solar heat gain through a window? How does the color of the surface of a shading device facing outside affect the solar heat gain?

12-105C   What is the effect of a low-e coating on the inner surface of a window glass on the (*a*) heat loss in winter and (*b*) heat gain in summer through the window?

12-106C   What is the effect of a reflective coating on the outer surface of a window glass on the (*a*) heat loss in winter and (*b*) heat gain in summer through the window?

12-107   A manufacturing facility located at 32° N latitude has a glazing area of 60 m^2 facing west that consists of double-pane windows made of clear glass (SHGC = 0.766). To reduce the solar heat gain in summer, a reflective film that will reduce the SHGC to 0.35 is considered. The cooling season consists of June, July, August, and September, and the heating season, October through April. The average daily solar heat fluxes incident on the west side at this latitude are 2.35, 3.03, 3.62, 4.00, 4.20, 4.24, 4.16, 3.93, 3.48, 2.94, 2.33, and 2.07 kWh/day · m^2 for January through December, respectively. Also, the unit costs of electricity and natural gas are $0.09/kWh and $0.45/therm, respectively. If the coefficient of performance of the cooling system is 3.2 and the efficiency of the furnace is 0.90, determine the net annual cost savings due to installing reflective coating on the windows. Also, determine the simple payback period if the installation cost of reflective film is $20/m^2.   *Answers:* $15,   27 years

12-108   A house located in Boulder, Colorado (40° N latitude), has ordinary double-pane windows with 6-mm-thick glasses and the total window areas are 8, 6, 6, and 4 m^2 on the south, west, east, and north walls. Determine the total solar heat gain of the house at 9:00, 12:00, and 15:00 solar time in July. Also, determine the total amount of solar heat gain per day for an average day in January.

12-109   Repeat Problem 12-108 for double-pane windows that are gray-tinted.

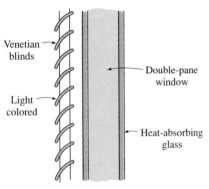

FIGURE P12-110

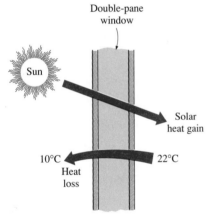

FIGURE P12-111

12-110 Consider a building in New York (40° N latitude) that has 200 m² of window area on its south wall. The windows are double-pane heat-absorbing type, and are equipped with light-colored venetian blinds with a shading coefficient of SC = 0.30. Determine the total solar heat gain of the building through the south windows at solar noon in April. What would your answer be if there were no blinds at the windows?

12-111 A typical winter day in Reno, Nevada (39° N latitude), is cold but sunny, and thus the solar heat gain through the windows can be more than the heat loss through them during daytime. Consider a house with double-door-type windows that are double paned with 3-mm-thick glasses and 6.4 mm of air space and have aluminum frames and spacers. The house is maintained at 22°C at all times. Determine if the house is losing more or less heat than it is gaining from the sun through an east window on a typical day in January for a 24-h period if the average outdoor temperature is 10°C.
Answer: less

12-112 Repeat Problem 12-111 for a south window.

12-113E Determine the rate of net heat gain (or loss) through a 9-ft-high, 15-ft-wide, fixed $\frac{1}{8}$-in. single-glass window with aluminum frames on the west wall at 3 PM solar time during a typical day in January at a location near 40° N latitude when the indoor and outdoor temperatures are 70°F and 45°F, respectively. *Answer:* 16,840 Btu/h gain

12-114 Consider a building located near 40° N latitude that has equal window areas on all four sides. The building owner is considering coating the south-facing windows with reflective film to reduce the solar heat gain and thus the cooling load. But someone suggests that the owner will reduce the cooling load even more if she coats the west-facing windows instead. What do you think?

Infiltration Heat Load and Weatherizing

12-115C What is infiltration? How does it differ from ventilation? How does infiltration affect the heating load in winter and the cooling load in summer? Explain.

12-116C Describe briefly two ways of measuring the infiltration rate of a building. Also, explain how the design infiltration rate differs from the seasonal average infiltration rate, and which is used when sizing a heating system.

12-117C How is the infiltration unit ACH (air changes per hour) defined? Why should too low and too high values of ACH be avoided?

12-118C How can the energy of the air vented out from the kitchens and bathrooms be saved?

12-119C Is latent heat load of infiltration necessarily zero when the relative humidity of the hot outside air in summer is the same as that of inside air? Explain.

12-120C Is latent heat load of infiltration necessarily zero when the humidity ratio (absolute humidity) of the hot outside air in summer is the same as that of inside air? Explain.

12-121C What are some practical ways of preventing infiltration in homes?

12-122C It is claimed that the infiltration rate and infiltration losses can be reduced by using radiant panel heaters since the air temperature can be lowered without sacrificing comfort. It is also claimed that radiant panels will increase the heat losses through the wall and the roof by conduction as a result of increased surface temperature. What do you think of these claims?

12-123E The average atmospheric pressure in Spokane, Washington (elevation = 2350 ft), is 13.5 psia, and the average winter temperature is 36.5°F. The pressurization test of a 9-ft-high, 3000-ft² older home revealed that the seasonal average infiltration rate of the house is 2.2 ACH. It is suggested that the infiltration rate of the house can be reduced by half to 1.1 ACH by winterizing the doors and the windows. If the house is heated by natural gas whose unit cost is $0.62/therm and the heating season can be taken to be six months, determine how much the home owner will save from the heating costs per year by this winterization project. Assume the house is maintained at 72°F at all times, and the efficiency of the furnace is 0.65. Also, assume the latent heat load during the heating season to be negligible.
Answer: $765

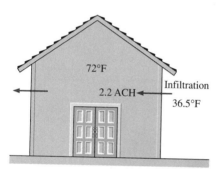

FIGURE P12-123E

12-124 Consider two identical buildings, one in Los Angeles, California, where the atmospheric pressure is 101 kPa and the other in Denver, Colorado, where the atmospheric pressure is 83 kPa. Both buildings are maintained at 21°C, and the infiltration rate for both buildings is 1.2 ACH on a day when the outdoor temperature at both locations is 10°C. Disregarding latent heat, determine the ratio of the heat losses by infiltration at the two cities.

12-125 Determine the rate of sensible heat loss from a building due to infiltration if outdoor air at −10°C and 90 kPa enters the building at a rate of 35 L/s when the indoors is maintained at 22°C. *Answer:* 1335 W

12-126 The ventilating fan of the bathroom of a building has a volume flow rate of 30 L/s and runs continuously. The building is located in San Francisco, California, where the average winter temperature is 12.2°C, and is maintained at 22°C at all times. The building is heated by electricity whose unit cost is $0.09/kWh. Determine the amount and cost of the heat "vented out" per month in winter. *Answers:* 225 kWh, $22.94

12-127 For an infiltration rate of 1.2 ACH, determine the sensible, latent, and total infiltration heat load, in kW, of a building at sea level, that is 20 m long, 13 m wide, and 3 m high when the outdoor air is at 32°C and 50 percent

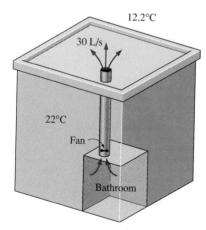

FIGURE P12-126

relative humidity. The building is maintained at 24°C and 50 percent relative humidity at all times.

Annual Energy Consumption

12-128C   Is it possible for a building in a city to have a higher peak heating load but a lower energy consumption for heating in winter than an identical building in another city? Explain.

12-129C   Can we determine the annual energy consumption of a building for heating by simply multiplying the design heating load of the building by the number of hours in the heating season? Explain.

12-130C   Considerable energy can be saved by lowering the thermostat setting in winter and raising it in summer by a few degrees. As the manager of a large commercial building, would you implement this measure to save energy and cut costs?

12-131C   What is the number of heating degree-days for a winter day during which the average outdoor temperature was 10°C, and it never went above 18°C?

12-132C   What is the number of heating degree-days for a winter month during which the average outdoor temperature was 12°C, and it never went above 18°C?

12-133C   Someone claims that the °C-days for a location can be converted to °F-days by simply multiplying °C-days by 1.8. But another person insists that 32 must be added to the result because of the formula $T(°F) = 1.8T(°C) + 32$. Which person do you think is right?

12-134C   What is balance-point temperature? Why is the balance-point temperature used in the determination of degree-days instead of the actual thermostat setting of a building?

12-135C   Under what conditions is it proper to use the degree-day method to determine the annual energy consumption of a building? What is the number of heating degree-days for a day during which the average outdoor temperature was 10°C, and it never went above 18°C?

12-136   Suppose you have moved to Syracuse, New York, in August, and your roommate, who is short of cash, offered to pay the heating bills during the upcoming year (starting January 1) if you pay the heating bills for the current calendar year until December 31. Is this a good offer for you?

12-137E   Using indoor and outdoor winter design temperatures of 70°F and −10°F, respectively, the design heat load of a 2500-ft² house in Billings, Montana, is determined to be 83,000 Btu/h. The house is to be heated by natural gas that will be burned in a 95 percent efficient, high-efficiency furnace. If the unit cost of natural gas is $0.65/therm, estimate the annual gas consumption of this house and its cost.   *Answers:* 1848 therms, $1201

12-138 Consider a building whose annual air-conditioning load is estimated to be 120,000 kWh in an area where the unit cost of electricity is $0.10/kWh. Two air conditioners are considered for the building. Air conditioner A has a seasonal average COP of 3.2 and costs $5500 to purchase and install. Air conditioner B has a seasonal average COP of 5.0 and costs $7000 to purchase and install. If all else are equal, determine which air conditioner is a better buy.

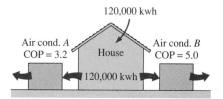

FIGURE P12-138

12-139 The lighting requirements of an industrial facility are being met by 700 40-W standard fluorescent lamps. The lamps are close to completing their service life, and are to be replaced by their 34-W high-efficiency counterparts that operate on the existing standard ballasts. The standard and high-efficiency fluorescent lamps can be purchased at quantity at a cost of $1.77 and $2.26 each, respectively. The facility operates 2800 hours a year and all of the lamps are kept on during operating hours. Taking the unit cost of electricity to be $0.08/kWh and the ballast factor to be 1.1, determine how much energy and money will be saved a year as a result of switching to the high-efficiency fluorescent lamps. Also, determine the simple payback period.   *Answers:* 12,936 kWh, $1035, 4 months

12-140 The lighting needs of a storage room are being met by six fluorescent light fixtures, each fixture containing four lamps rated at 60 W each. All the lamps are on during operating hours of the facility, which are 6 AM to 6 PM 365 days a year. The storage room is actually used for an average of three hours a day. If the price of the electricity is $0.08/kWh, determine the amount of energy and money that will be saved as a result of installing a motion sensor. Also, determine the simple payback period if the purchase price of the sensor is $32 and it takes 1 hour to install it at a cost of $40.

12-141 An office building in the Reno area (3346°C-days) is maintained at a temperature of 20°C at all times during the heating season, which is taken to be November 1 through April 30. The work hours of the office are 8 AM to 6 PM Monday through Friday, and there is no air-conditioning. An examination of the heating bills of the facility for the previous year reveals that the facility used 3530 therms of natural gas at an average price of $0.58/therm, and thus paid $2047 for heating. There are five thermostats in the facility to control the temperatures of various sections. The existing manual thermostats are to be replaced by programmable ones to reduce the heating costs. The new thermostats are to lower the temperature setting shortly before closing hour and raise it shortly before the opening hour. The thermostats will be set to 7.2°C during the off-work hours and to 20°C during the work hours. Determine the annual energy and cost savings as a result of installing five programmable thermostats. Also, determine the simple payback period if the installed cost of each programmable thermostat is $190.

12-142 A 75-hp motor with an efficiency of 91.0 percent is worn out and is to be replaced with a high-efficiency motor having an efficiency of 95.4 percent. The machine operates 4368 hours a year at a load factor of 0.75. Taking the cost of electricity to be $0.08/kWh, determine the amount of energy and money savings as a result of installing the high-efficiency motor

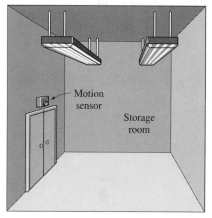

FIGURE P12-140

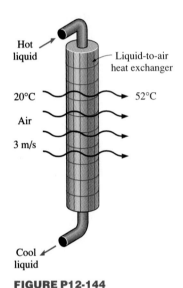

Hot
liquid

Liquid-to-air
heat exchanger

20°C   52°C

Air

3 m/s

Cool
liquid

FIGURE P12-144

instead of the standard one. Also, determine the simple payback period if the purchase prices of the standard and high-efficiency motors are $5449 and $5520, respectively.   *Answers:* 9289 kWh, $743, 1.1 months

12-143E   The steam requirements of a manufacturing facility are being met by a boiler of 3.8 million Btu/h input. The combustion efficiency of the boiler is measured to be 0.7 by a handheld flue gas analyzer. After tuning up the boiler, the combustion efficiency rises to 0.8. The boiler operates 1500 hours a year intermittently. Taking the unit cost of energy to be $4.35 per million Btu, determine the annual energy and cost savings as a result of tuning up the boiler.

12-144   The space heating of a facility is accomplished by natural gas heaters that are 80 percent efficient. The compressed air needs of the facility are met by a large liquid-cooled compressor. The coolant of the compressor is cooled by air in a liquid-to-air heat exchanger whose air flow section is 1.0 m high and 1.0 m wide. During typical operation, the air is heated from 20°C to 52°C as it flows through the heat exchanger. The average velocity of air on the inlet side is measured to be 3 m/s. The compressor operates 20 hours a day and 5 days a week throughout the year. Taking the heating season to be 6 months (26 weeks) and the cost of the natural gas to be $0.50/therm (1 therm = 105,500 kJ), determine how much money will be saved by diverting the compressor waste heat into the facility during the heating season.

12-145   Consider a family in Atlanta, Georgia, whose average heating bill has been $600 a year. The family now moves to an identical house in Denver, Colorado, where the fuel and electricity prices are also the same. How much do you expect the annual heating bill of this family to be in their new home?

12-146E   The design heating load of a new house in Cleveland, Ohio, is calculated to be 65,000 Btu/h for an indoor temperature of 72°F. If the house is to be heated by a natural gas furnace that is 90 percent efficient, predict the annual natural gas consumption of this house, in therms. Assume the entire house is maintained at indoor design conditions at all times.

12-147   For inside and outside design temperatures of 22°C and −12°C, respectively, a home located in Boise, Idaho, has a design heating load of 38 kW. The house is heated by electric resistance heaters, and the cost of electricity is $0.06/kWh. Determine how much money the home owner will save if she lowers the thermostat from 22°C to 14°C from 11 PM to 7 AM every night in December.

Review Problems

12-148   Consider a home owner who is replacing his 20-year-old natural gas furnace that has an efficiency of 60 percent. The home owner is considering a conventional furnace that has an efficiency of 82 percent and costs $1600, and a high-efficiency furnace that has an efficiency of 95 percent and costs $2700. The home owner would like to buy the high-efficiency furnace if the savings from the natural gas pay for the additional cost in less than eight years. If the home owner is presently paying $1100 a year for heating, determine if he should buy the conventional or high-efficiency model.

12-149 The convection heat transfer coefficient for a clothed person seated in air moving at a velocity of 0.2 to 4 m/s is given by $h = 8.3\mathcal{V}^{0.6}$ where $\mathcal{V}$ is in m/s and h is in W/m² · °C. The convection coefficient at lower velocities is 3.1 W/m² · °C. Plot the convection coefficient against the air velocity, and compare the average convection coefficient to the average radiation coefficient of about 5 W/m² · °C.

12-150 Workers in a casting facility are surrounded with hot surfaces, with an average temperature of 40°C. The activity level of the workers corresponds to a heat generation rate of 300 W, half of which is latent. If the air temperature is 22°C, determine the velocity of air needed to provide comfort for the workers. Take the average exposed surface area and temperature of the workers to be 1.8 m² and 30°C, respectively.

12-151 Replacing incandescent lights by energy-efficient fluorescent lights can reduce the lighting energy consumption to one-fourth of what it was before. The energy consumed by the lamps is eventually converted to heat, and thus switching to energy-efficient lighting also reduces the cooling load in summer but increases the heating load in winter. Consider a building that is heated by a natural gas furnace having an efficiency of 80 percent and cooled by an air conditioner with a COP of 3.5. If electricity costs $0.08/kWh and natural gas costs $0.70/therm, and the annual heating load of the building is roughly equal to the annual cooling load, determine if efficient lighting will increase or decrease the total heating and cooling cost of the building.   *Answer:* Decrease

12-152 A wall of a farmhouse is made of 200-mm common brick. Determine the rate of heat transfer through a 20-m² section of the wall when the indoor and outdoor temperatures are 20°C and −5°C, respectively, and the wall is exposed to 24 km/h winds.   *Answer:* 1280 W

12-153E Determine the R-value and the overall heat transfer coefficient (the U-factor) of a 45° pitched roof that is built around nominal 2-in. × 4-in. wood studs with a center-to-center distance of 24 in. The 3.5-in.-wide air space between the studs has a reflective surface and its effective emissivity is 0.05. The lower part of the roof is finished with $\frac{1}{2}$-in. gypsum wallboard and the upper part with $\frac{5}{8}$-in. plywood, building paper, and asphalt shingle roofing. The air space constitutes 80 percent of the heat transmission area, while the studs and headers constitute 20 percent.

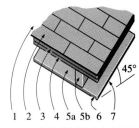

1 2 3 4 5a 5b 6 7

FIGURE P12-153E

12-154 The winter design heat load of a planned house is determined to be 32 kW. The original design involved aluminum-frame single-pane windows with a U-factor of 7.16 W/m² · °C, and heat losses through the windows accounted for 26 percent of the total. Alarmed by the projected high cost of annual energy usage, the owner decided to switch to vinyl-frame double-pane windows whose U-factor is 2.74 W/m² · °C. Determine the reduction in the heat load of the house as a result of switching to double-pane windows.

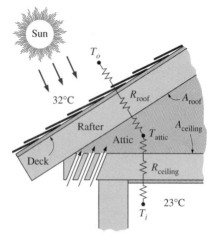

FIGURE P12-155

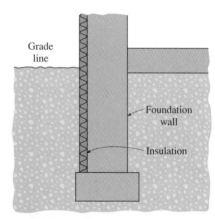

FIGURE P12-157

12-155 The attic of a 150-m^2 house in Thessaloniki, Greece (40° N latitude), is not vented in summer. The R-values of the roof and the ceiling are 1.4 m$^2 \cdot$ °C/W and 0.50 m$^2 \cdot$ °C/W, respectively, and the ratio of the roof area to ceiling area is 1.30. The indoors are maintained at 23°C at all times. Determine the rate of heat gain through the roof in late July at 16:00 (4 PM) solar time if the outdoor temperature is 32°C, assuming the roof is (*a*) light colored and (*b*) dark colored.   *Answers:* (*a*) 1700 W, (*b*) 2900 W

12-156 The dimensions of a house in Norfolk, Virginia, are 19 m × 10 m, and the house has a basement. The 1.8-m-high portion of the basement wall is below the ground level, and the top 0.6-m section of the wall below the grade is insulated with R-0.73 m$^2 \cdot$ °C/W insulation. If the basement is maintained at 17°C, determine the rate of design heat loss from the basement through its walls and floor.

12-157 A house in Anchorage, Alaska, has a concrete slab floor that sits directly on the ground at grade level. The house is 20 m long and 15 m wide, and the weather in Alaska is very severe. The walls of the house are made of 20-cm block wall with brick and are insulated from the edge to the footer with R-0.95 m$^2 \cdot$ °C/W insulation. If the house is maintained at 22°C, determine the heat loss from the floor at winter design conditions.

12-158 Repeat Problem 12-157 assuming the below-grade section of the wall is not insulated.   *Answer:* 3931 W

12-159 A university campus has 200 classrooms and 400 faculty offices. The classrooms are equipped with 12 fluorescent light bulbs, each consuming 110 W, including the electricity used by the ballasts. The faculty offices, on average, have half as many light bulbs. The campus is open 240 days a year. The classrooms and faculty offices are not occupied an average of 4 hours a day, but the lights are kept on. If the unit cost of electricity is $0.075/kWh, determine how much energy and money the campus will save a year if the lights in the classrooms and faculty offices are turned off during unoccupied periods.

12-160E For an infiltration rate of 0.8 ACH, determine the sensible, latent, and total infiltration heat load, in Btu/h, of a building at sea level that is 60 ft long, 50 ft wide, and 9 ft high when the outdoor air is at 82°F and 40 percent relative humidity. The building is maintained at 74°F and 40 percent relative humidity at all times.

12-161 It is believed that January is the coldest month in the Northern hemisphere. On the basis of Table 12-5, determine if this is true for all locations.

12-162 The December space heating bill of a fully occupied house in Louisville, Kentucky, that was kept at 22°C at all times was $110. How much do you think the December space heating bill of this house would be if it still were kept at 22°C but there were no people living in the house, no lights were on, no appliances were operating, and there was no solar heat gain?

12-163 For an indoor temperature of 22°C, the design heating load of a residential building in Charlotte, North Carolina, is determined to be 28 kW.

The house is to be heated by natural gas, which costs $0.70/therm. The efficiency of the gas furnace is 80 percent. Determine (*a*) the heat loss coefficient of the building, in kW/°C, and (*b*) the annual energy usage for heating and its cost.   *Answers:* (*a*) 1 kW/°C, (*b*) 1815 therms, $1271

Computer, Design, and Essay Problems

12-164   On average, superinsulated homes use just 15 percent of the fuel required to heat the same size conventional home built before the energy crises in the 1970s. Write an essay on superinsulated homes, and identify the features that make them so energy efficient as well as the problems associated with them. Do you think superinsulated homes will be economically attractive in your area?

12-165   Conduct the following experiment to determine the heat loss coefficient of your house or apartment in W/°C or Btu/h · °F. First make sure that the conditions in the house are steady and the house is at the set temperature of the thermostat. Use an outdoor thermometer to monitor outdoor temperature. One evening, using a watch or timer, determine how long the heater was on during a 3-h period and the average outdoor temperature during that period. Then using the heat output rating of your heater, determine the amount of heat supplied. Also, estimate the amount of heat generation in the house during that period by noting the number of people, the total wattage of lights that were on, and the heat generated by the appliances and equipment. Using that information, calculate the average rate of heat loss from the house and the heat loss coefficient.

12-166   Numerous professional software packages are available in the market for performing heat transfer analysis of buildings, and they are advertised in professional magazines such as the *ASHRAE Journal* magazine published by the American Society of Heating, Refrigerating, and Air-Conditioning Engineers (ASHRAE). Your company decides to purchase such a software package and asks you to prepare a report on the available packages, their costs, capabilities, ease of use, compatibility with the available hardware and other software, as well as the reputation of the software company, their history, financial health, customer support, training, and future prospects, among other things. After a preliminary investigation, select the top three packages and prepare a full report on them. Also, find out if there are any free software packages available that can be used for the same purpose.

12-167   A 1982 U.S. Department of Energy article (FS 204) states that a leak of one drip of hot water per second can cost $1.00 per month. Making reasonable assumptions about the drop size and the unit cost of energy, determine if this claim is reasonable.

12-168   Obtain the following weather data for the city in which you live: the elevation, the atmospheric pressure, the number of heating and cooling degree-days, the 97.5 percent level winter and summer design temperatures, the average annual and winter temperatures, the record high and low temperatures and when they occurred, and the unit cost of natural gas and electricity for residential buildings.

12-169 Your neighbor lives in a 2500-ft^2 (about 250-m^2) older house heated by natural gas. The current gas heater was installed in the early 1970s and has an efficiency (called the annual fuel utilization efficiency, or AFUE, rating of 65 percent. It is time to replace the furnace, and the neighbor is trying to decide between a conventional furnace that has an efficiency of 80 percent and costs $1500 and a high-efficiency furnace that has an efficiency of 95 percent and costs $2500. Your neighbor offered to pay you $100 if you help him make the right decision. Considering the weather data, typical heating loads, and the price of natural gas in your area, make a recommendation to your neighbor based on a convincing economic analysis.

12-170 The decision of whether to invest in an energy-saving measure is made on the basis of the length of time for it to pay for itself in projected energy (and thus cost) savings. The easiest way to reach a decision is to calculate the simple payback period by simply dividing the installed cost of the measure by the annual cost savings and comparing it to the lifetime of the installation. This approach is adequate for short payback periods (less than five years) in stable economies with low interest rates (under 10 percent) since the error involved is no larger than the uncertainties. However, if the payback period is long, it may be necessary to consider the *interest rate* if the money is to be borrowed, or the *rate of return* if the money is invested elsewhere instead of the energy conservation measure. For example, a simple payback period of five years corresponds to 5.0, 6.12, 6.64, 7.27, 8.09, 9.919, 10.84, and 13.91 for an interest rate (or return on investment) of 0, 6, 8, 10, 12, 14, 16, and 18 percent, respectively. Finding out the proper relations from engineering economics books, determine the payback periods for the interest rates given above corresponding to simple payback periods of 1 through 10 years.

12-171 A radiant heater mounted on the ceiling is to be designed to assist the heating of a 250-m^2 residence in Las Vegas, Nevada, whose walls are 4 m high. The indoor air is to be maintained at 15°C at all times, and the radiant heater is to provide thermal comfort for the sedentary occupants in light clothing. The system considered consists of hot water pipes buried in the ceiling and occupies the entire ceiling. Recommend a temperature for the ceiling, and estimate the energy consumption of the radiant heater. Does the radiant heater appear to save enough energy to justify its cost?

Refrigeration and Freezing of Foods

Refrigeration and freezing of perishable food products is an important and fascinating application area of heat transfer and thermodynamics. Refrigeration *slows down* the chemical and biological processes in foods and the accompanying deterioration and the loss of quality. The *storage life* of fresh perishable foods such as meats, fish, fruits, and vegetables can be extended by several days by cooling, and by several weeks or months by freezing. There are many considerations in the design and selection of proper refrigeration and heat transfer mechanisms, and this chapter demonstrates the importance of having a broad base and a good understanding of the processes involved when designing heat transfer equipment. For example, fruits and vegetables continue to *respire* and *generate heat* during storage; most foods freeze over a *range of temperatures* instead of a single temperature; the quality of frozen foods is greatly affected by the *rate of freezing;* the *velocity* of refrigerated air affects the rate of moisture loss from the products in addition to the rate of heat transfer, and so forth.

We start this chapter with an overview of *microorganisms* that are responsible for the spoilage of foods since the primary function of refrigeration is to retard the growth rate of microorganisms. Then we present the general considerations in the *refrigeration* and *freezing* of foods and describe the different methods of freezing. In the following sections we describe the distinctive features and refrigeration needs of *fresh fruits and vegetables, meats,* and *other food products.* Next we consider the heat transfer mechanisms in *refrigerated storage rooms.* Finally, we discuss the *transportation* of refrigerated foods since most refrigerated foods spend part of their life in transit in refrigerated trucks, railroad cars, ships, and even airplanes.

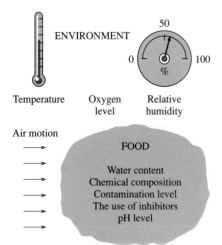

FIGURE 13-1

Typical growth curve of
microorganisms.

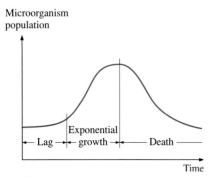

FIGURE 13-2

The factors that affect the rate of
growth of microorganisms.

13-1 ■ CONTROL OF MICROORGANISMS IN FOODS

Microorganisms such as *bacteria, yeasts, molds,* and *viruses* are widely encountered in air, water, soil, living organisms, and unprocessed food items, and cause *off-flavors* and *odors, slime production, changes* in the texture and appearances, and the eventual *spoilage* of foods. Holding perishable foods at warm temperatures is the primary cause of spoilage, and the prevention of food spoilage and the premature degradation of quality due to microorganisms is the largest application area of refrigeration. The first step in controlling microorganisms is to understand what they are and the factors that affect their transmission, growth, and destruction.

Of the various kinds of microorganisms, *bacteria* are the prime cause for the spoilage of foods, especially moist foods. Dry and acidic foods create an undesirable environment for the growth of bacteria, but not for the growth of yeasts and molds. *Molds* are also encountered on moist surfaces, cheese, and spoiled foods. Specific *viruses* are encountered in certain animals and humans, and poor sanitation practices such as keeping processed foods in the same area as the uncooked ones and being careless about handwashing can cause the contamination of food products.

When *contamination* occurs, the microorganisms start to adapt to the new environmental conditions. This initial slow or no-growth period is called the **lag phase,** and the shelf life of a food item is directly proportional to the length of this phase (Fig. 13-1). The adaptation period is followed by an *exponential growth* period during which the population of microorganisms can double two or more times every hour under favorable conditions unless drastic sanitation measures are taken. The depletion of nutrients and the accumulation of toxins slow down the growth and start the *death* period.

The *rate of growth* of microorganisms in a food item depends on the characteristics of the food itself such as the chemical structure, pH level, presence of inhibitors and competing microorganisms, and water activity as well as the environmental conditions such as the temperature and relative humidity of the environment and the air motion (Fig. 13-2).

Microorganisms need *food* to grow and multiply, and their nutritional needs are readily provided by the carbohydrates, proteins, minerals, and vitamins in a food. Different types of microorganisms have different nutritional needs, and the types of nutrients in a food determine the types of microorganisms that may dwell on them. The preservatives added to the food may also inhibit the growth of certain microorganisms. Different kinds of microorganisms that exist compete for the same food supply, and thus the composition of microorganisms in a food at any time depends on the *initial make-up* of the microorganisms.

All living organisms need *water* to grow, and microorganisms cannot grow in foods that are not sufficiently moist. Microbiological growth in refrigerated foods such as fresh fruits, vegetables, and meats starts at the *exposed surfaces* where contamination is most likely to occur. Fresh meat in a package left in a room will spoil quickly, as you may have noticed. A meat carcass hung in a controlled environment, on the other hand, will age healthily as a result of *dehydration* on the outer surface, which inhibits micro-biological growth there and protects the carcass.

Microorganism growth in a food item is governed by the combined effects of the *characteristics of the food* and the *environmental factors*. We cannot do much about the characteristics of the food, but we certainly can alter the environmental conditions to more desirable levels through *heating, cooling, ventilating, humidification, dehumidification,* and control of the *oxygen* levels. The growth rate of microorganisms in foods is a strong function of temperature, and temperature control is the single most effective mechanism for controlling the growth rate.

Microorganisms grow best at "warm" temperatures, usually between 20 and 60°C. The growth rate *declines* at high temperatures, and *death* occurs at still higher temperatures, usually above 70°C for most microorganisms. *Cooling* is an effective and practical way of reducing the growth rate of microorganisms and thus extending the *shelf life* of perishable foods. A temperature of 4°C or lower is considered to be a safe refrigeration temperature. Sometimes a small increase in refrigeration temperature may cause a large increase in the growth rate, and thus a considerable decrease in shelf life of the food (Fig. 13-3). The growth rate of some microorganisms, for example, doubles for each 3°C rise in temperature.

Another factor that affects microbiological growth and transmission is the *relative humidity* of the environment, which is a measure of the water content of the air. High humidity in *cold rooms* should be avoided since condensation that forms on the walls and ceiling creates the proper environment for *mold growth* and buildups. The drip of contaminated condensate onto food products in the room poses a potential health hazard.

Different microorganisms react differently to the presence of oxygen in the environment. Some microorganisms such as molds require oxygen for growth, while some others cannot grow in the presence of oxygen. Some grow best in low-oxygen environments, while others grow in environments regardless of the amount of oxygen. Therefore, the growth of certain microorganisms can be controlled by controlling the *amount of oxygen* in the environment. For example, vacuum packaging inhibits the growth of microorganisms that require oxygen. Also, the storage life of some fruits can be extended by reducing the oxygen level in the storage room.

Microorganisms in food products can be controlled by (1) *preventing* contamination by following strict sanitation practices, (2) *inhibiting* growth by altering the environmental conditions, and (3) *destroying* the organisms by heat treatment or chemicals. The best way to minimize contamination in food processing areas is to use fine air filters in ventilation systems to capture the *dust particles* that transport the bacteria in the air. Of course, the filters must remain dry since microorganisms can grow in wet filters. Also, the ventilation system must maintain a positive pressure in the food processing areas to prevent any airborne contaminants from entering inside by infiltration. The elimination of *condensation* on the walls and the ceiling of the facility and the diversion of *plumbing* condensation drip pans of refrigerators to the drain system are two other preventive measures against contamination. Drip systems must be cleaned regularly to prevent microbiological growth in

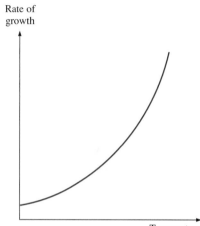

Rate of growth

Temperature

FIGURE 13-3

The rate of growth of microorganisms in a food product increases exponentially with increasing environmental temperature.

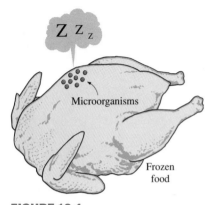

FIGURE 13-4

Freezing may stop the growth of microorganisms, but it may not necessarily kill them.

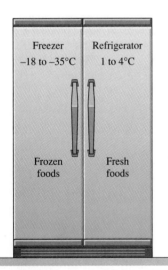

FIGURE 13-5

Recommended refrigeration and freezing temperatures for most perishable foods.

them. Also, any *contact* between raw and cooked food products should be minimized, and cooked products must be stored in rooms with positive pressures. Frozen foods must be kept at −18°C or below, and utmost care should be exercised when food products are packaged after they are frozen to avoid contamination during packaging.

The growth of microorganisms is best controlled by keeping the *temperature* and *relative humidity* of the environment in the desirable range. Keeping the relative humidity below 60 percent, for example, prevents the growth of all microorganisms on the surfaces. Microorganisms can be destroyed by *heating* the food product to high temperatures (usually above 70°C), by treating them with *chemicals,* or by exposing them to *ultraviolet light* or solar radiation.

Distinction should be made between *survival* and *growth* of microorganisms. A particular microorganism that may not grow at some low temperature may be able to survive at that temperature for a very long time (Fig. 13-4). Therefore, freezing is not an effective way of killing microorganisms. In fact, some microorganism cultures are preserved by freezing them at very low temperatures. The *rate of freezing* is also an important consideration in the refrigeration of foods since some microorganisms adapt to low temperatures and grow at those temperatures when the cooling rate is very low.

13-2 ■ REFRIGERATION AND FREEZING OF FOODS

The *storage life* of fresh perishable foods such as meats, fish, vegetables, and fruits can be extended by several days by storing them at temperatures just above freezing, usually between 1 and 4°C. The storage life of foods can be extended by several months by freezing and storing them at subfreezing temperatures, usually between −18 and −35°C, depending on the particular food (Fig. 13-5).

Refrigeration *slows down* the chemical and biological processes in foods, and the accompanying deterioration and loss of quality and nutrients. Sweet corn, for example, may lose half of its initial sugar content in one day at 21°C, but only 5 percent of it at 0°C. Fresh asparagus may lose 50 percent of its vitamin C content in one day at 20°C, but in 12 days at 0°C. Refrigeration also extends the shelf life of products. The first appearance of unsightly yellowing of broccoli, for example, may be delayed by three or more days by refrigeration.

Early attempts to freeze food items resulted in poor-quality products because of the large ice crystals that formed. It was determined that the *rate of freezing* has a major effect on the size of ice crystals and the quality, texture, and nutritional and sensory properties of many foods. During *slow freezing,* ice crystals can grow to a large size, whereas during *fast freezing* a large number of ice crystals start forming at once and are much smaller in size. Large ice crystals are not desirable since they can *puncture* the walls of the cells, causing a degradation of texture and a loss of natural juices during thawing. A *crust* forms rapidly on the outer layer of the product and seals in the juices, aromatics, and flavoring agents. The product quality is also affected adversely by temperature fluctuations of the storage room.

The ordinary refrigeration of foods involves *cooling* only without any phase change. The *freezing* of foods, on the other hand, involves three stages: *cooling* to the freezing point (removing the sensible heat), *freezing* (removing the latent heat), and *further cooling* to the desired subfreezing temperature (removing the sensible heat of frozen food), as shown in Figure 13-6.

Fresh fruits and vegetables are *live products,* and thus they continue giving off heat that adds to the refrigeration load of the cold storage room. The storage life of fruits and vegetables can be extended greatly by removing the field heat and cooling as soon after harvesting as possible. The optimum storage temperature of most fruits and vegetables is about 0.5 to 1°C above their freezing point. But this is not the case for some fruits and vegetables such as bananas and cucumbers that experience undesirable *physiological changes,* when exposed to low (but still above-freezing) temperatures, usually between 0 and 10°C. The resulting tissue damage is called the **chilling injury** and is characterized by internal discoloration, soft scald, skin blemishes, soggy breakdown, and failure to ripen. The severeness of the chilling injury depends on both the temperature and the length of storage at that temperature. The lower the temperature, the greater the damage in a given time. Therefore, products susceptible to chilling injury must be stored at higher temperatures. A list of vegetables susceptible to chilling injury and the lowest safe storage temperature are given in Table 13-1.

Chilling injury differs from **freezing injury,** which is caused by prolonged exposure of the fruits and vegetables to subfreezing temperatures and thus the actual *freezing* at the affected areas. The freezing injury is characterized by rubbery texture, browning, bruising, and drying due to rapid moisture loss. The freezing points of fruits and vegetables do not differ by much, but their susceptibility to freezing injury differs greatly. Some vegetables are frozen and thawed several times with no significant damage, but others such as tomatoes suffer severe tissue injury and are ruined after one freezing. Products near the refrigerator coils or at the bottom layers of refrigerator cars and trucks are most susceptible to freezing injury. To avoid freezing injury, the rail cars or trucks should be *heated* during transportation in subfreezing weather, and adequate air circulation must be provided in cold storage rooms. Damage also occurs during *thawing* if it is done too fast. It is recommended that thawing be done at 4°C.

Dehydration, or *moisture loss,* causes a product to shrivel or wrinkle and lose quality. Therefore, proper measures must be taken during cold storage of food items to minimize moisture loss, which also represents a direct loss of the salable amount. A fruit or vegetable that loses 5 percent moisture, for example, will weigh 5 percent less and will probably be sold at a lower unit price because of loss of quality.

The loss of moisture from fresh fruits and vegetables is also called **transpiration.** The amount of moisture lost from a fruit or vegetable per unit mass of the fruit or vegetable per unit time is called the *transpiration rate.* The transpiration rate varies with the environmental conditions such as the temperature, relative humidity, and air motion. Also, the transpiration rate is

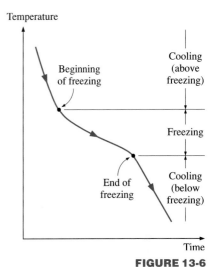

FIGURE 13-6

Typical freezing curve of a food item.

TABLE 13-1

Some vegetables susceptible to chilling injury and the lowest safe storage temperature (from ASHRAE *Handbook: Refrigeration*, Ref. 3, Chap. 17, Table 3)

Vegetable	Lowest safe temperature, °C
Cucumbers	10
Eggplants	7
Casaba melons	7 to 10
Watermelons	4
Okra	7
Sweet peppers	7
Potatoes	3 to 4
Pumpkins	10
Hard-shell squash	10
Sweet potatoes	13
Ripe tomatoes	7 to 10
Mature green tomatoes	13

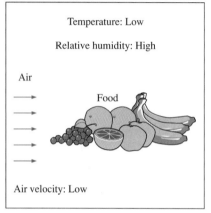

FIGURE 13-7

The proper environment for food storage to minimize moisture loss.

FIGURE 13-8

The cooling of food products by air involves heat transfer by convection, radiation, and evaporation.

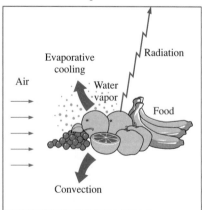

different for different fruits and vegetables. The tendency of a fruit or vegetable to transpire is characterized by the *transpiration coefficient,* which is the rate of transpiration per unit environmental vapor pressure deficit. The transpiration coefficient of apples, for example, is 58 ng/s · Pa · kg, whereas it is 1648 ng/s · Pa · kg for carrots and 8750 ng/s · Pa · kg for lettuce. This explains why the lettuce dehydrates quickly while the apples in the same environment maintain their fresh appearance for days.

Moisture loss can be minimized by (1) keeping the *storage temperature* of food as low as possible, (2) keeping the *relative humidity* of the storage room as high as possible, and (3) avoiding *high air velocities* (Fig. 13-7). However, air must be circulated continuously throughout the refrigerated storage room to keep it at a uniform temperature. To maintain high quality and product consistency, *temperature swings* of more than 1°C above or below the desired temperature in the storage room must be avoided. Air motion also minimizes the growth of molds on the surfaces of the wrapped or unwrapped food items and on the walls and ceiling of the storage room.

Waxing reduces moisture loss and thus slows down shriveling and maintains crispness in some products such as cucumbers, mature green tomatoes, peppers, and turnips. Waxing also gives the products an attractive glossy appearance. But a wax coating that is too thick may actually increase decay, especially when no fungicides are used.

Refrigeration is not necessary for all food items. For example, *canned foods* that are heat processed can be stored at room temperature for a few months without any noticeable change in flavor, color, texture, and nutritional value. Refrigeration should be considered for the storage of canned foods longer than two or three months to preserve quality and to avoid corrosion of the cans. *Dry foods* can last a long time, often more than a year, without refrigeration if they are protected against high temperatures and humidities. Dry foods that have been vacuum packed in water vapor–proof containers can maintain high quality and nutritional value for a long time. *Honey* can be stored at room temperature for about a year before any noticeable darkening or loss of flavor occurs. Cold storage below 10°C will extend the life of honey for several years. Storage of honey between 10 and 18° C is highly undesirable as it causes granulation.

The use of refrigeration is not limited to food items. It is commonly used in chemical and process industries to separate gases and solutions, to remove the heat of reaction, and to control pressure by maintaining low temperature. It is also used commonly in the beverage industry, in medicine, and even in the storage of furs and garments. Furs and wool products are commonly stored at 1 to 4° C to protect them against insect damage.

During cooling or freezing, heat is removed from the food usually by the combined mechanisms of *convection, radiation,* and *evaporation,* and the *rate of heat transfer* between the food and the surrounding medium at any time can be expressed as (Fig 13-8)

$$\dot{Q} = hA\Delta T = hA(T_{surface} - T_{ambient}) \qquad (W) \qquad (13\text{-}1)$$

where

h = average heat transfer coefficient for combined convection, radiation, and evaporation, W/m² · °C

A = exposed surface area of the food, m²

$T_{surface}$ = surface temperature of the food, °C

$T_{ambient}$ = temperature of the refrigerated fluid (air, brine, etc.) away from the food surface, °C

The *heat transfer coefficient h* is not a property of the food or refrigerated fluid. Its value depends on the *shape* of the food, the *surface roughness,* the *type* of cooling fluid, the *velocity* of the fluid, and the *flow regime.* The heat transfer coefficient is usually determined experimentally and is expressed in terms of the Reynolds and Prandtl numbers. Some experimentally determined values of the heat transfer coefficient are given in Table 13-2. The values of *h* include *convection* as well as other effects such as *radiation* and *evaporative cooling.*

Methods of Freezing

The method of freezing is an important consideration in the freezing of foods. Common freezing methods include *air-blast freezing,* where high-velocity air at about −30°C is blown over the food products; *contact freezing,* where packaged or unpackaged food is placed on or between cold metal plates and cooled by conduction; *immersion freezing,* where food is immersed in low-temperature brine; *cryogenic freezing,* where food is placed in a medium cooled by a cryogenic fluid such as liquid nitrogen or liquid or solid carbon dioxide; and the combination of the methods above.

In **air-blast freezers,** refrigerated air serves as the heat transfer medium and the heat transfer is primarily by convection. Perhaps the easiest method of freezing that comes to mind is to place the food items into a well-insulated *cold storage room* maintained at subfreezing temperatures. Heat transfer in this case is by natural convection, which is a rather slow process. The resulting low rates of freezing cause the growth of large ice crystals in the food and allow plenty of time for the flavors of different foods to mix. This hurts the quality of the product, and thus this simple method of freezing is usually avoided.

The next step is to use some large fans in the cold storage rooms to increase the convection heat transfer coefficient and thus the rate of freezing. These batch-type freezers, called *stationary blast cells* (Fig. 13-9), are still being used for many food products but are being replaced by conveyor-type air-blast freezers as appropriate since they allow the automation of the product flow and reduce labor costs. A simple version of mechanized freezers, called a *straight-belt freezer,* is very suitable for cooling fruits, vegetables, and uniform-sized products such as french fries. For smaller uniform-sized products such as peas and diced carrots, *fluidized-bed freezers* suspend the food items by a stream of cold air, usually at −40°C, as they travel on a conveyor belt. The high level of air motion and the large surface area result in high rates of heat transfer and rapid freezing of the food products. The floor space requirements for belt freezers can be minimized by using *multi-*

FIGURE 13-9

Schematic of a stationary blast cell freezer.

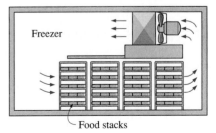

Freezer

Food stacks

TABLE 13-2

Surface heat transfer coefficients for food products cooled by air (adapted from ASHRAE *Handbook of Fundamentals*, Ref. 2, Chap. 30, Table 10)

Food	Heat transfer mechanisms	Size or characteristic length, L	Air velocity, m/s	Heat transfer coefficient, h, W/m$^2 \cdot$ °C[a]
Jonathan apples	Forced convection, radiation, and evaporation	52–62 mm	0.0	11.1–11.4
			0.40	15.9–17.0
			0.90	26.1–27.3
			2.0	39.2–45.3
			5.1	50.5–54.5
Red delicious apples	Forced convection, radiation, and evaporation	63–76 mm	1.5	14.2–27.3
			4.6	34.6–56.8
			Water[b], 0.27	55.7–90.9
Beef carcasses	Forced convection, radiation, and evaporation	64.5 kg 85 kg	1.8 0.3	21.8 10.0
Meat	Forced convection, radiation, and evaporation	23-mm-thick slabs	0.56	10.6
			1.4	20.0
			3.7	35.0
Perch fish	Forced convection, radiation, and evaporation	—	1.0–6.6	$h = 4.5 k_{air} \, Re^{0.28}/L$[c]
Potatoes	Forced convection, radiation, and evaporation	—	0.66	14.0
			4.0	19.1
			1.36	20.2
			1.73	24.4
Chicken Turkey	Forced convection and radiation	1.2–2.0 kg 5.4–9.5 kg	Agitated brine[d]	420–473
Eggs	Forced convection and radiation	34 mm 44 mm	2–8 2–8	$h = 0.46 k_{air} \, Re^{0.56}/L$ $h = 0.71 k_{air} \, Re^{0.55}/L$
Oranges, grapefruit, tangelos	Forced convection, radiation, and evaporation	53–80 mm	0.11–0.33	$h = 5.05 k_{air} \, Re^{0.333}/L$

[a]To convert to Btu/h · ft° · °F, multiply the given h values by 0.1761.

[b]The cooling medium is water instead of air.

[c]The L in this last column is the characteristic length given under the "size" column and $Re = \mathcal{V}L/\nu$ is the Reynolds number where $\mathcal{V}$ is the average air velocity and ν is the kinematic viscosity of air at the average temperature. The thermal conductivity and kinematic viscosity of air at 0°C are $k = 0.028$ W/m · °C and $\nu = 12.6 \times 10^{-6}$ m^2/s.

[d]The cooling medium is refrigerated brine instead of air.

pass straight *belt freezers* or *spiral belt freezers,* shown in Figure 13-10, which now dominate the frozen food industry. In another type of air-blast freezer, called the *impingement-style freezer,* cold air impinges upon the food product vertically from both sides of the conveyor belt at a high velocity, causing high heat transfer rates and very fast freezing.

In **contact freezers,** food products are sandwiched between two cold metal plates and are cooled by conduction. The plates are cooled by circulating cold refrigerant through the channels in the plates. Contact freezers are fast and efficient, but their use is limited to flat foods no thicker than about 8 cm with good thermal conductivity, such as meat patties, fish fillets, and chopped leafy vegetables. In **immersion freezing,** food products are immersed in brine or another fluid with a low-freezing point.

At atmospheric pressure, *liquid nitrogen* boils at $-195°C$ and absorbs 198 kJ/kg of heat during vaporization. Carbon dioxide is a solid at atmospheric pressure (called *dry ice*) and sublimes at $-79°C$ while absorbing 572 kJ/kg of heat. The saturated nitrogen and carbon dioxide vapors can further be used to precool the incoming food products before the vapors are purged into the atmosphere. The low boiling points and safety of these cryogenic substances make them very suitable for **cryogenic freezing** of food products. A common type of nitrogen freezer involves a long tunnel with a moving belt in it. Food products are frozen by nitrogen as they pass through the channel. Nitrogen provides extremely fast freezing because of the large temperature difference. Cryogenic cooling is used in limited applications because of its high cost. Sometimes cryogenic cooling is used in combination with air-blast freezing for improved quality and reduced cost. The food product is first crust-frozen in a bath of nitrogen to seal moisture and flavor in, and then transferred into the air-blast freezing section, where the freezing process is completed at a lower cost. This practice also reduces to negligible levels the dehydration losses, which can be as high as 4 percent for poorly designed and maintained systems.

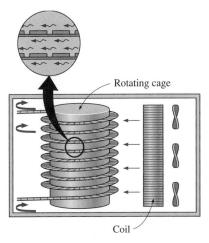

Rotating cage

Coil

FIGURE 13-10

Schematic of a horizontal airflow spiral freezer.

EXAMPLE 13-1   Cooling of Apples while Avoiding Freezing

Red delicious apples of 70 mm diameter and 85 percent water content initially at a uniform temperature of 30°C are to be cooled by refrigerated air at $-5°C$ flowing at a velocity of 1.5 m/s (Fig. 13-11). The average heat transfer coefficient between the apples and the air is given in Table 13-2 to be 21 W/m² · °C. Determine how long it will take for the center temperature of the apples to drop to 6°C. Also, determine if any part of the apples will freeze during this process.

Solution   The center temperature of apples drops to 6°C during cooling. The cooling time and if any part of the apples will freeze are to be determined.

Assumptions   **1** The apples are spherical in shape with a radius of $r_0 = 3.5$ cm. **2** The thermal properties and the heat transfer coefficient are constant.

Properties   The thermal conductivity and thermal diffusivity of apples are $k = 0.418$ W/m · °C and $\alpha = 0.13 \times 10^{-6}$ m²/s (Table A-7).

Analysis   Noting that the initial and the ambient temperatures are $T_i = 30°C$ and $T_\infty = -5°C$, the time required to cool the midsection of the apples to $T_0 = 6°C$ is

Air
$-5°C$
1.5 m/s

Apples
$D = 7$ cm
85% water
$T_i = 30°C$

FIGURE 13-11

Schematic for Example 13-1.

determined from the transient temperature charts in Chapter 4 for a sphere as follows:

$$\left.\begin{array}{l}\dfrac{1}{\text{Bi}} = \dfrac{k}{hr_o} = \dfrac{0.418 \text{ W/m} \cdot °\text{C}}{(21 \text{ W/m}^2 \cdot °\text{C})(0.035 \text{ m})} = 0.57 \\[2mm] \dfrac{T_0 - T_\infty}{T_i - T_\infty} = \dfrac{6 - (-5)}{30 - (-5)} = 0.314 \end{array}\right\} \tau = \dfrac{\alpha t}{r_o^2} = 0.46 \text{ (Fig. 4-15}a)$$

Therefore,

$$t = \frac{\tau r_o^2}{\alpha} = \frac{(0.46)(0.035 \text{ m})^2}{0.13 \times 10^{-6} \text{ m}^2/\text{s}} = 4335 \text{ s} \cong \mathbf{1.20 \text{ h}}$$

The lowest temperature during cooling will occur on the surface ($r/r_0 = 1$) of the apples that is in direct contact with refrigerated air and is determined from

$$\left.\begin{array}{l}\dfrac{1}{\text{Bi}} = \dfrac{k}{hr_o} = 0.57 \\[2mm] \dfrac{r}{r_o} = 1 \end{array}\right\} \dfrac{T(r) - T_\infty}{T_0 - T_\infty} = 0.50 \text{ (Fig. 4-15}b)$$

It gives

$$T_{\text{surface}} = T(r) = T_\infty + 0.50 \, (T_0 - T_\infty) = -5 + 0.5[6 - (-5)] = 0.5°\text{C}$$

which is above $-1.1°\text{C}$, the highest freezing temperature of apples. Therefore, **no part of the apples will freeze** during this cooling process.

13-3 ■ THERMAL PROPERTIES OF FOOD

Refrigeration of foods offers considerable challenges to engineers since the structure and composition of foods and their thermal and physical properties vary considerably. Furthermore, the properties of foods also change with *time* and *temperature*. Fruits and vegetables offer an additional challenge since they *generate heat* during storage as they consume oxygen and give off carbon dioxide, water vapor, and other gases.

The thermal properties of foods are dominated by their *water content*. In fact, the specific heat and the latent heat of foods are calculated with reasonable accuracy on the basis of their water content alone. The *specific heats* of foods can be expressed by **Siebel's formula** as

$$C_{p,\text{fresh}} = 3.35a + 0.84 \qquad (\text{kJ/kg} \cdot °\text{C}) \qquad (13\text{-}2)$$

$$C_{p,\text{frozen}} = 1.26a + 0.84 \qquad (\text{kJ/kg} \cdot °\text{C}) \qquad (13\text{-}3)$$

where $C_{p,\text{fresh}}$ and $C_{p,\text{frozen}}$ are the specific heats of the food before and after freezing, respectively; a is the fraction of water content of the food ($a = 0.65$ if the water content is 65 percent); and the constant 0.84 kJ/kg · °C represents the specific heat of the solid (nonwater) portion of the food. For example, the specific heats of fresh and frozen chicken whose water content is 74 percent are

$$C_{p,\text{fresh}} = 3.35a + 0.84 = 3.35 \times 0.74 + 0.84 = 3.32 \text{ kJ/kg} \cdot {}^\circ\text{C}$$
$$C_{p,\text{frozen}} = 1.26a + 0.84 = 1.26 \times 0.74 + 0.84 = 1.77 \text{ kJ/kg} \cdot {}^\circ\text{C}$$

Siebel's formulas are based on the specific heats of *water* and *ice* at 0°C of 4.19 and 2.10 kJ/kg · °C, respectively, and thus they result in the specific heat values of water and ice at 0°C for $a = 100$ (i.e., pure water). Therefore, Siebel's formulas give the specific heat values at 0°C. However, they can be used over a wide temperature range with reasonable accuracy.

The *latent heat* of a food product during freezing or thawing (the heat of fusion) also depends on its water content and is determined from (Fig. 13-12)

$$h_{\text{latent}} = 334a \qquad \text{(kJ/kg)} \qquad (13\text{-}4)$$

where a is again the fraction of water content and 334 kJ/kg is the *latent heat of water* during freezing at 0°C at atmospheric pressure. For example, the latent heat of chicken whose water content is 74 percent is

$$h_{\text{latent, chicken}} = 334a = 334 \times 0.74 = 247 \text{ kJ/kg}$$

Perishable foods are mostly water in content that turns to ice during freezing. Therefore, we may expect the food items to freeze at 0°C, which is the freezing point of pure water at atmospheric pressure. But the water in foods is far from being pure, and thus the freezing temperature of foods will be somewhat below 0°C, depending on the composition of a particular food. In general, food products freeze over a *range of temperatures* instead of a single temperature since the composition of the liquid in the food changes (becomes more concentrated in sugar) and its freezing point drops when some of the liquid water freezes (Table 13-3). Therefore, we often speak of the *average* freezing temperature, or, for foods like lettuce that are damaged by freezing, the temperature at which *freezing begins.* The freezing temperature of most foods is between −0.3 and −2.8°C. In the absence of the exact data, the freezing temperature can be assumed to be −2.0°C for meats and −1.0°C for vegetables and fruits.

The freezing temperature and specific and latent heats of common food products are given in Tables A-7 and A-7E. The *freezing temperature* in this table represents the temperature at which freezing starts for fruits and vegetables, and the average freezing temperature for other foods. The *water content* values are typical values for mature products and may exhibit some variation. The specific and latent heat values given are valid only for the *listed water content,* but they can be reevaluated easily for different water contents using the formulas above.

The accurate determination of heat transfer from the food during freezing to a certain temperature requires a knowledge of the unfrozen amount of water at that temperature. Therefore, it is convenient to present the enthalpies of foods at various temperatures in tabular or graphical form, as shown in Table 13-4 for sweet cherries and in Figure 13-13 for beef. Once the enthalpies are available, the heat transfer Q from the food can be determined from

$$Q = m(h_{\text{initial}} - h_{\text{final}}) = m(h_1 - h_2) \qquad \text{(kJ)} \qquad (13\text{-}5)$$

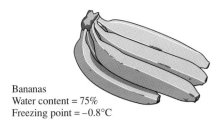

Bananas
Water content = 75%
Freezing point = −0.8°C

$C_{p,\text{ fresh}} = 3.35 \times 0.75 + 0.84 = 3.35 \text{ kJ/kg·°C}$

$C_{p,\text{ frozen}} = 1.26 \times 0.75 + 0.84 = 1.79 \text{ kJ/kg·°C}$

$h_{\text{latent}} = 334 \times 0.75 = 251 \text{ kJ/kg}$

FIGURE 13-12

The specific and latent heats of foods depend on their water content alone.

TABLE 13-3

The drop in the freezing point of typical ice cream whose composition is 12.5 percent milkfat, 10.5 percent serum solids, 15 percent sugar, 0.3 percent stabilizer, and 61.7 percent water (from ASHRAE *Handbook: Refrigeration,* **Ref. 3, Chap. 14, Table 9)**

Percent water unfrozen	Freezing point of unfrozen part, °C
0	−2.47
10	−2.75
20	−3.11
30	−3.50
40	−4.22
50	−5.21
60	−6.78
70	−9.45
80	−14.92
90	−30.36
100	−55.0

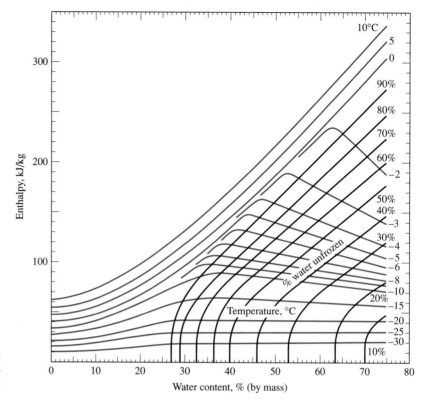

FIGURE 13.13

The enthalpy of beef
(from Riedel, Ref. 13).

where m is the mass of the food and $h_{initial}$ and h_{final} are the initial and final enthalpies of the food, respectively.

Other properties of foods such as density, thermal conductivity, and thermal diffusivity are listed in Tables A-7 and A-7E. Again, the exact values of these properties for a particular food depend on the composition, structure, and temperature of the food and may deviate from the listed values. However, sufficiently accurate heat transfer calculations can be made using the representative values in the tables.

The structures of some food products vary considerably in different directions, and thus the thermal conductivity and diffusivity of some food products can be somewhat different (usually about 10 percent) in directions parallel and perpendicular to the fiber structure. The lower values are listed in such cases to be conservative. Note that the thermal conductivity values for most foods are in the range of 0.2 to 1.0 W/m · °C, and the average thermal conductivity of foods is practically the same as the average thermal conductivity of water, which is 0.6 W/m · °C. In the absence of data, thermal diffusivity can be calculated from its definition, $\alpha = k/\rho C_p$.

EXAMPLE 13-2   Freezing of Beef

A 50-kg box of beef at 8°C having a water content of 72 percent is to be frozen to a temperature of −30°C in 4 h. Using data from Figure 13-13, determine (*a*) the total amount of heat that must be removed from the beef, (*b*) the amount of unfrozen water in beef at −30°C, and (*c*) the average rate of heat removal from the beef.

Solution A box of beef is to be frozen. The amount of heat removed, the remaining amount of unfrozen water, and the average rate of heat removal are to be determined.

Assumptions The beef is at uniform tempertures at the beginning and at the end of the process.

Properties At a water content of 72 percent, the enthalpies of beef at 8 and $-30°C$ are $h_1 = 312$ kJ/kg and $h_2 = 20$ kJ/kg (Fig. 13-13).

Analysis (a) The total heat transfer from the beef is determined from

$$Q = m(h_1 - h_2) = (50 \text{ kg})[(312 - 20) \text{ kJ/kg}] = \textbf{14,600kJ}$$

(b) The unfrozen water content at $-30°C$ and 72 percent water content is determined from Figure 13-13 to be about 10 percent. Therefore, the total amount of unfrozen water in the beef at $-30°C$ is

$$m_{\text{unfrozen}} = (m_{\text{total}})(\% \text{ unfrozen}) = (50 \text{ kg})(0.1) = 5 \text{ kg}$$

(c) Noting that 14,600 kJ of heat are removed from the beef in 4 h, the average rate of heat removal (or refrigeration) is

$$\dot{Q} = \frac{Q}{\Delta t} = \frac{14,600 \text{ kJ}}{(4 \times 3600 \text{ s})} = \textbf{1.01 kW}$$

Therefore, this facility must have a refrigeration capacity of at least 1.01 kW per box of beef.

This problem could also be solved by assuming the beef to be frozen completely at $-2°C$ by releasing its latent heat of 240 kJ/kg and using specific heat values of 3.25 kJ/kg · °C above freezing and 1.75 kJ/kg · °C below freezing. The total heat removal from the box of beef in this case would be 16,075 kJ. Note that the difference between the two results is $16,075 - 14,600 = 1475$ kJ, which is nearly equal to the latent heat released by 5 kg of water as it freezes.

EXAMPLE 13-3   Freezing of Sweet Cherries

A 40-kg box of sweet cherries at 10°C having a water content of 77 percent is to be frozen to a temperature of $-30°C$ (Fig. 13-14). Using enthalpy data from Table 13-4, determine the total amount of heat that must be removed from the cherries.

Solution A box of sweet cherries is to be frozen. The amount of heat that must be removed is to be determined.

Assumptions The cherries are at uniform temperatures at the beginning and at the end of the process.

Properties At a water content of 77 percent, the enthalpies of cherries at 0 and $-30°C$ are $h_1 = 324$ kJ/kg and $h_2 = 26$ kJ/kg and the specific heat of cherries is 3.24 kJ/kg · °C (Table 13-4).

Analysis The amount of heat removed as the cherries are cooled from 0°C to $-30°C$ is

$$Q_{\text{freezing}} = m(h_1 - h_2) = (40 \text{ kg})[(324 - 26) \text{ kJ/kg}] = 11,920 \text{ kJ}$$

TABLE 13-4

The variation of enthalpy and the unfrozen water content of sweet cherries (77 percent water content by mass, no stones) with temperature (from ASHRAE, *Handbook of Fundamentals,* **Chap. 30, Table 5)***

Tempera-ture, °C	Percent water unfrozen	Enthalpy, kJ/kg
0	100	324
−1	100	320
−2	100	317
−3	86	276
−4	67	225
−5	55	190
−6	47	166
−7	40	149
−8	36	133
−9	32	123
−10	29	114
−12	26	100
−14	21	87
−16	19	76
−18	17	66
−20	15	58
−30	9	26

*The specific heat of fresh (unfrozen) cherries can be taken to be 3.42 kJ/kg · °C.

FIGURE 13-14

Schematic for Example 13-3.

$\dot{Q} = mC_p(T_{\text{initial}} - T_{\text{final}})/\Delta t$

$= (0.14 \text{ kg})(3600 \text{ J/kg·°C})(23°C)/(3 \times 3600 \text{ s})$

$= 1.07 \text{ W}$ (per apple)

FIGURE 13-15

Refrigeration load of an apple-cooling
facility that cools apples from 25°C
to 2°C in 3 h.

The amount of heat removed as the cherries are cooled from 10°C to 0°C is

$$Q_{\text{cooling}} = mC_p\Delta T_{\text{cooling}} = (40 \text{ kg})(3.42 \text{ kJ/kg·°C})(10°C) = 1368 \text{ kJ}$$

Then the total heat removed as the cherries are cooled from 10°C to −30°C becomes

$$Q_{\text{total}} = Q_{\text{cooling}} + Q_{\text{freezing}} = 1368 + 11,920 = \mathbf{13,288 \text{ kJ}}$$

Also, the cherries at −30°C will contain 3.6 kg of water since 9 percent of the water in the cherries will still be unfrozen at −30°C.

13-4 ■ REFRIGERATION OF FRUITS AND VEGETABLES

Fruits and vegetables are frequently *cooled* to preserve preharvest freshness and flavor, and to extend storage and shelf life. Cooling at the field before the product is shipped to the market or storage warehouse is referred to as **precooling.** The cooling requirements of fruits and vegetables vary greatly, as do the cooling methods. Highly perishable products such as broccoli, ripe tomatoes, carrots, leafy vegetables, apricots, strawberries, peaches, and plums must be cooled as soon as possible after harvesting. Cooling is not necessary or as important for long-lasting fruits and vegetables such as potatoes, pumpkins, green tomatoes, and apples.

The refrigeration capacity of *precooling facilities* is usually much larger than that of cold storage facilities because of the requirement to cool the products to a specified temperature within a specified time. In applications where cooling is needed during daytime only, the size of the refrigeration equipment can be cut in half by utilizing cold storage facilities at night for activities such as ice making and during the day for precooling.

Fruits and vegetables are mostly water, and thus their properties are close in value to those of water. Initially, all of the heat removed from the product comes from the exterior of the products, causing a large temperature gradient within the product during fast cooling. But the *mass-average temperature,* which is the equivalent average temperature of the product at a given time, is used in calculations for simplicity.

The heat removed from the products accounts for the majority of the refrigeration load and is determined from

$$\dot{Q}_{\text{product}} = mC_p(T_{\text{initial}} - T_{\text{final}})/\Delta t \qquad \text{(W)} \qquad (13-6)$$

where $\dot{Q}_{\text{product}}$ is the *average rate of heat removal* from the fruits and vegetables, m is the total mass, C_p is the average specific heat, T_{initial} and T_{final} are the mass-average temperatures of the products before and after cooling, respectively, and Δt is the cooling time (Fig. 13-15). The heat of respiration is negligible when the cooling time is less than a few hours.

Fresh fruits and vegetables are *live products,* and they continue to *respire* at varying rates for days and even weeks after harvesting. During respiration, a sugar like glucose combines with O_2 to produce CO_2 and H_2O. **Heat of respiration** is released during this exothermic reaction, which adds to the refrigeration load during cooling of fruits and vegetables. The rate of respiration varies strongly with temperature. For example, the average heats of respiration of strawberries at 0, 10, and 20°C are 44, 213, and 442 mW/kg, respectively. The heat of respiration also varies greatly from one product to another. For example, at 20°C, it is 34 mW/kg for mature potatoes, 77 mW/kg for apples, 167 mW/kg for cabbage, and 913 mW/kg for broccoli. The heat of respiration of most vegetables decreases with time. The opposite is true for fruits that ripen in storage such as apples and peaches. For plums, for example, the heat of respiration increases from 12 mW/kg shortly after harvest to 21 mW/kg after 6 days and to 27 mW/kg after 18 days when stored at 5°C. Being infected with decay organisms also increases the respiration time.

The heat of respiration of some fruits and vegetables are given in Table 13-5. We should use the *initial rates* of respiration when calculating the cooling load of fruits and vegetables for the first day or two, and the *long-term equilibrium rates* when calculating the heat load for long-term cold storage. The *refrigeration load* due to *respiration* is determined from

$$\dot{Q}_{\text{respiration}} = \sum m_i \dot{q}_{\text{respiration}, i} \qquad (W) \qquad (13\text{-}7)$$

which is the sum of the mass times the heat of respiration for all the food products stored in the refrigerated space. Fresh fruits and vegetables with the highest rates of respiration are the most perishable, and refrigeration is the most effective way to slow down respiration and decay.

The primary precooling methods for fruits and vegetables are *hydrocooling,* where the products are cooled by immersing them into chilled water; *forced-air cooling,* where the products are cooled by forcing refrigerated air through them; *package icing,* where the products are cooled by placing crushed ice into the containers; and *vacuum cooling,* where the products are cooled by vaporizing some of the water content of the products under low pressure conditions.

Hydrocooling is a popular and effective method of precooling fruits and vegetables and is done by immersing or flooding products in chilled water or spraying chilled water over the products. Water is a good heat transfer medium compared to air, and the convection resistance at the product surface is usually negligible. This means the primary resistance to heat transfer is the internal resistance of the product, and internal heat is removed as soon as it arrives at the surface. The temperature difference between the product surface and the cooling water is normally less than 0.5°C. Under idealized conditions, the convection heat transfer coefficient and the cooling rate per unit surface area can be 680 W/m² · °C and 300 W/m², respectively.

The variation of the mass-average product temperature with time is given in Figure 13-16 for some fruits. Note that reducing the temperature difference between the fruit and the water to 10 percent of the initial value takes about 0.4 h for peaches but 0.7 h for citrus fruits. This is not much of a concern, however, since citrus fruits are often not cooled at all.

The water in a hydrocooling system is cooled by passing it through cooling coils in which a refrigerant flows at about −2°C. The water is normally

TABLE 13-5

Heat of respiration of some fresh fruits and vegetables at various temperatures (from ASHRAE *Handbook of Fundamentals,* Ref. 2, Chap 30, Table 2)

Product	Heat of respiration, mW/kg	
	5°C	20°C
Apples	13–36	44–167
Strawberries	48–98	303–581
Broccoli	102–475	825–1011
Cabbage	22–87	121–437
Carrots	20–58	64–117
Cherries	28–42	83–95
Lettuce	39–87	169–298
Watermelon	*	51–74
Mushrooms	211	782–939
Onions	10–20	50
Peaches	19–27	176–304
Plums	12–27	53–77
Potatoes	11–35	13–92
Tomatoes	*	71–120

*Indicates a chilling temperature.

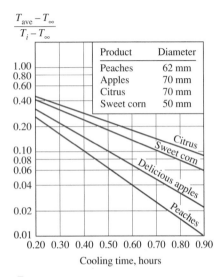

$$\frac{T_{ave} - T_\infty}{T_i - T_\infty}$$

FIGURE 13-16

The cooling time of some fruits under ideal conditions during hydrocooling (from ASHRAE, *Handbook: Refrigeration*, Ref. 3, Chap. 10, Fig. 2).

T_{ave} = average temperature

T_i = initial temperature

T_∞ = water temperature

FIGURE 13-17

The variation in the temperature of the fruits and vegetables with pressure during vacuum cooling from 25°C to 0°C.

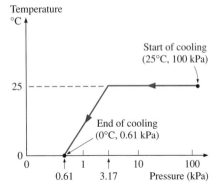

recirculated to save water and energy. But recirculation also saves the microorganisms that come in with fruits and vegetables. Chemicals such as chlorine are commonly added (usually at a rate of 50 to 100 mg/kg water) to reduce bacteria build-up in water.

Forced-air cooling systems have the advantage that they are *simple* and easy to maintain and do not have leakage problems. But they are not as effective as hydrocooling systems. The heat transfer coefficients encountered in forced-air cooling are considerably lower than those encountered in hydrocooling, and thus the cooling of fruits and vegetables by forced air usually takes much longer. The refrigerated air usually approaches the cooling section at a velocity of 1.5 to 2 m/s, and the convection heat transfer coefficients range from about 30 to 60 W/m² · °C. The rate of airflow ranges from about 1 to 3 L/s per kg of product. The cooling time can be reduced by spacing the product containers and allowing for airflow throughout instead of stacking them closely and disturbing the airflow. Various air cooling methods, such as batch cooling in refrigerated rooms and impingement air cooling as products continuously pass through a cooling tunnel on conveyor belts, are in common use today. Cut flowers are commonly cooled by forced air.

Vacuum cooling is a batch process, and it is based on *reducing the pressure* of the sealed cooling chamber to the saturation pressure at the desired low temperature and evaporating some water from the products to be cooled. The heat of vaporization during evaporation is absorbed from the products, which lowers the product temperature. The saturation pressure of water at 0°C is 0.61 kPa, and the products can be cooled to 0°C by lowering the pressure to this level. The cooling rate can be increased by lowering the pressure below 0.61 kPa. But this is not desirable because of the danger of freezing and the added cost.

In vacuum cooling, there are two distinct stages. In the first stage, the products at ambient temperature, say at 25°C, are loaded into the flash chamber and the operation begins. The temperature in the chamber remains constant until the *saturation pressure* is reached, which is 3.17 kPa at 25°C. In the second stage that follows, saturation conditions are maintained inside at progressively *lower pressures* and the corresponding *lower temperatures* until the desired temperature, usually slightly above 0°C, is reached (Fig. 13-17).

The amount of heat removed from the product is proportional to the mass of water evaporated, m_v, and the heat of vaporization of water at the average temperature, h_{fg}, and is determined from

$$Q_{vacuum} = m_v h_{fg} \qquad (kJ) \qquad (13\text{-}8)$$

If the product is cooled from 30°C to 0°C, the average heat of vaporization can be taken to 2466 kJ/kg, which corresponds to the *average temperature* of 15°C. Noting that the specific heat of products is about 4 kJ/kg · °C, the evaporation of 0.01 kg of water will lower the temperature of 1 kg of product by 24.66/4 = 6°C. That is, the vacuum-cooled products will lose 1 percent moisture for each 6°C drop in their temperature. This means the products will experience a *weight loss* of 4 percent for a temperature drop of about

24°C. To reduce the product moisture loss and enhance the effectiveness of vacuum cooling, the products are often *wetted* prior to cooling.

Vacuum cooling is usually more expensive than forced air or hydrocooling, and its use is limited to applications that result in much faster cooling. Products with large surface area per unit mass and a high tendency to release moisture such as *lettuce* and *spinach* are well suited for vacuum cooling. Products with low surface area to mass ratio are not suitable, especially those that have relatively impervious peel such as tomatoes and cucumbers. Some products such as mushrooms and green peas can be vacuum cooled successfully by wetting them first.

Package icing is used in small-scale applications to remove field heat and keep the product cool during transit, but its use is limited to products that are not harmed by contact with ice. Also, ice provides *moisture* as well as *refrigeration* (Fig. 13-18). Some shipping containers are cooled by pumping *slush ice* into them through a hose.

Fruits intended for long-term storage must be free of *mechanical injury* such as bruises and skin breaks since deterioration progresses rapidly in such areas. Also, the fruits must be harvested when *mature*. Undermature fruits will not ripen properly while overmature ones will deteriorate rapidly during storage. It is important that fruits be cooled as soon as possible after picking because at field temperatures some fruits deteriorate as much in one day as they do in one week in cold storage.

Fruits and vegetables are normally covered with *microorganisms* that cause decay and premature spoilage under favorable conditions. Refrigeration and the resulting low temperatures are the best protection against the damage caused by the growth of microorganisms. *High temperatures* increase the rate of ripening, decay, moisture loss, and loss of quality of food products during storage, transportation, and display. Precooling, cold storage, and refrigerated transportation have reduced these losses considerably and made it possible to move fruits and vegetables to distant markets at near-fresh conditions.

Fruits and vegetables require *oxygen* to respire, and studies have shown that the storage life of fruits and vegetables can be extended considerably by modifying the atmosphere in the cold storage rooms. This is usually done by reducing the oxygen level in the air to 1 to 5 percent and increasing the CO_2 level to retard the respiration rate and decay (Fig. 13-19). Therefore, many railcars and trucks that transport perishable products to distant cities are equipped with *modified atmosphere* systems. Lettuce is commonly shipped in a low oxygen atmosphere, but not all vegetables benefit from modified atmosphere. Certain vegetables benefit from a modified atmosphere for a limited time only. Excess CO_2 accumulation can be prevented in the cargo space by utilizing hydrated lime to absorb CO_2. Liquid nitrogen is often allowed to vaporize in the cargo room and force the regular air and thus oxygen out.

The modified atmosphere measure did not find much acceptance because of the complexities and the risks involved. For example, when the oxygen level is reduced too much, the products cannot respire and develop an alcoholic off-flavor. Also, very high levels of CO_2 harm the products. Modified

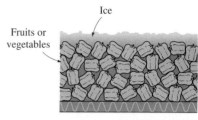

FIGURE 13-18

Ice is sometimes used to cool fruits and vegetables in the field and during short transit.

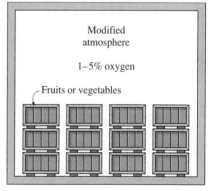

FIGURE 13-19

Reducing the oxygen level in storage rooms reduces respiration and extends storage life.

1°C
Oranges
Lemons
Grapefruits

14°C
Bananas

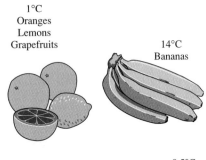

−1°C
Apples
Plums
Cherries
Grapes

−0.5°C
Peaches
Nectarines
Apricots

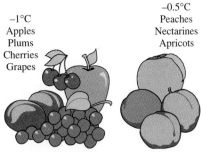

FIGURE 13-20

Recommended storage temperature
for various fruits.

atmosphere may *supplement* refrigeration, but it can never replace it. In fact, modified atmosphere may actually increase deterioration instead of decreasing it at temperatures above recommended values. An alternative way of extending the storage life is to *reduce* the oxygen level by reducing the atmospheric pressure below 0.2 atm in the storage room, accompanied by the removal of ethylene gas from the storage room to retard ripening.

Storage of fruits and vegetables in tightly sealed or tied *plastic bags* in order to keep the product moist is not recommended since the respiratory gases such as CO_2 and ethylene may reach harmful levels. Besides, the air packet inside acts as insulation that slows down the cooling of the product inside.

Most produce can be stored satisfactorily at 0°C and 90 to 95 percent relative humidity. An adequate refrigerator coil area should be provided to minimize the difference between the refrigerant and the air temperatures. A temperature difference of more than a few degrees will cause *excessive moisture removal* at the coils, and thus will make it difficult to maintain the storage room at the desired high humidity.

Different fruits and vegetables have different storage requirements, and they respond differently to prolonged cold storage. Therefore, the characteristics of a particular product must be known before an effective cooling and storage mechanism can be devised for it.

The recommended storage temperature for most varieties of *apples* is −1°C, which is 1°C above the highest freezing point of −2°C. But some apple varieties need to be stored at 2°C since they develop physiological disorders at −1°C. Some varieties susceptible to *chilling injury* are stored at about 4°C. *Pears* lose moisture faster than apples, and thus they should be maintained in humid environments. The recommended storage conditions for pears are −1°C and 90 to 90 percent relative humidity. Pears ripen best following storage at temperatures 16 to 21°C. *Plums* are usually stored at −1 to 0°C and 90 to 95 percent relative humidity. Most varieties of plums are not suitable for long-term storage, but some varieties can be stored for a few months with satisfactory results (Fig. 13-20).

Grapes are harvested after they are completely ripened but before being overripe since overripening may cause weakening of the stem attachment and increase the susceptibility to organisms causing decay. Grapes are normally precooled shortly after harvesting to minimize damage by the hot and dry field conditions. Care should be exercised during air cooling since the grapes have a large surface-to-volume ratio that makes them susceptible to moisture loss and drying. This is also the case for the stems of the grapes. Dry stems become brittle and break easily, causing a considerable loss of quality. Grapes are usually precooled by air supplied at 1.5°C or below a rate of at least 0.17 L/s per kg of grape and a minimum velocity of 0.5 m/s through the channels of the container. Sometimes the air is also humidified to minimize drying. It is also recommended that grapes be stored at −1°C and 90 to 95 percent relative humidity. After precooling, the air velocity is lowered to below 0.1 m/s in the container channels. Also, some varieties of grapes are fumigated with sulfur dioxide prior to storage to minimize decay and prevent new infections.

Peaches, especially the early season varieties, are not suitable for long-term storage. However, some late season varieties can be stored for up to six weeks without any noticeable deterioration in texture and flavor. If left unattended, peaches begin to soften and decay in a few hours. Therefore, it is important to cool peaches to 4°C as soon as possible after harvesting, usually by hydrocooling or forced air. Peaches are stored at −0.5°C and 90 to 95 percent relative humidity with low air velocities. *Nectarines* and *apricots* can be stored under similar storage conditions. *Sweet cherries* intended for storage must be picked with stems attached and precooled rapidly to −1°C by hydrocooling or forced-air cooling. Cherries retain their quality during storage for two weeks. Controlled atmosphere with 20 to 25 percent CO_2 or 0.5 to 2 percent O_2 helps preserve firmness and full vivid color during storage.

Citrus fruits such as *oranges, grapefruit,* and *lemons* do not undergo much changes in their composition after they are picked, and thus the product quality depends mostly on the degree of ripeness when the fruit is harvested. After harvesting, the fruits are transported to packinghouses where they are sorted, washed, and disinfected. The fruits are then polished and waxed after they are dried by warm air. Most varieties of oranges can be stored for 2 to 3 months at 0 to 1°C and 85 to 90 percent relative humidity. Maintaining high humidity in storage rooms is important since oranges tend to lose moisture rapidly. Storage temperatures up to 10°C can be used for shorter storage periods. Grapefruits and lemons can be stored at 10 to 15°C for more than a month.

Bananas are harvested when they are mature but unripe, as noticed by the green color of the peel. After they are washed and cut into retail-size clusters, they are boxed and transported from tropical climates to distant locations by refrigerated trucks or ships maintained at 14°C. The bananas are then moved to wholesale processing facilities by railcars or trucks again at 14°C. Product temperatures below 13°C can cause chilling injury, characterized by darkened areas of killed cells at the peel. Temperatures above about 16°C, on the other hand, may cause the bananas to ripen prematurely and must be avoided.

Unlike other produce storage facilities, bananas are ripened in carefully controlled wholesale processing facilities before they are shipped to retail stores. The ripening process takes four to eight days, depending on the temperatures used, which range between 14 and 18°C. The higher the storage temperature, the shorter the ripening time. The ripening process is initiated by the introduction of ethylene gas into the airtight storage air space at a mole fraction of 0.001 for one day. The refrigerant in the cooling system is maintained at a relatively high temperature (about 4.5° C) to avoid chilling injury.

The optimum temperatures and relative humidities for maximum storage at high quality of fruits and vegetables are well established and are available in handbooks (Fig. 13-21). The recommended storage temperatures and relative humidities are 0°C and 95 to 100 percent for artichokes, broccoli, brussels sprouts, leafy greens, lettuce, parsley, and radishes; 0°C and 98 to 100 percent for cabbage, carrots, celery, and parsnips; 0°C and 95 to 98 percent for sweet corn, green peas, and spinach; 0°C and 95 percent for cauliflower, mushrooms, and turnips; 0 to 2°C and 95 to 100 percent for

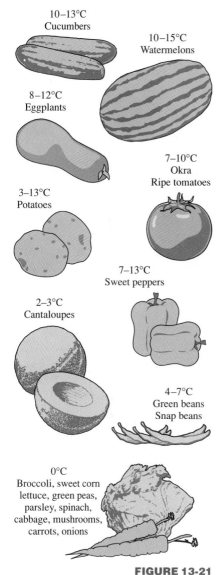

10–13°C
Cucumbers

10–15°C
Watermelons

8–12°C
Eggplants

7–10°C
Okra
Ripe tomatoes

3–13°C
Potatoes

7–13°C
Sweet peppers

2–3°C
Cantaloupes

4–7°C
Green beans
Snap beans

0°C
Broccoli, sweet corn
lettuce, green peas,
parsley, spinach,
cabbage, mushrooms,
carrots, onions

FIGURE 13-21

Recommended storage temperature
for various vegetables.

asparagus; 0°C and 65 to 70 percent for dry garlic and onions; 2 to 3°C and 90 to 95 percent for cantaloupes; 3 to 13°C and 90 to 95 percent for potatoes, 4 to 7°C and 95 percent for green or snap beans; 7 to 10°C and 90 to 95 percent for okra and ripe tomatoes; 7 to 13°C and 90 to 95 percent for sweet peppers; 8 to 12°C and 90 to 95 percent for eggplants; 10 to 13°C and 95 percent for cucumbers; and 10 to 15°C and 90 to 95 percent for watermelons.

Sprouting of onions, potatoes, and carrots becomes a problem in storage facilities that are not adequately refrigerated. The problem can be controlled by using sprout inhibitors. Gamma radiation also suppresses sprouting.

Heat treatment or *radiation* can be used to control decay and kill the insects and microorganisms on or near the fruit surfaces. For example, immersing the peaches in 55°C water for 1.5 minutes (or in 49°C water for 3 minutes) reduces brown rot considerably. Also, ultraviolet lamps are successfully used to kill bacteria and fungi on the exposed surfaces of the cold storage rooms. Gamma radiation is effectively used to control decay in some products, but consumer resistance to irradiated foods remains a concern.

EXAMPLE 13-4 Cooling of Bananas by Refrigerated Air

A typical one-half-carlot-capacity banana room contains 18 pallets of bananas. Each pallet consists of 24 boxes, and thus the room stores 432 boxes of bananas. A box holds an average of 19 kg of bananas and is made of 2.3 kg of fiberboard. The specific heats of banana and the fiberboard are 3.55 kJ/kg · °C and 1.7 kJ/kg · °C, respectively. The peak heat of respiration of bananas is 0.3 W/kg. The bananas are cooled at a rate of 0.2°C/h. If the temperature rise of refrigerated air is not to exceed 1.5°C as it flows through the room, determine the minimum flow rate of air needed. Take the density and specific heat of air to be 1.2 kg/m³ and 1.0 kJ/kg · °C.

Solution A banana cooling room is being analyzed. The minimum flow rate of air needed to cool bananas at a rate of 0.2°C/h is to be determined.

Assumptions **1** Heat transfer through the walls, floor, and ceiling of the banana room is negligible. **2** Thermal properties of air, bananas, and boxes are constant.

Properties The average properties air, bananas, and boxes are as specified in the problem statement.

Analysis A sketch of the banana room is given in Figure 13-22. Noting that the banana room holds 432 boxes, the total masses of the bananas and the boxes are determined to be

$$m_{banana} = \text{(Mass per box)(Number of boxes)} = (19 \text{ kg/box})(432 \text{ boxes})$$
$$= 8{,}208 \text{ kg}$$
$$m_{box} = \text{(Mass per box)(Number of boxes)} = (2.3 \text{ kg/box})(432 \text{ box})$$
$$= 993.6 \text{ kg}$$

The total refrigeration load in this case is due to the heat of respiration and the cooling of the bananas and the boxes and is determined from

$$\dot{Q}_{total} = \dot{Q}_{respiration} + \dot{Q}_{banana} + \dot{Q}_{box}$$

Banana cooling room

432 boxes
19 kg banana/box
2.3 kg fiberboard/box

Heat of respiration = 0.3 W/kg
Cooling rate = 0.2°C/h
$\Delta T_{air} = 1.5°C$

FIGURE 13-22

Schematic for Example 13-4.

where

$$\dot{Q}_{respiration} = m_{banana}\,\dot{q}_{respiration} = (8208\ kg)(0.3\ W/kg) = 2462\ W$$

$$\dot{Q}_{banana} = (mC_p\Delta T/\Delta t)_{banana}$$
$$= (8208\ kg)(3.55\ kJ/kg \cdot °C)(0.2°C/h) = 5828\ kJ/h$$
$$= 1619\ W \qquad (\text{since } 1\ W = 3.6\ kJ/h)$$

$$\dot{Q}_{box} = (mC_p\Delta T/\Delta t)_{box}$$
$$= (993.6\ kg)(1.7\ kJ/kg \cdot °C)(0.2°C/h) = 338\ kJ/h$$
$$= 94\ W \qquad (\text{since } 1\ W = 3.6\ kJ/h)$$

and the quantity $\Delta T/\Delta t$ is the rate of change of temperature of the products and is given to be 0.2°C/h. Then the total rate of cooling becomes

$$\dot{Q}_{total} = \dot{Q}_{respiration} + \dot{Q}_{banana} + \dot{Q}_{box} = 2462 + 1619 + 94 = 4175\ W$$

The temperature rise of air is limited to 1.5 K as it flows through the load. Noting that the air picks up heat at a rate of 4175 W, the minimum mass flow rate of air is determined to be

$$\dot{m}_{air} = \frac{\dot{Q}_{air}}{(C_p\Delta T)_{air}} = \frac{4175\ W}{(1000\ J/kg \cdot °C)(1.5°C)} = 2.78\ kg/s$$

Then the volume flow rate of air becomes

$$\dot{V}_{air} = \frac{\dot{m}_{air}}{\rho_{air}} = \frac{2.78\ kg/s}{1.2\ kg/m^3} = \mathbf{2.32\ m^3/s}$$

Therefore, the fan selected for the banana room must be large enough to circulate air at a rate of 2.32 m³/s.

13-5 ■ REFRIGERATION OF MEATS, POULTRY, AND FISH

Meat carcasses in slaughterhouses should be cooled *as fast as possible* to a uniform temperature of about 1.7°C to reduce the growth rate of micro-organisms that may be present on carcass surfaces, and thus minimize spoilage. The right level of *temperature, humidity,* and *air motion* should be selected to prevent excessive shrinkage, toughening, and discoloration.

The deep body temperature of an animal is about 39°C, but this temperature tends to rise a couple of degrees in the midsections after slaughter as a result of the *heat generated* during the biological reactions that occur in the cells. The temperature of the exposed surfaces, on the other hand, tends to drop as a result of heat losses. The thickest part of the carcass is the *round*, and the center of the round is the last place to cool during chilling. Therefore, the cooling of the carcass can best be monitored by inserting a thermometer deep into the central part of the round.

About 70 percent of the beef carcass is water, and the carcass is cooled mostly by *evaporative cooling* as a result of moisture migration toward the surface where evaporation occurs. But this shrinking translates into a loss of salable mass that can amount to 2 percent of the total mass during an overnight chilling. To prevent *excessive* loss of mass, carcasses are usually washed or sprayed with water prior to cooling. With adequate care, spray chilling can eliminate carcass cooling shrinkage almost entirely.

The average total mass of dressed beef, which is normally split into two sides, is about 300 kg, and the average specific heat of the carcass is about 3.14 kJ/kg · °C (Table 13-6). The *chilling room* must have a capacity equal to the daily kill of the slaughterhouse, which may be several hundred. A beef carcass is washed before it enters the chilling room and absorbs a large amount of water (about 3.6 kg) at its surface during the washing process. This does not represent a net mass gain, however, since it is lost by dripping or evaporation in the chilling room during cooling. Ideally, the carcass does not lose or gain any net weight as it is cooled in the chilling room. However, it does lose about 0.5 percent of the total mass in the *holding room* as it continues to cool. The actual product loss is determined by first weighing the dry carcass before washing and then weighing it again after it is cooled.

The refrigerated air temperature in the chilling room of beef carcasses must be sufficiently high to avoid *freezing* and *discoloration* on the outer surfaces of the carcass. This means a long residence time for the massive beef carcasses in the chilling room to cool to the desired temperature. Beef carcasses are only partially cooled at the end of an overnight stay in the chilling room. The temperature of a beef carcass drops to 1.7 to 7°C at the surface and to about 15°C in mid parts of the round in 10 h. It takes another day or two in the *holding room* maintained at 1 to 2°C to complete *chilling* and *temperature equalization*. But hog carcasses are fully chilled during that period because of their smaller size. The *air circulation* in the holding room is kept at minimum levels to avoid excessive moisture loss and discoloration. The refrigeration load of the holding room is much smaller than that of the chilling room, and thus it requires a smaller refrigeration system.

Beef carcasses intended for distant markets are shipped the day after slaughter in refrigerated trucks, where the rest of the cooling is done. This practice makes it possible to deliver fresh meat long distances in a timely manner.

The variation in temperature of the beef carcass during cooling is given in Figure 13-23. Initially, the cooling process is dominated by *sensible* heat transfer. Note that the average temperature of the carcass is reduced by about 28°C (from 36 to 8°C) in 20 h. The cooling rate of the carcass could be increased by *lowering* the refrigerated air temperature and *increasing* the air velocity, but such measures also increase the risk of *surface freezing*.

Most meats are judged on their **tenderness,** and the preservation of tenderness is an important consideration in the refrigeration and freezing of meats. Meat consists primarily of bundles of tiny *muscle fibers* bundled together inside long strings of *connective tissues* that hold it together. The tenderness of a certain cut of beef depends on the location of the cut, the age, and the activity level of the animal. Cuts from the relatively inactive mid-backbone section of the animal such as short loins, sirloin, and prime ribs are more tender than the cuts from the active parts such as the legs and the neck

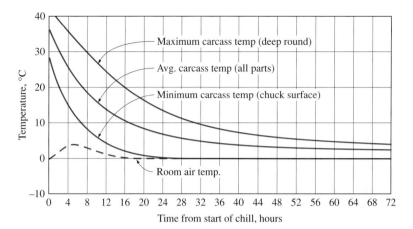

FIGURE 13-23

Typical cooling curve of a beef carcass in the chilling and holding rooms at an average temperature of 0°C (from ASHRAE, *Handbook: Refrigeration,* Ref. 3, Chap. 11, Fig. 2).

(Fig. 13-24). The more active the animal, the more the connective tissue, and the tougher the meat. The meat of an older animal is more flavorful, however, and is preferred for stewing since the toughness of the meat does not pose a problem for moist-heat cooking such as boiling. The protein *collagen,* which is the main component of the connective tissue, softens and dissolves in hot and moist environments and gradually transforms into *gelatin,* and tenderizes the meat.

The old saying "one should either cook an animal immediately after slaughter or wait at least two days" has a lot of truth in it. The biomechanical reactions in the muscle continue after the slaughter until the energy supplied to the muscle to do work diminishes. The muscle then stiffens and goes into *rigor mortis.* This process begins several hours after the animal is slaughtered and continues for 12 to 36 h until an enzymatic action sets in and tenderizes the connective tissue, as shown in Figure 13-25. It takes about seven days to complete tenderization naturally in storage facilities maintained at 2°C. Electrical stimulation also causes the meat to be tender. To avoid toughness, fresh meat should not be frozen before rigor mortis has passed.

You have probably noticed that steaks are tender and rather tasty when they are hot but toughen as they cool. This is because the gelatin that formed during cooking thickens as it cools, and meat loses its tenderness. So it is no surprise that first-class restaurants serve their steak on hot thick plates that keep the steaks warm for a long time. Also, cooking *softens* the connective tissue but *toughens* the tender muscle fibers. Therefore, barbecuing on *low heat* for a long time results in a tough steak.

Variety meats intended for long-term storage must be frozen rapidly to reduce spoilage and preserve quality. Perhaps the first thought that comes to mind to freeze meat is to place the meat packages into the *freezer* and wait. But the freezing time is *too long* in this case, especially for large boxes. For example, the core temperature of a 13-cm-deep box containing 32 kg of variety meat can be as high as 16°C 24 h after it is placed into a −30°C freezer.

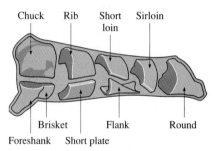

FIGURE 13-24

Various cuts of beef (from National Livestock and Meat Board).

FIGURE 13-25

Variation of tenderness of meat stored at 2°C with time after slaughter.

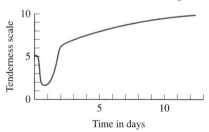

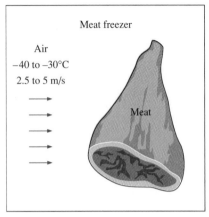

FIGURE 13-26

The freezing time of meat can be
reduced considerably by using low
temperature air at high velocity.

TABLE 13-7

**Storage life of frozen meat products
at different storage temperatures
(from ASHRAE** *Handbook:
Refrigeration,* **Ref. 3,
Chap. 10, Table 7)**

Product	Storage life, months		
	Temperature		
	−12°C	**−18°C**	**−23°C**
Beef	4–12	6–18	12–24
Lamb	3–8	6–16	12–18
Veal	3–4	4–14	8
Pork	2–6	4–12	8–15
Chopped beef	3–4	4–6	8
Cooked foods	2–3	2–4	

The freezing time of large boxes can be shortened considerably by adding
some *dry ice* into it.

A more effective method of freezing, called *quick chilling,* involves the
use of lower air temperatures, −40 to −30°C, with higher velocities of
2.5 m/s to 5 m/s over the product (Fig. 13-26). The internal temperature
should be lowered to −4°C for products to be transferred to a storage freezer
and to −18°C for products to be shipped immediately. The *rate of freezing*
depends on the *package material* and its insulating properties, the *thickness*
of the largest box, the *type* of meat, and the *capacity* of the refrigeration
system. Note that the air temperature will rise excessively during initial stages
of freezing and increase the freezing time if the capacity of the system is
inadequate. A smaller refrigeration system will be adequate if dry ice is to be
used in packages. Shrinkage during freezing varies from about 0.5 to
1 percent.

Although the average freezing point of lean meat can be taken to be −2°C
with a latent heat of 249 kJ/kg, it should be remembered that freezing occurs
over a *temperature range,* with most freezing occurring between −1 and
−4°C. Therefore, cooling the meat through this temperature range and removing the latent heat takes the most time during freezing.

Meat can be kept at an internal temperature of −2 to −1°C for local use
and storage for *under a week*. Meat must be frozen and stored at much lower
temperatures for *long-term storage*. The lower the storage temperature, the
longer the storage life of meat products, as shown in Table 13-7.

The *internal temperature* of carcasses entering the cooling sections
varies from 38 to 41°C for hogs and from 37 to 39°C for lambs and calves. It
takes about 15 h to cool the hogs and calves to the recommended temperature
of 3 to 4°C. The cooling-room temperature is maintained at −1 to 0°C and
the temperature difference between the refrigerant and the cooling air is kept
at about 6°C. Air is circulated at a rate of about 7 to 12 air changes per hour.
Lamb carcasses are cooled to an internal temperature of 1 to 2°C, which takes
about 12 to 14 h, and are held at that temperature with 85 to 90 percent
relative humidity until shipped or processed. The recommended rate of *air
circulation* is 50 to 60 air changes per hour during the first 4 to 6 h, which is
reduced to 10 to 12 changes per hour afterward.

Freezing does not seem to affect the *flavor* of meat much, but it affects
the *quality* in several ways. The *rate* and *temperature* of freezing may influence color, tenderness, and drip. Rapid freezing increases tenderness and
reduces the tissue damage and the amount of drip after thawing. Storage at
low freezing temperatures causes significant changes in *animal fat.* Frozen
pork experiences more undesirable changes during storage because of its fat
structure, and thus its acceptable storage period is shorter than that of beef,
veal, or lamb.

Meat storage facilities usually have a *refrigerated shipping dock* where
the orders are assembled and shipped out. Such docks save valuable storage
space from being used for shipping purposes and provide a more acceptable
working environment for the employees. Packing plants that ship whole or
half carcasses in bulk quantities may not need a shipping dock; a load-out
door is often adequate for such cases.

A refrigerated *shipping dock,* as shown in Figure 13-27, reduces the *refrigeration load* of freezers or coolers and prevents *temperature fluctuations* in the storage area. It is often adequate to maintain the shipping docks at 4 to 7°C for the coolers and about 1.5°C for the freezers. The dew point of the dock air should be below the product temperature to avoid condensation on the surface of the products and loss of quality. The rate of *airflow* through the loading doors and other openings is proportional to the *square root* of the temperature difference, and thus reducing the temperature difference at the opening by half by keeping the shipping dock at the average temperature reduces the rate of airflow into the dock and thus into the freezer by $1 - \sqrt{0.5} \cong 0.3$, or 30 percent. Also, the air that flows into the freezer is already cooled to about 1.5°C by the refrigeration unit of the dock, which represents about 50 percent of the cooling load of the incoming air. Thus, the net effect of the refrigerated shipping dock is a reduction of the *infiltration load* of the freezer by about 65 percent since $1 - 0.7 \times 0.5 = 0.65$. The net gain is equal to the difference between the reduction of the infiltration load of the freezer and the refrigeration load of the shipping dock. Note that the dock refrigerators operate at much higher temperatures (1.5°C instead of about −23°C), and thus they consume much less power for the same amount of cooling.

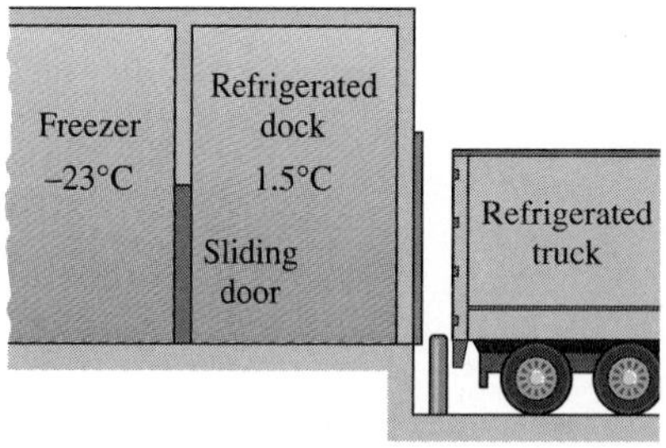

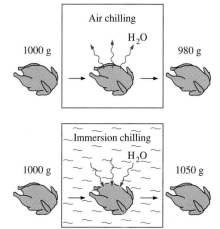

FIGURE 13-27

A refrigerated truck dock for loading frozen items to a refrigerated truck.

Poultry Products

Poultry products can be preserved by *ice-chilling* to 1 to 2°C or *deep chilling* to about −2°C for short-term storage, or by *freezing* them to −18°C or below for long-term storage. Poultry processing plants are completely *automated,* and the small size of the birds makes continuous conveyor line operation feasible.

The birds are first electrically stunned before cutting to prevent struggling, and to make the killing process more humane (!). Following a 90- to 120-s bleeding time, the birds are *scalded* by immersing them into a tank of warm water, usually at 51 to 55°C, for up to 120 s to loosen the feathers. Then the feathers are removed by feather-picking machines, and the eviscerated carcass is *washed* thoroughly before chilling. The internal temperature of the birds ranges from 24 to 35°C after washing, depending on the temperatures of the ambient air and the washing water as well as the extent of washing.

To control the microbial growth, the USDA regulations require that poultry be chilled to 4°C or below in less than 4 h for carcasses of less than 1.8 kg, in less than 6 h for carcasses of 1.8 to 3.6 kg. and in less than 8 h for carcasses more than 3.6 kg. Meeting these requirements today is not difficult since the slow *air chilling* is largely replaced by the rapid *immersion chilling* in tanks of slush ice. Immersion chilling has the added benefit that it not only prevents dehydration, but it causes a *net absorption of water* and thus increases the mass of salable product. Cool air chilling of unpacked poultry can cause a moisture loss of 1 to 2 percent, while water immersion chilling can cause a moisture absorption of 4 to 15 percent (Fig. 13-28). Water spray chilling can cause a moisture absorption of up to 4 percent. Most water absorbed is held

FIGURE 13-28

Air chilling causes dehydration and thus weight loss for poultry, whereas immersion chilling causes a weight gain as a result of water absorption.

between the flesh and the skin and the connective tissues in the skin. In immersion chilling, some soluble solids are lost from the carcass to the water, but the loss has no significant effect on flavor.

Many slush ice tank chillers today are replaced by *continuous* flow-type immersion slush ice chillers. Continuous slush ice-chillers can reduce the internal temperature of poultry from 32 to 4°C in about 30 minutes at a rate up to 10,000 birds per hour. Ice requirements depend on the inlet and exit temperatures of the carcass and the water, but 0.25 kg of ice per kg of carcass is usually adequate. However, *bacterial contamination* such as salmonella remains a concern with this method, and it may be necessary to chloride the water to control contamination.

Tenderness is an important consideration for poultry products just as it is for red meat, and preserving tenderness is an important consideration in the cooling and freezing of poultry. Birds cooked or frozen before passing through rigor mortis remain very tough. Natural tenderization begins soon after slaughter and is completed within 24 h when birds are held at 4°C. Tenderization is rapid during the first three hours and slows down thereafter. Immersion in hot water and cutting into the muscle adversely affect tenderization. Increasing the *scalding temperature* or the scalding time has been observed to increase toughness, and decreasing the scalding time has been observed to increase tenderness. The *beating action* of mechanical feather-picking machines causes considerable toughening. Therefore, it is recommended that any cutting be done after tenderization. *Cutting up* the bird into pieces before natural tenderization is completed reduces tenderness considerably. Therefore, it is recommended that any cutting be done after tenderization. *Rapid chilling* of poultry can also have a toughening effect. It is found that the tenderization process can be speeded up considerably by a patented *electrical stunning* process.

Poultry products are *highly perishable,* and thus they should be kept at the *lowest* possible temperature to maximize their shelf life. Studies have shown that the populations of certain bacteria double every 36 h at −2°C, 14 h at 0°C, 7 h at 5°C, and less than 1 h at 25°C (Fig. 13-29). Studies have also shown that the total bacterial counts on birds held at 2°C for 14 days are equivalent to those held at 10°C for 5 days or 24°C for 1 day. It has also been found that birds held at −1°C had 8 days of additional shelf life over those held at 4°C.

The growth of microorganisms on the *surfaces* of the poultry causes the development of an *off-odor* and *bacterial slime.* The higher the initial amount of bacterial contamination, the faster the sliming occurs. Therefore, good sanitation practices during processing such as cleaning the equipment frequently and washing the carcasses are as important as the storage temperature in extending shelf life.

Poultry must be frozen *rapidly* to assure a light, pleasing appearance. Poultry that is frozen slowly appears dark and develops large ice crystals that damage the tissue. The ice crystals formed during rapid freezing are small. Delaying freezing of poultry causes the ice crystals to become larger. Rapid freezing can be accomplished by forced air at temperatures of −23 to −40°C and velocities of 1.5 to 5 m/s in *air-blast tunnel freezers.* Most poultry is

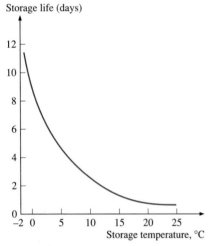

FIGURE 13-29

The storage life of fresh poultry decreases exponentially with increasing storage temperature.

frozen this way. Also, the packaged birds freeze much faster on open shelves than they do in boxes. If poultry packages must be frozen in boxes, then it is very desirable to leave the boxes open or to cut holes on the boxes in the direction of airflow during freezing. For best results, the blast tunnel should be fully loaded across its cross-section with even spacing between the products to assure uniform airflow around all sides of the packages. The freezing time of poultry as a function of refrigerated air temperature is given in Figure 13-30. Thermal properties of poultry are given in Table 13-8.

Other freezing methods for poultry include sandwiching between *cold plates, immersion* into a refrigerated liquid such as glycol or calcium chloride brine, and *cryogenic cooling* with liquid nitrogen. Poultry can be frozen in several hours by cold plates. Very high freezing rates can be obtained by *immersing* the packaged birds into a low-temperature brine. The freezing time of birds in −29°C brine can be as low as 20 min, depending on the size of the bird (Fig. 13-31). Also, immersion freezing produces a very appealing light appearance, and the high rates of heat transfer make continuous line operation feasible. It also has lower initial and maintenance costs than forced air, but *leaks* into the packages through some small holes or cracks remain a concern. The convection heat transfer coefficient is 17 W/m² · °C for air at −29°C and 2.5 m/s whereas it is 170 W/m² · °C for sodium chloride brine at −18°C and a velocity of 0.02 m/s. Sometimes liquid nitrogen is used to crust freeze the poultry products to −73°C. The freezing is then completed with air in a holding room at −23°C.

Properly packaged poultry products can be *stored* frozen for up to about a year at temperatures of −18°C or lower. The storage life drops considerably at higher (but still below-freezing) temperatures. Significant changes occur in flavor and juiciness when poultry is frozen for too long, and a stale rancid odor develops. Frozen poultry may become dehydrated and experience **freezer burn,** which may reduce the eye appeal of the product and cause toughening of the affected area. Dehydration and thus freezer burn can be controlled by *humidification, lowering* the storage temperature, and packaging the product in essentially *impermeable* film. The storage life can be extended by packing the poultry in an *oxygen-free* environment. The bacterial counts in precooked frozen products must be kept at safe levels since bacteria may not be destroyed completely during the reheating process at home.

Frozen poultry can be *thawed* in ambient air, water, refrigerator, or oven without any significant difference in taste. Big birds like turkey should be thawed safely by holding it in a refrigerator at 2 to 4°C for two to four days, depending on the size of the bird. They can also be thawed by immersing them into cool water in a large container for 4 to 6 h, or holding them in a paper bag. Care must be exercised to keep the bird's surface *cool* to minimize *microbiological growth* when thawing in air or water.

Fish

Fish is a *highly perishable* commodity, and the preservation of fish starts on the vessel as soon as it is caught. Fish deteriorates quickly because of bacterial and enzymatic activities, and *refrigeration* reduces these activities and

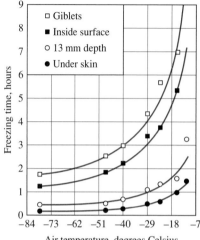

Note: Freezing time is the time required for temperature to fall from 0 to –4°C. The values are for 2.3 to 3.6 kg chickens with initial temperature of 0 to 2°C and with air velocity of 2.3 to 2.8 m/s.

FIGURE 13-30

The variation of freezing time of poultry with air temperature (from van der Berg and Lentz, Ref. 15).

TABLE 13-8

Thermal properties of poultry	
Quantity	Typical value
Average density:	
Muscle	1070 kg/m³
Skin	1030 kg/m³
Specific heat:	
Above freezing	2.94 kJ/kg · °C
Below freezing	1.55 kJ/kg · °C
Freezing point	−2.8°C
Latent heat of fusion	247 kJ/kg
Thermal conductivity:	(in W/m · °C)
Breast muscle	0.502 at 20°C
	1.384 at −20°C
	1.506 at −40°C
Dark muscle	1.557 at −40°C

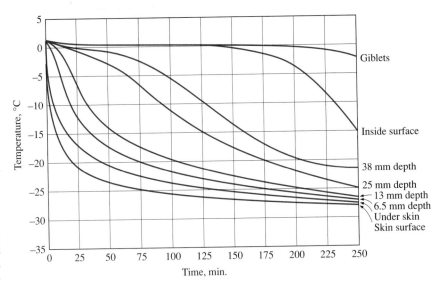

FIGURE 13-31

The variation of temperature of the breast of 6.8-kg turkeys initially at 1°C with depth during immersion cooling at −29°C (from van der Berg and Lentz, Ref. 15).

delays spoilage. Different species have different refrigeration requirements, and thus different practices exist for different species. *Large fish* are usually eviscerated, washed, and iced in the pens of the vessel's hold. *Smaller fish* such as ocean perch and flounder are iced directly without any processing. *Lobsters* and *crabs* are usually stored alive on the vessel without refrigeration. The fish caught in the arctic waters in winter are frozen by the cold weather and marketed as frozen fish. Salmon and halibut are usually stored in tanks of refrigerated sea water at −1°C. Fish raised in *aquaculture farms* are usually caught on demand and sold fresh in containers covered with ice. The ice used in the fishing industry can be a source of bacterial contamination itself, and thus ice should be made from chlorinated or potable water.

Fish are normally stored in chill rooms at 2°C as they await processing. They are packed in containers of 2- to 16-kg capacity in wet ice after processing and are transported to intended locations. It is desirable to keep fish *uniformly chilled* within 0 to 2°C during transit. At retail stores, fish should be displayed in special fish counters. Displaying in meat cases reduces the shelf life of fish considerably since the temperature of the meat case may be 4°C or higher.

The *maximum storage life* of fresh fish is 10 to 15 days, depending on the particular species, if it is properly iced and stored in refrigerated rooms at 0 to 2°C. Temperatures below 0°C should be avoided in storage rooms since they slow ice melting, which can result in high fish temperatures. Also, the humidity of the storage facility should be over 90 percent and the air velocity should be low to minimize dehydration. The shelf life of fresh fish can be doubled or tripled by treating the fish with *ionizing radiation*. Radiation has no adverse effect on quality, but consumer resistance has prevented its widespread use. The shelf life can also be extended by inhibiting the growth of bacteria by storing the fish in a *modified atmosphere* with high levels of carbon dioxide and low levels of oxygen.

Fish are commonly frozen for *long-term storage*. Some fatty fish species such as *mackerel* turn rancid during storage and are not suitable for long-term storage. But others such as *cod* respond well to freezing and have a much longer frozen storage life. The storage temperature of frozen fish should be as low as possible to maximize the storage life and to avoid oxidation of fish oils and the resulting off-flavor. Frozen fish should be stored at −26°C or below (Fig. 13-32). Storage at −29°C results in a shelf life of one year or more, but even occasional storage at −23°C or above results in a rapid loss of quality and shortened storage life. Low temperatures retard bacterial activity but do not stop deterioration entirely. As a result, fish that is stored too long exhibits a marked loss of quality. Poorly stored fish become opaque, dull, and spongy when thawed, and the flesh may lose integrity and break up. The overall quality of frozen fish depends on the *condition* of the fish before freezing, the *freezing method* and the *temperature* and *humidity* during storage and transportation.

The *packaging material* used should be thick enough to protect the product, but thin enough to allow rapid freezing while providing adequate protection against moisture loss during frozen storage. It is important to snug fit the product in the package to reduce the air space and its insulating effect and the resulting high freezing costs. Also, the freezing time of packaged fish fillets in plate freezers is proportional to the square of the package thickness. Therefore, packages that are too thick should be avoided to keep freezing time and cost at reasonable levels. Tightly fit packages also reduce moisture migration from the product to the inner surfaces of the package.

Air-blast freezers are commonly used to freeze fish with air velocities between 2.5 and 7.5 m/s. High velocities result in faster freezing but also higher unit freezing cost. They may also cause *freezer burn* and dehydration of unpacked fish. Typical freezing times of various fish packages are given in Figure 13-33.

Fish products packaged in 2.5- and 5-kg boxes are often cooled by *plate freezers* rapidly and efficiently. The air spaces in packages must be minimized for effective plate freezing. Fish fillets are sometimes frozen by combined conduction and convection by placing them on a slowly moving refrigerated stainless steel belt through an air-blast tunnel. Plate freezers are also used in large fishing vessels to freeze fish. *Immersion freezing* is not suitable for packaged products but is commonly used for freezing tuna at sea as well as shrimp and crab. Sodium chloride brine is commonly used for this purpose.

Frozen fish are *transported* in refrigerated trucks, railroad cars, or ships capable of maintaining −18°C over long distances under various weather conditions. The vehicles are precooled to at least −12°C before loading, and fish is stored at −18°C or below in retail stores. Frozen fish must be thawed in a refrigerator to avoid rapid deterioration.

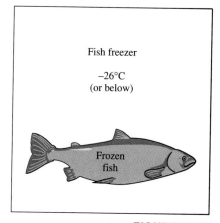

FIGURE 13-32
Frozen fish must be stored at −26°C or below to preserve high quality.

EXAMPLE 13-5 Chilling of Beef Carcasses in a Meat Plant

The chilling room of a meat plant is 18 m × 20 m × 5.5 m in size and has a capacity of 450 beef carcasses. The power consumed by the fans and the lights of the chilling room are 26 and 3 kW, respectively, and the room gains heat

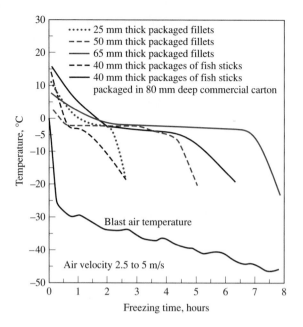

FIGURE 13-33

Freezing time of fish fillets and fish
sticks in a tunnel-type blast freezer
(from ASHRAE, *Handbook:
Refrigeration*, Ref. 3, Chap. 13, Fig. 2).

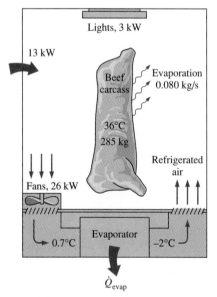

FIGURE 13-34

Schematic for Example 13.5.

through its envelope at a rate of 13 kW. The average mass of beef carcasses is 285 kg. The carcasses enter the chilling room at 36°C after they are washed to facilitate evaporative cooling and are cooled to 15°C in 10 h. The water is expected to evaporate at a rate of 0.080 kg/s. The air enters the evaporator section of the refrigeration system at 0.7°C and leaves at −2°C. The air side of the evaporator is heavily finned, and the overall heat transfer coefficient of the evaporator based on the air side is 20 W/m² · °C. Also, the average temperature difference between the air and the refrigerant in the evaporator is 5.5°C. Determine (*a*) the refrigeration load of the chilling room, (*b*) the volume flow rate of air, and (*c*) the heat transfer surface area of the evaporator on the air side, assuming all the vapor and the fog in the air freezes in the evaporator.

Solution The chilling room of a meat plant with a capacity of 450 beef carcasses is considered. The cooling load, the airflow rate, and the heat transfer area of the evaporator are to be determined.

Assumptions **1** Water evaporates at a rate of 0.080 kg/s. **2** All the moisture in the air freezes in the evaporator.

Properties The heat of fusion and the heat of vaporization of water at 0°C are 333.7 kJ/kg and 2501 kJ/kg (Table A-9). The density and specific heat or air at 0°C are 1.30 kg/m³ and 1.00 kJ/kg · °C (Table A-11). Also, the specific heat of beef carcass is determined from the relation in Table A-7b to be

$$C_p = 1.68 + 2.51 \times \text{(water content)} = 1.68 + 2.51 \times 0.58 = 3.14 \text{ kJ/kg} \cdot °C$$

Analysis (*a*) A sketch of the chilling room is given in Figure 13-34. The amount of beef mass that needs to be cooled per unit time is

$$\dot{m}_{beef} = \text{(Total beef mass cooled)/(Cooling time)}$$
$$= \text{(450 carcasses)(285 kg/carcass)/(10} \times \text{3600 s)} = 3.56 \text{ kg/s}$$

The product refrigeration load can be viewed as the energy that needs to be removed from the beef carcass as it is cooled from 36 to 15°C at a rate of 3.56 kg/s and is determined to be

$$\dot{Q}_{beef} = (\dot{m}C\Delta T)_{beef} = (3.56 \text{ kg/s})(3.14 \text{ kJ/kg} \cdot °C)(36 - 15)°C = 235 \text{ kW}$$

Then the total refrigeration load of the chilling room becomes

$$\dot{Q}_{total, chill room} = \dot{Q}_{beef} + \dot{Q}_{fan} + \dot{Q}_{lights} + \dot{Q}_{heat gain} = 235 + 26 + 3 + 13 = \textbf{277 kW}$$

The amount of carcass cooling due to evaporative cooling of water is

$$\dot{Q}_{beef, evaporative} = (\dot{m}h_{fg})_{water} = (0.080 \text{ kg/s})(2490 \text{ kJ/kg}) = 199 \text{ kW}$$

which is 199/235 = 85 percent of the total product cooling load. The remaining 15 percent of the heat is transferred by convection and radiation.

(*b*) Heat is transferred to air at the rate determined above, and the temperature of the air rises from −2°C to 0.7°C as a result. Therefore, the mass flow rate of air is

$$\dot{m}_{air} = \frac{\dot{Q}_{air}}{(C_p \Delta T_{air})} = \frac{277 \text{ kW}}{(1.0 \text{ kJ/kg} \cdot °C)[0.7 - (-2)°C]} = 102.6 \text{ kg/s}$$

Then the volume flow rate of air becomes

$$\dot{V}_{air} = \frac{\dot{m}_{air}}{\rho_{air}} = \frac{102.6 \text{ kg/s}}{1.30 \text{ kg/m}^3} = \textbf{78.9 m}^3\textbf{/s}$$

(*c*) Normally the heat transfer load of the evaporator is the same as the refrigeration load. But in this case the water that enters the evaporator as a liquid is frozen as the temperature drops to −2°C, and the evaporator must also remove the latent heat of freezing, which is determined from

$$\dot{Q}_{freezing} = (\dot{m}h_{latent})_{water} = (0.080 \text{ kg/s})(334 \text{ kJ/kg}) = 27 \text{ kW}$$

Therefore, the total rate of heat removal at the evaporator is

$$\dot{Q}_{evaporator} = \dot{Q}_{total, chill room} + \dot{Q}_{freezing} = 277 + 27 = \textbf{304 kW}$$

Then the heat transfer surface area of the evaporator on the air side is determined from $\dot{Q}_{evaporator} = (UA)_{air side} \Delta T$,

$$A = \frac{\dot{Q}_{evaporator}}{U\Delta T} = \frac{304,000 \text{ W}}{(20 \text{ W/m}^2 \cdot °C)(5.5°C)} = \textbf{2764 m}^2$$

Obviously, a finned surface must be used to provide such a large surface area on the air side.

13-6 ■ REFRIGERATION OF EGGS, MILK, AND BAKERY PRODUCTS

Eggs are important sources of nutrients, minerals, and vitamins, and they have a unique place in the diets of practically all cultures. Most eggs are consumed as *shell eggs,* while the rest are processed for use in other products

TABLE 13-9

Some properties of fresh chicken eggs

Quantity	Typical value
Average mass	60 g
Average density	1080 kg/m^3
Specific heat	3.23 kJ/kg · °C
Freezing point	−2.2°C
Water content	66 percent
Latent heat	220 kJ/kg

such as mayonnaise and dehydrated egg products. The production and consumption of eggs remain fairly steady throughout the year, and thus the freezing of eggs for long-term storage is not practical. However, eggs are highly perishable and must be *refrigerated* to preserve quality and to assure a reasonable shelf life.

Various properties of eggs are listed in Table 13-9. The average density of fresh eggs is 1080 kg/m^3, and thus fresh eggs settle at the bottom when they are put into a cup of water. The density decreases with time, however, as the egg loses moisture and the vacant space is filled with air. The egg shell contains about 17,000 pores through which air, water vapor, and microorganisms can pass. The *mass* of a chicken egg varies from about 35 to 80 g, with an average of 60 g. The white, the yolk, and the shell of an egg constitute 58, 31, and 11 percent of the egg mass, respectively.

In modern farms, eggs *laid* by the hens roll through the sloped floor onto a *conveyor* that transports the eggs to a *packing machine* that aligns the eggs into rows of 12 or 18. The dirt on eggs is a source of bacteria that may cause decomposition and spoilage. Therefore, the eggs are routed to the *washing section* where they are washed with warm detergent water at a temperature of 43 to 52°C by a series of sprays and brushes and are *rinsed* with warm water and sanitized with chlorine or another chemical. The temperature of eggs rises by about 3°C during washing. The eggs are then *dried* by air, *oiled,* and *checked* for internal and external defects under a high intensity light. The defective eggs are removed, and the remaining ones are *sized* automatically and are *packaged* into 12- or 18-egg cartons or flats.

Egg cartons are made of paper pulp or foam plastic, which are good insulating materials. Cartons that have *openings* on the top facilitate both viewing and cooling. Eggs in such stacks of cartons can be cooled by blasting cold air toward the openings. Cooling pallets of packed eggs by leaving them in cold storage may take two days or more since cold air will have difficulty reaching the mid section. Eggs are cooled most effectively on a *spiral belt cooler* after washing and before packaging, but incorporation of this cooling approach may require some costly modifications in most facilities. Cooling the eggs from about 32°C to about 7 to 13°C takes usually less than 1 h in this case. The eggs could also be cooled after they are packed provided that the boxing material does not prevent the air from flowing through the eggs, but the cooling time in this case would be considerably longer.

The most practical and effective way of preserving the quality of eggs and extending the storage life is *refrigeration,* which should start at the farm and continue through retail outlets. Shell eggs can be stored at 7 to 13°C and 75 to 80 percent relative humidity for *a few weeks*. Eggs lose about 1 percent of their mass per week as moisture. *Condensation* on the egg surfaces should be avoided since wet surfaces are conducive to bacterial and mold growth. Therefore, high moisture levels in storage rooms are undesirable. The chemical reactions that occur in eggs over time cause the yolk membrane to thin and the white to become watery. Absorption of *odors* from the environment also affects the flavor of eggs.

Milk

Milk is one of the most essential foods for humans, but also one of the most suitable environments for the *growth* of microorganisms (Fig. 13-35). Therefore, the production and processing of milk are heavily regulated. A dairy farmer must meet stringent federal and state *regulations* to produce Grade A milk, such as having healthy cows and adequate and sanitary facilities and equipment, and maintaining a bacteria count of less than 100,000 per mL. Also, the milk should not have objectionable flavors and odors.

The cows in a dairy farm are *mechanically milked,* and the milk flows into an insulated stainless steel *bulk tank* through sanitary tubes. The milk is *stirred* by an agitator and is *cooled* as milking continues. The raw milk is constantly agitated to maintain uniform milkfat distribution. The refrigeration capacity of the tank must be sufficient to cool 50 percent of its capacity from 32.2 to 10°C during the first hour, and from 10 to 4.4°C during the second hour. Milk is usually delivered to a dairy plant in a *stainless steel tank* on a truck. The tank is adequately *insulated* so that refrigeration is not required during transportation. The temperature rise of the tank filled with milk should not exceed 1.1°C in 18 h when the average temperature difference between the milk and the ambient air is 16.7°C. Horizontal storage tanks must also meet the same requirement.

A dairy plant receives, processes, and packages the milk. The minimum amounts of *milkfat* and *nonfat solids* in the milk are regulated. For example, the minimum legal milkfat requirements are 3.25 percent for whole milk, 0.5 percent for lowfat or skim milk (the milkfat contents of lowfat and skim milk in the market are about 2.0 and 0.5 percent, respectively), 18 percent for sour cream, and 36 percent for heavy cream (Fig. 13-36). The minimum legal requirement for nonfat solids in milk is 8.25 percent. The stored milk is first **standardized** to bring the milkfat content to desired levels using a *milk separator.* The separation of milkfat is easier and more efficient at higher temperatures. Therefore, milk is usually heated from the storage temperature of 4.4°C to about 20 to 33°C in warm milk separators. Only a portion of the milk needs to be separated. The remaining milk can be standardized by adding the required amount of skim milk or milkfat.

To kill the bacteria, milk is **pasteurized** in a batch or continuous flow-type system by heating the milk to a *minimum temperature* of 62.8°C normally by hot water or steam and holding it at that temperature for at least 30 min. Milk can also be pasteurized *quickly* by heating it to 71.7°C or above and holding it at that temperature at least 15 s (Fig. 13-37). Fast pasteurization is very suitable for continuous-type systems.

The pasteurized milk is first *cooled* to the environment temperature by the plant's *cold water,* and then it is cooled to 4.4°C or less by *refrigerated water* usually in plate-type heat exchangers. The entire cooling process should be completed in less than 1 h for sanitary reasons. The continuous-type pasteurizers are equipped with a *regenerator,* which is a counterflow heat exchanger in which the incoming cold raw milk is heated by the hot

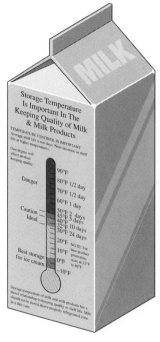

FIGURE 13-35

A label on a milk carton by Model Dairy that emphasizes the importance of storage temperature.

FIGURE 13-36

Typical milkfat contents of whole, lowfat, and skim milk.

| 3.5% milkfat | 2.0% milkfat | 0.5% milkfat |

Whole milk   Low-fat milk   Skim milk

FIGURE 13-37

The pasteurizing time of milk can be
minimized by raising the pasteurization
temperature.

pasteurized milk as it is cooled. The flow rates of both the pasteurized and
the raw milk are the same, and thus the temperature drop of pasteurized milk
will be practically equal to the temperature rise of the raw milk. The effectiveness of the regenerators are in the range of 80 to 90 percent.

The taste of milk depends on the feed of the cattle, and the milk is usually
vacuum processed after pasteurization to reduce the *undesirable flavors* and
odors in it. In this process, the temperature of pasteurized milk is raised to
82 to 93°C by injecting hot steam into it and then spraying the mixture into a
vacuum chamber where it is cooled by evaporation to the pasteurizing temperatures. The vapor as well as the noncondensible gases responsible for the
odors and off-flavor are removed by the vacuum pump. The unit is designed
such that the amount of moisture removed is equal to the amount of steam
added by the steam injector.

The milk is also **homogenized** before it is packaged to *break up* the large
fat globules into smaller ones, usually under 2 μm in size, to give it a "homogeneous" appearance. Homogenization distributes milkfat throughout the
body of milk and prevents the milkfat from collecting at the top because of
its lower density. This is done by pumping the warm milk at 56 to 82°C to
high pressure (usually 8 to 17 MPa) and shearing off the large globules by
forcing the milk through homogenizing valves.

The homogenized milk is *refrigerated* again to about 4°C and is *packaged* in the familiar paperboard, plastic, or glass containers for distribution
in refrigerated trucks. The paperboard carton is made of a 0.41-mm-thick
paper layer with 0.025- and 0.019-mm polyethylene film laminated on the
inside and outside, respectively. Plastic containers have a mass of about 60 g
and are made from polyethylene resin by blow-molding. Other milk products
such as yogurt, cheese, cream, and ice cream are produced by processing the
milk further.

Baked Products

Refrigeration plays an important part in all stages of modern bakery production from the preservation of raw materials such as flour and yeast to the
cooling of finished dough. *Flour* is usually stored in bins in air-conditioned
spaces at ambient temperatures. It is important to filter the *flour dust* in the
air since it may settle on the heat transfer surfaces of the air-conditioning
system and reduce heat transfer. *Yeast* is dormant at temperatures below 7°C,
and thus it is stored in refrigerated spaces below 7°C but above its freezing
point of −3.3°C since freezing kills most cells. Yeast is most active in the
temperature range of 27 to 38°C, especially in the presence of nutrients such
as *sugar* and *water,* but temperatures above 60°C must be avoided since yeast
cells cannot survive at those temperatures. Ingredients such as *corn syrup*
and *liquid sugar* are stored in heated tanks at about 52°C to prevent thickening
and crystallization.

The first step in the preparation of baked products is the *mixing* of flour,
yeast, water, and other ingredients, depending on the kind of product, into
dough. The temperature of the dough tends to rise during mixing as a result
of the *heat of hydration,* which is the heat released when dry flour absorbs

water, and the *heat of friction,* which is the mechanical energy supplied to the mixer by the motor. Therefore, *cooling* is needed during kneading of the dough. The easiest way to maintain the dough at the desired temperature of 26 to 27°C is to cool the water to 2 to 4°C before mixing it with the other ingredients. Another method is to cool the mixing chamber externally with water jackets.

After mixing, the dough is allowed to *ferment* in a room maintained at about 27°C and 75 percent relative humidity for a period of 3.5 to 5 h. The dough temperature rises by about 5°C during fermentation. Some water as well as other *fermentation gases* such as alcohols and esters evaporate from the dough surface, and the rate of evaporation depends on the relative humidity of the environment and the air velocity.

The dough is formed into *loaves* and is kept in an enclosure at 35 to 49°C for about 1 h. The fully developed loaves are then *baked,* usually in gas-fired ovens at 200 to 230°C for 18 to 30 min, depending on the type of bread. The hot air in the oven is circulated to ensure uniform baking. The exposed surfaces of the loaves are *sprayed with steam* to caramelize the sugars on the surface at early stages of baking to obtain a gold-colored crust. The temperature of the baked loaves is about 95°C inside and 230°C on the crust. The baked loaves are allowed to *cool* in racks to about 32 to 35°C, which takes about 1.5 to 3 h before they are bagged and shipped. The cooling time can be shortened by using forced airflow. The cooling rate is high initially because of the large temperature difference between the bread and the ambient air and the evaporative cooling due to the vaporization of the moisture in the bread (Fig. 13-38).

Bread starts losing its *freshness* shortly after baking as a result of the *starch crystallizing* and *losing moisture,* producing a crumbly texture. A tight wrap can slow the process by sealing the moisture in. Bread freezes at about −8°C. To preserve quality and to keep the cellular structure intact, the bread must be cooled as fast as possible. Wrapped bread loaves can be cooled from an internal temperature of 21°C to −10°C with cold air at −29°C and 3.5 m/s in about 3 h. Unwrapped bread will freeze faster, but the moisture loss in this case will be excessive. *Frozen bread* can be defrosted at room temperature with satisfactory results. The thermal properties of baked products are given in Table 13-10.

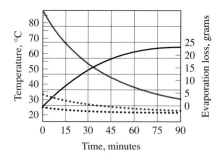

- Evaporation loss per kilogram of bread
- Temperature inside loaf
...... Air dry bulb temperature
...... Air wet bulb temperature

FIGURE 13-38

The variation of temperature and moisture loss of a hot bread with time in a counterflow bread-cooling tunnel (from ASHRAE, *Handbook: Refrigeration,* Ref. 3, Chap. 20, Fig. 1).

TABLE 13-10

Thermal properties of baked products and their ingredients (from ASHRAE *Handbook: Refrigeration,* Ref. 3, Chap. 20, Table 2)

Quantity	Typical value
Specific heat	
Dough	2.51 kJ/kg · °C
Flour	1.76 kJ/kg · °C
Milk	3.98 kJ/kg · °C
Ingredient mixture	1.68 kJ/kg · °C
Baked bread:	
Above freezing	2.93 kJ/kg · °C
Below freezing	1.42 kJ/kg · °C
Freezing point	−9 to −7°C
Latent heat of baked	
bread	109.3 kJ/kg
Heat of hydration of	
dough	15.1 kJ/kg

EXAMPLE 13-6   Retrofitting a Dairy Plant with a Regenerator

In a dairy plant, milk at 40°F is pasteurized continuously at 162°F at a rate of 2 gallons/s (Fig. 13-39). The milk is heated to pasteurizing temperature by hot water heated in a natural gas-fired boiler having an efficiency of 78 percent. The pasteurized milk is then cooled by cold water at 60°F before it is finally refrigerated back to 40°F. The plant is considering buying a regenerator with a certified effectiveness of 86 percent to save energy and water. The cost of the regenerator is $150,000 installed, and the plant is not interested in any retrofitting project that has a payback period of more than three years. If the cost of natural gas is $0.60/therm (1 therm = 100,000 Btu), determine if the dairy plant can justify purchasing this regenerator.

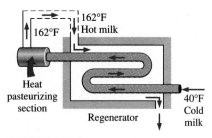

FIGURE 13-39
Schematic for Example 13-6.

Solution A regenerator is considered to save heat during the cooling of milk in a dairy plant. It is to be determined if such a generator will pay for itself from the fuel it saves in less than three years.

Assumptions The density and specific heat of milk are constant.

Properties The average density and specific heat of milk can be taken to be $\rho_{milk} \cong \rho_{water} = 62.4$ lbm/ft^3 = 8.35 lbm/gal and $C_{p,\,milk} = 0.928$ Btu/lbm · °F (Table A-7Eb).

Analysis The mass flow rate of the milk is

$$\dot{m}_{milk} = \rho \dot{V}_{milk} = (8.35 \text{ lbm/gal})(2 \text{ gal/s}) = 16.7 \text{ lbm/s} = 60{,}120 \text{ lbm/h}$$

To heat the milk from 40 to 162°F, as is being done currently, heat must be transferred to the milk at a rate of

$$\dot{Q}_{current} = [\dot{m}C_p(T_{pasteurization} - T_{refrigeration})]_{milk}$$
$$= (60{,}120 \text{ lbm/h})(0.928 \text{ Btu/lbm} \cdot \text{°F})(162 - 40) = 6{,}807{,}000 \text{ Btu/h}$$

The proposed regenerator has an effectiveness of $\varepsilon = 0.86$, and thus it will save 86 percent of this energy. Therefore,

$$\dot{Q}_{saved} = \varepsilon \dot{Q}_{current} = (0.86)(6{,}807{,}000 \text{ Btu/h}) = 5{,}854{,}000 \text{ Btu/h}$$

The boiler has an efficiency of $\eta_{boiler} = 0.78$, and thus the energy savings above correspond to fuel savings of

$$\text{Fuel savings} = \dot{Q}_{saved}/\eta_{boiler}$$
$$= (5{,}854{,}000 \text{ Btu/h})/0.78\left(\frac{1 \text{ therm}}{100{,}000 \text{ Btu}}\right) = 75.05 \text{ therm/h}$$

Noting that 1 year = 365 × 24 = 8760 h, the annual savings from the fuel cost will be

$$\text{Cost savings} = (\text{fuel saved}) (\text{Unit cost of fuel})$$
$$= (75.05 \text{ therm/h})(8760 \text{ h/yr})(\$0.60/\text{therm}) = \$394{,}500/\text{yr}$$

This gives the simple payback period of

$$\text{Payback period} = \frac{\text{Cost of equipment}}{\text{Annual cost savings}} = \frac{\$150{,}000}{\$394{,}500/\text{yr}} = 0.38 \text{ yr} = 4.6 \text{ months}$$

Therefore, the proposed regenerator will pay for itself in about 4.6 months from the natural gas cost it saves. The purchase and installation of the regenerator are definitely indicated in this case.

13-7 ■ REFRIGERATION LOAD OF COLD STORAGE ROOMS

Refrigerated spaces are maintained *below* the temperature of their surroundings, and thus there is always a driving force for heat flow toward the refrigerated space from the surroundings. As a result of this heat flow, the *temperature* of the refrigerated space will *rise* to the surrounding temperature unless the heat gained is promptly *removed* from the refrigerated space. A refrigeration system should obviously be large enough to remove the entire heat gain in order to maintain the refrigerated space at the desired low temperature. Therefore, the *size* of a refrigeration system for a specified refrigerated space is determined on the basis of the *rate of heat gain* of the refrigerated space.

The *total* rate of heat gain of a refrigerated space through all mechanisms under peak (or times of highest demand) conditions is called the **refrigeration load,** and it consists of (1) *transmission load,* which is heat conducted into the refrigerated space through its walls, floor, and ceiling; (2) *infiltration load,* which is due to surrounding warm air entering the refrigerated space through the cracks and open doors; (3) *product load,* which is the heat removed from the food products as they are cooled to refrigeration temperature; (4) *internal load,* which is heat generated by the lights, electric motors, and people in the refrigerated space; and (5) *refrigeration equipment load,* which is the heat generated by the refrigeration equipment as it performs certain tasks such as reheating and defrosting (Fig. 13-40).

The size of the refrigeration equipment must be based on *peak refrigeration load,* which usually occurs when the outside temperature is high and the maximum amount of products is brought into the cool storage room at field temperatures.

1 Transmission Load

The *transmission load* depends on the materials and construction of the walls, floor, and ceiling of the refrigerated space, the surface area, the air motion or wind conditions inside or outside, and the temperature difference between the refrigerated space and the ambient air. The *rate of heat transfer* through a particular wall, floor, or ceiling section can be determined from

$$\dot{Q}_{\text{transmission}} = UA_0\Delta T \quad \text{(W)} \tag{13-9}$$

where

A_0 = outside surface area of the section
ΔT = temperature difference between the outside air and the air inside the refrigerated space
U = overall heat transfer coefficient

Noting that the *thickness-to-thermal-conductivity* ratio of a layer represents its unit thermal resistance, the **overall heat transfer coefficient** is determined from (Fig. 13-41)

$$U = \frac{1}{\dfrac{1}{h_i} + \sum \dfrac{L}{k} = \dfrac{1}{h_0}} \quad \text{(W/m}^2 \cdot \text{°C)} \tag{13-10}$$

where

h_i = heat transfer coefficient at the inner surface of the refrigerated space

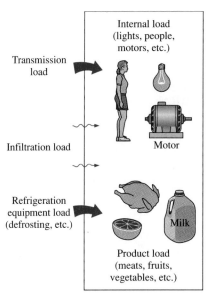

FIGURE 13-40

Various mechanisms of heat gain that make up the total refrigeration load of a refrigerated space.

FIGURE 13-41

The determination of thermal resistance and the overall heat transfer coefficient for a two-layer wall section for the calculation of transmission losses.

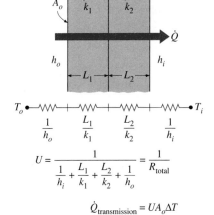

TABLE 13-11

Minimum recommended expanded polyurethane insulation thicknesses for refrigerated rooms (from ASHRAE, *Handbook: Refrigeration*, Ref. 3, Chap. 26, Table 2)

Storage temperature		Northern United States	Southern United States
10 to	16°C	25 mm	50 mm
4 to	10°C	50 mm	50 mm
−4 to	4°C	50 mm	75 mm
−9 to	−4°C	75 mm	75 mm
−18 to	−9°C	75 mm	100 mm
−26 to	−18°C	100 mm	100 mm
−40 to	−26°C	125 mm	125 mm

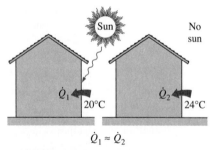

$$\dot{Q}_1 \approx \dot{Q}_2$$

FIGURE 13-42

Solar heating effect on the west or east wall can be compensated for by adding 4°C to the ambient temperature.

h_i = heat transfer coefficient at the outer surface of the refrigerated space

$\sum L/k$ = sum of the thickness-to-thermal-conductivity ratios of the layers that make up the section under consideration

For still air, it is common practice to take 10 W/m² · °C for h_i and h_0. A value of 20 or 30 W/m² · °C can be used for h_0 for low or moderate wind conditions outside, respectively.

The walls, floor, and ceiling of typical refrigerated rooms are well insulated, and the *unit thermal resistance L/k* of the *insulation layer* is usually much larger than the L/k of other layers such as the sheet metals and the convective resistance $1/h_i$ and $1/h_0$. Therefore, the thermal resistances of sheet metal layers can always be ignored. Also, the *convection resistances* $1/h_i$ and $1/h_0$ are often negligible, and thus having very accurate values of h_i and h_0 is usually not necessary.

When constructing refrigerated rooms, it is desirable to use *effective* insulation materials to *minimize* the refrigerated space for a fixed floor area. Minimum insulation thicknesses for expanded polyurethane ($k = 0.023$ W/m · °C) recommended by the refrigeration industry are given in Table 13-11. Note that the *larger* the temperature difference between the refrigerated space and the ambient air, the *thicker* the insulation required to reduce the transmission heat gain to reasonable levels. *Equivalent thicknesses* for other insulating materials can be obtained by multiplying the values in this table by k_{ins}/k_{poly} where k_{ins} and k_{poly} are the thermal conductivities of the insulation material and the polyurethane, respectively.

Direct exposure to the *sun* increases the refrigeration load of a refrigerated room as a result of the solar energy absorbed by the outer surface being conducted into the refrigerated space. The effect of *solar heating* can conveniently be accounted for by adding a few degrees to the ambient temperature. For example, the solar heating effect can be compensated for by adding 4°C to the ambient temperature for the east and west walls, 3°C for the south walls, and 9°C for flat rooms with medium-colored surfaces such as unpainted wood, brick and dark cement (Fig. 13-42). For dark- (or light-) colored surfaces, we should add (or subtract) 1°C to (or from) these values.

2 Infiltration Load

The heat gain due to the surrounding warm air entering the refrigerated space through the cracks and the open doors continues the *infiltration load* of the refrigeration system (Fig. 13-43). The infiltration load changes with time. We should consider the *maximum value* to properly size the refrigeration system, and the *average value* to property estimate the average energy consumption. In installations that require the doors to remain open for long periods of time, such as distribution warehouses, the infiltration load may amount to more than half of the total refrigeration load.

In the absence of any *winds,* the infiltration is due to the *density difference* between the cold air in the refrigerated space and the surrounding warmer air. When the average velocity of the air entering the refrigerated space under the influence of winds or pressure differentials is known, the rate of infiltration heat gain can be determined from (ASHRAE, *Handbook: Refrigeration*, Ref. 3, Chap. 26, p. 5):

$$\dot{Q}_{\text{infiltration, windy air}} = 6\mathcal{V}_{\text{air}}\, A_{\text{leak}}\, C_p (T_{\text{warm}} - T_{\text{cold}})\, \rho_{\text{cold}}\, D_{\text{time}} \quad (13\text{-}11)$$

where
$\mathcal{V}_{\text{air}}$ = average air velocity, usually between 0.3 and 1.5 m/s
A_{leak} = smaller of the inflow or outflow opening areas, m^2
$C_{p,\text{air}}$ = specific heat of air, 1.0 kJ/kg · °C
T_{warm} = temperature of the warm infiltration air, °C
T_{cold} = temperature of the cold refrigerated air, °C
ρ_{cold} = density of the cold refrigerated air, kg/m^3
D_{time} = doorway open-time factor

There is considerable uncertainty in the determination of the infiltration load, and field experience from similar installations can be a valuable aid in calculations. A practical way of determining the infiltration load is to estimate the rate of air infiltration in terms of **air changes per hour** (ACH), which is the number of times the entire air content of a room is replaced by the infiltrating air per hour. Once the number of air changes per hour is estimated, the *mass flow rate* of air infiltrating into the room is determined from

$$\dot{m}_{\text{air}} = \frac{V_{\text{room}}}{v_{\text{room}}}\, \text{ACH} \qquad (\text{kg/h}) \qquad (13\text{-}12)$$

where V_{room} is the volume of the room and v_{room} is the specific volume of the dry air in the room. Once $\dot{m}_{\text{air}}$ is available, the sensible and latent infiltration loads of the cold storage room can be determined from

$$\dot{Q}_{\text{infiltration, sensible}} = \dot{m}_{\text{air}}(h_{\text{ambient}} - h_{\text{room}}) \qquad (\text{kJ/h}) \qquad (13\text{-}13)$$

$$\dot{Q}_{\text{infiltration, latent}} = (w_{\text{ambient}} - w_{\text{room}})\, \dot{m}_{\text{air}}\, h_{fg} \qquad (\text{kJ/h}) \qquad (13\text{-}14)$$

where w is the humidity ratio of air (the mass of water vapor in 1 kg of dry air), h is the enthalpy of air, and h_{fg} is the heat of vaporization of water. The values of specific volume v, enthalpy h, and the humidity ratio w can be determined from the *psychrometric chart* when the temperature and relative humidity (or wet-bulb temperature) of air are specified. The total infiltration load is then the sum of the sensible and latent components.

The value of ACH should be determined under the most adverse operating conditions (such as during loading or unloading, and high winds) to ensure satisfactory performance under those conditions.

3 Product Load

The *heat removed* from the food products as they are cooled to refrigeration temperature and the heat released as the fresh fruits and vegetables respire in storage constitute the **product load** of the refrigeration system (Fig. 13-44).

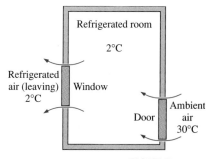

FIGURE 13-43

The warm air that enters a refrigerated space finds its way out after it is cooled, and the heat removed from the air constitutes the infiltration load.

FIGURE 13-44

The heat removed from the food products as they are cooled constitutes the product load of a refrigeration system.

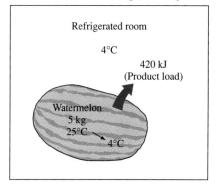

As we mentioned earlier, the *refrigeration* of foods involves cooling only, but the *freezing* of foods involves cooling to the freezing point (removing the sensible heat), freezing (removing the latent heat), and further cooling to the desired subfreezing temperature (removing the sensible heat of frozen food). The three components of the product load can be determined from

$$Q_{\text{cooling, fresh}} = mC_{p,\text{fresh}}\,(T_1 - T_{\text{freeze}}) \qquad \text{(kJ)} \qquad \text{(13-15)}$$

$$Q_{\text{freezing}} = mh_{\text{latent}} \qquad \text{(kJ)} \qquad \text{(13-16)}$$

$$Q_{\text{cooling, frozen}} = mC_{p,\text{frozen}}\,(T_{\text{freeze}} - T_2) \qquad \text{(kJ)} \qquad \text{(13-17)}$$

where

$$m = \text{mass of the food product}$$
$$C_{p,\text{fresh}} = \text{specific heat of the food before freezing}$$
$$C_{p,\text{frozen}} = \text{specific heat of the food after freezing}$$
$$h_{\text{latent}} = \text{latent heat of fusion of the food}$$
$$T_{\text{freeze}} = \text{freezing temperature of the food}$$
$$T_1 = \text{initial temperature of the food (before refrigeration)}$$
$$T_2 = \text{final temperature of the food (after freezing)}$$

The *rate of refrigeration* needed to cool a product from T_1 to T_2 during a time interval Δt can be determined from

$$\dot{Q}_{\text{product}} = \frac{Q_{\text{product, total}}}{\Delta t} = \frac{Q_{\text{cooling, fresh}} + Q_{\text{freezing}} + Q_{\text{cooling, frozen}}}{\Delta t} \qquad \text{(13-18)}$$

Note that it will take a relatively short time for a powerful refrigeration system to accomplish the desired cooling.

Food products are usually refrigerated in their *containers,* and the product load also includes cooling the containers. The amount of heat that needs to be removed from the container as it is cooled from T_1 to T_2 can be determined from

$$Q_{\text{container}} = mC_{p,\text{container}}(T_1 - T_2) \qquad \text{(kJ)} \qquad \text{(13-19)}$$

where m is the mass of the container and $C_{p,\text{container}}$ is the specific heat of the container, which is 0.50 kJ/kg · °C for stainless steel, 1.7 kJ/kg · °C for nylon, 1.9 kJ/kg · °C for polypropylene, and 2.3 kJ/kg · °C for polyethylene containers. The contribution of $Q_{\text{container}}$ to the refrigeration load is usually very small and can be neglected.

4 Internal Load

The heat generated by the *people, lights, electric motors,* and other *heat-dissipating equipment* in the refrigerated space constitutes the **internal load** of the refrigeration system (Fig. 13-45). The rate of heat dissipation by *people* depends on the size of the person, the temperature of the refrigerated space, the activity level, and the clothing, among other things. A person must generate more heat at lower temperatures to compensate for the increased rate of

FIGURE 13-45

The heat generated by the people, lights, electric motors, and other heat-dissipating equipment in a refrigerated space constitutes the internal load.

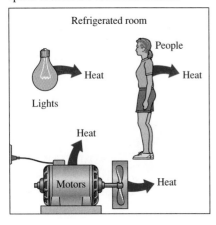

Refrigerated room

Lights

People

Heat

Heat

Heat

Motors

Heat

heat transfer at higher temperature differences. The *heat dissipated* by an *average person* in a refrigerated space maintained at temperature T in °C is expressed as (ASHRAE, *Handbook: Refrigeration*, Ref. 3, Chap. 26, p. 3)

$$\dot{Q}_{people} = 270 - 6T(°C) \qquad \text{(W/person)} \qquad (13\text{-}20)$$

Therefore, an average person will dissipate 210 W of heat in a space maintained at 10°C and 360 W in a space at −15°C.

The rate of heat dissipation from *lights* is determined by simply adding the wattage of the light bulbs and the fluorescent tubes. For example, five 100-W incandescent light bulbs and eight 40-W fluorescent tubes contribute 820 W to the refrigeration load when they all are on.

The calculation of the heat dissipation from the *motors* is more complicated because of the uncertainties involved in the operation such as the *motor efficiency,* the *load factor* (the fraction of the rated load at which the motor normally operates), and the *utilization factor* (the fraction of the time during which the motor actually runs), and whether the motor running a device such as a fan is located inside the refrigerated space. Noting that the motors are usually oversized, the rated power of a motor listed on its tag tells us little about its contribution to the refrigeration load. When the body of a motor running a device is housed *outside* the refrigerated space, then the internal heat load of this motor is simply the power consumed by the device in the refrigerated space. But when the motor is housed *inside* the refrigerated space, then the heat dissipated by the motor also becomes part of the internal heat load since this heat now must be removed from the refrigeration system. Keeping the motors inside the refrigerated space may increase the internal load due to motors by 30 to 80 percent.

5 Refrigeration Equipment Load

The *refrigeration equipment load* refers to the heat generated by the refrigeration equipment itself as it performs certain tasks such as *circulating* the cold air with a fan, *electric reheating* to prevent condensation on the surfaces of the refrigerator, and *defrosting* to prevent frost build-up and to evaporate moisture. Equipment load can be as little as 5 percent of the total refrigeration load for simple refrigeration systems or it may exceed 15 percent for systems maintained at very low temperatures. When the total refrigeration load is determined as described above, it is common practice to apply a safety factor of 10 percent to serve as a cushion to cover any unexpected situations.

EXAMPLE 13-7 Freezing of Chicken

A supply of 50 kg of chicken at 6°C contained in a box is to be frozen to −18°C in a freezer (Fig. 13-46). Determine the amount of heat that needs to be removed. The container box is 1.5 kg, and the specific heat of the box material is 1.4 kJ/kg · °C.

Solution   A box of chicken is to be frozen in a freezer. The amount of heat removal is to be determined.

FIGURE 13-46

Schematic for Example 13-7.

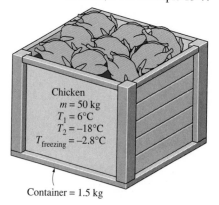

Chicken
$m = 50$ kg
$T_1 = 6°C$
$T_2 = -18°C$
$T_{freezing} = -2.8°C$

Container = 1.5 kg

Assumptions **1** The thermal properties of fresh and frozen chicken are constant. **2** The entire water content of chicken freezes during the process.

Properties For chicken, the freezing temperature is −2.8°C, the latent heat of fusion is 247 kJ/kg, the specific heat is 3.32 kJ/kg · °C above freezing and 1.77 kJ/kg · °C below freezing (Table A-7a).

Analysis The total amount of heat that needs to be removed (the cooling load of the freezer) is the sum of the latent heat and the sensible heats of the chicken before and after freezing as well as the sensible heat of the box, and is determined as follows:

Cooling fresh chicken from 6°C to −2.8°C:

$$Q_{cooling, fresh} = (mC\Delta T)_{fresh} = (50 \text{ kg})(3.32 \text{ kJ/kg} \cdot °C)[6 - (-2.8)]°C = 1461 \text{ kJ}$$

Freezing chicken at −2.8°C:

$$Q_{freezing} = mh_{latent} = (50 \text{ kg})(247 \text{ kJ/kg}) = 12{,}350 \text{ kJ}$$

Cooling frozen chicken from −2.8°C to −18°C:

$$Q_{cooling, frozen} = (mC\Delta T)_{frozen}$$
$$= (50 \text{ kg})(1.77 \text{ kJ/kg} \cdot °C)[-2.8 - (-18)]°C = 1345 \text{ kJ}$$

Cooling the box from 6°C to −18°C:

$$Q_{box} = (mC\Delta T)_{box} = (1.5 \text{ kg})(1.4 \text{ kJ/kg} \cdot °C)[6 - (-18)]°C = 50 \text{ kJ}$$

Therefore, the total amount of cooling that needs to be done is

$$Q_{total} = Q_{cooling, fresh} = Q_{freezing} + Q_{cooling, frozen} + Q_{box}$$
$$= 1461 + 12{,}350 + 1345 + 50 = \textbf{15{,}206 kJ}$$

Discussion Note that most of the cooling load (81 percent of it) is due to the removal of the latent heat during the phase change process. Also note that the cooling load due to the box is negligible (less than 1 percent), and can be ignored in calculations.

FIGURE 13-47

Schematic for Example 13-8.

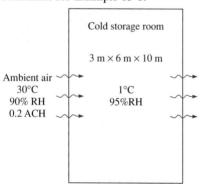

EXAMPLE 13-8 Infiltration Load of Cold Storage Rooms

A cold storage room whose internal dimensions are 3 m × 6 m × 10 m is maintained at 1°C and 95 percent relative humidity (Fig. 13-47). Under worst conditions, the amount of air infiltration is estimated to be 0.2 air changes per hour at the ambient conditions of 30°C and 90 percent relative humidity. Using the psychrometric chart, determine the total infiltration load of this room under these conditions.

Solution The infiltration rate of a cold storage room is given to be 0.2 ACH. The total infiltration load of the room is to be determined.

Assumptions The entire infiltrating air is cooled to 1°C before it exfiltrates.

Properties The heat of vaporization of water at the average temperature of 15°C is 2466 kJ/kg (Table A-9). The properties of the cold air in the room and the ambient air are determined from the psychrometric chart (Fig. A-13) to be

$$T_{ambient} = 30°C \atop \phi_{ambient} = 90\%} \quad {w_{ambient} = 0.024 \text{ kg/kg dry air} \atop h_{ambient} = 93 \text{ kJ/kg dry air}}$$

$$T_{room} = 1°C \atop \phi_{room} = 95\%} \quad {w_{room} = 0.0038 \text{ kg/kg dry air} \atop h_{room} = 11 \text{ kJ/kg dry air} \atop v_{room} = 0.780 \text{ m}^3/\text{kg dry air}}$$

Analysis Noting that the infiltration of ambient air will cause the air in the cold storage room to be changed 0.2 times every hour, the air will enter the room at a mass flow rate of

$$\dot{m}_{air} = \frac{V_{room}}{v_{room}} \text{ACH} = \frac{3 \times 6 \times 10 \text{ m}^3}{0.780 \text{ m}^3/\text{kg dry air}} (0.2 \text{ h}^{-1}) = 46.2 \text{ kg/h} = 0.0128 \text{ kg/s}$$

Then the sensible and latent infiltration heat gain of the room become

$$\dot{Q}_{infiltration, sensible} = \dot{m}_{air}(h_{ambient} - h_{room}) = (0.0128 \text{ kg/s})(93 - 11)\text{kJ/kg} = 1.05 \text{ kW}$$

$$\dot{Q}_{infiltration, latent} = (w_{ambient} - w_{room})\dot{m}_{air}h_{fg}$$
$$= (0.024 - 0.0038)(0.0128 \text{ kg/s})(2466 \text{ kJ/kg}) = 0.64 \text{ kW}$$

Therefore,

$$\dot{Q}_{infiltration} = \dot{Q}_{infiltration, sensible} + \dot{Q}_{infiltration, latent} = 1.05 + 0.64 = \mathbf{1.69 \text{ kW}}$$

That is, the refrigeration system of this cold storage room must be capable of removing heat at a rate of 1.69 kJ/s to meet the infiltration load. Of course, the total refrigeration capacity of the system will have to be larger to meet the transmission, product, and other loads as well.

13-8 ■ TRANSPORTATION OF REFRIGERATED FOODS

Perishable food products are transported by *trucks and trailers, railroad cars, ships, airplanes,* or a *combination* of them from production areas to distant markets. Transporting some farm products to a nearby market by a small truck may not require any special handling, but transporting large quantities over long distances usually requires strict climate control by refrigeration and adequate ventilation (Fig. 13-48).

Refrigerated *trucks* and *trailers* can be up to 17 m long, 2.6 m wide, and 4.3 m high, and they may have several compartments held at different temperatures. The body of the truck should be lightweight but stiff with good insulating characteristics. Using a *thick* insulation layer on walls reduces the rate of *heat gain* into the refrigerated space, but it also reduces the available *cargo space* since the outer dimensions of the trucks and trailers are limited. Urethane foam is commonly used as an insulating material because of its low thermal conductivity ($k = 0.026$ W/m · °C). The *thickness* of polyurethane insulation used is in the range of 7.5 to 10 cm for *freezer trucks* maintained at −18°C or lower, and 2.5 to 6.5 cm for *refrigerated trucks* maintained above

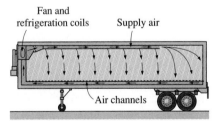

FIGURE 13-48

Airflow patterns in a refrigerated trailer.

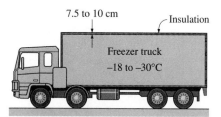

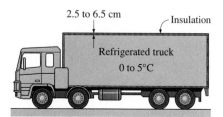

FIGURE 13-49

Proper thickness of polyurethane
insulation for the walls of refrigerated
and freezer trucks.

freezing temperatures (Fig. 13-49). The outer surface of the truck must be vapor proof to prevent water vapor from entering the insulation and condensing within the insulation.

Air entering inside the refrigerated space of the truck through cracks represents a significant heat gain since the *moisture* in the air will be cooled inside and condensed, releasing its latent heat. The *large temperature difference* between the ambient air and the interior of the refrigerated truck serves as a driving force for air infiltration. Air leakage is also caused by the *ram effect* of air during motion. It can be shown from Bernoulli's equation that a truck moving at a velocity of V will generate a ram air pressure of $\Delta P = \rho V^2/2$ at the front section (ρ is the density of air). This corresponds to a *pressure rise* of about 300 Pa on the surface of the truck at a velocity of 80 km/h. Road tests have shown that air leakage rates for large trailers can reach 12.5 L/s at 80 km/h. Further, the water vapor in the air that condenses inside the refrigerated space may cause *corrosion, damage to insulation, rotting,* and *odors.*

Products with a *large heat capacity* such as milk and orange juice can be transported safely without refrigeration by subcooling them before loading into the truck. For example, orange juice shipped from Florida to New York (a distance of 2300 km) warmed up only 5°C without any refrigeration. *Liquid nitrogen, dry ice,* or *ordinary ice* can also be used for refrigeration in trucks. The cargo space must be maintained at −18°C for most frozen products and −29°C for other deep-frozen products such as ice cream.

The rate of heat gain by properly insulated trailers that are 11 to 16 m long range from 40 to 130 W/°C under steady conditions. But on a sunny day the *solar radiation* can increase this number by more than 20 percent. The rate of heat gain can also increase significantly as a result of the product cooling load, air infiltration, and opening of the door during loading and unloading.

The interior of refrigerated trucks should be *precooled* to the desired temperature before loading. The fruits and vegetables should also be cooled to the proper storage temperature before they are loaded into the refrigerated railcars or trucks. This is because the refrigerated cars or trucks do not have the additional refrigeration capacity to cool the products fast enough. Also, the products must be protected from freezing in winter. Refrigerated cars or trucks provide *heating* in cold weather by electric heating or by operating the refrigeration system in reverse as a heat pump. Switching from cooling to heating is done automatically.

The heating and cooling processes in refrigerated cars and trucks are well automated, and all the operator needs to do is set the thermostat at the desired level. The system then provides the necessary refrigeration (or heating in cold weather) to maintain the interior at the set temperature. Thermostats must be calibrated regularly to ensure reliable operation since a deviation of a few degrees from the intended temperature setting may cause the products to be damaged by freezing or excessive decay.

Food products are normally loaded into cars or trucks in *containers* that can be handled by people or forklifts. The containers must allow for heat exchange while protecting the products. The containers must allow for air circulation through channels for uniform and effective cooling and ventilation.

The *railroad industry* also provides refrigerated car service for the transportation of perishable foods. The walls, floors, and ceiling of the cars are constructed with a minimum of 7.5-cm-thick insulation. Typical rates of heat gain are 132 W/°C for 15-m-long cars and 158 W/°C for 18-m-long cars (Fig. 13-50). Other aspects of transportation are similar to the trucks.

Perishable foods are also transported economically from one port to another by *refrigerated ships.* Some ships are designed specifically to transport fruits and vegetables, while others provide refrigerated sections for the transport of foods in various sizes of specially built containers that facilitate faster loading and unloading. Containers used to carry frozen food are insulated with at least 7.5-cm-thick urethane foam insulation. In fishing vessels, ice picked up from the port is commonly used to preserve fish and seafood for days with little or no additional refrigeration. Some fishing vessels cool the seawater to about 2°C and then use it to refrigerate the fish.

Perishable food products are also being transported *by air* in the cargo sections of passenger planes or in specifically designed cargo planes. As you would expect, air transport is more expensive, but the extra cost is justified for commodities that can command a high market price. Also, for some highly perishable products, the only means of transportation is air transportation. *Papaya* and *flowers,* for example, are transported from Hawaii to the mainland almost always by air. Likewise, *flowers* are transported to distant markets mostly by air. *Strawberries* are being transported anywhere in the world again by air at an increasing rate. *Fresh meats, seafood,* and *early season fruits and vegetables* are also commonly shipped by air. Many inland restaurants are able to serve high-quality fresh fish and seafood because of air transport. Air transportation may become economical for many perishable food items at times of short supply and thus higher prices (Fig. 13-51).

Air transport is completed in a matter of hours, and it is not practical to equip the airplanes with powerful but heavy refrigeration systems for such short periods. Airplane refrigeration is primarily based on *precooling* prior to loading, and the prevention of *temperature rise* during transit. This is achieved by precooling the products in a refrigerated storage facility; transporting them in refrigerated trucks; storing them in a refrigerated warehouse at the airport, if necessary; and transporting the products in well-insulated containers. Of course, quick loading and unloading are also important to minimize exposure to adverse conditions. Sometimes dry ice is used to provide temporary refrigeration to the container.

Air transport also offers some *challenges* to refrigeration engineers. For example, the outer surface temperature of an airplane cruising at a Mach number of 0.9 increases by about 28°C over the ambient. Also, the relative humidity of the cabin of an airplane cruising at high altitudes is very low, typically under 10 percent. However, the humidity level can still be maintained at about 90 percent without adding any water, thanks to the moisture released during respiration from fruits and vegetables.

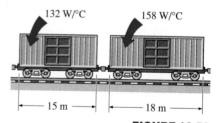

132 W/°C   158 W/°C

|← 15 m →| |← 18 m →|

FIGURE 13-50

Typical rates of heat gain by 15- and 18-m-long refrigerated railcars.

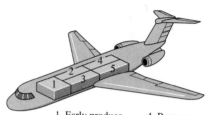

1. Early produce   4. Papayas
2. Seafood   5. Flowers
3. Strawberries

FIGURE 13-51

High-priced food products are commonly transported by air economically.

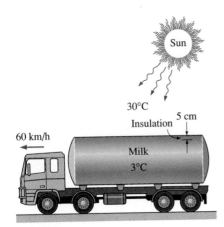

FIGURE 13-52

Schematic for Example 13-9.

EXAMPLE 13-9   Interstate Transport of Refrigerated Milk by Trucks

Milk is to be transported from Texas to California for a distance of 2100 km in a
7-m-long, 2-m-external-diameter cylindrical tank (Fig. 13-52). The walls of the tank
are constructed of 5-cm-thick urethane insulation ($k = 0.029$ W/m · °C) sandwiched between two metal sheets. The milk is precooled to 3°C before loading
and its temperature is not to exceed 7°C on arrival. The average ambient temperature can be as high as 30°C, and the effect of solar radiation and the radiation
from hot pavement surfaces can be taken to be equivalent to a 4°C rise in ambient
temperature. If the average velocity during transportation is 60 km/h, determine if
the milk can be transported without any refrigeration.

Solution   Milk is to be transported in a cylindrical tank by a truck. It is to be
determined if the milk can be transported without any refrigeration.

Assumptions   Thermal properties of milk and insulation are constant.

Properties   The thermal conductivity of urethane is given to be $k = 0.029$
W/m · °C. The density and specific heat of refrigerated milk at temperatures near
0°C are $\rho_{milk} \cong \rho_{water} \cong 1000$ kg/m³ and $C_{p, milk} \cong 3.79$ kJ/kg · °C (Table A-7).

Analysis   Problems of this kind that involve "checking" are best solved by performing the calculations under the worst conditions, with the understanding that
if the performance is satisfactory under those conditions, it will surely be satisfactory under any conditions.

We take the average ambient temperature to be 30°C, which is the highest
possible, and raise it to 34°C to account for the radiation from the sun and the
pavement. We also assume the metal sheets to offer no resistance to heat transfer and assume the convection resistances on the inner and outer sides of the
tank wall to be negligible. Under these conditions, the inner and outer surface
temperatures of the insulation will be equal to the milk and ambient temperatures,
respectively. Further, we take the milk temperature to be 3°C during heat transfer
calculations (to maximize the temperature difference) and the heat transfer area
to be the outer surface area of the tank (instead of the smaller inner surface
area), which is determined to be

$$A = 2A_{base} + A_{side} = 2(\pi D_o^2/4) + (\pi D_o)L_o$$
$$= 2\pi(2 \text{ m})^2/4 + \pi(2 \text{ m})(7 \text{ m}) = 50.3 \text{ m}^2$$

Then the rate of heat transfer through the insulation into the milk becomes

$$\dot{Q} = k_{ins}A\frac{\Delta T_{ins}}{L_{ins}} = (0.029 \text{ W/m} \cdot \text{°C})(50.3 \text{ m}^2)\frac{(34-3)\text{°C}}{0.05 \text{ m}} = 904.4 \text{ W}$$

This is the highest possible rate of heat transfer into the milk since it is determined
under the most favorable conditions for heat transfer.

At an average velocity of 60 km/h, transporting the milk 2100 km will take

$$\Delta t = \frac{\text{Distance traveled}}{\text{Average velocity}} = \frac{2100 \text{ km}}{60 \text{ km/h}} = 35 \text{ h}$$

Then the total amount of heat transfer to the milk during this long trip becomes

$$Q = \dot{Q}\Delta t = (904.4 \text{ J/s})(35 \times 3600 \text{ s}) = 113{,}950{,}000 \text{ J} = 113{,}950 \text{ kJ}$$

The volume and mass of the milk in a full tank is determined to be

$$V_{milk} = (\pi D_i^2/4)L_i = [\pi(1.9 \text{ m})^2/4](6.9 \text{ m}) = 19.56 \text{ m}^3$$
$$m_{milk} = \rho V_{milk} = (1000 \text{ kg/m}^3)(19.56 \text{ m}^3) = 19,560 \text{ kg}$$

The transfer of 113,950 kJ of heat into the milk will raise its temperature to

$$Q = mC_p(T_2 - T_1) \quad \rightarrow \quad T_2 = T_1 + \frac{Q}{mC_p}$$

$$= 3°C + \frac{113,950 \text{ kJ}}{(19,560 \text{ kg})(3.79 \text{ kJ/kg} \cdot °C)} = 4.5°C$$

That is, the temperature of the milk will rise from 3 to 4.5°C during this long trip under the most adverse conditions, which is well below the 7°C limit. Therefore, the milk can be transported even longer distances without any refrigeration.

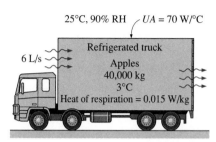

FIGURE 13-53

Schematic for Example 13-10.

EXAMPLE 13-10   Transport of Apples by Refrigerated Trucks

A large truck is to transport 40,000 kg of apples precooled to 3°C under average ambient conditions of 25°C and 90 percent relative humidity (Fig. 13-53). The structure of the walls of the truck is such that the rate of heat transmission is UA = 70 W per °C temperature difference between the ambient and the apples. From past experience, ambient air is estimated to enter the cargo space of the truck through the cracks at a rate of 6 L/s, and the average heat of respiration of the apples at 3°C is 0.015 W/kg for this particular load. Determine the refrigeration load of this truck and the amount of ice needed to meet the entire refrigeration need of the truck for a 20-h-long trip.

Solution   A large truck is to transport 40,000 kg of apples. The refrigeration load of the truck and the amount of ice needed to meet this need during a 20 h long trip is to be determined.

Assumptions   **1** Infiltrating air exits the truck saturated at 3°C. **2** The moisture in the air is condensed out at the exit temperature of 3°C.

Properties   The humidity ratio of air is 0.0180 kg water vapor/kg dry air at 25°C and 90 percent relative humidity, and 0.0047 at 3°C and 100 percent relative humidity (Fig. A-13). The latent heat of vaporization of water at 3°C is 2508 kJ/kg (Table A-9). The density of air at the ambient temperature of 25°C and 1 atm is 1.19 kg/m³, and its specific heat is C_p = 1.00 kJ/kg · °C (Table A-11). The latent heat of ice is 333.7 kJ/kg.

Analysis   The total refrigeration load of the truck is due to the heat gain by transmission, infiltration, and respiration. Noting that UA is given to be 70 W/°C, the rate of heat gain by transmission is determined to be

$$\dot{Q}_{transmission} = UA\Delta T = (70 \text{W/°C})(25 - 3)°C = 1540 \text{ W}$$

The rate of heat generation by the apples as a result of respiration is

$$\dot{Q}_{respiration} = mh_{respiration} = (40,000 \text{ kg})(0.015 \text{ W/kg}) = 600 \text{ W}$$

Ambient air enters the cargo space at a rate of 6 L/s, which corresponds to a mass flow rate of

$$\dot{m}_{air} = \rho \dot{V}_{air} = (1.19 \text{ kg/m}^3)(0.006 \text{ m}^3/\text{s}) = 0.00714 \text{ kg/s}$$

Noting that an equal amount of air at 3°C must leave the truck, the sensible heat gain due to infiltration is

$$\dot{Q}_{\text{infiltration, sensible}} = (\dot{m}C_p\Delta T)_{\text{air}}$$
$$= (0.00714 \text{ kg/s})(1.0 \text{ kJ/kg} \cdot °C)(25 - 3)°C = 0.157 \text{ kJ/s} = 157 \text{ W}$$

Ambient air enters with a humidity ratio of 0.0180 kg H_2O/kg air and leaves at of 0.0047 kg H_2O/kg air. The difference condenses in the truck and is drained out as a liquid, releasing $h_{fg} = 2508$ kJ/kg of latent heat. Then the latent heat gain of the truck becomes

$$\dot{Q}_{\text{infiltration, latent}} = \dot{m}_{\text{water}} h_{fg}$$
$$= (0.0180 - 0.0047)(0.00714 \text{ kg/s})(2508 \text{ kJ/kg}) = 238 \text{ W}$$

Then the refrigeration load, or the total rate of heat gain by the truck, becomes

$$\dot{Q}_{\text{total}} = \dot{Q}_{\text{transmission}} + \dot{Q}_{\text{respiration}} + \dot{Q}_{\text{infiltration}}$$
$$= 1540 + 600 + (157 + 238) = \textbf{2535 W}$$

The total amount of heat gain during the 20-h-long trip is

$$Q_{\text{total}} = \dot{Q}_{\text{total}}\Delta t = (2535 \text{ J/s})(20 \times 3600 \text{ s}) = 182,520,000 \text{ J} = 182,520 \text{ kJ}$$

Then the amount of ice needed to meet this refrigeration load is determined from

$$m_{\text{ice}} = \frac{Q_{\text{total}}}{h_{\text{latent, ice}}} = \frac{182,520 \text{ kJ}}{333.7 \text{ kJ/kg}} = \textbf{547 kg}$$

Discussion Note that about half a ton of ice (1.5 percent of the mass of the load) is sufficient in this case to maintain the apples at 3°C.

13-9 ■ SUMMARY

Refrigeration and freezing of perishable food products are major application areas of heat transfer and thermodynamics. *Microorganisms* such as bacteria, yeasts, molds, and viruses cause off-flavors and odors, slime production, changes in the texture and appearance, and eventual spoilage of foods. The rate of growth of microorganisms in a food item depends on the characteristics of the food itself as well as the environmental conditions, such as the temperature and relative humidity of the environment and the air motion. Refrigeration slows down the chemical and biological processes in foods and the accompanying deterioration. The storage life of fresh perishable foods such as meats, fish, vegetables and fruits can be extended by several days by storing them at temperatures just above freezing, and by several months by storing them at subfreezing temperatures.

Some fruits and vegetables experience undesirable physiological changes when exposed to low (but still above-freezing) temperatures, usually between 0 and 10°C. The resulting tissue damage is called the *chilling injury*. It differs from *freezing injury*, which is caused by prolonged exposure of the fruits and vegetables to subfreezing temperatures and thus the actual freezing at the affected areas. *Dehydration* or *moisture loss* causes a product to shrivel or wrinkle and lose quality. The loss of moisture from fresh fruits and vegetables is also called *transpiration*. Common freezing methods include *air-blast freezing*, where high-velocity air at about $-30°C$ is blown over the food products; *contact freezing*, where packaged or unpackaged food is placed on or between cold metal plates and cooled by conduction; *immersion freezing*, where the food is immersed in low-temperature brine; and *cryogenic freezing*, where food is placed in a medium cooled by a cryogenic fluid such as liquid nitrogen or solid carbon dioxide.

The thermal properties of foods are dominated by their water content. The *specific heats* of foods can be expressed by **Siebel's formula** as

$$C_{p,\text{fresh}} = 3.35a + 0.84 \qquad (\text{kJ/kg} \cdot °C)$$
$$C_{p,\text{frozen}} = 1.26a + 0.84 \qquad (\text{kJ/kg} \cdot °C)$$

where $C_{p,\text{fresh}}$ and $C_{p,\text{frozen}}$ are the specific heats of the food before and after freezing, respectively, and a is the fraction of water content of the food. The *latent heat* of a food product during freezing or thawing (the heat of fusion) also depends on its water content and is determined from

$$h_{\text{latent}} = 334a \qquad (\text{kJ/kg})$$

The heat removed from the food products accounts for the majority of the refrigeration load and is determined from

$$\dot{Q}_{\text{product}} = mC_p(T_{\text{initial}} - T_{\text{final}})/\Delta t \qquad (\text{W})$$

where $\dot{Q}_{\text{product}}$ is the *average rate of heat removal* from the fruits and vegetables; m is the total mass; C_p is the average specific heat; T_{initial} and T_{final} are the average temperatures of the products before and after cooling, respectively; and Δt is the cooling time. Fresh fruits and vegetables are live products, and they continue to respire after harvesting. *Heat of respiration* is released during this exothermic reaction. The refrigeration load due to respiration is determined from

$$\dot{Q}_{\text{respiration}} = \sum m_i \dot{q}_{\text{respiration},i} \qquad (\text{W})$$

which is the sum of the mass times the heat of respiration for all the food products stored in the refrigerated space.

The primary precooling methods for fruits and vegetables are *hydrocooling*, where the products are cooled by immersing them into chilled water; *forced-air cooling*, where the products are cooled by forcing refrigerated air through them; *package icing*, where the products are cooled by placing crushed ice into the containers; and *vacuum cooling*, where the products are cooled by vaporizing some of the water content of the products under low pressure conditions. Most produce can be stored satisfactorily at $0°C$ and 90 to 95 percent relative humidity.

Meat carcasses in slaughterhouses should be cooled as fast as possible to a uniform temperature of about $1.7°C$ to reduce the growth rate of microorganisms that may be present on carcass surfaces, and thus minimize spoilage. About 70 percent of the beef carcass is water, and the carcass is cooled mostly by *evaporative cooling* as a result of moisture migration toward the surface where evaporation occurs. Most meats are judged on their *tenderness,* and the preservation of tenderness is an important consideration in the refrigeration and freezing of meats. *Poultry products* can be preserved by *ice-chilling* to 1 to 2°C or *deep chilling* to about −2°C for short-term storage, or by *freezing* them to −18°C or below for long-term storage. Poultry processing plants are completely automated, and several processes involve strict temperature control. *Fish* is a highly perishable commodity, and the preservation of fish starts on the vessel as soon as it is caught.

Eggs must be refrigerated to preserve quality and to assure a reasonable shelf life. Refrigeration should start at the farm and continue through retail outlets. Shell eggs can be stored at 7 to 13°C and 75 to 80 percent relative humidity for a few weeks. *Milk* is one of the most essential foods for humans, but also one of the most suitable environments for the growth of microorganisms. The stored milk is first *standardized* to bring the milkfat content to desired levels using a milk separator. To kill the bacteria, milk is *pasteurized* in a batch or continuous flow-type system by heating the milk to a minimum temperature of 62.8°C and holding it at that temperature for at least 30 min. The milk is usually *vacuum processed* after pasteurization to reduce the undesirable flavors and odors in it. The milk is also *homogenized* before it is packaged to break up the large fat globules into smaller ones to give it a "homogeneous" appearance. Refrigeration plays an important part in all stages of modern *baker production* from the preservation of raw materials such as flour and yeast to the cooling of baked products.

The *total* rate of heat gain of a refrigerated space through all mechanisms under peak conditions is called the *refrigeration load,* and it consists of (1) *transmission load,* which is heat conducted into the refrigerated space through its walls, floor, and ceiling; (2) *infiltration load,* which is due to surrounding warm air entering the refrigerated space through the cracks and open doors; (3) *product load,* which is the heat removed from the food products as they are cooled to refrigeration temperature; (4) *internal load,* which is heat generated by the lights, electric motors, and people in the refrigerated space; and (5) *refrigeration equipment load,* which is the heat generated by the refrigeration equipment as it performs certain tasks such as reheating and defrosting.

The *rate of heat transfer* through a particular wall, floor, or ceiling section can be determined from

$$\dot{Q}_{\text{transmission}} = UA_0\Delta T \qquad \text{(W)}$$

where A_0 = the outside surface area of the section, ΔT = the temperature difference between the outside air and the air inside the refrigerated space, and U = the overall heat transfer coefficient. The infiltration load can be determined by estimating the rate of air infiltration in terms of *air changes per hour* (ACH), which is the number of times the entire air content of a room is replaced by the infiltrating air per hour. Then the *mass flow rate* of infiltrating air is determined from

$$\dot{m}_{\text{air}} = \frac{V_{\text{room}}}{v_{\text{room}}}\text{ACH} \qquad \text{(kg/h)}$$

where V_{room} is the volume of the room and v_{room} is the specific volume of the dry air in the room. Once $\dot{m}_{air}$ is available, the sensible and latent infiltration loads of the cold storage room can be determined from

$$\dot{Q}_{\text{infiltration, sensible}} = \dot{m}_{air}(h_{\text{ambient}} - h_{\text{room}}) \qquad \text{(kJ/h)}$$

$$\dot{Q}_{\text{infiltration, latent}} = (w_{\text{ambient}} - w_{\text{room}})\dot{m}_{air}h_{fg} \qquad \text{(kJ/h)}$$

where w is the humidity ratio of air (the mass of water vapor in 1 kg of dry air), h is the enthalpy of air, and h_{fg} is the heat of vaporization of water. The total infiltration load is then the sum of the sensible and latent components.

Perishable food products are transported by *trucks and trailers, railroad cars, ships, airplanes,* or a *combination* of them from production areas to distant markets. Transporting large quantities over long distances usually requires strict climate control by refrigeration and adequate ventilation, and adequate insulation to keep the heat transfer rates at reasonable levels.

REFERENCES AND SUGGESTED READING

1. H. B. Ashby, R. T. Hisch, L. A. Risse, W. G. Kindya, W. L. Craig, Jr., and M. T. Turezyn. "Protecting Perishable Foods during Transport by Truck." *Agricultural Handbook* No. 669. Washington, DC: U.S. Department of Agriculture, 1987.

2. ASHRAE, *Handbook of Fundamentals.* SI version. Atlanta, GA: American Society of Heating, Refrigerating, and Air-Conditioning Engineers, Inc., 1993.

3. ASHRAE, *Handbook of Refrigeration.* SI version. Atlanta, GA: American Society of Heating, Refrigerating, and Air-Conditioning Engineers, Inc., 1994.

4. J. C. Ayres, W. C. Ogilvy, and G. F. Stewart. "Postmortem Changes in Stored Meats. Part I, Microorganisms Associated with Development of Slime on Eviscerated and Cut-up Poultry." *Food Technology* 38 (1950), p. 275.

5. R. W. Burley and D. V. Vadehra. *The Avian Egg: Chemistry and Biology.* New York, NY: John Wiley & Sons, 1989.

6. M. D. Carpenter, D. M. Janky, A. S. Arafa, J. L. Oblinger, and J. A. Koburger. "The Effect of Salt Brine Chilling on Drip Loss of Icepacked Broiler Carcasses." *Poultry Science* 58 (1979), p. 369.

7. *Consumers Union. Consumer Reports.* February 94, pp. 80–85.

8. A. J. Farrel and E. M. Barnes. "Bacteriology of Chilling Procedures in Poultry Processing Plants." *British Poultry Science* 5 (1964), p. 89.

9. R. E. Hardenburg, A. E. Watada, and C. Y. Wang. "The Commercial Storage of Fruits, Vegetables, and Florist and Nursery Stocks." *Agricultural Handbook.* No. 66. Washington, DC: U.S. Department of Agriculture, 1986.

10. H. Hillman. *Kitchen Science.* Mount Vernon, NY: Consumers Union, 1981.

11. M. Peric, E. Rossmanith, and L. Leistner. "Verbesserung der Micro-biologischen Qualitat von Schlachthanchen Dursch die Sprunhkuhling." *Fleischwirstschaft,* April 1971, p. 574.

12. W. H. Redit. "Protection of Rail Shipments of Fruits and Vegetables." *Agricultural Handbook,* No. 195. Washington, DC: U.S. Department of Agriculture, 1969.

13. L. Riedel. "Calorimetric Investigation of the Meat Freezing Process." *Kaltetechnik* 9, no. 2 (1957), p. 89.

14. J. V. Spencer and W. J. Stadelman. "Effect of Certain Holding Conditions on Shelf Life of Fresh Poultry Meat." *Food Technology* 9 (1955).

15. L. van der Berg and C. P. Lentz. "Factors Affecting Freezing Rate and Appearance of Eviscerated Poultry Frozen in Air." *Food Technology* 12 (1958).

PROBLEMS*

Control of Microorganisms in Foods

13-1C   What are the common kinds of microorganisms? What undesirable changes do microorganisms cause in foods?

13-2C   How does refrigeration prevent or delay the spoilage of foods? Why does freezing extend the storage life of foods for months?

13-3C   What are the environmental factors that affect the growth rate of microorganisms in foods?

13-4C   What is the effect of cooking on the microorganisms in foods? Why is it important that the internal temperature of a roast in an oven be raised above 70°C?

13-5C   How can the contamination of foods with microorganisms be prevented or minimized? How can the growth of microorganisms in foods be retarded? How can the microorganisms in foods be destroyed?

13-6C   How does (*a*) the air motion and (*b*) the relative humidity of the environment affect the growth of microorganisms in foods?

Refrigeration and Freezing of Foods

13-7C   What is the difference between the freezing injury and the chilling injury of fruits and vegetables?

13-8C   How does the rate of freezing affect the size of the ice crystals that form during freezing and the quality of the frozen food products?

13-9C   What is transpiration? Does lettuce or an apple have a higher coefficient of transpiration? Why?

*Students are encouraged to answer *all* the concept "C" questions.

13-10C What are the mechanisms of heat transfer involved during the cooling of fruits and vegetables by refrigerated air?

13-11C What are the four primary methods of freezing foods?

13-12C What is air-blast freezing? How does it differ from contact freezing of foods?

13-13C What is cryogenic freezing? How does it differ from immersion freezing of foods?

13-14C Which type of freezing is more likely to cause dehydration in foods: air-blast freezing or cryogenic freezing?.

13-15 White potatoes ($k = 0.50$ W/m $\cdot$ °C and $\alpha = 0.13 \times 10^{-6}$ m^2/s) that are initially at a uniform temperature of 25°C and have an average diameter of 6 cm are to be cooled by refrigerated air at 2°C flowing at a velocity of 4 m/s. The average heat transfer coefficient between the potatoes and the air is experimentally determined to be 19 W/m^2 $\cdot$ °C. Determine how long it will take for the center temperature of the potatoes to drop to 6°C. Also, determine if any part of the potatoes will experience chilling injury during this process.

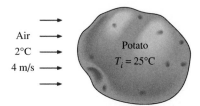

Air
2°C
4 m/s

Potato
$T_i = 25$°C

FIGURE P13-15

13-16E Oranges of 2.5-in. diameter ($k = 0.26$ Btu/h $\cdot$ ft $\cdot$ °F and $\alpha = 1.4 \times 10^{-6}$ ft^2/s) initially at a uniform temperature of 78°F are to be cooled by refrigerated air at 25°F flowing at a velocity of 1 ft/s. The average heat transfer coefficient between the oranges and the air is experimentally determined to be 4.6 Btu/h $\cdot$ ft^2 $\cdot$ °F. Determine how long it will take for the center temperature of the oranges to drop to 40°F. Also, determine if any part of the oranges will freeze during this process.

13-17 A 65-kg beef carcass ($k = 0.47$ W/m $\cdot$ °C and $\alpha = 0.13 \times 10^{-6}$ m^2/s) initially at a uniform temperature of 37°C is to be cooled by refrigerated air at -6°C flowing at a velocity of 1.8 m/s. The average heat transfer coefficient between the carcass and the air is 22 W/m^2 $\cdot$ °C. Treating the carcass as a cylinder of diameter 24 cm and height 1.4 m and disregarding heat transfer from the base and top surfaces, determine how long it will take for the center temperature of the carcass to drop to 4°C. Also, determine if any part of the carcass will freeze during this process. *Answer:* 14.0 h

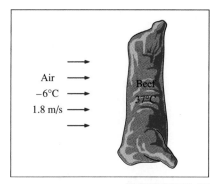

Air
-6°C
1.8 m/s

Beef
37°C

FIGURE P13-17

13-18 Layers of 23-cm-thick meat slabs ($k = 0.47$ W/m $\cdot$ °C and $\alpha = 0.13 \times 10^{-6}$ m^2/s) initially at a uniform temperature of 7°C are to be frozen by refrigerated air at -30°C flowing at a velocity of 1.4 m/s. The average heat transfer coefficient between the meat and the air is 20 W/m^2 $\cdot$ °C. Assuming the size of the meat slabs to be large relative to their thickness, determine how long it will take for the center temperature of the slabs to drop to -18°C. Also, determine the surface temperature of the meat slab at that time.

13-19E Layers of 6-in.-thick meat slabs ($k = 0.26$ Btu/h $\cdot$ ft $\cdot$ °F and $\alpha = 1.4 \times 10^{-6}$ ft^2/s) initially at a uniform temperature of 50°F are cooled by refrigerated air at 23°F to a temperature of 36°F at their center in 12 h. Estimate the average heat transfer coefficient during this cooling process. *Answer:* 1.5 Btu/h $\cdot$ ft^2 $\cdot$ °F

Chicken
1.7 kg

Brine –10°C

FIGURE P13-20

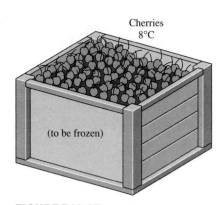

Cherries
8°C

(to be frozen)

FIGURE P13-27

13-20 Chickens with an average mass of 1.7 kg (k = 0.45 W/m · °C and α = 0.13 × 10^{-6} m^2/s) initially at a uniform temperature of 15°C are to be chilled in agitated brine at -10°C. The average heat transfer coefficient between the chicken and the brine is determined experimentally to be 440 W/m^2 · °C. Taking the average density of the chicken to be 0.95 g/cm^3 and treating the chicken as a spherical lump, determine the center and the surface temperatures of the chicken in 2 h and 30 min. Also, determine if any part of the chicken will freeze during this process.

Thermal Properties of Foods

13-21C Explain how to determine the latent heat of fusion of food products whose water content is known.

13-22C Whose specific heat is greater: apricots with a water content of 70 percent or apples with a water content of 82 percent?

13-23C Do carrots freeze at a fixed temperature or over a range of temperatures? Explain.

13-24C Consider 1 kg of cherries with a water content of 75 percent and 1 kg of roast beef also with a water content of 75 percent, both at 5°C. Now both the cherries and the beef are completely frozen to -40°C. The heat removed from the cherries will be (a) less than, (b) about equal to, or (c) greater than the heat removed from the beef.

13-25C Consider 1 kg of carrots with a water content of 65 percent and 1 kg of chicken also with a water content of 65 percent. Now both the carrots and the chicken are cooled from 12°C to 3°C. The heat removed from the carrots will be (a) less than, (b) about equal to, or (c) greater than the heat removed from the chicken.

13-26 A 35-kg box of beef at 6°C having a water content of 60 percent is to be frozen to a temperature of -20°C in 3 h. Using data from Figure 13-13, determine (a) the total amount of heat that must be removed from the beef, (b) the amount of unfrozen water in beef at -20°C, and (c) the average rate of heat removal from the beef.

13-27 A 50-kg box of sweet cherries at 8°C having a water content of 77 percent is to be frozen to a temperature of -20°C. Using enthalpy data from Table 13-4, determine (a) the total amount of heat that must be removed from the cherries and (b) the amount of unfrozen water in cherries at -20°C.

13-28 A 2-kg box made of polypropylene (C_p = 1.9 kJ/kg · °C) contains 40 kg of cod fish with a water content of 80.3 percent (by mass) at 18°C. The fish is to be frozen to an average temperature of -18°C in 2 h in its box. The enthalpy of the fish is given to be 323 kJ/kg at 0°C and 47 kJ/kg at -18°C. It is also given that the unfrozen water content of the fish at -18°C is 12 percent. Taking the average specific heat of the fish above freezing temperatures to be C_p = 3.69 kJ/kg · °C, determine (a) the total amount of heat that must be removed from the fish and its container, (b) the amount of

unfrozen water in fish at −18°C, and (c) the average rate of heat removal.
Answer: (a) 13,830 kJ, (b) 4.8 kg, (c) 1.92 kW

13-29E A 5-lbm container made of stainless steel (C_p = 0.12 Btu/lbm · °F) contains 90 lbm of applesauce with a water content of 82.8 percent (by mass) at 77°F. The applesauce is to be frozen to an average temperature of 7°F in its container. The enthalpy of the applesauce is given to be 147.5 Btu/lbm at 32°F and 31.4 Btu/lbm at 7°F. It is also given that the unfrozen water content of the applesauce at 7°F is 14 percent. Taking the average specific heat of the applesauce above freezing temperatures to be C_p = 0.89 Btu/lbm · °F, determine (a) the total amount of heat that must be removed from the applesauce and its container and (b) the amount of unfrozen water in applesauce at 7°F.

13-30 Fresh strawberries with a water content of 89.3 percent (by mass) at 26°C are stored in 0.8-kg boxes made of polyethylene (C_p = 2.3 kJ/kg · °C). Each box contains 25 kg of strawberries, and the strawberries are to be frozen to an average temperature of −16°C at a rate of 80 boxes per hour. The enthalpy of the strawberries is given to be 367 kJ/kg at 0°C and 54 kJ/kg at −16°C. Taking the average specific heat of the strawberries above freezing temperatures to be C_p = 3.94 kJ/kg · °C, determine the rate of heat removal from the strawberries and their boxes, in kJ/h. *Answer:* 232.5 kW

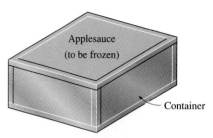

FIGURE P13-29E

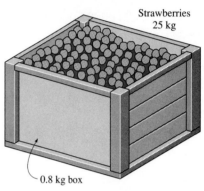

FIGURE P13-30

Refrigeration of Fruits, Vegetables, and Cut Flowers

13-31C What is precooling? Why is it commonly practiced for fruits and vegetables?

13-32C What are the primary cooling methods of fruits and vegetables?

13-33C What is the heat of respiration of fruits and vegetables?

13-34C How does hydrocooling differ from forced-air cooling of fruits and vegetables with respect to (a) cooling time and (b) moisture loss.

13-35C What is the operating principle of vacuum cooling of fruits and vegetables? How can moisture loss of fruits and vegetables be minimized during vacuum cooling?

13-36C What is modified atmosphere? How does modified atmosphere extend the storage life of some fruits and vegetables?

13-37C Why are apples, cucumbers, and tomatoes not suitable for vacuum cooling?

13-38C Why is it not recommended to store bananas below 13°C while it is highly recommended that apples be stored at −1°C?

13-39 A typical full-carlot-capacity banana room contains 36 pallets of bananas. Each pallet consists of 24 boxes, and thus the room stores 864 boxes of bananas. A box holds an average of 19 kg of bananas and is made of 2.3 kg of fiberboard. The specific heats of banana and the fiberboard are 3.55 kJ/kg · °C and 1.7 kJ/kg · °C, respectively. The peak heat of respiration

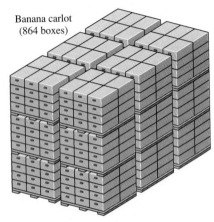

Banana carlot
(864 boxes)

FIGURE P13-39

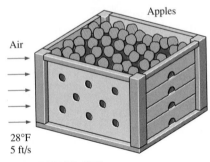

Apples

Air

28°F
5 ft/s

FIGURE P13-43E

of bananas is 0.3 W/kg. The bananas are cooled at a rate of 0.4°C/h. The rate of heat gain through the walls and other surfaces of the room is estimated to be 1800 kJ/h. If the temperature rise of refrigerated air is not to exceed 2.0°C as it flows through the room, determine the minimum flow rate of air needed. Take the density and specific heat of air to be 1.2 kg/m^3 and 1.0 kJ/kg · °C, respectively.

13-40   It is claimed that fruits and vegetables are cooled by 6°C for each percentage point of weight loss as moisture during vacuum cooling. Using calculations, demonstrate if this claim is reasonable.

13-41   Using Figure 13-16, determine how long it will take to cool 6-cm-diameter peaches from 30°C to an average temperature of 5°C by chilled water at 2°C. Compare your result with that obtained from transient one-term solutions in Chapter 4 using a heat transfer coefficient of 550 W/m^2 · °C. Also, determine how many metric tons of peaches can be cooled per day during a 10-h work day by a hydrocooling unit with a refrigeration capacity of 120 tons (1 ton of refrigeration = 211 kJ/min).

13-42   Using Figure 13-16, determine how long it will take to cool 7-cm-diameter apples from 28°C to an average temperature of 4°C by chilled water at 1.5°C. Compare your result with that obtained from transient one-term solutions in Chapter 4 using a heat transfer coefficient of 540 W/m^2 · °C. Also, determine how many metric tons of apples can be cooled per day during a 10-h work day by a hydrocooling unit with a refrigeration capacity of 80 tons (1 ton of refrigeration = 211 kJ/min).
       Answers: 0.39 h, 155.4 ton/day

13-43E   Apples (ρ = 52.4 lbm/ft^3, C_p = 0.91 Btu/lbm · °F, k = 0.242 Btu/h · ft · °F, and α = 1.47 × 10^{-6} ft^2/s) with a diameter of 2.5 in. and an initial temperature of 80°F are to be cooled to an average temperature of 38°F with air at 28°F that approaches the cooling section at a velocity of 5 ft/s. The average heat transfer coefficient between the apples and the air is estimated to be 7.8 Btu/h · ft^2 · °F, and it is recommended that air be supplied at a rate of 0.07 ft^3/s per kg of apples. The rectangular cooling section has a capacity of 10,000 lbm of apples with a porosity of 0.38 (i.e., the voids between the apples comprise 38 percent of the total volume). Disregarding the heat of respiration, determine (*a*) the volume of the cooling section where the apples are placed and its dimensions, (*b*) the amount of total heat transfer from a full load of apples to the cooling air, and (*c*) how long it will take for the center temperature of the apples to drop to 40°F.

13-44   Fresh strawberries with a water content of 88 percent (by mass) at 30°C are stored in 0.8-kg boxes made of nylon (C_p = 1.7 kJ/kg · °C). Each box contains 23 kg of strawberries, and the strawberries are to be cooled to an average temperature of 4°C at a rate of 60 boxes per hour. Taking the average specific heat of the strawberries to be C_p = 3.89 kJ/kg · °C and the average rate of heat of respiration to be 210 mW/kg, determine the rate of heat removal from the strawberries and their boxes, in kJ/h. What would be the percent error involved if the strawberry boxes were ignored in calculations?

13-45 Lettuce is to be vacuum cooled from the environment temperature of 24°C to a temperature of 2°C in 45 min in a 4-m outer diameter insulated spherical vacuum chamber whose walls consist of 3-cm-thick urethane insulation ($k = 0.020$ W/m · °C) sandwiched between metal plates. The vacuum chamber contains 5000 kg of lettuce when loaded. Disregarding any heat transfer through the walls of the vacuum chamber, determine (*a*) the final pressure in the vacuum chamber and (*b*) the amount of moisture removed from the lettuce, in kg. It is claimed that the error involved in (*b*) due to neglecting heat transfer through the wall of the chamber is less than 2 percent. Is this a reasonable claim?   *Answers:* (*a*) 0.714 kPa, (*b*) 179 kg

13-46 Spinach is to be vacuum cooled from the environment temperature of 27°C to a temperature of 3°C in 50 min in a 5-m outer diameter insulated spherical vacuum chamber whose walls consist of 2.5-cm-thick urethane insulation ($k = 0.020$ W/m · °C) sandwiched between metal sheets. The vacuum chamber contains 6500 kg of spinach when loaded. Disregarding any heat transfer through the walls of the vacuum chamber, determine (*a*) the final pressure in the vacuum chamber and (*b*) the amount of moisture removed from the spinach, in kg. It is claimed that the error involved in (*b*) due to neglecting heat transfer through the wall of the chamber is less than 2 percent. Is this a reasonable claim?

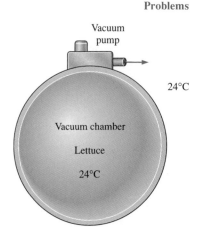

FIGURE P13-45

FIGURE P13-46

Refrigeration of Meats, Poultry, and Fish

13-47C Why does a beef carcass lose up to 2 percent of its weight as it is cooled in the chilling room? How can this weight loss be minimized?

13-48C The cooling of a beef carcass from 37°C to 5°C with refrigerated air at 0°C in a chilling room takes about 48 h. To reduce the cooling time, it is proposed to cool the carcass with refrigerated air at −10°C. How would you evaluate this proposal?

13-49C Consider the freezing of packaged meat in boxes with refrigerated air. How do (*a*) the temperature of air, (*b*) the velocity of air, (*c*) the capacity of the refrigeration system, and (*d*) the size of the meat boxes affect the freezing time?

13-50C How does the rate of freezing affect the tenderness, color, and the drip of meat during thawing?

13-51C It is claimed that beef can be stored for up to two years at −23°C but no more than one year at −12°C. Is this claim reasonable? Explain.

13-52C What is a refrigerated shipping dock? How does it reduce the refrigeration load of the cold storage rooms?

13-53C How does immersion chilling of poultry compare to forced-air chilling with respect to (*a*) cooling time, (*b*) moisture loss of poultry, and (*c*) microbial growth.

13-54C What is the proper storage temperature of frozen poultry? What are the primary methods of freezing for poultry?

13-55C What are the factors that affect the quality of frozen fish?

13-56 The chilling room of a meat plant is 15 m × 18 m × 5.5 m in size and has a capacity of 350 beef carcasses. The power consumed by the fans and the lights in the chilling room are 22 and 2 kW, respectively, and the room gains heat through its envelope at a rate of 11 kW. The average mass of beef carcasses is 280 kg. The carcasses enter the chilling room at 35°C, after they are washed to facilitate evaporative cooling, and are cooled to 16°C in 12 h. The air enters the chilling room at −2.2°C and leaves at 0.5°C. Determine (*a*) the refrigeration load of the chilling room and (*b*) the volume flow rate of air. The average specific heats of beef carcasses and air are 3.14 and 1.0 kJ/kg · °C, respectively, and the density of air can be taken to be 1.28 kg/m³.

13-57 Turkeys with a water content of 64 percent that are initially at 1°C and have a mass of about 7 kg are to be frozen by submerging them into brine at −29°C. Using Figure 13-31, determine how long it will take to reduce the temperature of the turkey breast at a depth of 3.8 cm to −18°C. If the temperature at a depth of 3.8 cm in the breast represents the average temperature of the turkey, determine the amount of heat transfer per turkey assuming (*a*) the entire water content of the turkey is frozen and (*b*) only 90 percent of the water content of the turkey is frozen at −18°C. Take the specific heats of turkey to be 2.98 and 1.65 kJ/kg · °C above and below the freezing point of −2.8°C, respectively, and the latent heat of fusion of turkey to be 214 kJ/kg.   *Answers:* (*a*) 1753 kJ, (*b*) 1617 kJ

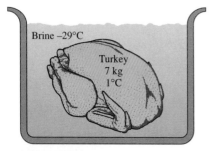

Brine −29°C

Turkey
7 kg
1°C

FIGURE P13-57

13-58E Chickens with a water content of 74 percent, an initial temperature of 32°F, and a mass of about 6 lbm are to be frozen by refrigerated air at −40°F. Using Figure 13-30, determine how long it will take to reduce the inner surface temperature of chickens to 25°F. What would your answer be if the air temperature were −80°F?

13-59 Chickens with an average mass of 2.2 kg and average specific heat of 3.54 kJ/kg · °C are to be cooled by chilled water that enters a continuous-flow-type immersion chiller at 0.5°C. Chickens are dropped into the chiller at a uniform temperature of 15°C at a rate of 500 chickens per hour and are cooled to an average temperature of 3°C before they are taken out. The chiller gains heat from the surroundings at a rate of 210 kJ/min. Determine (*a*) the rate of heat removal from the chicken, in kW, and (*b*) the mass flow rate of water, in kg/s, if the temperature rise of water is not to exceed 2°C.

13-60 In a meat processing plant, 10-cm-thick beef slabs (ρ = 1090 kg/m³, C_p = 3.54 kJ/kg · °C, k = 0.47 W/m · °C, and α = 0.13 × 10⁻⁶ m²/s) initially at 15°C are to be cooled in the racks of a large freezer that is maintained at −12°C. The meat slabs are placed close to each other so that heat transfer from the 10-cm-thick edges is negligible. The entire slab is to be cooled below 5°C, but the temperature of the steak is not to drop below −1°C anywhere during refrigeration to avoid "frost bite." The convection heat transfer coefficient and thus the rate of heat transfer from the steak can be controlled by varying the speed of a circulating fan inside. Using the transient tempera-

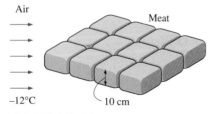

Air

Meat

−12°C   10 cm

FIGURE P13-60

ture charts, determine the heat transfer coefficient h that will enable us to meet both temperature constraints while keeping the refrigeration time to a minimum.   *Answer:* 9.9 W/m$^2 \cdot$ °C

Refrigeration of Eggs, Milk, and Bakery Products

13-61C   Why does a fresh egg settle at the bottom of a cup full of water while an old egg rises to the top?

13-62C   What is standardization of milk? How is milk standardized?

13-63C   Why and how is milk pasteurized? How can the pasteurization time be minimized? How does a regenerator reduce the energy cost of pasteurization?

13-64C   What is the homogenization of milk? How is the milk homogenized?

13-65C   Why is yeast usually stored between 1 and 7°C? Why is the range of optimum dough temperature from 27 to 38°C?

13-66C   What is heat of hydration? How can the temperature rise of dough due to the heat of hydration be prevented during kneading?

13-67   In a dairy plant, milk at 4°C is pasteurized continuously at 72°C at a rate of 12 L/s for 24 hours a day and 365 days a year. The milk is heated to the pasteurizing temperature by hot water heated in a natural gas–fired boiler that has an efficiency of 82 percent. The pasteurized milk is then cooled by cold water at 18°C before it is finally refrigerated back to 4°C. To save energy and money, the plant installs a regenerator that has an effectiveness of 82 percent. If the cost of natural gas is $0.52/therm (1 therm = 105,500 kJ), determine how much fuel and money the regenerator will save this company per year.

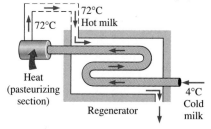

FIGURE P13-67

13-68   A 1.5-kg box made of polypropylene (C_p = 1.9 kJ/kg $\cdot$ °C) contains 30 loaves of white bread, each 0.45 kg, with a water content of 37.3 percent (by mass) at 20°C. The breads are to be frozen to an average temperature of -12°C in 3 h in their box by refrigerated air. The enthalpy of bread is given to be 137 kJ/kg at 0°C and 56 kJ/kg at -12°C. Taking the average specific heat of the bread above 0°C to be C_p = 2.60 kJ/kg $\cdot$ °C, determine (*a*) the total amount of heat that must be removed from a box of breads and (*b*) the average rate of heat removal from the breads and their box to the air per box.

13-69   Eggs (ρ = 1.08 g/cm^3, k = 0.56 W/m $\cdot$ °C, C_p = 3.34 kJ/kg $\cdot$ °C, and α = 0.14 $\times$ 10^{-6} m^2/s) with an average mass of 70 g and initially at a uniform temperature of 32°C are to be cooled by refrigerated air at 2°C flowing at a velocity of 3 m/s with an average heat transfer coefficient between the eggs and the air of 45 W/m$^2 \cdot$ °C. Determine how long it will take for the center temperature of the eggs to drop to 12°C. Also, determine the temperature difference between the center and the surface of the eggs and the amount of heat transfer per egg.   *Answers:* 26.7 min, 5.6°C, 5.50 kJ

FIGURE P13-69

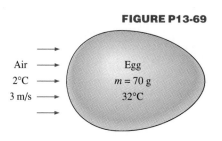

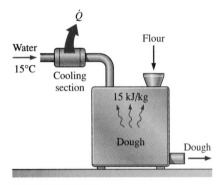

FIGURE P13-71

FIGURE P13-78

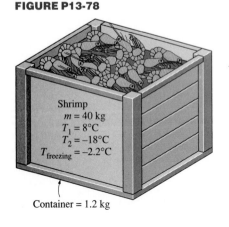

13-70E A refrigeration system is being designed to cool eggs ($\rho = 67.4$ lbm/ft^3, $k = 0.32$ Btu/h · ft · °F, $C_p = 0.80$ Btu/lbm · °F, and $\alpha = 1.5 \times 10^{-6}$ ft^2/s) with an average mass of 0.14 lbm from an initial temperature of 90°F to a final average temperature of 50°F by air at 34°F at a rate of 10,000 eggs per hour. Determine (a) the rate of heat removal from the eggs, in Btu/h; (b) the required volume flow rate of air, in ft^3/h, if the temperature rise of air is not to exceed 10°F; and (c) the size of the compressor of the refrigeration system, in kW, for a COP of 3.5 for the refrigeration system.

13-71 The heat of hydration of dough, 15 kJ/kg, will raise the dough's temperature to undesirable levels unless some cooling mechanism is utilized. A practical way of absorbing the heat of hydration is to use refrigerated water when kneading the dough. If a recipe calls for mixing 2 kg of flour with 1 kg of water, and the temperature of the city water is 15°C, determine the temperature to which the city water must be cooled before mixing it with flour in order for the water to absorb the entire heat of absorption when the water temperature rises to 15°C. Take the specific heats of the flour and the water to be 1.76 and 4.18 kJ/kg · °C, respectively.   *Answer:* 4.2°C

Refrigeration Load of Cold Storage Rooms

13-72C What does the refrigeration load of a cold storage room represent? What does the refrigeration load consist of?

13-73C Define the transmission load of a cold storage room, and explain how it is determined. Discuss how the transmission load can be minimized.

13-74C What is infiltration heat gain for cold storage rooms? How can it be minimized?

13-75C Define the product load of a frozen food storage room, and explain how it is determined.

13-76C Define the internal load of a refrigeration room. Name the factors that constitute the internal load.

13-77C Consider the motors that power the fans of a refrigerated room. How does locating the motors inside the refrigeration room or outside of it affect the refrigeration load of the room?

13-78 A supply of 40 kg of shrimp at 8°C contained in a box is to be frozen to −18°C in a freezer. Determine the amount of heat that needs to be removed. The latent heat of the shrimp is 277 kJ/kg, and its specific heat is 3.62 kJ/kg · °C above freezing and 1.89 kJ/kg · °C below freezing. The container box is 1.2 kg and is made up of polyethylene, whose specific heat is 2.3 kJ/kg · °C. Also, the freezing temperature of shrimp is −2.2°C.

13-79E A cold storage room whose internal dimensions are 12 ft × 15 ft × 30 ft is maintained at 35°F and 95 percent relative humidity. Under worst conditions, the amount of air infiltration is estimated to be 0.4 air change per hour at the ambient conditions of 90°F and 90 percent relative humidity.

Using the psychrometric chart, determine the total infiltration load of this room under these conditions. _Answer:_ 92,940 Btu/h

13-80 A holding freezer whose outer dimensions are 80 m × 25 m × 7 is maintained at −25°C at a location where the annual mean ambient temperature is 15°C. The construction of the walls is such that it has an _R_-value of 3 m² · °C/W (equivalent to _R_-17 in English units of h · ft² · °F/Btu), which is well below the recommended value of _R_-6.5. The refrigeration system has a COP of 1.3, and the cost of electricity is $0.10/kWh. Taking the heat transfer coefficients on the inner and outer surfaces of the walls to be 10 and 20 W/m² · °C, respectively, determine the amount of electrical energy and money this facility will save per year by increasing the insulation value of the walls to the recommended level of _R_-6.5. Assume continuous operation for the entire year. What is the percent error involved in the total thermal resistance of the wall in neglecting the convection resistances on both sides?

13-81 A chilling room whose outer dimensions are 60 m × 30 m × 6 m is maintained at −2°C at a location where the annual mean ambient temperature is 18°C. The construction of the roof of the room is such that it has an _R_-value of 2 m² · °C/W (equivalent to _R_-11.4 in English units of h · ft² · °F/Btu), which is well below the recommended value of _R_-7. The refrigeration system has a COP of 2.4, and the cost of electricity is $0.09/kWh. Taking the heat transfer coefficients on the inner and outer surfaces of the walls to be 10 and 25 W/m² · °C, respectively, determine the amount of electrical energy and money this facility will save per year by increasing the insulation value of the walls to the recommended level of _R_-7.
 Answers: 42,997 kWh/yr, $3870/yr

13-82 A cold storage room whose inner dimensions are 50 m × 30 m × 7 m is maintained at 6°C. At any given time, 15 people are working in the room, and lighting for the facility is provided by 150 light bulbs, each consuming 100 W. There are also several electric motors in the facility that consume a total of 25 kW of electricity. It is estimated that the ambient air that enters the room through the cracks and openings is equivalent to 0.2 air change per hour. Taking the ambient air temperature to be 20°C and disregarding any condensation of moisture, determine (_a_) the internal load of this cold storage room and (_b_) the infiltration load, both in kW.

In order to save energy, it is proposed that the incandescent lights be replaced by 40 high-efficiency fluorescent tubes, each consuming 110 W. If the lights are on for an average of 15 h a day, every day, determine the amount of electrical energy and money this facility will save per year as a result of switching to fluorescent lighting. Assume the refrigeration system has a COP of 2.8, and the cost of electricity is $0.09/kWh.

Transportation of Refrigerated Foods

13-83C Why are the refrigerated trucks insulated? What are the advantages and disadvantages of having a very thick layer of insulation in the walls of a refrigerated truck?

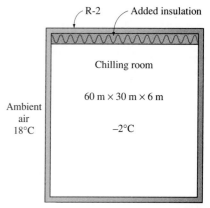

FIGURE P13-81

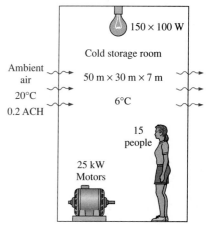

FIGURE P13-82

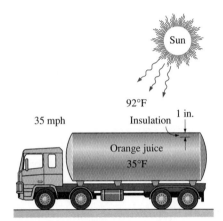

FIGURE P13-87E

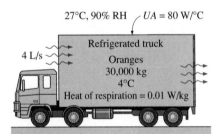

FIGURE P13-88

13-84C Under what conditions is it worthwhile to transport food items by airplanes?

13-85C Why are the trucks precooled before they are loaded? Also, do refrigerated trucks need to be equipped with heating systems? Why? Explain.

13-86C A truck driver claims to have transported orange juice from Florida to New York in warm weather without any refrigeration, and that the temperature of the orange juice rose by only 4°C during this long trip. Is this claim reasonable?

13-87E Orange juice ($\rho = 62.4$ lbm/ft^3 and $C_p = 0.90$ Btu/lbm $\cdot$ °F) is to be transported from Florida to New York for a distance of 1250 miles in a 27-ft-long, 6.3-ft-external-diameter cylindrical tank. The walls of the tank are constructed of 1-in.-thick urethane insulation ($k = 0.017$ Btu/h $\cdot$ ft $\cdot$ °F) sandwiched between two metal sheets. The juice is precooled to 35°F before loading, and its temperature is not to exceed 46°F on arrival. The average ambient temperature can be as high as 92°F, and the effect of solar radiation and the radiation from hot pavement surfaces can be taken to be equivalent to a 12°F rise in ambient temperature. If the average velocity during transportation is 35 mph, determine if the orange juice can be transported without any refrigeration.

13-88 A large truck is to transport 30,000 kg of oranges precooled to 4°C under average ambient conditions of 27°C and 90 percent relative humidity, which corresponds to a humidity ratio of 0.0205 kg water vapor/kg dry air. The structure of the walls of the truck is such that the rate of heat transmission is $UA = 80$ W per °C temperature difference between the ambient and the oranges. From past experience, ambient air is estimated to enter the cargo space of the truck through the cracks at a rate of 4 L/s. The moisture in the air is condensed out at an average temperature of 15°C, at which the latent heat of vaporization of water is 2466 kJ/kg. Also, the average heat of respiration of the oranges at 4°C is 0.017 W/kg for this particular load. Taking the density of air to be 1.15 kg/m^3, determine the refrigeration load of this truck and the amount of ice needed to meet the entire refrigeration need of the truck for a 15-h-long trip.

13-89 Consider a large freezer truck whose outer dimensions are 14 m × 2.5 m × 4 m that is maintained at −18°C at a location where the ambient temperature is 25°C. The walls of the truck are constructed of 2.5-cm-thick urethane insulation ($k = 0.026$ W/m $\cdot$ °C) sandwiched between thin metal plates. The recommended minimum thickness of insulation in this case is 8 cm. Taking the heat transfer coefficients on the inner and outer surfaces of the walls to be 8 and 4 W/m^2 $\cdot$ °C, respectively, determine the reduction in the transmission part of the refrigeration load if the insulation thickness is upgraded to the recommended value. Also, taking the gross density of the load to be 600 kg/m^3, determine the reduction in the cargo load of the truck, in kg, as a result of increasing the thickness of insulation while holding the outer dimensions constant.

13-90 The cargo space of a refrigerated truck whose inner dimensions are 12 m × 2.3 m × 3.5 m is to be precooled from 25°C to an average temperature of 5°C. The construction of the truck is such that a transmission heat gain occurs at a rate of 80 W/°C. If the ambient temperature is 25°C, determine how long it will take for a system with a refrigeration capacity of 8 kW to precool this truck. Take the density of air to be 1.20 kg/m³.
Answer: 5.4 min

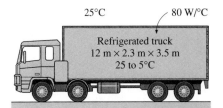

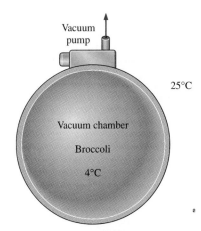

FIGURE P13-90

Review Problems

13-91 Wet broccoli is to be vacuum cooled from the environment temperature of 25°C to a temperature of 4°C in 1 h in a 4-m outer diameter insulated spherical vacuum chamber whose walls consist of 3-cm-thick fiberglass insulation ($k = 0.026$ W/m·°C) sandwiched between metal plates. The vacuum chamber contains 6000 kg of broccoli when loaded, but 2 percent of this mass is due to water that sticks to the surface of the broccoli during wetting. Disregarding any heat transfer through the walls of the vacuum chamber, determine the final mass of the broccoli after cooling. It is claimed that the error involved in the predicted final mass of broccoli due to neglecting heat transfer through the wall of the chamber is less than 5 percent. Is this a reasonable claim?   *Answer:* 5801 kg

13-92 Red Delicious apples of 65-mm diameter and 85 percent water content ($k = 0.42$ W/m·°C and $\alpha = 0.14 \times 10^{-6}$ m²/s) initially at a uniform temperature of 22°C are to be cooled by (*a*) refrigerated air at 1°C flowing at a velocity of 1.5 m/s with an average heat transfer coefficient of 45 W/m²·°C between the apples and the air and (*b*) chilled water at 1°C flowing at a velocity of 0.3 m/s with an average heat transfer coefficient of 80 W/m²·°C between the apples and the water. Determine how long it will take for the center temperature of the apples to drop to 8°C and the temperature difference between the center and the surface of the apples for each case.

13-93E A 3.5-lbm box made of polypropylene ($C_p = 0.45$ Btu/lbm·°F) contains 70 lbm of haddock fish with a water content of 83.6 percent (by mass) at 60°F. The fish is to be frozen to an average temperature of −4°F in 4 h in its box. The enthalpy of the fish is given to be 145 Btu/lbm at 32°F and 18 Btu/lbm at −4°F. It is also given that the unfrozen water content of the fish at −4°F is 9 percent. Taking the average specific heat of the fish above freezing temperatures to be $C_p = 0.89$ Btu/lbm·°F, determine (*a*) the total amount of heat that must be removed from the fish, (*b*) the amount of unfrozen water in fish at −4°F, and (*c*) the average rate of heat removal from the fish.

13-94 Fresh carrots with a water content of 87.5 percent (by mass) at 22°C are stored in 1.4-kg boxes made of polyethylene ($C_p = 2.3$ kJ/kg·°C). Each box contains 30 kg of carrots, and the carrots are to be frozen to an average temperature of −18°C at a rate of 50 boxes per hour. The enthalpy of the carrots is given to be 361 kJ/kg at 0°C and 51 kJ/kg at −18°C. It is also given that the unfrozen water content of the carrots at −18°C is 7 percent. Taking

FIGURE P13-91E

FIGURE P13-93

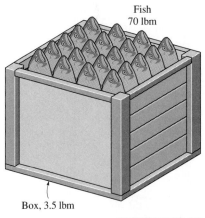

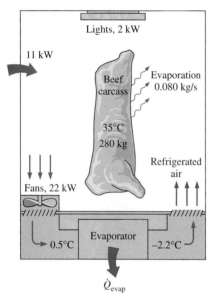

FIGURE P13-95

FIGURE P13-97E

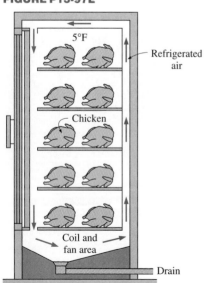

the average specific heat of the carrots above freezing temperatures to be $C_p = 3.90$ kJ/kg · °C, determine (a) the rate of heat removal from the carrots and their boxes, in kJ/h, and (b) the rate at which the water in carrots freezes during this process, in kg/h.   *Answers:* (a) 166.7 kW, (b) 0.336 kh

13-95   The chilling room of a meat plant is 15 m × 18 m × 5.5 m in size and has a capacity of 350 beef carcasses. The power consumed by the fans and the lights of the chilling room are 22 and 2 kW, respectively, and the room gains heat through its envelope at a rate of 11 kW. The average mass of beef carcasses is 280 kg. The carcasses enter the chilling room at 35°C, after they are washed to facilitate evaporative cooling, and are cooled to 16°C in 10 h. The water is expected to evaporate at a rate of 0.080 kg/s. The air enters the evaporator section of the refrigeration system at 0.5°C and leaves at −2.2°C. The air side of the evaporator is heavily finned, and the overall heat transfer coefficient of the evaporator based on the air side is 22 W/m² · °C. Also, the average temperature difference between the air and the refrigerant in the evaporator is 5.5°C. Determine (a) the refrigeration load of the chilling room, (b) the volume flow rate of air, and (c) the heat transfer surface area of the evaporator on the air side, assuming all the vapor and the fog in the air freeze in the evaporator. The average specific heats of beef carcasses and air are 3.14 and 1.0 kJ/kg · °C, respectively. Also, the heat of fusion of water is 334 kJ/kg and the heat of vaporization of water and the density of air can be taken to be 2490 kJ/kg and 1.28 kg/m³, respectively.

13-96   Turkeys with an average mass of 7.5 kg and average specific heat of 3.28 kJ/kg · °C are to be cooled by chilled water that enters a continuous-flow-type immersion chiller at 0.6°C. Turkeys are dropped into the chiller at a uniform temperature of 14°C at a rate of 200 turkeys per hour and are cooled to an average temperature of 4°C before they are taken out. The chiller gains heat from the surroundings at a rate of 120 kJ/h. Determine (a) the rate of heat removal from the turkeys, in kW, and (b) the mass flow rate of water, in kg/s, if the temperature rise of water is not to exceed 2.5°C.

13-97E   In a chicken processing plant, whole chickens ($\rho = 65.5$ lbm/ft³, $C_p = 0.85$ Btu/lbm · °F, $k = 0.27$ Btu/h · ft · °F, and $\alpha = 1.4 \times 10^{-6}$ ft²/s) averaging five pounds each and initially at 60°F are to be cooled in the racks of a large refrigerator that is maintained at 5°F. The entire chicken is to be cooled below 40°F, but the temperature of the chicken is not to drop below 33°F anywhere during refrigeration. The convection heat transfer coefficient and thus the rate of heat transfer from the chicken can be controlled by varying the speed of a circulating fan inside. Treating the chicken as a homogeneous spherical object, determine the heat transfer coefficient that will enable us to meet both temperature constraints while keeping the refrigeration time to a minimum.

13-98   A person puts a few peaches ($\rho = 960$ kg/m³, $C_p = 3.9$ kJ/kg · °C, $k = 0.53$ W/m · °C, and $\alpha = 0.14 \times 10^{-6}$ m²/s) into the freezer at −18°C to cool them quickly for the guests who are about to arrive. Initially, the peaches are at a uniform temperature of 20°C, and the heat transfer coefficient on the surfaces is 18 W/m² · °C. Treating the peaches as 8-cm-diameter spheres,

determine the center and surface temperatures of the peaches in 45 minutes. Also, determine the amount of heat transfer from each peach.
Answers: 7.2°C, −3.7°C, 19.7 kJ

13-99 A refrigeration system is to cool bread loaves with an average mass of 450 g from 22°C to −10°C at a rate of 500 loaves per hour by refrigerated air at −30°C and 1 atm. Taking the average specific and latent heats of bread to be 2.93 kJ/kg · °C and 109.3 kJ/kg, respectively, determine (a) the rate of heat removal from the breads, in kJ/h; (b) the required volume flow rate of air, in m³/h, if the temperature rise of air is not to exceed 8°C; and (c) the size of the compressor of the refrigeration system, in kW, for a COP of 1.2 for the refrigeration system.

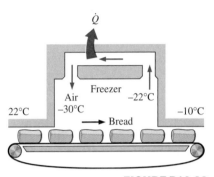

FIGURE P13-99

13-100 A holding freezer whose outer dimensions are 70 m × 22 m × 6 m is maintained at −23°C at a location where the annual mean ambient temperature is 14°C. The construction of the roof is such that it has an *R*-value of 4 m² · °C/W (equivalent to *R*-23 in English units of h · ft² · °F/Btu), which is well below the recommended value of *R*-8.5. The refrigeration system has a COP of 1.25, and the cost of electricity is $0.07/kWh. Taking the heat transfer coefficients on the inner and outer surfaces of the roof to be 12 and 30 W/m² · °C, respectively, determine the amount of electrical energy and money this facility will save per year by increasing the insulation value of the walls to the recommended level of *R*-8.5. What is the percent error involved in the total thermal resistance of the roof in neglecting the convection resistances on both sides?   *Answers:* 50,650 kWh/yr, $3546/yr, 2.9 percent

13-101 Milk is to be transported from Wyoming to Chicago for a distance of 1600 km in an 8-m-long, 2.2-m-external-diameter cylindrical tank. The walls of the tank are constructed of 3-cm-thick urethane insulation (*k* = 0.029 W/m · °C) sandwiched between two metal sheets. The milk is precooled to 2°C before loading, and its temperature is not to exceed 5°C on arrival. The average ambient temperature can be as high as 32°C, and the effect of solar radiation and the radiation from hot pavement surfaces can be taken to be equivalent to a 5°C rise in ambient temperature. If the average velocity during transportation is 60 km/h, determine if the milk can be transported without any refrigeration.

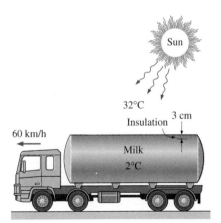

FIGURE P13-101

Computer, Design, and Essay Problems

13-102 Write an essay on the operating principle of thermoelectric refrigerators, and discuss their advantages and disadvantages.

13-103 Using a thermometer, measure the temperature of the main food compartment of your refrigerator, and check if it is between 1 and 4°C. Also, measure the temperature of the freezer compartment, and check if it is at the recommended value of −18°C.

13-104 Using a timer (or watch) and a thermometer, conduct the following experiment to determine the predictable heat load of your refrigerator. First, make sure that the door of the refrigerator is not opened for at least a few hours to make sure that steady operating conditions are established. Start the timer when the refrigerator stops running and measure the time Δt_1 it stays off before it kicks in. Then measure the time Δt_2 it stays on. Noting that the heat removed during Δt_2 is equal to the heat gain of the refrigerator during $\Delta t_1 + \Delta t_2$ and using the power consumed by the refrigerator when it is running, determine the average rate of heat gain for your refrigerator, in W. Take the COP (coefficient of performance) of your refrigerator to be 1.3 if it is not available.

Now, clean the condenser coils of the refrigerator and remove any obstacles on the way of airflow through the coils. By repeating the measurements above, determine the improvement in the COP of the refrigerator.

13-105 Design a hydrocooling unit that can cool fruits and vegetables from 30°C to 5°C at a rate of 20,000 kg/h under the following conditions:

The unit will be of flood type that will cool the products as they are conveyed into the channel filled with water. The products will be dropped into the channel filled with water at one end and picked up at the other end. The channel can be as wide as 3 m and as high as 90 cm. The water is to be circulated and cooled by the evaporator section of a refrigeration system. The refrigerant temperature inside the coils is to be −2°C, and the water temperature is not to drop below 1°C and not to exceed 6°C.

Assuming reasonable values for the average product density, specific heat, and porosity (the fraction of air volume in a box), recommend reasonable values for the quantities related to the thermal aspects of the hydrocooler, including (a) how long the fruits and vegetables need to remain in the channel, (b) the length of the channel, (c) the water velocity through the channel, (d) the velocity of the conveyor and thus the fruits and vegetables through the channel, (e) the refrigeration capacity of the refrigeration system, and (f) the type of heat exchanger for the evaporator and the surface area on the water side.

13-106 Design a scalding unit for slaughtered chicken to loosen their feathers before they are routed to feather-picking machines with a capacity of 1200 chickens per hour under the following conditions:

The unit will be of immersion type filled with hot water at an average temperature of 53°C at all times. Chicken with an average mass of 2.2 kg and an average temperature of 36°C will be dipped into the tank, held in the water for 1.5 min, and taken out by a slow-moving conveyor. The chicken is expected to leave the tank 15 percent heavier as a result of the water that sticks to its surface. The center-to-center distance between chickens in any direction will be at least 30 cm. The tank can be as wide as 3 m and as high as 60 cm. The water is to be circulated through and heated by a natural gas furnace, but the temperature rise of water will not exceed 5°C as it passes through the furnace. The water loss is to be made up by the city water at an average temperature of 16°C. The ambient air temperature can be taken to be 20°C. The walls and the floor of the tank are to be insulated with a 2.5-cm-thick urethane layer. The unit operates 24 h a day and 6 days a week.

Assuming reasonable values for the average properties, recommend reasonable values for the quantities related to the thermal aspects of the scalding tank, including (*a*) the mass flow rate of the make-up water that must be supplied to the tank; (*b*) the length of the tank; (*c*) the rate of heat transfer from the water to the chicken, in kW; (*d*) the velocity of the conveyor and thus the chickens through the tank; (*e*) the rate of heat loss from the exposed surfaces of the tank and if it is significant; (*f*) the size of the heating system in kJ/h; (*g*) the type of heat exchanger for heating the water with flue gases of the furnace and the surface area on the water side; and (*h*) the operating cost of the scalding unit per month for a unit cost of $0.56 therm of natural gas (1 therm = 105,000 kJ).

13-107   A company owns a refrigeration system whose refrigeration capacity is 200 tons (1 ton of refrigeration = 211 kJ/min), and you are to design a forced-air cooling system for fruits whose diameters do not exceed 7 cm under the following conditions:

The fruits are to be cooled from 28°C to an average temperature of 8°C. The air temperature is to remain above −2°C and below 10°C at all times, and the velocity of air approaching the fruits must remain under 2 m/s. The cooling section can be as wide as 3.5 m and as high as 2 m.

Assuming reasonable values for the average fruit density, specific heat, and porosity (the fraction of air volume in a box), recommend reasonable values for the quantities related to the thermal aspects of the forced-air cooling, including (*a*) how long the fruits need to remain in the cooling section; (*b*) the length of the cooling section; (*c*) the air velocity approaching the cooling section; (*d*) the product cooling capacity of the system, in kg · fruit/h; (*e*) the volume flow rate of air; and (*f*) the type of heat exchanger for the evaporator and the surface area on the air side.

Cooling of Electronic Equipment

Electronic equipment has made its way into practically every aspect of modern life, from toys and appliances to high-power computers. The reliability of the electronics of a system is a major factor in the overall reliability of the system. Electronic components depend on the passage of electric current to perform their duties, and they become potential sites for excessive heating, since the current flow through a resistance is accompanied by heat generation. Continued miniaturization of electronic systems has resulted in a dramatic increase in the amount of heat generated per unit volume, comparable in magnitude to those encountered at nuclear reactors and the surface of the sun. Unless properly designed and controlled, high rates of heat generation result in high operating temperatures for electronic equipment, which jeopardizes its safety and reliability.

The failure rate of electronic equipment increases exponentially with temperature. Also, the high thermal stresses in the solder joints of electronic components mounted on circuit boards resulting from temperature variations are major causes of failure. Therefore, thermal control has become increasingly important in the design and operation of electronic equipment.

In this chapter, we discuss several cooling techniques commonly used in electronic equipment such as conduction cooling, natural convection and radiation cooling, forced air cooling, liquid cooling, immersion cooling, and heat pipes. This chapter is intended to familiarize the reader with these techniques and put them into perspective. The reader interested in an in-depth coverage of any of these topics can consult numerous other sources available, such as those listed in the references.

14-1 ■ INTRODUCTION AND HISTORY

The field of electronics deals with the construction and utilization of devices that involve current flow through a vacuum, a gas, or a semiconductor. This exciting field of science and engineering dates back to 1883, when Thomas Edison discovered the vacuum diode. The **vacuum tube** served as the foundation of the electronics industry until the 1950s, and played a central role in the development of radio, TV, radar, and the digital computer. Of the several computers developed in this era, the largest and best known is the ENIAC (Electronic Numerical Integrator and Computer), which was built at the University of Pennsylvania in 1946. It had over 18,000 vacuum tubes and occupied a room 7 m × 14 m in size. It consumed a large amount of power, and its reliability was poor because of the high failure rate of the vacuum tubes.

The invention of the bipolar **transistor** in 1948 marked the beginning of a new era in the electronics industry and the obsolescence of vacuum tube technology. Transistor circuits performed the functions of the vacuum tubes with greater reliability, while occupying negligible space and consuming negligible power compared with vacuum tubes. The first transistors were made from germanium, which could not function properly at temperatures above 100°C. Soon they were replaced by silicon transistors, which could operate at much higher temperatures.

The next turning point in electronics occurred in 1959 with the introduction of the **integrated circuits** (IC), where several components such as diodes, transistors, resistors, and capacitors are placed in a single chip. The number of components packed in a single chip has been increasing steadily since then at an amazing rate, as shown in Figure 14-1. The continued miniaturization of electronic components has resulted in *medium-scale integration* (MSI) in the 1960s with 50–1000 components per chip, *large-scale integration* (LSI) in the 1970s with 1000–100,000 components per chip, and *very large-scale integration* (VLSI) in the 1980s with 100,000–10,000,000 components per chip. Today it is not unusual to have a chip 3 cm × 3 cm in size with several million components on it.

The development of the **microprocessor** in the early 1970s by the Intel Corporation marked yet another beginning in the electronics industry. The accompanying rapid development of large-capacity memory chips in this decade made it possible to introduce capable personal computers for use at work or at home at an affordable price. Electronics has made its way into practically everything from watches to household appliances to automobiles. Today it is difficult to imagine a new product that does not involve any electronic parts.

The current flow through a resistance is always accompanied by *heat generation* in the amount of I^2R, where I is the electric current and R is the resistance. When the transistor was first introduced, it was touted in the newspapers as a device that "produces no heat." This certainly was a fair statement, considering the huge amount of heat generated by vacuum tubes. Obviously, the little heat generated in the transistor was no match to that generated in its predecessor. But when thousands or even millions of such components are packed in a small volume, the heat generated increases to

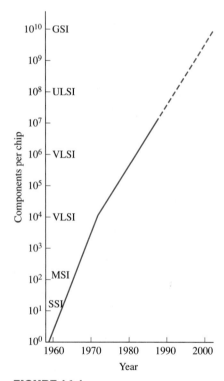

FIGURE 14-1

The increase in the number of components packed on a chip over the years.

such high levels that its removal becomes a formidable task and a major concern for the safety and reliability of the electronic devices. The heat fluxes encountered in electronic devices range from less than 1 W/cm² to more than 100 W/cm².

Heat is generated in a resistive element for as long as current continues to flow through it. This creates a *heat build-up* and a subsequent *temperature rise* at and around the component. The temperature of the component will continue rising until the component is destroyed unless heat is transferred away from it. The temperature of the component will remain constant when the rate of heat removal from it equals the rate of heat generation.

Individual electronic components have *no moving parts,* and thus nothing to wear out with time. Therefore, they are inherently reliable, and it seems as if they can operate safely for many years. Indeed, this would be the case if components operated at room temperature. But electronic components are observed to fail under prolonged use at high temperatures. Possible causes of failure are *diffusion* in semiconductor materials, *chemical reactions,* and *creep* in the bonding materials, among other things. The failure rate of electronic devices increases almost *exponentially* with the operating temperature, as shown in Figure 14-2. The cooler the electronic device operates, the more reliable it is. A rule of thumb is that the failure rate of electronic components is halved for each 10°C reduction in their junction temperature.

14-2 ■ MANUFACTURING OF ELECTRONIC EQUIPMENT

The narrow band where two different regions of a semiconductor (such as the p-type and n-type regions) come in contact is called a **junction.** A transistor, for example, involves two such junctions, and a diode, which is the simplest semiconductor device, is based on a single p-n junction. In heat transfer analysis, the circuitry of an electronic component through which electrons flow and thus heat is generated is also referred to as the junction. That is, junctions are the sites of heat generation and thus the hottest spots in a component. In silicon-based semiconductor devices, the junction temperature is limited to 125°C for safe operation. However, lower junction temperatures are desirable for extended life and lower maintenance costs. In a typical application, numerous electronic components, some smaller than 1 μm in size, are formed from a silicon wafer into a chip.

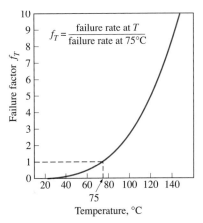

$$f_T = \frac{\text{failure rate at } T}{\text{failure rate at } 75°C}$$

FIGURE 14-2

The increase in the failure rate of bipolar digital devices with temperature (from Ref. 17).

The Chip Carrier

The chip is housed in a **chip carrier** or substrate made of ceramic, plastic, or glass in order to protect its delicate circuitry from the detrimental effects of the environment. The chip carrier provides a rugged housing for the safe handling of the chip during the manufacturing process, as well as the connectors between the chip and the circuit board. The various components of the chip carrier are shown in Figure 14-3. The chip is secured in the carrier by bonding it to the bottom surface. The thermal expansion coefficient of the *plastic* is about 20 times that of silicon. Therefore, bonding the silicon chip directly to the plastic case would result in such large thermal stresses that the

FIGURE 14-3

The components of a chip carrier (from Dally, Ref. 4).

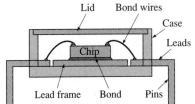

reliability would be seriously jeopardized. To avoid this problem, a *lead frame* made of a copper alloy with a thermal expansion coefficient close to that of silicon is used as the bonding surface.

The design of the chip carrier is the *first level* in the thermal control of electronic devices, since the transfer of heat from the chip to the chip carrier is the first step in the dissipation of the heat generated on the chip. The heat generated on the chip is transferred to the case of the chip carrier by a combination of conduction, convection, and radiation. However, it is obvious from the figure that the common chip carrier is designed with the *electrical aspects* in mind, and little consideration is given to the *thermal aspects*. First of all, the cavity of the chip carrier is filled with a gas, which is a poor heat conductor, and the case is often made of materials that are also poor conductors of heat. This results in a relatively large thermal resistance between the chip and the case, called the **junction-to-case resistance,** and thus a large temperature difference. As a result, the temperature of the chip will be much higher than that of the case for a specified heat dissipation rate. The junction-to-case thermal resistance depends on the *geometry* and the *size* of the chip and the chip carrier as well as the material properties of the *bonding* and the *case*. It varies considerably from one device to another and ranges from about 10°C/W to more than 100°C/W.

Moisture in the cavity of the chip carrier is highly undesirable, since it causes corrosion on the wiring. Therefore, chip carriers are made of materials that prevent the entry of moisture by diffusion and are *hermetically sealed* in order to prevent the direct entry of moisture through cracks. Materials that outgas are also not permitted in the chip cavity, because such gases can also cause corrosion. In products with strict hermeticity requirements, the more expensive ceramic cases are used instead of the plastic ones.

A common type of chip carrier for high-power transistors is shown in Figure 14-4. The transistor is formed on a small silicon chip housed in the disk-shaped cavity, and the I/O pins come out from the bottom. The case of the transistor carrier is usually attached directly to a flange, which provides a large surface area for heat dissipation and reduces the junction-to-case thermal resistance.

It is often desirable to house more than one chip in a single chip carrier. The result is a *hybrid* or *multichip package*. Hybrid packages house several chips, individual electronic components, and ordinary circuit elements connected to each other in a single chip carrier. The result is improved performance due to the shortening of the wiring lengths, and enhanced reliability. Lower cost would be an added benefit of multichip packages if they are produced in sufficiently large quantity.

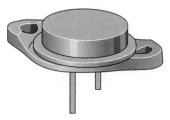

FIGURE 14-4

The chip carrier for a high-power transistor attached to a flange for enhanced heat transfer.

EXAMPLE 14-1 Predicting the Junction Temperature of a Transistor

The temperature of the case of a power transistor that is dissipating 3 W is measured to be 50°C. If the junction-to-case resistance of this transistor is specified by the manufacturer to be 15°C/W, determine the temperature at the junction of the transistor.

Solution The case temperature of a power transistor and the junction-to-case resistance are given. The junction temperature is to be determined.

Assumptions Steady operating conditions exist.

Analysis The schematic of the transistor is given in Figure 14-5. The rate of heat transfer between the junction and the case in steady operation can be expressed as

$$\dot{Q} = \left(\frac{\Delta T}{R}\right)_{\text{junction-case}} = \frac{T_{\text{junction}} - T_{\text{case}}}{R_{\text{junction-case}}}$$

Then the junction temperature becomes

$$T_{\text{junction}} = T_{\text{case}} + \dot{Q}R_{\text{junction-case}}$$
$$= 50°C + (3\ W)(15°C/W)$$
$$= \textbf{95°C}$$

Therefore, the temperature of the transistor junction will be 95°C when its case is at 50°C.

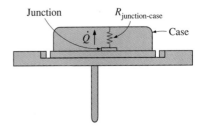

FIGURE 14-5
Schematic for Example 14-1.

EXAMPLE 14-2 Determining the Junction-to-Case Thermal Resistance

The following experiment is conducted to determine the junction-to-case thermal resistance of an electronic component. Power is supplied to the component from a 15-V source, and the variations in the electric current and in the junction and the case temperatures with time are observed. When things are stabilized, the current is observed to be 0.1 A and the temperatures to be 80°C and 55°C at the junction and the case, respectively. Calculate the junction-to-case resistance of this component.

Solution The power dissipated by an electronic component as well as the junction and case temperatures are measured. The junction-to-case resistance is to be determined.

Assumptions Steady operating conditions exist

Analysis The schematic of the component is given in Figure 14-6. The electric power consumed by this electronic component is

$$\dot{W}_e = VI = (15\ V)(0.1\ A) = 1.5\ W$$

In steady operation, this is equivalent to the heat dissipated by the component. That is,

$$\dot{Q} = \left(\frac{\Delta T}{R}\right)_{\text{junction-case}} = \frac{T_{\text{junction}} - T_{\text{case}}}{R_{\text{junction-case}}} = 1.5\ W$$

Then the junction-to-case resistance is determined to be

$$R_{\text{junction-case}} = \frac{T_{\text{junction}} - T_{\text{case}}}{\dot{Q}} = \frac{(80 - 55)°C}{1.5\ W} = \textbf{16.7°C/W}$$

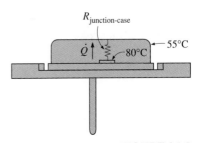

FIGURE 14-6
Schematic for Example 14-2.

Discussion   Note that a temperature difference of 16.7°C will occur between the electronic circuitry and the case of the chip carrier for each W of power consumed by the component.

Printed Circuit Boards

A **printed circuit board** (PCB) is a properly wired plane board made of polymers and glass–epoxy materials on which various electronic components such as the ICs, diodes, transistors, resistors, and capacitors are mounted to perform a certain task, as shown in Figure 14-7. The PCBs are commonly called *cards,* and they can be replaced easily during a repair. The PCBs are plane boards, usually 10 cm wide and 15 cm long and only a few millimeters thick, and they are not suitable for heavy components such as transformers. Usually a copper cladding is added on one or both sides of the board. The cladding on one side is subjected to an etching process to form wiring strips and attachment pads for the components. The power dissipated by a PCB usually ranges from 5 W to about 30 W.

A typical electronic system involves several layers of PCBs. The PCBs are usually cooled by direct contact with a fluid such as air flowing between the boards. But when the boards are placed in a hermetically sealed enclosure, they must be cooled by a cold plate (a heat exchanger) in contact with

FIGURE 14-7

A printed circuit board (PCB) with a variety of components on it (courtesy of Litton Systems, Inc.).

the edge of the boards. The *device-to-board edge thermal resistance* of a PCB is usually high (about 20–60°C/W) because of the small thickness of the board and the low thermal conductivity of the board material. In such cases, even a thin layer of copper cladding on one side of the board can decrease the device-to-board edge thermal resistance in the plane of the board and enhance heat transfer in that direction drastically.

In the thermal design of a PCB, it is important to pay particular attention to the components that are not tolerant of high temperatures, such as certain high-performance capacitors, and to ensure their safe operation. Often when one component on a PCB fails, the whole board fails and must be replaced.

Printed circuit boards come in three types: *single-sided, double-sided,* and *multilayer* boards. Each type has its own strengths and weaknesses. Single-sided PCBs have circuitry lines on one side of the board only and are suitable for low-density electronic devices (10–20 components). Double-sided PCBs have circuits on both sides and are best suited for intermediate-density devices. Multilayer PCBs contain several layers of circuitry and are suitable for high-density devices. They are equivalent to several PCBs sandwiched together.

The single-sided PCB has the lowest cost, as expected, and it is easy to maintain, but it occupies a lot of space. The multilayer PCB, on the other hand, allows the placement of a large number of components in a three-dimensional configuration, but it has the highest initial cost and is difficult to repair. Also, temperatures are most likely to be the highest in multilayer PCBs.

In critical applications, the electronic components are placed on boards attached to a conductive metal, called the *heat frame,* that serves as a conduction path to the edge of the circuit board and thus to the cold plate for the heat generated in the components. Such boards are said to be *conduction-cooled.* The temperature of the components in this case will depend on the location of the components on the boards: it will be highest for the components in the middle and lowest for those near the edge, as shown in Figure 14-8.

Materials used in the fabrication of circuit boards should be (1) effective electrical insulators to prevent electrical breakdown and (2) good heat conductors to conduct away the heat generated. They should also have (3) high material strength to withstand forces and to maintain dimensional stability; (4) thermal expansion coefficients that closely match that of copper, to prevent cracking in the copper cladding during thermal cycling; (5) resistance to moisture absorption, since moisture can affect both mechanical and electrical properties and degrade performance; (6) stability in properties at temperature levels encountered in electronic applications; (7) ready availability and manufacturability; and, of course, (8) low cost. As you might have already guessed, no existing material has all of these desirable characteristics.

Glass–epoxy laminates made of an epoxy or polymide matrix reinforced by several layers of woven glass cloth are commonly used in the production of circuit boards. Polymide matrices are more expensive than epoxy but can withstand much higher temperatures. Polymer or polymide films are also used without reinforcement for flexible circuits.

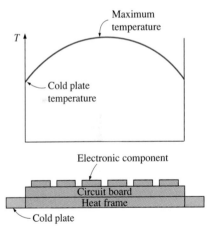

FIGURE 14-8

The path of heat flow in a conduction-cooled PCB and the temperature distribution.

FIGURE 14-9

A cabinet-style enclosure.

The Enclosure

An electronic system is not complete without a rugged enclosure (a case or a cabinet) that will *house* the circuit boards and the necessary peripheral equipment and connectors, *protect* them from the detrimental effects of the environment, and *provide* a cooling mechanism (Fig. 14-9). In a small electronic system such a personal computer, the enclosure can simply be an inexpensive box made of sheet metal with proper connectors and a small fan. But for a large system with several hundred PCBs, the design and construction of the enclosure are challenges for both electronic and thermal designers. An enclosure must provide *easy access* for service personnel so that they can identify and replace any defective parts easily and quickly in order to minimize down time, which can be very costly. But, at the same time, the enclosure must *prevent* any easy access by unauthorized people in order to protect the sensitive electronics from them as well as the people from possible electrical hazards. Electronic circuits are powered by low voltages (usually under ± 15 V), but the currents involved may be very high (sometimes a few hundred amperes).

Plug-in-type circuit boards make it very easy to replace a defective board and are commonly used in low-power electronic equipment. High-power circuit boards in large systems, however, are tightly attached to the racks of the cabinet with special brackets. A well-designed enclosure also includes switches, indicator lights, a screen to display messages and present information about the operation, and a key pad for user interface.

The printed circuit boards (PCBs) in a large system are plugged into a *back panel* through their edge connectors. The back panel supplies power to the PCBs and interconnects them to facilitate the passage of current from one board to another. The PCBs are assembled in an orderly manner in card racks or chassis. One or more such assemblies are housed in a *cabinet,* as shown in Figure 14-10.

Electronic enclosures come in a wide variety of sizes and shapes. Sheet metals such as thin-gauge aluminum or steel sheets are commonly used in the production of enclosures. The thickness of the enclosure walls depends on the shock and vibration requirements. Enclosures made of thick metal sheets or by casting can meet these requirements, but at the expense of increased weight and cost.

Electronic boxes are sometimes sealed to prevent the fluid inside (usually air) from leaking out and the water vapor outside from leaking in. Sealing against moisture migration is very difficult because of the small size of the water molecule and the large vapor pressure outside the box relative to that within the box. Sealing adds to the size, weight, and cost of an electronic box, especially in space or high-altitude operation, since the box in this case must withstand the larger forces due to the higher pressure differential between inside and outside the box.

14-3 ■ COOLING LOAD OF ELECTRONIC EQUIPMENT

The first step in the selection and design of a cooling system is the determination of the heat dissipation, which constitutes the *cooling load*. The easiest way to determine the power dissipation of electronic equipment is to measure

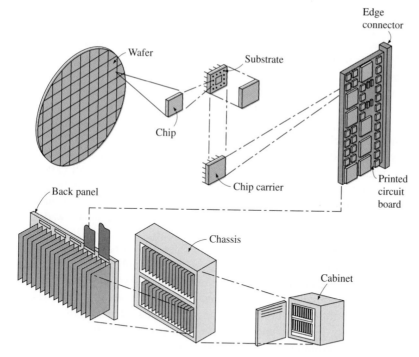

Wafer

Substrate

Chip

Chip carrier

Edge
connector

Printed
circuit
board

Back panel

Chassis

Cabinet

FIGURE 14-10

Different stages involved in the
production of an electronic system
(from Dally, Ref. 4).

the voltage applied V and the electric current I at the entrance of the electronic device under full-load conditions and to substitute them into the relation

$$\dot{W}_e = VI = I^2R \qquad (W) \qquad (14\text{-}1)$$

where $\dot{W}_e$ is the electric power consumption of the electronic device, which constitutes the *energy input* to the device.

The first law of thermodynamics requires that in *steady* operation the energy input into a system be equal to the energy output from the system. Considering that the only form of energy leaving the electronic device is heat generated as the current flows through resistive elements, we conclude that the heat dissipation or cooling load of an electronic device is equal to its power consumption. That is, $\dot{Q} = \dot{W}_e$, as shown in Figure 14-11. The exception to this rule is equipment that outputs other forms of energy as well, such as the emitter tubes of a radar, radio, or TV installation emitting radiofrequency (RF) electromagnetic radiation. In such cases, the cooling load will be equal to the difference between the power consumption and the RF power emission. An equivalent but cumbersome way of determining the cooling load of an electronic device is to determine the heat dissipated by each component in the device and then to add them up.

The discovery of *superconductor* materials that can operate at room temperature will cause drastic changes in the design of electronic devices and cooling techniques, since such devices will generate hardly any heat. As a result, more components can be packed into a smaller volume, resulting in

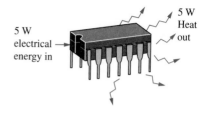

5 W
electrical
energy in

5 W
Heat
out

FIGURE 14-11

In the absence of other energy
interactions, the heat output of an
electronic device in steady operation is
equal to the power input to the device.

enhanced speed and reliability without having to resort to exotic cooling techniques.

Once the cooling load has been determined, it is common practice to inflate this number to leave some *safety margin*, or a "cushion," and to make some allowance for future growth. It is not uncommon to add another card to an existing system (such as adding a fax/modem card to a PC) to perform an additional task. But we should not go overboard in being conservative, since an *oversized* cooling system will cost more, occupy more space, be heavier, and consume more power. For example, there is no need to install a large and noisy fan in an electronic system just to be "safe" when a smaller one will do. For the same reason, there is no need to use an expensive and failure-prone liquid cooling system when air cooling is adequate. We should always keep in mind that the most desirable form of cooling is *natural convection cooling,* since it does not require any moving parts, and thus it is inherently reliable, quiet, and, best of all, free.

The cooling system of an electronic device must be designed considering the actual *field* operating conditions. In critical applications such as those in the military, the electronic device must undergo extensive testing to satisfy stringent requirements for safety and reliability. Several such codes exist to specify the minimum standards to be met in some applications.

The *duty cycle* is another important consideration in the design and selection of a cooling technique. The actual power dissipated by a device can be considerably less than the rated power, depending on its duty cycle (the fraction of time it is on). A 5-W power transistor, for example, will dissipate an average of 2 W of power if it is active only 40 percent of the time. If the chip of this transistor is 1.5 mm wide, 1.5 mm high, and 0.1 mm thick, then the heat flux on the chip will be $(2 \text{ W})/(0.15 \text{ cm})^2 = 89 \text{ W/cm}^2$.

An electronic device that is not running is in *thermal equilibrium* with its surroundings, and thus is at the temperature of the surrounding medium. When the device is turned on, the temperature of the components and thus the device starts rising as a result of absorbing the heat generated. The temperature of the device stabilizes at some value when the heat generated equals the heat removed by the cooling mechanism. At this point, the device is said to have reached *steady* operating conditions. The warming-up period during which the component temperature rises is called the *transient* operation stage (Fig. 14-12).

Another thermal factor that undermines the reliability of electronic devices is the thermal stresses caused by *temperature cycling*. In an experimental study (see Hilbert and Kube, Ref. 11), the failure rate of electronic devices subjected to deliberate temperature cycling of more than 20°C is observed to increase eightfold. *Shock* and *vibration* ore other common causes of failure for electronic devices and should be considered in the design and manufacturing process for increased reliability.

Most electronic devices operate for long periods of time, and thus their cooling mechanism is designed for steady operation. But electronic devices in some applications never run long enough to reach steady operation. In such cases, it may be sufficient to use a limited cooling technique, such as thermal storage for a short period, or not to use one at all. Transient operation can

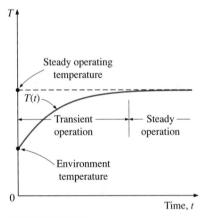

FIGURE 14-12

The temperature change of an electronic component with time as it reaches steady operating temperature after it is turned on.

also be caused by large swings in the environmental conditions. A common cooling technique for transient operation is to use a double-wall construction for the enclosure of the electronic equipment, with the space between the walls filled with a wax with a suitable melting temperature. As the wax melts, it absorbs a large amount of heat and thus delays overheating of the electronic components considerably. During off periods, the wax solidifies by rejecting heat to the environment.

14-4 ■ THERMAL ENVIRONMENT

An important consideration in the selection of a cooling technique is the *environment* in which the electronic equipment is to operate. Simple *ventilation holes* on the case may be all we need for the cooling of low-power-density electronics such as a TV or a VCR in a room, and a *fan* may be adequate for the safe operation of a home computer (Fig. 14-13). But the thermal control of the electronics of an *aircraft* will challenge thermal designers, since the environmental conditions in this case will swing from one extreme to another in a matter of minutes. The expected *duration of operation* in a hostile environment is also an important consideration in the design process. The thermal design of the electronics for an aircraft that cruises for hours each time it takes off will be quite different than that of a missile that has an operation time of a few minutes.

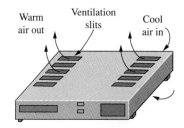

FIGURE 14-13

Strategically located ventilation holes are adequate to cool low-power electronics such as a TV or VCR.

The thermal environment in *marine applications* is relatively stable, since the ultimate heat sink in this case is water with a temperature range of 0°C to 30°C. For *ground applications,* however, the ultimate heat sink is the atmospheric air, whose temperature varies from −50°C at polar regions to +50°C in desert climates, and whose pressure ranges from about 70 kPa (0.7 atm) at 3000 m elevation to 107 kPa (1.08 atm) at 500 m below sea level. The combined convection and radiation heat transfer coefficient can range from 10 W/m² · °C in calm weather to 80 W/m² · °C in 100 km/h (62 mph) winds. Also, the surfaces of the devices facing the sun directly can be subjected to solar radiation heat flux of 1000 W/m² on a clear day.

In *airborne applications,* the thermal environment can change from 1 atm and 35°C on the ground to 19 kPa (0.2 atm) and −60°C at a typical cruising altitude of 12,000 m in minutes (Fig. 14-14). At supersonic velocities, the surface temperature of some part of the aircraft may rise 200°C above the environment temperature.

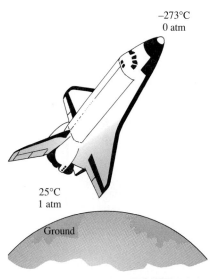

FIGURE 14-14

The thermal environment of a spacecraft changes drastically in a short time, and this complicates the thermal control of the electronics.

Electronic devices are rarely exposed to uncontrolled environmental conditions directly because of the wide variations in the environmental variables. Instead, a conditioned fluid such as air, water, or a dielectric fluid is used to serve as a local heat sink and as an intermediary between the electronic equipment and the environment, just like the air-conditioned air in a building providing thermal comfort to the human body. Conditioned air is the preferred cooling medium, since it is benign, readily available, and not prone to leakage. But its use is limited to equipment with low power densities, because of the low thermal conductivity of air. The thermal design of electronic equipment in military applications must comply with strict military standards in order to satisfy the utmost reliability requirements.

14-5 ■ ELECTRONICS COOLING IN DIFFERENT APPLICATIONS

The cooling techniques used in the cooling of electronic equipment vary widely, depending on the particular application. Electronic equipment designed for *airborne applications* such as airplanes, satellites, space vehicles, and missiles offers challenges to designers because it must fit into odd-shaped spaces because of the curved shape of the bodies, yet be able to provide adequate paths for the flow of fluid and heat. Most such electronic equipment are cooled by forced convection using pressurized air bled off a compressor. This compressed air is usually at a high temperature, and thus it is cooled first by expanding it through a turbine. The moisture in the air is also removed before the air is routed to the electronic boxes. But the removal process may not be adequate under rainy conditions. Therefore, electronics in some cases are placed in sealed finned boxes that are externally cooled to eliminate any direct contact with electronic components.

The electronics of short-range *missiles* do not need any cooling because of their short cruising times (Fig. 14-15). The missiles reach their destinations before the electronics reach unsafe temperatures. Long-range missiles such as cruise missiles, however, may have a flight time of several hours. Therefore, they must utilize some form of cooling mechanism. The first thing that comes to mind is to use forced convection with the air that rams the missile by utilizing its large *dynamic pressure*. However, the *dynamic temperature* of air, which is the rise in the temperature of the air as a result of the ramming effect, may be more than 50°C at speeds close to the speed of sound (Fig. 14-16). For example, at a speed of 320 m/s, the dynamic temperature of air is

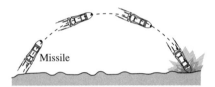

$$T_{\text{dynamic}} = \frac{\mathcal{V}^2}{2C_p} = \frac{(320 \text{ m/s})^2}{2(1005 \text{ J/kg} \cdot °C)} \left(\frac{1 \text{ J/kg}}{1 \text{ m}^2/\text{s}^2} \right) = 51°C \qquad (14\text{-}2)$$

Therefore, the temperature of air at a velocity of 320 m/s and a temperature of 30°C will rise to 81°C as a result of the conversion of kinetic energy to internal energy. Air at such high temperatures is not suitable for use as a cooling medium. Instead, cruise missiles are often cooled by taking advantage of the cooling capacity of the large quantities of *liquid fuel* they carry. The electronics in this case are cooled by passing the fuel through the cold plate of the electronic enclosure as it flows toward the combustion chamber.

Electronic equipment in *space vehicles* is usually cooled by a liquid circulated through the components, where heat is picked up, and then through a space radiator, where the waste heat is radiated into deep space at 0 K. Note that radiation is the only heat transfer mechanism for rejecting heat to the vacuum environment of space, and radiation exchange depends strongly on surface properties. Desirable radiation properties on surfaces can be obtained by special coatings and surface treatments. When electronics in sealed boxes are cooled by a liquid flowing through the outer surface of the electronics box, it is important to run a fan in the box to circulate the air, since there are no natural convection currents in space because of the absence of a gravity field.

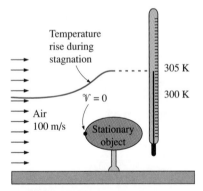

FIGURE 14-15

The electronics of short-range missiles may not need any cooling because of the short flight time involved.

FIGURE 14-16

The temperature of a gas having a specific heat C_p flowing at a velocity of $\mathcal{V}$ rises by $\mathcal{V}^2/2C_p$ when it is brought to a complete stop.

Electronic equipment in *ships* and *submarines* is usually housed in rugged cabinets to protect it from vibrations and shock during stormy weather. Because of easy access to water, water-cooled heat exchangers are commonly used to cool seaborn electronics. This is usually done by cooling air in a closed- or open-loop air-to-water heat exchanger and forcing the cool air to the electronic cabinet by a fan. When forced-air cooling is used, it is important to establish a flow path for air such that no trapped hot-air pockets will be formed in the cabinets.

Communication systems located at remote locations offer challenges to thermal designers because of the extreme conditions under which they operate. These electronic systems operate for long periods of time under adverse conditions such as rain, snow, high winds, solar radiation, high altitude, high humidity, and extremely high or low temperatures. Large communication systems are housed in specially built shelters. sometimes it is necessary to air-condition these shelters to safely dissipate the large quantities of heat dissipated by the electronics of communication systems.

Electronic components used in high-power *microwave equipment* such as radars generate enormous amounts of heat because of the low conversion efficiency of electrical energy to microwave energy. Klystron tubes of high-power radar systems where radiofrequency (RF) energy is generated can yield local heat fluxes as high as 2000 W/cm^2, which is close to one-third of the heat flux on the sun's surface. The safe and reliable dissipation of these high heat fluxes usually requires the immersion of such equipment in a suitable dielectric fluid that can remove large quantities of heat by boiling.

The manufacturers of electronic devices usually specify the *rate of heat dissipation* and the *maximum allowable component temperature* for reliable operation. These two numbers help us determine the cooling techniques that are suitable for the device under consideration.

The heat fluxes attainable at specified temperature differences are plotted in Figure 14-17 for some common heat transfer mechanisms. When the power rating of a device or component is given, the heat flux is determined by dividing the power rating by the exposed surface area of the device or component. Then suitable heat transfer mechanisms can be determined from Figure 14-17 from the requirement that the temperature difference between the surface of the device and the surrounding medium not exceed the allowable maximum value. For example, a heat flux of 0.5 W/cm^2 for an electronic component would result in a temperature difference of about 500°C between the component surface and the surrounding air if natural convection in air is used. Considering that the maximum allowable temperature difference is typically under 80°C, the natural convection cooling of this component in air is out of question. But forced convection with air is a viable option if using a fan is acceptable. Note that at heat fluxes greater than 1 W/cm^2, even forced convection with air will be inadequate, and we must use a sufficiently large heat sink or switch to a different cooling fluid such as water. Forced convection with water can be used effectively for cooling electronic components with high heat fluxes. Also note that dielectric liquids such as fluorochemicals can remove high heat fluxes by immersing the component directly in them.

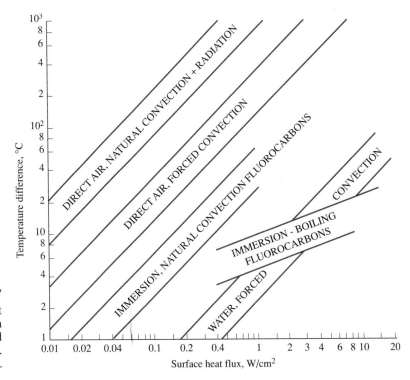

FIGURE 14-17

Heat fluxes that can be attained at specified temperature differences with various heat transfer mechanisms and fluids (from Kraus and Bar-Cohen, Ref. 16, p. 22; reproduced with permission).

FIGURE 14-18

The thermal resistance of a medium is proportional to its length in the direction of heat transfer, and inversely proportional to its heat transfer surface area and thermal conductivity.

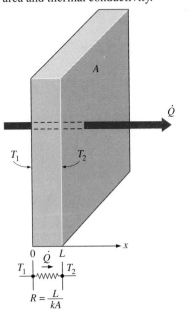

14-6 ■ CONDUCTION COOLING

Heat is *generated* in electronic components whenever electric current flows through them. The generated heat causes the temperature of the components to rise, and the resulting temperature difference drives the heat away from the components through a path of *least thermal resistance*. The temperature of the components stabilizes when the heat dissipated equals the heat generated. In order to minimize the temperature rise of the components, *effective heat transfer paths* must be established between the components and the ultimate heat sink, which is usually the atmospheric air.

The selection of a cooling mechanism for electronic equipment depends on the *magnitude* of the heat generated, *reliability* requirements, *environmental* conditions, and *cost*. For low-cost electronic equipment, inexpensive cooling mechanisms such as natural or forced convection with air as the cooling medium are commonly used. For high-cost, high-performance electronic equipment, however, it is often necessary to resort to expensive and complicated cooling techniques.

Conduction cooling is based on the *diffusion* of heat through a solid, liquid, or gas as a result of molecular interactions in the absence of any bulk fluid motion. Steady *one-dimensional* heat conduction through a plane medium of thickness L, heat transfer surface area A, and thermal conductivity k is given by (Fig. 14-18).

$$\dot{Q} = kA\frac{\Delta T}{L} = \frac{\Delta T}{R} \qquad \text{(W)} \qquad (14\text{-}3)$$

where

$$R = \frac{L}{kA} = \frac{\text{Length}}{\text{Thermal conductivity} \times \text{Heat transfer area}} \qquad (14\text{-}4)$$

is the *thermal resistance* of the medium and ΔT is the temperature difference across the medium. Note that this is analogous to the *electric current* being equal to the potential difference divided by the electrical resistance.

The thermal resistance concept enables us to solve heat transfer problems in an analogous manner to electric circuit problems using the thermal resistance network, as discussed in Chapter 3. When the rate of heat conduction $\dot{Q}$ is known, the temperature drop along a medium whose thermal resistance is R is simply determined from

$$\Delta T = \dot{Q}R \qquad \text{(°C)} \qquad (14\text{-}5)$$

Therefore, the greatest temperature drops along the path of heat conduction will occur across portions of the heat flow path with the largest thermal resistances.

Conduction in Chip Carriers

The conduction analysis of an electronic device starts with the *circuitry* or *junction* of a chip, which is the site of *heat generation*. In order to understand the heat transfer mechanisms at the chip level, consider the DIP (dual in-line package) type chip carrier shown in Figure 14-19.

The heat generated at the junction spreads throughout the chip and is conducted across the *thickness* of the chip. The spread of heat from the junction to the body of the chip is three-dimensional in nature, but can be

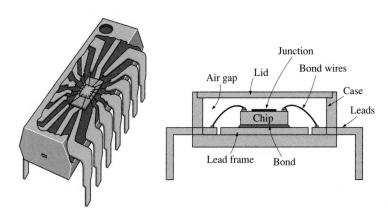

FIGURE 14-19

The schematic for the internal geometry and the cross-sectional view of a DIP (dual in-line package) type electronic device with 14 leads (from Dally, Ref. 4).

approximated as one-dimensional by adding a *constriction thermal resistance* to the thermal resistance network. For a small heat generation area of diameter d on a considerably larger body, the constriction resistance is given by

$$R_{\text{constriction}} = \frac{1}{2\sqrt{\pi dk}} \quad (°\text{C/W}) \tag{14-6}$$

where k is the thermal conductivity of the larger body.

The chip is attached to the lead frame with a highly conductive *bonding material* that provides a low-resistance path for heat flow from the chip to the lead frame. There is no metal connection between the *lead frame* and the *leads,* since this would short-circuit the entire chip. Therefore, heat flow from the lead frame to the leads is through the dielectric case material such as *plastic* or *ceramic.* Heat is then transported outside the electronic device through the leads.

When solving a heat transfer problem, it is often necessary to make some simplifying assumptions regarding the *primary* heat flow path and the magnitudes of heat transfer in other directions (Fig. 14-20). In the chip carrier discussed above, for example, heat transfer through the top is disregarded since it is very small because of the large thermal resistance of the stagnant air space between the chip and the lid. Heat transfer from the base of the electronic device is also considered to be negligible because of the low thermal conductivity of the case material and the lack of effective convection on the base surface.

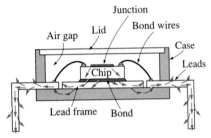

FIGURE 14-20

Heat generated at the junction of an electronic device flows through the path of least resistance.

EXAMPLE 14-3 Analysis of Heat Conduction in a Chip

A chip is dissipating 0.6 W of power in a DIP with 12 pin leads. The materials and the dimensions of various sections of this electronic device are as given in the table below. If the temperature of the leads is 40°C, estimate the temperature at the junction of the chip.

Section and material	Thermal conductivity, W/m · °C	Thickness, mm	Heat transfer surface area
Junction constriction	—	—	diameter 0.4 mm
Silicon chip	120[†]	0.4	3 mm × 3 mm
Eutectic bond	296	0.03	3 mm × 3 mm
Copper lead frame	386	0.25	3 mm × 3 mm
Plastic separator	1	0.2	12 × 1 mm × 0.25 mm
Copper leads	386	5	12 × 1 mm × 0.25 mm

[†]The thermal conductivity of silicon varies greatly with temperature from 153.5 W/m · °C at 27°C to 113.7 W/m · °C at 100°C, and the value 120 W/m · °C reflects the anticipation that the temperature of the silicon chip will be close to 100°C.

Solution The dimensions and power dissipation of a chip are given. The junction temperature of the chip is to be determined.

Assumptions **1** Steady operating conditions exist. **2** Heat transfer through various components is one-dimensional. **3** Heat transfer through the air gap and the lid on top of the chip is negligible because of the very large thermal resistance involved along this path.

Analysis The geometry of the device is as shown in Figure 14-20. We take the primary heat flow path to be the chip, the eutectic bond, the lead frame, the plastic insulator, and the 12 leads. When the constriction resistance between the junction and the chip is considered, the thermal resistance network for this problem becomes as shown in Figure 14-21.

The various thermal resistances on the path of primary heat flow are determined as follows.

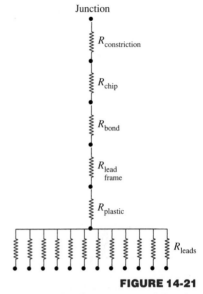

$$R_{constriction} = \frac{1}{2\sqrt{\pi}dk} = \frac{1}{2\sqrt{\pi}(0.4 \times 10^{-3} \text{ m})(120 \text{ W/m} \cdot °C)} = 5.88°C/W$$

$$R_{chip} = \left(\frac{L}{kA}\right)_{chip} = \frac{0.4 \times 10^{-3} \text{ m}}{(120 \text{ W/m} \cdot °C)(9 \times 10^{-6} \text{ m}^2)} = 0.37°C/W$$

$$R_{bond} = \left(\frac{L}{kA}\right)_{bond} = \frac{0.03 \times 10^{-3} \text{ m}}{(296 \text{ W/m} \cdot °C)(9 \times 10^{-6} \text{ m}^2)} = 0.01°C/W$$

$$R_{lead\ frame} = \left(\frac{L}{kA}\right)_{lead\ frame} = \frac{0.25 \times 10^{-3}\text{m}}{(386 \text{ W/m} \cdot °C)(9 \times 10^{-6} \text{ m}^2)} = 0.07°C/W$$

$$R_{plastic} = \left(\frac{L}{kA}\right)_{plastic} = \frac{0.2 \times 10^{-3}\text{m}}{(1 \text{ W/m} \cdot °C)(12 \times 0.25 \times 10^{-6} \text{ m}^2)} = 66.67°C/W$$

$$R_{leads} = \left(\frac{L}{kA}\right)_{leads} = \frac{5 \times 10^{-3}\text{m}}{(386 \text{ W/m} \cdot °C)(12 \times 0.25 \times 10^{-6} \text{ m}^2)} = 4.32°C/W$$

FIGURE 14-21

Thermal resistance network for the electronic device considered in Example 14-3.

Note that for heat transfer purposes, all 12 leads can be considered as a single lead whose cross-sectional area is 12 times as large. The alternative is to find the resistance of a single lead and to calculate the equivalent resistance for 12 such resistances connected in parallel. Both approaches give the same result.

All the resistances determined above are in series. Thus the total thermal resistance between the junction and the leads is determined by simply adding them up:

$$R_{total} = R_{junction\text{-}lead} = R_{constriction} + R_{chip} + R_{bond} + R_{lead\ frame} + R_{plastic} + R_{leads}$$
$$= (5.88 + 0.37 + 0.01 + 0.07 + 66.67 + 4.32)°C/W$$
$$= 77.32°C/W$$

Heat transfer through the chip can be expressed as

$$\dot{Q} = \left(\frac{\Delta T}{R}\right)_{junction\text{-}leads} = \frac{T_{junction} - T_{leads}}{R_{junction\text{-}leads}}$$

Solving for $T_{junction}$ and substituting the given values, the junction temperature is determined to be

$$T_{junction} = T_{leads} + \dot{Q}R_{junction\text{-}leads} = 40°C + (0.6 \text{ W})(77.32°C/W) = \textbf{86.4°C}$$

Note that the plastic layer between the lead frame and the leads accounts for 66.67/77.32 = 86 percent of the total thermal resistance and thus the 86 percent of the temperature drop ($0.6 \times 66.67 = 40°C$) between the junction and the leads. In other words, the temperature of the junction would be just $86.5 - 40 = 46.5°C$ if the thermal resistance of the plastic was eliminated.

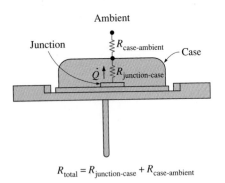

$$R_{total} = R_{junction-case} + R_{case-ambient}$$

$$T_{junction} = T_{ambient} + \dot{Q}R_{total}$$

FIGURE 14-22

The junction temperature of a chip depends on the external case-to-ambient thermal resistance as well as the internal junction-to-case resistance.

Discussion The simplified analysis given above points out that any attempt to reduce the thermal resistance in the chip carrier and thus improve the heat flow path should start with the plastic layer. We also notice from the magnitudes of individual resistances that some sections, such as the eutectic bond and the lead frame, have negligible thermal resistances, and any attempt to improve them further will have practically no effect on the junction temperature of the chip.

The analytical determination of the junction-to-case thermal resistance of an electronic device can be rather complicated and can involve considerable uncertainty, as shown above. Therefore, the manufacturers of electronic devices usually determine this value *experimentally* and list it as part of their product description. When the thermal resistance is known, the *temperature difference* between the junction and the outer surface of the device can be determined from

$$\Delta T_{junction-case} = T_{junction} - T_{case} = \dot{Q}R_{junction-case} \qquad (°C) \qquad (14\text{-}7)$$

where $\dot{Q}$ is the power consumed by the device.

The determination of the *actual junction temperature* depends on the ambient temperature $T_{ambient}$ as well as the thermal resistance $R_{case-ambient}$ between the case and the ambient (Fig. 14-22). The magnitude of this resistance depends on the type of ambient (such as air or water) and the fluid velocity. The two thermal resistances discussed above are in series, and the total resistance between the junction and the ambient is determined by simply adding them up:

$$R_{total} = R_{junction-ambient} = R_{junction-case} + R_{case-ambient} \qquad (°C/W) \qquad (14\text{-}8)$$

Many manufacturers of electronic devices go the extra step and list the *total resistance* between the junction and the ambient for various chip configurations and ambient conditions likely to be encountered. Once the total thermal resistance is available, the *junction temperature* corresponding to the specified power consumption (or heat dissipation rate) of $\dot{Q}$ is determined from

$$T_{junction} = T_{ambient} + \dot{Q}R_{junction-ambient} \qquad (°C) \qquad (14\text{-}9)$$

A typical chart for the total junction-to-ambient thermal resistance for a single DIP-type electronic device mounted on a circuit board is given in Figure 14-23 for various air velocities and lead numbers. The values at the intersections of the curves and the vertical axis represent the thermal resistances corresponding to *natural convection* conditions (zero air velocity). Note that the thermal resistance and thus the junction temperature decrease with increasing air velocity and the number of leads extending from the electronic device, as expected.

EXAMPLE 14-4 Predicting the Junction Temperature of a Device

A fan blows air at 30°C and a velocity of 200 m/min over a 1.2-W plastic DIP with 16 leads mounted on a PCB, as shown in Figure 14-24. Using data from Figure 14-23, determine the junction temperature of the electronic device. What would the junction temperature be if the fan were to fail?

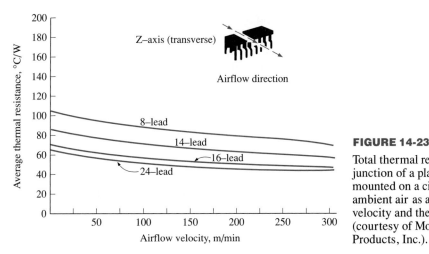

FIGURE 14-23

Total thermal resistance between the junction of a plastic DIP device mounted on a circuit board and the ambient air as a function of the air velocity and the number of leads (courtesy of Motorola Semiconductor Products, Inc.).

Solution A plastic DIP with 16 leads is cooled by forced air. Using data supplied by the manufacturer, the junction temperature is to be determined.

Assumptions Steady operating conditions exist

Analysis The junction-to-ambient thermal resistance of the device with 16 leads corresponding to an air velocity of 200 m/min is determined from Figure 14-23 to be

$$R_{\text{junction-ambient}} = 55°C/W$$

Then the junction temperature can be determined from Equation 14-9 to be

$$T_{\text{junction}} = T_{\text{ambient}} + \dot{Q}R_{\text{junction-ambient}} = 30°C + (1.2\ W)(55°C/W) = \textbf{96°C}$$

When the fan fails, the airflow velocity over the device will be zero. The total thermal resistance in this case is determined from the same chart by reading the value at the intersection of the curve and the vertical axis to be

$$R_{\text{junction-ambient}} = 70°C/W$$

which gives

$$T_{\text{junction}} = T_{\text{ambient}} + \dot{Q}R_{\text{junction-ambient}} = 30°C + (1.2\ W)(70°C/W) = \textbf{114°C}$$

Discussion Note that the temperature of the junction will rise by 18°C when the fan fails. Of course, this analysis assumes the temperature of the surrounding air still to be 30°C, which may no longer be the case. Any increase in the ambient temperature as a result of inadequate airflow will reflect on the junction temperature, which will seriously jeopardize the safety of the electronic device.

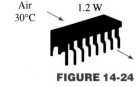

FIGURE 14-24

Schematic for Example 14-4.

Conduction in Printed Circuit Boards (PCBs)

Heat-generating electronic devices are commonly mounted on thin rectangular *boards,* usually 10 cm × 15 cm in size, made of electrically insulating materials such as *glass–epoxy laminates,* which are also poor conductors of heat. The resulting printed circuit boards (PCBs) are usually cooled by blowing air or passing a dielectric liquid through them. In such cases, the components on the PCBs are cooled directly, and we are not concerned about heat conduction along the PCBs. But in some critical applications such as those encountered in the military, the PCBs are contained in *sealed enclosures,* and the boards provide the only effective heat path between the components and the heat sink attached to the sealed enclosure. In such cases, heat transfer from the side faces of the PCBs is negligible, and the heat generated in the components must be *conducted* along the PCB toward its edges, which are clamped to cold plates for removing the heat externally.

Heat transfer along a PCB is complicated in nature because of the multidimensional effects and nonuniform heat generation on the surfaces. We can still obtain sufficiently accurate results by using the thermal resistance network in one or more dimensions.

Copper or aluminum *cladding, heat frames,* or *cores* are commonly used to enhance heat conduction along the PCBs. The thickness of the copper cladding on the PCB is usually expressed in terms of *ounces of copper,* which is the thickness of 1-ft^2 copper sheet made of one ounce of copper. An ounce of copper is equivalent to 0.03556-mm (1.4-mil) thickness of a copper layer.

When analyzing heat conduction along a PCB with *copper* (or aluminum) *cladding* on one or both sides, often the question arises whether heat transfer along the epoxy laminate can be ignored relative to that along the copper layer, since the thermal conductivity of copper is about 1500 times that of epoxy. The answer depends on the relative cross-sectional areas of each layer, since heat conduction is proportional to the cross-sectional area as well as the thermal conductivity.

Consider a copper-cladded PCB of width w and length L, across which the temperature difference is ΔT, as shown in Figure 14-25. Assuming heat conduction is along the length L only and heat conduction in other directions is negligible, the rate of heat conduction along this PCB is the sum of the heat conduction along the *epoxy board* and the *copper* layer and is expressed as

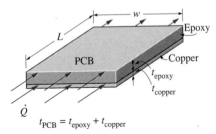

FIGURE 14-25

Schematic of a copper-cladded epoxy board and heat conduction along it.

$$\dot{Q}_{PCB} = \dot{Q}_{epoxy} + \dot{Q}_{copper} = \left(kA\frac{\Delta T}{L}\right)_{epoxy} + \left(kA\frac{\Delta T}{L}\right)_{copper}$$

$$= [(kA)_{epoxy} + (kA)_{copper}]\frac{\Delta T}{L} \qquad (14\text{-}10)$$

$$= [(kt)_{epoxy} + (kt)_{copper}]\frac{w\Delta T}{L}$$

where t denotes the thickness. Therefore, the relative magnitudes of heat conduction along the two layers depend on the relative magnitudes of the thermal conductivity–thickness product kt of the layer. Therefore, if the kt product of the copper is 100 times that of epoxy, then neglecting heat conduction along the epoxy board will involve an error of just 1 percent, which is negligible.

We can also define an **effective thermal conductivity** for metal-cladded PCBs as

$$k_{\text{eff}} = \frac{(kt)_{\text{epoxy}} + (kt)_{\text{copper}}}{t_{\text{epoxy}} + t_{\text{copper}}} \quad (\text{W/m} \cdot {}^\circ\text{C}) \quad (14\text{-}11)$$

so that the rate of heat conduction along the PCB can be expressed as

$$\dot{Q}_{\text{PCB}} = k_{\text{eff}}(t_{\text{epoxy}} + t_{\text{copper}})\frac{w\Delta T}{L} = k_{\text{eff}} A_{\text{PCB}} \frac{\Delta T}{L} \quad (\text{W}) \quad (14\text{-}12)$$

where $A_{\text{PCB}} = w(t_{\text{epoxy}} + t_{\text{copper}})$ is the area normal to the direction of heat transfer. When there are holes or discontinuities along the copper cladding, the above analysis needs to be modified to account for their effect.

EXAMPLE 14-5   Heat Conduction along a PCB with Copper Cladding

Heat is to be conducted along a PCB with copper cladding on one side. The PCB is 10-cm long and 10-cm wide, and the thickness of the copper and epoxy layers are 0.04 mm and 0.16 mm, respectively, as shown in Figure 14-26. Disregarding heat transfer from side surfaces, determine the percentages of heat conduction along the copper ($k = 386$ W/m · °C) and epoxy ($k = 0.26$ W/m · °C) layers. Also, determine the effective thermal conductivity of the PCB.

Solution   A PCB with copper cladding is given. The percentages of heat conduction along the copper and epoxy layers as well as the effective thermal conductivity of the PCB are to be determined.

Assumptions   **1** Heat conduction along the PCB is one-dimensional since heat transfer from side surfaces is negligible. **2** The thermal properties of epoxy and copper layers are constant.

Analysis   The length and width of both layers are the same, and so is the temperature difference across each layer. Heat conduction along a layer is proportional to the thermal conductivity–thickness product kt, which is determined for each layer and the entire PCB to be

$(kt)_{\text{copper}} = (386\ \text{W/m} \cdot {}^\circ\text{C})(0.04 \times 10^{-3}\ \text{m}) = 15.44 \times 10^{-3}\ \text{W/}{}^\circ\text{C}$
$(kt)_{\text{epoxy}} = (0.26\ \text{W/m} \cdot {}^\circ\text{C})(0.16 \times 10^{-3}\ \text{m}) = 0.04 \times 10^{-3}\ \text{W/}{}^\circ\text{C}$
$(kt)_{\text{PCB}} = (kt)_{\text{copper}} + (kt)_{\text{epoxy}} = (15.44 + 0.04) \times 10^{-3}\ \text{m} = 15.48 \times 10^{-3}\ \text{W/}{}^\circ\text{C}$

Therefore, heat conduction along the epoxy board will constitute

$$f = \frac{(kt)_{\text{epoxy}}}{(kt)_{\text{PCB}}} = \frac{0.04 \times 10^{-3}\ \text{W/}{}^\circ\text{C}}{15.48 \times 10^{-3}\ \text{W/}{}^\circ\text{C}} = 0.0026$$

or 0.26 percent of the thermal conduction along the PCB, which is negligible. Therefore, heat conduction along the epoxy layer in this case can be disregarded without any reservations.

The effective thermal conductivity of the board is determined from Equation 14-11 to be

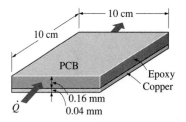

FIGURE 14-26
Schematic for Example 14-5.

$$k_{\text{eff}} = \frac{(kt)_{\text{epoxy}} + (kt)_{\text{copper}}}{t_{\text{epoxy}} + t_{\text{copper}}} = \frac{(15.44 + 0.04) \times 10^{-3} \text{ W/°C}}{(0.16 + 0.04) \times 10^{-3} \text{ m}} = \textbf{77.4 W/m} \cdot \textbf{°C}$$

That is, the entire PCB can be treated as a 0.20-mm-thick single homogeneous layer whose thermal conductivity is 77.4 W/m · °C for heat transfer along its length.

Discussion Note that a very thin layer of copper cladding on a PCB improves heat conduction along the PCB drastically, and thus it is commonly used in conduction-cooled electronic devices.

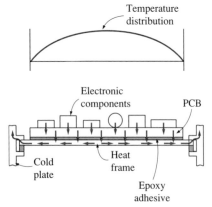

FIGURE 14-27

Conduction cooling of a printed circuit board with a heat frame, and the typical temperature distribution along the frame.

FIGURE 14-28

Planting the epoxy board with copper fillings decreases the thermal resistance across its thickness considerably.

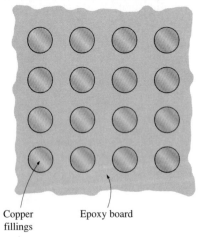

Heat Frames

In applications where direct cooling of circuit boards by passing air or a dielectric liquid over the electronic components is not allowed, and the junction temperatures are to be maintained relatively low to meet strict safety requirements, a thick *heat frame* is used instead of a thin layer of copper cladding. This is especially the case for multilayer PCBs that are packed with high-power output chips.

The schematic of a PCB that is conduction-cooled via a heat frame is shown in Figure 14-27. Heat generated in the chips is conducted through the circuit board, through the epoxy adhesive, to the center of the heat frame, along the heat frame, and to a heat sink or cold plate, where heat is externally removed.

The heat frame provides a low-resistance path for the flow of heat from the circuit board to the heat sink. The thicker the heat frame, the lower the thermal resistance, and thus the smaller the temperature difference between the center and the ends of the heat frame. When the heat load is evenly distributed on the PCB, there will be thermal symmetry about the centerline, and the temperature distribution along the heat frame and the PCB will be *parabolic* in nature, with the chips in the middle of the PCB (furthest away from the edges) operating at the highest temperatures and the chips near the edges operating at the lowest temperatures. Also, when the PCB is cooled from two edges, heat generated in the left half of the PCB will flow toward the left edge and heat generated in the right half will flow toward the right edge of the heat frame. But when the PCB is cooled from all four edges, the heat transfer along the heat frame as well as the resistance network will be two-dimensional.

When a heat frame is used, heat conduction in the epoxy layer of the PCB is through its *thickness* instead of along its length. The epoxy layer in this case offers a much smaller resistance to heat flow because of the short distance involved. This resistance can be made even smaller by drilling holes in the epoxy and filling them with copper, as shown in Figure 14-28. These copper fillings are usually 1 mm in diameter and their centers are a few millimeters apart. Such highly conductive fillings provide easy passageways for heat from one side of the PCB to the other and result in considerable reduction in the thermal resistance of the board along its thickness, as shown in the following examples.

EXAMPLE 14-6   Thermal Resistance of an Epoxy Glass Board

Consider a 10-cm × 15-cm glass–epoxy laminate ($k = 0.26$ W/m · °C) whose thickness is 0.8 mm, as shown in Figure 14-29. Determine the thermal resistance of this epoxy layer for heat flow (a) along the 15 cm long side and (b) across its thickness.

Solution   The dimensions of an epoxy–glass laminate are given. The thermal resistances for heat flow along the layers and across the thickness are to be determined.

Assumptions   **1** Heat conduction in the laminate is one-dimensional in either case. **2** Thermal properties of the laminate are constant.

Analysis   The thermal resistance of a plane parallel medium in the direction of heat conduction is given by

$$R = \frac{L}{kA}$$

where L is the length in the direction of heat flow, k is the thermal conductivity, and A is the area normal to the direction of heat conduction. Substituting the given values, the thermal resistances of the board for both cases are determined to be

(a)

$$R_{along\ length} = \left(\frac{L}{kA}\right)_{along\ length}$$

$$= \frac{0.15\ m}{(0.26\ W/m \cdot °C)(0.1\ m)(0.8 \times 10^{-3}\ m)} = 7212°C/W$$

(b)

$$R_{across\ thickness} = \left(\frac{L}{kA}\right)_{across\ thickness}$$

$$= \frac{0.8 \times 10^{-3}\ m}{(0.26\ W/m \cdot °C)(0.1\ m)(0.15\ m)} = 0.21°C/W$$

Discussion   Note that heat conduction at a rate of 1 W along this PCB would cause a temperature difference of 7212°C across a length of 15 cm. But the same rate of heat conduction would cause a temperature difference of only 0.21°C across the thickness of the epoxy board.

EXAMPLE 14-7   Planting Cylindrical Copper Fillings in an Epoxy Board

Reconsider the 10-cm × 15-cm glass–epoxy laminate ($k = 0.26$ W/m · °C) of thickness 0.8 mm discussed in Example 14-6. In order to reduce the thermal resistance across its thickness from the current value of 0.21°C/W, cylindrical copper fillings ($k = 386$ W/m · °C) of 1-mm diameter are to be planted throughout the board with a center-to-center distance of 2.5 mm, as shown in Figure 14-30. Determine the new value of the thermal resistance of the epoxy board for heat conduction across its thickness as a result of this modification.

Solution   Cylindrical copper fillings are planted throughout an epoxy glass board. The thermal resistance of the board across its thickness is to be determined.

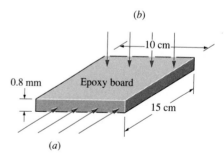

FIGURE 14-29
Schematic for Example 14-6.

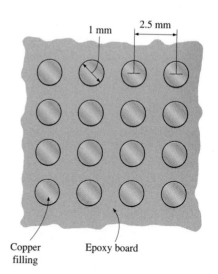

Copper   Epoxy board
filling

FIGURE 14-30
Schematic for Example 14-7.

Assumptions **1** Heat conduction along the board is one-dimensional. **2** Thermal properties of the board are constant.

Analysis Heat flow through the thickness of the board in this case will take place partly through the copper fillings and partly through the epoxy in parallel paths. The thickness of both materials is the same and is given to be 0.8 mm. But we also need to know the surface area of each material before we can determine the thermal resistances.

It is stated that the distance between the centers of the copper fillings is 2.5 mm. That is, there is only one 1-mm-diameter copper filling in every 2.5 mm × 2.5 mm square section of the board. The number of such squares and thus the number of copper fillings on the board are

$$n = \frac{\text{Area of the board}}{\text{Area of one square}} = \frac{(100 \text{ mm})(150 \text{ mm})}{(2.5 \text{ mm})(2.5 \text{ mm})} = 2400$$

Then the surface areas of the copper fillings and the remaining epoxy layer become

$$A_{copper} = n\frac{\pi D^2}{4} = (2400)\frac{\pi(1 \times 10^{-3} \text{ m})^2}{4} = 0.001885 \text{ m}^2$$

$$A_{total} = (\text{Length})(\text{Width}) = (0.1 \text{ m})(0.15 \text{ m}) = 0.015 \text{ m}^2$$

$$A_{epoxy} = A_{total} - A_{copper} = (0.015 - 0.001885) \text{ m}^2 = 0.013115 \text{ m}^2$$

The thermal resistance of each material is

$$R_{copper} = \left(\frac{L}{kA}\right)_{copper} = \frac{0.8 \times 10^{-3} \text{ m}}{(386 \text{ W/m} \cdot °\text{C})(0.001885 \text{ m}^2)} = 0.0011°\text{C/W}$$

$$R_{epoxy} = \left(\frac{L}{kA}\right)_{epoxy} = \frac{0.8 \times 10^{-3} \text{ m}}{(0.26 \text{ W/m} \cdot °\text{C})(0.013115 \text{ m}^2)} = 0.2346°\text{C/W}$$

Noting that these two resistances are in parallel, the equivalent thermal resistance of the entire board is determined from

$$\frac{1}{R_{board}} = \frac{1}{R_{copper}} + \frac{1}{R_{epoxy}} = \frac{1}{0.0011°\text{C/W}} + \frac{1}{0.2346°\text{C/W}}$$

which gives

$$R_{board} = \mathbf{0.00109°\text{C/W}}$$

Discussion Note that the thermal resistance of the epoxy board has dropped from 0.21°C/W by a factor of almost 200 to just 0.00109°C/W as a result of implanting 1-mm-diameter copper fillings into it. Therefore, implanting copper pins into the epoxy laminate has virtually eliminated the thermal resistance of the epoxy across its thickness.

EXAMPLE 14-8   Conduction Cooling of PCBs by a Heat Frame

A 10-cm × 12-cm circuit board dissipating 24 W of heat is to be conduction-cooled by a 1.2-mm-thick copper heat frame ($k = 386$ W/m · °C) 10 cm × 14 cm in size. The epoxy laminate ($k = 0.26$ W/m · °C) has a thickness of 0.8 mm and is attached to the heat frame with conductive epoxy adhesive ($k = 1.8$ W/m · °C) of 0.13-mm thickness, as shown in Figure 14-31. The PCB is attached to a heat sink by clamping a 5-mm-wide portion of the edge to the heat sink from both ends.

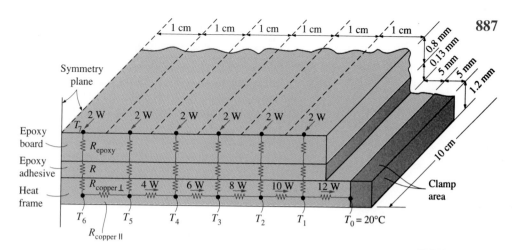

FIGURE 14-31

The schematic and thermal resistance network for **Example 14-8.**

The temperature of the heat frame at this point is 20°C. Heat is uniformly generated on the PCB at a rate of 2 W per 1-cm × 10-cm strip. Considering only one-half of the PCB board because of symmetry, determine the maximum temperature on the PCB and the temperature distribution along the heat frame.

Solution A circuit board with uniform heat generation is to be conduction-cooled by a copper heat frame. Temperature distribution along the heat frame and the maximum temperature in the PCB are to be determined.

Assumptions **1** Steady operating conditions exist. **2** Thermal properties are constant. **3** There is no direct heat dissipation from the surface of the PCB, and thus all the heat generated is conducted by the heat frame to the heat sink.

Analysis The PCB under consideration possesses thermal symmetry about the centerline. Therefore, the heat generated on the left half of the PCB is conducted to the left heat sink, and the heat generated on the right half is conducted to the right heat sink. Thus we need to consider only half of the board in the analysis.

The maximum temperature will occur at a location furthest away from the heat sinks, which is the symmetry line. Therefore, the temperature of the electronic components located at the center of the PCB will be the highest, and their reliability will be the lowest.

Heat generated in the components on each strip is conducted through the epoxy layer underneath. Heat is then conducted across the epoxy adhesive and to the middle of the copper heat frame. Finally, heat is conducted along the heat frame to the heat sink.

The thermal resistance network associated with heat flow in the right half of the PCB is also shown in Figure 14-31. Note that all vertical resistances are identical and are equal to the sum of the three resistances in series. Also note that heat conduction toward the heat sink is assumed to be predominantly along the heat frame, and conduction along the epoxy adhesive is considered to be negligible. This assumption is quite reasonable, since the conductivity–thickness product of the heat frame is much larger than those of the other two layers.

The properties and dimensions of various sections of the PCB are summarized in the following table.

Section and material	Thermal conductivity, W/m · °C	Thickness, mm	Heat transfer surface area
Epoxy board	0.26	0.8	10 mm × 100 mm
Epoxy adhesive	1.8	0.13	10 mm × 100 mm
Copper heat frame, ⊥ (normal to frame)	386	0.6	10 mm × 100 mm
Copper heat frame, ∥ (along the frame)	386	10	1.2 mm × 100 mm

Using the values in the table, the various thermal resistances are determined to be

$$R_{epoxy} = \left(\frac{L}{kA}\right)_{epoxy} = \frac{0.8 \times 10^{-3}\,m}{(0.26\,W/m \cdot °C)(0.01\,m \times 0.1\,m)} = 3.077°C/W$$

$$R_{adhesive} = \left(\frac{L}{kA}\right)_{adhesive} = \frac{0.13 \times 10^{-3}\,m}{(1.8\,W/(m \cdot °C)(0.01\,m \times 0.1\,m)} = 0.072°C/W$$

$$R_{copper, \perp} = \left(\frac{L}{kA}\right)_{copper, \perp} = \frac{0.6 \times 10^{-3}\,m}{(386\,W/m \cdot °C)(0.01\,m \times 0.1\,m)} = 0.002°C/W$$

$$R_{frame} = R_{copper, \parallel} = \left(\frac{L}{kA}\right)_{copper, \parallel} = \frac{0.01\,m}{(386\,W/m \cdot °C)(0.0012\,m \times 0.1\,m)}$$

$$= 0.216°C/W$$

The combined resistance between the electronic components on each strip and the heat frame can be determined, by adding the three resistances in series, to be

$$R_{vertical} = R_{epoxy} + R_{adhesive} + R_{copper, \perp}$$
$$= (3.077 + 0.072 + 0.002)°C/W$$
$$= 3.151°C/W$$

The various temperatures along the heat frame can be determined from the relation

$$\Delta T = T_{high} - T_{low} = \dot{Q}R$$

where R is the thermal resistance between two specified points, $\dot{Q}$ is the heat transfer rate through that resistance, and ΔT is the temperature difference across that resistance.

The temperature at the location where the heat frame is clamped to the heat sink is given as $T_0 = 20°C$. Noting that the entire 12 W of heat generated on the right half of the PCB must pass through the last thermal resistance adjacent to the heat sink, the temperature T_1 can be determined from

$$T_1 = T_0 + \dot{Q}_{1-0} R_{1-0} = 20°C + (12\,W)(0.216°C/W) = \mathbf{22.59°C}$$

Following the same line of reasoning, the temperatures at specified locations along the heat frame are determined to be

$$T_2 = T_1 + \dot{Q}_{2-1}R_{2-1} = 22.59°C + (10\ W)(0.216°C/W) = \mathbf{24.75°C}$$

$$T_3 = T_2 + \dot{Q}_{3-2}R_{3-2} = 24.75°C + (8\ W)(0.216°C/W) = \mathbf{26.48°C}$$

$$T_4 = T_3 + \dot{Q}_{4-3}R_{4-3} = 26.48°C + (6\ W)(0.216°C/W) = \mathbf{27.78°C}$$

$$T_5 = T_4 + \dot{Q}_{5-4}R_{5-4} = 27.78°C + (4\ W)(0.216°C/W) = \mathbf{28.64°C}$$

$$T_6 = T_5 + \dot{Q}_{6-5}R_{6-5} = 28.64°C + (2\ W)(0.216°C/W) = \mathbf{29.07°C}$$

Finally, T_7, which is the maximum temperature on the PCB, is determined from

$$T_7 = T_6 + \dot{Q}_{\text{vertical}}R_{\text{vertical}} = 29.07°C + (2\ W)(3.151°C/W) = \mathbf{35.37°C}$$

Discussion The maximum temperature difference between the PCB and the heat sink is only 15.37°C, which is very impressive considering that the PCB has no direct contact with the cooling medium. The junction temperatures in this case can be determined by calculating the temperature difference between the junction and the leads of the chip carrier at the point of contact to the PCB and adding 35.37°C to it. The maximum temperature rise of 15.37°C can be reduced, if necessary, by using a thicker heat frame.

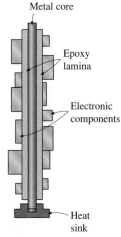

FIGURE 14-32

A two-sided printed circuit board with a metal core for conduction cooling.

Conduction cooling can also be used when electronic components are mounted on *both sides* of the PCB by using a copper or aluminum core plate in the middle of the PCB, as shown in Figure 14-32. The heat load in this case will be twice that of a PCB that has components on one side only. Again, heat generated in the components will be conducted through the thickness of the epoxy layer to the *metal core,* which serves as a *channel* for effective heat removal. The thickness of the core is selected such that the maximum component temperatures remain below specified values to meet a prescribed reliability criterion.

The thermal expansion coefficients of aluminum and copper are about twice as large as that of the glass–epoxy. This large difference in *thermal expansion coefficients* can cause *warping* on the PCBs if the epoxy and the metal are not bonded properly. One way of avoiding warping is to use PCBs with components on both sides, as discussed above. Extreme care should be exercised during the bonding and curing process when components are mounted on only one side of the PCB.

The Thermal Conduction Module (TCM)

The heat flux for logic chips has been increasing steadily as a result of the increasing circuit density in the chips. For example, the peak flux at the chip level has increased from 2 W/cm^2 on IBM System 370 to 20 W/cm^2 on IBM System 3081, which was introduced in the early 1980s. The conventional forced-air cooling technique used in earlier machines was inadequate for removing such high heat fluxes, and it was necessary to develop a new and more effective cooling technique. The result was the **thermal conduction**

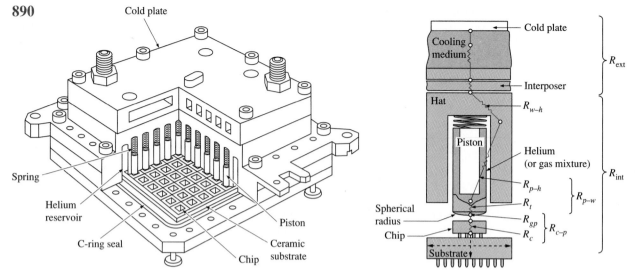

FIGURE 14-33

Cutaway view of the thermal conduction module (TCM), and the thermal resistance network between a single chip and the cooling fluid (courtesy of IBM Corporation).

module (TCM), shown in Figure 14-33. The TCM was different from previous chip packaging designs in that it incorporated both electrical and thermal considerations in early stages of chip design. Previously, a chip would be designed primarily by electrical designers, and the thermal designer would be told to come up with a cooling scheme for the chip. That approach resulted in unnecessarily high junction temperatures, and reduced reliability, since the thermal designer had no direct access to the chip. The TCM reflects a new philosophy in electronic packaging in that the thermal and electrical aspects are given equal treatment in the design process, and a successful thermal design starts at the chip level.

In the TCM, one side of the chip is reserved for electrical connections and the other side for heat rejection. The chip is cooled by direct contact to the cooling system to minimize the junction-to-case thermal resistance.

The TCM houses 100–118 logic chips, which are bonded to a multilayer ceramic substrate 90 mm × 90 mm in size with solder balls, which also provide the electrical connections between the chips and the substrate. Each chip dissipates about 4 W of power. The heat flow path from the chip to the metal casing is provided by a *piston,* which is pressed against the back surface of the chip by a spring. The tip of the piston is slightly curved to ensure good thermal contact even when the chip is tilted or misaligned.

Heat conduction between the chip and the piston occurs primarily through the gas space between the chip and the piston because of the limited contact area between them. To maximize heat conduction through the gas, the air in the TCM cavity is evacuated and is replaced by helium gas, whose thermal conductivity is about six times that of air. Heat is then conducted through the piston, across the surrounding helium gas layer, through the module housing, and finally to the cooling water circulating through the cold plate attached to the top surface of the TCM.

The total *internal thermal resistance* R_{int} of the TCM is about 8°C/W, which is rather impressive. This means that the temperature difference between the chip surface and the outer surface of the housing of the module will be only 24°C for a 3-W chip. The *external thermal resistance* R_{ext} between the housing of the module and the cooling fluid is usually comparable in magnitude to R_{int}. Also, the thermal resistance between the junction and the surface of the chip can be taken to be 1°C/W.

The compact design of the TCM significantly reduces the *distance* between the chips, and thus the signal transmission time between the chips. This, in turn, increases the *operating speed* of the electronic device.

EXAMPLE 14-9   Cooling of Chips by the Thermal Conduction Module

Consider a thermal conduction module with 100 chips, each dissipating 3 W of power. The module is cooled by water at 25°C flowing through the cold plate on top of the module. The thermal resistances in the path of heat flow are R_{chip} = 1°C/W between the junction and the surface of the chip, R_{int} = 8°C/W between the surface of the chip and the outer surface of the thermal conduction module, and R_{ext} = 6°C/W between the outer surface of the module and the cooling water. Determine the junction temperature of the chip.

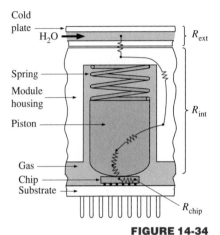

FIGURE 14-34

Thermal resistance network for Example 14-9.

Solution   A thermal conduction module TCM with 100 chips is cooled by water. The junction temperature of the chip is to be determined.

Assumptions   **1** Steady operating conditions exist. **2** Heat transfer through various components is one-dimensional.

Analysis   Because of symmetry, we will consider only one of the chips in our analysis. The thermal resistance network for heat flow is given in Figure 14-34. Noting that all resistances are in series, the total thermal resistance between the junction and the cooling water is

$$R_{total} = R_{junction-water} = R_{chip} + R_{int} + R_{ext} = (1 + 8 + 6)°C/W = 15°C/W$$

Noting that the total power dissipated by the chip is 3 W and the water temperature is 25°C, the junction temperature of the chip in steady operation can be determined from

$$\dot{Q} = \left(\frac{\Delta T}{R}\right)_{junction-water} = \frac{T_{junction} - T_{water}}{R_{junction-water}}$$

Solving for $T_{junction}$ and substituting the specified values gives

$$T_{junction} = T_{water} + \dot{Q}R_{junction-water} = 25°C + (3 \text{ W})(15°C/W) = \textbf{70°C}$$

Therefore, the circuits of the chip will operate at about 70°C, which is considered to be a safe operating temperature for silicon chips.

Cold plates are usually made of metal plates with fluid channels running through them, or copper tubes attached to them by brazing. Heat transferred to the cold plate is conducted to the tubes, and from the tubes to the fluid

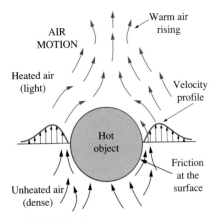

FIGURE 14-35

Natural convection currents around a
hot object in air.

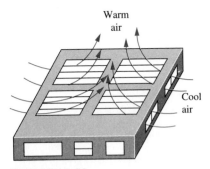

FIGURE 14-36

Natural convection cooling of electronic
components in an enclosure with air
vents.

flowing through them. The heat carried away by the fluid is finally dissipated
to the ambient in a heat exchanger.

14-7 ■ AIR COOLING: NATURAL CONVECTION AND RADIATION

Low-power electronic systems are conveniently cooled by *natural convection*
and *radiation*. Natural convection cooling is very desirable, since it does not
involve any fans that may break down.

Natural convection is based on the *fluid motion* caused by the density
differences in a fluid due to a temperature difference. A fluid expands when
heated and becomes less dense. In a gravitational field, this lighter fluid rises
and initiates a motion in the fluid called *natural convection currents* (Fig.
14-35). Natural convection cooling is most effective when the path of the
fluid is relatively free of obstacles, which tend to slow down the fluid, and is
least effective when the fluid has to pass through narrow flow passages and
over many obstacles.

The magnitude of the natural convection heat transfer between a surface
and a fluid is directly related to the *flow rate* of the fluid. The higher the flow
rate, the higher the heat transfer rate. In natural convection, no blowers are
used and therefore the flow rate cannot be controlled externally. The flow rate
in this case is established by the dynamic balance of *buoyancy* and *friction*.
The *larger* the temperature difference between the fluid adjacent to a hot
surface and the fluid away from it, the *larger* the buoyancy force, and the
stronger the natural convection currents, and thus the *higher* the heat transfer
rate. Also, whenever two bodies in contact move relative to each other, a
friction force develops at the contact surface in the direction opposite to that
of the motion. This opposing force slows down the fluid, and thus reduces
the flow rate of the fluid. Under steady conditions, the airflow rate driven by
buoyancy is established at the point where these two effects *balance* each
other. The friction force increases as more and more solid surfaces are introduced, seriously disrupting the fluid flow and heat transfer.

Electronic components or PCBs placed in *enclosures* such as a TV or
VCR are cooled by natural convection by providing a sufficient number of
vents on the case to enable the cool air to enter and the heated air to leave
the case freely, as shown in Figure 14-36. From the heat transfer point of
view, the vents should be as large as possible to minimize the flow resistance
and should be located at the bottom of the case for air entering and at the top
for air leaving. But equipment and human safety requirements dictate that the
vents should be quite narrow to discourage unintended entry into the box.
Also, concern about human habits such as putting a cup of coffee on the
closest flat surface make it very risky to place vents on the top surface. The
narrow clearance allowed under the case also offers resistance to airflow.
Therefore, vents on the enclosures of natural convection–cooled electronic
equipment are usually placed at the *lower section* of the side or back surfaces
for air inlet and at the *upper section* of those surfaces for air exit.

The heat transfer from a surface at temperature T_s to a fluid at temperature T_{fluid} by convection is expressed as

$$\dot{Q}_{\text{conv}} = h_{\text{conv}} A \Delta T = h_{\text{conv}} A(T_s - T_{\text{fluid}}) \qquad \text{(W)} \qquad (14\text{-}13)$$

where h_{conv} is the convection heat transfer coefficient and A is the heat transfer surface area. The value of h_{conv} depends on the geometry of the surface and the type of fluid flow, among other things.

Natural convection currents start out as *laminar* (smooth and orderly) and turn *turbulent* when the dimension of the body and the temperature difference between the hot surface and the fluid are large. For air, the flow remains laminar when the temperature differences involved are less than 100°C and the characteristic length of the body is less than 0.5 m, which is almost always the case in electronic equipment. Therefore, the airflow in the analysis of electronic equipment can be assumed to be laminar.

The natural convection heat transfer coefficient for laminar flow of *air* at atmospheric pressure is given by a simplified relation of the form

$$h_{\text{conv}} = K\left(\frac{\Delta T}{L}\right)^{0.25} \qquad \text{(W/m}^2 \cdot \text{°C)} \qquad (14\text{-}14)$$

where $\Delta T = T_s - T_{\text{fluid}}$ is the temperature difference between the surface and the fluid, L is the characteristic length (the length of the body along the heat flow path), and K is a constant whose value depends on the *geometry* and *orientation* of the body.

The heat transfer coefficient relations are given in Table 14-1 for some common geometries encountered in electronic equipment in both SI and English unit systems. Once h_{conv} has been determined from one of these relations, the rate of heat transfer can be determined from Equation 14-13. The relations in Table 14-1 can also be used at pressures other than 1 atm by multiplying them by $\sqrt{P}$, where P is the *air pressure in atm* (1 atm = 101.325 kPa = 14.696 psia). That is,

$$h_{\text{conv}, P\,\text{atm}} = h_{\text{conv}, 1\,\text{atm}} \sqrt{P} \qquad \text{(W/m}^2 \cdot \text{°C)} \qquad (14\text{-}15)$$

When hot surfaces are surrounded by cooler surfaces such as the walls and ceilings of a room or just the sky, the surfaces are also cooled by *radiation,* as shown in Figure 14-37. The magnitude of radiation heat transfer, in general, is comparable to the magnitude of natural convection heat transfer. This is especially the case for surfaces whose emissivity is close to *unity,* such as plastics and painted surfaces (regardless of color). Radiation heat transfer is negligible for *polished metals* because of their very low emissivity and for bodies surrounded by surfaces at about the same temperature.

Radiation heat transfer between a surface at temperature T_s completely surrounded by a much larger surface at temperature T_{surr} can be expressed as

$$\dot{Q}_{\text{rad}} = \varepsilon A \sigma (T_s^4 - T_{\text{surr}}^4) \qquad \text{(W)} \qquad (14\text{-}16)$$

where ε is the emissivity of the surface, A is the heat transfer surface area, and σ is the Stefan–Boltzmann constant, whose value is $\sigma = 5.67 \times 10^{-8}$ W/

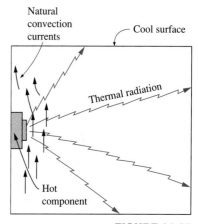

FIGURE 14-37

Simultaneous natural convection heat transfer to air and radiation heat transfer to the surrounding surfaces from a hot electronic component mounted on the wall of an enclosure.

TABLE 14-1

Simplified relations for natural convection heat transfer coefficients for various geometries in air at atmospheric pressure for laminar flow conditions

Geometry	Natural convection heat transfer coefficient	
	$W/m^2 \cdot °C$ (ΔT in °C, L or D in m)	$Btu/h \cdot ft^2 \cdot °F$ (ΔT in °F, L or D in ft)
Vertical plate or cylinder 	$h_{conv} = 1.42 \left(\dfrac{\Delta T}{L} \right)^{0.25}$	$h_{conv} = 0.29 \left(\dfrac{\Delta T}{L} \right)^{0.25}$
Horizontal cylinder 	$h_{conv} = 1.32 \left(\dfrac{\Delta T}{D} \right)^{0.25}$	$h_{conv} = 0.27 \left(\dfrac{\Delta T}{D} \right)^{0.25}$
Horizontal plate ($L = 4A/p$, where A is surface area and p is perimeter) (a) Hot surface facing up	$h_{conv} = 1.32 \left(\dfrac{\Delta T}{L} \right)^{0.25}$	$h_{conv} = 0.27 \left(\dfrac{\Delta T}{L} \right)^{0.25}$
 (b) Hot surface facing down	$h_{conv} = 0.59 \left(\dfrac{\Delta T}{L} \right)^{0.25}$	$h_{conv} = 0.12 \left(\dfrac{\Delta T}{L} \right)^{0.25}$
Components on a circuit board 	$h_{conv} = 2.44 \left(\dfrac{\Delta T}{L} \right)^{0.25}$	$h_{conv} = 0.50 \left(\dfrac{\Delta T}{L} \right)^{0.25}$
Small components or wires in free air 	$h_{conv} = 3.53 \left(\dfrac{\Delta T}{L} \right)^{0.25}$	$h_{conv} = 0.72 \left(\dfrac{\Delta T}{L} \right)^{0.25}$
Sphere 	$h_{conv} = 1.92 \left(\dfrac{\Delta T}{D} \right)^{0.25}$	$h_{conv} = 0.39 \left(\dfrac{\Delta T}{D} \right)^{0.25}$

Sources: Refs. 5 and 6.

$m^2 \cdot K^4 = 0.1714 \times 10^{-8}$ Btu/h $\cdot$ ft^2 $\cdot$ R^4. Here, both temperatures must be expressed in K or R. Also, if the hot surface analyzed has only a partial view of the surrounding cooler surface at T_{surr}, the result obtained from Equation 14-16 must be multiplied by a *view factor*, which is the fraction of the view of the hot surface blocked by the cooler surface. The value of the view factor ranges from 0 (the hot surface has no direct view of the cooler surface) to 1 (the hot surface is completely surrounded by the cooler surface). In preliminary analysis, the surface is usually assumed to be completely surrounded by a single hypothetical surface whose temperature is the equivalent average temperature of the surrounding surfaces.

Arrays of low-power PCBs are often cooled by natural convection by mounting them within a chassis with adequate openings at the top and at the bottom to facilitate airflow, as shown in Figure 14-38. The air between the PCBs rises when heated by the electronic components and is replaced by the cooler air entering from below. This initiates the natural convection flow through the parallel flow passages formed by the PCBs. The PCBs must be placed *vertically* to take advantage of natural convection currents and to minimize trapped air pockets (Fig. 14-39). Placing the PCBs too far from each other wastes valuable cabinet space, and placing them too close tends to "choke" the flow because of the increased resistance. Therefore, there should be an optimum spacing between the PCBs. It turns out that a distance of about 2 cm between the PCBs provides adequate air flow for effective natural convection cooling.

In the heat transfer analysis of PCBs, *radiation* heat transfer is *disregarded,* since the view of the components is large *blocked* by other heat-generating components. As a result, hot components face other hot surfaces instead of a cooler surface. The exceptions are the two PCBs at the ends of the chassis that view the cooler side surfaces. Therefore, it is wise to mount any high-power components on the PCBs facing the walls of the chassis to take advantage of the additional cooling provided by radiation.

Circuit boards that dissipate up to about 5 W of power (or that have a power density of about 0.02 W/cm^2) can be cooled effectively by natural convection. Heat transfer from PCBs can be analyzed by treating them as rectangular plates with uniformly distributed heat sources on one side, and insulated on the other side, since heat transfer from the back surfaces of the PCBs is usually small. For PCBs with electronic components mounted on both sides, the rate of heat transfer and the heat transfer surface area will be twice as large.

It should be remembered that natural convection currents occur only in *gravitational fields.* Therefore, there can be no heat transfer in space by natural convection. This will also be the case when the air passageways are blocked and hot air cannot rise. In such cases, there will be no air motion, and heat transfer through the air will be by convection.

The heat transfer from hot surfaces by natural convection and radiation can be enhanced by attaching fins to the surfaces. The heat transfer in this case can best be determined by using the data supplied by the manufacturers, as discussed in Chapter 3, especially for complex geometries.

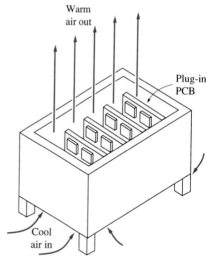

FIGURE 14-38

A chassis with an array of vertically oriented PCBs cooled by natural convection.

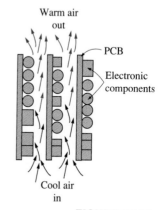

FIGURE 14-39

The PCBs in a chassis must be oriented vertically and spaced adequately to maximize heat transfer by natural convection.

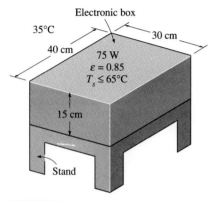

35°C

Electronic box

30 cm

40 cm

75 W
$\varepsilon = 0.85$
$T_s \le 65°C$

15 cm

Stand

FIGURE 14-40

Schematic for Example 14-10.

EXAMPLE 14-10   Cooling of a Sealed Electronic Box

Consider a sealed electronic box whose dimensions are 15 cm × 30 cm × 40 cm placed on top of a stand in a room at 35°C, as shown in Figure 14-40. The box is painted, and the emissivity of its outer surface is 0.85. If the electronic components in the box dissipate 75 W of power and the outer surface temperature of the box is not to exceed 65°C, determine if this box can be cooled by natural convection and radiation alone. Assume the heat transfer from the bottom surface of the box to the stand to be negligible.

Solution   The surface temperature of a sealed electronic box placed on top of a stand is not to exceed 65°C. It is to be determined if this box can be cooled by natural convection and radiation alone.

Assumptions   **1** The box is located at sea level so that the local atmospheric pressure is 1 atm. **2** The temperature of the surrounding surfaces is the same as the air temperature in the room.

Analysis   The sealed electronic box will lose heat from the top and the side surfaces by natural convection and radiation. All four side surfaces of the box can be treated as 0.15-m-high vertical surfaces. Then the natural convection heat transfer from these surfaces is determined to be

$$L = 0.15 \, \text{m}$$
$$A_{\text{side}} = (2 \times 0.4 \, \text{m} + 2 \times 0.3 \, \text{m})(0.15 \, \text{m}) = 0.21 \, \text{m}^2$$
$$h_{\text{conv, side}} = 1.42\left(\frac{\Delta T}{L}\right)^{0.25} = 1.42\left(\frac{65-35}{0.15}\right)^{0.25} = 5.34 \, \text{W/m}^2 \cdot °\text{C}$$
$$\dot{Q}_{\text{conv, side}} = h_{\text{conv, side}} A_{\text{side}}(T_s - T_{\text{fluid}})$$
$$= (5.34 \, \text{W/m}^2 \cdot °\text{C})(0.21 \, \text{m}^2)(65 - 35)°\text{C}$$
$$= 33.6 \, \text{W}$$

Similarly, heat transfer from the horizontal top surface by natural convection is determined to be

$$L = \frac{4A}{p} = \frac{4(0.3 \, \text{m})(0.4 \, \text{m})}{2(0.3 + 0.4) \, \text{m}} = 0.34 \, \text{m}$$
$$A_{\text{top}} = (0.3 \, \text{m})(0.4 \, \text{m}) = 0.12 \, \text{m}^2$$
$$h_{\text{conv, top}} = 1.32\left(\frac{\Delta T}{L}\right)^{0.25} = 1.32\left(\frac{65-35}{0.34}\right)^{0.25} = 4.05 \, \text{W/m}^2 \cdot °\text{C}$$
$$\dot{Q}_{\text{conv, top}} = h_{\text{conv, top}} A_{\text{top}}(T_s - T_{\text{fluid}})$$
$$= (4.05 \, \text{W/m}^2 \cdot °\text{C})(0.12 \, \text{m}^2)(65 - 35)°\text{C}$$
$$= 14.6 \, \text{W}$$

Therefore, the natural convection heat transfer from the entire box is

$$\dot{Q}_{\text{conv}} = \dot{Q}_{\text{conv, side}} + \dot{Q}_{\text{conv, top}} = 33.6 + 14.6 = 48.2 \, \text{W}$$

The box is completely surrounded by the surfaces of the room, and it is stated that the temperature of the surfaces facing the box is equal to the air temperature in the room. Then the rate of heat transfer from the box by radiation can be determined from

$$\dot{Q}_{rad} = \varepsilon A \sigma (T_s^4 - T_{surr}^4)$$
$$= 0.85[(0.21 + 0.12) \text{ m}^2](5.67 \times 10^{-8} \text{ W/m}^2 \times \text{K}^4)$$
$$\times [(65 + 273)^4 - (35 + 273)^4]\text{K}^4$$
$$= 64.5 \text{ W}$$

Note that we must use absolute temperatures in radiation calculations. Then the total heat transfer from the box is simply

$$\dot{Q}_{total} = \dot{Q}_{conv} + \dot{Q}_{rad} = 48.2 + 64.5 = \textbf{112.7 W}$$

which is greater than 75 W. Therefore, this box can be cooled by combined natural convection and radiation, and there is no need to install any fans. There is even some safety margin left for occasions when the air temperature rises above 35°C.

EXAMPLE 14-11   Cooling of a Component by Natural Convection

A 0.2-W small cylindrical resistor mounted on a PCB is 1 cm long and has a diameter of 0.3 cm, as shown in Figure 14-41. The view of the resistor is largely blocked by the PCB facing it, and the heat transfer from the connecting wires is negligible. The air is free to flow through the parallel flow passages between the PCBs. If the air temperature at the vicinity of the resistor is 50°C, determine the surface temperature of the resistor.

Solution   A small cylindrical resistor mounted on a PCB is being cooled by natural convection and radiation. The surface temperature of the resistor is to be determined.

Assumptions   **1** Steady operating conditions exist. **2** The device is located at sea level so that the local atmospheric pressure is 1 atm. **3** Radiation is negligible in this case since the resistor is surrounded by surfaces that are at about the same temperature, and the net radiation heat transfer between two surfaces at the same temperature is zero. This leaves natural convection as the only mechanism of heat transfer from the resistor.

Analysis   Using the relation for components on a circuit board from Table 14-1, the natural convection heat transfer coefficient for this cylindrical component can be determined from

$$h_{conv} = 2.44 \left(\frac{T_s - T_{fluid}}{D} \right)^{0.25}$$

where the diameter $D = 0.003$ m, which is the length in the heat flow path, is the characteristic length. We cannot determine h_{conv} yet since we do not know the surface temperature of the component and thus ΔT. But we can substitute this relation into the heat transfer relation to get

$$\dot{Q}_{conv} = h_{conv}A(T_s - T_{fluid}) = 2.44 \left(\frac{T_s - T_{fluid}}{D} \right)^{0.25} A(T_s - T_{fluid})$$
$$= 2.44 A \frac{(T_s - T_{fluid})^{1.25}}{D^{0.25}}$$

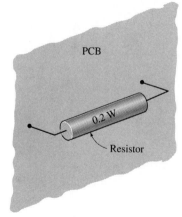

FIGURE 14-41
Schematic for Example 14-11.

The heat transfer surface area of the component is

$$A = 2 \times \tfrac{1}{4}\pi D^2 + \pi DL = 2 \times \tfrac{1}{4}\pi(0.3 \text{ cm})^2 + \pi(0.3 \text{ cm})(1 \text{ cm}) = 1.084 \text{ cm}^2$$

Substituting this and other known quantities in proper units (W for $\dot{Q}$, °C for T, m^2 for A, and m for D) into this equation and solving for T_s yields

$$0.2 = 2.44(1.084 \times 10^{-4})\frac{(T_s - 50)^{1.25}}{0.003^{0.25}} \longrightarrow \quad T_s = \textbf{113°C}$$

Therefore, the surface temperature of the resistor on the PCB will be 113°C, which is considered to be a safe operating temperature for the resistors. Note that blowing air to the circuit board will lower this temperature considerably as a result of increasing the convection heat transfer coefficient and decreasing the air temperature at the vicinity of the components due to the larger flow rate of air.

EXAMPLE 14-12   Cooling of a PCB in a Box by Natural Convection

A 15-cm × 20-cm PCB has electronic components on one side, dissipating a total of 7 W, as shown in Figure 14-42. The PCB is mounted in a rack vertically together with other PCBs. If the surface temperature of the components is not to exceed 100°C, determine the maximum temperature of the environment in which this PCB can operate safely at sea level. What would your answer be if this rack is located at a location at 4000 m altitude where the atmospheric pressure is 61.66 kPa?

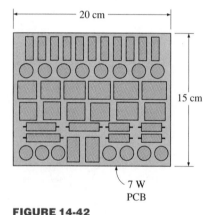

FIGURE 14-42
Schematic for Example 14-12.

Solution   The surface temperature of a PCB is not to exceed 100°C. The maximum environment temperatures for safe operaton at sea level and at 4000 m altitude are to be determined.

Assumptions   **1** Steady operating conditions exist. **2** Radiation heat transfer is negligible since the PCB is surrounded by other PCBs at about the same temperature. **3** Heat transfer from the back surface of the PCB will be very small and thus negligible.

Analysis   The entire heat load of the PCB is dissipated to the ambient air by natural convection from its front surface, which can be treated as a vertical flat plate.

   Using the simplified relation for a vertical surface from Table 14-1, the natural convection heat transfer coefficient for this PCB can be determined from

$$h_{\text{conv}} = 1.42\left(\frac{\Delta T}{L}\right)^{0.25} = 1.42\left(\frac{T_s - T_{\text{fluid}}}{L}\right)^{0.25}$$

The characteristic length in this case is the height ($L = 0.15$ m) of the PCB, which is the length in the path of heat flow. We cannot determine h_{conv} yet, since we do not know the ambient temperature and thus ΔT. But we can substitute this relation into the heat transfer relation to get

$$\dot{Q}_{\text{conv}} = h_{\text{conv}}A(T_s - T_{\text{fluid}}) = 1.42\left(\frac{T_s - T_{\text{fluid}}}{L}\right)^{0.25}A(T_s - T_{\text{fluid}})$$

$$= 1.42A\frac{(T_s - T_{\text{fluid}})^{1.25}}{L^{0.25}}$$

The heat transfer surface area of the PCB is

$$A = (\text{Width})(\text{Height}) = (0.2\,\text{m})(0.15\,\text{m}) = 0.03\,\text{m}^2$$

Substituting this and other known quantities in proper units (W for $\dot{Q}$, °C for T, m^2 for A, and m for L) into this equation and solving for T_{fluid} yields

$$7 = 1.42(0.03)\frac{(100 - T_{\text{fluid}})^{1.25}}{0.15^{0.25}} \longrightarrow T_{\text{fluid}} = \mathbf{59.5°C}$$

Therefore, the PCB will operate safely in environments with temperatures up to 59.4°C by relying solely on natural convection.

At an altitude of 4000 m, the atmospheric pressure is 61.66 kPa, which is equivalent to

$$P = (61.66\,\text{kPa})\frac{1\,\text{atm}}{101.325\,\text{kPa}} = 0.609\,\text{atm}$$

The heat transfer coefficient in this case is obtained by multiplying the value at sea level by $\sqrt{P}$, where P is in atm. Substituting

$$7 = 1.42(0.03)\frac{(100 - T_{\text{fluid}})^{1.25}}{0.15^{0.25}}\sqrt{0.609} \longrightarrow T_{\text{fluid}} = \mathbf{50.6°C}$$

which is about 10°C lower than the value obtained at 1 atm pressure. Therefore, the effect of altitude on convection should be considered in high-altitude applications.

14-8 ■ AIR COOLING: FORCED CONVECTION

We mentioned earlier that convection heat transfer between a solid surface and a fluid is proportional to the velocity of the fluid. The *higher* the velocity, the *larger* the flow rate and the *higher* the heat transfer rate. The fluid velocities associated with natural convection currents are naturally low, and thus natural convection cooling is limited to low-power electronic systems.

When natural convection cooling is not adequate, we simply add a *fan* and blow air through the enclosure that houses the electronic components. In other words, we resort to *forced convection* in order to enhance the velocity and thus the flow rate of the fluid as well as the heat transfer. By doing so, we can increase the heat transfer coefficient by a factor of up to about 10, depending on the size of the fan. This means we can remove heat at much higher rates for a specified temperature difference between the components and the air, or we can reduce the surface temperature of the components considerably for a specified power dissipation.

The *radiation* heat transfer in forced-convection-cooled electronic systems is usually disregarded for two reasons. First, forced convection heat transfer is usually much larger than that due to radiation, and the consideration of radiation causes no significant change in the results. Second, the electronic components and circuit boards in convection-cooled systems are mounted so close to each other that a component is almost entirely surrounded by other components at about the same high temperature. That is, the components have hardly any direct view of a cooler surface. This results in little or no radiation heat transfer from the components. The components near the edges of circuit boards with a large view of a cooler surface may benefit somewhat from the additional cooling by radiation, and it is a good design practice to reserve those spots for high-power components to have a thermally balanced system.

When heat transfer from the outer surface of the enclosure of the electronic equipment is negligible, the amount of heat absorbed by the air becomes equal to the amount of heat given up (or power dissipated) by the electronic components in the enclosure, and can be expressed as (Fig. 14-43)

$$\dot{Q} = \dot{m}C_p(T_{out} - T_{in}) \qquad \text{(W)} \qquad (14\text{-}17)$$

where $\dot{Q}$ is the rate of heat transfer to the air; C_p is the specific heat of air; T_{in} and T_{out} are the average temperatures of air at the inlet and exit of the enclosure, respectively; and $\dot{m}$ is the mass flow rate of air.

Note that for a specified mass flow rate and power dissipation, the *temperature rise* of air, $T_{out} - T_{in}$, remains constant as it flows through the enclosure. Therefore, the higher the inlet temperature of the air, the higher the exit temperature, and thus the higher the surface temperature of the components. It is considered a good design practice to limit the temperature rise of air to 10°C and the maximum exit temperature of air to 70°C. In a properly designed forced-air-cooled system, this results in a maximum component surface temperature of under 100°C.

The mass flow rate of air required for cooling an electronic box depends on the temperature of air available for cooling. In cool environments, such as an air-conditioned room, a smaller flow rate will be adequate. However, in hot environments, we may need to use a larger flow rate to avoid overheating the components and the potential problems associated with it.

Forced convection is covered in detail in Chapter 6. For those who skipped that chapter because of time limitations, below we present a brief review of basic concepts and relations.

The fluid flow over a body such as a transistor is called **external flow,** and flow through a confined space such as inside a tube or through the parallel passage area between two circuit boards in an enclosure is called **internal flow** (Fig. 14-44). Both types of flow are encountered in a typical electronic system.

Fluid flow is also categorized as being **laminar** (smooth and streamlined) or **turbulent** (intense eddy currents and random motion of chunks of fluid). Turbulent flow is desirable in heat transfer applications since it results in a much larger heat transfer coefficient. But is also comes with a much larger friction coefficient, which requires a much larger fan (or pump for liquids).

Numerous experimental studies have shown that turbulence tends to occur at larger velocities, during flow over larger bodies or flow through larger channels, and with fluids having smaller viscosities. These effects are combined into the dimensionless **Reynolds number,** defined as

$$\text{Re} = \frac{\mathcal{V}D}{\nu} \qquad (14\text{-}18)$$

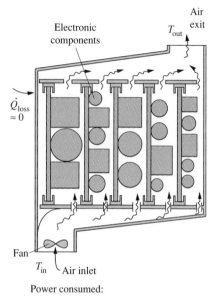

Power consumed:
$$\dot{W}_{electric} = \dot{Q}_{absorbed} = \dot{m}C_p(T_{out} - T_{in})$$

FIGURE 14-43

In steady operation, the heat absorbed by air per unit time as it flows through an electronic box is equal to the power consumed by the electronic components in the box.

FIGURE 14-44

Internal flow through a circular tube and external flow over it.

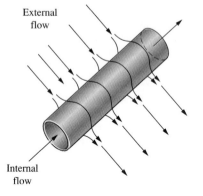

where

    $\mathcal{V}$ = velocity of the fluid (*free-stream* velocity for external flow and *average* velocity for internal flow), m/s

    D = characteristic length of the geometry (the length the fluid flows over in external flow, and the equivalent diameter in internal flow), m

    $\nu = \mu/\rho$ = kinematic viscosity of the fluid, m²/s.

The Reynolds number at which the flow changes from laminar to turbulent is called the *critical Reynolds number,* whose value is 2300 for internal flow, 500,000 for flow over a flat plate, and 200,000 for flow over a cylinder or sphere.

    The *equivalent* (or *hydraulic*) *diameter* for internal flow is defined as

$$D_h = \frac{4A_c}{p} \quad \text{(m)} \tag{14-19}$$

where A_c is the cross-sectional area of the flow passage and p is the perimeter. Note that for a circular pipe, the hydraulic diameter is equivalent to the ordinary diameter.

    The convection heat transfer is expressed by *Newton's law of cooling* as

$$\dot{Q}_{\text{conv}} = hA(T_s - T_{\text{fluid}}) \quad \text{(W)} \tag{14-20}$$

where

    h = average convection heat transfer coefficient, W/m² · °C

    A = heat transfer surface area, m²

    T_s = temperature of the surface, °C

    T_{fluid} = temperature of the fluid sufficiently far from the surface for *external* flow, and average temperature of the fluid at a specified location in *internal* flow, °C

When the heat load is distributed uniformly on the surfaces with a constant heat flux $\dot{q}$, the total rate of heat transfer can also be expressed as $\dot{Q} = \dot{q}A$.

    In *fully developed flow* through a pipe or duct (i.e., when the entrance effects are negligible) subjected to constant heat flux on the surfaces, the convection heat transfer coefficient h remains *constant*. In this case, both the surface temperature T_s and the fluid temperature T_{fluid} increase *linearly,* as shown in Figure 14-45, but the difference between them, $T_s - T_{\text{fluid}}$, remains *constant*. Then the *temperature rise* of the surface above the fluid temperature can be determined from Equation 14-20 to be

$$\Delta T_{\text{rise, surface}} = T_s - T_{\text{fluid}} = \frac{\dot{Q}_{\text{conv}}}{hA} \quad \text{(°C)} \tag{14-21}$$

Note that the temperature rise of the surface is inversely proportional to the convection heat transfer coefficient. Therefore, the greater the convection coefficient, the lower the surface temperature of the electronic components.

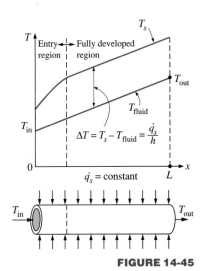

FIGURE 14-45

Under constant heat flux conditions, the surface and fluid temperatures increase linearly, but their difference remains constant in the fully developed region.

When the exit temperature of the fluid, T_{out}, is known, the highest surface temperature that will occur at the end of the flow channel can be determined from Equation 14-21 to be

$$T_{s,max} = T_{fluid,max} + \frac{\dot{Q}}{hA} = T_{out} + \frac{\dot{Q}}{hA} \qquad (^\circ C) \qquad (14\text{-}22)$$

If this temperature is within the safe range, then we don't need to worry about temperatures at other locations. But if it is not, it may be necessary to use a larger fan to increase the flow rate of the fluid.

In convection analysis, the convection heat transfer coefficient h is usually expressed in terms of the dimensionless **Nusselt number** Nu as

$$h = \frac{k}{D}\text{Nu} \qquad (\text{W/m}^2 \cdot {}^\circ C) \qquad (14\text{-}23)$$

where k is the thermal conductivity of the fluid and D is the characteristic length of the geometry. Relations for the average Nusselt number based on experimental data are given in Table 14-2 for external flow and in Table 14-3 for laminar (Re < 2300) internal flow under a uniform heat flux condition, which is closely approximated by electronic equipment. For *turbulent flow* (Re > 2300) through smooth tubes and channels, the Nusselt number can be determined from

$$\text{Nu} = 0.023 \, \text{Re}^{0.8}\text{Pr}^{0.4} \qquad (14\text{-}24)$$

for any geometry. Here Pr is the dimensionless **Prandtl number,** and its value is about 0.7 for air at room temperature.

The fluid properties in the above relations are to be evaluated at the *bulk mean fluid temperature* $T_{ave} = \frac{1}{2}(T_{in} + T_{out})$ for internal flow, which is the arithmetic average of the mean fluid temperatures at the inlet and the exit of the tube, and at the *film temperature* $T_{film} = \frac{1}{2}(T_s + T_{fluid})$ for external flow, which is the arithmetic average of the surface temperature and free-stream temperature of the fluid.

The relations in Table 14-3 for internal flow assume fully developed flow over the entire flow section, and disregard the heat transfer enhancement effects of the development region at the entrance. Therefore, the results obtained from these relations are on the conservative side. We don't mind this much, however, since it is common practice in engineering design to have some safety margin to fall back to "just in case," as long as it does not result in a grossly overdesigned system. Also, it may sometimes be necessary to do some local analysis for critical components with small surface areas to assure reliability and to incorporate solutions to local problems such as attaching heat sinks to high power components.

Fan Selection

Air is supplied to electronic equipment by one or several fans. Although the air is free and abundant, the fans are not. Therefore, a few words about the fan selection are in order.

TABLE 14-2

903

Air Cooling: Forced
Convection

Empirical correlations for the average Nusselt number for forced convection
over a flat plate and circular and noncircular cylinders in cross flow

Cross-section of the cylinder	Fluid	Range of Re	Nusselt number
Circle	Gas or liquid	0.4 – 4 4 – 40 40 – 4000 4000 – 40,000 40,000 – 400,000	$Nu = 0.989\,Re^{0.330}\,Pr^{1/3}$ $Nu = 0.911\,Re^{0.385}\,Pr^{1/3}$ $Nu = 0.683\,Re^{0.466}\,Pr^{1/3}$ $Nu = 0.193\,Re^{0.618}\,Pr^{1/3}$ $Nu = 0.027\,Re^{0.805}\,Pr^{1/3}$
Square	Gas	5000 – 100,000	$Nu = 0.102\,Re^{0.675}\,Pr^{1/3}$
Square (tilted 45°)	Gas	5000 – 100,000	$Nu = 0.246\,Re^{0.588}\,Pr^{1/3}$
Flat plate	Gas or liquid	$0 - 5 \times 10^5$ $5 \times 10^5 - 10^7$	$Nu = 0.664\,Re^{1/2}\,Pr^{1/3}$ $Nu = (0.037\,Re^{4/5} - 871)\,Pr^{1/3}$
Vertical plate	Gas	4000 – 15,000	$Nu = 0.228\,Re^{0.731}\,Pr^{1/3}$

Sources: Jakob, Ref. 13, and Zhukauskas, Ref. 19.

TABLE 14-3

Nusselt number of fully developed
laminar flow in circular tubes and
rectangular channels

Cross-section of the tube	Aspect ratio	Nusselt number
Circle	—	4.36
Square	—	3.61
Rectangle	a/b 1 2 3 4 6 8 ∞	 3.61 4.12 4.79 5.33 6.05 6.49 8.24

A fan at a fixed speed (or fixed rpm) will deliver a fixed volume of air regardless of the altitude and pressure. But the mass flow rate of air will be less at *high altitude* as a result of the lower density of air. For example, the atmospheric pressure of air drops by more than 50 percent at an altitude of 6000 m from its value at sea level. This means that the fan will deliver *half* as much air mass at this altitude at the same rpm and temperature, and thus the temperature rise of air cooling will double. This may create serious reliability problems and catastrophic failures of electronic equipment if proper precautions are not taken. *Variable-speed fans* that automatically increase speed when the air density decreases are available to avoid such problems. Expensive electronic systems are usually equipped with thermal cutoff

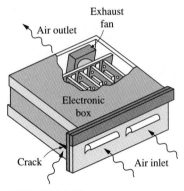

FIGURE 14-46

A fan placed at the exit of an electronic box draws in air as well as contaminants in the air through cracks.

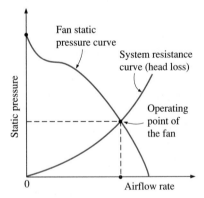

FIGURE 14-47

The airflow rate a fan delivers into an electronic enclosure depends on the flow resistance of that system as well as the variation of the static head of the fan with flow rate.

switches to prevent overheating due to inadequate airflow rate or the failure of the cooling fan.

Fans draw in not only cooling air but also all kinds of *contaminants* that are present in the air, such as lint, dust, moisture, and even oil. If unattended, these contaminants can pile up on the components and plug up narrow passageways, causing overheating. It should be remembered that the dust that settles on the electronic components acts as an insulation layer that makes it very difficult for the heat generated in the component to escape. To minimize the contamination problem, air filters are commonly used. It is good practice to use the largest air filter practical to minimize the pressure drop of air and to maximize the dust capacity.

Often the question arises about whether to place the fan at the inlet or the exit of an electronic box. The generally preferred location is the *inlet*. A fan placed at the inlet draws air in and pressurizes the electronic box and prevents air infiltration into the box from cracks or other openings. Having only one location for air inlet makes it practical to install a filter at the inlet to clean the air from all the dust and dirt before they enter the box. This allows the electronic system to operate in a *clean* environment. Also, a fan placed at the inlet handles cooler and thus *denser* air, which results in a higher mass flow rate for the same volume flow rate or rpm. Since the fan is always subjected to cool air, this has the added benefit that it increased the reliability and extends the life of the fan. The major disadvantage associated with having a fan mounted at the inlet is that the *heat* generated by the fan and its motor is picked up by air on its way into the box, which adds to the heat load of the system.

When the fan is placed at the *exit,* the heat generated by the fan and its motor is immediately discarded to the atmosphere without getting blown first into the electronic box. However, a fan at the exit creates a *vacuum* inside the box, which draws air into the box through inlet vents as well as any cracks and openings (Fig. 14-46). Therefore, the air is difficult to filter, and the dirt and dust that collect on the components undermine the reliability of the system.

There are several types of fans available on the market for cooling electronic equipment, and the right choice depends on the situation on hand. There are two primary considerations in the selection of the fan: the *static pressure head* of the system, which is the total resistance an electronic system offers to air as it passes through, and the *volume flow rate* of the air. *Axial fans* are simple, small, light, and inexpensive, and they can deliver a large flow rate. However, they are suitable for systems with relatively small pressure heads. Also, axial fans usually run at very high speeds, and thus they are noisy. The *radial* or *centrifugal fans,* on the other hand, can delivery moderate flow rates to systems with high static pressure heads at relatively low speeds. But they are larger, heavier, more complex, and more expensive than axial fans.

The performance of a fan is represented by a set off curves called the *characteristic curves,* which are provided by fan manufacturers to help engineers with a selection of fans. A typical static pressure head curve for a fan is given in Figure 14-47 together with a typical system flow resistance curve plotted against the flow rate of air. Note that a fan creates the *highest* pressure head at *zero* flow rate. This corresponds to the limiting case of blocked exit vents of the enclosure. The flow rate increases with decreasing static head and reaches its maximum value when the fan meets no flow resistance.

Any electronic enclosure will offer some resistance to flow. The system resistance curve is *parabolic* in shape, and the pressure or head loss due to this resistance is nearly proportional to the *square* of the flow rate. The fan must overcome this resistance to maintain flow through the enclosure. The design of a forced convection cooling system requires the determination of the total system resistance characteristic curve. This curve can be generated accurately by measuring the static pressure drop at different flow rates. It can also be determined approximately by evaluating the pressure drops.

A fan will operate at the *point* where the fan static head curve and the system resistance curve *intersect*. Note that a fan will deliver a higher flow rate to a system with a low flow resistance. The required airflow rate for a system can be determined from heat transfer requirements alone, using the *design heat load* of the system and the *allowable temperature rise* of air. Then the flow resistance of the system at this flow rate can be determined analytically or experimentally. Knowing the flow rate and the needed pressure head, it is easy to select a fan from manufacturers' catalogs that will meet both of these requirements.

Below we present some general guidelines associated with the forced-air cooling of electronic systems.

1 Before deciding on forced-air cooling, check to see if *natural convection* cooling is adequate. If it is, which may be the case for low-power systems, incorporate it and avoid all the problems associated with fans such as cost, power consumption, noise, complexity, maintenance, and possible failure.

2 Select a fan that is neither too small nor too large. An *undersized* fan may cause the electronic system to overhead and fail. An *oversized* fan will definitely provide adequate cooling, but it will needlessly be larger and more expensive and will consume more power.

3 If the temperature rise of air due to the power consumed by the motor of the fan is acceptable, mount the fan at the *inlet* of the box to pressurize the box and filter the air to keep dirt and dust out (Fig. 14-48).

4 Position and size the air exit vents so that there is *adequate airflow* throughout the entire box. More air can be directed to a certain area by enlarging the size of the vent at that area. The total exit areas should be at least as large as the inlet flow area to avoid the choking of the airflow, which may result in a reduced airflow rate.

5 Place the most critical electronic components near the *entrance,* where the air is coolest. Place the components that consume a lot of power near the exit (Fig. 14-49).

6 Arrange the circuit boards and the electronic components in the box such that the *resistance* of the box to airflow is *minimized* and thus the flow rate of

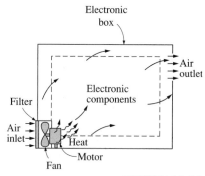

FIGURE 14-48

Installing the fan at the inlet keeps the dirt and dust out, but the heat generated by the fan motor in.

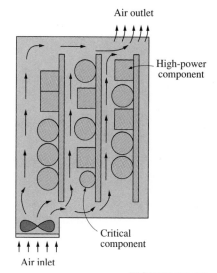

FIGURE 14-49

Sensitive components should be located near the inlet and high-power components near the exit.

air through the box is maximized for the same fan speed. Make sure that no hot air pockets are formed during operation.

7 Consider the effect of *altitude* in high-altitude applications.

8 Try to avoid any flow sections that increase the flow resistance of the systems, such as unnecessary corners, sharp turns, sudden expansions and contractions, and very high velocities (greater than 7 m/s), since the flow resistance is nearly proportional to the flow rate. Also, avoid very low velocities since these result in a poor heat transfer performance and allow the dirt and the dust in the air to settle on the components.

9 Arrange the system such that *natural convection* helps forced convection instead of hurting it. For example, mount the PCBs vertically, and blow the air from the bottom toward the top instead of the other way around.

10 When the design calls for the use of two or more fans, a decision needs to be made about mounting the fans in parallel or in series. Fans mounted in *series* will boost the pressure head available and are best suited for systems with a high flow resistance. Fans connected in *parallel* will increase the flow rate of air and are best suited for systems with small flow resistance.

Cooling Personal Computers

The introduction of the 4004 chip, the first general-purpose microprocessor, by the Intel Corporation in the early 1970s marked the beginning of the electronics era in consumer goods, from calculators and washing machines to personal computers. The microprocessor, which is the "brain" of the personal computer, is basically a DIP-type LSE package that incorporates a central processing unit (CPU), memory, and some input/output capabilities.

A typical desktop personal computer consists of a few circuit boards plugged into a mother board, which houses the microprocessor and the memory chips, as well as the network of interconnections enclosed in a formed sheet metal chassis, which also houses the disk and CD-ROM drives. Connected to this "magic" box are the monitor, a keyboard, a printer, and other auxiliary equipment (Fig. 14-50). The PCBs are normally mounted vertically on a mother board, since this facilitates better cooling.

A small and quite fan is usually mounted to the rear or side of the chassis to cool the electronic components. There are also louvers and openings on the side surfaces to facilitate air circulation. Such openings are not placed on the top surface, since many users would block them by putting books or other things there, which will jeopardize safety, and a coffee or soda spill can cause major damage to the system.

FIGURE 14-50

A desktop personal computer with monitor and keyboard.

EXAMPLE 14-13 Forced-Air Cooling of a Hollow-Core PCB

Some strict specifications of electronic equipment require that the cooling air not come into direct contact with the electronic components, in order to protect them from exposure to the contaminants in the air. In such cases, heat generated in the components on a PCB must be conducted a long way to the walls of the enclosure through a metal core strip or a heat frame attached to the PCB. An alternative solution is the *hollow-core PCB,* which is basically a narrow duct of rectangular cross-section made of thin glass–epoxy board with electronic components mounted on both sides, as shown in Figure 14-51. Heat generated in the components is conducted to the hollow core through a thin layer of epoxy board and is then removed by the cooling air flowing through the core. Effective sealing is provided to prevent air leakage into the component chamber.

Consider a hollow-core PCB 12 cm high and 18 cm long, dissipating a total of 40 W. The width of the air gap between the two sides of the PCB is 0.3 cm. The cooling air enters the core at 20°C at a rate of 0.72 L/s. Assuming the heat generated to be uniformly distributed over the two side surfaces of the PCB, determine (*a*) the temperature at which the air leaves the hollow core and (*b*) the highest temperature on the inner surface of the core.

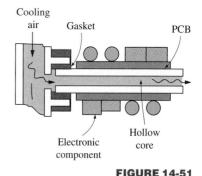

FIGURE 14-51

The hollow-core PCB discussed in Example 14-13.

Solution A hollow-core PCB is cooled by forced air. The outlet temperature of air and the highest surface temperature are to be determined.

Assumptions **1** Steady operating conditions exist. **2** The inner surfaces of the duct are smooth. **3** Air is an ideal gas. **4** Operation is at sea level and thus the atmospheric pressure is 1 atm. **5** The entire heat generated in electronic components is removed by the air flowing through the hollow core.

Properties The temperature of air varies as it flows through the core, and so do its properties. We will perform the calculations using property values at 27°C from Table A-11 since the air enters at 20°C and its temperature will increase.

$$\rho = 1.177 \ \text{kg/m}^3$$
$$C_p = 1005 \ \text{J/kg} \cdot °\text{C}$$
$$\text{Pr} = 0.712$$

$$k = 0.0261 \ \text{W/m} \cdot °\text{C}$$
$$\nu = 1.57 \times 10^{-5} \ \text{m}^2/\text{s}$$

After we calculate the exit temperature of air, we can repeat the calculations, if necessary, using properties at the average temperature.

Analysis The cross-sectional area of the channel and its hydraulic diameter are

$$A_c = (\text{Height})(\text{Width}) = (0.12 \ \text{m})(0.003 \ \text{m}) = 3.6 \times 10^{-4} \ \text{m}^2$$
$$D_h = \frac{4A_c}{p} = \frac{4 \times (3.6 \times 10^{-4} \ \text{m}^2)}{2 \times (0.12 + 0.003) \ \text{m}} = 0.00585 \ \text{m}$$

The average velocity and the mass flow rate of air are

$$\mathcal{V} = \frac{\dot{V}}{A_c} = \frac{0.72 \times 10^{-3} \ \text{m}^3/\text{s}}{3.6 \times 10^{-4} \ \text{m}^2} = 2.0 \ \text{m/s}$$
$$\dot{m} = \rho\dot{V} = (1.177 \ \text{kg/m}^3)(0.72 \times 10^{-3} \ \text{m}^3/\text{s}) = 0.847 \times 10^{-3} \ \text{kg/s}$$

(*a*) The temperature of air at the exit of the hollow core can be determined from

$$\dot{Q} = \dot{m}C_p(T_\text{out} - T_\text{in})$$

Solving for T_out and substituting the given values, we obtain

$$T_\text{out} = T_\text{in} + \frac{\dot{Q}}{\dot{m}C_p} = 20°\text{C} + \frac{40 \ \text{J/s}}{(0.847 \times 10^{-3} \ \text{kg/s})(1005 \ \text{J/kg} \cdot °\text{C})} = \textbf{67°C}$$

(b) The surface temperature of the channel at any location can be determined from

$$\dot{Q}_{conv} = hA(T_s - T_{fluid})$$

where the heat transfer surface area is

$$A = 2A_{side} = 2(\text{Height})(\text{Length}) = 2(0.12\ \text{m})(0.18\ \text{m}) = 0.0432\ \text{m}^2$$

To determine the convection heat transfer coefficient, we first need to calculate the Reynolds number:

$$\text{Re} = \frac{\mathcal{V}D_h}{\nu} = \frac{(2\ \text{m/s})(0.00585\ \text{m})}{1.57 \times 10^{-5}\ \text{m}^2/\text{s}} = 745 < 2300$$

Therefore, the flow is laminar, and, assuming fully developed flow, the Nusselt number for the airflow in this rectangular cross-section corresponding to the aspect ratio $a/b = (12\ \text{cm})/(0.3\ \text{cm}) = 40 \approx \infty$ is determined from Table 14-3 to be

$$\text{Nu} = 8.24$$

and thus

$$h = \frac{k}{D_h}\text{Nu} = \frac{0.0261\ \text{W/m} \cdot {}^\circ\text{C}}{0.00585\ \text{m}}(8.24) = 36.7\ \text{W/m}^2 \cdot {}^\circ\text{C}$$

Then the surface temperature of the hollow core near the exit is determined to be

$$T_{s,max} = T_{out} + \frac{\dot{Q}}{hA} = 67^\circ\text{C} + \frac{40\ \text{W}}{(36.7\ \text{W/m}^2 \cdot {}^\circ\text{C})(0.0432\ \text{m}^2)} = \mathbf{92.2^\circ C}$$

Discussion Note that the temperature difference between the surface and the air at the exit of the hollow core is 25.2°C. This temperature difference between the air and the surface remains at that value throughout the core, since the heat generated on the side surfaces is uniform and the convection heat transfer coefficient is constant. Therefore, the surface temperature of the core at the inlet will be 20°C + 25.2°C = 45.2°C. In reality, however, this temperature will be somewhat lower because of the entrance effects, which affect heat transfer favorably. The fully developed flow assumption gives somewhat conservative results but is commonly used in practice because it provides considerable simplification in calculations.

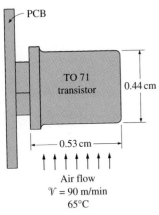

PCB

TO 71
transistor

0.44 cm

0.53 cm

Air flow
$\mathcal{V}$ = 90 m/min
65°C

FIGURE 14-52

Schematic for Example 14-14.

EXAMPLE 14-14   Forced-Air Cooling of a Transistor Mounted on a PCB

A TO 71 transistor with a height of 0.53 cm and a diameter of 0.44 cm is mounted on a circuit board, as shown in Figure 14-52. The transistor is cooled by air flowing over it at a velocity of 90 m/min. If the air temperature is 65°C and the transistor case temperature is not to exceed 90°C, determine the amount of power this transistor can dissipate safely.

Solution A transistor mounted on a circuit board is cooled by air flowing over it. The power dissipated when its case temperature is 90°C is to be determined.

Assumptions **1** Steady operating conditions exit. **2** Air is an ideal gas. **3** Operation is at sea level and thus the atmospheric pressure is 1 atm.

Properties The properties of air at 1 atm pressure and the film temperature of $T_f = (T_s + T_{\text{fluid}})/2 = (90 + 65)/2 = 77.5°C \cong 350$ K are (Table A-11)

$\rho = 1.009$ kg/m³

$C_p = 1008$ J/kg · °C

Pr = 0.706

$k = 0.0297$ W/m · °C

$\nu = 2.06 \times 10^{-5}$ m²/s

Analysis The transistor is cooled by forced convection through its cylindrical surface as well as its flat top and bottom surfaces. The characteristic length for flow over a cylinder is the diameter $D = 0.0044$ m. Then the Reynolds number becomes

$$\text{Re} = \frac{\mathcal{V}D}{\nu} = \frac{(\frac{90}{60}\ \text{m/s})(0.0044\ \text{m})}{2.06 \times 10^{-5}\ \text{m}^2/\text{s}} = 320$$

which falls into the range 40–4000. Using the corresponding relation from Table 14-2 for the Nusselt number, we obtain

$$\text{Nu} = 0.683\ \text{Re}^{0.466}\text{Pr}^{1/3} = 0.683(320)^{0.466}(0.706)^{1/3} = 8.94$$

and

$$h = \frac{k}{D}\text{Nu} = \frac{0.0297\ \text{W/m} \cdot °\text{C}}{0.0044\ \text{m}}(8.94) = 60.4\ \text{W/m}^2 \cdot °\text{C}$$

Also,

$$A_{\text{cyl}} = \pi DL = \pi(0.0044\ \text{m})(0.0053\ \text{m}) = 0.733 \times 10^{-4}\ \text{m}^2$$

Then the rate of heat transfer from the cylindrical surface becomes

$$\dot{Q}_{\text{cyl}} = hA_{\text{cyl}}(T_s - T_{\text{fluid}})$$
$$= (60.4\ \text{W/m}^2 \cdot °\text{C})(0.733 \times 10^{-4}\ \text{m}^2)(90 - 65)°\text{C} = 0.11\ \text{W}$$

We now repeat the calculations for the top and bottom surfaces of the transistor, which can be treated as flat plates of length $L = 0.0044$ m in the flow direction (which is the diameter), and, using the proper relation from Table 14-2,

$$\text{Re} = \frac{\mathcal{V}L}{\nu} = \frac{(\frac{90}{60}\ \text{m/s})(0.0044\ \text{m})}{2.06 \times 10^{-5}\ \text{m}^2/\text{s}} = 320$$

$$\text{Nu} = 0.664\ \text{Re}^{1/2}\text{Pr}^{1/3} = 0.664(320)^{1/2}(0.706)^{1/3} = 10.6$$

and

$$h = \frac{k}{D}\text{Nu} = \frac{0.0297\ \text{W/m} \cdot °\text{C}}{0.0044\ \text{m}}(10.6) = 76.1\ \text{W/m}^2 \cdot °\text{C}$$

Also,

$$A_{\text{flat}} = A_{\text{top}} + A_{\text{bottom}} = 2 \times \tfrac{1}{4}\pi D^2 = 2 \times \tfrac{1}{4}\pi(0.0044\ \text{m})^2 = 0.30 \times 10^{-4}\ \text{m}^2$$
$$\dot{Q}_{\text{flat}} = hA_{\text{flat}}(T_s - T_{\text{fluid}}) = (71.6\ \text{W/m}^2 \cdot °\text{C})(0.30 \times 10^{-4}\ \text{m}^2)(90 - 65)°\text{C}$$
$$= 0.05\ \text{W}$$

Therefore, the total rate of heat that can be dissipated from all surfaces of the transistor is

$$\dot{Q}_{\text{total}} = \dot{Q}_{\text{cyl}} + \dot{Q}_{\text{flat}} = (0.11 + 0.05)\ \text{W} = \mathbf{0.16\ W}$$

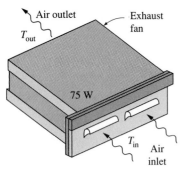

FIGURE 14-53
Schematic for Example 14-15.

which seems to be low. This value can be increased considerably by attaching a heat sink to the transistor to enhance the heat transfer surface area and thus heat transfer, or by increasing the air velocity, which will increase the heat transfer coefficient.

EXAMPLE 14-15   Choosing a Fan to Cool a Computer

The desktop computer shown in Figure 14-53 is to be cooled by a fan. The electronics of the computer consume 75 W of power under full-load conditions. The computer is to operate in environments at temperatures up to 40°C and at elevations up to 2000 m, where the atmospheric pressure is 79.50 kPa. The exit temperature of air is not to exceed 70°C to meet reliability requirements. Also, the average velocity of air is not to exceed 75 m/min at the exit of the computer case, where the fan is installed, to keep the noise level down. Determine the flow rate of the fan that needs to be installed and the diameter of the casing of the fan.

Solution   A desktop computer is to be cooled by a fan safely in hot environments and high elevations. The airflow rate of the fan and the diameter of the casing are to be determined.

Assumptions   **1** Steady operation under worst conditions is considered. **2** Air is an ideal gas.

Properties   The specific heat of air at the average temperature of (40 + 70)/2 = 55°C = 328 K is 1006 J/kg · °C (Table A-11).

Analysis   We need to determine the flow rate of air for the worst-case scenario. Therefore, we assume the inlet temperature of air to be 40°C and the atmospheric pressure to be 79.50 kPa and disregard any heat transfer from the outer surfaces of the computer case. Note that any direct heat loss from the computer case will provide a safety margin in the design.

Noting that all the heat dissipated by the electronic components is absorbed by the air, the required mass flow rate of air to absorb heat at a rate of 75 W can be determined from

$$\dot{Q} = \dot{m}C_p(T_{out} - T_{in})$$

Solving for $\dot{m}$ and substituting the given values, we obtain

$$\dot{m} = \frac{\dot{Q}}{C_p(T_{out} - T_{in})} = \frac{75 \text{ J/s}}{(1006 \text{ J/kg} \cdot °\text{C})(70 - 40)°\text{C}}$$
$$= 0.00249 \text{ kg/s} = 0.149 \text{ kg/min}$$

In the worst case, the exhaust fan will handle air at 70°C. Then the density of air entering the fan and the volume flow rate become

$$\rho = \frac{P}{RT} = \frac{79.50 \text{ kPa}}{(0.287 \text{ kPa} \cdot \text{m}^3/\text{kg} \cdot \text{K})(70 + 273) \text{ K}} = 0.8076 \text{ kg/m}^3$$

$$\dot{V} = \frac{\dot{m}}{\rho} = \frac{0.149 \text{ kg/min}}{0.8076 \text{ kg/m}^3} = \textbf{0.184 m}^3\textbf{/min}$$

Therefore, the fan must be able to provide a flow rate of 0.184 m³/min or 6.5 cfm (cubic feet per minute). Note that if the fan were installed at the inlet instead of the exit, then we would need to determine the flow rate using the density of air at the inlet temperature of 40°C, and we would need to add the power consumed by the motor of the fan to the heat load of 75 W. The result may be a slightly smaller or larger fan, depending on which effect dominates.

For an average velocity of 75 m/min, the diameter of the duct in which the fan is installed can be determined from

$$\dot{V} = A_c \mathscr{V} = \frac{1}{4}\pi D^2 \mathscr{V}$$

Solving for D and substituting the known values, we obtain

$$D = \sqrt{\frac{4\dot{V}}{\pi \mathscr{V}}} = \sqrt{\frac{4 \times (0.184 \text{ m}^3/\text{min})}{\pi(75 \text{ m/min})}} = 0.056 \text{ m} = \textbf{5.6 cm}$$

Therefore, a fan with a casing diameter of 5.6 cm and a flow rate of 0.184 m³/min will meet the design requirements.

EXAMPLE 14-16   Cooling of a Computer by a Fan

A computer cooled by a fan contains 6 PCBs, each dissipating 15 W of power, as shown in Figure 14-54. The height of the PCBs is 15 cm and the length is 20 cm. The clearance between the tips of the components on the PCB and the back surface of the adjacent PCB is 0.4 cm. The cooling air is supplied by a 20-W fan mounted at the inlet. If the temperature rise of air as it flows through the case of the computer is not to exceed 10°C, determine (a) the flow rate of the air the fan needs to deliver, (b) the fraction of the temperature rise of air due to the heat generated by the fan and its motor, and (c) the highest allowable inlet air temperature if the surface temperature of the components is not to exceed 90°C anywhere in the system.

Solution   A computer is to be cooled by a fan, and the temperature rise of air is limited to 10°C. The flow rate of the air, the fraction of the temperature rise of air caused by the fan and its motor, and maximum allowable air inlet temperature are to be determined.

Assumptions   1 Steady operating conditions exist. 2 Air is an ideal gas. 3 The computer is located at sea level so that the local atmospheric pressure is 1 atm. 4 The entire heat generated by the electronic components is removed by air flowing through the opening between the PCBs. 5 The entire power consumed by the fan motor is transferred as heat to the cooling air. This is a conservative approach since the fan and its motor are usually mounted to the chassis of the electronic system, and some of the heat generated in the motor may be conducted to the chassis through the mounting brackets.

Properties   We use properties of air at 27°C since the air enters at room temperature, and the temperature rise of air is limited to just 10°C (Table A-11)

$$\rho = 1.177 \text{ kg/m}^3$$
$$C_p = 1005 \text{ J/kg} \cdot °\text{C}$$
$$\text{Pr} = 0.712$$

$$k = 0.0261 \text{ W/m} \cdot °\text{C}$$
$$\nu = 1.57 \times 10^{-5} \text{ m}^2/\text{s}$$

Analysis   Because of symmetry, we consider the flow area between the two adjacent PCBs only. We assume the flow rate of air through all six channels to be identical, and to be equal to one-sixth of the total flow rate.

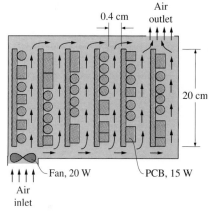

FIGURE 14-54

Schematic of the computer discussed in Example 14-16.

(a) Noting that the temperature rise of air is limited to 10°C and that the power consumed by the fan is also absorbed by the air, the total mass flow rate of air through the computer can be determined from

$$\dot{Q} = \dot{m}C_p(T_{out} - T_{in})$$

Solving for $\dot{m}$ and substituting the given values, we obtain

$$\dot{m} = \frac{\dot{Q}}{C_p(T_{out} - T_{in})} = \frac{(6 \times 15 + 20) \text{ J/s}}{(1005 \text{ J/kg} \cdot °\text{C})(10°\text{C})} = 0.0109 \text{ kg/s}$$

Then the volume flow rate of air and the air velocity become

$$\dot{V} = \frac{\dot{m}}{\rho} = \frac{0.0109 \text{ kg/s}}{1.177 \text{ kg/m}^3} = 0.00926 \text{ m}^3/\text{s} = \textbf{0.556 m}^3\textbf{/min}$$

$$\mathcal{V} = \frac{\dot{V}}{A_c} = \frac{\frac{1}{6}(0.00926 \text{ m}^3/\text{s})}{6 \times 10^{-4} \text{ m}^2} = 2.57 \text{ m/s}$$

Therefore, the fan needs to supply air at a rate of 0.556 m³/min, or about 20 cfm.

(b) The temperature rise of air due to the power consumed by the fan motor can be determined by assuming the entire 20 W of power consumed by the motor to be transferred to air as heat:

$$\Delta T_{air, rise} = \frac{\dot{Q}_{fan}}{\dot{m}C_p} = \frac{20 \text{ J/s}}{(0.0109 \text{ kg/s})(1005 \text{ J/kg} \cdot °\text{C})} = \textbf{1.8°C}$$

Therefore, 18 percent of the temperature rise of air is due to the heat generated by the fan motor. Note that the fraction of the power consumed by the fan is also 18 percent of the total, as expected.

(c) The surface temperature of the channel at any location can be determined from

$$\dot{Q}_{conv} = \dot{Q}_{conv}/A = h(T_s - T_{fluid})$$

where the heat transfer surface area is

$$A = A_{side} = (\text{Height})(\text{Length}) = (0.15 \text{ m})(0.20 \text{ m}) = 0.03 \text{ m}^2$$

To determine the convection heat transfer coefficient, we first need to calculate the Reynolds number. The cross-sectional area of the channel and its hydraulic diameter are

$$A_c = (\text{Height})(\text{Width}) = (0.15 \text{ m})(0.004 \text{ m}) = 6 \times 10^{-4} \text{ m}^2$$

$$D_h = \frac{4A_c}{p} = \frac{4 \times (6 \times 10^{-4} \text{ m}^2)}{2 \times (0.15 + 0.004) \text{ m}} = 0.0078 \text{ m}$$

Then the Reynolds number becomes

$$\text{Re} = \frac{\mathcal{V} D_h}{\nu} = \frac{(2.57 \text{ m/s})(0.0078 \text{ m})}{1.57 \times 10^{-5} \text{ m}^2/\text{s}} = 1277 < 2300$$

Therefore, the flow is laminar, and, assuming fully developed flow, the Nusselt number for the airflow in this rectangular cross-section corresponding to the aspect ratio $a/b = (15 \text{ cm})/(0.4 \text{ cm}) = 37.5 \approx \infty$ is determined from Table 14-3 to be

$$\text{Nu} = 8.24$$

and thus

$$h = \frac{k}{D_h} \text{Nu} = \frac{0.0261 \text{ W/m} \cdot \text{°C}}{0.0078 \text{ m}} (8.24) = 27.6 \text{ W/m}^2 \cdot \text{°C}$$

Disregarding the entrance effects, the temperature difference between the surface of the PCB and the air anywhere along the channel is determined to be

$$T_s - T_{\text{fluid}} = \left(\frac{\dot{q}}{h}\right)_{\text{PCB}} = \frac{(15 \text{ W})/(0.03 \text{ m}^2)}{27.6 \text{ W/m}^2 \cdot \text{°C}} = 18.1\text{°C}$$

That is, the surface temperature of the components on the PCB will be 18.1°C higher than the temperature of air passing by.

The highest air and component temperatures will occur at the exit. Therefore, in the limiting case, the component surface temperature at the exit will be 90°C. The air temperature at the exit in this case will be

$$T_{\text{out, max}} = T_{s, \text{max}} - \Delta T_{\text{rise}} = 90\text{°C} - 18.1\text{°C} = 71.9\text{°C}$$

Noting that the air experiences a temperature rise of 10°C between the inlet and the exit, the inlet temperature of air is

$$T_{\text{in, max}} = T_{\text{out, max}} - 10\text{°C} = (71.9 - 10)\text{°C} = \textbf{61.9°C}$$

This is the highest allowable air inlet temperature if the surface temperature of the components is not to exceed 90°C anywhere in the system.

It should be noted that the analysis presented above is approximate since we have made some simplifying assumptions. However, the accuracy of the results obtained is adequate for most engineering purposes.

14-9 ■ LIQUID COOLING

Liquids normally have much higher thermal conductivities than gases, and thus much higher heat transfer coefficients associated with them. Therefore, liquid cooling is far more effective than gas cooling. However, liquid cooling comes with its own risks and potential problems, such as leakage, corrosion, extra weight, and condensation. Therefore, liquid cooling is reserved for applications involving power densities that are too high for safe dissipation by air cooling.

Liquid cooling systems can be classified as **direct cooling** and **indirect cooling** systems. In *direct cooling* systems, the electronic components are in direct contact with the liquid, and thus the heat generated in the components is transferred directly to the liquid. In *indirect cooling* systems, however, there is no direct contact with the components. The heat generated in this case is first transferred to a medium such as a *cold plate* before it is carried away by the liquid. Liquid cooling systems are also classified as *closed-loop* and *open-loop* systems, depending on whether the liquid is discarded or recirculated after it is heated. In open-loop systems, tap water flows through the cooling system and is discarded into a drain after it is heated. The heated liquid in closed-loop systems is cooled in a heat exchanger and recirculated through the system. Closed-loop systems facilitate better temperature control while conserving water.

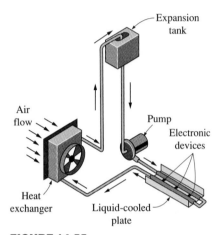

FIGURE 14-55

Schematic of an indirect liquid cooling system.

The electronic components in direct cooling systems are usually completely immersed in the liquid. The heat transfer from the components to the liquid can be by natural or forced convection or *boiling,* depending on the temperature levels involved and the properties of the fluids. Immersion cooling of electronic devices usually involves boiling and thus very high heat transfer coefficients, as discussed in the next section. Note that only *dielectric fluids* can be used in immersion or direct liquid cooling. This limitation immediately excludes water from consideration as a prospective fluid in immersion cooling. Fluorocarbon fluids such as FC75 are well suited for direct cooling and are commonly used in such applications.

Indirect liquid cooling systems of electronic devices operate just like the cooling system of a *car engine,* where the water (actually a mixture of water and ethylene glycol) circulates through the passages around the cylinders of the engine block, picking up heat generated in the cylinders by combustion. The heated water is then routed by the water pump to the car radiator, where it is cooled by air blown through the radiator coils by the cooling fan. The cooled water is then rerouted to the engine to pick up more heat. To appreciate the effectiveness of the cooling system of a car engine, it will suffice to say that the temperatures encountered in the engine cylinders are typically much higher than the melting temperatures of the engine blocks.

In an electronic system, the heat is generated in the *components* instead of the combustion chambers. The components in this case are mounted on a metal plate made of a highly conducting material such as copper or aluminum. The metal plate is cooled by circulating a cooling fluid through tubes attached to it, as shown in Figure 14-55. The heated liquid is then cooled in a heat exchanger, usually by air (or sea water in marine applications), and is recirculated by a pump through the tubes. The expansion and storage tank accommodates any expansions and contractions of the cooling liquid due to temperature variations while acting as a liquid reservoir.

The liquids used in the cooling of electronic equipment must meet several requirements, depending on the specific application. Desirable characteristics of cooling liquids include *high thermal conductivity* (yields high heat transfer coefficients), *high specific heat* (requires smaller mass flow rate), *low viscosity* (causes a smaller pressure drop, and thus requires a smaller pump), *high surface tension* (less likely to cause leakage problems), *high dielectric strength* (a must in direct liquid cooling), *chemical inertness* (does not react with surfaces with which it comes into contact), *chemical stability* (does not decompose under prolonged use), *nontoxic* (safe for personnel to handle), *low freezing and high boiling points* (extends the useful temperature range), and *low cost.* Different fluids may be selected in different applications because of the different priorities set in the selection process.

The heat sinks or cold plates of an electronic enclosure are usually cooled by *water* by passing it through channels made for this purpose or through tubes attached to the cold plate. High heat removal rates can be achieved by circulating water through these channels or tubes. In high-performance systems, a *refrigerant* can be used in place of water to keep the temperature of the heat sink at subzero temperatures and thus reduce the junction temperatures of the electronic components proportionally. The heat transfer and

FIGURE 14-56
Liquid cooling of TO-3 packages placed on top of the coolant line (courtesy of Wakefield Engineering).

pressure drop calculations in liquid cooling systems can be performed as described in Chapter 6.

Liquid cooling can be used effectively to cool clusters of electronic devices attached to a tubed metal plate (or heat sink), as shown in Figure 14-56. Here 12 TO-3 cases, each dissipating up to 150 W of power, are mounted on a heat sink equipped with tubes on the back side through which a liquid flows. The thermal resistance between the case of the devices and the liquid is minimized in this case, since the electronic devices are mounted directly over the cooling lines. The case-to-liquid thermal resistance depends on the spacing between the devices, the quality of the thermal contact between the devices and the plate, the thickness of the plate, and the flow rate of the liquid, among other things. The tubed metal plate shown is 15.2 cm $\times$ 18 cm $\times$ 2.5 cm in size and is capable of dissipating up to 2 kW of power.

The thermal resistance network of a liquid cooling system is shown in Figure 14-57. The junction temperatures of silicon-based electronic devices are usually limited to 125°C. The *junction-to-case* thermal resistance of a device is provided by the manufacturer. The *case-to-liquid* thermal resistance can be determined experimentally by measuring the temperatures of the case and the liquid, and dividing the difference by the total power dissipated. The *liquid-to-air* thermal resistance at the heat exchanger can be determined in a similar manner. That is,

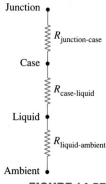

FIGURE 14-57
Thermal resistance network for a liquid-cooled electronic device.

$$R_{\text{case-liquid}} = \frac{T_{\text{case}} - T_{\text{liquid, device}}}{\dot{Q}},$$

$$R_{\text{liquid-air}} = \frac{T_{\text{liquid, hx}} - T_{\text{air, in}}}{\dot{Q}}$$

(°C/W) (14-25)

where $T_{\text{liquid, device}}$ and $T_{\text{liquid, hx}}$ are the inlet temperatures of the liquid to the electronic device and the heat exchanger, respectively. The required mass flow rate of the liquid corresponding to a specified temperature rise of the liquid as it flows through the electronic systems can be determined from Equation 14-17.

Electronic components mounted on liquid-cooled metal plates should be provided with good thermal contact in order to minimize the thermal resistance between the components and the plate. The thermal resistance can be minimized by applying silicone grease or beryllium oxide to the contact surfaces and fastening the components tightly to the metal plate. The liquid cooling of a cold plate with a large number of high-power components attached to it is illustrated in the following example.

EXAMPLE 14-17 Cooling of Power Transistors on a Cold Plate by Water

A cold plate that supports 20 power transistors, each dissipating 40 W, is to be cooled with water, as shown in Figure 14-58. Half of the transistors are attached to the back side of the cold plate. It is specified that the temperature rise of the water is not to exceed 3°C and the velocity of water is to remain under 1 m/s. Assuming that 20 percent of the heat generated is dissipated from the components to the surroundings by convection and radiation, and the remaining 80 percent is removed by the cooling water, determine the mass flow rate of water needed and the diameter of the pipe to satisfy the restriction imposed on the flow velocity. Also, determine the case temperature of the devices if the total case-to-liquid thermal resistance is 0.030°C/W and the water enters the cold plate at 35°C.

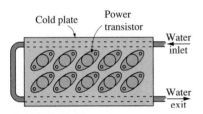

FIGURE 14-58

Schematic for Example 14-17.

Solution   A cold plate is to be cooled by water. The mass flow rate of water, the diameter of the pipe, and the case temperature of the transistors are to be determined.

Assumptions   **1** Steady operating conditions exist. **2** About 20 percent of the heat generated is dissipated from the components to the surroundings by convection and radiation.

Properties   The properties of water at room temperature are $\rho = 1000 \text{ kg/m}^3$ and $C_p = 4180 \text{ J/kg} \cdot °C$ (Table A-9).

Analysis   Noting that each of the 20 transistors dissipates 40 W of power and 80 percent of this power must be removed by the water, the amount of heat that must be removed by the water is

$$\dot{Q} = (20 \text{ transistors})(40 \text{ W/transistor})(0.80) = 640 \text{ W}$$

In order to limit the temperature rise of the water to 3°C, the mass flow rate of water must be no less than

$$\dot{m} = \frac{\dot{Q}}{C_p \Delta T_{\text{rise}}} = \frac{640 \text{ J/s}}{(4180 \text{ J/kg} \cdot °C)(3°C)} = 0.051 \text{ kg/s} = \textbf{3.06 kg/min}$$

The mass flow rate of a fluid through a circular pipe can be expressed as

$$\dot{m} = \rho A \mathcal{V} = \rho \frac{\pi D^2}{4} \mathcal{V}$$

Then the diameter of the pipe to maintain the velocity of water under 1 m/s is determined to be

$$D = \sqrt{\frac{4\dot{m}}{\pi\rho\mathcal{V}}} = \sqrt{\frac{4(0.051 \text{ kg/s})}{\pi(1000 \text{ kg/m}^3)(1 \text{ m/s})}} = 0.0081 \text{ m} = \mathbf{0.81 \text{ cm}}$$

Noting that the total case-to-liquid thermal resistance is 0.030°C/W and the water enters the cold plate at 35°C, the case temperature of the devices is determined from Equation 14-25 to be

$$T_{case} = T_{liquid, \, device} + \dot{Q}R_{case\text{-}liquid} = 35°C + (640 \text{ W})(0.03°C/W) = \mathbf{54.2°C}$$

The junction temperature of the device can be determined similarly by using the junction-to-case thermal resistance of the device supplied by the manufacturer.

14-10 ■ IMMERSION COOLING

High-power electronic components can be cooled effectively by immersing them in a *dielectric liquid* and taking advantage of the very high heat transfer coefficients associated with boiling. Immersion cooling has been used since the 1940s in the cooling of electronic equipment, but for many years its use was largely limited to the electronics of high-power radar systems. The miniaturization of electronic equipment and the resulting high heat fluxes brought about renewed interest in immersion cooling, which had been largely viewed as an exotic cooling technique.

You will recall from thermodynamics that, at a specified pressure, a fluid boils isothermally at the corresponding saturation temperature. A large amount of heat is absorbed during the boiling process, essentially in an isothermal manner. Therefore, immersion cooling also provides a constant-temperature bath for the electronic components and eliminates hot spots effectively.

The simplest type of immersion cooling system involves an *external reservoir* that supplies liquid continually to the electronic enclosure. The vapor generated inside is simply allowed to escape to the atmosphere, as shown in Figure 14-59. A pressure relief valve on the vapor vent line keeps the pressure and thus the temperature inside at the preset value, just like the petcock of a pressure cooker. Note that without a pressure relief valve, the pressure inside the enclosure would be atmospheric pressure and the temperature would have to be the boiling temperature of the fluid at the atmospheric pressure.

The **open-loop**-type immersion cooling system described above is simple, but there are several impracticalities associated with it. First of all, it is heavy and bulky because of the presence of an external liquid reservoir, and the fluid lost through evaporation needs to be replenished continually, which adds to the cost. Further, the release of the vapor into the atmosphere greatly limits the fluids that can be used in such a system. Therefore, the use of open-loop immersion systems is limited to applications that involve occasional use and thus have a light duty cycle.

More sophisticated immersion cooling systems operating in a **closed loop** in that the vapor is condensed and returned to the electronic enclosure instead of being purged to the atmosphere. Schematics of two such systems

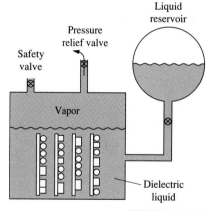

FIGURE 14-59

A simple open-loop immersion cooling system.

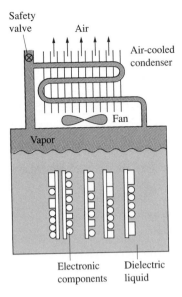

(a) System with external condenser

(b) System with internal condenser

FIGURE 14-60

The schematics of two closed-loop immersion cooling systems.

are given in Figure 14-60. The first system involves a condenser *external* to the electronics enclosure, and the vapor leaving the enclosure is cooled by a cooling fluid such as air or water outside the enclosure. The condensate is returned to the enclosure for reuse. The condenser in the second system is actually *submerged* in the electronic enclosure and is part of the electronic system. The cooling fluid in this case circulates through the condenser tube, removing heat from the vapor. The vapor that condenses drips on top of the liquid in the enclosure and continues to recirculate.

The performance of closed-loop immersion cooling systems is most susceptible to the presence of *noncondensable gases* such as air in the vapor space. An increase of 0.5 percent of air by mass in the vapor can cause the condensation heat transfer coefficient to drop by a factor of up to 5. Therefore, the fluid used in immersion cooling systems should be degassed as much as practical, and care should be taken during the filling process to avoid introducing any air into the system.

The problems associated with the condensation process and noncondensable gases can be avoided by *submerging* the condenser (actually, heat exchanger tubes in this case) in the *liquid* instead of the vapor in the electronic enclosure, as shown in Figure 14-61(a). The cooling fluid, such as water, circulating through the tubes absorbs heat from the dielectric liquid at the top portion of the enclosure and subcools it. The liquid in contact with the electronic components is heated and may even be vaporized as a result of absorbing heat from the components. But these vapor bubbles collapse as they move up, as a result of transferring heat to the cooler liquid with which they come in contact. This system can still remove heat at high rates from the surfaces of electronic components in an isothermal manner by utilizing the

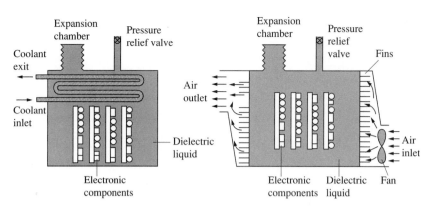

(a) System with internal cooling   (b) System with external cooling

FIGURE 14-61

The schematics of two all-liquid immersion cooling systems.

boiling process, but its overall capacity is limited by the rate of heat that can be removed by the external cooling fluid in a *liquid-to-liquid* heat exchanger. Noting that the heat transfer coefficients associated with forced convection are far less than those associated with condensation, this all-liquid immersion cooling system is not suitable for electronic boxes with very high power dissipation rates per unit volume.

A step further in the all-liquid immersion cooling systems is to remove the heat from the dielectric liquid directly from the *outer surface* of the electronics enclosure, as shown in Figure 14-61(*b*). In this case, the dielectric liquid inside the sealed enclosure is heated as a result of absorbing the heat dissipated by the electronic components. The heat is then transferred to the walls of the enclosure, where it is removed by external means. This immersion cooling technique is the most reliable of all since it does not involve any penetration into the electronics enclosure and the components reside in a completely sealed liquid environment. However, the use of this system is limited to applications that involve moderate power dissipation rates. The heat dissipation is limited by the ability of the system to reject the heat from the outer surface of the enclosure. To enhance this ability, the outer surfaces of the enclosures are often *finned,* especially when the enclosure is cooled by air.

Typical ranges of heat transfer coefficients for various dielectric fluids suitable for use in the cooling of electronic equipment are given in Figure 14-62 for natural convection, forced convection, and boiling. Note that extremely high heat transfer coefficients (from about 1500 to 6000 W/m² · °C) can be attained with the boiling of *fluorocarbon fluids* such as FC78 and FC86 manufactured by the 3M company. Fluorocarbon fluids, not to be confused with the ozone-destroying fluorochloro fluids, are found to be very suitable for immersion cooling of electronic equipment. They have boiling points ranging from 30°C to 174°C and freezing points below −50°C. They are nonflammable, chemically inert, and highly compatible with materials used in electronic equipment.

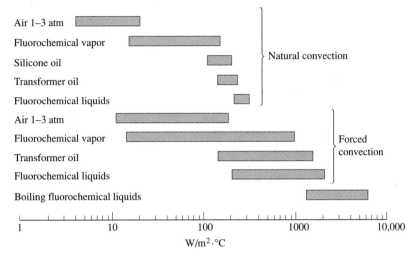

FIGURE 14-62

Typical heat transfer coefficients for various dielectric fluids (from Hwang and Moran, Ref. 12).

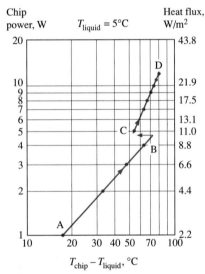

FIGURE 14-63

Heat transfer from a chip immersed in the fluorocarbon fluid FC86 (from Hwang and Moran, Ref. 12).

Experimental results for the power dissipation of a chip having a heat transfer area of 0.457 cm^2 and its substrate during immersion cooling in an FC86 bath (boiling point $-57°C$) are given in Figure 14-63. The FC86 liquid is maintained at a bulk temperature of 5°C during the experiments by the use of a heat exchanger. Heat transfer from the chip is by *natural convection* in regime A–B, and *bubble formation* and thus boiling begins in regime B–C. Note that the chip surface temperature suddenly drops with the onset of *boiling* because of the high heat transfer coefficients associated with boiling. Heat transfer is by nucleate boiling in regime C–D, and very high heat transfer rates can be achieved in this regime with relatively small temperature differences.

EXAMPLE 14-18   Immersion Cooling of a Logic Chip

A logic chip used in an IBM 3081 computer dissipates 4 W of power and has a heat transfer surface area of 0.279 cm^2, as shown in Figure 14-64. If the surface of the chip is to be maintained at 80°C while being cooled by immersion in a dielectric fluid at 20°C, determine the necessary heat transfer coefficient and the type of cooling mechanism that needs to be used to achieve that heat transfer coefficient.

Solution   A logic chip is to be cooled by immersion in a dielectric fluid. The minimum heat transfer coefficient and the type of cooling mechanism are to be determined.

Assumptions   Steady operating conditions exist.

Analysis   The average heat transfer coefficient over the surface of the chip can be determined from Newton's law of cooling, expressed as

$$\dot{Q} = hA(T_{chip} - T_{fluid})$$

Solving for h and substituting the given values, the convection heat transfer coefficient is determined to be

$$h = \frac{\dot{Q}}{A(T_{chip} - T_{fluid})} = \frac{4 \text{ W}}{(0.279 \times 10^{-4} \text{ m}^2)(80 - 20)^\circ\text{C}} = \textbf{2390 W/m}^2 \cdot {}^\circ\textbf{C}$$

which is rather high. Examination of Figure 14-62 reveals that we can obtain such high heat transfer coefficients with the boiling of fluorocarbon fluids. Therefore, a suitable cooling technique in this case is immersion cooling in such a fluid. A viable alternative to immersion cooling is the thermal conduction module discussed earlier.

EXAMPLE 14-19 Cooling of a Chip by Boiling

An 8-W chip having a surface area of 0.6 cm^2 is cooled by immersing it into FC86 liquid that is maintained at a temperature of 15°C, as shown in Figure 14-65. Using the boiling curve in Figure 14-63, estimate the temperature of the chip surface.

Solution A chip is cooled by boiling in a dielectric fluid. The surface temperature of the chip is to be determined.

Assumptions The boiling curve in Figure 14-63 is prepared for a chip having a surface area of 0.457 cm^2 being cooled in FC86 maintained at 5°C. The chart can be used for similar cases with reasonable accuracy.

Analysis The heat flux is

$$\dot{q} = \frac{\dot{Q}}{A} = \frac{8 \text{ W}}{0.6 \text{ cm}^2} = 13.3 \text{ W/cm}^2$$

Corresponding to this value on the chart is $T_{chip} - T_{fluid} = 60°C$. Therefore,

$$T_{chip} = T_{fluid} + 60 = 15 + 60 = \textbf{75°C}$$

That is, the surface of this 8-W chip will be at 75°C as it is cooled by boiling in the dielectric fluid FC86.

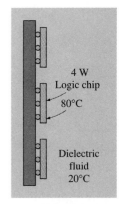

FIGURE 14-64
Schematic for Example 14-18.

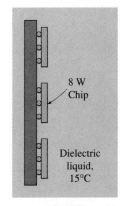

FIGURE 14-65
Schematic for Example 14-19.

A liquid-based cooling system brings with it the possibility of leakage and associated reliability concerns. Therefore, the consideration of immersion cooling should be limited to applications that require precise temperature control and those that involve heat dissipation rates that are too high for effective removal by conduction or air cooling.

14-11 ■ HEAT PIPES

A **heat pipe** is a simple device with no moving parts that can transfer large quantities of heat over fairly large distances essentially at a constant temperature without requiring any power input. A heat pipe is basically a sealed slender tube containing a wick structure lined on the inner surface and a small amount of fluid such as water at the saturated state, as shown in Figure 14-66. It is composed of three sections: the *evaporator* section at one end, where heat is absorbed and the fluid is vaporized; a *condenser* section at the other end, where the vapor is condensed and heat is rejected; and the *adiabatic* section in between, where the vapor and the liquid phases of the

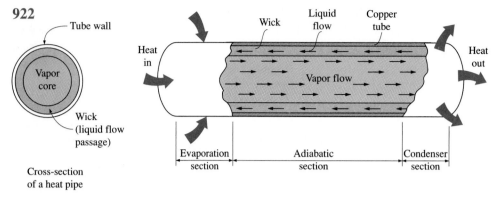

Cross-section
of a heat pipe

FIGURE 14-66

Schematic and operation of a heat pipe.

fluid flow in opposite directions through the core and the wick, respectively, to complete the cycle with no significant heat transfer between the fluid and the surrounding medium.

The *type of fluid* and the *operating pressure* inside the heat pipe depend on the *operating temperature* of the heat pipe. For example, the critical- and triple-point temperatures of water are 0.01°C and 374.1°C, respectively. Therefore, water can undergo a liquid-to-vapor or vapor-to-liquid phase change process in this temperature range only, and thus it will not be a suitable fluid for applications involving temperatures beyond this range. Furthermore, water will undergo a phase-change process at a specified temperature only if its pressure equals the saturation pressure at that temperature. For example, if a heat pipe with water as the working fluid is designed to remove heat at 70°C, the pressure inside the heat pipe must be maintained at 31.2 kPa, which is the boiling pressure of water at this temperature. Note that this value is well below the atmospheric pressure of 101 kPa, and thus the heat pipe operates in a vacuum environment in this case. If the pressure inside is maintained at atmospheric pressure instead, heat transfer would result in an increase in the temperature of the water instead of evaporation.

Although water is a suitable fluid to use in the moderate temperature range encountered in electronic equipment, several other fluids can be used in the construction of heat pipes to enable them to be used in cryogenic as well as high-temperature applications. The suitable temperature ranges for some common heat pipe fluids are given in Table 14-4. Note that the overall temperature range extends from almost absolute zero for cryogenic fluids such as helium to over 1600°C for liquid metals such as lithium. The ultimate temperature limits for a fluid are the *triple-* and *critical-point* temperatures. However, a narrower temperature range is used in practice to avoid the extreme pressures and low heats of vaporization that occur near the critical point. Other desirable characteristics of the candidate fluids are having a high surface tension to enhance the capillary effect and being compatible with the wick material, as well as being readily available, chemically stable, nontoxic, and inexpensive.

TABLE 14-4

Suitable temperature ranges for some fluids used in heat pipes

Fluid	Temperature range, °C
Helium	−271 to −268
Hydrogen	−259 to −240
Neon	−248 to −230
Nitrogen	−210 to −150
Methane	−182 to −82
Ammonia	−78 to −130
Water	5 to 230
Mercury	200 to 500
Cesium	400 to 1000
Sodium	500 to 1200
Lithium	850 to 1600

The concept of heat pipe was originally conceived by R. S. Gaugler of the General Motors Corporation, who filed a patent application for it in 1942. However, it did not receive much attention until 1962, when it was suggested for use in space applications. Since then, heat pipes have found a wide range of applications, including the cooling of electronic equipment.

The Operation of a Heat Pipe

The operation of a heat pipe is based on the following physical principles:

- At a specified pressure, a liquid will vaporize or a vapor will condense at a certain temperature, called the *saturation temperature*. Thus, fixing the pressure inside a heat pipe fixes the temperature at which phase change will take place.
- At a specified pressure or temperature, the amount of heat *absorbed* as a unit mass of liquid vaporizes is equal to the amount of heat *rejected* as that vapor condenses.
- The capillary pressure developed in a wick will move a liquid in the wick even *against* the gravitational field as a result of the capillary effect.
- A fluid in a channel flows in the direction of *decreasing pressure.*

Initially, the *wick* of the heat pipe is saturated with liquid and the *core section* is filled with vapor. When the evaporator end of the heat pipe is brought into contact with a hot surface or is placed into a hot environment, heat will flow into the heat pipe. Being at a saturated state, the liquid in the evaporator end of the heat pipe will *vaporize* as a result of this heat transfer, causing the vapor pressure there to rise. This resulting pressure difference drives the vapor through the core of the heat pipe from the evaporator toward the condenser section. The condenser end of the heat pipe is in a cooler environment, and thus its surface is slightly cooler. The vapor that comes into contact with this cooler surface *condenses,* releasing the heat a vaporization, which is rejected to the surrounding medium. The liquid then returns to the evaporator end of the heat pipe through the wick as a result of *capillary action* in the wick, completing the cycle. As a result, heat is absorbed at one end of the heat pipe and is rejected at the other end, with the fluid inside serving as a transport medium for heat.

The boiling and condensation processes are associated with extremely high heat transfer coefficients, and thus it is natural to expect the heat pipe to be an extremely effective heat transfer device, since its operation is based on alternate boiling and condensation of the working fluid. Indeed, heat pipes have effective conductivities *several hundred times* that of copper or silver. That is, replacing a copper bar between two mediums at different temperatures by a heat pipe of equal size can increase the rate of heat transfer between those two mediums by several hundred times. A simple heat pipe with water as the working fluid has an effective thermal conductivity of the order of 100,000 W/m · °C compared with about 400 W/m · °C for copper. For a heat pipe, it is not unusual to have an effective conductivity of 400,000 W/m · °C, which is 1000 times that of copper. A 15-cm-long, 0.6-cm-diameter horizontal cylindrical heat pipe with water inside, for example, can transfer heat at a rate of 300 W. Therefore, heat pipes are preferred in some critical applications, despite their high initial cost.

There is a small pressure difference between the evaporator and condenser ends, and thus a small temperature difference between the two ends of the heat pipe. This temperature difference is usually between 1°C and 5°C.

The Construction of a Heat Pipe

The wick of a heat pipe provides the means for the return of the liquid to the evaporator. Therefore, the structure of the wick has a strong effect on the performance of a heat pipe, and the design and construction of the wick are the most critical aspects of the manufacturing process.

The wicks are often made of porous ceramic or woven stainless wire mesh. They can also be made together with the tube by extruding axial grooves along its inner surface, but this approach presents manufacturing difficulties.

The performance of a wick depends on its structure. The characteristics of a wick can be changed by changing the *size* and the *number* of the pores per unit volume and the *continuity* of the passageway. Liquid motion in the wick depends on the dynamic balance between two opposing effects: the *capillary pressure,* which creates the suction effect to draw the liquid, and the *internal resistance to flow* as a result of friction between the mesh surfaces and the liquid. A small pore size increases the capillary action, since the capillary pressure is inversely proportional to the effective capillary radius of the mesh. But decreasing the pore size and thus the capillary radius also increases the friction force opposing the motion. Therefore, the core size of the mesh should be reduced so long as the increase in capillary force is greater than the increase in the friction force.

Note that the *optimum pore size* will be different for different fluids and different orientations of the heat pipe. An improperly designed wick will result in an inadequate liquid supply and eventual failure of the heat pipe.

Capillary action permits the heat pipe to operate in any orientation in a gravity field. However, the performance of a heat pipe will be best when the capillary and gravity forces act in the same direction (evaporator end down) and will be worst when these two forces act in opposite directions (evaporator end up). Gravity does not affect the capillary force when the heat pipe is in the horizontal position. The heat removal capacity of a horizontal heat pipe can be *doubled* by installing it vertically with evaporator end down so that gravity helps the capillary action. In the opposite case, vertical orientation with evaporator end up, the performance declines considerably relative the horizontal case since the capillary force in this case must work against the gravity force.

Most heat pipes are cylindrical in shape. However, they can be manufactured in a variety of shapes involving 90° bends, S-turns, or spirals. They can also be made as a flat layer with a thickness of about 0.3 cm. Flat heat pipes are very suitable for cooling high-power-output (say, 50 W or greater) PCBs. In this case, flat heat pipes are attached directly to the back surface of the

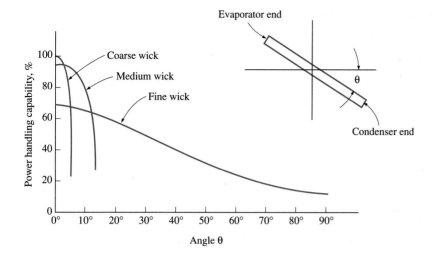

FIGURE 14-67

Variation of the heat removal capacity of a heat pipe with tilt angle from the horizontal when the liquid flows in the wick against gravity (from Steinberg, Ref. 18).

PCB, and they absorb and transfer the heat to the edges. Cooling fins are usually attached to the condenser end of the heat pipe to improve its effectiveness and to eliminate a bottleneck in the path of heat flow from the components to the environment when the ultimate heat sink is the ambient air.

The decline in the performance of a 122-cm-long water heat pipe with the tilt angle from the horizontal is shown in Figure 14-67 for heat pipes with coarse, medium, and fine wicks. Note that for the horizontal case, the heat pipe with a coarse wick performs best, but the performance drops off sharply as the evaporator end is raised form the horizontal. The heat pipe with a fine wick does not perform as well in the horizontal position but maintains its level of performance greatly at tilted positions. It is clear from this figure that heat pipes that work against gravity must be equipped with *fine* wicks. The heat removal capacities of various heat pipes are given in Table 14-5.

A major concern about the performance of a heat pipe is degradation with time. Some heat pipes have failed within just a few months after they are put into operation. The major cause of degradation appears to be *contamination* that occurs during the sealing of the ends of the heat pipe tube and affects the vapor pressure. This form of contamination has been minimized by electron beam welding in clean rooms. Contamination of the wick prior to installation in the tube is another cause of degradation. Cleanliness of the wick is essential for its reliable operation for a long time. Heat pipes usually undergo extensive testing and quality control process before they are put into actual use.

An important consideration in the design of heat pipes is the compatibility of the materials used for the tube, wick, and fluid. Otherwise, reaction between the incompatible materials produces noncondensable gases, which degrade the performance of the heat pipe. For example,the reaction between stainless steel and water in some early heat pipes generated hydrogen gas, which destroyed the heat pipe.

TABLE 14-5

Typical heat removal capacity of various heat pipes

Outside diameter, cm (in.)	Length, cm (in.)	Heat removal rate, W
0.635($\frac{1}{4}$)	15.2(6)	300
	30.5(12)	175
	45.7(18)	150
0.95($\frac{3}{8}$)	15.2(6)	500
	30.5(12)	375
	45.7(18)	350
1.27($\frac{1}{2}$)	15.2(6)	700
	30.5(12)	575
	45.7(18)	550

EXAMPLE 14-20   Replacing a Heat Pipe by a Copper Rod

A 30-cm-long cylindrical heat pipe having a diameter of 0.6 cm is dissipating heat at a rate of 180 W, with a temperature difference of 3°C across the heat pipe, as shown in Figure 14-68. If we were to use a 30-cm-long copper rod instead to remove heat at the same rate, determine the diameter and the mass of the copper rod that needs to be installed.

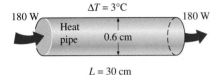

FIGURE 14-68

Schematic for Example 14-20.

Solution   A 30-cm long cylindrical heat pipe dissipates heat at a rate of 180 W. The diameter and mass of a 30-cm long copper rod that can conduct heat at the same rate are to be determined.

Assumptions   Steady operating conditions exist.

Properties   The properties of copper at room temperature are $\rho = 8950$ kg/m^3 and $k = 386$ W/m · °C (Table A-3).

Analysis   The rate of heat transfer $\dot{Q}$ through the copper rod can be expressed as

$$\dot{Q} = kA\frac{\Delta T}{L}$$

where k is the thermal conductivity, L is the length, and ΔT is the temperature difference across the copper bar. Solving for the cross-sectional area A and substituting the specified values gives

$$A = \frac{L}{k\Delta T}\dot{Q} = \frac{0.3\text{ m}}{(386\text{ W/m} \cdot \text{°C})(3\text{°C})}(180\text{ W}) = 0.04663\text{ m}^2 = 466.3\text{ cm}^2$$

Then the diameter and the mass of the copper rod become

$$A = \frac{1}{4}\pi D^2 \longrightarrow D = \sqrt{4A/\pi} = \sqrt{4(466.3\text{ cm}^2)/\pi} = \textbf{24.4 cm}$$

$$m = \rho V = \rho AL = (8590\text{ kg/m}^3)(0.04663\text{ m}^2)(0.3\text{ m}) = \textbf{125.2 kg}$$

Therefore, the diameter of the copper rod needs to be almost 25 times that of the heat pipe to transfer heat at the same rate. Also, the rod would have a mass of 125.2 kg, which is impossible for an average person to lift.

14-12 ■ SUMMARY

Electric current flow through a resistance is always accompanied by heat generation, and the essence of thermal design is the safe removal of this internally generated heat by providing an effective path for heat flow from electronic components to the surrounding medium. In this chapter, we have discussed several *cooling techniques* commonly used in electronic equipment, such as conduction cooling, natural convection and radiation cooling, forced-air convection cooling, liquid cooling, immersion cooling, and heat pipes.

In a chip carrier, heat generated at the junction is conducted through the thickness of the chip, the bonding material, the lead frame, the case material, and the leads. The *junction-to-case thermal resistance* $R_{\text{junction-case}}$ represents the total resistance to heat transfer between the junction of a component and its case. This resistance should be as low as possible to minimize the temperature rise of the junction above the case temperature. The epoxy board used in PCBs is a poor heat conductor, and so it is necessary to use copper cladding or to attach the PCB to a heat frame in conduction-cooled systems.

Low-power electronic systems can be cooled effectively with natural convection and radiation. The heat transfer from a surface at temperature T_s to a fluid at temperature T_{fluid} by *convection* is expressed as

$$\dot{Q}_{conv} = hA(T_s - T_{fluid}) \qquad \text{(W)}$$

where h is the convection heat transfer coefficient and A is the heat transfer surface area. The natural convection heat transfer coefficient for laminar flow of air at atmospheric pressure is given by a simplified relation of the form

$$h = K\left(\frac{\Delta T}{L}\right)^{0.25} \qquad \text{(W/m}^2 \cdot {}^\circ\text{C)}$$

where $\Delta T = T_s - T_{fluid}$ is the temperature difference between the surface and the fluid, L is the characteristic length (the length of the body along the heat flow path), and K is a constant, whose value is given in Table 14-1. The relations in Table 14-1 can also be used at pressures other than 1 atm by multiplying them by $\sqrt{P}$, where P is the air pressure in atm.

Radiation heat transfer between a surface at temperature T_s completely surrounded by a much larger surface at temperature T_{surr} can be expressed as

$$\dot{Q}_{rad} = \varepsilon A \sigma (T_s^4 - T_{surr}^4) \qquad \text{(W)}$$

where ε is the emissivity of the surface, A is the heat transfer surface area, and σ is the Stefan–Boltzmann constant, whose value is $\sigma = 5.67 \times 10^{-8}$ W/m$^2 \cdot$ K$^4 = 0.1714 \times 10^{-8}$ Btu/h $\cdot$ ft$^2 \cdot$ R^4.

Fluid flow over a body such as a transistor is called *external flow,* and flow through a confined space such as inside a tube or through the parallel passage area between two circuit boards in an enclosure is called *internal flow.* Fluid flow is also categorized as being *laminar* or *turbulent,* depending on the value of the Reynolds number. In convection analysis, the convection heat transfer coefficient is usually expressed in terms of the dimensionless *Nusselt number* Nu as

$$h = \frac{k}{D}\text{Nu} \qquad \text{(W/m}^2 \cdot {}^\circ\text{C)}$$

where k is thermal conductivity of the fluid and D is the characteristic length of the geometry. Relations for the average Nusselt number based on experimental data are given in Table 14-2 for external flow and in Table 14-3 for laminar internal flow under the uniform heat flux condition, which is closely approximated by electronic equipment. In forced-air-cooled systems, the heat transfer can also be expressed as

$$\dot{Q} = \dot{m}C_p(T_{out} - T_{in}) \qquad \text{(W)}$$

where $\dot{Q}$ is the rate of heat transfer to the air; C_p is the specific heat of air; T_{in} and T_{out} are the average temperatures of air at the inlet and exit of the enclosure, respectively; and $\dot{m}$ is the mass flow rate of air.

The heat transfer coefficients associated with liquids are usually an order of magnitude higher than those associated with gases. Liquid cooling systems can be classified as *direct cooling* and *indirect cooling* systems. In direct cooling systems, the electronic components are in direct contact with the liquid, and thus the heat generated in the components is transferred directly to the liquid. In indirect cooling systems, however, there is no direct contact with the components. Liquid cooling systems are also classified as *closed-loop* and *open-loop* systems, depending on whether the liquid is discharged or recirculated after it is heated. Only *dielectric fluids* can be used in immersion or direct liquid cooling.

High-power electronic components can be cooled effectively by immersing them in a *dielectric liquid* and taking advantage of the very high heat transfer coefficients associated with boiling. The simplest type of immersion cooling system involves an *external reservoir* that supplies liquid continually to the electronic enclosure. This *open-loop*-type immersion cooling system is simple but often impractical. Immersion cooling systems usually operate in a *closed loop,* in that the vapor is condensed and returned to the electronic enclosure instead of being purged to the atmosphere.

A *heat pipe* is basically a sealed slender tube containing a wick structure lined on the inner surface, which can transfer large quantities of heat over fairly large distances essentially at constant temperature without requiring any power input. The type of fluid and the operating pressure inside the heat pipe depend on the operating temperature of the heat pipe.

REFERENCES AND SUGGESTED READING

1 E. P. Black and E. M. Daley. "Thermal Design Considerations for Electronic Components." ASME Paper No. 70-DE-17, 1970.

2 S. W. Chi. *Heat Pipe Theory and Practice.* Washington, DC: Hemisphere, 1976.

3 R. A. Colclaser, D. A. Neaman, and C. F. Hawkins. *Electronic Circuit Analysis.* New York: John Wiley & Sons, 1984.

4 J. W. Dally. *Packaging of Electronic Systems.* New York: McGraw-Hill, 1990.

5 *Design Manual of Cooling Methods for Electronic Equipment.* NAVSHIPS 900-190. Department of the Navy, Bureau of Ships, March 1955.

6 *Design Manual of Natural Methods of Cooling Electronic Equipment.* NAVSHIPS 900-192. Department of the Navy, Bureau of Ships, November 1956.

7 G. N. Ellison. *Thermal Computations for Electronic Equipment.* New York: Van Nostrand Reinhold, 1984.

8 J. A. Gardner. "Liquid Cooling Safeguards High-Power Semiconductors." *Electronics,* February 24, 1974, p. 103.

9 R. S. Gaugler. "Heat Transfer Devices." U.S. Patent 2350348, 1944.

10 G. M. Grover, T. P. Cotter, and G. F. Erickson. "Structures of Very High Thermal Conductivity." *Journal of Applied Physics* 35 (1964), pp. 1190–91.

11 W. F. Hilbert and F. H. Kube. "Effects on Electronic Equipment Reliability of Temperature Cycling in Equipment." Report No. EC-69-400. Final Report, Grumman Aircraft Engineering Corporation, Bethpage, NY, February 1969.

12 U. P. Hwang and K. P. Moran. "Boiling Heat Transfer of Silicon Integrated Circuits Chip Mounted on a Substrate." *ASME HTD* 20 (1981), pp. 53–59.

13 M. Jakob. *Heat Transfer.* Vol. 1. New York: John Wiley & Sons, 1949.

14 R. D. Johnson. "Enclosures—State of the Art." *New Electronics,* July 29, 1983, p. 29.

15 R. Kemp. "The Heat Pipe—A New Tune on an Old Pipe." *Electronics and Power,* August 9, 1973, p. 326.

16 A. D. Kraus and A. Bar-Cohen. *Thermal Analysis and Control of Electronic Equipment.* New York: McGraw-Hill/Hemisphere, 1983.

17 *Reliability Prediction of Electronic Equipment.* U.S. Department of Defense, MIL-HDBK-2178B, NTIS, Springfield, VA, 1974.

18 D. S. Steinberg. *Cooling Techniques for Electronic Equipment.* New York: John Wiley & Sons, 1980.

19 A. Zhukauskas. "Heat Transfer from Tubes in Cross Flow." In *Advances in Heat Transfer,* vol. 8, ed. J. P. Hartnett and T. F. Irvine, Jr. New York: Academic Press, 1972.

PROBLEMS*

Introduction and History

14-1C What invention started the electronic age? Why did the invention of the transistor mark the beginning of a revolution in that age?

14-2C What is an integrated circuit? What is its significance in the electronics era? What do the initials MSI, LSI, and VLSI stand for?

14-3C When electric current i passes through an electrical element having a resistance $R,$ why is heat generated in the element? How is the amount of heat generation determined?

14-4C Consider a TV that is wrapped in the blankets from all sides except its screen. Explain what will happen when the TV is turned on and kept on for a long time, and why. What will happen if the TV is kept on for a few minutes only?

*Students are encouraged to answer *all* the concept "C" questions.

14-5C Consider an incandescent light bulb that is completely wrapped in a towel. Explain what will happen when the light is turned on and kept on. (P.S. Do not try this at home!)

14-6C A businessman ties a large cloth advertisement banner in front of his car such that it completely blocks the airflow to the radiator. What do you think will happen when he starts the car and goes on a trip?

14-7C Which is more likely to break: a car or a TV? Why?

14-8C Why do electronic components fail under prolonged use at high temperatures?

14-9 The temperature of the case of a power transistor that is dissipating 12 W is measured to be 60°C. If the junction-to-case thermal resistance of this transistor is specified by the manufacturer to be 5°C/W, determine the junction temperature of the transistor.   *Answer:* 120°C

14-10 Power is supplied to an electronic component from a 12-V source, and the variation in the electric current, the junction temperature, and the case temperatures with time are observed. When everything is stabilized, the current is observed to be 0.15 A and the temperatures to be 90°C and 65°C at the junction and the case, respectively. Calculate the junction-to-case thermal resistance of this component.

14-11 A logic chip used in a computer dissipates 3.5 W of power in an environment at 55°C and has a heat transfer surface area of 0.32 cm². Assuming the heat transfer from the surface to be uniform, determine (*a*) the amount of heat this chip dissipates during an eight-hour work day, in kWh, and (*b*) the heat flux on the surface of the chip, in W/cm².

14-12 A 15-cm × 20-cm circuit board houses 90 closely spaced logic chips, each dissipating 0.1 W, on its surface. If the heat transfer from the back surface of the board is negligible, determine (*a*) the amount of heat this circuit board dissipates during a 10-h period, in kWh, and (*b*) the heat flux on the surface of the circuit board, in W/cm².

14-13E A resistor on a circuit board has a total thermal resistance of 130°F/W. If the temperature of the resistor is not to exceed 360°F, determine the power at which it can operate safely in an ambient at 120°F.

14-14 Consider a 0.1-kΩ resistor whose surface-to-ambient thermal resistance is 300°C/W. If the voltage drop across the resistor is 7.5 V and its surface temperature is not to exceed 150°C, determine the power at which it can operate safely in an ambient at 30°C.   *Answer:* 0.4 W

Manufacturing of Electronic Equipment

14-15C Why is a chip in a chip carrier bonded to a lead frame instead of the plastic case of the chip carrier?

14-16C Draw a schematic of a chip carrier, and explain how heat is transferred from the chip to the medium outside of the chip carrier.

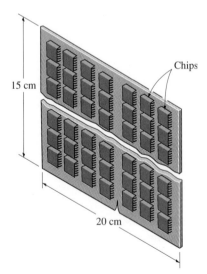

15 cm

Chips

20 cm

FIGURE P14-12

14-17C   What does the junction-to-case thermal resistance represent? On what does it depend for a chip carrier?

14-18C   What is a hybrid chip carrier? What are the advantages of hybrid electronic packages?

14-19C   What is a PCB? Of what is the board of a PCB made? What does the "device-to-PCB edge" thermal resistance in conduction-cooled systems represent? Why is this resistance relatively high?

14-20C   What are the three types of printed circuit boards? What are the advantages and disadvantages of each type?

14-21C   What are the desirable characteristics of the materials used in the fabrication of the circuit boards?

14-22C   What is an electronic enclosure? What is the primary function of the enclosure of an electronic system? Of what materials are the enclosures made?

Cooling Load of Electronic Equipment and Thermal Environment

14-23C   Consider an electronics box that consumes 120 W of power when plugged in. How is the heating load of this box determined?

14-24C   Why is the development of superconducting materials generating so much excitement among designers of electronic equipment?

14-25C   How is the duty cycle of an electronic system defined? How does the duty cycle affect the design and selection of a cooling technique for a system?

14-26C   What is temperature cycling? How does it affect the reliability of electronic equipment?

14-27C   What is the ultimate heat sink for (*a*) a TV, (*b*) an airplane, and (*c*) a ship? For each case, what is the range of temperature variation of the ambient?

14-28C   What is the ultimate heat sink for (*a*) a VCR, (*b*) a spacecraft, and (*c*) a communication system on top of a mountain? For each case, what is the range of temperature variation of the ambient?

Electronics Cooling in Different Applications

14-29C   How are the electronics of short-range and long-range missiles cooled?

14-30C   What is dynamic temperature? What causes it? How is it determined? At what velocities is it significant?

14-31C   How are the electronics of a ship or submarine cooled?

14-32C   How are the electronics of the communication systems at remote areas cooled?

14-33C How are the electronics of high-power microwave equipment such as radars cooled?

14-34C How are the electronics of a space vehicle cooled?

14-35 Consider an airplane cruising in the air at a temperature of $-10°C$ at a velocity of 850 km/h. Determine the temperature rise of air at the nose of the airplane as a result of the ramming effect of the air.

850 km/h

FIGURE P14-35

14-36 The temperature of air in high winds is measured by a thermometer to be 12°C. Determine the true temperature of air if the wind velocity is 90 km/h.   *Answer:* 11.7°C

14-37 Air at 25°C is flowing in a channel at a velocity of (*a*) 1, (*b*) 10, (*c*) 100, and (*d*) 1000 m/s. Determine the temperature that a stationary probe inserted into the channel will read for each case.

14-38 An electronic device dissipates 2 W of power and has a surface area of 5 cm^2. If the surface temperature of the device is not to exceed the ambient temperature by more than 50°C, determine a suitable cooling technique for this device. Use Figure 14-17.

14-39E A stand-alone circuit board, 6 in. $\times$ 8 in. in size, dissipates 20 W of power. The surface temperature of the board is not to exceed 170°F in an 85°F environment. Using Figure 14-17 as a guide, select a suitable cooling mechanism.

Conduction Cooling

14-40C What are the major considerations in the selection of a cooling technique for electronic equipment?

14-41C What is thermal resistance? To what is it analogous in electrical circuits? Can thermal resistance networks be analyzed like electrical circuits? Explain.

14-42C If the rate of heat conduction through a medium and the thermal resistance of the medium are known, how can the temperature difference between the two sides of the medium be determined?

14-43C Consider a wire of electrical resistance R, length L, and cross-sectional area A through which electric current I is flowing. How is the voltage drop across the wire determined? What happens to the voltage drop when L is doubled while I is held constant?

Now consider heat conduction at a rate of $\dot{Q}$ through the same wire having a thermal resistance of R. How is the temperature drop across the wire determined? What happens to the temperature drop when L is doubled while $\dot{Q}$ is held constant?

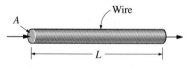

A   Wire

L

FIGURE P14-43C

14-44C What is a heat frame? How does it enhance heat transfer along a PCB? Which components on a PCB attached to a heat frame operate at the highest temperatures: those at the middle of the PCB or those near the edge?

14-45C What is constriction resistance in heat flow? To what is it analogous in fluid flow through tubes and electric current flow in wires?

14-46C What does the junction-to-case thermal resistance of an electronic component represent? In practice, how is this value determined? How can the junction temperature of a component be determined when the case temperature, the power dissipation of the component, and the junction-to-case thermal resistance are known?

14-47C What does the case-to-ambient thermal resistance of an electronic component represent? In practice, how is this value determined? How can the case temperature of a component be determined when the ambient temperature, the power dissipation of the component, and the case-to-ambient thermal resistance are known?

14-48C Consider an electronic component whose junction-to-case thermal resistance $R_{\text{junction-case}}$ is provided by the manufacturer and whose case-to-ambient thermal resistance $R_{\text{case-ambient}}$ is determined by the thermal designer. When the power dissipation of the component and the ambient temperature are known, explain how the junction temperature can be determined. When $R_{\text{junction-case}}$ is greater than $R_{\text{case-ambient}}$, will the case temperature be closer to the junction or ambient temperature?

14-49C Why is the rate of heat conduction along a PCB very low? How can heat conduction from the mid-parts of a PCB to its outer edges be improved? How can heat conduction across the thickness of the PCB be improved?

14-50C Why is the warping of epoxy boards that are copper-cladded on one side a major concern? What is the cause of this warping? How can the warping of PCBs be avoided?

14-51C Why did the thermal conduction module receive so much attention from thermal designers of electronic equipment? How does the design of TCM differ from traditional chip carrier design? Why is the cavity in the TCM filled with helium instead of air?

14-52 Consider a chip dissipating 0.8 W of power in a DIP with 18 pin leads. The materials and the dimensions of various sections of this electronic device are given in the table below. If the temperature of the leads is 50°C, estimate the temperature at the junction of the chip.

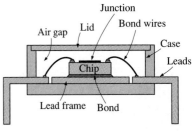

FIGURE P14-52

Section and material	Thermal conductivity, W/m·°C	Thickness, mm	Heat transfer surface area
Junction constriction	—	—	Diameter 0.5 mm
Silicon chip	120	0.5	4 mm × 4 mm
Eutectic bond	296	0.05	4 mm × 4 mm
Copper lead frame	386	0.25	4 mm × 4 mm
Plastic separator	1	0.3	18 × 1 mm × 0.25 mm
Copper leads	386	6	18 × 1 mm × 0.25 mm

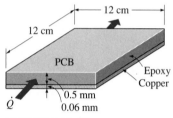

FIGURE P14-54

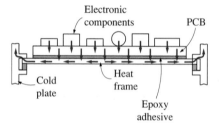

FIGURE P14-58

FIGURE P14-59

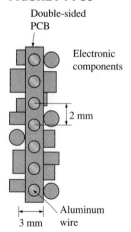

14-53 A fan blows air at 35°C over a 2-W plastic DIP with 16 leads mounted on a PCB at a velocity of 300 m/min. Using data from Figure 14-23, determine the junction temperature of the electronic device. What would the junction temperature be if the fan were to fail?

14-54 Heat is to be conducted along a PCB with copper cladding on one side. The PCB is 12 cm long and 12 cm wide, and the thicknesses of the copper and epoxy layers are 0.06 mm and 0.5 mm, respectively. Disregarding heat transfer from the side surfaces, determine the percentages of heat conduction along the copper ($k = 386$ W/m · °C) and epoxy ($k = 0.26$ W/m · °C) layers. Also, determine the effective thermal conductivity of the PCB.
Answers: 0.6 percent, 99.4 percent, 41.6 W/m · °C

14-55 The heat generated in the circuitry on the surface of a silicon chip ($k = 130$ W/m · °C) is conducted to the ceramic substrate to which it is attached. The chip is 6 mm × 6 mm in size and 0.5-mm thick and dissipates 2 W of power. Determine the temperature difference between the front and back surfaces of the chip in steady operation.

14-56E Consider a 6-in. × 7-in. glass–epoxy laminate ($k = 0.15$ Btu/h · ft · °F) whose thickness is 0.05 in. Determine the thermal resistance of this epoxy layer for heat flow (a) along the 7-in.-long side and (b) across its thickness.

14-57 Consider a 15-cm × 18-cm glass–epoxy laminate ($k = 0.26$ W/m · °C) whose thickness is 1.4 mm. In order to reduce the thermal resistance across its thickness, cylindrical copper fillings ($k = 386$ W/m · °C) of diameter 1 mm are to be planted throughout the board with a center-to-center distance of 3 mm. Determine the new value of the thermal resistance of the epoxy board for heat conduction across its thickness as a result of this modification.

14-58 A 12-cm × 15-cm circuit board dissipating 45 W of heat is to be conduction-cooled by a 1.5-mm-thick copper heat frame ($k = 386$ W/m · °C) 12 cm × 17 cm in size. The epoxy laminate ($k = 0.26$ W/m · °C) has a thickness of 2 mm and is attached to the heat frame with conductive epoxy adhesive ($k = 1.8$ W/m · °C) of thickness 0.12 mm. The PCB is attached to a heat sink by clamping a 5-mm-wide portion of the edge to the heat sink from both ends. The temperature of the heat frame at this point is 30°C. Heat is uniformly generated on the PCB at a rate of 3 W per 1-cm × 12-cm strip. Considering only one-half of the PCB board because of symmetry, determine the maximum surface temperature on the PCB and the temperature distribution along the heat frame.

14-59 Consider a 15-cm × 20-cm double-sided circuit board dissipating a total of 30 W of heat. The board consists of a 3-mm-thick epoxy layer ($k = 0.26$ W/m · °C) with 1-mm-diameter aluminum wires ($k = 237$ W/m · °C) inserted along the 20-cm-long direction, as shown in Figure P14-59. The distance between the centers of the aluminum wires is 2 mm. The circuit board is attached to a heat sink from both ends, and the temperature of the board at both ends is 30°C. Heat is considered to be uniformly generated on both sides of the epoxy layer of the PCB. Considering only a portion of the PCB because of symmetry, determine the magnitude and location of the maximum temperature that occurs in the PCB. *Answer:* 136.7°C

14-60 Repeat Problem 14-59, replacing the aluminum wires by copper wires ($k = 386$ W/m · °C).

14-61 Repeat Problem 14-59 for a center-to-center distance of 4 mm instead of 2 mm between the wires.

14-62 Consider a thermal conduction module with 80 chips, each dissipating 4 W of power. The module is cooled by water at 22°C flowing through the cold plate on top of the module. The thermal resistances in the path of heat flow are $R_{chip} = 12$°C/W between the junction and the surface of the chip, $R_{int} = 9$°C/W between the surface of the chip and the outer surface of the thermal conduction module, and $R_{ext} = 7$°C/W between the outer surface of the module and the cooling water. Determine the junction temperature of the chip.

14-63 Consider a 0.3-mm-thick epoxy board ($k = 0.26$ W/m · °C) that is 15 cm × 20 cm in size. Now a 0.1-mm-thick layer of copper ($k = 386$ W/m · °C) is attached to the back surface of the PCB. Determine the effective thermal conductivity of the PCB along its 20-cm-long side. What fraction of the heat conducted along that side is conducted through copper?

14-64 A 0.5-mm-thick copper plate ($k = 386$ W/m · °C) is sandwiched between two 3-mm-thick epoxy boards ($k = 0.26$ W/m · °C) that are 12 cm × 18 cm in size. Determine the effective thermal conductivity of the PCB along its 18-cm-long side. What fraction of the heat conducted along that side is conducted through copper?

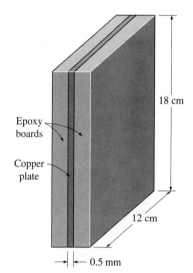

FIGURE P14-64

14-65E A 6-in. × 8-in. × 0.07-in. copper heat frame is used to conduct 20 W of heat generated in a PCB along the 8-in.-long side toward the ends. Determine the temperature difference between the midsection and either end of the heat frame. *Answer:* 34.8°F

14-66 A 12-W power transistor is cooled by mounting it on an aluminum bracket ($k = 237$ W/m · °C) that is attached to a liquid-cooled plate by 0.2-mm-thick epoxy adhesive ($k = 1.8$ W/m · °C), as shown in Figure P14-66. The thermal resistance of the plastic washer is given as 2.5°C/W. Preliminary calculations show that about 20 percent of the heat is dissipated by convection and radiation, and the rest is conducted to the liquid-cooled plate. If the temperature of the cold plate is 50°C, determine the temperature of the transistor case.

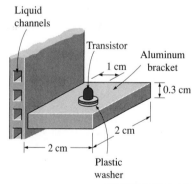

FIGURE P14-66

Air Cooling: Natural Convection and Radiation

14-67C A student puts his books on top of a VCR, completely blocking the air vents on the top surface. What do you think will happen as the student watches a rented movie played by that VCR?

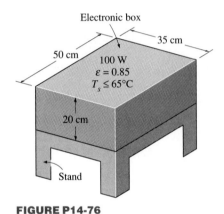

Electronic box
50 cm
35 cm
100 W
$\varepsilon = 0.85$
$T_s \leq 65°C$
20 cm
Stand

FIGURE P14-76

FIGURE P14-79

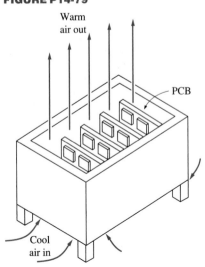

Warm
air out
PCB
Cool
air in

14-68C Can a low-power electronic system in space be cooled by natural convection? Can it be cooled by radiation? Explain.

14-69C Why are there several openings on the various surfaces of a TV, VCR, and other electronic enclosures? What happens if a TV or VCR is enclosed in a cabinet with no free air space around?

14-70C Why should radiation heat transfer always be considered in the analysis of natural convection–cooled electronic equipment?

14-71C How does atmospheric pressure affect natural convection heat transfer? What are the best and worst orientations for heat transfer from a square surface?

14-72C What is view factor? How does it affect radiation heat transfer between two surfaces?

14-73C What is emissivity? How does it affect radiation heat transfer between two surfaces?

14-74C For most effective natural convection cooling of a PCB array, should the PCBs be placed horizontally or vertically? Should they be placed close to each other or far from each other?

14-75C Why is radiation heat transfer from the components on the PCBs in an enclosure negligible?

14-76 Consider a sealed 20-cm-high electronic box whose base dimensions are 35 cm × 50 cm that is placed on top of a stand in a room at 30°C. The emissivity of the outer surface of the box is 0.85. If the electronic components in the box dissipate a total of 100 W of power and the outer surface temperature of the box is not to exceed 65°C, determine if this box can be cooled by natural convection and radiation alone. Assume the heat transfer from the bottom surface of the box to the stand to be negligible, and the temperature of the surrounding surfaces to be the same as the air temperature of the room.

14-77 Repeat Problem 14-76, assuming the box is mounted on a wall instead of a stand such that it is 0.5 m high. Again, assume heat transfer from the bottom surface to the wall to be negligible.

14-78E A 0.15-W small cylindrical resistor mounted on a circuit board is 0.5-in. long and has a diameter of 0.15 in. The view of the resistor is largely blocked by the circuit board facing it, and the heat transfer from the connecting wires is negligible. The air is free to flow through the parallel flow passages between the PCBs as a result of natural convection currents. If the air temperature in the vicinity of the resistor is 130°F, determine the surface temperature of the resistor.   *Answer:* 194°F

14-79 A 14-cm × 20-cm PCB has electronic components on one side, dissipating a total of 7 W. The PCB is mounted in a rack vertically (height 14 cm) together with other PCBs. If the surface temperature of the components is not to exceed 90°C, determine the maximum temperature of the environment in which this PCB can operate safely at sea level. What would your answer be if this rack is located at a location at 3000 m altitude where the atmospheric pressure is 70.12 kPa?

14-80 A cylindrical electronic component whose diameter is 2 cm and length is 4 cm is mounted on a board with its axis in the vertical direction and is dissipating 3 W of power. The emissivity of the surface of the component is 0.8, and the temperature of the ambient air is 30°C. Assuming the temperature of the surrounding surfaces to be 20°C, determine the average surface temperature of the component under combined natural convection and radiation cooling.

14-81 Repeat Problem 14-80, assuming the component is oriented horizontally.

14-82 Consider a power transistor that dissipates 0.1 W of power in an environment at 30°C. The transistor is 0.4 cm long and has a diameter of 0.4 cm. Assuming heat to be transferred uniformly from all surfaces, determine (a) the heat flux on the surface of the transistor, in W/cm^2, and (b) the surface temperature of the transistor for a combined convection and radiation heat transfer coefficient of 12 $W/m^2 \cdot °C$.

14-83 The components of an electronic system dissipating 150 W are located in a 1-m-long horizontal duct whose cross-section is 15 cm × 15 cm. The components in the duct are cooled by forced air, which enters at 30°C at a rate of 0.4 m^3/min and leaves at 45°C. The surfaces of the sheet metal duct are not painted, and thus radiation heat transfer from the outer surfaces is negligible. If the ambient air temperature is 25°C, determine (a) the heat transfer from the outer surfaces of the duct to the ambient air by natural convection and (b) the average temperature of the duct.
    Answers: (a) 31.7 W, (b) 40°C

14-84 Repeat Problem 14-83 for a circular horizontal duct of diameter 10 cm.

14-85 Repeat Problem 14-83, assuming that the fan fails and thus the entire heat generated inside the duct must be rejected to the ambient air by natural convection from the outer surfaces of the duct.

14-86 A 20-cm × 20-cm circuit board containing 81 square chips on one side is to be cooled by combined natural convection and radiation by mounting it on a vertical surface in a room at 25°C. Each chip dissipates 0.1 W of power, and the emissivity of the chip surfaces is 0.7. Assuming the heat transfer from the back side of the circuit board to be negligible and the temperature of the surrounding surfaces to be the same as the air temperature of the room, determine the surface temperature of the chips.

14-87 Repeat Problem 14-86, assuming the circuit board to be positioned horizontally with (a) chips facing up and (b) chips facing down.

Air Cooling: Forced Convection

14-88C Why is radiation heat transfer in forced-air-cooled systems disregarded?

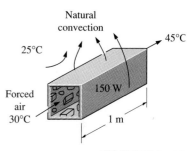

FIGURE P14-83

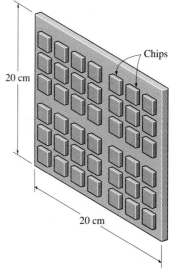

FIGURE P14-86

14-89C If an electronic system can be cooled adequately by either natural convection or forced-air convection, which would you prefer? Why?

14-90C Why is forced convection cooling much more effective than natural convection cooling?

14-91C Consider a forced-air-cooled electronic system dissipating a fixed amount of power. How will increasing the flow rate of air affect the surface temperature of the components? Explain. How will it affect the exit temperature of the air?

14-92C To what do internal and external flow refer in forced convection cooling? Give an example of a forced-air-cooled electronic system that involves both types of flow.

14-93C For a specified power dissipation and air inlet temperature, how does the convection heat transfer coefficient affect the surface temperature of the electronic components? Explain.

14-94C How does high altitude affect forced convection heat transfer? How would you modify your forced-air cooling system to operate at high altitudes safely?

14-95C What are the advantages and disadvantages of placing the cooling fan at the inlet or at the exit of an electronic box?

14-96C How is the volume flow rate of air in a forced-air-cooled electronic system that has a constant-speed fan established? If a few more PCBs are added to the box while keeping the fan speed constant, will the flow rate of air through the system change? Explain.

14-97C What happens if we attempt to cool an electronic system with an undersized fan? What about if we do that with an oversized fan?

14-98 Consider a hollow core PCB that is 15 cm high and 20 cm long, dissipating a total of 30 W. The width of the air gap in the middle of the PCB is 0.25 cm. The cooling air enters the core at 30°C at a rate of 1 L/s. Assuming the heat generated to be uniformly distributed over the two side surfaces of the PCB, determine (*a*) the temperature at which the air leaves the hollow core and (*b*) the highest temperature on the inner surface of the core.
  Answers: (*a*) 55°C, (*b*) 66.4°C

14-99 Repeat Problem 14-98 for a hollow core PCB dissipating 45 W.

14-100E A transistor with a height of 0.25 in. and a diameter of 0.2 in. is mounted on a circuit board. The transistor is cooled by air flowing over it at a velocity of 400 ft/min. If the air temperature is 140°F and the transistor case temperature is not to exceed 175°F, determine the amount of power this transistor can dissipate safely. *Answer:* 0.15 W

14-101 A desktop computer is to be cooled by a fan. The electronic components of the computer consume 60 W of power under full-load conditions. The computer is to operate in environments at temperatures up to 45°C and

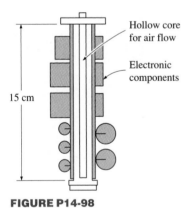

Hollow core
for air flow

Electronic
components

15 cm

FIGURE P14-98

at elevations up to 3400 m where the atmospheric pressure is 66.63 kPa. The exit temperature of air is not to exceed 60°C to meet reliability requirements. Also, the average velocity of air is not to exceed 110 m/min at the exit of the computer case, where the fan is installed to keep the noise level down. Determine the flow rate of the fan that needs to be installed and the diameter of the casing of the fan.

14-102 Repeat Problem 14-101 for a computer that consumes 100 W of power.

14-103 A computer cooled by a fan contains eight PCBs, each dissipating 12 W of power. The height of the PCBs is 12 cm and the length is 18 cm. The clearance between the tips of the components on the PCB and the back surface of the adjacent PCB is 0.3 cm. The cooling air is supplied by a 25-W fan mounted at the inlet. If the temperature rise of air as it flows through the case of the computer is not to exceed 15°C, determine (a) the flow rate of the air that the fan needs to deliver, (b) the fraction of the temperature rise of air due to the heat generated by the fan and its motor, and (c) the highest allowable inlet air temperature if the surface temperature of the components is not to exceed 80°C anywhere in the system.

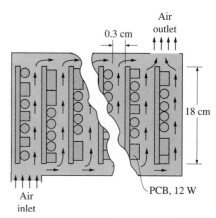

FIGURE P14-103

14-104 An array of power transistors, each dissipating 2 W of power, is to be cooled by mounting them on a 20-cm × 20-cm square aluminum plate and blowing air over the plate with a fan at 30°C with a velocity of 3 m/s. The average temperature of the plate is not to exceed 60°C. Assuming the heat transfer from the back side of the plate to be negligible, determine the number of transistors that can be placed on this plate. *Answer:* 9

14-105 Repeat Problem 14-104 for a location at an elevation of 1610 m where the atmospheric pressure is 83.4 kPa.

14-106 An enclosure contains an array of circuit boards, 15 cm high and 20 cm long. The clearance between the tips of the components on the PCB and the back surface of the adjacent PCB is 0.3 cm. Each circuit board contains 75 square chips on one side, each dissipating 0.15 W of power. Air enters the space between the boards through the 0.3-cm × 15-cm cross-section at 40°C with a velocity of 300 m/min. Assuming the heat transfer from the back side of the circuit board to be negligible, determine the exit temperature of the air and the highest surface temperature of the chips.

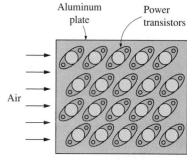

FIGURE P14-104

14-107 The components of an electronic system dissipating 120 W are located in a 1-m-long horizontal duct whose cross-section is 15 cm × 15 cm. The components in the duct are cooled by forced air, which enters at 30°C at a rate of 0.5 m³/min. Assuming 80 percent of the heat generated inside is transferred to air flowing through the duct and the remaining 20 percent is lost through the outer surfaces of the duct, determine (a) the exit temperature of air and (b) the highest component surface temperature in the duct.

14-108 Repeat Problem 14-107 for a circular horizontal duct of diameter 10 cm.

Liquid Cooling

14-109C If an electronic system can be cooled adequately by either forced-air cooling or liquid cooling, which one would you prefer? Why?

14-110C Explain how direct and indirect liquid cooling systems differ from each other.

14-111C Explain how closed-loop and open-loop liquid cooling systems operate.

14-112C What are the properties of a liquid ideally suited for cooling electronic equipment?

14-113 A cold plate that supports 10 power transistors, each dissipating 40 W, is to be cooled with water. It is specified that the temperature rise of the water not exceed 4°C and the velocity of water remain under 0.5 m/s. Assuming 25 percent of the heat generated is dissipated from the components to the surroundings by convection and radiation, and the remaining 75 percent is removed by the cooling water, determine the mass flow rate of water needed and the diameter of the pipe to satisfy the restriction imposed on the flow velocity. Also, determine the case temperature of the devices if the total case-to-liquid thermal resistance is 0.04°C/W and the water enters the cold plate at 25°C.

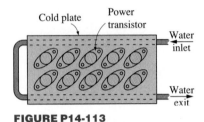

FIGURE P14-113

14-114E Water enters the tubes of a cold plate at 95°F with an average velocity of 60 ft/min and leaves at 105°F. The diameter of the tubes is 0.25 in. Assuming 15 percent of the heat generated is dissipated from the components to the surroundings by convection and radiation, and the remaining 85 percent is removed by the cooling water, determine the amount of heat generated by the electronic devices mounted on the cold plate.
   Answer: 263 W

14-115 A sealed electronic box is to be cooled by tap water flowing through channels on two of its sides. It is specified that the temperature rise of the water not exceed 4°C. The power dissipation of the box is 2 kW, which is removed entirely by water. If the box operates 24 h a day, 365 days a year, determine the mass flow rate of water flowing through the box and the amount of cooling water used per year.

14-116 Repeat Problem 14-115 for a power dissipation of 3 kW.

Immersion Cooling

14-117C What are the desirable characteristics of a liquid used in immersion cooling of electronic devices?

14-118C How does an open-loop immersion cooling system operate? How does it differ from closed-loop cooling systems?

14-119C How do immersion cooling systems with internal and external cooling differ? Why are externally cooled systems limited to relatively low-power applications?

14-120C Why is boiling heat transfer used in the cooling of very high-power electronic devices instead of forced air or liquid cooling?

14-121 A logic chip used in a computer dissipates 3 W of power and has a heat transfer surface area of 0.3 cm^2. If the surface of the chip is to be maintained at 70°C while being cooled by immersion in a dielectric fluid at 20°C, determine the necessary heat transfer coefficient and the type of cooling mechanism that needs to be used to achieve that heat transfer coefficient.

14-122 A 6-W chip having a surface area of 0.5 cm^2 is cooled by immersing it into FC86 liquid that is maintained at a temperature of 25°C. Using the boiling curve in Figure 14-63, estimate the temperature of the chip surface.
  Answer: 82°C

14-123 A logic chip cooled by immersing it in a dielectric liquid dissipates 3.5 W of power in an environment at 50°C and has a heat transfer surface area of 0.8 cm^2. The surface temperature of the chip is measured to be 95°C. Assuming the heat transfer from the surface to be uniform, determine (*a*) the heat flux on the surface of the chip, in W/cm^2; (*b*) the heat transfer coefficient on the surface of the chip, in W/m$^2 \cdot$ °C; and (*c*) the thermal resistance between the surface of the chip and the cooling medium, in °C/W.

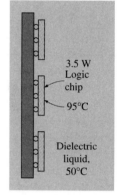

3.5 W Logic chip

95°C

Dielectric liquid, 50°C

FIGURE P14-123

14-124 A computer chip dissipates 5 W of power and has a heat transfer surface area of 0.4 cm^2. If the surface of the chip is to be maintained at 55°C while being cooled by immersion in a dielectric fluid at 10°C, determine the necessary heat transfer coefficient and the type of cooling mechanism that needs to be used to achieve that heat transfer coefficient.

14-125 A 3-W chip having a surface area of 0.2 cm^2 is cooled by immersing it into FC86 liquid that is maintained at a temperature of 40°C. Using the boiling curve in Figure 14-63, estimate the temperature of the chip surface.
  Answer: 103°C

14-126 A logic chip having a surface area of 0.3 cm^2 is to be cooled by immersing it into FC86 liquid that is maintained at a temperature of 35°C. The surface temperature of the chip is not to exceed 60°C. Using the boiling curve in Figure 14-63, estimate the maximum power that the chip can dissipate safely.

FIGURE P14-127

14-127 A 2-kW electronic device that has a surface area of 120 cm^2 is to be cooled by immersing it in a dielectric fluid with a boiling temperature of 60°C contained in a 1-m × 1-m × 1-m cubic enclosure. Noting that the combined natural convection and the radiation heat transfer coefficients in air are typically about 10 W/m$^2 \cdot$ °C, determine if the heat generated inside can be dissipated to the ambient air at 20°C by natural convection and radiation. If it cannot, explain what modification you could make to allow natural convection cooling.

  Also, determine the heat transfer coefficient at the surface of the electronic device for a surface temperature of 80°C. Assume the liquid temperature remains constant at 60°C throughout the enclosure.

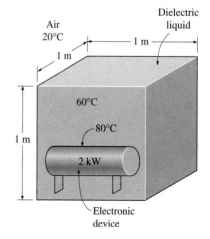

Air 20°C

Dielectric liquid

1 m

1 m

60°C

80°C

2 kW

1 m

Electronic device

Heat Pipes

14-128C What is a heat pipe? How does it operate? Does it have any moving parts?

14-129C A heat pipe with water as the working fluid is said to have an effective thermal conductivity of 100,000 W/m · °C, which is more than 100,000 times the conductivity of water. How can this happen?

14-130C What is the effect of a small amount of noncondensable gas such as air on the performance of a heat pipe?

14-131C Why do water-based heat pipes used in the cooling of electronic equipment operate below atmospheric pressure?

14-132C What happens when the wick of a heat pipe is too coarse or too fine?

14-133C Does the orientation of a heat pipe affect its performance? Does it matter if the evaporator end of the heat pipe is up or down? Explain.

14-134C How can the liquid in a heat pipe move up against gravity without a pump? For heat pipes that work against gravity, is it better to have coarse or fine wicks? Why?

14-135C What are the important considerations in the design and manufacture of heat pipes?

14-136C What is the major cause for the premature degradation of the performance of some heat pipes?

14-137 A 40-cm-long cylindrical heat pipe having a diameter of 0.5 cm is dissipating heat at a rate of 150 W, with a temperature difference of 4°C across the heat pipe. If we were to use a 40-cm-long copper rod ($k = 386$ W/m · °C and $\rho = 8950$ kg/m^3) instead to remove heat at the same rate, determine the diameter and the mass of the copper rod that needs to be installed.

14-138 Repeat Problem 14-137 for an aluminum rod instead of copper.

14-139E A plate that supports 10 power transistors, each dissipating 35 W, is to be cooled with 1-ft-long heat pipes having a diameter of $\frac{1}{4}$ in. Using Table 14-5, determine how many pipes need to be attached to this plate.
Answer: 2

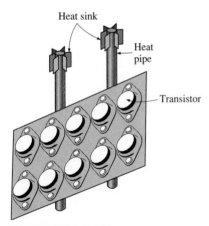

Heat sink

Heat pipe

Transistor

FIGURE P14-139E

Review Problems

14-140C Several power transistors are to be cooled by mounting them on a water-cooled metal plate. The total power dissipation, the mass flow rate of water through the tube, and the water inlet temperature are fixed. Explain what you would do for the most effective cooling of the transistors.

14-141C Consider heat conduction along a vertical copper bar whose sides are insulated. One person claims that the bar should be oriented such that the hot end is at the bottom and the cold end is at the top for better heat transfer, since heat rises. Another person claims that it makes no differences to heat conduction whether heat is conducted downward or upward, and thus the orientation of the bar is irrelevant. With which person do you agree?

14-142 Consider a 15-cm × 15-cm multilayer circuit board dissipating 22.5 W of heat. The board consists of four layers of 0.1-mm-thick copper ($k = 386$ W/m · °C) and three layers of 0.5-mm-thick glass–epoxy ($k = 0.26$ W/m · °C) sandwiched together, as shown in Figure P14-142. The circuit board is attached to a heat sink from both ends, and the temperature of the board at those ends is 35°C. Heat is considered to be uniformly generated in the epoxy layers of the PCB at a rate of 0.5 W per 1-cm × 15-cm epoxy laminate strip (or 1.5 W per 1-cm × 15-cm strip of the board). Considering only a portion of the PCB because of symmetry, determine the magnitude and location of the maximum temperature that occurs in the PCB. Assume heat transfer from the top and bottom faces of the PCB to be negligible.
Answer: 108°C

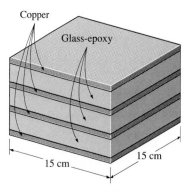

FIGURE P14-142

14-143 Repeat Problem 14-142, assuming that the board consists of a single 1.5-mm-thick layer of glass–epoxy, with no copper layers.

14-144 The components of an electronic system that is dissipating 200 W are located in a 1-m-long horizontal duct whose cross-section is 10 cm × 10 cm. The components in the duct are cooled by forced air, which enters at 30°C and 50 m/min and leaves at 45°C. The surfaces of the sheet metal duct are not painted, and so radiation heat transfer from the outer surfaces is negligible. If the ambient air temperature is 30°C, determine (*a*) the heat transfer from the outer surfaces of the duct to the ambient air by natural convection, (*b*) the average temperature of the duct, and (*c*) the highest component surface temperature in the duct.

14-145 Two 10-W power transistors are cooled by mounting them on the two sides of an aluminum bracket ($k = 237$ W/m · °C) that is attached to a liquid-cooled plate by 0.2-mm-thick epoxy adhesive ($k = 1.8$ W/m · °C), as shown in Figure P14-145. The thermal resistance of each plastic washer is given as 2°C/W, and the temperature of the cold plate is 40°C. The surface of the aluminum plate is untreated, and thus radiation heat transfer from it is negligible because of the low emissivity of aluminum surfaces. Disregarding heat transfer from the 0.3-cm-wide edges of the aluminum plate, determine the surface temperature of the transistor case. Also, determine the fraction of heat dissipation to the ambient air by natural convection and to the cold plate by conduction. Take the ambient temperature to be 25°C.

14-146E A fan blows air at 85°F and a velocity of 500 ft/min over a 1.5-W plastic DIP with 24 leads mounted on a PCB. Using data from Figure 14-23, determine the junction temperature of the electronic device. What would the junction temperature be if the fan were to fail?

14-147 A 15-cm × 18-cm double-sided circuit board dissipating a total of 18 W of heat is to be conduction-cooled by a 1.2-mm-thick aluminum core plate ($k = 237$ W/m · °C) sandwiched between two epoxy laminates ($k = 0.26$ W/m · °C). Each epoxy layer has a thickness of 0.5 mm and is attached to the aluminum core plate with conductive epoxy adhesive ($k = 1.8$ W/m · °C) of

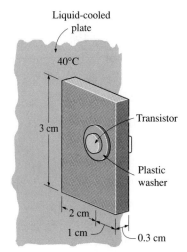

FIGURE P14-145

thickness 0.1 mm. Heat is uniformly generated on each side of the PCB at a rate of 0.5 W per 1-cm × 15-cm epoxy laminate strip. All of the heat is conducted along the 18-cm side since the PCB is cooled along the two 15-cm-long edges. Considering only part of the PCB board because of symmetry, determine the maximum temperature rise across the 9-cm distance between the center and the sides of the PCB. *Answer:* 10.1°C

14-148 Ten power transistors, each dissipating 2 W, are attached to a 7-cm × 7-cm × 0.2-cm aluminum plate with a square cutout in the middle in a symmetrical arrangement, as shown in Figure P14-148. The aluminum plate is cooled from two sides by liquid. If 85 percent of the heat generated by the transistors is estimated to be conducted through the aluminum plate, determine the temperature rise across the 1-cm-wide section of the aluminum plate between the transistors and the heat sink.

14-149 The components of an electronic system are located in a 1.2-m-long horizontal duct whose cross-section is 10 cm × 20 cm. The components in the duct are not allowed to come into direct contact with cooling air, and so are cooled by air flowing over the duct at 30°C with a velocity of 250 m/min. The duct is oriented such that air strikes the 10-cm-high side of the duct normally. If the surface temperature of the duct is not to exceed 60°C, determine the total power rating of the electronic devices that can be mounted in the duct. What would your answer be if the duct is oriented such that air strikes the 20-cm-high side normally? *Answers:* 487 W, 389 W

14-150 Repeat Problem 14-149 for a location at an altitude of 5000 m, where the atmospheric pressure is 54.05 kPa.

14-151E A computer that consumes 80 W of power is cooled by a fan blowing air into the computer enclosure. The dimensions of the computer case are 6 in. × 20 in. × 24 in., and all surfaces of the case are exposed to the ambient, except for the base surface. Temperature measurements indicate that the case is at an average temperature of 95°F when the ambient temperature and the temperature of the surrounding walls are 80°F. If the emissivity of the outer surface of the case is 0.85, determine the fraction of heat lost from the outer surfaces of the computer case.

Computer, Design, and Essay Problems

14-152 Bring an electronic device that is cooled by heat sinking to class and discuss how the heat sink enhances heat transfer.

14-153 Obtain a catalog from a heat sink manufacturer and select a heat sink model that can cool a 10-W power transistor safely under (*a*) natural convection and radiation and (*b*) forced convection conditions.

14-154 Take the cover off a PC or another electronic box and identify the individual sections and components. Also, identify the cooling mechanisms involved and discuss how the current design facilitates effective cooling.

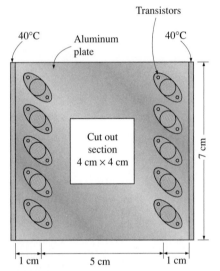

FIGURE P14-148

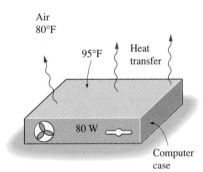

FIGURE P14-151E

Property Tables and Charts (SI Units)

TABLE A-1

Molar mass, gas constant, and critical-point properties

Substance	Formula	Molar mass, M kg/kmol	Gas constant, R kJ/kg · K*	Critical-point properties		
				Temperature, K	Pressure, MPa	Volume, m^3/kmol
Air	—	28.97	0.2870	132.5	3.77	0.0883
Ammonia	NH_3	17.03	0.4882	405.5	11.28	0.0724
Argon	Ar	39.948	0.2081	151	4.86	0.0749
Benzene	C_6H_6	78.115	0.1064	562	4.92	0.2603
Bromine	Br_2	159.808	0.0520	584	10.34	0.1355
n-Butane	C_4H_{10}	58.124	0.1430	425.2	3.80	0.2547
Carbon dioxide	CO_2	44.01	0.1889	304.2	7.39	0.0943
Carbon monoxide	CO	28.011	0.2968	133	3.50	0.0930
Carbon tetrachloride	CCl_4	153.82	0.05405	556.4	4.56	0.2759
Chlorine	Cl_2	70.906	0.1173	417	7.71	0.1242
Chloroform	$CHCl_3$	119.38	0.06964	536.6	5.47	0.2403
Dichlorodifluoromethane (R-12)	CCl_2F_2	120.91	0.06876	384.7	4.01	0.2179
Dichlorofluoromethane (R-21)	$CHCl_2F$	102.92	0.08078	451.7	5.17	0.1973
Ethane	C_2H_6	30.070	0.2765	305.5	4.48	0.1480
Ethyl alcohol	C_2H_5OH	46.07	0.1805	516	6.38	0.1673
Ethylene	C_2H_4	28.054	0.2964	282.4	5.12	0.1242
Helium	He	4.003	2.0769	5.3	0.23	0.0578
n-Hexane	C_6H_{14}	86.179	0.09647	507.9	3.03	0.3677
Hydrogen (normal)	H_2	2.016	4.1240	33.3	1.30	0.0649
Krypton	Kr	83.80	0.09921	209.4	5.50	0.0924
Methane	CH_4	16.043	0.5182	191.1	4.64	0.0993
Methyl alcohol	CH_3OH	32.042	0.2595	513.2	7.95	0.1180
Methyl chloride	CH_3Cl	50.488	0.1647	416.3	6.68	0.1430
Neon	Ne	20.183	0.4119	44.5	2.73	0.0417
Nitrogen	N_2	28.013	0.2968	126.2	3.39	0.0899
Nitrous oxide	N_2O	44.013	0.1889	309.7	7.27	0.0961
Oxygen	O_2	31.999	0.2598	154.8	5.08	0.0780
Propane	C_3H_8	44.097	0.1885	370	4.26	0.1998
Propylene	C_3H_6	42.081	0.1976	365	4.62	0.1810
Sulfur dioxide	SO_2	64.063	0.1298	430.7	7.88	0.1217
Tetrafluoroethane (R-134a)	CF_3CH_2F	102.03	0.08149	374.3	4.067	0.1847
Trichlorofluoromethane (R-11)	CCl_3F	137.37	0.06052	471.2	4.38	0.2478
Water	H_2O	18.015	0.4615	647.3	22.09	0.0568
Xenon	Xe	131.30	0.06332	289.8	5.88	0.1186

*The unit kJ/kg · K is equivalent to kPa · m^3/kg · K. The gas constant is calculated from $R = R_u/M$, where R_u = 8.314 kJ/kmol · K and M is the molar mass.

Source: K. A. Kobe and R. E. Lynn, Jr., Chemical Review 52 (1953), pp. 117–236; and ASHRAE, Handbook of Fundamentals, (Atlanta, GA: American Society of Heating, Refrigerating and Air-Conditioning Engineers, Inc., 1993), pp. 16.4 and 36.1.

TABLE A-2

Boiling- and freezing-point properties

Substance	Boiling data at 1 atm		Freezing data		Liquid properties		
	Normal boiling point, °C	Latent heat of vaporization, h_{fg} kJ/kg	Freezing point, °C	Latent heat of fusion, h_{if} kJ/kg	Temp., °C	Density, ρ kg/m³	Specific heat, C_p kJ/kg · °C
Ammonia	−33.3	1357	−77.7	322.4	−33.3	682	4.43
					−20	665	4.52
					0	639	4.60
					25	602	4.80
Argon	−185.9	161.6	−189.3	28	−185.6	1394	1.14
Benzene	80.2	394	5.5	126	20	879	1.72
Brine (20% sodium chloride by mass)	103.9	—	−17.4	—	20	1150	3.11
n-Butane	−0.5	385.2	−138.5	80.3	−0.5	601	2.31
Carbon dioxide	−78.4*	230.5 (at 0°C)	−56.6		0	298	0.59
Ethanol	78.2	838.3	−114.2	109	25	783	2.46
Ethylene glycol	198.1	800.1	−10.8	181.1	20	1109	2.84
Ethyl alcohol	78.6	855	−156	108	20	789	2.84
Glycerine	179.9	974	18.9	200.6	20	1261	2.32
Helium	−268.9	22.8	—	—	−268.9	146.2	22.8
Hydrogen	−252.8	445.7	−259.2	59.5	−252.8	70.7	10.0
Isobutane	−11.7	367.1	−160	105.7	−11.7	593.8	2.28
Kerosene	204–293	251	−24.9	—	20	820	2.00
Mercury	356.7	294.7	−38.9	11.4	25	13,560	0.139
Methane	−161.5	510.4	−182.2	58.4	−161.5	423	3.49
					−100	301	5.79
Methanol	64.5	1100	−97.7	99.2	25	787	2.55
Nitrogen	−195.8	198.6	−210	25.3	−195.8	809	2.06
					−160	596	2.97
Octane	124.8	306.3	−57.5	180.7	20	703	2.10
Oil (light)					25	910	1.80
Oxygen	−183	212.7	−218.8	13.7	−183	1141	1.71
Petroleum	—	230–384			20	640	2.0
Propane	−42.1	427.8	−187.7	80.0	−42.1	581	2.25
					0	529	2.53
					50	449	3.13
Refrigerant-134a	−26.1	216.8	−96.6	—	−50	1443	1.23
					−26.1	1374	1.27
					0	1294	1.34
					25	1206	1.42
Water	100	2257	0.0	333.7	0	1000	4.23
					25	997	4.18
					50	988	4.18
					75	975	4.19
					100	958	4.22

*Sublimation temperature. (At pressures below the triple-point pressure of 518 kPa, carbon dioxide exists as a solid or gas. Also, the freezing-point temperature of carbon dioxide is the triple-point temperature of −56.5°C.)

947

TABLE A-3

Properties of solid metals

Composition	Melting point, K	ρ kg/m³	c_p J/kg·K	k W/m·K	$\alpha \times 10^6$ m²/s	100	200	400	600	800	1000
Aluminum:											
Pure	933	2702	903	237	97.1	302	237	240	231	218	
						482	798	949	1033	1146	
Alloy 2024-T6	775	2770	875	177	73.0	65	163	186	186		
(4.5% Cu, 1.5% Mg, 0.6% Mn)						473	787	925	1042		
Alloy 195, Cast (4.5% Cu)		2790	883	168	68.2			174	185		
Beryllium	1550	1850	1825	200	59.2	990	301	161	126	106	90.8
						203	1114	2191	2604	2823	3018
Bismuth	545	9780	122	7.86	6.59	16.5	9.69	7.04			
						112	120	127			
Boron	2573	2500	1107	27.0	9.76	190	55.5	16.8	10.6	9.60	9.85
						128	600	1463	1892	2160	2338
Cadmium	594	8650	231	96.8	48.4	203	99.3	94.7			
						198	222	242			
Chromium	2118	7160	449	93.7	29.1	159	111	90.9	80.7	71.3	65.4
						192	384	484	542	581	616
Cobalt	1769	8862	421	99.2	26.6	167	122	85.4	67.4	58.2	52.1
						236	379	450	503	550	628
Copper:											
Pure	1358	8933	385	401	117	482	413	393	379	366	352
						252	356	397	417	433	451
Commercial bronze	1293	8800	420	52	14		42	52	59		
(90% Cu, 10% Al)							785	160	545		
Phosphor gear bronze	1104	8780	355	54	17		41	65	74		
(89% Cu, 11% Sn)							—	—	—		
Cartridge brass	1188	8530	380	110	33.9	75	95	137	149		
(70% Cu, 30% Zn)							360	395	425		
Constantan	1493	8920	384	23	6.71	17	19				
(55% Cu, 45% Ni)						237	362				
Germanium	1211	5360	322	59.9	34.7	232	96.8	43.2	27.3	19.8	17.4
						190	290	337	348	357	375
Gold	1336	19,300	129	317	127	327	323	311	298	284	270
						109	124	131	135	140	145
Iridium	2720	22,500	130	147	50.3	172	153	144	138	132	126
						90	122	133	138	144	153
Iron:											
Pure	1810	7870	447	80.2	23.1	134	94.0	69.5	54.7	43.3	32.8
						216	384	490	574	680	975
Armco		7870	447	72.7	20.7	95.6	80.6	65.7	53.1	42.2	32.3
(99.75% pure)						215	384	490	574	680	975
Carbon steels:											
Plain carbon (Mn ≤ 1%,		7854	434	60.5	17.7			56.7	48.0	39.2	30.0
Si ≤ 0.1%)								487	559	685	1169

TABLE A-3

Properties of solid metals (*Continued*)

Composition	Melting point, K	ρ kg/m³	c_p J/kg·K	k W/m·K	$\alpha \times 10^6$ m²/s	100	200	400	600	800	1000
AISI 1010		7832	434	63.9	18.8			58.7	48.8	39.2	31.3
								487	559	685	1168
Carbon–silicon (Mn ≤ 1%, 0.1% < Si ≤ 0.6%)		7817	446	51.9	14.9			49.8	44.0	37.4	29.3
								501	582	699	971
Carbon–manganese–silicon (1% < Mn ≤ 1.65% 0.1% < Si ≤ 0.6%)		8131	434	41.0	11.6			42.2	39.7	35.0	27.6
								487	559	685	1090
Chromium (low) steels:											
$\frac{1}{2}$Cr–$\frac{1}{4}$Mo–Si (0.18% C, 0.65% Cr, 0.23% Mo, 0.6% Si)		7822	444	37.7	10.9			38.2	36.7	33.3	26.9
								492	575	688	969
1Cr–$\frac{1}{2}$Mo (0.16% C, 1% Cr, 0.54% Mo, 0.39% Si)		7858	442	42.3	12.2			42.0	39.1	34.5	27.4
								492	575	688	969
1Cr–V (0.2% C, 1.02% Cr, 0.15% V)		7836	443	48.9	14.1			46.8	42.1	36.3	28.2
								492	575	688	969
Stainless steels:											
AISI 302		8055	480	15.1	3.91			17.3	20.0	22.8	25.4
								512	559	585	606
AISI 304	1670	7900	477	14.9	3.95	9.2	12.6	16.6	19.8	22.6	25.4
						272	402	515	557	582	611
AISI 316		8238	468	13.4	3.48			15.2	18.3	21.3	24.2
								504	550	576	602
AISI 347		7978	480	14.2	3.71			15.8	18.9	21.9	24.7
								513	559	585	606
Lead	601	11,340	129	35.3	24.1	39.7	36.7	34.0	31.4		
						118	125	132	142		
Magnesium	923	1740	1024	156	87.6	169	159	153	149	146	
						649	934	1074	1170	1267	
Molybdenum	2894	10,240	251	138	53.7	179	143	134	126	118	112
						141	224	261	275	285	295
Nickel:											
Pure	1728	8900	444	90.7	23.0	164	107	80.2	65.6	67.6	71.8
						232	383	485	592	530	562
Nichrome (80% Ni, 20% Cr)	1672	8400	420	12	3.4			14	16	21	
								480	525	545	
Inconel X-750 (73% Ni, 15% Cr, 6.7% Fe)	1665	8510	439	11.7	3.1	8.7	10.3	13.5	17.0	20.5	24.0
						—	372	473	510	546	626
Niobium	2741	8570	265	53.7	23.6	55.2	52.6	55.2	58.2	61.3	64.4
						188	249	274	283	292	301
Palladium	1827	12,020	244	71.8	24.5	76.5	71.6	73.6	79.7	86.9	94.2
						168	227	251	261	271	281
Platinum:											
Pure	2045	21,450	133	71.6	25.1	77.5	72.6	71.8	73.2	75.6	78.7
						100	125	136	141	146	152
Alloy 60Pt–40Rh (60% Pt, 40% Rh)	1800	16,630	162	47	17.4			52	59	65	69
								—	—	—	—

TABLE A-3

Properties of solid metals (*Concluded*)

Composition	Melting point, K	Properties at 300 K				Properties at various temperatures (K), k(W/m · K)/c_p(J/kg · K)					
		ρ kg/m³	c_p J/kg · K	k W/m · K	$\alpha \times 10^6$ m²/s	100	200	400	600	800	1000
Rhenium	3453	21,100	136	47.9	16.7	58.9	51.0	46.1	44.2	44.1	44.6
						97	127	139	145	151	156
Rhodium	2236	12,450	243	150	49.6	186	154	146	136	127	121
						147	220	253	274	293	311
Silicon	1685	2330	712	148	89.2	884	264	98.9	61.9	42.4	31.2
						259	556	790	867	913	946
Silver	1235	10,500	235	429	174	444	430	425	412	396	379
						187	225	239	250	262	277
Tantalum	3269	16,600	140	57.5	24.7	59.2	57.5	57.8	58.6	59.4	60.2
						110	133	144	146	149	152
Thorium	2023	11,700	118	54.0	39.1	59.8	54.6	54.5	55.8	56.9	56.9
						99	112	124	134	145	156
Tin	505	7310	227	66.6	40.1	85.2	73.3	62.2			
						188	215	243			
Titanium	1953	4500	522	21.9	9.32	30.5	24.5	20.4	19.4	19.7	20.7
						300	465	551	591	633	675
Tungsten	3660	19,300	132	174	68.3	208	186	159	137	125	118
						87	122	137	142	146	148
Uranium	1406	19,070	116	27.6	12.5	21.7	25.1	29.6	34.0	38.8	43.9
						94	108	125	146	176	180
Vanadium	2192	6100	489	30.7	10.3	35.8	31.3	31.3	33.3	35.7	38.2
						258	430	515	540	563	597
Zinc	693	7140	389	116	41.8	117	118	111	103		
						297	367	402	436		
Zirconium	2125	6570	278	22.7	12.4	33.2	25.2	21.6	20.7	21.6	23.7
						205	264	300	332	342	362

Source for Tables A-3 and A-4: Frank P . Incropera and David P . DeWitt, *Fundamentals of Heat and Mass Transfer*, 3d ed. (New York: John Wiley & Sons, 1990), pp. A3–A8. Originally compiled from various sources. Reprinted by permission of John Wiley & Sons, Inc.

TABLE A-4

Properties of solid nonmetals

Composition	Melting point, K	ρ kg/m³	c_p J/kg·K	k W/m·K	$\alpha \times 10^6$ m²/s	100	200	400	600	800	1000
		Properties at 300 K				**Properties at various temperatures (K), k(W/m·K)/c_p(J/kg·K)**					
Aluminum oxide, sapphire	2323	3970	765	46	15.1	450	82	32.4	18.9	13.0	10.5
						—	—	940	1110	1180	1225
Aluminum oxide, polycrystalline	2323	3970	765	36.0	11.9	133	55	26.4	15.8	10.4	7.85
						—	—	940	1110	1180	1225
Beryllium oxide	2725	3000	1030	272	88.0			196	111	70	47
								1350	1690	1865	1975
Boron	2573	2500	1105	27.6	9.99	190	52.5	18.7	11.3	8.1	6.3
						—	—	1490	1880	2135	2350
Boron fiber epoxy (30% vol.) composite	590	2080									
k, ∥ to fibers				2.29		2.10	2.23	2.28			
k, ⊥ to fibers				0.59		0.37	0.49	0.60			
c_p			1122			364	757	1431			
Carbon											
Amorphous	1500	1950	—	1.60	—	0.67	1.18	1.89	21.9	2.37	2.53
						—	—	—	—	—	—
Diamond, type IIa insulator	—	3500	509	2300		10,000	4000	1540			
						21	194	853			
Graphite, pyrolytic	2273	2210									
k, ∥ to layers				1950		4970	3230	1390	892	667	534
k, ⊥ to layers				5.70		16.8	9.23	4.09	2.68	2.01	1.60
c_p			709			136	411	992	1406	1650	1793
Graphite fiber epoxy (25% vol.) composite	450	1400									
k, heat flow to fibers				11.1		5.7	8.7	13.0			
k, heat flow to fibers				0.87		0.46	0.68	1.1			
c_p			935			337	642	1216			
Pyroceram, Corning 9606	1623	2600	808	3.98	1.89	5.25	4.78	3.64	3.28	3.08	2.96
						—	—	908	1038	1122	1197
Silicon carbide	3100	3160	675	490	230			—	—	—	87
								880	1050	1135	1195
Silicon dioxide, crystalline (quartz)	1883	2650									
k, ∥ to c-axis				10.4		39	16.4	7.6	5.0	4.2	
k, ⊥ to c-axis				6.21		20.8	9.5	4.70	3.4	3.1	
c_p			745			—	—	885	1075	1250	
Silicon dioxide, polycrystalline (fused silica)	1883	2220	745	1.38	0.834	0.69	1.14	1.51	1.75	2.17	2.87
						—	—	905	1040	1105	1155
Silicon nitride	2173	2400	691	16.0	9.65	—	—	13.9	11.3	9.88	8.76
						—	578	778	937	1063	1155
Sulfur	392	2070	708	0.206	0.141	0.165	0.185				
						403	606				
Thorium dioxide	3573	9110	235	13	6.1			10.2	6.6	4.7	3.68
								255	274	285	295
Titanium dioxide, polycrystalline	2133	4157	710	8.4	2.8			7.01	5.02	3.94	3.46
								805	880	910	930

TABLE A-5

Properties of building materials
(at a mean temperature of 24°C)

Material	Thickness, L mm	Density, ρ kg/m³	Thermal conductivity, k W/m·°C	Specific heat, C_p kJ/kg·°C	R-value (for listed thickness, L/k), °C·m²/W
Building Boards					
Asbestos–cement board	6 mm	1922	—	1.00	0.011
Gypsum of plaster board	10 mm	800	—	1.09	0.057
	13 mm	800	—	—	0.078
Plywood (Douglas fir)	—	545	0.12	1.21	—
	6 mm	545	—	1.21	0.055
	10 mm	545	—	1.21	0.083
	13 mm	545	—	1.21	0.110
	20 mm	545	—	1.21	0.165
Insulated board and sheating	13 mm	288	—	1.30	0.232
(regular density)	20 mm	288	—	1.30	0.359
Hardboard (high density, standard tempered)	—	1010	0.14	1.34	—
Particle board:					
Medium density	—	800	0.14	1.30	—
Underlayment	16 mm	640	—	1.21	0.144
Wood subfloor	20 mm	—	—	1.38	0.166
Building Membrane					
Vapor-permeable felt	—	—	—	—	0.011
Vapor-seal (2 layers of mopped 0.73 kg/m² felt)	—	—	—	—	0.021
Flooring Materials					
Carpet and fibrous pad	—	—	—	1.42	0.367
Carpet and rubber pad	—	—	—	1.38	0.217
Tile (asphalt, linoleum, vinyl)	—	—	—	1.26	0.009
Masonry Materials					
Masonry units:					
Brick, common		1922	0.72	—	—
Brick, face		2082	1.30	—	—
Brick, fire clay		2400	1.34	—	—
		1920	0.90	0.79	—
		1120	0.41	—	—
Concrete blocks (3 oval cores, sand and gravel aggregate)	100 mm	—	0.77	—	0.13
	200 mm	—	1.0	—	0.20
	300 mm	—	1.30	—	0.23
Concretes:					
Lightweight aggregates (including expanded shale, clay, or slate; expanded slags; cinders; pumice; and scoria)		1920	1.1	—	—
		1600	0.79	0.84	—
		1280	0.54	0.84	—
		960	0.33	—	—
		940	0.18	—	—

TABLE A-5

Properties of building materials (*Concluded*)
(at a mean temperature of 24°C)

Material	Thickness, L mm	Density, ρ kg/m³	Thermal conductivity, k W/m·°C	Specific heat, C_p kJ/kg·°C	R-value (for listed thickness, L/k), °C·m²/W
Cement/lime, mortar, and stucco		1920	1.40	—	—
		1280	0.65	—	—
Stucco		1857	0.72	—	—
Roofing					
Asbestos-cement shingles		1900	—	1.00	0.037
Asphalt roll roofing		1100	—	1.51	0.026
Asphalt shingles		1100	—	1.26	0.077
Built-in roofing	10 mm	1100	—	1.46	0.058
Slate	13 mm	—	—	1.26	0.009
Wood shingles (plain and plastic/film faced)		—	—	1.30	0.166
Plastering Materials					
Cement plaster, sand aggregate	19 mm	1860	0.72	0.84	0.026
Gypsum plaster:					
Lightweight aggregate	13 mm	720	—	—	0.055
Sand aggregate	13 mm	1680	0.81	0.84	0.016
Perlite aggregate	—	720	0.22	1.34	—
Siding Material (on flat surfaces)					
Asbestos-cement shingles	—	1900	—	—	0.037
Hardboard siding	11 mm	—	—	1.17	0.12
Wood (drop) siding	25 mm	—	—	1.30	0.139
Wood (plywood) siding, lapped	10 mm	—	—	1.21	0.111
Aluminum of steel siding (over sheating):					
Hollow backed	10 mm	—	—	1.22	0.11
Insulating-board backed	10 mm	—	—	1.34	0.32
Architectural glass	—	2530	1.0	0.84	0.018
Woods					
Hardwoods (maple, oak, etc.)	—	721	0.159	1.26	—
Softwoods (fir, pine, etc.)	—	513	0.115	1.38	—
Metals					
Aluminum (1100)	—	2739	222	0.896	—
Steel, mild	—	7833	45.3	0.502	—
Steel, Stainless	—	7913	15.6	0.456	—

Source: Tables A-5 and A-6 are adapted from ASHRAE, *Handbook of Fundamentals* (Atlanta, GA: American Society of Heating, Refrigerating, and Air-Conditioning Engineers, 1993), Chap. 22, Table 4. Used with permission.

TABLE A-6

Properties of insulating materials
(at a mean temperature of 24°C)

Material	Thickness, L mm	Density, ρ kg/m³	Thermal conductivity, k W/m · °C	Specific heat, C_p kJ/kg · °C	R-value (for listed thickness, L/k) °C · m²/W
Blanket and Batt					
Mineral fiber (fibrous form	50 to 70 mm	4.8–32	—	0.71–0.96	1.23
processed from rock, slag,	75 to 90 mm	4.8–32	—	0.71–0.96	1.94
or glass)	135 to 165 mm	4.8–32	—	0.71–0.96	3.32
Board and Slab					
Cellular glass		136	0.055	1.0	—
Glass fiber (organic bonded)		64–144	0.036	0.96	—
Expanded polystyrene (molded beads)		16	0.040	1.2	—
Expanded polyurethane (R-11 expanded)		24	0.023	1.6	—
Expanded perlite (organic bonded)		16	0.052	1.26	—
Expanded rubber (rigid)		72	0.032	1.68	—
Mineral fiber with resin binder		240	0.042	0.71	—
Cork		120	0.039	1.80	—
Sprayed or Formed in Place					
Polyurethane foam		24–40	0.023–0.026	—	—
Glass fiber		56–72	0.038–0.039	—	—
Urethane, two-part mixture (rigid foam)		70	0.026	1.045	—
Mineral wool granules with asbestos/inorganic					
binders (sprayed)		190	0.046	—	—
Loose Fill					
Mineral fiber (rock, slag, or glass)	~ 75 to 125 mm	9.6–32	—	0.71	1.94
	~165 to 222 mm	9.6–32	—	0.71	3.35
	~191 to 254 mm	—	—	0.71	3.87
	~185 mm	—	—	0.71	5.28
Silica aerogel		122	0.025	—	—
Vermiculite (expanded)		122	0.068	—	—
Perlite, expanded		32–66	0.039–0.045	1.09	—
Sawdust or shavings		128–240	0.065	1.38	—
Cellulosic insulation (milled paper or wood pulp)		37–51	0.039–0.046	—	—
Roof Insulation					
Cellular glass	—	144	0.058	1.0	—
Preformed, for use above deck	13 mm	—	—	1.0	0.24
	25 mm	—	—	2.1	0.49
	50 mm	—	—	3.9	0.93
Reflective Insulation					
Silica powder (evacuated)		160	0.0017	—	—
Aluminum foil separating fluffy glass mats; 10–12 layers					
(evacuated); for cryogenic applications (150 K)		40	0.00016	—	—
Aluminum foil and glass paper laminate; 75–150					
layers (evacuated); for cryogenic applications (150 K)		120	0.000017	—	—

TABLE A-7

Properties of common foods

(a) Specific heats and freezing-point properties

Food	Water content,[a] % (mass)	Freezing point,[a] °C	Specific heat,[b] kJ/kg·°C Above freezing	Specific heat,[b] kJ/kg·°C Below freezing	Latent heat of fusion,[c] kJ/kg
Vegetables					
Artichokes	84	−1.2	3.65	1.90	281
Asparagus	93	−0.6	3.96	2.01	311
Beans, snap	89	−0.7	3.82	1.96	297
Broccoli	90	−0.6	3.86	1.97	301
Cabbage	92	−0.9	3.92	2.00	307
Carrots	88	−1.4	3.79	1.95	294
Cauliflower	92	−0.8	3.92	2.00	307
Celery	94	−0.5	3.99	2.02	314
Corn, sweet	74	−0.6	3.32	1.77	247
Cucumbers	96	−0.5	4.06	2.05	321
Eggplant	93	−0.8	3.96	2.01	311
Horseradish	75	−1.8	3.35	1.78	251
Leeks	85	−0.7	3.69	1.91	284
Lettuce	95	−0.2	4.02	2.04	317
Mushrooms	91	−0.9	3.89	1.99	304
Okra	90	−1.8	3.86	1.97	301
Onions, green	89	−0.8	3.82	1.96	297
Onions, dry	88	−0.8	3.79	1.95	294
Parsley	85	−1.1	3.69	1.91	284
Peas, green	74	−0.6	3.32	1.77	247
Peppers, sweet	92	−0.7	3.92	2.00	307
Potatoes	78	−0.6	3.45	1.82	261
Pumpkins	91	−0.8	3.89	1.99	304
Spinach	93	−0.3	3.96	2.01	311
Tomatos, ripe	94	−0.5	3.99	2.02	314
Turnips	92	−1.1	3.92	2.00	307
Fruits					
Apples	84	−1.1	3.65	1.90	281
Apricots	85	−1.1	3.69	1.91	284
Avocados	65	−0.3	3.02	1.66	217
Bananas	75	−0.8	3.35	1.78	251
Blueberries	82	−1.6	3.59	1.87	274
Cantaloupes	92	−1.2	3.92	2.00	307
Cherries, sour	84	−1.7	3.65	1.90	281
Cherries, sweet	80	−1.8	3.52	1.85	267
Figs, dried	23	—	—	1.13	77
Figs, fresh	78	−2.4	3.45	1.82	261
Grapefruit	89	−1.1	3.82	1.96	297
Grapes	82	−1.1	3.59	1.87	274
Lemons	89	−1.4	3.82	1.96	297
Olives	75	−1.4	3.35	1.78	251
Oranges	87	−0.8	3.75	1.94	291
Peaches	89	−0.9	3.82	1.96	297
Pears	83	−1.6	3.62	1.89	277
Pineapples	85	−1.0	3.69	1.91	284
Plums	86	−0.8	3.72	1.92	287
Quinces	85	−2.0	3.69	1.91	284
Raisins	18	—		1.07	60
Strawberries	90	−0.8	3.86	1.97	301
Tangerines	87	−1.1	3.75	1.94	291
Watermelon	93	−0.4	3.96	2.01	311
Fish/Seafood					
Cod, whole	78	−2.2	3.45	1.82	261
Halibut, whole	75	−2.2	3.35	1.78	251
Lobster	79	−2.2	3.49	1.84	264
Mackerel	57	−2.2	2.75	1.56	190
Salmon, whole	64	−2.2	2.98	1.65	214
Shrimp	83	−2.2	3.62	1.89	277
Meats					
Beef carcass	49	−1.7	2.48	1.46	164
Liver	70	−1.7	3.18	1.72	234
Round, beef	67	—	3.08	1.68	224
Sirloin, beef	56	—	2.72	1.55	187
Chicken	74	−2.8	3.32	1.77	247
Lamb leg	65	—	3.02	1.66	217
Pork carcass	37	—	2.08	1.31	124
Ham	56	−1.7	2.72	1.55	187
Pork sausage	38	—	2.11	1.32	127
Turkey	64	—	2.98	1.65	214
Other					
Almonds	5	—	—	0.89	17
Butter	16	—		1.04	53
Cheese, cheddar	37	−12.9	2.08	1.31	124
Cheese, Swiss	39	−10.0	2.15	1.33	130
Chocolate, milk	1	—	—	0.85	3
Eggs, whole	74	−0.6	3.32	1.77	247
Honey	17	—	—	1.05	57
Ice cream	63	−5.6	2.95	1.63	210
Milk, whole	88	−0.6	3.79	1.95	294
Peanuts	6	—	—	0.92	20
Peanuts, roasted	2	—	—	0.87	7
Pecans	3	—	—	0.87	10
Walnuts	4	—	—	0.88	13

Sources: [a]Water content and freezing-point data are from ASHRAE, *Handbook of Fundamentals*, SI version (Atlanta, GA: American Society of Heating, Refrigerating and Air-Conditioning Engineers, Inc., 1993), Chap. 30, Table 1. Used with permission. Freezing point is the temperature at which freezing starts for fruits and vegetables, and the average freezing temperature for other foods.

[b]Specific heat data are based on the specific heat values of water and ice at 0°C and are determined from Siebel's formulas: $C_{p,\, fresh} = 3.35 \times (\text{Water content}) + 0.84$, above freezing, and $C_{p,\, frozen} = 1.26 \times (\text{Water content}) + 0.84$, below freezing.

[c]The latent heat of fusion is determined by multiplying the heat of fusion of water (334 kJ/kg) by the water content of the food.

TABLE A-7

Properties of common foods (*Concluded*)
(*b*) Other properties

Food	Water content, % (mass)	Temperature, $T\,°C$	Density, $\rho\,kg/m^3$	Thermal conductivity, $k\,W/m\cdot°C$	Thermal diffusivity, $\alpha\,m^2/s$	Specific heat, C_p kJ/kg · °C
Fruits/Vegetables						
Apple juice	87	20	1000	0.559	0.14×10^{-6}	3.86
Apples	85	8	840	0.418	0.13×10^{-6}	3.81
Apples, dried	41.6	23	856	0.219	0.096×10^{-6}	2.72
Apricots, dried	43.6	23	1320	0.375	0.11×10^{-6}	2.77
Bananas, fresh	76	27	980	0.481	0.14×10^{-6}	3.59
Broccoli	—	−6	560	0.385	—	—
Cherries, fresh	92	0–30	1050	0.545	0.13×10^{-6}	3.99
Figs	40.4	23	1241	0.310	0.096×10^{-6}	2.69
Grape juice	89	20	1000	0.567	0.14×10^{-6}	3.91
Peaches	89	2–32	960	0.526	0.14×10^{-6}	3.91
Plums	—	−16	610	0.247	—	—
Potatoes	78	0–70	1055	0.498	0.13×10^{-6}	3.64
Raisins	32	23	1380	0.376	0.11×10^{-6}	2.48
Meats						
Beef, ground	67	6	950	0.406	0.13×10^{-6}	3.36
Beef, lean	74	3	1090	0.471	0.13×10^{-6}	3.54
Beef fat	0	35	810	0.190	—	—
Beef liver	72	35	—	0.448	—	3.49
Cat food	39.7	23	1140	0.326	0.11×10^{-6}	2.68
Chicken breast	75	0	1050	0.476	0.13×10^{-6}	3.56
Dog food	30.6	23	1240	0.319	0.11×10^{-6}	2.45
Fish, cod	81	3	1180	0.534	0.12×10^{-6}	3.71
Fish, salmon	67	3	—	0.531	—	3.36
Ham	71.8	20	1030	0.480	0.14×10^{-6}	3.48
Lamb	72	20	1030	0.456	0.13×10^{-6}	3.49
Pork, lean	72	4	1030	0.456	0.13×10^{-6}	3.49
Turkey breast	74	3	1050	0.496	0.13×10^{-6}	3.54
Veal	75	20	1060	0.470	0.13×10^{-6}	3.56
Other						
Butter	16	4	—	0.197	—	2.08
Chocolate cake	31.9	23	340	0.106	0.12×10^{-6}	2.48
Margarine	16	5	1000	0.233	0.11×10^{-6}	2.08
Milk, skimmed	91	20	—	0.566	—	3.96
Milk, whole	88	28	—	0.580	—	3.89
Olive oil	0	32	910	0.168	—	—
Peanut oil	0	4	920	0.168	—	—
Water	100	0	1000	0.569	0.14×10^{-6}	4.217
	100	30	995	0.618	0.15×10^{-6}	4.178
White cake	32.3	23	450	0.082	0.10×10^{-6}	2.49

Source: Data obtained primarily from ASHRAE, *Handbook of Fundamentals*, SI version (Atlanta, GA: American Society of Heating, Refrigerating and Air-Conditioning Engineers, Inc., 1993), Chap. 30, Tables 7 and 9. Used with permission.

Most specific heats are calculated from $C_p = 1.68 + 2.51 \times$ (Water content), which is a good approximation in the temperature range of 3 to 32°C. Most thermal diffusivities are calculated from $\alpha = k/\rho C_p$. Property values given above are valid for the specified water content.

TABLE A-8

Properties of miscellaneous materials
(Values are at 300 K unless indicated otherwise)

Material	Density, ρ kg/m³	Thermal conductivity k, W/m·K	Specific heat, C_p J/kg·K	Material	Density, ρ kg/m³	Thermal conductivity k, W/m·K	Specific heat, C_p J/kg·K
Asphalt	2115	0.062	920	Ice			
Bakelite	1300	1.4	1465	273 K	920	1.88	2040
Brick, refractory				253 K	922	2.03	1945
Chrome brick				173 K	928	3.49	1460
473 K	3010	2.3	835	Leather, sole	998	0.159	—
823 K	—	2.5	—	Linoleum	535	0.081	—
1173 K	—	2.0	—		1180	0.186	—
Fire clay, burnt				Mica	2900	0.523	—
1600 K				Paper	930	0.180	1340
773 K	2050	1.0	960	Plastics			
1073 K	—	1.1	—	Plexiglass	1190	0.19	1465
1373 K	—	1.1	—	Teflon			
Fire clay, burnt				300 K	2200	0.35	1050
1725 K				400 K	—	0.45	—
773 K	2325	1.3	960	Lexan	1200	0.19	1260
1073 K	—	1.4	—	Nylon	1145	0.29	—
1373 K	—	1.4	—	Polypropylene	910	0.12	1925
Fire clay brick				Polyester	1395	0.15	1170
478 K	2645	1.0	960	PVC, vinyl	1470	0.1	840
922 K	—	1.5	—	Porcelain	2300	1.5	—
1478 K	—	1.8	—	Rubber, natural	1150	0.28	—
Magnesite				Rubber, vulcanized			
478 K	—	3.8	1130	Soft	1100	0.13	2010
922 K	—	2.8	—	Hard	1190	0.16	—
1478 K	—	1.9	—	Sand	1515	0.2–1.0	800
Chicken meat,				Snow, fresh	100	0.60	—
white (74.4%				Snow, 273 K	500	2.2	—
water content)				Soil, dry	1500	1.0	1900
198 K	—	1.60	—	Soil, wet	1900	2.0	2200
233 K	—	1.49	—	Sugar	1600	0.58	—
253 K	—	1.35	—	Tissue, human			
273 K	—	0.48	—	Skin	—	0.37	—
293 K	—	0.49	—	Fat layer	—	0.2	—
Clay, dry	1550	0.930	—	Muscle	—	0.41	—
Clay, wet	1495	1.675	—	Vaseline	—	0.17	—
Coal, anthracite	1350	0.26	1260	Wood, cross-grain			
Concrete (stone				Balsa	140	0.055	—
mix)	2300	1.4	880	Fir	415	0.11	2720
Cork	86	0.048	2030	Oak	545	0.17	2385
Cotton	80	0.06	1300	Yellow pine	640	0.15	2805
Fat	—	0.17	—	White pine	435	0.11	—
Glass				Wood, radial			
Window	2800	0.7	750	Oak	545	0.19	2385
Pyrex	2225	1–1.4	835	Fir	420	0.14	2720
Crown	2500	1.05	—	Wool, ship	145	0.05	—
Lead	3400	0.85	—				

Source: Compiled from various sources.

TABLE A-9

Properties of saturated water

Temperature, $T\,°C$	Saturation pressure, P kPa	Density, ρ kg/m^3 Liquid	Density, ρ kg/m^3 Vapor	Enthalpy of vaporization, h_{fg} kJ/kg	Specific heat, C_p J/kg·°C Liquid	Specific heat, C_p J/kg·°C Vapor	Thermal conductivity, k W/m·°C Liquid	Thermal conductivity, k W/m·°C Vapor	Dynamic viscosity, μ kg/m·s Liquid	Dynamic viscosity, μ kg/m·s Vapor	Prandtl number, Pr Liquid	Prandtl number, Pr Vapor	Volume expansion coefficient, β 1/K Liquid
0.01	0.6113	999.8	0.0048	2501	4217	1854	0.561	0.0171	1.792×10^{-3}	0.922×10^{-5}	13.5	1.00	-0.068×10^{-3}
5	0.8721	999.9	0.0068	2490	4205	1857	0.571	0.0173	1.519×10^{-3}	0.934×10^{-5}	11.2	1.00	0.015×10^{-3}
10	1.2276	999.7	0.0094	2478	4194	1862	0.580	0.0176	1.307×10^{-3}	0.946×10^{-5}	9.45	1.00	0.733×10^{-3}
15	1.7051	999.1	0.0128	2466	4186	1863	0.589	0.0179	1.138×10^{-3}	0.959×10^{-5}	8.09	1.00	0.138×10^{-3}
20	2.339	998.0	0.0173	2454	4182	1867	0.598	0.0182	1.002×10^{-3}	0.973×10^{-5}	7.01	1.00	0.195×10^{-3}
25	3.169	997.0	0.0231	2442	4180	1870	0.607	0.0186	0.891×10^{-3}	0.987×10^{-5}	6.14	1.00	0.247×10^{-3}
30	4.246	996.0	0.0304	2431	4178	1875	0.615	0.0189	0.798×10^{-3}	1.001×10^{-5}	5.42	1.00	0.294×10^{-3}
35	5.628	994.0	0.0397	2419	4178	1880	0.623	0.0192	0.720×10^{-3}	1.016×10^{-5}	4.83	1.00	0.337×10^{-3}
40	7.384	992.1	0.0512	2407	4179	1885	0.631	0.0196	0.653×10^{-3}	1.031×10^{-5}	4.32	1.00	0.377×10^{-3}
45	9.593	990.1	0.0655	2395	4180	1892	0.637	0.0200	0.596×10^{-3}	1.046×10^{-5}	3.91	1.00	0.415×10^{-3}
50	12.35	988.1	0.0831	2383	4181	1900	0.644	0.0204	0.547×10^{-3}	1.062×10^{-5}	3.55	1.00	0.451×10^{-3}
55	15.76	985.2	0.1045	2371	4183	1908	0.649	0.0208	0.504×10^{-3}	1.077×10^{-5}	3.25	1.00	0.484×10^{-3}
60	19.94	983.3	0.1304	2359	4185	1916	0.654	0.0212	0.467×10^{-3}	1.093×10^{-5}	2.99	1.00	0.517×10^{-3}
65	25.03	980.4	0.1614	2346	4187	1926	0.659	0.0216	0.433×10^{-3}	1.110×10^{-5}	2.75	1.00	0.548×10^{-3}
70	31.19	977.5	0.1983	2334	4190	1936	0.663	0.0221	0.404×10^{-3}	1.126×10^{-5}	2.55	1.00	0.578×10^{-3}
75	38.58	974.7	0.2421	2321	4193	1948	0.667	0.0225	0.378×10^{-3}	1.142×10^{-5}	2.38	1.00	0.607×10^{-3}
80	47.39	971.8	0.2935	2309	4197	1962	0.670	0.0230	0.355×10^{-3}	1.159×10^{-5}	2.22	1.00	0.653×10^{-3}
85	57.83	968.1	0.3536	2296	4201	1977	0.673	0.0235	0.333×10^{-3}	1.176×10^{-5}	2.08	1.00	0.670×10^{-3}
90	70.14	965.3	0.4235	2283	4206	1993	0.675	0.0240	0.315×10^{-3}	1.193×10^{-5}	1.96	1.00	0.702×10^{-3}
95	84.55	961.5	0.5045	2270	4212	2010	0.677	0.0246	0.297×10^{-3}	1.210×10^{-5}	1.85	1.00	0.716×10^{-3}
100	101.33	957.9	0.5978	2257	4217	2029	0.679	0.0251	0.282×10^{-3}	1.227×10^{-5}	1.75	1.00	0.750×10^{-3}

Temp.													
110	143.27	950.6	0.8263	2230	4229	2071	0.682	0.0262	0.255×10^{-3}	1.261×10^{-5}	1.58	1.00	0.798×10^{-3}
120	198.53	943.4	1.121	2203	4244	2120	0.683	0.0275	0.232×10^{-3}	1.296×10^{-5}	1.44	1.00	0.858×10^{-3}
130	270.1	934.6	1.496	2174	4263	2177	0.684	0.0288	0.213×10^{-3}	1.330×10^{-5}	1.33	1.01	0.913×10^{-3}
140	361.3	921.7	1.965	2145	4286	2244	0.683	0.0301	0.197×10^{-3}	1.365×10^{-5}	1.24	1.02	0.970×10^{-3}
150	475.8	916.6	2.546	2114	4311	2314	0.682	0.0316	0.183×10^{-3}	1.399×10^{-5}	1.16	1.02	1.025×10^{-3}
160	617.8	907.4	3.256	2083	4340	2420	0.680	0.0331	0.170×10^{-3}	1.434×10^{-5}	1.09	1.05	1.145×10^{-3}
170	791.7	897.7	4.119	2050	4370	2490	0.677	0.0347	0.160×10^{-3}	1.468×10^{-5}	1.03	1.05	1.178×10^{-3}
180	1002.1	887.3	5.153	2015	4410	2590	0.673	0.0364	0.150×10^{-3}	1.502×10^{-5}	0.983	1.07	1.210×10^{-3}
190	1254.4	876.4	6.388	1979	4460	2710	0.669	0.0382	0.142×10^{-3}	1.537×10^{-5}	0.947	1.09	1.280×10^{-3}
200	1553.8	864.3	7.852	1941	4500	2840	0.663	0.0401	0.134×10^{-3}	1.571×10^{-5}	0.910	1.11	1.350×10^{-3}
220	2318	840.3	11.60	1859	4610	3110	0.650	0.0442	0.122×10^{-3}	1.641×10^{-5}	0.865	1.15	1.520×10^{-3}
240	3344	813.7	16.73	1767	4760	3520	0.632	0.0487	0.111×10^{-3}	1.712×10^{-5}	0.836	1.24	1.720×10^{-3}
260	4688	783.7	23.69	1663	4970	4070	0.609	0.0540	0.102×10^{-3}	1.788×10^{-5}	0.832	1.35	2.000×10^{-3}
280	6412	750.8	33.15	1544	5280	4835	0.581	0.0605	0.094×10^{-3}	1.870×10^{-5}	0.854	1.49	2.380×10^{-3}
300	8581	713.8	46.15	1405	5750	5980	0.548	0.0695	0.086×10^{-3}	1.965×10^{-5}	0.902	1.69	2.950×10^{-3}
320	11,274	667.1	64.57	1239	6540	7900	0.509	0.0836	0.078×10^{-3}	2.084×10^{-5}	1.00	1.97	—
340	14,586	610.5	92.62	1028	8240	11870	0.469	0.110	0.070×10^{-3}	2.255×10^{-5}	1.23	2.43	—
360	18,651	528.3	144.0	720	14,690	25,800	0.427	0.178	0.060×10^{-3}	2.571×10^{-5}	2.06	3.73	—
374.14	22,090	317.0	317.0	0	∞	∞	∞	∞	0.043×10^{-3}	4.313×10^{-5}	—	—	—

Source: Viscosity and thermal conductivity data are from J. V. Sengers and J. T. R. Watson, *Journal of Physical and Chemical Reference Data* 15 (1986), pp. 1291–1322. Other data are obtained from various sources or calculated.

Note: Kinematic viscosity ν and thermal diffusivity α can be calculated from their definitions, $\nu = \mu/\rho$ and $\alpha = k/\rho C_p = \nu/Pr$. The temperatures 0.01°C, 100°C, and 374.14°C are the triple-, boiling-, and critical-point temperatures of water, respectively. The properties listed above (except the vapor density) can be used at any pressure with negligible error except at temperatures near the critical-point value.

TABLE A-10

Properties of liquids

Temperature, $T\,°C$	Density, ρ kg/m³	Specific heat, C_p J/kg·°C	Thermal conductivity, k W/m·°C	Thermal diffusivity, α m²/s	Dynamic viscosity, μ kg/m·s	Kinematic viscosity, ν m²/s	Prandtl number, Pr
				Ammonia			
−40	692	4467	0.546	1.78×10^{-7}	2.81×10^{-4}	4.06×10^{-7}	2.28
−20	667	4509	0.546	1.82×10^{-7}	2.54×10^{-4}	3.81×10^{-7}	2.09
0	640	4635	0.540	1.82×10^{-7}	2.39×10^{-4}	3.73×10^{-7}	2.05
20	612	4798	0.521	1.78×10^{-7}	2.20×10^{-4}	3.59×10^{-7}	2.02
40	581	4999	0.493	1.70×10^{-7}	1.98×10^{-4}	3.40×10^{-7}	2.00
				Ethyl alcohol (C_2H_6O)			
−40	823	2037	0.186	1.11×10^{-7}	4.81×10^{-3}	5.84×10^{-6}	52.7
−20	815	2124	0.179	1.03×10^{-7}	2.83×10^{-3}	3.47×10^{-6}	33.6
0	806	2249	0.174	0.960×10^{-7}	1.77×10^{-3}	2.20×10^{-6}	22.9
20	789	2395	0.168	0.889×10^{-7}	1.20×10^{-3}	1.52×10^{-6}	17.0
40	772	2572	0.162	0.816×10^{-7}	0.834×10^{-3}	1.08×10^{-6}	13.2
60	755	2781	0.156	0.743×10^{-7}	0.592×10^{-3}	0.784×10^{-6}	10.6
80	738	3026	0.150	0.672×10^{-7}	0.430×10^{-3}	0.583×10^{-6}	8.7
				Ethylene glycol ($C_2H_6O_2$)			
0	1131	2295	0.254	9.79×10^{-8}	65.1×10^{-3}	57.5×10^{-6}	588
20	1117	2386	0.257	9.64×10^{-8}	21.4×10^{-3}	19.2×10^{-6}	199
40	1101	2476	0.259	9.50×10^{-8}	9.57×10^{-3}	8.69×10^{-6}	91
60	1088	2565	0.262	9.39×10^{-8}	5.17×10^{-3}	4.75×10^{-6}	51
80	1078	2656	0.265	9.26×10^{-8}	3.21×10^{-3}	2.98×10^{-6}	32
100	1059	2750	0.267	9.17×10^{-8}	2.15×10^{-3}	2.03×10^{-6}	22
				Freon-12 refrigerant (CCl_2F_2)			
−40	1515	885	0.069	5.14×10^{-8}	4.24×10^{-4}	2.80×10^{-7}	5.4
−20	1457	907	0.071	5.38×10^{-8}	3.43×10^{-4}	2.35×10^{-7}	4.4
0	1393	935	0.073	5.59×10^{-8}	2.98×10^{-4}	2.14×10^{-7}	3.8
20	1327	966	0.073	5.66×10^{-8}	2.62×10^{-4}	1.97×10^{-7}	3.5
40	1254	1002	0.069	5.46×10^{-8}	2.40×10^{-4}	1.91×10^{-7}	3.5
				Glycerin			
−20	1288	2143	0.282	1.02×10^{-7}	134	104×10^{-3}	$1020 \times 10^3 \times 10^6$
0	1276	2261	0.284	0.98×10^{-7}	12.1	9.5×10^{-3}	96×10^3
20	1264	2386	0.287	0.95×10^{-7}	1.49	1.2×10^{-3}	12.4×10^3
40	1252	2513	0.290	0.92×10^{-7}	0.27	0.2×10^{-3}	2.3×10^3
				Lead			
601*	10,588	161	15.5	0.91×10^{-5}	2.62×10^{-3}	2.47×10^{-7}	0.0272
700	10,476	157	17.4	1.06×10^{-5}	2.15×10^{-3}	2.05×10^{-7}	0.0194
800	10,359	153	19.0	1.20×10^{-5}	2.05×10^{-3}	1.98×10^{-7}	0.0165
900	10,237	149	20.3	1.33×10^{-5}	1.54×10^{-3}	1.50×10^{-7}	0.0113
1000	10,111	145	21.5	1.47×10^{-5}	1.32×10^{-3}	1.30×10^{-7}	0.0089
				Mercury			
234*	13,723	142	7.3	3.8×10^{-6}	2.00×10^{-3}	1.46×10^{-7}	0.0389
273	13,628	140	8.2	4.3×10^{-6}	1.69×10^{-3}	1.24×10^{-7}	0.0289
300	13,562	139	8.9	4.7×10^{-6}	1.51×10^{-3}	1.11×10^{-7}	0.0237
350	13,441	138	10.0	5.4×10^{-6}	1.31×10^{-3}	0.98×10^{-7}	0.0181
400	13,320	137	11.0	6.1×10^{-6}	1.18×10^{-3}	0.89×10^{-7}	0.0147
500	13,081	136	12.7	7.1×10^{-6}	1.02×10^{-3}	0.78×10^{-7}	0.0109
600	12,816	134	14.2	8.3×10^{-6}	0.84×10^{-3}	0.66×10^{-7}	0.0080
				Unused engine oil			
0	899	1796	0.147	9.11×10^{-8}	3850×10^{-3}	4280×10^{-6}	47,100
20	888	1880	0.145	8.72×10^{-8}	800×10^{-3}	901×10^{-6}	10,400
40	876	1964	0.144	8.34×10^{-8}	212×10^{-3}	242×10^{-6}	2870
60	864	2047	0.140	8.00×10^{-8}	72.5×10^{-3}	83.9×10^{-6}	1050
80	852	2131	0.138	7.69×10^{-8}	32.0×10^{-3}	37.5×10^{-6}	490
100	840	2219	0.137	7.38×10^{-8}	17.1×10^{-3}	20.3×10^{-6}	276
120	829	2307	0.135	7.10×10^{-8}	10.2×10^{-3}	12.4×10^{-6}	175
140	817	2395	0.133	6.86×10^{-8}	6.53×10^{-3}	8.0×10^{-6}	116
160	806	2483	0.132	6.63×10^{-8}	4.49×10^{-3}	5.6×10^{-6}	84

*Melting point.

Source: Tables A-10 and A-11 are adapted from Frank M. White, *Heat and Mass Transfer* (Reading, MA: Addison-Wesley, 1988), pp. 677–88 and 692–94. Originally compiled from various sources. Reprinted by permission of Addison-Wesley Longman Publishing Company, Inc.

TABLE A-11

Properties of gases at 1 atm pressure

Temperature, T K	Density, ρ kg/m³	Specific heat, C_p J/kg · °C	Thermal conductivity, k W/m · °C	Thermal diffusivity, α m²/s	Dynamic viscosity, μ kg/m · s	Kinematic viscosity, ν m²/s	Prandtl number, Pr
				Air			
200	1.766	1003	0.0181	1.02×10^{-5}	1.34×10^{-5}	0.76×10^{-5}	0.740
250	1.413	1003	0.0223	1.57×10^{-5}	1.61×10^{-5}	1.14×10^{-5}	0.724
280	1.271	1004	0.0246	1.95×10^{-5}	1.75×10^{-5}	1.40×10^{-5}	0.717
290	1.224	1005	0.0253	2.08×10^{-5}	1.80×10^{-5}	1.48×10^{-5}	0.714
298	1.186	1005	0.0259	2.18×10^{-5}	1.84×10^{-5}	1.55×10^{-5}	0.712
300	1.177	1005	0.0261	2.21×10^{-5}	1.85×10^{-5}	1.57×10^{-5}	0.712
310	1.143	1006	0.0268	2.35×10^{-5}	1.90×10^{-5}	1.67×10^{-5}	0.711
320	1.110	1006	0.0275	2.49×10^{-5}	1.94×10^{-5}	1.77×10^{-5}	0.710
330	1.076	1007	0.0283	2.64×10^{-5}	1.99×10^{-5}	1.86×10^{-5}	0.708
340	1.043	1007	0.0290	2.78×10^{-5}	2.03×10^{-5}	1.96×10^{-5}	0.707
350	1.009	1008	0.0297	2.92×10^{-5}	2.08×10^{-5}	2.06×10^{-5}	0.706
400	0.883	1013	0.0331	3.70×10^{-5}	2.29×10^{-5}	2.60×10^{-5}	0.703
450	0.785	1020	0.0363	4.54×10^{-5}	2.49×10^{-5}	3.18×10^{-5}	0.700
500	0.706	1029	0.0395	5.44×10^{-5}	2.68×10^{-5}	3.80×10^{-5}	0.699
550	0.642	1039	0.0426	6.39×10^{-5}	2.86×10^{-5}	4.45×10^{-5}	0.698
600	0.589	1051	0.0456	7.37×10^{-5}	3.03×10^{-5}	5.15×10^{-5}	0.698
700	0.504	1075	0.0513	9.46×10^{-5}	3.35×10^{-5}	6.64×10^{-5}	0.702
800	0.441	1099	0.0569	11.7×10^{-5}	3.64×10^{-5}	8.25×10^{-5}	0.704
900	0.392	1120	0.0625	14.2×10^{-5}	3.92×10^{-5}	9.99×10^{-5}	0.705
1000	0.353	1141	0.0672	16.7×10^{-5}	4.18×10^{-5}	11.8×10^{-5}	0.709
1200	0.294	1175	0.0759	22.2×10^{-5}	4.65×10^{-5}	15.8×10^{-5}	0.720
1400	0.252	1201	0.0835	27.6×10^{-5}	5.09×10^{-5}	20.2×10^{-5}	0.732
1600	0.221	1240	0.0904	33.0×10^{-5}	5.49×10^{-5}	24.9×10^{-5}	0.753
1800	0.196	1276	0.0970	38.3×10^{-5}	5.87×10^{-5}	29.9×10^{-5}	0.772
2000	0.177	1327	0.1032	44.1×10^{-5}	6.23×10^{-5}	35.3×10^{-5}	0.801
				Ammonia (NH_3)			
200	1.038	2199	0.0153	0.67×10^{-5}	6.89×10^{-6}	0.66×10^{-5}	0.990
250	0.831	2248	0.0197	1.05×10^{-5}	8.53×10^{-6}	1.03×10^{-5}	0.973
300	0.692	2298	0.0246	1.55×10^{-5}	10.27×10^{-6}	1.48×10^{-5}	0.959
350	0.593	2349	0.0302	2.17×10^{-5}	12.06×10^{-6}	2.03×10^{-5}	0.938
400	0.519	2402	0.0364	2.92×10^{-5}	13.90×10^{-6}	2.68×10^{-5}	0.917
450	0.461	2455	0.0433	3.82×10^{-5}	15.76×10^{-6}	3.42×10^{-5}	0.894
500	0.415	2507	0.0506	4.86×10^{-5}	17.63×10^{-6}	4.25×10^{-5}	0.873
550	0.378	2559	0.0580	6.00×10^{-5}	19.5×10^{-6}	5.16×10^{-5}	0.860
600	0.346	2611	0.0656	7.26×10^{-5}	21.4×10^{-6}	6.18×10^{-5}	0.852
700	0.297	2710	0.0811	10.1×10^{-5}	25.1×10^{-6}	8.45×10^{-5}	0.839
800	0.260	2810	0.0977	13.4×10^{-5}	28.8×10^{-6}	11.1×10^{-5}	0.828
				Argon			
200	2.435	523.6	0.0124	0.98×10^{-5}	1.60×10^{-5}	0.66×10^{-5}	0.674
250	1.948	522.2	0.0152	1.49×10^{-5}	1.95×10^{-5}	1.00×10^{-5}	0.672
300	1.623	521.6	0.0177	2.09×10^{-5}	2.27×10^{-5}	1.40×10^{-5}	0.669
350	1.392	521.2	0.0201	2.78×10^{-5}	2.57×10^{-5}	1.85×10^{-5}	0.666
400	1.218	521.0	0.0223	3.52×10^{-5}	2.85×10^{-5}	2.34×10^{-5}	0.665
450	1.082	520.9	0.0244	4.33×10^{-5}	3.12×10^{-5}	2.88×10^{-5}	0.665

TABLE A-11

Properties of gases at 1 atm pressure (*Continued*)

Tempera-ture, T K	Density, ρ kg/m^3	Specific heat, C_p J/kg · °C	Thermal conductivity, k W/m · °C	Thermal diffusivity, α m^2/s	Dynamic viscosity, μ kg/m · s	Kinematic viscosity, ν m^2/s	Prandtl number, Pr
500	0.974	520.8	0.0264	5.20×10^{-5}	3.37×10^{-5}	3.45×10^{-5}	0.664
550	0.886	520.7	0.0283	6.14×10^{-5}	3.60×10^{-5}	4.07×10^{-5}	0.662
600	0.812	520.6	0.0301	7.12×10^{-5}	3.83×10^{-5}	4.72×10^{-5}	0.662
700	0.696	520.6	0.0336	9.28×10^{-5}	4.25×10^{-5}	6.11×10^{-5}	0.658
800	0.609	520.5	0.0369	11.6×10^{-5}	4.64×10^{-5}	7.62×10^{-5}	0.655
900	0.541	520.5	0.0398	14.1×10^{-5}	5.01×10^{-5}	9.26×10^{-5}	0.654
1000	0.487	520.5	0.0427	16.8×10^{-5}	5.35×10^{-5}	11.0×10^{-5}	0.652
1200	0.406	520.5	0.0481	22.8×10^{-5}	5.99×10^{-5}	14.8×10^{-5}	0.648
1400	0.348	520.4	0.0535	29.6×10^{-5}	6.56×10^{-5}	18.9×10^{-5}	0.638
			Carbon dioxide (CO_2)				
200	2.683	759	0.0095	0.47×10^{-5}	1.02×10^{-5}	0.38×10^{-5}	0.814
250	2.146	806	0.0129	0.75×10^{-5}	1.26×10^{-5}	0.59×10^{-5}	0.790
300	1.789	852	0.0166	1.09×10^{-5}	1.50×10^{-5}	0.84×10^{-5}	0.768
350	1.533	897	0.0205	1.49×10^{-5}	1.73×10^{-5}	1.13×10^{-5}	0.755
400	1.341	939	0.0244	1.94×10^{-5}	1.94×10^{-5}	1.45×10^{-5}	0.747
450	1.192	979	0.0283	2.43×10^{-5}	2.15×10^{-5}	1.80×10^{-5}	0.743
500	1.073	1017	0.0323	2.96×10^{-5}	2.35×10^{-5}	2.19×10^{-5}	0.740
550	0.976	1049	0.0363	3.55×10^{-5}	2.54×10^{-5}	2.60×10^{-5}	0.734
600	0.894	1077	0.0403	4.18×10^{-5}	2.72×10^{-5}	3.04×10^{-5}	0.727
700	0.767	1126	0.0487	5.64×10^{-5}	3.06×10^{-5}	3.99×10^{-5}	0.708
800	0.671	1169	0.0560	7.14×10^{-5}	3.39×10^{-5}	5.05×10^{-5}	0.708
900	0.596	1205	0.0621	8.65×10^{-5}	3.69×10^{-5}	6.19×10^{-5}	0.716
1000	0.537	1235	0.0680	10.25×10^{-5}	3.97×10^{-5}	7.40×10^{-5}	0.721
1200	0.447	1283	0.0780	13.6×10^{-5}	4.49×10^{-5}	10.04×10^{-5}	0.739
1400	0.383	1315	0.0867	17.2×10^{-5}	4.97×10^{-5}	13.0×10^{-5}	0.754
			Carbon monoxide (CO)				
200	1.708	1045	0.0175	0.98×10^{-5}	1.27×10^{-5}	0.75×10^{-5}	0.763
250	1.366	1048	0.0214	1.50×10^{-5}	1.54×10^{-5}	1.13×10^{-5}	0.753
300	1.138	1051	0.0252	2.11×10^{-5}	1.78×10^{-5}	1.56×10^{-5}	0.743
350	0.976	1056	0.0288	2.80×10^{-5}	2.01×10^{-5}	2.05×10^{-5}	0.735
400	0.854	1060	0.0323	3.57×10^{-5}	2.21×10^{-5}	2.59×10^{-5}	0.727
450	0.759	1065	0.0355	4.39×10^{-5}	2.41×10^{-5}	3.18×10^{-5}	0.723
500	0.683	1071	0.0386	5.28×10^{-5}	2.60×10^{-5}	3.80×10^{-5}	0.720
550	0.621	1077	0.0416	6.22×10^{-5}	2.77×10^{-5}	4.46×10^{-5}	0.717
600	0.569	1084	0.0444	7.20×10^{-5}	2.94×10^{-5}	5.17×10^{-5}	0.718
700	0.488	1099	0.0497	9.27×10^{-5}	3.25×10^{-5}	6.66×10^{-5}	0.718
800	0.427	1114	0.0549	11.5×10^{-5}	3.54×10^{-5}	8.29×10^{-5}	0.718
900	0.379	1128	0.0596	13.9×10^{-5}	3.81×10^{-5}	10.04×10^{-5}	0.721
1000	0.342	1142	0.0644	16.5×10^{-5}	4.06×10^{-5}	11.9×10^{-5}	0.720
1100	0.310	1155	0.0692	19.3×10^{-5}	4.30×10^{-5}	13.9×10^{-5}	0.718
1200	0.285	1168	0.0738	22.2×10^{-5}	4.53×10^{-5}	15.9×10^{-5}	0.717
			Helium				
200	0.2440	5197	0.115	0.91×10^{-4}	1.50×10^{-5}	0.61×10^{-4}	0.676
250	0.1952	5197	0.134	1.54×10^{-4}	1.75×10^{-5}	0.90×10^{-4}	0.680

TABLE A-11

Properties of gases at 1 atm pressure (*Continued*)

Tempera-ture, T K	Density, ρ kg/m^3	Specific heat, C_p J/kg · °C	Thermal conductivity, k W/m · °C	Thermal diffusivity, α m^2/s	Dynamic viscosity, μ kg/m · s	Kinematic viscosity, ν m^2/s	Prandtl number, Pr
300	0.1627	5197	0.150	1.77×10^{-4}	1.99×10^{-5}	1.22×10^{-4}	0.690
350	0.1394	5197	0.165	2.28×10^{-4}	2.21×10^{-5}	1.59×10^{-4}	0.698
400	0.1220	5197	0.180	2.83×10^{-4}	2.43×10^{-5}	1.99×10^{-4}	0.703
450	0.1085	5197	0.195	3.45×10^{-4}	2.63×10^{-5}	2.43×10^{-4}	0.702
500	0.0976	5197	0.211	4.17×10^{-4}	2.83×10^{-5}	2.90×10^{-4}	0.695
550	0.0887	5197	0.229	4.97×10^{-4}	3.02×10^{-5}	3.40×10^{-4}	0.684
600	0.0813	5197	0.247	5.84×10^{-4}	3.20×10^{-5}	3.93×10^{-4}	0.673
700	0.0697	5197	0.278	7.67×10^{-4}	3.55×10^{-5}	5.09×10^{-4}	0.663
800	0.0610	5197	0.307	9.68×10^{-4}	3.88×10^{-5}	6.37×10^{-4}	0.657
900	0.0542	5197	0.335	11.9×10^{-4}	4.20×10^{-5}	7.75×10^{-4}	0.652
1000	0.0488	5197	0.363	14.3×10^{-4}	4.50×10^{-5}	9.23×10^{-4}	0.645
1200	0.0407	5197	0.416	19.7×10^{-4}	5.08×10^{-5}	12.5×10^{-4}	0.635
1400	0.0349	5197	0.469	25.9×10^{-4}	5.61×10^{-5}	16.1×10^{-4}	0.622
1600	0.0305	5197	0.521	32.9×10^{-4}	6.10×10^{-5}	20.0×10^{-4}	0.608
1800	0.0271	5197	0.570	40.4×10^{-4}	6.57×10^{-5}	24.2×10^{-4}	0.599
2000	0.0244	5197	0.620	48.9×10^{-4}	7.00×10^{-5}	28.7×10^{-4}	0.587
			Hydrogen				
200	0.1299	13,540	0.128	0.77×10^{-4}	0.68×10^{-5}	0.55×10^{-4}	0.717
250	0.0983	14,070	0.156	1.13×10^{-4}	0.79×10^{-5}	0.80×10^{-4}	0.713
300	0.0819	14,320	0.182	1.55×10^{-4}	0.89×10^{-5}	1.09×10^{-4}	0.705
350	0.0702	14,420	0.203	2.01×10^{-4}	0.99×10^{-5}	1.42×10^{-4}	0.705
400	0.0614	14,480	0.221	2.49×10^{-4}	1.09×10^{-5}	1.78×10^{-4}	0.714
450	0.0546	14,500	0.239	3.02×10^{-4}	1.18×10^{-5}	2.17×10^{-4}	0.719
500	0.0492	14,510	0.256	3.59×10^{-4}	1.27×10^{-5}	2.59×10^{-4}	0.721
550	0.0447	14,520	0.274	4.22×10^{-4}	1.36×10^{-5}	3.04×10^{-4}	0.722
600	0.0410	14,540	0.291	4.89×10^{-4}	1.45×10^{-5}	3.54×10^{-4}	0.724
700	0.0351	14,610	0.325	6.34×10^{-4}	1.61×10^{-5}	4.59×10^{-4}	0.724
800	0.0307	14,710	0.360	7.97×10^{-4}	1.77×10^{-5}	5.76×10^{-4}	0.723
900	0.0273	14,840	0.394	10.8×10^{-4}	1.92×10^{-5}	7.03×10^{-4}	0.723
1000	0.0246	14,990	0.428	11.6×10^{-4}	2.07×10^{-5}	8.42×10^{-4}	0.724
1200	0.0205	15,370	0.495	15.7×10^{-4}	2.36×10^{-5}	11.5×10^{-4}	0.733
			Nitrogen				
200	1.708	1043	0.0183	1.02×10^{-5}	1.29×10^{-5}	0.75×10^{-5}	0.734
250	1.367	1042	0.0222	1.56×10^{-5}	1.55×10^{-5}	1.13×10^{-5}	0.725
300	1.139	1040	0.0260	2.19×10^{-5}	1.79×10^{-5}	1.57×10^{-5}	0.715
350	0.967	1041	0.0294	2.92×10^{-5}	2.01×10^{-5}	2.08×10^{-5}	0.711
400	0.854	1045	0.0325	3.64×10^{-5}	2.21×10^{-5}	2.59×10^{-5}	0.710
450	0.759	1050	0.0356	4.47×10^{-5}	2.41×10^{-5}	3.17×10^{-5}	0.709
500	0.683	1057	0.0387	5.36×10^{-5}	2.59×10^{-5}	3.79×10^{-5}	0.708
550	0.621	1065	0.0414	6.26×10^{-5}	2.76×10^{-5}	4.45×10^{-5}	0.711
600	0.569	1075	0.0441	7.20×10^{-5}	2.93×10^{-5}	5.14×10^{-5}	0.713
700	0.488	1098	0.0493	9.20×10^{-5}	3.24×10^{-5}	6.63×10^{-5}	0.720
800	0.427	1122	0.0541	11.3×10^{-5}	3.52×10^{-5}	8.24×10^{-5}	0.730
900	0.380	1146	0.0587	13.5×10^{-5}	3.79×10^{-5}	9.97×10^{-5}	0.739
1000	0.342	1168	0.0631	15.8×10^{-5}	4.04×10^{-5}	11.8×10^{-5}	0.747

Properties of gases at 1 atm pressure (*Concluded*)

Tempera-ture, T K	Density, ρ kg/m³	Specific heat, C_p J/kg·°C	Thermal conductivity, k W/m·°C	Thermal diffusivity, α m²/s	Dynamic viscosity, μ kg/m·s	Kinematic viscosity, ν m²/s	Prandtl number, Pr
1200	0.285	1205	0.0713	20.8×10^{-5}	4.50×10^{-5}	15.8×10^{-5}	0.761
1400	0.244	1233	0.0797	26.5×10^{-5}	4.92×10^{-5}	20.2×10^{-5}	0.761
Oxygen							
200	1.951	906	0.0182	1.03×10^{-5}	1.47×10^{-5}	0.75×10^{-5}	0.728
250	1.561	914	0.0225	1.58×10^{-5}	1.78×10^{-5}	1.14×10^{-5}	0.721
300	1.301	920	0.0267	2.23×10^{-5}	2.07×10^{-5}	1.59×10^{-5}	0.711
350	1.115	929	0.0306	2.95×10^{-5}	2.34×10^{-5}	2.10×10^{-5}	0.710
400	0.976	942	0.0342	3.72×10^{-5}	2.59×10^{-5}	2.65×10^{-5}	0.713
450	0.867	956	0.0377	4.55×10^{-5}	2.83×10^{-5}	3.26×10^{-5}	0.717
500	0.780	971	0.0412	5.44×10^{-5}	3.05×10^{-5}	3.91×10^{-5}	0.720
550	0.709	987	0.0447	6.38×10^{-5}	3.27×10^{-5}	4.61×10^{-5}	0.722
600	0.650	1003	0.0480	7.36×10^{-5}	3.47×10^{-5}	5.34×10^{-5}	0.725
700	0.557	1032	0.0544	9.46×10^{-5}	3.85×10^{-5}	6.91×10^{-5}	0.730
800	0.488	1054	0.0603	11.7×10^{-5}	4.21×10^{-5}	8.63×10^{-5}	0.736
900	0.434	1074	0.0661	14.2×10^{-5}	4.54×10^{-5}	10.5×10^{-5}	0.738
1000	0.390	1091	0.0717	16.8×10^{-5}	4.85×10^{-5}	12.4×10^{-5}	0.738
1200	0.325	1116	0.0821	22.6×10^{-5}	5.42×10^{-5}	16.7×10^{-5}	0.737
1400	0.278	1136	0.0921	29.1×10^{-5}	5.95×10^{-5}	21.3×10^{-5}	0.734
Water vapor (steam)							
300	0.0253*	2041	0.0181	$35.1 \times 10^{-5*}$	0.91×10^{-5}	$36.1 \times 10^{-5*}$	1.03
350	0.258*	2037	0.0222	$4.22 \times 10^{-5*}$	1.12×10^{-5}	$4.33 \times 10^{-5*}$	1.02
400	0.555	2000	0.0264	2.38×10^{-5}	1.32×10^{-5}	2.38×10^{-5}	1.00
450	0.491	1968	0.0307	3.17×10^{-5}	1.52×10^{-5}	3.10×10^{-5}	0.98
500	0.441	1977	0.0357	4.09×10^{-5}	1.73×10^{-5}	3.92×10^{-5}	0.96
550	0.401	1994	0.0411	5.15×10^{-5}	1.93×10^{-5}	4.82×10^{-5}	0.94
600	0.367	2022	0.0464	6.25×10^{-5}	2.13×10^{-5}	5.82×10^{-5}	0.93
700	0.314	2083	0.0572	8.74×10^{-5}	2.54×10^{-5}	8.09×10^{-5}	0.93
800	0.275	2148	0.0686	11.6×10^{-5}	2.95×10^{-5}	10.7×10^{-5}	0.92
900	0.244	2217	0.078	14.4×10^{-5}	3.36×10^{-5}	13.7×10^{-5}	0.95
1000	0.220	2288	0.087	17.3×10^{-5}	3.76×10^{-5}	17.1×10^{-5}	0.99

*At saturation pressure (less than 1 atm).

For ideal gases, the properties C_p, k, μ, and Pr are independent of pressure. The properties ρ, ν, and α at a pressure P other than 1 atm are determined by multiplying the values of ρ at the given temperature by P and by dividing the values of ν and α at the given temperature by P, where P is in atm (1 atm = 101.325 kPa = 14.696 psi).

TABLE A-12

Properties of the atmosphere at high altitude

Altitude, m	Temperature, °C	Pressure, kPa	Gravity, g m/s²	Speed of sound, m/s	Density, kg/m³	Viscosity, μ kg/m·s	Thermal conductivity, W/m·°C
0	15.00	101.33	9.807	340.3	1.225	1.789×10^{-5}	0.0253
200	13.70	98.95	9.806	339.5	1.202	1.783×10^{-5}	0.0252
400	12.40	96.61	9.805	338.8	1.179	1.777×10^{-5}	0.0252
600	11.10	94.32	9.805	338.0	1.156	1.771×10^{-5}	0.0251
800	9.80	92.08	9.804	337.2	1.134	1.764×10^{-5}	0.0250
1000	8.50	89.88	9.804	336.4	1.112	1.758×10^{-5}	0.0249
1200	7.20	87.72	9.803	335.7	1.090	1.752×10^{-5}	0.0248
1400	5.90	85.60	9.802	334.9	1.069	1.745×10^{-5}	0.0247
1600	4.60	83.53	9.802	334.1	1.048	1.739×10^{-5}	0.0245
1800	3.30	81.49	9.801	333.3	1.027	1.732×10^{-5}	0.0244
2000	2.00	79.50	9.800	332.5	1.007	1.726×10^{-5}	0.0243
2200	0.70	77.55	9.800	331.7	0.987	1.720×10^{-5}	0.0242
2400	−0.59	75.63	9.799	331.0	0.967	1.713×10^{-5}	0.0241
2600	−1.89	73.76	9.799	330.2	0.947	1.707×10^{-5}	0.0240
2800	−3.19	71.92	9.798	329.4	0.928	1.700×10^{-5}	0.0239
3000	−4.49	70.12	9.797	328.6	0.909	1.694×10^{-5}	0.0238
3200	−5.79	68.36	9.797	327.8	0.891	1.687×10^{-5}	0.0237
3400	−7.09	66.63	9.796	327.0	0.872	1.681×10^{-5}	0.0236
3600	−8.39	64.94	9.796	326.2	0.854	1.674×10^{-5}	0.0235
3800	−9.69	63.28	9.795	325.4	0.837	1.668×10^{-5}	0.0234
4000	−10.98	61.66	9.794	324.6	0.819	1.661×10^{-5}	0.0233
4200	−12.3	60.07	9.794	323.8	0.802	1.655×10^{-5}	0.0232
4400	−13.6	58.52	9.793	323.0	0.785	1.648×10^{-5}	0.0231
4600	−14.9	57.00	9.793	322.2	0.769	1.642×10^{-5}	0.0230
4800	−16.2	55.51	9.792	321.4	0.752	1.635×10^{-5}	0.0229
5000	−17.5	54.05	9.791	320.5	0.736	1.628×10^{-5}	0.0228
5200	−18.8	52.62	9.791	319.7	0.721	1.622×10^{-5}	0.0227
5400	−20.1	51.23	9.790	318.9	0.705	1.615×10^{-5}	0.0226
5600	−21.4	49.86	9.789	318.1	0.690	1.608×10^{-5}	0.0224
5800	−22.7	48.52	9.785	317.3	0.675	1.602×10^{-5}	0.0223
6000	−24.0	47.22	9.788	316.5	0.660	1.595×10^{-5}	0.0222
6200	−25.3	45.94	9.788	315.6	0.646	1.588×10^{-5}	0.0221
6400	−26.6	44.69	9.787	314.8	0.631	1.582×10^{-5}	0.0220
6600	−27.9	43.47	9.786	314.0	0.617	1.575×10^{-5}	0.0219
6800	−29.2	42.27	9.785	313.1	0.604	1.568×10^{-5}	0.0218
7000	−30.5	41.11	9.785	312.3	0.590	1.561×10^{-5}	0.0217
8000	−36.9	35.65	9.782	308.1	0.526	1.527×10^{-5}	0.0212
9000	−43.4	30.80	9.779	303.8	0.467	1.493×10^{-5}	0.0206
10,000	−49.9	26.50	9.776	299.5	0.414	1.458×10^{-5}	0.0201
12,000	−56.5	19.40	9.770	295.1	0.312	1.422×10^{-5}	0.0195
14,000	−56.5	14.17	9.764	295.1	0.228	1.422×10^{-5}	0.0195
16,000	−56.5	10.53	9.758	295.1	0.166	1.422×10^{-5}	0.0195
18,000	−56.5	7.57	9.751	295.1	0.122	1.422×10^{-5}	0.0195

Source: U.S. Standard Atmosphere Supplements, U.S. Government Printing Office, 1966. Based on year-round mean conditions at 45° latitude and varies with the time of the year and the weather patterns. The conditions at sea level ($z = 0$) are taken to be P = 101.325 kPa, T = 15°C, ρ = 1.2250 kg/m³, g = 9.80665 m²/s.

FIGURE A-13

Psychrometric chart at 1 atm total pressure. (Reprinted by permission of the American Society of Heating, Refrigerating and Air-Conditioning Engineers, Inc., Atlanta.)

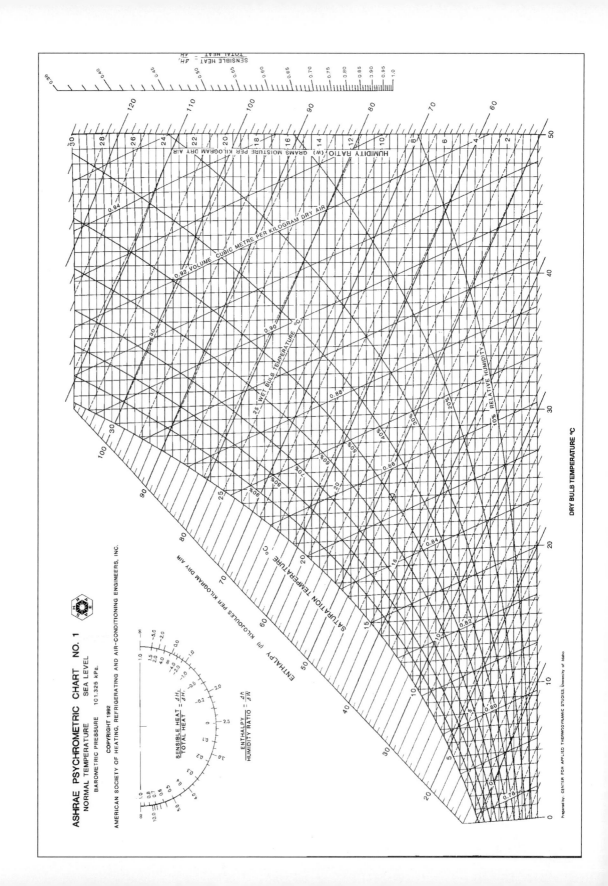

TABLE A-14

Emissivities of surfaces

(a) Metals

Material	Temperature, K	Emissivity, ε	Material	Temperature, K	Emissivity, ε
Aluminum			Magnesium, polished	300–500	0.07–0.13
Polished	300–900	0.04–0.06	Mercury	300–400	0.09–0.12
Commercial sheet	400	0.09	Molybdenum		
Heavily oxidized	400–800	0.20–0.33	Polished	300–2000	0.05–0.21
Anodized	300	0.8	Oxidized	600–800	0.80–0.82
Bismuth, bright	350	0.34	Nickel		
Brass			Polished	500–1200	0.07–0.17
Highly polished	500–650	0.03–0.04	Oxidized	450–1000	0.37–0.57
Polished	350	0.09	Platinum, polished	500–1500	0.06–0.18
Dull plate	300–600	0.22	Silver, polished	300–1000	0.02–0.07
Oxidized	450–800	0.6	Stainless steel		
Chromium, polished	300–1400	0.08–0.40	Polished	300–1000	0.17–0.30
Copper			Lightly oxidized	600–1000	0.30–0.40
Highly polished	300	0.02	Highly oxidized	600–1000	0.70–0.80
Polished	300–500	0.04–0.05	Steel		
Commercial sheet	300	0.15	Polished sheet	300–500	0.08–0.14
Oxidized	600–1000	0.5–0.8	Commercial sheet	500–1200	0.20–0.32
Black oxidized	300	0.78	Heavily oxidized	300	0.81
Gold			Tin, polished	300	0.05
Highly polished	300–1000	0.03–0.06	Tungsten		
Bright foil	300	0.07	Polished	300–2500	0.03–0.29
Iron			Filament	3500	0.39
Highly polished	300–500	0.05–0.07	Zinc		
Case iron	300	0.44	Polished	300–800	0.02–0.05
Wrought iron	300–500	0.28	Oxidized	300	0.25
Rusted	300	0.61			
Oxidized	500–900	0.64–0.78			
Lead					
Polished	300–500	0.06–0.08			
Unoxidized, rough	300	0.43			
Oxidized	300	0.63			

TABLE A-14

Emissivities of surfaces (*Concluded*)

(*b*) Nonmetals

Material	Temperature, K	Emissivity, ε	Material	Temperature, K	Emissivity, ε
Alumina	800–1400	0.65–0.45	Paper, white	300	0.90
Aluminum oxide	600–1500	0.69–0.41	Plaster, white	300	0.93
Asbestos	300	0.96	Porcelain, glazed	300	0.92
Asphalt pavement	300	0.85–0.93	Quartz, rough, fused	300	0.93
Brick			Rubber		
Common	300	0.93–0.96	Hard	300	0.93
Fireclay	1200	0.75	Soft	300	0.86
Carbon filament	2000	0.53	Sand	300	0.90
Cloth	300	0.75–0.90	Silicon carbide	600–1500	0.87–0.85
Concrete	300	0.88–0.94	Skin, human	300	0.95
Glass			Snow	273	0.80–0.90
Window	300	0.90–0.95	Soil, earth	300	0.93–0.96
Pyrex	300–1200	0.82–0.62	Soot	300–500	0.95
Pyroceram	300–1500	0.85–0.57	Teflon	300–500	0.85–0.92
Ice	273	0.95–0.99	Water, deep	273–373	0.95–0.96
Magnesium oxide	400–800	0.69–0.55	Wood		
Masonry	300	0.80	Beech	300	0.94
Paints			Oak	300	0.90
Aluminum	300	0.40–0.50			
Black, lacquer, shiny	300	0.88			
Oils, all colors	300	0.92–0.96			
Red primer	300	0.93			
White acrylic	300	0.90			
White enamel	300	0.90			

TABLE A-15

Solar radiative properties of materials

Description/Composition	Solar absorptivity, α_s	Emissivity, ε, at 300 K	Ratio, α_s/ε	Solar transmissivity, τ_s
Aluminum				
Polished	0.09	0.03	3.0	
Anodized	0.14	0.84	0.17	
Quartz-overcoated	0.11	0.37	0.30	
Foil	0.15	0.05	3.0	
Brick, red (Purdue)	0.63	0.93	0.68	
Concrete	0.60	0.88	0.68	
Galvanized sheet metal				
Clean, new	0.65	0.13	5.0	
Oxidized, weathered	0.80	0.28	2.9	
Glass, 3.2-mm thickness				
Float or tempered				0.79
Low iron oxide type				0.88
Marble, slightly off-white (nonreflective)	0.40	0.88	0.45	
Metal, plated				
Black sulfide	0.92	0.10	9.2	
Black cobalt oxide	0.93	0.30	3.1	
Black nickel oxide	0.92	0.08	11	
Black chrome	0.87	0.09	9.7	
Mylar, 0.13-mm thickness				0.87
Paints				
Black (Parsons)	0.98	0.98	1.0	
White, acrylic	0.26	0.90	0.29	
White, zinc oxide	0.16	0.93	0.17	
Paper, white	0.27	0.83	0.32	
Plexiglas, 3.2-mm thickness				0.90
Porcelain tiles, white (reflective glazed surface)	0.26	0.85	0.30	
Roofing tiles, bright red				
Dry surface	0.65	0.85	0.76	
Wet surface	0.88	0.91	0.96	
Sand, dry				
Off-white	0.52	0.82	0.63	
Dull red	0.73	0.86	0.82	
Snow				
Fine particles, fresh	0.13	0.82	0.16	
Ice granules	0.33	0.89	0.37	
Steel				
Mirror-finish	0.41	0.05	8.2	
Heavily rusted	0.89	0.92	0.96	
Stone (light pink)	0.65	0.87	0.74	
Tedlar, 0.10-mm thickness				0.92
Teflon, 0.13-mm thickness				0.92
Wood	0.59	0.90	0.66	

Source: V. C. Sharma and A. Sharma, "Solar Properties of Some Building Elements," *Energy* 14 (1989), pp. 805–10, and other sources.

Property Tables and Charts (English Units)

2

TABLE A-1E

Molar mass, gas constant, and critical-point properties

| Substance | Formula | Molar mass, M lbm/lbmol | Gas constant, R | | Critical-point properties | | |
			Btu/ lbm · R*	psia · ft³/ lbm · R*	Tempera-ture, R	Pres-sure, psia	Volume, ft³/lbmol
Air	—	28.97	0.06855	0.3704	238.5	547	1.41
Ammonia	NH_3	17.03	0.1166	0.6301	729.8	1636	1.16
Argon	Ar	39.948	0.04971	0.2686	272	705	1.20
Benzene	C_6H_6	78.115	0.02542	0.1374	1012	714	4.17
Bromine	Br_2	159.808	0.01243	0.06714	1052	1500	2.17
n-Butane	C_4H_{10}	58.124	0.03417	0.1846	765.2	551	4.08
Carbon dioxide	CO_2	44.01	0.04513	0.2438	547.5	1071	1.51
Carbon monoxide	CO	28.011	0.07090	0.3831	240	507	1.49
Carbon tetrachloride	CCl_4	153.82	0.01291	0.06976	1001.5	661	4.42
Chlorine	Cl_2	70.906	0.02801	0.1517	751	1120	1.99
Chloroform	$CHCl_3$	119.38	0.01664	0.08988	965.8	794	3.85
Dichlorodifluoromethane (R-12)	CCl_2F_2	120.91	0.01643	0.08874	692.4	582	3.49
Dichlorofluoromethane (R-21)	$CHCl_2F$	102.92	0.01930	0.1043	813.0	749	3.16
Ethane	C_2H_6	30.020	0.06616	0.3574	549.8	708	2.37
Ethyl alcohol	C_2H_5OH	46.07	0.04311	0.2329	929.0	926	2.68
Ethylene	C_2H_4	28.054	0.07079	0.3825	508.3	742	1.99
Helium	He	4.003	0.4961	2.6805	9.5	33.2	0.926
n-Hexane	C_6H_{14}	86.178	0.02305	0.1245	914.2	439	5.89
Hydrogen (normal)	H_2	2.016	0.9851	5.3224	59.9	188.1	1.04
Krypton	Kr	83.80	0.02370	0.1280	376.9	798	1.48
Methane	CH_4	16.043	0.1238	0.6688	343.9	673	1.59
Methyl alcohol	CH_3OH	32.042	0.06198	0.3349	923.7	1154	1.89
Methyl chloride	CH_3Cl	50.488	0.03934	0.2125	749.3	968	2.29
Neon	Ne	20.183	0.09840	0.5316	80.1	395	0.668
Nitrogen	N_2	28.013	0.07090	0.3830	227.1	492	1.44
Nitrous oxide	N_2O	44.013	0.04512	0.2438	557.4	1054	1.54
Oxygen	O_2	31.999	0.06206	0.3353	278.6	736	1.25
Propane	C_3H_8	44.097	0.04504	0.2433	665.9	617	3.20
Propylene	C_3H_6	42.081	0.04719	0.2550	656.9	670	2.90
Sulfur dioxide	SO_2	64.063	0.03100	1.1675	775.2	1143	1.95
Tetrafluoroethane (R-134a)	CF_3CH_2F	102.03	0.01946	0.1052	673.7	589.9	2.96
Trichlorofluoromethane (R-11)	CCl_3F	137.37	0.01446	0.07811	848.1	635	3.97
Water	H_2O	18.015	0.1102	0.5956	1165.3	3204	0.90
Xenon	Xe	131.30	0.01513	0.08172	521.55	852	1.90

*Calculated from $R = R_u/M$, where $R_u = 1.986$ Btu/lbmol · R = 10.73 psia · ft³/lbmol · R and M is the molar mass.

Source: K. A. Kobe and R. E. Lynn, Jr., Chemical Review 52 (1953), pp. 117–236, and ASHRAE, Handbook of Fundamentals (Atlanta, GA: American Society of Heating, Refrigerating and Air-Conditioning Engineers, Inc., 1993), pp. 16.4 and 36.1.

TABLE A-2E

Boiling- and freezing-point properties

Substance	Boiling data at 1 atm		Freezing data		Liquid properties		
	Normal boiling point, °F	Latent heat of vaporization, h_{fg} Btu/lbm	Freezing point, °F	Latent heat of fusion, h_{if} Btu/lbm	Temperature, °F	Density, ρ lbm/ft³	Specific heat, C_p Btu/lbm·°F
Ammonia	−27.9	24.54	−107.9	138.6	−27.9	42.6	1.06
					0	41.3	1.083
					40	39.5	1.103
					80	37.5	1.135
Argon	−302.6	69.5	−308.7	12.0	−302.6	87.0	0.272
Benzene	176.4	169.4	41.9	54.2	68	54.9	0.411
Brine (20% sodium chloride by mass)	219.0	—	0.7	—	68	71.8	0.743
n-Butane	31.1	165.6	−217.3	34.5	31.1	37.5	0.552
Carbon dioxide	−109.2*	99.6 (at 32°C)	−69.8	—	32	57.8	0.583
Ethanol	172.8	360.5	−173.6	46.9	77	48.9	0.588
Ethylene glycol	388.6	344.0	12.6	77.9	68	69.2	0.678
Ethyl alcohol	173.5	368	−248.8	46.4	68	49.3	0.678
Glycerine	355.8	419	66.0	86.3	68	78.7	0.554
Helium	−452.1	9.80	—	—	−452.1	9.13	5.45
Hydrogen	−423.0	191.7	−434.5	25.6	−423.0	4.41	2.39
Isobutane	10.9	157.8	−255.5	45.5	10.9	37.1	0.545
Kerosene	399 to 559	108	−12.8	—	68	51.2	0.478
Mercury	674.1	126.7	−38.0	4.90	77	847	0.033
Methane	−258.7	219.6	296.0	25.1	−258.7	26.4	0.834
					−160	20.0	1.074
Methanol	148.1	473	−143.9	42.7	77	49.1	0.609
Nitrogen	−320.4	85.4	−346.0	10.9	−320.4	50.5	0.492
					−260	38.2	0.643
Octane	256.6	131.7	−71.5	77.9	68	43.9	0.502
Oil (light)	—	—			77	56.8	0.430
Oxygen	−297.3	91.5	−361.8	5.9	−297.3	71.2	0.408
Petroleum	—	99–165			68	40.0	0.478
Propane	−43.7	184.0	−305.8	34.4	−43.7	36.3	0.538
					32	33.0	0.604
					100	29.4	0.673
Refrigerant-134a	−15.0	93.2	−141.9	—	−40	88.5	0.283
					−15	86.0	0.294
					32	80.9	0.318
					90	73.6	0.348
Water	212	970.5	32	143.5	32	62.4	1.01
					90	62.1	1.00
					150	61.2	1.00
					212	59.8	1.01

*Sublimation temperature. (At pressures below the triple-point pressure of 75.1 psia, carbon dioxide exists as a solid or gas. Also, the freezing-point temperature of carbon dioxide is the triple-point temperature of −69.8°F.)

Properties of solid metals

Composition	Melting point, R	ρ lbm/ft³	c_p Btu/ lbm·R	k Btu/ h·ft·R	$\alpha \times 10^6$ ft²/s	180	360	720	1080	1440	1800
Aluminum:	1679	168	0.216	137	1045	174.5	137	138.6	133.4	126	
Pure						0.115	0.191	0.226	0.246	0.273	
Alloy 2024-T6	1395	173	0.209	102.3	785.8	37.6	94.2	107.5	107.5		
(4.5% Cu, 1.5% Mg, 0.6% Mn)						0.113	0.188	0.22	0.249		
Alloy 195, Cast (4.5% Cu)		174.2	0.211	97	734			100.5	106.9		
								—	—		
Beryllium	2790	115.5	0.436	115.6	637.2	572	174	93	72.8	61.3	52.5
						0.048	0.266	0.523	0.621	0.624	0.72
Bismuth	981	610.5	0.029	4.6	71	9.5	5.6	4.06			
						0.026	0.028	0.03			
Boron	4631	156	0.264	15.6	105	109.7	32.06	9.7	6.1	5.5	5.7
						0.03	0.143	0.349	0.451	0.515	0.558
Cadmium	1069	540	0.055	55.6	521	117.3	57.4	54.7			
						0.047	0.053	0.057			
Chromium	3812	447	0.107	54.1	313.2	91.9	64.1	52.5	46.6	41.2	37.8
						0.045	0.091	0.115	0.129	0.138	0.147
Cobalt	3184	553.2	0.101	57.3	286.3	96.5	70.5	49.3	39	33.6	30.1
						0.056	0.09	0.107	0.12	0.131	0.145
Copper:	2445	559	0.092	231.7	1259.3	278.5	238.6	227.07	219	212	203.4
Pure						0.06	0.085	0.094	0.01	0.103	0.107
Commercial bronze	2328	550	0.1	30	150.7		24.3	30	34		
(90% Cu, 10% Al)							0.187	0.109	0.130		
Phosphor gear bronze	1987	548.1	0.084	31.2	183		23.7	37.6	42.8		
(89% Cu, 11% Sn)							—	—	—		
Cartridge brass	2139	532.5	0.09	63.6	364.9	43.3	54.9	79.2	86.0		
(70% Cu, 30% Zn)							0.09	0.09	0.101		
Constantan	2687	557	0.092	13.3	72.3	9.8	1.1				
(55% Cu, 45% Ni)						0.06	0.09				
Germanium	2180	334.6	0.08	34.6	373.5	134	56	25	15.7	11.4	10.05
						0.045	0.069	0.08	0.083	0.085	0.089
Gold	2180	334.6	0.08	34.6	373.5	189	186.6	179.7	172.2	164.09	156
	2405	1205	0.03	183.2	1367	0.026	0.029	0.031	0.032	0.033	0.034
Iridium	4896	1404.6	0.031	85	541.4	99.4	88.4	83.2	79.7	76.3	72.8
						0.021	0.029	0.031	0.032	0.034	0.036
Iron:	3258	491.3	0.106	46.4	248.6	77.4	54.3	40.2	31.6	25.01	19
Pure						0.051	0.091	0.117	0.137	0.162	0.232
Armco		491.3	0.106	42	222.8	55.2	46.6	38	30.7	24.4	18.7
(99.75% pure)						0.051	0.091	0.117	0.137	0.162	0.233
Carbon steels:		490.3	0.103	35	190.6			32.8	27.7	22.7	17.4
Plain carbon (Mn ≤ 1%, Si ≤ 0.1%)								0.116	0.113	0.163	0.279
AISI 1010		489	0.103	37	202.4			33.9	28.2	22.7	18
								0.116	0.133	0.163	0.278
Carbon–silicon		488	0.106	30	160.4			28.8	25.4	21.6	17
(Mn ≤ 1%, 0.1% < Si ≤ 0.6%)								0.119	0.139	0.166	0.231

Properties of solid metals (*Continued*)

Composition	Melting point, R	ρ lbm/ft³	Properties at 540 R c_p Btu/ lbm·R	k Btu/ h·ft·R	$\alpha \times 10^6$ ft²/s	Properties at various temperatures (R), k(Btu/h·ft·R)/c_p(Btu/lbm·R) 180	360	720	1080	1440	1800	2160
Carbon–manganese–silicon (1% < Mn ≤ 1.65% 0.1% < Si ≤ 0.6%)			508	0.104	23.7	125			24.4 0.116	23 0.133	20.2 0.163	16 0.260
Chromium (low) steels: $\frac{1}{2}$Cr–$\frac{1}{4}$Mo–Si (0.18% C, 0.65% Cr, 0.23% Mo, 0.6% Si)			488.3	0.106	21.8	117.4			22 0.117	21.2 0.137	19.3 0.164	15.6 0.231
1 Cr–$\frac{1}{2}$Mo (0.16% C, 1% Cr, 0.54% Mo, 0.39% Si)			490.6	0.106	24.5	131.3			24.3 0.117	22.6 0.137	20 0.164	15.8 0.231
1 Cr–V (0.2% C, 1.02% Cr, 0.15% V)			489.2	0.106	28.3	151.8			27.0 0.117	24.3 0.137	21 0.164	16.3 0.231
Stainless steels: AISI 302			503	0.114	8.7	42			10 0.122	11.6 0.133	13.2 0.140	14.7 0.144
AISI 304	3006		493.2	0.114	8.6	42.5	5.31 0.064	7.3 0.096	9.6 0.123	11.5 0.133	13 0.139	14.7 0.145
AISI 316			514.3	0.111	7.8	37.5			8.8 0.12	10.6 0.131	12.3 0.137	14 0.143
AISI 347			498	0.114	8.2	40			9.1 0.122	1.1 0.133	12.7 0.14	14.3 0.144
Lead	1082		708	0.03	20.4	259.4	23 0.028	21.2 0.029	19.7 0.031	18.1 0.034		
Magnesium	1661		109	0.245	90.2	943	87.9 0.155	91.9 0.223	88.4 0.256	86.0 0.279	84.4 0.302	
Molybdenum	5209		639.3	0.06	79.7	578	1034 0.033	82.6 0.053	77.4 0.062	72.8 0.065	68.2 0.068	64.7 0.070
Nickel: Pure	3110		555.6	0.106	52.4	247.6	94.8 0.055	61.8 0.091	46.3 0.115	37.9 0.141	39 0.126	41.4 0.134
Nichrome (80% Ni, 20% Cr)	3010		524.4	0.1	6.9	36.6			8.0 0.114	9.3 0.125	12.2 0.130	
Inconel X-750 (73% Ni, 15% Cr, 6.7% Fe)	2997		531.3	0.104	6.8	33.4	5 —	5.9 0.088	7.8 0.112	9.8 0.121	11.8 0.13	13.9 0.149
Niobium	4934		535	0.063	31	254	31.9 0.044	30.4 0.059	32 0.065	33.6 0.067	35.4 0.069	32.2 0.071
Palladium	3289		750.4	0.058	41.5	263.7	44.2 0.04	41.4 0.054	42.5 0.059	46 0.062	50 0.064	54.4 0.067
Platinum: Pure	3681		1339	0.031	41.4	270	44.7 0.024	42 0.03	41.5 0.032	42.3 0.034	43.7 0.035	45.5 0.036
Alloy 60Pt–40Rh (60% Pt, 40% Rh)	3240		1038.2	0.038	27.2	187.3			30 —	34 —	37.5 —	40 —
Rbenium	6215		1317.2	0.032	27.7	180	34 0.023	30 0.03	26.6 0.033	25.5 0.034	25.4 0.036	25.8 0.037
Rhodium	4025		777.2	0.058	86.7	534	107.5 0.035	89 0.052	84.3 0.06	78.5 0.065	73.4 0.069	70 0.074
Silicon	3033		145.5	0.17	85.5	960.2	510.8 0.061	152.5 0.132	57.2 0.189	35.8 0.207	24.4 0.218	18.0 0.226

TABLE A-3E

Properties of solid metals (*Concluded*)

Composition	Melting point, R	ρ lbm/ft³	c_p Btu/ lbm·R	k Btu/ h·ft·R	$\alpha \times 10^6$ ft²/s	Properties at various temperatures (R), k(Btu/h·ft·R)/c_p(Btu/lbm·R) 180	360	720	1080	1440	1800	2160
Silver	2223	656	0.056	248	1873	257	248.4	245.5	238	228.8	219	
						0.044	0.053	0.057	0.059	0.062	0.066	
Tantalum	5884	1036.3	0.033	33.2	266	34.2	33.2	33.4	34	34.3	34.8	
						0.026	0.031	0.034	0.035	0.036	0.036	
Thorium	3641	730.4	0.028	31.2	420.9	34.6	31.5	31.4	32.2	32.9	32.9	
						0.024	0.027	0.029	0.032	0.035	0.037	
Tin	909	456.3	0.054	38.5	431.6	49.2	42.4	35.9				
						0.044	0.051	0.058				
Titanium	3515	281	0.013	12.7	100.3	17.6	14.2	11.8	11.2	11.4	12	
						0.071	0.111	0.131	0.141	0.151	0.161	
Tungsten	6588	1204.9	0.031	100.5	735.2	120.2	107.5	92	79.2	72.2	68.2	
						0.020	0.029	0.032	0.033	0.034	0.035	
Uranium	2531	1190.5	0.027	16	134.5	12.5	14.5	17.1	19.6	22.4	25.4	
						0.022	0.026	0.029	0.035	0.042	0.043	
Vanadium	3946	381	0.117	17.7	110.9	20.7	18	18	19.3	20.6	22.0	
						0.061	0.102	0.123	0.128	0.134	0.142	
Zinc	1247	445.7	0.093	67	450	67.6	68.2	64.1	59.5			
						0.07	0.087	0.096	0.104			
Zirconium	3825	410.2	0.067	13.1	133.5	19.2	14.6	12.5	12	12.5	13.7	
						0.049	0.063	0.072	0.77	0.082	0.087	

Source: Tables A-3E and A-4E are obtained from the respective tables in SI units in Appendix 1 using proper conversion factors.

TABLE A-4E

Properties of solid nonmetals

Composition	Melting point, R	ρ lbm/ft³	c_p Btu/ lbm·R	k Btu/ h·ft·R	$\alpha \times 10^6$ ft²/s	Properties at various temperatures (R), k(Btu/h·ft·R)/c_p(Btu/lbm·R) 180	360	720	1080	1440	1800
Aluminum oxide,	4181	247.8	0.182	26.6	162.5	260	47.4	18.7	11	7.5	6
sapphire						—	—	0.224	0.265	0.281	0.293
Aluminum oxide,	4181	247.8	0.182	20.8	128	76.8	31.7	15.3	9.3	6	4.5
polycrystalline						—	—	0.244	0.265	0.281	0.293
Beryllium oxide	4905	187.3	0.246	157.2	947.3			113.2	64.2	40.4	27.2
								0.322	0.40	0.44	0.459
Boron	4631	156	0.264	16	107.5	109.8	30.3	10.8	6.5	4.6	3.6
								0.355	0.445	0.509	0.561
Boron fiber epoxy	1062	130									
(30% vol.) composite											
k, ∥ to fibers				1.3		1.2	1.3	1.31			
k, ⊥ to fibers				0.34		0.21	0.28	0.34			
c_p			0.268			0.086	0.18	0.34			
Carbon	2700	121.7	—	0.92	—	0.38	0.68	1.09	1.26	1.36	1.46
Amorphous											
Diamond,	—	219	0.121	1329	—	5778	2311.2	889.8			
type IIa insulator								0.005	0.046	0.203	
Graphite, pyrolytic	4091	138									
k, ∥ to layers				1126.7		2871.6	1866.3	803.2	515.4	385.4	308.5
k, ⊥ to layers				3.3		9.7	5.3	2.4	1.5	1.16	0.92
c_p			0.169			0.032	0.098	0.236	0.335	0.394	0.428
Graphite fiber epoxy (25% vol.)	810	87.4									
composite											
k, ∥ heat flow to fibers				6.4		3.3	5.0	7.5			
k, heat flow ⊥ to fibers				0.5	5	0.4	0.63				
c_p			0.223			0.08	0.153	0.29			
Pyroceram, Corning 9606	2921	162.3	0.193	2.3	20.3	3.0	2.3	2.1	1.9	1.7	1.7
Silicon carbide	5580	197.3	0.161	283.1	2475.7			—	—	—	50.3
								0.210	0.25	0.27	0.285
Silicon dioxide, crystalline (quartz)	3389	165.4									
k, ∥ to c-axis				6		22.5	9.5	4.4	2.9	2.4	
k, ⊥ to c-axis				3.6		12.0	5.9	2.7	2	1.8	
c_p			0.177			—	—	0.211	0.256	0.298	
Silicon dioxide,	3389	138.6	0.177	0.79	9	0.4	0.65	0.87	1.01	1.25	1.65
polycrystalline (fused silica)						—	—	0.216	0.248	0.264	0.276
Silicon nitride	3911	150	0.165	9.2	104	—	—	8.0	6.5	5.7	5.0
						—	0.138	0.185	0.223	0.253	0.275
Sulfur	706	130	0.169	0.1	1.51	0.095	0.1				
						0.962	0.144				
Thorium dioxide	6431	568.7	0.561	7.5	65.7			5.9	3.8	2.7	2.12
								0.609	0.654	0.680	0.704
Titanium dioxide,	3840	259.5	0.170	4.9	30.1			4.0	2.9	2.3	2
polycrystalline								0.192	0.210	0.217	0.222

TABLE A-5E

Properties of building materials
(at a mean temperature of 75°F)

Material	Thickness, L in.	Density, ρ lbm/ft³	Thermal conductivity, k Btu-in./ft. · °F	Specific heat, C_p Btu/lbm · °F	R-value (for listed thickness, L/k), °F · h · ft²/Btu
Building Boards					
Asbestos–cement board	$\frac{1}{4}$ in.	120	—	0.24	0.06
Gypsum of plaster board	$\frac{3}{8}$ in.	50	—	0.26	0.32
	$\frac{1}{2}$ in.	50	—	—	0.45
Plywood (Douglas fir)	—	34	0.80	0.29	—
	$\frac{1}{4}$ in.	34	—	0.29	0.31
	$\frac{3}{8}$ in.	34	—	0.29	0.47
	$\frac{1}{2}$ in.	34	—	0.29	0.62
	$\frac{3}{4}$ in.	34	—	0.29	0.93
Insulated board and sheating	$\frac{1}{2}$ in.	18	—	0.31	1.32
(regular density)	$\frac{25}{32}$ in.	18	—	0.31	2.06
Hardboard (high density, standard tempered)	—	63	1.00	0.32	—
Particle board:					
Medium density	—	50	0.94	0.31	—
Underlayment	$\frac{5}{8}$ in.	40	—	0.29	0.82
Wood subfloor	$\frac{3}{4}$ in.	—	—	0.33	0.94
Building Membranes					
Vapor-permeable felt	—	—	—	—	0.06
Vapor-seal (2 layers of mopped 0.73 kg/m² felt)	—	—	—	—	0.12
Flooring Materials					
Carpet and fibrous pad	—	—	—	0.34	2.08
Carpet and rubber pad	—	—	—	0.33	1.23
Tile (asphalt, linoleum, vinyl)	—	—	—	0.30	0.05
Masonry Materials					
Masonry units:					
Brick, common		120	5.0	—	—
Brick, face		130	9.0	—	—
Brick, fire clay		150	9.3	—	—
		120	6.2	0.19	—
		70	2.8	—	—
Concrete blocks (3 oval cores, sand and gravel aggregate)	4 in.	—	5.34	—	0.71
	8 in.	—	6.94	—	1.11
	12 in.	—	9.02	—	1.28
Concretes:					
Lightweight aggregates		120	5.2	—	—
(including expanded shale,		100	3.6	0.2	—
clay, or slate; expanded slags;		80	2.5	0.2	—
cinders; pumice; and scoria)		60	1.7	—	—
		40	1.15	—	—

TABLE A-5E

Properties of building materials (*Concluded*)
(at a mean temperature of 75°F)

Material	Thickness, L in.	Density, ρ lbm/ft³	Thermal conductivity, k Btu-in./ft·°F	Specific heat, C_p Btu/lbm·°F	R-value (for listed thickness, L/k), °F·h·ft²/Btu
Cement/lime, mortar, and stucco		120	9.7	—	—
		80	4.5	—	—
Stucco		116	5.0	—	—
Roofing					
Asbestos-cement shingles		120	—	0.24	0.21
Asphalt roll roofing		70	—	0.36	0.15
Asphalt shingles		70	—	0.30	0.44
Built-in roofing	$\frac{3}{8}$ in.	70	—	0.35	0.33
Slate	$\frac{1}{2}$ in.	—	—	0.30	0.05
Wood shingles (plain and plastic film faced)		—	—	0.31	0.94
Plastering Materials					
Cement plaster, sand aggregate	$\frac{3}{4}$ in.	116	5.0	0.20	0.15
Gypsum plaster:					
Lightweight aggregate	$\frac{1}{2}$ in.	45	—	—	0.32
Sand aggregate	$\frac{1}{2}$ in.	105	5.6	0.20	0.09
Perlite aggregate	—	45	1.5	0.32	—
Siding Material (on flat surfaces)					
Asbestos-cement shingles	—	120	—	—	0.21
Hardboard siding	$\frac{7}{16}$ in.	—	—	0.28	0.67
Wood (drop) siding	1 in.	—	—	0.31	0.79
Wood (plywood) siding, lapped	$\frac{3}{8}$ in.	—	—	0.29	0.59
Aluminum or steel siding (over sheeting):					
Hollow backed	$\frac{3}{8}$ in.	—	—	0.29	0.61
Insulating-board backed	$\frac{3}{8}$ in.	—	—	0.32	1.82
Architectural glass	—	158	6.9	0.21	0.10
Woods					
Hardwoods (maple, oak, etc.)	—	45	1.10	0.30	—
Softwoods (fir, pine, etc.)	—	32	0.80	0.33	—
Metals					
Aluminum (1100)	—	171	1536	0.214	—
Steel, mild	—	489	314	0.120	—
Steel, Stainless	—	494	108	0.109	—

Source: Tables A-5E and A-6E are adapted from ASHRAE, *Handbook of Fundamentals* (Atlanta, GA: American Society of Heating, Refrigerating, and Air-Conditioning Engineers, 1993), Chap. 22, Table 4. Used with permission.

TABLE A-6E

Properties of insulating materials
(at a mean temperature of 75°F)

Material	Thickness, L inches	Density, ρ lbm/ft³	Thermal conductivity, k Btu-in./ft · °F	Specific heat, C_p Btu/lbm · °F	R-value (for listed thickness, L/k) °F · h ft²/Btu
Blanket and Batt					
Mineral fiber (fibrous form	~2 to $2\frac{3}{4}$ in.	0.3–2.0	—	0.17–0.23	7
processed from rock, slag,	~3 to $3\frac{1}{2}$ in.	0.3–2.0	—	0.17–0.23	11
or glass)	~$5\frac{1}{4}$ to $6\frac{1}{2}$ in.	0.3–2.0	—	0.17–0.23	19
Board and Slab					
Cellular glass		8.5	0.38	0.24	—
Glass fiber (organic bonded)		4–9	0.25	0.23	—
Expanded polystyrene (molded beads)		1.0	0.28	0.29	—
Expanded polyurethane (R-11 expanded)		1.5	0.16	0.38	—
Expanded perlite (organic bonded)		1.0	0.36	0.30	—
Expanded rubber (rigid)		4.5	0.22	0.40	—
Mineral fiber with resin binder		15	0.29	0.17	—
Cork		7.5	0.27	0.43	—
Sprayed or Formed in Place					
Polyurethane foam		1.5–2.5	0.16–0.18	—	—
Glass fiber		3.5–4.5	0.26–0.27	—	—
Urethane, two-part mixture (rigid foam)		4.4	0.18	0.25	—
Mineral wool granules with asbestos/inorganic binders (sprayed)		12	0.32	—	—
Loose Fill					
Mineral fiber (rock, slag,	~3.75 to 5 in.	0.6–0.20	—	0.17	11
or glass)	~6.5 to 8.75 in.	0.6–0.20	—	0.17	19
	~7.5 to 10 in.	—	—	0.17	22
	~7.25 in.	—	—	0.17	30
Silica aerogel		7.6	0.17	—	—
Vermiculite (expanded)		7–8	0.47	—	—
Perlite, expanded		2–4.1	0.27–0.31	—	—
Sawdust or shavings		8–15	0.45	—	—
Cellulosic insulation (milled paper or wood pulp)		0.3–3.2	0.27–0.32	—	—
Cork, granulated		10	0.31	—	—
Roof Insulation					
Cellular glass	—	9	0.4	0.24	—
Preformed, for use above deck	$\frac{1}{2}$ in.	—	—	0.24	1.39
	1 in.	—	—	0.50	2.78
	2 in.	—	—	0.94	5.56
Reflective Insulation					
Silica powder (evaculated)		10	0.0118	—	—
Aluminum foil separating fluffy glass mats; 10–12 layers (evacuated); for cryogenic applications (270R)		2.5	0.0011	—	—
Aluminum foil and glass paper laminate; 75–150 layers (evacuated); for cryogenic applications (270R)		7.5	0.00012	—	—

TABLE A-7E

Properties of common foods
(a) Specific heats and freezing-point properties

Food	Water content,[a] % (mass)	Freezing point,[a] °F	Specific heat,[b] Btu/lbm·°F Above freezing	Specific heat,[b] Btu/lbm·°F Below freezing	Latent heat of fusion,[c] Btu/lb
Vegetables					
Artichokes	84	30	0.873	0.453	121
Asparagus	93	31	0.945	0.481	134
Beans, snap	89	31	0.913	0.468	128
Broccoli	90	31	0.921	0.471	129
Cabbage	92	30	0.937	0.478	132
Carrots	88	29	0.905	0.465	126
Cauliflower	92	31	0.937	0.478	132
Celery	94	31	0.953	0.484	135
Corn, sweet	74	31	0.793	0.423	106
Cucumbers	96	31	0.969	0.490	138
Eggplant	93	31	0.945	0.481	134
Horseradish	75	29	0.801	0.426	108
Leeks	85	31	0.881	0.456	122
Lettuce	95	32	0.961	0.487	136
Mushrooms	91	30	0.929	0.474	131
Okra	90	29	0.921	0.471	129
Onions, green	89	30	0.913	0.468	128
Onions, dry	88	31	0.905	0.465	126
Parsley	85	30	0.881	0.456	122
Peas, green	74	31	0.793	0.423	106
Peppers, sweet	92	31	0.937	0.478	132
Potatoes	78	31	0.825	0.435	112
Pumpkins	91	31	0.929	0.474	131
Spinach	93	31	0.945	0.481	134
Tomatos, ripe	94	31	0.953	0.484	135
Turnips	92	30	0.937	0.478	132
Fruits					
Apples	84	30	0.873	0.453	121
Apricots	85	30	0.881	0.456	122
Avocados	65	31	0.721	0.396	93
Bananas	75	31	0.801	0.426	108
Blueberries	82	29	0.857	0.447	118
Cantaloupes	92	30	0.937	0.478	132
Cherries, sour	84	29	0.873	0.453	121
Cherries, sweet	80	29	0.841	0.441	115
Figs, dried	23	—	—	0.270	33
Figs, fresh	78	28	0.825	0.435	112
Grapefruit	89	30	0.913	0.468	128
Grapes	82	29	0.857	0.447	118
Lemons	89	29	0.913	0.468	128
Olives	75	29	0.801	0.426	108
Oranges	87	31	0.897	0.462	125
Peaches	89	30	0.913	0.468	128
Pears	83	29	0.865	0.450	119
Pineapples	85	30	0.881	0.456	122
Plums	86	31	0.889	0.459	124
Quinces	85	28	0.881	0.456	122
Raisins	18	—	—	0.255	26
Strawberries	90	31	0.921	0.471	129
Tangerines	87	30	0.897	0.462	125
Watermelon	93	31	0.945	0.481	134
Fish/Seafood					
Cod, whole	78	28	0.825	0.435	112
Halibut, whole	75	28	0.801	0.426	108
Lobster	79	28	0.833	0.438	113
Mackerel	57	28	0.657	0.372	82
Salmon, whole	64	28	0.713	0.393	92
Shrimp	83	28	0.865	0.450	119
Meats					
Beef carcass	49	29	0.593	0.348	70
Liver	70	29	0.761	0.411	101
Round, beef	67	—	0.737	0.402	96
Sirloin, beef	56	—	0.649	0.369	80
Chicken	74	27	0.793	0.423	106
Lamb leg	65	—	0.721	0.396	93
Pork carcass	37	—	0.497	0.312	53
Ham	56	29	0.649	0.369	80
Pork sausage	38	—	0.505	0.315	55
Turkey	64	—	0.713	0.393	92
Other					
Almonds	5	—	—	0.216	7
Butter	16	—	—	0.249	23
Cheese, cheddar	37	9	0.497	0.312	53
Cheese, Swiss	39	14	0.513	0.318	56
Chocolate, milk	1	—	—	0.204	1
Eggs, whole	74	31	0.793	0.423	106
Honey	17	—	—	0.252	24
Ice cream	63	22	0.705	0.390	90
Milk, whole	88	31	0.905	0.465	126
Peanuts	6	—	—	0.219	9
Peanuts, roasted	2	—	—	0.207	3
Pecans	3	—	—	0.210	4
Walnuts	4	—	—	0.213	6

Source: [a]Water content and freezing point data are from ASHRAE, *Handbook of Fundamentals*, I-P version (Atlanta, GA: American Society of Heating, Refrigerating and Air-Conditioning Engineers, Inc., 1993), Chap. 30, Table 1. Used with permission. Freezing point is the temperature at which freezing starts for fruits and vegetables, and the average freezing temperature for other foods.

[b]Specific heat data are based on the specific heat values of water and ice at 32°F and are determined from Siebel's formulas: $C_{p, \text{fresh}} = 0.800 \times$ (Water content) + 0.200, above freezing, and $C_{p, \text{frozen}} = 0.300 \times$ (Water content) + 0.200, below freezing.

[c]The latent heat of fusion is determined by multiplying the heat of fusion of water (143 Btu/lbm) by the water content of the food.

TABLE A-7E

Properties of common foods (*Concluded*)
(*b*) Other properties

Food	Water content, % (mass)	Temperature, T°F	Density, ρ lbm/ft^3	Thermal conductivity, k Btu/h·ft·°F	Thermal diffusivity, α ft^2/s	Specific heat, C_p Btu/lbm·°F
Fruits/Vegetables						
Apple juice	87	68	62.4	0.323	1.51×10^{-6}	0.922
Apples	85	32–86	52.4	0.242	1.47×10^{-6}	0.910
Apples, dried	41.6	73	53.4	0.127	1.03×10^{-6}	0.650
Apricots, dried	43.6	73	82.4	0.217	1.22×10^{-6}	0.662
Bananas, fresh	76	41	61.2	0.278	1.51×10^{-6}	0.856
Broccoli	—	21	35.0	0.223	—	—
Cherries, fresh	92	32–86	65.5	0.315	1.42×10^{-6}	0.952
Figs	40.4	73	77.5	0.179	1.03×10^{-6}	0.642
Grape juice	89	68	62.4	0.328	1.51×10^{-6}	0.934
Peaches	36–90	2–32	59.9	0.304	1.51×10^{-6}	0.934
Plums	—	3	38.1	0.143	—	—
Potatoes	32–158	0–70	65.7	0.288	1.40×10^{-6}	0.868
Raisins	32	73	86.2	0.217	1.18×10^{-6}	0.592
Meats						
Beef, ground	67	43	59.3	0.235	1.40×10^{-6}	0.802
Beef, lean	74	37	68.0	0.272	1.40×10^{-6}	0.844
Beef fat	0	95	50.5	0.110	—	—
Beef liver	72	95	—	0.259	—	0.832
Cat food	39.7	73	71.2	0.188	1.18×10^{-6}	0.638
Chicken breast	75	32	65.5	0.275	1.40×10^{-6}	0.850
Dog food	30.6	73	77.4	0.184	1.18×10^{-6}	0.584
Fish, cod	81	37	73.7	0.309	1.29×10^{-6}	0.886
Fish, salmon	67	37	—	0.307	—	0.802
Ham	71.8	72	64.3	0.277	1.51×10^{-6}	0.831
Lamb	72	72	64.3	0.263	1.40×10^{-6}	0.832
Pork, lean	72	39	64.3	0.263	1.40×10^{-6}	0.832
Turkey breast	74	37	65.5	0.287	1.40×10^{-6}	0.844
Veal	75	72	66.2	0.272	1.40×10^{-6}	0.850
Other						
Butter	16	39	—	0.114	—	0.496
Chocolate cake	31.9	73	21.2	0.061	1.29×10^{-6}	0.591
Margarine	16	40	62.4	0.135	1.18×10^{-6}	0.496
Milk, skimmed	91	72	—	0.327	—	0.946
Milk, whole	88	82	—	0.335	—	0.928
Olive oil	0	90	56.8	0.097	—	—
Peanut oil	0	39	57.4	0.097	—	—
Water	100	0	62.4	0.329	1.51×10^{-6}	1.000
	100	30	59.6	0.357	1.61×10^{-6}	1.000
White cake	32.3	73	28.1	0.047	1.08×10^{-6}	0.594

Source: Data obtained primarily from ASHRAE, *Handbook of Fundamentals*, I-P version (Atlanta, GA: American Society of Heating, Refrigerating and Air-Conditioning Engineers, Inc., 1993), Chap. 30, Tables 7 and 9. Used with permission.

Most specific heats are calculated from $C_p = 0.4 + 0.6 \times$ (Water content), which is a good approximation in the temperature range of 40 to 90°F. Most thermal diffusivities are calculated from $\alpha = k/\rho C_p$. Property values given above are valid for the specified water content.

TABLE A-8E

Properties of miscellaneous materials
(Values are at 540 R unless indicated otherwise)

Material	Density, ρ lbm/ft³	Thermal conductivity, k, Btu/h·ft·R	Specific heat, C_p Btu/lbm·R	Material	Density, ρ lbm/ft³	Thermal conductivity, k, Btu/h·ft·R	Specific heat, C_p Btu/lbm·R
Asphalt	132.0	0.036	0.220	Ice			
Bakelite	81.2	0.81	0.350	492 R	57.4	1.09	0.487
Brick, refractory				455 R	57.6	1.17	0.465
Chrome brick				311 R	57.9	2.02	0.349
851 R	187.9	1.33	0.199	Leather, sole	62.3	0.092	—
1481 R	—	1.44	—	Linoleum	33.4	0.047	—
2111 R	—	1.16	—		73.7	0.11	—
Fire clay, burnt				Mica	181.0	0.30	—
2880 R				Paper	58.1	0.10	0.320
1391 R	128.0	0.58	0.229	Plastics			
1931 R	—	0.64	—	Plexiglass	74.3	0.11	0.350
2471 R	—	0.64	—	Teflon			
Fire clay, burnt				540 R	137.3	0.20	0.251
3105 R				720 R	—	0.26	—
1391 R	145.1	0.75	0.229	Lexan	74.9	0.11	0.301
1931 R	—	0.81	—	Nylon	71.5	0.17	—
2471 R	—	0.81	—	Polypropylene	56.8	0.069	0.388
Fire clay brick				Polyester	87.1	0.087	0.279
860 R	165.1	0.58	0.229	PVC, vinyl	91.8	0.058	0.201
1660 R	—	0.87	—	Porcelain	143.6	0.87	—
2660 R	—	1.04	—	Rubber, natural	71.8	0.16	—
Magnesite				Rubber, vulcanized			
860 R	—	2.20	0.270	Soft	68.7	0.075	0.480
1660 R	—	1.62	—	Hard	74.3	0.092	—
2660 R	—	1.10	—	Sand	94.6	0.1–0.6	0.191
Chicken meat, white (74.4% water content)				Snow, fresh	6.24	0.35	—
				Snow, 492 R	31.2	1.27	—
356 R	—	0.92	—	Soil, dry	93.6	0.58	0.454
419 R	—	0.86	—	Soil, wet	118.6	1.16	0.525
455 R	—	0.78	—	Sugar	99.9	0.34	—
492 R	—	0.28	—	Tissue, human			
527 R	—	0.28	—	Skin	—	0.21	—
Clay, dry	96.8	0.54	—	Fat layer	—	0.12	—
Clay, wet	93.3	0.97	—	Muscle	—	0.24	—
Coal, anthracite	84.3	0.15	0.301	Vaseline	—	0.098	—
Concrete (stone mix)	143.6	0.81	0.210	Wood, cross-grain			
Cork	5.37	0.028	0.485	Balsa	8.74	0.032	—
Cotton	5.0	0.035	0.311	Fir	25.9	0.064	0.650
Fat	—	0.10	—	Oak	34.0	0.098	0.570
Glass				Yellow pine	40.0	0.087	0.670
Window	174.8	0.40	0.179	White pine	27.2	0.064	—
Pyrex	138.9	0.6–0.8	0.199	Wood, radial			
Crown	156.1	0.61	—	Oak	34.0	0.11	0.570
Lead	212.2	0.49	—	Fir	26.2	0.081	0.650
				Wool, ship	9.05	0.029	—

TABLE A-9E

Properties of saturated water

Temperature, T °F	Saturation pressure, P, psia	Density, ρ lbm/ft³ Liquid	Density, ρ lbm/ft³ Vapor	Enthalpy of vaporization, h_{fg} Btu/lbm	Specific heat, C_p Btu/lbm·°F Liquid	Specific heat, C_p Btu/lbm·°F Vapor	Thermal conductivity, k Btu/h·ft·°F Liquid	Thermal conductivity, k Btu/h·ft·°F Vapor	Dynamic viscosity, μ lbm/ft·h Liquid	Dynamic viscosity, μ lbm/ft·h Vapor	Prandtl number, Pr Liquid	Prandtl number, Pr Vapor	Volume expansion coefficient, β 1/R Liquid
32.02	0.0887	62.41	0.00030	1075	1.010	0.446	0.324	0.0099	4.336	0.0223	13.5	1.00	-0.038×10^{-3}
40	0.1217	62.42	0.00034	1071	1.004	0.447	0.329	0.0100	3.740	0.0226	11.4	1.01	0.003×10^{-3}
50	0.1780	62.41	0.00059	1065	1.000	0.448	0.335	0.0102	3.161	0.0229	9.44	1.01	0.047×10^{-3}
60	0.2563	62.36	0.00083	1060	0.999	0.449	0.341	0.0104	2.713	0.0232	7.95	1.00	0.080×10^{-3}
70	0.3632	62.30	0.00115	1054	0.999	0.450	0.347	0.0106	2.360	0.0236	6.79	1.00	0.115×10^{-3}
80	0.5073	62.22	0.00158	1048	0.999	0.451	0.352	0.0108	2.075	0.0240	5.89	1.00	0.145×10^{-3}
90	0.6988	62.12	0.00214	1043	0.999	0.453	0.358	0.0110	1.842	0.0244	5.14	1.00	0.174×10^{-3}
100	0.9503	62.00	0.00286	1037	0.999	0.454	0.363	0.0112	1.648	0.0248	4.54	1.01	0.200×10^{-3}
110	1.2763	61.86	0.00377	1031	0.999	0.456	0.367	0.0115	1.486	0.0252	4.05	1.00	0.224×10^{-3}
120	1.6945	61.71	0.00493	1026	0.999	0.458	0.371	0.0117	1.348	0.0256	3.63	1.00	0.246×10^{-3}
130	2.225	61.55	0.00636	1020	0.999	0.460	0.375	0.0120	1.230	0.0260	3.28	1.00	0.267×10^{-3}
140	2.892	61.38	0.00814	1014	0.999	0.463	0.378	0.0122	1.129	0.0264	2.98	1.00	0.287×10^{-3}
150	3.722	61.19	0.0103	1008	1.000	0.465	0.381	0.0125	1.040	0.0269	2.73	1.00	0.306×10^{-3}
160	4.745	60.99	0.0129	1002	1.000	0.468	0.384	0.0128	0.963	0.0273	2.51	1.00	0.325×10^{-3}
170	5.996	60.79	0.0161	996	1.001	0.472	0.386	0.0131	0.894	0.0278	2.90	1.00	0.346×10^{-3}
180	7.515	60.57	0.0199	990	1.002	0.475	0.388	0.0134	0.834	0.0282	2.15	1.00	0.367×10^{-3}
190	9.343	60.35	0.0244	984	1.004	0.479	0.390	0.0137	0.781	0.0287	2.01	1.00	0.382×10^{-3}
200	11.53	60.12	0.0297	978	1.005	0.483	0.391	0.0141	0.733	0.0291	1.88	1.00	0.395×10^{-3}

210	14.125	59.87	0.0359	972	1.007	0.487	0.392	0.0144	0.690	0.0296	1.77	1.00	0.412×10^{-3}
212	14.698	59.82	0.0373	970	1.007	0.488	0.392	0.0145	0.682	0.0297	1.75	1.00	0.417×10^{-3}
220	17.19	59.62	0.0432	965	1.009	0.492	0.393	0.0148	0.651	0.0300	1.67	1.00	0.429×10^{-3}
230	20.78	59.36	0.0516	959	1.011	0.497	0.394	0.0152	0.616	0.0305	1.58	1.00	0.443×10^{-3}
240	24.97	59.09	0.0612	952	1.013	0.503	0.394	0.0156	0.585	0.0310	1.50	1.00	0.462×10^{-3}
250	29.82	58.82	0.0723	946	1.015	0.509	0.395	0.0160	0.556	0.0310	1.43	1.00	0.480×10^{-3}
260	35.42	58.53	0.0850	939	1.018	0.516	0.395	0.0164	0.530	0.0319	1.37	1.00	0.497×10^{-3}
270	41.85	58.24	0.0993	932	1.020	0.523	0.395	0.0168	0.506	0.0324	1.31	1.01	0.514×10^{-3}
280	49.18	57.94	0.1156	925	1.023	0.530	0.395	0.0172	0.484	0.0328	1.25	1.01	0.532×10^{-3}
290	57.53	57.63	0.3390	918	1.026	0.538	0.395	0.0177	0.464	0.0333	1.21	1.01	0.549×10^{-3}
300	66.98	57.31	0.1545	910	1.029	0.547	0.394	0.0182	0.445	0.0338	1.16	1.02	0.566×10^{-3}
320	89.60	56.65	0.2033	895	1.036	0.567	0.393	0.0191	0.412	0.0347	1.09	1.03	0.636×10^{-3}
340	117.93	55.95	0.2637	880	1.044	0.590	0.391	0.0202	0.383	0.0356	1.02	1.04	0.656×10^{-3}
360	152.92	55.22	0.3377	863	1.054	0.617	0.389	0.0213	0.359	0.0365	0.973	1.06	0.681×10^{-3}
380	195.60	54.46	0.4275	845	1.065	0.647	0.385	0.0224	0.337	0.0375	0.932	1.08	0.720×10^{-3}
400	241.1	53.65	0.5359	827	1.078	0.683	0.382	0.0237	0.318	0.0384	0.893	1.11	0.771×10^{-3}
450	422.1	51.46	0.9082	775	1.121	0.799	0.370	0.0271	0.278	0.0407	0.842	1.20	0.912×10^{-3}
500	680.0	48.95	1.479	715	1.188	0.972	0.352	0.0312	0.246	0.0432	0.830	1.35	1.111×10^{-3}
550	1046.7	45.96	4.268	641	1.298	1.247	0.329	0.0368	0.219	0.0461	0.864	1.56	1.445×10^{-3}
600	1541	42.32	3.736	550	1.509	1.759	0.299	0.0461	0.194	0.0497	0.979	1.90	1.885×10^{-3}
650	2210	37.31	6.152	422	2.086	3.103	0.267	0.0677	0.167	0.0555	1.30	2.54	—
700	3090	27.28	13.44	168	13.80	25.90	0.254	0.1964	0.123	0.0736	6.68	9.71	—
705.44	3204	19.79	19.79	0	∞	∞	∞	∞	0.104	0.1043	—	—	—

Source: Viscosity and thermal conductivity data are from J. V. Sengers and J. T. R. Watson, *Journal of Physical and Chemical Reference Data* 15 (1986), pp. 1291–1322. Other data are obtained from various sources or calculated.

Note: Kinematic viscosity ν and thermal diffusivity α can be calculated from their definitions, $\nu = \mu/\rho$ and $\alpha = k/\rho C_p = \nu/Pr$. The temperatures 32.02°F, 212°F, and 705.44°F are the triple-, boiling-, and critical-point temperatures of water, respectively. All properties listed above (except the vapor density) can be used at any pressure with negligible error except at temperatures near the critical-point value.

TABLE A-10E

Properties of liquids

Tempera-ture, T °F	Density, ρ lbm/ft³	Specific heat, C_p Btu/lbm ·°F	Thermal conduc-tivity, k Btu/h·ft ·°F	Dynamic viscosity, μ lbm/ft·s	Kinematic viscosity, ν ft²/s	Thermal diffusiv-ity, α ft²/h	Volume expansiv-ity, β 1/R	Prandtl number, Pr
				Ammonia				
−20	42.4	1.07	0.317	17.6×10^{-5}	0.417×10^{-5}	6.94×10^{-3}		2.15
0	41.6	1.08	0.316	17.1×10^{-5}	0.410×10^{-5}	7.04×10^{-3}		2.09
10	40.8	1.09	0.314	16.6×10^{-5}	0.407×10^{-5}	7.08×10^{-3}		2.07
32	40.0	1.11	0.312	16.1×10^{-5}	0.402×10^{-5}	7.03×10^{-3}	1.2×10^{-3}	2.05
50	39.1	1.13	0.307	15.5×10^{-5}	0.396×10^{-5}	6.95×10^{-3}	1.3×10^{-3}	2.04
80	37.2	1.17	0.293	14.5×10^{-5}	0.386×10^{-5}	6.73×10^{-3}		2.01
120	35.2	1.22	0.275	13.0×10^{-5}	0.355×10^{-5}	6.40×10^{-3}		1.99
				Benzene				
60	55.1	0.40	0.093	46.0×10^{-5}	0.835×10^{-5}	4.22×10^{-3}	0.60×10^{-3}	7.2
80	54.6	0.42	0.092	39.6×10^{-5}	0.725×10^{-5}	4.01×10^{-3}		6.5
100	54.0	0.44	0.087	35.1×10^{-5}	0.650×10^{-5}	3.53×10^{-3}		5.1
150	53.5	0.46		26.0×10^{-5}	0.480×10^{-5}			4.5
200				20.3×10^{-5}				4.0
				n-Butyl alcohol				
60	50.5	0.55	0.097	226×10^{-5}	4.48×10^{-5}	3.49×10^{-3}		46.6
100	49.7	0.61	0.096	129×10^{-5}	2.60×10^{-5}	3.16×10^{-3}	0.45×10^{-3}	29.5
150	48.5	0.68	0.095	67.5×10^{-5}	1.39×10^{-5}	2.88×10^{-3}	0.48×10^{-3}	17.4
200	47.2	0.77	0.094	38.6×10^{-5}	0.815×10^{-5}	2.58×10^{-3}		11.3
				Glycerin				
50	79.3	0.554	0.165	2.56	0.0323	3.76×10^{-3}		31,000
70	78.9	0.570	0.165	1.0	0.0127	3.67×10^{-3}	0.28×10^{-3}	12,500
85	78.5	0.584	0.164	0.424	0.0054	3.58×10^{-3}	0.30×10^{-3}	5,400
100	78.2	0.600	0.163	0.188	0.0024	3.45×10^{-3}		2,500
120	77.7	0.617		0.124	0.0016			1,600
				Light oil				
60	57.0	0.43	0.077	5820×10^{-5}	102×10^{-5}	3.14×10^{-3}	0.38×10^{-3}	1170
80	56.8	0.44	0.077	2780×10^{-5}	49×10^{-5}	3.09×10^{-3}	0.38×10^{-3}	570
100	56.0	0.46	0.076	1530×10^{-5}	27.4×10^{-5}	2.95×10^{-3}	0.39×10^{-3}	340
150	54.3	0.48	0.075	530×10^{-5}	9.8×10^{-5}	2.88×10^{-3}	0.40×10^{-3}	122
200	54.0	0.51	0.074	250×10^{-5}	4.6×10^{-5}	2.69×10^{-3}	0.42×10^{-3}	62
250	53.0	0.52	0.074	139×10^{-5}	2.6×10^{-5}	2.67×10^{-3}	0.44×10^{-3}	35
300	51.8	0.54	0.073	830×10^{-5}	1.6×10^{-5}	2.62×10^{-3}	0.45×10^{-3}	22
				Mercury				
50	847	0.033	4.7	1.07×10^{-3}	1.2×10^{-6}	0.17	1.01×10^{-4}	0.027
200	834	0.033	6.0	0.84×10^{-3}	1.0×10^{-6}	0.22	1.01×10^{-6}	0.016
300	826	0.033	6.7	0.74×10^{-3}	0.90×10^{-6}	0.25	1.01×10^{-6}	0.012
400	817	0.032	7.2	0.67×10^{-3}	0.82×10^{-6}	0.27	1.01×10^{-6}	0.011
600	802	0.032	8.1	0.58×10^{-3}	0.72×10^{-6}	0.31	1.03×10^{-6}	0.0084
				Sodium				
200	58.0	0.33	49.8	0.47×10^{-3}	8.1×10^{-6}	2.6		0.011
400	56.3	0.32	46.4	0.29×10^{-3}	5.1×10^{-6}	2.6		0.0072
700	53.7	0.31	41.8	0.19×10^{-3}	3.5×10^{-6}	2.5		0.0050
1000	51.2	0.30	37.8	0.14×10^{-3}	2.7×10^{-6}	2.4		0.0040
1300	48.6	0.30	34.5	0.12×10^{-3}	2.5×10^{-6}	2.4		0.0038

Source: F. Kreith, *Principles of Heat Transfer,* 3d ed., Addison-Wesley Educational Publishers, Inc., New York, 1973. Originally compiled from various sources. Reprinted by permission of Addison-Wesley Educational Publishers, Inc.

TABLE A-11E

Properties of gases at 1 atm pressure*

Tempera-ture, $T\,°F$	Density, ρ lbm/ft³	Specific heat, C_p Btu/lbm $·°F$	Thermal conduc-tivity, k Btu/h·ft $·°F$	Dynamic viscosity, μ lbm/ft·s	Kinematic viscosity, ν ft²/s	Thermal diffusiv-ity, α ft²/h	Volume expansiv-ity, β 1/R	Prandtl number, Pr
				Air				
0	0.086	0.239	0.0133	1.110×10^{-5}	0.130×10^{-3}	0.646	2.18×10^{-3}	0.73
32	0.081	0.240	0.0140	1.165×10^{-5}	0.145×10^{-3}	0.720	2.03×10^{-3}	0.72
60	0.077	0.240	0.0146	1.214×10^{-5}	0.159×10^{-3}	0.796	1.93×10^{-3}	0.72
80	0.074	0.240	0.0150	1.250×10^{-5}	0.170×10^{-3}	0.851	1.86×10^{-3}	0.72
100	0.071	0.240	0.0154	1.285×10^{-5}	0.180×10^{-3}	0.905	1.79×10^{-3}	0.72
120	0.069	0.240	0.0158	1.316×10^{-5}	0.192×10^{-3}	0.964	1.74×10^{-3}	0.72
140	0.067	0.241	0.0162	1.347×10^{-5}	0.204×10^{-3}	1.023	1.68×10^{-3}	0.72
160	0.064	0.241	0.0166	1.378×10^{-5}	0.215×10^{-3}	1.082	1.63×10^{-3}	0.72
180	0.062	0.241	0.0170	1.409×10^{-5}	0.227×10^{-3}	1.141	1.57×10^{-3}	0.72
200	0.060	0.241	0.0174	1.440×10^{-5}	0.239×10^{-3}	1.20	1.52×10^{-3}	0.72
300	0.052	0.243	0.0193	1.610×10^{-5}	0.306×10^{-3}	1.53	1.32×10^{-3}	0.71
400	0.046	0.245	0.0212	1.750×10^{-5}	0.378×10^{-3}	1.88	1.16×10^{-3}	0.689
500	0.0412	0.247	0.0231	1.890×10^{-5}	0.455×10^{-3}	2.27	1.04×10^{-3}	0.683
600	0.0373	0.250	0.0250	2.000×10^{-5}	0.540×10^{-3}	2.68	0.943×10^{-3}	0.685
700	0.0341	0.253	0.0268	2.14×10^{-5}	0.625×10^{-3}	3.10	0.862×10^{-3}	0.690
800	0.0314	0.256	0.0286	2.25×10^{-5}	0.717×10^{-3}	3.56	0.794×10^{-3}	0.697
900	0.0291	0.259	0.0303	2.36×10^{-5}	0.815×10^{-3}	4.02	0.735×10^{-3}	0.705
1000	0.0271	0.262	0.0319	2.47×10^{-5}	0.917×10^{-3}	4.50	0.685×10^{-3}	0.713
1500	0.0202	0.276	0.0400	3.00×10^{-5}	1.47×10^{-3}	7.19	0.510×10^{-3}	0.739
2000	0.0161	0.286	0.0471	3.45×10^{-5}	2.14×10^{-3}	10.2	0.406×10^{-3}	0.753
				Carbon dioxide				
0	0.132	0.184	0.0076	0.88×10^{-5}	0.067×10^{-3}	0.313	2.18×10^{-3}	0.77
100	0.108	0.203	0.0100	1.05×10^{-5}	0.098×10^{-3}	0.455	1.79×10^{-3}	0.77
200	0.092	0.216	0.0125	1.22×10^{-5}	0.133×10^{-3}	0.63	1.52×10^{-3}	0.76
500	0.063	0.247	0.0198	1.67×10^{-5}	0.266×10^{-3}	1.27	1.04×10^{-3}	0.75
1000	0.0414	0.280	0.0318	2.30×10^{-5}	0.558×10^{-3}	2.75	0.685×10^{-3}	0.73
1500	0.0308	0.298	0.0420	2.86×10^{-5}	0.925×10^{-3}	4.58	0.510×10^{-3}	0.73
2000	0.0247	0.309	0.050	3.30×10^{-5}	1.34×10^{-3}	6.55	0.406×10^{-3}	0.735
3000	0.0175	0.322	0.061	3.92×10^{-5}	2.25×10^{-3}	10.8	0.289×10^{-3}	0.745
				Carbon monoxide				
0	0.0835	0.2482	0.0129	1.065×10^{-5}	0.128×10^{-3}	0.621	2.18×10^{-3}	0.75
200	0.0582	0.2496	0.0169	1.390×10^{-5}	0.239×10^{-3}	1.16	1.52×10^{-3}	0.74
400	0.0446	0.2532	0.0208	1.670×10^{-5}	0.374×10^{-3}	1.84	1.16×10^{-3}	0.73
600	0.0362	0.2592	0.0246	1.910×10^{-5}	0.527×10^{-3}	2.62	0.943×10^{-3}	0.725
800	0.0305	0.2662	0.0285	2.134×10^{-5}	0.700×10^{-3}	3.50	0.794×10^{-3}	0.72
1000	0.0263	0.2730	0.0322	2.336×10^{-5}	0.887×10^{-3}	4.50	0.685×10^{-3}	0.71
1500	0.0196	0.2878	0.0414	2.783×10^{-5}	1.420×10^{-3}	7.33	0.510×10^{-3}	0.70
				Helium				
0	0.012	1.24	0.078	1.140×10^{-5}	0.950×10^{-3}	5.25	2.18×10^{-3}	0.67
200	0.00835	1.24	0.097	1.480×10^{-5}	1.77×10^{-3}	9.36	1.52×10^{-3}	0.686
400	0.0064	1.24	0.115	1.780×10^{-5}	2.78×10^{-3}	14.5	1.16×10^{-3}	0.70
600	0.0052	1.24	0.129	2.02×10^{-5}	3.89×10^{-3}	20.0	0.943×10^{-3}	0.715
800	0.00436	1.24	0.138	2.285×10^{-5}	5.24×10^{-3}	25.5	0.794×10^{-3}	0.73
1000	0.00377	1.24	—	2.520×10^{-5}	6.69×10^{-3}	—	0.685×10^{-3}	—
1500	0.0028	1.24	—	3.160×10^{-5}	11.10×10^{-3}	—	0.510×10^{-3}	—

TABLE A-11E

Properties of gases at 1 atm pressure* (*Concluded*)

Tempera-ture, T°F	Density, ρ lbm/ft³	Specific heat, C_p Btu/lbm ·°F	Thermal conduc-tivity, k Btu/h·ft ·°F	Dynamic viscosity, μ lbm/ft·s	Kinematic viscosity, ν ft²/s	Thermal diffusiv-ity, α ft²/h	Volume expansiv-ity, β 1/R	Prandtl number, Pr
Hydrogen								
0	0.0060	3.39	0.094	0.540×10^{-5}	0.89×10^{-3}	4.62	2.18×10^{-3}	0.70
100	0.0049	3.42	0.110	0.620×10^{-5}	1.26×10^{-3}	6.56	1.79×10^{-3}	0.695
200	0.0042	3.44	0.122	0.692×10^{-5}	1.65×10^{-3}	8.45	1.52×10^{-3}	0.69
500	0.0028	3.47	0.160	0.884×10^{-5}	3.12×10^{-3}	16.5	1.04×10^{-3}	0.69
1000	0.0019	3.51	0.208	1.160×10^{-5}	6.2×10^{-3}	31.2	0.685×10^{-3}	0.705
1500	0.0014	3.62	0.260	1.415×10^{-5}	10.2×10^{-3}	51.4	0.510×10^{-3}	0.71
2000	0.0011	3.76	0.307	1.64×10^{-5}	14.4×10^{-3}	74.2	0.406×10^{-3}	0.72
3000	0.0008	4.02	0.380	1.72×10^{-5}	24.2×10^{-3}	118.0	0.289×10^{-3}	0.66
Nitrogen								
0	0.0840	0.2478	0.0132	1.055×10^{-5}	0.125×10^{-3}	0.635	2.18×10^{-3}	0.713
100	0.0690	0.2484	0.0154	1.222×10^{-5}	0.177×10^{-3}	0.898	1.79×10^{-3}	0.71
200	0.0585	0.2490	0.0174	1.380×10^{-5}	0.236×10^{-3}	1.20	1.52×10^{-3}	0.71
400	0.0449	0.2515	0.0212	1.660×10^{-5}	0.370×10^{-3}	1.88	1.16×10^{-3}	0.71
600	0.0364	0.2564	0.0252	1.915×10^{-5}	0.526×10^{-3}	2.70	0.943×10^{-3}	0.70
800	0.0306	0.2623	0.0291	2.145×10^{-5}	0.702×10^{-3}	3.62	0.794×10^{-3}	0.70
1000	0.0264	0.2689	0.0330	2.355×10^{-5}	0.891×10^{-3}	4.65	0.685×10^{-3}	0.69
1500	0.0197	0.2835	0.0423	2.800×10^{-5}	1.420×10^{-3}	7.58	0.500×10^{-3}	0.676
Oxygen								
0	0.0955	0.2185	0.0131	1.215×10^{-5}	0.127×10^{-3}	0.627	2.18×10^{-3}	0.73
100	0.0785	0.2200	0.0159	1.420×10^{-5}	0.181×10^{-3}	0.880	1.79×10^{-3}	0.71
200	0.0666	0.2228	0.0179	1.610×10^{-5}	0.242×10^{-3}	1.20	1.52×10^{-3}	0.722
400	0.0511	0.2305	0.0228	1.955×10^{-5}	0.382×10^{-3}	1.94	1.16×10^{-3}	0.710
600	0.0415	0.2390	0.0277	2.26×10^{-5}	0.545×10^{-3}	2.79	0.943×10^{-3}	0.704
800	0.0349	0.2465	0.0324	2.53×10^{-5}	0.725×10^{-3}	3.76	0.794×10^{-3}	0.695
1000	0.0301	0.2528	0.0366	2.78×10^{-5}	0.924×10^{-3}	4.80	0.685×10^{-3}	0.690
1500	0.0224	0.2635	0.0465	3.32×10^{-5}	1.480×10^{-3}	7.88	0.510×10^{-3}	0.677
Water vapor								
212	0.0372	0.451	0.0145	0.870×10^{-5}	0.234×10^{-3}	0.864	1.49×10^{-3}	0.96
300	0.0328	0.456	0.0171	1.000×10^{-5}	0.303×10^{-3}	1.14	1.32×10^{-3}	0.95
400	0.0288	0.462	0.0200	1.130×10^{-5}	0.395×10^{-3}	1.50	1.16×10^{-3}	0.94
500	0.0258	0.470	0.0228	1.265×10^{-5}	0.490×10^{-3}	1.88	1.04×10^{-3}	0.94
600	0.0233	0.477	0.0257	1.420×10^{-5}	0.610×10^{-3}	2.31	0.943×10^{-3}	0.94
700	0.0213	0.485	0.0288	1.555×10^{-5}	0.725×10^{-3}	2.79	0.862×10^{-3}	0.93
800	0.0196	0.494	0.0321	1.700×10^{-5}	0.855×10^{-3}	3.32	0.794×10^{-3}	0.92
900	0.0181	0.50	0.0355	1.810×10^{-5}	0.987×10^{-3}	3.93	0.735×10^{-3}	0.91
1000	0.0169	0.51	0.0388	1.920×10^{-5}	1.13×10^{-3}	4.50	0.685×10^{-3}	0.91
1200	0.0149	0.53	0.0457	2.14×10^{-5}	1.44×10^{-3}	5.80	0.603×10^{-3}	0.88
1400	0.0133	0.55	0.053	2.36×10^{-5}	1.78×10^{-3}	7.25	0.537×10^{-3}	0.87
1600	0.0120	0.56	0.061	2.58×10^{-5}	2.14×10^{-3}	9.07	0.485×10^{-3}	0.87
1800	0.0109	0.58	0.068	2.81×10^{-5}	2.58×10^{-3}	10.8	0.442×10^{-3}	0.87
2000	0.0100	0.60	0.076	3.03×10^{-5}	3.03×10^{-3}	12.7	0.406×10^{-3}	0.86

*For ideal gases, the properties C_p, k, μ, and Pr are independent of pressure. The properties ρ, ν, and α at a pressure P other than 1 atm are determined by multiplying the values of ρ at the given temperature by P and by dividing the values of ν and α at the given temperature by P, where P is in atm (1 atm = 101.325 kPa = 14.696 psi).

Source: F. Kreith, *Principles of Heat Transfer*, 3d ed. (New York: Addison-Wesley Educational Publishers, Inc. 1973). Originally compiled from various sources. Reprinted by permission of Addison-Wesley Educational Publishers, Inc.

TABLE A-12E

Properties of the atmosphere at high altitude

Altitude, ft	Temperature, °F	Pressure, psia	Gravity, g ft/s^2	Speed of sound, ft/s	Density, lbm/ft^3	Viscosity, μ lbm/ft · s	Thermal conductivity, Btu/h · ft · °F
0	59.00	14.7	32.174	1116	0.07647	1.202×10^{-5}	0.0146
500	57.22	14.4	32.173	1115	0.07536	1.199×10^{-5}	0.0146
1000	55.43	14.2	32.171	1113	0.07426	1.196×10^{-5}	0.0146
1500	53.65	13.9	32.169	1111	0.07317	1.193×10^{-5}	0.0145
2000	51.87	13.7	32.168	1109	0.07210	1.190×10^{-5}	0.0145
2500	50.09	13.4	32.166	1107	0.07104	1.186×10^{-5}	0.0144
3000	48.30	13.2	32.165	1105	0.06998	1.183×10^{-5}	0.0144
3500	46.52	12.9	32.163	1103	0.06985	1.180×10^{-5}	0.0143
4000	44.74	12.7	32.162	1101	0.06792	1.177×10^{-5}	0.0143
4500	42.96	12.5	32.160	1099	0.06690	1.173×10^{-5}	0.0142
5000	41.17	12.2	32.159	1097	0.06590	1.170×10^{-5}	0.0142
5500	39.39	12.0	32.157	1095	0.06491	1.167×10^{-5}	0.0141
6000	37.61	11.8	32.156	1093	0.06393	1.164×10^{-5}	0.0141
6500	35.83	11.6	32.154	1091	0.06296	1.160×10^{-5}	0.0141
7000	34.05	11.3	32.152	1089	0.06200	1.157×10^{-5}	0.0140
7500	32.26	11.1	32.151	1087	0.06105	1.154×10^{-5}	0.0140
8000	30.48	10.9	32.149	1085	0.06012	1.150×10^{-5}	0.0139
8500	28.70	10.7	32.148	1083	0.05919	1.147×10^{-5}	0.0139
9000	26.92	10.5	32.146	1081	0.05828	1.144×10^{-5}	0.0138
9500	25.14	10.3	32.145	1079	0.05738	1.140×10^{-5}	0.0138
10,000	23.36	10.1	32.145	1077	0.05648	1.137×10^{-5}	0.0137
11,000	19.79	9.72	32.140	1073	0.05473	1.130×10^{-5}	0.0136
12,000	16.23	9.34	32.137	1069	0.05302	1.124×10^{-5}	0.0136
13,000	12.67	8.99	32.134	1065	0.05135	1.117×10^{-5}	0.0135
14,000	9.12	8.63	32.131	1061	0.04973	1.110×10^{-5}	0.0134
15,000	5.55	8.29	32.128	1057	0.04814	1.104×10^{-5}	0.0133
16,000	+1.99	7.97	32.125	1053	0.04659	1.097×10^{-5}	0.0132
17,000	−1.58	7.65	32.122	1049	0.04508	1.090×10^{-5}	0.0132
18,000	−5.14	7.34	32.119	1045	0.04361	1.083×10^{-5}	0.0130
19,000	−8.70	7.05	32.115	1041	0.04217	1.076×10^{-5}	0.0129
20,000	−12.2	6.76	32.112	1037	0.04077	1.070×10^{-5}	0.0128
22,000	−19.4	6.21	32.106	1029	0.03808	1.056×10^{-5}	0.0126
24,000	−26.5	5.70	32.100	1020	0.03553	1.042×10^{-5}	0.0124
26,000	−33.6	5.22	32.094	1012	0.03311	1.028×10^{-5}	0.0122
28,000	−40.7	4.78	32.088	1003	0.03082	1.014×10^{-5}	0.0121
30,000	−47.8	4.37	32.082	995	0.02866	1.000×10^{-5}	0.0119
32,000	−54.9	3.99	32.08	987	0.02661	0.986×10^{-5}	0.0117
34,000	−62.0	3.63	32.07	978	0.02468	0.971×10^{-5}	0.0115
36,000	−69.2	3.30	32.06	969	0.02285	0.956×10^{-5}	0.0113
38,000	−69.7	3.05	32.06	968	0.02079	0.955×10^{-5}	0.0113
40,000	−69.7	2.73	32.05	968	0.01890	0.955×10^{-5}	0.0113
45,000	−69.7	2.148	32.04	968	0.01487	0.955×10^{-5}	0.0113
50,000	−69.7	1.691	32.02	968	0.01171	0.955×10^{-5}	0.0113
55,000	−69.7	1.332	32.00	968	0.00922	0.955×10^{-5}	0.0113
60,000	−69.7	1.048	31.99	968	0.00726	0.955×10^{-5}	0.0113

Source: U.S. Standard Atmosphere Supplements, U.S. Government Printing Office, 1966. Based on year-round mean conditions at 45° latitude and varies with the time of the year and the weather patterns. The conditions at sea level ($z = 0$) are taken to be P = 14.696 psia, T = 59°F, ρ = 0.076474 lbm/ft^3, g = 32.1741 ft^2/s.

FIGURE A-13E

Psychrometric chart at 1 atm total pressure. (From the American Society of Heating, Refrigerating and Air-Conditioning Engineers; used with permission.)

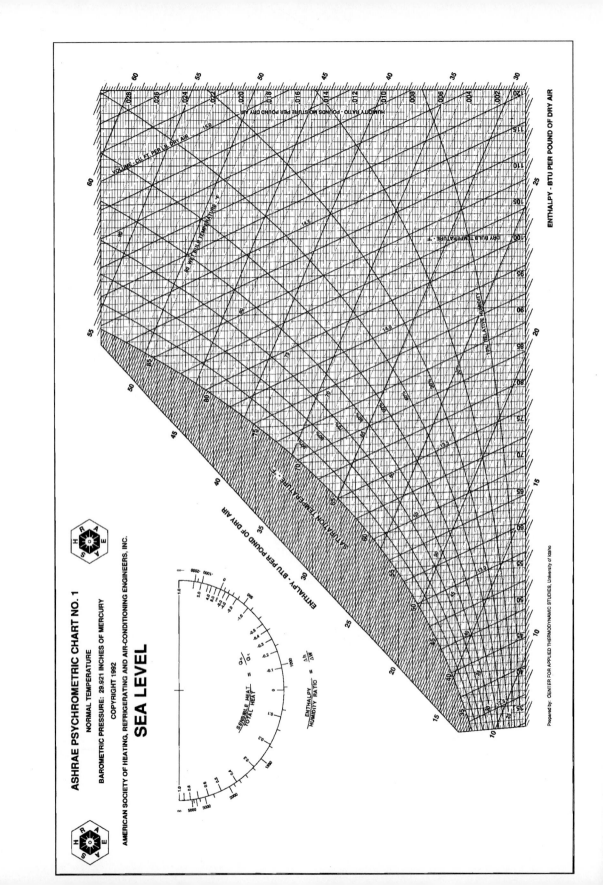

Introduction to EES

OVERVIEW

EES (pronounced "ease") is an acronym for Engineering Equation Solver. The basic function provided by EES is the numerical solution of nonlinear algebraic and differential equations. In addition, EES provides built-in thermodynamic and transport property functions for many fluids, including water, dry and moist air, refrigerants, combustion gases, and others. Additional property data can be added by the user. The combination of equation-solving capability and engineering property data makes EES a very powerful tool.

A license for EES is provided by WCB/McGraw-Hill to departments of educational institutions that adopt this text. If you need more information, contact your local WCB/McGraw-Hill representative, call 1-800-338-3987, or visit our website at www.mhhe.com. A commercial version of EES can be obtained from:

F-Chart Software
4406 Fox Bluff Road
Middleton, WI 53562
Phone: (608) 836-8531
Fax: (608) 836-8536
http://www.fchart.com

BACKGROUND INFORMATION

The EES program is probably installed on your departmental computer. In addition, the license agreement for EES allows students and faculty in a participating educational department to copy the program for educational use on their personal computer systems. Ask your instructor for details.

To start EES from the Windows File Manager or Explorer, double-click on the EES program icon or on any file created by EES. You can also start EES from the Windows Run command in the Start menu. EES begins by displaying a dialog window that shows registration information, the version number, and other information. Click the OK button to dismiss the dialog window.

Detailed help is available at any point in EES. Pressing the F1 key will bring up a Help window relating to the foremost window. Clicking the Contents button will present the Help index shown in Fig. 1. Clicking on an underlined word (shown in green on color monitors) will provide help relating to that subject.

EES commands are distributed among nine pull-down menus. A brief summary of their functions follows. (A tenth pull-down menu, which is made

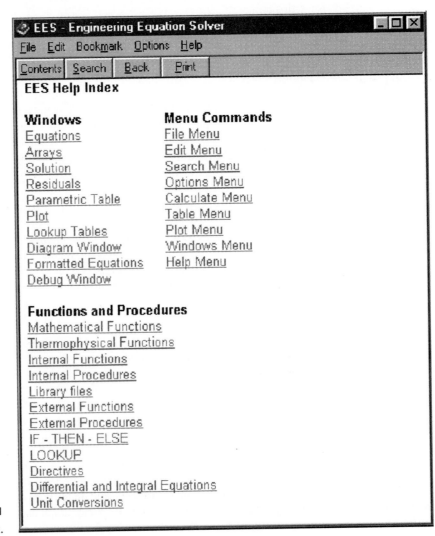

FIGURE 1
EES Help Index.

visible with the `Load Textbook` command described below, provides access to problems from this text.)

The `System` menu appears above the `File` menu. The `System` menu is not part of EES, but rather a feature of the Windows Operating System. It holds commands that allow window moving, resizing, and switching to other applications.

The `File` menu provides commands for loading, merging, and saving work files and libraries, and printing. The `Load Textbook` command in this menu reads the problem disk developed for this text and creates a new menu to the right of the `Help` menu for easy access to EES problems accompanying this text.

The `Edit` menu provides the editing commands to cut, copy, and paste information.

The `Search` menu provides `Find` and `Replace` commands for use in the Equations window.

The `Options` menu provides commands for setting the guess values and bounds of variables, the unit system, default information, and program preferences. A command is also provided for displaying information on built-in and user-supplied functions.

The `Calculate` menu contains the commands to check, format, and solve the equation set.

The `Tables` menu contains commands to set up and alter the contents of the Parametric and Lookup Tables and to do linear regression on the data in these tables. The Parametric Table, which is similar to a spreadsheet, allows the equation set to be solved repeatedly while varying the values of one or more variables. The Lookup Table holds user-supplied data that can be interpolated and used in the solution of the equation set.

The `Plot` menu provides commands to modify an existing plot or prepare a new plot of data in the Parametric, Lookup, or Array tables. Curve-fitting capability is also provided.

The `Windows` menu provides a convenient method of bringing any of the EES windows to the front or to organize the windows.

The `Help` menu provides commands for accessing the online help documentation.

A basic capability provided by EES is the solution of a set of nonlinear algebraic equations. To demonstrate this capability, start EES and enter this simple example problem in the Equations window.

Text is entered in the same manner as for any word processor. Formatting rules are as follows:

1 Upper- and lowercase letters are not distinguished. EES will (optionally) change the case of all variables to match the manner in which they first appear.

2 Blank lines and spaces may be entered as desired since they are ignored.

3 Comments must be enclosed within braces { } or within quote marks " ". Comments may span as many lines as needed. Comments within braces may be nested, in which case only the outermost set of braces is recognized. Comments within quotes will also be displayed in the Formatted Equations window.

4 Variable names must start with a letter and consist of any keyboard characters except ( ) ' | * / + - ^ { } : " or ;. Array variables are identified with square brackets around the array index or indices, for example, X[5,3]. The maximum variable length is 30 characters.

5 Multiple equations may be entered on one line if they are separated by a semi-colon (;). The maximum line length is 255 characters.

6 The caret symbol (^) or ** is used to indicate raising to a power.

7 The order in which the equations are entered does not matter.

8 The position of knowns and unknowns in the equation does not matter.

If you wish, you may view the equations in mathematical notation by selecting the `Formatted Equations` command from the `Windows` menu.

Select the `Solve` command from the `Calculate` menu. A dialog window will appear indicating the progress of the solution. When the calculations are completed, the button will change from Abort to Continue (Fig. 2).

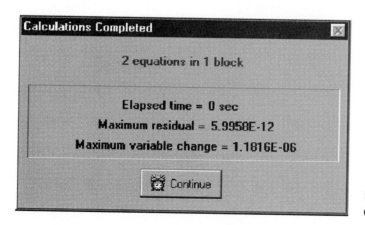

FIGURE 2

Calculations Completed window.

Click the Continue button. The solution to this equation set will then be displayed.

A HEAT TRANSFER EXAMPLE PROBLEM

In this section, problem 1-106 from the text is worked from start to finish illustrating some of the capabilities of the EES program. EES is particularly appropriate for this problem since a trial-and-error solution would be required if the problem were done by hand. The problem to be solved is stated as follows.

The roof of a house consists of a 15-cm-thick concrete slab (k = 2 W/m·°C) that is 15 m wide and 20 m long. The emissivity of the outer surface of the roof is 0.9 and the convection heat transfer coefficient on that surface is estimated to be 15 W/m²·°C. The inner surface of the roof is maintained at 15°C. On a clear winter night, the ambient air is reported to be at 10°C while the night sky temperature for radiation heat transfer is 255 K. Considering both radiation and convection heat transfer, determine the outer surface temperature and the rate of heat transfer through the roof.

If the house is heated by a furnace burning natural gas with an efficiency of 85 percent and the unit cost of natural gas is $0.60/therm, determine the money lost through the roof that night during a 14-hour period.

This problem involves conduction, convection, and radiation. The problem is facilitated by the energy-flow diagram in Fig. 3.

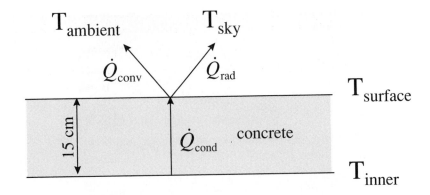

FIGURE 3

Energy-flow diagram for
example problem.

To solve this problem, it is necessary to equate the conduction heat transfer rate through the concrete roof to the combined convective and radiative heat transfer rates from the roof surface.

$$\dot{Q}_{\text{cond}} = k\,A(T_{\text{inner}} - T_{\text{surface}})/L \tag{1}$$

$$\dot{Q}_{\text{conv}} = h\,A(T_{\text{surface}} - T_{\text{ambient}}) \tag{2}$$

$$\dot{Q}_{\text{rad}} = \sigma\,\varepsilon\,A(T_{\text{surface}}^4 - T_{\text{sky}}^4) \tag{3}$$

$$\dot{Q}_{\text{cond}} = \dot{Q}_{\text{conv}} + \dot{Q}_{\text{rad}} \tag{4}$$

where

A	= roof area (m^2)
k	= thermal conductivity (W/m·K)
h	= convection coefficient (W/m²·K)
T_{inner}	= temperature of the inner surface of the roof
T_{surface}	= temperature of the outer surface of the roof
T_{ambient}	= temperature of the outdoor air
T_{sky}	= effective radiative temperature of the sky
σ	= Stefan–Boltzman constant (5.67E-8 W/m²·K⁴)
ε	= emissivity of the roof surface
L	= thickness of the concrete slab

There are a total of 13 variables in this problem. Of these, 9 (A, k, h, T_{inner}, T_{ambient}, T_{sky}, σ, ε, L) are specified in the problem statement, leaving 4 unknown variables. There are four equations relating these variables, so the problem is completely defined. However, the equations are nonlinear and, as a result, cumbersome to solve. This is where EES can help.

Start EES or select the New command from the File menu if you have already been using the program. A blank Equations window will appear. Since this problem does not require any of the thermophysical property functions built into EES, it is not necessary to specify the unit system. However, it is good practice to set the unit system at the start of a problem. To view or change the unit system, select Unit System from the Options menu (Fig. 4).

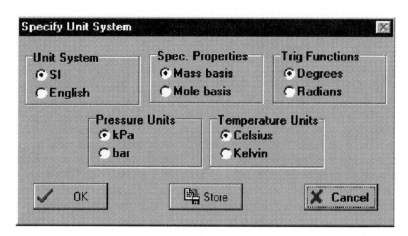

FIGURE 4

Unit System dialog.

After entering the equations for this problem and (optionally) checking the syntax using the Check/Format command in the Calculate menu, the Equations window will appear as shown in Fig. 5. Comments are normally displayed in blue on a color monitor. Other formatting options are set with the Preferences command in the Options menu.

Many of the variable names include an underscore. The underscore is a special formatting character that indicates the start of a subscript. Using underscores is optional but it improves the display of the Formatted Equations and Solutions window.

Only one built-in function has been used in this problem: the Convert function, which is used to convert W-hrs into therms. The Convert function should be able to provide any unit conversions you encounter. There are many other useful functions in EES, including functions that provide thermodynamic and transport-property data. Select the Function Information command in the Options menu. This command will bring up the Function Information dialog window. Click on the radio buttons at the top of the dialog window to view the available functions. The Info button in this dialog provides additional information relating to the selected function.

It is usually a good idea to set the guess values and (possibly) the lower and upper bounds for the variables before attempting to solve the equations. This is done with the Variable Information command in the Options menu. Before displaying the Variable Information dialog, EES checks syntax and compiles newly entered and/or changed equations, and then solves all equations with one unknown. The Variable Information dialog will then appear.

The Variable Information dialog (Fig. 6) contains a line for each variable appearing in the Equations window. By default, each variable has a guess value of 1.0 with lower and upper bounds of negative and positive infinity. (The lower and upper bounds are shown in italics if EES has previously calculated the value of the variable. In this case, the Guess value column displays the calculated value. The italicized values may still be edited.)

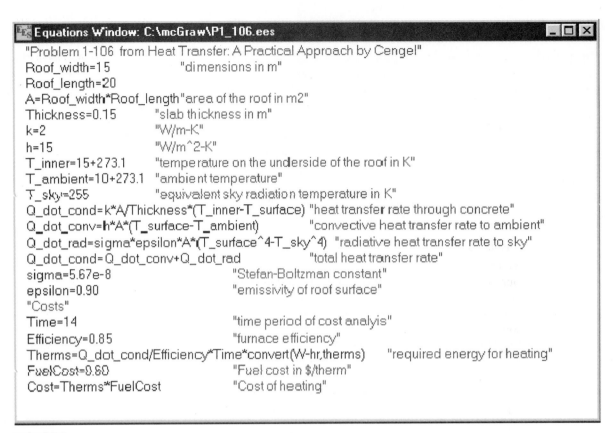

Equations Window: C:\mcGraw\P1_106.ees

```
"Problem 1-106  from Heat Transfer: A Practical Approach by Cengel"
Roof_width=15              "dimensions in m"
Roof_length=20
A=Roof_width*Roof_length"area of the roof in m2"
Thickness=0.15          "slab thickness in m"
k=2                        "W/m-K"
h=15                       "W/m^2-K"
T_inner=15+273.1          "temperature on the underside of the roof in K"
T_ambient=10+273.1    "ambient temperature"
T_sky=255                  "equivalent sky radiation temperature in K"
Q_dot_cond=k*A/Thickness*(T_inner-T_surface) "heat transfer rate through concrete"
Q_dot_conv=h*A*(T_surface-T_ambient)          "convective heat transfer rate to ambient"
Q_dot_rad=sigma*epsilon*A*(T_surface^4-T_sky^4)  "radiative heat transfer rate to sky"
Q_dot_cond=Q_dot_conv+Q_dot_rad          "total heat transfer rate"
sigma=5.67e-8                    "Stefan-Boltzman constant"
epsilon=0.90                     "emissivity of roof surface"
"Costs"
Time=14                          "time period of cost analyis"
Efficiency=0.85                  "furnace efficiency"
Therms=Q_dot_cond/Efficiency*Time*convert(W-hr,therms)      "required energy for heating"
FuelCost=0.60                    "Fuel cost in $/therm"
Cost=Therms*FuelCost             "Cost of heating"
```

FIGURE 5

Equations window.

FIGURE 6

Variable Information dialog.

Variable Information

Variable	Guess	Lower	Upper	Display			Units
A	300	-infinity	infinity	A	3	N	m^2
Cost	1	-infinity	infinity	A	3	N	$
Efficiency	0.85	-infinity	infinity	A	3	N	
epsilon	0.9	-infinity	infinity	A	3	N	
h	15	-infinity	infinity	A	3	N	W/m^2-K
k	2	-infinity	infinity	A	3	N	W/m-K
Q_dot_cond	1	-infinity	infinity	A	3	N	W
Q_dot_conv	1	-infinity	infinity	A	3	N	W
Q_dot_rad	1	-infinity	infinity	A	3	N	W
Roof_length	20	-infinity	infinity	A	3	N	m

| ✓ OK | 🖨 Print | 📋 Update | ✗ Cancel |

The A in the Display options column indicates that EES will automatically determine the display format for the numerical value of the variable when it is displayed in the Solution window. In this case, EES will select an appropriate number of digits, so the digits column to the right of the A is disabled. Automatic formatting is the default. Alternative display options are F (for fixed number of digits to the right of the decimal point) and E (for exponential format). The display and other defaults can easily be changed with the `Default Information` command in the `Options` menu. The third Display options column controls the highlighting effects such as normal (default), bold, and boxed. The units of the variables can be specified, if desired. The units will be displayed with the variable in the Solution window and/or in the Parametric Table. The units can also be set directly in the Solution window. EES does not automatically do unit conversions but it can provide unit conversions using the `Convert` function. The units information entered here is only for display purposes.

With nonlinear equations, it is sometimes necessary to provide reasonable guess values and bounds in order to determine the desired solution. (It is not necessary for this problem.) The bounds of some variables are known from the physics of the problem. In the example problem, the roof surface temperature should be somewhere between 255 and 285 K. It would be good practice to set the guess value for `T_surface` to something reasonable, for example, 270 K.

To solve the equation set, select the `Solve` command from the `Calculate` menu. An information dialog will appear indicating the elapsed time, the maximum residual (i.e., the difference between the left-hand side and right-hand side of an equation), and the maximum change in the values of the variables since the last iteration. When the calculations are completed, EES displays the total number of equations in the problem and the number of blocks. A block is a subset of equations that can be solved independently. EES automatically blocks the equation set, whenever possible, to improve the calculation efficiency. When the calculations are completed, the button will change from Abort to Continue.

By default, the calculations are stopped when 100 iterations have occurred, the elapsed time exceeds 3600 sec, the maximum residual is less than 10^{-6}, or the maximum variable change is less than 10^{-9}. These defaults can be changed with the `Stop Criteria` command in the `Options` menu. If the maximum residual is larger than the value set for the stopping criteria, the equations were not correctly solved, possibly because the bounds on one or more variables constrained the solution. Clicking the Continue button will remove the information dialog and display the Solution window (Fig. 7). The problem is now completed since the values of `T_surface`, `Q_dot_cond`, `Q_dot_conv`, and `Q_dot_rad` are determined.

An interesting and unexpected result is evident in the solution. Because of the very low sky temperature, the radiant losses cause the roof surface temperature to be lower than the ambient temperature and thus the roof is actually warmed by the convection from the air.

One of the most useful features of EES is its ability to provide parametric studies. For example, in this problem, it may be of interest to see how the

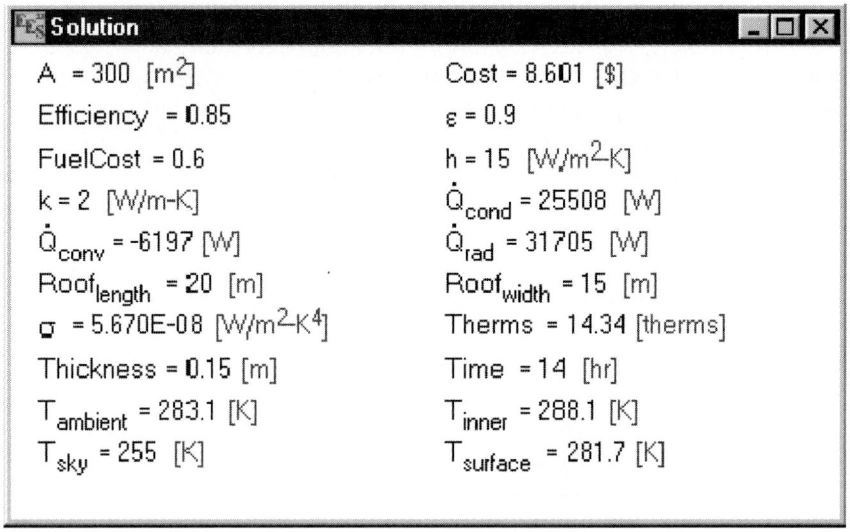

FIGURE 7
Solution window.

heating cost varies with the thermal conductivity of the roof material. A series of calculations can be automated and plotted in EES.

Select the New Table command in the Tables menu. A dialog will be displayed listing the variables appearing in the Equations window. In this case, we will construct a table containing the variables k, Q_dot_cond, and Cost. Click on k from the variable list on the left. This will cause k to be highlighted and the Add button will become active (Fig. 8). Repeat for Q_dot_cond and Cost, using the scroll bar to bring the variable into view if necessary. (As a shortcut, you can double-click on the variable name in the list on the left to move it to the list on the right.) The table setup dialog should now appear as shown in Fig. 8. Click the Add button to move the selected variables into the list on the right and then click the OK button to create the table.

The Parametric Table works much like a spreadsheet. You can type numbers directly into the cells. Numbers that you enter are shown in black and produce the same effect as if you set the variable to that value with an equation in the Equations window. Delete the k = 2 equation currently in the Equations window or enclose it in comment braces { }. This equation will not be needed because the value of k will be set in the table. Now enter values of k in the table for which Q_dot_cond and Cost are to be determined. Values of k between 2 and 0.2 have been chosen for this example. (The values could also be entered automatically using Alter Values in the Tables menu or by using the Alter Values control at the upper right of each table column header.) The Parametric Table should now appear as in Fig. 9. Now, select Solve Table from the Calculate menu. The Solve Table dialog window (Fig. 10) will appear, allowing you to choose the runs for which calculations will be done.

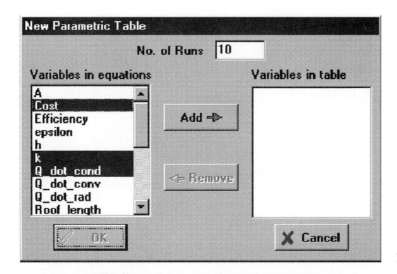

FIGURE 8
New Parametric Table dialog.

	1 $\dot{Q}_{cond}$ [W]	2 k [W/m-K]	3 Cost [$]
Run 1		2	
Run 2		1.8	
Run 3		1.6	
Run 4		1.4	
Run 5		1.2	
Run 6		1	
Run 7		0.8	
Run 8		0.6	
Run 9		0.4	
Run 10		0.2	

FIGURE 9
Parametric Table window.

When the Update Guess Values control is selected, as shown, the solution for the last run will provide guess values for the current run. Click the OK button. A status window will be displayed, indicating the progress of the solution. When the calculations are completed (Fig. 11), the values of calculated values of Q_dot_cond and Cost will be entered into the table. The values calculated by EES will be displayed in blue, bold, or italic type depending on the setting made in the Screen Display tab of the Preferences command in the Options menu.

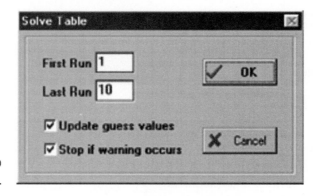

FIGURE 10

Solve Table dialog.

	$\dot{Q}_{cond}$ [W]	k [W/m-K]	Cost [$]
Run 1	25508	2	8.601
Run 2	23827	1.8	8.068
Run 3	22207	1.6	7.488
Run 4	20329	1.4	6.855
Run 5	18269	1.2	6.16
Run 6	16000	1	5.395
Run 7	13488	0.8	4.548
Run 8	10691	0.6	3.605
Run 9	7558	0.4	2.548
Run 10	4022	0.2	1.356

FIGURE 11

Parametric Table window after
calculations have been completed.

The relationship between variables such as Cost and k is now apparent, but it can be seen more clearly with a plot. Select New Plot Window from the Plot menu. The New Plot Window dialog (Fig. 12) will appear. Choose k to be the *x*-axis by clicking on k in the X-Axis list. Click on Cost in the Y-Axis list. You may wish to adjust the scale limits or add grid lines. When you click the OK button, the plot will be constructed and the plot window will appear (Fig. 13). The plot shows how the Cost varies with the thermal conductivity, indicating why most people do not have roofs made of concrete.

Once created, there are a variety of ways in which the appearance of the plot can be changed. Double-click the mouse in the plot rectangle or on the plot axis to see some of these options.

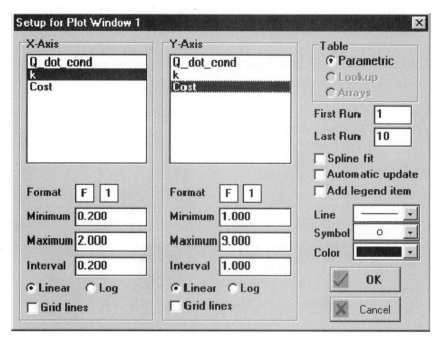

FIGURE 12
New Plot Window dialog.

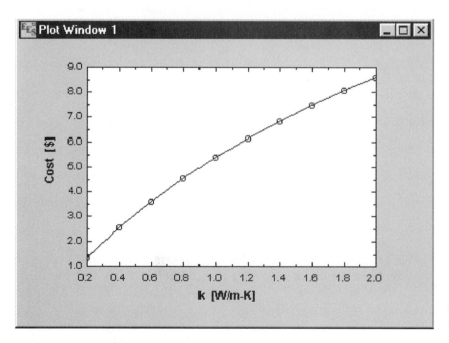

FIGURE 13
Plot window.

LOADING A TEXTBOOK FILE

A Problems disk developed for EES has been included with this textbook. Place the disk in the disk drive and then select the Load Textbook command in the File menu. Use the Windows Open File command to open the textbook problem index file, which, for this book, is named CENGEL.TXB. A new menu called Cengel Heat Transfer will appear to the right of the Help menu. This menu will provide access to all of the EES problem solutions developed for this book organized by chapter. As an example, select Chapter 1 from the Cengel Heat Transfer menu. A dialog window will appear listing the problems in Chapter 1. Select Problem 1-106 Combined Heat Transfer Mechanisms. This problem is a modification of the problem you just entered. It provides a Diagram window in which you can enter the sky temperature and other information. Enter values and then select the Solve command in the Calculate menu to see their effect on the heating cost.

At this point, you should explore. Try whatever you wish. You can't hurt anything. The online help (invoked by pressing F1) will provide details for the EES commands. EES is a powerful tool that you will find very useful in your studies.

Index

1006

Some Physical Constants

Universal gas constant	$R_u = 8.31434$ kJ/kmol $\cdot$ K
	$= 8.31434$ kPa $\cdot$ m^3/kmol $\cdot$ K
	$= 0.0831434$ bar $\cdot$ m^3/kmol $\cdot$ K
	$= 82.05$ L $\cdot$ atm/kmol $\cdot$ K
	$= 1.9858$ Btu/lbmol $\cdot$ R
	$= 1545.35$ ft $\cdot$ lbf/lbmol $\cdot$ R
	$= 10.73$ psia $\cdot$ ft^3/lbmol $\cdot$ R
Standard acceleration of gravity	$g = 9.80665$ m/s^2
	$= 32.174$ ft/s^2
Standard atmospheric pressure	1 atm $= 101.325$ kPa
	$= 1.01325$ bar
	$= 14.696$ psia
	$= 760$ mmHg ($0°$C)
	$= 29.9213$ inHg ($32°$F)
	$= 10.3323$ mH$_2$O ($4°$C)
Stefan–Boltzmann constant	$\sigma = 5.66961 \times 10^{-8}$ W/m^2 $\cdot$ K^4
	$= 0.1714 \times 10^{-8}$ Btu/h $\cdot$ ft^2 $\cdot$ R^4
Boltzmann's constant	$k = 1.380622 \times 10^{-23}$ kJ/kmol $\cdot$ K
Speed of light in vacuum	$c = 2.9979 \times 10^8$ m/s
	$= 9.836 \times 10^8$ ft/s
Speed of sound in dry air at $0°$C and 1 atm	$C = 331.36$ m/s
	$= 1089$ ft/s
Heat of fusion of water at 1 am	$h_{if} = 333.7$ kJ/kg
	$= 143.5$ Btu/lbm
Heat of vaporization of water at 1 atm	$h_{fg} = 2257.1$ kJ/kg
	$= 970.4$ Btu/lbm